Non-Linear Electromagnetic Systems

Advanced Techniques and Mathematical Methods

Edited by

V. Kose

*Physikalisch-Technische Bundesanstalt,
Braunschweig, Germany*

and

J. Sievert

*Physikalisch-Technische Bundesanstalt,
Braunschweig, Germany*

Amsterdam • Berlin • Oxford • Tokyo • Washington, DC

ISBN 90 5199 381 1 (IOS Press)
ISBN 4 274 90190 4 C3042 (Ohmsha)
Library of Congress Catalog Card Number 97-75193

Publisher
IOS Press
Van Diemenstraat 94
1013 CN Amsterdam
Netherlands
fax: +31 20 620 3419
e-mail: order@iospress.nl

Distributor in the UK and Ireland
IOS Press/Lavis Marketing
73 Lime Walk
Headington
Oxford OX3 7AD
England
fax: +44 1865 75 0079

Distributor in the USA and Canada
IOS Press, Inc.
5795-G Burke Center Parkway
Burke, VA 22015
USA
fax: +1 703 323 3668
e-mail: iosbooks@iospress.com

Distributor in Germany
IOS Press
Spandauer Strasse 2
D-10178 Berlin
Germany
fax: +49 30 242 3113

Distributor in Japan
Ohmsha, Ltd.
3-1 Kanda Nishiki-cho
Chiyoda-ku, Tokyo 101
Japan
fax: +81 3 3233 2426

Preface

The 8th International Symposium on Non-Linear Electromagnetic Systems (ISEM Braunschweig) was held in Braunschweig, Germany, from 12th to 14th May 1997. The Symposium was organized by the Electricity Division, in particular by the Magnetic Measurements Laboratory, of the Physikalisch-Technische Bundesanstalt under the chairmanship of Professor V. Kose, with Professor H. Bachmair and Dr. J. Sievert as co-chairmen.

We would like to express our thanks to the sponsors of the Symposium and to our co-workers engaged in the organization.

Previous ISEMs were held in Japan, Korea and the United Kingdom. The next Symposium will be organized under the chairmanship of Professor A. Savini in Pavia, Italy, from 10 to 12 May, 1999.

ISEM Braunschweig was of particular interdisciplinary character evident in this publication. A few tutorial papers presented by recognized experts of the respective fields of research are intended to help introduce the readers to subjects of related disciplines. This book comprises about 220 papers which cover all the ISEM Braunschweig topics:

Biomagnetism and biomagnetic applications
Inverse problems and non-destructive testing
Advanced mathematical methods, computational techniques
Micro-electromechanics, sensors and actuators, intelligent materials
Material properties and their modeling
Design of magnetic devices using FEM and BEM techniques
MAGLEV Systems

The subjects range from micro- to macro-systems (e.g. from nanowires to high-speed trains) and also from the very cold (superconducting materials) to the very hot state of matter (plasma). The understanding of biomagnetic signals in the femto-tesla range, as well as magnetic field densities which are 15 orders of magnitude higher, were and still are of great interest and importance. Seemingly different specializations such as non-destructive testing and biomagnetic research proved to be closely related to each other.

We hope that the readers will have great benefit from this book, which presents an excellent review of the state of the art of current research in non-linear electromagnetic systems.

Finally we should like to express our sincere thanks to all those have contributed to the publication of the ISEM Proceedings, particularly to Brenda Moses, Christine Dukaczewski, Ingrid Rieger and all the reviewers of the papers.

Volkmar Kose
Johannes Sievert

Braunschweig, April 1998

Contents

Inverse Problems, NDT

Microelectromechanics, Sensors and Actuators, Intelligent Materials

Advanced Mathematical and Computational Techniques

Material Properties, Design and Modeling

FEM and BEM Calculations

Magnetic Devices, Levitation

Miscellaneous

Biomagnetism, Biomagnetic Applications

Non-Linear Electromagnetic Systems
V. Kose and J. Sievert (Eds.)
IOS Press, 1998

3

Functional Imaging in Medicine:

Application of SQUIDs in Biomagnetism

T. KATILA

Laboratory of Biomedical Engineering, Helsinki University of Technology,
02150 Espoo, and
BioMag Research Centre, Helsinki University Central Hospital, 00290 Helsinki,
Finland
E-mail: toivo.katila@hut.fi

Abstract. The methods used for medical imaging can roughly be divided into two classes. Methods such as conventional X-ray imaging, X-ray computed tomography, magnetic resonance imaging and ultrasound imaging produce anatomic and geometric information on the human body. On the other hand, functional information can be obtained by using methods such as single photon emission tomography, positron emission tomography, functional magnetic resonance imaging and optical imaging. The methods most often used in functional studies of the bioelectric sources inside the human brain and heart are the electroencephalography (EEG) and the electrocardiography (ECG). Here we discuss the magnetic counterparts of the bioelectric studies, the functional imaging using the magnetoencephalography (MEG) and the magnetocardiography (MCG). So far, most biomagnetic studies have been close to basic research. Several research groups have also started clinical studies. For a few years, commercial multichannel instruments have been available both for MEG- and MCG-studies. Presently, most instrumentations utilize low temperature superconducting quantum interference devices (SQUIDs) in order to reach the high sensitivity, needed in biomagnetic measurements. The high-temperature superconductors may make the biomagnetic instrumentation in the future simpler, smaller and cheaper.

1. Introduction

One megatrend in health care today is the trend towards noninvasive, or at least minimally invasive methods. This trend requires the use of new technologies. The role of various medical imaging methods is emphasized. However, health care is conservative. Any new technology must be approved, and it must be better that the old ones. Every new technology must therefore be carefully assessed.

In industrialized countries, around 6 - 12 % of the gross national product is used for health. World wide, the health care was using yearly during the first half of the '90s around 2000 billion USD. Several governments have tried to cut down, but not very successfully, the cost of health care. Instead, better cost-efficiency is sought. Indeed, another megatrend in health care today is the requirement of increasing efficiency.

2. Medical imaging

In the last twenty years, in vivo imaging of humans and the computed image processing have strongly influenced studies of anatomy and physiology. Medical imaging has made it possible to monitor e.g. brain and cardiac anatomy and function completely noninvasively. It has also in many ways changed the clinical practise.

X-ray and magnetic resonance imaging produce anatomic pictures of the subject studied. X-ray images are based on the different absorption of radiation in different tissues. Computed tomography (CT) utilizes images taken at different angles to produce volume data. Magnetic resonance (MR) imaging uses the nuclear magnetic resonance signal, in most cases the signal of protons situated in human tissue. With the aid of magnetic field gradients, three-dimensional volume data are collected.

Both CT and MR imaging show structural and geometrical information in humans. The images are often studied as two-dimensional slices, but three-dimensional images can be computed as well. However, in general, these images do not give information on functional activity of the tissue.

There are other methods which are able to produce functional images. In positron emission tomography (PET), the intensity in the images is proportional to the local consumption of the radioactively marked sugar. The PET techniques can reveal the sites in the human brain in which e.g. hearing, reading, spelling a word or even thinking take place, but it cannot reveal the thoughts themselves. In addition to PET, single photon emission tomography (SPET) is used in hospitals for functional imaging. A recent development is the use of MRI to functional (f) studies as well. In fMRI, the small differences in the MR-images, caused by changes in metabolism, are studied. Another recent method is the use of optical imaging for functional studies.

The main topics of this report are the biomagnetic and bioelectric methods in studying the bioelectric activity of humans. The bioelectromagnetic fields appear at such low frequencies, mostly below 1 kHz, that the electric and magnetic fields can be measured separately. The fields are produced by the bioelectric activity of the human tissue. Thus no illumination with electromagnetic radiation is needed, and the measurements are totally noninvasive.

The study of the heart with the aid of the electric surface potential measurements is called the electrocardiography (ECG), and the corresponding biomagnetic study the magnetocardiography (MCG)[1]. Similarly, the bioelectric measurements of the brain are called the electroencephalography (EEG), and the magnetic measurements the magnetoencephalography (MEG) [2]. The examples of measurements chosen for this paper have been performed in the BioMag Research Centre of the Helsinki University Central Hospital [3].

3. Electromagnetic source imaging

The bioelectromagnetic methods for functional imaging are based on the measurements of the electric and magnetic fields produced by humans. These fields are generated by the bioelectric activity of the excitable cells, e.g. in the heart, brain, eye, muscles etc. The fields can be studied in millisecond time scale, much faster than using other methods mentioned above.

In imaging methods, in general, an inversion must be made to obtain the image from the measured data. The disadvantage of the biomagnetic and bioelectric studies is that the inversion does not have a unique solution. And sometimes, the inversion does not even work in practice.

The task of the electromagnetic inverse problem is to estimate the sources of the fields. We assume that we know the (measured) electric or magnetic fields, or both, e.g. at the outer surface of the subject studied. Further, the geometric structures and their electromagnetic properties inside the subject must be known. The geometry of the subject measured can be obtained e.g. from MR-images, taken separately.

We can make use of the solution of the forward problem. Assuming that the source current distribution $\vec{j}_p(\vec{r}')$ is located within the piecewise (i) homogeneous and isotropic volume conductor $v'(\vec{r}')$, the magnetic field $\vec{B} = \nabla \times \vec{A}$, where

$$\vec{A} = \vec{A}_\infty + \vec{A}_S \tag{1}$$

$$= \frac{\mu_0}{4\pi} \int_{v''} \frac{\vec{j}_p(\vec{r}')}{|\vec{r} - \vec{r}'|} dv' + \frac{\mu_0}{4\pi} \int_{v'} \frac{\vec{j}_S(\vec{r}')}{|\vec{r} - \vec{r}'|} dv' \tag{2}$$

and

$$\vec{j}_S dv' = -\sum_i \phi_{S_i}(\sigma_i'' - \sigma_i')d\vec{S}_i. \tag{3}$$

The terms $\vec{A}_\infty$ and $\vec{A}_S$ refer to an approximation of an infinite homogeneous volume conductor and to the influence of the conductivity boundaries, correspondingly. The primary source currents are assumed to be confined within the subvolume v''. The parameters σ_i'' and σ_i' are the electric conductivities outside and inside the surface S_i.

Equations similar to Eqs. (1 - 3) can also be found to the electric potential ϕ. The equations are integral equations and as the surface potential appears inside the integrals, the solutions must be found iteratively.

Several different methods have been used to solve the bioelectromagnetic inverse problem. The actual sources can be modelled as simplified equivalent sources, which then can be solved uniquely. The simplest source current model is a current dipole. Multiple dipoles and current multipoles have been used as well [4]. To obtain an estimate of the actual current source distribution, e.g. the minimum norm approach, the probabilistic entropy concept, linear estimation and spatial Fourier filtering have been used.

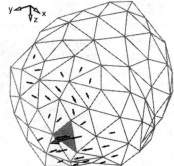

Fig. 1. For the same experimental data from a study of the cardiac evoked biomagnetic field, the large arrow in the middle of the shaded triangle indicates the current-dipole solution of the source current, the set of the smaller arrows depicts the solution for the current distribution on the heart's surface.[5]

The lead field of the magnetic detector (i) can be defined as the electric current field in the source volume conductor, energized by feeding (alternating) unit current to the detector. Let us assume that the lead fields $\vec{L}_i(\vec{r}')$, where i = 1 ... N, are known. Then the minimum norm estimate $\vec{j}^\star$ for the source currents can be calculated as,

$$\vec{j}^\star(\vec{r}') = \sum_{i=1}^{N} \omega_i \vec{L}_i(r'). \tag{4}$$

The coefficients ω_i are calculated from the actually measured biomagnetic field values. However, in general the inverse solution of the bioelectromagnetic problems is not unique. Therefore, regularization techniques must be used for successful convergence of the solution.

Fig. 1 demonstrates the importance of selecting the appropriate method for solving the inverse problem. Due to the fact that the biomagnetic inverse problem is ill-posed, the selection influences the result obtained. Thus the methods must be tailored for the specific purposes.

4. Bioelectromagnetic studies

Due to the extreme weakness of the biomagnetic fields, their measurements must usually be performed in a magnetically shielded room. The walls of such a room consist of ferromagnetic and/or aluminum layers. Modern neuromagnetometers consist of one hundred to several hundred channels. The size of such instrumentation is today comparable to CT X-ray and MR imaging equipment. The size is mostly determined by the helium cryostat, needed to cool the superconducting quantum interference devices (SQUID-detectors) to 4 K [6]. The detector coils are placed at the bottom of the cryogenic dewar. In brain studies, a helmet type design offers a good coverage to the cortical sources. In cardiomagnetometers, detector setups close to planar are used and so far the number of channels is smaller, but the area covered larger than in neuromagnetometers. In the BioMag Centre, the neuromagnetometer now has 122 and the cardiomagnetometer 67 magnetic channels.[3]

The SQUIDs are wired to corresponding electronic detection and amplification units. In addition, there are devices to indicate the positioning of the subject, to produce various stimuli needed and measure simultaneously the corresponding electric potentials.

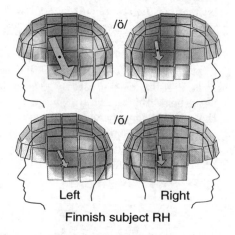

Fig. 2. Magnetic source imaging of language processing in human brain. The squares indicate the individual sensor modules of the neuromagnetometer. The arrows indicate the estimated current dipoles, the interpretation is that they show locations of language-dependent memory traces in the brain (from ref. [7], reprinted with permission from Nature).

In the BioMag Centre, the accuracy of the source current localization has been studied both for cardiac and neural fields. In order to test the accuracy in cardiomagnetic studies, the cardiac evoked magnetic fields, produced in vivo using a non-magnetic catheter, were measured.[5] The catheter was specially designed to produce an elementary current dipole around the tip of it (the large arrow in Fig. 1). In five patients, the location error was 4 ± 1 mm. The consistency of the source localizations obtained from fMRI- and MEG-studies was tested using an electric stimulus on the median nerve at the wrist. In MEG, the cortical source was calculated 35 ms later. In this case as well, the accuracy of the results was a few mm.

Biomagnetic measurements have been made in a number of different fields. To give just one recent example, Näätänen et al. [7] studied at the BioMag Centre the language-dependent memory trace in the human brain (Fig. 2). Finnish and Estonian languages were studied, the languages are related, but the presentations e.g. of the Finnish phoneme /ö/ and the Estonian /õ/ differ considerably. Note especially the functional difference between the hemispheres of the brain.

5. Clinical applications

Presently, several biomagnetism research centres are developing clinical applications. Most applications are based on the ability of the biomagnetic studies to localize the centres of the bioelectric current sources more accurately than the bioelectric measurements. In BioMag, the main clinical projects in brain studies have been the localization of the epileptic foci, as well as e.g. the localization of the central sulcus as a presurgical evaluation of the patients. In addition, several other patient groups have been studied using the MEG.

In cardiac studies in the BioMag, the localization of sources of life-threatening arrhythmias has been one of the major clinical projects. The localization studies include both congenital and acquired arrhythmogeneity. The localization procedure needs advanced source and thorax models. The localization of the onset site of tachyarrhythmias is used in guiding the curative catheter treatment. Another large project is the identification of the patients prone to malignant arrhythmia. Several patient groups are studied for this purpose [3]. Multichannel ECG mappings have also been reported [8].

6. Conclusion

The combination of the noninvasive bioelectromagnetic measurements with other medical imaging methods is a very powerful research tool for functional studies of the heart and brain, both in basic research and in clinical applications. For bioelectric and biomagnetic studies, advanced data processing and analysis are needed. Although the methods are useful for clinical applications as well, they still have to be developed for more easy and reliable use.

The main obstacle in the extensive use of the biomagnetic method today is the high cost of the biomagnetic instrumentation. A smaller size of the instrumentation would also be preferable. With the advent of the new high temperature superconducting SQUIDs this situation may change. Thereafter, the biomagnetic measurement method will compete even more successfully.

7. Acknowledgement

The author is grateful to the researchers of the Biomedical Engineering Laboratory and the BioMag Research Centre for their help during the preparation of the manuscript.

References

[1] G. Stroink, M.J.R. Lamothe and M.J. Gardner, "Magnetocardiographic and electrocardiographic mapping studies", H. Weinstock (ed.), SQUID Sensors: Fundamentals, Fabrication and Applications, 413 - 444, Kluwer Academic Publishers (Netherlands, 1996)

[2] M. Hämäläinen, R. Hari, R.J. Ilmoniemi, J. Knuutila and O.V. Lounasmaa, "Magnetoencephalography - theory, instrumentation and applications to noninvasive studies of the working human brain", Rev. Mod. Phys. 65, 413 - 497 (1993)

[3] R. J. Ilmoniemi, A.Ahonen, K. Alho, H. J. Aronen, J. Huttunen, J. Karhu, P. Karp, M. Mäkijärvi, R. Näätänen, R. Paetau, E. Pekkonen, C.-G. Standertskjöld-Nordenstam, J. Tiihonen, L. Toivonen and T. Katila, "MEG and MCG in a clinical environment", Proc. 10th Int. Conf. on Biomagnetism (1996), to be publ.

[4] M. Burghoff, W. Haberkorn, B.-M. Mackert and G. Curio, "Multipole Source Analysis of the Magnetic Field from Stimulated Peripheral Nerves", this Conference

[5] K. Grönros, J. Nenonen, R. Fenici and T. Katila, "Comparison of regularization methods when applied to epicardial minimum norm estimates", to be publ.

[6] H. Koch, "SQUIDs - from Basic Circuits to Complex Systems", this Conference

[7] R. Näätänen, A. Lehtokoski, M. Lennes, M. Cheour, M. Huotilainen, A. Iivonen, M. Vainio, P. Alku, R.J. Ilmoniemi, A. Luuk, J. Allik, J. Sinkkonen and K. Alho, "Language-specific phoneme representations revealed by electric and magnetic brain responses", Nature 385, 432-4 (1997)

[8] L. Reinhardt, M. Mäkijärvi, T. Fetch, J. Montonen, G. Sierra, A. Martinez-Rubio, T. Katila, M. Morggrefe, G. Breithardt, "Predictive Value of Wavelet Correlation Functions of Signal-Averaged Electrocardiogram in Patients After Anterior versus Inferior Myocardial Infarction", J. Am. Coll. Cardiol. 27(1), 53-59 (1996)

Non-Linear Electromagnetic Systems
V. Kose and J. Sievert (Eds.)
IOS Press, 1998

SQUIDs

From Basic Circuits to Complex Systems

Hans Koch

Physikalisch-Technische Bundesanstalt, Fachbereich „Biosignale",
Abbestr. 2 - 12, 10587 Berlin, Germany[#]

Abstract. Although the basic circuits of SQUIDs (Superconducting QUantum Interference Devices) are very simple, their non-linear character makes them quite complex and this in turn leads to a very rich variety of designs, functions, and applications.
The aim of this contribution is to present a few instructive examples on the role of non-linearities and the methods employed to improve the performance of different components of SQUID-systems, like the sensors themselves, the read-out electronics, and the impact of gradiometer configurations on the output signals.

1. Introduction

Two main categories of SQUIDs exist. They are termed according to their mode of signal read-out: rf-SQUIDs and dc-SQUIDs.

The basic circuit of an rf-SQUID is a superconducting ring with inductance L_s interrupted by one Josephson junction (c.f. Fig. 1). The non-linear relationship between externally applied magnetic flux ϕ_a and the internal flux inside the SQUID-ring allows the signal read-out by a circuit that acts as a parametric amplifier. By coupling the SQUID to a tank circuit (L_T, C_T) pumped by an rf-bias, the weak measured signal will be mixed, amplified parametrically and then detected as a side-band signal. The rectified output signal amplitude $\hat{V}_{rf}$ depends periodically on the value of the applied magnetic field. The period is given by the flux quantum $\phi_0 = h / 2e$.

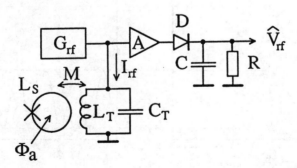

Fig. 1a Basic circuit of an rf-SQUID coupled to a read-out electronic

---

[#] Dedicated to Prof. Dr. V. Kose, Vice-President of Physikalisch-Technische Bundesanstalt (PTB), on the occasion of his retirement

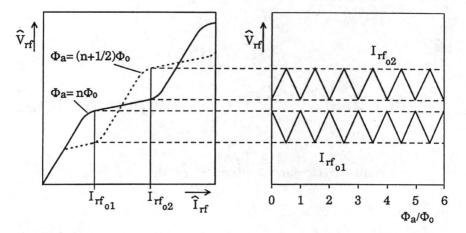

Fig. 1b Schematic characteristics of an rf-SQUID

The basic circuit of a dc-SQUID is a superconducting ring with inductance L_s interrupted by two Josephson junctions (c.f. Fig. 2). That set-up allows a dc current bias generating an output voltage $\langle V \rangle$ that again depends periodically on the applied flux ϕ_a.

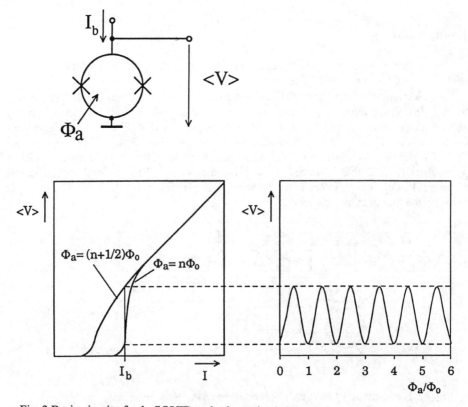

Fig. 2 Basic circuit of a dc-SQUID and schematic characteristics.

Periodic, non-linear input-output-characteristics are not very convenient. They are multi-valued and offer a small dynamic range. Thus, it is common practice to employ so-called flux-locked-loop circuits to linearize the output (c.f. Fig. 3): the output voltage of the SQUID is amplified and the integrated signal V_{out} is used to generate a flux in a feed-back coil coupled to the SQUID-loop, thus locking the flux inside the SQUID-ring to a fixed working point. V_{out} is then linearly dependent on the applied flux ϕ_a.

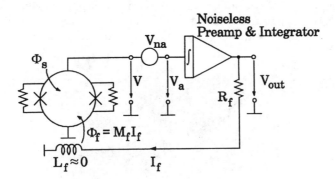

Fig. 3 Idealized SQUID flux locked loop electronic

Flux resolution in the order of $10^{-6}\ \phi_0\ /\ \sqrt{Hz}$ and a slew rate of $> 10^7\ \phi_0\ /\ s$ are achieved with modern read-out electronics. An excellent and up to date overview on SQUIDs and read-out electronics is given in [1].

2. Real SQUID Sensors

In practice, real SQUID sensors look quite different and are more complex than their basic circuits. Fig. 4 shows a microscope image of a modern Niobium thin film dc-SQUID and its schematic circuit. Two features are obvious: i) instead of the simple ring the SQUID inductance is designed as a multiloop structure, ii) several additional circuit elements have been added to the device.

The multiloop structure, i.e. the parallel arrangement of several inductances, is only one of several possible solutions to the following dilemma: the output voltage modulation depth ($\Delta V_{out} = \Delta V_{out,max} - \Delta V_{out,min}$) is only sufficient for SQUIDs with a very low loop inductance L_s in the order of 100 pH. On the other hand, for practical applications SQUIDs should not be flux sensitive but sensitive to magnetic induction $B = \phi / A$. Thus a large pick-up loop area A would be beneficial for field resolution but contradicts the need for a small inductance L_s. That led to SQUID designs with parallel inductances, flux focussers, or flux transformers; all of them are methods to achieve small inductances and large effective areas.

The additional circuit elements are required to improve the performance of the SQUID by optimizing the circuit. The term dc-SQUID is misleading for characterizing the SQUID itself, in reality it is a microwave device! The Josephson relationships which connect the current density j and voltage v of the junction via the phase difference δ:

$$j = J_0 \sin \delta$$
$$v = (\hbar / 2e)\ \partial \delta / \partial t$$

govern the non-linear properties of superconducting electronic devices. Thus the junctions are subject to an alternate current with frequency $\omega = (2e/\hbar)\,v$. The frequencies of the two dc-SQUID junctions interfere („Superconducting Quantum *Interference* Device") and the output voltage V_{out} alternates with the beating frequency accordingly. With read-out amplifiers only the averaged voltage $\langle V_{out} \rangle$ is detected. But the SQUID structure itself with its striplines and distributed inductances and capacitances is a complicated microwave circuit easily leading to resonances affecting the output characteristic. Without measures like damping resistors and other circuit elements only mediocre SQUID performance will be achieved. Several papers describe the effect of microwave resonances in SQUID structures [2]. Theory and experimental verification of optimized SQUID design may be found in [3].

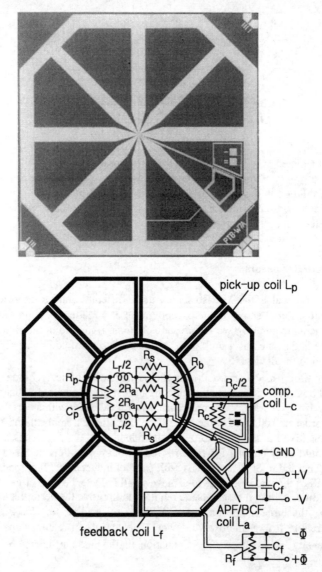

Fig. 4 Microscopic image of a dc-SQUID chip (7 mm x 7 mm) and schematic circuit (right).

3. Dc-SQUID Read-Out with Additional Positive Feedback

The schematic flux locked loop circuit shown in Fig. 3 is of course over-simplified. The SQUID voltage modulation is usually so small, that the voltage noise of even the best pre-amplifiers would seriously limit the signal to noise ratio. Sophisticated lock-in techniques are usually necessary to obtain a sufficiently low white noise level of the output signal. Such an electronic is quite complex, expensive and bulky. Particularly for multichannel SQUID-systems that approach is inappropriate.

However, a few years ago a new scheme was introduced utilizing an additional positive feedback (APF): the SQUID voltage drop over resistor R_{APF} generates a current that couples flux into the SQUID-loop L_s via the APF-coil L_{APF}, see Fig. 5. The effect is a steepening of the V-ϕ-characteristics at the working point. Due to the increased dV/dϕ the voltage noise of the preamplifier has a reduced effect on the composition of the flux noise. Therefore a direct read-out of the SQUID with the preamplifier is possible now, making lock-in techniques obsolete. Details of the APF-technique may be found in [4].

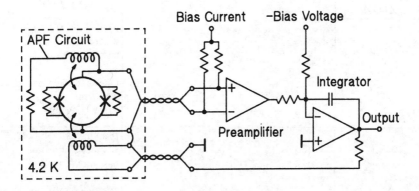

Fig. 5 Dc-SQUID with direct coupled APF electronic

4. The Effect of Different Gradiometer Configurations

SQUIDs are extremely sensitive sensors for magnetic fields. Thus they are applied just to detect extremely small signals like in biomagnetic applications (see accompanying article [5]). Unwanted signals from interfering sources of our industrialized environment like powerline hum, subway systems etc. are often much stronger than the measurement signal. Heavy magnetic shielding is one means to decrease the influence of interference, another means are gradiometers i.e. pick-up coil configurations that discriminate between distant and adjacent sources. Several different gradiometer configurations have been employed in different brands of biomagnetic SQUID-systems. Fig. 6 shows some of them and the MCG-signal of the same heart beat (MCG is the magnetic equivalent of an electrocardiogram (ECG)). It is obvious that all signals have quite different shapes, amplitudes and character-istic time intervals (e.g. duration of the QRS-structure).

This clearly shows how a comparability of results obtained with different SQUID-systems with different gradiometer configurations is no longer possible. This is a critical draw-back if multicentered clinical studies with different SQUID-systems are performed.

Recently a signal transformation method has been proposed that allows a regained comparability of results of such different systems [6].

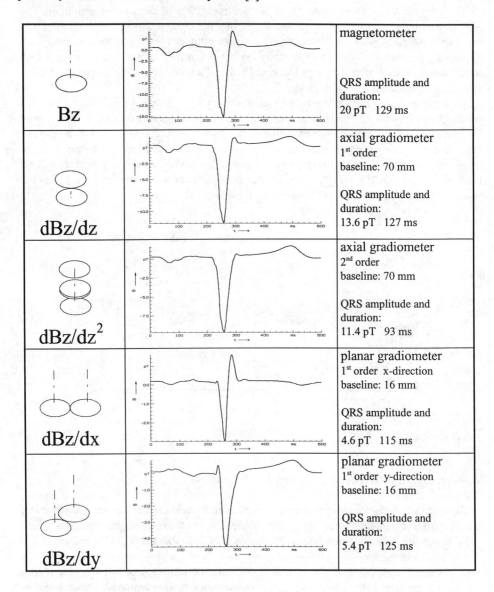

Fig. 6 Influence of different gradiometer configurations on the resulting output signals

5. Complex Systems

In order to become a versatile instrument in the hands of e.g. physicians a wide range of additional supplies have to be added and should interact with the SQUID-system without restricting the operation. The following list might offer a taste of what has to be considered when developing a fully functioning system: dewars fitting to application (helmet shape for brain research, flat bottom for cardiac applications), liquid helium management, magnetically shielded room, multichannel data processing, acquisition and storage in appropriate data bases, and very complex software for retrieving results relevant to the physician. The accompanying paper [5] will present a further overview on this very difficult task.

Summary

SQUID-systems are good examples to demonstrate the role non-linearities may play and how one is able to tame and exploit these non-linearities.

Nowadays SQUID-systems became very robust and versatile and thus open a broad spectrum of interesting applications of which biomagnetism is the most spectacular.

References

[1] H. Weinstock (ed.), SQUID Sensors: Fundamentals, Fabrication and Applications, ISBN 0-7923-4350-6, Kluwer Academic Publishers, 1996.
[2] K. Enpuku *et al.*, Modeling the DC Superconducting Quantum Interference Device Coupled to the Multi-turn Input Coil II, *J. Appl. Phys.* **71** (1992) 2338-2346, and part III, *J. Appl. Phys.* **72** (1992) 1000-1006.
[3] D. Drung *et al.*, Theory of the Multiloop SQUID and Experimental Verification, *J. Appl. Phys.* **77** (1995) 4088-4098.
[4] D. Drung *et al.*, Low-Noise High-Speed DC Superconducting Quantum Interference Device Magnetometer with Simplified Feedback Electronics, *Appl. Phys.* **57** (1990) 406-408.
[5] T. Katila, these proceedings.
[6] M. Burghoff *et al.*, Comparability of Measurement Results Obtained with Multi-SQUID-Systems of Different Sensor Configurations, *IEEE Trans. Appl. Supercond.* **7** (1997) 3465-3468.

16   *Non-Linear Electromagnetic Systems*
V. Kose and J. Sievert (Eds.)
IOS Press, 1998

Inverse Problem Aspects in the Field of Biomagnetic Applications

Shoogo UENO

Department of Biomedical Engineering, Graduate School of Medicine, University of Tokyo, Tokyo 113 Japan

Abstract. Biomagnetics is an interdisciplinary field where magnetics, biology and medicine overlap. This paper focuses on some topics in biomagnetics, especially, transcranial magnetic stimulation, neuromagnetic measurements by super-conducting quantum interference device(SQUID) systems, and functional magnetic resonance imaging (fMRI). The inverse problem is essential for the estimation of the brain electrical activities from the measured data over the surface of the head, where the importance of the inhomogeneities in the head is emphasized. A new method of impedance imaging based on MRI techniques is proposed to get imaging of electrical properties in the body.

1. Introduction

Biomagnetics is an interdisciplinary field where magnetics, biology and medicine overlap. Biomagnetics includes four main subjects: biomagnetic stimulation, study of biological effects of magnetic fields, measurement of biomagnetic fields using superconducting quantum interference devise(SQUID) systems, and magnetic resonance imaging (MRI). These fields have expanded rapidly in recent years. Among them, transcranial magnetic stimulation, neuromagnetic imaging by SQUIDs, and functional MRI have become important tools in functional brain research and clinical diagnosis. The inverse problem is essential for imaging and estimating localized area of brain electrical activities.

The present study is focused on source estimation of neuronal electrical activities in the brain associated with motor evoked potentials and higher brain function where the importance of the inhomogeneities in the head is emphasized. The limitation of source estimation based on dipole models is also discussed. In order to get imaging of electrical properties in the body, a new method of impedance MRI is proposed.

2. Magnetic Stimulation of the Human Brain

Magnetic nerve stimulation has been widely used in neurophysiological research and clinical diagnosis [1]. We have developed a method of localized magnetic stimulation of the human brain [2]. The basic idea is to concentrate induced eddy currents locally in the vicinity of a target by a pair of opposingly pulsed magnetic fields which can be produced by a figure-eight coil. We have calculated spatial distributions of eddy currents induced in cubical and spherical volume conductor models using a finite element method. The computer simulation has shown that the current density at the target under the intersection of the figure-eight coil attains a peak which is 3 times higher than current densities in the surrounding regions. Based on this principle, we were able to stimulate the human motor cortex within a 5mm resolution [3].

Functional maps of the cortex varied with the orientation of the stimulating current. To explain the mechanism responsible for producing this anistropic response to brain stimulation, models of neural excitation elicited by magnetic stimulation were proposed [4] [5]. A model was developed to examine how the threshold for nerve excitation changes with the depth and length of the nerve fibers and the bending angle of the axon. In this study, a model is proposed to examine the effect of inhomogeneities on nerve excitation. This model shows that the activating function, a parameter for nerve excitation, is enhanced at the boundary between media with different electrical conductivities. The existence of the vectorial characteristics of stimulating currents for neural excitation reflects both the functional and anatomical organization of neural fibers in the brain.

In biomagnetic stimulation, the electric fields arise from two sources, time-varying magnetic fields and charge distributions at tissue boundaries. This can be expressed as

$$E = E_A + E_\Phi = -\partial A/\partial t - \nabla\Phi \qquad (1)$$

where E is the net applied electric field, A is the magnetic vector potential, and is the electrostatic potential [6].

The vector potential A can be calculated from the coil geometry by

$$E_A = -\frac{\partial A}{\partial t} = -\frac{\mu_0 N}{4\pi}\frac{\partial I}{\partial t}\int\frac{d\,l'}{R} \qquad (2)$$

where N is the number of turns in the coil, i is the coil current, dl' is a vector representing a small length of the coil (pointing in the direction of the current), and R is the distance from dl' to the position where A is calculated. EA can be calculated by approximating the stimulating coil as a polygon and the sum of the induced electric field produced by each segment.

We consider only the excitation of a long straight nerve fiber. For convenience, we assume here that a nerve fiber runs in the direction parallel to the x axis. The first derivative of the electric field component parallel to the nerve fiber, $\partial Ex/\partial x$, called the activating function, contributes mostly to the depolarization or hyperpolarization of the membrane [7] [8]. The site of excitation is a point at which the electric field gradient is the largest. The membrane at the place where the activating function is negative is to be depolarized, while the membrane is to be hyperpolarized if the activating function is positive. In other words, the outward currents out of the membrane, which contribute to the membrane, can be caused at the place where the activating function is negative.

The inward currents from the outside of the membrane, which give rise to the hyperpolarization of the membrane, can be caused at the place where the activating function is positive.

No nerve excitation occurs where the induced electric field is most intense, because uniform electric fields along the nerve axon cause no outward currents.

The contribution of the radial components of the electric field to the membrane excitation is negligibly small, compared with the contribution of the gradient of axial components of the electric field.

For instance, in an inhomogeneous medium, as shown in Fig. 1, the activating function shows a negative peak at the boundary between the tissues with different conductivity, σ_1 and σ_2, where $\sigma_1/\sigma_2 < 1$, when the induced electric field E_x is in the direction from σ_2 to σ_1.   In contrast, the activating function shows a positive peak when the induced electric field E_x is in the direction from σ_2 to σ_1, where $\sigma_1/\sigma_2 < 1$.

This effect is derived by the conservation law of currents under quasi-static electric field assumption in magnetic stimulation.   That is, div $i = 0$ in the whole region.

In order to calculate spatial distributions of the activating function in an inhomogeneous volume conductor, we use a computing model based on a finite element method. E_A is easily obtained by Bio-Savart law. $E\phi$ is the negative gradient of potential ϕ, and ϕ is obtained by solving the Laplace's equation

$$\nabla^2 \Phi = 0$$

(3)

with the boundary condition on the outer boundary between conductor and outer space

$$\frac{\partial \Phi}{\partial n} = n \cdot E_A$$

(4)

and on the inner boundary between subregions of different conductivities

$$\sigma_1 \frac{\partial \Phi_1}{\partial n} = \sigma_2 \frac{\partial \Phi_2}{\partial n}$$

(5)

where ϕ_1 and ϕ_2 are scalar potentials in σ_1 and σ_2 respectively.

In this study, we introduce a cubical model with conductivity ϕ_1 surrounding a cubical subspace of conductivity ϕ_2, as shown in Fig. 2.

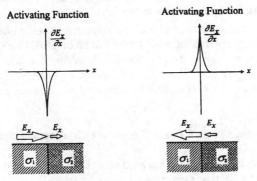

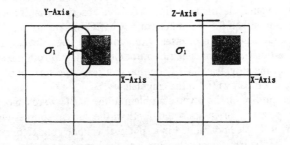

Fig.1 activating function at the boundary

Fig.2 Cubical volume conductor model

A figure-eight coil was positioned over ϕ_2 near the boundary between $\sigma1$and $\sigma2$. We calculated the spatial distributions of the activating function $\partial Ex / \partial x$ over the plane z = 0.8 in order to determine the excitation sites and optimal direction for nerve excitation.

Fig. 3 shows an equipotential contour map of the activating function in cubical volume mode when $\sigma_1/\sigma_2 = 0.1$. The peak appears at x=0.3, y=0.5 on the boundary between σ_1 and σ_2. If the targeted neurons lie in that area, the nerve can be easily excited. The absolute value of $\partial Ex / \partial x$ at the boundary is significantly higher than in a homogeneous volume conductor. It is noted that the optimal direction of stimulating current for nerve excitation is the direction of the +x-axis in this case.

To demonstrate the usefulness of the model, the transcranial magnetic stimulation was carried out using a figure-eight coil which was driven by discharged currents through a capacitor which was driven by discharged currents through a capacitor band. The magneticstimulation of the motor cortex with pulsed magnetic fields of 0.5T with a duration of 80 ms was able to evoke muscle contractions of the hand. The motor evoked potentials at the hand and the arm muscles were measured, changing the direction of the stimulating currents.

It was difficult to stimulate the target neurons located near the edema by the vectors which orient from the posterior to the anterior. In contrast, the reversed vector stimulation of the neurons located near the edema was easily performed. This phenomena is explained by the results in Fig. 2.

The limitations found using the model of the long straight nerve fibers in inhomogenous media were that the neuron fibers of brain tissues are complex and do not exist in straight lines as suggested by the model. However, this model is useful for understanding the importance of the effect of the inhomogeneities on the excitation site in magnetic stimulation of the brain and the peripheral nervous systems.

3. Neuromagnetic Imaging of Brain Function by Magnetoencephalography

Magnetoencephalography (MEG) has become a useful tool to study higher brain function as well as to diagnose neurophysiological disorders [9] [10] [11]. One of the advantages of neural magnetic measurement using a dc-SQUID machine is to measure magnetic fields associated with the direct current or the extremely low frequency components of neural electric activities. We have observed that the extremely slow magnetic fields less than 0.5 Hz appear on the MEG recordings in subjects during short-term memory and cognitive processes with a four tone task. Inverse problems are important in estimating neural electric sources in the brain. From the spatial distributions of the slow wave MEG components over the surface of the hand, we estimated the electrical source in the cingulate gyrus.

Most of the studies on the inverse problem in the MEG were concerned with estimation of position, orientation and depth of the current sources which were modeled as single current dipoles positioned in a spherical volume conductor with homogenous conductivity. However, the neural sources in the brain are complex, and the head as a volume conductor consists of different materials with different electrical conductivities. These include brain tissues, skull, ventricles, brain lesions, etc. Brain lesions themselves can be also expected to have different conductivities, depending upon their pathology such as edema and calcification.

The present study focuses on the influence of these Inhomogeneities in the head on the MEG topographic patterns, modeling the neural sources as single or multiple dipoles located in an inhomogeneous volume conductor. Electrical potentials and magnetic fields over the surface of volume conductors are calculated, changing the conductivity of the inhomogeneous regions [12] [13].

One of the most important aspects of computer simulation studies is that flux reversal phenomena can be observed in MEG topographic patterns in special cases where conductivities of the inhomogeneous regions change with brain pathology.

For example, we consider a brain lesion model as shown in Fig. 5. The electrical conductivity of brain lesions varies with pathology such as edema and calcification. Edema gives a high conductivity, whereas calcification gives a low conductivity. Edema gives a high conductivity, whereas calcification is expected to give a low conductivity. A brain lesion is assumed to be a sphere of radius 0.03m and positioned at $(x,y,z) =$ (0.025,0.025,0.03545). A current dipole is located $(x,y,z) = (0.025,0.025,0.005)$, parallel to the X axis.

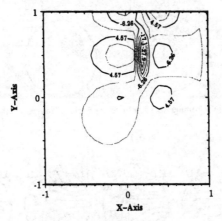

Fig3. Contour map of $\partial E_x / \partial x$ over the plane z=0.8 in the case of inhomogeneous volume conductor with $\sigma_1 / \sigma_2 = 0.1$. A nerve axon runs along the x-axis at y=0.5. The negative peak of $\partial E_x / \partial x$ appears at the place (x,y) = (0.2,0.5) where the nerve can be excited

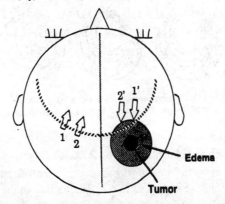

Fig.4. Schematic diagram of vectorial characteristics in magnetic stimulation of the brain with a tumor surrounded by edema. The arrows 1 and 1' show points which innervate the thenar muscles, and the arrows 2 and 2' show points which innervate the brachioradial muscles. The optimal directions of stimulating currents for nerve excitation are reversed in the right and left hemispheres.

Fig. 4 shows the experimental results of magnetic brain stimulation of a patient with a metastatic brain tumor in the right parietal region. The tumor itself is small, but the large surrounding edema has a high electric conductivity. The arrows how the optimal directions for excitation of neurons which innervate the thenar and brachioradial muscles. The dashed line roughly coincides with the central sulcus. The left hemisphere is normal. In the left hemisphere, the optimal directions for neural excitation orient from the posterior to anterior region. In contrast, the optimal direction in the abnormal right hemisphere is in the direction opposite to that in the normal left hemisphere. While the output of our stimulator was fixed, the direction of the stimulating current was changed, and motor evoked potentials were measured. The potential at optimal direction was 1.5 mV. A clear phenomena between the optimal direction and opposite directions were observed. This reversal phenomena can be explained, in part, by the peak of the activating function at the boundary of the lesion.

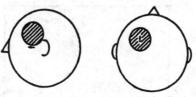

Fig. 5 Current dipole in inhomogeneous volume conductor. Brain lesion is located in a sphere.

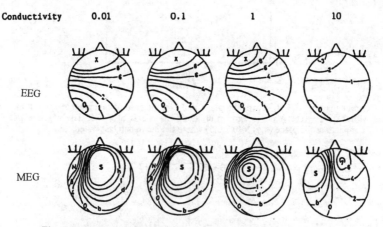

Fig. 6 Spatial distributions of EEGs and MEGs over the surface of the hemi-spherical model when the conductivity ratio varies from 0.01 to 10

Figure 6 shows calculated results. Maps in cases of $\sigma_2/\sigma_1 = 10$ correspond to edema, and maps in cases of $\sigma_2/\sigma_1 = 0.1, 0.01$ correspond to calcification. The patterns of potential distributions are analogous to each other although the amplitudes decrease with the increase of σ_2. In contrast with the potential distribution, the magnetic isofields contour maps are influenced dramatically by the inhomogeneities. The poles of the North and South are reversed in the range between $\sigma_2/\sigma_1 = 10$ and $\sigma_2/\sigma_1 = 0.1$. The results suggest that magnetic flux reversal phenomena can appear in isofield contour maps of evoked magnetic fields when pathological situations of brain lesions are changed from edema to calcification.

4. Impedance Magnetic Resonance Imaging

Magnetic Resonance Imaging (MRI) techniques have become important tools in medicine and biology. Conventional MRI, however, produces no information about the electrical properties of the body. We have proposed a new method for imaging electrical properties such as conductivity and admittance or impedance based on MRI techniques. The basic idea is to use the shielding effects of induced eddy currents in the body upon spin precessions. Two types of methods are introduced ; i) a large flip angle method, and ii) a third coil method. Phantom experiments have been carried out to verify this concept using an MRI system of 7.05 T.

i) Large flip angle method

The basic idea of this method is to use very large flip angles ($\alpha > \pi/2$, π or 2π) to enhance the shielding effects of induced eddy currents in conductive tissues. First, two images with identical scan parameters but different flip angles are obtained. When the 1st and 2nd RF pulses give different flip angles a and p + a respectively to different tissues of a different conductivity. Since a transversal magnetization is proportional to the sine function of the flip angle, the two images yield an image reflecting the conductivity distribution. Images of rats have been obtained with strong contrasts of conductivity distributions.

Experiments were first performed on a columnar phantom which has a water compartment and a saline compartment. A 7.05 T, 18.3 cm bore system was used for imaging. Two images were obtained with flip angles 231 to the water and simultaneously 210 to the saline and 264 to the water and simultaneously 240 to the saline. The images were then divided. Similarly, low and high angle images were obtained to yield calculated images of conductivity distributions.

Fig. 7 shows the results of impedance magnetic resonance imaging of a rat brain. The MR image at the left side is an image obtained with a flip angle of 18. The image of the right side shows an image obtained with a flip angle of 120. The substraction of the left side image from the right side image gives the image of the brain which gives the impedance distribution of the brain.

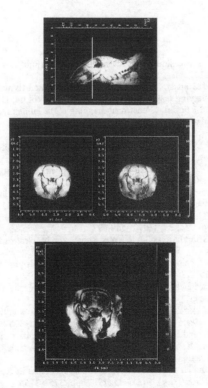

Fig. 7. Impedance MR imaging of a rat brain. The upper figure is saggital imaging. The middle left and right figures are images obtained with a flip angle of 18 ° and 120°, separately. The lower figure is the subtraction image of the left side from the right side.

ii) Third Coil Method

The method utilizes a third electrical coil to produce a time-varying magnetic field Bc parallel to the main static magnetic field B0. The additional magnetic field Bc induces eddy currents in the body. The induced eddy currents give rise to shielding effects which influence spin precessions. If the shielding effects are reflected by a medium with different electrical properties, such as conductivity and dielectric constant, the signal reflects the local admittance of the medium. Changing the frequency of additional magnetic fields Bc, imaging of conductivity or admittance distributions at different frequencies can be obtained.

Experiments were first performed on a 5-cm diametric columnar phantom which has a water compartment and a saline compartment. A 7.05 T , 18.3 cm bore system was used for imaging. The phantoms was positioned through a four turn solenoidal Bc coil 6 cm in diameter. The applied frequency of the Bc field was low (1-2 kHz) and varied independently of the magnetic resonance frequency w0. Rats were also subjected to this admittance MRI.

Spin precessions were obviously affected by the Bc field and image intensities were reduced. High conductivity objects were less sensitive to the signal reduction by the Bc field. Imaging of rats were obtained, the intensity of which strongly reflects the conductance distribution independently of the magnetic resonance frequency w0, because Bc frequency was able to be low and varied independently.

References

[1] Barker, R.Jalinous and I.L.Freeston, "Non-invasive magnetic stimulation of human motor cortex," *Lancet* ii, (1985) pp. 1106-1107.
[2] S.Ueno, T.Tashiro and K.Harada, "Localized stimulation of neural tissues in the brain by means of paired configuration of time-varying magnetic fields", *J.Appl.Phys*, Vol 64, (1988)5862-5864.
[3] S.Ueno, T.Matsuda and M.Fujiki, "Functional mapping of the human motor cortex obtained by focal and vectorial magnetic stimulation of the brain", *IEEE Trans. on Magnetics*. Vol.26, (1990) pp. 1539-1544.
[4] S.Ueno, T.Matsuda and M.Hiwaki, "Estimation of structures of neural fibers in the human brain by vectorial magnetic stimulation", *IEEE Trans on Magnetics*, Vol 27, (1991), pp. 5387-5389.
[5] Hyodo, S.Ueno, "Nerve excitation model for localized magnetic stimulation of finite neuronal structures", *IEEE Trans on Magnetics* Vol 32, (1996), pp. 5112-5114.
[6] P.J.Basser, "Magnetic stimulation of peripheral axon: models and experiments" in *Biomagnetic Stimulation* (Edited by S.Ueno), Plenum Press, New York, (1994) pp.119-129.
[7] B.J.Roth, P.J.Basser, "The model of the stimulation of a nerve fiber by electromagnetic induction", *IEEE Trans. on Biomed. Eng.*, Vol 37, (1990) 588-597.
[8] P.J.Basser, B.J.Roth, "Stimulation of myelinated nerve axon by electromagnetic induction", *Med.Biol.Eng.Comput.*, Vol 29 (1991):261-268.
[9] Cohen, D. "Magnetoencephalography: Evidence of magnetic fields produced by alpha-rythm currents." *Science*, (1968), 161:784-786.
[10] Cohen, D. "Magnetoencephalography: Detection of the brain's electrical activity with a superconducting magnetometer." *Science* (1972), 175: 664-666.
[11] Williamson, S.J. and Kaufman, L. "Application of SQUID sensors to the investigation of neural activity in the human brain." *IEEE Trans Magn.*, MAG-19, (1983)pp. 835-844.
[12] Ueno, S., Wakisako, H. and Harada, K. "Flux reversal phenomena in spatial distributions of magnetoencephalograms." In: H. Weinberg, G. Stroink and T. Katila (Eds.), Biomagnetism; Application and Theory, Pergamon Press, New York, (1985) 289-293.
[13] Ueno, S., Iramina, K. and Harada, K. "Effects of inhomogeneities in cerebral modeling for magnetoencephalography." *IEEE Trans.* MAG-23:(1987) pp. 3753-3755.

Non-Linear Electromagnetic Systems
V. Kose and J. Sievert (Eds.)
IOS Press, 1998

Reconstruction of 3D Current Distributions - A Biomagnetic Inverse Problem

**H. Brauer[1], U. Tenner[1], H. Wiechmann[1], A. Arlt[1], M. Ziolkowski[1],
J. Haueisen[2], H. Nowak[2], and U. Leder[3]**

[1] *Technical University of Ilmenau, P.O.Box 10 0565, D-98684 Ilmenau, Germany*
[2] *Friedrich-Schiller-University of Jena, Biomagnetic Center, D-07740 Jena, Germany*
[3] *Friedrich-Schiller-University of Jena, Clinic of Internal Medicine III, D-07740 Jena, Germany*

Abstract- Both biomagnetic simulation studies and experimental investigations of
the source reconstruction techniques (inverse problem) require detailed
descriptions of field source models. We developed a physical thorax phantom
and propose a technique for physical modelling of extended current sources to
use for biomagnetic studies. The new source models were applied in physical
phantoms of the human thorax. The Boundary-Element-Method (BEM) was
used for volume conductor modelling. The impressed current distributions were
reconstructed applying different minimum-norm solution methods. It was found
that it is very difficult to apply the well-proved methods of dipole localization to
the reconstruction of extended sources. The reconstruction of 3-D current
distributions requires development of much more improved methods.

1. Introduction

In the last ten years more and more biomagnetic measurement systems became available to
many clinical centers. Biomagnetic measurements can provide magnetic data sets for functional
localization of heart or brain activity non-invasively. But the interpretation of measured, very weak
magnetic fields generated by electrically active organs requires special algorithms for localization
or reconstruction of the sources.

We developed a physical thorax phantom which will help us to validate the numerical
techniques for reconstruction of dipolar and extended sources. Physical modelling of extended
biomagnetic sources is still a widely unsolved problem. Both in neuromagnetism and
cardiomagnetism detailed descriptions of totally non-magnetic source models are needed, for
computer simulation studies as well as for experimental validation of inverse methods applied to
reconstruction of the current sources.

We propose a new technique of physical modelling of extended current sources using platinum
electrodes as well as special polymer membranes. These source models were applied in physical
phantoms. We used a tank model and a realistically shaped phantom of the human thorax which
was prepared by reproducing the body surface of an adult male volunteer. The biomagnetic
measurements were performed at the *Biomagnetic Center of the Friedrich-Schiller-University of
Jena* applying the Philips twin-dewar SQUID system (62 channels).

2. Forward Modelling of Biomagnetic Fields

We consider biomagnetic signals generated by electric currents due to transmembrane ionic flow
in excitable tissues. These magnetic fields are linked to bioelectric potentials and can be measured
with magnetometer or gradiometer systems in magnetically shielded rooms. Since the great number

of elementary sources are activated simultaneously, it is convenient to represent them as a continuous macroscopic primary current distribution. The total current density incorporates the impressed current density and the effect of microscopic changes in conductivity due to cellular structure, which are represented by the ohmic volume current density. The primary current distribution in a bounded volume can be approximated by a macroscopic equivalent current dipole. Because of the relatively slow variations of variables in time (< 1 kHz), treatment of sources and fields in a quasi-static approximation is allowed. If an inhomogeneous volume conductor consisting of homogeneous subvolumes with constant conductivity is assumed, the surface potential is defined by an integral equation. The external magnetic field can be found from Biot-Savart law if the sources and the properties of the volume conductor are known. This calculation is called the solution of the forward problem, which is important in verifying the validity of the source and volume conductor models. Because of the irregular shape of the body, calculations can be performed only numerically, using the Finite Element (FEM), or the Boundary Element Method (BEM). We applied the BEM because it is easier to perform on workstations [1].

The volume conductor models were obtained from Magnetic Resonance Images (MRI) using a specific mesh generation technique [2]. The phantom surface was modelled by reproducing the body surface of an adult man (Fig. 1). The phantom was made of epoxy resin and has been filled up for magnetic field measurements with saline solution (0.1 %) which corresponds to a conductivity of 0.21 S/m. Further, the phantom contains an universal support for fixing the generator models. This module has a defined position on the back inside the phantom and can be exchanged to choose the proper generator setup. By this way it can be easily adapted to any source model, either dipolar or extended sources.

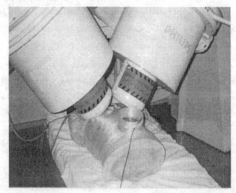

Fig. 1. Realistically shaped thorax phantom of an adult man, with 138 surface electrodes

Fig.2. Magnetic field measurement of a thorax phantom in the shielded room of the Biomagnetic Center of Jena using the Philips SQUID system

Dipolar sources were made of twisted silver or platinum wires. For modelling of extended sources at the tips of such dipole different plates made of platinum or aluminium were mounted and isolated against each other (Fig. 3). Furthermore, there was found another new method to model extended sources. We applied a certain polymer membrane (which is usually used for desalinating sea water) coated with platinum electrodes on both sides. One version of such an extended source where the current flows in the membrane sheet is shown in Fig. 4.

The MRI data of the phantom were used to construct the BEM model. The magnetic field measurements were performed using the Philips twin-dewar SQUID system (2 x 31 channels) in the shielded room of the Biomagnetic Center Jena (Fig. 2). The study of source localization and reconstruction techniques, respectively, was performed with a simple tank model. A glass tank was filled with 0.1 % saline solution (corresponding to a mean conductivity of 0.21 S/m), and then the extended source shown in Fig. 3 was placed about 5 cm below the water surface. In Fig. 5 is shown the related magnetic field pattern measured in central position about 45 mm above the tank surface.

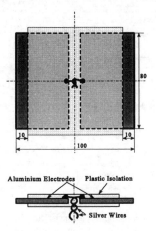

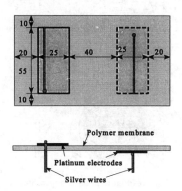

Fig. 3. Extended source working as an
expanded, 'Butterfly' dipole

Fig. 4. Extended membrane source model
working as a thin current sheet

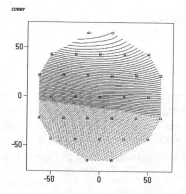

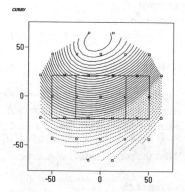

Fig. 5. Magnetic field pattern of the extended
'Butterfly' source in the glass tank

Fig. 6. Magnetic field pattern of the membrane source
model in a spherical tank, with membrane contour

The magnetic field pattern of the membrane source model measured with one dewar including 31 channels is shown in Fig. 6. It looks like a magnetic field caused by a nonhomogeneous current distribution across the membrane which is due to an unsufficient contact of the feed wires and the platinum electrodes on the membrane. But the main reason for such field pattern is the equivalent dipole which is due to the closed current loop arising from the feed wires and the membrane. To avoid this effect the 'Butterfly' dipole (Fig. 3) was prepared and finally used for verification of reconstruction methods.

3. Results

The phantoms were developed to validate the numerical solution methods of source reconstruction. We applied the above described extended source models. The reconstruction of the current sources was performed applying the minimum-norm methods which are included in the Philips software CURRY. We found the L1-norm solution much better estimates the depth than it does the L2-norm (see Fig. 8). But any method was not able to reconstruct the current we assumed based on the measurements shown in Fig. 5 or 7. Because we measure only one component of the magnetic field the reconstructed current dipoles are usually oriented more or less normal to the source plane.

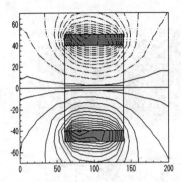

Fig. 7. Electrical potential field pattern measured in a plane 5 mm above the `Butterfly` dipole

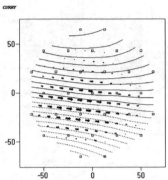

Fig. 8. Reconstructed current of the `Butterfly` dipole applying the L1-norm solution of CURRY

It was found that the reconstruction of a plane current distribution applying the widely used minimum-norm solutions provides acceptable results only if the L1-norm was chosen. The results of reconstruction of volume sources are so far widely unsatisfactory. New approaches are still required for appropriate reconstruction of current distributions in 3D.

4. Conclusions

We realized physical models of extended current sources applicable to biomagnetic measurements. Further experiments are still necessary to guarantee optimal impressed currents and to minimize the distortion of the measured magnetic field due to the feed wires. The proposed new technique of extended source modelling offers new possibilities for validating biomagnetic inverse methods but the methods for reconstructing extended current sources must be further improved.

Acknowledgment

This work was supported by „Deutsche Forschungsgemeinschaft (DFG)" with Grant Br 1195/1-3 and by the Thuringian Ministry of Science, Research and Culture with Grant B 511-95003.

References

[1] H. Brauer, O. Kosch, U. Tenner, H. Wiechmann and A. Arlt, "A Modified Linear Estimation Approach for Solving Biomagnetic Inverse Problems", IEEE Transactions on Magnetics,vol. 32(1996)3, pp. 1298-1301
[2] M. Ziolkowski and H. Brauer, "Methods of Mesh Generation for Biomagnetic Problems", IEEE Transactions on Magnetics, vol. 32(1996)3, pp.1345-1348
[3] H. Brauer, O. Kosch, A. Arlt, U. Tenner and O. Michelsson, "Inverse Calculation of Biomagnetic Fields Using Realistically Shaped Boundary Element Models", Proc. 6th Internat. IGTE-Symposium, Graz 1994, pp. 270-275
[4] H. Brauer, U. Tenner, H. Wiechmann, A. Arlt, U. Leder, J. Haueisen, H. Nowak, L. Trahms and M Burghoff, "Modellierung eines Thoraxphantoms für die Validierung der biomagnetischen Quellenlokalisation", Biomedizinische Technik, Band 41, Ergänzungsband 1, 1996, S. 294-295
[5] U. Tenner, H. Brauer, H. Wiechmann, A. Arlt, J. Haueisen, H. Nowak and U. Leder, "Reconstruction of Dipolar and Extended Biomagnetic Sources in a Physical Thorax Phantom", Proc. 18th Annual International Conference of IEEE Engineering in Medicine and Biology Society, Amsterdam, 1996
[6] H. Brauer, U. Tenner, H. Wiechmann, A.Arlt, J. Haueisen and H. Nowak, "A New Field Source Modelling Technique and its Application to Validation of Biomagnetic Source Reconstruction Methods", Proc. 7th International IGTE-Symposium, Graz, 1996, pp. 290 - 295

Non-Linear Electromagnetic Systems
V. Kose and J. Sievert (Eds.)
IOS Press, 1998

Biomagnetic Inverse Solutions in Terms of Multipoles

Wolfgang Haberkorn

Physikalisch-Technische Bundesanstalt, Abbestraße 2-12, 10587 Berlin, Germany

Abstract. A spherical harmonic multipole expansion of the magnetic induction due to a primary current density is performed and compared with a tensor formulation based upon the Taylor series expansion. The effects of a bounded volume conductor with conductivity interfaces on the multipole fields are treated. The results provide the basis for the determination of the multipole moments from measurements of the fields. For the magnetic heart vector the inverse solution is presented. Furthermore, it is shown that simple biomagnetic sources can be reconstructed from the first few multipole terms.

1. Introduction

Electric currents in tissues like the heart, the brain, and the nerves produce magnetic fields around the human body. The calculation of the fields from a known current distribution is called the forward problem. Of greater interest is the solution of the inverse problem, i.e., the determination of the sources from the measured fields. In contrast to the forward problem the inverse problem does not possess a mathematically unique solution. This difficulty is circumvented by using an equivalent source description in terms of multipoles.

For the magnetic field generated by a primary current density a spherical harmonic multipole expansion is performed. Furthermore, a multipole expansion in Cartesian coordinates is considered to study the tensor properties of the multipole fields. The relationships between both expansions are investigated. In addition, the effects of the volume conductor on the multipole source fields are treated.

The multipole source is described by two types of multipole coefficients, which can be classified as electric and magnetic moments. The determination of the electric moments from body surface potentials is a well-known result [1]. The corresponding inverse solution for the magnetic multipole moments is presented in this contribution. The result includes the important case of the magnetic heart vector, i.e., the solution for the magnetic dipole moments of the human heart.

The multipole moments reflect the features of the underlying current source. For cardio- and neuromagnetic source models the multipole moments are calculated to study the information content and the meaning of higher-order terms. It is shown that for simple cases the source parameters can be obtained from the first few multipole terms.

2. Method

The starting point of the multipole expansion is the vector potential $\vec{A}$ due to a primary current density representing the biological sources. The effects of the body are taken into account by a finite, linear, and isotropic volume conductor with conductivity interfaces between homogeneous regions. The solution of the Maxwell equations under quasi-static conditions by taking $\nabla \vec{A} = 0$ (Coulomb gauge) is [2]

$$\vec{A}(\vec{r}) = \vec{A}^{\mathrm{p}}(\vec{r}) + \vec{A}^{\mathrm{v}}(\vec{r}).$$

The source term is given by

$$\frac{4\pi}{\mu_0} \vec{A}^{\mathrm{p}}(\vec{r}) = \int_G \frac{\vec{J}^{\mathrm{p}}(\vec{r}')}{|\vec{r} - \vec{r}'|} \, \mathrm{d}V',$$

where $\vec{J}^{\mathrm{p}}$ is the primary current density, G is the source volume, and μ_0 is the permeability of free space. The contribution of the volume conductor of the volume V is

$$\frac{4\pi}{\mu_0} \vec{A}^{\mathrm{v}}(\vec{r}) = -\sum_k \int_{S_k} (\sigma_i^k - \sigma_o^k) \frac{\Phi(\vec{r}')}{|\vec{r} - \vec{r}'|} \, \mathrm{d}\vec{S}'_k + \int_V \sigma(\vec{r}')\Phi(\vec{r}') \frac{(\vec{r} - \vec{r}')}{|\vec{r} - \vec{r}'|^3} \, \mathrm{d}V'.$$

In this equation, S_k are the internal surfaces and the external boundary of the volume conductor; σ_i^k, σ_o^k are the values of the conductivity σ on the inside and outside of S_k, where $\sigma = 0$ outside V; Φ is the electric potential.

The expression for $\vec{A}^{\mathrm{p}}$ in series of spherical functions is obtained by using the expansion of the dyadic Green's function [3]. Furthermore, the expansion of $\vec{A}^{\mathrm{p}}$ in Cartesian coordinates is found by Taylor series expansion of $1/|\vec{r} - \vec{r}'|$. Finally, the curl of the vector potential $\vec{A}$ is taken to obtain the magnetic induction $\vec{B}$.

3. Results

The magnetic induction $\vec{B}$ is the sum of the source field $\vec{B}^{\mathrm{p}}$ and the contribution of the volume conductor $\vec{B}^{\mathrm{v}}$. The spherical multipole expansion of $\vec{B}^{\mathrm{p}}$ is found to be

$$\frac{4\pi}{\mu_0} \vec{B}^{\mathrm{p}}(\vec{r}) = \mathrm{Re} \sum_{n=0}^{\infty} \sum_{m=0}^{n+1} \frac{A_{n+1,m} + \mathrm{i}B_{n+1,m}}{n+1} \nabla\left[\frac{P_{n+1}^m(\cos\vartheta)\mathrm{e}^{-\mathrm{i}m\varphi}}{r^{n+2}}\right] \times \vec{r}$$
$$- \mathrm{Re} \sum_{n=1}^{\infty} \sum_{m=0}^{n} (\alpha_{nm} + \mathrm{i}\beta_{nm}) \nabla\left[\frac{P_n^m(\cos\vartheta)\mathrm{e}^{-\mathrm{i}m\varphi}}{r^{n+1}}\right],$$

where Re denotes the real part, and r, ϑ, φ are the spherical coordinates. $P_n^m(\cos\vartheta)$ are the associated Legendre functions of the first kind, i represents the imaginary unit, and ∇ is the del operator. There are two types of multipole moments of the primary current density [4]. The electric multipole coefficients are given by

$$A_{n+1,m} + \mathrm{i}B_{n+1,m} = g_{n+1,m} \int_G \vec{J}^{\mathrm{p}}(\vec{r}') \nabla'\left[r'^{n+1} P_{n+1}^m(\cos\vartheta')\mathrm{e}^{\mathrm{i}m\varphi'}\right] \mathrm{d}V',$$

and the magnetic multipole coefficients are

$$\alpha_{nm} + \mathrm{i}\beta_{nm} = \frac{g_{nm}}{n+1} \int_G [\vec{r}' \times \vec{J}^{\mathrm{p}}(\vec{r}')] \nabla'\left[r'^n P_n^m(\cos\vartheta')\mathrm{e}^{\mathrm{i}m\varphi'}\right] \mathrm{d}V'.$$

Here, $g_{nm} = (2 - \delta_{m0})(n-m)!/(n+m)!$, where δ_{m0} is the Kronecker delta, which is unity for $m = 0$ and zero otherwise. Furthermore, the field contribution of the volume conductor $\vec{B}^{\mathrm{v}}$ is given by

$$\frac{4\pi}{\mu_0} \vec{B}^{\mathrm{v}}(\vec{r}) = \sum_k \int_{S_k} (\sigma_i^k - \sigma_o^k) \Phi(\vec{r}') \frac{(\vec{r} - \vec{r}')}{|\vec{r} - \vec{r}'|^3} \times \mathrm{d}\vec{S}'_k.$$

This expression depends on the electric potential, for which an equation can be found by using Green's theorem [1]. The same result can be easily obtained from $\nabla \vec{A} = 0$, namely,

$$4\pi\sigma(\vec{r})\Phi(\vec{r}) = \mathrm{Re}\sum_{n=0}^{\infty}\sum_{m=0}^{n+1}(A_{n+1,m}+\mathrm{i}B_{n+1,m})\frac{P_{n+1}^{m}(\cos\vartheta)\,\mathrm{e}^{-\mathrm{i}m\varphi}}{r^{n+2}}$$

$$-\sum_{k}\int_{S_k}(\sigma_i^k-\sigma_o^k)\Phi(\vec{r}')\frac{(\vec{r}-\vec{r}')}{|\vec{r}-\vec{r}'|^3}\,\mathrm{d}\vec{S}_k'.$$

These results show that the magnetic field is described by both types of multipole moments, whereas the electric potential is only determined by the electric moments. The magnetic field contributions due to the magnetic moments are not affected by the volume conductor.

Performing Taylor series expansion, the source magnetic field is given by

$$\frac{4\pi}{\mu_0}B_l^{\mathrm{p}} = -\varepsilon_{lmn}\frac{p_n x_m}{r^3} - \varepsilon_{lmn}Q_{nk}\frac{3x_k x_m - r^2\delta_{km}}{r^5} + \cdots,$$

where x_m are the Cartesian coordinates $x_1 = x, x_2 = y, x_3 = z$, and ε_{lmn} equals $+1$ (-1) for l, m, n an even (odd) permutation of 1, 2, 3 and zero otherwise. In the lowest order contribution p_n is the dipole moment

$$p_n = \int_G J_n^{\mathrm{p}}(\vec{r}')\,\mathrm{d}V'.$$

In the next order term Q_{nk} represents the traceless quadrupole tensor

$$Q_{nk} = \frac{1}{3}\int_G [3J_n^{\mathrm{p}}(\vec{r}')x_k' - \vec{J}^{\mathrm{p}}(\vec{r}')\vec{r}'\delta_{nk}]\,\mathrm{d}V'$$

with $Q_{kk} = 0$. The tensor Q_{nk} can be written as the sum of symmetric and antisymmetric tensors. The symmetric part Q_{nk}^s contains five independent components, whereas the antisymmetric part Q_{nk}^a has three independent elements.

Comparing both types of multipole expansions, the relations between the electric dipole coefficients of the spherical expansion and the coordinates p_n of the electric dipole vector $\vec{p}$ are given by $A_{11} = p_x$, $B_{11} = p_y$, and $A_{10} = p_z$. The electric quadrupole coefficients are related to the quadrupole tensor by the equations

$$A_{20} = -3(Q_{11}^s + Q_{22}^s),\ A_{21} = 2Q_{13}^s,\ B_{21} = 2Q_{23}^s,\ A_{22} = (Q_{11}^s - Q_{22}^s)/2,\ B_{22} = Q_{12}^s.$$

The relationship between the magnetic dipole coefficients and the quadrupole tensor is found to be $m_i = \varepsilon_{ijk}Q_{kj}^a/2$, where the coordinates of the magnetic dipole vector $\vec{m}$ are given by $m_x = \alpha_{11}, m_y = \beta_{11}$, and $m_z = \alpha_{10}$.

4. Discussion

The results can be applied for the inverse determination of the multipole moments from measurements of the fields. The calculation of the electric multipole moments from body surface potentials by surface integration is a well-known procedure [1]. The inverse solution for the magnetic multipole coefficients can be obtained by using the orthogonality of the spherical harmonics. Considering a finite homogeneous volume conductor, the magnetic moments are found to be

$$\alpha_{nm} + \mathrm{i}\beta_{nm} = \frac{2n+1}{n+1}\frac{g_{nm}}{\mu_0}\int_{S_s} r^n P_n^m(\cos\vartheta)\,\mathrm{e}^{\mathrm{i}m\varphi}[\vec{B}(\vec{r}) - \vec{B}^{\mathrm{v}}(\vec{r})]\,\mathrm{d}\vec{S},$$

where S_S is a spherical surface enclosing the volume conductor, and the center of the sphere is at the location of the multipole. The expression for $\vec{B}^v$ is given by

$$\vec{B}^v(\vec{r}) = \frac{\mu_0 \sigma}{4\pi} \int_{S_0} \Phi(\vec{r}') \frac{(\vec{r} - \vec{r}')}{|\vec{r} - \vec{r}'|^3} \times d\vec{S}'.$$

In this equation, σ is the homogeneous conductivity, and S_0 denotes the external surface of the volume conductor. The expression for the magnetic moments includes the solution for the magnetic heart vector, which can be shown to be

$$\vec{m} = \frac{3}{2\mu_0} \int_{S_S} \{\vec{r}[\vec{B}(\vec{r}) - \vec{B}^v(\vec{r})]\} d\vec{S}.$$

Applying this result, the magnetic heart vector can be obtained and compared with the electric heart vector, which is given by [1]

$$\vec{p} = \sigma \int_{S_0} \Phi(\vec{r}) d\vec{S}.$$

To calculate the multipole moments for the case of inhomogeneous volume conductors, an equivalent source, which includes the effect of the internal inhomogeneities, can be defined [1]. Note that an alternative approach to determine the multipole moments from magnetic field measurements is the least-squares error method [1], which needs the solution of the forward problem.

Furthermore, the above results can be applied for the characterization of biomagnetic sources. First, a line source positioned on the x axis is considered. The end points are at $\pm a/2$, and the current strength is I. This model can be used to describe nerve activity. The multipole moments up to the order $n = 2$ are given by the electric dipole coefficient $A_{11} = aI$ and by the electric octupole moments $A_{31} = -A_{11}a^2/16$ and $A_{33} = A_{11}a^2/96$. Note that all magnetic moments vanish. The source extent and the current strength can be determined from the electric dipole and octupole moments. Moreover, a sheet source in the form of a planar rectangular layer is considered. The sheet lies in the $x = 0$ plane and has a continuous distribution of perpendicular current dipoles pointing in the x direction. The center of the layer is at the coordinate origin, the constant dipole density is q, and the dimensions of the source in y and z directions are b and c, respectively. Note that the sheet source can be used to model cardiac and brain activity. The multipole moments up to the order $n = 2$ are given by the electric moments $A_{11} = qbc$, $A_{31} = A_{11}(4c^2 - b^2)/48$, and $A_{33} = -A_{11}b^2/96$, and by the magnetic quadrupole coefficient $\beta_{21} = A_{11}(c^2 - b^2)/36$. It is found that $A_{31} + 6A_{33} - 3\beta_{21} = 0$. The source parameters can be determined from the electric dipole and octupole moments. These examples demonstrate that for simple configurations the source parameters can be obtained from the first few multipole moments.

References

[1]   R. M. Gulrajani, P. Savard, and F. A. Roberge, The Inverse Problem in Electrocardiography: Solutions in Terms of Equivalent Sources, *CRC Crit. Rev. Biomed. Eng.* **16** (1988) 171-214.

[2]   D. B. Geselowitz, On the Magnetic Field Generated Outside an Inhomogeneous Volume Conductor by Internal Current Sources, *IEEE Trans. Magn.* MAG-6 (1970) 346-347.

[3]   F. Grynszpan and D. B. Geselowitz, Model Studies of the Magnetocardiogram, *Biophys. J.* **13** (1973) 911-925.

[4]   W. Haberkorn, M. Burghoff, and L. Trahms, Multipolquellenanalyse biomagnetischer Felder, *Biomed. Eng.* **37** (suppl. 2) (1992) 156-157.

Non-Linear Electromagnetic Systems
V. Kose and J. Sievert (Eds.)
IOS Press, 1998

Considerations on an Electrical Model of the Human Cardiac Oscillator

Monica COSMA°, Viorel Mihail COSMA*, Gabriel S. POPESCU*,
Danut BURDIA*; Thierry BUCLIN°

° - *Division of Clinical Pharmacology, Centre Hospitalier Universitaire Vaudois,*
1011 Lausanne, Switzerland
* - *Faculty of Electrotechnics and Faculty of Electronics and Telecommunications,*
Technical University "Gh. Asachi ", 6600 Iasi, Romania

Abstract : In the normal human heart, each beat originates in the sino-atrial (SA) node, a specialized structure that comprises "pacemaker" cells. Impulses generated in the SA node pass through the atrio-ventricular (AV) node to the ventricles. But the AV node also possesses discharging cells, with an intrinsic rhythm of about 2/3 of that of the SA node [6,7,8]. By modifying a recent model developed by West et al. [3,4] using the nonlinear dynamics [5], we reproduced the activity of the SA and AV nodes under normal and some pathological conditions and we described the possibility of synchronised and nonsynchronised interactions.

1. Introduction

According to the classic view, the SA node, having a greater firing rate, discharges the excitable membranes of the AV node cells, which function only as a passive conduit [6,7,8]. In contrast to the above mentioned model, a new concept [5] was introduced by West et al. [3,4], suggesting that the AV node may function as an active oscillator. It also predicts various arhythmias and conduction defects when certain parameters are modified, which are not unlike clinical patterns observed in practice.

2. Method

In order to simulate the interaction between the SA and the AV node, we initially considered the electrical model proposed by J. P. Gollub [1,2]. Our model differs from the original in that these 2 nonlinear resistors have different characteristics. The 2 relaxation oscillators are linearly coupled by means of 2 linear resistors. The circuit is as follows :

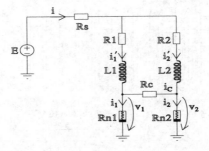

Fig. 1 : The electrical model of the circuit modelling the human cardiac oscillator.

The values of its components are: E = 5V or 10V; R_s, R_c = variable; R_1 = 1 Ohm; R_2 = 1 Ohm; L_1 = 2 H; L_2 = 2 H; R_{n1} and R_{n2} are the nonlinear resistors, their characteristics I(U) are presented in the following figure :

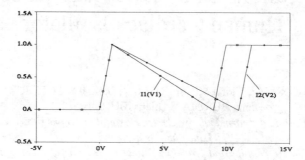

Fig. 2 : The characteristics of nonlinear resistors used in the model :
intensity as a function of applied tension.

These values were chosen to provide a normal cardiac frequency (60 - 90 beats/min). The application of Kirchhoff's laws gave the following equations :

$$E = L_1\ di_1'/dt + (R_1 + R_s)\ i_1' + R_s\ i_2' + v_1$$

$$E = L_2\ di_2'/dt + (R_2 + R_s)\ i_2' + R_s\ i_1' + v_2$$

$$v_1 = R_c\ i_c + v_2$$

(i_1', i_2'=intensity of the currents corresponding to the 2 relaxation oscillators; i_1, i_2=intensity of the currents corresponding to the 2 nonlinear resistors; v_1, v_2=tensions corresponding to the 2 nonlinear resistors)

The proposed electric model was simulated by with SPICE, a specialized electronic simulation software.

3. Results

We first attempted to reproduce the normal functioning of a healthy heart, where the SA node drives the AV node at the same frequency. By varying the values of the coupling resistances R_s and R_c, we were able to obtain a synchronised functioning of the two cardiac pacemakers (fig. 3). By modifying the values of E, the SA node could be blocked (v_1 = cst.), and the AV node became active, with a discharging rate of two-thirds of the SA rate, reproducing a phenomenon encountered in certain pathological situations.

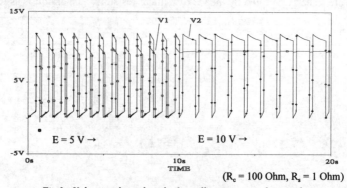

(Rc = 100 Ohm, Rs = 1 Ohm)

Fig 3 : Voltage pulses when the 2 oscillators are synchronised .

Another setting for R_s and Rc produced a nonsynchronised functioning, the SA node again becoming silent at high E value.

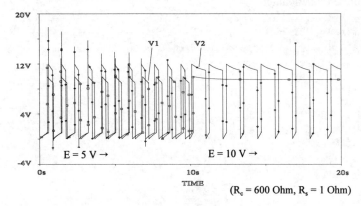

Fig 4 : Voltage pulses when the 2 oscillators are not synchronised.

The relationship between i_1 and i_2, corresponding to the synchronised and to the nonsynchronised functioning of the 2 oscillators (the SA and the AV node) is depicted in Figs. 5 and 6.

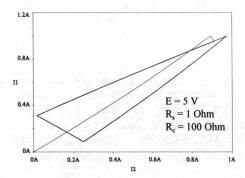

Fig 5 : The relationship of i_1 and i_2 when the 2 oscillators are synchronised.

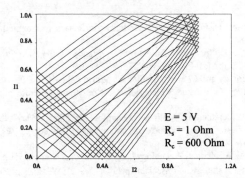

Fig 6 : The relationship of i_1 and i_2 when the 2 nodes oscillate with different frequencies (nonsynchronised functioning).

In the latter situation, the chaotic behaviour of the 2 cardiac oscillators is obvious.

Finally, we explored the variation of the oscillation periods of the SA and the AV nodes, $T(v_1)$, respectively $T(v_2)$, in the case of the synchronised and of the nonsynchronised functioning:

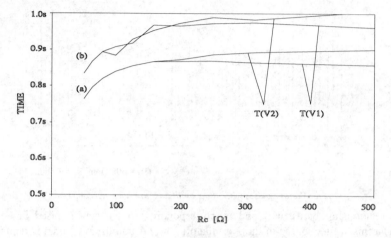

Fig. 7. Period length of the two oscillators as a function of coupling (R_C).
The (a) curve corresponds to R_S = 0 Ohm, the (b) curve for R_S = 0.5 Ohm.

4. Conclusions

By reproducing the activity of the SA and the AV nodes with the above-mentioned electronic model, we describe the possibility of synchronised and nonsynchronised interactions. The AV node plays an active role both in the generation of a normal cardiac rhythm and in the appearance of SA-AV dissociations. Particular conditions can render the SA node silent, while the AV node generates a slower rhythm. This simple model is thus able to reproduce some patterns of physiological and pathological heart rhythms.

References

1. P. Gollub, T. O. Brunner; B. G. Danly - "Periodicity and Chaos in Coupled Nonlinear Oscillators". Science : vol. 200, 1978.
2. P. Gollub, E. I. Romer, J. E. Socolar - "Trajectory Divergence for Coupled Oscillators : Measurements and Models". Journal of Statistical Physics, vol. 23, n° 3, 1980.
3. J. West - "Fractal Physiology and Chaos in Medicine", World Scientific, p. 93-105, 1980.
4. J. West, A. L. Goldberger, G. Rovner, V. Bhargava - "Nonlinear Dynamics of the Heartbeat". Physica 17D, 1985.
5. D. Kaplan, L. Glass - "Understanding Nonlinear Dynamics". Springer-Verlag, New York Inc, 1995.
6. E. Braunwald - "Heart Disease - A Textbook of Cardiovascular Medicine". W. B. Saunders Company, 4th Ed., p. 588-628, 667-726, 1992.
7. A. C Guyton - "Textbook of Medical Physiology", W. B. Saunders Company, 6th Ed., p. 165-176, 1981.
8. A. C Guyton - "Human Physiology and Mechanisms of Disease", W. B. Saunders Company, 4th Ed., p. 90-98, 1987.

Non-Linear Electromagnetic Systems
V. Kose and J. Sievert (Eds.)
IOS Press, 1998

Magnetocardiographic Spatio-Temporal Analysis of Abnormal Ventricular Activation

K. Czerski [1,3], M. Oeff [1], R. Agrawal [1], M. Burghoff [2], U. Steinhoff [2]

[1] *Dept. of Cardiology, Klinikum Benjamin Franklin, Free University of Berlin, Germany*
[2] *Physikalisch-Technische Bundesanstalt Berlin, Germany*
[3] *Institut für Atomare und Analytische Physik, Technical University of Berlin, Germany*

Abstract. In the frame of our new spatio-temporal turbulence analysis, the beat-to-beat variability of the heart activity has been investigated by means of magnetocardiographic mapping in 10 patients with ventricular tachycardia (VT) and with coronary heart disease, and compared to 8 normal persons. In all VT-patients at the last phase of ventricular depolarization (RS-interval) we have found a strong enhancement of the parameter Circulation, which is sensitive to local disturbance of the electric signal propagation. In addition, a significantly larger fluctuation of the angle between the current dipole vector and its velocity has been observed throughout the QRS-complex in VT-patients than in healthy normals.

1. Introduction

The cardiovascular system has been shown to behave like a low dimensional chaotic system [1,2]. The nonlinearity of its dynamics orginates from the behaviour of single cardiac cells as well as from the complex interaction with external events (for instance breathing process or blood pressure) controlled by the autonomic nervous system. The mentioned leads not only to fluctuations in the heart rate but also to significant spatial beat-to-beat differences in the electric heart activity. The latter can provide an important information for diagnosis of different heart diseases and their courses.

An ideal tool to study such effects is magnetocardiographic mapping (MCG) which enables, mainly due to recent improvement of the signal to noise ratio, a single beat analysis of the magnetic heart activity. Consequently, by solving the inverse problem the current source can be localized and an investigation of the spatio-temporal beat-to-beat variability can be carried out. In order to test our new approach we have investigated patients with ventricular tachycardia (VT) and coronary heart disease comparing to a collective of healthy persons. The VT-patients are known for their disturbed conduction of electric signals (slow conduction areas) manifesting electrically by fragmentation of the QRS-complex or by "late potentials". Thus, this patient group should be a very good test for our new analysis.

2. Magnetocardiographic Measurements

The magnetocardiographic mapping was carried out in 10 VT-patients and in 8 healthy persons, applying a 49-channel 1st order gradiometer in a magnetically shielded hospital environment [3]. The white noise level was better than 2.5 fT Hz$^{-1/2}$. The sensor measured the z-component of the magnetic field in a plane parallel to the anterior chest wall of the subject, at a distance of about 2 cm apart. Simultaneously to magnetic signals, the 3 lead vector ECG was also recorded. Acquisation time amounted to 100s at a sampling rate of 1000 Hz. Hardware band pass filtering from 0.16-250 Hz was applied for both magnetic and electric channels.

3. Spatio-Temporal Turbulence-Analysis of MCG

To study spatio-temporal beat-to-beat variability, we applied the equivalent current dipole model in a half plane volume conductor to parameterize an actual location and strength of the current source:

$$\vec{B} = \frac{\mu_0}{4\pi} \cdot \frac{\vec{J} \cdot \vec{R}}{\left|\vec{R}\right|^3} \tag{1}$$

where $\vec{B}$, $\vec{J}$ and $\vec{R}$ are the magnetic flux density, equivalent current dipole and location vector, respectively. The vectors $\vec{J}$ and $\vec{R}$ are calculated for every time point of a measurement. To compare different heart beats in time domain to each other, we have applied R-peak triggering. Thus, it is possible to calculate the mean values of $\vec{J}$ and $\vec{R}$ vectors and simultaneously to investigate the spatial distribution of $\vec{J}$ vectors for each time point of the heart cycle separately. By means of scalar and vector products the current dipole vectors can be decomposed, relative to the mean location vector $\vec{R}_0(t)$, in two orthogonal components: tangential and radial. These two vector components form new variables: Source S and Circulation $\vec{C}$, defined as follows

$$S(t) = \frac{1}{N\sigma(\vec{R}_0(t))} \sum_{i=1}^{N} \vec{J}_i \cdot \vec{R}_i' \,, \qquad\qquad \vec{R}_i' = \vec{R}_i - \vec{R}_0$$

$$\vec{C}(t) = \frac{1}{N\sigma(\vec{R}_0(t))} \sum_{i=1}^{N} \vec{J}_i \times \vec{R}_i' \tag{2}$$

where N, $\sigma(\vec{R}_0(t))$ are the number of heart beats (typically N=100) and the standard deviation of the location vector, respectively.. The parameters Source and Circulation reflect the spatio-temporal beat-to-beat variability and describe the interdependence between the current dipole vectors and their locations.

To complete the spatio-temporal description of the electric signal propagation we have defined an additional parameter: the mean angle between the current dipole vector and its velocity $\dot{\vec{R}}$

$$\alpha(t) = \frac{1}{N} \sum_{i=1}^{N} \arccos\left(\frac{\dot{\vec{R}}_i(t) \cdot \vec{J}_i(t)}{\left|\dot{\vec{R}}_i(t)\right| \cdot \left|\vec{J}_i(t)\right|} \right) \tag{3}$$

4. Results

For all VT-patients a strong enhancement of Circulation (see Fig. 1) in the last phase of the QRS-complex (RS-interval) is characteristic. Additionally, a prominent maximum of Circulation can be observed for both groups at the R-peak.

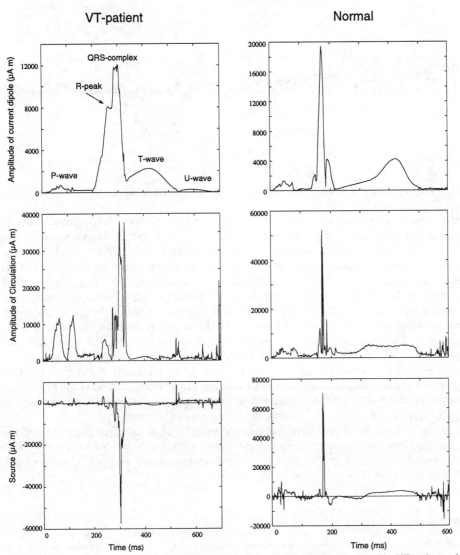

Fig.1 Time dependence of the mean current dipole, Circulation and Source for an exemplary VT-patient and a normal person

A large variation of the angle parameter (between 0 and 180 degrees) has been found for VT-patients as well as for the control group (Fig.2). However, there is a difference in the number of extrema throughout the QRS-complex between both groups, as follows

N=18±3 for VT-patients
N=12±3 for healthy persons

5. Discussion

The method proposed is based on solving the inverse problem by means of the equivalent current dipole model. It ensures that calculated parameters are independent of the sensor

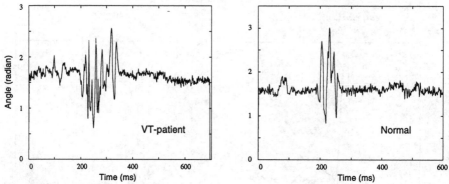

Fig.2 Time dependence of the mean angle between the current dipole and its velocity.

and measurement geometry, but on the other hand provides only global, integral information about the instantaneous electric heart activity. In spite of this we argue that the analysis of the beat-to-beat variability of the current dipole can provide a more detailed insight not only into electric signal propagation but also into the structure of its medium i.e. the myocardium. The instantaneous Circulation parameter corresponds to a spatial shift of current dipole vectors, correlated to the change of their amplitude. Therefore, an increase of Circulation could be due to slow conduction areas which contribute to the spatial dispersion of the electric signal conduction. The physiological meaning of the Source component is not yet clear. The prominent structures in the time dependence of Source and Circulation parameters, which are different for VT-patients and healthy persons, support our hypothesis. Moreover, applying the same method we have recently found for Long-QT-patients a characteristic enhancement of Circulation during repolarization of the myocardial tissue [4]. We can therefore conclude that our new approach provides a useful tool for the noninvasive diagnosis. Further investigations are however necessary.

In the present work we have additionally established a significant increase of the variation of the angle parameter throughout the QRS-complex for VT-patients, compared to the controls. This finding is analogical to signal fragmentation observed in the averaged, filtered MCG-signals.

References

[1] D.R. Chialvo and J. Jalive, Nonlinear dynamics of cardiac excitation and impulse propagation, *Nature* **330** (1987) 749-753

[2] R. Femat, J. Alvarez-Ramirez and M. Zarazua, Chaotic behavior from a human biological signal, *Phys. Lett.* **A 214** (1996) 175-179

[3] D. Drung, The PTB 83-SQUID System for Biomagnetic Applications in a Clinic, *IEEE Trans Appl Supercond* **5** (1995) 1-6.

[4] L. Schmitz, K. Czerski, K. Brockmeier, R. Agrawal, U. Steinhoff, L. Trahms and M. Oeff, Magnetocardiographic Turbulence Analysis in Patients with the Long QT Syndrom, accepted for publication in *Journal of Electrocardiology*

Non-Linear Electromagnetic Systems
V. Kose and J. Sievert (Eds.)
IOS Press, 1998

Modeling Spatially Extended Brain Currents by Multipole Expansion

G. Nolte, G. Curio.

Neurophysics Group, Dept. of Neurology, University Klinikum Benjamin Franklin; Berlin

Abstract. Spatially extended current sources may adequately be described by a multipole expansion significantly reducing the number of degrees of freedom compared to the usual dipole field approach. After implementing the respective inverse calculation we simulated the reconstruction of the spatial extent for a one-dimensionally distributed current source using parameters of a realistic MEG-measurement. We found that (a) the presently available signal-to-noise ratio allows for a satisfactory reconstruction of the width of the current distribution and (b) the approximation of a realistic head by a spherical volume conductor is too poor to give reliable results.

1. Introduction

Low-noise low-T_C SQUID measurements of magnetic fields induced by brain currents [1] allow for a reconstruction of the current source which goes far beyond the equivalent current dipole approximation. While the latter merely indicates 'center' and magnitude of activity the multipole expansion is adequate to model further source parameters like spatial extent. Truncating a multipole series results in a small number of degrees of freedom and is justified if the extent of the source is reasonably small (in the order of a few centimeters). This is in contrast to the dipole field approach which needs (biasing) regularization terms to have a unique solution of the inverse calculation.

Here we simulate the estimation of the spatial extent of a one-dimensionally extended current source under conditions as realistic as possible. First, we choose a spherical volume conductor for the forward and inverse calculation, add noise to the forward calculated field and analyze the accuracy of the estimation for realistic data. Second, by choosing a realistic volume conductor for the forward and a fitted sphere for the inverse calculation we quantify the misinterpretation due to the idealized volume conductor.

2. Model

We modeled the generator of the brain's "N20" response to electric nerve stimulation by a current $\bar{J}(\xi)$ distributed along a straight line which is parametrized by a coordinate ξ [1]. While two-dimensional approximations based on biasing regularisation techniques are possible [2], additional calculations have shown that a one-dimensional approximation is valid to first order if lateral source extent is estimated using multipole expansion. We assumed the straight line to be tangential at $\xi = 0$ in the coordinates of a spherical volume conductor. $\vec{J}$ is assumed to have constant direction orthogonal to the straight line and to be tangential at $\xi = 0$. For this effectively one-dimensional model there is only one multipole coefficient a_n for each order n

$$\vec{J}_0 a_n = \int_{-\infty}^{+\infty} \vec{J}(\xi) \xi^n d\xi \qquad (1)$$

$\vec{J}_0$ is a unit vector in direction of $\vec{J}$. The magnetic field can be written in closed form to all orders of the multipole expansion as

$$\vec{B}(\vec{r}) = \sum_{n=0}^{\infty} \frac{a_n}{n!} \frac{d^n}{d\xi^n} \vec{B}_{dip}(\vec{r}, \vec{r}_0(\xi)) \bigg|_{\xi=0} \qquad (2)$$

where $\vec{B}_{dip}(\vec{r}, \vec{r}_0(\xi))$ is the magnetic field at point $\vec{r}$ due to a dipole placed at $\vec{r}_0$ with unit magnitude and direction $\vec{J}$. Hence, the forward calculation is reduced to the calculation of the dipole field with a well known solution in case of a spherical volume conductor [1].

Assuming that $\vec{J}(\xi)$ does not change sign it may be regarded as a probability density. The quadrupole coefficient is set to zero, implicitly defining the expansion point $\vec{r}_0$ to be the 'center of mass' of the current distribution. The spatial extent σ may now reasonably be defined as the standard deviation of ξ resulting in $\sigma = \sqrt{a_2/a_0}$. Truncating the series after the octapole term (n=2) leaves 6 degrees of freedom (Fig. 1) describing position of center, magnitude of dipole and octapole, (a_0 and a_2) and orientation angle ω in the plane which is tangential at $\vec{r}_0(\xi = 0)$.

3. Simulation

For all simulations we assumed a first order gradiometer (7 cm baseline) consisting of 49 planar channels measuring the z component of the magnetic field.

Source Model

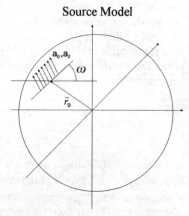

Fig.1 An N20-generator is modeled by a one-dimensionally distributed current source described by its center $\vec{r}_0$, its dipole and octapole components a_0 and a_2, and its orientation ω.

Though for the inverse calculation we will not try to reconstruct the details of the source distribution this has to be specified for the simulated forward calculations: here we will always use a distribution which is constant along a line of length $l=2$ cm with a total dipole moment of 20 nAm corresponding to the strength of a typical N20-generator.

All reconstruction results are given in terms of l (or sometimes l^2) and not in terms of the standard deviation σ, which is related to l by $l = \sqrt{12}\ \sigma$. We want to emphasize, that for general (one-dimensional) source distributions only σ has a definite meaning, which we merely scale to make the figures easier to read.

For the noise analysis we assumed a spherical volume conductor with its center located 10.5 cm below the measuring plane. The center of the current distribution was placed at $x=y=0$ and $z=5$ cm in the coordinate system of the sphere. The resulting magnetic field was then of the order of ± 100 fT at its minimum/maximum. To this field we added spatially uncorrelated noise with standard deviation in the range from 0 fT to 2.5 fT. After band pass filtering (20 Hz - 400 Hz) and averaging $N \approx 10.000$ measurements the standard deviation of the real data will be of the order 1 fT.

In Fig.2 we show the mean and standard deviation of the reconstructed spatial extent l as a function of the noise level. Since there is almost no systematic error detectable (a significant deviation of the mean from the correct value of 2 cm) the multipole expansion is capable of bias free reconstruction of spatial extent. In practice one has to cope with an error of about 5 mm for one individual measurement.

To analyze the effect of volume conductor idealization we made a forward calculation (using *CURRY* software) for a 'realistic' head model, a one compartment volume conductor comprising brain and liquor found in an MRI-measurement of a real head. For the inverse calculation a sphere was fitted to the boundary of the realistic volume conductor resulting in a geometry very similar to the previous case.

Noise Analysis

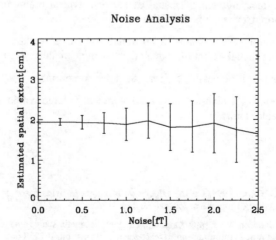

Fig.2 Estimated spatial extent as a function of noise level. For each noise value the mean and standard deviation was calculated for 100 simulated measurements.

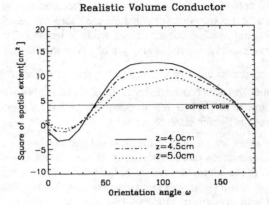

Fig.3 Estimated square of spatial extent as a function of orientation of the current distribution entering the forward calculation

For this realistic head we simulated the effects of various head shapes by rotating the direction of the current source for the forward calculation suppressing or enhancing various aspects of the deformed surface. In Fig.3 we show the reconstructed square of spatial extent (taking the root eventually results in an imaginary estimate for the extent) as a function of the orientation angle ω. Since the estimated l^2 oscillates with a large amplitude, any individual measurement may contain an error larger than the parameter which is to be reconstructed.

4. Conclusion

We simulated the measurement of spatial extent of a typical brain current by means of a multipole expansion. Our result is threefold:

- The multipole expansion of a current source is a method which is capable of measuring spatial extent of brain activity with negligible systematic error (2%).

- Spatial extent is detectable for realistic noise conditions.

- Approximating a realistic head by a spherical volume conductor is insufficient to give reliable results.

References

[1] Nolte, G., and G. Curio. 1997. On the calculation of magnetic fields based on multipole modeling of focal biological sources. Biophysical Journal 73: 1253-1262.

[2] Wang, J.-Z., Williamson, W. J., and Kaufman, L. 1992. Magnetic source images determined by a lead-field analysis: the unique minimum-norm least-squares estimation. IEEE Trans. Biomed. Eng. 39: 665-675.

Supported by DFG Ma 1782/1-1.

Non-Linear Electromagnetic Systems
V. Kose and J. Sievert (Eds.)
IOS Press, 1998

Clinical and Theoretical Research on Transcranial Magnetic Stimulation in Antidepressive Treatment

A. KRAWCZYK, T. ZYSS*, and S. WIAK‡

Institute of Electrical Engineering, Warsaw, Poland
*Institute of Psychiatry, Jagiellonian University, Cracow, Poland
‡ Technical University of Lódź, Lódź, Poland

Abstract. The paper shows the recent developments in the transcranial magnetic stimulation (TMS) that is applied in the treatment of deep depression of an endogenous nature. The therapy that uses an electromagnetic field is believed to substitute electroconvulsive shocks (ECS) used widely in psychiatry some years ago. The magnetic stimulation, which has to generate eddy currents in undercortical layers in order to excite neural cells, is based on a low frequency pulse magnetic field. To penetrate deep layers of brain by magnetic field one needs the exciting field to be of high value, up to 2 T. There is a real technical problem in reaching parameters of stimulation that are required. Some experiments, already made on rats and humans, are described in the paper. The results of the experiments are very promising for further research and applications. The mathematical modelling is also discussed in the paper.

1. Introduction

The application of transcranial magnetic stimulation (TMS) to medicine is relatively new and for neurological purpose was introduced in 1985 [1]. For therapeutical application to psychiatry, it has been developed in recent years, among others, by the authors [2]. This is the use of TMS in the treatment of deep depression of an endogenous nature. The basis of the treatment is given elsewhere [2,3], thus here only a short description is presented. The therapy is based on the hypothesis formulated by Zyss [4], that the medical effect of electroconvulsive shock (ECS) in treating depression is the same as the effect caused by magnetic stimulation and the latter is deprived of all negative aspects which are associated with the electroshock treatment (pain, convulsions and the like). This therapeutic effect is due to excitation of neural cells by an electric current (eddy currents) which is generated by an AC magnetic field. The method is being intensively investigated and all the results obtained look very promising [5].

2. Experimental Research

2.1 Medical generators of magnetic field

Until 1985, magnetic stimulators existed for physicotherapeutic treatment (orthopedic), at which the so called "time parameters" are appropriate: they are able to work for a long time (t_{total} < 99min.) with relative high frequency (up to 50-60 Hz). Their drawback is a low value of generated magnetic field (up to 0.01-0.015 T).

The time and strength parameters of stimulators, introduced after 1985 and used in neurological diagnostics, are just of opposite features. They are able to generate the magnetic field of 1-2 T. On the other hand, their time parameters are insufficient for psychiatric purpose. The stimulus repetition rates available from commonly used magnetic stimulators are low (0.25-1 Hz) with the maximum output. This frequency is unable to stimulate the neural cells in an optimal way.

TMS studies on animals and humans have been performed by means of two units:

a) the apparatus for magnetic therapy Magnetronic MF-10, accepted for use by Polish Ministry of Health and Social Care; the magnetic field of rectangular time shape produced by the stimulator possesses the following parameters: $B = 0.01$ T, $t_{rise} = 5$ ms, $f = 50$ Hz,

b) the prototype magnetic stimulator MS-2, constructed at the Department of Electronics of the Academy of Mining and Metallurgy in Cracow, generates the impulse magnetic field of the following parameters: $B = 0.1$ T, $t_{rise} = 200$ μs, $f = 50$ Hz.

2.2 Experimental investigation on animals

The experiments were aimed at comparing in rats some of behavioral and biochemical effects of multiple TMS and ECS that are regarded as relevant for the clinical effects of antidepressant treatments, and some others which are regarded as signs of adverse effects of ECS.

The experiments were carried out on male Wistar rats with standard conditions [6]. The treatment was applied once daily for 10 consecutive days. Each group, *i.e.* controlled, TMS and ECS, contained 10 animals. In the TMS examination a magnetic stimulator MS-2 with the parameters listed above, was used. The stimulation time reached 5 min. The initial and predictive calculations indicate that the conditions generated in the rat brain are equivalent to the conditions generated in the human head by the magnetic field of 1-2 T. ECS was induced using the ZK-2 apparatus. The current ($I = 150$ mA, $f = 50$ Hz) was applied for 0.5 s. The control animals were handled once daily.

For estimation of the stimulation effects several tests typical for neurological examination (the Porsolt forced swimming test, the measurement of exploration, basal locomotor activity, the tail flick test) have been used. Omitting the details, one can deduce from statistical analysis that the antidepressant effects of magnetic and electric stimulations are almost the same.

2.3 Clinical investigations on humans

To assess possible cerebral activation by prolonged rapid-rate TMS, we acquired EEG and measured serum prolactin level. The determined cortisol values may be treated as stress indicator during the stimulation session. It should be stressed here that all the experiments with humans aimed at showing that the method is safe and they did not have to confirm the antidepressant character of the method. The volunteer group[1] consisted of 10 healthy adults. The magnetic field from Magnetronic MF-10 (B= 0.01 T, $t_{rise} = 5$ ms, $f = 50$ Hz) was applied over the left temporal region for 15 min. For the safety, the second stimulation was carried out on the other part of brain with a larger stimulator and the time of stimulation decreased. Stimulation with the magnetic stimulator MS-2 ($B = 0.1$ T, $t_{rise} = 200$ μs, $f = 50$ Hz) was carried out over the right temporal region for 5 min. Subjects underwent successive stimulation with

---

[1]The experiment was approved by the local ethics committee. All subjects gave their consent. None was receiving medication.

LAYER	OUTER RADIUS (mm)	CONDUCTIVITY (S/m)
scalp	92	0.2
scull	85	.02
brain tissue	80	.4

Table 1: Parameters of model

both stimulators; the first was the MF-10 stimulator (of lower power), while the second stimulator was the MS-2 unit. The interval between each session was 30 min.

The detailed procedure of the investigation is described elsewhere [5]. As mentioned above, the results of examination are of threefold purpose: the data of EEG, serum prolactin level and cortisol level.

Open examination of all EEGs obtained before and after stimulation revealed no abnormal transients and neither focal nor generalized slowing. The numerical analysis detected a right shift of the frequency spectrum; it seems the dominant frequency was accelerated. This activation of 0.4 - 1.3 Hz was greater on the contralateral side of the stimulated hemisphere. These changes are, however, not significant from a statistical point of view.

The statistical analysis shows that there was no significant difference between the mean prolactin concentration before and after the stimulation. A similar conclusion can be stated in the case of the cortisol level. It is clear that the level of cortisol in the blood is not greatly influenced by TMS; it shows, however, the level of stress in the subject during TMS procedure.

3. Mathematical Modelling

Regarding all the material and structural problems occurring in mathematical modelling of the biomagnetic system, it seems to be impossible to make such a model at a realistic cost and time. For this reason, many simplifications are introduced just in order to find the first approximation.

The problem of eddy currents in the layers of human brain has been solved by FEM with $A - \varphi$ formulation and externally prescribed boundary conditions [6]. The model analysed here has been assembled as a three-layer sphere subjected to a magnetic field generated by a coil that was assumed to be a turn placed over the head (Fig.1). The coil is supplied with sinusoidal current of frequency 50 Hz[2]. Numerical results were obtained by means of the commercial package PC-OPERA 3D. Because of symmetry 1/8 of the whole sphere was under examination.

The material parameters that were used in calculation are given in Table 1. The numerical results are presented in Fig.1. The numerical results show how eddy currents are distributed among particular layers — it is the direct problem. There is also an indirect problem in finding the magnetic flux density which induces eddy currents of density that is required from therapeutical point of view. As the model is, for the time being linear the indirect problem may be solved easily by re-scaling. The first attempt led to the value of magnetic flux density about 2T which was somehow anticipated [5].

---

[2]In real treatment, we have AC supplying current, but it is of the short pulse form - the time of pulse is much shorter than 10 ms.

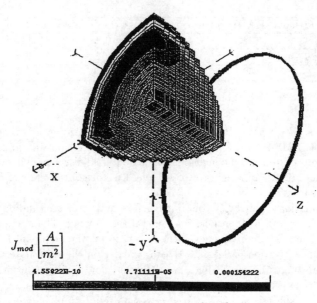

Figure 1: Eddy current distribution

4. Conclusions

The paper shows the novel medical application of electromagnetism. The usage of eddy currents in antidepressive treatment helps to solve some common problems in psychiatry. The experimental results of animal examinations suggest that TMS shows some features regarded as preclinical signs of antidepressive action. The investigations on humans show that the method gives no bad side effects. Thus, TMS may be considered as a possible alternative for the technique based on electric stimulation (ECS), but deprived of all negative aspects of the latter. The results of mathematical modelling confirm the intuition as to eddy current distribution. As these investigations are in their initial state, the results, both experimental and numerical, are also of an introductory nature. Nevertheless, they confirm the validity of the method in question as well as the need for further research.

References

[1] A.T. Barker *et al.*: Non-invasive magnetic stimulation of human motor cortex, *Lancet*, No.1, 1985, pp.1106-7 (1985)

[2] A. Krawczyk *et al.*: Magnetic Stimulation in Antidepressive Treatment — Theory and Experiment, *JSAEM Studies in Applied Electromagnetics*. No.4, 1996, pp.204-210

[3] A. Krawczyk *et al.*: Magnetotherapy due to eddy currents induced in neural cells by magnetic stimulation, *COMPEL*, vol.14, No.4, 1996, pp.283-286

[4] T. Zyss: Deep Magnetic Brain Stimulation - The End of Psychiatric Electroshock Therapy?, *Medical Hypotheses*, No.43, 1994, pp.68-74

[5] T. Zyss, A. Krawczyk: The magnetic brain stimulation in treatment of depression: the search for the perfect stimuli, Psychiatria Polska (Polish Psychiatry), vol. XXX, No.4, 1996, pp.611-628

[6] A. Krawczyk and J.A. Tegopoulos, Numerical Modelling of Eddy Currents, Clarendon Press, Oxford, 1993

Non-Linear Electromagnetic Systems
V. Kose and J. Sievert (Eds.)
IOS Press, 1998

Source Analysis of Somatosensory Evoked Response to Electrical Stimulation of the Fingers Using MEG and Functional MR Images

Keiji IRAMINA, Masato YUMOTO*, Kohki YOSHIKAWA**,
Hirotake KAMEI and Shoogo UENO
Department of Biomedical Engineering, Graduate School of Medicine,
* Department of Laboratory Medicine,
**Institute of Medical Science, University of Tokyo, Tokyo 113, Japan

Abstract. This study focuses on the comparison of characteristics of functional magnetic resonance imaging (fMRI) and magnetoencepalography (MEG). Activations of the primary somatosensory cortex were investigated using MEG and fMRI. A right thumb and a ring finger were stimulated by the electrical current pulse. Both MEG and fMRI identified expected anatomical regions of primary somatosensory cortex. When the thumb and the ring finger were stimulated simultaneously, it was possible in fMRI to discriminate the area of the thumb and the ring finger in the primary somatosensory cortex. In MEG, a single current dipole model might fit with a sufficiently large correlation coefficient because two sources are closely located within 10mm. It is difficult to discriminate two closely located dipoles, if no initial information is given.

1. Introduction

In recent years, several non-invasive techniques for imaging brain function, such as MEG (magnetoencephalography), fMRI(functional magnetic resonance imaging)[1] and differential NMR(nuclear magnetic resonance)[2], have advanced significantly. Since the successful measurement of fMRI[1], the relationships between MEG and fMRI have been studied[3]-[7] in order to clarify the origin of brain electrical activities. The fMRI provides high spatial resolution (less than a few mm), but it provides poor temporal resolution because this technique observes local changes in celebral blood flow. The MEG is able to measure electrical activities of populations of neurons and to provide good temporal resolution. In this study, we compared the characteristics of fMRI and MEG. We estimated the source of the somatosensory response to electrical stimuli of the fingers using MEG and evaluated the results by comparison with the localization provided by fMRI.

2. Methods

Somatosensory evoked magnetic fields were measured using a 37-channel SQUID magnetometer. The neuromagnetic field pattern was recorded over the parietotemporal cortex contralateral to the stimulation. A right thumb and the ring finger were stimulated with rectangular 0.2 msec constant current pulses at a strength of four times the sensory threshold. The stimulation rate was 3 Hz. MEG data were averaged 1000 times and filtered at 1.0-1200 Hz. To solve the inverse problem for localization of sources of somatosensory evoked magnetic responses, a single current dipole in a spherical volume conductor was used as a theoretical model. The best fitting dipole for the measured magnetic fields was calculated by means of an iterative least square method. In order to estimate the accuracy of dipole fitting, goodness-of-fit was calculated using correlation between the magnetic fields derived from a model dipole and measured fields.

The fMRI was performed with echo-planar imaging using a 1.5T whole body system (TE=50msec, TR=4000msec, flip angle 90deg). Seven 6.5mm contiguous slices (FOV = 240mm) in the region of the sensory cortex were obtained. Seven cycles of activity/rest periods (40 seconds per period) were used. fMRI activation images were produced using pixel-by-pixel t-test processing(threshold=99% probability). The stimulation rate of electric stimuli was 10Hz.

3. Result and Discussion

We measured somatosensory magnetic fields evoked by stimulation of the thumb. Two short-latency magnetic components and two middle-latency components were recorded. These components were referred to as N20m, P30m, N40m and P60m respectively. We estimated the source localizations of these components and superimposed them on the MRI scan as shown in Fig.1. Dipoles were obtained in the contralateral primary somatosensory cortex with the goodness of fit better than 99%. Although the exact positions of the four dipoles for each component differed, they were located within a 15 mm area. The fMRI were also obtained. The activated area was observed in the contralateral primary somatosensory cortex and was distributed along the central sulcus in the left hemisphere. The distances between the nearest point of distributed fMRI activation and the MEG location of each component were 5.3 mm, 4.1 mm, 5.5 mm and 6.0 mm respectively. These differences mainly appeared in z direction. The differences on the x-y plane were only one or two mm. Distributed fMRI area can be explained by the plots of the dipoles which are estimated in each component of the MEG. These results show that fMRI provide temporal information over relatively long periods of activation, while the MEG provides more accurate temporal information.

Fig. 2 shows the superposition of all 37 curves and topographies of the N20m, P30m, N40m and P55m. When the thumb and the ring finger were stimulated simultaneously, topographies of N20m, P30m and P55m are analogous to patterns of the thumb stimulation. It is reasonable that these topographies are explained by the single dipole model. Each value of goodness of fit is better than the case of stimulation of the thumb or the ring finger, and each value is 0.997, 0.996, 0.947, and 0.986 respectively. Fig. 3 shows the location of estimated dipoles of N20m and P30m in the stimulation of the thumb, the ring finger and both the

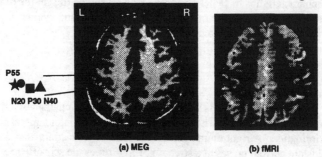

Fig. 1 Dipole estimation and fMRI of sensory evoked response.
(a) Estimated positions of current dipoles for N20m, P30m, N40m and P60m are superimposed on the MRI scan. (b)fMRI. Sources are located in the primary somatosensory cortex in the left hemisphere. The activated area is shown by black point.

thumb and the ring finger. The dipole in the ring finger stimulation exists 10 mm superior to the location of thumb's area along the central sulcus. Dipole locations of N20m and P30m in the stimulation of both fingers are estimated deeper than the location of the thumb's dipole. Because spread areas, including regions of the thumb and the ring finger in the somatosensory cortex are active at the same time, it is not adequate to fit the single dipole model. However, the value of goodness-of-fit indicates a much higher value than 0.99.

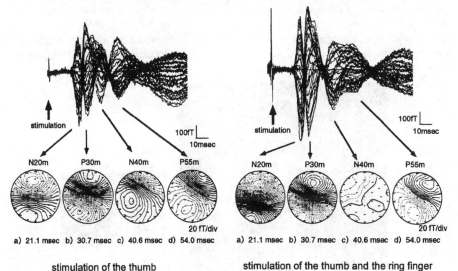

Fig. 2 The superposition of all 37 curves and topographies of magnetic field distributions for the N20m, P30m, N40m and P55m. Left: stimulation of the thumb Right: stimulation of the thumb and the ring finger

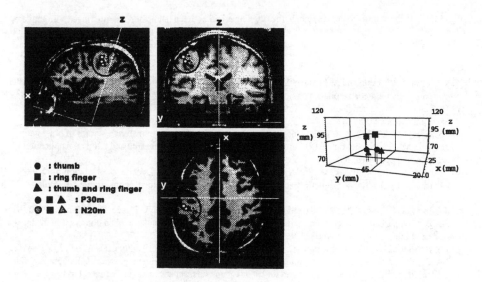

Fig.3 Dipole estimation of N20m and P30m components in the stimulation of the thumb, the ring finger and both the thumb and the ring finger.

Fig. 4 shows the fMRI in the stimulation of the thumb, the ring finger and both the thumb and the ring finger. When the thumb or the ring finger was stimulated, the activated areas of fMRI were observed in each expected anatomical region of the sensory cortex separately. These areas are in the same sulcus that is identified by MEG. In the stimulation of both fingers, activated areas are separated clearly to the thumb's and the ring finger's.

These results indicate that the fMRI are able to resolve somatosensory cortex with a high accuracy better than 10mm. For the dipole estimation in the MEG, if anatomical information about the initial condition is given, it would be possible to discriminate the closely located sources. However, a single current dipole model might fit with a sufficiently large correlation coefficient. It is difficult to discriminate two closely located dipoles, if no initial information is given.

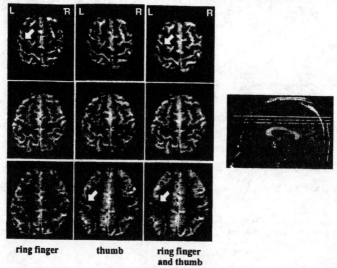

ring finger   thumb   ring finger
                                    and thumb

Fig. 4 fMRI in the stimulation of the thumb, the ring finger and both the thumb and the ring finger. The activated area as determined by t-test processing is shown by black point. Three 6.5 mm contiguous slices (FOV = 240 mm) are shown.

References
[1] S. Ogawa, D.W. Tank, and R. Menon, "Intrinsic signal changes accompanying stimulation: Functional brain mapping with magnetic resonance imaging" Proc. Natl. Acad. Sci. USA 89(1992) 5951
[2] H. Kamei, Y. Katayama, and H. Yokoyama, in Microcirculation, An Updated, ed. M. Tsuchiya et al. 1, Excerpta Medica (1989) 417
[3] J.S. George, J.A. Sanders, J. D. Lewine, A. Caprihan and C.J. Aine, "Comparative studies of brain activation with MEG and functional MRI" in Biomagnetism: Fundamental Research and Clinical Application C. Baumgartner et al. Eds. IOS Press (1995) 60-65
[4] R. Beisteiner, G. Gomiscek, M. Erdler, C. Teichtmeister, E. Moser, L. Deecke, "Correlation of results of localization by functional magnetic resonance tomography with magnetoencephalography" Radiologe. 35 (1995) 290-293
[5] T. Morioka, T. Yamamoto, A. Mizushima, S. Tombimatsu, H. Shigeto, K. Hasuo, S. Nishio, K. Fujii, M. Fukui, "Comparison of magnetoencephalography, functional MRI, and motor evoked potentials in the localization of the sensory-motor cortex" Neurological Research. 17 (1995) 361-367
[6] J.S. George, C.J. Aine, J.C. Mosher , D.M. Schmidt, D.M. Ranken, H.A. Schlitt, C.C. Wood , J.D. Lewine, J.A. Sanders, J.W. Belliveau, "Mapping function in the human brain with magnetoencephalography, anatomical magnetic resonance imaging, and functional magnetic resonance imaging", Journal of Clinical Neurophysiology. 12 (1995)406-431
[7] J.A. Sanders, J.D.Lewine, and W.W. Orrison,Jr., "Comparison of primary motor cortex localization using functional magnetic resonance imaging and magnetoencephalography", Human Brain Mapping 4 (1996) 47-57

Non-Linear Electromagnetic Systems
V. Kose and J. Sievert (Eds.)
IOS Press, 1998

Surface Charge Moving Process on a Bioimplant

Peter GRABNER
Department of High Voltage Engineering and Equipment
Technical University of Budapest
H-1521 Budapest, Hungary

Abstract. After frequent ophthalmologic operations involving the implantation of intraocular lenses (IOLs), mild or rarely serious cellular reactions were observed around the IOLs. These postoperative reactions called the author's attention to the electrostatic attitudes of intraocular lenses and other instruments.

In order to determine how a non-toxic implant made of highly insulating materials can carry harmful contamination to the eye on their surface, the electrostatic properties of the implant materials were examined. After previous studies a problem oriented finite-difference numerical model was developed in order to examine the influence of changing dielectric properties of the IOL's environment and the transient effect of the accidental initial charge distribution.

1. Introduction

The development of medical techniques and wide application of new materials for bioimplantation were induced by the industrial application of the results of organic chemistry. Nowadays these new materials showing good chemical stability and good biocompatibility are used in routine microsurgery operations.

Fig. 1. The usual application of the
investigated intraocular lens
(IOL)

The intraocular lens, or IOL, is a permanent plastic lens implanted inside the eye to replace the crystalline lens (Fig. 1). Previous studies have examined the effects of manufacturing, packing and implantation. Based on laboratory measurements of the electric field, the surface charge density, and field experiments, significant charging of the surface of the IOLs has been detected just before the implantation [1], [2], [3].

The usual materials of IOLs are Polymethylmetacrilate (PMMA), Polypropylene and Silicone but the most popular is the PMMA lens. These materials (different types of PMMA) are industrially the most important materials of the ester polymers of acrylic acid. The material is characterised by a dielectric constant of 3.5 (3.5 – 4.5). In basic state the charge carriers of both positive and negative signs are distributed in the materials uniformly, so materials seem to be electrically neutral. Transitional and definite disintegration of this neutral state results in an electrostatically charged state of materials.

During the manufacturing processes all types of the IOLs (Fig. 2.) become highly charged due to the manufacturing technology. At the same time most of the package materials are a very good insulators. All of the packing materials (except the paper box) became highly charged while opening regardless of the type and origin.

2. Governing equations

The different types of IOLs have relatively complicated geometric form. The basic condition was that the process could be described in a special co-ordinate system [7], [5]. The authors found that the shape of a typical IOL fits on a spheroid that is defined in the oblate spheroid co-ordinate system. The handles of the lens were therefore eliminated and we examined a cylindrical symmetrical problem. The conclusions of the other computations were that this simplification does not definitely influence the results.

In order to compute the charge moving phenomena on the IOL's surface we had to use the well-known postulate on the behaviour of charges (1).

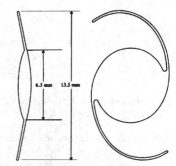

Fig. 2. Typical intraocular lens (IOL) used in routine implantation.

$$div\mathbf{J} + \frac{\partial \rho(\mathbf{r}, t)}{\partial t} = 0 \qquad (1)$$

$\rho(\mathbf{r},t)$ represents the charge density and $\mathbf{J}$ represents the current density. The current density is defined by two orthogonal vector (tangential and normal) components [1].
After using (2), (3) equations and divergence theorem formula can be written in the mentioned co-ordinate system. As a trial solution (4) was substituted into the previous equation. After regulation of equation the result may be written as (5.1), (5.2):

$$\mathbf{J} = -\langle\gamma\rangle \cdot grad\ \varphi , \qquad (2)$$

$$\rho = -\langle\varepsilon\rangle \cdot div(grad\ \varphi), \qquad (3)$$

where $\langle\gamma\rangle$, $\langle\varepsilon\rangle$ represent the different permittivities and conductivities on the surface of the lens and in the nearby space.

$$\varphi(\eta,\vartheta) = \frac{1}{\sqrt{ch\eta \cdot sin\vartheta}} \Phi(\eta,\vartheta) \qquad (4)$$

The surface conductivity of insulating materials and the present discussion IOLs are usually strongly influenced by the local physical and electrical environment of the materials but in this case γ_ϑ conductivity is constant across the surface and the space conductivity is constant in the full region. It means that in this region γ_η is equivalent with γ_ϑ. The operation with ε_ϑ and ε_η permittivities was similar to the previous process. The structure of the time dependent term is equivalent with (5.2), in this case conductivities were substituted by permittivities.

$$\gamma_\eta \frac{\partial^2 \Phi}{\partial \eta^2} + \gamma_\vartheta \frac{\partial^2 \Phi}{\partial \vartheta^2} + \Phi \cdot S(\eta,\vartheta,\gamma_\eta,\gamma_\vartheta) - \frac{\partial}{\partial t}\left(\varepsilon_\eta \frac{\partial^2 \Phi}{\partial \eta^2} + \varepsilon_\vartheta \frac{\partial^2 \Phi}{\partial \vartheta^2} + \Phi \cdot S(\eta,\vartheta)\right) = 0 \qquad (5.1)$$

$$S(\eta,\vartheta,\gamma_\eta,\gamma_\vartheta) = \frac{1}{4}\left(\gamma_\eta th\eta + \gamma_\vartheta ctg\,\vartheta\right) + \frac{\gamma_\eta}{ch^2\eta} \qquad (5.2)$$

For the investigation Neumann and Dirichlet ($\Phi \rightarrow 0|\ \eta = \eta_{max}$; $\Phi = \Phi_0$) boundary conditions were used. Homogenous Neumann conditions were defined on ϑ_0 and ϑ_{imax} edges, which mean $\mathbf{J}_t = 0$. These boundary conditions were defined for the basic equation (5.1).

An efficient numerical solution method was used for time-dependent simulation. After previous studies the approximate solution was created by the *finite-difference* method [4]. In replacing the equation (5.1) by a set of finite-difference equations that connect values of the potential function at discrete points, any planar distributions of the grid points can be used. In this problem a regular grid was used which is defined by η_j and ϑ_i parameters.

The computation was worked with the well-known Crank - Nicolson scheme (other schemes can also be used) within the time range. During the simulation of the above equations the initial condition are satisfied (6):

$$\{\dot{\Phi}\} = \{0\}, \qquad \sigma_i \rightarrow \{G\} \rightarrow \{\Phi\}_{t=0} = \{\Phi_0\}. \tag{6}$$

3. Numerical results

In the present study three different types of effects were investigated. The first simulation represents the changing of surface potential if the boundary condition contains homogenous surface charge density on the edge. This time variation is illustrated in Fig. 3. The potential decays strictly monotonically with arc in each time steps. The shape of the function represents the discharge process of the surface.

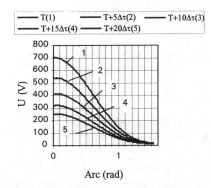

Fig. 3. Potential distribution on the surface in different time steps
$(\gamma_\eta = 1e^{-12} \, \text{S}; \gamma_\vartheta = 0,5e^{-12} \, \text{S}; \varepsilon_\eta = 1;$
$\varepsilon_\vartheta = 3,5; E_{i=imax, j=0} = 5,5e^{+4} \, \text{V/m})$

The second simulation represents the effect of the IOL's shape, the time variation of the potential in single points. The boundary condition on the surface of the lens was also homogenous surface charge density. Three different points were followed up ($\vartheta = 0$ and $\vartheta = \pi/2$ are singular points considering the trial function, $\Delta\vartheta = 0,6°$). This variation is illustrated in Fig. 4. Remarkable experience that the velocity of the process is relatively high in the selected circumstances. The third simulation presents the effect of changing surface conductivity (Fig. 5). The results prove that after increasing surface conductivity, the velocity of the potential time, variation was also increased in $\vartheta = \pi/4$ point.

4. Conclusions

The natural and general solution to prevent bacterial contamination due to electrostatic forces is the dissipation of static charges. The best practical method would be adding antistatic additives to the materials of plastic bioimplants or using hydrophilic coating materials [6]. Another way of eliminating electrostatic charging is by using eliminators. In order to design an efficient eliminator, the movement of accidental charges on the surface of a typical single IOL must be cleared. The simulation method described in the paper is suitable to detect the transient effect of the initial charge distribution, too. Basic equations for the problem orient method have been worked out and the first numerical results were presented. This simulation provides adequate possibilities for further accurate investigation, different surface charge densities and dielectric properties and this method is an effective tool to examine the movement of the charges.

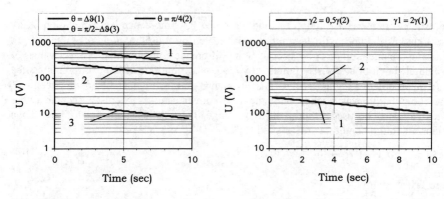

Fig. 4. Time variation of potential
function in three characteristic
points
$(\gamma_\eta = 1e^{-12} S; \gamma_9 = 1e^{-12} S;$
$\varepsilon_r = 1; \varepsilon_9 = 3,5;$
$E_{i=imax, j=0} = 5,5e^{+4} V/m)$

Fig. 5. The effect of changing conductivity (γ_9)
$(\gamma_\eta = 1e^{-12} S; \gamma_{92} = 0,5e^{-12} S; \gamma_{91} = 2e^{-12} S;$
$\varepsilon_r = 1; \varepsilon_9 = 3,5)$

5. Acknowledgement

The author wishes to thank A. Iványi, I. Berta, I. Kiss and Gy. Czvikovszky, for their co-operation and many valuable discussions.

References

[1]   Berta I, Czvikovszky Gy, Grabner P, Electrostatic Charging of Plastic Bioimplants, 3rd Japan-Hungary Seminar on Applied Electromagnetics in Material and Computational Technology, Budapest TUB 1994.

[2]   Berta I, Static Control: Modelling and Application, Journal of Electrostatics, 30 (1993), pp365-380

[3]   Czvikovszky G, Vogt G, Berta I, Electrostatic Behaviour of PMMA Implants: a New Approach to "Toxic Lens Syndrome", Congress Ophthalmology, Brussels, 1992.

[4]   Gisbert S, Galina T, Numerical methods, ELTE - TypoTEX, Budapest (1993), pp31-107.

[5]   Grabner P, Surface phenomena investigation due to a frequent ophthatmologic bioimplant, 4th Japan-Hungary Seminar on Applied Electromagnetics in Material and Computational Technology, Fukuyama University, Japan 1996.

[6]   Hogt AH, Dankert J, Feijen J. Adhesion of coagulase-negative staphylococci with different surface characteristics onto a hydrophobic biomaterial. Antoine van Leeuwenhoek 1985: 51, pp510-512.

[7]   Moon P, Spencer D.E, Field theory for engineers, D.Van Nostrand Company, Inc. (1961), pp267-290.

Non-Linear Electromagnetic Systems
V. Kose and J. Sievert (Eds.)
IOS Press, 1998

Enzymatic Processes of Oxidation-Reduction Systems in Magnetic Fields

Masakazu IWASAKA and Shoogo UENO
*Department of Biomedical Engineering, Graduate School of Medicine, University of Tokyo,
Tokyo 113, Japan*

Abstract. This study focuses on whether enzymatic activities can be affected by magnetic fields. We examined the effect of magnetic fields of up to 14T on SOD, peroxidase, xanthine oxidase, and catalase. We obtained clear negative results in SOD, peroxidase, and xanthine oxidase under magnetic fields. In the case of catalase, it was observed that the initial rate of the reaction which was exposed to magnetic fields of up to 8T was 50 – 85 % lower than the control data. However, this magnetic field-effect was not observed when the reaction mixture was bubbled with nitrogen gas to remove the dissolved oxygen molecules produced in the solution. We also carried out the experiments by changing the gas-pressure around the optical cell in the superconducting magnet. When the oxygen gas pressure around the optical cell increased by 2 mmH_2O, it was observed that the initial reaction rate of the mixture significantly decreased in a magnetic field at 14T. The results of this study indicate that magnetic fields affect dynamic movement of oxygen bubbles produced in the reaction mixture by the decomposition of hydrogen peroxide, but not the catalytic activity of catalase.

1. Introduction

The question of whether magnetic fields affect enzymatic activity is of considerable interest in biomagnetics and biochemistry. We have investigated the effects of magnetic fields on the fibrinolytic process[1]. We have observed that dissolutions of fibrin gels were enhanced in gradient magnetic fields up to 8 T [2]. We have also investigated changes in concentrations of fibrin in magnetic fields of up to 8 T. We measured levels of fibrin degradation products by colorimetric determination and concluded that changes in concentrations of fibrin polymers occurred in gradient magnetic fields of up to 8 T [3].

This study focuses on whether magnetic fields affect the activity of enzymes such as superoxide-dismutase (SOD), peroxidase, xanthine oxidase, and catalase.

During the oxidation of xanthine, the enzyme releases superoxide anion radicals as intermediates which reduce the ferricytochrome c (Fe^{3+}). We examined a possible effect of magnetic fields up to 1.0 T on the biochemical reaction catalyzed by xanthine oxidase, and obtained negative results[4].

SOD neutralizes the activity of superoxide anion. Catalase and peroxidase contain a ferric state (Fe^{3+}) iron atom, and decompose toxic hydrogen peroxide (H_2O_2).

It has been reported that magnetic fields of up to 6 T enhance the activity of catalase 4.9 ~ 52% [5]. However, we have observed that magnetic fields of up to 1.0 T did not alter the rate of decomposition of hydrogen peroxide catalyzed by catalase [4].

In this study, we examined the effects of magnetic fields of up to 14 T on SOD, peroxidase, xanthine oxidase, and catalase (Fig. 1).

2. Methods

We used a spectrophotometric system with an external optical cell room in a superconducting magnet as shown in Fig. 2.

Substrates for the enzymatic reactions of SOD, peroxidase, and xanthine oxidase, SOD-525 (5,6,6a,11b-tetrahydro-3,9,10-trihydroxybenzo[c]fluorene), hydrogen peroxide (0.02%), and xanthine were used respectively.

In the case of catalase, the rate of disappearance of hydrogen peroxide was measured by observing the decrease in absorbance at 240 nm, with and without nitrogen-gas-bubbling.

The final concentrations of catalase and hydrogen peroxide were 0.03 – 0.3 μ Mol and 2.5 – 50 m Mol, respectively.

3. Results and discussion

We observed no effects of magnetic fields of up to 14T on SOD, peroxidase, and xanthine oxidase, as shown in Fig. 3.

On the other hand, we obtained an effect of magnetic fields on the reaction of catalase with 50 mM of hydrogen peroxide, as shown in Fig. 4 (a). We observed changes in the absorbance of the reaction mixture of hydrogen peroxide and catalase at 240 nm, both during and after 8T magnetic field exposures.

When the reaction mixture was not treated with nitrogen-gas-bubbling, we observed that the initial reaction rates of the mixture which were exposed to magnetic fields was 50 – 85 % lower than the control data (Fig. 4 (a)).

However, this magnetic field-effect was not observed when the reaction mixture was bubbled with nitrogen gas to remove the dissolved oxygen molecules produced in the solution (Fig. 4 (b)).

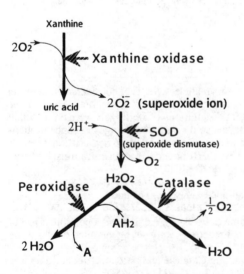

Fig. 1 Enzymatic reaction system of superoxide-dismutase (SOD), peroxidase, xanthine oxidase, and catalase.

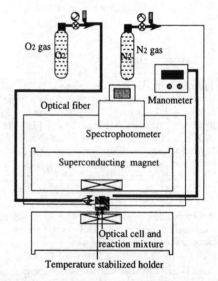

Fig. 2 A spectrophotometric system with an external optical cell room in a superconducting magnet.

Furthermore, we carried out the same experiments by changing the gas-pressure around the optical cell in the superconducting magnet. Oxygen gas was introduced into the bore of the magnet, and the gas-pressure was measured by a differential gas-pressure meter (Fig. 2). When the oxygen gas pressure around the optical cell increased by 2 mmH2O, we observed that the initial reaction rates of the mixtures significantly decreased in a magnetic field at 14T, as shown in Fig. 5.

The results of the present study indicate that magnetic fields affect the dynamic movement of oxygen bubbles produced in the reaction mixture by the decomposition of hydrogen peroxide, but not the catalytic activity of catalase itself.

4. Conclusion

We conclude that magnetic fields of up to 14 T have no effects on the enzymatic activities of SOD, peroxidase, and xanthine oxidase.

In the case of catalase, the initial rate of the reaction was inhibited by magnetic fields of up to 14 T. This magnetic field-effect was changed depending on the pressures of oxygen molecules inside and outside the reaction mixture. The results of the present study indicate that magnetic fields affect dynamic movement of oxygen molecules, but not the catalytic activity of catalase.

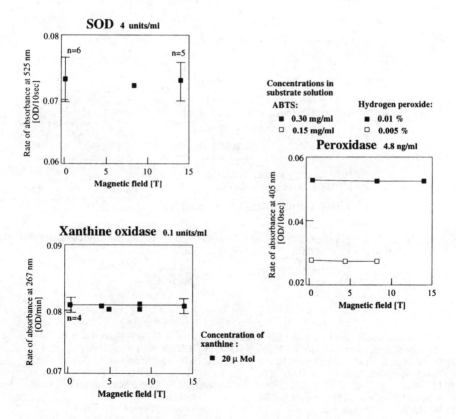

Fig. 3 Enzyme activities of SOD, peroxidase, and xanthine oxidase under magnetic fields of up to 14T.

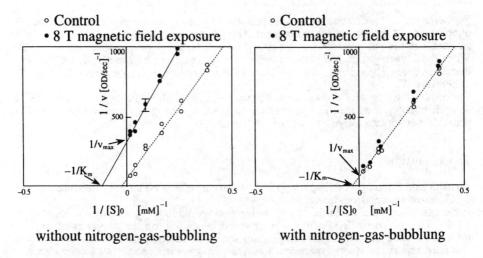

Fig. 4 Changes in the absorbance of the reaction mixture of hydrogen peroxide and catalase at 240 nm, with and without an 8T magnetic field exposure.

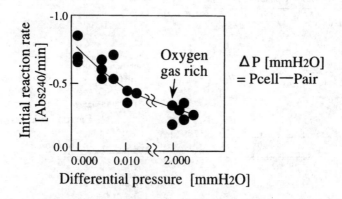

Fig. 5 Effects of the gas-pressure around the optical cell in the superconducting magnet on the initial reaction rates of catalase with hydrogen peroxide.

References

[1] S. Ueno, M. Iwasaka, and H. Tsuda: Effects of Magnetic Fields on Fibrin Polymerization and Fibrinolysis, IEEE Trans. Magn., vol.29, No.6, 1993, pp. 3352-3354.
[2] M. Iwasaka, S. Ueno, and H. Tsuda: Effects of Magnetic Fields on Fibrinolysis, J.Appl. Phys., vol.75, No.10, 1994, pp.7162-7164.
[3] M. Iwasaka, S. Ueno, and H. Tsuda: Diamagnetic Properties of Fibrin and Fibrinogen, IEEE Trans. Magn., vol.30, No.6, 1994, pp. 4695-4697.
[4] W. Haberditzl: Enzyme Activity in High Magnetic Fields, Nature, 213,pp.72-73, 1967.
[5] S. Ueno and K. Harada: Experimental Difficulties in Observing the Effects of Magnetic Fields on Biological and Chemical Processes, IEEE Trans. Magn., MAG-22, 5, 1986, pp.868-873.

Superconductivity, Non-Linear Dynamical Concepts

Non-Linear Electromagnetic Systems
V. Kose and J. Sievert (Eds.)
IOS Press, 1998

Prediction of Mesoscopic Phenomena of Superconductors with Fluxoid Dynamics Method

Kazuyuki Demachi, Kenzo Miya and Kentaro Takase

Faculty of Engineering, The University of Tokyo,

22-2 Shirane Shirakata Tokai-mura Naka-gun Ibaraki 319-11, Japan

Abstract. The Fluxoid Dynamics method was developed in Miya's laboratory
to simulate the behavior of the fluxoids in the type-II superconductor. The
macroscopic electromagnetic phenomena were predicted by this method, and
the results were compared with the well-known empirical ones.

1.Introduction

Recently, applications of superconductors have been explored in various fields such as
flywheels, magnetic bearings, magnetic levitation etc, and numerical analyses of their
nonlinear phenomena are becoming more important for their realization. Mesoscopic
behavior of the fluxoids is sufficient for the purpose and hence has to be analyzed
aiming at establishing a new correct macroscopic model for several physical phenomena
like $E-J$ and $B-J_c$ relationships.

Study of superconductivity and superconductor can be categorized into three aspects, *i.e.* microscopic, mesoscopic and macroscopic phenomena. They are characterized respectively by the BCS theory, the Ginzburg-Landau (G-L) theory and the
macroscopic physical constants relating to the critical state model and flux flow–creep
one. As long as we are concerned with application of superconductors in the form of
flux and electromagnetic force utilizations, we can do not need to consult with the BCS
theory. On the other hand, we have to be familiar with the mesoscopic electromagnetic
phenomena of superconductivity for the following two reasons;

(1) The mesoscopic phenomena represented by the behavior of fluxoids can lead to
prediction of macroscopic phenomena specified by the critical current, the flux
flow resistivity and anisotropy of superconductors.

(2) The macroscopic approach has limitation with prediction of the anisotropic nature of crystal axis orientation dependency on the critical current and so on.

Solution of distributed fluxoids in the mesoscopic level can be obtained theoretically
by solving the G-L equations[1]. However, it takes a great deal of computational
time. Practical aspects of fluxoid dynamics require development of a simulator with
significantly less CPU time compared with the direct solution of G-L equations. We
have already developed the 2 and 3-dimensional fluxoid dynamics simulation methods
of Type II superconductor[2],[3] based on the Molecular Dynamics method combined
with the G-L theory. These methods were named the Fluxoid Dynamics (FD) methods
and they made it possible to simulate the mesoscopic electromagnetic phenomena with
the practical CPU time.

2. Fluxoid Dynamics Methods

The three models of FD method have been already developed and verified in our
laboratory; 1) 2-dimensional one for Low Temperature SuperConductor (LTSC), 2)

3-dimensional one for LTSC and 3) 3-dimensional one for High Temperature Super-Conductor (HTSC) introducing the pancake fluxoid model. These models simulate each fluxoid's movements by several kinds of forces in the superconductor.

2.1. 2-dimensional FD method for LTSC (2D–L–FD method)

The fluxoids exist in the form of continuing flux line quantized into Φ_0. In the 2D–L–FD method, it is assumed that the fluxoids are parallel to the z- axis and are uniform in the z- direction. The fluxoids are affected by 5 kinds of forces as shown in Fig. 1 and they move until the equilibrium of the force is attained.

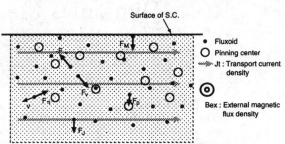

Fig. 1 : 5 kinds of forces in 2D–L–FD method

The kinetic equation of i-th fluxoid movement is written as,

$$\eta_f \frac{\partial r_i}{\partial t} = \sum_j F_p(r_{ij}) + \sum_k F_f(r_{ik}) + F_M(r_i) + F_J(r_i), \qquad (1)$$

where the inertial term was neglected because the mass of electrons around a fluxoid is negligibly small. r_i is a position vector of i-th fluxoid and r_{ij} and r_{ik} are relative position vectors of i-th fluxoid with j-th pinning center and k-th fluxoids. η_f is the viscous coefficient. F_p and F_f are pinning and repulsive forces between two fluxoids. F_M and F_J are Lorentz forces due to Meissner and transport current, respectively. The left side of Eq. (1) corresponds to the viscous force $-F_\eta$. This equation was discretized to calculate the location of i-th fluxoid, r_i. The value of the time increment is chosen to be small enough so that each fluxoid does not skip over any pinning center.

2.2. 3-dimensional FD method for LTSC (3D–L–FD method)

In the 3D–L–FD method, the fluxoids are affected by 6 kinds of forces and torques because the tension due to the curvature of fluxoid must also be considered.

A fluxoid is discretized with the N circular arc elements as shown in Fig. 2. r_i and r_{i+1} are the location vectors of the nodes which are both edges of the i-th element and ℓ_i is the length of the i-th element. o_i and R_i are the location vectors of the center of curvature and the radius of curvature of the i-th element. n_i is the normal vector which directs from the center of the i-th element to o_i. F_i and C_i are the force and torque acting on the i-th element.

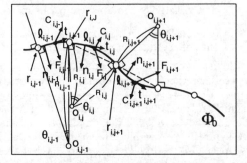

Fig. 2 : Discretization of a fluxoid with arc elements in 3D–L–FD method

The governing equations of the 3D–L–FD method concerning the force and torque are written as follows,

$$\eta_F v_i = -F_i^\eta = \sum_j F_{i,j}^p + \sum_{k,k\neq i} F_{i,k}^f + F_i^M + F_i^J + F_i^T, \qquad (2)$$

$$-C_i^\eta \;=\; \sum_j C_{i,j}^p + \sum_{k,k\neq i} C_{i,k}^f + C_i^M + C_i^J + F_i^T, \tag{3}$$

where the super- and subscripts are the same as the 2D–L–FD method. The forces and torques acting on i-th element are calculated as

$$F_i = \int_{-\frac{t_i}{2}}^{\frac{t_i}{2}} df_i \, d\ell' \qquad and \qquad C_i = \int_{-\frac{t_i}{2}}^{\frac{t_i}{2}} (r - r_{g,i}) \times df_i \, d\ell', \tag{4}$$

where df_i is the force per unit length of the i-th element according to the 2D–L–FD method and $r_{g,i}$ is the location vector of the center of gravity of the i-th element.

2.3. 3-dimensional FD method for HTSC (3D–H–FD method)

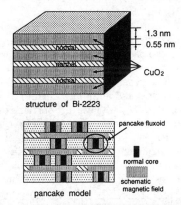

1.3 nm
0.55 nm
CuO$_2$

structure of Bi-2223

pancake fluxoid

normal core

schematic
magnetic field

pancake model

Fig. 3 : Pancake fluxoids in CuO$_2$ planes

We extended this simulation method to the high temperature superconductor Bi-2223 which has the strong 2-dimensionality. In this model, it was assumed that the fluxoids exist only in the CuO$_2$ planes in the form of "pancake" fluxoids, there is no fluxoid string between the CuO$_2$ planes and the fluxoids in different CuO$_2$ planes are affected only by the electromagnetic interaction among them. Then the fluxoids can move in the each CuO$_2$ plane as shown in Fig. 3. The kinetic equation of each fluxoid can be expressed in the same way as Eq. (1).

In this model, the shielding current density of a pancake fluxoid was numerically calculated by solving the following equation[4]

$$\mu_0 j(r) = \mu_0 j(r) e_\theta = \frac{1}{\lambda^2}\left(\frac{\phi_0}{2\pi r}\delta_{n0} e_\theta - A(r)\right), \tag{5}$$

where A is the vector potential, and n is the number of CuO$_2$ planes on which the normal core of fluxoid exists. The gauge of A is taken to satisfy

$$A(r) = \frac{1}{4\pi}\int \frac{\mu_0 j(r')}{|r - r'|} dv'. \tag{6}$$

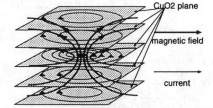

CuO2 plane

magnetic field

current

Fig. 4 : J and B_z around a pancake fluxoid

From this it is revealed that the current flows counterclockwise in the CuO$_2$ plane on which the core of a pancake fluxoid exists, when B_z generated by that pancake fluxoid is positive. In other planes, the current flows clockwise as shown in Fig. 4. This means that the force between two pancake fluxoids with the same sign of B_z, is repulsive when they are in the same plane, and attractive when they are in different planes from each other.

3. Numerical results from three models of the FD method

3.1. Distribution of fluxoids in NbTi

Fig. 5 shows the numerical results of the distribution of fluxoids in NbTi by the 3D–L–FD method in equilibrium before the transport current was induced in the case where $B=0.1361$T and its angle is $\theta=45°$. The rectangular prisms correspond to the α-Ti pinning centers arranged parallel to the x-axis. Some fluxoids are trapped strongly by the pinning centers and knuckled like the stairs.

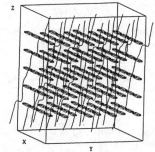

Fig. 5 : Distribution of fluxoid lines in NbTi

3.2. Prediction of critical current density : J_c

When the fluxoids move with speed v, the electric field is yielded in the superconductor by the electromotive force acting upon the fluxoids. Fig. 6 shows the numerical result of the constitutive relationship between the transport current densities J_t and the electric field E for several external magnetic flux density B_{ex} by the 3D–H–FD method. E keeps low values when J_t is small, but it rises suddenly at a certain J_t and the gradient of the E–J_t curve is larger for the higher B_{ex}. These properties correspond to the well-known "Critical State Model" and the critical current density J_c can be predicted from these numerical results. Fig. 7 shows the numerical results of the relationship between B_{ex} and J_c. J_c decreases monotonously while B_{ex} increases and this feature corresponds to the empirical one. Thus it can be said that this method is useful for prediction of electromagnetic properties of the type II superconductor.

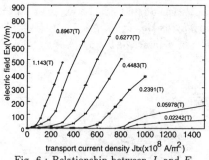

Fig. 6 : Relationship between J_t and E

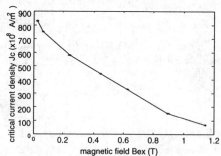

Fig. 7 : Relationship between B and J_c

4. Conclusion

The three models based on the FD method were developed to simulate the behavior of fluxoids and to predict its electromagnetic properties, two being for the LTSC and the other being for the HTSC. The validity of this simulation method was verified with the empirical properties concerning E–J_t and B–J_c relationships.

References

[1] K. Demachi, Y. Yoshida, H. Asakura and K. Miya, "Numerical analysis of Magnetization Processes in Type II Superconductors Based on Ginzburg-Landau Theory", *IEEE Trans. Magn.*, Vol. 32, No. 3, pp. 1156–1159, 1996.

[2] K. Demachi, H. Tsumori and K. Miya, "Numerical Analysis of the Behaviors of Fluxoids in Tpye II Superconductor Based on the Fluxoid Dynamics Method Combined with the Ginzburg-Landau Theory", *Proc. 8th Int. Work. Crit. Curr. Supercon.*, pp. 263-266, 1996.

[3] K. Demachi, H. Tsumori and K. Miya, "Fluxoid Dynamics Method of Type II Superconductor ", *Proc. Asian Joint Semin. Appl. Electromag.*, pp. 87-92, 1996.

[4] K. Takase, K. Demachi and K. Miya, *Cryogenics*, to be published.

Non-Linear Electromagnetic Systems
V. Kose and J. Sievert (Eds.)
IOS Press, 1998

Topological Defects in a Large Josephson Junction: Analytical Description

A.B. BORISOV and V.V. KISELIEV

Institute of Metal Physics, S.Kovalevskaya 18, GSP-170,
Ekaterinburg 620219, Russia

Abstract. Large Josephson junction in an external magnetic field applied along the junction plane is analyzed. Non-linear defects induced by penetration of an Abricosov vortex filament into the tunnel junction are predicted. Two-dimensional deformations of the flux lattice and Josephson current density near defects have been analytically described.

1. Introduction

Josephson tunnel junctions have a variety of applications: Josephson interferometer, SQUID's, and high- speed logic and memory devices. In Josephson junctions with dimensions greater λ_J (see equation (1)) 2D vortex dipoles have been experimentally observed [1]. Such topological defects emerge in junction plane as a result of the penetration of Abricosov vortex filaments from the superconducting electrodes into the tunnel junction (see [1]). Theoretical description of vortex dipoles involves great difficulties. Indeed, two-dimensional non-linear elliptic sine-Gordon equation

$$\left(\partial_x^2 + \partial_y^2\right)\varphi = \lambda_J^{-2}\sin\varphi, \qquad \lambda_J = \left(\hbar c^2/8\pi e d J_1\right)^{1/2} \tag{1}$$

with the proper boundary conditions must be integrated to determine the Josephson current. In the general form this problem has not yet been solved. Here $\varphi(x,y)$ denotes the phase difference of the macroscopic wave functions of the upper and the lower superconducting electrodes. The Josephson current density $J(x,y)$ and the magnetic field in the plane of the contact H are expressed in terms of the field $\varphi(x,y)$:

$$J(x,y) = J_1\sin\varphi(x,y), \qquad H = \hbar c(2ed)^{-1}[\mathbf{n} \times \boldsymbol{\nabla}\varphi]$$

where d denotes the effective thickness of the tunnel barrier, $\mathbf{n} = (0,0,1)$ is the vector normal to the contact plane. Fur further treatment it is convenient to introduce the dimensionless variables $x\lambda_J^{-1} \to x$, $y\lambda_J^{-1} \to y$, then equation (1) reduced to:

$$\left(\partial_x^2 + \partial_y^2\right)\varphi = \sin\varphi. \tag{2}$$

If an external magnetic field is applied in the plane of the junction ($\mathbf{H_0} \uparrow\downarrow 0Y$), then at certain field magnitude a one-dimensional magnetic flux lattice is formed in the tunnel junction:

$$\varphi_0(x) = -2\operatorname{am}(\chi + \chi_0, k) + \pi \tag{3}$$

where $\chi = x/k$, $\chi_0 = const$ and $am\chi$ is the Jacobian amplitude χ function with the modulus k. The function $\varphi_0(x)$ represents the sequence of identical 2π-kinks separated by extensive regions of length $2Kk$, within which the function $\varphi_0(\chi)$ change more slowly ($\varphi_0(\chi) \simeq 2\pi s$, where s is integer). Here $k = K(k)$ is the complete elliptic integral of the first kind. Each 2π-kink corresponds to a magnetic flux in the junction plane. In terms of $\varphi(x, y)$ the vortex dipole represents a string configuration from the 2π-kink section against the background of a one-dimensional 2π-kink lattice. In this work we have used the inverse scattering transform to calculate field $\varphi(x, y)$ caused by the vortex dipoles against the background the magnetic flux lattice.

2. Analytical description of a vortex dipole on the flux lattice background

Let us consider the vortex dipole which occupies a symmetric position between two neighboring 2π- kinks of the lattice (3). The dipole central line coordinates have the following form: $x = 0$, $|y| \leq l$ ($\chi_0 = K$). The field φ of a vortex dipole with topological charges

$$Q = \frac{1}{2\pi} \oint_\gamma \nabla\varphi \cdot d\mathbf{r} = \pm 1 \qquad (4)$$

at points $(0, \mp l)$ corresponds to a solution of equation (2) with the singular sources:

$$\left(\partial_x^2 + \partial_y^2\right)\varphi = \sin\varphi + J^{(0)}; \qquad J^{(0)} = \partial_x q^{(0)}(x, y) + f^{(0)}(x, y);$$

$$q^{(0)}(x, y) = 2\pi \int_{-l}^{l} \delta(y - y')dy'\delta(x); \quad f^{(0)}(x, y) = [f_+^{(0)}\delta(y - l) + f_-^{(0)}\delta(y + l)]\delta(x);$$

$$f_\pm^{(0)} = const; \qquad \varphi(x, y) \rightarrow \varphi_0(\chi), \qquad x^2 + y^2 \equiv r^2 \rightarrow \infty. \qquad (5)$$

The source $q^{(0)}(x, y)$ takes into account an ambiguous (4) of the field $\varphi(x, y)$ in circulating around the vortex centers $(0, \pm l)$. The source $f^{(0)}(x, y)$ simulated injection currents, which are localized in the points $(0, \pm l)$.

At the present time the inverse scattering transform is the only possible method which can find an analytical solution to the essentially nonlinear problem (5). In this paper we apply a modification of the procedure, which was used in works [2], [3] for the analytical description of vortex dipoles. We found, that the vortex dipole field is calculated from the integral equations of the Marchenko type:

$$\sin\frac{1}{4}(\varphi - \varphi_0) = \int_{-\infty}^{y} dy'[a(y, y')\cos\frac{\varphi_0}{2} + b(y, y')\sin\frac{\varphi_0}{2}]\exp\frac{1}{4}\partial_x\varphi_0(x)(y - y');$$

$$8a(y, y'') - \sin\frac{1}{4}(\varphi + \varphi_0)\Phi_{1,1}(y + y'') + \cos\frac{1}{4}(\varphi + \varphi_0)\Phi_{0,2}(y + y'') +$$
$$+ \int_{-\infty}^{y} dy'[a(y, y')\Phi_{1,1}(y' + y'') + b(y, y')\Phi_{0,2}(y' + y'')] = 0 ; \qquad (6)$$

$$8b(y, y'') - \sin\frac{1}{4}(\varphi + \varphi_0)\Phi_{2,0}(y + y'') + \cos\frac{1}{4}(\varphi + \varphi_0)\Phi_{1,1}(y + y'') +$$
$$+ \int_{-\infty}^{y} dy'[a(y, y')\Phi_{2,0}(y' + y'') + b(y, y')\Phi_{1,1}(y' + y'')] = 0; \qquad (y'' < y).$$

The kernels $\Phi_{m,n}(y)$ are given:

$$\Phi_{m,n}(y) = \frac{1}{2\pi i}\int_{\Gamma}(-\frac{2d\alpha}{\alpha\tau})w_1^m(\alpha)w_2^n(\alpha)\rho(\alpha,\tau,x)\exp(-\frac{i\tau y}{4}); \qquad (7)$$

$$\rho(\alpha,\tau,x) = \left[\frac{i\tau + \partial_x\varphi_0}{-i\tau + \partial_x\varphi_0}\right]^{1/2}\exp\left\{-\frac{i}{2}w_1 w_2\tau\int_0^x dx'[w_1^2 - \cos^2\frac{\varphi_0}{2}]^{-1}\right\}\rho_0(\alpha,\tau).$$

Here $w_1 = (\alpha + \alpha^{-1})/2$, $w_2 = (\alpha - \alpha^{-1})/2i$, $\tau = 2(w_1^2 - k^{-2})^{1/2}$ are the functions of spectral parameter α. The set $\Gamma = \{\alpha \mid \mathrm{Im}\,\tau = 0\}$ forms a continuous spectrum of the scattering problem and in parametrization $\alpha = \mathrm{sn}(\lambda, k) + i\,\mathrm{cn}(\lambda, k)$ has a simple form: $\Gamma = \{\lambda \mid \lambda = u \pm iK'\}$ where $-2K \leq u \leq 2K$. We found that the function $\rho_0(\alpha, \tau)$ is equal to zero on pieces of contour Γ which are given: 1) $-2K \leq u \leq -K$, $0 \leq u \leq K$·for $x > 0$ and·2) $-K \leq u \leq 0, K \leq u \leq 2K$·for $x < 0$. The nonzero coefficient $\rho_0(\alpha, \tau)$ has the following form:

$$\rho_0(\alpha,\tau) = -\frac{2\pi}{\tau}w_1 w_2 m^{(\infty)}\,\mathrm{sign}\,x\left[(\tau/2)^2 + (k'/k)^2\right]^{-\frac{1}{2}}\sin\frac{\tau l}{2} - \qquad (8)$$
$$-\frac{i}{2\tau}\,\mathrm{sign}\,x\left[(\tau/2)^2 + (k'/k)^2\right]^{\frac{1}{2}}\left\{f_+^{(\infty)}\exp\frac{i\tau l}{2} + f_-^{(\infty)}\exp\left(-\frac{i\tau l}{2}\right)\right\}$$

The constants $m^{(\infty)}$, $f_{\pm}^{(\infty)}$ are expressed via $f_{\pm}^{(0)}$. In particular, $m^{(\infty)} \approx 1$ at $f_{\pm}^{(0)} = 0$, $f_{\pm}^{(0)} = 8\sinh\frac{f_{\pm}^{(0)}}{8}$ at $l \to \infty$, $k \to 1$, $q^{(0)} = 0$. Since the kernels $\Phi_{m,n}(y)$ are known, the two-dimensional nonlinear problem (5) reduces to the solution of a one-dimensional linear system (6).

3. Asymptotic behavior of the vortex dipole field far from a defect center

We find effective formulae for the main terms of the dipole field $\varphi(x,y)$ expansion as $r \to \infty$, which are convenient for experimental checking of the result. Using saddle-point and Laplace methods we obtain

$$\varphi \cong \varphi_0(x + \alpha_1(x,y) + \alpha_2(x,y)) + \varphi_p(x,y), \qquad r \to \infty; \qquad (9)$$

$$\alpha_1(x,y) = \frac{km^{(\infty)}}{2k'}\left\{\arctan\frac{y_+}{x_0} - \arctan\frac{y_-}{x_0}\right\}, \qquad x_0 = \frac{x}{\sqrt{K_1}}, \qquad y_{\pm} = \frac{y \mp l}{\sqrt{K_2}},$$

$$\alpha_2(x,y) = -\frac{k}{4\pi}\left\{f_+^{(\infty)}\ln\frac{r_+}{R} + f_-^{(\infty)}\ln\frac{r_-}{R}\right\}, \qquad K_1 = \frac{4Kk'^2}{k^2 E}, \qquad K_2 = \frac{4E}{k^2 K}$$

where R is an external truncation radius (size of the junction), $E = E(k)$ is a complete elliptic integral of the second kind and $r_{\pm}^2 = x_0^2 + y_{\pm}^2$. The parameters K_1, K_2 are the effective elastic constants of the flux lattice.

The first term in (9) gives an "adiabatic" approximation of $\varphi(x)$ and describes a two-dimensional deformation of the one-dimensional flux lattice (3) due to the presence of a vortex dipole. The scheme of deformation of 2π-kink lines of field $\varphi(x,y)$ is presented in the Fig.1. It is of interest that a vortex dipole is taking features of logarithmic spirals, while the Abricosov filament penetration is attended by the injection currents $(f^{(0)}(x,y), f_{\pm}^{(\infty)} \neq 0, \quad \alpha_2(x,y) \neq 0)$.

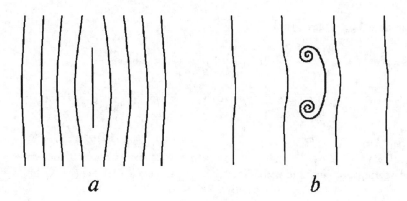

Figure 1: Magnetic vortex dipole on a flux lattice background (scheme of the level lines $\varphi = \pi(2n+1)$, n is an integer number): a) $f_{\pm}^{(0)} = 0$, b) $f_{\pm}^{(0)} < 0$.

The second term $\varphi_p(x, y)$ describes a proper field of vortex dipole. In the limit $k \to \infty$, when the period of flux lattice tends to infinity, $\varphi_p(x, y)$ has a simple form:

$$\partial_x \varphi_p(x, y) = \left(m^{(\infty)} \partial_x - \frac{f_+^{(\infty)}}{2\pi} \partial_y \right) K_0(r_+) - \left(m^{(\infty)} \partial_x + \frac{f_-^{(\infty)}}{2\pi} \partial_y \right) K_0(r_-), \quad r \to \infty$$

which at $f_{\pm}^{(\infty)} = 0$ agreed with result obtained in the work [2]. Here $K_0(r)$ is zero-order Mcdonald function.

4. Conclusion

We have succeeded to provide a qualitative description of new non-linear defects in a large Josephson junction. These defects deform the flux lattice and significantly change the Josephson current. We believe that topological defects are essential elements of intrinsic structure of a large Josephson junction, which determine the electrical properties of the tunnel barrier.

Work supported by the Russian Fundamental Research Foundation under Grant N 93-02-2011.

References

[1]   J. Manhart *et al.*, Spatial Distribution of the Maximum Josephson Current in Superconducting Tunnel Junction, *J. Low. Temp. Phys.* **70** (1988) 459-484.

[2]   A.B. Borisov and S.N. Ionov, Vortices and Vortex Dipoles in 2D sine-Gordon Model, *Physica D* **99** (1996) 18-34.

[3]   A.B. Borisov and V.V. Kiseliev, Vortex Dipole on a Soliton Lattice Background. Solution of the Boundary-Value Problem by the Inverse Spectral Transform, *Physica D* (1997) (in press).

Non-Linear Electromagnetic Systems
V. Kose and J. Sievert (Eds.)
IOS Press, 1998

Frequency Characteristic of Magnetic Flux Invasion on YBCO Superconductor Ring Specimen

Katsuhiro FUKUOKA[a] and Mitsuo HASHIMOTO[b]

[a]*Dept. of Electrical Technique, Tokyo Polytechnic College,*
2-32-1 Ogawanishimachi, Kodaira, Tokyo 187, Japan
[b]*Dept. of Electrical Engineering, Polytechnic University,*
4-1-1 Hashimotodai, Sagamihara, Kanagawa 229, Japan

Abstract. Because the bulk type YBCO superconductor prepared by the MPMG process characterizes high critical current density J_c, it is expected to be applied to various fields. It is important to develop measurement methods of magnetic properties of the high T_c superconductor (HTSC). It is necessary to value not only DC magnetic properties but also AC magnetic properties. So we determined the magnetic properties of the HTSC under an AC magnetic field (5~500Hz). From the experimental results we can understand that there is the first order lag element caused by the flux flow effect, and the magnetic properties in low frequency are ruled by the flux flow.

1. Introduction

In recent years, for the purpose of adding many pinning points in high T_c superconductor (HTSC) materials, the half melt solidification methods by QMG (Quench and Melt Growth) and MPMG (Melt Powder Melt Growth) methods have been developed [1]. YBCO superconductors, which have the characteristic of critical current density over 1×10^8 A/m² in liquid nitrogen temperature (77K) on 1T [2], can be produced. Therefore, it is expected to be applied to various fields such as magnetic bearing [3], fly wheel for electricity storage, magnetic levitation [4], etc.. In these applications, evaluation of magnetic characteristics of the HTSC is an important problem. We have evaluated magnetic characteristics of the HTSC in a DC magnetic field [5]. On the other hand, in practical applications of the HTSC, ripples occur to the applied magnetic field from permanent magnets which are not uniform. The ripples cause energy loss. Therefore, evaluation of magnetic properties in an AC magnetic field is important, too. So we evaluated magnetic properties of the HTSC in an AC magnetic field on 5~500Hz in this study.

2. Experimental Method

We used a YBCO superconductor made by the MPMG method for this experiment. The test specimen is processed as a ring shape (the outside diameter is 34mm, the inside diameter is 24mm and the thickness is 10mm), because the magnetic flux inside the ring specimen can easily be evaluated using pickup coils. AC magnetic properties are evaluated

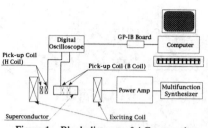

Figure 1   Block diagram of AC magnetic property measurement system.

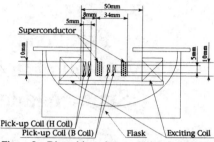

Figure 2   Disposition of superconductor and coil.

by measuring the inner magnetic flux of the ring specimen and the applied AC magnetic field. A system of measuring AC magnetic properties is shown in Figure 1. The AC magnetic field is generated by an exciting coil to which the AC current is controlled by an amplifier. For the measurements of magnetic flux pickup coils are used. Signals of pickup coils are measured by a digitizing oscilloscope, and sent to a computer through a GP-IB interface. The pickup coils are a H coil (5mm × 40mm × 10mm, 100 turns) to measure the outside (applied) magnetic field and a B coil (the inside diameter : 18mm, 100 turns) to measure the magnetic field which invades inside the HTSC ring specimen. Figure 2 shows the arrangement of the HTSC specimen in a low temperature flask, the exciting coil and the pickup coils. The exciting coil is expected to generate a high AC magnetic field but small in size, because it is set in the low temperature flask. So we used a square coil which surrounded the HTSC specimen. The exciting coil, in which copper wire of 0.65mm² is wound 690 turns, is capable of generating a magnetic field of 7.8×10^3A/m at an exciting current of 1A.

3. Experimental Results

When a magnetic field greater than the lower part of the critical magnetic field is applied to a YBCO superconductor, magnetic field invades inside of the HTSC ring specimen. Figure 3 shows the relationship between the applied magnetic field measured with the H coil and the invading magnetic flux Φ measured with the B coil, when an AC magnetic field (10Hz) is applied. In this figure there is a phase difference between the applied magnetic field H and the invading flux Φ. Figure 4 shows hysteresis curves on each

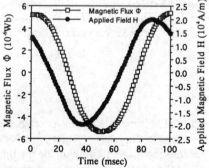

Figure 3   Magnetic flux of inside of ring and applied magnetic field.

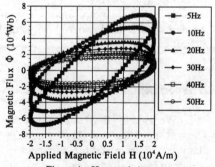

Figure 4   Hysteresis loop.

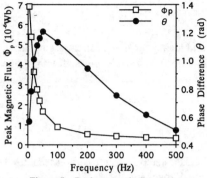

Figure 5   Peak magnetic flux and phase difference.

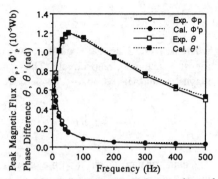

Figure 6   Comparison of survey data and calculation data.

frequency (5~50Hz). In this case, the applied magnetic fields are sine waves which have a peak value of 2.0×10^4A/m. This figure shows, when frequency of the applied AC magnetic field becomes low, the peak value of the magnetic flux Φ invading the inside of the ring becomes large (peak value of magnetic flux Φ at 5Hz is about 4.2 times as large as at 50Hz).

Figure 5 shows the frequency characteristic of peak magnetic flux Φ_p (maximum value of Φ) and phase difference θ (between H and Φ). The peak magnetic flux Φ_p decreases rapidly in the region of less than 50Hz, but decreases slowly in the high frequency of more than 50Hz. The phase difference θ increases where frequency is less than 50Hz, but decreases in the high frequency of more than 50Hz. Because the time change of the outside magnetic field is slow in the low frequency, a lot of magnetic flux is able to invade the ring specimen. This phenomenon is generally called the flux flow. Therefore, in the frequency range of 5~50Hz, the invading magnetic flux is decreasing and its phase difference is increasing at higher frequency. In other words, its property is directed by the flux flow in this range.

On the other hand, in the frequency region greater than 50Hz, the magnetic flux is hard to invade into the ring specimen. At a frequency of 500Hz, the inside magnetic flux of the ring specimen is slightly measured. Its phase difference approaches about zero. In this region, we think that measured magnetic flux includes that passing around from the top and bottom sides of the ring specimen, and it is mainly measured by the B coil. The thickness of the ring specimen used in this study is relatively thin, and magnetic flux passing around is the same phase as the applied magnetic field.

4. AC Magnetic Shield Model

We have already stated that the magnetic flux invading the inside of the ring specimen was the sum of the magnetic flux by the flux flow and that passing around from the top and bottom sides of the ring specimen. In this case, we think that some constant time is necessary for the magnetic flux to invade the inside of the ring specimen by the flux flow, because the magnetic flux must move through each pinning point in the HTSC and there is the viscosity to movement of the magnetic flux. So we propose the AC magnetic shield model assuming that the flux flow uses first order lag element. The transfer function G of the AC magnetic shield model is expressed as follows:

$$G = \frac{K}{1 + sT} + L \tag{1}$$

where K, T and L are gain constant, time constant and leakage magnetic flux passing around the ring specimen, respectively. The invading magnetic flux Φ' is expressed with applied magnetic field H.

$$\Phi' = \left(\frac{K}{1 + j\omega T} + L \right) H \tag{2}$$

The gain constant K and the time constant T can be calculated from measuring response of the invading magnetic flux using step wise applied magnetic field. The element L is calculated from Φ_p at 500Hz in Figure 5. The amplitude of the invading magnetic flux Φ'_p and its phase difference θ' can be calculated by equation (2) using each given constant. Figure 6 shows a comparison between calculated and experimental results concerning the characteristics of the magnetic shield. The calculation values of the magnetic field and the phase difference agree with experiment values. It is confirmed that the proposed model of the AC magnetic shield is useful.

5. Summary

We evaluated the AC magnetic characteristics of the YBCO superconductor bulk material using the ring specimen in the regions of frequency 5~500Hz. From this study we conclude as follows:
(1) The invading magnetic flux has hysteresis and phase lag.
(2) It is confirmed that the invading magnetic flux gives expression to the vector sum of the magnetic flux by the flux flow and that passing around from the top and bottom sides of the ring specimen.
(3) We proposed the AC magnetic shield model used the first order lag element. The invading magnetic flux behavior under the AC magnetic field could be evaluated by this model.

References
[1] M.Murakami, A.Kondoh, H.Fujimoto, N.Sakai, N.Koshizuka and S.Tanaka, "Melt processing of bulk YBaCuO superconductors with high J_c", *J. Eng. Mat. Tec.*, **114**, 2, p.189 (1992)
[2] H.Fujimoto, M.Murakami, S.Gotoh, N.Koshizuka, T.Oyama, Y.Shiohara and S.Tanaka, "Melt processing of YBaCuO oxide superconductors", *Advances in Superconductivity II*, p.285 (Springer-Verlag, Tokyo, 1990)
[3] F.C.Moon, M.Yanoviak and R.Ware, "Hysteretic levitation forces in superconducting ceramics", *Appl. Phys. Lett.*, **52**, p.1534 (1988)
[4] M.Murakami, T.Oyama, H.Fujimoto, T.Taguchi, S.Goto, Y.Shiohara, N.Koshizuka and S.Tanaka, "Large levitation force due to flux pinning in YBaCuO superconductors fabricated by MPMG process", *Jpn. J. Appl. Phy.*, **29**, p.L1991 (1990)
[5] M.Hashimoto and K.Fukuoka, "Magnetic property observation of high temperature superconducting bulk material by magnetic flux distribution and flux creep measurement", *Elsevier Studies in Appl. Electromagn. in Mater.*, **6**, pp.399-402 (1995)

Non-Linear Electromagnetic Systems
V. Kose and J. Sievert (Eds.)
IOS Press, 1998

Evaluation of Anisotropy of Critical Current Density of Bi-based High Tc Superconductors

Y. Yoshida, M. Rabara, R. Shinagawa and K. Miya
Nuclear Engineering Research Laboratory, The University of Tokyo,
Tokai-mura, Ibaraki 319-11, Japan

Abstract. The angular dependence of the critical current density of BSCCO-2223 superconducting tape is investigated theoretically and experimentally. The model proposed is based on the effective coherence tensor model which can deal with the anisotropy of the coherence length. The resulting mathematical model shows that the critical current is dependent only on the normal component of the field to the tape surface. The measurements of the critical current were made by the four-point DC method and their results show the validity of the theoretical considerations.

1. Introduction

Recent enhancement of properties of high Tc superconductors promises their applications in various fields of technology, e.g., energy storage, magnet, cable, current lead and SQUID. To realize such applications of copper-oxide superconductors, the electromagnetic and mechanical properties of the superconductors must be investigated and characterized. From this point of view, the anisotropy due to the crystal structure is one of the significant features of the copper-oxide superconductors.

In the copper-oxide high Tc superconductor, the CuO_2 plane, which is parallel to the ab-plane of its crystal structure, has good electrical conductivity but it is sandwiched between the insulation layers, hence the anisotropy of the coherence length. This anisotropy results in the anisotropy of the upper critical field, the critical current density, the irreversibility field and so on. In this paper, the dependence of the critical current density upon the direction of the magnetic field is investigated theoretically and experimentally.

2. Theory

Let us consider the coordinate system depicted in Fig. 1, where the z-axis and xy plane are respectively parallel to the c-axis and ab-plane. Denoting the coherence lengths in the ab-plane as ξ_{ab} and in the c-axis as ξ_c, the effective coherence length tensor $\bar{\xi}$ is given by $\bar{\xi} = \xi_{ab}\hat{x}\hat{x} + \xi_{ab}\hat{y}\hat{y} + \xi_c\hat{z}\hat{z}$, [1] where $\hat{x}$, $\hat{y}$ and $\hat{z}$ are the base vectors in the Cartesian coordinate system. From this, one can readily obtain the following expression of the coherence length in the r-direction (for the definition of r, see Fig. 1):

$$\xi = (\xi_{ab}^2 \sin^2 \theta' + \xi_c^2 \cos^2 \theta')^{1/2}. \tag{1}$$

By letting the angle between the z-axis and the applied field be θ and assuming the transport current to be x-directed, Eq. (1) gives the radius of the normal core of a fluxoid in the following manner (see Fig. 2)

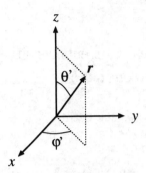

Fig. 1 Coordinate system

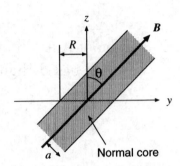

Fig. 2 Normal core in superconductor

$$a = (\xi_{ab}^2 \cos^2 \theta + \xi_c^2 \sin^2 \theta)^{1/2}, \tag{2}$$

which readily leads to the size of the pancake normal core in the ab-plane,

$$R = a/\cos \theta = (\xi_{ab}^2 \cos^2 \theta + \xi_c^2 \sin^2 \theta)^{1/2}/\cos \theta. \tag{3}$$

By defining a function g as $a = \xi_{ab} g(\theta)$, the elementary pinning force f_p is then written

$$f_p = \frac{U_p}{R} = \frac{U_p \cos \theta}{\xi_{ab} g(\theta)}, \tag{4}$$

where U_p is the pinning potential. Since the pinning force density is a function of the fluxoid density and the elementary pinning force, it can be written as

$$F_p = F_p(f_p, B \cos \theta). \tag{5}$$

Furthermore, the following balance equation of the pinning and Lorentz forces holds in the critical state:

$$J_c B \cos \theta = F_p. \tag{6}$$

Finally, the combination of Eqs. (4) to (6) gives

$$J_c = \frac{F_p(U_p \cos \theta/\xi_{ab} g(\theta), B \cos \theta)}{B \cos \theta}. \tag{7}$$

The above equation means that the critical current is a function of $\cos \theta/g(\theta)$ and $B \cos \theta$. In the case of the BSCCO superconductor ($\xi_{ab} = 2.57$ nm, $\xi_c = 0.13$ nm [2]), $g(\theta)$ coincides with $\cos \theta$ in most regions of angle θ ($0 \sim 80$ degree) as shown in Fig. 3. This implies that the critical current density depends upon $B \cos \theta$, in other words, the c-axis directed component of the magnetic field determines the critical current density.

3. Experiment

The measurement of the critical current density was made based on the four-point DC method. The superconductor samples used in the measurement are Ag-sheathed BSCCO-2223 superconducting tapes fabricated by the powder-in-tube method [3], which has a cross section of 3.417 mm × 0.24 mm. The silver ratio of the sample is 2.5 and there are 61 superconducting filaments in the tape. As shown in Fig. 4, the angle between the applied magnetic field and the normal direction to the tape surface (i.e., c-axis) is defined as θ.

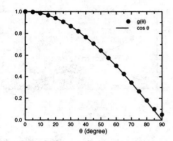

Fig. 3 Comparison of $g(\theta)$ with $\cos\theta$.

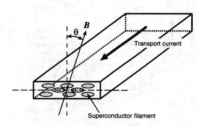

Fig. 4 Schematic of BSCCO-2223 tape.

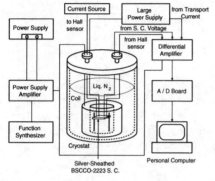

Fig. 5 Experimental set-up.

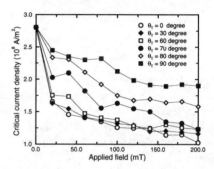

Fig. 6 Measured relationship between J_c and B.

The applied magnetic field is generated by the copper winding and its maximum magnitude is 0.2 T. The measurement was made at the liquid nitrogen temperature in such a way that the sample which is fixed to the rotatory support and the exciting coil is immersed in the liquid nitrogen. The critical current is determined by the off-set method.

4. Results and Discussion

Fig. 6 shows the measured results of the critical current density as a function of the applied magnetic field in the six cases of direction of the applied field. Here the horizontal axis shows the absolute value of the applied field. In the case of the angle being less than 60 degrees, the critical current density is found not to be sensitive to the angle of the field. Otherwise, the strong angular dependence of the critical current is observed in these results.

In Fig. 7(a), the critical current density as a function of the magnetic field component normal to the ab plane is shown, i.e., the magnetic field parallel to the c-axis. The critical current density is clearly found to depend only on the c-axis oriented component of the magnetic field. We should note that the misalignment of crystal growth is taken into account in the same manner in Ref. [4] so that the c-axis directed component of the magnetic field is not zero even in the case of $\theta=90$ degrees in this figure.

As for the scaling law for the BSCCO-2223 describing the dependence of the critical current density on the magnetic field , the following relationship is proposed in Ref. [5]:

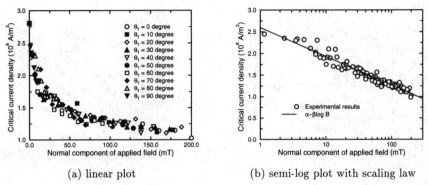

(a) linear plot   (b) semi-log plot with scaling law

Fig. 7 Relation between the critical current density and
the normal component of the applied magnetic field.

$$J_c = \alpha - \beta \log B.$$

From the experimental results, Fig. 7(a), it is possible to put the normal component
of the magnetic field as B in the above equation. Adopting this scaling law, unknown
parameters α and β are determined so as to fit the scaling law to the measurements.
In this way, we obtain $\alpha = 0.602$, $\beta = 0.665$ and Fig. 7(b) shows the critical current
density dependence together with the predicted line based on the scaling law.

5. Conclusion

A mathematical model which describes the angular dependence of the critical current density of the BSCCO-2223 superconducting tape is proposed accounting for the
anisotropy of the coherence length. The model indicates that the critical current density
depends only on the c-axis oriented component of the magnetic field and the measurements of the critical current density based on the four-point DC method validated this
theoretical consideration.

Acknowledgment

The authors would like to express their thanks to Sumitomo Electric Industries for
providing the BSCCO-2223 tape sample.

References

[1] R. C. Morris, R. V. Coleman and R. Bhandari, Superconductivity and Magnetoresistance
    in NbSe$_2$, *Phys. Rev. B* 5 (1972) 895-901.
[2] T. Matsushita, Flux Pinning and Electromagnetic Phenomena, Sangyou-tosho, 1994 (in
    Japanese).
[3] H. Mukai, K. Ookura, N. Shibuta, T. Hikata et al., Development of A Multicore Silver-
    Coated Bismuth Superconducting Wire, *Sumitomo-Denki*, 143 (1993) 67-70 (in Japanese).
[4] T. Kaneko, Y. Torii, H. Takei et al., Magnetic Field Angle Dependence of The Critical
    Current Density in Bi-2223 Superconducting Tape, *Adv. Supercond.* VI (1993) 761-764.
[5] S. Kobayashi, T. Kaneko, T. Kato et al., A Novel Scaling of Field Dependencies of Critical
    Currents for Ag-sheathed Bi-2223 Superconducting Tape, *Physica C* 258 (1996) 336-340.

Non-Linear Electromagnetic Systems
V. Kose and J. Sievert (Eds.)
IOS Press, 1998

Current Distribution within HTSC's and its Identification by Field Measurements

W.-R. Canders, H. May and R. Palka
Institut für Elektrische Maschinen, Antriebe und Bahnen,
Technische Universität Braunschweig, Hans Sommer-Str. 66,
38106 Braunschweig, Germany

Abstract. The quality of high temperature bulk superconductors is characterised by the number, dimension, location and orientation of the subdomains and their individual value of the critical current density J_c. This paper proposes different methods to determine overall quality criteria of bulk superconductors using measurements of the magnetic fields. Furthermore a dynamic field calculation model (virtual displacement) is proposed to determine the J_c distribution within HTSC's at arbitrary external field exposures by a nonlinear iteration process.

1. Introduction

For monolithic High Temperature Superconductors the value and distribution of the induced currents are the most important properties. They determine the quality and performance of HTSC's if applied as magnetic bearings, magnetic shields and excitation units ("HTSC-permanent magnets") for energy converters and they are the key values for the optimisation of the HTSC layout.

Real HTSC's are composed of some single weak linked domains with nearly unpredictable orientations which react individually with their own critical current densities (J_c) on external fields. For the quality control after fabrication the HTSC's are inserted e.g. into the warm bore of a high field magnet and cooled below the critical temperature T_c. In this magnetically saturated state the magnetic flux is trapped into the HTSC's due to pinning centres and the associated currents. Based on measurements of the external magnetic field the value of J_c and its distribution within the HTSC can be determined in a nondestructive manner (NDT) by solving it as an inverse problem.

2. Analytical determination of the J_c within HTSC's

A single domain cylindrically shaped HTSC (Fig. 1) reacts to an external axis symmetric 1-dimensional magnetic field in the z-direction with shielding currents of equal symmetry. If the magnetising field is strong enough the superconductor becomes saturated - it carries the current in the whole volume ($R_1=0$). In the nonsaturated state the superconductor is exposed to a magnetic field of reduced magnitude that only a part of its cross section is needed to suppress the flux in the centre completely ($R_1>0$).

As all components of the magnetic field at any point outside a superconductor are defined by the current distribution within the superconductor they can be calculated by using classical formulas (elliptic integrals of the first and second kind) describing the magnetic field of axis symmetric coils [2,3]. Using these formulas it is possible to calculate the spatial distribution of all magnetic flux components for different values of R_1, R_2 and J_c. Fig. 2 shows the axial flux density distribution for the saturated and non saturated cylindrical superconductor at the top surface for the same current density J_c in a normalised form.

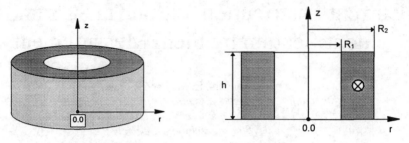

Fig. 1. Cylindrically shaped non-saturated HTSC.

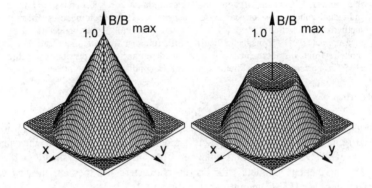

Fig. 2. Axial field component of a saturated $(R_1=0)$ and non-saturated $(R_1>0)$ HTSC.

In the case of saturation the value of the critical current density J_c is a direct function of any outer components of the magnetic flux density. To ascertain this state of the HTSC it is necessary to measure the axial flux density along the radius. Only if the magnetic field decreases linearly as a function of r, the complete saturation is secured (Bean's cone in Fig. 2).

For nonsaturated HTSC's there are two unknown parameters: the critical current density J_c and the inner radius R_1. If the critical current J_c is known from previous measurements or calculations the inner radius R_1 can be obtained directly by using only a single value of the measured flux density on the axis. If both the distribution and the value of the current density are unknown they can also be determined by measuring appropriate magnetic flux components because of their definite relationship. In Reference [2] the way of determining the J_c and its distribution is described for various shapes of HTSC's (cylindrical, quadratic quaders and rectangular quaders).

The above mentioned investigations can be used to determine the critical current density J_c of HTSCs and to control uniformity of superconducting bulks.

3. 2-D finite element model of superconductivity

Single domain superconductors react in the zero field frozen state on nonuniform external magnetic fields with a J_c distribution which is a function of both coordinates r and z. To calculate its distribution under these nonuniform conditions it is necessary to build a physical model of superconductivity describing the macroscopic behaviour of HTSC's. The electromagnetic field outside a superconductor is determined by the classical Maxwell's equations. The proposed model here is based on the analogy between the superconductivity and the eddy current effects in conventional conductors. It describes the electromagnetic phenomena of superconductivity by additional nonlinear flux-current density relationships.

The calculation of the complete magnetic field and the current density distribution within the HTSC is obtained by an iterative process. The starting point is given by the calculation of the current density distribution in HTSC caused by a virtual displacement of the HTSC assuming a high conductivity κ of the superconducting area. This approach allows iterative corrections of the conductivities under the following assumptions:

$$\kappa_{new} = \kappa_{old} \frac{J_c}{J}, \qquad \kappa_{new} \rightarrow 0 \text{ if } J << J_c \qquad (1)$$

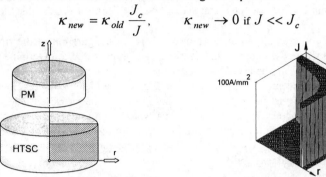

Fig.3: HTSC-permanent magnet configuration

Fig. 4: Current density distribution within HTSC (J_c=100A/mm²)

By this nonlinear process (optimisation with constraints) the determination of the current density distribution within HTSC's is possible. Due to the rotational symmetry of the investigated structures a 2D-FE axis symmetric method was applied for the calculations.

Fig. 3 shows the typical axis symmetric configuration used for examination and comparison of different HTSC's.

As can be seen from Fig. 4 the superconductor is non-saturated as only a part of its cross section leads the critical current J_c. The corresponding field plot is shown in Fig. 5.

The algorithm from Section 2 together with this nonlinear calculation allows the effective determination of the magnetic field and the forces of single domain HTSC's.

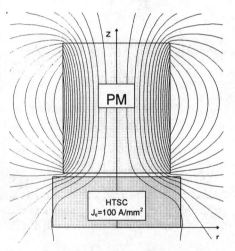

Fig.5: Field plot for J_c=100A/mm²

4. Identification of the J_c distribution

Industrially manufactured HTSCs usually exhibit an inhomogeneous structure with unknown positions, dimensions and orientations of superconducting subdomains with individually unknown J_c's. Because the extrapolation of flux measurements leads usually to the ill-posed problem [4], the determination of these parameters is not possible by the above mentioned methods. Ill-posed problems means that their solution, even if it exists, may not be unique and may not depend continuously upon the data.

This task can be solved by the method presented in [1,4]. The basic idea of this method depends on the extension of the set of usually known finite element equations by those equations describing the magnetic field distribution in measurement areas. This leads to an over determined set of linear equations with unknown vector potential values in the whole region and unknown J_c values within the HTSC. The so-called minimal least squares solution of this system of equations can be obtained from the singular values decomposition (SVD) of the equation matrix [4]. This method makes the determination of the unknown J_c distribution within HTSC's possible and allows a quality control even under these difficult conditions.

Fig. 6 shows for example an axis symmetric HTSC configuration in the trapped state (J_c=100 A/mm^2) where the external flux densities can be measured to determine the current density distribution within the HTSC.

The results of the identification are shown in Fig. 7. The identified J_c distribution differs from the initial one (J_c=const. in subregions), but it excites the measured field with a high accuracy. Such nonuniqueness of the solution is a typical feature of inverse problems (IFP). The precondition necessary for a successful identification of the unknown J_c distribution is the exact measurements of the magnetic field components at many points in the vicinity of the current carrying regions (Fig. 6).

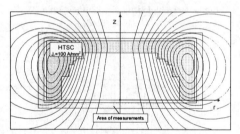

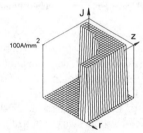

Fig. 6: HTSC with the area of measurement and arbitrary assumed current densities

Fig. 7: Identified current density distribution within the HTSC

5. Conclusions

Methods for the quality control of HTSC's are presented. One is based on the definite dependencies of the magnetic field on its sources, a second one uses the inverse problems methodology. Furthermore a nonlinear dynamic field calculation model (virtual displacement) is proposed to determine the J_c distribution at arbitrary external field exposures.

References

[1] W.-R. Canders, H. May, R. Palka, Identification of the current density distribution of monolithic superconductors, ISTET'97, Palermo
[2] Th. Klupsch *et al.*, Field Mapping Characterisation for Axially Magnetised, Superconducting Cylinders in the Remanent State: Theory and Experiment, J. Appl. Phys. (to be published)
[3] H. May, R. Palka, Identifikation und Berechnung der Stromdichteverteilung in zylindrischen und quaderförmigen, monolithischen Supraleitern. Technical Report, TU Braunschweig 1996
[4] R. Palka, Synthesis of magnetic fields by optimisation of the shape of areas and source distribution, Arch. Elektr. 75 (1991)
[5] M. Tsuchimoto, T. Kojima, H. Takeuchi, T. Honma, Numerical Analysis of Levitation Force and Flux Creep on High T$_c$ Superconductors, IEEE Trans. on Magnetics, VOL. 29, NO 6, Nov. 1993

Non-Linear Electromagnetic Systems
V. Kose and J. Sievert (Eds.)
IOS Press, 1998

Estimation of Shielding Current Density
of A Thin Film High Tc Superconductor

Masanori TSUCHIMOTO

Hokkaido University, Kita 13, Nishi 8, Kita-ku, Sapporo 060, Japan

Abstract. Hysteresis of the levitation force with a permanent magnet and a high Tc superconducting (HTS) thin film is discussed from calculated shielding current distribution obtained using a macroscopic numerical simulation code. Shielding current density of the thin film can be roughly estimated from the measurement of the levitation force. The levitation force of a stack of the thin films is also compared with that of bulk HTS. Numerical evaluation is useful to explain the experimental results and to estimate the shielding current density of the HTS thin film.

1. Introduction

Hysteresis of the levitation force is obtained when a permanent magnet is moved towards and moved away from a high Tc superconductor (HTS) [1]. The experimental hysteresis of the levitation force was explained well by using macroscopic numerical simulation codes [2-4]. Experimental levitation forces with HTS thin film were reported by Lucke et.al. [5]. The thin film has a high shielding current density compared with bulk material, but its total shielding current is small because of the thick substructure. In contrast to levitation force of the HTS bulk, almost symmetric hysteresis is obtained in the experiments .

In the present paper, the numerical code based on the critical state model is applied to the analysis of the levitation with the HTS thin film. The reported symmetric hysteresis of the HTS thin film is discussed from the calculated shielding current distribution [6]. Application of the force measurement to the estimation of the shielding current density of the HTS thin film is discussed based on the numerical results.

2. Numerical Formulation

Numerical formulation of the critical state model was explained precisely in the previous papers [2-4]. Macroscopic electromagnetic phenomena in the type-II superconductors are described by the Maxwell equations :

$$\nabla \times E = -\frac{\partial B}{\partial t}, \quad \nabla \times B = \mu_0 J, \quad \nabla \cdot B = 0, \quad J = J_0 + J_{SC}, \tag{1}$$

where μ_0, E, B are magnetic permeability in air, the electric and magnetic field, respectively. The magnetic field is caused by the shielding current J_{SC} and external current J_0. The standard critical state model is applied to the present analysis of the quasi-static levitation force. Constitutive relationships between the shielding current J_{SC} and the electric field E are obtained from force balance on the fluxoid [2] :

$$J_{SC} = J_C \frac{E}{|E|} \Delta S \quad (\text{if } |E| \neq 0), \qquad \frac{\partial J_{SC}}{\partial t} = 0 \quad (\text{if } |E| = 0), \tag{2}$$

where J_C and ΔS are the critical current density and an area of a numerical element. In the cylindrical coordinates (r, θ, z), the following axisymmetric scalar equation is obtained with the magnetic vector potential ($B = \nabla \times A$, $A = (0, A, 0)$) [3]:

$$(\nabla^2 - \frac{1}{r^2}) A_{SC} = -\mu_0 J_{SC}, \quad \nabla^2 = \frac{1}{r} \frac{\partial}{\partial r} (r \frac{\partial}{\partial r}) + \frac{\partial^2}{\partial z^2}, \tag{3}$$

$$E = -(\frac{\partial A_{SC}}{\partial t} + \frac{\partial A_0}{\partial t}), \tag{4}$$

where A_{SC} and A_0 are caused by the shielding current J_{SC} and the external current J_0. Equation (3) is transformed to the integral equation and is solved by the boundary element method [3]. Selfconsistent solutions, which satisfy nonlinear Eq. (2) at each time step, are obtained by using numerical techniques in iterative calculations [3, 4]. The shielding current J_n on each element in the n-th step is corrected as follows with the electric field E_n in Eq.(4) under the condition of $J_n \leq J_c \Delta S$:

$$J_n = J_{n-1} + \delta J \cdot \text{sign}(E_n) \quad \text{if } |E_n| > \varepsilon, \tag{5}$$

$$J_n = J_{n-1} \quad \text{if } |E_n| \leq \varepsilon,$$

where ε is the convergence condition and δJ is set to a small value according to the convergence of the solution.

3. Numerical Results and Discussion

The permanent magnet has 1.0 T magnetization, and a diameter and thickness of 10.0 mm and 5.0 mm. The HTS is 20.0 mm in diameter, and its thickness d and critical current density J_c are changed to simulate a bulk material and a thin film. The hysteresis of the levitation force is evaluated in the zero-field-cooling case, where the permanent magnet is slowly moved towards the HTS from a sufficiently long distance. When the gap between the HTS and the magnet is 1.0 mm, the magnet is moved away from the HTS. The experimental hysteresis of the levitation force, obtained in the bulk HTS, was explained from the shielding current distribution calculated by the macroscopic numerical simulation codes [2, 3]. Though the critical current density J_c strongly depends on the magnetic field, the Bean model (J_c=constant) is applied to the present numerical analysis for simple comparisons with the reported experimental results [1].

Solid lines in Fig. 1 show numerical levitation force of a typical HTS thin film ($J_c = 2 \times 10^{10}$ A/m^2, $d = 1 \times 10^{-7}$ m.). The reported symmetric hysteresis [5] is also obtained from the numerical analysis. Since the total induced currents are small to shield the applied field by the magnet, the currents flow in the whole volume of the film when the magnet is approached. When the magnet is moved away, reverse currents flow in the whole volume. The same absolute values of the attractive and the repulsion forces are obtained at the same gap length. Since the film is thin enough compared with the gap length, the obtained levitation forces are equal for the same $J_c d$ cases. The shielding current density of the HTS thin film can be roughly estimated from the measurement of the levitation force from its thickness [6].

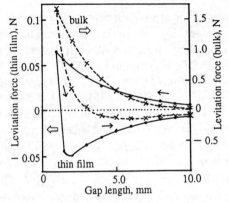

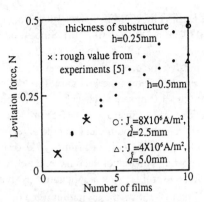

Fig.1 Numerical levitation force of a thin film and a bulk HTS

Fig.2 Numerical levitation force of a filmstack

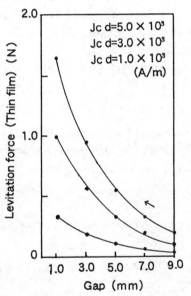

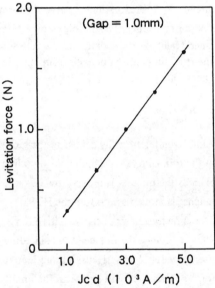

Fig.3 Numerical levitation force of a thin film for several $J_c d$

Fig.4 Numerical levitation force of a thin film at 1.0mm gap

Dashed lines in Fig.1 show numerical levitation force of a typical bulk material ($J_c = 5 \times$ 10^7 A/m², $d = 5 \times 10^{-3}$ m). Since total currents are large, not all the induced currents are reversed when the magnet is moved away. The attractive force is small compared with the repulsive force, and the hysteresis of the levitation force is not symmetrical [6].

Figure 2 shows the levitation force of a stack with films ($J_c d = 2 \times 10^3$ A/m) at the gap length =1.0 mm. Rough value of the reported experimental results [5] for 1 and 3 films (thickness of substructure h=0.5mm) are also shown by $\times$. The effect of multi films agrees well in both the numerical and the experimental results. Numerical results of a bulk material with equivalent J_c for 10 films are also shown by $\bigcirc$ and $\triangle$. The equivalent J_c is still small compared with that of the bulk material. The filmstack with a thin substructure and higher $J_c d$ will be useful for such applications as small actuators.

Figure 3 shows the levitation force of a thin film for different $J_c d$. The levitation force at the gap length =1.0 mm is also shown in Fig.4. Since the shielding currents flow in the whole volume of the HTS, the linear relationship between the levitation force and the $J_c d$ is obtained from analysis with the Bean model. From this figure, the shielding current density of a thin film can be roughly estimated given information of the thickness d and levitation force measurement. Since the shielding current density J_c of a thin film strongly depends on the magnetic field, the dependence must be considered with the Kim model [7], etc. The estimation of the shielding current density from experimental measurement data will be carried out in the near future.

4. Conclusion

1. Almost symmetrical hysteresis of the levitation force is obtained in the numerical analysis of the thin film HTS. The whole shielding currents, induced in the whole volume of the HTS as the magnet is moved towards to the HTS, changes to reverse currents when the magnet is moved away from the HTS.

2. The same hysteresis of the levitation force is obtained for the thin film HTS with the same value $J_c d$, since the induced total shielding currents are equal in each cases. It is useful to use the value $J_c d$ in discussing the levitation force of the thin film HTS. The levitation force of filmstack will be useful for small sized applications.

3. The shielding current density of the thin film HTS can be roughly estimated from the measurement of the levitation force using information of its thickness.

References

[1] M. Murakami, Melt Processed High-Temperature Superconductors, World Scientific, 1992.

[2] T. Sugiura, H. Hashizume and K. Miya, Int. J. Appl. Electromagn. in Materials, 2, 183, 1991.

[3] M. Tsuchimoto, et.al., IEE Japan, 114-D, 741, 1994.

[4] M. Tsuchimoto, et.al., Appl.Supercond. 2, 549, 1994.

[5] B. Lucke, H.G. Kurschner, B. Lehndorff, M. Lenkens and H. Piel, Physica C, 259, 151, 1996 .

[6] M.Tsuchimoto, et.al.,ISS96,Springer-Verlag, 1997.

[7] Y. B. Kim, et.al.,Phys. Rev. Lett., 9, 306, 1962.

Non-Linear Electromagnetic Systems
V. Kose and J. Sievert (Eds.)
IOS Press, 1998

MOCVD of superconducting $RBa_2Cu_3O_{7-\delta}$ thin films for microwave applications

Sergey V. SAMOYLENKOV, Oleg Yu. GORBENKO, Andrey R. KAUL

Chemistry Department, Moscow State University, 119899 Moscow, Russia

Asan R. KUZHAKHMETOV, Sergey A. ZHGOON

Moscow Power Engineering Institute, 111250 Moscow, Russia

Georg WAHL

IOPW, TU Braunschweig, Bienroder Weg 53, D-38108 Braunschweig, Germany

Abstract. $RBa_2Cu_3O_{7-\delta}$ (R=Lu,Ho,Gd) thin films with high superconducting characteristics (T_c=87-92 K, j_c(77 K) up to 2.7×10^6 A/cm^2) were prepared by MOCVD technique on $LaAlO_3$ and CeO_2-buffered Al_2O_3 substrates. The dependence of structural and superconducting characteristics of the films on R^{3+} radius is briefly discussed taking into account the difference in thermodynamics of $RBa_2Cu_3O_{7-\delta}$ phases. Microstrip and coplanar microwave resonators were fabricated from $LuBa_2Cu_3O_{7-\delta}$/$LaAlO_3$ thin films and for the coplanar resonators the loaded Q(67 K, 6.6 GHz) factor values up to 5500 were obtained. The microstrip resonators with a superconducting ground plane were fabricated from double-sided $RBa_2Cu_3O_{7-\delta}$ (R=Lu,Ho) films on $LaAlO_3$ substrates. The value of Q(63 K, 8.8 GHz) equal to 8000 was observed for $HoBa_2Cu_3O_{7-\delta}$ resonator.

1. Introduction

The family of $RBa_2Cu_3O_{7-\delta}$ (R-rare earth element) superconducting oxides with T_c of about 90 K belongs to one of the most attractive materials to be used for the fabrication of microwave devices operating at 77 K. The contemporary film deposition techniques, such as metallorganic chemical vapour deposition (MOCVD) [1], provide reproducible preparation of $RBa_2Cu_3O_{7-\delta}$ films making possible the design and fabrication of high-performance microwave circuits [2].

The properties of $RBa_2Cu_3O_{7-\delta}$ phases are known to vary considerably with the variation of R^{3+} ionic radius. However, until now most of the studies in the field of thin film deposition were performed for $YBa_2Cu_3O_{7-\delta}$ and the comparative systematic studies of $RBa_2Cu_3O_{7-\delta}$ films with different R are still lacking. It is known that the optimum $p(O_2)$-T deposition conditions of $YBa_2Cu_3O_{7-\delta}$ films correspond roughly to those of CuO/Cu_2O equilibrium line [3], the same behaviour has been recently observed for $LuBa_2Cu_3O_{7-\delta}$ films as well [4]. The superconducting properties and structural characteristics of $RBa_2Cu_3O_{7-\delta}$ films prepared under identical deposition conditions were shown to depend strongly on R^{3+} ionic radius and the behaviour was understood taking into account the difference in thermodynamics of $RBa_2Cu_3O_{7-\delta}$ phases [5]. Thus, a surface mobility during the film growth must be highest in the case of $LuBa_2Cu_3O_{7-\delta}$ phase which possesses the lowest peritectic melting temperature among $RBa_2Cu_3O_{7-\delta}$.

Recently, "fully oxygenated" $RBa_2Cu_3O_{7-\delta}$ phases (R= Y,Sm-Lu; $\delta\approx0$) were found to be overdoped (i.e. the hole concentration in CuO_2 planes of $RBa_2Cu_3O_7$ structure is higher than the optimum value) [6-8]. Furthermore, the overdoping effect was found to increase with the decrease of R^{3+} ionic radius reaching its maximum level for $LuBa_2Cu_3O_7$ [6,8]. Thus, an increase of T_c by 3 K was observed after the optimization of the oxygen content in $LuBa_2Cu_3O_{7-\delta}$ films (T_c=90 K for δ=0.15-0.20) [7]; an increase of T_c by about 2 K has been reported previously for $YBa_2Cu_3O_{7-\delta}$ (T_c≈93 K for δ=0.10) [6].

Here we present the study of $RBa_2Cu_3O_{7-\delta}$ films deposited by MOCVD on various substrates and the fabrication of microwave resonators from some of these films.

2. Experimental

$RBa_2Cu_3O_{7-\delta}$ (R=Lu,Ho,Gd) thin films of 300-700 nm thickness were prepared by powder flash evaporation MOCVD technique [9] on $LaAlO_3$ (001) – in a cubic system and ceria-buffered sapphire – CeO_2(001)/Al_2O_3 ($1\bar{1}02$) – single crystalline substrates. 60 nm-thick films of CeO_2 were prepared by band flash evaporation MOCVD to prevent the chemical interaction between the $RBa_2Cu_3O_{7-\delta}$ film and Al_2O_3 [10]. The films prepared were characterized by EDX and XRD analyses. The superconducting transition temperature T_c, the width of the superconducting transition ΔT_c and the temperature dependences of critical current density j_c(T) [11] were derived from AC magnetic susceptibility χ(T) measurements.

The microwave resonators were patterned by a standard photolithography process with a positive photoresist and wet chemical etching in NH_4F-buffered glycerol solution of H_3PO_4 (25 %) and HCl (25 %). The substrate dimensions were 10 mm × 10 mm × 0.5 mm. The dimensions of the resonators transmission lines were usually about 5 mm × 0.5 mm that corresponded to a resonator frequency in the range of 6-9 GHz. The microwave characteristics measurements were performed in a vacuum chamber of a special cryostat that allowed the sample to cool down to 63 K [12,13]. For the study of the oxygen content effect, the $LuBa_2Cu_3O_{7-\delta}$/$LaAlO_3$ microstrip resonators were annealed at 460 °C and $p(O_2)$=0.1 bar for 2 hours as described in [7].

3. Results and discussion

The properties of $RBa_2Cu_3O_{7-\delta}$ films are known to depend strongly on deposition conditions. Previously we have studied the influence of $p(O_2)$ on the characteristics of $LuBa_2Cu_3O_{7-\delta}$/$LaAlO_3$ and $LuBa_2Cu_3O_{7-\delta}$/$SrTiO_3$ films prepared at 800 °C [4]. The optimum $p(O_2)$ value was found to be about 1.3 mbar, that corresponded to the $p(O_2)$ value of $CuO \Leftrightarrow Cu_2O + 0.5\ O_2$ equilibrium. Besides, the comparative study of $RBa_2Cu_3O_{7-\delta}$/$LaAlO_3$ (R=Lu,Ho,Y,Gd) thin films deposited at 800 °C and $p(O_2)$=1.7 mbar was carried out [5]. It was shown that j_c(77 K) of $RBa_2Cu_3O_{7-\delta}$ films decreases remarkably with the increase of R^{3+} ionic radius. The behaviour was regarded to be a result of the decrease of surface mobility during the film growth with the increase of R^{3+} radius due to the increase of the peritectic melting temperature of $RBa_2Cu_3O_{7-\delta}$ phases. Accordingly, the tendency to c-axis oriented growth of $RBa_2Cu_3O_{7-\delta}$ films, which is known to be more preferable than the a-axis oriented growth due to thermodynamic factors [14], increases with the decrease of R^{3+} radius [5]. In order to avoid the formation of additional a-axis orientation and to improve the quality of $HoBa_2Cu_3O_{7-\delta}$ and $GdBa_2Cu_3O_{7-\delta}$ films, the deposition temperature was increased to 820 °C. Consequently, c-axis epitaxial $HoBa_2Cu_3O_{7-\delta}$ films with high superconducting characteristics were prepared (Table 1); the deposition conditions of $GdBa_2Cu_3O_{7-\delta}$ films should be further optimized.

Table 1. The deposition conditions and superconducting characteristics of "fully oxygenated" $RBa_2Cu_3O_{7-\delta}$ films prepared on $LaAlO_3$ and CeO_2 - buffered sapphire substrates

Film	Deposition temperature, °C	Deposition $p(O_2)$, mbar	T_c, K	ΔT_c, K	$j_c(77 \text{ K})$, $10^6 A/cm^2$	Film thickness, nm
LuBCO/LaAlO₃	800	1.3	87.0	1.0	2.7	300
LuBCO/CeO₂/Al₂O₃	800	1.3	89.0*	2.5	1.5	300
HoBCO/LaAlO₃	820	1.7	92.0	0.5	2.5	500
GdBCO/LaAlO₃	820	1.7	91.1	7.0	0.3	500

* the higher T_c value of this film is caused probably by a small oxygen deficiency of the $LuBa_2Cu_3O_{7-\delta}$ phase

The temperature dependences of the loaded quality factor Q of microstrip and coplanar resonators fabricated from $LuBa_2Cu_3O_{7-\delta}/LaAlO_3$ films are shown in Fig.1. The Q factor of the microstrip resonators at temperatures lower than 75 K was limited to the value of 3000, which was supposed to be due to losses in a copper ground plane. Its substitution by the silver one led to an increase in the value of Q by 5 %, confirming the supposition. The use of the coplanar resonator geometry, though it should provide lower Q factor values, allows to avoid the appearance of the upper limit of Q, since the grounded part of the circuit is made from the same superconducting film in this case. The Q factors achieved for both types of the resonators corresponded to $R_s(77 \text{ K}, 10 \text{ GHz})$ values of about 1 mOhm. The optimization of the oxygen content of $LuBa_2Cu_3O_{7-\delta}$ phase was shown to enhance $Q(77 \text{ K})$ of $LuBa_2Cu_3O_{7-\delta}/LaAlO_3$ microstrip resonator by 15% of the value and to lower $R_s(77 \text{ K}, 10 \text{ GHz})$ down to 700 µOhm.

Sapphire is usually considered as one of the most attractive substrates for the fabrication of superconducting microwave devices due to its unique dielectric characteristics ($\varepsilon_r \approx 10$ and $tg\delta \approx 4 \times 10^{-8}$; for $LaAlO_3$ $\varepsilon_r \approx 25$ and $tg\delta \approx 10^{-5}$) [15]. However, because of the strong chemical interaction of the $RBa_2Cu_3O_{7-\delta}$ phase with Al_2O_3 during the deposition, the use of the buffer layer is necessary. Epitaxial $LuBa_2Cu_3O_{7-\delta}/CeO_2/Al_2O_3$ structures were prepared at 800 °C and $p(O_2)=1.3$ mbar with quite high superconducting characteristics (Table 1). However, for the coplanar $LuBa_2Cu_3O_{7-\delta}/CeO_2/Al_2O_3$ resonator a $Q(66 \text{ K}, 9.2 \text{ GHz})$ value equal to 1700 was found, so further optimization of the film quality is necessary.

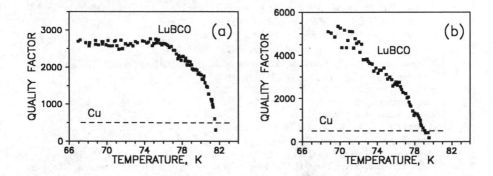

Figure 1. The temperature dependences of the loaded Q factor for the microstrip 8.9 GHz (a) and coplanar 6.6 GHz (b) resonators, fabricated from $LuBa_2Cu_3O_{7-\delta}/LaAlO_3$ films. Dashed line - the same data for identical copper resonators.

The resonators with the superconducting ground plane made from the films deposited on both sides of the substrate (Fig. 2) should provide higher Q factors as compared to the resonators of microstrip or coplanar geometry [16]. However, the degradation of the layer, deposited first, during the deposition of the second layer, due to a chemical interaction with a substrate holder, makes the preparation of the double-sided films difficult. In this work, in order to prevent the interaction, a 150 μm-thick plate of single crystalline $NdGaO_3$ was placed between the sample and the substrate holder. Using this technique, $LuBa_2Cu_3O_{7-\delta}$ and $HoBa_2Cu_3O_{7-\delta}$ double-sided films on single crystalline $LaAlO_3$ substrates were prepared. According to XRD and EDX analyses, the films deposited on the opposite sides of the substrate were nearly identical in their quality. The magnetic measurements revealed the sharp (ΔT_c=1.5 K) transition with T_c above 86 K for the $LuBa_2Cu_3O_{7-\delta}$ sample. These facts show that the chemical interaction was successfully eliminated. The temperature dependence of the Q factor for the microstrip resonator prepared from a $HoBa_2Cu_3O_{7-\delta}$ sample is given in Fig. 2. A Q(63 K, 8.8 GHz) value as high as 8000 was achieved for the resonator. For the $LuBa_2Cu_3O_{7-\delta}$ sample a Q(63 K, 8.8 GHz)=5500 was observed.

This study was supported by BMBF 13N6947/5 project, RFBR (No.96-03-33027) and Russian National Programm on HTSC (No.93-458). S.V.S. acknowledges the support of ISSEP.

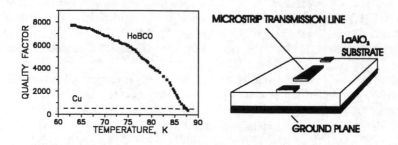

Figure 2. The Q(T) dependence of a 8.8 GHz microstrip resonator, fabricated from double-sided $HoBa_2Cu_3O_{7-\delta}$ film on $LaAlO_3$ substrate as shown in the scheme. The dashed line represents the dependence for an identical copper resonator.

References

[1]   M.Leskela et al., Supercond. Sci. Technol. 6 (1993) 627.
[2]   N.Newman and W.G.Lyons, J. Supercond. 6 (1993) 119.
[3]   R.Feenstra et al., J.Appl.Phys. 69 (1991) 6569.
[4]   S.V.Samoylenkov et al., J. Alloys and Compounds 251 (1997) 342.
[5]   S.V.Samoylenkov et al., in: Chemical Vapor Deposition, ed. M.D.Allendorf and C.Bernard, The Electrochemical Society, Inc., NJ, 97-25, 1997, p.990.
[6]   G.V.M.Williams and J.L.Tallon, Physica C 258 (1996) 41.
[7]   S.V.Samoylenkov et al., Physica C 267 (1996) 74.
[8]   S.V.Samoylenkov et al., Physica C 278 (1997) 49.
[9]   I.S.Chuprakov and A.R.Kaul, J.CVD 2 (1993) 123.
[10] I.E.Graboy et al., J. Alloys and Compounds 251 (1997) 347.
[11] J.Z.Sun et al., Phys. Rev. B 44 (1991) 5275.
[12] A.R. Kuzhakhmetov et al., Proc. XIII Int. Conf. Microwave Ferrites, Busteni, Romania, 1996, p.267.
[13] R.A.Pucel et al., IEEE Trans. Microwave Theory and Technology No.6 (1968) 342.
[14] S.J.Pennycook et al., Physica C 202 (1992) 1.
[15] Z.-Y.Shen, High-Temperature Superconducting Microwave Circuits, Artech House, Boston, London, 1994, p.11.
[16] S.Miura et al., J. Appl. Phys. 76 (1994) 4440.

Non-Linear Electromagnetic Systems
V. Kose and J. Sievert (Eds.)
IOS Press, 1998

High-Tc SQUIDs With a Ferromagnetic Antenna for a Magnetic Microscope

S.I.Bondarenko, A.A.Shablo

Special Research and Development Bureau of B.Verkin Institute for Low Temperature Physics and Engineering of National Academy of Sciences of Ukraine, 47, Lenin ave., Kharkov, 310164, Ukraine

Abstract. A new possibility of extending the information output of squid-based magnetic measurements on «warm» (T=300 K) microobjects is described. The possibility is afforded by a combination of a SQUID and a ferromagnetic antenna with a small-radius end-part. The "SQUID-ferromagnetic antenna" system promises a higher spatial resolution.

1. Justification

High-sensitivity measurements of magnetic fields from sources at distances much longer than the linear size of the SQUID (far sources) and sources at distances much shorter than SQUID size (near sources) call for various approaches to SQUID system design.

Typical near sources include magnetic defects of solid surfaces, magnetic inclusions in machine members, magnetic components of biological structures and living organisms. A high spatial-resolution record of the magnetic field is the aim of magnetic field testing, one of the non - destructive methods of control of materials and other objects.

So far, high-Tc (HTc) SQUIDs were used [1] to control near "warm" (T=300K) objects when no ferromagnetic antennae (FMA) were applied in the space between the SQUID detector and the object.

We believe that this arrangement needs further improvement.

2. Ferromagnetic Antennae of HTc SQUID for Measuring Magnetic Fields of Near Sources.

Until recently the HTc SQUIDs used to study "warm" (T≅300 K) sources suffered the following disadvantages:

- the HTc SQUID detector and the warm object can be brought together no less than 2-3 mm;

- the HTc SQUID detector integrates the magnetic field induction from the source within the detector area (hundreds of micrometers and more);

-the HTc SQUID detector cannot operate in the environment of a dc magnetic fields over$\cdot 10^{-4}$ T [1-4].

The use of FMA with a small-radius end-part permits detection and location of microsources, which is important to the present-day physics of magnetism [2]. On the

other hand, small magnetic moments of such sources require   high sensitivity of
measurement , which is ensured by SQUIDs.

It is essential that the «SQUID-FMA» system is designed so that the FMA effect
on the magnetic noise and the inductivity of the SQUID should be at a minimum. We
propose two methods of reducing this effect: (i) - a decrease in the magnetic SQUID-
FMA coupling or (ii) fulfilment of the condition $S_2 > > S_1$, where S_1 is the cross section
of a uniform linear FMA and S_2 is the area of the receiving loop of the superconducting
SQUID antenna. Both methods promise higher sensitivity of magnetic measurements of
near source parameters.

The methods have been studied experimentally using the "SQUID - FMA"
system. In the first case the HT_c SQUID detector was a cylindrical RF - interferometer
made of the bulk ceramic $YBa_2Cu_3O_{7-\delta}$ (Fig.1).

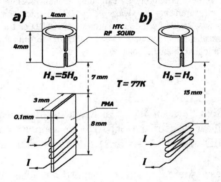

Figure 1. Sketch of experiment with HT$_c$-SQUID and   FMA

The magnetic field microsource was simulated with a small-size solenoid carrying
the current I. With a FMA between the SQUID and the source (solenoid), the SQUID
measures the induction $B_a = 5 B_0$, where B_0 is the magnetic field induction induced in
the SQUID region by the same solenoid when there is no FMA (Fig.1b). In this case the
FMA is a rectangular piece of permalloy foil.

The five times enhancement of the magnetic field in the SQUID region is
essentially higher than the slight (1.5 - 1.8 times) increase in the magnetic noise which
occurs in the presence of a FMA near the SQUID (Fig.2).

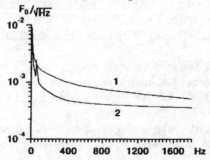

Figure 2. SQUID noise spectral density: 1- with FMA;   2- without FMA.

In the other case we used a SQUID interferometer and a superconducting antenna with the receiving loop area S_2 (Fig.3). The condition $S_2 >> S_1$ was obeyed. The FMA was a permalloy foil rolled to form a tube. The magnetic moment (M) of a field source was simulated with the same solenoid placed at the distant end of the FMA (the distance from the SQUID is like in Fig.1.

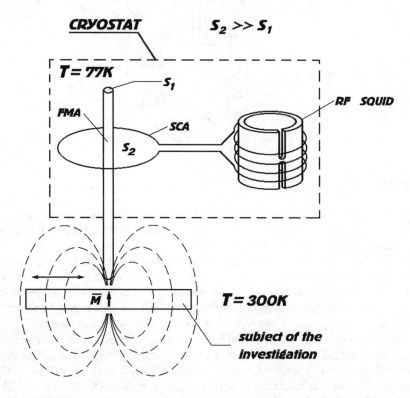

Figure 3. Sketch of magnetic microscope ""SQUID-FMA" with magnetic coupling FMA with superconducting antennae (SCA) of RF-SQUID . M- magnetic moment of subject of investigation.

With this arrangement, the enhancement of the magnetic field in the SQUID region is 16 times higher than in the absence of a FMA between the source and the SQUID. With the presence of a FMA near the superconducting antenna, there is no appreciable increase in the magnetic noise of the SQUID.

Let us estimate the expected magnetic noise induction produced by a FMA. It is known experimentally, that the magnetic noise of permalloy-based ferroprobe detectors does not exceed $B_m = 10^{-9}$ T/Hz$^{1/2}$ at T=77 K. As a result, the magnetic noise flux in the linear FMA is $\Phi = B_{n1} * S_1$. In the worst limiting case (no allowance for scattering) this flux goes through the superconducting antenna (SCA). We can find the corresponding induction B_{n2} of this SCA noise. Taking into account that $B_{n1} * S_1 = B_{n2} * S_2$, we obtain $B_{n2} = B_{n1} / (S_2 / S_1)$. If we take $S_2 = 1$ cm^2 and $S_1 = 10^{-4}$ cm^2, then $S_2 / S_1 = 10^4$ and $B_{n2} = 10^{-13}$ T/Hz$^{1/2}$. It is seen that the FMA - generated noise is at the level of the intrinsic noise of the SQUID without a FMA. This suggests an extremely small contribution of a FMA to the SQUID noise.

To avoid a loss of the cryogen through a FMA, we propose an antenna design with a gap(δ) which is in the vacuum «jacket» of the cryostat (Fig.4).

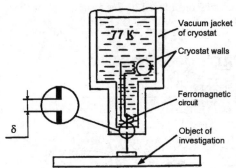

Fig.4 Schematical view of a SQUID-containing cryostat and a ferromagnetic antenna with the gapδ.

3. Conclusions

The magnetic field acting on the SQUID can be increased considerably if a ferromagnetic antenna is placed between the field source and the SQUID. This will bring only a slight increase in the magnetic noise of the SQUID.

If necessary, the FMA end diameter may be reduced to less then 10^{-5} cm. This FMA end can be brought very close to the object, even to the extent of an immediate contact. The location of a magnetic field microsource can therefore be much more accurate than it is permissible with the current SQUID detectors having linear sizes $10^{-2}...10^{-1}$ cm and being 2-3 cm away from the "warm" objects to be measured.

The combination of a SQUID and a FMA with a small-radius end can be used to develop a new version of a high-sensitivity magnetic microscope, which will cope with the physical problems typical for current magnetic force microscopes but will be advantageous in considerably larger scanned area. Since a FMA can operate both at low and high temperatures, the microscope will be efficient with ''warm'' and ''cold'' objects. Finally, the use of FMA will permit us to extend the local external magnetic field in the region of a FMA and an object to $10^{-3}...10^{-1}$T which will contribute to further improvement of the spatial resolution against current SQUID-microscopes operating under field not more than $(1..3) \cdot 10^{-4}$ T [4].

References

[1] Fan C.-X., Lu D.F., Wong K.W. et al High temperature RF SQUID gradiometer applied to non-destructive testing, Cryogenics (1994) , 34, 667-670
[2] Gibson G.A., Schultz S. A high - sensitivity alternating gradient magnetometer for use in quantifying magnetic force microscopy. Journal of Appl. Phys., 15 April (1991), 5880-5885.
[3] Edelman V.S. Scanning tunnelling microscopes. PTE (Pribory i technika experimenta) (1989) N5, 25-49.
[4] Gudoshnikov S.A., Vengrus I.I., Andreev K.E., Snigirev O.V. Magnetic microscope based YBaCuO bicrital thin film DC SQUID operating at 77 K, Cryogenics (1994), 34, 907-909.

Non-Linear Electromagnetic Systems
V. Kose and J. Sievert (Eds.)
IOS Press, 1998

Magnetic Force Acting on a Soft Magnetic Material Using the Flux Pinning of Bulk Superconductors

H. OHSAKI, M. TAKABATAKE and E. MASADA
The University of Tokyo, Bunkyo-ku, Tokyo 113, Japan

Abstract. The mixed-μ levitation system and the magnetic gradient levitation system are experimentally studied, which use bulk superconductors for a stable levitation of a soft magnetic material like iron. The generation of a restoring force against the displacement of the iron is demonstrated and the electromagnetic force characteristics are discussed.

1. Introduction

Stable levitation systems utilizing strong flux pinning of bulk superconductors (bulk-SC) have been recently investigated. The authors have proposed the magnetic gradient levitation system, which uses bulk superconductors both as a flux source and as a field shaping material to generate stable levitation force on an iron rail [1]. The mixed-μ levitation system [2-4] and the levitation system proposed by Tsutsui, et al. [5] can be also classified into the same category as the magnetic gradient levitation system because they all form a flux distribution around the superconductor which generates restoring force against displacement without active control [6]. This paper deals with the mixed-μ levitation system (permanent magnet + bulk-SC + Iron) and the magnetic gradient levitation system (bulk-SC + Iron) shown in Fig. 1. These levitation systems only need an iron rail with which to interact. Therefore, an economical levitation system can be realized, in particular the application to a transportation system over relatively long distances.

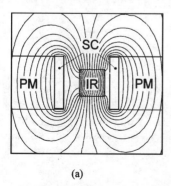 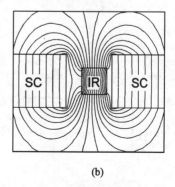

(a)　　　　　　　　　　　　　　　(b)

Fig. 1. Fundamental structure and magnetic flux distribution of the levitation systems using bulk superconductors: (a) PM-type mixed-μ levitation system, (b) magnetic gradient levitation system. (PM: permanent magnet, SC: superconductor, IR: iron rail.)

Both levitation systems have an essential stability against a displacement of the iron rail in the y-direction, even if there is no superconductor in the mixed-μ system. However, the stability against a displacement in the x-direction depends on the system design and materials used. The finite element analysis has shown that the levitation characteristics depend strongly upon the superconducting properties, such as a critical current density and its dependence on the magnetic field, etc. [7]. Bulk superconductors with a lower critical current density cannot realize a stable levitation. However, the recent development of the melt processes of bulk superconductors has enabled good Y-Ba-Cu-O and RE-Ba-Cu-O bulk materials with a high critical current density, strong pinning force and very few weak links to be manufactured. These are expected to realize a stable levitation.

In this paper the mixed-μ levitation system and the magnetic gradient levitation system, which can support a soft magnetic material like iron without active control, are experimentally studied. The generation of a restoring force against the displacement of the iron rail is demonstrated and the electromagnetic force characteristics are discussed.

2. Experimental Setup

Figure 2 shows the configurations of the mixed-μ and the magnetic gradient levitation systems for electromagnetic force measurement. The c axis of the bulk superconductor coincided with the x-axis in the mixed-μ system and with the y-axis in the magnetic gradient levitation system, respectively. SmCo permanent magnets were used in the mixed-μ system. The flux density on the surface of the permanent magnet was about 0.34T. YBCO bulk superconductors produced by Nippon Steel Corp. were used in the experiment. Due to the restriction of the size and shape of available bulk superconductors the x-z cross-section of the superconductors used in the magnetic gradient levitation was a half circle. All the components shown in Fig. 2 were cooled down to 77K in liquid nitrogen.

Figure 3 shows the flux trapping characteristics of the superconductor used in the magnetic gradient levitation system. After the field cooling of the superconductor in the magnetic field of about 1T the flux density distribution in the plane 1.5mm above the superconductor surface was measured with a Hall sensor. The maximum flux density was 0.41T. From the flux density profile the critical current density J_C of the superconductor was calculated: $J_C=7 \times 10^7 A/m^2$. The superconductor used in the mixed-μ levitation system has almost the same value of J_C.

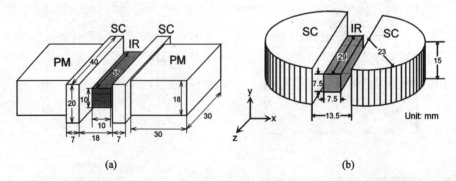

(a)   (b)

Fig. 2. System configuration for force measurement: (a) mixed-μ levitation system, (b) magnetic gradient levitation system.

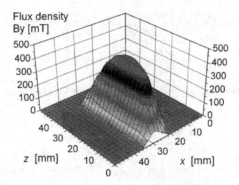

Fig. 3. Flux density distribution of the bulk superconductor used for the magnetic gradient levitation system. (Field cooling, 1.5mm above the SC surface, B_{max}=0.41T)

In the experiment of the mixed-μ system the zero-field cooling was applied to the superconductors with the permanent magnets kept at a distance. Then the permanent magnets were moved to the right position shown in Fig. 2. The iron rail was supported by linear bearings and could move only in the x direction. The electromagnetic force acting on the iron rail in the x direction was measured with load cells. In the experiment of the magnetic gradient levitation system the bulk superconductors were cooled down in the uniform magnetic field and after the switch-off of the externally applied field the iron rail was inserted between the superconductors as shown in Fig. 2. The force measurement method was the same as that for the mixed-μ system.

3. Results and Discussion

Figure 4 (a) shows the influence of the bulk superconductors on the characteristics of the x-directional electromagnetic force acting on the iron rail in the mixed-μ levitation system. By placing bulk superconductors between the permanent magnets and the iron rail as shown in Fig. 2 (a) the electromagnetic force characteristics was stabilized, that is, a restoring force was obtained.

Figure 4 (b) shows the dependence of the x-directional electromagnetic force characteristics on the magnetic field applied during the field cooling in the magnetic gradient levitation system shown in Fig. 2 (b). When the applied field was lower than 0.2T, a restoring force was produced. When the applied field was larger than 0.25T, the force became unstable. It is considered that the magnetic flux required for a stable levitation could not be kept in the bulk superconductor due to its finite critical current density. The critical point existed between 0.2 and 0.25T.

The two-dimensional analysis of the influence of the critical current density J_C of the superconductors on the electromagnetic force characteristic of the magnetic gradient levitation system showed that $J_C=1 \times 10^8 A/m^2$ was the marginal value for a stable levitation when the flux density in the superconductor was 0.37T [7]. In the measurements, the superconductors had a critical current density of $J_C=7 \times 10^7 A/m^2$ and three-dimensional effects would deteriorate the levitation stability to some extent. We consider, therefore, such actual conditions reduced the critical value of the applied field during field cooling and it was between 0.2 and 0.25T.

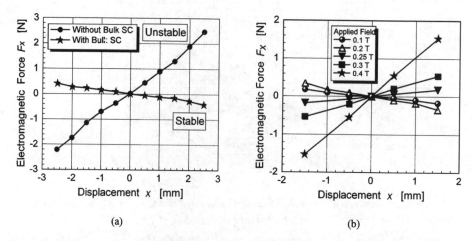

(a)   (b)

Fig. 4. Characteristics of the x-directional electromagnetic force acting on the iron rail as a function of the displacement of the iron rail: (a) influence of the bulk superconductors in the mixed-μ levitation system, (b) dependence on the applied magnetic field during field cooling in the magnetic gradient levitation system.

4. Conclusion

We studied experimentally the mixed-μ system and the magnetic gradient levitation system, which use bulk superconductors for a levitation stability. From the measurement of electromagnetic force acting on the iron rail the generation of a restoring force against the displacement of the iron rail has been confirmed in both systems. In the magnetic gradient levitation system there was the critical value of the applied field during field cooling: above this value the electromagnetic force became unstable due to the finite critical current density of the superconductor.

References

[1] H. Ohsaki, H. Kitahara, and E. Masada, Magnetic levitation systems using a high-Tc superconducting bulk magnet, Proc. 14th Int. Conf. on Magnetically Levitated Systems (Maglev '95), VDE, Germany, pp.203-208, 1995.

[2] G. J. Homer, T. C. Randle, C. R. Walter, M. N. Wilson and M. K. Bevir, A New Method for Stable Levitation of an Iron Body using Superconductors, J. Phys. D: Appl. Phys., vol.10, pp.879-886, 1977.

[3] H. Joyce, J. T. Williams, R. J. A. Paul, G. Gallagher-Daggitt and J. Brown, Some Initial Experimental Results for a New Form of Magnetic Suspension, IEEE Trans. Magn., vol.17, pp.2154-2157, 1981.

[4] H. Weh, H. May and H. Hupe, Magnet concepts with one- and two-dimensional stable suspension characteristics, Proc. 11th Int. Conf. on Magnetically Levitated Systems and Linear Drives (Maglev '89), IEE of Japan, pp.251-256, 1989.

[5] Y. Tsutsui, A. Yamamoto and T. Higuchi, Magnetic suspension using soft magnetic materials and high-Tc superconductors, Proc. Int. Symp. on Linear Drives for Industry Applications (LDIA '95), IEE of Japan, pp.259-262, 1995.

[6] H. Ohsaki, M. Takabatake and E. Masada, Stable Magnetic Levitation of Soft Ferromagnetic Materials by Flux Pinning of Bulk Superconductors, INTERMAG97, HD-07, New Orleans, U.S.A.

[7] H. Ohsaki, M. Takabatake and E. Masada, Magnetic Gradient Levitation using High-Tc Bulk Superconductors, IEEE Applied Superconductivity Conference 1996, LSC-5, Pittsburgh, U.S.A.

Non-Linear Electromagnetic Systems
V. Kose and J. Sievert (Eds.)
IOS Press, 1998

Magnetic Shielding Analysis of Superconducting Plates by Using 2D Current Vector Potential Method

Atsushi KAMITANI* and Takafumi YOKONO**

* Yamagata University, Johnan 4-3-16, Yonezawa, Yamagata 992, Japan
**University of Tsukuba, Tennoudai 1-1-1, Tsukuba, Ibaraki 305, Japan

Abstract. The magnetic shielding effect of the superconducting plate is investigated numerically. By assuming that the shielding current does not flow along the thickness direction, the superconducting plate is modelized into an assembly of multiple thin layers. Under these assumptions, the governing equation of the shielding current is expressed in terms of a scalar function. A numerical code to integrate the equation has been developed and the magnetic shielding effect is analyzed by use of the code. The results of computations show that the magnetic shielding ability of the superconductor does not depend on the frequency of the applied magnetic field.

1. Introduction

It has recently been demonstrated that high-Tc superconductors can shield the magnetic field over a wide frequency range[1,2]. Although high-conductivity or high-permeability materials have already been used to construct magnetic shielding apparatuses, they cannot cut off magnetic fields with a low frequency below 10 Hz. In this sense, high Tc superconductors are superior to the high-conductivity and the high-permeability materials in the magnetic shielding. In addition, high Tc superconductors are easy to keep in the superconducting state by use of liquid nitrogen or a mini-refrigerator. On account of these advantages, high Tc superconductors have recently attracted great attention as materials for shielding magnetic fields.

The purpose of the present study is to investigate the magnetic shielding effect of superconducting plates by means of numerical simulation.

2. Mathematical Formulation

In this section, we introduce the governing equation of the shielding current flowing in a superconducting plate. Throughout the present study, we assume that the superconducting plate has the same cross section along its thickness direction. Let us use the Cartesian coordinate system, the z axis of which is taken along the thickness direction. In the following, the cross-section of the superconducting plate and its boundary are denoted by Ω and Γ, respectively.

Recently it has been observed that the critical current density along the crystallographic c axis of the MPMG-YBCO superconductor is about one third of that in the a-b plane and that the c axis is parallel to the thickness direction. Thus we may assume that the superconducting plate consists of multiple thin layers and that the shielding current density has no z component in every layer. The central plane of the i-th layer is placed on $z = z_i$ and its thickness is denoted by $2\varepsilon_i$. By using these notations, the interior of the i-th layer can be expressed as { (x, y, z): $(x, y) \in \Omega, |z - z_i| \le \varepsilon_i$}. We further assume that each layer is thin so that the shielding current density does not change its magnitude through thickness of each layer. In addition, the shielding current density is a divergence-free vector if frequency of the externally applied magnetic field is low enough and, therefore, the displacement current term can be neglected. In this case, there exists a scalar function $T_i(x, y, t)$ such that

$$j_i = \nabla T_i \times e_z \ . \tag{1}$$

Here j_i denotes the shielding current density flowing in the i-th layer.

Under the above assumptions, the temporal evolution of the shielding current density can be expressed in terms of $T_i(x, y, t)$ as follows:

$$-\nabla \cdot \left(\frac{1}{\sigma_i} \nabla T_i \right) + \frac{\partial \langle B_{0z} \rangle_i}{\partial t} + \mu_0 \frac{\partial T_i}{\partial t} + \mu_0 \frac{\partial}{\partial t} \sum_{k=1}^{K} \iint_{\Omega} T_k(x', y', t) \, q_{ki} \, dx' \, dy' = 0 \ . \tag{2}$$

Here K is the number of layers and μ_0 represents permeability of vacuum. Further $\langle B_{0z} \rangle_i$ denotes the average of the z component of the externally applied magnetic field along thickness of the i-th layer. The fourth term in the right-hand side of Eq. (2) represents the self and the mutual induction of the shielding current and, hence, q_{ki} corresponds to the mutual inductance between the k-th and the i-th layer. The explicit form of q_{ki} is given by

$$q_{ki} \equiv -\frac{1}{8\pi\varepsilon_i} \sum_{m=1}^{2} \sum_{n=1}^{2} \frac{(-1)^{m+n}}{\sqrt{r^2 + \{ [z_k + (-1)^m \varepsilon_k] - [z_i + (-1)^n \varepsilon_i] \}^2}} \ , \tag{3}$$

where $r^2 \equiv (x - x')^2 + (y - y')^2$. In addition, σ_i denotes a macroscopic electric conductivity of the i-th layer. Although the electric conductivity of the superconductor is actually infinite, the spatial distribution of σ_i is determined iteratively at each time. The method for the evaluation of σ_i is described later. The initial and the boundary conditions of the shielding current j_i can be written as $j_i \cdot n = 0$ on the layer surface and $j_i = 0$ at $t = 0$, respectively. Rewriting these conditions in terms of $T_i(x, y, t)$, we get the initial and the boundary conditions for Eq. (2) as follows: $T_i = 0$ on Γ and $T_i = 0$ at $t = 0$.

Now let us explain the method for evaluating the macroscopic electric conductivity σ_i by use of the modified Bean model[3]. In using this model as a critical state model, σ_i is calculated by employing an iterative procedure. At first, the initial value $\sigma_i^{(0)}$ of the electric conductivity is assumed to be sufficiently large, e.g., $\sigma_{cu} \times 10^8$. Here σ_{cu} denotes an electric conductivity of copper. In the n-th cycle, $\sigma_i^{(n)}$ is substituted into σ_i and, subsequently, Eq. (2) is solved with the boundary conditions to determine $T_i(x, y, t)$. Next, the spatial distribution $j_i^{(n)}(x, t)$ of the shielding current is determined by using Eq. (1). Finally, the electric conductivity $\sigma_i^{(n)}$ is corrected by

$$\sigma_i^{(n+1)}(x, t) = \sigma_i^{(n)}(x, t) \frac{j_C}{\text{Max}(| j_i^{(n)}(x, t) |, j_C)} \ , \tag{4}$$

where j_C is the critical current density of the superconductor. The above cycle is iterated until the current density becomes lower than j_C all over the superconducting plate.

Equation (2) is discretized by use of the finite element method and the resulting discretized equation is advanced in time by employing the complete implicit scheme. The numerical code to solve Eq. (2) by this method has been developed. By using the code, we can investigate the magnetic shielding effect of the superconducting plate. As the measure of the magnetic shielding effect, we define the time-averaged shielding coefficient α by

$$\alpha \equiv 20 \log_{10} \left(\langle B^2 \rangle / \langle B_0^2 \rangle \right)^{1/2} . \tag{5}$$

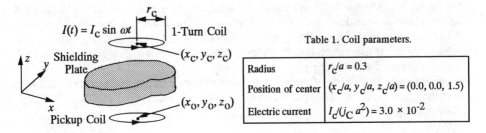

Fig. 1. Schematic view of the magnetic shielding experiments.

Table 1. Coil parameters.

Radius	$r_c/a = 0.3$
Position of center	$(x_c/a, y_c/a, z_c/a) = (0.0, 0.0, 1.5)$
Electric current	$I_c/(j_C\,a^2) = 3.0 \times 10^{-2}$

Here B and B_0 denote the total and the applied magnetic flux density, respectively, and the bracket denotes the time average.

3. Magnetic Shielding Analysis

In this section, let us investigate the magnetic shielding effect of superconducting plates by using the code explained in the previous section. Figure 1 shows the schematic view of the magnetic shielding experiments. In order to reproduce the experimental conditions exactly, the 1-turn coil is placed above the plate and the electric current $I(t) = I_c \sin \omega t$ is applied. The specifications of the coil used in the present study are shown in Table 1. Throughout the present analysis, the shielding plate is assumed to have a square cross section and its side length is denoted by a. Further, the observation point is fixed as $(x_0, y_0, z_0) = (0, 0, -0.2a)$.

First, we investigate the influence of frequency ω of the applied magnetic field on the shielding effect. The time-averaged shielding coefficients are calculated for various values of ω and are depicted in Fig. 2(a). This figure shows that the magnetic shielding effect of the superconductor is independent of field frequency. On the other hand, the shielding effect of copper is negligibly small while the inequality $\omega\tau_{Cu} < 10^1$ is satisfied. However, once the applied magnetic field frequency exceeds $10^1/\tau_{Cu}$, the shielding effect of copper is considerably improved with the increase of field frequency. Figure 2(a) also indicates that the shielding ability of the normal conductor is enhanced with the increase of its conductivity. This is mainly due to the phase shift δ between the applied field and the shielding current. In Fig. 2(b), we show the values of δ as functions of $\omega\tau_{Cu}$. We see from this figure that the phase shift

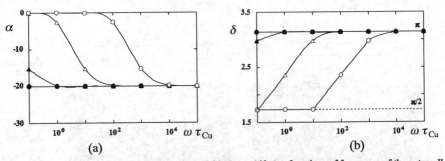

Fig. 2. (a) Time-averaged shielding coefficient α and (b) phase shift δ as functions of frequency of the externally applied magnetic field for single-layer plate with $\varepsilon_1/a = 0.05$ and $z_1 = 0$. The symbols, $\bigcirc$, $\triangle$ and $\blacktriangle$, denote the values of $\bar{\alpha}$ and δ for the normal conductor whereas the symbol $\bullet$ indicates those for the superconductor. The electric conductivity for the symbols, $\bigcirc$, $\triangle$ and $\blacktriangle$, is given by $\sigma_1/\sigma_{Cu} = 1$, 10^2 and 10^4, respectively. Here τ_{Cu} represents the magnetic diffusion time of copper.

δ for the superconductor is always equal to π and, therefore, the magnetic field does not penetrate inside the shielding plate. On the other hand, the phase shift for the normal conductor increases monotonously from $\pi/2$ to π with the increase of ω. This tendency explains the low shielding ability of the normal conductor under low-frequency magnetic field and its high ability under high-frequency field.

Next, the shielding ability of the multiple-layer superconducting plate is compared with that of the single-layer one. The spatial distribution of the time-averaged shielding coefficient is shown in Fig. 3.

4. Conclusions

We have developed a numerical code for analyzing the time evolution of the shielding current and have investigated the magnetic shielding effect of a superconductor and of a normal conductor by means of the code. Conclusions obtained in the present study are summarized as follows. 1) The magnetic shielding ability of the superconductor has no dependency on ω whereas that of the normal conductor depends strongly on ω. With the increase of ω, the magnetic shielding effect of the normal conductor is enhanced remarkably. 2) The spatial distribution of α for the double-layer superconducting plate is almost the same as that for the single-layer one. This is because the shielding current is almost entirely distributed in the upper layer of the double-layer plate.

Since the finite element method is applied to the discretization of the governing equation, the resulting matrix equation includes a dense matrix due to the integral term in the left-hand side of Eq.(2). Hence, it costs us much CPU time and computer memory to solve the matrix equation numerically if the layer number K is increased. In order to save the computer resources, we have to develop the fast algorithm for the solution of Eq. (2). This remains to us as a future problem.

References

[1] S. Ohshima and K. Okuyama, Magnetic Shielding Effect of $Ba_2YCu_3O_{7-\delta}$ Plates, *Jpn. J. Appl. Phys.* **29** (1990) 2403-2406.
[2] S. Ohshima, H. Ohtsu, A. Kamitani, S. Kambe and K. Okuyama, Magnetic Shielding Effect of Overlapped BPSCCO Plates. In: Y. Bando and H. Yamauchi (eds.), *Proc. 5th Int. Symp. Superconductivity (ISS '92), Kobe, 1992*. Springer, Tokyo, 1993, pp. 1273-1276.
[3] H. Hashizume, T. Sugiura, K. Miya, Y. Ando, S. Akita, S. Torii, Y. Kubota and T. Ogasawara, Numerical analysis of a.c. losses in superconductors, *Cryogenics* **31** (1991) 601-606.

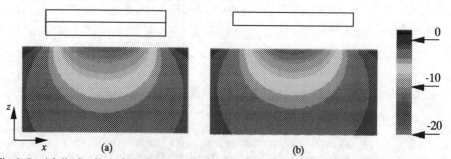

Fig. 3. Spatial distributions of the time-averaged shielding coefficient for (a) the double-layer and (b) the single-layer superconducting plates for the case $\omega\tau_{Cu} = 0.1$. Here, values of z_i are fixed as $z_1 = 0$ and $z_2 = -0.1$, and thickness of every layer is equal to $0.05a$.

Non-Linear Electromagnetic Systems
V. Kose and J. Sievert (Eds.)
IOS Press, 1998

Quench Stability Analysis for a New Type Forced-Flow Cooled Superconducting Magnet

Masahiro MIYAUCHI

Yamagata University, Johnan 4-3-16, Yonezawa, Yamagata 992, Japan

Abstract. A simple one-dimensional modeling method for a multi-channel forced-flow cooled superconducting coil is proposed. For the consideration of the non-linear thermal properties of coolant and conductor at very low temperature such as 4K, the properties are given by the measurements. By use of this method, quenching effect is numerically reproduced, then the transient behavior of the supercritical helium coolant is calculated.

1. Introduction

The superconductor (SC) cable used for SC magnet is made of many thin Nb_3-Sn or Nb-Ti wires with stabilizing material, such as copper or pure aluminum. For generating very high magnetic field, large current is operated to the cable, cooled by very low temperature helium. In the SC cable, a small thermal disturbance sometimes occurs by flux jump and displacement of the materials from electromagnetic force. This small amount of heating can change a part of the SC material into a normal conducting state. Then, Joule heating would be caused by the operating current, and propagate the normal conducting region. Hence, the cooling system of the SC should be designed to avoid this instability, so called quench. In this sense, a forced-flow cooled (FFC) SC is developed to enhance the stability.

However, multiple stability of the FFC-SC was detected by J.W.Lue, J.R.Miller, L.Dresner [1,2], and the same kind of phenomenon was also numerically reproduced by T.Amano, A.Kamitani et al.[3]. For avoiding this complex stability margin, multi-channel type FFC-SC is proposed. Thus, the purpose of the present study is to make a simple numerical model of multi-channel FFC-SC and to estimate its stability.

2. Numerical model and scheme for analysis

A multi-channel FFC-SC having subchannels beside the main-channel, is shown in Fig.1, compared with the ordinary type. In the present section, first, modeling and formulation of the coolant (supercritical helium) and conductor (SC and copper) in each channel are shown. Next, the boundary condition and closs flow modeling between the channels are described.

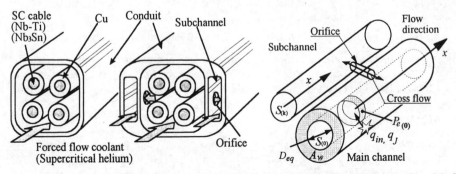

(a) Ordinary type   (b) Subchannel type
Fig.1 Forced flow cooled SC

Fig.2 Multi-channel one-dimensional model of forced-flow cooled SC

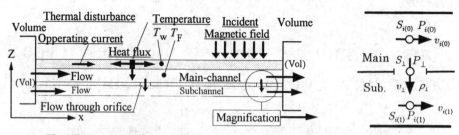

Fig.3 Present one-dimensional modeling for flow passages with subchannel

2.1 One-dimensional passage modeling for each channel

Each channel is assumed as an one-dimensional passage, and coordinate x is located along the passage. Section S and perimeter Pe are not changed with time. Each channel is connected by orifices at several points, and coolant can flow between the channels, shown in Fig.2 and Fig.3. There are sufficiently large volumes at both ends of the passage. Flow velocity in the volumes are assumed to be zero. A_w in Fig.2 represents the section of the conductor including SC and stabilizing conductor (Cu in present case). Governing equations of fluid flow along the x-direction of each channel are represented as follows.

$$\frac{\partial}{\partial t}\begin{bmatrix} \rho \\ m \\ e \end{bmatrix} + \frac{\partial}{\partial x}\begin{bmatrix} m \\ mv+P \\ hv \end{bmatrix} = \begin{bmatrix} -\dfrac{m}{S}\dfrac{dS}{dx} \\ -\dfrac{mv}{S}\dfrac{dS}{dx} + \rho(F_B - F_F) \\ -\dfrac{mh}{S}\dfrac{dS}{dx} + mF_B + q \end{bmatrix}, \quad \text{for each channel} \tag{1}$$

Here, ρ, v, P, m, e and h are density, velocity, pressure, mass flow rate through unit section, internal energy and enthalpy in unit volume, respectively. Mass flow rate and enthalpy are also written as $m = \rho v$ and $h = e + P$. Friction force F_F and body force F_B in Eq.(1) for unit mass are represented by,

$$F_F = f v|v|/(2D_{eq}), \quad F_B = g_0\, dZ/dx, \text{ (Equivalent diameter } D_{eq} = 4S/Pe). \tag{2}$$

Where, g_0 is acceleration of gravity. Friction coefficient $f(x,t)$ is defined as a function of local Reynolds number Re, by using Eqs.(3) and (4).

$$f = 64/Re \quad (Re \le Re_c), \qquad f^{-1/2} = 0.87\ln(Re\, f^{1/2}) - 2.4 \quad (Re > Re_c) \tag{3}$$

$$Re = D_{eq}|v|/v \tag{4}$$

Here, v is local kinetic viscosity. Critical Reynolds number Re_c is fixed as 2000 in the present study. Eq.(3) is derived by Nikuradse's equation. In Eq.(1), q means heat flux into unit volume at unit time, and is evaluated by,

$$q = 4h_{wF}(T_F - T_w)/\rho D_{eq} \quad \text{for the main-channel}, \qquad q{=}0 \quad \text{for the subchannel.} \tag{5}$$

In the above equation, T_F and T_w are temperature of fluid and conductor wall, respectively. Suffices "F" and "w" are used to distinguish fluid and conductor wall. h_{wF} is a local coefficient of heat transfer between the fluid and wall, and is evaluated by the following Dittus-Boelter's equation.

$$h_{wF} = 0.023Re^{0.8}\, Pr^{0.4}\, \lambda/D_{eq}, \quad Pr = \rho v C_P/\lambda = \mu C_P/\lambda. \tag{6}$$

Here, μ, v, λ and C_P are viscosity, kinetic viscosity, coefficient of heat conduction and specific heat of the coolant, respectively. They and P, e are evaluated by measured thermal property data of helium as functions of density ρ and temperature T_F. In addition, gas constant R and specific heat ratio γ are also given by Eq.(7). Non-linear property of C_P is shown in Fig.4, as an example.

$$R = P /(\rho\, T_{\text{F}}), \qquad \gamma = (e + P)/e. \tag{7}$$

On the other hand, heat barnacle in the conductor is written as follows.

$$\begin{cases} C_{\text{w}} \rho_{\text{w}} A_{\text{w}} \dfrac{\partial T_{\text{w}}}{\partial t} = \dfrac{\partial}{\partial x}\left(A_{\text{w}} \lambda_{\text{w}} \dfrac{\partial T_{\text{w}}}{\partial x} \right) - P e\, h_{\text{wF}}\,(T_{\text{w}} - T_{\text{F}}) + q_{\text{in}} + q_{J}, & \text{(for the main-channel)} \qquad (8) \\[2mm] A_{\text{w}} = \sum_i A_i, \quad \lambda_{\text{w}} A_{\text{w}} = \sum_i \lambda_i A_i, \quad C_{\text{w}} \rho_{\text{w}} A_{\text{w}} = \sum_i C_i \rho_i A_i \end{cases} \tag{9}$$

$A_i, C_i, \lambda_i, \rho_i$ are sections of conductor area, specific heat, heat transfer coefficient, density, of i-th material, respectively. q_{in} and q_{J} is initial incident heat and Joule heating, respectively. Characteristics of the SC material are determined by the operated current density J, temperature T_{w} and incident magnetic field H_{inc} (or magnetic induction B_{inc}). Critical points of them are denoted by J_{c}, T_{c}, H_{c} (or B_{c}). In the present model, current distribution is regarded as shown in Fig.5, presenting a cut surface of the $B = B_{\text{inc}}$ plane. Here, the operating current I_{op} is divided into a SC part I_{sc} and Cu part I_{Cu}. Joule heating is assumed to be generated only by I_{Cu}. Hence,

$$q_{J} = \rho_{\text{Cu}} I_{\text{op}}^{\,2} F_{J} / A_{\text{Cu}}, \tag{10}$$

$$F_{J} = \begin{bmatrix} 0 & \text{for } T_{\text{w}} < T_{\text{cs}}(B_{\text{inc}}) \\ 1 & \text{for } T_{\text{c}}(B_{\text{inc}}) < T_{\text{w}} \end{bmatrix}, \quad F_{J} = \frac{T_{\text{w}} - T_{\text{cs}}(B_{\text{inc}})}{T_{\text{c}} - T_{\text{cs}}(B_{\text{inc}})} \text{ for } T_{\text{cs}}(B_{\text{inc}}) < T_{\text{w}} < T_{\text{c}}(B_{\text{inc}}) \tag{11}$$

In Eq.(10), admittance of copper ρ_{Cu} is given by the following. Unit of admittance is $[\Omega \cdot \text{m}]$.

$$\begin{cases} \rho_{\text{Cu}} = \rho_0 \left[1 + \dfrac{1.6 \times 10^4 (B_g)^2}{1 + 7.5 \times 10^2 (B_g)^2} \right], \quad B_g = B_{\text{inc}}\, \rho_g / \rho_0, \quad \rho_g = 1.99 \times 10^{-8}, \quad \rho_0 = \rho_L + 2 \times 10^{-10}, \\[2mm] \rho_L = \rho_\theta\, 425.2\, T_{\text{w}\theta}^{\,5} / \left(1 - 5.494\, T_{\text{w}\theta} + 36.41\, T_{\text{w}\theta}^{\,2} + 0.92\, T_{\text{w}\theta}^{\,3} + 394.2\, T_{\text{w}\theta}^{\,4}\right), \quad T_{\text{w}\theta} \equiv T_{\text{w}} / \theta, \quad \theta = 333 [\text{K}]. \end{cases} \tag{12}$$

Eq.(1) to Eq.(12) are calculated by using the two-step Lax-Wendroff method, then time dependent thermo-hydro dynamic values are evaluated for each channel.

2.2 Modeling of cross flow between channels

In the present section, suffices (0), (1) and $\perp$ refer to values in the main-channel, subchannel and orifice part, respectively. Flow passing the orifice is assumed to be driven by difference of static pressure between the channels. Now we describe a case of main-channel to subchannel flow. Just after the n-th step of the Lax-Wendroff, (n+1)-th pressure field in each channel is determined by the method in previous section.

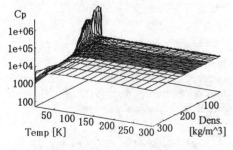

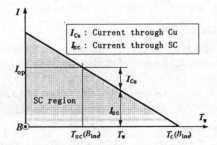

Fig.4 Cp of helium as function of T_{F} and ρ.   Fig.5 Current distribution model in a section

Then the following equations are solved to evaluate the density and velocity at the orifice.

$$P_{i(1)} = P_\perp, \quad P_\perp\, \rho_\perp^{-\gamma} = P_{i(0)}\, \rho_{i(0)}^{-\gamma}, \quad \frac{C_{P\,i(0)}}{R_{i(0)}} \frac{P_{i(0)}}{\rho_{i(0)}} - \frac{C_{P\perp}}{R_\perp} \frac{P_\perp}{\rho_\perp} + \frac{(v_\perp^{n+1})^2}{2}. \tag{13}$$

Next, mass flow rate in each channels are corrected by the density and velocity evaluated through Eq.(14).

$$\bar{m}_{i(0)} = m_{i(0)}^{n+1} - \rho_\perp v_\perp (S_\perp / S_{i(0)}), \quad \bar{m}_{i(1)} = m_{i(1)}^{n+1} + \rho_\perp v_\perp (S_\perp / S_{i(0)}).$$ (14)

Here, $\bar{m}_{i(0)}$ and $\bar{m}_{i(1)}$ are corrected mass flow rates through the main-channel and subchannel at the i-th point and (n+1)-th time step.

3. Results

Stability margins of incident thermal disturbance Q_{-in} were investigated for some example any cases shown in Table 1. The margin is obviously improved in case B2 compared with case A as shown in Fig.6 presenting the relation between Q_{-in} and operating current, under 7.0 T magnetic induction and 5.0 g/sec mass flow rate of coolant. By monitoring the flux through the orifice, the amount of heat passing from the main channel to subchannel is very low compared with that of the main channel. However, the stagnation in the main channel at the upwind of the heating point is successfully decreased. This effect might contribute the enhancement of stability margin.

Table 1 Design parameters of example cases

Example case	Case A (Ordinary)	Case B1 (Subchannel)	Case B2 (Subchannel)
Number of orifice	0 (non-subchannel)	10 (1.0m interval)	20 (0.5m interval)
Section, Main channel	$S_{(0)}$= 400 mm^2	$S_{(0)}$= 300 mm^2	$S_{(0)}$= 300 mm^2
Section, Subchannel	$S_{(1)}$= 0 mm^2	$S_{(1)}$= 100 mm^2	$S_{(1)}$= 100 mm^2

Other conditions: SC material Nb$_3$-Sn, Cable length 10m, Initial temp.(fluid) 4.2K, Initial pressure 10 atm, Magnetic induction 7 T, Perimeter 100 mm, Total conductor section A_w= 400 mm^2, Stabilizer section A_{Cu}= 200 mm^2, Crossing passage section $S_\perp$ = 50 mm^2, Mass flow rate 5.0 g/sec.

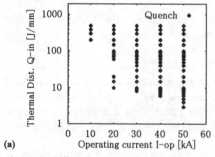
(a)

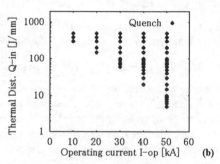
(b)

Fig. 6. Stability characteristics of FFC-SC in present example cases. (a) case A , (b) case B2.

4. Conclusion

The stability of FFC-SC with a subchannel is investigated using a multi-channel one-dimensional numerical model. The obtained results from some examples are summarized as follows. 1) The subchannel is very effective for avoiding multiple stability in FFC-SC, and 2) The cross flow through orifice may contribute not only to transport the heat from the main channel but also to avoid temporary stagnation of coolant flow in main channel.

References

[1] J.W.Lue, J.R.Miller and L.Dresner, Stability of cable-in-conduit superconductors, J. Appl. Phys., Vol.51, pp.772-782, 1980.

[2] J.R.Miller, Experimental investigation of factors affecting the stability of cable-in-conduit superconductors, Cryogenics, Vol.25, pp.552-557, 1985.

[3] T.Amano, S.Nakamura, T.Yamada, S.Watabe and A.Kamitani, Comparison of Measured Value and Calculated Value of Multiple Stability in Forced-Flow Cooled Superconductors, IEEE Trans. Appl. Superconductor., Vol.2, No.4, pp.205-213, 1992.

Non-Linear Electromagnetic Systems
V. Kose and J. Sievert (Eds.)
IOS Press, 1998

Fundamental Study of a Motor
Using High-Temperature Bulk Superconductors
in the Rotor

Noboru ARAKI, Hiroyuki OHSAKI, Eisuke MASADA
Department of Electrical Engineering, Faculty of Engineering,
The University of Tokyo. 7-3-1, Hongo, Bunkyo-ku. Tokyo 113, JAPAN

Abstract. We have studied the motor of an axial flux type using high-Tc bulk superconductors in the rotor. The characteristics of this kind of motor are considered to be greatly dependent on the electromagnetic properties of the superconductors. In this paper, a rotating magnetic field parallel with the rotation axis is applied to the yttrium-based superconductors, and torque characteristics are investigated for the synchronous mode operation. The results show that torque is larger when the field cooling method is used than when zero field cooling is used, and that the hysteresis characteristics of the superconductors influence the motor performance.

1. Introduction

Since the discovery of high-temperature (high-Tc) superconducting oxides in 1986, various kinds of superconducting applications have been investigated, while the development of the melt processes enabled the production of YBCO bulk material with a strong pinning force. This superconductor can produce a higher magnetic field than a conventional permanent magnet. However, studies on motors using high-Tc bulk superconductors have been so far limited to electromagnetic phenomena [1,2].

We have analyzed the electromagnetic phenomena in the bulk superconductor under the influence of an applied traveling field using the finite element method, in order to investigate AC motors with high-Tc bulk superconductors used on the secondary side [3]. In this paper we investigate the electromagnetic characteristics of the motor which is of an axial field type with 4 YBCO bulk superconductors in the rotor. In the present experimental machine, permanent magnets on a rotating disk create an axial magnetic field, which rotates parellel with the rotation axis.

2. Applications of superconductors to motors

Superconductors have hysteresis characteristics due to the pinning force, and they can be used as a magnetic screening material or a flux source. Such a magnetic behavior can improve the performance of motors. Fig. 1 shows one of the possible motor configurations with bulk superconductors in the rotor. Bulk superconductors can be used as a screening material to form a variable reluctance (VR) type motor, or as a flux source to play the role of a field magnet of a synchronous motor. In addition, as for the motor of an axial field type like this, we can easily realize the structure with a relatively small air gap. The strong pinning force of the bulk superconductors and this structure improve the power of the motor.

3. Experimental machine

To investigate the electromagnetic characteristics of a motor like the one shown in Fig. 1, we have constructed an experimental machine to measure torque characteristics.

Fig. 2 shows the structure of the experimental machine, and the upper view of the rotor and locations of permanent magnets are shown in Fig. 3. Four permanent magnets on the upper rotating disk (PM disk), which is controlled by a stepping motor, apply an axial magnetic field on four YBCO bulk superconductors in the rotor. Liquid nitrogen keeps the superconductors in a superconducting state. Generated electromagnetic force is measured by the structure of clamp screws, a linear bearing and load cells, and the torque is calculated.

Main parameters of the permanent magnet and the bulk superconductor are shown in Table 1. We have estimated the critical current density by comparison between the result of a numerical analysis and measured flux distribution of a magnetized bulk superconductor.

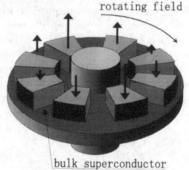

Fig. 1   Motor using bulk superconductors in the rotor.

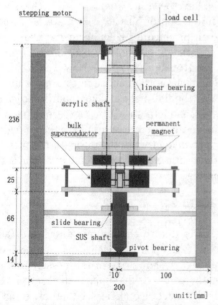

Fig. 2   Experimental machine.

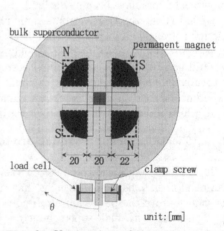

Fig. 3   Upper view of the rotor and locations of permanent magnets.

Table 1 : Main parameters of the permanent magnet and the bulk superconductor.

permanent magnet	size	$20 \times 20 \times 10 [\text{mm}^3]$
	type	SmCo
	residual induction	$0.85 \sim 0.95 [\text{T}]$
	coercive force	$630 \sim 760 [\text{kA/m}]$
	maximum energy product	$143 \sim 175 [\text{kJ/m}^3]$
bulk superconductor	cross section	sector form, $\frac{1}{4} 22^2 \pi [\text{mm}^2]$
	thickness	18.8[mm] (mean value)
	type	$YBa_2Cu_3O_{7-x}$
	critical current density	$4.5 \times 10^7 [\text{A/m}^2]$ (estimated)

4. Results

4.1 Experimental conditions

Synchronous torques were measured by holding the PM disk still, which assumes the motor was running in the synchronous mode operation. When the permanent magnets are in the locations shown in Fig. 3, the angle of rotation is defined as 0 degree.

Bulk superconductors were refrigerated into a superconducting state in different ways, and accordingly the mechanisms of torque generation were considered different. When the PM disk is kept away in the process of cooling (zero field cooling), it is assumed that reluctance torque based on the pinning force is generated. If the PM disk is kept at the same height (field cooling), superconductors are expected to act as flux sources. Here both methods were used for cooling. The gap length was fixed at 5 mm.

4.2 Field cooling

Fig. 4 shows the dependence of the rotation angle on the torque when the superconductors are cooled by field cooling. Three sets of data are adopted for this graph, in which both ends of the error bars indicate maximum and minimum values and the curve mean values.

The PM disk rests stably at 0 degree, and unstably at around 90 degrees. In Fig. 4 the value at 90 degrees is not 0, due to the hysteresis of the superconductors and the change of the applied magnetic field in the process of the measurement. Torque is smaller than expected from overall tendency at around 45 degrees, where the permanent magnets are farther from the superconductors and electromagnetic interaction is supposed to be weaker. Fig. 5 shows the distribution of the z component of the flux density applied on the superconductor, along the circle with radius r=27.9mm which contains the centers of gravity of the superconductors. Due to the third harmonic component of 29.8% of the fundamental component, values around 0 and 90 degrees are small.

4.3 Zero field cooling

Fig. 6 shows the torque characteristics when the superconductors are refrigerated by zero field cooling. Two sets of data are used for this graph.

The PM disk rests stably at 45 degrees, and unstably at around 90 degrees. The peak value is smaller than the one in Fig. 4, because the invasion of flux is limited and the total amount of the Lorentz force is smaller.

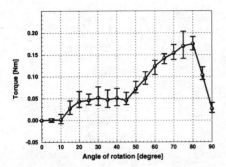

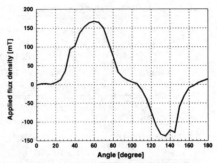

Fig. 4 Torque characteristics (field cooling).

Fig. 5 Distribution of the flux density applied on superconductors.

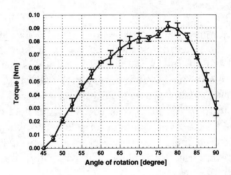

Fig. 6 Torque characteristics (zero field cooling).

5. Conclusions

We have studied the axial field type motor using high-Tc bulk superconductors in the rotor. A rotating magnetic field was applied to the superconductors and torque characteristics in synchronous mode operation were measured. Results show that both field cooling and zero field cooling of the superconductor can make the motor generate torque, and torque is larger when the field cooling method is used.

References

[1] A.D.Crapo, J.D.Lloyd, "Homopolar DC motor and trapped flux brushless DC motor using high temperature superconductor materials", IEEE Transactions on Magnetics, Vol.27, No.2, pp. 2244-2247, March 1991.
[2] P.Görnert, W.Gawalek, "High-Temperature Superconductors for Innovative Applications ", In Proceedings of Maglev '95, pp. 483-488, 1995.
[3] H.Ohsaki, M.Takabatake, N.Araki, E.Masada, "Numerical Analysis of Electromagnetic Phenomena in Bulk Superconductors under the Influence of an Applied Traveling Field", 9th International Symposium on Superconductivity, SAP-27, Sapporo, 1996.

Non-Linear Electromagnetic Systems
V. Kose and J. Sievert (Eds.)
IOS Press, 1998

Multicriterion Designing of Additional Windings of the Cylindrical Cryomagnet

Paweł SURDACKI and Jerzy MONTUSIEWICZ*

*Lublin Technical University, Dept. of Fund. Electr. Eng., *Dept. of Techn. Fund.,
Nadbystrzycka 38a, 20-618 Lublin, Poland*

Abstract. In this paper, the FEM analysis of the magnetic field of the cryomagnet has been performed for the shape parameters and current density of the additional windings. In order to obtain the optimum values of magnetic flux density in selected points of the main winding bore, the multicriterion strategy has been applied, based on the representative solution method. The optimum solutions show considerable improvement of the conditions of the winding stability and the HGMS separator operation.

1. Introduction

Effective operation of the cryomagnet as an energy source for the HGMS magnetic separator requires large values and considerable homogeneity of magnetic flux density in the bore of a cylindrical coil [1]. Moreover, in order to assure stable operation of the winding superconductor, it is essential to minimize the magnetic field value in *hot spots*, where the strongest values occur. Exceeding the critical values of magnetic flux density as well as temperature at the definite local current density leads to a local superconductivity loss and initiation of the resistive zone propagation in the winding [2].

In order to obtain a desirable distribution of the magnetic flux density in the coil bore, the additional windings are wound. The proper choice of their dimensions and of a working current is required to fulfil satisfactorily all the imposed conditions of effective and stable operation.

2. Design model of the cryomagnet

The main winding of the cylindrical cryomagnet (Fig. 1) is made of composite wire which consists of several NbTi superconductor filaments placed in a Cu matrix and which is cooled by liquid helium. The following dimensions of the main winding are assumed [1]: internal radius $a_1 = 0.0815$ m, external radius $a_{11} = 0.09247$ m, a half of length b = 0.157 m. The working space of the separator matrix, which is to collect ferromagnetic particles from a flowing suspension, has diameter d=0.09 m and length l=0.2 m. The additional windings with internal radius a_{11}, external radius a_2 and central incision length 2b' are wound around the main winding edges. The additional winding dimensions a_2 and b' are varied in order to investigate their influence on the magnetic field distribution in the cryomagnet. The dimensionless shape coefficients of the main (α and β) and additional (α' and β') windings, respectively, are introduced to generalize the considerations

$$\alpha = a_{11}/a_1 \quad \beta = b/a_1 \quad \alpha' = a_2/a_{11} \quad \beta' = b'/a_{11}. \tag{1}$$

In accordance with the accepted notation, the shape coefficients of the main winding are: $\alpha=1.1346$ and $\beta=1.926$. The invariable current density value $J_1=10^8$ A/m^2 is assumed in the main winding and the current density in the additional winding J_2 is to be changed from zero to $2*10^8$ A/m².

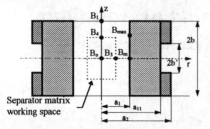

Fig. 1. Cross-section of the cryomagnet with the additional windings and separator matrix: B_o, B_m - magnetic flux density modules in the middle of the bore and in the weak point, respectively, B_{max} - maximum flux density.

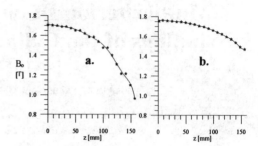

Fig. 2. Magnetic flux density of the main winding: a) along axis, b) along internal surface of the winding; the main winding current density is 10^8 A/m².

The design process is aimed at finding a set of design variable vectors $\mathbf{x} = [\alpha', \beta', J_2]^T$ that assure correct operation of the separator during a matrix exchange and superconducting winding stability. In order to obtain the above requirements, the following evaluation criteria are accepted: $F_1=B_o$, $F_2=B_{max}/B_o$, $F_3=|B_3/B_o-1|$, $F_4=|B_4/B_o-1|$, $F_5=|B_1/B_o-1|$, where B_{max} is a maximum value of the magnetic field density distribution, and the others are the magnetic field density values in: the geometrical centre of the bore (B_o), at the separator space edge along radial coordinate (B_3), axial coordinate (B_4), and at the bore edge (B_1), respectively (Fig. 1). The first criterion should be maximized and the remaining ones should be minimized. The specific i-th solution $\mathbf{x}^i$ corresponds to the vector $\mathbf{F}^i[F_1(\mathbf{x}^i), F_2(\mathbf{x}^i), F_3(\mathbf{x}^i), F_4(\mathbf{x}^i), F_5(\mathbf{x}^i)]^T$ of the evaluation criteria.

3. Computations of the magnetic field distribution

Multiple magnetic flux density distributions have been computed for various design variable vectors using FLUX2D (Cedrat) software based on FEM, in order to determine the additional winding influence on the accepted operation and stability evaluation criteria.

The distributions obtained for the cryomagnet without additional windings (Fig. 2) exhibit substantial nonhomogeneity along both internal surface and z-axis, where the values decrease nearly two times. The additional windings may improve this nonuniformity but cause a maximum magnetic flux density shift from the point B_m (main winding only) to the point B_{max} (the additional windings present) (Fig. 1). The magnetic flux density module at these points approaches the critical values that increase the possibility of a superconductivity disturbance.

The additional winding influence for value $\beta'=1$ (Fig. 3) is clearest in extreme parts of the main winding. The magnetic field increase in the central part of the bore is much smaller. Together with additional winding thickness, the maximum value of magnetic flux density increases, making worse the field homogeneity along the internal surface, but simultaneously increases it in the central part of the bore. For the value $\alpha'=1.04$ the increase of maximum field is the smallest but, at the same time, the distribution homogeneity along a z-axis is better than without additional windings.

The distribution near the internal surface edges has been improved for $\alpha'=1.04$ and $\beta'=1.5$ (the smallest of the analysed additional thicknesses) (Fig. 4c) while along the winding axis (Fig. 4b,c) it has remained practically unchanged. An initial decrease of β' makes the range of homogeneous field expand and then it is decreased and made worse near the coil edges.

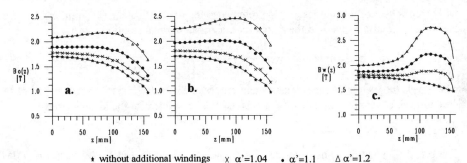

* without additional windings   × α'=1.04   • α'=1.1   △ α'=1.2

Fig. 3. Magnetic flux density along the bore axis at current density J_2 = a) $0.7*10^8$ A/m^2, b) 10^8 A/m^2, and c) along the winding internal surface at $J_2 = 0.7*10^8$ A/m^2 for constant value β' = 1 and changing α'.

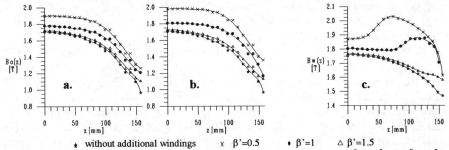

* without additional windings   × β'=0.5   • β'=1   △ β'=1.5

Fig. 4. Magnetic flux density along the bore axis at current density J_2 = a) $0.7*10^8$ A/m^2, b) 10^8 A/m^2, and c) along the winding internal surface at $J_2 = 0.7*10^8$ A/m^2 for constant value α' = 1.04 and changing β'.

A straightforward analysis of the obtained distributions does not allow the choice of the best vectors of the design variable. The assumed criteria are opposing (maximization of B_o, minimization of B_{max} and homogeneity along rotation and symmetry axes) because the improvement of one of them worsens the other. Therefore, the set of 92 obtained solutions should be analysed by means of a polyoptimal method.

4. Multicriterion analysis

The strategy of multicriterion optimization [3] is based on a simultaneous analysis of designer selected characteristics, computed in the process of solving the formulated problem. It consists of the following stages: 1) adopt a mesh of design variable values α', β', J_2, 2) compute the magnetic field distributions and extract the required evaluation characteristics B_o, B_1, B_3, B_4, B_{max}, 3) carry out a polyoptimal analysis in a criterion space with the accepted set of the evaluation criteria F_1 - F_5: a) delimit a subset of nondominated solutions, b) delimit a subset of representative solutions, 4) carry out an analysis in design variable space, 5) repeat the above points for the improved design variable mesh. In order to determine the nondominated solution subset in point a), the optimal solutions in Pareto sense [4] are evaluated.

The solution is optimal in Pareto sense if no criterion $F_1(x)$, $F_2(x)$, .., $F_5(x)$ can be improved without worsening at least one other criterion. In point b) the method of searching for representative solutions from the finite nondominated solution set has been used, based on a weighted min-max metric [4]. The method is composed of two phases. In the first phase, only one min-max solution is obtained. In the second phase we obtain several solutions depending on the choice of the completion formula of the search procedure. The weights are assigned to the relative deviations from the minimal solutions; hence they are a fairly true reflection of the designer's preferences for the particular evaluation criteria. For the weighted min-max method the computer assigns an optimal solution:

$$\xi(x) = \min_{x \in X} \max_{i \in I} \left\{ w_i \frac{\left| F_i^0(x) - F_i(x) \right|}{\left| F_i^0(x) \right|} \right\}, \tag{2}$$

where w_i is the weight of the i-th assesment criterion, $\Sigma w_i = 1$, $I = \{1,...,m\}$ is the set of the criterion vector component indices, $F_i^0(x)$ is the i-th component of the ideal vector [4].

5. Results of design analysis

The set of 92 solutions was generated for the various design variables by means of FEM. The set of 64 nondominated solutions was obtained using a multicriterion analysis. Subsequently, the set of 11 representative solutions was extracted. As a result of the above strategy, the ranges of design variables have been obtained for α' from 1.04 to 1.14, β' from 0.6 to 1, and J_2 from 0.7 to $1.5*10^8$ A/m^2, which have the nondominating solutions. The obtained design variable space consists of a few disjoint subsets.

The magnetic flux module distributions B(r,z) for exemplary design variable sets have been shown for a dominated (Fig. 5a) and an optimum (Fig 5b) solution. The maximum value B$_{max}$ in the vicinity of the main winding internal surface has been reduced from 3.5 to 2.17 T, therefore the stability of the superconducting winding has been considerably improved. In the separator matrix space the magnetic flux density distribution varies from 2.19 T to 2.65 T for a dominated solution (Fig. 5a), whereas the field has much smaller variations around 1.93 T for an optimum solution (Fig. 5b). That makes both a separator stationary operation and a matrix removal from the bore more effective.

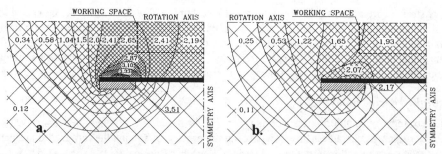

Fig. 5. Distributions of the magnetic flux module (in teslas) in the left quarter of the cryomagnet space for: a) dominated solution ($\alpha'= 1.07$, $\beta'=1.1$, $J_2=2.5*10^8$A/m^2) and b) optimum solution ($\alpha'= 1.07$, $\beta'=1$, $J_2=10^8$A/m^2).

References

[1]   A. Cieśla and H. Malinowski, Analysis of the Magnetic Field Distribution in the Working Space of a Superconductor Magnetic Filter, *Proc. of the 19-th SPETO*, Gliwice-Ustroń (Poland), 1996, pp. 213-216.

[2]   P. Surdacki and T. Janowski, Computation of Resistive Zone Propagation in Superconducting Winding, *IEEE Trans. Magn.*, **31** (1995) no. 3, pp. 1829-1832.

[3]   P. Surdacki, J. Montusiewicz and R. Kuchcewicz, An Approach to Multicriterion Shape Optimization of Superconducting Solenoid Windings, *Proc. of Int. Conf. ELMECO*, Lublin (Poland), 1994, pp. 295-300.

[4]   P. Surdacki and J. Montusiewicz, Approach to Multicriterion Optimization of Quench Performance of Superconducting Winding, *IEEE Trans. Magn.*, **32** (1996) no. 3, pp. 1266-1269.

Non-Linear Electromagnetic Systems
V. Kose and J. Sievert (Eds.)
IOS Press, 1998

Experiments for Selective Detection of Magnetic Dipoles Using Superconducting and Ferromagnetic Cylinders

K. Sakasai and K. Ara

Japan Atomic Energy Research Institute,
2-4, Tokai-mura, Naka-gun, Ibaraki-ken, 319-11 Japan

Abstract. The authors have studied the system, consisting of superconducting, ferromagnetic cylinders and a magnetic sensor, for selectively detecting a magnetic dipole parallel to the extension of the axis of the cylinders. The superconducting cylinder has an open slit in its side so that the magnetic fields from the dipole can be measured selectively according to its diamagnetic characteristics. The experiments using Bi-Sr-Ca-Cu-O based superconducting cylinder were carried out to confirm the effectiveness of the system. It may be possible to use the system for positioning of the dipoles. The system will be useful for non-destructive testing of metallic materials.

1. Introduction

Noninvasive techniques are essentially important in measuring magnetic fields produced by ionic currents accompanied with electrical activities of various human tissues. Since Baule and McFee first measured the magnetic fields from the human heart[1], the measurement of such weak magnetic fields from the human body is of great interest in biomedical study. More weaker magnetic fields from the human brain were first measured by Cohen[2]-[3] using the SQUID. One of the most important purposes of the biomagnetic field measurement is to localize the source for diagnostic purposes in the human body[4]-[6]. The same situation occurs in non-destructive testing of metallic substance. When cracks or defects exist in metallic substance, they can be considered as magnetic sources or magnetic dipoles by magnetizing the substance or letting currents flow in it. However, localization of the magnetic sources, so called the inverse problem in magnetic field measurement, is known as a problem which has no unique solution[6]. The non-uniqueness does not come from the fact that the number of measurable data is insufficient. The difficulty cannot be overcome without knowing the internal characteristics of human tissues or metallic substance. Therefore, many attempts have been made to detect the sources introducing a simple source model such as current dipoles or magnetic dipoles, or using special methods[6]-[11].

On the other hand, the authors have introduced a selective detection system for magnetic dipoles using superconducting and ferromagnetic cylinders for non-destructive testing of metallic substance[12]-[13]. The effectiveness of the system was confirmed by the magnetic field analysis[14]-[15]. The results of experiments using Bi-Sr-Ca-Cu-O based superconducting and ferromagnetic cylinders are presented in this paper.

2. Detection System

The detection system has been already reported[12]-[15], and therefore will be only briefly described here. The proposed method is schematically shown in Fig.1. The magnetometer was installed in a superconducting cylinder. The superconducting cylinder has an electric insulation line (i.e., an open slit) in its side to interrupt the flow of circumferential shielding currents. The sensitive region which can be searched selectively with the system is the area being cut out by the axial direction of the cylinder. The magnetometer can detect selectively magnetic fields produced by magnetic dipoles which exist in the sensitive region. In the figure, the authors assumed that the dipoles have their directions parallel to the axis of the cylinder, because the dipoles can be considered

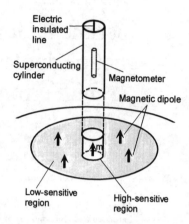

Fig. 1   Selective detection of magnetic dipoles

to have an certain direction in many cases by magnetizing the metallic substance or letting currents flow in it. Since the system must be cooled down below the critical temperature of the superconductor to obtain the diamagnetic characteristics of the superconductor, the magnetometer must be used at low temperature with a high sensitivity. This type of magnetometer was developed and reported as a flux-gate type magnetometer by the authors[16]. It is also expected by magnetostatic analysis that a ferromagnetic cylinder is also useful for the system when it is placed outside the superconducting cylinder, since the magnetic fields through the slit will be shielded by the ferromagnetic cylinder[14],[15].

3. Experimental

In order to confirm the effectiveness of the system with both superconducting and ferromagnetic cylinders, some fundamental experiments have been carried out by using a high-T_c Bi-Sr-Ca-Cu-O based superconducting and iron cylinders. Figure 2 shows the experimental set-up. The superconducting cylinder is made by depositing a superconducting layer on the surface of a silver cylinder. The thickness of the layer is about 0.2 mm. The critical temperature and critical current density of the cylinder are around 90 K and 800 A/cm^2, respectively. The magnetometer used in the experiments was developed for low temperature use and has a sensitivity of 4.26 V/G at 77K. It has a good linearity in a magnetic field range between -2 G and +2 G. The outside diameter and the sensitive length of the magnetometer are about 1mm and 20 mm, respectively. In the experiments, the system was directly immersed in

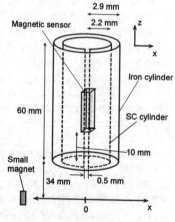

Fig. 2   Experimental set-up

liquid nitrogen and the z-component of the magnetic fields was measured as a function of a small magnet that equivalently works as a magnetic dipole, as shown in Fig.2.

4. Results and Discussion

The experimental results are shown in Fig.3, where the measured magnetic fields are normalized with the peak values. The curve of the case with superconducting and iron cylinders has a sharper peak than that of the case without cylinders. When we define the spatial resolution of the sensor as the full width of the curve at a half value of the measured peak, the spatial resolution with superconducting and iron cylinders and without cylinders are 3.84 cm and 5.82 cm, respectively. This means that the spatial resolution of the system is improved by using superconducting and iron cylinders.

The improved spatial resolution suggests that positioning of dipoles will be possible with the system. From the experimental results the authors estimated the magnetic fields when two parallel dipoles exist. Figure 4 shows the magnetic fields in the case of two parallel dipoles, where the distance between the two dipoles is 40 mm. In the figure, two arrows show the positions and directions of the dipoles. As seen in the figure, two clear peaks exist in the case with cylinders but only one peak in the case without cylinders. The positions of peaks correspond to those of the two dipoles. This means that the system can be used to position the dipoles.

To investigate the characteristics of the system, the authors calculated an effective transfer function from the obtained data. Since the Fourier transform of the sensor output with cylinders, $B_{with}(k)$, is written as

$$B_{with}(k) = B_{without}(k)T(k), \qquad (1)$$

where $B_{without}(k)$ is the Fourier transform of the output without cylinders, $T(k)$ the effective transfer function, and k the spatial frequency. The transfer function can be obtained by

$$T(k) = B_{with}(k) / B_{without}(k). \qquad (2)$$

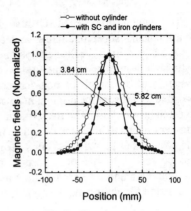

Fig. 3   Experimental results

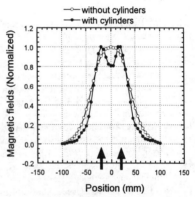

Fig. 4   Magnetic fields by two parallel magnets

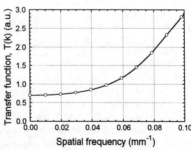

Fig.5   Effective transfer function

The calculated transfer function is shown in Fig.5. The transfer function has low values at low spatial frequencies and high values at high frequencies. In other words, the low frequency component of the output without cylinders is decreased but the high frequency component is increased by the transfer function. Therefore, the system has characteristics of a so-called "high-pass filter" and hence the spatial resolution of the sensor may be improved with the system.

5. Summary

The authors proposed a selective detection system for magnetic dipoles as magnetic sources using superconducting and ferromagnetic cylinders. The effectiveness of the system is confirmed with the experiments using a Bi-Sr-Ca-Cu-O based superconducting cylinder. It may be possible to detect magnetic dipoles selectively by the system. This system will be useful for non-destructive testing of metallic materials. Since it is not possible to measure the original magnetic fields directly with our system, one of the ways to estimate the original sources is to use a normal sensor and apply reconstruction methods (an iterative method or a least-square method, etc.) to the measured data after determining the positions of sources with our system. It may be also possible to do so by calibration of our sensor. On the other hands, the effectiveness greatly depends on the system configuration and they are planning to perform extensive experiments as well as computer simulation for optimizing the system.

References

[1] G. M. Baule and R. McFee, Detection of the magnetic fields of the heart, *Am. Heart J.* **66** (1963) 95-96.

[2] D. Cohen, Magnetoencephalography: Evidence of magnetic fields produced by a alpha-rhythm currents, *Science*, **161** (1968) 784-786.

[3] D. Cohen, E. Edelsack, and J. E. Zimmerman, Magnetocardiograms taken inside a shielded room with a superconducting point-contact interferometer, *Appl. Phys. Lett.*, **16** (1970) 278-280.

[4] S. Ueno, Recent Advances in Biomagnetics, *Trans. IEE Jpn.*, **116-A** (1996) 141-144.

[5] Y. Uchikawa and M. Kotani, Tracing of the equivalent source localization in the brain to somatosensory evoked magnetic field using equivalent current dipole technique, *Trans. IEE Jpn.*,**112-A** (1992)127-132.

[6] S. J. Williamson and L. Kaufmann, Biomagnetism, *J. Magn. Magn. Mat.*, **22** (1981) 129-201.

[7] H. Saotome, K. Kutsuta, S. Hayano, and Y. Saito, A Neural Behavior Estimation by the Generalized Correlative Analysis, *IEEE Trans. Magn.*, **Mag-29** *(1993) 1389-1394.*

[8] M. Kishimoto, K. Sakasai, and K. Ara, Solution of electromagnetic inverse problem using combinational method of Hopfield neural network and genetic algorithm, *J. Appl. Phys.*, **79** (1996) 1-7.

[9] W.J.Dallas, Fourier space solution to the magnetostatic imaging problem, *Appl.Opt.*,**24** (1985) 4543-4546.

[10] W. Kullman and W. J. Dallas, Fourier Imaging of Electrical currents in the Human Brain from Their Magnetic Fields, *IEEE Trans. Biomed. Eng.*, **BME-34** (1987) 837-842

[11] B. J. Roth, N. G. Sepulveda, and J. P. Wikswo, Jr., Using a magnetometer to image a two-dimensional current distribution, *J. Appl. Phys.*, **65** (1988) 361-372.

[12] K. Sakasai, M. Kishimoto, and K. Ara, Detection of Magnetic-flux Sources by a System of Perfect Diamagnetic Substance and Magnetic Sensor, *J. Mag. Soc. Jpn.*, **16** (1992) 337-342.

[13] K. Sakasai, M. Kishimoto, and K. Ara, Experiments for Selective Detection of Magnetic Flux Sources Using a System of High-T_c Superconductors and a Magnetic Sensor, *J.Mag. Soc.Jpn.*,**18**(1994)709-714.

[14] K. Sakasai, M. Kishimoto, and K. Ara, Method for Selective Detection of Magnetic Flux Sources, Using a System of Superconductors, a Ferromagnetic Substance, and a Magnetic Sensor, *J. Mag. Soc. Jpn.*, **19** (1995) 593-596.

[15] K. Sakasai and K. Ara, A possible device for selective detection of magnetic field sources using superconducting and ferromagnetic cylinders, *Rev. Sci. Instrum.*, **68**(1997) 1739-1742.

[16] K. Sakasai *et al.*, Electric and magnetic characteristics of a Co-Fe-Si-B-based amorphous wire and its application to multivibrator-type magnetometer at low temperature, *Rev. Sci. Instrum.*, **65**(1994)1657-1662.

Non-Linear Electromagnetic Systems
V. Kose and J. Sievert (Eds.)
IOS Press, 1998

Spatio-Temporal Stochastic Resonance of Domain Walls in Inhomogeneous Magnets

I. Dikshtein, A. Neiman*, and L. Schimansky-Geier**

Institute of Radioengineering and Electronics, 103907 Moscow, Russia
*Department of Physics, Saratov State University, 410071 Saratov, Russia
**Institut für Physik, Humboldt-Universität, D-10115 Berlin, Germany

Abstract. We study the motion of a small-angle domain wall (DW) in a rhombic ferromagnet with localized attracting inhomogeneities in the vicinity of the reorientational phase transiton induced by temperature. Usually a DW can be trapped by the inhomogeneities. If the system is additionally driven by a noisy magnetic field the DW is able to perform stochastic motion and hence achieves probability to jump between several attracting inhomogeneities. With a small periodic field applied to the system we will observe stochastic resonance in the motion of the DW. At an optimal noise level the hopping dynamics of the DW become most coherent and the response of the system to the periodic force is maximal.

1. Introduction

In the last decade the phenomenon of stochastic resonance (SR) has attracted much attention. This phenomenon underlies a "positive", constructive role which noise might play in nonlinear systems: with an increase of noise applied to a nonlinear system a response of the system to a small periodic signal can be amplified. Nowadays, SR in nonmagnetic spatially-distributed systems has attracted particular interest. For example, the phenomenon of array enhanced SR and spatiotemporal synchronization was found for the linearly coupled bistable elements [1]. Despite a great deal of SR studies in a wide class of systems there are only a few studies dealing with SR in ferromagnets (FM). In particular, SR in a zero-dimensional Ising-like single-domain particle was investigated in [2]. SR in the spatially-distributed Ising model has been studied in [3].

As distinct from the previous studies our goal here is to consider a drift of a small-angle domain wall (DW) in a rhombic FM with a double-well shaped inhogeneity perturbed by noise. SR predicts that the transitions between localized attracting inhomogeneity can be synchronized to an external periodic field if the noise is tuned to an optimal value.

2. The Model and Numerical Investigation

The starting point for the analysis is the energy density of a rhombic FM in an external magnetic field $\mathbf{H} = H_y \mathbf{e}_y + H_z \mathbf{e}_z$

$$W = \frac{1}{2}\alpha(\frac{\partial \mathbf{M}}{\partial x})^2 - \frac{\beta_z}{2}M_z^2 - \frac{\beta_y}{2}M_y^2 + \frac{1}{4M^2}(b_z M_z^4 - 2b_{yz}M_y^2 M_z^2 - b_y M_y^4) - M_y H_y(x) - M_z H_z(t),$$

$$(1)$$

Here $\mathbf{M}$ is the magnetization vector, α is the exchange constant, $\beta_i > 0$ and $b_i > 0$ are the magnetic anisotropy constants $(\beta_z = \beta_{0z} + \beta_{1z}(x))$, β_{0z} and β_{1z} are the homogeneous and inhomogeneous parts of the anisotropy constant β_z (the rest of the anisotropy constants do not depend on x), $H_{y,z}$ are external magnetic fields which stand for the introduced inhomogeneity and for the random and periodic perturbation, respectively.

In the angular variables θ and ψ parametrizing the vector $\mathbf{M}(\mathbf{M} = \mathbf{M}(\cos\theta\sin\psi, \cos\theta\cos\psi, \sin\theta)$ the Lagrange function of the magnet is given by

$$L = (M/g)(1 - \sin\theta)\partial\psi/\partial t - W(\theta, \psi) \tag{2}$$

with g, the gyromagnetic ratio. Dissipative processes are included in the model through a dissipative function $Q = \tilde{\lambda}\dot{\mathbf{M}}^2/(2gM)$ with $\tilde{\lambda}$, the Gilbert relaxation constant.

For $0 < \tilde{\beta} < \tilde{b}$ in the ground state of the unperturbed magnet ($\mathbf{H} = \beta_{1z} = \mathbf{0}$) there are two equilibrium orientations of the magnetization vector $\mathbf{M}$: $\theta_0 = \pm\arcsin(\sqrt{\tilde{\beta}/\tilde{b}}$, $\psi_0 = 0$ with $\tilde{\beta} = \beta_{0z} - \beta_y + b_{yz} - b_y$ and $\tilde{b} = b_z + 2b_{yz} - b_y$. Therefore, for $\nu^2 = \tilde{\beta}/\tilde{b} \ll 1$ a small-angle DW ($2\theta_0$- DW) can exist in the magnet, dividing the latter into domains with different orientations of $\mathbf{M}$. If the effective anisotropy constant $\tilde{\beta}$ changes sign with a temperature T a reorientational phase transition (PT) from the angular phase ($\theta_0 \neq 0$) for $\tilde{\beta}(T) > 0$ to the phase ($\theta_0 = 0$) for $\tilde{\beta}(T) < 0$ occurs.

In the general case the solutions of the dynamical equations (Landau Lifshitz equations) are very cumbersome. However, near PT when the parameter ν is small the problem is substantially simplified and in the main approximation with respect to the small parameter ν the angle variable ψ can be excluded from the equations of motion and the dynamical equation for another variable θ can be reduced to the form characteristic of the ϕ^4 model

$$\omega_0^{-2}\phi_{tt} + (\lambda/\omega_0)\phi_t = x_0^2\phi_{xx} + 2\phi(1 - \phi^2) - h_1(x,t) - 2h_2(x,t)\phi \tag{3}$$

with $\phi = \theta/\nu$, $x_0^2 = 2\alpha/(\nu^2\tilde{b})$, $\omega_0^2 = (1/2)g^2\nu^2(\beta_y + b_y)\tilde{b}M^2$, $\lambda = (\sqrt{2}\tilde{\lambda})/\nu)\sqrt{(\beta_y + b_y)/\tilde{b}}$, $h_1(t) = -2H_z(t)/(\nu^3\tilde{b}M)$, $h_2(x) = [H_y(x) - \beta_{1z}M]/(\nu^2\tilde{b}M)$.

The unperturbed system (3) ($h_1 = h_2 = 0$) possesses the stationary solitary solution (kink) $\phi_0(x) = \tanh(x/x_0)$, satisfying the boundary conditions $\phi(\pm\infty, t) = \pm 1$.

Now we specify the external fields $h_{1,2}(x,t)$ in the model (3). $h_2(x)$ will describe a stationary inhomogeneity of the medium. We introduce a localized double-well

$$h_2(x) = \epsilon\left(\delta^2 x^2/2 - x^4/4\right) \text{ for } |x| < 2\delta, \qquad -2\epsilon\delta^4 \text{otherwise}, \tag{4}$$

The distance of the two wells δ should be large compared with the width x_0 of the interface ($\delta \gg x_0$). To suffer stability of the front profile the strength of the inhomogeneities $\epsilon\delta^2$ should be small with respect to the depth of the potential for the unperturbed ϕ^4 dynamics, i.e. ($\epsilon\delta^2 \ll 1$). Using the parameters of typical ferromagnets, $g \sim 2.5\times10^5 ms^{-1}A^{-1}$, $\sqrt{\alpha/\tilde{b}} \sim 10^{-8}m$, $\tilde{\lambda} \sim 10^{-2}, \tilde{b} \sim 10-100, M \sim 8\times10^3 A\times m^{-1}$ and considering $\nu = 10^{-1}$, one gets the numerical estimate of the characteristic length $x_0 \sim 10^{-7}m$. The inhomogeneity of a magnetic field and a magnetic anisotropy with the characteristic wavelength $\delta \sim 10^{-5}$ m can be easily set up by a system of permanent magnets and by the method of ionic implantation or by the laser annealing method.

The temporal perturbation $h_1(t)$ will refer to the external periodic with random initial phase Θ and noisy fields:

$$h_1 = h_1(t) = A\cos(\Omega t + \Theta) + \sqrt{2D}\eta(t), \tag{5}$$

The noise term $\eta(t)$ is taken to be Gaussian δ-correlated with zero mean and intensity D. The amplitude of periodic excitation A is always assumed to be sub-threshold, i.e. there will be no deterministic transitions of the interface over the barrier between the two inhomogeneities in the presence of the periodic signal alone. Free boundary conditions are assumed throughout. For the low noise intensity $D \ll E_0$ with $E_0 = 4x_0/3$, being the unperturbed rest energy of the kink, we will ignore an opportunity of the kink-antikink nucleation inside the double-well potential during a time of observation in numerical simulations.

The finite difference analogue of (3) in the overdamped case $\lambda \rightarrow \infty$ is a system of coupled stochastic differential equations

$$\dot{\phi}_n = (\omega_0/\lambda)[2\phi_n(1-\phi_n^2)+x_0^2\frac{\phi_{n+1}-2\phi_n+\phi_{n-1}}{(\Delta x)^2}-2h_{2n}-A\cos(\Omega t+\Theta)+\sqrt{2D}\eta(t)], \quad (6)$$

where $\phi(x,t) \rightarrow \phi(n\Delta x,t) \equiv \phi_n(t)$, $n = 0,...,N$, $\Delta x = 2L/N$, L is the size of the system and N is the number of partitions. We note, that (6) can be considered indeed as an array of coupled bistable element [1], however, with the inhomogeneous term h_{2n}. The stochastic differential equations (6) have been simulated for $N = 100$, $L = 20$ using an algorithm of [4]. We use the unperturbed kink solution as the initial state of the system. In order to quantifity SR we introduce a new dichotomic process $u(t)$ following for the motion of the center of the kink $X(t)$: $u(t) = +1$, if $X(t) > 0$ and $u(t) = -1$, otherwise. By calculating the power spectrum of this process we calculate signal-to noise ratio (SNR). The results of the calculation of SNR versus noise intensity D are presented in Fig. 1 for increasing values of the coupling strength x_0 and exhibit the bell-shaped maximum. The output SNR increases with the growth of x_0.

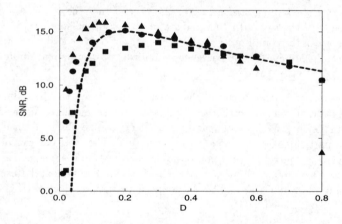

Fig.1: Numerical determined SNR vs D for different values of coupling strength $x_0 = p * a$: $p = 1.5$ (triangles), $p = 1$ (circles), $p = 0.5$ (squares). The dashed line presents theoretical estimation from adiabatic theory for $p = 1$. Other parameters $a = 10^{-5}$, $D = 0.2$, $\delta = 4a$, $\epsilon = 0.0005/a^2$, $A = 0.01$, $\Omega = 0.1$, $\lambda = 0.005$.

Now we will apply perturbative methods (collective variables approach) from soliton theory to our problem. We seek a solution of (3) in the form $\phi = \phi_0(x) + \tilde{\phi}(x,t)$, with $\phi_0(x)$, the stationary kink of an unperturbed problem and $\tilde{\phi}(x,t)$, the distortion of the stationary profile.

An expansion of $\tilde{\phi}(\xi, t)$ in orthogonal eigenfunctions of the linearized operator of (3) a projection gives in first order perturbation theory $\phi = \phi_0(\xi)$ where $\xi = x - X(t)$ with $X(t)$ being the co-ordinate of the interface at time t. Thereby, the velocity of the interface $V(t) = \dot{X}(t)$ is determined by the inhomogeneity and time dependent fields. $X(t)$ obeys the stochastic differential equation

$$\dot{X} = -dU(X)/dX + \tilde{A}\cos(\Omega t + \Theta) + \sqrt{2\tilde{D}}\eta(t). \tag{7}$$

The rescaled noise intensity and amplitude of the periodic force read

$$\tilde{A} = 3x_0\omega_0 A/(2\lambda), \qquad \tilde{D} = 9x_0^2\omega_0^2 D/(4\lambda^2), \tag{8}$$

and the potential $\tilde{U}(X)$ is defined as

$$-dU(X)/dX = (x_0\lambda/\omega_0)^{-1}\int_{-\infty}^{+\infty} h_2(x)\text{sech}^3(\xi/x_0)\sinh(\xi/x_0)dx.$$

To discuss SR phenomena we first estimate the mean escape rate from a metastable state of the potential $U(X)$ in the absence of a periodic field. For weak noise we use Kramers formula for the mean escape rate in the overdamped case and get

$$r_0 = \pi^{-1}\sqrt{|U"(X_s)U"(X_u)|}\exp(-\Delta U/D) \tag{9}$$

with $\Delta U = U(X_u) - U(X_s)$, where X_s and X_u are stable and unstable points of the potential, respectively. Further, the adiabatic theory of McNamara and Wiesenfeld [5] is employed and the dependence of the SNR versus D is depicted in Fig.1 (dashed line). It follows from Fig. 1 that the collective coordinate approach is in a rough qualitative correspondence with the numerical data.

From (9), (8) one immediately concludes that the rescaled noise strength becomes stronger with the growing couplings x_0. Therefore, the mean escape rate increases with the increase of the coupling strength. On the other hand, the rescaled amplitude of periodic force (8) increases with the increase of x_0 as well. Therefore, SNR increases with the increase of the coupling strength. Thus, the collective coordinate method turns out to predict phenomena whose existence is confirmed by numerical simulations.

3. Conclusions

In conclusion, we generalise the phenomenon of SR for the periodically driven stochastic motion of coherent structures, kinks, for the ϕ^4 model with the double-well shaped inhomogeneity. We expect that similar effects can be observed for other model equations, in particular for the Sine-Gordon equation. Therefore SR can be used as a new mechanism to control propagation of nonlinear waves in inhomogeneous media.

This work was supported, in part, by the Russian Fundamental Research Fund (RFRF) (grant 96-02-16082-a) and by the common research project of DFG and RFRF (grant 436 RUS 113/334/0 (R)).

References

[1] J. Lindner *et al.*, Array enhanced SR and spatiotemporal synchronization, *Phys. Rev. Lett.* **75** (1995) 3-6.

[2] J. Brey and A. Prados, SR in a One-Dimensional Ising model, *Phys. Lett.* **A216** (1996) 240-246.

[3] A.N.Grigorenko, *et al.*, Magnetostochastic resonance, *JETP Lett.* **52** (1990) 593-596.

[4] L. Schimansky-Geier and Ch. Zülicke, Kink Motion Induced by Multiplicative Noise, *Z.Phys.B* **82** (1991) 157-162.

[5] B. McNamara and K. Wiesenfeld, Theory of SR, *Phys. Rev.* **A39** (1989) 4854-4869.

Non-Linear Electromagnetic Systems
V. Kose and J. Sievert (Eds.)
IOS Press, 1998

Shock Waves on Quasi-Homogeneous Transmission Lines with Magnetic Non-Linearities

Oldrich BENDA and Július ORAVEC

Slovak University of Technology, Ilkovičova 3, 81219 Bratislava, Slovakia

Abstract. A new construction of a non-linear transmission line exhibiting low dispersion is described. Quasi-homogeneous distribution of a ferrite filler allows the formation of shock waves with rise time $\cong$ 1ns and instantaneous power $\cong$ 1MW without creation of undesirable solitons. The methods for experimental estimation of the dynamic inductance, non-linear impedance in shock wave regime and the matching conditions at the junctions of non-linear and linear lines are presented.

1. Introduction

As it is well known the non-linear transmission lines (NTL) make possible the formation of electromagnetic shock waves (ESW). They found a large application in pulse generators for magnetron modulation in radar technology, in high power pulse sources used in nuclear research and in other applications of sub-microsecond, or nanosecond pulse technology. The necessary condition of the creation of stable ESW is the sufficiently strong non-linearity of the energy accumulating elements, either $L(i)$ or $C(u)$. At the same time the NTL should exhibit a low dispersion to avoid the immediate change of ordinary waves into lattice solitons, as it is in the case of ladder-lines with discrete structure due to their pronounced intrinsic dispersion [1].

In the present paper a new construction of quasi-homogeneous coaxial NTL fulfilling both requirements is described. It exploits the non-linear current dependence of the ferrite filler

$$L_d(i) = L(i) + i\frac{\partial L(i)}{\partial i} \tag{1}$$

When the ordinary wave with the input voltage U_{1max} up to 10^4 V, input current I_{1max} up to 10^2 A and with a front in the order of a few hundred nanoseconds propagates in NTL, for the total differential of the current $i(x,t)$ it holds [2]

$$di(x,t) = \frac{\partial i(x,t)}{\partial x}[dx - v(i)dt] \tag{2}$$

Points with different *i*-values propagate with different velocities *v(i)*, being function of instantaneous current values

$$v(i) = \frac{1}{\sqrt{C_o L_d(i)}} \tag{3}$$

Due to the monotonic decreasing function $L_d(i)$ the steepness of the wave front increases but the time intervals between points with the same *i*-values are preserved. In the distance x^* a finite discontinuity zone at the front of the ordinary wave appears. The derivative $\partial i(x,t)/\partial x$ increases theoretically to infinity. The formation of the ESW begins.

$$x^* = v^2(i)\left[\frac{\partial v(i)}{\partial i} \cdot \frac{\partial i(x,t)}{\partial t}\right]^{-1} \qquad (4)$$

The zone seizes the whole front and successively narrows until the process is stopped by dissipation joined with the prevalent coherent rotation of magnetization M in the vicinity of ferromagnetic resonance, described by a non-linear relation

$$\frac{\partial M}{\partial t} = \frac{\alpha\gamma M_s}{1+\alpha^2}\left(1-\frac{M^2}{M_s^2}\right)(H-H_0) \qquad (5)$$

where α is the damping constant. It occurs when $dH/dt \cong 10^9$ $Acm^{-1}s^{-1}$. For the velocity v_z of the propagation of discontinuity holds then

$$I_2 - I_1 = v_z C(U_2 - U_1) \qquad (6)$$

where subscripts 1, 2 indicate voltage and current values in front or behind the discontinuity of the zone.

2. Experimental Set-up

A schematic picture of a new quasi-homogeneous NTL is shown in Fig. 1.

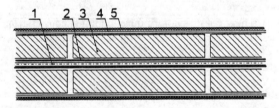

Fig.1 Schematic cut of the new non-linear coaxial line with ferrite filler. The indices are explained in the text.

The inner line (1) is made of a multifilament high-frequency conductor with diameter 1.5 mm and rubber insulation (2), the outer line (5) consists of the standard coaxial copper shielding separated by teflon foil (4) from a ferrite filler (3). The filler consists of several identical sections of NiZn ferrite toroidal cores 50/6/2 mm, type H11, having H_c=30 A/m, B_m(2kA/m) = 0.38 T, B_r = 0.15 T and $\varepsilon_r \cong 12$. The space between adjacent toroids is less than 0.1 mm. With respect to the spectrum of the observed waves the dispersion of ε_r can be neglected. The length of the line was 4 m.

3. Experimental Results

The current dependence of the dynamic inductance $L_d(i)$ has been experimentally estimated by measurement of the velocity $v(i)$ for various i-levels along the NTL and afterward calculated using equation (1). The results are shown in Fig. 2 (curve 1) [3]. The experimentally obtained function can be well fitted by expression

$$L_d(i) = L_d(0)\exp[-(qi)^2] = \frac{d\Phi(i)}{di}$$

(7)

where $L_d(0) = 4.8\times10^{-6}$ H, $q = 0.12$ A^{-1} (Fig. 2, curve 2). The magnetic flux in the ferrite filler $\Phi(i)$ is of essential importance for analysis of the fast magnetization processes.

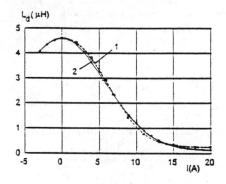 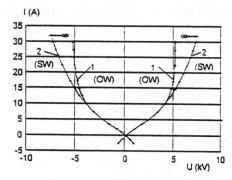

Fig. 2 Experimentally estimated and by equation (9) fitted $L_d(i)$ function of NTL with ferrite H11 and $dH/dt > 10^9$ Acm^{-1}s^{-1}.

Fig.3 The interaction characteristics of NTL. 1-ordinary wave, 2-shock wave

From equation (7) by integration of $L_d(i)$ we obtain

$$\Phi(i) = L_d(0)\frac{\sqrt{\pi}}{2q}\operatorname{erf}(qi)$$

(8)

For the points of the ordinary wave profile beyond its slope holds

$$u_2 - u_1 = \int_{i_1}^{i_2}\sqrt{\frac{L_d(i)}{C_0}}\,di = \sqrt{\frac{\pi L_d(0)}{2C_0}}\frac{1}{q}\left[\operatorname{erf}(qi_2/2) - \operatorname{erf}(qi_1/2)\right]$$

(9)

and for ESW in the case of the passive line ($U_1 = 0$, $I_1 = 0$; $U_2 = U$, $I_2 = I$) it holds

$$U = \sqrt{\frac{\pi L_d(0)}{2C_0 q}}I\cdot\operatorname{erf}(qI)$$

(10)

Both interaction characteristics are shown in Fig. 3.

4. Transition Between Non-Linear Line and Linear Load

On the junction of NTL with a linear line or in the case of NTL loaded with resistor R a very complicated non-linear effects between incident and reflected waves occur. This problem can be solved in the first approximation by introduction of the notion of non-linear impedance $Z_F(I)$ of the ferrite coaxial line working in the ESW regime, taking in consideration equation (10)

$$Z_F(I) = \sqrt{\frac{\pi L_d(0)}{2C_0 qI}} \, \mathrm{erf}(qI) \tag{11}$$

The method allowing experimental estimation of $Z_F(I)$ we proposed in [4] uses the double reflection of the ESW between the resistor load $R \ll Z_0$ and the NTL, connected through a linear line with $Z_0 \ll Z_F$, as it is shown in Fig. 4.

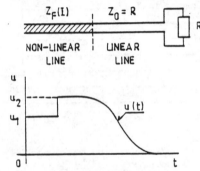

Fig. 4 Voltage u(t) on the resistor load R used to determination of $Z_F(I)$

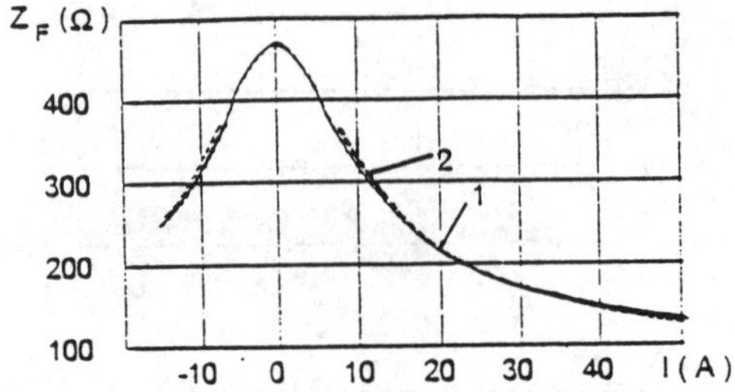

Fig. 5 Non-linear impedance $Z_F(I)$ of NTL in shock wave regime. 1-calculated using equation (11), 2-estimated from experiments using equation (12)

The measured difference $\Delta u = (u_2 - u_1)$ allows to determine $Z_F(I)$

$$Z_F(I) = Z_0 \frac{u_2 + \Delta u(1+\alpha)}{u_2 - \Delta u(1+\alpha)} \tag{12}$$

where $\alpha = 2R/(Z_0 - R)$.

The measured and calculated dependence $Z_F(I)$ is shown in Fig. 5. An optimal matching with minimum reflection and maximum power transfer is achieved when the usual condition $R = Z_0 = Z_F(I)$ for required maximum current I_{max} of ESW is fulfilled.

References

[1] O. Benda, J. Červeň, J. Oravec, Shock waves and solitons in applied electromagnetics, *Journal of Technical Physics* **35** (1994) 13-24
[2] O. Benda, J. Oravec, Non-linear transmission line, a powerful tool to pulse forming, *Proceedings of the 23. Int. Kolloqium. TH Ilmenau* (1978) 25-28
[3] J. Oravec, Types of interactive waves in a non-linear coaxial transmission line with a homogeneous ferrite filler, *Journal of Electrical Engineering* **47** (1996) 306-311
[4] J. Oravec, M. Oravec, Impedance matching of non-linear coaxial cable with a homogeneous ferrite filler, *Journal of Electrical Engineering* **47** (1996) 143-146

Non-Linear Electromagnetic Systems
V. Kose and J. Sievert (Eds.)
IOS Press, 1998

Non-local Dynamics and the Soliton Regimes in an Antiferromagnetic Film

V.V. KISELIEV and A.P. TANKEYEV
*Institute of Metal Physics, Urals Branch of the Academy of Sciences
Ekaterinburg, 620219 Russia*

Abstract. An effective integro-differential equation describing the interaction of quasi-one-dimensional exchange-dipole Goldstone spin waves is derived. New types of solitons and multi-soliton regimes in films have been predicted. The conditions for soliton existence are investigated as a function of film thickness and magnitude of the magnetic field perpendicular to the film surface. The dynamical properties of solitons, their interaction and peculiarities of relaxation have been described.

1. Introduction

Non-linear properties of antiferromagnetic films have attracted interest. This interest is caused by fundamental physical properties of these materials, as well as by the progress in preparation of high-quality films and their potential applications. The peculiarities in the propagation of the activation spin waves in antiferromagnets were studied in [1]. For a wide variety of two-sublattice antiferromagnets with an "easy-plane" anisotropy the zero-gap (Goldstone) modes play a more important role since they have a lower excitation energy. When describing Goldstone modes in thin films the non-local part of magnetostatic spin wave dispersion should be taken into account. We propose a version of the perturbation theory, which overcomes the difficulties associated with the non-linear and non-local nature of the problem.

2. Problem formulation

Let us consider an antiferromagnetic film with an "easy- plane" anisotropy. The constant magnetic field H_0 and anisotropy axis are directed along the z-axis, which is perpendicular to the surface of the film. The energy density has the form:

$$W = \frac{1}{2}\alpha\left[(\nabla M_1)^2 + (\nabla M_2)^2\right] + \alpha'\nabla M_1 \cdot \nabla M_2 + \delta M_1 \cdot M_2 - \beta'(M_1 \cdot n)(M_2 \cdot n) - \frac{1}{2}\beta\left[(M_1 \cdot n)^2 + (M_2 \cdot n)^2\right] - \left[nH_0 + \frac{1}{2}H^{(m)}\right] \cdot (M_1 + M_2), \tag{1}$$

where α, α', δ are the exchange interaction constants, β, β' are the magnetic anisotropy constants $(\beta' - \beta > 0)$, $n=(0, 0, 1)$, $H^{(m)}$ is the field defined by magnetostatic equations. Landau-Lifshitz and magnetostatic equations are solved in case of free spins at the surface of the slab. We used the following parameterization for the magnetization vectors of the sublattices:

$$\mathbf{M}_i = M_0(\cos\theta_i \cos\varphi_i, \cos\theta_i \sin\varphi_i, \sin\theta_i), \quad i = 1\,2.$$

The equilibrium values of the angles are defined by

$$\varphi_1^0 - \varphi_2^0 = \pi, \quad \theta_1^0 = \theta_1^0 = \theta^0, \quad \sin\theta^0 = H_0 \Big/ \Big[2M_0(\delta + 4\pi - \beta_2)\Big], \quad \beta_2 = \frac{1}{2}(\beta' + \beta).$$

3. Effective equations of motion

Our problem involves two characteristic space scales: the slab thickness d and the size Δ of a magnetic inhomogeneity. We shall deal with the weakly nonlinear waves under the following conditions:

$$\Delta \gg \max\Big\{(\alpha + \alpha'' \tan^2\theta^0/2\pi d, \, d\Big\}, \quad d/\delta\Delta \ll \Big[H_0/2M_0\big(\delta + 4\pi - \beta_2\big)\Big]^2 \ll 1, \quad (2)$$

where $\alpha = \frac{1}{2}(\alpha + \alpha')$, $\alpha'' = \frac{1}{2}(\alpha - \alpha')$. It can be shown that the spin-wave spectrum of an antiferromagnetic film has activation branches and a Goldstone branch, which in range (2) is exchange-magnetostatic with a dispersion law

$$\omega^2 = (2gM_0)^2 \alpha'' k_x^2 \cos^2\theta^0 \Big(\delta + 4\pi - \beta_2 - 2\pi d|k_x|\Big).$$

Here g is the magnetomechanical ratio, the waves propagate along the x-axis.

To derive the effective equations for Goldstone modes, we use a nonlinear perturbation theory [2], which can be generalized to films with non-uniform material parameters along the normal to film surface and to arbitrary exchange boundary conditions. A similar procedure was used earlier to describe the propagation waves in stratified fluid [3]. Inside the slab the angles θ_i, φ_i and the magnetostatic potential φ ($\mathbf{H}^{(m)} = -\nabla\varphi$) have the following form:

$$\theta_i = \theta^0 + \sum_{n=1}^{\infty} \varepsilon^n \theta_i^{(n)}(\xi, \tau, z), \quad s = 2M_0 g\cos\theta^0\Big[\alpha''\big(\delta + 4\pi - \beta_2\big)\Big]^{\frac{1}{2}},$$

$$\varphi_i = \varphi_i^0(\xi, \tau, z) + \sum_{n=1}^{\infty} \varepsilon^n \varphi_i^{(n)}(\xi, \tau, z), \quad \xi = \varepsilon(x + st), \quad \tau = \varepsilon^2 t, \quad i = 1,2,$$

$$\varphi = \varphi^{(0)}(\xi, \tau, z) + \sum_{n=1}^{\infty} \varepsilon^n \varphi^{(n)}(\xi, \tau, z).$$

Outside the film we look for a solution of the magnetostatic equation in the form:

$$\varphi = \varphi^{(0)}(\eta, \tau, z) + \sum_{n=1}^{\infty} \varepsilon^n \varphi^{(n)}(\eta, \tau, z), \quad \eta = x + st, \quad \tau = \varepsilon^2 t.$$

Here ε is a small parameter, s is the phase velocity of spin waves. Magnetostatic boundary-value problems are solved in explicit form and the variable ξ and z are separated in each order of such perturbation theory. The long-range character of dipole-dipole forces is manifested in a non-locality of the field $\mathbf{H}^{(m)}$. The effective equations for Goldstone modes are obtained as a result of the matching of two versions of the perturbation theory at the boundary of the film. The functions $\theta_1^{(1)}$, φ_1^0 satisfy the following resultant equations:

$$\partial_\tau \theta_1^{(1)} - v \partial_\xi^2 \hat{H} \ \theta_1^{(1)} + q \ \partial_\xi \left(\theta_1^{(1)}\right)^2 = 0, \qquad \theta_1^{(1)} = \theta_2^{(1)} = \gamma \partial_\xi \varphi_1^0, \qquad (3)$$

$$\gamma = \left[\alpha''/(\delta + 4\pi - \beta_2)\right]^{\frac{1}{2}}, \ v = 2\pi d M_{0} \gamma \cos\theta^0, \ q = \frac{3}{2}\, s\tan\theta^0,$$

which are equivalent to the Benjamin-Ono equation. Here $\hat{H} u$ denotes Hilbert

transform of u: $\hat{H} u \equiv \dfrac{1}{\pi} P \displaystyle\int\limits_{-\infty}^{+\infty} \dfrac{dx' u(x')}{x' - x}$.

The Benjamin-Ono model acknowledges the multi-soliton regimes and can be investigated in detail using the inverse scattering method (see references in [2]).

We have shown that in weak magnetic fields $\left(H_0 / 2M_0 \delta < \theta_1^{(1)}\right)$ the interaction of Goldstone modes is less intensive and defined by the cubic (rather than quadratic as in (3)) term in field $\theta_1^{(1)}$ in the resultant equation. Effective equations differ from (3) and do not have soliton solutions.

4. Multi-soliton excitations in an antiferromagnetic film

The exchange-magnetostatic soliton is described by the "algebraic" wave:

$$\theta_1^{(1)} = -\sigma\Delta / \left[(\xi + v\tau)^2 + \Delta^2\right], \qquad \varphi_1^0 = -(\sigma/\gamma) \arctan\left[(\xi + v\tau)/\Delta\right], \qquad (4)$$

$$\Delta = 2\pi d \gamma g v^{-1} M_0 \cos\theta^0 > 0, \qquad \sigma = \left(4\pi d \cot \theta^0\right) / 3\left[\delta + 4\pi - \beta_2\right],$$

where v is a positive real parameter $\left(v/s \propto d/\delta\Delta \ll 1\right)$. The soliton velocity v+s is above the phase velocity s of the exchange-dipole spin waves. The solitons arise in a threshold way as a result of the increase in amplitude of the spin waves and are realized only in dynamics. The system tends to reduce its energy by creation of solitons. We have shown, that the soliton energy is, by factor

$N \propto \pi \left[8\pi d M_0 / 3aH_0\right]^2 \gg 1$, smaller than the magnon energy with the same momentum (a is lattice constant).

In terms of $\theta_1^{(1)}$ and φ_1^0 the vectors ferro- and antiferromagnetism have the following form:

$$\mathbf{M} \equiv \mathbf{M}_1 + \mathbf{M}_2 = 2M_0(0;\ 0;\ \sin\theta^0 + \theta_1^{(1)} \cos\theta^0) + O(\varepsilon^2),$$

$$\mathbf{L} \equiv \mathbf{M}_1 - \mathbf{M}_2 = 2M_0(\cos\theta^0 \cos\varphi_1^0;\ \cos\theta^0 \sin\varphi_1^0;\ 0) + O(\varepsilon).$$

In the localization region of a soliton the component M_3 is smaller than at infinity; so M_3 describes a "dark" soliton. Vector $\mathbf{L}$ lies in the plane of the slab being

turned by an angle $\Delta\varphi_1^{(0)} \propto \left(8\pi^2 d M_0 / 3aH_0\right) \gg \pi$ in the region of localization of the soliton i.e. completing many revolutions.

Multi-soliton solutions of (3) describe the elastic collisions between these "algebraic" solitons. The "algebraic" solitons in principle are forming the surface magnetic charges and are absent in an infinite antiferromagnet. In the infinite sample the magnetostatic interaction contribution is of secondary importance in comparison with the short-range exchange interaction contribution and non-linear small-amplitude waves are described by local equations of the Korteveg -de Vries type. A weak spatial localization of the soliton (4) is connected with the non-local long-range dipole-dipole forces. Certainly in the case of a thick film the strongly localized "exponential" solitons are realized rather than "algebraic" solitons.

The model considered above is sufficiently simple and consistent with real antiferromagnets. Specific evaluations can be performed using the material parameters for $MnCO_3$ (Neel point $T_N=325K$), α-Fe_2O_3 with $T_N=950K$ having an "easy-plane" anisotropy above the Morin point $T_M=260K$, and $FeBO_3$ with $T_N=348K$. In these systems it is easy to realize conditions (2) when $d \propto 5-10$ μm. Considering that for $FeBO_3$ $\delta \propto 10^3$ and $M_0=828$ A/cm, we obtain a soliton velocity close to the velocity of sound propagation in these materials: $v+s\sim10^4$-10^5 ms^{-1}. Evidently, the presence of solitons can be revealed through their resonance interaction with the elastic subsystem of the antiferromagnet.

The soliton relaxation due to the exchange interaction can be taken into account in a phenomenological manner by adding the term $-\mu\,\partial_\xi^2\theta_1^{(1)}$ to the left - hand side of equation (3):

$$\partial_\tau\theta_1^{(1)} - v\partial_\xi^2\hat{H}\,\theta_1^{(1)} - \mu\partial_\xi^2\theta_1^{(1)} + q\,\partial_\xi\left(\theta_1^{(1)}\right)^2 = 0.$$

The obtained equation acknowledges a wide class of exact solutions. The simplest of these has the form:

$$\theta_1^{(1)} = -\sigma\left\{\left[\xi+\xi_0(\tau)\right]^2 + \Delta^2(\tau)\right\}^{-1}\left\{\Delta(\tau) + \frac{\mu}{v}\left[\xi+\xi_0(\tau)\right]\right\}, \tag{5}$$

$$\xi_0(\tau) = (v/\mu)\left[\Delta(\tau) - \delta_0\right], \quad \Delta(\tau) = \left[2\mu\tau + \delta_0^2\right]^{1/2}, \quad \mu, \delta_0 = \text{const} > 0$$

and in the $\mu\rightarrow\infty$ limit reduces to (4). From equation (5), it follows that the localized excitation becomes blurred in time and extends in width, while the amplitude and velocity are diminished.

References

[1]   A. Boardman et al., Existence of Spin-wave Solitons in an Antiferromagnetic Film, Phys. Rev. B **48** (1993) 13602-13606.
[2]   V. Kiseliev and A. Tankeyev, Weakly Non-linear Excitations in an Antiferromagnetic Film, J. Phys.: Condens. Matter **7** (1995) 2087-2094.
[3]   H. Ono, Algebraic Solitary Waves in Stratified Fluids, J. Phys. Soc. Japan **39** (1975) 108-1091.

Non-Linear Electromagnetic Systems
V. Kose and J. Sievert (Eds.)
IOS Press, 1998

Non-linear Dynamics of Exchange-Dipole Spin Waves in Ferromagnetic Films

A.P. Tankeyev, V.V. Kiseliev
*Institute of Metal Physics, Urals Branch of the Academy
of Sciences, Ekaterinburg 620219 Russia*

Abstract. The non-linear dynamics of exchange-dipole spin waves in ferromagnetic films has been investigated for various space-time scales. It is shown that in the range of small wavenumbers **k** the effective equation of evolution is of an integro-differential type and not reducible to the local non-linear Schrodinger equation (NSE). It is also shown that this equation permits the existence of new localized states of the algebraic soliton type. The conditions of their existence and their dynamic properties are investigated depending on the film thickness and the magnitude and orientation of the external magnetic field. It is found that for sufficiently thick films (slabs) at moderate and large wavenumbers the description of weakly non-linear dynamics for each activation branch of the spin wave spectrum can be reduced to a completely integrable model within the NSE. This equation permits the existence of exponential solitons. The possibilities of experimental detection of these new states are discussed.

It is well known that the main peculiarities of linear spin dynamics in ferromagnetic films are determined by the interactions, which form the lowest-energy (ground) state of the system. The long-range magnetostatic interaction has no significant effect on the ground state. Nevertheless, this interaction plays an important role in the spin dynamics. The long-range dipole-dipole interactions determine in the main the spatial dispersion of the long-wave part of the linear spectrum and the analytical properties of the dispersion law. These properties have a dominant role in the formation of a non-linear dynamic regime. The competition between short-range exchange and long-range dipole-dipole interactions proves to be of importance in the formation of this regime.

At present a series of papers is known where the dipole-exchange spectrum of spin waves in ferromagnetic films was studied in detail (see [1] and references therein). The principal result of this work lies in the fact that the dipole-exchange spectrum is quantized because of the spatial quantization of the spin-wave vector along the normal to the surface of the film. In this case the dipole-exchange spectrum of spin excitations is described by a set of curves each of which corresponds to its intrinsic wave propagating along the film. As indicated in [2], the dispersion law $\omega(\mathbf{k})$ is a differentiable function in all domains of its definition except for the interval of small wavenumbers **k** ($|kd| \ll 1$, d is the thickness of the film). The additional peculiar interval of the dispersion law is the domain of the interaction of the various branches of spin-wave spectrum (the domain of the anomalous dispersion) , $|kd| \geq 1$. In addition, the competition between exchange and magnetostatic interactions gives rise to the inflection on the $\omega(\mathbf{k})$ curve for thin ferromagnetic films. The description of the weakly nonlinear dynamics of the spin excitations has intrinsic peculiarities in all regions mentioned above. The present paper is devoted to the analysis of these peculiarities of the spin dynamics over a wide range of variation of wavenumber $\mathbf{k}$.

Consider an isotropic ferromagnetic film of thickness d along the z-axis magnetized by a uniform magnetic field H_0 directed along the normal to the surface (namely, along the z- axis) or tangentialy (along the y- axis). The equation of motion of the magnetization M has the form

$$\partial_t \, \boldsymbol{M} = -|\gamma| \Big[\boldsymbol{M} \times (\boldsymbol{H}_0 + \boldsymbol{H}^{(m)} + \alpha \Delta \, \boldsymbol{M}) \Big] \tag{1}$$

where $\partial_t \equiv \partial/\partial t$; $M^2 = M_0^2$, M_0 is the saturation magnetization, γ is the magnetomechanical ratio, α is the exchange interaction constant and Δ is the Laplacian operator. Below the modulus sign around γ is omitted. The analysis is performed for the case of free spins at the surface of the film, that is

$$\partial_z \, \boldsymbol{M} \big|_{z=\pm d/2} = 0. \tag{2}$$

The demagnetizing field $\boldsymbol{H}^{(m)}$ satisfies the equation of the magnetostatics

$$\text{curl } \boldsymbol{H}^{(m)} = 0 \qquad\qquad \text{div}\Big(\boldsymbol{H}^{(m)} + 4\pi \, \boldsymbol{M} \Big) = 0 \tag{3}$$

and continuity conditions of tangential components of the magnetic field vector $\boldsymbol{H}^{(m)}$ and normal component of the induction vector $\boldsymbol{B}^{(m)} = \boldsymbol{H}^{(m)} + 4\pi \boldsymbol{M}$ at the boundary of a ferromagnet:

$$\Big(\boldsymbol{H}_+^{(m)} \Big)_\tau = \Big(\boldsymbol{H}_-^{(m)} \Big)_\tau \qquad\qquad \Big(\boldsymbol{H}_+^{(m)} \Big)_v + 4\pi \, \boldsymbol{M}_v = \Big(\boldsymbol{H}_-^{(m)} \Big)_v \tag{4}$$

where the indices $(+)$ and $(-)$ denote the fields inside and outside the ferromagnetic film, respectively, and the indices τ and v correspond to tangential and normal components of the vectors $\boldsymbol{H}^{(m)}$ and M at the surface of a film. In order to obtain the spin- wave spectrum it is necessary to linearize the equations (1), (3) in combination with the boundary conditions (2), (4) in terms of small deviations of magnetization from the ground state $M = M_0 \, \boldsymbol{n}$:

$$M = M_0 \boldsymbol{n} + \boldsymbol{m} \qquad\qquad m_z = (2M_0)^{-1} \Big(m_x^2 + m_y^2 \Big). \tag{5}$$

The expressions for the lowest type branches of the spin-wave spectrum are given below.

If the uniform magnetic field is directed along the normal to the surface of the film, that is, along the z-axis, we have the following expression for the linear mode spectrum (in the long-wave limit $|kd| \langle\langle 1$)

$$\omega^2 \cong \omega_H^2 + \omega_H \omega_M \frac{|kd|}{2} + \omega_H k^2 (2\gamma \alpha M_0 - \frac{\omega_M d^2}{6}). \tag{6}$$

Here

$$\omega_H = \omega_H^0 - \omega_M \qquad\qquad \omega_H^0 = \gamma H_0 \qquad\qquad \omega_M = \gamma \, 4\pi M_0 \; .$$

Let us note that in this geometry the spectrum does not depend on the direction of wave propagation in the film plane. If the field is directed along the y-axis, then the simple calculation results in the following expression for the spectrum (in the limit $|kd| \langle\langle 1$)

$$\omega^2 \cong \omega_H^2(\omega_H^0 + \omega_M) + \frac{|kd|}{2}\,\omega_M(\omega_M cos^2\zeta - \omega_H^0 sin^2\zeta) + k^2(\gamma\alpha M_0(2\omega_H^0 + \omega_M) -$$
$$\frac{5}{12}(\omega_H dcos\zeta)^2 + \frac{1}{6}\omega_M\omega_H^0(dsin\zeta)^2) \tag{7}$$

where ζ is the angle between the Ox- and $O\xi$- axes ($O\xi$- axis coincides with the direction of wave propagation). It should be noted that the dispersion relationships in both cases depend on $|\mathbf{k}| \equiv k$. As a result the functions $\omega(\mathbf{k})$ become non-differentiable at $|\mathbf{k}| \to 0$.

The presence of the modulus of wave vector $\mathbf{k}$ in the dispersion law rather than its projection on the direction of wave propagation changes substantially the effective equation of weakly nonlinear dynamics of these waves, which becomes non-local and irreducible to the NSE. The effective evolution equation of weakly non-linear exchange-dipole waves assuming that these waves propagate along $O\xi$- axis and that $|kd|\langle\langle 1$ takes the form [3]:

$$\partial_t^2\varphi + \omega_0^2\varphi + a\partial_\xi^2\hat{H}\varphi + b\partial_\xi^2\varphi + g|\varphi|^2\varphi = 0. \tag{8}$$

Here

$$\varphi = M_0^{-1}(m_x + im_y), \quad \omega_0^2 = \omega_H^2, \quad a = -\omega_H\omega_m d / 2, \quad b = \omega_H\left[(\omega_M d^2 / 6) - 2\gamma\alpha M_0\right],$$

$$g = \omega_H\omega_M, \qquad \hat{H}u = \pi^{-1}P\int_{-\infty}^{+\infty}\frac{d\xi' u(\xi')}{\xi' - \xi} \quad \text{is a Hilbert operator.}$$

When the external magnetic field lies in the plane of the film (for definition, along the y-axis), the effective equation of evolution coincides in form with (8), where φ, ω_0^2, a, b and g should be now replaced by the new expressions [3]. Equation (8) is not completely integrable, but it permits exact localized solutions like solitons. Provided that $ag \rangle 0$, $bg \rangle 0$ the following solution is possible:

$$\varphi = \frac{2i\Delta|C|exp(i\Omega t + i\varphi_0)}{\xi^2 + \Delta^2} \qquad \Omega^2 = \omega_0^2 + \frac{a^2}{3b} \qquad |C|^2 = \frac{2b}{g} \qquad \Delta = \frac{3b}{a}. \tag{9}$$

It corresponds to a precessing soliton. Here φ_0 is the real parameter. The analysis shows that if the film is magnetized normally to its boundaries, the soliton regime is impossible. The conditions of a soliton existence are more easily satisfied if the film is magnetized tangentially. The formation of the precessing algebraic soliton leads to a resonance energy absorption at a frequency smaller than the frequency of the uniform resonance (uniform ferromagnetic resonance). The experimental detection of this new absorption might be conclusive evidence for the occurrence of this soliton-like state. .

It is well known that for sufficiently thick films (slabs) at moderate wavenumbers the effective equations are reduced to the local NSE [4]. This equation admits the existence of exponential solitons. The authors have extended these results to the exchange-dipole spectrum using a specially developed perturbation theory [4]. This theory allows us to take into account the key aspects of the problem: the non-linearity of the magnetic subsystem, the nonuniform distribution of magnetization along the normal to the film plane, electromagnetic and exchange boundary conditions, the long-range character of the magnetic dipole-dipole interaction which gives rise to the non-local spatial dispersion. The effective non-linear evolution equation in "slow" variables X_1

and τ for the function ψ ($\psi \approx (m_x + imy)$) takes the form   coinciding with NSE for an arbitrary branch of the dipole-exchange spectrum

$$i\partial_\tau \psi - \frac{1}{2}\left(\partial_k^2 \omega(k)\right)\partial_{x_1}^2 \psi + g(k)|\psi|^2 \psi = 0.\tag{10}$$

For the lowest branch of this spectrum the dispersion law $\omega(k)$ and the spin wave interaction constant g(k) are presented in [4]. It is well known that equation (10) admits the soliton regimes, if its parameters satisfy the Lighthill criterion

$$g(k)\partial_k^2 \omega(k) \langle 0.\tag{11}$$

As a numerical analysis shows, in the case of sufficiently thick films this criterion will be satisfied over a fairly broad range of $|kd|$ up to $|kd| \approx 10$ (see [4]). In the case of thin films, the condition   (11) will be violated even at $|kd| \approx 1$. Near this point $\partial_k^2 \omega$ vanishes and   changes its sign afterwards. In this range of wavenumbers, the dispersion term of equation (10) vanishes. It is evident that the soliton regime of spin waves propagation are changed within an indicated range. It is interesting to note that the possibility of a change in the soliton regime of spin waves propagation owing to a change of sign of $\partial_k^2 \omega$ is mentioned by Kalinikos et al. [5]. They, however, have   associated this occurrence with the existence of an anomalous dispersion region that is due to the interaction between the various modes of the dipole-exchange spectrum. In this paper, we associate the change of sign with the competition between the exchange and dipole dispersion for the lowest branch of the spectrum.

Let us note in conclusion that in the present paper the fairly complete space-time picture of weakly non-linear dynamics taking into account the interaction of quasi-one-dimensional exchange-dipole spin excitations is constructed for a thin ferromagnetic film. The peculiarities of the behaviour for small-amplitude spin waves has been analyzed over a wide range of wavenumbers **k**. It is found that the non-local part of the magnetostatic dispersion determines to a large degree the characteristic properties of the non-linear excitations. The results of the investigation are of basic interest for an understanding of the character of the non-linear phenomena in thin ferromagnetic films.

References

[1]   B.A. Kalinikos, The Spectrum and Linear Excitation of Spin Waves in Ferromagnetic Films, *Izvestiya Vysshykh Uchebnykh Zavedenii, Fizika* **8** (1981) 42-56.

[2]   R.E. De Wames and T. Wolfram, Dipole-Exchange Spin Waves in Ferromagnetci Films, *Journal of Applied Physics* **41** (1970) 987-993.

[3]   V.V. Kiseliev and A.P. Tankeyev, Non-local Dynamics of Weakly Non-linear Spin Excitations in Thin Ferromagnetic Films, *Journal of Physics: Condensed Matter,* **8** (1996) 10219-10229.

[4]   V.V. Kiseliev and A.P.Tankeyev, Weakly Non-Linear Dynamics of Dipole-Exchange Spin Waves in Ferromagnetic Plates of Finite Thickness, *Fizika Metallov i Metallovedenie,* **82**(1996) 38-58.

[5]   B.A. Kalinikos et al., Solitons of the Envelope and Modulation Instability of Dipole-Exchange Magnetization Waves in Yttrium Iron Garnet Films, *Zhurnal Eksperimentalnoii i Teoreticheskoii Fiziki,* **94** (1988) 159-176.

Non-Linear Electromagnetic Systems
V. Kose and J. Sievert (Eds.)
IOS Press, 1998

Modeling of Chaotic Behavior in Discharge Plasmas

Andrzej J. TURSKI

Institute of Fundamental Technological Research-PAS,PL-00-049 WARSAW,
Ul.Swiętokrzyska 21

Abstract. Plasma discharges produced by electric current flows and related oscillations (Hopf bifurcations), saddle-node and period-doubling bifurcations and chaos are considered. If wavelengths of the wave phenomena are much greater than the physical size of the system, we consider it as a nonlinear circuit of lumped elements. A space-extended plasma system is also considered. This is a system of a dimension, which is comparable or longer than wavelength of travelling waves. Plasma equations are reduced to a set of hyperbolic PDE for electric charge and current densities with nonlinear boundary. The boundary-value problem for the equations is reduced to a difference equation, which plays a similar role to the Poincaré map of the first model.

1. Modeling Principles

The study of chaotic plasma systems may be performed by direct measurements of laboratory plasma discharge systems,see [1] and [2], or by analyzing experimental data recorded, e.g. in space, as a series of measurements in time of pertinent and easily accessible state variables. In most cases such variables describe a global or averaged properties of the system. Although, a vast amount of literature already exists describing experimental results concerning bifurcation, chaos, quasiperiodicity, intermittency and 1/f-noise in plasma discharge and turbulent systems, coherent discussions and a theory derived from plasma equations are still lacking. Two models are put forward. One is based on an assumption that the dimensions of the system in relation to wavelengths are small and lumped circuit simulations are justified. It is known that the direct current discharge system leads to selfexcitation of the system as the regulating discharge parameters reach the Hopf bifurcation point of a lumped electrical circuit with a nonlinear conductance simulating gas discharges. Next, the nonlinear capacity of a double layer (DL) plays a main role in the processes of period-doubling and saddle-node bifurcations, intermittency and chaos. These phenomena are related to charge separation, which takes place in the case of electric current flow through the plasma system. The plasma DL appearance is related to Bohm conditions for electron thermal velocities at the cathode side and for ion velocities at the anode side. Periodically excited R-L-Capacity Diode, first investigated by Tanaka et al. [3], exhibits fractal and chaotic properties due to nonlinear capacity. We note that the nonlinearity is of the same origin in plasma DL and in semicoductor charge separation. This is a main assumption of our simulation model. Thus, we obtain a 2-D Poincaré map, which is fundamental for investigations of bifurcation trees, 1/f noise and intermittencies related to plasma discharge systems, see [4]. A space-extended plasma system is also considered. This is a system of a dimension,which is comparable or longer than the wavelength of travelling waves. Plasma equations are reduced to a set of hyperbolic PDE for electric charge and current densities with a nonlinear boundary. The boundary-condition problem for the equations is reduced to a *difference* equation, which plays a similar role to the Poincaré map of the first model.

2. Lumped Circuit Model

In view of the well-known electrical circuit of glow discharges, e.g. see[5], one can write the following equations describing the circuit:

$$E = RI + V_c, \quad V_c = L\frac{di}{dt} + \Psi(i) \quad and \quad I = i + C\frac{dV_c}{dt}$$

The equations can be reduced to the second order ODE:

$$\frac{d^2i}{dt^2} + \frac{di}{dt}\frac{RC\Psi'(i) + L}{RLC} + \frac{i}{LC} = \frac{E - \Psi(i)}{RLC}.$$

Following the well known method of determination of Hopf bifurcation point, we linearize the last equation to obtain

$$\frac{d^2i^*}{dt^2} + \frac{di^*}{dt}\frac{R\rho C + L}{RLC} + i^*\frac{R + \rho}{RLC} = 0,$$

where $i = i_0 + i^*$, $V_c = V_0 + v^*$, $\Psi(i_0 + i^*) \approx \Psi(i_0) + i^*\Psi'(i_0)$, $\Psi'(i_0) \triangleq \rho$ and we assume that i_0, V_0 and ρ are constants and the linearization is related to the point (V_0, i_0).

By use of the last equation, one can obtain the characteristic equation to find, e.g. in (R, ρ) plane and for arbitrary but fixed L and C,the bifurcation point

$M(\sqrt{L/C}, -\sqrt{l/C})$ and selfoscillation frequency $\omega = \sqrt{1/LC - 1/R^2C^2}$.

Next, we consider the R-L-Diode circuit, which exhibits the nonlinear complexity of a DL system. We note that the voltage-charge plot of a diode, see[4], as well as the respective plot of a plasma DL are similar. We approximate the experimental characteristic by a related piece-wise linear characteristic. It seems to be a drastic approximation but it preserves crucial features of the observed nonlinear processes. By virtue of the assumptions, the circuit equations are as follows:

$$\frac{dq}{dt} = i,$$

$$L\frac{di}{dt} = -Ri - \begin{cases} q/C_1 & for \ q \geq 0 \\ q/C_2 & for \ q < 0 \end{cases} - E_0 + \begin{cases} E & for \ nT \leq t < (n + 1/2)T \\ -E & for \ (n + 1/2)T \leq t < (n + 1)T \end{cases}$$

where $C(q) = C_1 \ if \ q \geq 0$ and $C_2 \ as \ q < 0$.

A square wave source of period T=1/f replaced the sinusoidal excitation of period T. Introducing the following dimensionless parameters

$$q \Longrightarrow (Lf^2/E)q, \ i \Longrightarrow (Lf/E)i, \ t \Longrightarrow tf, \ k = R/Lf, \ \alpha = 1/LC_1f^2$$

and $\beta = 1/LC_2f^2$,

we have the following circuit equations

$$\frac{dq}{dt} = i \ , \ \frac{di}{dt} = -ki - \{ \begin{array}{l} \alpha q \ \ for \ \ q \geq 0 \\ \beta q \ \ for \ \ q < 0 \end{array} - \frac{E}{E_0} + \{ \begin{array}{l} 1 \ for \ n \leq t < (n+1/2) \\ -1 \ for \ (n+1/2) \leq t < (n+1) \end{array}$$

and the 2-D Poincaré map, takes the form

$$x_{n+1=y_n} - 1 + \{ \begin{array}{l} a_1 x_n \ \ for \ \ x_n \geq 0 \\ -a x_n \ \ for \ \ x_n < 0 \end{array}$$

$$y_{n+1} = b x_n$$

where $a_1 = e^{\lambda_1} + e^{\lambda_2}$, $b = -e^{(\lambda_1+\lambda_2)} = -e^{-R/Lf}$

and $\lambda_{1,2} = -R/2Lf \pm (1/2f)\sqrt{(R/L)^2 - 4/LC_1}$

We note,that a_2 is a solution dependent constant and it can be determined numerically. By use of the Poincaré map, we can obtain nearly all the complex properties of the chaotic system.

3. Reduction of Spatially Extended Plasma DL System to Difference Equations of Poincaré Map Type.

Let us consider plasma system with electric current flow. The system consists of a cathode and anode immersed in discharge plasmas. At the cathode we have a DL, which plays the role of a nonlinear boundary condition. We assume that the current-charge characteristic $J(\rho)$ can be reconstructed. We note that the distance between electrodes is comparable or larger than the wavelength of travelling waves. Electric field $E(x,t)$,current density $J(x,t)$ and charge density $\rho(x,t)$ satisfy the following equations :

$$\partial_t \rho + \partial_x J = 0, \quad \epsilon_0 \partial_x E - \rho = 0, \quad \epsilon_0 \partial_t E + J = 0. \tag{1}$$

Travelling wave approximation is accepted, that is

$$E(x,t) \rightarrow E(x - Ut,t) \sim E(x - Ut), \tag{2}$$

where U is wave velocity. By virtue of Eq. 1, we have

$$\partial_t J + U^2 \partial_x \rho = \epsilon_0 (\partial_{tt} E - U^2 \partial_{xx} E) = 0. \tag{3}$$

Eqs. 1 are reduced to the following hyperbolic p.d. equations:

$$\partial_t \rho + \partial_x J = 0, \ \partial_t J + U^2 \partial_x \rho = 0. \tag{4}$$

The equations are linear but the boundary relation for $J(\rho)$ is nonlinear at $x = l$. Assuming suitable initial- and boundary- value conditions, we proceed with the following transformations of variables:

$$J = 1/2(u+v), \ x = l\xi, \ u = u(\xi,\tau), \rho = 1/2(u-v), \tag{5}$$
$$t = (l/U)\tau, \ v = v(\xi,\tau)$$

and we have

$$\partial_\tau u + \partial_\xi u = 0, \ \partial_\tau v + \partial_\xi v = 0, \tag{6}$$

where $(\xi, \tau) \in [0, 1) \times R^{+}$. The boundary–value conditions take the form

$$u(0,\tau) = i_0 + U r_0 , \; i_0 = i_0(\frac{l}{U}\tau)$$

$$v(0,\tau) = i_0 - U r_0 , \; r_0 = r_0(\frac{l}{U}\tau), v(1,\tau) = f[u(1,\tau)]. \tag{7}$$

The function $f(u)$ is given indirectly by the following relation:

$$(u + f)/2 = g(\frac{u - f}{2v}). \tag{8}$$

We note that

$$u(\xi,\tau) = y(\tau - \xi) \; and \; v(\xi,\tau) = y(\tau + \xi),$$

where $u(\xi, 0) = y(-\xi)$ and $v(\xi, 0) = y(\xi)$. By virtue of (7) we have:

$$v(1,\tau + 1) = f[u(1, \tau + 1)] \; and \; y(\tau + 2) = f[y(\tau)]. \tag{9}$$

Discretization of the last equation leads to the 2–D map of Poincaré type, that is

$$y_{n+2} = f(y_n). \tag{10}$$

Investigation of the map leads to interesting results of selfsimilarity.

4.Conclusions

The dynamical systems, considered here, are advantageous as they may be easily measured and computed. In the first model there are three parameters a_1, b and a_2 which allow applications and simulations of different dynamical processes. The algorithm of difference equations can serve for demonstration of many interesting features of chaotic dynamics and fractal trajectories as well as low–dimensional noises.

References

1.   T.Braun, J.A.Lisboa, R.E.Francke and J.A.Gallas, Observation of deterministic chaos in electrical discharges in gases,Phys.Rev.Lett.,**59**,613,1987

2.   D.Weixing, H.Wei, W.Xiaodong and C.Yu, Quasiperiodicity transition to chaos in a plasma, Phys.Rev.Lett. **70**,170.1993

3.   S.Tanaka, T.Matsumoto and Chua, Bifurcation Scenario in a Driven R–L–Diode Circuit, Physica **28D**, 317– 344, 1987.

4.   B.Atamaniuk, Intermittent chaos and 1/f noise in electrostatic double layer system., ISEM, Braunschweig, 1997.

5.   N.Minorsky, Nonlinear Oscillations, Van Nostrand Co. Inc. ,Princenton, N. Jersey, 1962

Modelling of Chaos and Noise Conversion in Magnetic Systems by Means of Maps with Nonlinear Rotations

Vladimir V. ZVEREV

Physics-Technical Department
Ural State Technical University, 620002 Ekaterinburg, Russia

Abstract. It is shown that the behavior of magnetization in ferromagnets may be described by nonlinear maps with rotating transformations. To investigate dynamics under random perturbations the Kolmogorov-Chapman equation is used. The stationary (asymptotic) solution in the long-time limit is obtained as a power expansion "near" the state with complete angular randomization. The conditions of convergence of power series are founded.

1. Introduction

In this paper attention is focused on a class of nonlinear multidimensional maps with nonlinearities caused by rotations. Typically a map under discussion may be represented in the form:

$$x_{N+1} = a + \xi + \hat{R}\hat{U}(\lambda \cdot f(x_N))x_N , \tag{1}$$

where $\hat{U}$ is an element of the rotation (orthogonal, unitary) group and $\hat{R}$ denotes the matrix of contractive linear transformation associated with the proper mechanism of relaxation (energy dissipation). The angle of rotation $\phi = \lambda \cdot f(x_N)$ is a function of the original vector x_N ; a and ξ are constant and fluctuating vectors respectively. A well-known example of the map (1) arises from the Ikeda model of the optical ring cavity [1,2]. Here we show that maps belonging to the above-mentioned class can be obtained as a result of conversion of differential equations of motion into the difference equations: [3] - for a nuclear magnetization in ferromagnets, [4] - for parametrically driven spin waves in ferromagnets.

As has been shown in [2], the control parameter λ in Eq.(1) assigns *the degree of rotational mixing* (for the infinitely large λ we have *the perfect angular randomization*). The basic idea of our approach is to obtain the stationary solution of the Kolmogorov - Chapman equation at first, for the infinitely large value of λ, and further, for a finite λ, *in the form of a power expansion*.

2. Model 1: NMR in Ferromagnets (the Uniform Mode)

Consider 3D dynamics of the nuclear magnetization under action of a periodic sequence of radio-frequency (rf) pulses (further, $T_p = T_{rf} + T_{prec}$ is the pulse repetition period, where T_{rf} is the duration of rf field action). We are interested in nonlinear regimes caused by the indirect interaction of nuclear spins via electron spin waves (Sulh-Nakamura mechanism [5]). Neglecting the nonlinear and relaxation effects during rf field action one can integrate the equation of motion

$$\frac{d}{dt}M = \gamma_n\left[M \times \left\{A\left(M_e^0 e_z + \chi h_{rf}\right) + A^2\chi M_\perp\right\}\right] - \frac{1}{T_2}M_\perp - \frac{1}{T_1}\left(M_\| - M_n^0 e_z\right), \quad (2)$$

obtaining the evolutionary map [3]:

$$m_{N+1} = F(m_N) = e_z + \hat{R}\hat{U}_{prec}\left(\alpha + \lambda\left(e_z \cdot \hat{U}_{rf}m_N\right)\right)\hat{U}_{rf}m_N, \quad (3)$$

where

$$\hat{U}_{prec}(\phi) = \exp\left(i\phi\,\hat{S}_z\right), \quad \hat{U}_{rf} = \exp\left(iT_{rf}\left(\omega_1\hat{S}_x + \Delta\omega\hat{S}_z\right)\right), \quad \hat{R} = \exp\left(-T_{prec}\cdot\mathrm{diag}\left(\frac{1}{T_2},\frac{1}{T_2},\frac{1}{T_1}\right)\right),$$

$$\alpha = -\Delta\omega T_{prec} + A^2\chi M_n^0\left(T_{prec} - T_1(1-\gamma)\right), \quad \lambda = A^2\chi M_n^0 T_1(1-\gamma)^2, \quad \gamma = \exp\left(-T_{prec} / T_1\right),$$

$m_N = M\left(t_0 + NT_p\right) / M_n^0(1-\gamma)$, γ_n is the nuclear gyromagnetic ratio, A is the constant of electron-nuclear interaction, χ is a linear susceptibility of the electron spin system, $h_{rf} = 2e_x H_{rf}\cos\omega t$ is an external rf field, $\omega_1 = \gamma_n A\chi H_{rf}$, $T_{1,2}$ are nuclear spin relaxation times, $\Delta\omega = \gamma_n AM_e^0 - \omega$ is a frequency detuning, M_n^0, M_e^0 are equilibrium magnetizations of spin subsystems, $\hat{S}_\alpha$ are the spin 1 matrices (in the Cartesian coordinates with the basis e_α, $\alpha = x, y, z$). For nuclear systems in some real ferromagnets and antiferromagnets $\lambda \sim 50 - 5\cdot10^5$.

3. Model 2: Parametrically Excited Spin Waves

This section is concerned with nonlinear dynamics of spin-wave modes parametrically excited through microwave absorption beyond the Suhl instability threshold. We examine the "S-theory" theoretical model [6]

$$\frac{d}{dt}\theta = 2\Gamma\left(-b\sin\theta + \Delta - n\right), \qquad \frac{d}{dt}n = 2\Gamma n(b\cos\theta - 1), \quad (4)$$

where n is the number of quasiparticles, θ is the angle of phase divergence for the spin-wave pair, $b = h / h_{thresh}$ is the normed amplitude of rf field, Γ is the damping parameter, Δ is a detuning (in normalized units). Let the detuning modulation represents a periodic sequence of rectangular pulses: $\Delta = \Delta_+$ for $NT_p < t < NT_p + T_+$, $\Delta = \Delta_-$ for $NT_p + T_+ < t < (N+1)T_p$, $T_p = T_+ + T_-$, $N = 0,1,2...$. Using the method of slowly varying amplitudes for $b \gg 1$, one can obtain the 2D (complex) map [4]

$$v_{N+1} = F_-\left(F_+(v_N)\right), \quad F_\pm(v) = \pm1 + \kappa_\pm v \exp\left\{2i\Gamma b\Omega_\pm T_\pm - ig_\pm|v|^2\right\}, \quad (5)$$

where

$$g_\pm = \left(\Omega_0 - \Omega_0^{-1}\right)\left(1 - \kappa_\pm^2\right)b\Omega_1^2, \quad \Omega_\pm = \Omega_0 \pm\Omega_1 = \left(1 + \Delta_\pm / b\right)^{1/2}, \quad b\Omega_1^2 \sim 1,$$

$$\kappa_\pm = \exp(-\Gamma T_\pm), \qquad v_N = (4i\Omega_1)^{-1}\left\{\delta\theta\left(NT_p + 0\right) + i\delta n\left(NT_p + 0\right) / b\Omega_+\right\};$$

$\delta\theta$ and δn are the deviations from the static solution of Equation (4) with $\Delta = \Delta_+$.

4. Statistical Theory and Asymptotic Expansions

To investigate dynamics under external noise action (the additional fluctuating term is supplemented phenomenologically) , we use the Kolmogorov-Chapman equation for probability distribution functions:

$$P_{N+1}(x) = \iint dy\, dz\, P_{fl}(x - y)\,\delta\!\left(y - F(z)\right) P_N(z)\ . \tag{6}$$

Here we consider Model 1 in detail; the proper analysis for Model 2 may be fulfilled similarly. We obtain the stationary (asymptotic) solution in the long time limit $N \to \infty$ as a power expansion "near" the state with perfect angular randomization ("phase mixing"). It is convenient to rewrite Eq. (6) in the form

$$\Psi_{N+1}(u) = \left(K_0 + \varepsilon K_1\right)\Psi_N(u)\ , \tag{7}$$

where Ψ_N is the Fourier transform of the probability function, and ε is a formal "small parameter" (as it was shown in [3], $\varepsilon K_1 \Psi_N \to 0$ as $\lambda \to \infty$). As well,

$$K_0 f(u) = h(u)\left\langle f\!\left(\hat{Q}(\beta)u\right)\right\rangle_\beta\ , \tag{8}$$

$$K_1 f(u) = h(u)\sum_{v \neq 0}\left\langle \exp(-iv\beta) f\!\left(\hat{Q}(\beta)u - v\lambda l\right)\right\rangle_\beta\ , \tag{9}$$

$$h(u) = \exp\!\left[-i\left(u \cdot e_z\right)\right]\Lambda_{fl}(u), \quad \hat{Q}(\beta) = \hat{U}_{rf}^{-1}\hat{R}\hat{U}_{prec}^{-1}(\alpha + \beta)\ ; \quad \Lambda_{rf}(u) = \exp\!\left(-\tfrac{1}{2}\sigma^2|u|^2\right) \text{ for}$$

the Gaussian noise, $l = \hat{U}_{rf}^{-1}e_z$. Averaging over the angles $\beta \in [0,2\pi]$ in Eqs. (8), (9) is denoted by angular brackets. Fulfilling the iterations with Eq. (7), one can obtain:

$$K_0^N f(u) = h(u)\left\langle \left\{\prod_{r=1}^{N-1} h\!\left(\prod_{s=1}^{r}\hat{Q}(\beta_s)u\right)\right\} f\!\left(\prod_{s=1}^{N}\hat{Q}(\beta_s)u\right)\right\rangle_{\text{all }\beta\text{'s}} \tag{10}$$

(in the products of noncommutative matrices $\hat{Q}(\beta_s)$ the indices increase from the right to the left). The matrices $\hat{R}$ and $\hat{Q}(\beta_s)$ represent contractive transformations. It follow that

$$\prod_{s=1}^{N}\hat{Q}(\beta_s)u \xrightarrow{\ N \to \infty\ } 0\ , \qquad K_0^N f(u) \xrightarrow{\ N \to \infty\ } f(0)\cdot \lim_{N \to \infty} K_0^N \cdot 1\ ,$$

and there exists *the stationary asymptotic solution ($N, N+1 \to$ st), corresponding to the limiting case $\lambda \to \infty$ (intense phase mixing):*

$$\Psi_{st}^{(0)} = \lim_{N \to \infty} K_0^N \cdot 1, \qquad \Psi_{st}^{(0)} = K_0 \Psi_{st}^{(0)}\ . \tag{11}$$

Further, we seek the stationary solution of Eq. (7) in the form of power series in ε . Taking into account the normalization conditions $\Psi_{st}^{(0)} = \Psi_{st} = 1$, we obtain the formal expansion

$$\Psi_{st} = \Psi_{st}^{(0)} + \sum_{p=1}^{\infty} \varepsilon^p \left[\left(\sum_{m=0}^{\infty} K_0^m\right)K_1\right]^p \Psi_{st}^{(0)}\ . \tag{12}$$

The expression (11) may be treated as a "stationary point" of a proper operator:

$$\Psi_{st} = A\Psi_{st}\ , \qquad Af = \Psi_{st}^{(0)} + \varepsilon\!\left(\sum_{m=0}^{\infty} K_0^m\right)K_1 f\ . \tag{13}$$

Now our goal is to prove that A is *a contraction operator in an appropriate complete metric space*. In this work we use a metric induced by the norm

$$\|f\| = \sup_{u \in R^3, \alpha}\left|\frac{1}{\alpha!}\frac{\partial^{|\alpha|}}{\partial u^\alpha}\!\left(\Lambda^{-1/2}(u)f(u)\right)\right|\ ; \tag{14}$$

one can prove the completeness of the corresponding space (in the definition (14) the standard conventions for multi-indices $\alpha = (\alpha_1, \alpha_2, \alpha_3)$, $|\alpha| = \alpha_1 + \alpha_2 + \alpha_3$, $\alpha! = \alpha_1!\,\alpha_2!\,\alpha_3!$, $u^\alpha = u_x^{\alpha_1} u_y^{\alpha_2} u_z^{\alpha_3}$ are used). We have proved the condition of contraction

$$\|Af_2 - Af_1\| < \eta\|f_2 - f_1\|, \qquad \eta < 1, \tag{15}$$

for a certain domain of parameters. Precisely, we obtained the estimation

$$\eta = \frac{27\sqrt{3}\sigma\kappa_m}{2\pi^{3/2}} \sum_{p=0}^{\infty} \left\{ \prod_{q=0}^{p} G_q \exp\left(3\sqrt{3}\kappa_m^q\right) \right\} \frac{\kappa_m^p}{1 - \kappa_m^{2(p+1)}} \frac{1 + 6\sqrt{3}\kappa_m^{p+1}}{\left(1 - 3\sqrt{3}\kappa_m^{p+1}\right)^4} \frac{a_p^3}{\left(2 - a_p^2\right)^{3/2}}$$

$$\times I^2\left(\tfrac{1}{2}\sigma a_p\right)\left\{2a_p\left(a_p I\left(\tfrac{1}{2}\sigma a_p\right) + 1\right)L_{0p} + \kappa_m^{p+1} I\left(\tfrac{1}{2}\sigma a_p\right)L_{1p}\right\}, \tag{16}$$

where

$$G_0 = 1, \quad G_q = \frac{6\sigma\kappa_m^q}{\sqrt{2\pi}} + \left(1 + 9\sigma^2\kappa_m^{2q}\right)\exp\left(\tfrac{9}{2}\sigma^2\kappa_m^{2q}\right)\left(1 + \mathrm{erf}\left(\frac{3\sigma\kappa_m^q}{\sqrt{2}}\right)\right), \quad q > 0,$$

$$L_{dp} = \sum_{v=1}^{\infty}(v\lambda\sigma)^d \exp\left[-\frac{1}{4}(v\lambda\sigma)^2 \frac{1 - 2\kappa_m^{2(p+1)}}{1 - \kappa_m^{2(p+1)}}\right], \quad \kappa_m = \max_{i=1,2}\left\{\exp\left(-\frac{T_{prec}}{T_i}\right)\right\},$$

$$I(x) = \sqrt{\pi}\exp\left(x^2\right)\left(1 + \mathrm{erf}(x)\right), \qquad\qquad a_p = \sqrt{1 + \sqrt{3}\kappa_m^{2(p+1)}}.$$

The condition $\eta < 1$ is satisfied in a wide region of parameters, for sufficienly small κ_m, σ and sufficiently large λ. For example, $\eta < 0.8$; $1.6 \cdot 10^{-4}$; $2.5 \cdot 10^{-9}$ for $\kappa_m = 0.15$, $\sigma = 0.45$, $\lambda = 15$; 20; 25. According to *the contraction mapping principle, in this case the unique stationary solution (12) exists.* Notice in conclusion, that as it was shown in [7], the limiting stationary distribution (the Fourier transform of $\Psi_{st}^{(0)}$) may have a hidden fractal structure.

Acknowledgements

Work has been partially supported by RF State Committee on High Education (Grant No.95-0-8.3-14) and Russian Foundation of Basic Research (Grant No. 97-02-26727).

References

[1]   K. Ikeda, H. Daido, O. Akimoto, Optical Turbulence: Chaotic Behavior of Transmitted Light from a Ring Cavity, *Physical Review Letters*, **45** (1980) 709-712.

[2]   V.V. Zverev, B.Ja. Rubinstein, Chaotic Oscillations and Noise Transformations in a Simple Dissipative System with Delayed Feedback, *Journal of Statistical Physics*, **63** (1991) 221-239.

[3]   V.V. Zverev, On an Origin of a Chaotic Attractor in a Motion of Nuclear Spins, *Pricladnaja Nelinejnaja Dinamika (Applied Nonlinear Dynamics)*, **1** (1993) 1-72 (Saratov, in Russian).

[4]   V.L. Safonov, V.V. Zverev, Dynamical Chaos of Collective Oscillations in Magnets, *Fizika Tverdogo Tela (Solid State Physics)*, **36** (1994) 1939-1949.

[5]   M.I. Kurkin, E.A. Turov, NMR in Magnetically Ordered Materials and Its Applications. Nauka, Moscow, 1990.

[6]   V.E. Zakharov, V.S. L'vov, S.S.Starobinets, Spin Waves Turbulence Beyond the Threshold of Its Parametrical Exitations, *Uspehi Fizicheskih Nauk (Sov. Phys. - Usp.)*, **114** (1974) 609-654.

[7]   V.V. Zverev, On the Conditions for the Existence of Fractal Domain Integrals, *Theoretical and Mathematical Physics* , **107** (1996) 419-426.

Inverse Problems, NDT

Non-Linear Electromagnetic Systems
V. Kose and J. Sievert (Eds.)
IOS Press, 1998

An Introduction To Electromagnetic Nondestructive Testing

Yushi Sun

Materials Characterization Research Group
Department of Electrical and Computer Engineering
Iowa State University, Ames, IA 50011, USA

Abstract. In this tutorial paper some fundamental knowledge of electromagnetic non-destructive evaluation is introduced. Three examples using three different techniques based on different electromagnetic phenomena are described. Both theoretical and practical sides of the three techniques are provided.

1. Introduction

Nondestructive Testing (NDT) plays an increasely important role in diverse areas of modern society ranging from aerospace, nuclear power plants to oil and gas transportation. NDT techniques help promote safety, environmental protection and economy in production.

The goal of most NDT techniques is to reveal anomalies in a test object without destroying the test specimen. Such inspection can avoid unnecessary damage to the objects and eliminate or minimize the number of catastrophic accidents associated with its use. NDT techniques may also be used for monitoring the growth in the size of defects until they reach a critical level. Thus the service time of some expensive facilities can be maximized. A new area of NDE applications is process control. NDE techniques may be able to provide real time information of the status and the quality of a product during the manufacturing process. The information can be used as a basis for final product sorting or as feed-back for system adjustments.

Most physical principles, mechanics, acoustics, thermotics, electromagnetics, optics, nuclear physics, etc., have been utilized in NDT techniques. Each principle possesses certain advantages for certain application areas.

A generic diagram describing the principle of a NDE system is shown in Fig. 1. A NDE probe excited by an appropriate energy source sets up a field in a local area of the material under

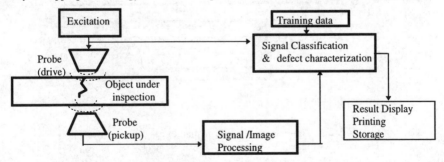

Fig. 1 Generic NDE system

inspection. Signals representing the field/anomaly interaction are then measured. The signal processing unit improves the signals quality. In most cases, the signals are sent for display, hard copy printing or recorded on an appropriate medium. Operators can also analyze the signals in real time or afterwards from the records.

Defect characterization involves either signal classification and/or characterization. When the signals are applied to a classification unit, they are classified into certain classes, e.g. good and bad, or normal signals caused by known objects and abnormal signals that need to be analyzed or defect signals that need to be characterized. Finally, the defect parameters or profiles are obtained after the abnormal signals are analyzed by the characterization unit.

Traditional algorithms for classification and characterization are mainly empirical, based on comparison of the processed signals with experimental data measured from a set of standard samples called calibration gauges. In recent years, advanced approaches, e.g. those used in pattern recognition, are used for solving NDE inverse problems.

Due to limitation of space, only electromagnetic NDT techniques are used as examples. Electromagnetic phenomena, governed by the Maxwell Equations, are classified by their frequencies as magnetostatic, magneto-stationary (when f = 0 Hz), quasi-static or diffusion (when $\sigma^2 >> \omega^2\varepsilon^2$) and wave (when $\sigma^2 << \omega^2\varepsilon^2$). All of these techniques are widely utilized. Three typical electromagnetic NDT methods, their working principles and examples of their applications, are given below. They are:

1. Magnetic flux leakage (MFL) method and its application in gas pipeline inspection [1].
2. Eddy current (EC) method and its application in stream generator tubing inspection [2].
3. Remote field eddy current (RFEC) method and its application to metallic tube and plate inspection [3], [4].

2. MFL method and its application in gas pipeline inspection

A magnetostatic field, or a magneto-stationary field in a ferromagnetic material is perturbed by the existence of an anomaly. The perturbation field is measured for anomaly detection. The so-called MFL method is based on this principle. Fig. 2 shows a typical example of applying the MFL method to natural gas pipeline inspection. The tool consists of two pieces of permanent magnets, a back iron, steel brushes, etc., providing a magnetic path for the flux to pass through the pipe wall under inspection. A Hall element array is placed at the center of the probe and measures the flux leakage variations due to anomalies in the pipe wall.

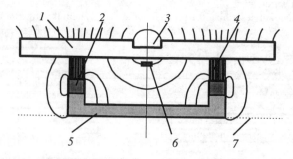

Fig. 2   Application of MFL method to gas pipeline inspection
1 - pipe wall, 2 - magnet, 3 - defect, 4 - brush, 5 - back iron, 6 - sensor, 7 - pipe axis.

Anomalies and their detection

Different anomalies are encountered in NDE practice. They can be classified, from a modeling point of view, as follows:

1. metal loss, when a certain volume of metal is removed from the object. A typical example is a corrosion pit.

2. tight cracks; where broken layers of very limited thickness appear in the object material which restrict or prevent magnetic flux or induced current from passing through these layers. A typical example is inter-granular stress corrosion cracking, IGSCC.
3. changes in the material property, μ or v. Typical examples are heat burns and mechanical damage.
4. changes in the material properties, B_r or H_c. Typical examples are also heat burn and mechanical damage.

The perturbation signals caused by different anomalies have different features, hence different techniques, probe models and parameters are used for their detection to maximize the sensitivity of inspection. Fig. 3 suggests some of these applications.

A reluctance probe (see sensor location #1 in Fig. 3) measures flux density between one of probe magnet poles and the object under inspection. Hence, it is sensitive to near side metal loss.

Active MFL method (sensor location #2 in Fig. 3) measures the flux leakage due to the existence of an anomaly at the near side or the far side of the object. The tangential component of the measured field represents H_z inside the object according to Ampere's law. Hence, it is sensitive to near side and, to a certain extent, far side metal loss and permeability variation.

Residual MFL method (sensor location #3 in Fig. 3) measures the residual magnetic field in the object after the sample is magnetized by the MFL probe. Hence it is sensitive to variation in B_r and H_c of the object material.

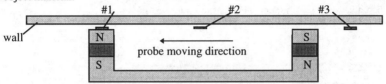

Fig. 3 Some suggested sensor positions to maximize sensitivity of inspection

MFL methods inherently have very limited sensitivity to tight cracks. However, in high velocity situations the velocity induced eddy current establishes an additional field. Studies show [6] there is an effect called velocity induced, or motion induced, remote field eddy current effect (VIRFEC or MIRFEC), occurring in the object under inspection. Sensor location #3 measures the VIRFEC signals that have penetrated through the object twice and hence brought rich information about the object far side. It is quite possible to detect tight cracks by making use of the VIRFEC effect.

Signal classification and defect characterization

Interpreting the obtained MFL information is an inverse problem that involves signal invariance transformation [7], signal classification and defect characterization [8].

Signals obtained by an MFL sensor from a given defect vary with many factors, such as material property represented by its *B-H* curve and the probe velocity. The objective of invariance transformation is to get rid of these unwanted variations. Suppose a signal, B_z, is a function of both a defect parameter d and an unwanted parameter u, or $B_z = f_1(d,u)$. If we can obtain one more signal, say $B_r = f_2(d,u)$, from the same defect measurement, in theory, we will be able to get a signal, $S = f_3(d)$, that is invariant to variation in u.

All the three processes, invariance transformation, signal classification, and defect characterization, are inverse problems that require a large amount of data obtained either by analytical or numerical modeling of the problem, or from experimental measurements.

3. EC techniques and steam generator tubing inspection

Traditional EC methods are based on the measurement of a probe coil impedance, *Z*, which is defined as

$$Z = R + j\omega L \tag{1}$$

where R is resistance of the coil, ω is angular frequency, L is self-inductance of the coil and is defined as

$$L = \Sigma\phi/I \qquad (2)$$

where $\Sigma\phi$ is the total flux linked by each turn of the coil and I is the current in the coil.

When an alternating current of certain frequency is applied to a probe coil placed close to a conducting object under inspection, the induced eddy current in the object alters the magnetic field and the total flux, ϕ, linked by the coil. Therefore, the coil impedance of an EC probe is a function not only of the coil parameters, but also the object conductivity and geometries. In other words, the impedance of an EC probe is related to the material properties of the object under inspection. The probe consists of a single coil and is called an absolute EC probe.

Anomalies on the surface of a conducting object also alter the induced current and change the probe coil impedance. EC probes used for detecting anomalies of an object usually consist of two coils connected in differential mode to enhance its sensitivity and are called differential EC probes. Differential probes have found wide applications for inspecting surface and subsurface anomalies of conducting objects. A typical application of EC techniques is the steam generation tubing inspection as shown in Fig. 4a. Differential EC probes are inserted into inconel tubes. An alternating current of certain frequency is applied to the probe coils. Signals reflecting the probe coil impedance variation are displayed, recorded and analyzed. A typical signal collected from the tube when it passes a steel support plate is shown in Fig. 4b.

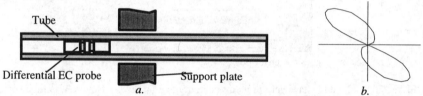

Fig. 4 Application of EC method to steam generation tubing inspection
    a.   A differential EC probe is operating in a steam generator tube with support plate;
    b.   A typical impedance plane trajectory signal of a differential probe.

Impedance plane trajectory

A simplified approach to describe the impedance variation relations is the equivalent circuit of a transformer shown in Fig. 5. The probe coil can be regarded as the primary side of the 'transformer', while the conducting object is regarded as its secondary side. At first, assume an ideal coupling between the probe coil and the object and ignore the leakage inductance, L_s, and coil wire resistance, R_0. The actual impedance, Z, and its two components, R and ωL, are as follows:

$$|Z| = \omega rL_0/\sqrt{[r^2 + (\omega L_0)^2]}, \qquad (3)$$
$$R = \omega L_0\{\omega rL_0/[r^2 + (\omega L_0)^2]\}, \qquad (4)$$
$$\omega L = r\{\omega rL_0/[r^2 + (\omega L_0)^2]\}, \qquad (5)$$

where L_0 is the coil inductance when it is away from the object; r is the conducting object's equivalent resistance normalized with respect to the primary side.

The existence of an anomaly in the object alters the resistance, r, and, consequently, changes the coil impedance. The eddy current method typically display changes in impedance on an impedance plane.

Impedance plane trajectories are often normalized by dividing Z, R and ωL by ωL_0, and ignoring the coil wire resistance, R_0. Thus, we have

$$|Z_n| = r/\sqrt{[r^2 + (\omega L_0)^2]}, \qquad (6)$$
$$R_n = \omega rL_0/[r^2 + (\omega L_0)^2], \qquad (7)$$
$$\omega L_n = r^2/[r^2 + (\omega L_0)^2]. \qquad (8)$$

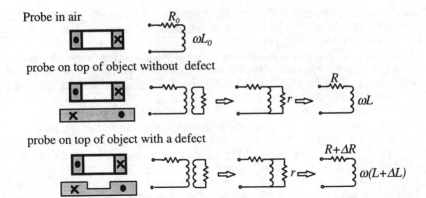

Probe in air

probe on top of object without defect

probe on top of object with a defect

Fig. 5 Equivalent circuit and circuit parameters of an EC probe

We see that the operating point of an EC probe traces a half-circle as shown in Fig. 6a. The top point where $|Z_n| = 1$ corresponds to the situation where the probe is away from the object, in other words the lift-off is infinite. The operating point of a probe moves downward when either r decreases, which implies the object is more conducting, or ω increases. Apparently, the existence of a defect in the object intends to increase the r and moves the operating point up tangentially along the trace.

In real cases the coupling is not perfect and the L_S is not negligible. The half-circle trace of an impedance plane moves up as shown in Fig. 6b. Because L_S increases with the increase in lift-off, the bigger the lift-off of the probe is, the higher and smaller the half-circle trace will be. If we take a probe gradually away from an object, we increase its lift-off, hence the operating point of the probe moves from the biggest trace to the smaller traces until we reach the top point that is common to all traces.

Notice, the direction of movement of the operating point caused by r variations is different from those caused by lift-off. Usually, lift-off cannot be controlled perfectly during inspection and causes

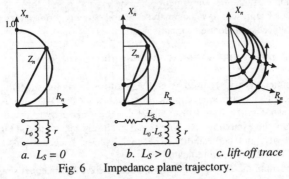

a. $L_S = 0$ b. $L_S > 0$ c. lift-off trace

Fig. 6   Impedance plane trajectory.

unwanted signals to be generated. The difference in the directions of movement of an operating point due to r changes and lift-off variation is often used to minimize the lift-off noise from defect signals.

Different types of EC probes

The probe described earlier is an absolute EC probe. It is used mainly for sorting conducting materials. Impedance variation of such a probe due to an anomaly is usually less than one percent of the total impedance value.

The differential probe measures impedance difference of two closely placed coils and gives much higher sensitivity in defect detection. A typical impedance plane trajectory obtained using a differential probe in shown in Fig. 4a.

The reflection probe consists of two coils. One of them called the drive coil establishes a magnetic field in the object material. The other coil called the pickup coil senses the field variations due to defects. In reflection probes the mutual inductance between the drive and pickup coils is exploited. The existence of a defect, changes flux linkage or mutual inductance of the coils, and consequently changes the induced voltage measured by the pickup coil.

In recent years, ferromagnetic cores and magnetic/nonmagnetic shields are often used in EC probe designs. Multiple-coil probes and probes with multiple semiconductor sensors, replacing bigger sized pickup coils, are often used in NDE. These enhance the sensitivity of probes' detection capability of detection, as well as resolution.

The major advantage of the EC technique is that it is simple and is sensitive to tight cracks.

The major disadvantage of the EC technique is that it is insensitive to defects that are deeply embedded below the object surface due to the skin-depth effect.

Different EC techniques

The EC technique described above utilizes a single frequency alternating current and is called the single frequency method. Other EC techniques include the multi-frequency technique [9] and pulsed EC technique [10].

In the multi-frequency technique excitation signals at a variety of frequencies are used to eliminate artifacts in the signal. The artifacts interfere with signal interpretation. Example of such artifacts include support plate signals encountered in steam generator inspection.

In the pulsed EC technique, time domain signals are used to obtain information related to deeper anomalies and to locate a defect with respect to its distance from the object surface.

Signal interpretation

Different signal processing and pattern recognition algorithms are used for EC signal classification and characterization. A typical example [11] is first to express the closed loop of a trajectory as a complex function. The function can be expanded in a form of a Fourier series to obtain a feature vector representing an anomaly. A neural network, trained by data obtained from a number of known defects, can be used for classification.

4. RFEC method and its application to metallic plate inspections

The RFEC technique is widely used as a NDE tool for inspecting metallic pipes and tubing. Essentially, the RFEC phenomenon can be observed when an AC coil is excited inside a conducting tube (see Fig. 7). The RFEC signal can be sensed by a pickup coil located 2-3 diameters away from the excitation coil. The signal is closely related to the tube wall conditions, thickness, conductivity, and permeability. The signal phase, especially, has an approximately linear relationship with the tube wall thickness.

For tubing inspection, the RFEC technique is characterized by its equal sensitivity to an inner diameter (ID) or an outer diameter (OD) defect, its insensitivity to probe wobble or lift-off, and not being limited by the penetration depth, which has traditionally been a major disadvantage of the conventional EC technique.

Studies show that the RFEC effect results when a portion of the energy released from the excitation coil passes twice through the tube wall (from the tube interior towards the exterior, say in the near field, and from tube exterior towards the exterior in the remote field). Fig. 8 is a Poynting vector plot [12] showing the energy movement as described.

This technique has been extended to metallic plate inspection [4] in recent years. Fig, 9 shows the schematic of an RFEC probe for inspecting up to 25.4" thick aluminum plate. Fig. 10 gives a typical experimental measurement results from such an inspection.

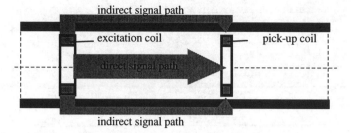

Fig. 7. Schematic of an RFEC probe for tube inspection and two signal paths between the excitation coil and the pick-up coil

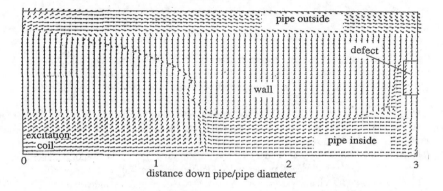

Fig. 8 Poynting vector showing energy flow around an RFEC probe

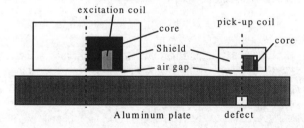

Fig. 9 Schematic of the probe developed for inspecting aluminum plates

Brief Summary

1. NDE plays an increasingly important role in modern society, it promotes safety, environmental protection and economy.
   NDE involves most physical principles and utilizes a lot of advanced technology.
2. In the MFL method, perturbation to an magnetostatic or a magneto-stationary field is measured for anomaly detection. Invariance transformation, signal classification and characterization techniques are used to analyze detected signals.
   The MFL method can be used for detecting ID and OD metal loss, heat burn and mechanical damage. VIRFEC effect may possible be used to detect IGSCC.
3. The eddy current methods are based on the measurement of probe coil self-impedance or mutual-impedance. Changes in impedance are usually displayed in impedance plane. Impedance trajectory loops can be used to attract feature vectors representing anomalies.

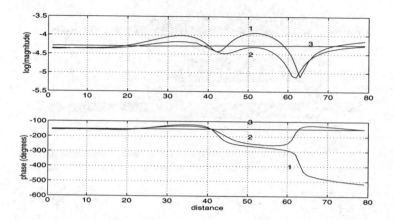

Fig. 10 Experimental measurement results from an RFEC probe inspecting a 12.7 mm thick aluminum plate. All defects are on the far side of the plate:1 - 25.7 mm long, 2 mm wide and 50% deep defect; 2 - 25.7 mm long, 2 mm wide and 30% deep defect; 3 - no defect.

The major advantages of the EC method are its simplicity and its sensitivity to tight cracks. The major disadvantage of the EC method is its insensitivity to defects that are deeply embedded below the surfaces of a test object.
4.   The RFEC method is based on the RFEC effect that results when a portion of the energy released from the excitation coil passes twice through the tube wall under inspection.
      The RFEC method is characterized by its sensitivity to deep defects and equal sensitivity to an ID or OD defect of a tube.
      The RFEC method has recently been successfully extended to thick metallic plate inspection, as well as to other application areas.

References

[1] Udpa, L., Mandayam, S., Udpa, S., et al, *Materials Evaluation/54/April 1996*, pp.467-472.
[2] Lord, W., *Electromagnetomechanical Interaction in Deformable Solids and Structures - Proceedings of the IUTAM Symposium held in Tokyo, Japan, 12-17 October, 1986, edited by Y. Yamamoto and K. Miya, Elsevier Science PublisherB.V.(North-Holland), 1987.*
[3] Schmidt, T.R., *Materials Evaluation/42/Feb. 1984*, pp. 225-230.
[4] Sun, Y. et al, *Review of Progress in QNDE, Vol. 15A*, pp. 1137-1144, Edited by D.O. Thompson and D.E. Chimenti, Plenum Press, New York, 1996.
[5]. Y.S. Sun, S.S. Udpa abd W. Lord, *Review of Progress in QNDE, Vol. 6B*, pp.1651-1658, 1986.
[6]. Y.S. Sun, *IEEE Trans. On Magnetics, Vol. 30, No. 5*, pp. 3304-3307, September 1994.
[7]. S. Mandayam, L. Udpa, S.S. Udpa, W. Lord, *IEEE Trans. On Magnetics, Vol. 32, No. 3*, pp. 1577-1580, may 1994.
[8]. G. Xie, M. Chao, C. Yeoh, S. Mandayam, S.S.. Udpa, L. Udpa, W. Lord, *Review of Progress in QNDE, Vol. 15A*, pp.813-820, 1996.
[9]. J. Stolte, L. Udpa, W. Lord, *Review of Progress in QNDE, Vol. 7A*, pp.821-830, 1988.
[10]. J. C. Moulder, J. A. Bieber, S. K. Shaligram, W. W. Ward III, and J. H. Rose, <u>The First US-Japan Symposium on Advances in NDT</u>, June 24-28,1996, Kahuku (Island of Oahu) Hawaii.
[11]. S.S. Udpa and W. Lord, *Materials Evaluation/42/Aug. 1984*, pp.1136-1141.
[12].W. Lord, Y.S. Sun, S.S. Udpa and S. Nath, *IEEE Trans. On Magnetics, Vol. 24, No. 1*, pp. 435-438, January 1988.

Non-Linear Electromagnetic Systems
V. Kose and J. Sievert (Eds.)
IOS Press, 1998

Fractal Dimensions of Magnetic Noises: A Diagnosis Tool for Non-Destructive Testing

Koji YAMADA, Sinnichi SHOJI , Katsuhiko YAMAGUCHI, Yoshinori TANAKA,

Dept. .Materials Science, Saitama University, Urawa ,338 Saitama, Japan,

Roland Groessinger

Inst. for Experimental Physics, 1040 Hauptstrasse, 1100 Wien, Austria

Abstract. The magnetic noises in iron based material caused by magnetic domain jumps, so called Barkhausen jumps, were investigated on various residual strains by static stress, rapid quenching or mechanical roll. The transient magnetic noises were observed by using a sensitive pulsed magnetometer, and Fourier power spectra were calculated with respect to the magnetic field. We found several times smaller power amplitudes in a frequency range between 0.012 [m/A] and 0.12 [m/A] for strained samples by static stress more than 200 MP for a yield of 600 [MP] of A533B samples. Much higher Fourier amplitudes in the easy axis than those in the hard axis were also found for Supermalloy thin films sputtered in a magnetic field of 1.6×10^4 [A/m]. Fractal dimensions of these power spectra were discussed on the residual strains, in conjunction with NDE of these material.

1. Introduction

The magnetic noises caused by domain jumps in iron based materials are of great interest in the physical origin and the relationship between the noise spectra and their lattice imperfections. In this paper, we investigated the basic nature of these noises and their spectra for differently strained materials, e.g. rolled metallic thin ribbon, sputtered thin film, in connection with the microscopic imperfections inside the materials [1,2]. For these purposes, we have been developed a sensitive pulsed magnetometer available down to 10^{-4} [emu] for transient magnetization of samples smaller than 1 [μg] [3]. By using this pulsed magnetometer, the spectra of these noises were analyzed by Fourier transform with respect to magnetic fields, and the noises in the strained samples were compared with those in as-annealed samples. We found that the fractal dimension of the noises in an iron based rolled thin ribbon was around 1.66 [1,4], which coincided with a theoretical simulation by Geoffroy et al.[5]. Further, in this study, the noise spectral amplitudes and the fractal dimensions decreased in a specific frequency range for strained samples quenched by pinning centers. We propose the application of this analysis for the nondestructive testing of iron based material by evaluating the Fourier spectra and the fractal dimensions of the magnetic noises.

2. Experimental

The magnetic noises in iron based material were observed at room temperature for rolled iron based thin ribbon[2], ultra thin Supermalloy prepared by sputtering method and strained A533B (see note in Fig. 3). The sample shape of A533B was a wire of $1mm^2$ cross section and 150 mm long, and the Supermalloy samples of ultra thin films up to 1 [μm] on a quartz substrate. A coil pair for magnetic noise detection was inserted in the center of the main coil, connected in reverse to cancel the electromotive force as shown in Fig.1. The magnetic field was linearly increased up to 1000 [A/m] to obtain the linear relationship between dM/dH and dM/dt. Here, M, H and t denote the magnetic field, the magnetization and time. The noises were recorded by a 12 bits analogue to digital converter at every 1 [μs] up to 17 [ms].

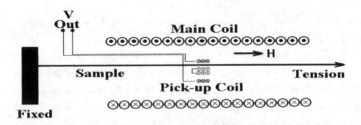

Fig. 1   Experimental set-up of magnetic noise observation

It corresponds to the noise observations up to 1000 [A/m] with every 5.9×10^{-2} [A/m], a low enough speed for magnetization in these material. Fig.2(a) shows the direct trace of the generated magnetic noises observed for $Ni_{81}Fe_{91}$ (Supermalloy) sample parallel to the hard axis. The magnetic noise differed for the observation direction

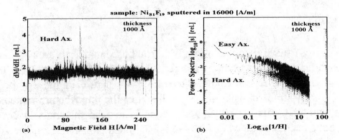

Fig.2   The magnetic noises observed in Supermalloy sample and the spectra. (a) The original traces obtained by ADC.   (b) Fourier spectra for the original trace of (a) parallel to the hard axis and to the easy axis in the lower trace.

between the easy and hard axis as shown in Fig. 2(b). Apparently, the noise observed for easy axis were larger than those for the hard axis. The differential magnetizations of A533B material were observed for differently strained conditions in zero stress after the removal of applied stresses up to 520 [MP] . Note here, that the yield of this material is around 600 [MP] and the coercivity H_c

was increased by 180 [A/m] at P=500 [MP], from H_c=544[A/m] at P=0.

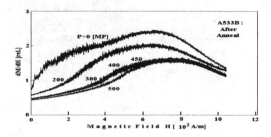

Fig. 3   The differential permeability of A533B for variously strained conditions.
[A533B composition:C(0.17),Si(0.24),Mn(1.37), Cu(0.02),Ni(0.6),Cr(0.07),Mo(0.46),S(0.003),P]

3. Analyses and Discussions

The giant magnetic noises, so called Barkhausen jumps (BJ), were observed by the electromotive force (emf) in the pick coil. The magnetic fields were increased at a slow speed both for
enough magnetization and the electronic devices to observe every jump with enough
resolutions for the ADC in this experiment. The analysis of BJ was performed by Fourier
transform of dM/dH with respect to H as

$$\frac{dM}{dH} = \frac{dM}{dt}\left(\frac{dH}{dt}\right)^{-1} = \frac{1}{c}\frac{dM}{dt} \quad , \tag{1}$$

$$F\left(\frac{dM}{dH}\right)_H = \frac{1}{\sqrt{2\pi}}\int\frac{dM}{dH}e^{-i\omega H} - i\omega dH = i\omega F\left(M(H)\right)_H \approx f^{-(\beta+1)} \quad . \tag{2}$$

Here, M and H denote magnetization and field. $F(..)_H$ denotes the Fourier transform with respect to H and
$\omega = 2\pi f = 2\pi / H$ representing the periodicity of the spectra at $f = 1/H$.   In Fig.2(b), the Fourier power
spectra S are shown for the original traces in Fig. 2(a) for the field directions in the easy axis.   As was
pointed out, the spectra obtained for the easy axis are larger than those for the hard axis, reflecting the
jump suppressions by strong pinning centers caused by anisotropy energies in differently strained
directions.   The spectrum power index β as $S=(1/H)^{-\beta}$ can be evaluated by inserting best fitting lines as
in Fig. 4. by using a well known relationship between the power index β and the fractal dimension D [5]
as $2D=5-\beta$.   The fractal dimensions of the broken lines are D=1.5, obtained by $F(M(H))$ spectra. It is
evident that the dimensions decrease with increasing stress P as D=1.57 (P=0, H_c=544 [A/m])
and D=1.4 (P=500 [MP], H_c=704 [A/m]) at A and B points, respectively in Fig. 4.   In other
words, an as- annealed sample shows a large fractal dimension than those for the strained,
due to the larger numbers of the smaller scale pinning centers in an as-annealed sample than
those observed in the strained samples by stresses more than 200 [MP].   These are reflected
by the spectral amplitude decreases in a low frequency range between 1x10^{-3} and 1 x10^{-2}
[m/A]   observed at P=0 and P=200 [MP], respectively.   The spectral intensity at point
"A"   for P=0 in Fig. 4, is 10 times larger than that at point "B" for P=300 [MP]. Indicated by

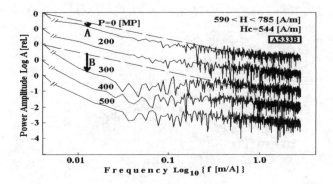

Fig. 4   The spectral dependence of the Barkhausen jumps for A533B at variously strained conditions after removal of the stresses P=0-500, indicated in this figure.

the arrow lengths in this figure. It is worth noting here that in the range larger than f>0.25 [m/A]. the spectral amplitudes were unchanged by strains, reflecting the nature of the barriers for magnetic domain jumps. This phenomenon is plausibly understood by a physical mechanism that the high frequency magnetic noises are generated by the non-magnetic atoms as the pinning centers in the alloy of A533B material.   On the other hands, the dislocations by strains causes the stronger pinning centers and the higher barriers than those · originated by the non-magnetic atoms. In this way, the strains in iron based material were visualized by the Fourier spectra and the fractal dimensions.

4. Conclusions

The fractal theory approach to the BJ analysis seems valid for our purposes. However, it is very important to assign the various pinning centers for the domain wall jumps to the specific lattice imperfections as Kronmuller has developed [7] for crystals. This work should be further developed in the future.

References

[1] K. Yamada et. al. Proc. Int. Conf. High Magnetic Fields (1990)

[2] K. Yamada et. al. J. Magn. Mag. Mater. **104-107** (1992)341

[3] T. Saitoh, March 1994, Master Thesis, Saitama University, Japan

[4] K. Yamada et. al. Supplement of Int. J. Appl. Electromagnetics in material, **6**(1995)533

[5] O. Geoffroy, et. al. J. Magn. Mag. Mater. **104-107**(1992)379

[6] H. Takahashi, Fractal(in Japanese, Asakura Publishing Company) (1988)

[7] H. Kronmueller, Int. J. Nondestructive Test. **3**(1972)315

Non-Linear Electromagnetic Systems
V. Kose and J. Sievert (Eds.)
IOS Press, 1998

Wavefront Reconstruction from
Shearing Interferograms: an Inverse Problem

Stefan LOHEIDE

Physikalisch-Technische Bundesanstalt Braunschweig,
Bundesallee 100, D-38116 Braunschweig, Germany

Abstract. A new method for quantitatively determining the shape of a wavefront from an interferogram recorded with a shearing interferometer will be presented. The reconstruction is subdivided into two steps, each of them representing an inverse problem. At first, the information about the wavefront, which is coded in the form of the difference between the wavefront and a laterally sheared copy, has to be extracted from the interferogram. To determine the wavefront from the difference a new procedure will be presented which is based on filtering in the frequency domain with a transfer function specifically developed for this purpose. This techniuqe requires hardly pre-information about the shape of the wavefront. It reconstructs the wavefront inside the whole aperture and furnishes accurate results with high lateral resolution for shears up to 20% of the aperture.

1 Introduction

Shearing interferometry was devised by Ronchi [1] and can essentially be characterized as a self-referencing wavefront detection method. The wavefront, after having been influenced by the object under test, is coherently superposed with a laterally sheared copy. Shearing interferometry can be performed either by reflection from the front and the rear face of a plate whose surfaces are known [2] or by diffraction at a combination of gratings [3]. After splitting, two waves with the same aberration are available, which are laterally sheared with respect to each other.

The intensity distribution $I(x, y)$ of the interference pattern created by two plane waves with the same amplitude, disturbed by the wave aberration $w(x, y)$ and laterally sheared with respect to each other in the x-direction by the shear s, is given by:

$$I(x, y) = I_0 \{1 + \cos[(\mathbf{k}_2 - \mathbf{k}_1)\mathbf{r} + w(x + s/2, y) - w(x - s/2, y)]\},$$

(1)

where I_0 is the average intensity, $\mathbf{k}_{1,2}$ the wave vectors of the two interacting beams and $\mathbf{r} = (x, y, z)$ the bound vector. The grating vector of the fringe pattern, which results when the interacting beams are not oriented parallel to each other is denoted by $\mathbf{K} = \mathbf{k}_2 - \mathbf{k}_1$. From the interference pattern formed by superposition, the wave aberration is to be determined, which is coded in the form of its difference.

The reconstruction of the shape of the wavefront from a shearing interferogram can be divided into two steps. First, the wavefront difference $\Delta w(x, y) = w(x + s/2, y) - w(x - s/2, y)$, has to be determined from the interferogram. For this, it is to be taken into account that the cosine function is not monotonic and that the inverse function thus is ambiguous. Solutions of this fundamental problem of interferometry are described as "phase unwrapping" and represent a typical inverse problem. A special technique for extracting the information about the wavefront difference from the shearing interferogram will be presented in section 2.

After this the wavefront has to be reconstructed, which can also be interpreted as an inverse problem, because the wave aberration is not recorded directly but is coded in the interference pattern in the form of its difference. A point in the interferogram thus furnishes only a comparison between two points of the wavefront spaced by the lateral shear, and not a comparison with a defined and well-known reference. Although several different methods have been available up to now [4][5][6][7] this is the first time that a procedure for wavefront reconstruction from shearing interferograms with a comparatively high accuracy of evaluation is presented. This evaluation technique is based on Fourier filtering in the frequency domain and allows the wavefront to be reconstructed with high lateral resolution and without preconditions for the aberrations. In addition, the method provides information about the reconstructed wavefront over its whole aperture for relatively large shears. An representative example for this technique of evaluation is published elsewhere [8].

2 Phase Unwrapping

The basis of the technique of extracting the information about the wavefront difference from the shearing interferogram is the availability of the wavefront difference in the form of a complex signal [9]. This can be performed by tilting the interacting beams with respect to each other, which leads to a fringe pattern with grating vector K (see equation (1)). If the frequency of the fringe dominates all contributions of the wavefront difference $\Delta w(x, y)$, a Fourier transformation of the interference pattern enables the separation of the complex signal, shifted by $K/2$ in the frequency domain from the conjugate complex of this signal shifted by $-K/2$. When the complex signal is reshifted by $-K/2$ into the center of the frequency domain, an inverse Fourier transformation yields the desired complex signal containing the wavefront difference.

Phase unwrapping is a process by which the absolute value of the phase angle of the complex signal, that extends over a range of more than 2π, relative to a predefined starting point and which directly represents the wavefront difference, is recovered. This absolute value is lost when the phase term is wrapped upon itself with a repeat distance of 2π due to the fundamental sinusoidal nature of the wave functions used in interferometric measurements.

The basic principle of phase unwrapping is to integrate the wrapped phase along the path through the data [10]. At each pixel, the phase gradient is calculated by differentiation. If the gradient exceeds a certain threshold such as π, a phase fringe edge or 2π discontinuity is assumed. The phase jump is corrected by adding or subtracting 2π, according to the sign of the gradient, to all points which have not yet been taken into account.

3 New evaluation technique for wavefront reconstruction

It is assumed that the wavefront difference $\Delta w(x, y)$ is known and available for the evaluation described in the following. To further elucidate the new evaluation method, it is sufficient to restrict ourselves to that dimension in which the lateral shear takes place. The method can, however, be easily extended to two dimensions as is necessary for practical applications in the field of interferometry. Up to now, infinitely extended signals have been considered. The wavefronts to be investigated are, however, always limited by an aperture of width p which is taken into account by using $W(x)$ instead of $w(x)$ where $W(x) = w(x)$ for $|x| \leq p/2$ and $W(x) = 0$ otherwise. The reconstruction of the wave aberration $W(x)$ from the difference $\Delta W(x) = W(x + s/2) - W(x - s/2)$ is described in terms of Fourier optics using Fourier integrals. With the aid of shift operations and additional integral calculations, the following relationship is obtained [11]:

$$F[W](v) = T(v)\, F[\Delta W](v) \qquad \text{with} \qquad T(v) = \frac{-i}{2\,\sin(2\pi vs/2)}.$$

$$(2)$$

Here $F[\Delta W](v)$ and $F[W](v)$ denote the Fourier transforms of $\Delta W(x)$ and $W(x)$, respectively, x the spatial coordinates, and v the spatial frequencies. With equation (2) we have introduced a transfer function $T(v)$ with which filtering of the difference wavefront ΔW in the frequency domain is performed to obtain $W(x)$. Equation (2) can also serve to show a basic limitation of shearing interferometry. Periodic contributions to the wave aberration cannot be reconstructed if the shear s corresponds to an integer multiple of the period length as the transfer function $T(v)$ has a singularity at this frequency v. This is the fundamental difficulty of shearing interferometry but not of the evaluation method described here. In practice, when discrete measurement points are used, it is, however, seen that a suitable choice of the shear allows errors to be avoided.

To reconstruct the wave aberration $W(x)$ using equation (2), $\Delta W(x) = W(x + s/2) - W(x - s/2)$ must be known. However, a shearing interferogram contains information only about the difference between the wave aberrations laterally sheared where the two waves interfere. Thus, not $\Delta W(x)$ is available for evaluation but $\Delta W^{st}(x) = win^{st}_{p-s}(x)\, \Delta W(x)$, where $win^{st}_{p-s}(x)$ denotes a window function of width $p - s$ ($win^{st}_{p-s}(x) = 1$ for $|x| \leq (p - s)/2$, $win^{st}_{p-s}(x) = 0$ otherwise) according to the width of the interference pattern. As the flanks of the window function $win^{st}_{p-s}(x)$ at $|x| \leq (p - s)/2$ are steep, the corresponding symbols are denoted by the superscript st to distinguish them from the symbols used in the case of another window function described below.

To extract the information contained in $\Delta W^{st}(x)$ through $W(x)$, equation (2) will be further used. However, $\Delta W(x)$ is replaced by the expression $\Delta W^{st}(x)$ which contains the finite size of the interferogram. Applying the evaluation technique, described by equation (2) to $\Delta W^{st}(x)$, we obtain $W^{st}(x)$ instead of $W(x)$. As a solution, equation (3) furnishes the wave aberration $W^{st}(x)$ for which the following condition is valid:

$$W^{st}(x + s/2) - W^{st}(x - s/2) = win^{st}_{p-s}(x)\, \Delta W(x).$$

$$(3)$$

What is the relationship between $W^{st}(x)$ and $W(x)$ according to equation (3)? Equation (3) can be fulfilled if $W^{st}(x)$ has the form $W^{st}(x) = W(x) + o(x)$ inside the region $|x| \leq p/2$, $o(x)$ being a periodic function at first unknown. The period length corresponds to the shear s ($o(x) = o(x + s)$) so that $o(x)$ can be omitted when $\Delta W^{st}(x)$ is calculated. This is a condition necessary for the modifying function $o(x)$. Restrictions of the form of $o(x)$ also result from the ratio of the width p of the pupil to the shear s, which generally is non-integer, as well as from the gradients of the function $W(x)$ at the edge $|x| = p/2$ of the pupil. The problem of evaluation has now been reduced to the task of determining the periodic error function $o(x)$.

To select $o(x)$, changes will be discussed, which result from the use of another window function; for this window function, $win^{fl}_{p-s}(x) = 0$ remains valid for $|x| \geq (p - s)/2$

and $win^{fl}_{p-s}(x) = 1$ for $|x| \le s$, but in the region $s < |x| < (p - s)/2$ the function is continuous and flat (superscript ^{fl}). For a continuous window function $win^{fl}_{p-s}(x)$, the result $W^{fl}(x)$, after application of the transfer function (2), is also continuous, and the influence of the periodic function $o(x)$ decreases with decreasing shear s. Furthermore, the difference between $W^{fl}(x)$ and $W(x)$ increases towards the edge of the pupil. In the region $|x| \le s/2$, however, $W^{fl}(x) = W(x)$ is valid for $s \ll p$.

Now, the periodic function $o(x) = W^{st}(x) - W^{fl}(x)$ can be selected for $|x| \le s/2$. Because of the periodicity of $o(x)$ the extension to the region $|x| \le p/2$ and thus the calculation of $W(x)$ from $W^{st}(x)$ for the range $|x| \le p/2$ of values are possible. This is the central idea of the evaluation method described here. This also shows how the loss of information makes itself felt. The larger s, the steeper the flanks of the window function $win^{fl}_{p-s}(x)$ and the poorer the approximation $s \ll p$. Then $W^{fl}(x)$ is affected by periodic contributions $o(x)$ which falsify the reconstructed function. With the aid of this method it is possible to very accurately determine the wave aberration from the wavefront difference for shears up to 0.2 of the aperture. This determination comprises two evaluations with two different superposed window functions which are combined with each other.

Acknowledgment: Financial support by the Bundesministerium für Bildung, Wissenschaft, Forschung und Technologie (FKZ: 13N6754) is gratefully acknowledged.

References

[1] V. Ronchi, Forty years of history of a grating interferometer, *Appl. Opt.* **3** (1964) 437

[2] M. V. R. K. Murty, The use of a single plane parallel plate as a lateral shearing interferometer with a visible gas laser source, *Appl. Opt.* **3** (1964) 531

[3] J. Schwider, Continuous lateral shearing interferometer, *Appl. Opt.* **23** (1984) 4403

[4] G. Harbers, P.J. Kunst, and G.W.R. Leibbrandt, Analysis of lateral shearing interferograms by use of Zernike polynomials, *Appl. Opt.* **35** (1996) 6162

[5] M. P. Rimmer, Method for Evaluating Lateral Shearing Interferometer, *Appl. Opt.* **13** (1974) 623

[6] B. R. Hunt, Matrix formulation of the reconstruction of phase values from phase differences, *J. Opt. Soc. Am.* **69** (1979) 393

[7] W. H. Southwell, Wave-front estimation from wave-front slope measurements, *J. Opt. Soc. Am.* **70** (1980) 998

[8] S. Loheide and I. Weingärtner, New Procedure for Wavefront Reconstruction from Shearing Interferograms, *Optik,* accepted for publication: June 12, 1997

[9] M. Takeda, H. Ina, and S. Kobayashi, Fourier-transform method of fringe pattern analysis for computer-based topography and interferometry, *J. Opt. Soc. Am.* **72** (1982) 156

[10] D. J. Bone, Fourier fringe analysis: the two-dimensional phase unwrapping problem, *Appl. Opt.* **30** (1991) 3627

[11] K. R. Freischlad and C. L. Koliopoulos, Modal estimation of a wave front from difference measurements using the discrete Fourier transform, *J. Opt. Soc. Am. A* **3** (1986) 1852

Non-Linear Electromagnetic Systems
V. Kose and J. Sievert (Eds.)
IOS Press, 1998

Regularization of the Eddy Current Tomography Problem : First and Second Order Linearized Approach

Claude Cohen-Bacrie [(1)], Yves Goussard[*] and Riadh Zorgati
EDF/Research & Development Division - BP 49 - 78401 Chatou Cedex, France
[*]École Polytechnique de Montréal - Institut de Génie Biomédical
BP 6079 - Sucursalle "Centre Ville", Montréal (Québec), H3C317 Canada
(Email : riadh.zorgati@der.edfgdf.fr, yves@grbb.polymtl.ca)

Abstract. One of the major difficulties of Eddy Current Tomography (ECT) lies in the severe depth attenuation of the electromagnetic field inside a metal. This paper proposes an extension to ECT of the *variance uniformization* regularization method originally developed for Electrical Impedance Tomography (EIT). This regularization, which can be viewed as a heuristic way of accounting for the attenuation, is applied to both first and second order linearized approximations of the forward problem. The results obtained with simulated data representing a crack in a homogeneous medium show significant improvement over Tikhonov regularization, particularly when the data are corrupted by noise.

1. Introduction

In nuclear power plants, steam generator tubes testing is mainly performed by an Eddy Current (EC) Non-Destructive Technique (NDT). With a view to obtaining ever more accurate information on equipment integrity, we attempt to image flaws affecting tubes. Flaws show up like localized spatial variations of conductivity with respect to its standard value in sound metal. Thus, the Eddy Current Tomography consists of estimating the inner conductivity distribution x of a flawed object excited by an incident electromagnetic field, from the measurements of the anomalous electric field y generated outside [1], [2]. This inverse problem is ill-posed and non-linear with respect to conductivity. In addition, in EC regime, attenuation phenomena considerably impairs the quality of the images currently reconstructed [3]. This paper investigates the variance uniformization regularization (VUR) technique [4] as a possible improvement to the ECT attenuation issue. Section 2 introduces two models of the forward problem that respectively consist of first- and second-order linear approximations of the actual relationship between conductivity and measurements [2]. In Section 3, the VUR method is applied and compared to classical Tikhonov regularization. Conclusions are presented in Section 4.

[(1)] now with Professionnal Imaging Systems - Laboratoire d'Electronique Philips S.A.S
22, avenue Descartes - 94453 Limeil-Brevannes Cedex. Email : bacrie@lep-philips.fr.

2. Forward Problem Modelling

The EC-NDT configuration is as follows. In the air, an EC probe produces an electromagnetic excitation and collects the result of its interaction with a flaw, occupying area Ω, embedded in the metal.

This defect is described by the contrast function $x(\vec{r}) = \sigma(\vec{r}) - \sigma_0$. The model linking x and the anomalous field measured in the air y is derived from Maxwell's equations by applying Green's theorem. After discretization, the continuous model, made up of a pair of Fredholm integral equations of the first and of the second kind, becomes a nonlinear matrix system :

$$\begin{cases} G_1 X e_2 = y \\ (I - G_2 X)e_2 = e_{02} \end{cases} \Leftrightarrow \begin{cases} G_1 E_2 x = y \\ (I - G_2 X)e_2 = e_{02} \end{cases} \quad \begin{array}{l} X = \text{diag}(x = \sigma - \sigma_0) \\ E_2 = \text{diag}(e_2) \\ y = e_1 - e_{01} = G_1 X e_2 = G_1 X(I - G_2 X)^{-1} e_{02} \end{array} \quad (1)$$

G_1 and G_2 denote the dyadic Green's functions for air-metal and metal-metal medium, while e_2 and e_{02} are the internal electric fields within the flawed and sound areas respectively. This general model is approximated by the two following linearized models.

First-Order linearization. A very rough linearization of the problem is performed by approximating the total electrical field E_2 (unknown) by the known incident electrical field: $\hat{E}_2 = E_{02}$ (strong Born approximation). Provided this approximation is valid, (1) becomes linear with respect to x. First-order linearization yields a relationship of the form : $(G_1 E_{02})x + b = e_1 - e_{01} = y = Sx + b$ (2). The noise term b corresponds to measurement and modelling errors. Matrix S expresses underlying physical properties, and particularly the strong attenuation, which greatly contributes to the ill-posed nature of the problem.

Second-Order linearization. From the nonlinear model $(G_1 E_2) x = G_1 z = y$ (1a) and $(I - G_2 X)e_2 = e_{02}$ or $e_2 = e_{02} - G_2 z$ (1b), we estimate x in two steps. Equation (1a) is first solved for z. Through (1b), this yields an estimate of e2 (second-order linearization), which is used in the second stage to solve (1a) for x.

3. Inversion

The goal of the inversion is to estimate x from (1). With both first and second order linearization, one must solve *ill-posed* linear inverse problems of the form $y = Sx + b$ where y, x and b respectively denote the measured, unknown and noise vectors and where S is the system matrix. In order to obtain acceptable solutions, the problem should be regularized by introducing prior information on the unknown quantity x into the inversion process. A simple form of regularization consists of modeling x as an independent identically distributed Gaussian field. Its covariance matrix K is proportional to identity and the solution is given by $\hat{x} = (S^t S + K)^{-1} S^t y$ In order to obtain more accurate results, we propose to select K in the above equation according to the contribution of the corresponding pixel to the measured data. This can be achieved heuristically by *uniformizing* the estimated variance of the reconstructed object [4]. Assuming that $S^t S$ and K can be diagonalized on the same basis, the variance uniformization condition

$COV(\hat{x}) = c = \text{cste}$ yields a closed-form expression of the eigenvalues of K from which the exact expression of K can be easily derived. Such an approach was successfully applied to EIT which presents attenuation characteristics that are similar to ECT. The technique can be readily transposed to inversion of the first-order linearized problem, or sequentially applied to both equations of the second-order linearized problem in a sequential manner. We now present a simulation example where the object to retrieve is a 1 mm depth, 1/9 mm width crack embedded in a 1mm x 1mm homogeneous medium (9 x 9 pixels). y represents the simulated anomalous field at 200 kHz.

With the first-order linearized model, TR and VUR reconstructions are plotted in Figure 1. When data are noiseless, Tikhonov Regularization technique (TR) only reconstructs the superficial part of the crack. This can be explained by the attenuation phenomena. On the other hand, VUR provides adequate reconstruction of the deeper parts of the crack. When noise with a 20 dB signal-to-noise ratio is added to the simulated data, only VUR provides acceptable reconstruction. The same robustness was observed for EIT.

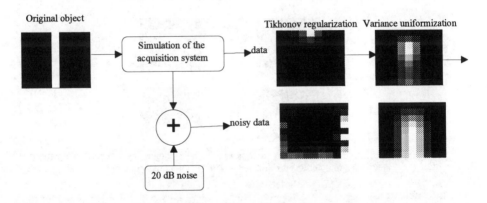

Figure 1: Reconstruction of the crack with classical Tikhonov regularization and with variance uniformization technique, using simulated data at 200kHz with the first-order linearized model.

With the second order linearized model. Reconstruction consists of inverting both equations (1a) and (1b) sequentially.The first step is an estimation of the induced EC, $z=E_2x$, inside the unknown object. In the second step, the estimate is integrated into equation (1b) and the resulting relationship is inverted so as to estimate the conductivity x. Since two inversions are required, we tried a combination of TR (T) and VUR (U) for both steps. Four combinations are possible: T-T, T-U, U-T and U-U. Figure 2 shows reconstructions obtained with and without noise. Reconstruction on both noisy and noiseless data with VUR are more complete than with classical TR. As opposed to observations made with the first-order approximation, TR provides acceptable results on the superficial region. The precision of the crack localization decreases with TR when the data are noise-corrupted. VUR shows superior robustness with respect to noise.

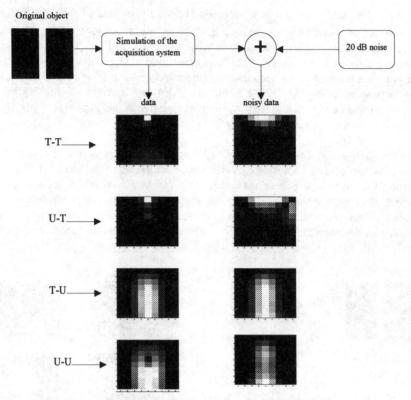

Figure 2: Reconstruction of the crack combining classical Tikhonov regularization and variance uniformization, with data simulated at 200kHz with the second order-linearized model.

4. Conclusions

In order to improve the quality of the EC images currently reconstructed, attenuation phenomena should be taken into account during inversion. An heuristic way of accounting for this issue, the variance uniformization regularization, has proven its efficiency in Electrical Impedance Tomography and was successfully adapted for Eddy Current Tomography. In addition to the correct reconstructions obtained, this kind of regularization seems very promising for inverting real noisy data.

References

[1]   H. A. Sabbagh, L. D. Sabbagh, "An eddy Current Model for Three Dimensional Inversion" IEEE Trans. Magn., vol. 22, n° 4, pp. 282-291, 1986.
[2]   R. Zorgati, V. Monebhurrun, P.O. Gros, B. Duchêne, D. Lesselier, C. Chavant, "INES: 3D Eddy Current Imaging for a Nondestructive Evaluation System Applied to Steam Generator Tubes", Review of Progress in Quant. Non-Destructive Evaluation, Brunswick (USA), July 28-August 2, 1996.
[3]   R. Zorgati, "Eddy Current Imaging: Approaches,Formulations and Problems", Proceedings of the Second International Symposium on Inverse Problems - ISIP'94, Paris, 2-4 Nov. 1994, 393-400, Inverse Problems in Engineering Mechanics, H. D. Bui Ed.,.Balkema Books, Rotterdam, 1994.
[4]   C. Cohen-Bacrie, Y. Goussard, "Electrical Impedance Tomography: Regularized Reconstruction Using a Variance Uniformization Constraint", Proc. IEEE-EMBS Conf., 569-570, Montréal, Québec, Canada, 1995.

Non-Linear Electromagnetic Systems
V. Kose and J. Sievert (Eds.)
IOS Press, 1998

Experiments on Eddy Current NDE with HTS RF SQUIDs

M. v. KREUTZBRUCK , M. MÜCK, U. BABY and C. HEIDEN

Institut für Angewandte Physik der JLU Giessen, Heinrich-Buff-Ring 16,
35392 Giessen, Germany

Abstract High Temperature Superconductor (HTS) Superconducting Quantum Interference Devices (SQUIDs) are promising sensors for applications in eddy current nondestructive evaluation (NDE). Due to their high field sensitivity at low frequencies, they are especially suitable for applications, where a large penetration depth is required. We have investigated two different SQUID-based NDE systems, one of which is optimised for testing felloes of aircraft wheels. The second system allows for testing planar structures using a motorised x-y-stage, which moves the cryostat above the planar samples. As sensors 3 GHz rf SQUIDs made from YBCO were used, having a field noise of about 1 pT/√Hz. This results in a dynamic range of our SQUID system of about 155 dB/√Hz. In most cases, the SQUIDs have been cooled by immersing them in liquid nitrogen. We have however also developed a cryosystem, which allows the cooling of the sensors by a Ne-gas flow. In planar test structures we could detect flaws with lengths of 10 mm, having a height of 0.6 mm at a depth of 13 mm. In aircraft felloes, flaws located at the inner surface of the felloe (thickness 8 mm) were easily detectable despite a high static background field of up to 0.5 G caused by ferromagnetic structures inside the felloe. For flaws at a depth of 5 mm, the spatial resolution of both systems was about 8 mm without applying image postprocessing.
*This work is supported by the German Federal Ministry of Education and Research (BMBF) under Contract No. 13N6677/8 and in part by the Prize of the Justus-Liebig-Universität Giessen.

1. Introduction

The high field sensitivity of SQUIDs at low frequencies makes them ideally suitable for nondestructive evaluation [1-6] of relatively thick conductive objects, like layered stacks of aluminium sheets or felloes. In contrast to a direct injection of dc or ac currents into the samples to be tested, the induction of eddy currents by an external excitation coil generates a locally limited current distribution. Since no electrical connection of the sample is required, eddy current NDE is easier to use from a practical point of view, however, the choice of the optimum measurement parameters like e.g. the excitation frequency is more critical. Furthermore, the calculation of the current flow in the sample from the measured field distribution tends to be more difficult than in the case of a direct current injection.

2. Experimental configuration

We have used HTS rf SQUIDs [7] in our experiments, which have been operated at 3 GHz. The field noise of the sensors was about 1 pT/√Hz at signal frequencies above 1 Hz, and the field to flux transfer coefficient was about 50 nT/Φ_0. The SQUID readout system had a slewrate in excess of $5 \cdot 10^4$ Φ_0/s and a dynamic range of ± 500 Φ_0. This made the operation of the system in an unshielded environment possible. To minimise the excitation field at the location of the SQUID, we used either a gradiometric excitation coil shaped like a double D, or a circular field coil with local compensation of the excitation field at the SQUID by a second coil. In both cases, the excitation field at the sample can be as high as 5 mT. With the gradiometric excitation coil, the excitation field at the location of the SQUID could be reduced by a factor of about 200 (less than 20 Φ_0 at the SQUID). With a local compensation of the field at the SQUID by a second coil, the reduction could be as high as 5000.

We have set up two prototypes for NDE testing. The first system makes use of a motorised x-y-stage (scanning range 1 m × 1 m) which is used for two dimensional scanning of flat samples. A simple glass dewar is used for cooling the SQUIDs. The distance between SQUID and room temperature sample is about 7 mm. The excitation coil is attached to the bottom of the dewar and can be shifted horizontally by means of two set screws to adjust for minimum excitation field at the SQUID. The dewar is mounted to the moveable part of the x-y-stage and is thus scanned above the fixed test object. The stage is controlled by a PC, which moves the stage with an accuracy of better than 100 μm in both directions.

The z-component of the magnetic field above the test specimen is sensed by the SQUID and further processed by a lock-in amplifier. A 16-bit analogue to digital converter is used to record the position of the SQUID and the output signal of the lock-in amplifier with a PC.

In addition to the liquid nitrogen dewar, a cryocooler was used in some cases to cool the SQUIDs [8]. In order to have a larger separation between SQUIDs and cooler, we have developed a closed cycle gas flow cryosystem. This system uses a miniature Stirling cooler, which cools neon gas down to about 50 K. A transfer system consisting of two flexible coaxial tubes with a length of 2 m transports the cold Ne gas to the SQUID mount and cools the SQUIDs to about 65K. The flexible transfer lines allow for a 3D movement of the SQUIDs.

For testing aircraft felloes we used a different set-up, where the felloe is rotated by an electric motor with a speed of 6-20 rpm. The SQUID is placed inside a conventional dewar, which is mounted close to the outer surface of the rotating felloe.

3. Results

Most of the investigated test objects consisted of aluminium alloy with an electrical conductivity of 15 to 30 MS/m. With a thickness of the samples of up to 25 mm, an optimum excitation frequency of the order of a few hundred Hz is calculated. When measuring the external noise in our laboratory, it turned out that at these frequencies the external noise was somewhat lower than the intrinsic noise of our SQUIDs, which is in the order of 1 pT/√Hz. At least in case of these high excitation frequencies, the use of an electronic gradiometer is not necessary. We think that in the case of large test samples, a field noise of 1 pT/√Hz is sufficient for all measurements, since the thermal magnetic noise of the test samples at room temperature can easily be as high as a few pT [9].

Without any mathematical post-processing, the spatial resolution of the system was found to be about 8 mm, which corresponds nicely to the separation between sample and SQUID. The spatial resolution can be improved further by using suitable differentiating filters (calculating dB_x/dx^2), however, on the expense of a slight reduction in the signal to noise ratio.

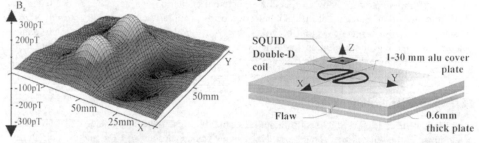

Fig.1 An artificial slot (40x1.2x0.15mm) below 13mm plate of aluminium. Double-D exitation coil: 120Hz

Fig. 1 shows the result of a measurement of an aluminium sheet with an artificial slot (40 mm long, 0.15 mm wide and 1.2 mm deep) covered by a 12.7 mm thick aluminium plate. This measurement was performed using a double-D coil and an excitation frequency of 120 Hz. Since we measured the z-component of the eddy current field, we mainly detect effects from currents flowing in the x-y plane of the sample. The distortion of the eddy current distribution is maximum at the outer edges of the slot, leading to a corresponding high field strength at these points (see Fig. 1).

We then measured various test objects representing original aircraft parts. Fig. 2 shows the two dimensional field distribution above a test object consisting of three aluminium sheets bolted together by aluminium rivets. Close to some rivets, artificial cracks of 3-5 mm length had been introduced in the bottom aluminium layer. A gradiometric coil with a diameter of 9 mm was used here. Due to the depths of the cracks of about 3-7 mm, the optimum excitation frequency was about 1kHz. A periodic pattern can be observed in Fig. 2, which is due to the distortion of the eddy current distribution by the aluminium rivets. The cracks lead to a further distortion of the eddy current flow, which results in an increase in the magnetic field amplitude at positions close to the cracks (white spots in Fig. 2). As a guide, the position of the rivets (black circles) and the cracks (black lines adjacent to circles) are also given in Fig. 2. The figure shows the raw data, and no further image processing was used. The measurements discussed above could be carried out with both SQUIDs cooled by liquid nitrogen or cooled by the gas cooling.

For more detailed measurements on stacked aluminium sheets (3 layers) with titanium bolts, we have developed a rotating coil system, where the SQUID is placed above the rotation axis of two circular coils (Fig. 3, top). Due to the gradiometric design of the coils, the excitation field of the SQUID is reduced by a factor of 200. Fig. 3, bottom, shows the eddy current field above the bolt for a flawless stack (0mm) and for cracks of different length (5mm, 10mm, 15mm) in the second aluminium layer. The relatively large signal even in the case of a flawless sample is due to an asymmetry in the system. If either the SQUID or bolt are not exactly placed in the centre of the rotating coils, such an asymmetry will result. Subtracting the signal of a flawless plate from the measurements can further enhance the sensitivity of the system.

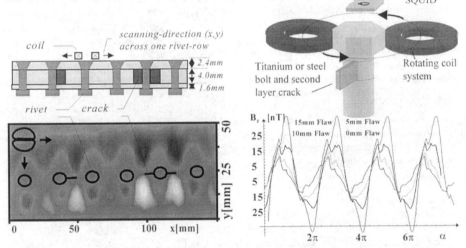

Fig. 2   Response of a rivet plate of a A-320
          Double D: ∅=9mm, 1kHz, 150mA

Fig.3   Signals of a rotating system for bolt testing
        for flaws with different lengths in the second layer.

4. Comparison with conventional systems

Finally we have performed some measurements to compare the microwave SQUID system with an eddy current system based on conventional induction coils. For these tests we used a sensitive conventional system (Elotest B1) with a pick up coil without ferrite core (PLA-44). The test sample had a small flaw at a depth of 13 mm. Although the maximum amplification was used (pre: 24dB, post: 60dB) the crack was not visible in a bandwidth of 30 Hz. We therefore performed an A-scan in a bandwidth of 2 Hz, and achieved a better result with a signal to noise ratio still smaller than 3. Figure 4, right, shows a measurement of the SQUID system. Although the bandwidth of nearly 50Hz was used the result of the signal to noise ratio is better than a factor of 100. In order to detect small defects like the one in Figure 4 with a conventional system large pick up coils have to be used which result in a decreased spatial resolution. The spatial resolution of the SQUID system is limited by the distance between sensor and sample or the diameter of the excitation coil, which can be adapted to the depth of the sample without appreciable loss of sensitivity.

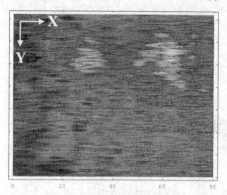 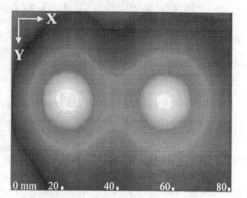

Fig.4 An artificial slot (40x1.2x0.15mm) below a 13mm thick plate of aluminium.
Left: Conventional system, induction coil with a   right: SQUID system, circular exitation coil with a
diameter of 44mm and 170Hz.   diameter of 20mm and 90Hz.

Up to now cracks with a length of 15mm and a height of 0.6 mm (6 times smaller cross sectionnal area than in Figure 4) can easily be found in even samples. At the moment the detection of smaller defects is limited by the noise caused by vibrations while scanning the sample. Furthermore effects like variation in conductivity and the changing distance between SQUID and sample (spacing effect) or edge effects caused signals prevent the detection of smaller defects.

5. Conclusion

We have demonstrated a SQUID based prototype for eddy current NDE, which can be operated in an unshielded environment. A cryocooler operated version could also be realised which allows a three dimensional moving and tilting of the SQUIDs. Various measurements on original aircraft parts could be carried out, and technically relevant cracks could easily be detected. A comparison with a conventional system shows the advantages of the SQUID system with a higher dynamic range and a higher field sensitivity.

Acknowledgments

The authors are grateful to W.B.Klemmt of DASA, F.Schur of Lufthansa, M.Junger and J.Rohmann of Rohmann GmbH and Y.Zhang, H.Bousack and H.J.Krause of Research Centre Jülich for supplying the various test samples and for helpful discussions.

References

[1] H.Weinstock, IEEE Trans.Mag. MAG-27, 3231 (1991)
[2] RR.Gans and R.M.Rose, Journal of Nondestructive Evaluation 12, 199 (1993)
[3] A.D.Hibbs, Review of Progress in Quantitative NDE 12, 1129 (1993)
[4] R.Hohmann, H. -J.Krause, H. Soltner, Y. Zhang, C.A.Copetti, H.Bousack and A.I.Braginski, M.I.Faley, accepted for publication in IEEE Trans. Appl.Supercond. AS-7 (1997)
[5] A.Cochran, J.C.Macfarlane, L.N.C.Morgan, J.Kuznik, R.Weston, R.Hao, R.M.Bowman and G.B.Donaldson, IEEE Trans.Appl.Supercond. AS-4, 128 (1994)
[6] Y.P.Ma and J.P.Wickswo, Review of Progress in Quantitative NDE 13, 303 (1994)
[7] The rf SQUIDs were supplied by Y. Zhang of Research Centre Jülich
[8] H. J. Tröll, Dissertation, JLU Gießen 1997
[9] L.Vant-Hull, R.A.Simpkins and J.T.Harding, Phys.Lett. A 24, 736 (1967)

Non-Linear Electromagnetic Systems
V. Kose and J. Sievert (Eds.)
IOS Press, 1998

New Inversion Procedure for the Separate Measurement of Conductivity and Thickness of Metallic Foils using Eddy Currents

Frank Röper

Institut für Elektrische Meßtechnik, TU Braunschweig, Hans-Sommer-Str. 66, 38106 Braunschweig, Germany

Abstract. An eddy-current through-transmission set-up is analysed analytically. By discussing the complex-voltage plane, it is shown that a separation of thickness and conductivity of a metallic foil is possible at relatively high frequencies. An inversion procedure based on frequency adjustment is presented. Finally the design of the coils is optimised to improve the amplitude of the output-voltage.

1. Introduction

An important application of eddy-current metrology is the noncontact measuring of electrical conductivity and thickness of metallic foils or films on isolating base material. One major advantage of the relatively inexpensive eddy-current technique is the insensitivity against dust, steam or oil covering the test material's surface. Two principle coil configurations exit:

- Reflection type: one or more coils are located on the same side of the test object,
- Through-transmission type: the primary coil is placed on one side and the signal pick-up coil is placed on the opposite side, ensuring that both sides of the specimen are accessible.

The exact solution of the forward problem, calculation of the coil impedance or the induced voltage from given test-foil parameters, for both coil set-ups was presented by Dodd and Deeds [1]. So far only through-transmission tests of flat metallic sheets have been reported which determine the product of conductivity and thickness [2]. A separate determination has been done for reflection type coils placed in contact with the test foils by Moulder et al. [3]. The measured data were compared with the exact solution using a least-squares norm and the conductivity and thickness were extracted by minimising the norm. The measurement of the impedance of a reflection-type coil is extremely sensitive to small changes in the lift-off, hence the coil must be adjusted very carefully.

2. Analytical modelling

The geometry of the through-transmission set-up is shown in Fig. 1. The primary coil generates the exciting field which induces eddy-currents within the metallic sheet. The reaction field weakens the exciting field and reduces the magnetic coupling between the primary and pick-up coil. Both the resistance and inductance of the primary coil and the amplitude and phase lag of the induced voltage in the pick-up coil are affected by this reaction. Considering a primary coil excited by a constant ac-current I_{ex} of frequency f, the induced voltage is given by the following integral equation according to [1]:

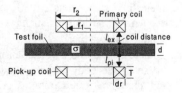

Fig. 1. Geometry of the through-transmission set-up

$$U_t = \frac{j\,8\pi^2 f\,\mu_0\,I_{ex}\,n_{ex}\,n_{pi}}{A_{ex}\,A_{pi}} \int_0^\infty \frac{J\,(r_{2ex},r_{1ex})\;J\,(r_{2pi},r_{1pi})}{\alpha^5}\,H_1\,(\alpha,T_{ex},T_{pi})\;Exp[-\alpha\,(l_{ex}+l_{pi}+d)]\,H_2[\alpha,d,\sigma\cdot f]\,d\alpha$$

Only air-core coils and non-magnetic materials are considered and the displacement current is neglected. The ohmic resistance of the coil winding is not included in this formalism. The subscripts „ex" and „pi" denote the primary and pick-up coil respectively, n is the number of turns and α the separation constant. The geometry parameters are shown in Fig. 1. The radial dimensions of the coils are incorporated via the function J:

$$J = \alpha^2 \int_{r_1}^{r_2} x\,\text{BesselJ}_1\,(\alpha x)\,dx$$

The expressions H_1, H_2 are combinations of exponential functions.
From this exact analytical solution two important conclusions can be derived:
- the magnetic field is determined by the product: $\sigma \cdot f$ and not by the two parameters separately,
- the sum $(l_{ex} + l_{pi} + d)$ indicates that the output signal U_t is not influenced by the foil position between the coils.

Because of the position-insensitivity continuous inspection of moving foils is possible with a through-transmission set-up.

The spatial distribution and the magnitude of the eddy-currents are determined by the thickness (d) and electrical conductivity (σ) of the test foil. If the penetration depth is much larger than d, the amplitude and the phase lag of the eddy currents remain constant in the z-direction. Changes of d or σ result in the same changes of phase and the time-averaged power loss of the eddy currents in the test sheet. Thus in the low-frequency regime only the product: $\sigma \cdot d$ is accessible.

Of course only σ can be measured with reflection coils, if the penetration depth is much smaller than the materials thickness. Hence separation of σ and d is only possible, if penetration depth and thickness lie in the same order of magnitude.

3. Complex voltage plane

The calculated loci for frequency variation for two different values of ($\sigma \cdot d$) are shown in Fig. 2. Additionally the thickness and conductivity loci are presented. It can be seen, that a separation angle Φ occurs in the fourth quadrant of the complex plane, making possible the separation of the two interesting parameters. At lower frequencies the loci for changes in σ and d fall together.

Increasing d and σ yields an increase in the eddy-current density and therefore a reduction of the amplitude of U_t. In fact, calcula-

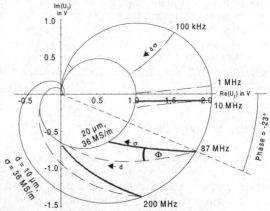

Fig. 2. Complex plane diagram of the voltage induced in the pick-up coil for two values of ($\sigma \cdot d$), phase-angle with respect to I_{ex} ; r_1=17.5mm, r_2=22.5mm, T=5mm, coil-distance=20mm, n=100, I_{ex}=100mA

tions show that the amplitudes of voltage phasors with the same phase-angle are inversely proportional to ($\sigma \cdot d$) in the whole frequency range.

If the eddy-current distribution is constant in the z-direction, the phase-angle of U_t is a function of the parameter product: $f \cdot \sigma \cdot d \cdot A_{eff}$ [2], where the factor A_{eff} represents the geometry of the coil configuration. In the fourth quadrant (phase-angles $< -10°$) the phase is only a function of the quotient of penetration depth and thickness:

$$\frac{d}{\delta} \sim d \cdot \sqrt{f \cdot \sigma} \quad ,$$

the coil geometry has no more influence on the phase-angle of U_t.

This is confirmed by the graphs in Fig.3, showing the separation angle vs. test frequency. The maximum separation angle occurs at frequencies determined by: $d \cdot \sqrt{f \cdot \sigma} = 560\sqrt{Hz\,Sm}$, corresponding to a phase angle of $-23°$. The maximum value of Φ is $12.5°$. Calculations for other coil configurations lead to exactly the same graphs.

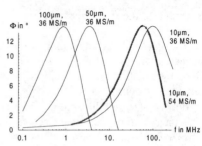

Fig. 3. Separation angle vs. frequency for three foil thicknesses and two conductivities, parameters see Fig. 2.

4. Inversion procedure

From the above discussion of the complex voltage an inversion procedure can be derived. Fig. 4 shows the block diagram of the corresponding circuit.

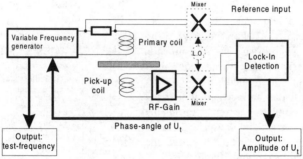

Fig. 4. Circuit for measuring d and σ separately

The test frequency is manipulated in a closed loop to balance the phase-angle of the output voltage to a specific, constant value. The frequency is determined by the equation: $d \cdot \sqrt{\sigma \cdot f} = C_1$. C_1 is fixed by the value of the phase-angle and independent of the coil configuration. The maximum separation angle occurs at $C_1 = 560\sqrt{Hz\,Sm}$, but an phase angle below this maximum might be chosen to decrease the needed frequency range. The resulting amplitude is now given by: $|U_t(d\sqrt{\sigma \cdot f} = C_1)| = (C_2 \cdot \sigma \cdot d)^{-1}$. The coefficient C_2 is dependent on the coil set-up, because better coupling of the primary and pick-up coil increases the output voltage. Thus, C_2 must be derived from a calibration measurement of a foil with known thickness and conductivity. Once the coefficients are fixed, σ and d can be inferred from the frequency and output voltage amplitude:

$$d = \frac{C_1^2 C_2 |U_t|}{f} \quad , \qquad \sigma = \frac{f}{(C_1 C_2 |U_t|)^2}$$

Modern lock-in amplifiers use DSP-technology. Because at frequency ranges above 1 MHz the use of pure digital techniques is not feasible, an analog down-conversion is necessary for thin foils (see Fig. 3).

5. Design of the transmission coils

The amplitude of the output-voltage depends on the coupling of the coils. The coupling increases with decreasing coil distance and increasing coil radius. If the coil distance is fixed by installation requirements, only the coil radius, cross-section and the number of turns can be optimised.

Between the individual turns of wire of the coil winding a distributed capacitance occurs. This interwinding capacitance acts as a lumped capacitance in parallel with a series inductance and resistance, caused by the ohmic resistance of the wire. Therefore a technical coil behaves as a parallel resonant circuit. To obtain reliable measurement results the coils must be operated well below their resonant frequency: $1 / (2\pi\sqrt{LC})$

The capacitance can be reduced by increasing the spacing between the individual turns or by using special winding techniques like honeycomb winding, but the value of the capacitance will still lie in the picofarad range. Thus because of the high frequencies needed for the separation of d and σ only low inductance coils can be used. The coil inductance is mainly proportional to the mean radius and to the number of turns. Therefore strong coupling and low inductance represent opposite design goals.

In Fig. 5 the dependence of the output voltage: $|U(\text{phase} = -23°)|$ on the coils' mean radii is shown. The primary and pick-up coils have identical shapes and mean radii, their distance is 5mm. The inductance of the coils in air is calculated and set to 1 µH by adjusting the number of turns. From the graphs it can be concluded that pancake coils generate the greatest output voltage and the optimum coil radius is about 10 mm.

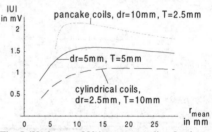

Fig. 5. $|U(\text{phase} = -23°)|$ vs. mean radius, L_{air} is adjusted to 1 µH. Coil distance =5mm, f=850kHz, d=100µm, σ=36 MS/m

In the above discussion the influence of the lateral dimensions of the test foil has been neglected. Measurements with quadratic foils with different edge-lengths were performed to investigate this parameter. The measurements showed, that for foils with edges larger than five times the coil mean diameter, the foil shape has no influence on the output-voltage.

References

[1]   C.V. Dodd, et al., Solutions to some Eddy Current Problems, *International Journal of Nondestructive Testing* **1** (969) 29-90

[2]   F. Förster and H.L. Libby, Analysis of Probe and Through-Transmission Test of Sheets and Foils. In: R.C. McMaster (ed), Nondestructive Testing Handbook, Vol.4, The American Society for Nondestructive Testing, Colombus OH, 1986

[3]   J.C. Moulder et al., Thickness and conductivity of metallic layers from eddy-current measurements, *Rev. Sci. Instrum.* **63** (1992) 3455--3465

Prediction Algorithm for Non-Destructive On-Line Evaluation of Mechanical Hardness of Electrical Steels

A. J. Moses, H. J. Stanbury[+,] Z. S. Soghomonian [++]

++ *Now with A.B.B. Kent Meters Limited, Pondwicks Road, Luton, Beds., LU1 3LJ, U.K.*
+ *European Electrical Steels, Orb Works, P.O. Box 30, Newport, Gwent, NP9 0XT, U.K.*
Wolfson Centre for Magnetics Technology, School of Engineering , University of Wales Cardiff, P.O. Box 917, Newport Road, Cardiff, CF2 1XH, U.K.

Abstract   Mechanical hardness of electrical steels is generally measured destructively. A computerised system based on a generic prediction algorithm has been developed for non-destructive determination of mechanical hardness in real time. Discrete data-bases of magnetic and mechanical hardness characteristics are compiled then subjected to a grouping analysis and quantised with respect to the desired resolution of the hardness predictions. Combinations of the magnetic parameters are used to perform a synchronous field matching process between various discrete magnetic-hardness quantised data bases. The results of the field mapping operation allows a series of specific coincidence patterns to emerge. The final Vickers Pyramidal Numbers (VPN) have uncertainty levels of ± 5 VPN. The deduction technique has the potential to predict mechanical hardness of other continuous strip materials.

1. Introduction

A non-destructive technique is introduced for predicting on-line mechanical hardness of non-oriented electrical steels during production. It is based upon a statistical super-positioning algorithm, which is applied to pre-selected magnetic/hardness data-bases during the deduction procedure. These data bases may be regularly updated in order to improve the resolution of the predicted Vickers hardness numbers. By accounting for alloying composition, density and gauge variations, the technique can provide the range as well as the statistical mean and distribution of the localised mechanical hardness (VPN10) of a region of steel strip. On-line magnetic measurements and the predicted hardness can be customised to deduce hardness values corresponding to a specific processing stage of the strip and thus be used to monitor the uniformity and the consistency of the mechanical hardness in real-time.

2. Description of the technique

The prediction algorithm links two separately quantised fields where the mechanical hardness and magnetic data are stored. The link between the two separate fields is found without the use of any direct mathematical or correlated relationships.

General mathematical modelling procedures carried out between the magnetic and mechanical properties give inconclusive results. In simple or polynomial multi-variable models, the magnetic parameters could be grouped or discretely included within a selected mathematical model structure, but the practical results of the model tends to be ambiguous.   Therefore, the independent (magnetic) variables such as coercive field strength and power loss and the dependent (hardness number) variable are treated statistically based on a pre-defined analytical and quantisation structure, in order to realise their finite symmetrical properties.   The coherent symmetry eliminates many problems generally encountered with the complex multi-variable mathematical models normally regressed for the prediction of the dependent parameter.

A series of magnetic and hardness measurements were made on grades of non-oriented electrical steel. For each composition a large number of transversal Epstein samples (30 mm x 300 mm) were tested in the cold rolled condition. These samples were annealed in a decarburising atmosphere at 800 to 830 °C allowing gradual hardness changes to take place in the samples similar to that achieved in a given production processing stage.

The hardness and the magnetic properties of the treated samples were measured using Vickers hardness indentation method and conventional off-line testing methods, respectively.   The power loss (W_{spec}), relative permeability (μ_{rel}) , remanence flux density B_{rem} and the ac coercive field strength Hc were measured at 0.5 to 1.5 T (50Hz).

The materials were categorised according to their compositions. Further structuring of magnetic and hardness data allowed the entire magnetic and hardness variation spectrum to be separated into smaller groups separated by 5VPN increments.   To maintain the symmetries between the hardness and magnetic variable fields within their respective subsets, a number of groups were created in the magnetic subsets corresponding to the same incremental variation in the "hardness subsets".   Each numbered group in both magnetic and hardness subsets were analysed statistically to obtain their separate normal distributions.   The properties of these small normal distributions such as the population mean and variances, were used to characterise each group. The population minimum and maximum measurement values were used to separate the leading and lagging boundaries of the neighbouring groups. This quantisation process depicted in Figure: 1, was carried out separately for each magnetic parameter.

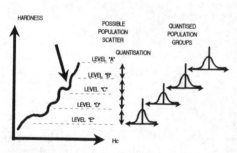

FIGURE: 1
SCHEMATIC VISUALISATION OF THE
QUANTISATION PROCESS ON EACH
COMPOSITION MAGNETIC & HARDNESS DATABASE

The statistical results obtained from this quantisation process is shown in a generic form in Table: 1 for a given magnetic parameter, denoted by "Q". The number of data points within each composition data-set, and consequently the associated magnetic and hardness data-sets, were allowed to be increased continually. Such accumulation and expansion of the reference magnetic and hardness data improves the characteristics of each quantised group for all magnetic parameters.   Prior to the deduction process, the corresponding quantisation look up table of a material grade is either recompiled or updated and saved in a format similar to that shown in Table: 1.

The magnetic properties of steel strip are measured continuously on-line, scaled and compared with the corresponding magnetic/hardness look-up tables. The corresponding composition data-set is looked-up, followed by the magnetic subsets and the common reference hardness subset. The given magnetic parameter, is looked-up and scanned against the minimum and maximum levels of its quantised subsets, as shown in Table: 1 (column 4). This scanning operation is carried out in the first group, until the magnitude of the parameter satisfies the upper and the lower conditions of one group or two to three successive quantisation groups. These groups are stored in a temporary dynamic memory

Table 1 Quantisation "look-up " tables for four generic quantised population groups, created symmetrically for the magnetic parameters with respect to the quantised hardness data-groups.

Quantised groups	Mean hardness (vpn_{10})	Minimum & maximum levels of hardness (vpn_{10})	Minimum & maximum levels of a magnetic parameter	Population mean the magnetic parameter	Population variance the magnetic parameter
Group A	H_{VPN_A}	Hv1 <--< Hv2	Q1 < -.- < Q2	Q (MEAN A)	$\eth_1$ (VAR. A)
Group B	H_{VPN_B}	Hv2 <--< Hv3	Q2 < -.- < Q3	Q (MEAN B)	$\eth_2$ (VAR. B)
Group C	H_{VPN_C}	Hv3 <--< Hv4	Q3 < -.- < Q4	Q (MEAN C)	$\eth_3$ (VAR. C)
Group D	H_{VPN_D}	Hv4 <--< Hv5	Q4 < -.- < Q5	Q (MEAN D)	$\eth_4$ (VAR. D)

location. The magnitude of the parameter is compared with the mean and the distribution levels of those stored. The group with the closest mean and the smallest relative distribution spread is selected. The same scan and comparison procedures are carried out sequentially for each magnetic parameter. Upon completion of magnetic data-set scanning processes, the mapped quantisation groups corresponding to magnetic properties of the steel strip, are statistically processed further to extract the most probable limits within which the corresponding mechanical hardness can lie.

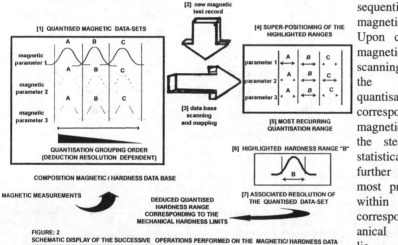

FIGURE: 2
SCHEMATIC DISPLAY OF THE SUCCESSIVE OPERATIONS PERFORMED ON THE MAGNETIC/ HARDNESS DATA USED FOR DEDUCING MECHANICAL HARDNESS VALUES

All the necessary quantisation groups which have been separately extracted from the reference magnetic data-sets corresponding to the new magnetic measurements are analysed further to deduce the most probable quantised hardness range. In the mapping operation, the magnetic propeerties are treated independently with respect to the quantised hardness range. The super-positioning in the final deduction procedures of the mechanical hardness is schematically shown in Figure: 2.

After the synchronous scanning of each magnetic parameter across the pre-calculated magnetic data-sets, the corresponding quantised hardness ranges are highlighted and subsequently analysed for their spread and frequency of occurrence in relation to the other highlighted hardness quantisation groups. In order to filter through different coincidence patterns which could occur for any of the highlighted quantisation groups.

It was found experimentally that the way in which the highlighted quantisation groupings were selected by the discrete consideration of the magnetic parameters tended to reproduce coincidence patterns. Each pattern could have different permutations depending on the gradual changes in the metallurgical condition of the material and the fluctuations in the related magnetic parameters. The reoccurrence of such patterns at this stage of the deduction process was not influenced by any dominance of a given magnetic parameter. The hardness number may either be expressed as a quantised hardness range or as a mean hardness value along with its associated uncertainty limits .

Figure: 3 shows output plots of the traced magnetic parameters and the deduced hardness along the length of a coil of non-oriented steel. The real-time on-line trace typifies an over-annealing and under-annealing effects. The fluctuations in the magnetic parameters matches the hardness trace. The sudden drop in the magnetic and mechanical hardness are typical of cases where the natural flow of the annealing line is interrupted manually causing the material to be momentarily exposed to the annealing conditions longer than necessary, thus resulting in a drop of mechanical hardness.

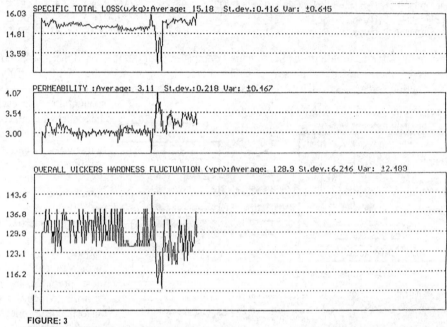

FIGURE: 3
THE ON-LINE OUTPUT OF THE THREE SELECTED MAGNETIC PARAMETERS MEASURED AT 1.5 Hz AT EVERY 1.5 METERS ALONG THE STRIP WITH THE CORRESPONDING FLUCTUATIONS IN THE PREDICTED VICKERS HARDNESS NUMBERS (VPN$_{10}$) DETERMINED BY THE DEDUCTION PROGRAMMES. (MATERIAL GRADE: 0.1 SILICON SEMI-PROCESSED NON-ORIENTED STEEL, GAUGE 0.5 mm)

3. Conclusions

In the deduction technique, the mechanical hardness and magnetic parameters were inter-linked directly and discretely , without any intermediate mathematical relationships. The pre-conditioning, the quantisation data-base scanning and the super-positioning or pattern coincident recognition operations, when combined together could successfully produce hardness predictions which closely agree with the known variations in hardness of a continuous steel strip. The overall correlation between the destructive and non-destructively predicted on-line hardness values were calculated to be between 97.1% and 99.5%. The entire deduction technique has been implemented in the form of multi-tasking computer software and used successfully on a commercial production line.

Acknowledgements

The work was carried out as part of an EPSRC/DTI Teaching Company project and the authors are duly grateful to Orb Electrical Steels and the TCD for funding.

Non-Linear Electromagnetic Systems
V. Kose and J. Sievert (Eds.)
IOS Press, 1998

Investigation of the Magnetic Anisotropy of Electrical Steel

Slawomir Tumanski, Bernard Fryskowski
Warsaw University of Technology, Koszykowa 75, 00-661 Warsaw, Poland

Abstract. Two methods of investigations of electrical steel anisotropy by means of MR Permalloy sensor have been presented. The first method is based on mapping the magnetic field for the sheet magnetised parallel and perpendicularly to the rolling direction. The second one is based on measurement of the magnetic field with rotation of the sensor-yoke assembly.

1. Introduction

It is possible to determine the stray magnetic field component directly above the investigated steel sheet with the aid of thin film Permalloy magnetoresistors [1,2]. If the sensor is placed above the sheet, the output signal of the sensor is proportional to the local value of magnetic field strength in the steel sheet. In this way we can determine the local value of magnetic field strength for different magnetising directions.

The main idea of Permalloy MR sensor application for the determination of the magnetic anisotropy has been pointed in the earlier work of the author [1]. In this paper two variants of the measuring system are tested. The basic structures of these systems are presented in Fig. 1.

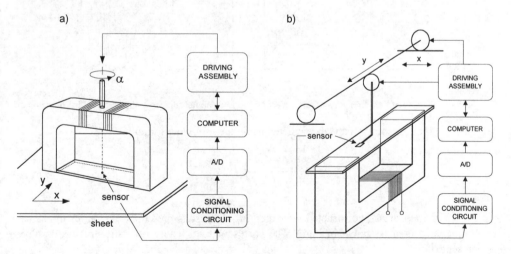

Fig. 1. The basic structure of two measuring systems

In the measuring system presented in Fig.1a, the sensor is placed in the centre of the space between the poles of the magnetising yoke. The yoke-sensor assembly may be rotated with respect to the investigated electrical sheet. The magnetic field strength in the sheet is measured for a fixed value of magnetic flux density in the yoke and for different values of the angle α between the magnetising field and the rolling direction. In the system presented in Fig. 1b, the MR sensor is moved above the sheet magnetised by C-yoke (or double-C yoke). The sheet can be magnetised with different angles between the rolling direction and magnetising field. For determination of the magnetic field the KMZ10B sensor with dimensions 1mm by 1 mm has been used.

2. The map of magnetic anisotropy

Using an Epstein frame the anisotropy can be determined by investigating two kinds of samples. One kind of sample is cut parallel to the rolling direction and the second one perpendicular to the rolling direction [3]. We can also get information about anisotropy by measuring magnetic field in the sheet for two directions of magnetisation - parallel and perpendicular to the rolling direction. In Fig. 2 two maps for the same area 2.5 cm by 2.5 cm are presented.

H_0 H_{90}

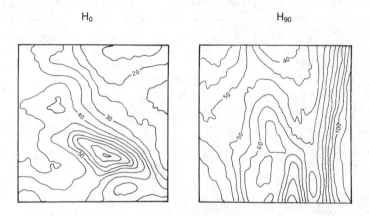

Fig. 2. The map of field strength above the sheet magnetised parallel and perpendicular to rolling direction

The magnetic anisotropy can be calculated using the following formula:

$$\delta H = \frac{H_{90} - H_0}{H_{90} + H_0}$$

(2)

The map of δH, constructed after merging the results presented above, for the same investigated area is shown in Fig.3. The corresponding picture of grain structure is also presented.

The mean value of magnetic field anisotropy calculated for the whole area was equal to 0.25. It means that H_0 was about 60% of H_{90}. There is no direct correlation between the picture of grain structure and the map of H_0, H_{90} or δH. However it is possible to find the area and individual grains responsible for reduction of anisotropy.

δH   grain structure

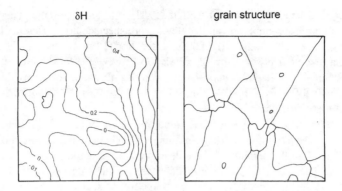

Fig. 3. The map of magnetic anisotropy and corresponding picture of the grain structure

3. Measurement of the local value of anisotropy

We can get more information about magnetic field anisotropy by measuring the changes of magnetic field strength for different directions of magnetising field (using the system with rotating yoke-sensor assembly presented in Fig.1a). The magnetic field strength is measured for different values of the angle α (angle with respect to the rolling direction) and the same value of flux density.

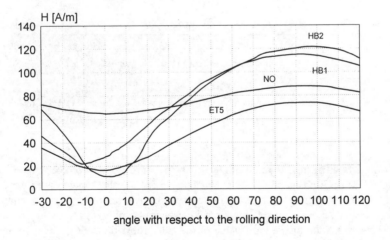

Fig. 4. The change of magnetic field strength for different directions of magnetising field

The relations $H = f(\alpha)$ for different materials: GO SiFe steel (ET5), NO SiFe (NO) and Hi-B steel (HB) are presented in Fig.4. The anisotropy, calculated according to formula (2), for typical GO SiFe steel is approximately 0.6. The same value determined for NO steel is equal to 0.12, and for Hi-B reaches the value of 0.8.

The lines HB1 and HB2 represent the relation H = f (α) for the same material but for two different points. Both amplitude and angle α_{min} are spectacularly different. The spread of the measured values of the angle α_{min} for the tested area (10 cm^2) was about ± 20°. It can be observed that there exists some sort of correlation between the local value of H and the angle α_{min} . When the measured value of H is significantly larger it means that also the value of α_{min} will differ from 0°. We can, therefore, draw instant conclusions about the quality of anisotropy from the map.

The characteristics H = f (α), for the same measuring point and different values of flux density, are presented in Fig. 5. The minimum of the characteristics around the 0° point is rather flat. Therefore the resolution of the determination of angle α_{min} is not better than several degrees. The magnetic anisotropy depends on the value of flux density. The anisotropy should be measured for a small value of flux density when the state of magnetisation depends mainly on the anisotropy energy.

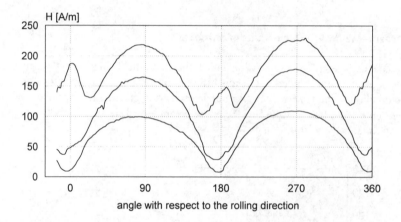

Fig. 5. The characteristics H = f (α) for the same measuring point and different values of flux density

4. Conclusions

Two methods of investigation of electrical steel anisotropy have been presented. The first method is based on mapping the magnetic field for the sheet magnetised parallel and perpendicular to the rolling direction. More comprehensive, but more complex and slower, is the second method based on rotating the yoke-sensor assembly.

For determination of the local value of the anisotropy, for typical GO SiFe steel, the sensor with dimensions 1mm by 1mm seems to be rather too large. Such a sensor can average the field value above several neighbouring grains.

References
[1]   S. Tumanski, The application of Permalloy magnetoresistive sensors for nondestructive testing of electrical steel sheets, *J.Magn.Magn.Mat.*, v.75 (1988) 266-272
[2]   S. Tumanski and M. Stabrowski, Magnetovision system: new method of investigating steel sheets, *J.Magn.Magn.Mat.*, v.160 (1996) 165-166
[3]   M. Soinski and A.J. Moses , Anisotropy in iron-based soft magnetic materials, Handbook of Magnetic Materials, Chapter 4, v.8 (1995), Elsevier Sc.

Usage of Magnetic Induction Higher Harmonic Components for Hardness Nondestructive Testing of Details after Surface Hardening

N.Gusak, A.Chernyshev, S.Murlin

*Institute of Applied Physics of Academy of Sciences of Belarus,
16, F.Scorina str., 220072, Minsk, Belarus.*

Abstract. The dependence of the third harmonic amplitude of the superimposed transducer emf on the coercive force and sample hardness after surface hardening has been determined. The influence of the various interfering factors (initial structures of samples before hardening, residual magnetization, lift-off (distance) between the transducer and testing surface) on E_{3m} was also investigated. A possibility of the hardness nondestructive testing of the complex form details after surface hardening by the higher harmonic method is shown.

1. Introduction

The modern industrial enterprises make great use of complex geometrical form details (cog-wheels, crankshafts, distributive shafts and others) which are exposed to surface hardening by high frequency currents, by laser hardening and others. One of the parameters, describing quality of surface hardening is hardness. Nowadays, for the hardness control of hardened layers the standard methods (Rockwell, Vickers and others), and also devices based on use of shock, ultrasonic, resonant, magnetic and other methods are applied. However, in many cases these devices cannot be applied for the hardness control, as for their work a high standard of surface cleanliness of a controllable detail site is required. Besides, transducers used in these devices do not allow measurement of hardness in difficult-to-achieve places in detail, for example, on teeth of cog-wheels, in apertures, slots and others, and also on little details of the complex form. In many cases the usage of the multiparameter system [1], based on the measurement of the various characteristics of a material, is not justified economically for the hardness control of hardened layers. The industrial enterprises prefer, in some cases, simpler and cheaper devices with higher accuracy of measurement, than complex and expensive ones.

2. Experimental Results

We have investigated a possibility of the hardness nondestructive control of hardened layers by measurements of the higher harmonic components (3rd, 5th and others) of the superimposed transducer emf, caused by the nonlinearity of magnetic characteristics of ferromagnetic materials at magnetic reversal by sinusoidal field. It is known [1-2], that for some steels, subjected for thermal processing, a correlation between the hardness and magnetic characteristics, in particular, the coercitivity *Hc*, exists. In turn, the higher harmonics also depend on *Hc*; for example, at magnetic reversal of a ferromagnet in the Rayleigh area by sinusoidal magnetic field of intensity $H=H_m\sin(\omega t)$ in a periodic signal of a magnetic induction the third harmonic component B_3 will be present, which amplitude is equal [3]

$$B_{3m} = \mu_0 \tfrac{4}{15\pi} bH_m^2 \qquad (1)$$

where b - coefficient at the square member of the Rayleigh's law, $\mu_0 = 4\pi \cdot 10^{-7}$ H/m. It is valid for mordern steels which remagnetization minor loop is described by Rayleigh formula. It follows from [4-5], that in some cases the product bH_c^2 is equal to a constant k or close to it. Hence, $b = k / H_c^2$ and the expression (1) takes the form

$$B_{3m} = \mu_0 \tfrac{4}{15\pi} \cdot kH_m^2 / H_c^2 \qquad (2)$$

The given expression relates to the quasistatic magnetic reversal case. On account of eddy currents, and also magnetization of a controllable site by a non-uniform field of the superimposed transducer, resulting in non-uniform distribution of the magnetization, the character of dependence (2) can essentially change. First of all the result is that correlation of B_{3m} with initial magnetic permeability μ_{in} occurs. However, experimental researches show, that in areas of rather high hardnesses the dependence of the third harmonic amplitude of the transducer output emf E_{3m}, proportional B_{3m}, on H_c is close to (2). It can form the basis for the hardness nondestructive control by measurements E_{3m}. As an example, in Fig.1 the dependence E_{3m} on H_c is given, received by measurements on samples from 45 steel (0,45 wt% C, 0,6 wt% Mn) at the exciting field frequency 2000 Hz.

As the hardness control of the complex form details must be carried out on a local site, for research realization miniature electromagnetic transducers with a diameter of 5 mm and height of 7 mm were developed. With the help of these transducers we have investigated dependences E_{3m} on the sample hardness after various kinds of surface hardening. The frequency and intensity amplitude of an exciting field have been elected to receive the information on hardness of a surface layer and to exclude influence of its depth over the E_{3m} magnitude. In Fig.2 the dependence E_{3m} on the hardness of samples made from the various steels after an induction hardening by high frequency currents is presented. The same character of the dependence E_{3m} on the hardness was observed for samples after a laser hardening, chemical-thermal processing and others. The sensitivity E_{3m} to hardness is constant with change of exciting field frequency from tens Hz up to tens kHz.

The influence of the various interfering factors on E_{3m} was also investigated: initial structure of the samples before hardening, residual magnetization, lift-off between the transducer and controllable surface and others. Various initial structures have been received by various thermal processing of samples. The investigation results of the samples from 40H steel are submitted in table 1.

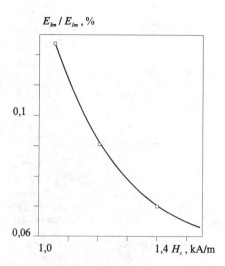

Fig.1. E_{3m} dependence (in % against the first harmonic amplitude of the transducer output e.m.f. E_{1m}) on H_c for the 45 steel.

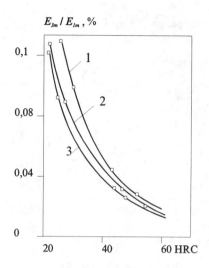

Fig.2. E_{3m} dependence on the sample hardness after hardening by high frequency currents. 1 - 45 steel, 2 - 40H steel (0,4 wt% C, ~1,0 wt% Cr), 3 - 65G steel (0,65 wt% C, ~ 1 wt% Mn)

Table 1. The third harmonic amplitude E_{3m} at various initial conditions and after an individual hardening of the samples from 40 H steel at the same regime.

Sample №	Initial condition				After hardening		
	Thermal processing kind	Structure	Hardness, HRC	E_{3m}/E_{1m} ,%	Thermal processing kind	Hardness, HRC	E_{3m}/E_{1m},%
1	Without thermal processing	perlite+ ferrite	<20	0,09		52	0,016
2	Annealing 860°C	perlite+ ferrite	<20	0,079	Induction hardening	51	0,017
3	Annealing+ hardening 850°C + tempering 150°C	beinite	49÷50	0,018	+ tempe- ring	52	0,015
4	Annealing + hardening 850°C tempering 350°C	beinite - trostite	44÷45	0,026	200°C	51,5	0,015
5	Annealing+ hardening 850°C+ tempering 500°C	sorbite	34÷35	0,032		52	0,016

Residual magnetization of the samples, as our researches have shown, can influence essentially on the third harmonic amplitude (Fig.3). With the residual magnetization field increase, E_{3m} decreases. The fastest fall is observed at H_r increase from 0 up to 10 A/cm. The E_{3m} reduction can reach 30 % from its initial meaning on the demagnetized sample. It is necessary to take this into account at the hardness nondestructive control by the higher harmonic method .

One of the main factors, which is necessary to take into account at the hardness control by the higher harmonic method using superimposed transducers, is the influence of the lift-off δ between the transducer and product. With the δ increase the amplitude of the third harmonic E_{3m} decreases by the nonlinear law (Fig.4, curves 1 and 2). The E_{3m} decrease with the δ increase is caused by reduction of a variable magnetic field intensity, influencing a sample at the exciting coil removal, as by reduction of a signal

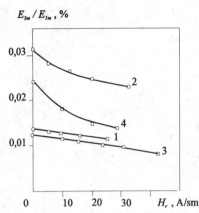

E_{3m}/E_{1m}, %

0,03

0,02

2

4

0,01

1

3

0   10   20   30 H_r, A/sm

Fig.3. E_{3m} dependence on the residual magnetization field strength H_r:
1,2 - 45 steel, T_{temp} = 200°C(1), 400°C(2);
3,4 - 13HN2MA steel (0,13 wt % C,
~2 wt% Ni), T_{hard} = 850°C (3);
and without thermal processing (4) .

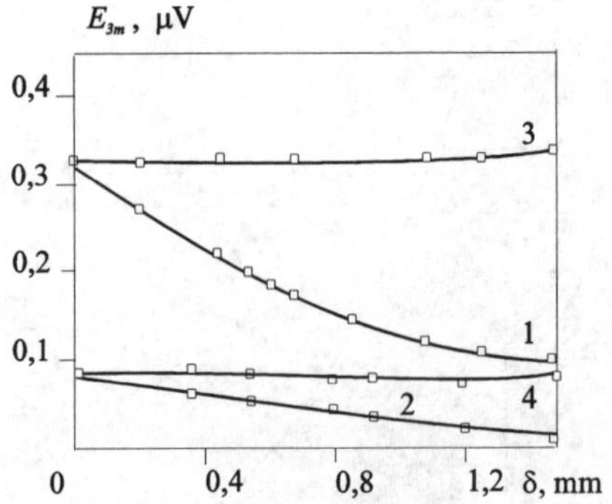

E_{3m}, µV

0,4

3

0,3

0,2

0,1

1

2   4

0   0,4   0,8   1,2 δ, mm

Fig.4. E_{3m} dependence on the lift - off , measured on the sample made from the 45 steel :
1,2 - before tuning off and 3,4 - after tuning
off the lift - off;
1,3 - samples without thermal processing,
2,4 - after hardening by high frequency
currents

directly in the measuring coil. To compensate the δ influence on E_{3m} the change of a transducer magnetization current by given algorithm was used. As a parameter sensitive to the lift-off, the total signal effective meaning of the transducer measuring coil was chosen. As a result, it became possible to tune off the lift-off up to 1,5 mm with an error no more than 10 % (Fig. 4, curves 3 and 4).

On the basis of research the devices for the hardness nondestructive control of the complex form details after surface hardening were developed.

3. Conclusions

The following results were found:
(1) The third harmonic amplitude decreases with increase of coercive force and sample hardness after surface hardening.
(2) A change of the sample initial structure before hardening does not influence essentially on E_{3m} measured after hardening. E_{3m} is decreased up to 30% , approximatly, by residual magnetization.
(3) The tuning off the lift-off up to 1,5 mm with an error no more than 10% was carried out by magnetizing current change according to the given algorithm.

References

[1]   W.A.Theiner et. al., Process Integrated Nondestructive Testing for Evaluation of Hardness. 14th World Conference on Non-Destructive Testing, New-Delhi, India, December 8-13, 1996, v.2, p. 573-576.
[2]   M.N.Miheev, E.S.Gorkunov, Magnetic Methods of Structure Analysis and Nondestructive Testing. M., Nauka, 1993.
[3]   R. Bozort, Ferromagnetizm, IIL, 1956.
[4]   H.Kronmuller, Statistical Theory of Rayleigh's Law. Z. angew Physik, 1970, 30, No.1, p. 9-13.
[5]   B. Astie et. al., Prediction of the Random Potential Energy Models of Domain Wall Motion: an Experimental Investigation on High-Purity Iron, J. Magn. Magn. Mat., 1982, 28. p. 149-153.

Non-Linear Electromagnetic Systems
V. Kose and J. Sievert (Eds.)
IOS Press, 1998

Non-Destructive Testing
Using 2D Infinite Elements

R. Sikora, J. Sikora[§], B. Pańczyk[*] and A. Kamińska[*]
Szczecin Technical University, Al. Piastów 19, 70-310, Szczecin, Poland
[§] *Warsaw University of Technology, ul. Koszykowa 75, 00-662 Warsaw, Poland*
[*] *Lublin Technical University, ul. Nadbystrzycka 36b, 20-618 Lublin, Poland*

Abstract. The non-destructive testing (NDT) problems can be solved as an inverse problem for the electromagnetic field with conductivity γ as the unknown material coefficient. This method could be adopted for the routine testing of nuclear reactor cooling tubes. To calculate such problem efficiently the infinite elements are used. The inverse problem is calculated for some types of cracks. The results are introduced and discussed.

1. Introduction

Using the Electrical Impedance Tomography (EIT) technique the unknown interior of the object can be reconstructed. The Eddy Current Tomography (ECT), a kind of EIT, may be adopted to the non-destructive testing (NDT) of the reactor's cooling tubes.

In this case the NDT problem is defined as the inverse problem of the electromagnetic field where the material coefficients (conductivity γ) in partial differential equations are unknown. The basis for the non-destructive tests is the measured impedance of differential coils. The probe is put inside the tube (Fig. 1) and its impedance is measured at a few different positions close to the supposed crack location. The impedance trajectory is the data base of the ECT image reconstruction algorithm. The forward problem can be solved using the Finite Element Method (FEM) and the inverse problem using the optimisation approach.

The following assumption have been made: the source current density $\vec{J}_s$ and the magnetic vector potential $\vec{A}$ vary sinusoidally with time, eddy currents within the current source can be neglected; magnetic permeability of materials involved in the modelling are

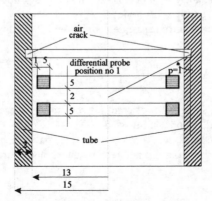

Fig. 1. Geometry of the NDT example

single valued and the displacement currents are negligible. The problem can be described by the Helmholtz equation:

$$\nabla^2 \underline{\vec{A}} - k^2 \underline{\vec{A}} = -\mu \underline{\vec{J}}_s, \tag{1}$$

where $k^2 = j\omega\mu\gamma$ ($j=\sqrt{-1}$, ω - angular frequency, μ - permeability, γ - conductivity).

2. Hybrid Finite Element Method

In the eddy current analysis in some problems the non-standard elements are used [1]. This is one of the methods to make EIT more efficient. The Hybrid Finite Element Method (HFEM) means classical FEM that uses non-standard elements.

Some FEM problems can be considered as open-boundary ones. In this case infinite elements can be used [2,3]. Infinite elements are easy to include in existing finite element programs. It is also more economical to use them because this method needs much less time and space for solving the inverse problem solution than the standard Finite Element Method.

Figure 1 shows three types of infinite elements used in FEM-algorithms. The shape functions of each kind of infinite elements are defined as:

IE1: $N_{1,2} = \mp \left(\dfrac{d}{r}\right)^m \dfrac{z - z_c \mp l}{2l}$, $\qquad$ (2)

IE2: $N_1 = \left(\dfrac{cd}{rz}\right)^m$, $\qquad$ (3)

IE3: $N_{1,2} = \mp \left(\dfrac{c}{z}\right)^m \dfrac{r - r_c \mp h}{2h}$, $\qquad$ (4)

where r_c, z_c, h, l, c, d are values shown in Fig. 2 and m - the decay factor (we assume that the magnetic vector potential $\vec{A}$ decays towards infinity as $1/r^m$ function).

The matrix coefficients of the infinite element are given by:

$$\underline{k}_{ij} = 2 \iint_\Omega \left(\frac{\partial N_i}{\partial x} \frac{\partial N_j}{\partial x} + \frac{\partial N_i}{\partial y} \frac{\partial N_j}{\partial y} + k^2 N_i N_j \right) \frac{1}{r} dr dz \tag{5}$$

The discretisation of the analysed model is shown in Fig. 3. The region is divided into 828 finite and infinite elements and has 893 nodes.

As the optimisation routine the steepest descent method (SDM) with momentum term has been used.

3. Image reconstruction

Using the EIT method some types of cracks described in [4] are reconstructed. Figure 4 shows part of an analysed region with a window (12x4 elements) in which the cracks are modelled. Notch shapes treated here are: elliptical (Fig. 5a) and stepwise (Fig. 5b). In the Figure 6 the crack shape discretisations and reconstructed images are presented.

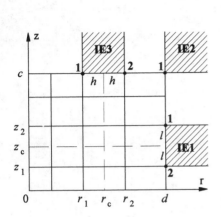

Fig. 2. Infinite elements

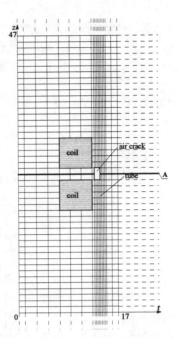

Fig. 3. Finite and infinite element mesh

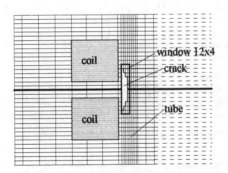

Fig. 4. Window with supposed crack

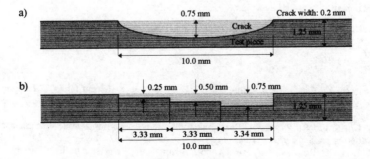

Fig. 5. Crack shapes [4] : elliptical (a) and stepwise (b)

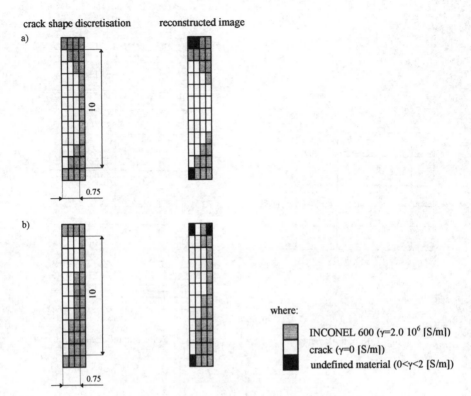

Fig. 6. Reconstructed images of cracks: elliptical (a) and stepwise (b)

4. Conclusions

In this paper a new approach to the Eddy Current NDT using the Hybrid Finite Element Method is discussed. The computing time for image reconstruction could be remarkably reduced by introducing the infinite elements. The developed algorithm has been applied to some examples, and good images were obtained. Through the obtained results it is found that the method can give information concerning the shape of the flaws. The suggested algorithm can be extended to other inverse problems.

Acknowledgement

This work was supported in part by the KBN Poland under the Grant No. 8 T10A 001 11

References

[1]   Ishibashi K., Eddy current analysis by BEM utilizing crack element for ECT. *Proc. of the Fourth Japanese and Polish Joint Seminar on Electromagnetic Phenomena Applied to Technology,* Oita, Japan, pp. 36-43, 1997.
[2]   S. Gratkowski, L. Pichon, A. Razek: New infinite elements for a finite element analysis of 2D scattering problems. *IEEE Transactions on Magnetics,* Vol. 32, No. 3, pp 882-885, 1996.
[3]   R. Sikora, B. Pańczyk, A. Kamińska and J. Sikora, Infinite elements in Eddy Current Tomography, *Proc. of 7th International Field Calculations in Electrical Engineering,* Graz, September 1996.
[4]   T. Takagi, J. Bowler, N. Nakagawa and J. Pávó, *E'NDE Benchmark Problems,* Ver. 1.0, Tokyo, 1996.

Non-Linear Electromagnetic Systems
V. Kose and J. Sievert (Eds.)
IOS Press, 1998

Dual-Frequency Eddy Current NDT Measuring Technique Improved Depth Sensitivity in Metallic Plates

[a]Cs. S. Daróczi and [b]A. Gasparics

[a]*Hungarian Academy of Sciences, Research Institute for Materials Science (ATKI),
H-1525 Budapest, P.O. Box 49, Hungary*
[b]*Technical University of Budapest, Department of Electromagnetic Theory (BME),
H-1521 Budapest, Hungary*

Abstract. Eddy Current Testing (ECT) can be used to obtain information on the inhomogeneity of the electrical conductivity of different materials along the sample surface. Due to the skin-effect the penetration depth of the exciting electromagnetic field decreases with the frequency, therefore the electromagnetic field induced by the surface eddy currents are usually much stronger than the field of the eddy currents flowing in the deeper regions. If the detector is placed on the same side of the sample as the exciting coil (as usually), then the shielding-effect will also appear as it rushes down the signal originated from the deeper regions. We have applied a simultaneous dual-frequency excitation to the sample that enables the surface part of the signal to be subtracted and to obtain better depth sensitivity. The subtraction can be done electronically in real time to avoid the need of repetitive measuring of (exactly) the same area of the sample at different frequencies.

1. Introduction

ECT is widely used for the investigation of defects in metals, alloys and other conducting materials. There are many ways to improve the sensitivity of ECT methods, such as applying increased excitation/measuring frequency, smaller lift-off (between the sample and the detector), using coils that have relatively large effective cross-section, separated excitation and detector coils, lock-in technique, more sensitive magnetic field sensors (SQUID, FluxGate, FluxSet) and extended measuring time. Unfortunately, the sensitivity of the whole system is usually controlled by the unwanted part of the induced signal itself and not by the thermal noise or other signals [1]. This leads to much smaller signal/disturb ratio than signal/noise. When we are observing defects well under the sample surface, the signal induced in the deeper regions will be overshadowed by the much stronger signal induced in the surface near to the exciting coil. One way to avoid this kind of signal degradation is to lower the excitation frequency, because according to the *skin-effect* the contribution of the deeper regions in the induced signal will be higher. However, in this case the absolute value of the induced signal is smaller, that demands more sensitive detectors especially in the low frequency domain (down to DC), like *fluxset*. In this paper we present the results of a modified setup of the more conventional ECT measurement that has been developed recently for the investigation of internal discontinuities (cracks) in metallic specimens [2] made of ICONEL-600.

2. Sample and Detector

For the measuring we have used the same sample as in earlier investigations [2]. There are four artificially made cracks on it, with the same width (0.2 mm) and length (10 mm), but with different height (1, 2, 3 and 4 mm high, respectively), *Fig. 1.*

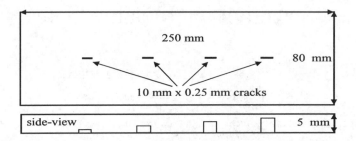

Figure 1. Reference sample made of ICONEL-600 with four artificially made "cracks" on it.

According to the naming conventions introduced in [1] our sensor is the Dipole-Dipole type. This means that for excitation and for detection orthogonal coils are used which are placed over the sample surface so that the induced signal of a point-like defect results in a dipole type image, *Fig. 2.* Regarding the original design, we have decreased the height of the exciting coil in order to decrease the effective distance between the sample and the probe. This enables stronger excitation without a large exciting current.   Another improvement is to centralize the detector coil inside the exciting coil. Along the axis of the detector coil the exciting field component is almost zero (in the same direction). Because of the dual-frequency excitation we need a core material which behaves linearly in order to avoid unwanted harmonics, therefore ferrite material was applied instead of the metallic glass frequently used in *FluxSet* sensors [3], for example. The amplification of the induced signal is performed by a lock-in technique which is phase sensitive, consequently under the spots "A" and "B" the sign of the rectified signal is just the opposite.

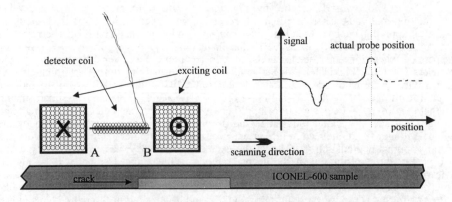

Figure 2. Inductive ECT sensor with optimized shape. The detector is sensitive to the difference of the electrical conductivity near to positions "A" and "B".

3. Measuring System

As it can be seen in *Fig. 3.* we have applied two independent lock-in amplifiers. The first is used to measure the high-frequency part of the signal, while the second is to measure the lower frequency component. Because the subtraction of the surface component of the signal is done after the lock-in amplifiers (but before the digitizing board), the ratio of the two exciting frequencies can be arbitrary between the ranges of 10 kHz ... 100 kHz, and 1 kHz ... 10 kHz, respectively. In our case the lock-in amplifiers act just as very narrow band filters which enable the wanted frequency component to be selected. For the dual-frequency excitation, we have applied also two independent function generators and a single power current source which is fast and highly linear up to the frequency of 100 kHz. During the measuring the probe is scanned over the sample surface and 2D images acquired by the controlling computer. Later 1D line-cuts can be made by software.

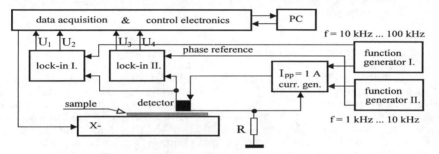

Figure 3. Block diagram of the ECT measuring system.

4. Results

The sample was always scanned from the back side, or in other words the cracks were covered and electromagnetically shielded by a 1, 2, 3 and 4 mm thick conducting layer, respectively. This measuring configuration simulates the so called inside cracks, *Fig. 4.*

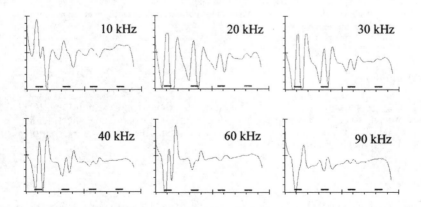

Figure 4. Line scans over the sample surface with the ECT probe at different exciting frequencies. (The broken lines show the positions of the cracks.)

Figure 5. 2D scans over the sample surface with the ECT probe using exciting frequencies of 10, 20, 30, 40, 50, 60, 70, 80 and 90 kHz respectively. Note the change in the shape of the images of the same crack at different frequencies and the change in the contrast.

In *Fig. 5*. 2D representations of 9 consecutive image scans are shown. Many scans were done up to 100 kHz exciting frequency. The images of the 4 and 3 mm high cracks are always visible, while the 2 mm high crack can be seen only below 90 kHz, and the 1 mm high crack only below 30 kHz. When we use dual-frequency excitation, the smallest crack can be seen with improved contrast. However, for this purpose properly chosen frequencies are needed. For example in *Fig. 6*. the result of a dual-frequency measurement can be seen which was done at 20 kHz / 40 kHz. In this case, the relative improvement of the contrast of the smallest crack (1 mm) to that of the largest (4 mm) is almost 2. If we increase the frequency difference, further improvement can be obtained. However, the apparent shape of the cracks becomes more complicated. To solve the reverse problem (to determine the real shape of the cracks) we plane to utilize Neural Network simulations.

Figure 6. 20 kHz / 40 kHz dual-frequency image of the ICONEL-600 sample from the back side. Note the improved contrast of the smallest crack on the right side of the image.

Acknowledgments

This work was partially supported by the Hungarian Scientific Research Fund (OTKA) under the project No. T-023559 and the MANODET INCO Copernicus Project, under No. ERBIC-15CT-960703 The authors would also like to thank the Organizing Committee of ISEM '97 for the financial contribution.

References

[1]   Cs. S. Daróczi et al., Nondestructive Testing of Materials. Amsterdam: IOS Press, pp.75-86, 1995
[2]   Cs. S. Daróczi et al., Nonlinear Electromagnetic Systems. Amsterdam: IOS Press, pp.748-751, 1996
[3]   G. Vértesy et al., Electronic Horizon, Vol.53 (1992) pp.105-107

Non-Linear Electromagnetic Systems
V. Kose and J. Sievert (Eds.)
IOS Press, 1998

Approximate Numerical Analysis Method for Alternating Magnetic Flux Leakage Testing

Y.Gotoh and M.Hashimoto

Department of Electrical Engineering, Polytechnic University,

4-1-1 Hashimotodai, Sagamihara, Kanagawa 229-11, Japan

Abstract. Alternating magnetic flux leakage testing is one of the surface inspection methods for iron and steel in iron works. This method consists of detecting leakage flux from cracks on ferromagnetic material magnetized by alternating magnetic flux by an exciting coil. But an evaluation method for this testing is not always established. Therefore we propose an approximate numerical analysis for alternating magnetic flux leakage testing. We clarify the phenomenon of alternating magnetic flux leakage testing using the approximate numerical analysis method.

1. Introduction

Alternating magnetic flux leakage testing[1,2] is one surface inspection method for steel material. The exciting frequency for this method becomes higher and higher, because precision detection of very small surface cracks in iron and steel materials has been demanded. But an evaluation method of this test is not established. In order to clarify this phenomenon, development of a non-linear alternating magnetic field numerical analysis is useful. Generally speaking, this numerical analysis needs time series analysis. But the analysis needs many time steps in a cycle and long calculation time. Therefore we propose an approximate numerical analysis method for alternating magnetic flux leakage testing. In this method, the magnetic flux is dealt with assuming that the flux is usually sine wave without distortion, and the relative permeability in the material is dealt with as a non-linear property. The developed approximate numerical analysis[3,4] is useful to clarify the phenomenon of alternating magnetic flux leakage testing.

2. Analysis Method

Equation (1) shows the basic equation of non-linear alternating magnetic field numerical analysis.

$$rot(\frac{1}{\mu}rotA) = J_0 - J_e \qquad (1)$$

Where **A** , J_0, J_e and $\frac{1}{\mu}$ are vector potential, current density, eddy current density and resistivity, respectively. Eddy current density J_e is expressed as following.

$$J_e = -j\omega\sigma A \qquad (2)$$

Where σ and ω are electric conductivity and angular frequency. The equation (1) is calculated as a non-linear solution in this method, where the magnetic flux is dealt with assuming an alternating sine wave without distortion. Variation of relative permeability in the material is accounted for the non-linear property of ferromagnetic material. Figure 1 shows magnetization curve measured in static magnetic field. We assume that initial magnetic relative permeability is 10% of the maximum. The used ferromagnetic material is SUS430. The electric conductivity of SUS430 is 1.533×10^5 S/m. Its maximum relative permeability is 782. Figure 2 shows a 1/4 region of a tube model used for the numerical analysis. Binary exciting coils (1000turns) are established on the upper and lower sides of a crack. The leakage magnetic flux is measured on the surface of the tube of the center region of the exciting coils.

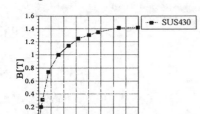

Fig.1   B-H curve in static magnetic field

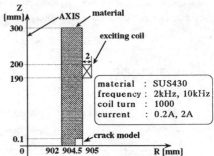

Fig.2   Model of numerical analysis

3. Result and Consideration

Figures 3 and 4 show distribution of the magnetic flux density using contour and vector displays in the material near the crack. From these figures, we can understand that there are convergence and distribution of the magnetic flux by the crack. Figures 5 and 6 show the distribution of relative permeability calculated by the non-linear numerical analysis, when the exciting current is 0.2A of frequencies of 2kHz and 10kHz. In both figures relative permeability on the surface area of the tube changes to about maximum permeability, and its domains become narrow when frequency is higher. We understand that a part with lower permeability exists at corners near the surface and the crack. Figure 7 also shows the distribution of relative permeability when the exciting current is 2A. From this figure, when the exciting magnetic flux is high, the relative permeability on the surface without a crack (Z=1.0mm) and the bottom region of the crack saturate, and the relative permeability of the internal part of the tube is about maximum. Figure 8 shows a distribution of magnetic flux by the non-linear alternating magnetic field numerical analysis by vector display. Figure 9 also shows one by linear numerical analysis (relative permeability is 100 fixed). From those results distribution of the magnetic flux in the material near the crack are quite different.

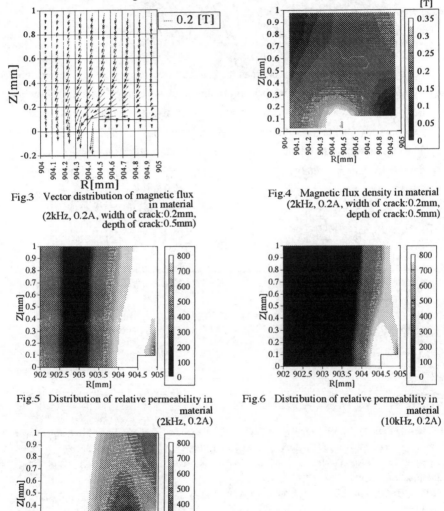

Fig.3 Vector distribution of magnetic flux in material
(2kHz, 0.2A, width of crack:0.2mm, depth of crack:0.5mm)

Fig.4 Magnetic flux density in material
(2kHz, 0.2A, width of crack:0.2mm, depth of crack:0.5mm)

Fig.5 Distribution of relative permeability in material
(2kHz, 0.2A)

Fig.6 Distribution of relative permeability in material
(10kHz, 0.2A)

Fig.7 Distribution of relative permeability in material
(2kHz, 2.0A)

At the result by the non-linear field numerical analysis (Figure 8), the magnetic flux behaves naturally as the flow passing around the crack, and a maximum magnetic flux value in the material is reasonable. But from the results by the linear numerical analysis (Figure 9), the magnetic flux vectors do not change continuously, and   magnetic flux values in the material are too high. We can understand the non-linear numerical analysis is useful for alternating magnetic flux leakage testing because it is important that the distribution of the magnetic flux leakage on the surface of the tube can be estimated.

4. Conclusions

We developed the non-linear alternating magnetic field numerical analysis method for alternating magnetic flux leakage testing . From this study we can conclude the following.
 (1) We developed non-linear alternating magnetic field numerical analysis method using
      sine wave approximation without distortion for alternating magnetic flux leakage testing.
 (2) The phenomena of the alternating magnetic flux leakage testing are made clear by the
      developed numerical analysis method.

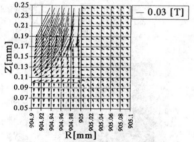
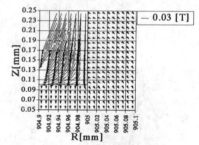

Fig.8   Non-linear alternating magnetic
field numerical analysis
(2kHz, 0.2A, width of crack:0.2mm,
depth of crack:0.5mm)

Fig.9   Linear alternating magnetic
field numerical analysis
(2kHz, 0.2A, width of crack:0.2mm,
depth of crack:0.5mm,
constant relative permeability:100)

References

[1] Toshiyuki Suzuma, Hirotsugu Fujiwara: Application of 2 Dimensional Non-linear Elector-magnetic Field
    Simulation For Alternative Current Magnetic Leakage Flux Testing, The Japanese Society for
    Non-Destructive Inspection Research & Technical Committee on Surface Method, pp.113-118, Tokyo,
    Japan, January (1997)
[2] Y.Gotoh, M.Hashimoto: Investigation of Alternating Magnetic Flux Leakage Testing Analysis.
    The Japanese Society for Non-Destructive Inspection, pp.87-90, Tokyo, Japan, March (1996)
[3] Yuji Gotoh, Mitsuo Hashimoto: Simulation of Alternating Magnetic Flux Leakage Testing, The Japanese
    Society for Non-Destructive Inspection Research & Technical Committee on Surface Method, pp.119-124,
    Tokyo, Japan, January (1997)
[4] Yuji Gotoh, Mitsuo Hashimoto: Equivalent Sine Wave Non-Linear Numerical Analysis Method for
    Alternating Magnetic Flux Leakage Testing, MAGDA Conference in Shinshu, pp.29-34 , Japan, March
    (1997)

Non-Linear Electromagnetic Systems
V. Kose and J. Sievert (Eds.)
IOS Press, 1998

Non - Destructive Testing Method for Spark Plug Insulators

Radu BURLICA, Mihai CHIRUTA, Mihai DUCA
ICPE-Trafil S.A., bd. Mangeron nr. 49, IASI, 6600, Romania

Abstract. The paper presents a non-destructive testing method for ceramic insulators in order to provide information about the electric strength of ceramic material.

The non-destructive testing method for spark-plug insulators is based on the leakage currents through the microcracks, or/and non-homogeneity of the dielectric material, submitted to a high intensity electric field.

In order to separate the currents that are generated by surface discharges and corona effects, which are not a measure of the volume insulation quality, the testing device has the possibility of measuring only the volume leakage currents determined by the microcracks and non-homogeneities of the dielectric material. The insulator testing system was tested and improved during the laboratory research to meet international standards requirements.

1. Introduction

The testing system consists of a testing cell with a special geometry of electrodes, which provides a signal on the insulator's electric strength, and an electronic measuring system.

The principle of the method is to place the insulator in an electric field of high intensity, created by a special system of electrodes.

If there are microcracks or non-homogeneities in the dielectric material of the insulators, an electric discharge will occur through the volume of ceramic material. The measurement and detection of the discharge current provides information about damaged insulators.

Under the circumstances, due to the high value of the electric field between electrodes, a surface discharge will occur.

Therefore, the electrode system of the testing cell is designed in a manner to permit the separation between the surface and volume discharge currents and also to make possible the detection and measurement of the leakage current through the dielectric material.

The fault signal value depends mainly on the geometrical shape of the high voltage electrode, as well as on the electrode position against the probe insulator.

The above mentioned conditions provide the testing sensitivity along the insulator and also the fault signal accuracy provided by the testing cell in the case of a damaged insulator.

The analog signals from the probes are detected, processed by the electronic measuring system, converted into digital signals and the data obtained are collected and stored by a computer.

The computer can also control the whole testing process.

2.   The testing cell

The testing cell is a structure of electrodes, which create different paths of discharge for surface and volume currents.

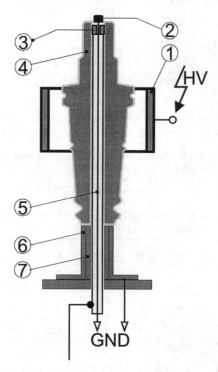

Fig.1 The testing cell

The high voltage electrode (1) is a circular ring surrounding the plug insulator (Fig.1). This is connected to the testing high voltage source (about 22 kV, 50 Hz).

The electrodes (2) and (6), are connected directly to the ground and with the high voltage electrode make up the surface discharge circuit.

Both electrodes (2) and (6) are insulated from the testing electrode (5) by the insulators (3) and (7). This one provides the signal for detection of the leakage currents through the material volume and is connected to the ground by a resistor which converts the current signal into a voltage one.

The spark plug insulator (4) will be placed on the electrodes system (5) and (6).

When the electrode (1) is connected to the high voltage power supply, the high intensity electric field determines also a surface discharge which is connected to the ground by electrodes (2) and (6), and volume discharge through the insulator's microcracks is settled between the ring electrode and testing electrode (5).

The geometric dimensions and shape of the high voltage electrode are essential.

The shape and position of this electrode determines the sensitivity of the testing cell along the insulator length and also minimises the energy and electromagnetic disturbances produced by the surface discharges.

3. The testing system

The information on the status of the insulator dielectric material is provided by the measuring resistor (8) as a voltage level, proportional to the value of the volume discharge current settled through the material (Fig.2).

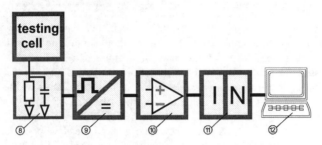

Fig. 2 The testing system. Block diagram

The signal must be processed by a radiofrequency filter due to the electromagnetic disturbances induced by the surface discharges in the measuring system.

After the voltage signal is rectified by the rectifier (9), it is compared with a reference voltage by the comparator (10).

The selection of the spark plug insulators is made by the threshold reference voltage, because a permanent leakage current through the insulator material is present, due to corona discharge on the surface of the insulator. A signal lower than the reference threshold voltage corresponds to a high quality insulator. If the signals caused by discharge through the insulator material exceed the level of the reference voltage, the insulator is damaged.

The analog output signal of the comparator is converted into a digital one by the interface (11) and it is stored and processed by the computer (12).

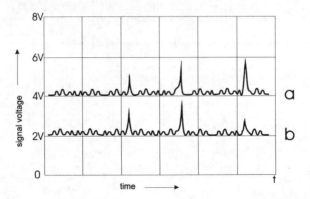

Fig.3 Signals diagram
a-   The form of the signal for a damaged insulator.
b-   The form of the signal for a good insulator.

Fig.3 shows the volume discharge current signal for a good insulator, trace (b), and the signal for an insulator with microcracks, trace (a).

The electromagnetic disturbances due to the surface discharges are pointed out in both traces (a) and (b) by intermittent jumps in the value of the voltage signal.

4. Conclusion

Using the method presented, based on measuring the leakage currents through the insulator's microcracks, experimented and tested in our laboratories, good results were obtained and industrial application of this method should be considered in order to improve the quality of the spark plugs, used in the automotive industry.

The testing system is designed to allow full automation and computer control of the testing process. Therefore it is possible to test a lot of insulators at the same time. The testing conditions meet the Romanian quality standards (STAS 5581/1-80) requirements. The system can also be used for testing other types of bushing insulators.

Afterwards, based on this non-destructive method we developed testing equipment, fully automated, which provides high productivity in accordance with the requirements of spark plugs industrial production.

Acknowledgements

Mircea SLININI, PhD is gratefully acknowledged for his help, encouragement and advice throughout this work. We are grateful to Braunschweig Technical University for financial support given for participation in the conference.

References

[1] F. H. Kreuger, Discharge detection in high voltage equipment, London, Temple Press Books Ltd., 1964
[2] N. H. Miller, Non-destructive high-potential testing, London, Illife Books Ltd., 1964.
[3] L.L. Alston, High Voltage Technology, Oxford University Press, London, 1968
[4] I.H. Moson, Dielectric Breakdown in Solid Insulation Progress in Dielectrics, Vol.I, Haywood, London, 1979
[5] T. Horvath, Incercarea izolatiei electrice, Ed. Tehnica, Bucuresti, 1982

Non-Linear Electromagnetic Systems
V. Kose and J. Sievert (Eds.)
IOS Press, 1998

Defect Identification on the Opposite Side of the Conducting Material by Means of Genetic Algorithm

Masato ENOKIZONO and Takeshi KAI
Dept. of Electrical and Electric Engineering, Faculty of Engineering,
Oita University, 700 Dannoharu, Oita 870-11, Japan

Abstract. This paper presents a new method to identify defects on the opposite side of the conducting material by means of Genetic Algorithm(GA). GA is a kind of optimization technique based on the evolutional process such as natural selection, crossover and mutation in biological systems. But we obtained different solutions with each change in the mutation rate. We propose changing the mutation rate in every generation.

1. Introduction

In recent years, the numerical techniques such as the FEM and BEM have been developed, so that we have been able to analyze various physical conditions in the material accurately. However, many of these methods have been developed for direct analysis where the material property and shape are given previously. Today, inverse analysis to obtain information on a specimen from outer magnetic field becomes very important for many industrial systems such as the nuclear power generation system and has being studied all over the world. In general, it is difficult to solve the inverse problem that has more unknown quantities than the number of simultaneous equations. In this case, we can solve the inverse problem by transforming it into a combinatorial optimization problem.

This paper presents a new defect characterization method to identify the defect on the opposite side of the conducting material by means of GA(Genetic Algorithm). Such identification problem is transformed into an optimization problem with discrete combination and GA is applied to the problem[1]. But we obtained different solutions with each change of the mutation rate. A practical identification method is realized by changing the mutation rate automatically in every generation.

2. Transforming magnetic inverse-problems into a combinatorial optimization problem

In this paper, we investigate the identification of the defect of the conducting material as shown in Fig. 1. When alternating current is flowing in the two exciting coils, we can measure the perpendicular and parallel components of the magnetic field intensity on the surface of the conducting material at each sensing point indicated with "*." In this paper, at first we assumed distribution of the defect as shown in Fig. 1, and we used the magnetic field intensities calculated at sensing points by means of the finite element method (FEM) as objective values. To transform the above-mentioned problem into a combinatorial optimization problem, we proceeded as follows:

2.1. Segmentation

The region of the conducting material is divided into the same-sized square cells as shown in Fig. 1. Each cell is usually called a segment. If perpendicular components of a segment of the conducting material contain the defect, then the state is assumed to vary from "1" to "4" depending on the depth of the defect, then and if the defect does not exist, the state is assumed to be "0,"

2.2. Evaluation

When the number of sensing points is equal to n, the perpendicular component H_P^O and the parallel component H_H^O at sensing points, can be written as follows:

$$H_P^O = \left[H_{P1}^O, H_{P2}^O, \cdots, H_{Pn}^O\right]^T, \quad H_H^O = \left[H_{H1}^O, H_{H2}^O, \cdots, H_{Hn}^O\right]^T \tag{1}$$

We call these column vectors "observed vector". Those elements are calculated with FEM for the defect location assumed as shown in Fig. 1. These values are also objective values.

The estimation of the defect location is started from an assumed candidate for solutions. The candidate can be initialed randomly as shown in Fig. 2. In the same way, the following perpendicular component H_P^A and parallel component H_H^A at sensing points in the case of the candidate, can be calculated.

$$H_P^A = \left[H_{P1}^A, H_{P2}^A, \cdots, H_{Pn}^A\right]^T, \quad H_H^A = \left[H_{H1}^A, H_{H2}^A, \cdots, H_{Hn}^A\right]^T \tag{2}$$

We call these column vectors "assumed vector". The candidate is iteratively improved in the process of minimizing the square error between the object values and the calculated values. The function used in this evaluation is as follows:

$$F\left(H_P^O, H_H^O, H_P^A, H_H^A\right) = \frac{\left(H_P^O - H_P^A\right)^2 + \left(H_H^O - H_H^A\right)^2}{\left(H_P^{O2} + H_H^{O2}\right)} \tag{3}$$

From the above, it is evident that this identification poses a problem of locating the defect segments to minimize the evaluation function. Thus, the problem was transformed into a combinatorial optimization problem. However when the total number of the segments is equal to m, then there are 5^m candidates for solutions. Thus the number of the combinations increases exponentially, as m gets larger. This kind of problem is called a "Nondesterministic Polynomial Complete Problem," and finding an optimum solution with deterministic approaches becomes difficult, as the value m gets larger. To overcome the difficulty, we introduced the Genetic Algorithm.

3. Estimating method by means of Genetic Algorithm

3.1. Improved Genetic Algorithm

Genetic Algorithm (GA) is a kind of optimization technique based on the evolutional process such as natural selection, crossover and mutation in biological systems. In GA, we set up an imaginary life population with a chromosome in a computer cord, and perform the generative simulation. The imaginary life population and chromosomes are evolved so that the new individual in the next generation has a larger adaptation probability than the one adjusting itself to an environment given beforehand. Figure 4 shows the fundamental structure of GA. GA has the following advantages: (1) We never encounter a local minimum, because the finding of an optimum solution is started from a population including many searching points (individuals or chromosomes) and performed at the same time in the mutual cooperation and competition of individuals. (2) We can also apply the algorithm to problems for which the evaluation function can't be clearly defined, because it needs only the evaluation criterion (fitness) at the time of searching, and the derivative and continuity of fitness are needless. Therefore, GA is very useful for identification problems such as the above-mentioned problem. Fitness used in this problem is defined by

$$fitness = \exp\left(-F\left(H_P^O, H_H^O, H_P^A, H_H^A\right)\right) \tag{4}$$

The minimum value and the maximum value of Eq. (3) are transformed into 1 and 0, respectively. Probability that an individual i is to be the elite in considering population P, can be expressed as follows:

$$select_i = fitness_i \bigg/ \sum_{j=1}^{P} fitness_i \tag{5}$$

Figure 5 shows the flowchart of crossover. As shown in this figure, a new chromosome is made by splicing parent's chromosomes at random location. The couples are chosen by using the values of *select* to cross two individuals in many pairs selected randomly. Natural selection based on *select* is also performed in this process. Mutation as shown in Fig. 6 is an operation that randomly changes one gene in a chromosome. If we change the mutation rate, we obtain different solutions.

In this paper, we propose a new method to determine the mutation rate by the number of groups in population P. Plural number of the same individuals exist in a certain generation, we regard the same individuals as a group. When the number of groups in a certain generation is Ng, we written function of the mutation rate,

$$mutation\ \ rate = 0.5 \times \frac{P - Ng}{P - 1} \tag{6}$$

where 0.5 is the maximum value of the mutation rate. The mutation rate is decided automatically.

3.2. Coding method into chromosome

To use GA in the above problem, it is necessary to code a candidate for solution as a chromosome. In this paper, the depth of the defect corresponds to one gene as shown in Fig. 6, the candidate is coded into a $\{0,1,2,3,4\}$ bit-string chromosome. Therefore the one gene denotes the existence of the depth of the defect and the chromosome denotes the defect distribution in the cross-sectional area of the conducting material under consideration.

4. Simulation Method

The initial candidates, the P chromosomes in which $\{0,1,2,3,4\}$ are randomly located, are used in this analysis. The principle of our new method is to evolve the initial population toward an optimum solution with genetic operators such as crossover and mutation. A next generative chromosome is decoded as a candidate of the solution, and the magnetic field intensities at sensing points are numerically calculated by using FEM to obtain the assumed vector. The candidate is iteratively improved by means of GA to minimize the square error between the objective vector and the assumed vector with Eq. (3). Table 1 shows the parameters used in this analysis.

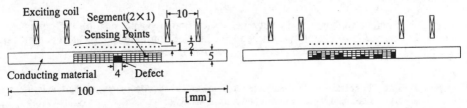

Fig. 1 Definition of an identification problem of defect on the opposite side of the conducting material.

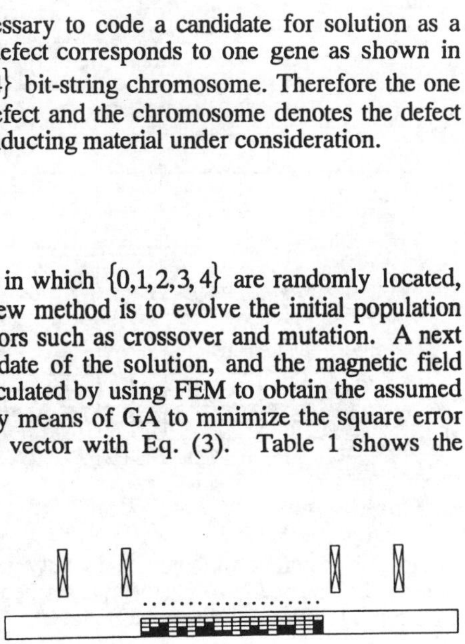

Fig. 2 A candidate for solution.

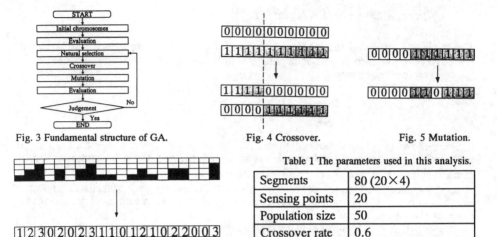

Fig. 3 Fundamental structure of GA.   Fig. 4 Crossover.   Fig. 5 Mutation.

Fig. 6 Coding method into chromosome.

Table 1 The parameters used in this analysis.

Segments	80 (20×4)
Sensing points	20
Population size	50
Crossover rate	0.6
Mutation rate	0.1, 0.5, our method

5. Results and Discussion

Figure 7 shows the transition of the error (generation vs. error). Figure 8 shows the expansion of Fig. 7 between the 100th and 500th generations. Figure 9 shows the estimated defective location at the 500th generation by using our method, and black segments are the defects. In Fig. 8, we can observe that our method requires a short interval to find out the optimum solution in comparison with the other methods. In Fig. 9, the defects of edge exist, presumably influenced by the proximity effect of the exciting coils. The advantage of our method keep the same number of groups, so our method is hard to be thrown into a local minima.

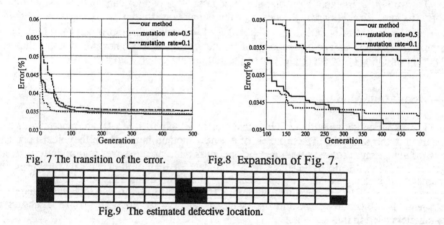

Fig. 7 The transition of the error.   Fig.8 Expansion of Fig. 7.

Fig.9 The estimated defective location.

6. Conclusion

In this paper, the method to identify defects on the opposite side of the conducting material by means of GA has been described. We have demonstrated the great importance of the proper selection of the mutation rate for the estimation process.

References
[1]   A. Ishiguro, Y. Tanaka and Y. Uchikawa, A Study of Genetic Algorithms' Application to Inverse Problems in Electromagnetics, ISEM-Nagoya, pp.57-60, 1992
[2]   M. Enokizono and Y. Akinari, Defect identification in magnetic materials by using genetic algorithm, ISEM-Seoul, pp.117-120, 1994

Non-Linear Electromagnetic Systems
V. Kose and J. Sievert (Eds.)
IOS Press, 1998

Magneto-Optical Non-Destructive Testing of Near-Surface Layers of Thin CoNi Films with the Depth Depended Properties

Y.B.DONG, V.E.ZUBOV

*Department of Physics, Moscow State University, Vorobyevy Gory, Moscow 119899,
Russia*

By magneto-optical investigation of hysteresis loops on two surfaces of
thin CoNi films prepared under the variation of the sputtering angle it
is established that magnetic properties of the films are depth
dependent. Magnetization curves measured on the films' surfaces with
the help of the equatorial Kerr effect depend on the light wave length.
The observed effect is explained by the depth dependence of films'
magnetic properties and by the dependence of magneto-optical effects
formation depth on the light wave length

1. Introduction

Thin CoNi films prepared by oblique incidence deposition on a moving substrate are
considered to have excellent characteristics for a high density recording medium [1]. The
variation of sputtering angle during deposition results in the bending of deposited
microcrystals [2]. Dependence on depth of the films crystal structure must result in depth
dependence of magnetic properties (coercive force, magnetic anisotropy, etc.), which
depend on shape and orientation of deposited microcrystals. In this paper the investigation
of depth distribution of magnetic properties of CoNi films is conducted with the help of the
new method of hysteresis loops measurement on the ferromagnets surface, based on the
employment of meridional intensity magneto-optical effect (MIE) [3]. Besides the
magnetization curves of films near-surface layers are measured with the help of the
equatorial Kerr effect (EKE) for different magnitudes of light wave length (λ) by
modulation method in ac magnetic field [4].

2. Samples and Experimental Methods

The investigated samples are Co-30%Ni films of 30-300 nm thick prepared by oblique
incidence deposition on a poliethylene terephtalate substrate. Two groups of samples are
investigated: in the first group the sputtering angle changed from a large value to a small
one, and in the second - from a small value to a large one. The sputtering angle (α) is the
angle between the molecular beam and perpendicular to the substrate. The initial sputtering
angle in the first group of films was 90^0 , the final angle was 57^0-70^0 . The initial angle in the

second group of films was 55^0 , the final angle was 90^0. The measurement of hysteresis loops with the help of MIE was fulfilled at $\lambda \approx 0.6$ μm. The measuring of magnetization curves with the help of the EKE was fulfilled at different λ in the range of 0.33 μm $\leq \lambda \leq$ 0.83 μm. Magnetic measurements in the volume of the films were performed using a vibrating sample magnetometer. Refraction (n) and absorption (k) indices of light of CoNi films were received by magneto-optical techniques in self-consistent determination of optical and magneto-optical constants of magnetics [5].

3. Depth Dependent Magnetic Properties of Films

The hysteresis loops investigation on both surfaces of films showed that there was uniaxial magnetic anisotropy with easy axis (EA), parallel to the plane of deposition. Fig.1 shows hysteresis loops measured in the field H ‖ EA on the free film surface, on the surface adjacent to the substrate and in the volume of the CoNi film 70 nm thick (film №1) from the first group. One can see from the figure that coercive force (H_c) on the surface adjacent to the substrate is bigger than H_c on the free surface of the film. With increasing film thickness the differences of H_c magnitudes increase on the various surfaces of films.

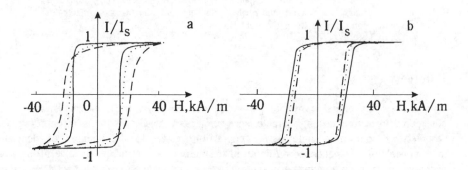

Fig.1. Hysteresis loops in CoNi films measured in the field H ‖ EA: a - in the film 70 nm thick prepared under the variation of the sputtering angle from a large value to a small one; b - in the film 65 nm thick prepared under the variation of the sputtering angle from a small value to a large one; solid curves - for the free film surfaces, dash lines - for surfaces adjacent to the substrate; the hysteresis loops in the volume are presented by dotted lines; I, I_s - magnetization and saturation magnetization of films, correspondingly.

Fig.1b shows hysteresis loops measured on both surfaces and in the volume of film 65 nm thick (film №2) from the second group. One can see that it is opposite for this film: on the free surface H_c is bigger than on the surface adjacent to the substrate. This situation is typical for all the films of the second group. The observed distinction in magnetic properties on different surfaces of the films and difference in these properties in the films from the two groups may be explained by special features of the films microstructure. The angle between the deposited columnar microcrystals and the perpendicular to the film increases, if the angle α increases [2,6]. Therefore the slope of microcrystals on the surface adjacent to the substrate is bigger than on the free surface of the films of the first group. EA is situated in deposition plane for $\alpha \geq 60^0$, and H_c increases if α increases [1]. The mentioned facts explain the distinction H_c on the different surfaces of the films and the qualitative distinction in magnetic properties in the films of the two groups.

4. Formation Depth of Magneto-Optical Effects and Magnetization Curves of Near-Surface Layers of Films

The measuring of magnetization curves of the near-surface layers of CoNi films with the help of EKE showed that these curves depended on λ. Fig.2 shows the magnetization curves measured on film №1 (Fig.2a) and film №2 (Fig.2b) free surfaces at different λ. From Fig.2 one can see that the magnetization curves of film №1 become more sloping and the magnetization curves of film №2 more abrupt when λ increases.

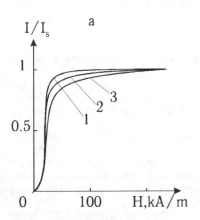 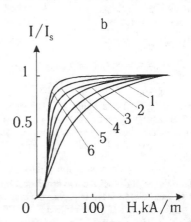

Fig.2. Magnetization curves measured with the help of EKE on the free films surfaces for different λ. a - in the film 70 nm thick; λ (μm): 1 - 0.33, 2 - 0.50, 3 - 0.83. b - in the film 65 nm thick; λ (μm): 1 - 0.33, 2 - 0.35, 3 - 0.41, 4 - 0.50, 5 - 0.62, 6 - 0.83.

The observed effect can be explained if one takes into consideration that the magnetic properties of the investigated films are depth dependent and the thickness of the near-surface layers under investigation in the magneto-optic experiment is determined by the formation depth of magneto-optical signal (l). In accordance with [7], l is determined by that of the two parameters, $L=\lambda/4\pi k$ or $Z_0=\lambda/8n$, the magnitude of which is lower. The measuring of optical constants n and k showed that l in CoNi films determined by L, which depended on λ. The magnitude of L increases from 22 to 31,5 nm when λ changes in the range of 0.33-0.83 μm. One can see that l and thickness of films have the same order of magnitude.

Consequently, the increase of λ leads to the growth of thickness of the near-surface layers under investigation. The deeper the layer on the free surface of film № 1 the bigger is the magnitude of H_c, therefore the saturation field must grow with increasing thickness of the layer under investigation. This conclusion is consistent with the experimental results. In the case of film №2 the situation must be opposite, and this conclusion is also in accordance with the experiment.

5. Conclusion

CoNi films prepared under the variation of the sputtering angle have a depth dependent microcryctal structure, and for this reason their magnetic properties are also depth dependent. The coercive force is always stronger on that surface of film where the deposition angle is bigger.

The formation depth of magneto-optical effects in reflected light depends on λ. Therefore the magnetization curves measured with the help of the magneto-optical Kerr effect in the films with depth dependent magnetic properties also depend on λ. This effect becomes more pronounced when l and the film thickness have the same order of magnitude. The dependence of magneto-optical effects formation depth on λ can be used for non-destructive testing of the magnetic properties of near-surface layers with different thickness.

References

[1] S.L.Zeder, J.F.Silvain, M.E.Re et al, Magnetic and structural properties of Co-Ni thin film prepared by oblique incidence deposition, J.Appl.Phys. **61** (1987) 3804-3806.
[2] K.Nakamura, Y.Ohta, A.Itoh et al, Magnetic properties of thin film prepared by continuous vapor deposition, IEEE Trans. Magn. **18** (1982) 1077-1079.
[3] Y.B.Dong, V.E.Zubov, A.D.Kudakov, A new method of measuring hysteresis loops on local areas of the surface of a ferromagnet, J.Magn.Magn.Mat. **160** (1996) 157-158.
[4] V.E.Zubov, G.S.Krinchik, V.A.Lyskov, Magnetooptical properties of hematite, Sov.Phys.JETP **54** (1981) 789-793.
[5] V.E.Zubov, G.S.Krinchik, A.S.Tablin, and A.A.Kostyurin, Magnetooptical techniques in self-consistent determination of optical and magnetooptical constants of magnetics, Phys.Stat.Sol.(a) **119** (1990) 297-306.
[6] A.Feuerstein, M.Mayr, High vacuum evaporation ferromagnetic materials - a new production technology for magnetic tapes, ,IEEE Trans.Magn. **20** (1984) 51-56.
[7] G.Traeger, L.Wenzel, A.Hubert, Computer experiments on the information depth and the figure of merit in magnetooptics, Phys.Stat.Sol.(a) **131** (1992) 201-227.

Non-Linear Electromagnetic Systems
V. Kose and J. Sievert (Eds.)
IOS Press, 1998

A Knowledge Based Approach for Fast Computation of ECT Signals

Zhenmao CHEN and Kenzo MIYA

Nuclear Engineering Research Laboratory, The University of Tokyo
2-22 Shirakata-shirane, Tokai-mura, Naka-gun, Ibaraki, 319-11 Japan

M. Kurokawa

Takasago Research & Development Center
Mitsubishi Heavy Industries, Ltd.
2-1-1 Shinahama Arai-cho, Takasago, Hyogo, 676, Japan

Abstract. A new way for the fast evaluation of ECT signals was proposed by using data bases and an FEM-BEM hybrid approach. The calculation of flaw signals was localized to the flaw region with the use of a newly developed scheme based on the $A - \phi$ formulae. A great deal of computational time was reduced by using this method compared with the conventional FEM codes. The fast and accurate features of the new approach make the FEM-BEM method a powerful forward solver for the reconstruction of flaw profiles, one of the typical and important problems of ECT inversion.

1. Introduction

In the optimization approach for ECT inversion, a fast forward solver is necessary because the ECT signals need to be calculated accurately at the every step of iteration. Some fast ways were developed based on the analytical solutions of the dyadic Green's function[1],[2]. Unfortunately these methods are only efficient for some special geometries of conductor and exciting coils. Z.Badics et al[3] have proposed a new approach, where all nine components of the dyadic Green's functions were calculated numerically *a priori* by FEM through applying a uniformly distributed unit current in a small cubic cell which is located along the radial direction of a tube. Even this new approach also needs a big element number for a large crack since the cells used for the data bases must be very small. Besides, it is only applicable for a crack in the shape being made up of the cells used for the data base.

   In this paper, the drawbacks of the approach of Ref.[3] were overcome by introducing a scheme which does not require the electric-electric dyadic Green's function to be calculated. A small part of the inverse matrix of the coefficient matrix for an unflawed conductor, which was formed during the FEM-BEM discretization, was memorized as the data base instead. This makes the interpolation of eddy current become a continuous piece-wise linear function, and consequently, makes the fast calculation of large cracks possible. In order to evaluate cracks with an arbitrary shape, a new element containing both conductor and air media was also introduced in unmagnetic material conditions. The results of the present approach agree quite well with the conventional FEM-BEM hybrid method, further the CPU time was reduced significantly.

2. Basic Theory

Through subtracting governing equations about potentials A, Φ of a conductor with flaw present from those about the potentials A^u, Φ^u of the unflawed conductor, the reduced governing equations for the disturbed vector potential and scalar potential can be written as following for an AC problem,

$$\frac{1}{\mu_0}\nabla^2 \mathbf{A}^f - j\omega\sigma_0(\mathbf{A}^f + \nabla\Phi^f) = -j\omega[\sigma_0 - \sigma(\mathbf{r})](\mathbf{A} + \nabla\Phi), \tag{1}$$

$$j\omega\nabla \cdot \sigma_0(\mathbf{A}^f + \nabla\Phi^f) = j\omega\nabla \cdot [\sigma_0 - \sigma(\mathbf{r})](\mathbf{A} + \nabla\Phi), \qquad (in\ conductor) \tag{2}$$

$$\frac{1}{\mu_0}\nabla^2 \mathbf{A}^f = 0, \qquad\qquad\qquad (in\ air) \tag{3}$$

where, $A^f = A - A^u$, $\Phi^f = \Phi - \Phi^u$ are the perturbation of the potentials due to the presence of a crack. Comparing the equations (1)~(3) with the equations of the unflawed conductor, one can find that they are in the same form besides the term with $[\sigma_0 - \sigma(\mathbf{r})]$. This means that the left hand of the equations discretized from the equations (1)~(3) by the FEM-BEM hybrid approach are the same as those of the unflawed situation. Discretizing the equation (1) and (2) with FEM and equation (3) with BEM and putting the term with $[\sigma_0 - \sigma(\mathbf{r})]$ in the right hand, the following linear equations are obtained,

$$[P + j\omega Q + j\omega R + K] \left\{ \begin{array}{c} A^f \\ \Phi^f \end{array} \right\} = [j\omega Q' + j\omega R'] \left\{ \begin{array}{c} A \\ \Phi \end{array} \right\}, \tag{4}$$

where matrix P are the coefficients derived from the first term in left hand of equation (1) and Q the coefficient matrix of the second term. R is the discretizing result of the left hand of equation (2) with FEM and K the coefficient matrix of equation (3) by BEM. Q' and R' are corresponding to the terms with $[\sigma_0 - \sigma(\mathbf{r})]$, which can be derived similarly with the Q and R. But the integration is localized in the flaw region because of the vanished $[\sigma_0 - \sigma(\mathbf{r})]$ in the unflawed area. Based on this consideration, most elements in the matrix Q' and R' are zero. Therefore, the equation (4) can be rewritten as the following submatrix form,

$$\left[\begin{array}{cc} \bar{K}_{11} & \bar{K}_{12} \\ \bar{K}_{21} & \bar{K}_{22} \end{array} \right] \left\{ \begin{array}{c} q_1^f \\ q_2^f \end{array} \right\} = \left[\begin{array}{cc} \tilde{K}_{11} & 0 \\ 0 & 0 \end{array} \right] \left\{ \begin{array}{c} q_1^f + q_1^u \\ q_2^f + q_1^u \end{array} \right\}, \tag{5}$$

where, $\bar{K} = P + j\omega Q + j\omega R + K, \tilde{K} = j\omega Q' + j\omega R'$, $q^f = \{A^f, \Phi^f\}^T, q^u = \{A^u, \Phi^u\}^T$. The potential vector $\{q\}$ was divided into two parts, $\{q_1\}$ is the value at the nodes of crack element, $\{q_2\}$ is the potential at the other nodes.

Multipling H, the inverse matrix of $\bar{K}$, to the equation (5), we find,

$$\left\{ \begin{array}{c} q_1^f \\ q_2^f \end{array} \right\} = \left[\begin{array}{cc} H_{11} & H_{12} \\ H_{21} & H_{22} \end{array} \right] \left[\begin{array}{cc} \tilde{K}_{11} & 0 \\ 0 & 0 \end{array} \right] \left\{ \begin{array}{c} q_1^f + q_1^u \\ q_2^f + q_2^u \end{array} \right\}. \tag{6}$$

The equations related to the unknowns $\{q_1^f\}$ can be separated from the equation above simply. It reads,

$$\{q_1^f\} = [H_{11}][\tilde{K}_{11}][\{q_1^f\} + \{q_1^u\}] \tag{7}$$

Rearrange this equation and let $[G] = [H_{11}][\tilde{K}_{11}]$, the relation between the field of the unflawed conductor and the perturbation due to a crack is finally obtained as follows,

$$[I - G]\{q_1^f\} = [G]\{q_1^u\}. \tag{8}$$

From equation (8), the A^f, Φ^f, and further, the electric field in the flaw region can be calculated from the unflawed field directly. As these equations are only related to the nodes of a crack, only a little computational time is necessary. The information about the conductor and exciting coils was contained in the coefficients $[H_{11}]$ and $\{q^u\}$, which can be calculated a priori since they are independent of the crack.

For the inverse problem recognizing a crack profile, the position and depth of the crack usually can be predicted roughly by some classification method. Therefore, the area where a crack possibly exists, can be decided before the inversion. The unflawed field and the inverse matrix H_{11} over this area can be calculated previously through discretizing this region into small cells. Every column of the inverse matrix can be determined by solving the governing equations of the unflawed conductor through applying a unit potential to the corresponding node and memorizing the induced field at the possible crack region. Considering the fact that the tube is of infinite length compared with the dimensions of the possible crack for SG tubing problem, these data bases are small because shifting technique is applicable.

The following equation based on the reciprocity theorem is adopted for impedance calculation [2],

$$\triangle Z = \frac{1}{I^2} \int_{coil} \mathbf{E}^f \cdot \mathbf{J_0} dv = \frac{1}{I^2} \int_{flaw} \mathbf{E}^u \cdot (\mathbf{E}^f + \mathbf{E}^u)\sigma_0 dv, \tag{9}$$

where $\sigma_0(\mathbf{E}^f + \mathbf{E}^u)$ is corresponding to the current dipoles in the flaw region. Calculating impedance with equation (9) and FEM can save most of the CPU time compared with using the Biot-Svart's law, because the integration can be realized simply by using the element matrix.

3. Consideration of cracks with arbitrary shapes

As the data bases have been established based on given cells a priori, only the cracks consisting of these cells can be considered with these data bases if only one type of conductivity is allowed in one element. However, since the permeability and dielectric constant of the conducting material are just the same as the free space for SG tubing, the electromagnetic fields are continuous and the potentials have one order differentialability at the crack boundary. Together with the fact that the cells discretized for the data base are usually very small, it is reasonable to apply linear interpolation in the whole cell even it has different media. Introducing such an element will enable the calculation of cracks with arbitrary shapes.In actual discretization, the shape function of the new element only needs to be chosen as same as the normal FEM element. The difference of the material is taken into account in the integration over an element during the calculation of the element coefficient matrix.

4. Numerical results

The numerical results of this new approach were calculated and compared with the conventional FEM-BEM hybrid code. A zone of 10mm × 1.27mm × 0.2mm in a tube was considered as the possible region of cracks and was divided into 21 × 8 cells. A self induction pancake coil and the tube geometry are considered in the calculation. Reasonable agreements were obtained between the two methods while the calculation of the conventional method needs several hundred or several thousand times more computational time than the present one. Consequently, this verified the validity of using this method in the inverse problem.

Fig.1 shows the axial scan signals calculated by the present method and by the conventional FEM-BEM code. Where 50% inner crack with 0.2 mm width and 6 mm length was considered, for 41 points of the axial scan, the new approach only needs 10 seconds of CPU time while the FEM-BEM code needs several hours to get results with the same accuracy.

Fig.2 shows the results of a crack which cannot be made up of the given cells shown in the figure, where the points are the results of the FEM-BEM code using meshes with 4 and 8 layers respectively, and the lines are the results of the present method. It is obvious that the FEM-BEM results of the 8 layer mesh agrees better with the present method though both results have good accuracy. This shows us that the impedance due to any crack can be evaluated accurately in a very short CPU time by using the present scheme and new element. The impedance calculations of an elliptic crack and many other cases have also been performed. All the results agreed to a large extent with those of the conventional FEM-BEM code.

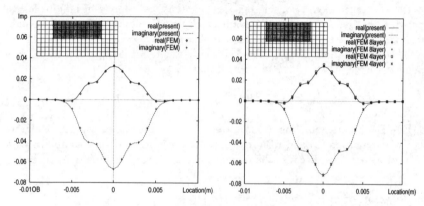

Fig.1 Results of crack ID50%   Fig.2 Results due to an ID56% crack

5. Conclusions

A knowledge based approach for the fast evaluation of impedance signals was proposed and validated in this work. Because a one order interpolation function was adopted in the new method, not only bigger elements can be chosen for data base calculation, but new elements with different media can also be introduced. This enables the impedances due to an arbitrary crack to be evaluated accurately in a short time. Therefore, it is advantageous to apply this new approach to the ECT inversion.

References

[1] J.R.Bowler, Eddy current interaction with an idea crack, I. the forward problem, J. Appl. Phys. 75, (1994) 8128-8137.

[2] H.A.Sabbagh and L.D.Sabbagh, An eddy current model for three-dimensional inversion, IEEE Trans, MAG-22, (1986) 282-291.

[3] Z.Badics et.al, Rapid flaw reconstruction scheme for 3-d inverse problems in eddy current NDE, In: Proceedings of E'NDE Tokyo, IOS press, 1996 (in press).

Non-Linear Electromagnetic Systems
V. Kose and J. Sievert (Eds.)
IOS Press, 1998

A Second Order Approximation for Conductivity Imaging

G. RUBINACCI, A. TAMBURRINO, S. VENTRE, F. VILLONE

Dipartimento di Ingegneria Industriale, Università di Cassino, I-03043 Cassino, Italy

Abstract. A non-conventional conductivity retrieval method, based on a second order approximation, is used in this paper. A deep understanding both of the characteristics of this method and of the intrinsic limits of the inverse problem here considered allows us to extract as much information as possible about the unknown conductivity profile from the external field measurements.

1. Introduction

Conductivity imaging consists of reconstructing an unknown conductivity profile using a number of noisy measurements of the scattered field due to a known excitation field; it exploits the capability of eddy currents to penetrate physical objects. This reconstruction gives rise to a non-linear inverse and ill-posed problem. The non-destructive evaluation of conductivity has a wide range of application in industry [1], medicine [2], geophysics [3] and many other fields.

The linearization of the mapping between the conductivity and measurements (the widely used Born approximation) sets the problem into the classic frame of the theory of the linear ill-posed problems. Unfortunately, the range of validity of this approximation is limited to weakly scattering bodies, whose dimensions are small as compared to the skin depth and whose spatial variations are smooth enough. A quadratic approximation has been recently proposed for phase retrieval and permittivity identification problems [4],[5], and analysed also in the conductivity imaging field in order to enlarge the set of retrievable profiles [6],[7].

Up to now, a study of how much information about the unknown conductivity is available from a given set of measurements and which part of it is retrievable with a given method of solution, is still lacking. In this paper, we aim to give some indications in this sense, in order to be able to understand how to set up both the experimental configuration (i.e. number and position of the measurement points) and the numerical methods.

This paper is organised as follows. In section 2, we recall the basic aspects of this quadratic approach, with reference to a three-dimensional numerical integral formulation of the eddy current problem. Then, in section 3, we will discuss the main features of the quadratic approach describing its advantages and drawbacks as compared to the linear one and present a numerical example of application of the method.

2. Formulation of the problem

The model is described by a magnetoquasi-static integral formulation in terms of the 2-components vector potential **T** [8]. After discretization, using the Galerkin method, the following linear system of algebraic equations is obtained:

$$(j\omega\{L\}+\{R\})[I]=[V] \tag{1}$$

where [I] is the vector of the unknown currents, [V] is the vector of the excitations and {R} is a sparse matrix that depends linearly on the electric resistivity $\eta(r)$. The induced flux density $B_k = B_s(r_k)$ at the probe locations r_k can be easily computed in terms of the vector [I] as [B]={Q}[I] where Q is a suitable matrix arising from the Biot-Savart law.

The inverse problem here considered consists of finding the unknown resistivity profile $\eta(r)$. The excitation is prescribed; B_s is the scattered magnetic field resulting from a given excitation at a set of measurement points in the space-frequency domain. Hence, the problem is described by the scattering operator $S(\eta)$ which maps the excitation into the scattered magnetic field, and depends on the unknown resistivity η in a non-linear way:

$$[B]=\{Q\}(j\omega\{L\}+\{R\})^{-1}[V]=S(\eta)[V] \tag{2}$$

Therefore, the solution is defined as the minimum with respect to $\eta(r)$ of the error functional $\Phi(\eta)=\sum_{k=1}^{M}\left\|S_k(\eta)[V]-\tilde{B}_k\right\|^2$, where $\left\{\tilde{B}_1,...,\tilde{B}_M\right\}$ is the set of M noisy measurements of the induced field at the given points $(r_1,\omega_1),...,(r_M,\omega_M)$ respectively, and $S_k(\eta)[V]$ is the value of $S(\eta)[V]$ at the point (r_k,ω_k).

Let us assume that the resistivity profile η can be supposed as consisting of a small perturbation $\Delta\eta$ around a known background value η_0, like $\eta = \eta_0 + \Delta\eta$; then, we expand the non-linear operator $S(\eta)$ around η_0 [3], obtaining:

$$[B]=[B_0]+\{Q\}([I_1]+[I_2]+...+[I_n]+...) \tag{3}$$

where, having called $\{R_0\}$ and $\{\Delta R\}$ the resistivity matrices related to η_0 and $\Delta\eta$ respectively, it results: $[B_0]=\{Q\}[I_0]$, $[I_0]=\{Y_0\}[V]$, $[I_n]=-\{Y_0\}\{\Delta R\}[I_{n-1}]$, $\{Y_0\}=(\{R_0\}+j\omega\{L\})^{-1}$.

The usual linear (Born) approximation consists of stopping the expansion at the first order. Conversely, the quadratic approach basically retains the first two terms of the expansion; it has been demonstrated [7] that this can give substantial advantages, as will be discussed afterwards.

3. Results and discussion

Now, we discuss the main features of the quadratic approach, as compared to the Born approximation. First of all, it must be noted that the quadratic approximation has a wider range of validity than the linear one. In particular, the quadratic method here proposed is by far superior to the Born approach in the situations where the effects of the mutual interactions between different parts of the conducting body are dominant. As a matter of fact, the linear approximation does not take into account these effects at all. This happens, for instance, when the resistivity perturbation is rapidly spatially varying; in these cases, in the linear approximation the equivalent current gives rise to a negligible scattered field, so that we cannot distinguish the actual resistivity profile from the unperturbed one, giving rise to the so-called "invisible profiles" problem. The ability to distinguish rapidly varying profiles allows to enhance the spatial resolution of the reconstruction.

Secondly, the main drawback of the quadratic approach is the fact that a non-linear inverse problem must be solved, as compared to a linear one in the Born approach. This gives rise to additional difficulties from both the numerical and the theoretical point of view. The most serious problem that must be tackled is the occurrence of local minima, since, due to the non-linearity of the operators involved, their presence in the error functional cannot be excluded *a priori*. Fortunately, it is possible to control the occurrence of these local minima by acting on the ratio between the number of equations (i.e., of measurements) and the number of unknowns (i.e., of parameters describing the conductivity profile) [9].

Now we will describe all the steps that must be tackled when dealing with a conductivity retrieval problem. For the sake of simplicity, we will refer to the following configuration. The object to be reconstructed is a $L \times L$ square conducting plate (L=10 cm), with a thickness D=3 mm. The incident field is produced by a circular filamentary coil with radius r=5 cm, placed at 2 mm above the surface, carrying a sinusoidal current with normalised amplitude and frequency f=1 kHz. The flux density samples are taken on a regular grid placed 1 mm above the plate. The resistivity is expanded as

$$\Delta\eta(x, y, z) = \sum_{i=1}^{N_x} \sum_{j=1}^{N_y} \Delta\eta_{ij} P(x - x_i, y - y_i, z), \text{ where } x_i=(i\text{-}1/2)L/N_x, y_j=(j\text{-}1/2)L/N_y, \text{ and P}(\cdot) \text{ is}$$

a rectangular pulse function (Fig. 1a). In general, the basis functions should be chosen in order to fit the expected resistivity profiles, on the basis of a rough knowledge of the geometry of the configuration.

The first problem to solve is to find an estimate of the number N_d of degrees of freedom of the problem, i.e. the number of independent measurements of the scattered field. For this purpose, we compute the singular value decomposition (SVD) of the operator arising from the linear approximation. The number of singular values exceeding a given threshold will provide the required estimate. In principle, this analysis should be carried out on the continuous operator; in this application, we confine ourselves to studying the discrete operator resulting from a (18x18x1) finite elements subdivision of the conductive plate (Fig. 1b), and from assuming a (9x9) resistivity grid. The resulting number of independent measurements is around N_d=60, if a 10% threshold is chosen. We have checked that this result does not depend on the number or the position of the measurement points, since moving the measurement points from a (15x15) grid to a (31x31) grid does not result in any significant change of the singular values pattern. Once N_d is known, we have to choose the number $N = N_x \times N_y$ of unknowns, i.e. of independent coefficients $\Delta\eta_{ij}$ describing the resistivity profile. It must result $N \leq N_d/2$, in order to avoid the occurrence of local minima, as discussed in [9]. The electromagnetic mesh is chosen in order that the current patterns resulting from the class of conductivity profiles previously chosen is correctly represented. In our test case, we subdivide each resistivity grid element into 3 sub-elements, so that the electromagnetic grid will be made again of 18x18x1 hexahedral elements. Figure 3a shows the actual resistivity profile. In Fig. 2 the real and the imaginary part of the induced flux density z component at the measurement points is shown. While the linear approximation error is around 62%, and 47% for the real and the imaginary part respectively, the quadratic approach provides an error of 36% and 22%. In Figures 3b and 3c the reconstructed

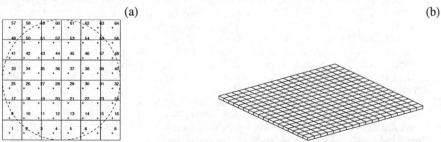

(a)   (b)

Figure 1. (a) The resistivity grid together with the excitation coil (dashed line) and the field probes (+).
(b) The finite element mesh.

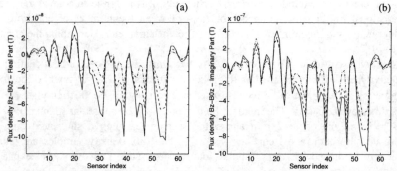

Figure 2. Real (a) and imaginary (b) part of the measured magnetic flux density B_z-B_{0z} (solid line) together with its linear (dash-dot) and quadratic (dashed) approximations.

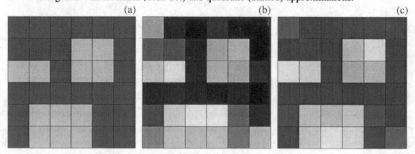

Figure 3. Grey scale representations of the actual resistivity profile (a); dark grey stays for $2 \cdot 10^{-6}$ Ω m and light grey for $6 \cdot 10^{-6}$ Ω m. The reconstructions obtained by linear (b) and quadratic (c) approaches.

resistivity profiles are shown. The square mean error for the reconstructed profiles is 98% (linear approach) and 26% (quadratic approach).

Acknowledgements

This work was supported in part by MURST of Italy and by the INCO-COPERNICUS project PL964037 of the European Commission.

References

[1] R. Zorgati, B. Duchêne, D. Lesselier, F. Pons, Eddy Current Testing of Anomalies in Conductive Materials, Part I: Qualitative Imaging via Diffraction Tomography Techniques, *IEEE Trans. Mag.*, **27** (1991) 4416-4437.

[2] A. K. Louis, Medical Imaging: State of the Art and Future Development, *Inverse Problems,* **8** (1992) 709.

[3] K. A. Dines, R. J. Lytle, Analysis of Electrical Conductivity Imaging, *Geophysics,* **46** (1981) 1025.

[4] T. Isernia, G. Leone, R. Pierri, Phase Retrieval of Radiated Fields, *Inverse Problems,* **11** (1995) 183-203.

[5] A. Brancaccio, V. Pascazio, R. Pierri, A quadratic model for inverse profiling: the one-dimensional case, *Journal of Electromagnetic Waves and Applications*, **9** (1995) 673-696.

[6] R. Pierri, G. Rubinacci, A. Tamburrino, A numerical algorithm for the reconstruction of 3D conductiong profiles using eddy currents testing, *Nondestructive Testing of Materials*, R. Collins et al. (Eds.) 235-244, IOS Press, 1995.

[7] R. Pierri, G. Rubinacci, A. Tamburrino, A quadratic approach for the reconstruction of conductivity profiles using eddy currents, *IEEE Trans. Mag.*, **32**, (1996) 1310-1313.

[8] R. Albanese and G. Rubinacci, Integral Formulation for 3D Eddy Current Computation Using Edge Elements, *IEE Proc.*, **135A**, (1988) 457-462.

[9] R. Pierri and A. Tamburrino, On the local minima problem in conductivity imaging via a quadratic approach, *Inverse Problems*, **13** (1997) 1547-1568.

Microelectromechanics, Sensors and Actuators, Intelligent Materials

Non-Linear Electromagnetic Systems
V. Kose and J. Sievert (Eds.)
IOS Press, 1998

Microsystem Technology:
A Survey of the State of the Art

Stephanus BÜTTGENBACH

Institute for Microtechnology, Technical University of Braunschweig
Langer Kamp 8, D-38106 Braunschweig, Germany

Abstract. Microsystem technology opens up a new way of integration of sensors, actuators, and information processing components resulting in multifunctional, adaptive, and intelligent systems similar to those created by nature. In this paper a general idea of what is feasible with this technology will be given. After a short introduction the key technologies for the fabrication of microsystems are described and illustrated by several examples of current applications.

1. Introduction

The great advances of microelectronics technology during the last three decades have led to a large demand for sensors and actuators which are compatible to integrated circuits with respect to driving voltages, power consumption, size, weight, complexity, and price, and which can be integrated with microelectronic functions into one device. Compatibility of transducers and microelectronic devices is achieved by the new concept "micromachining", which stands for the application of the highly developed batch processing techniques of integrated circuits to the fabrication of sensors and actuators with very small dimensions [1]. The development of complete miniaturized systems is expected to be the next step of integration after the integrated circuit.

The enormous potential of microsystems results from particular features of this new technology. The utilization of materials and sophisticated processes already developed in semiconductor technology leads to low production costs. In addition, new materials, such as biological receptors used in biosensors or shape memory alloys and ferroelectric compounds applied in microactuators, extend the range of application. The extremely small size and weight of microsystems as well as their low power consumption are essential for stand-alone, portable or implantable systems, and due to the reduced number of external connections and to new concepts of control architectures and signal processing a large increase in reliability can be expected.

2. Key technologies for the fabrication of microsystems

Microelectronics technology usually employs planar processes only, whereas in micromachining three-dimensional and movable elements with mechanical functions have to be

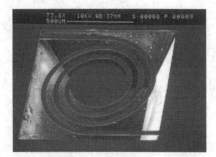 

Fig. 1. SEM micrographs of silicon micromachined structures

fabricated. For this purpose special processes and materials have to be added to the technological basis borrowed from microelectronics.

2.1 Bulk micromachining

Bulk micromachining utilizes chemical anisotropic etching as well as deep dry etching
to fabricate three-dimensional microstructures mainly of silicon. Single crystal silicon
plays a dominant role in micromachining because of its excellent mechanical properties
and compatibility with microelectronics technology. However, for special applications also
other materials such as single crystal quartz, gallium-arsenide or photosensitive glass are
used.

Basic structures that can be fabricated by anisotropic etching of silicon are pits, grooves, membranes, cantilevers and bridge-like structures [2]. These structures are the starting
points for the fabrication of miniaturized sensors and actuators. A membrane, for example,
can be used as the movable part of a valve, cantilevers may act as spring-mass systems in
accelerometers (Fig. 1).

In dry etching processes, which are widely applied in microelectronics technology, a
plasma instead of a liquid is used as the source of chemical reagents. The main advantage
when used in micromachining is the fact, that the shape of the resulting microstructures is
not dictated by the crystallographic orientation of the substrate as in chemical etching. For
deep etching of silicon a high-rate cryogenic reactive ion etching process using sulfur he-
xafluoride and a Ni/Al mask shows an almost perfect anisotropy. Etch depths of 200 μm
have been obtained [3].

2.2 Surface micromachining

In surface micromachining the mechanical devices are machined in thin layers that have been deposited on the surface of the substrate. This technology is based on the sacrificial layer method which makes use of the selectivity of isotropic etchants to different materials. One starts from sandwich layers made, for example, of silicon dioxide and polycrystalline silicon, and deposited on standard silicon substrates. The poly-silicon is used as
the mechanical material and is structured by lithography and etching. The sacrificial silicon dioxide layer is etched away completely with a high selective hydrogen fluoride etch-
ant leaving free standing poly-silicon structures or releasing movable parts. In order to
overcome problems due to sticking effects and to constraints in structure height some re-

cent developments [4] use plasma release techniques instead of wet chemical etching for the fabrication of movable elements.

2.3 High aspect ratio technologies

Micromechanical structures of metals and plastics with high aspect ratios can be fabricated by deep X-ray lithography. In the LIGA process [5] polymethyl-methacrylate type resist up to 600 μm thickness is irradiated by synchrotron radiation using special masks. The developed resist structure is filled up by metal deposited by electroforming. If parts of the microstructure shall be movable they have to be applied on a sacrificial layer which is etched selectively after electroforming. The metallic microstructures can also be used as a mould insert for the fabrication of plastic replicas.

A low-cost alternative to the LIGA process uses photosensitive polyimide as a resist material, ordinary masks and ultraviolet light exposure [6]. Although the resolution of this process is inferior to the LIGA process it has the advantage that it is simple and can be carried out using commercially available equipment.

2.4 Laser micromachining and microcutting

The technologies based on integrated circuit processing techniques are complemented by techniques borrowed from precision engineering such as laser micromachining, microcutting, and micro electro discharge machining. These methods extend the variety of available three-dimensional microstructures. In Fig. 2 vertical-walled shafts fabricated by laser micromachining combined with anisotropic etching [7] and silicon pins (15 μm x 15 μm x 400 μm) produced by microcutting demonstrate the capability of these techniques.

3. Examples of current microsystems

Today, the integration of micromachined sensors and actuators and microelectronic functions as single-chip or hybrid microsystems has been demonstrated at the laboratory as well as at the industrial level. Some examples illustrate the state of the art.

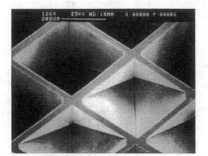 

Fig. 2. SEM micrographs of silicon structures fabricated by laser micromachining and microcutting

3.1 Automotive applications

Due to the large number of units needed automotive applications are one of the biggest market segments of microsystem technology. The available space in cars decreases, whereas the number of additional functions is growing rapidly because of more stringent safety, environmental and economic demands. These reverse developments require the use of micromachined components. Therefore, the first silicon chips, in which microelectronic and micromachined functions have been integrated monolithically, have been developed for automotive applications. Examples are monolithic accelerometers for airbag release manufactured by bulk as well as by surface micromachining, micromechanical gyroscopes, which are needed in active chassis development and for inertial navigation systems, and mass-flow sensors for fuel mixture management.

3.2 Resonant microsensors

Rigorous safety demands in automotive and other applications require the development of microsensors with an integrated self-test function. Resonant microsensors, which change their output frequency as a function of the quantity to be measured, meet this requirement. They exhibit a wide field of applications and further benefits such as high sensitivity, high resolution, and semi-digital output [8].

As an example, Fig. 3 shows the scheme of a resonant pressure sensor mounted on a glass plate coated with a patterned conductive ITO layer, which is glued to an Al carrier. The sensor is based on a novel design of a quartz membrane realized with an AT crystal cut. The resonator consists of a full-thickness bossed membrane of about 4 mm diameter, which is monolithically attached to the bulk frame. Applied pressure induces a deformation of the membrane. Extensive finite-element modelling has been carried out to determine the electrode configuration and the shape of the structure for exciting a low-frequency bending mode in the 30-50 kHz range. A very sensitive and stable frequency shift response of about 20 Hz/kPa has been measured [9].

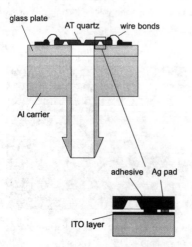

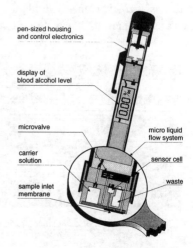

Fig. 3. Scheme of a resonant quartz pressure sensor

Fig. 4. Scheme of a pen-sized alcohol meter

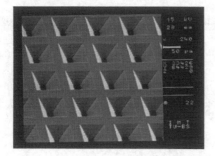 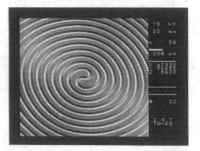

Fig. 5. SEM micrographs of a silicon porous diaphragm and a flow channel system

3.3 Microflow devices and systems

An attractive field of increasing interest is the use of microsystems for chemical analysis and for accurate delivery of small amounts of liquids or gases. There are possible applications in a broad range of the market, for example in industrial process control, biotechnology, environmental control, and medical applications. Fig. 4 shows the scheme of an alcohol meter presently under development [10]. The portable instrument, whose operation is based on the principle of flow analysis and that measures the alcohol transmitted through the skin, consists of two parts, a disposable measuring head containing a micro liquid flow system and a pen-sized housing containing the control electronics. Components of microflow systems are, for example, miniaturized valves, flow channel systems and silicon porous diaphragms (Fig. 5).

3.4 Applications in precision machining

Precision machining is one of the areas in which the application of microsystem technology is of particular interest. Due to the increasing competition in the production industry monitoring and adaptive readjustment of CNC-machine tools are of increasing importance for cost reduction and quality management. Fig. 6 shows the layout of a microsystem that is part of a tool monitoring system and that is integrated near to the cutting process [11]. It contains several strain gauge sensors to measure the metal-cutting forces, a vibration sensor to analyse the spectrum of the tool oscillations, a temperature sensor to correct for the thermal influence on the sensors, and integrated circuits for signal preprocessing. The system is completed by a transponder board and a coupling coil for wireless transmission of data and power.

3.5 Light modulators and deflectors

Applications for optical microactuators include displays, printing and optical scanning. Different technologies and actuation principles have been used to fabricate light modulators and deflectors. A device that is produced in series for commercial application is the digital micromirror device (DMD) with 864 x 576 highly reflective mechanical mirrors with an area of 16 x 16 μm^2 [12]. Each mirror is suspended by two torsion hinges and built over a pair of address electrodes that are connected to an underlying SRAM cell. The

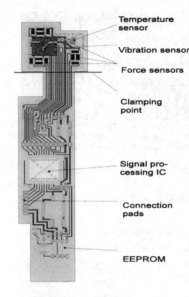

Temperature sensor

Vibration sensor

Force sensors

Clamping point

Signal processing IC

Connection pads

EEPROM

Fig. 6. Layout of the tool monitoring microsystem

mirrors can be tilted electrostatically about their torsional axis.

The DMD is used in conjunction with darkfield optics to provide a highly efficient projection display. It is suitable for all applications that require the modulation or directional switching of light, and it provides a promising alternative to LCD technology in projection television applications.

4. Conclusion

The state of the art of microsystem technology can be summarized as follows. Microsystem technology provides a broad technology basis to fabricate miniaturized multifunctional systems in which sensors and actuators are combined with microelectronic devices. Integrated sensors for the measurement of pressure, acceleration and other variables are already in volume production, and significant progress has been made to develop microactuators such as microvalves, pumps and deflectable mirrors using micromachining techniques. There is no doubt that microsystem technology will trigger a new cycle of innovation and that this will bear the key to future technological progress.

References

[1] S. Büttgenbach, Mikromechanik. Teubner, Stuttgart, 2. ed., 1994

[2] K.E. Petersen, Silicon as a Mechanical Material, Proc. IEEE **70** (1982) 420-457

[3] M. Esashi et al., High-rate directional deep dry etching for bulk silicon micromachining, J. Micromech. Microeng. **5** (1995) 5-10

[4] M. de Boer et al., The black silicon method V: a study of the fabrication of movable structures for micro electromechanical systems, Proc. 8th Int. Conf. on Solid-State Sensors and Actuators, Stockholm, 1995, Vol. I, pp. 565-568

[5] E.-W. Becker et al., Fabrication of microstructures with high aspect ratio and great structural heights by synchrotron radiation lithography, galvanoforming and plastic moulding (LIGA-process), Microelectronic Engineering **4** (1986) 35-56

[6] A.B. Frazier and M.G. Allen, High aspect ratio electroplated microstructures using a photosensitive polyimide process, Proc. Micro Electro Mechanical Systems, Travemünde, 1992, pp. 87-92

[7] M. Alavi et al., Fabrication of microchannels by laser machining and anisotropic etching of silicon, Sensors and Actuators A **32** (1992) 299-302

[8] S. Büttgenbach et al., Resonant force and pressure microsensors, Proc. SENSOR 95, Nürnberg, 1995, pp. 27-32

[9] H.-J. Wagner et al., Design and fabrication of resonating AT-quartz diaphragms as pressure transducers, Sensors and Actuators A **41-42** (1994) 389-393

[10] S. Büttgenbach et al., Pen-sized alcohol meter, mst news 19 (1997) p. 17

[11] K. Feldmann et al., Tool Monitoring of an automated lathe with microsystem technology, Proc. Microsystem Technologies 96, Potsdam, 1996, pp. 163-167

[12] J.M. Younse, Mirrors on a chip, IEEE Spectrum **30** (1993) No. 11, pp. 27-31

Non-Linear Electromagnetic Systems
V. Kose and J. Sievert (Eds.)
IOS Press, 1998

Compact High-Tc Superconducting Actuator System Design

Tadashi Kitamura

Kyushu Institute of Technology, Kawatsu, Iizuka-city, Fukuoka 820, Japan

Abstract: This paper describes the system design of centimeter-sized YBCO superconducting linear and rotary actuators and their appropriate bearings that have been developed since 1989. An MPMG-YBCO superconductor tile of the linear actuator driven by two solenoids is guided and levitated over two magnetic bearings. Its position, measured by a laser sensor, is satisfactorily well controlled by linear feedback with biased currents applied to the solenoids. A disk-shaped rotor of the same superconductor for the rotary actuator with two poles magnetized by field cooling was stably driven by four solenoids arranged evenly around the rotor up to about 400rpm, levitating over a disk-shaped permanent magnet. The rotation angle is detected by the laser sensor and a lab made absolute encoder. The control system of both types of actuator were designed by computer simulations with a simple repulsion force model, developed by the author. Damped oscillations can occur dependent on feedback gains for the rotary actuator, but this drawback will be improved by generating uniform torque based on current control depending on the position of the actuator. The hysteretic torque dependent on the angle for this actuator should be reduced in future work.

1. Introduction

Our project of high-Tc superconducting (HTS) mechatronics design was started with hopes of biomedical applications of HTS actuators and bearings available at room temperature, such as an artificial heart actuator and artificial joint, [1] right after the HTS fever in 1987 and 88. Since then, although stable superconductivity is not yet realized at room temperature, YBCO stable bulk superconductors processed by melt-power-melt-growth (MPMG) have been manufactured, and showed to generate strong magnetic forces [2]. Industrial applications, such as bearings for fly wheel, magnets, and actuators, have been investigated [3-4]. We have developed linear actuators, rotary actuators, and their appropriate bearings [5-6], and also built a simple model of repulsion force for control system design by computer simulation [7]. This paper reports the system design of centimeter-sized YBCO superconducting actuators and discusses experimental results of their control performance in comparison with computer simulations.

2. Linear Actuators

We have developed three versions of high-Tc superconducting linear actuators, all of which have the same electromechanical structure such that a moving part of the superconductor is driven back and forth by two solenoids placed on both sides of the superconductor, and are guided and levitated over two magnetic bearings. For the first version, YBCO superconductor tiles, which were made by a sintering method in our lab, were used. The next version used an MPMG-YBCO supercoductor tile made by the Japan Steel Company, and the latest version used ones supplied by the Superconductivity Research Laboratory with simpler assembly and a more sophisticated drive system than the second one. The bearings for both second and latest versions of linear actuaotrs are made of rare earth metal permanent magnet bars arranged in the alternating polarity pattern. It was shown by our previous work [5] that the alternating polarity pattern of permanent magnets, as illustrated in Fig.1, guarantees stable levitation of a superconductor tile, and that there is the optimal width of bar magnets providing the maximal repulsion force for a given gap between the superconductor and magnets.

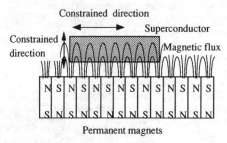

Fig.1   Magnetic Bearing with Alternating Polarity Pattern of Permanent Magnets: Motion of the superconductor is restricted in the direction of the alternating polarity pattern, but free in the direction vertical to this page if the magnets length is infinite, i.e., the magnetic flux is uniform in this direction.

The schematic diagram of the latest version is shown in Fig.2. This actuator is under undesirable constraints in the moving direction by a position dependent spring-like force due to flux gradients of the bearing magnets at both ends. Therefore, the actuator is modeled by a second order linear differential equation of mass, damper and spring that are identified by

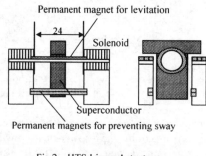

Fig.2   HTS Linear Actuator

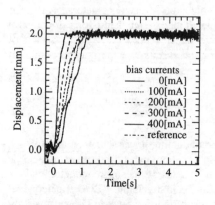

Fig. 3   Position Control of Linear Actuator

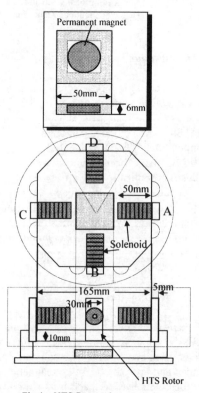

Fig.4   HTS Rotary Actuator

a free response of the actuator. The position of the superconductor of the latest version of actuators, measured by a laser position sensor, is controlled with the accuracy of 0.04mm with feedback gains determined by a standard pole assignment technique, where a linear model of the repulsion force with the current is used. The step response is improved with the biased currents applied to the solenoids because a stiffer spring can be produced than with no biased current. Fig.3 shows experimental results of the position control with four biased currents applied to both solenoids. More details of control design with the spring constant adjusted in the control law are reported by the author elsewhere in this Proceedings.

3. Rotary Actuator

The schematic diagram of the rotary actuator and a photograph of the real actuator system are shown in Fig.4 and Fig.5, respectively. Its cylinder-shaped rotor consists of two MPMG-YBCO superconductor disks and a plastic disk in-between; the upper one for rotation and the lower one for levitation over a disk-shaped permanent magnet with which the lower superconductor is field-cooled. The upper superconductor is magnetized with the poles of N and S on the two opposite sides of the superconductor with field cooling. The four solenoids, evenly arranged around the rotor at every 90°, stably drive the rotor up to about 370rpm by alternately switching the polarities of each solenoid in the fashion that each pole of the rotor receives repulsion by the one solenoid and attraction by the next one at the same time.

Fig.5   Photograph of Rotary Actuator System

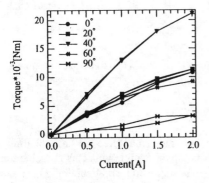

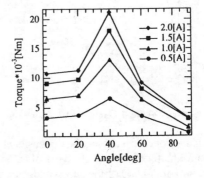

Fig.6-a   Relation of Torque to Current   Fig.6 -b   Relation of Torque to Angle

The rotation angle is detected by the laser sensor and a lab made conic spiral-shaped absolute encoder attached on the upper superconductor, where the angle corresponds to the
height of the continuous spiral slope. Improvement of the timing of switching the solenoids
may make it possible to rotate over 400rpm.

The static characteristics of the torque versus the driving current and angle was
measured by a load cell for control design. Fig.6-a and -b show the relationship of the
torque to the current with the angle fixed and to the angle with the fixed current,
respectively. In these experiments, the angle was measured from the standard position
where the two poles of the upper superconducting magnet for rotation and a solenoid are
arranged in a straight line, and therefore, the torque is maximized at about $45°$ as shown in
Fig6-b. Ignoring the fact that as Fig.6-b shows the torque is not linear with the angle, it is
modeled as a linear function of the current for the control algorithm of the angle, where the
proportional coefficient K is identified with the gradient of the tangent at the origin for the
angle of about $45°$ shown in Fig.6-a. Then, the equation of the rotor is given by,

$$Jd^2\theta/dt^2 + cd\theta/dt = K\,i \qquad\qquad (1)$$

where θ, i, J, c, and K are the angle, current, moment of inertia, friction coefficient and
torque constant, respectively. Then, the control i is given by $i=k_1(\theta_r-\theta)+k_2d\theta/dt$ for the reference θ_r, where k_1 and k_2 are computed by a standard pole assignment technique, with the
initial conditions of $\theta(0)=0$, and $d\theta/dt(0)=0$. Fig.7 shows the experimental results of control performance with an assigned double pole of s=-8 for the reference angles of $20°$, $40°$,
and $60°$. Fig.8 shows the results for the four sorts of double poles assigned at -7, -8, -9, and
-11, where the angle settles down to the reference with no oscillation for pole s=-7, but the
other three responses are oscillatory. These oscillations should occur mainly due to the remarkable dependency of the output torque on the angle which is shown in Fig.5, and maybe
partially due to hysterisis of the torque, which was observed to be small. For smaller poles
than s=-11, oscillations in responses were not damped, i.e., the control did not work at all.
From the above, uniform torque independent of the angle should improve the control performance, and such torque will be obtained by controlling the solenoids' currents in future
work.

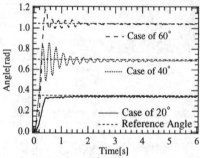
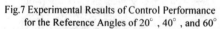
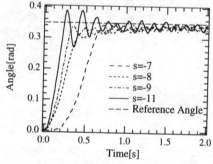

Fig.7 Experimental Results of Control Performance
for the Reference Angles of $20°$, $40°$, and $60°$

Fig. 8 Experimental Results of Control Performance
for Four Different Assigned Double Poles.

4. Control System Design and Analysis

We developed a simple one-dimensional model of repulsion force for investigating the
control performance of both types of actuator by computer simulation. This model assumes
that in a thin plate of the superconductor there exist superconducting shielding currents of

the only ϕ component of the cylindrical coordinate system, based on the mixed state model generating the repulsion force by the Bean model. The repulsion force is computed by the Lorentz force, i.e., the vector product of the superconducting shielding current and the external magnetic field that is assumed to be the one produced by a minute magnetic dipole. Then, the repulsion force in the driving direction F_z to a thin, minute disk-shaped superconductor is given by,

$$F_z = \int_V J_\phi B_{Er} dV \tag{2}$$

$$B_{Er} = \frac{3\mu_0 MV}{4\pi} \frac{(z_0)r}{\left\{(z_0)^2 + r^2\right\}^{\frac{5}{2}}} \tag{3}$$

where B_{Er} is the external magnetic field, J_ϕ the ϕ component of the superconducting shielding current i, $M = \alpha i$ the intensity of magnetization of the solenoid, V the solenoid's volume, z_0 the distance between the superconductor and solenoid, and μ_0 the permeability of vacuum. The radius of the disk is determined first so that the shape of static hysteretic loop data of the repulsion force versus the solenoid current agrees with that of experimental data, and then the repulsion force is scaled-up to the experimental data of the force by multiplying by a scale factor, the ratio of the original superconductor's volume to the minute disk. This modeling technique means that the shape and sizes of the superconductor are reduced to two parameters, radius and scale factor, to be adjusted for agreement with experimental data.

Computer simulations with this simple model of repulsion force show that the feedback controller employed for both linear and rotary actuators provides satisfactory performances, which agree with experimental results. Fig.9 shows the computer simulation of control performance for the rotary actuator which corresponds to the experimental results shown in Fig.7. However, this computer simulation with the above model cannot represent oscillatory responses. A higher order model including the mechanism of flux creep and flux flow may provide better fittings with actual performances of those two actuators. The computer simulation for the linear actuator also showed that biased currents applied to the solenoids on both sides reduce hysterisis as shown in Fig.10.

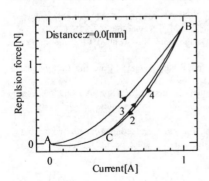
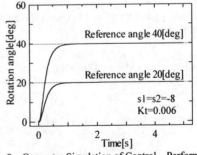

Fig.9   Computer Simulation of Control   Performance
for   Rotary Actuator

Fig.10   Reduction of Hysterisis by Biased Current: When the current is decreased from A to B, and then decreased back to A, the force draws the major hysteretic loop (path 1 to 2).   If the current is increased at C of 0.4A on the way back from B to A, then the force reaches B again through the path 3.   Decreasing the current from B to C makes the same path 4 on 2.   Thus, the minor loop 3 and 4, much less hysteretic than the major one appears by the biased current of 0.4A

The use of the biased currents in the experiment with the linear actuator gave the stiffer spring constant resulting in smaller oscillations discussed in Paragraph 2. As stability analysis by the Popov theory suggests, a wider domain of the feedback gains guarantees asymptotic stability of the control system when using the biased currents [8].

5. Conclusion

The position of the latest version of the linear actuator, driven by two solenoids, guided and levitated over two magnetic bearings, was satisfactorily controlled by linear feedback with the biased currents applied to the solenoids.   The current-dependency of the spring constant should be canceled by the control law for better performance.   A disk-shaped rotor of the rotary actuator with two poles magnetized by field cooling was stably driven by four solenoids arranged evenly around the rotor up to about 400rpm, levitating over a disk-shaped permanent magnet. Damped oscillations dependent on feedback gains should be improved by generating uniform torque based on current control depending on the angle in future work.

Acknowledgements

The author is thankful to my Doctor Course student, Tetsuya Nakao, and Master Course students, Satoshi, Uryu, Yoshinari Ito and Kouji Okamoto, for conducting their experiments with the actuators and helping to organize this paper.

References

[1] T.Kitamura: The Challenge to Superconducting Mechatronics--A Dream of Artificial Heart Technologists, Journal of Robotics and Mechatronics, vol.3, no.1, 1992.
[2] M.Murakami, et al: Large Levitation Force Due to Flux Pinning in MPMG Processed YbaCuO Superconductors with Ag Doping, in Advances in Superconductivity III (Eds: Kkajimura and Hayakawa) pp.753-756, 1991
[3] F.C.Moon and P.Z.Chang: High-speed rotation of Magnets on High Tc Superconducting bearings, Appl. Phys. Lett. 56, pp.397-399, 1990
[4] B.Takahata, H.Ueyama, and T.Yotsuya: Load Carrying of Superconducting Magnetic Bearings, in Electromagnetic Forces and Applications (Eds. J.Tani and T.Takagi), Elsevier, pp.347-351, 1992
[5] M.Komori and T.Kitamura: Superconducting Actuator Design, Int. J. Applied Electromagnetics in Materials, vol.2, pp. 243-252, 1991
[6] S.Uryu and T.Kitamura: Design of a High-Tc Superconducting Rotary Actuator, Proc. the 8th Symposium on Electromagnetics and Dynamics, pp.311-314, 1996 (in Japanese)
[7] S.Uryu, T.Nakao, T.Kitamura, and M.Murakami: Computer Simulation of High Tc Superconducting Linear Actuator, Proc. the 7th Symposium on Electromagnetics and Dynamics, pp.11-16, 1995 (in Japanese).
[8] T.Kitamura: Stability Analysis to a Fuzzy Control System of a Superconducting Actuator Including two Nonlinear Blocks, Proc. 2nd. IEEE Int. Conf.Fuzzy Sys., San Francisco, vol.2, pp.1321-1326, 1993.

Non-Linear Electromagnetic Systems
V. Kose and J. Sievert (Eds.)
IOS Press, 1998

Magneto-Impedance Effect and Application to Micro Magnetic Sensors

L. V. Panina and K. Mohri

Department of Electrical Engineering, Nagoya University, Nagoya 464, Japan

Abstract. A review on magneto-impedance (MI) in high permeability magnetic materials is presented with the emphasis on the application of this effect to micro magnetic sensor development. MI utilizes the skin effect at which both the skin depth and the transverse permeability can be controlled with an external magnetic field. Since the impedance change is caused by the magnetization in a thin surface layer the demagnetization effect is reduced allowing the MI head to be made in micro-size without a sufficient drop in sensitivity. In 0.5 - 2 mm long CoFeSiB near zero magnetostrictive amorphous wires the impedance change sensitivity to the longitudinal field can reach up to 100 % / (10^{-4} T) for a sinusoidal current excitation with a frequency of several MHz. MI depends considerably upon a particular magnetic structure, a frequency range and an excitation method, providing various types of impedance vs. field behaviors. This suggests a great technological potential of MI in a wide range of sensor applications. Various sensitive and quick response MI sensors are realized by incorporating a tiny MI element in a self-oscillations circuit, which demonstrate a field resolution of 10^{-10} T for an ac field detection.

1. Introduction

Progress in miniaturization technology applied to such fields as high density magnetic recording , industrial measurements and control, and medical electronics requires new sensitive and quick - response micro magnetic sensors to detect localized weak magnetic fields of 10^{-7} to 10^{-10} T in the frequency range up to 10 MHz. Widely used Hall elements, magnetoresistive and giant magnetoresistive elements, and fluxgate sensors are expected to be insufficient in the near future due to the limitation in sensitivity, response time and size. The recently discovered giant magneto-impedance (MI) effect [1-5] is a promising phenomenon to develop sensor with required parameters. Various MI based magnetic sensors have been constructed by incorporating small MI heads (amorphous wire or thin film) in self oscillations circuits [6].

This paper presents a review on the MI effect in some prospective materials such as high permeability amorphous wires, single-layer films and sandwich films and outlines the application of this effect to micro magnetic sensor development. The MI effect includes a very large and sensitive change in a high frequency voltage or a pulse voltage across a high permeability magnetic material under the application of a dc (or lower frequency) magnetic field. The high field sensitivity is preserved even for specimens of few mm in length. In CoFeSiB near-zero magnetostrictive amorphous wires 0.5-2 mm long, the voltage sensitivity to the dc longitudinal external field can be 10-100 % / 10^{-4} T for a frequency of several MHz. The largest values of MI ratio of about 600 % are obtained with sandwich films having two outer magnetic layers and a non-magnetic low resistance inner layer (for instance, made of Cu) separated by insulator materials [7]. Such structures also give very high MI sensitivity of 30-60 % / 10^{-4} T. The voltage change is caused by the impedance vs. field behavior resulting from the   skin effect in conjunction with the transverse magnetization (with respect to the ac current) which is very sensitive to the applied field for certain magnetic structures.   Since the magnetization of a thin surface layer determines the dependence of impedance on field,   the effect of demagnetizing fields is reduced allowing the MI heads to be made in micro size without a drop in sensitivity. Thus, the MI effect gives a sufficient basis for construction of sensitive and quick-response micro magnetic sensors.

The giant MI effect can be considered as a high frequency analogy of giant magnetoresistance. In the latter case, although the nominal values of the resistance change can be very large (~ 50 % in multilayers [8]), the sensitivity to the field is relatively low and does not exceed 1.5 % [9] due to high saturation fields. Besides, there are hysteresis and temperature instability problems. A flux gate element which has also a very high sensitivity [10], for many applications does not satisfy requirements in size (more than 10 mm) and response speed (less than 10 kHz).

2. Impedance calculation

In homogeneous and uniform materials (for both magnetic and conducting properties), the MI effect is based on the dependence of the penetration depth $\delta = 1/\sqrt{\omega\sigma\mu/2}$ on the magnetic permeability μ. With increasing frequency ω, an ac current I tends to concentrate near the surface changing the material impedance Z, and hence, the voltage $V = ZI$. For some simple geometries as in the case of a wire or a film, in a linear approximation (over ac magnetic H^{ac} and electric E^{ac} fields) the relationship between the impedance and the skin depth has a simple form:

$$Z_w = R_{dc}\, kaJ_0(ka)/2J_1(ka), \qquad Z_f = R_{dc} \cdot jka\coth(jka) \qquad (1)$$

respectively for a wire and a film. Here $k = (1+j)/\delta$, $2a$ is the cross section size (diameter or thickness), R_{dc} is the dc resistance, J_0, J_1 are Bessel functions, σ is the conductivity. At high frequencies when the skin effect is strong ($a/\delta \gg 1$), expressions (1) give: $Z_{w(f)} \propto \sqrt{\omega\mu}$. Then, in this frequency region the impedance behavior reflects the dependence of the permeability on the external magnetic field, which can be extremely sensitive for certain magnetic structures as in the case of a circular domain structure typical of negative magnetostrictive amorphous wires.

Expressions (1) have a rather limited field of applications. First, the magnetic properties are not homogeneous due to the existence of a domain structure. Second, in the range of ac magnetic fields higher than the coercivity, the magnetic behavior is not linear. Third, the permeability has a tensor form. However, with some modifications expressions (1) can be used for some cases including a practically important case of a transverse domain structure (circular domains in a wire and stripe domains in a film), when I and an external field H_{ex} are applied in the length direction, as shown in Fig. 1. In this case μ corresponds to the linearized total transverse permeability averaged over domains with the opposite magnetization [2,5].

This simplification is possible since the averaged dc magnetization M_0 lies in the longitudinal direction. On the contrary, for many cases (a helical anisotropy, or if an ac current is biased with a dc one) M_0 has some angle ψ with the longitudinal axis, z. For such configurations the tensor of surface impedance $\zeta_{\alpha\beta}$ can be introduced according to:

$$E_\alpha^{ac} = \zeta_{\alpha\beta}\,(H^{ac} \times n)_\beta\,,$$ where the electric and magnetic fields are taken on the

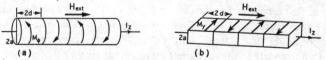

Fig. 1. Magnetic structure and geometry for MI in a wire and a film.

specimen surface, n is a unit vector directed inside it. If a specimen is excited by the longitudinal current I_z the diagonal component ζ_{zz} determines the voltage V_z between its ends (for a wire $V_z = (l/2a)\zeta_{zz}I_z$, l is the wire length) whereas the off-diagonal component ζ_{zy} determines the voltage V_c induced in the coil, if any, mounted on it. It should be noted that at low frequencies the off-diagonal components are related to the well known Matteucci and inverse Wiedemann effects. The impedance tensor has a simple general form in the case of a sufficiently strong skin effect when $a/\delta \gg 1$, $a/\delta_0 \gg 1$, δ_0 is the skin depth for a unit permeability:

$$\zeta = \zeta^{(dc)} x_0 \begin{pmatrix} \mu^{1/2}\cos^2\psi + \sin^2\psi & (\mu^{1/2}-1)\sin\psi\cos\psi \\ (\mu^{1/2}-1)\sin\psi\cos\psi & \mu^{1/2}\sin^2\psi + \cos^2\psi \end{pmatrix} \quad (2)$$

where $\zeta^{(dc)} = 1/\sigma a$, $x_0 = j(1+j)a/\delta_0$, μ is composed of the components of the permeability tensor found in the coordinate system with the axis z' along M_0 [11]. All the components $\zeta_{\alpha\beta}$ are of the same order and have a similar field dependence, but the off-diagonal terms $\zeta_{zy} = \zeta_{yz}$ can be even more sensitive to the field exhibiting a maximum at lower external fields. They can add an essential contribution to V_z if the current I_z is applied in the presence of a longitudinal ac magnetic field [12].

3. Magneto-impedance characteristics

A variety of MI characteristics has been obtained with $(Co_{1-x}Fe_x)_{72.5}Si_{12.5}B_{15}$ amorphous wires with x < 0.06 possessing a negative magnetostriction. A circumferential anisotropy existing in the outer layer of the wire results in an alternate left and right handed circular domain structure. Annealing under tensile stress enhances the circumferential anisotropy, and the circular domains exist almost in the entire wire. The current flowing along the wire creates an easy axis driving field, whereas the external axial field H_{ex} being a hard axis field greatly affects this easy axis magnetization. As seen from the circumferential hysteresis loops given in Fig. 2a, the permeability caused by the irreversible wall movements is strongly suppressed by H_{ex}. In the case of amorphous wires with x=0.06 (magnetostriction is -10^{-7}) which are cold drawn and then annealed with tensions of 2-8 kg/mm^2, having 30 μm in diameter and 5 mm in length, the quasi dc circumferential permeability is $1.3\cdot10^4$ for $H_{ex}= 0$ and it decreases 8 times after the application of a field of 2 Oe which is of the order of the anisotropy field, H_K. On the other hand, the rotational contribution which is dominant at high frequencies increases with $H_{ex} < H_K$. Consequently, the impedance behavior determined by the ac circumferential permeability changes its behavior with increasing frequencies.

Fig. 2b shows plots of the voltage amplitude V_0 across the wire vs. external field for different frequencies [5]. At relatively low frequencies less than 1 MHz, V_0 decreases with H_{ex}. The sensitivity of the voltage drop defined as $\xi = [V_0(H_{ex})/V_0(H_{ex}=0)-1]/H_{ex}$ 100% increases with frequency up to 25 %/10^{-4} T for $f=1$ MHz. At higher frequencies V_0 has a minimum at $H_{ex} = 0$. The tendency of the voltage to increase with the field becomes more pronounced as frequency is increased and $\xi = 50\%$ /10^{-4} T at $f= 10$ MHz. The theoretical MI characteristics are shown in Fig. 2b by dashed lines. The calculation is made on the basis of expression (1) for the wire impedance with the linearized ac permeability μ associated with both the domain wall movement and magnetization rotation.

Theoretical curves reasonably describe the impedance vs. field behavior for all frequencies.

In the case of a helical anisotropy which can be established by annealing under a

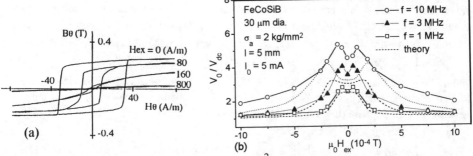

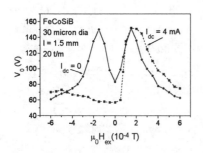

Fig. 2. For CoFeSiB tension annealed ($\sigma_a = 2\,kg\,/\,mm^2$) amorphous wires, in (a) circumferential hysteresis loops in the presence of H_{ex} and in (b) voltage amplitude vs. H_{ex} for various current frequencies. Theoretical curves are shown as dashed lines.

Fig. 3. V_0 vs. H_{ex} for a wire from Fig. 2 under a torsion of 20 turns/m.

Fig. 4. Bistable MI in CoSiB amorphous wire under a torsion of 2 turns/m for different I_0 .

torsion or applying a torsion to a wire, MI characteristics have a minimum at zero field for all frequencies since the domain wall permeability increases with H_{ex} in this case. The helical anisotropy allows asymmetrical MI characteristics to be obtained (see Fig. 3, the dashed curve), if the ac current is biased with a dc current. The field sensitivity in this case is especially impressive 120 %/10^{-4} T [13].

The MI effect does not exhibit a hysteresis for the variation of H_{ex}, even in the case of a helical anisotropy. This seems to be   associated with a non linear transverse magnetization depending on both fields: H_{ex} and H_{ac} induced by the ac current. The situation changes in the case of very high frequencies, at which the ac magnetization *m* is small and the corresponding permeability is almost independent of H_{ac}. The MI effect with a sharp hysteresis (named as a bistable MI [14]) has been reported for a twisted CoFeSiB amorphous wire excited by a short current pulse. The bistability is more pronounced for CoSiB wires having much higher negative magnetostriction of $-3 \cdot 10^{-6}$. Figure 4 shows the voltage response peak value $_0$ vs. H_{ex}, for different current peaks, I_0 . With increasing I_0, the voltage jump and the hysteresis area initially change little but further increase in I_0 leads to an abrupt shift of the hysteresis to the positive fields with its area shrunk considerably and voltage jump increased. The critical field can be increased by annealing of a wire at a higher tension. In the present case H^* can be varied within (2-15) 10^{-4} T. The bistable MI effect is satisfactorily described within a linear model by applying   the Fourier integral transform method to the Maxwell's equations [14]. It has been demonstrated that if the characteristic frequency of the current pulse $\omega_0 = 1/\Delta t$ corresponds to a strong skin effect ($a/\delta(\omega_0) \gg 1$), the voltage peak vs. the external field resembles the field dependence of the real part of $\mu_{ef} = \mu \cos^2 \psi$ (see expression (2)) taken at $\omega \cong \omega_0$, which has a hysteresis.

MI ratio in soft ferromagnetic films is of the same order as that in wires. In CoFeB sputtered films of 1-4 μm thickness a large MI occurs at frequencies higher than 50 MHz with the impedance sharply increasing with the field and having a maximum at $H_{ex} \sim H_K$. For frequencies of 80-100 MHz this maximum reaches 40-80 %. However, due to a high transverse anisotropy induced by annealing in the presence of a strong magnetic field the sensitivity does not exceed 8 %/10^{-4} T [15].

A very large change in impedance can be obtained for two magnetic films sandwiching a Cu-layer, in which the MI ratio can be up to 10 times larger than that in a single layer [16]. Recent experimental results [10] on MI in CoSiB/Cu/CoSiB multilayers of several micrometers in thickness have demonstrated the maximum of impedance change of about 300 % for $f = 10$ MHz, with a sensitivity of about 50 %/10^{-4} T. On the other hand, for a single CoSiB magnetic layer of the same thickness the MI effect is very small (< 5 %) at such frequencies. In a sandwich film, the contribution of magnetic layers to the total impedance corresponds to the external inductance, which becomes much larger than the resistance defined by the inner conductor. This means that the condition of a strong skin effect is not required in this case. If the resistance is decreased by separating the magnetic and conducting layers with insulator material, for example SiO_2, the MI ratio increases up to 600%. However, for a sandwich film with a much smaller total thickness of the order of 0.1 μm the MI effect decreases down to about 10 - 15 %, as it is reported for $Ni_{80}Fe_{20}/Cu/Ni_{80}Fe_{20}$ films of 0.2 μm thickness[16], yet in the case of a similar single layer film the MI effect would not be noticeable. The sandwich structure is important to further reduce the MI element size.

4. Principles of MI based sensors

Advantages of MI based sensors are associated with the possibility to simultaneously realize high sensitivity, micro size and quick response, which can not be reached with other convential sensors. On the other hand, a high frequency excitation in the sensor circuit involves many problems like instability in lead wires connecting an MI element with an ac source, impedance mismatching, and the existence of reflected signals. To avoid these problems, self-oscillation circuits such as a Colpitts oscillator and a resonance multivibrator operating with a dc voltage source and a power consumption less than 10 mW can be used, as shown in Fig. 5. In the resonant circuits the rate of the voltage change becomes even several times higher than that for a constant amplitude current case, as can be seen from the comparison presented in Fig. 6 , where a thin film is used as an MI element [15]. A strong negative feedback loop has to be added to obtain a linear field sensing. The MI field sensor using the Colpitts oscillator with an amorphous CoFeSiB wire 1-3 mm long produces a field-sensing resolution of 10^{-10} T for ac fields and 10^{-9} T for dc fields, non-linearity less than 0.2% FS(±1 Oe), a quick response operation up to 1 MHz with a Colpitts oscillation frequency of 50 -100 MHz [16]. A gradient-field sensor can be constructed by using a pair of MI elements installed in a multivibrator circuit [17]. Such a sensor circuit can be modified to detect a uniform field by applying a dc bias field inversely to the two MI heads. A linear field sensor without applying an additional dc bias field can be constructed on the basis of asymmetrical MI characteristics obtained in twisted amorphous wires. However, applying a torsion involves such engineering problems as stress stabilization. Utilizing the non-diagonal components of the impedance tensor, asymmetrical MI characteristics are found in CoFeSiB amorphous wire excited by a pulse current flowing through a wire and a coil mounted on it. Such current induces a spiral

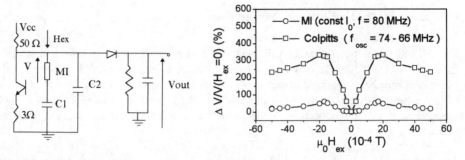

Fig. 5 MI- Colpitts oscillator type field sensor

Fig. 6. MI vs. Hex characteristics for a constant current case and obtained with Colpitts oscillator.

magnetization for both the dc and ac components, although the easy anisotropy axis in the wire is along the circumferential direction.   The bistable MI effect is applicable to construct a micro proximity sensor and a neural network element.

References

[1]   K. Mohri, Application of Amorphous Magnetic Wires to Computer Peripherals, *Materials Science and Engineering* **A185** (1995) 141.

[2]   L. V. Panina, K. Mohri, K. Bushida, M. Noda, Giant Magneto-Inductive and Magneto-Impedance Effects in Amorphous Alloys, *J.Appl. Phys* . **76** (1994) 6198.

[3]   L. V. Panina and K. Mohri, Magneto-Impedance Effect in Amorphous Wires, *Appl. Phys. Lett.* **65** (1994) 1189.

[4]   R. S. Beach and A. E. Berkowitz, Giant Magnetic Field Dependent Impedance of Amorphous FeCoSiB Wires, *Appl. Phys. Lett.* **64** (1994) 3652.

[5]   L.V. Panina, K. Mohri, T. Uchiyama, K. Bushida, M. Noda, Giant Magneto-Impedance in Co-Rich Amorphous Wires and Films, *IEEE Trans. Magn.* **31** (1995) 1249.

[6]   K. Mohri et al,   Sensitive and Quick-Response Micro Magnetic Sensor Utilizing Magneto-Impedance in Co-Rich Amorphous Wires, *IEEE Trans. Magn.* **31** (1995) 1266.

[7]   T. Morikawa et al , Thin-Film GMI Element with High Sensitivity, *J.Mag.Soc.Jpn* **20** (1996) 553.

[8]   S. S. P. Parkin, Giant Magnetoresistance in Antiferromagnetic Co/Cu Multilayers, *Appl. Phys. Lett.* **60** (1992) 512.

[9]   T.C. Anthony, J.A. Brag, Shufeng Zhang , Magnetoresistance of Symmetric Spin Valve Structures, *IEEE Trans. Magn.* **30** (1994) 3819.

[10] F. Primdahl, *J.Phys.E: Sci. Instrum* **12** (1979) 241.

[11] L.V. Panina and K. Mohri, Bistable Magneto-Impedance Effect in Twisted Amorphous Wires, *J.Mag.Soc.Japan* **20** (1996) 625.

[12] T. Gunji, L.V. Panina andK. Mohri, Asymmetrical Magneto-Impedance Effect in Amorphous Wire Spirally Magnetized with Pulse Current, *J.Mag.Soc.Japan* **21** (1997).

[13] T. Kitoh, K. Mohri, T. Uchiyama, Asymmetrical Magneto-Impedance Effect in Twisted Amorphous Wire, *IEEE Trans. Magn.* **31**   (1995)3137.

[14] M. Noda, L.V.Panina and K. Mohri, Pulse Response Bistable Magneto-Impedance Effect in Amorphous Wires, *IEEE Trans. Magn* **31** (1995) 3167.

[15] T. Uchiyama, K. Mohri, L.V. Panina, K.Furuno, Magneto-Impedance in Sputtered Amorphous Films for Micro-Magnetic Sensor, *IEEE Trans. Magn.* **31** (1995) 3182.

[16] K. Hika, L.V. Panina and K. Mohri, Magneto-Impedance in Sandwich Films for Magnetic Sensor Heads, *IEEE Trans. Magn.***32** (1996) 4594.

[17] K. Bushida, K. Mohri, T. Uchiyama,   Sensitive and Quick-Response Micro Magnetic Sensor Using Amorphous Wire MI Element Colpitts Oscillator, *IEEE Trans. Magn.***31** (1995) 4594.

[18] K. Bushida, K. Mohri, T.Kanno, D. Katoh, A. Kobayashi. Amorphous Wire MI Micro Magnetic Sensor for Gradient Field Detection, *IEEE Trans. Magn.***31** (1995) 4594.

Non-Linear Electromagnetic Systems
V. Kose and J. Sievert (Eds.)
IOS Press, 1998

Recent Research on Intelligent Material Systems

Junji Tani
Institute of Fluid Science, Tohoku University
2-1-1 Katahira, Aoba-ku, Sendai 980, Japan

Abstract. "Intelligent material systems" is a newly created concept. This concept aims to create an artificially designed material system. This material system has several functions in itself as a sensor, an actuator, and controller functions in combination with the properties and functions of the material. Hence, the system can change the strength, rigidity, configuration and dynamic response. The intelligent material systems can be very important and useful in many fields and their interdisciplinary fields such as mechanics, electronics, electricity, control and material engineering. In this paper, a composite cylindrical shell without vibration is shown as an intelligent material system. Piezoelectric films are used as a sensor and an actuator. The forced vibration of the cylindrical shell is suppressed by the combination of feedback and feedforward controls. This method has wide application in the vibration control of structures.

1. Introduction

It is well known that materials occupy key positions all along human history and evolution. Successive expertise in the processes of elementary materials are milestones which marked the early stages of mankind development, that is stone-age, bronze-age, etc. For a very long time, the traditional use of structural or functional materials has been related to only one of their properties. It is only since the beginning of the century that materials became multifunctional and required optimization of their different physical, mechanical, and/or chemical properties. The evolution is going on toward the concept and the process of composite or heterogeneous material where somewhat opposite properties are often associated. Nowadays, another step is to be undertaken by introducing the concept of intelligence. The analogy with biological systems is obvious and led to the concept of intelligent or smart materials which are able to adjust their behavior to the changes of external or, maybe internal parameters. In intelligent materials, sensors, actuators and controllers are integrated at the macroscopic level and the concept of material appears again if these components are incorporated into a classical composite material. Electromagnetic materials such as piezoelectric material, ferromagnetic material, magnetic shape memory alloy, electro-rheological fluid etc have functions of sensor and actuator. Therefore, when the sensor and actuator are incorporated into a classical composite material and a controller is added, an intelligent material system appears.

This paper describes an intelligent composite cylindrical shell without vibration as an example of the intelligent material system. The vibration control of cylindrical shells has become an active area of research in the last few years. In recent studies, small patches of PZT actuators and distributed PVDF actuators were used, since spillover due to distributed actuators can be reduced by the optimization of size, position and output of the actuators [1,2].

Several control methods have been used in past research. These methods can be divided into two categories: disturbance cancellation and feedback control. The disturbance cancellation method(DCM) can only be used in the control of vibration induced by predetermined disturbances or disturbances which can be directly measured, but is more effective than the feedback control method [3, 4]. The feedback control method is suitable for the control of vibration induced by both predetermined and unknown disturbances [1, 2].

This paper demonstrates the vibration control of a multi-layered composite piezo-electric shell using the integrated distributed sensor and actuators and a hybrid control method which is a combination of the feedback control method and disturbance cancellation method. It is expected that the combination of the two control methods will give better control effect than either method used independently. The μ-synthesis control theory is applied in the implementation of feedback control.

2. Model and Analysis

Figure 1 shows the configuration of the cylindrical shell used in this study. The cylindrical shell is made of two layers of piezoelectric films and one layer of polyester film, which are bounded together. The electrodes on the surface of the piezoelectric films are divided into several regions so that one region of them can be used as a sensor and another two can be used as actuators as shown in Figure 1. The lower end of the shell is fixed to a vibration table and the upper end is free. The shell is excited by the horizontal motion of the vibration table.

The coordinate system is fixed to the vibration table as shown in Figure 1, and the displacements of point (x, y) in x, y and z directions are denoted by $U(x,y,t)$, $V(x,y,t)$ and $W(x,y,t)$, respectively. For convenience, the coordinates and displacements are non-dimensionalized as follows:

$$\xi = \frac{x}{L}, \quad \eta = \frac{y}{L}, \quad \bar{\nabla}^2 = \frac{\partial^2}{\partial \xi^2} + \frac{\partial^2}{\partial \eta^2}, \quad u = \frac{RU}{Lh}, \quad v = \frac{RV}{Lh}, \quad w = \frac{W}{h}, \quad \tau = \Omega t \quad (1)$$

where R, L and h are radius, length and thickness of the shell. Using the dimensionless coordinates and displacements, the equation of motion can be expressed in the following form:

$$\left.\begin{array}{l} u_{,\xi\xi} + \nu_1(1 + k)u_{,\eta\eta} + (1 - \nu_1)v_{,\xi\eta} - \nu w_{,\xi} + k\kappa \cdot (w_{,\xi\xi\xi} - \nu_1 w_{,\xi\eta\eta}) - \dfrac{1}{\kappa}u_{,\tau\tau} = -\bar{f}_x \\[2mm] (1 - \nu_1)u_{,\xi\eta} + \nu_1(1 + 3k)v_{,\xi\xi} + v_{,\eta\eta} - w_{,\eta} - \dfrac{1}{\kappa}v_{,\tau\tau} = -\bar{f}_y \\[2mm] -\nu u_{,\xi} - v_{,\eta} + k\kappa^2\bar{\nabla}^4 w + k\kappa(u_{,\xi\xi\xi} - \nu_1 u_{,\xi\eta\eta}) + (1 + \nu_1)k\kappa v_{,\xi\xi\eta} + (1 + k)w \\[2mm] \qquad + 2k\kappa w_{,\eta\eta} + w_{,\tau\tau} = \bar{f}_z \end{array}\right\} \quad (2)$$

in which $\bar{f}_x$, $\bar{f}_y$ and $\bar{f}_z$ represent distributed forces generated by excitation and piezo-electric actuators. The parameters k, κ, C and Ω are given in the following expressions:

$$k = \frac{h^2}{12R^2}, \quad C = \frac{Eh}{1 - \nu^2}, \quad \kappa = \frac{R^2}{L^2}, \quad \Omega = \frac{E}{\rho R^2(1 - \nu^2)} \quad (3)$$

where E, ν and ρ are the Young's modulus, Poisson's ration and density of the shell. The values of material properties and geometric parameters are shown in Table 1.

The lower end of the shell is clamped and the upper end is free. These boundary conditions can be expressed in the following form:

$$w = w_{,\xi} = u = v = 0 \quad (\xi = 0) \quad \text{and} \quad S_x = M_x = N_x = T_x = 0 \quad (\xi = 1). \quad (4)$$

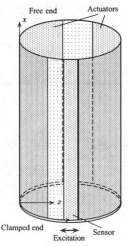

Figure 1: Configuration of the shell

Table 1: Parameter of the model

Length of shell L (mm)	600
Radius of shell R (mm)	50
Young's mod. of shell E (N/m²)	2.3×10^9
Young's mod. of PVDF E_p (N/m²)	2.0×10^9
Internal damping coefficient ζ	0.02
Density of shell ρ (kg/m³)	1.78×10^3
Thickness of PVDF h_p (mm)	0.052
Thickness of shell h (mm)	0.19
Width of actuators (mm)	$2\pi R/3$
Width of sensor (mm)	$\pi R/6$

Table 2: Natural frequency

Mode (i, j)	$(1, 1)$	$(1, 2)$	$(1, 5)$
Sim.(Hz)	75.3	27.5	87.9
Exp.(Hz)	73.5	28.6	–

where S_x, M_x, N_x and T_x are lateral force, bending moment, in-plane force and shear force, respectively.

By solving the equation of motion (2) with the boundary condition (4), a second order ordinary equation can be obtained for each mode. Ignoring the unimportant modes, we can obtain the following state equation:

$$\dot{X}(t) = A_c X(t) + B_1 u_c(t) + B_2 w_d(t) \tag{5}$$

where X is the state vector, and A_c, B_1 and B_2 are matrices depending on the geometrical and physical parameters of the shell and actuators. The output voltage of the piezoelectric sensor is a function of the state vector and the output equation can be expressed in the following form:

$$y_c = C_c X \tag{6}$$

where C_c is a vector depending on the size and position of the sensor.

3. Design of Controller

In this study, the feedback control and the disturbance cancellation method are combined in the calculation of the control input, that is, the control input u_c is the sum of the control inputs obtained from the two control methods and can be expressed in the following form:

$$u_c = u_a + u_p \tag{7}$$

where u_a and u_p are inputs of feedback control and disturbance cancellation control. If the transfer function of the feedback controller and disturbance cancellation controller are denoted by K_a and K_p, the control inputs u_a and u_p can be expressed in the following form:

$$u_a(s) = K_a(s)y_c(s) \tag{8}$$

$$u_p(s) = K_p(s)w_d(s) \tag{9}$$

The input of feedback control is calculated from the output y_c and the input of disturbance cancellation is calculated from the external disturbance w_d. The feedback controller K_a is designed using the μ-synthesis method as described in reference [2]. The disturbance cancellation controller K_p is designed as described in reference [5].

4. Experimental Set-up

The experimental set-up is shown in Figure 2. The shell was vertically mounted on a vibration table. The upper end of the shell was free and the lower end clamped. The vibration table was excited horizontally by an electromagnetic shaker and the shell was excited by the base movement. The motion of the table was measured by an acceleration sensor and the vibration of the shell was measured by the integrated piezoelectric sensor. Figure 3 shows the relationship between the output voltage of the sensor and the amplitude of excitation. When the amplitude is small, the relationship can be considered to be linear.

The outputs of the acceleration sensor and piezoelectric sensor were converted to digital signals and sent to the personal computer. The control input was calculated from the two sensor outputs by using the designed feedback controller and disturbance cancellation controller. The control input was converted to an analog signal, amplified and applied to the piezoelectric actuators.

5. Results and Discussion

The experimentally measured and analytically predicted natural frequencies of modes (1, 1) and (1, 2) are shown in Table 2. In the control experiments, the shell was excited at a frequency of 75 Hz which approximately equals the natural frequency of mode (1, 1). The exciting amplitude of the table was set to 30μm. The forced vibration of the shell was controlled by using the μ-synthesis method, disturbance cancellation method and the combination of the two methods, respectively, and the results are compared.

5.1 Results of Feedback Control

Figure 4(a) shows the simulation result of piezoelectric sensor output when the μ-synthesis control method is used. The amplitude of vibration is reduced to about 1/4 in a few cycles after control input is applied. Figure 4(b) shows the experimental result obtained under the same condition as the simulation. The simulation and experimental results are in good agreement.

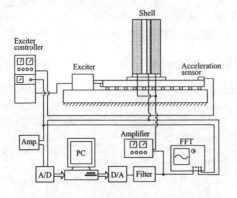

Figure 2: Experimental set-up

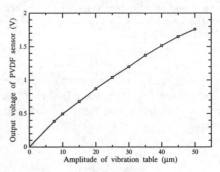

Figure 3: Property of sensor

5.2 Results of Disturbance Cancellation

Figure 5(a) shows the simulation results when the disturbance cancellation method is used. The amplitude of the shell vibration decays with time and becomes very small. The experimental results are shown in Figure 5(b). The amplitude of vibration is reduced only to about one half of the uncontrolled vibration. There are two main factors which lead to the difference. In the simulation the exact disturbance is known and used in the computation of the control input, but in the experiment it is unknown and the measured one is used in the computation of control input. The measurement error is one of the main factors. The second main factor is the phase delay of control input due to A/D and D/A conversion and the low-pass filter used to remove the high frequency components introduced by D/A conversion. By comparing the sensor output and the control input after D/A conversion, the phase delay due to A/D and D/A conversion is estimated to be about 12°. The phase delay due to the filter is about 10°. Hence the total phase delay is 22°. When this phase delay is included in the simulation, the simulation and experimental results are in good agreement.

5.3 Results of Hybrid Control

To achieve better performance, the two control methods are combined in the calculation of the control input as shown in Equation (7). Figure 6(a) shows the simulation result of the hybrid control when the phase delay of control input is considered. The

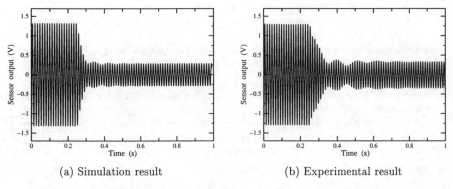

     (a) Simulation result   (b) Experimental result

Figure 4: Simulation and experimental result of response: μ-synthesis only

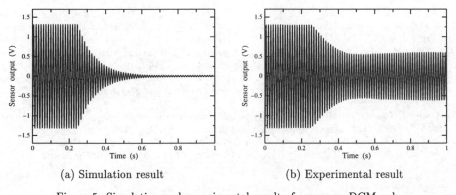

     (a) Simulation result   (b) Experimental result

Figure 5: Simulation and experimental result of response: DCM only

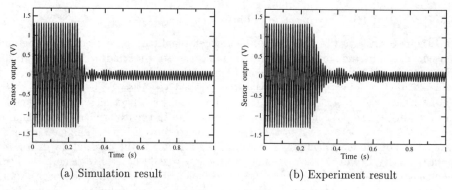

(a) Simulation result   (b) Experiment result

Figure 6: Simulation and experimental result of response: hybrid control

amplitude of the shell vibration is reduced to a very small level within a few cycles after the control input is applied. It can be seen that the decay rate with the hybrid control is much faster than with the disturbance cancellation method, but the amplitude of residual vibration is smaller than with the μ-synthesis control. The corresponding experimental result is shown in Figure 6(b).

6. Conclusion

Simulations and experiments were used to demonstrate the control of forced vibration of a cantilever piezoelectric composite cylindrical shell excited by a horizontal base movement. The integrated piezoelectric films are used as distributed actuators. A hybrid control is used and its performance is compared with that of the feedback control method and the disturbance cancellation method. From the simulation and experimental results, the following conclusions can be obtained:

(1) The integrated piezoelectric actuators are effective in the control of the shell vibration excited by base movement.

(2) The hybrid control method gives a better performance than the feedback control method and the disturbance cancellation method when they are used individually. It is expected that hybrid control will have wide application in the control of engineering structures since the external disturbance can be measured in many cases.

References

[1] J. Tani, J. Qiu and H. Miura, "Vibration Control of Cylindrical Shell System Using Piezoelectric Film", *Trans. of JSME*(C), Vol.60, No.570, pp.443-449, 1994.

[2] J. Qiu and J. Tani, "Vibration Control of a Cylindrical Shell Used in MRI equipment", *Journal of Smart Materials and Structures*, Vol.4, Supplement A, pp.75-81, 1995.

[3] J. Qiu and J. Tani, "Vibration Control of a Cylindrical Shell Using Distributed Piezoelectric Sensors and Actuators", *Journal of Intelligent Material Systems and Structures* Vol.6, No.6, pp.474-481, 1995.

[4] J. Qiu, J. Tani and K. Ohtomo, "Intelligent Coil Drum with Electromagnetic Force Cancellation for MRI Equipment", *Trans. of JSME*(C), Vol.61, No.581, pp.2412-2417, 1994.

[5] J. Qiu and J. Tani, "Vibration Control of a Cylindrical Shell Using the Hybrid Control Method", *Journal of Intelligent Material Systems and Structures*, 1996(Accepted for publication).

Non-Linear Electromagnetic Systems
V. Kose and J. Sievert (Eds.)
IOS Press, 1998

Magnetic Amorphous Glass-Covered Wires

Horia CHIRIAC and Tibor-Adrian ÓVÁRI

National Institute of R&D for Technical Physics, 47 Mangeron Blvd., 6600 Iasi, Romania

Abstract. A review of the basic magnetic properties of amorphous glass-covered wires with positive, negative, and nearly zero magnetostriction is presented. The application possibilities of these wires based on their specific magnetic behavior are also discussed. Recent results concerning the influence of Sn, Cr, and Mn additions on the magnetic properties of amorphous glass-covered wires from the Fe-Si-B system, and the magnetic behavior of nanocrystalline glass-covered wires prepared from amorphous precursors are also presented.

1. Introduction

Amorphous glass-covered wires (AGCW) produced by glass-coated melt spinning are relatively new materials which present a special interest for basic research and potential applications in magnetic sensors. These wires have a specific magnetic behavior which originates in their very small diameters - 3 to 25 μm - and in the large internal stresses induced during preparation [1].

In this paper we review recent results on the basic magnetic behavior of AGCW of three important classes related to their potential applications: AGCW with positive, negative, and nearly zero magnetostriction, having the compositions $Fe_{77.5}Si_{7.5}B_{15}$ ($\lambda = 35 \times 10^{-6}$), $Co_{80}Si_{10}B_{10}$ ($\lambda = -4 \times 10^{-6}$), and $Co_{68.15}Fe_{4.35}Si_{12.5}B_{15}$ ($\lambda = -1 \times 10^{-7}$) respectively. The potential applications of these wires, together with their practical advantages over 'conventional' water-quenched ones are also discussed. Recent results concerning the influence of Sn, Cr, and Mn additions on the magnetic properties of AGCW from the Fe-Si-B system, and the magnetic behavior of nanocrystalline glass-covered wires prepared from amorphous precursors are also presented.

The reported results are based on the interpretation of the experimental results on the grounds of internal stresses induced during preparation, that strongly influence the domain structure formation in these wires.

2. Basic magnetic behavior of amorphous glass-covered wires

The most outstanding feature of the magnetic behavior of AGCW is the appearance of an axial large Barkhausen effect (LBE). LBE appears in $Fe_{77.5}Si_{7.5}B_{15}$ amorphous wires before and after glass removal, while in the $Co_{80}Si_{10}B_{10}$ and $Co_{68.15}Fe_{4.35}Si_{12.5}B_{15}$ amorphous wires it appears only after glass removal, this behavior being explained by considering the changes in the domain structure determined by the glass cover removal. This effect is very appropriate for applications in magnetic sensors.

Fig. 1 illustrates the LBE in an $Fe_{77.5}Si_{7.5}B_{15}$ AGCW (A) and in the same wire after glass removal (B). Fig. 2 shows the hysteresis loops for an as-cast $Co_{80}Si_{10}B_{10}$ AGCW (A), for the same wire after glass removal (B), and for the wire after glass removal subjected to an external tensile stress of 50 MPa (C).

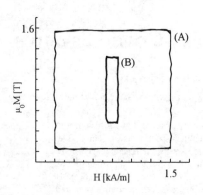

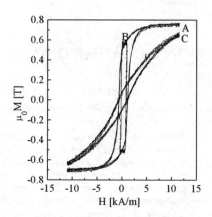

Fig. 1. LBE in an $Fe_{77.5}Si_{7.5}B_{15}$ amorphous glass-covered wire with 7.30 μm in diameter and a glass cover thickness of 7.50 μm (A), and in the same wire after glass removal (B).

Fig. 2. M-H loops for a $Co_{80}Si_{10}B_{10}$ glass-coated wire 15 μm diameter and a glass cover of 6.6 μm (A), for the wire without glass (B), and for the wire without glass subjected to a 50 MPa tensile stress (C).

The LBE - a one-step magnetization reversal process that occurs at a given value of the axially applied field, called the switching field H^* - appears in AGCW with $\lambda > 0$ and it is maintained in such wires after glass removal, but it does not appear in AGCW with $\lambda < 0$ and $\lambda \approx 0$. LBE appears in wires with $\lambda < 0$ and $\lambda \approx 0$ obtained from AGCW after glass removal, and it vanishes when such wires are subjected to a tensile stress of a certain value. The values of H^* and of the reduced remanence M^*/M_S - the ratio between the remanent magnetization M^* that corresponds to H^* and the saturation magnetization M_S - depend on the values of λ and on the wire dimensions [2].

The LBE is related to the existence of an appropriate magnetic structure in a sample, as a necessary condition for its appearance. For AGCW with $\lambda > 0$, the minimization of the magnetoelastic energy arising from the coupling between internal stresses induced during preparation and λ leads to the formation of a domain structure that consists of a cylindrical inner core (CIC) with axial easy axis and an outer shell (OS) with radial easy axis. This domain structure favors the appearance of the LBE, and previous studies demonstrated that the CIC is the effective region in which nucleation and propagation occur. The volume of the CIC is more than 90% of the wire volume, a fact supported by the very large value of the reduced remanence ($M^*/M_S \approx 1$). H^* is also high due to the large internal stresses. After glass removal, stress relief occurs, and the volume of the CIC decreases, a fact proved by the decrease of M^*/M_S. However, the domain structure remains qualitatively similar.

For AGCW with $\lambda < 0$, the strong magnetoelastic coupling overcomes the increase of the magnetostatic and exchange terms - unlike in the case of water-quenched 'conventional' amorphous wires with $\lambda < 0$ [3] - and the resulting domain structure consists of a radially magnetized CIC and a circumferentially magnetized OS, the appearance of the LBE being obstructed. After glass removal, the strength of the magnetoelastic interactions decreases due to stress relief, and the domain structure changes, i.e. the easy axis from the CIC becomes axial. Accordingly, LBE appears in such wires after glass removal. If a sample after glass removal is subjected to a tensile stress with a given value, LBE vanishes. This behavior indicates a return to the initial situation when the CIC is radially magnetized, and also the existence of a fine balance between the magnetoelastic energy on one hand, and the exchange and magnetostatic ones on the other.

A similar situation exists in AGCW with $\lambda \approx 0$ ($\lambda = -1 \times 10^{-7}$), in which, even if λ is one order of magnitude smaller, internal stresses are still large, and the resulting magnetoelastic coupling is strong [4]. Nevertheless, in this case, the competition between the magnetoelastic energy and the exchange and magnetostatic ones is stronger. This fact was expected due to the lower magnetostriction, and it is proved by the suppressing of LBE in such wires after glass removal by a smaller tensile stress.

3. Effect of Sn, Mn, and Cr additions on the magnetic properties of amorphous glass-covered wires from the Fe-Si-B alloy system

We have previously shown that the magnetic properties of AGCW can be controlled starting from the preparation process through their dimensions, or by different post-production treatments [2, 4]. Another method that allows controlled modifications of the magnetic properties together with improvements in other properties - thermal, chemical - is the addition of certain elements to the basic alloy system of which AGCW are prepared. Additions of Mn, Sn, and Cr in small amounts to FeSiB amorphous alloys result in the improvement of thermal stability and reduce the hysteresis losses, improving the corrosion resistance as well (Cr). On the other hand, a study of the changes induced by their presence in the good soft magnetic properties of FeSiB AGCW is required. We investigated $Fe_{79}Si_5B_{16}$, $Fe_{75}Si_5B_{16}Mn_4$, $Fe_{77}Si_5B_{16}Sn_2$, and $Fe_{77}Si_5B_{16}Cr_2$ AGCW. The saturation magnetization M_S of the $Fe_{79}Si_5B_{16}$ samples (1.58 T) is changed to 1.35 T, 1.39 T, and 1.41 T for the $Fe_{75}Si_5B_{16}Mn_4$, $Fe_{77}Si_5B_{16}Sn_2$, and $Fe_{77}Si_5B_{16}Cr_2$ wires, respectively.

Positive magnetostrictive $Fe_{79}Si_5B_{16}$ AGCW ($\lambda \approx 27 \times 10^{-6}$) present LBE at relatively low axial fields, depending on their dimensional characteristics. M^*/M_S is close to 1. LBE is maintained after glass removal, but H^* decreases one order of magnitude, and M^*/M_S decreases to 50%. Similar behavior was found for the $Fe_{75}Si_5B_{16}Mn_4$, $Fe_{77}Si_5B_{16}Sn_2$, and $Fe_{77}Si_5B_{16}Cr_2$ AGCW and wires after glass removal, but H^* is one order of magnitude smaller than that of $Fe_{79}Si_5B_{16}$ ones. M^*/M_S is very large for all AGCW, and it decreases after glass removal, reaching its initial value (for the AGCW) or slightly higher ones when the samples are subjected to an external tensile stress σ_{ext}. After glass removal, the external stresses required for M^*/M_S to reach its initial value are one order of magnitude smaller for $Fe_{75}Si_5B_{16}Mn_4$, $Fe_{77}Si_5B_{16}Sn_2$, and $Fe_{77}Si_5B_{16}Cr_2$ wires than for $Fe_{79}Si_5B_{16}$ ones.

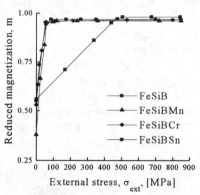

Fig. 3. Stress dependence of M^*/M_S for FeSiB, FeSiBMn, FeSiBSn, and FeSiBCr amorphous wires after glass removal.

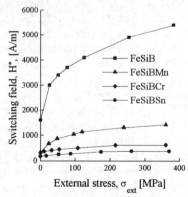

Fig. 4. Stress dependence of H^* for FeSiB, FeSiBMn, FeSiBSn, and FeSiBCr amorphous glass-covered wires.

Fig. 3 illustrates the stress dependence of M^*/M_S for $Fe_{79}Si_5B_{16}$, $Fe_{75}Si_5B_{16}Mn_4$, $Fe_{77}Si_5B_{16}Sn_2$, and $Fe_{77}Si_5B_{16}Cr_2$ amorphous wires after glass removal. FeSiBMn, FeSiBSn, and FeSiBCr wires have almost similar behavior from this point of view, while the FeSiB one is clearly detached. This fact occurs since after glass removal, the OS has two regions: an inner one where the radial easy axis is given by the coupling between small positive radial tensile stresses and $\lambda > 0$, and an outer one where the radial easy axis is given by the coupling between $\lambda > 0$ and high negative circumferential and axial stresses [5]. The behavior of M^*/M_S for wires after glass removal is affected by the changes that occur in the critical region from the vicinity of the OS's inner limit, where domination of axial tensile stresses ceases, and radial ones take over the role of determining the easy axis. An external tensile stress σ_{ext} applied on a sample obtained after glass removal determines transverse Poisson contractions on the radial and circumferential directions, diminishing the radial tensile stresses that determine the radial easy axis in the inner region of the OS. The axial tensile stress from the CIC increases simultaneously, and the CIC increases to the detriment of the OS. In this process, the magnitude of λ has a decisive role: it gives the strength of the magnetoelastic coupling that determines the easy axes distribution. Hence, the higher is λ, the higher is the strength of this coupling. Thus, it is easier for the CIC to increase to the detriment of the OS in FeSiBMn, FeSiBSn, and FeSiBCr wires, since the radial anisotropy that has to be overcome by σ_{ext} is smaller in this case due to the reduction of λ by the additions of Mn, Sn, and Cr. Consequently, smaller values of σ_{ext} are required for M^*/M_S to reach its initial value. M^*/M_S does not change with σ_{ext} for AGCW with $\lambda > 0$, since it determines transverse contractions that reinforce the radial easy axis from the OS determined in this case by the coupling between internal circumferential compressive stresses and λ, inhibiting the increase of the CIC to the detriment of the OS.

Fig. 4 shows the stress dependence of H^* for FeSiB, FeSiBMn, FeSiBSn, and FeSiBCr AGCW. H^* increases with σ_{ext} for all the wires, typical behavior for wires with $\lambda > 0$ [2]. One observes that FeSiB AGCW have the highest values of H^*, and they are the most sensitive to σ_{ext}. FeSiBMn, FeSiBSn, and FeSiBCr AGCW have much smaller values of H^*, and the slope of its stress dependence is also smaller. Since the magnitude and stress sensitivity of H^* is determined by the coupling between stresses - applied and internal - and λ, this behavior suggests a decrease of λ for the FeSiB alloy, determined by the Mn, Sn, and Cr additions. Indeed, we measured that the initial magnetostriction of the FeSiB alloy decreases from 27×10^{-6} to 18×10^{-6}, 19×10^{-6}, and 20×10^{-6}, for the FeSiBMn, FeSiBSn, and FeSiBCr alloys, respectively. However, the reduction of λ does not fully explain the results presented in Fig. 4. It is plausible to attribute the higher sensitivity of H^* to σ_{ext} for FeSiBMn AGCW - having the smallest λ - to the higher values of internal stresses induced in these wires due to changes produced in the material parameters - thermal conductivity, specific heat, thermal expansion coefficient, etc. - by the Mn addition. In this way, the magnetoelastic coupling is large enough and the differences in the values of λ are compensated by the other factor - internal stresses. This hypothesis also explains the initial value of H^* for AGCW prepared from this alloy, which is larger than those of FeSiBSn and FeSiBCr AGCW. Our hypothesis is well supported by the larger values of the coercive force found in the FeSiBMn amorphous alloy as compared to the FeSiBCr one, being known that the magnitude of the coercive force is directly related to mechanical stresses [6].

Summarizing, these additions improve other physical properties of Fe-Si-B wires without causing damage to their soft magnetic properties: LBE is maintained with a large M^*/M_S, and M_S decreases very slightly. On the other hand, they offer a convenient tool for tailoring the magnetic properties of these wires: reduction of H^* and λ can be controlled through additions of appropriate elements.

4. Magnetic behavior of nanocrystalline glass-covered wires

Let us discuss the evolution of the magnetic behavior of $Fe_{73.5}Cu_1Nb_3Si_{13.5}B_9$ glass-covered wires starting from the amorphous state after different stages of annealing for 1 h at temperatures up to 600°C, in order to analyze the changes induced by the appearance of the α-FeSi crystalline grains, and also by stress relief. FeCuNbSiB AGCW present LBE at low fields due to the axially magnetized CIC, arising from the coupling between λ (25×10^{-6}) and internal stresses. M^*/M_S is close to 1. Wires after glass removal also present LBE, but H^* and M^*/M_S are smaller due to the stress relief determined by glass removal [5].

Fig. 5 illustrates the dependence of H^* and M^*/M_S on the annealing temperature T_a. One observes that both quantities decrease when T_a increases up to 550°C, when LBE vanishes. H^* and M^*/M_S decrease due to the stress relief that occurs during annealing. At 550°C, LBE disappears due to the appearance of the nanocrystalline phase [7], i.e. since the positive magnetostriction of the residual amorphous matrix and the negative one of α-FeSi nanocrystals are averaged out, resulting in a nearly zero macroscopic magnetostriction.

Additional information about thermal relaxation of internal stresses and growth of crystalline grains can be obtained by studying the dependencies of the coercive force H_C on T_a for a glass-covered wire and for the same wire after glass removal, illustrated in Fig. 6. H_C decreases when T_a increases up to 550°C, when a minimum value is reached in both cases. The minimum corresponds to the nanocrystalline phase formation, which is characterized by good soft magnetic properties, and especially by low coercivity. The decrease of H_C is determined by stress relief in both cases for $T_a < 550°C$, while for $T_a = 550°C$, a major contribution is also given by the decrease of the magnetoelastic anisotropy due to the reduction of λ. H_C increases for $T_a > 550°C$ for both wires, indicating a dimensional increase of the crystalline grains, that also determines an increase of the role of magnetocrystalline anisotropy. The larger values of internal stresses in these thin wires as compared to the 'conventional' ones, influence their behavior up to the appearance of the nanocrystals, determining larger values of H^* and M^*/M_S. Once the nanocrystalline phase is formed, the influence of these stresses becomes less important from the point of view of the magnetoelastic anisotropy, since λ is strongly diminished, but internal stresses are still important for H_C.

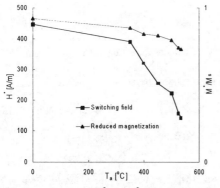

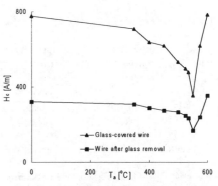

Fig. 5. Dependencies of H^* and M^*/M_S on the annealing temperature for FeCuNbSiB glass-covered wires.

Fig. 6. Dependence of H_C on the annealing temperature for FeCuNbSiB glass-covered wires and wires after glass removal.

5. Potential applications and advantages of amorphous glass-covered wires

The basic operating principles that can be used to develop magnetic sensors using AGCW as sensitive elements are:

- LBE in positive magnetostrictive amorphous wires with and without glass cover and in negative and nearly zero magnetostrictive amorphous wires after glass removal;
- Matteucci effect in highly positive magnetostrictive amorphous glass-covered wires and wires after glass removal;
- Other dependencies of magnetic quantities like M^*/M_S and H^* on σ_{ext}, as well as of the circumferential magnetization on torsional stresses;
- Giant magneto-impedance and magneto-inductive effects in nearly zero magnetostrictive amorphous and nanocrystalline glass-covered wires and wires after glass removal.

These operating principles allow the achievement of high sensitivity magnetic field, current, torque, and stress sensors. Sensor prototypes with AGCW or wires after glass removal as sensitive elements have been already developed in our laboratory at the National Institute of Research and Development for Technical Physics, Iaşi, Romania.

The magnetic properties of these kinds of wires are easily tailored through an accurate control of their dimensional characteristics - wire diameter, glass cover thickness - as well as through different types of post-production treatments (heat treatments, field annealing, stress annealing, field-stress annealing).

The utilization of AGCW and wires after glass removal in applications offers several advantages:

- They are obtained in a one-step process (two-step for wires after glass removal) at smaller diameters in comparison with the thin amorphous wires obtained by cold drawing in several steps starting from 'conventional' amorphous wires;
- They offer miniaturization opportunities for sensor devices working with 'conventional' amorphous wires - e.g. the critical length at which the LBE still appears in such wires is about 1-2 mm [5];
- They have more degrees of freedom for tailoring their magnetic properties due to the existence of the glass cover, and thus, one can adjust their basic magnetic properties - coercive force, switching field, permeability, remanence - in a wide range of values.

AGCW are appropriate for sensor applications even if the sensitive element is working under applied stress due to their very good mechanical properties.

References

[1] H. Chiriac, T.A. Óvári and Gh. Pop, Internal Stress Distribution in Glass-Covered Magnetic Amorphous Wires, *Physical Review B* **52** (1995) 10104-10113.

[2] H. Chiriac, T.A. Óvári and Gh. Pop, Magnetic Behavior of Glass-Covered Amorphous Wires, *Journal of Magnetism and Magnetic Materials* **157/158** (1996) 227-228.

[3] M. Vázquez and D.-X. Chen, The Magnetization Reversal Process in Amorphous Wires, *IEEE Transactions on Magnetics* **31** (1995) 1229-1238.

[4] H. Chiriac, Gh. Pop, T.A. Óvári and F. Barariu, Magnetic Behavior of Negative and Nearly Zero Magnetostrictive Glass-Covered Amorphous Wires, *IEEE Transactions on Magnetics* **32** (1996) 4872-4874.

[5] H. Chiriac, T.A. Óvári, Gh. Pop and F. Barariu, Effect of Glass Removal on the Magnetic Behavior of FeSiB Glass-Covered Wires, *IEEE Transactions on Magnetics* **33** (1997) 782-787.

[6] S. Jiansheng, Z. Hongru, Q. Dirong, W.H. Qin, S. Songyao and X. Qingzheng, Magnetic Properties of FeMnSiB Amorphous Alloy, *IEEE Transactions on Magnetics* **23** (1987) 2146-2148.

[7] H. Chiriac, T.A. Óvári, Gh. Pop and F. Barariu, Magnetic Behavior of Nanostructured Glass-Covered Metallic Wires, *Journal of Applied Physics* **81** (1997) 5817-5819.

Non-Linear Electromagnetic Systems
V. Kose and J. Sievert (Eds.)
IOS Press, 1998

Optimum Design of a Permanent Magnet Linear Actuator for Reciprocating Electro-Mechanical Systems

Steve EVANS

University of Newcastle Upon Tyne, Newcastle Upon Tyne, UK

Ivor SMITH, Gordon KETTLEBOROUGH

Loughborough University, Leicestershire, UK

Abstract. With the rapid progress in the past two decades in permanent magnet technology, through the use of high energy rare-earth materials, a range of compact high performance short-stroke cylindrical linear actuators is now available. This paper describes the design and optimization of such a device for use in reciprocating electro-mechanical systems. The actuator dimensions most significant in affecting device performance are identified through a parametric study, and an optimum design is thus established. Simulation results are compared with the experimental performance of two prototype actuators, to illustrate the accuracy of the modelling technique adopted.

1. Introduction

Permanent magnet linear actuators are replacing conventional electrical and hydraulic systems in a wide range of applications, such as Stirling cycle cryogenic refrigerators and artificial heart motors. The actuator under investigation here is essentially a linear version of the Laws' Relay [1]. It takes the form of a bi-directional moving-iron or variable air-gap device and operates by balancing the two forces that act on a soft-iron armature, positioned on the central axis of two opposing rare-earth ring magnets producing a constant polarizing flux [2, 3]. The control flux produced by a solenoidal coil situated between the ring magnets disturbs the symmetrical distribution of the polarizing flux, and thereby unbalances the forces acting on the armature. This then moves until the magnetic force produced by the energised coil is counter-balanced by that due to the permanent magnets (a reluctance force or magnetic stiffness) and any external load. The magnetic co-energy is thereby minimised and the flux is maximised. A typical device is shown in Figure 1.

2. Optimum Design Specification

An optimized actuator design should possess a linear coil current/armature displacement characteristic (both loaded and unloaded), produce the maximum possible force on the armature and have a fast dynamic response. A restoring axial magnetic stiffness is an essential requirement, to ensure that the armature returns to its central position in the absence of any coil current, and eliminate the need for mechanical springs. An optimum force-displacement characteristic for various coil currents is shown in Figure 2. Ideally, the actuator should operate at its mechanical

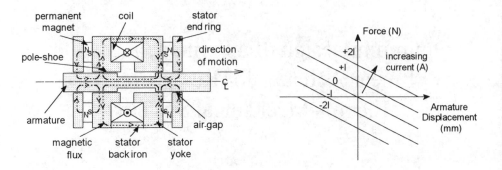

Figure 1 Permanent magnet linear actuator   Figure 2 Ideal actuator force characteristics

resonant frequency, to produce a highly efficient drive system, and a design needs to be under-taken with this criterion in mind. It should also be extremely rugged and reliable with very few components. The armature can be supported by either linear or diaphragm spring bearings.

3. Modelling Technique

Due to the complex geometry of the actuator, and the saturation characteristics of the magnetic materials involved, a non-linear axi-symmetric finite element analysis (MEGA) was used to predict the force-displacement and the coil flux linkage characteristics. The force was calculated using Maxwell's stress tensor method and the flux linkage was determined from energy considerations. These results were then included into a state-variable model containing the differential equations for the actuator and drive, and which can be employed to determine the overall dynamic performance [3].

4. Prototype Actuator Designs

Two prototype actuators were produced, both being 57mm long and 46mm outer diameters (excluding housing and bearings), but with different armature diameters of 10mm and 15mm. The number of turns on the excitation coil was 720. The magnetic material used was Radiometal (50% NiFe), chosen for its high saturation, high permeability and corrosion resistance, with the high saturation level leading to a high air-gap flux density and a large force. The permanent magnet material was Neodymium-iron-boron (NdFeB). The high remanence and coercive force of this material produce an actuator with a potentially large force/volume ratio.

The force-displacement characteristics for the prototype actuators were similar, with the 15mm diameter design producing the larger force. Experimental and finite element characteristics for this design are shown in Figure 3. The two sets of results compare favourably and highlight the problem that the prototype designs have a poor stiffness (force characteristic at zero current). They show that there is very little force attempting to return the armature to its central position at displacements up to ±4.0mm from this position. This arises from armature saturation and is further illustrated by the other force curves not being parallel and equi-distant and producing a non-linear movement of the armature. The device simply acts as a bistable actuator, with two stable positions at ±4.2mm from the central position.

Figure 4 shows the dynamic response of both actuators excited with a 0.6A peak sinusoidal current at frequencies up to 50Hz. The correlation between the theoretical and experimental

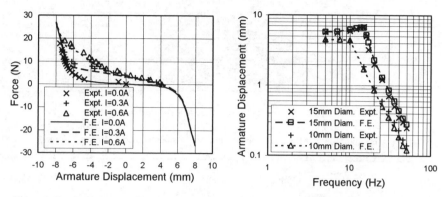

Figure 3 Prototype actuator force characteristics

Figure 4 Prototype actuator dynamic responses

results are good even though the simulations use flux linkage data obtained from magnetostatic field solutions that neglect the effects of eddy currents in the magnetic circuit which would inhibit the flux and possibly reduce the armature displacement.

5. Optimum Actuator Design

The armature outer diameter was increased to 23mm to reduce the saturation and improve the device linearity, but this significantly increased its mass and degraded the dynamic response. A hollow armature with a 6mm diameter hole was therefore employed to reduce the mass, and although this improved the dynamic performance it, as expected, reduced the linearity. The 50% NiFe used for the hollow armature was then replaced by Permendur (50% CoFe), which has a higher saturation level and restored the linearity whilst keeping the mass approximately unchanged. A detailed parametric study was then undertaken [3], which led to a further improvement being achieved by a reduction in the stator yoke/armature pole-face overlap. This was achieved by increasing the narrow centre-section of the armature and reducing the stator yoke pole-shoes, to produce a much stiffer system with further improved linearity. Figures 5 and 6 illustrate the optimum actuator static and dynamic characteristics, and the resonant frequency can be clearly seen in Figure 6 to be about 20Hz. A theoretical analysis determined the frequency to be about 21Hz. A flux plot of the optimum design is shown in Figure 7. The specification of the optimum design is given in Table 1 below.

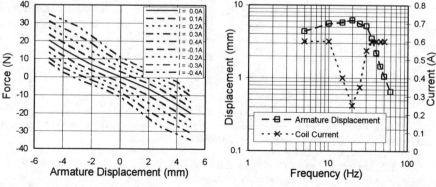

Figure 5 Optimum actuator force characteristics

Figure 6 Optimum actuator dynamic response

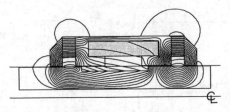

Figure 7 Flux plot of the optimum actuator design

Table 1 Optimum Actuator Specification (Parameter units not explicitly stated are mm)

Parameter	Value	Parameter	Value
Stator (50% NiFe):		Permanent Magnets (NdFeB):	
Overall length	57.0	Length	5.00
Overall outer diameter	46.0	Volume (cm^3)	2 × 4.66
Inner pole-shoe length	11.4	Coil (Copper):	
End-ring and Yoke length	5.00	No. of turns	825
Back-iron length	27.0	Resistance (Ω)	13.4
Armature (50% CoFe):		Maximum coil current (A dc)	0.70
Outer diameter	23.2	Performance:	
Inner diameter	6.00	Stiffness (N/mm)	4.00
Narrow centre-section length	25.7	Peak Force (N)	49.4
Narrow centre-section diameter	19.8	Motor Constant (N/A)	28.0
Mass (g)	213	Operating range	dc (±4.4mm) - 50Hz (±1.1mm)
Air-gap length	0.31	Resonant frequency	21Hz (±6.2mm)

6. Conclusions

The paper has described the optimum design of a permanent magnet linear actuator for use in reciprocating electro-mechanical systems. The design was achieved using a magnetostatic finite element analysis, coupled with a mathematical model consisting of the motor and drive system to determine the dynamic performance of the actuator. Predicted results compare favourably with the experimental performance of two prototype devices.

Acknowledgements

The authors wish to thank the Department of Electronic & Electrical Engineering, Loughborough University, UK for the research studentship awarded to Dr S A Evans, and for the facilities available for the conduct of the research. Thanks are also due to Coercive Systems, Rochester, UK, especially to Mr S K Jim, for providing the hardware used in the research.

References

[1]   A. E. Laws, An electro-mechanical transducer with permanent magnet polarisation, Technical Note No. G.W.202, Royal Aircraft Establishment, Farnborough, UK, 1952.

[2]   S. A. Evans *et al.*, Design and analysis of a permanent magnet linear actuator, Proc. of the 7th Power Electronics and Motion Control Conference, ISBN: 963 420 488 0, Budapest (1996) **3** 266-270.

[3]   S. A. Evans, Design and optimization of a permanent magnet linear reluctance motor for reciprocating electro-mechanical systems, Loughborough University Ph.D. Thesis, Loughborough, UK, 1996.

Non-Linear Electromagnetic Systems
V. Kose and J. Sievert (Eds.)
IOS Press, 1998

Static Testing of
Microelectromechanical Sensors

Erwin PEINER, Dirk SCHOLZ, Klaus FRICKE, Andreas SCHLACHETZKI
*Institut für Halbleitertechnik, Technische Universität Carolo-Wilhelmina zu
Braunschweig, Hans-Sommer-Str. 66, D-38106 Braunschweig, Germany*

Abstract. In this study a novel load-deflection technique for testing
microelectromechanical sensors is described which utilizes a commercial surface
tracer common to most IC laboratories. It is used to characterize a spring-mass
system in Si based on a cantilever comprising a piezoresistive bridge for
mechanical-electrical signal conversion. Due to buckling the cantilever exhibits a
bending stiffness which is strongly increased with respect to the limit of plane
strain. The buckling which is indicated by the deflection of the unloaded cantilever
is caused by biaxial stress due to a thin SiO_2 coating.

1. Introduction

Quality and reliability verification is of major importance for the transfer of prototype
microelectromechanical systems (MEMS) to marketable products [1]. In MEMS, cantilever
structures are used for sensing physical parameters (acceleration, forces, vibrations) and as
actuators (switch). Their response to mechanical load which is important with respect to
device calibration, reliability, aging and failure depends on bending stiffness, residual stress
and signal conversion efficiency. For static characterization of MEMS conventionally high-
speed centrifuges are employed [2,3] which are not optimal for routine inspection. In a
recent study, a novel load-deflection technique for the characterization of thin films on Si
microcantilevers was presented which utilizes a commercial mechanical surface tracer [4].
In this study, we describe the application of this technique to the measurement of the
electromechanical response of a micromachined vibration sensor [5].

2. Theory

Figure 1 shows a schematic of a one-side
clamped cantilever comprising a seismic
mass m_{sm} and, at a distance x_b to the
clamping position, resistors arranged in a
symmetric Wheatstone full bridge for
conversion of the cantilever deflection
into an electrical signal. It is deflected by
a stylus as a transverse load F. The given
notation is employed in the following
where we consider the deflection curve
$\delta_c(x)$ of the beam due to F and

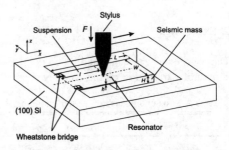

Figure 1. Schematic of the sensor chip.

$F_{sm} = m_{sm}g$ which act at $(L + l)/2$ and x', respectively, as well as due to internal stress by a thin SiO_2 coating. g denotes the acceleration due to gravity. In the following the film and substrate parameters are marked by the indices f and s, respectively. Restricting our considerations to elastic stresses and bending deformation (i.e., shear deformation is neglected) we find for $h_f << h$:

$$\delta_c(x) = \frac{x^2}{2D}\left[F\left(x' - \frac{x}{3}\right) + F_{sm}\left(\frac{L+l}{2} - \frac{x}{3}\right)\right] + x^2 \frac{Wh_s h_f}{4E_s'I}\sigma_f. \tag{1}$$

where the bending stiffness is $D \approx E_s''I$ with the geometrical moment of inertia $I \approx Wh_s^3/12$. $E' = E/(1 - v)$ and $E'' = E/(1 - v^2)$ denote the biaxial modulus and the plate modulus, respectively, assuming that the cantilever is under plane strain.

The stress σ_f of the unloaded film comprises an internal component σ_i and a thermal component σ_{th} generated during cool-down from the coating temperature T_1 to measurement temperature T_0:

$$\sigma_{th} = E_f'(T_1 - T_0)(\alpha_f - \alpha_s), \tag{2}$$

where α denotes the thermal expansion coefficient. Values of the material parameters are given in Table 1. During the load-deflection scan the position x of the deflection measurement equals the position x' where F is applied and eq. (1) simplifies to:

$$\delta(x) = \frac{x^3}{3D}F \tag{3}$$

Table 1. Material parameters [4,6,7]

Parameter	Si	SiO_2
E (GPa)	169[a]	75
v	0.064[a]	0.16
$\alpha(10^{-6}\ K^{-1})$	2.55	0.55
π_{eff} ($10^{-10}\ m^2/N$)	3.7	-

[a] (100) crystal plane, [011] direction

for the deflection due to F and:

$$\delta_0(x) = \delta_c(x) - \delta(x) - \frac{x^2}{4D}\left(\frac{L+l}{2} - \frac{x}{3}\right)F_{sm} = x^2 \frac{Wh_s h_f}{4E_s'I}\sigma_f = \frac{x^2}{2\rho}. \tag{4}$$

for the deflection at $F = 0$. ρ denotes the radius of curvature. The stress σ generated at the cantilever surface acts on the piezoresistors and leads to a detuning of the Wheatstone bridge. The contribution of the external load to the output signal is given by:

$$U = \pi_{eff}\sigma U_b = \frac{\pi_{eff}h_s}{2I}(x - x_b)U_b F \tag{5}$$

with $x = x'$, $x_b = 350\ \mu m$, π_{eff} as the effective piezoresistive coefficient and U_b as the supply voltage of the bridge.

3. Experiment

The cantilever was fabricated by etching of n-type (100)-Si in KOH (30 %) at 60°C. The structural parameters of the cantilever are listed in Table 2. Boron diffusion at 1100°C

Table 2. Structural parameters of the cantilever

L(mm)	l(mm)	H(mm)	W(mm)	$h_s(\mu m)$	$h_f(\mu m)$	x_b(mm)	$F_{sm}(\mu N)$
7.0	3.5	0.54	5.0	15.5	0.5	0.35	200

from a Emulsitone Borofilm 100 emulsion was employed to realize the p^+-regions for the piezoresistors. The resistivity was 0.01 Ωcm. A SiO_2 coating fabricated by thermal oxidation at 1100°C was used as the mask material for etching as well as for diffusion. Electrical contacts with a specific resistance of 10^{-3} Ωcm^2 were formed by Au/Cr which is only slightly removed by KOH at a rate of 30 nm/h. The most critical dimension, the cantilever height h_s, could be defined within bounds of ± 0.5 µm. For the static cantilever deflection we used a commercial mechanical surface tracer (Dektak 3ST).

4. Results

In Figs. 2 and 3 the deflection δ of the SiO_2-coated Si cantilever and the output voltage of the Wheatstone bridge are displayed in dependence of the applied external load F. The observed linear dependences as well as $\delta^{1/3} \propto x$ and $U \propto x$ (not shown here) are expected according to eqs. (3) and (5). By evaluation of these curves values of D and π_{eff} were determined. The results are displayed in Fig. 4 where a deviation of D from the theoretically expected value of 2.6×10^{-7} Nm^2 by a factor of 1.2 to 4.2 is visible. For π_{eff} we found values ranging from 2.2 to 3.3×10^{-10} m^2/N which is in reasonable agreement with 3.7×10^{-10} m^2/N reported for piezoresistors fabricated by boron implantation [3]. The sensitivity $S = U/U_b g^{-1}$ of the sensor calculated using eq. (5) with $x = (L + l)/2 - x_b$ and $F = F_{sm}$, amounts to 0.4 mV/Vg^{-1}.

A reduced value of D at small x can be attributed to shear deformation of the cantilever at the not ideally fixed clamping [6]. The increase of D observed in Fig. 4 is related to a curling of the cantilever into a spherical shape caused by the biaxial film stress. While inhibited at the clamping position, a buckling of the cantilever occurs at large x, i. e., the condition of plane strain fails. The geometrical moment of inertia increases with respect to I_0 at plane strain:

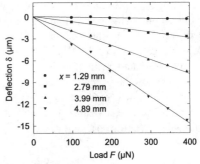

Figure 2. Load-deflection curves with F applied at x.

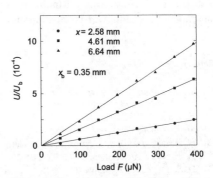

Figure 3. Output voltage of the Wheatstone bridge vs. load.

$$I = I_0\left(1 + \frac{W^4}{60h^2\rho_y^2}\right),\qquad(6)$$

where for the radius of curvature ρ_y in y direction $|\rho_y| \gg W$ is assumed. For biaxial stress, $\rho_y = \rho$ can be determined from the measured $\delta_0(x)$ curve according to eq. (4) employing σ_f (or ρ) as the fitting parameter. We find $\sigma_f = -400$ MPa corresponding to $|\rho_y| = 0.14$ m thus confirming $W \ll |\rho_y|$. Using eq. (2) we calculate $\sigma_{th} = -190$ MPa leading to $\sigma_i = -210$ MPa which is in reasonable agreement with a reported value of

160 MPa [7]. Using Eq. (6) and the measured value of ρ_y we obtain $I/I_0 = 3.2$. This result is well within the measured range of 1.2 to 4.2 (cf. Fig. 4).

σ_f causes a stress response in the Si surface of $-4h_f/h_s\sigma_f = 50$ MPa. Close to the cantilever clamping the x-component of σ_f is much greater than the y-component. Thus using eq. (5) an offset voltage ratio $(U/U_b)_{off}$ at the output of the Wheatstone bridge of 16 mV/V can be estimated which is close to the measured value of 15 mV/V. Furthermore, using eqs. (2) and (5) an offset drift $(U/U_b\Delta T^{-1})$ due to the temperature dependence of σ_{th} of 7 μV/V°C^{-1} is calculated (cf. ref. [8]).

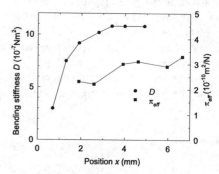

Figure 4. Bending stiffness and effective piezoresistive coefficient vs. x.

The presented results show that a comprehensive characterization of micromachined cantilever structures can be performed by static load-deflection measurements using a commercial surface tracer. We expect that the described technique is of considerable potential for routine testing and calibration of microelectromechanical sensors on-wafer or immediately before packaging.

Acknowledgements

We are indebted to Doris Rümmler, Christine Pabsch and Sören Sturm for technical support during cantilever fabrication and stylus measurements.

References

[1] T. Olbrich et al., Integrating testability into microsystems, *Microsystem Technologies* **3** (1997) 72-79. H. Baltes, CMOS as sensor technology, *Sensors and Actuators A* **37-38** (1993) 51-56.
[2] Y. Ning et al., Fabrication and characterization of high g-force, silicon piezoresistive accelerometers, *Sensors and Actuators A* **48** (1995) 55-61.
[3] H. Crazzolara et al., Piezoresistive accelerometer with overload protection and low cross-sensitivity, *Sensors and Actuators A* **39** (1993) 201-207.
[4] D. Scholz et al., A simple load-deflection technique for the determination of Young's modulus of thin films, Proc. MicroMat '95, DVM, Berlin, 1995, 557-562.
[5] D. Scholz et al., An Online Vibration-Control System for Rotating Machinery based on Smart Sensors Fabricated by Si-MST, Proc. Micro System Technologies '96. ISBN: 3-8007-2200-3. VDE-Verlag, Berlin, 1996, 157-162, A smart resonant sensor fabricated by Si-MST for online vibration control of rotating machinery, Proc. ESSDERC '96. ISBN: 2-86332-196-X. Editions Frontiéres, Gif-sur-Yvette, France, 1996, 721-724.
[6] M. P. Knauß et al., Microbeam bending experiments for determination of the mechanical properties of thin films, Proc. MicroMat '95, DVM, Berlin, 1995, 653-658.
[7] L. M. Mack et al., Stress Measurements of Thermally Grown Thin Oxides on (100) Si Substrates, *J. Electrochem. Soc.* **136** (1989) 3433-3437.
[8] H. Muro et al., Stress analysis of SiO$_2$/Si bi-metal effect in silicon accelerometers and its compensation, *Sensors and Actuators A* **34** (1992) 43-49.

Dynamics of Force Measuring Devices with Respect to Electromechanical Systems

Rolf Kumme

Physikalisch-Technische Bundesanstalt, Bundesallee 100, D-38116 Braunschweig, Germany

Abstract. The dynamic properties of force measuring devices are described and experimental results presented. The results demonstrate that the dynamic properties are influenced by different mechanical and electrical factors because force measuring devices are electromechanical systems.

1. Introduction

In many applications, dynamic forces are measured using statically calibrated force measuring devices. The unknown dynamic properties can result in systematic deviations. New methods and calibration facilities have, therefore, been developed for the dynamic calibration of force measuring devices [1,2,3,4]. The dynamic calibration is traceable to the definition of force through the generation of well-defined inertia forces.

2. Dynamic calibration of force measuring devices

2.1. The principle of dynamic calibration

For the dynamic calibration of force measuring devices a calibration procedure is used which is based on a defined realisation of mass forces. The force transducer to be calibrated is mounted on an electrodynamic shaker, and a load mass m_1 is screwed on the force transducer as shown in Figure 1. Excitation by the shaker results in a dynamic force F acting on the force transducer:

$$F = \left(m_1 + m_e \right) \cdot \ddot{x}_1 \tag{1}$$

where $\ddot{x}_1$ is the acceleration of the load mass m_1 and m_e the end mass of the force transducer. The end mass is the part of the transducer mass "situated between the strain gauges and the point of force introduction". It therefore results in an additional dynamic force. The simple equation (1) does not take into account the influences of relative motions of the load mass and the influences of side force which must be considered because the force is a vectorial quantity. Side forces can be reduced by using air bearings as described in [1,2]. To allow for the influence of relative motions, the dynamic force must be determined from the acceleration distribution $\ddot{u}(x,t)$ and the mass distribution with density ρ according to

$$F = \int_V \rho \cdot \ddot{u}(x,t) \cdot dV . \tag{2}$$

For the determination of the acceleration distribution, multicomponent acceleration measurements are necessary as shown in Figure 1, and the theory presented in [1,4] must be used to calculate the dynamic force. According to Equation (1) or, more accurately, according to Equation (2), the dynamic force is traceable to the definition of the force according to Newton's law.

2.2. Description of force measuring devices

Electrical force measuring devices consist of a force transducer and a force indication unit as shown in Figure 2. In dynamic applications the output signal U_v of the measuring amplifier is measured, for example, with a signal analyser. The dynamic sensitivity of a force measuring device is defined by the ratio of the output signal of the measuring amplifier U_v to the acting dynamic force F according to Equation (3):

$$S = \frac{U_v}{F} .\qquad\qquad(3)$$

The dynamic sensitivity of the whole force measuring device is determined using the calibration facility described above; it therefore includes the frequency response of force transducer and measuring amplifier.

In most applications strain gauge force transducers of different geometrical design are used. The dynamic behaviour of the transducer is mainly determined by the mechanical properties such as elasticity modulus and geometrical design. The dynamics of the measuring amplifier depends on the amplifier type and principle, for example direct voltage (DC) or carrier frequency (CF) measuring amplifier.

Piezoelectric force measuring devices are often used for dynamic measurements because of the high stiffness of the piezoelectric material. The charge output signal of the piezoelectric force transducer is measured with a charge amplifier. The dynamic response is mainly determined by the selected filter, e.g. low-pass and / or high-pass filter.

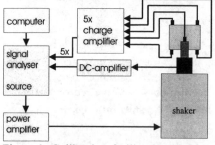

Figure 1: Calibration facility for dynamic force measurement.

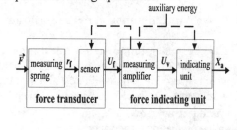

Figure 2: Block diagram of an electrical force measuring device.

3. Results of the calibration of force measuring devices

3.1. Mechanical influences

The main mechanical influences on the dynamic properties of force measuring devices are those from the mechanical structure of the force transducer and from force introduction. As shown in Figure 3, the two transducers A and B, both of high static quality and with the same static capacity of 2 kN, may exhibit a completely different dynamic behaviour. These differences can be explained by the mechanical structure of the transducer. Transducer A is a shear-type transducer of higher stiffness than transducer B which is a double-beam-type transducer. The lower stiffness of transducer B results in a higher decrease of sensitivity. Furthermore, transducer B shows more irregularities than transducer A, because transducer B is more sensitive to side forces. Figure 3 also shows that force introduction can lead to a significant change in sensitivity. If the force is introduced by a load button with loading pad, the stiffness of the force introduction is reduced compared with a screwed joint.

3.2. Electrical influences

The sensitivity of a force measuring device consisting of a strain gauge force transducer and a direct voltage measuring amplifier was determined. The measurements were repeated with the same force transducer and a 5 kHz carrier frequency measuring amplifier. The results obtained for the amplitude and phase response of the sensitivity are represented in Figure 4. As cross resonances occur at higher frequencies, the measurements were carried out only up to 900 Hz. In this frequency range, the measuring amplifiers lead to amplitude deviations of 3.5% and to phase differences of up to 160 degrees. The best results have been achieved by means of the direct voltage measuring amplifier. The greater deviations of the measurements made with the carrier frequency measuring amplifier are due to internal filters which are to limit the bandwidth, and to filters which are to smoothen the carrier frequency in the output signal of the measuring amplifier.

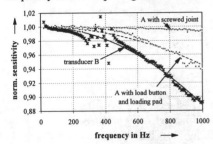

Figure 3: Normalized dynamic sensitivity of a double-beam-type transducer B and of a shear-type transducer A with different types of force introduction.

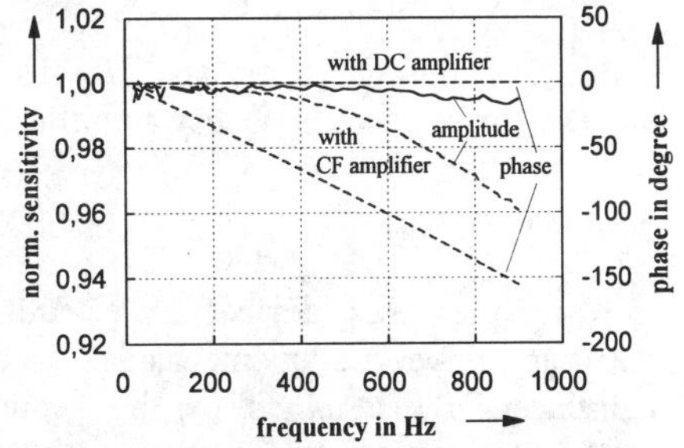

Figure 4: Amplitude and phase response of a force measuring device with direct voltage measuring amplifier and carrier frequency measuring amplifier.

In many dynamic applications piezoelectric force measuring devices are used. The investigations have shown that the dynamic sensitivity of the piezoelectric force measuring devices is mainly influenced by the frequency response of the charge amplifier.

4. Dynamic calibration of the amplifier

It was demonstrated that the dynamic sensitivity of a force measuring device is determined by the dynamic properties of the force transducer and the measuring amplifier. To determine the dynamic properties of a single force transducer, the dynamic properties of the amplifier must be determined.

4.1. DC and carrier frequency amplifier

DC and carrier frequency amplifiers are used to measure the detuning of a Wheatstone measuring bridge. In dynamic applications, the resistances of strain gauge force transducers are amplitude-modulated. For a dynamic calibration of the amplifiers it is necessary to simulate well-defined bridge detuning. With the bridges described in another paper [5] it is possible to investigate the dynamic properties of amplifiers used in strain gauge force transducers. Figure 5 shows the amplitude and phase response of a DC and a 5 kHz carrier frequency amplifier. The differences in section 3.2 can be explained on the basis of these results.

4.2. Charge amplifier

In the case of piezoelectric force measuring devices the frequency response is mainly determined by the charge amplifier. The dynamic sensitivity of the charge amplifier was determined with a calibrated capacitor. The results obtained with piezoelectric force measuring devices can be explained by the amplitude and phase response of a charge amplifier shown in Figure 6.

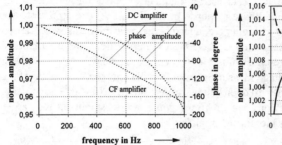

Figure 5: Amplitude and phase response of a DC and a carrier frequency amplifier.

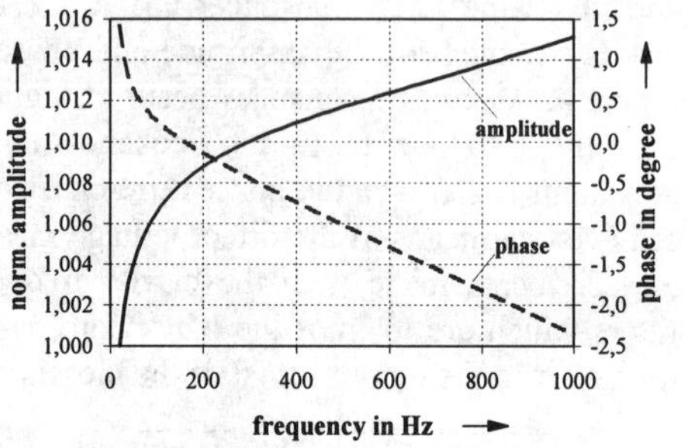

Figure 6: Amplitude and phase response of a charge amplifier.

5. Conclusions

Force measuring devices are electromechanical systems. The dynamic properties are therefore influenced by mechanical and electrical factors. The stiffness of the force transducer is determined by the geometrical design of the spring element and the mechanical properties of the material. Furthermore, the force transducer interacts with the surrounding mechanical structure. However, the electrical components of a force measuring device can also influence the dynamic properties. In the case of strain gauge force transducers, the frequency response of the measuring amplifier must be taken into consideration. The dynamic properties are limited by the amplifier principle (carrier frequency or DC) and the filter used in the amplifier. In the case of piezoelectric force transducers, the frequency response of the charge amplifier must be taken into account. The results show that the dynamic properties of a force measuring device can only be approximated by a simple spring-mass model. In a more complicated model the force introduction can be described by an additional spring element, and the frequency response of the amplifier must be considered separately. In dynamic calibration, the validity of the spring-mass model must, therefore, be proved and the differences between the static and dynamic sensitivity of the force measuring device determined.

References
[1] Kumme, R.: Untersuchungen eines direkten Verfahrens zur dynamischen Kalibrierung von Kraftmeßgeräten - ein Beitrag zur Verringerung der Meßunsicherheit. Doctoral Thesis, Braunschweig Technical University, 1996 (PTB-Bericht MA-48), ISBN 3-89429-744-1.
[2] Kumme, R.: The principle of a new testing and calibration facility for dynamic force measurement. In: Proceedings of the 14th IMEKO TC3 Conference on Measurement of Force and Mass, Warszawa (Poland), 12-15 September, 1995, pp.191-196.
[3] Kumme, R.; Peters, M.; Sawla, A.: Improvements of dynamic force calibration. Part 2: contains task 2.2 and 2.3. Brussels, Luxembourg: Commission of the European Communities, 1995 (BCR information, Report EUR No. 16497EN), ISBN 92-827-5344-1.
[4] Kumme, R.: The determination of the effective dynamic force for the calibration of force transducers, with due regard to the distribution of mass and acceleration. In: Proceedings of the 15th IMEKO TC3 Conference, Madrid (Spain), 7-11 October, 1996, pp.129-138.
[5] Kumme, R.: Influence of measuring amplifiers on dynamic force measurement. In: Proceedings of the 13th International Conference on Force and Mass Measurement, Helsinki (Finland), 10-14 May, 1993, pp.25-31.

Non-Linear Electromagnetic Systems
V. Kose and J. Sievert (Eds.)
IOS Press, 1998

Modelling and Simulation of an Integrated Electromechanical Multicoordinate Drive

Olaf Enge, Heiko Freudenberg, Gerald Kielau, Peter Maißer

Institute of Mechatronics at the Technical University Chemnitz,
Reichenhainer Str. 88, D-09126 Chemnitz, Germany

Abstract. Electromechanical systems are characterized by interactions of electromagnetic fields with inertial bodies. Electromechanical interactions can be described by so-called constitutive equations. A unified approach based on the principle of virtual work is used to obtain the motion equations. The example of a planar two-dimensional linear motor driven directly by electromagnetic forces is given.

1. Introduction

Electromechanical systems (EMS) are characterized by interactions between electromagnetic fields and inertial bodies. These interactions can be expressed by constitutive equations (generalized force laws). Using a coupling of multibody dynamics with Kirchhoff's theory, these constitutive equations define discrete EMS, i.e. systems with a finite degree of freedom. In order to use a unified approach, the motion equations are obtained as Lagrange's equations in explicit form based on the principle of virtual work. These motion equations can be generated automatically. The mathematical description follows the classical analytical mechanics completed by some basic concepts and methods of graph theory [5] to characterize topological properties of electrical networks [1, 2, 3].

2. Electromechanical Systems

A discrete electrical system is understood as a finite network of electrical components with lumped parameters (multipoles). It can be represented by a network graph of abstract 2-poles. Let G be an arbitrary (but fixed) spanning tree of the representing graph Γ and H the corresponding cotree. Then, using the fundamental loop matrix A and the fundamental cut matrix Q, Kirchhoff's laws read

$$\sum_{i\in\Gamma} A^i{}_j V_i = 0, \ j \in H$$

$$\sum_{i\in\Gamma} Q_i{}^j I^i = 0, \ j \in G,$$

where V_i denotes the branch voltages and I^i are the branch currents.

In the following, the charges in the fundamental loops (mesh charges) are used as generalized coordinates q^μ of the electrical system. If the spanning tree does not contain current generators (q_0), each admissible configuration can be described by

$$\bar{q}^j = A^j{}_\mu q^\mu + A^j{}_\xi q_0^\xi, \tag{1}$$

with $\mu \in H^*$ (= set of cotree branches not containing current generators) and $\xi \in H_0 = H \setminus H^*$. Hence, every branch charge is a linear combination of mesh charges completed by a source part.

The electromechanical interaction is described by constitutive equations. In the following, electrically linear constitutive equations are assumed:

$$
\begin{aligned}
V_i^{(L)} &= \dot{\Psi}_i, \quad \Psi_i = L_{ij}(q^\kappa, t)\dot{\bar{q}}^j + \Psi_{i0}(q^\kappa, t), \quad L_{ij} = L_{ji}, \\
V_i^{(C)} &= C_{ij}(q^\kappa, t)\bar{q}^j + C_{i0}(q^\kappa, t), \qquad\qquad C_{ij} = C_{ji}, \\
V_i^{(R)} &= R_{ij}(q^\kappa, t)\dot{\bar{q}}^j, \qquad\qquad\qquad\qquad R_{[ij]} = R_{[ij]}(t), \\
V_{i0} &= V_{i0}(t).
\end{aligned}
\tag{2}
$$

Hence, the integrability conditions for the existence of state function (Lagrange formalism) are valid. The constitutive equations (2) respresent the electrical components inductor (L_{ij}), capacitor (C_{ij}), resistor (R_{ij}) and voltage source (V_{i0}).

The actual motion of an EMS is characterized by the vanishing of the virtual work

$$\delta'W := -V_i \delta \bar{q}^i + S\delta\mathfrak{r}(d\mathfrak{k} - \ddot{\mathfrak{r}}dm) = 0 \qquad \forall \delta\bar{q}^i, \delta\mathfrak{r} \text{ virtual} \tag{3}$$

at any time t. Generalized coordinates of EMS will be denoted by q^κ, q^μ (κ, λ, ϱ - mechanical, μ, ν, ω - electrical; $a, b \in \{\kappa\}\cup\{\mu\}$ and with $q^0 = t$: $\alpha, \beta \in \{\kappa\}\cup\{\mu\}\cup\{0\}$).

Lagrange's motion equations follow from (3)

$$(\partial_a \Lambda)\dot{} - \partial_a \Lambda + \dot{\partial}_a D = Q_a, \tag{4}$$

where $\Lambda := T + \Psi - V$ is the Lagrangian of the EMS ($T(\dot{q}^\lambda, q^a, t)$ - kinetic coenergy, $\Psi(\dot{q}^\nu, q^a, t)$ - magnetomechanical copotential, $V(\dot{q}, q, t)$ - generalized electromechanical potential), $D(\dot{q}, q, t)$ denotes the dissipation function and Q_a are generalized forces/voltages. The state functions Λ and D have the following form:

$$\Lambda = \tfrac{1}{2}g_{\alpha\beta}(q, t)\dot{q}^\alpha \dot{q}^\beta, \quad D = \tfrac{1}{2}s_{\alpha\beta}(q, t)\dot{q}^\alpha \dot{q}^\beta.$$

The metric of the configuration space of the EMS defined by $g_{ab} = \dot{\partial}_a \dot{\partial}_b \Lambda$ is a direct sum of the mass matrix and the matrix of inductivities. It will be Riemannian if the matrix of inductivities is positive definite because the constitutive equations are electrically linear. Then, the explicit Lagrange's equations follow from (4)

$$
\begin{aligned}
g_{\mu\nu}\ddot{q}^\nu &&+\partial_\lambda g_{\mu\nu}\dot{q}^\lambda\dot{q}^\nu &&+(2\Gamma_{\mu b0} + s_{\mu b})\dot{q}^b &&+\Gamma_{\mu 00} &&+s_{\mu 0} &&= Q_\mu \\
g_{\kappa\lambda}\ddot{q}^\lambda &+\Gamma_{\kappa\lambda\varrho}\dot{q}^\lambda\dot{q}^\varrho &-\tfrac{1}{2}\partial_\kappa g_{\nu\omega}\dot{q}^\nu\dot{q}^\omega &&+(2\Gamma_{\kappa b0} + s_{\kappa b})\dot{q}^b &&+\Gamma_{\kappa 00} &&+s_{\kappa 0} &&= Q_\kappa.
\end{aligned}
\tag{5}
$$

For dynamical investigation of EMS with the simulation system **alaska**, one has to define (extra to the mechanical subsystem) the topology of the electrical system (network graph), the constitutive parameters (L_{ij}, Ψ_{i0}, C_{ij}, C_{i0}, R_{ij}, V_{i0}), their partial derivations to the generalized mechanical coordinates, and the current sources q_0^j. The motion equations (5) are generated and solved automatically.

3. Example of an Integrated Electromechanical Multicoordinate Drive

Phenomenology of the model:

The multicoordinate drive or planar motor (Fig. 1, see also [4]) consists of a stator and an aerostatical supported slide. It works as a DC-motor with commutation. The inductances are fixed at the stator and the permanent magnets at the slide.

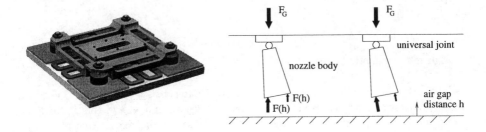

Figure 1: Planar motor   Figure 2: Modelling of the air bearings

The slide consists of a main body and four nozzles of the air bearings which are coupled to the slide by universal joints. The areal loads produced by the air bearings are represented by a number of single forces (Fig. 2), the magnitude of which depends on the air gap (about 10 μm) in a nonlinear way.

The drive of the slide is realized by eight coils interacting with the permanent magnets, i.e. the generation of motion works contactlessly. Four coils are used for each direction. The currents through each inductance are controlled separately. For continuous slide movement, the coils are supplied by commutative currents using suitable time dependent functions. Using different current intensities for coils located side by side on the carrier, the yaw angle φ of the slide can be controlled to be zero.

Mathematical modelling:

To compute the electromagnetic force acting on the slide, the variation of the magnetic flux Φ depending on the coil position is needed. Starting with Maxwell's equations, the distribution of the magnetic field between permanent magnets and stator yoke has been calculated

$$\mathfrak{B}_x = \frac{2B\sqrt{2}}{\pi} \sum_{k=0} \frac{(-1)^k}{(2k+1)} \frac{e^{\lambda(2k+1)z_p} - e^{-\lambda(2k+1)z_p}}{e^{\lambda(2k+1)z_s} + e^{-\lambda(2k+1)z_s}} \cos(2k+1)\lambda x$$

$$\mathfrak{B}_z = \frac{2B\sqrt{2}}{\pi} \sum_{k=0} \frac{(-1)^k}{(2k+1)} \frac{e^{\lambda(2k+1)z_p} + e^{-\lambda(2k+1)z_p}}{e^{\lambda(2k+1)z_s} + e^{-\lambda(2k+1)z_s}} \sin(2k+1)\lambda x.$$

where $\mathfrak{B}_x$ denotes the magnetic field in driving direction ($\mathfrak{B}_z$ - vertical), B is a permanent magnet constant, z_p denotes the distance between magnets and yoke, z_s that between coils and yoke, $\lambda = \pi/l$ with the distance l between the magnets.

Figs. 3 and 4 show the distribution of the magnetic field in different directions parameterized by z_p.

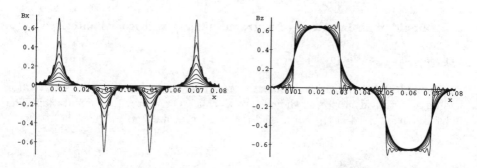

Figure 3: Magnetic field (x-direction)   Figure 4: Magnetic field (z-direction)

The simplest model of the mechanical substructure consists of five rigid bodies: the moving platform and four nozzles. For a more profound modelling, the platform can be replaced by four corner-bodies and four edge-bodies connected to each other by 3-dimensional spring-damper-systems. Stiffness and damping parameters are computed using an FEM model. The representing graph of the electrical network consists of eight loops, each containing one branch with a current generator and one branch with an inductance. The position control in the x-y-plane and also the retention of the slide orientation (yaw angle φ) have been realized with state controllers.

4. Results

Different kinds of air bearing approximations (variable number of single forces and variable directions of force action with respect to the nozzle body) have been simulated. Based on the results, remarks about the quality of the air bearing approximation can be made. The influence of separate model components on the system behaviour has been evaluated by simulation of various models differing in complexity (e.g. 5-body-model, 12-body-model, restricted freedom of slide motion).

References

[1] Enge, O., Kielau, G., Maißer, P.: *Dynamiksimulation elektromechanischer Systeme.* VDI-Fortschritt-Berichte, Reihe 20, Nr. 165. VDI-Verlag, Düsseldorf, 1995.

[2] Maißer, P., Enge, O., Freudenberg, H., Kielau, G.: Electromechanical Interactions in Multibody Systems Containing Electromechanical Drives. *Journal Multibody System Dynamics,* 1(3):281–302, 1997.

[3] Maißer, P., Steigenberger, J.: Lagrange-Formalismus für diskrete elektromechanische Systeme. *ZAMM,* 59:717–730, 1979.

[4] Schäffel, C., Saffert, E., Kallenbach, E.: Integrated Electrodynamic Multicoordinate Drive – Modern Components for Intelligent Motions. In *ASPE Annual Meeting,* pages 456–461. American Society for Precision Engineering, 1996.

[5] Seshu, S., Reed, M.B.: *Linear Graphs and Electrical Networks.* Addison-Wesley Publishing Company, Reading, MA, 1961.

Non-Linear Electromagnetic Systems
V. Kose and J. Sievert (Eds.)
IOS Press, 1998

On Possible Chaotic Behaviour in Magnetic Actuator

B. Z. Kaplan and G. Sarafian
Ben-Gurion University of The Negev
Department of Electrical and Computer Engineering
P. O. Box 653, Beer-Sheva 84105, Israel

Abstract. A simple electromagnetic actuator is investigated. The system dynamic behaviour consists of a series of period doublings that leads to a chaotic-like motion. The resulting broad band vibrations are successfully tamed by an OPF controller. Due to the similarity of the system to a tuned circuit magnetic levitator, the present results are likely to contribute to the stabilization of the related magnetic levitator.

1. Introduction

The present work is devoted to the investigation of a magnetic actuator, whose structure resembles a device, which is used for magnetic levitation. The device employs a method, which utilizes a simple tuned circuit [1-3]. The electromagnet that is used in the system for providing the magnetic force for suspension is the inductive part of a tuned circuit driven by an ac source detuned above resonance (see Fig. 1a). The movement of a ferromagnetic object suspended in the electromagnet field causes the resonant circuit current to vary in such a manner that the overall force-distance characteristic of the electromagnet-object system is similar to that of a mechanical spring, and a statically stable suspension can be achieved. However, the tuned circuit on its own is usually not sufficient for providing the dynamic stability. The suspended object in most cases tends to exhibit diverging mechanical oscillations [1-3]. The first author's own experience and some hints in the literature [3] indicate that the oscillations are in some circumstances more complicated, and this may suggest chaos. Furthermore, the nonlinearity in the system is due mainly to the square-law relationship between the current and the magnetic force. The system is at least of fourth order and is nonautonomous. Hence, the degree of complexity of the system permits chaotic behaviour. One of the aims of the present work is to examine the complicated system dynamics (chaotic) and to demonstrate the feasibility of a chaos control method in stabilizing the system dynamics. The experimental search for chaos becomes difficult because of the many possible geometrical degrees of freedom of movement of the suspended object. This may mask possible chaotic behaviour. As a result, we have developed a simple electromechanical device (actuator) whose mechanical movement is limited to one (translational) degree of freedom [4]. Furthermore, the moving member of the electromagnet system is now part of a simple mechanical resonator. This enables a simplified treatment of the initially complicated electromagnetic device.

2. Description of the experimental system

The mechanically variable inductor of the present system is implemented by employing a simple commercial vibrating shaving machine [5]. The system is described schematically in Fig. 1b. The U core and the windings are stationary, while the armature is movable. The movable member is connected to a stationary frame by means of a spring. Damping D should also be taken into account in parallel with the spring. The main damping component in a shaving machine is due to the machine's regular performance and it is therefore usually characterized by Coulomb friction. The blades have been removed in our experiment. Hence, the small remaining friction is mainly viscous. The original electrical system of the shaving machine was not altered. The machine was, however, connected to an ac power source not directly as usual but via a series capacitor which is needed for achieving an electric resonant circuit. A small resistor of 0.1 ohm is also connected in series. It is used for probing the resonator current. Its losses are relatively negligible when compared to the coil losses. It is interesting that the ac power source employed by us in the experiment is not the mains supply. We employed a powerful variable voltage source, whose amplitude and frequency can be controlled manually.

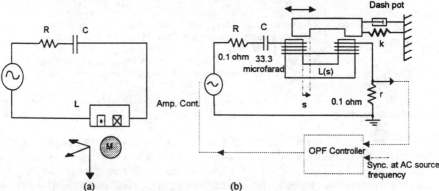

(a)   (b)

Fig. 1 (a) Magnetic levitation by LC tuned circuit. (b) The magnetic actuator. The dotted lines indicate the OPF controller, (s represents the distance coordinate).

The equation describing the mechanical resonator in Fig. 1 (b) is as follows,

$$M\ddot{s} + D\dot{s} + K \cdot (s - s_0) = \frac{1}{2}\Gamma(s)\dot{q}^2 \qquad (1)$$

where M,D, and K are the movable mass ($5 \cdot 10^{-3}$ [Kg]), the viscous damping (0.65 [N$\cdot$Sec/m]) and the spring function respectively. s_0 is a constant ($3.5 \cdot 10^{-4}$ [m]). The right hand side of (1) is the magnetic force, where $\Gamma(s) = -\dfrac{dL}{ds}$, and L(s) is the inductance-distance characteristic. $L(s) = 3.1 + \dfrac{1.2}{0.5 + 120 \cdot s}$ where L in [Hy] and s in [m], see [2]. K=K(s) is also a nonlinear function. The nonlinearity contributed by the spring coefficient K is intended to describe the fact that free motion of the moving member is limited by hitting the rigid frame of the machine. This is modelled by making the spring coefficient relatively large (it becomes almost rigid) at the hitting points.(K(s)=1800 where $-0.002 \le s \le 0.015$ and 12000 elswhere). $\dot{q}$ is the current in the circuit. The second order electrical part of the system can be described by two state variables q, and λ, which represent the charge in the capacitor and the flux linkage respectively. Hence

$$\dot{q} = \frac{\lambda}{L(s)} \quad , \quad \dot{\lambda} = -\frac{\lambda}{L(s)}r_e - \frac{q}{C} + A\cos(\omega t) , \qquad (2)$$

where A is the amplitude of the forcing function implemented by the ac voltage source. r_e - is the equivalent resistor that represents the combined electrical dissipation sources. The nonlinearity of the coupled electromechanical system as a whole is due mainly to the square law behaviour of the magnetic force, as shown in (1), and to the nonlinearity of the spring.

3. The simulation and measurement results

The measurements are accomplished by probing the resonator current and by sensing the armature displacement. The displacement measurement was conducted by a capacitive displacement sensor. The purpose of the experiments was mainly to demonstrate a series of period doublings which leads to a dynamics with a broadband spectrum. The demonstration was accomplished by examining the system behaviour in several cases. In each case, the ac source level was adjusted to a certain value (the frequency of the ac source was constant at 75.5 Hz). The selected values were determined such that the spectrum of the measured voltage and the spectrum of the armature displacement signal had typically distinct characteristics. The measured spectrum for 141 Vrms indicates a line spectrum where subharmonic of order 1/2 can be clearly identified. By increasing the level of the ac source to 161 Vrms, the subharmonic of order 1/4 starts to emerge. When the ac source level was increased to 179 Vrms, the measured spectrum revealed a wide band chaotic like behaviour. Both the probed current and the mechanical displacement signal have the same type of characteristics, e.g. either a line spectrum or a broad band spectrum. It is interesting to see whether chaos can be obtained without the mechanical hitting phenomenon associated with the actuator structure. We have simulated a similar system, where the spring is linear and L(s) characteristic is steeper, and we found that chaos appears also in this case.

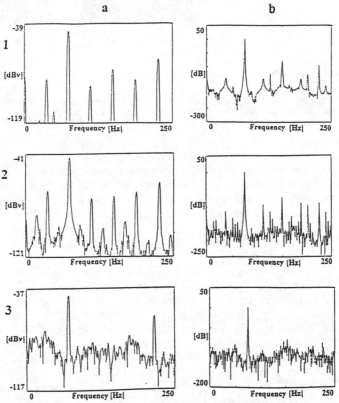

Fig. 2 Measurement results (a), and the related simulation results (b), of the circuit current spectrum for various values of the ac source amplitude. (The frequency of the ac source is 75.5 Hz.) (a1) 141 Vrms, (a2) 161 Vrms, (a3) 179 Vrms. (b1) 130 Vrms, (b2) 147 Vrms, (b3) 180 Vrms.

4. Control of the system vibrations

The present actuator dynamics is similar to that of an unstabilized tuned circuit levitator. Hence, controlling the actuator vibrations appears to possess implications regarding the levitation stabilization. The control method adopted here is OPF (Occasional Proportional Feedback) [5]. This method is usually employed in chaotic systems, and it appears presently suitable, because of the chaotic-like features of the present system. The OPF controller is shown in Fig. 1a (the dotted lines). The controller output stage generates short impulses that occasionally act to increase or decrease the amplitude of the ac source. The stabilization effect is experimentally demonstrated and presented by measuring the Poincare map. The chaotic-like situation is demonstrated here by the area filling feature of the dots in the map, while the controlled state reveals a finite number of points in the map.

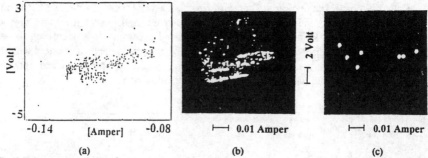

| (a) | (b) | (c) |

Fig. 3 Poincare maps in the q̇ . Vc plane: (a) The poincare map as evaluated from the simulation results (for the case where the amplitude of the ac source is 169 Vrms) (b) The measured poincare map in the experimental system where the ac source is 180 Vrms. (c) The measured poincare map of a motion stabilized by using OPF controller (period 7).

5. Conclusions

The present work demonstrates that the experimental magnetic actuator exhibits a series of period doublings which leads to chaos. Simulation results support the observations. We have experimented with a chaos control method (OPF controller). It has been found possible to stabilize the vibrations and tame them into a regularly stabilized motion.

References

[1]   B. Z. Kaplan, Topological considerations of parametric electromechanical devices and their parametric analysis, *IEEE Trans. Magnetics*, **12** (1976), 373-380.

[2]   B. Z. Kaplan, Estimation of mechanical transients in "Tuned-Circuit" levitators by employing steady state impedances, *J. Appl. Phys.*, **47** (1976), 78-84.

[3]   R. B. Parente, Stability of magnetic suspension device, *IEEE Trans. Aerospace Elect. Syst.*, **5** (1969), 475-485.

[4]   B. Z. Kaplan and G. Sarafian, An RC-Electromechanically Variable L Resonator: The Chaotic Shaver, *Elect. Let.* **33,** (1997), 111-113.

[5]   E. R. Hunt, Stabilizing high-period orbits in a chaotic system: the diode resonator, *Phys. Rev. Lett.*, **67,** (1991), 1953-1955.

Non-Linear Electromagnetic Systems
V. Kose and J. Sievert (Eds.)
IOS Press, 1998

Drive and Control of a High-Tc Superconducting Linear Actuator

Tetsuya Nakao and Tadashi Kitamura

Kyushu Institute of Technology, Kawatsu, Iizuka-city, Fukuoka 820, Japan

Abstract. This paper describes the development of the drive and control system of a compact High-Tc superconducting linear actuator. An MPMG-YBCO super-conductor tile of the linear actuator driven by two solenoids is guided and levitated over two magnetic bearings. Its position, measured by a laser sensor, was satisfactorily controlled by a linear state feedback with biased currents applied to the solenoids. Experimental investigations show that the motion of the actuator is governed by the second order linear system of mass, damper and spring linearly forced by the current, but the spring constant remarkably becomes larger with increase in the currents to the solenoids. For better control performance, another position feedback term should be added to the original feedback control law so that the dependency of the spring constant on the driving current is cancelled.

1. Introduction

YBCO stable bulk superconductors processed by melt-power-melt-growth (MPMG) have been manufactured, and shown to generate strong magnetic forces [1]. Industrial applications, such as bearings for fly wheels, magnets, and actuators, have been investigated [2-4]. We have developed compact actuators and their appropriate magnetic bearings to get basic design information toward low temperature fluid gas pumps and superconducting micromachines [5]. The major advantage of the use of high-Tc superconductors for actuators is that the superconductor can be used for drive as well as non-contact bearing at a lower cost than conventional actuators. This paper reports both static and dynamic characteristics of a centimeter-sized YBCO superconducting linear actuator and also discusses its position control design and experimental results in comparison with computer simulation.

2. Description of the Actuator

Fig.1 and 2 show a schematic diagram of the developed high-Tc superconducting linear actuator and a photograph of the actuator, respectively. The YBCO-MPMG superconductortile, supplied by SRL, is driven back and forth by two solenoids placed on both sides of the superconductor, and is guided and levitated over two magnetic bearings. The bearings are made of rare earth permanent magnet bars arranged in the alternating polarity pattern that makes it possible to stably levitate the superconductor. But this actuator undergoes undesirable constraints in the moving direction by a position dependent spring-like force due to flux gradients of the bearing magnets at both ends. Both solenoids receive the equal and constant biased currents so that the superconductor is equally pressed from both sides for a stiff damping effect, and the driving current superimposed on to the biased currents is switched from one solenoid to another depending on the control law. The solenoids are controlled with PWM amps by a personal computer with a sampling interval of 10msec which is larger than the electric constant, 0.8msec, of the solenoids. The position of the superconductor is measured by a laser position sensor for control.

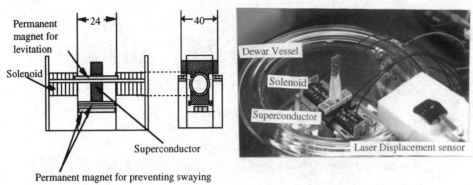

Permanent magnet for levitation

Solenoid

Superconductor

Permanent magnet for preventing swaying

Fig.1 Schematic Diagram of Actuator

Dewar Vessel

Solenoid

Superconductor

Laser Displacement sensor

Fig.2 Photograph of Actuator

3. Basic Characteristics of the Actuator

The static repulsion force is measured by a load cell when the superconductor is kept in touch with the solenoid with the current changed in the range from 0 to 1.7A. As Fig.3 shows, the force is hysteretic, but is modelled as a linear function of the current for a simple control system design. Although the static repulsion force also depends on the distance between the superconductor and the solenoid, this dependence is neglected in this model because the flux density of the two solenoids can be seen to be almost constant in the moving range of the superconductor. Thus, the equation of motion of the superconductor is given by,

$$m\ddot{x} + c\dot{x} + kx = K_f(i + i_b) \qquad (1)$$

where i and i_b are the driving and biased currents, respectively, K_f the motor constant, m the mass of the superconductor of 61g, c the coefficient depending on the viscosity of liquid nitrogen, and k the spring constant due to the magnets for levitation. The back emf in the solenoids is neglected because it is much smaller than the voltage difference across the solenoid's resistance.

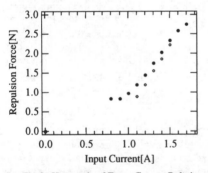

Fig.3 Hysteresis of Force-Current Relation:
● and ○ denote increase and decrease in the current, respectively.

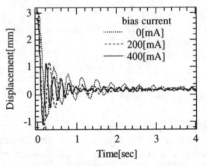

Fig.4 Experimental Results of Free Responses

Dynamic characteristics of the actuator are investigated by experiments of free response, step response and frequency response. The free responses with the initial position of 3mm are shown for three biased currents equally applied to the two solenoids in Fig.4. The parameters of c and k are identified with these free response data, which show that c, k and the frequency of oscillations all increase with increase of the biased currents. The remarkable dependency of k on the current, as shown in Fig.5, is caused by interactions between the solenoids' flux and that of the bearing magnets.

Fig 6 shows step responses to the driving current of 200mA for three biased currents. The motor constant K_p, identified by each step response as kx_s/i_d for the step response of displacement x_s and the driving current i_d, also increases with increase of the biased current. The parameters K_f and c are fixed for each given biased current for control design. Fig.7 shows computer simulation compared with the experimental result for the step response with the biased current of 400mA in Fig.6. This figure shows that for the first three periods there is a good agreement between the experimental data and those obtained by computer simulation.

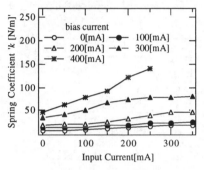

Fig.5 Dependence of k on Biased Current

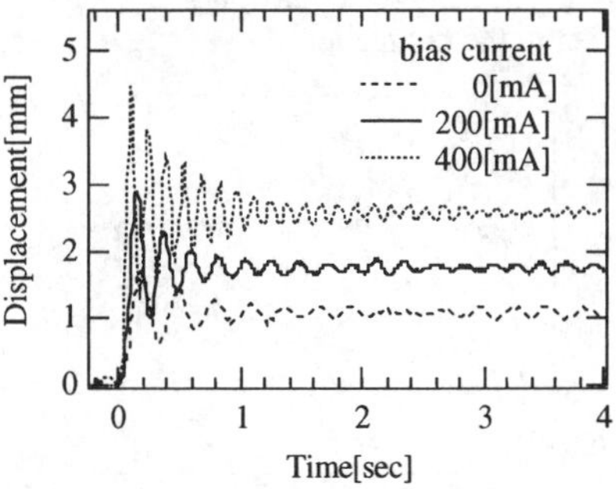

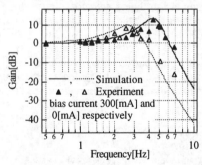

Fig.6 Experimental Results of Step Responses

Fig.8 shows the Bode diagram of the actuator with the amplitude of the driving current of 100mA for the biased current of 0 and 300mA. The resonant frequency becomes higher when the biased current is changed from 0 to 300mA mainly because the spring constant increases. Two resonant frequencies exist in the experimental data for each biased current, which shows characteristics of a higher order system including flux creep and flux flow in the mechanism generating the repulsion force.

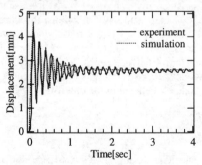

Fig.7 Computer Simulation for Step Response

Fig.8 Bode Diagram of Open-Loop Characteristics

4. Control of the Actuator

Position control is investigated with standard state feedback given by,

$$i = \frac{1}{T_I}\int (x_r - x)dt - k_1 x - k_2 \dot{x} \tag{2}$$

where x_r is the reference position, the feedback gains T_I, k_1, and k_2 are determined by a standard pole assign technique, i.e., $T_l = -K_f /(ms_1 s_2 s_3)$, $k_1 = \{m(s_1 s_2 + s_2 s_3 + s_3 s_1) - k\}/K_f$, $k_2 = \{-m(s_1 + s_2 + s_3) - c\}/K_f$ for assigned poles, s_1, s_2, and s_3. Control performance of the real actuator is much more dependent on k than it follows from linear model with the same controller. Fig.9 shows the remarkable k-dependency of control performance of the real actuator with the assigned triple pole of s=-12 and three k values of 43, 55, and 60 N/m for the biased current of 300mA, c=0.40Ns/m, and K_f=0.45N/A. Another example of the k-dependency is that the control response overshoots when k is changed from 43 to 55 with the biased current of 400mA. As Fig.10 shows, however, this overshoot is removed and the response changes back to the non-oscillatory one with k=43 by assigning the complex poles that cancel the overshoot. The k-dependency of control performance is actually caused by the strong dependency of k on the driving current, and therefore, this fact suggests that better performances should be obtained by adding a current-dependent position feedback term such as $k_a ix$ to the original control law so that the linear dependency of k on the driving current is cancelled in the controller.

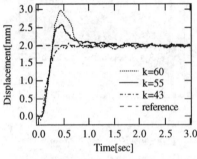   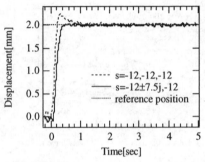

Fig.9 Control Performance Depending on k Fig.10 Cancellation of Overshoot by Complex Poles

5. Conclusion

This paper describes the development of the drive and control system of a compact High-Tc superconducting linear actuator. Its position was satisfactorily controlled by a linear state feedback with biased currents applied to the solenoids. Experimental investigations show that the motion of the actuator is governed by the second order linear system of mass, damper and spring linearly forced by the current, but the spring constant remarkably becomes larger with increase in the current to the solenoids. For better control performances, another position feedback term should be added to the original feedback control law so that the linear dependency of the spring constant on the driving current is cancelled in the controller.

References

[1] M.Murakami, et al: Large Levitation Force Due to Flux Pinning in MPMG Processed YbaCuO Superconductors with Ag Doping, in Advances in Superconductivity III (Eds: Kajimura and Hayakawa) pp.753-756, 1991
[2] F.C.Moon and P.Z.Chang: High-speed rotation of Magnets on High Tc Superconducting bearings, Appl. Phys. Lett. 56, pp.397-399, 1990
[3] B.Takahata, H.Ueyama, and T.Yotsuya: Load Carrying of Superconducting Magnetic Bearings, in Electromagnetic Forces and Applications (Eds. J.Tani and T.Takagi), Elsevier, pp.347-351, 1992
[4] M.Komori and T.Kitamura: Superconducting Actuator Design, Int. J. Applied Electromagnetics in Materials, vol.2, pp. 243-252, 1991
[5] T.Nakao and T.Kitamura: Drive and Control of a High-Tc Superconducting Actuator(2nd Report), Proc. the 8th Symposium on Electromagnetics and Dynamics, pp.315-318, 1996 (in Japanese)

Non-Linear Electromagnetic Systems
V. Kose and J. Sievert (Eds.)
IOS Press, 1998

273

Torsion Sensor Based on the Large Matteucci Effect in FeSiB Amorphous Glass-Covered Wires

Horia CHIRIAC, Sandrino Catalin MARINESCU, Tibor Adrian OVARI and
Valentin NAGACEVSCHI
National Institute of Research & Development for Technical Physics, 6600 Iasi 3, Romania

Abstract. The large Matteucci effect in $Fe_{77.5}Si_{7.5}B_{15}$ amorphous glass-covered wires allows the use of these materials in torsion sensors. The torque sensitivity of the Matteucci voltage induced in these wires is strongly improved after glass removal by chemical etching reaching about 2.5mV/deg. A new sensing device based on $Fe_{77.5}Si_{7.5}B_{15}$ wire, which allows a direct and simultaneous determination of the torsional angle and its direction is proposed. The sensor's characteristic shows a good linearity of the response in the range of $\pm45°$. The operating principle, construction, and characteristics of such a device are presented.

1. Introduction

The remarkable magnetic properties presented by amorphous magnetic materials can lead to their application in various types of sensor and transducers. Soft magnetic amorphous wires with positive magnetostriction present a particular effect, consisting in the generation of a induced voltage across the ends of a twisted wire in the presence of an alternating magnetic field, known as the Matteucci effect [1].

Recently, amorphous glass-covered magnetic wires (AGCMW) have been obtained by the glass-coated melt spinning method [2], with the diameter of the metallic core ranging between 2 and 25 μm and the thickness of the glass cover ranging between 2 and 15 μm. These materials present high internal stresses induced during preparation [3]. The coupling between internal stresses and magnetostriction plays a significant role in the domain structure formation [4]. In the case of $Fe_{77.5}Si_{7.5}B_{15}$ AGCMW, the magnetoelastic coupling between the high positive magnetostriction ($\lambda = 35 \times 10^{-6}$) and internal stresses leads, in a first approximation, to a domain structure consisting of an inner core with axial magnetization and an outer shell with radial magnetization [5]. The existence of a helical anisotropy, i.e. of an inclination of the easy axis of magnetization from the axial direction toward the azimuthal one, was recently reported [6].

Applied torque contributes to the increase of the easy axis inclination and leads to an increase of the azimuthal magnetization, determining an increase of the Matteucci voltage induced in these wires. Due to the helical magnetic anisotropy, $Fe_{77.5}Si_{7.5}B_{15}$ AGCMW present a large Matteucci effect. After the glass removal, this induced voltage increases up to several hundreds of mV [6]. The torque sensitivity of the voltage induced in these wires allows the use of these materials in torsion sensors.

A new sensing device using an $Fe_{77.5}Si_{7.5}B_{15}$ AGCMW, that allows a direct and simultaneous determination of the torsional angle and its direction, is presented.

2. Principle of the Torsion Sensor and Construction

The Matteucci effect was studied by measuring the induced voltage V_M across the wire ends in an alternating axial magnetic field with a maximum value of 2000 A/m at the frequency value of 10 kHz, subjected to applied torque.

The tests were performed on $Fe_{77.5}Si_{7.5}B_{15}$ AGCMW with the diameter of the metallic core 16 μm and the thickness of the glass cover 3.65 μm, having 100 mm in length. The glass removal was achieved by chemical etching with a hydrofluoric acid solution the concentration of which was gradually diminished to avoid the etching of the metal [7].

The determination of the torsional angle requires the detection of the V_M magnitude. With this aim, we analyzed the Matteucci signal synchronic with the square exciting field, as shown in Figure 1 (a).

From the synchronic diagram, one observes that, on the positive alternation of the square exciting field $V_M > 0$ for "positive" torsional direction and $V_M < 0$ for "negative" torsional direction, respectively.

Because the magnitude of the induced voltage V_M is related only to the torsional angle, the determination of the torsional direction was accomplished by means of electronic synchronic detection.

After the synchronic detection performed by an excitation signal controlled electronic switch, the sense of the detected signal allows to discern the torsional direction as it can be observed from Figure 1 (b).

The sensing device proposed, uses a $Fe_{77.5}Si_{7.5}B_{15}$ amorphous wire after glass removal, 16 μm in diameter and 10 mm in length, which allows a direct and simultaneous determination of the torsional angle and its direction by means of the synchronic-detection technique.

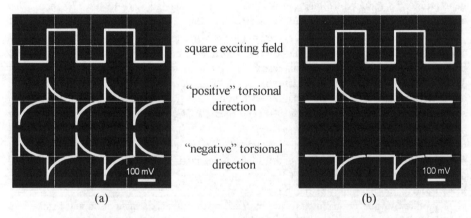

square exciting field

"positive" torsional direction

"negative" torsional direction

(a)   (b)

Figure 1. Synchronic diagram of the square exciting field and Matteucci voltage for "positive" and "negative" torsional direction, respectively; f=10 kHz, torque angle=50 rad/m; (a) normal detection, (b) synchronic-detection.

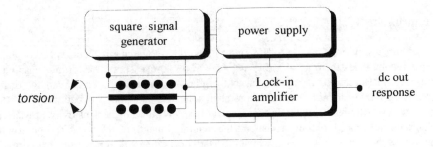

Figure 2. Block-diagram of the torsion sensor based on Matteucci effect in FeSiB glass-covered amorphous wires obtained by the glass-coated melt spinning method after glass removal.

The torsion sensor presented in block-diagram in Figure 2 consists of sensitive element, square signal generator (10 kHz) and synchronic-detection amplifier.

The sensor's characteristic presented in Figure 3, shows a good linearity of the response in the range ±45°.

In the torque linearity range of V_M (± 50 rad/m), the dc signal obtained after peak-detection is proportional with the torsional angle and does not require any correction.

The value of the torque sensitivity of V_M, about 2.5 mV/deg, assures an angular precision better than 20''.

Due to the relatively high value of the torque sensitivity and the good linearity of the Matteucci voltage induced in $Fe_{77.5}Si_{7.5}B_{15}$ wires after glass removal, a torsion sensor based on this effect requires very simple electronics. This fact allows the miniaturization and cost reduction of the device. A possible application of this torsion sensor is a sensitive control of mechanical arms movement.

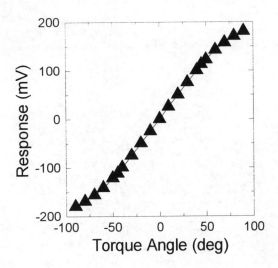

Figure 3. The response of torsion sensor versus the torsional angle.

3. Conclusion

The sensitivity of $Fe_{77.5}Si_{7.5}B_{15}$ amorphous glass-covered magnetic wires to applied torque recommends them as sensitive elements in torsion sensors. The torque sensitivity of wires after glass removal is higher than that of glass-covered wires. The advantage of these wires is that they do not require stress-relief annealing like the conventional amorphous wires. It is worthwhile to mention that sensors based on the torque sensitivity of Matteucci voltage do not require pick-up coils, and the minimum length of the wires can be around 2 mm, that is the minimum length at which the Matteucci effect still appears in such wires. The relatively high torsion sensitivity and the good linearity allows the miniaturization and cost reduction of this sensor.

Such a device was designed and tested, and was found to work well on viscosity measurement apparatus based on torque pendulum method. Another possible application of this torsion sensor is a sensitive control of mechanical arms movement.

References

[1]   P. T. Squire, D Atkinson, M. R. J. Gibbs and S. Atalay, Amorphous Wires and Their Applications. *Journal of Magnetism and Magnetic Materials* **132** (1994) 10-21.

[2]   M. Hagiwara and A. Inoue, Production Techniques of Alloy Wire by Rapid Solidification. In: H. H. Liebermann (ed.), Rapidly Solidified Alloys. Marcel Dekker Inc., New York, 1993, pp. 139-155.

[3]   H. Chiriac, T.A. Ovari, and Gh. Pop, Internal Stress Distribution in Glass-Covered Magnetic Amorphous Wires, *Physical Revue B* **52** (1995) 10104-10113.

[4]   H. Chiriac, T.A. Ovari, Gh. Pop. and F. Barariu, Internal Stresses in Highly Magnetostrictive Glass-Covered Amorphous Wires, *Journal of Magnetism and Magnetic Materials* **160** (1996) 237-238.

[5]   H. Chiriac, T.A. Ovari, and Gh. Pop. Magnetic Behavior of Negative and Zero Magnetostrictive Glass-Covered Amorphous Wires, *Journal of Magnetism and Magnetic Materials* **157/158** (1996) 227-228.

[6]   H. Chiriac, T.A. Ovari, S.C. Marinescu, and V. Nagacevschi, Magnetic Anisotropy in FeSiB Amorphous Glass-Covered Wire, *IEEE Transactions on Magnetics* **32** (1996) 4755-4757.

[7]   H. Chiriac, T. A. Ovari, Gh. Pop, and F. Barariu, *IEEE Transactions on Magnetics* **33** (1997) 782-787.

Non-Linear Electromagnetic Systems
V. Kose and J. Sievert (Eds.)
IOS Press, 1998

Simulation of Impedance Characteristics of a Meander Coil for a Linear Displacement Sensor

H. Wakiwaka, M. Nishizawa, K. Tsuchimichi, K. Masaki* and T. Suganuma*
Shinshu University, 500 Wakasato, Nagano 380-8553, Japan
*Tamagawa Co.,Ltd., 1879 Ohyasumi, Iida 395-0068, Japan

Abstract. The present paper examines the recently developed meander coil type linear displacement sensor. When the detection coil is moved along the scale coil, mutual inductance fluctuates according to the movement, which causes the impedance of the detection coil to change. If the mutual inductance changes as an ideal sine waveform, a highly accurate sensor can be developed. Thus, the displacement characteristics of the mutual inductance were calculated by Neumann's equation and the distortion factor of these characteristics was calculated. When the scale coil width was 1.1 mm, the distortion factor was minimal over the total width of the detection coil.

1. Introduction

We have previously reported the development of a linear displacement sensor that uses meander coils[1]. This sensor was referred to as the Meander coil type Linear displacement Sensor (MLS). Because the excitation frequency of the meander coil is several MHz, a high-speed response can be expected. Furthermore, this sensor is not influenced by dust because the displacement can be detected according to the impedance change that is resistant to environmental influences. In addition, an extremely compact sensor can be constructed by combining the displacement sensor with the meander coil. MLS does not need any connecting line to excite the scale coil and, in so far, differs from Inductosyn[2]. This paper describes the following points:

(1) Comparison of the calculated and measured characteristics of mutual inductance vs. displacement,

(2) Determination and simulation of the distortion factor of mutual inductance vs. coil width,

(3) Simulation of the impedance characteristics vs. displacement relationship using equivalent circuit.

2. Principle operation of MLS

Fig. 1 shows a meander coil above another meander coil. Fig. 2 shows their equivalent circuit. Impedance Z observed at the terminal A-B end of Fig. 2 is expressed by (1). When a high

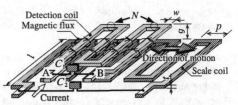

Fig. 1 A meander coil above another meander coil.

$$Z=\cfrac{\cfrac{1}{j\omega C_1}\left[R_1+\cfrac{\omega^2 M^2 R_2}{R_2^2+\left(\omega L_2-\cfrac{1}{\omega C_2}\right)^2}+j\omega\left[L_1-\cfrac{\omega M^2\left(\omega L_2-\cfrac{1}{\omega C_2}\right)}{R_2^2+\left(\omega L_2-\cfrac{1}{\omega C_2}\right)^2}\right]\right]}{R_1+\cfrac{\omega^2 M^2 R_2}{R_2^2+\left(\omega L_2-\cfrac{1}{\omega C_2}\right)^2}+j\omega\left[L_1-\cfrac{\omega M^2\left(\omega L_2-\cfrac{1}{\omega C_2}\right)}{R_2^2+\left(\omega L_2-\cfrac{1}{\omega C_2}\right)^2}\right]-\cfrac{1}{\omega^2 C_1}} \quad (\Omega) \quad (1)$$

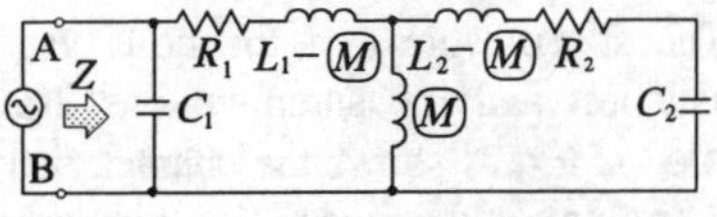

R_1, R_2: Resistance of detection and scale coil (Ω),
L_1, L_2: Self inductance of detection and scale coil (H),
C_1, C_2: Capacitance which is conected detection and scale coil (F),
M: Mutual inductance (H)

Fig. 2 Equivalent circuit of 2 facing meander coils.

frequency excitation current flows through the detection coil, the induced current flows through the scale coil as a result of mutual inductance between the detection coil and scale coil. As a result, the impedance of the detection coil fluctuates. The basic structure of the MLS is as follows: the detection coil moves in the direction of motion; the mutual inductance between the detection coil and the cale coil fluctuates periodically and the impedance of the detection coil varies sinusoidally according to displacement.

Fig. 3 shows the basic structure of the MLS. The MLS is composed of the sensor unit circuit, the position detection circuit, and the counter. The sensor unit circuit is

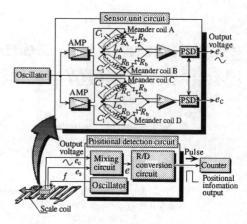

Fig. 3 MLS structure.

composed of phase sensitive detectors (PSD) and two bridge circuits that consist of four meander coils (A-D) and four resonance capacitors. Coil A is displaced by half a pitch with respect to coil B (electrical angle is 180 degree), and coil C is arranged similarly to coil D. Coils A and B are displaced by a quarter pitch to coils C and D (electrical angle is 90 degree) resulting in an output voltage that has a sine waveform and a cosine waveform, respectively. The output voltage is then converted into a pulse that corresponds to a displacement of 12.5 μm, which is 1/128 of the coil pitch of 1.6 mm. This enables the position information to be detected using a counter. If the mutual inductance variation is large, the output voltage from the sensor also becomes large. Moreover, if the mutual inductance changes sinusoidally, the output voltage from the sensor also changes sinusoidally. As the result, a highly accurate sensor can be developed.

3. Mutual inductance characteristics of MLS

In this section the displacement characteristics of the mutual inductance is calculated using Neumann's equation[3] as well as the distortion factor of these characteristics.

3.1 Comparison of the calculated and measured characteristics of mutual inductance vs. displacement relationship according to differences in the scale coil width

The specifications of the coils used for calculation and measurement are listed in Table. 1. Fig. 4 shows the characteristics of the mutual inductance vs. displacement according to the differences in the scale coil width. The detection coil was meander coil No. 4. The scale coil consisted of meander coils No. 1, No. 2 and No. 3. The gap between the detection coil and the scale coil

Table 1   Specifications of the meander coil

Item		Meander coil			
		No.1	No.2	No.3	No.4
Coil line width	w (mm)	0.3	0.5	0.7	0.3
Turn number	N	50	50	50	8
Coil line length	l (mm)	160	160	160	42
Coil line thickness	t (μm)	15	15	15	35
Coil pitch	p (mm)	1.6	1.6	1.6	1.6

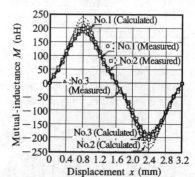

Fig. 4 Characteristics of mutual inductance vs. displacement relationship according to difference in scale coil width (Gap $g = 0.1$ (mm), $f = 2$ (MHz), Detection : No. 4).

was 0.1 mm. When the meander coil conductors of the detection coil and the scale coil are perfectly in the same position the displacement is 0 mm. We moved the detection coil in 0.1 mm increments for a total of 3.2 mm in the direction of motion. From this, the mutual inductance was calculated and measured. The excitation frequency was 2 MHz. The mutual inductance of the meander coil changes in one sine-wave period when the detection coil moves by two pitches. The maximum mutual inductance decreased from 210.0 nH to 185.0 nH (approximately 11.7%) as a result of changing the scale coil width from 0.3 mm to 0.7 mm. Therefore, the scale coil appears to be under the influence of the flux generated by the twin detection coils when the scale coil width is narrow. In contrast, the scale coil appears to be under the influence of the flux which flows back from the adjacent detection coil when the scale coil width is great.

3.2 Determination and simulation of the distortion factor of mutual inductance vs. coil width

The mutual inductance waveform was calculated using the Fast Fourier Transformation (FFT) [4] and the distortion factor of its waveform [5] was determined as follows.

$$d = \frac{\sqrt{|M_2|^2 + |M_3|^2 + \cdots |M_n|^2 + \cdots}}{|M_1|} \times 100 \quad \% \tag{2}$$

where,

$|M_1|$: fundamental element of the mutual inductance waveform (H),

$|M_n|$: nth higher harmonic effective value of the mutual inductance waveform (H)
$(2 \leqq n \leqq 20)$

Fig. 5 shows the characteristics of the distortion factor of mutual inductance vs. coil width. The gap was 0.1 mm and the excitation frequency was 2.0 MHz. The correlation between the detection coil and the scale coil becomes stronger as the coil width decreases. Thus, mutual inductance appears to vary dramatically with the displacement, such that when the coil width is narrow, the distortion factor is great. Furthermore, the distortion factor decreases remarkably as the coil width increases.

When the scale coil width was 1.1 mm, the distortion factor was minimal over the total width of the detection coil. Therefore, a scale coil width of 1.1 mm appears to be optimal for constructing a highly accurate sensor.

Fig. 6 shows the characteristics of the distortion factor of mutual inductance vs. scale coil width when the detection coil width is 0.3, 0.4, and 0.5 mm. When the detection coil

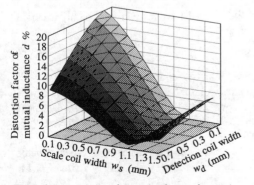

Fig. 5 Characteristics of distortion factor of mutual inductance vs. coil width
(Gap $g = 0.1$ (mm), $f = 2.0$ (MHz), Detection coil : Turn 8, Scale coil : Turn 50).

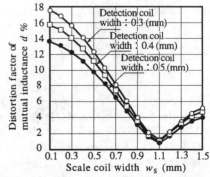

Fig. 6 Characteristics of distortion factor of mutual inductance vs. scale coil width
(Gap $g = 0.1$ (mm), $f = 2$ (MHz)).

width is 0.5 mm, the distortion factor decreases from 13.50% to 0.86% (approximately 94%) as a result of changing the scale coil width from 0.1 mm to 1.1 mm.

4. Simulation of the impedance vs. displacement characteristics using equivalent circuit

Fig. 7 shows the distortion factor of the characteristics of the impedance vs. displacement using an equivalent circuit. The gap was 0.1 mm and the measurement frequency was 2.0 MHz. The scale coil has been short-circuited.

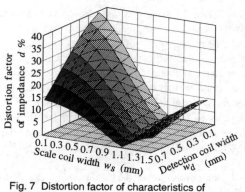

Fig. 7 Distortion factor of characteristics of impedance vs. displacement using equivalent circuit (Gap $g = 0.1$ (mm), $f = 2.0$ (MHz), Detection coil : Turn 8, Scale coil : Turn 50).

The distortion factors of both the impedance vs. displacement characteristics and the mutual inductance vs. displacement characteristics change, such that when the coil width is narrow, the distortion factor is great. Furthermore, the distortion factor decreases remarkably as the coil width increases. When the scale coil width is 1.1 mm, the distortion factor is minimal (less than approximately 4%) over the total width of the detection coil. However, the distortion factor appears to be influenced by the measurement frequency, gap and condition of the associated scale coil.

5. Summary

The following results were found:

(1) Mutual inductance of the meander coil changed in a sine waveform on 2 pitches. The maximum mutual inductance decreased from 210.0 nH to 185.0 nH (approximately 11.7%) as a result of changing the scale coil width from 0.3 mm to 0.7 mm.
(2) When the scale coil width was 1.1 mm, the distortion factor of the mutual inductance characteristics was minimal over the total width of the detection coil.
(3) When the scale coil width was 1.1 mm, the distortion factor of the impedance characteristics was minimal (less than approximately 4%) over the total width of the detection coil.

Thus, a scale coil width of 1.1 mm appears to be optimal for constructing a highly accurate sensor.

References

[1]   H. Wakiwaka, M. Nishizawa and S. Yanase, Analysis of Impedance Characteristics of Meander Coil, *IEEE Transactions on Magnetics*, Vol.32, No.5, 1996, pp.4332-4334.
[2]   LI C, GAO R X, Error Compensation Techniques for a Linear Inductosyn Displacement Measurement System, *B0689B Conf Proc IEEE Instrum Meas Technol Conf*, Vol.1995, 1995, pp.364-369.
[3]   O. Oshiro, H. Tsujimoto and K.Shirae, High frequency characteristics of a planar inductor and a magnetic coupling control device, *Journal of the Magnetic Society of Japan*, Vol.14, No.2, 1990, pp.411-414.
[4]   S. Koike, Calculation of scientific techniques by C, *CQ Publishing*, 1987, pp.263-265.
[5]   S. Ohsita, Electric circuit, *Kyoritsu Shuppan*, 1987, p.128.

Non-Linear Electromagnetic Systems
V. Kose and J. Sievert (Eds.)
IOS Press, 1998

Physical Principles of Impact Energy Measurements Using a Contactless Magnetostrictive Sensor

Leszek MAŁKIŃSKI
Institute of Physics of the Polish Academy of Sciences,
al. Lotników 32/46, PL-02-668 Warszawa, Poland

Abstract. A prototype of a magnetoelastic sensor for impact energy measurements has been studied in order to determine the physical phenomena responsible for the sensor operation. It was found that the shape and amplitude of a signal in the pick-up coil produced by a magnetoelastic shock wave strongly depend on the geometry of the cross-section of the magnetostrictive rod subjected to the impact. This was due to eddy current shielding of the inner core of the rod. The nonlinear characteristics of the sensor are presented. The effect of thermomechanical treatment has also been studied. The best repeatability of the impact energy measurements was obtained for magnetic biasing fields between 15 and 30 kA/m. It was concluded that linear magnetostriction and rotations of magnetization vectors in domains located close to the surface of the rod are responsible for generation of magnetoelastic shock waves

1. Introduction

Strain gauge sensors are commonly used to measure impact energy produced by drilling or chiseling machines. A certain disadvantage of this technique is frequent damage of lead wires, which can be easily avoided by applying a contactless sensor based on magneto-elastic principle. The technical description of two designs of magnetoelastic sensors for impact energy measurements have been published by Hecker [1] and Rahf et al. [2].

The aim of this paper is to discuss the physical phenomena which relate to the principle of magnetoelastic sensors operation and therefore are important for design of such sensors.

2. Experimental

The energy of impact was determined by the kinetic energy of a hammer which was shot with adjustable speed by a pneumatic gun and hit one end of a cylindrical steel rod. The magnetoelastic wave propagating along the rod was detected by a system of three coaxial coils. One of the coils was used to produce a biasing magnetic field and the two other coils (sensing coil and trigger coil) measured changes in the magnetic flux caused by the magnetoelastic wave. The sensing coil was placed in the middle of the biasing coil, like in the Hecker's sensor [1], whereas the position of the trigger coil was movable in order to change the delay time between the voltage pulses in the two coils.

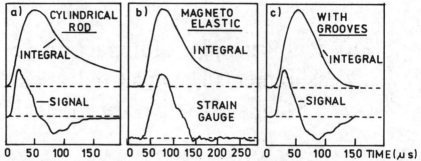

Fig.1. Signals and their integrals resulting from the magnetoelastic wave propagation: a) proper shape of the signal measured in the cylindrical rod, b) comparison of integrated output signal of the magneto-elastic sensor with strain gauge measurements, c) signal measured in cylindrical rod with grooves

3. Magnetization Processes and Hysteretic Effects

The proper shape of the signal induced in the pick-up coil, presented in Fig.1a), consists of positive and negative parts. An integrated signal reflects behaviour of the magnetization in the rod caused by the shock wave. The shape of the ascending part of this signal agrees well with the waveforms measured by strain gauges (Fig.1b). However, the decay of the signal in the pick-up coil is much slower due to eddy currents. The effect of the eddy currents was revealed when 8 thin grooves were made along the rod. The magnitude of the signal in such a rod increased by more than 2 times and the magnetization decay was much steeper, as demonstrated in fig.1c. This is the evidence that the main contribution to the signal originates from the outer shell of the rod, because the magnetic flux from the inner part of the rod is shielded by the eddy currents. Moreover, the component of the magnetization change transverse to the rod surface showed a much faster decay than the axial one due to the different configuration of eddy currents.

The signals observed in a longer time scale contain waves reflected from the ends of the rod. They may also include reflections from the hammer. If the biasing field in the vicinity of the detection coil is inhomogeneous the displacements of the whole rod may also induce voltage pulses. The shielding of the detection coil is important to minimize stray noise. A noise to signal ratio in a properly designed magnetoelastic sensor is better than in the equivalent strain gauge sensors. In order to detect the whole magnetoelastic pulse as shown in Fig. 1 the distance between the pick-up coil and the working end of the rod must be larger than 0.4 m, otherwise overlapping of reflected signals would occur.

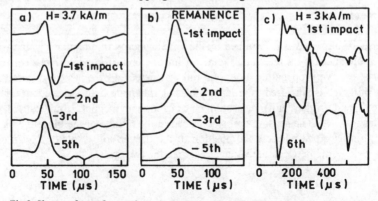

Fig.2. Shapes of waveforms after subsequent impacts for different initial magnetic states.

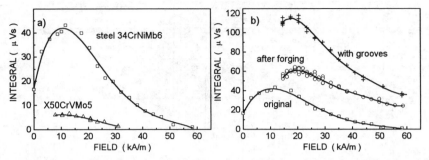

Fig.3. The dependence of the integrated sensor signal on magnetic bias in rods made from different steels a) and the effects of the forging and grooving on the magnetic field dependence of the sensor signal b)

It is important to mention that the amplitude and the shape of the output signal of the sensor depend on the biasing field. The signal vanishes at sufficiently strong fields close to the magnetic saturation of the rod. On the other hand, at low magnetic fields comparable with the coercivity of the steel rod, the signal exhibits irreversible behaviour dependent on the magnetic history of the rod. It has been shown in Fig.2 that not only the peak value of the pulse can change after subsequent impacts but also its shape and sign. In this range of the fields a maximum of the Barkhausen noise has been detected, which originates from irreversible displacements of domain walls.

All these properties indicate that linear magnetostricion is responsible for the generation of magnetoelastic waves. Unlike volume magnetostriction, the linear magnetostriction vanishes with the approach to magnetic saturation and is sensitive to magnetization processes at lower magnetic fields. Therefore, it can demonstrate both anisotropic and hysteretic behaviours.

The effect of forging, presented in Fig.3b, which changes domain structure and magnetic anisotropy of the rod, is an additional confirmation of the fact that the magnetoelastic waves are strongly dependent on domain processes. Fig.3a shows the dependences of the sensor signal on the magnetic biasing field for two kinds of steels with different magnetostriction coefficients. In order to well define the magnetic state before each impact measurement the impacts the rod was magnetically saturated and then the bias field was reduced. For negative biasing fields the signal decreases to zero at the coercive field, reaches minimum and then approaches zero at higher negative fields. This behaviour becomes clear if the signal is compared with the magnetostrictive strain (Fig.4). The slope of the hysteresis loop of the magnetostricive strain determines the sensitivity of the steel to stresses: it is small at the saturation and coercive field and has its maximum at the fields corresponding to highest slope of the magnetostriction curve.

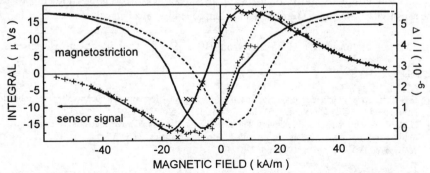

Fig.4. A comparison of the sensor signal for different bias fields with the magnetostrictive strain of the rod

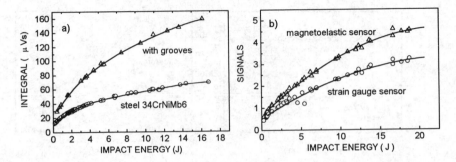

Fig.5. a) Impact sensor characteristics for cylindrical and grooved rods made of 34CrNiMo6 steel
and b) A comparison between characteristics of the magnetoelasic and strain gauge impact sensors

This interpretation explains also the complicated shapes of the pulses at the coercive field
where the derivative of the magnetostriction changes its sign.

Thus, from the point of view of sensor applications two ranges of biasing fields are
excluded: weak fields for which hysteretic and irreversible phenomena occur and strong
fields, because of the weak sensitivity. These stray effects are easily avoided by adjusting
the magnetic bias in the range from 15 to 30 kA/m, which ensures simultaneously high
level of the sensor signal and excellent repeatability of the measurements. The reversible
rotations of magnetization vectors are commonly considered to be the predominant
magnetization mechanism in this case.

4. Sensor Characteristics

The dependences of the sensor signal on the impact energy for the rods with different
geometry of the cross-sections are presented in Fig.5a. In spite of certain nonlinearity of
these characteristics the magnetoelastic sensors exhibit better accuracy (below 1%) of the
measurements, a faster dynamical response, as well as higher signal to noise ratio than the
strain gauge ones. A comparison between the two kinds of sensors is given in Fig.5b. It is
worth mentioning that the shapes of the magnetoelastic pulses remain almost unchanged for
the impact energies up to 15 J, though their heights change more than one order of
magnitude.

Thus it can be concluded that it is possible to make a contactless magnetoelastic sensor
for impact energy testing which can compete with strain gauges. The optimum conditions
of the sensor operation are achievable for magnetic bias fields around 25 kA/m which
eliminates undesirable effects of magnetic hysteresis and irreversible domain processes.
The sensitivity of the measurements can be improved by mechanical treatments of the rod.

The author is grateful to Dr. H. Kousek from HILTI Co. for enabling the measurements.

References

[1] R.Hecker, Anwendung des magnetoelstischen Effekts zur Messung von Dehnwellen in stabförmigen
     Körpern schlagender Maschinen", Techn. Messen, 55, no.5 (1988) 132
[2] L. Rahf, T. Kiesewetter and J.Sievert, The measurement of shock wave energy by means of magneto-
     elastic sensor, IOS Studies in Applied Electromagnetics and Mechanics 10 (A.Moses and
     A. Basakeditas) Amsterdam 1996, pp. 776-779

Non-Linear Electromagnetic Systems
V. Kose and J. Sievert (Eds.)
IOS Press, 1998

Non-Linear Effect in the Switching
Mechanism of Fe-Based Amorphous Wires

Horia CHIRIAC, Tibor-Adrian ÓVÁRI, Cătălin-Sandrino MARINESCU, Firuța BARARIU
National Institute of R&D for Technical Physics, 47 Mangeron Blvd., 6600 Iași, Romania

Manuel VÁZQUEZ, Antonio HERNANDO
Inst. de Magnetismo Aplicado UCM-RENFE and ICM-CSIC, Las Rozas, Madrid, Spain

Abstract. Results on the changes in the characteristics of the large Barkhausen effect in amorphous Fe-based wires due to modifications of the wire ends' relative position with respect to the ends of the magnetizing coil are reported. These studies are important as regarding the exact value of the switching field at which this effect occurs, as well as the applications of these wires in magnetic sensors based on it.

1. Introduction

One of the most peculiar aspects of the magnetic behavior of amorphous wires is the appearance of a large Barkhausen effect (LBE). This effect consists of a one-step magnetization reversal at a certain value of an axially applied magnetic field, called the switching field. The presence of the LBE was reported in water-quenched amorphous wires - the so-called 'conventional' amorphous wires (CAW) with diameters of 80 to 300 μm - but also in amorphous glass-covered wires (AGCW) with diameters of 3 to 25 μm [1, 2]. LBE was observed in positive and negative magnetostrictive CAW, in positive magnetostrictive AGCW and in such wires after glass removal, and also in negative and nearly zero magnetostrictive AGCW after glass removal [1-3].

The aim of this paper is to analyze the switching mechanism from the point of view of the influence of the wire ends relative position with respect to the ends of the magnetizing coil, since changes in this position alter the magnitude of the switching field, reflecting modifications of the well-known nucleation-propagation process. The results are important from the point of view of the experiments concerning the LBE, i.e. of the exact values of the switching field that are reported in numerous papers, as well as from the point of view of the applications of these wires, i.e. of the practical aspects of LBE-based magnetic sensors.

2. Experimental Results

We prepared positive magnetostrictive $Fe_{77.5}Si_{7.5}B_{15}$ CAW by in-rotating-water spinning and AGCW by glass-coated melt spinning. For our experiments we have chosen a CAW with a diameter, Φ, of 110 μm, and three AGCW with a metallic wire diameter, Φ_m, of 2.1, 7.5, and 13.8 μm, and a glass cover thickness, t_g, of 10, 7.3, and 2.1 μm, respectively. Magnetic measurements were performed by a fluxmetric method, using an alternating field with the maximum value of 3500 A·m⁻¹, at a frequency of 500 Hz. We measured the switching field at which the LBE occurs for all the samples, for different positions of the wire ends with respect to the ends of the magnetizing coil, with a 2.5 cm long pick-up coil placed in the middle of the magnetizing one. All the measured samples were 60 cm long, the length of the magnetizing coil being 10 cm. We mention that the reported results hold irrespective of these dimensions, as long as the samples are longer than the critical length for which LBE appears [1, 2].

Fig. 1 shows the dependence of the switching field H^* on the distance L between one end of the wire and the corresponding end of the magnetizing coil for the $Fe_{77.5}Si_{7.5}B_{15}$ CAW. The other end of the wire is always left outside the coil. One observes that for $L > 10$ cm, H^* is constant, having a value of 240 A·m^{-1}. For $0 < L \leq 10$ cm, H^* decreases as the wire end comes closer toward the end of the magnetizing coil, reaching to about 230 A·m^{-1} for $L = 0$. This value of H^* remains constant if the wire end is placed inside the magnetizing coil ($L < 0$), and even if both wire ends are inside this coil.

Fig. 2 illustrates the same dependence for the three $Fe_{77.5}Si_{7.5}B_{15}$ AGCW. One observes that H^* has much larger values for AGCW than for CAW, and it increases with the decrease of the wire diameter Φ_m and the increase of the glass cover thickness t_g, as has been previously reported [2]. For $L > 5$ cm, H^* is constant, being 2070 A·m^{-1} for the AGCW with $\Phi_m = 2.1$ µm and $t_g = 10$ µm, 1280 A·m^{-1} for the AGCW with $\Phi_m = 7.5$ µm and $t_g = 7.3$ µm, and 100 A·m^{-1} for the AGCW with $\Phi_m = 13.8$ µm and $t_g = 2.1$ µm. For $0 < L \leq 5$ cm, for the AGCW with $\Phi_m = 13.8$ µm and $t_g = 2.1$ µm, H^* decreases as the wire end comes closer toward the end of the magnetizing coil, reaching to about 57 A·m^{-1} when $L = 0$. This value of H^* remains constant even if the AGCW is placed with one end or entirely inside the coil. For the AGCW with $\Phi_m = 2.1$ µm and $t_g = 10$ µm, as well as for the one with $\Phi_m = 7.5$ µm and $t_g = 7.3$ µm, one observes an increase of H^* prior to the decrease, which still occurs before the wire end is placed inside the coil. This effect - that can be called 'proximity effect' - is more prevalent for AGCW with small Φ_m and large t_g. For both AGCWs, H^* remains constant after the wire end is placed inside the coil ($L \leq 0$): 200 A·m^{-1} for the one with $\Phi_m = 7.3$ µm and 400 A·m^{-1} for the one with $\Phi_m = 2.1$ µm. These values of H^* remain constant even if the AGCWs are placed entirely in the magnetizing coil (with both ends).

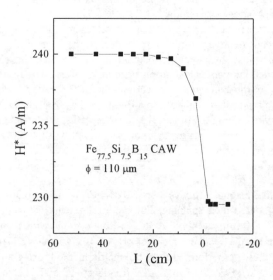

Fig. 1. Dependence of the switching field H^* on the distance L between one end of the wire and the corresponding end of the magnetizing coil for the $Fe_{77.5}Si_{7.5}B_{15}$ 'conventional' amorphous wire (CAW).

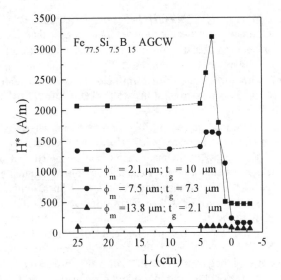

Fig. 2. Dependence of the switching field H^* on the distance L between one end of the wire and the corresponding end of the magnetizing coil for three $Fe_{77.5}Si_{7.5}B_{15}$ amorphous glass-covered wires (AGCW).

3. Discussion

It is meaningful to begin with a brief review of the current status of the domain structures in $Fe_{77.5}Si_{7.5}B_{15}$ CAW and AGCW. The domain structure of $Fe_{77.5}Si_{7.5}B_{15}$ CAW consists of an axially magnetized inner core (IC), that lies on about 70% of the wire volume, and a radially magnetized outer shell (OS) [1]. The domain structure of $Fe_{77.5}Si_{7.5}B_{15}$ AGCW is qualitatively similar, but the IC occupies about 90% of the wire volume [2]. Previous studies emphasized a direct relationship between domain structure and LBE, i.e. that LBE occurs in the IC. Studies on the switching mechanism in CAW showed that at each wire end there is an axially magnetized domain with reversed magnetization, the 180° domain wall between this domain and the IC being the wall that propagates on the wire length when the applied field is equal to H^* [1]. These 'end' domains with reversed magnetization are pre-existent in the wires and they appear to diminish the overall demagnetization energy of the samples.

Let us discuss our results by considering the above mentioned domain structures, and also their correlation with the LBE. We begin with AGCW, since the dependence of H^* on L is more complex in such wires, and we expect its characteristics in CAW to be a particular case of the same effect in AGCW. The most complex dependencies of H^* on L are observed for the AGCWs with Φ_m of 2.1 and 7.5 µm. We can separate three regions in Fig. 2: (i) - $L > 5$ cm, where H^* is constant, (ii) - $0 < L \leq 5$ cm, where H^* increases and then decreases (the 'proximity' effect), and (iii) - $L \leq 0$, where H^* is also constant.

In region (i), both wire ends are well outside the magnetizing coil, and the pre-existent 'end' domains are not affected by the field of the magnetizing coil. Therefore, the 'end' domains do not participate in the LBE, and it is necessary to nucleate a new domain with reversed magnetization in the wire, somewhere inside the coil ($L \leq 0$), in order to attain the LBE. The newly formed 180° wall will then propagate on the wire length. In this case, the

LBE consists of *nucleation and propagation*, H^* being the nucleation field H_N, according to the switching mechanism. Since the propagation field H_P is smaller than H_N ($H_N > H_P$), the 180° wall propagates immediately after the domain with reversed magnetization is nucleated.

In region (iii), one of the wire ends is inside the coil ($L \leq 0$), and the corresponding pre-existent 'end' domain is within the range of uniform field. Thus, no nucleation is required, but only *propagation* of the already existent 180° wall. In this case, H^* is the propagation field H_P, smaller than H_N - that is H^* in region (i).

The most interesting things occur in region (ii) - $0 < L \leq 5$ cm, where H^* increases over the value of H_N, and subsequently decreases to H_P ('proximity' effect). In this region, the 'end' domain is placed in the vicinity of the magnetizing coil's end. Thus, it 'feels' the field, but in this region the field is not uniform, and its axial component is much smaller than its value inside the coil, due to the strong dispersion of the field lines. When the field is H^* inside the coil, in its outside vicinity (less than 5 cm distance from coil's end), the axial component is smaller, but if the field from the outside vicinity reaches the value that corresponds to H^* (i.e. H_P), the 180° wall propagates, and LBE can be detected. At the same time, the field inside the coil is larger than H_N. While the 'end' domain comes closer to the zone of uniform field, and even begins to enter this zone, the 'proximity' effect is damped, since the difference between the field measured inside the magnetizing coil (with the pick-up one) and the one from its immediate outer vicinity decreases as L decreases. For larger wire diameters, the increase of H^* in region (ii) is smaller, since the 'end' domains are longer due to larger demagnetizing effect, and thus, the 'end' domain reaches faster the zone of uniform field. Hence, over a given diameter, the 'proximity' effect vanishes. This is the case of the AGCW with $\Phi_m = 13.8$ μm and of CAW. The 'proximity' effect does not imply changes in the switching mechanism, the LBE being a simple *propagation* of the pre-existent 180° wall between the 'end' domain and the IC, but it is only a result of the spatial difference between the place where H^* is measured and the place where LBE begins.

4. Conclusion

The measured value of H^* at which LBE occurs in CAW and AGCW, together with the switching mechanism itself, depend on the position of the wire ends with respect to the ends of the magnetizing coil. If both wire ends are well outside the coil's end, LBE consists of *nucleation and propagation*, and H^* is the nucleation field H_N. If one or both wire ends are inside the coil, LBE consists of *propagation* only, and H^* is the propagation field H_P. For AGCWs with small Φ_m and large t_g, one observes the 'proximity' effect, i.e. that H^* increases over H_N, before its decrease to H_P, if one of the wire ends is in the immediate outer vicinity of the coil end, and the other one is well outside. The results are of practical importance, related to the possibility of adjusting the value of H^* in LBE-based sensors.

References

[1] M. Vázquez and D.-X. Chen, The Magnetization Reversal Process in Amorphous Wires, *IEEE Transactions on Magnetics* **31** (1995) 1229-1238.

[2] H. Chiriac, T.A. Óvári, Gh. Pop and F. Barariu, Effect of Glass Removal on the Magnetic Behavior of FeSiB Glass-Covered Wires, *IEEE Transactions on Magnetics* **33** (1997) 782-787.

[3] H. Chiriac, Gh. Pop, T.A. Óvári and F. Barariu, Magnetic Behavior of Negative and Nearly Zero Magnetostrictive Glass-Covered Amorphous Wires, *IEEE Transactions on Magnetics* **32** (1996) 4872-4874.

Non-Linear Electromagnetic Systems
V. Kose and J. Sievert (Eds.)
IOS Press, 1998

Measurement and Valuation of Touch Sensation
(AE Sensor Readings Compared with Tactile Perception of Forefinger)

Mami TANAKA,   Seiji CHONAN,   Zhong-Wei JIANG,   Tomohiro HIKITA
Department of Mechatronics and Precision Engineering, Tohoku University, Sendai, Miyagi 980-77, Japan

Abstract. This paper is concerned with the measurement and evaluation of the human touch sensation. First, the fabrics such as crepe, velvet and corduroy are laid inside a blind box and touched and stroked by the forefingers of examiners and their feelings of touch are obtained and classified by means of questionnaires. Next, the AE sensor is pressed and slid over the same samples and the output signals from the sensor are recorded. The features on the collected data are then extracted by using the auto-correlation function and the FFT analysis and compared with the valuation of forefingers. It is seen that the AE sensor readings partially correspond with the tactile perception of the forefingers.

1.Introduction

Of all our senses, the sense of touch is the most intimately connected to our feeling. In daily life, various kinds of things are touched by the fingers and their physical as well as geometrical features are extracted and evaluated. Thus, the sense of touch is indispensable to our life like the sense of sight. In our tactile function of fingers, the digital pulp is pressed against the object and then the stroke or rubbing action is started over the surface of an object to feel its quality. It seems that most of the force sensors developed so far have been functional to the measurement of applied force magnitude and no sensors are available for the evaluation of the texture of the object materials.

This paper is a study on the comparison of the human tactile perception with the output of the piezoceramic pressure transducer. First the fabrics such as silk crepe, velvet or corduroy are laid inside the blind box and touched and stroked by the forefingers of examiners and the feelings of touch on the objects are questioned and classified. Next, the AE sensor is pressed and slid over the same samples and the output voltage signals are collected. The characteristics of the sensor's measurements are then extracted by using the auto-correlation function and the FFT analysis and the obtained results are compared with the tactile perception of the forefinger. Further, the surface condition of the objects is observed by using a microscope and the correspondence of the sense of touch with that of sight are compared .

2.Investigation on Touch of Sample Fabrics(Experiment 1)

In this investigation, five fabrics were picked up as the sample objects to sense the texture. They were wool, velvet, corduroy, silk crepe and polyethylene. The examiners were asked how the materials felt and their opinions were classified. First, twenty examiners (10 female and 10 male) were asked to touch and rub small pieces of fabrics, which were laid inside a blind box so that the examiners could evaluate the textures only by the sense of touch. The direction of rubbing over the object was one way since the texture is dependent on the direction of the fibers. The examiners were asked how the objects felt and whether the touch was comfortable or not. The opinions on the wool and the polyethylene were quite variable and no general conclusion could be drawn. Because of this, in the following, the results are presented only for velvet, corduroy and silk crepe. Table 1 presents the feelings of touch and the corresponding number of examiners' answers for the fabrics (a)velvet, (b)corduroy and (c)silk crepe, respectively. The opinion given by the majority of examiners are considered to be the major feeling on the fabric. It is seen that the velvet and corduroy give similar results. Further, the overall evaluation of touch is comfortable for the velvet and the corduroy, while it is not for the silk crepe as given in Table 2.

3. Measurement by Piezoceramic AE Sensor (Experiment 2)

Next, the output voltage signals from the piezoceramic sensor[FN Electric Instrument AE-900S-WB] were obtained   by sliding the sensor over the fabrics. When the fingers are slid over the object, the force magnitude is adjusted depending on the information to be collected. In general, the quality of surface is sensed by the stroke over the fabric, while the stiffness and the elasticity is evaluated by rubbing or pushing the object strongly. Considering those functions of the fingers, the output voltage of the sensor was measured both for the stroke and the hard rubbing of the fabrics.

The output voltage signals from the piezoceramic sensor are presented in Figs.1—3. In each figure, (a) shows the voltage for the fabric stroked   gently and (b) the output when it is rubbed strongly. It is seen that the output signals have different features depending on the fabric. Table 3 presents the average peak-to-peak amplitude obtained from three measurements. It is seen that the amplitude becomes greater in the order of the velvet, corduroy and silk crepe both for (a) soft stroke and (b) hard rubbing. Next, the power spectrum densities were obtained for the output of Figs.1—3 and are presented in Figs. 4—6. It is seen that the high frequency signals are comparatively significant for the velvet and corduroy, while the low frequency components are remarkable for the silk crepe. It is further noted that the high frequency components are significant for the soft stroke while the low frequency signals are remarkable for the hard rubbing, which is similar to the human tactile functions where the qualitative and quantitative information is collected selectively by changing the force magnitude of the finger to the object.

Table 1   Feelings on fabrics

(a) Velvet

Feeling	No. of Answers
Light	8
Soft	7
Thick	4
Fine	2

(b) Corduroy

Feeling	No. of Answers
Soft	5
Thick	4
Light	3
Fine	3

(C) Silk Crepe

Feeling	No. of Answers
Rough	18
Thin	3

Table 2 Overall evaluation of touch.

Fabric	Evaluation
Velvet	Comfortable
Corduroy	Comfortable
Silk Crepe	Not Comfortable

The comparison of Table 3 and Figs.4—6 with Table 1 leads to an understanding that the sensor readings of small amplitude and high frequency correspond to the feeling " fine" and the readings of large amplitude and low frequency to the feeling "rough" in the case of human perception.

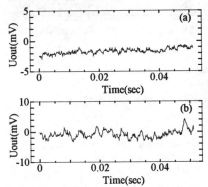

Fig.1 Output of sensor when the velvet is (a) stroked gently and (b) rubbed strongly.

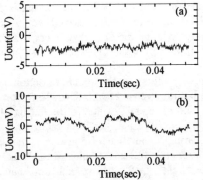

Fig.2 Output of sensor when the corduroy is (a) stroked gently and (b) rubbed strongly.

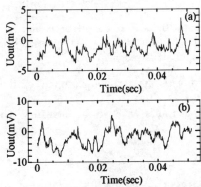

Fig.3 Output of sensor when the silk crepe is (a) stroked gently and (b) rubbed strongly.

Table 3 Average peak—to—peak amplitude

Fabric	Soft Stroke	Hard Rubbing
Velvet	1.2mV	1.9mV
Corduroy	1.6mV	3.5mV
Silk Crepe	3.0mV	8.2mV

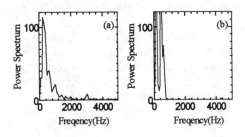

Fig.4 Power spectrum on sensor's measurement when the velvet was (a) stroked gently and (b) rubbed strongly.

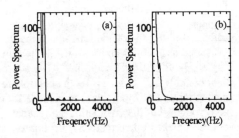

Fig.5 Power spectrum on sensor's measurement when the corduroy was (a) stroked gently and (b) rubbed strongly.

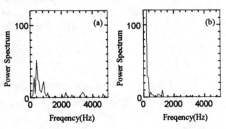

Fig.6 Power spectrum on sensor's measurement when the silk crepe was (a) stroked gently and (b) rubbed strongly.

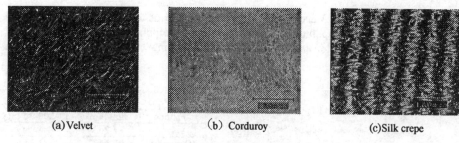

(a) Velvet   (b) Corduroy   (c) Silk crepe

Fig.10 Surface conditions of fabrics observed with a microscope.

4. Comparison of Human Tactile Perception with AE Sensor Measurements

The feeling of the forefinger (Exp.1) is compared with the output of the ceramic sensor (Exp.2). It is seen from Table 3 and Figs. 4—5 that the voltage amplitudes are small but the frequency components are significant at frequencies higher than 400Hz both for the velvet and for the corduroy, which correspond to the feeling "fine". As for the silk crepe, the voltage amplitude is greater than that of the velvet and corduroy but the signals are not distributed at low frequencies range significant compared with the output on the velvet and corduroy, which would result in the feeling "rough" in the case of human perception.

To confirm the surface conditions, the fabrics were observed with a microscope and are presented in Fig.7. It is seen that the surfaces of velvet and corduroy are covered with long fibers while the surface of silk crepe has fine periodic wrinkles but no extending fibers, which will result in the feelings "light", "soft" and "thick" on the velvet and the corduroy, and the feeling "rough" on the silk crepe.

Finally, the discrimination ability of the AE sensor relating to the touch sensation is summarized. By the tactile function of the forefinger, both the velvet and corduroy have been evaluated as "comfortable", which is the total impression of the fabrics due to the feeling of "light", "soft", "thick" and "fine". The silk crepe, on the other hand, has been estimated as "not comfortable", which is by the multiple feeling of "rough" and "thin". Compared with the perception of the forefinger, the piezoceramic AE sensor feels the surface qualities as "fine" and "rough", but the volumetic features represented by "light", "soft", "thick" and "thin" have not been picked up cerrectively, the improvement of which is beyond the scope of this paper and will be discussed in future work.

5. Conclusions

A comparison of   the human tactile perception with the output   of the piezoeceramic AE sensor has been presented when both the sensor and the human forefinger were pressed and slid over a variety of fabrics. The obtained results can be summarized as follows.

1. The AE sensor readings partially correspond with the tactile perception of the forefinger.
2. The active sensing using the piezoceramic AE sensor is effective for simulating the tactile functions of human fingers.

References

[1] K. Mori "Tribosensor –the Basic Concept of a New Tactile Sensor", 37-2(1992),91-95.
[2] S.Chonan, Z.W.Jiang, K.Mori and Y.Munekata " Development of Tribosensor System Using Neural Networks (Tribosensor Using Piezoelectric Ceramics)", Transactions of the JSME C, 61(1995),2996—3003.

Non-Linear Electromagnetic Systems
V. Kose and J. Sievert (Eds.)
IOS Press, 1998

A New Touch-Stress Sensor Based on Amorphous Wires

Leszek MAŁKIŃSKI and Johannes SIEVERT[*]
Institute of Physics of the Polish Academy of Sciences,
al. Lotników 32/46, PL-02-668 Warszawa, Poland
[*]Physikalisch-Technische Bundesanstalt,
Bundesallee 100, D-38116 Braunschweig, Germany

Abstract. The idea of the resonant-antiresonant method is used to design new touch sensors based on amorphous wires. Ultrasonic resonant vibrations of the wire which is fixed at one end or in the middle of its length are excited by ac magnetic field due to magnetoelastic properties of the wire. Changes of impedance or permeability resulting from applied stresses are detected. The main advantages of the sensor are: low cost, security, high sensitivity, small cross-section, fast frequency response and a large temperature range of operation

1. Introduction

Metallic glass ribbons have been successfully applied in a number of magnetoelastic sensors for last two decades [1-4]. The stress or strain sensors based on amorphous ribbons use the excellent soft magnetic properties (high magnetic permeability and small hysteresis) and relatively large magnetostrictions (several times higher than those of conventional, polycrystalline alloys), resulting in a huge magnetoelastic coupling, e.g. [5,6]. An increasing number of applications of amorphous wires have been noted for last few years. Though the magnetostriction of the wires remains the same as that of the ribbons, the cylindrical shape of the wires gives rise to differences in the domain structure, shape anisotropy, eddy current losses, as well as residual stress distribution.

2. Principle of Operation

The idea of the resonance-antiresonance method [7,8] was used to design a very sensitive touch-stress sensor. In this method, longitudinal, resonant vibrations of a sample are excited by an alternating magnetic field at ultrasonic frequencies. Due to magnetoelastic coupling the magnetic permeability of the sample is sensitive to applied stress. Consequently, a touch can change the impedance of an excitation coil which includes the amorphous wire.

The magnetomechanical coupling coefficient k, which describes a ratio of a magnetic field energy being transformed to kinetic energy of vibrations (or vice versa) to the total energy delivered to the system, is one of the fundamental quantities characterizing dynamical properties of magnetostrictive materials. The coupling coefficient can be determined experimentally by two methods referring to alternative versions of the touch sensor.

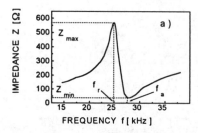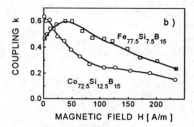

Fig.1. The frequency dependence of the impedance Z of an amorphous wire a) and the relations between the magnetomechanical coupling coefficient k and the magnetic bias H in Fe- and Co-rich amorphous wires b)

In the resonant-antiresonant method the coefficient k is evaluated from the resonant, f_R , and antiresonant, f_A , frequencies according to the following formula [7,8]:

$$k \approx \frac{\pi}{\sqrt{8}} \sqrt{1 - \frac{f_R^2}{f_A^2}} \quad , \tag{1}$$

where the frequencies f_R and f_A correspond, respectively, to the maximum and minimum dynamical impedance of the excitation coil with the amorphous wire. An example of the resonant curve and the dependences of the magnetomechanical coupling coefficient on magnetic biasing field are presented in Fig.1.

Another method is based on dynamic permeability measurements of the magnetostrictive sample when it is vibrating freely (i.e. without external stress) or when the ends of the samples are fixed (i.e. the sample length is constant) and k is given by the following approximate relation [7]:

$$k \approx \frac{\pi}{\sqrt{8}} \sqrt{\frac{\mu_\sigma - \mu_\varepsilon}{\mu_\sigma - 1}} \quad , \tag{2}$$

where μ_σ and μ_ε denote permeabilities of free vibrating and fixed samples, respectively.

A touch of the free end of the sensor is equivalent to clamping of the wire. Applied stress affects both the real and imaginary components of the permeability because it introduces stress-induced anisotropy and increases external damping (load). Therefore, the permeability is reduced and the resonance peak of the impedance drops out under touch.

3. Sensor Design

Schematic diagrams of two types of magnetoelastic sensors are presented in Fig.2. A piece of an amorphous wire is placed inside a glass capillary with an excitation coil wound around it and the wire is fixed at one end (Fig.2a) or in the middle (Fig.2b). Typical length of the wire ranged from 10 to 40 mm, depending on the required resonant frequency of operation and the active part of the wire could stick from 1 to 5 mm out of the coil. The resistor R which is much larger than the impedance of the coil ensures constant ac current conditions in the coil. In both cases, the magnetic biasing field is produced by a dc offset voltage of the oscillator. The main difference between two types of the sensors consists in detection system. In the simplest version (Fig. 2a) the impedance of the excitation coil is measured by a voltmeter or a voltage level detector. In the modified sensor, the voltage

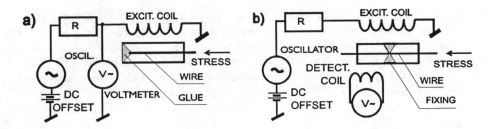

Fig. 2. Block diagrams of touch-sensors using magnetoelastic properties of amorphous wires - a) the sensor with the wire fixed at the end, b) the sensor with the wire fixed in the middle of its length

induced in an additional pick-up coil wound around inactive half of the wire core, corresponds directly to the magnetic permeability of the wire. The advantage of this solution is that several wires with independent detection coils can be placed inside a common excitation coil and the active ends can be positioned very close to each other. Such sensors can be applied for instance in robotics for recognition of fine element shapes.

4. Sensor Properties and Potential Applications

Amorphous wires of compositions $Fe_{77.5}Si_{7.5}B_{15}$ and $Co_{72.5}Si_{12.5}B_{15}$, kindly supplied by UNITICA Co. from Japan, have been annealed in order to improve their magnetoelastic properties. The best results (i.e. k=0.6) for the sensor with fixed wire end have been obtained by annealing Fe-based wire at 630 K for 4 hours in the presence of a transverse magnetic field of 300 kA/m. In the case of Co-rich alloys the highest magnetomechanical coupling coefficient (k=0.53) was obtained with an annealing of an initially twisted wire at 690 K for 90 minutes. In both cases the impedance of the sensors decreased about 4 times during touching (compressive force above 0.1 N). On the other hand, an axial stress of 0.1 MPa caused a drop of the impedance more than two times. It is worth to mention that the magnetomechanical coupling in the sensor from Fig.2b, which represents typical system for magnetoelastic measurements, is usually better than that in the sensor with the fixed wire end because reflections of the magnetoelastic wave from the fixing point are not perfect. The magnetic permeability of the sensor from Fig.2b decreased almost of one order of magnitude under axial compressive stress.

The frequency of the first mode of longitudinal resonant vibrations was changing from 29 to 66 kHz for the wire lengths from 35 to 16 mm, respectively. The frequency range of the sensor operation is limited by the length of the sample and it is typically from 20 to 120 kHz. Thus the response of the sensor to touch can be as fast as 17 μs.

The optimum biasing field corresponds to maximum magnetomechanical coupling coefficient and is related to the magnetic anisotropy of the wire and it is usually in the range from 50 to 300 A/m. The amplitude of the driving field was equal to 10 A/m. The sensors cannot work properly near by hard magnets because the soft magnetic wires of the sensors can be magnetically saturated by the strong external fields of the magnets, however, magnetic shielding or differential sensor system with reference (dummy) sensor can considerably reduce the sensitivity of the sensors to stray magnetic fields.

The sensor shows excellent properties (both mechanical and magnetoelastic) at low temperatures but it can also work at elevated temperatures up to 500 K.

Security of the sensor is ensured by electrical isolation of the active element of the sensor from the electric circuit which also operates at low voltage levels.

Though the length of the sensor cannot be shorter than few millimeters the diameter of the active wire can be as small as 0.1 mm and the diameter of the coil can be well below 1 mm. It can be minimized by using glass coated wires [9] instead of amorphous wires placed inside glass tube. Then the diameter of the sensor is expected to be smaller than 0.2 mm. Due to small cross-section of the wire small forces of order of single mN can easily be detected.

Some possible fields of application of the touch sensor can be proposed as follows:
- fast, sensitive, secure and long-life switching element (mechanical fatigue of the amorphous wire is minimized by small amplitude of ultrasonic vibrations as well as small applied stresses),
- elements of precise positioning systems operating in special conditions, e.g. low temperatures, elevated temperatures or in vacuum,
- measurements of depth of narrow gaps, cracks and small cavities (when the touch-sensor is combined with any displacement sensor),
- measurements of small forces (e.g. surface tension or viscosity of liquids)
- pattern recognition (a set of sensors with common excitation coil can detect shapes of fine elements).

5. Conclusions

A new magnetoelastic touch-stress sensor has been design using the excellent magneto-mechanical properties of amorphous wires. The low cost sensor is characterized by a simple construction - the basic element consists of an amorphous wire and an excitation coil. The main features of the sensor are its high sensitivity to applied stresses, relatively fast response, wide temperature range of operation and small diameter.

The authors gratefully acknowledge partial support from the European Community Commission under grant ERB3510PL920388

References

[1] G. Hinz, H. Viogt, Magnetoelastic sensors, in Sensors vol.5 - Magnetic Sensors, ed. R. Boll, K.J. Overshott, VCH Veralgsgeselschaft, Germany 1989, pp.98-153
[2] M.R.J. Gibbs, P.T. Squire, Applications, in Sensors, vol.5 Magnetic Sensors. Ed. R. Boll, K.J.Overshott, VCH Verlagsgeselschaft, Germany 1989, pp.448-476
[3] K. Mohri, Review of recent advances in the field of amorphous-metal sensors and transducers, IEEE Trans. on Magn. MAG-20, 5 (1984) 942-7
[4] H.R. Hilzinger, Applicatioins of metallic glasses in electronic industry, IEEE Trans. On Magn. MAG-21, 5 (1985) 2020-5
[5] B.S. Berry, Elastic and anelastic behaviour, in: Metallic galsses, ed. American Society for Metals, Metal Park, Ohio (1978) pp.160-200
[6] K.I. Arai, N. Tsuya, Magnetomechanical coupling and variable delay characteristics by means a giant ΔE-effect in Iron-rich amorphous ribbon. J.Appl. Phys. 49, 3 (1978) 1718-20
[7] C.M. Van der Burgt, Dynamical physical parameters of the magnetostrictive excitations of the extensional and torsional vibrations in ferrites, Philips res. Rep. 8,2 (1953) 91-133
[8] D.A. Berlincourt, D.R. Curran and H. Jaffe, Piezoelectric and piezomagnetic materials and their function in transducers, Physical Acoustics 1A, (1964) pp 169-270
[9] H.Chiriac and T.Ovari, Magnetic behaviour of glass-covered amorphous wires, J. Magn. Magn. Mater., vol. 157-158 (1996) 227-8

Magnetic Properties of CoMnSiB Amorphous Glass-Covered Wires

Horia CHIRIAC, Firuţa BARARIU, Tibor Adrian ÓVÁRI, and Gheorghe POP
National Institute of R&D for Technical Physics, 47 Mangeron Blvd., Iaşi, Romania

Abstract. Results on the magnetic behavior and properties of the small positive magnetostrictive $Co_{67}Mn_9Si_9B_{15}$ glass-covered wires and on their changes determined by geometrical dimensions, glass removal, and applied tensile stresses are reported. The experimental data explain their specific behavior resulting from the changes produced in the domain structure by the Mn addition to the CoSiB negative magnetostrictive alloy and the other mentioned factors. The Mn addition lead to an improvement in the thermal stability of the wires, the appearance of a large Barkhausen effect and also to a decrease of the coercive force and switching field, the wires behaving like typical positive magnetostrictive ones.

1. Introduction

The intensive studies carried out in recent years on the magnetic behavior and properties of glass-covered amorphous wires (AGCW) have confirmed their suitability for sensing applications. Recently, new results have been published on the magnetic properties of positive, negative and nearly zero magnetostrictive AGCW [1,2]. The magnetic properties of these wires strongly depend on the composition, but also on their dimensional characteristics such as: diameter of the metallic core, thickness of the glass cover, and their ratio.

For the purpose of applications our studies were extended to obtain improved properties: increased thermal stability, reduced coercive force, and nearly zero but positive magnetostriction.

In this paper we report our results on the specific magnetic behavior and properties of $Co_{67}Mn_9Si_9B_{15}$ AGCW. The aim of this paper is to analyse the influence of the replacement of 9 at.% Co with Mn on the magnetic behavior of the negative magnetostrictive amorphous CoSiB wires and to show the changes produced by different factors such as: ratio metallic core-glass cover, glass removal, and external tensile stresses.

2. Experiment

We have prepared by the glass-coated melt spinning method [3] $Co_{67}Mn_9Si_9B_{15}$ AGCW with diameters of the metallic core (d_m) ranging between 2 and 25 μm and thicknesses of the glass cover (t_g) ranging between 1 and 6 μm. The diameters and the thicknesses of the glass cover were measured using an optical microscope coupled to a computer.

The amorphous state was checked by differential thermal analysis (DTA), derivative-thermo-magneto-gravimetry (DTMG) and thermomagnetic measurements.

We performed magnetic measurements by a fluxmetric method in an alternating sinusoidal field having a maximum value of 12,000 A/m, at different frequencies.

The saturation magnetization of the sample was determined with a vibrating sample magnetometer, and was found to be 0.96 T. The saturation magnetostriction was determined by the Small-Angle Magnetization Rotation method [4] and was found to be $\lambda = 10^{-7}$. The glass removal was achieved by a chemical method [5]. The experimental set-up also allows the determination of these quantities when the samples are subjected to external tensile stresses.

3. Results and discussion

We ascertained that all as-cast samples are amorphous, the Curie temperature being $T_c = 415°C$. The alloy presents two picks of crystallization $T_{x1} = 430°C$ and $T_{x2} = 570°C$. The higher values of T_c and T_x as compared to the same quantities for the negative magnetostrictive CoSiB wires ($T_c = 360°C$, $T_x = 520°C$) or nearly zero magnetostrictive CoFeSiB wires ($T_c = 310°C$, $T_x = 540°C$) indicate an increased thermal stability of this alloy [2].

We studied the dependence of the coercive force H_c on the ratio between the radius of the metallic core R_m and the thickness of the glass cover t_g ($k = R_m/t_g$) at 50 Hz and 1KHz at a measuring field of 2,000 A/m (Figure 1). It can be observed that for $k < 1$ H_c strongly depends on this ratio, while for $k > 1$ the dependence becomes insignificant. This behavior is due to the fact that for $k < 1$ the glass cover strongly influences the domain structure by inducing high internal stresses during the preparation process while for $k > 1$ the induced internal stresses have smaller values.

Amorphous CoMnSiB glass-covered wires present switching of the magnetization (large Barkhausen effect which consists of a one-step magnetization at a certain value of an axially applied magnetic field, called the switching field) at relatively low values of the applied magnetic field due to their axially magnetized inner core arising from the coupling between their small positive magnetostriction and high internal stresses induced during preparation (several GPa). The reduced magnetization M^*/M_s, i.e. the ratio between the magnetization M^* at the switching field H^* and the saturation magnetization M_s is about 0.9, this indicating that the volume of the inner core is very large as compared to the volume of the outer shell.

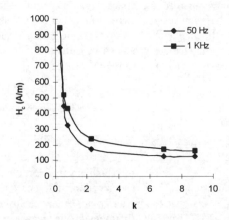

Figure 1. Dependencies of the coercive force on the k factor for $Co_{67}Mn_9Si_9B_{15}$ AGCW having different radius of the metallic core and thickness of the glass cover

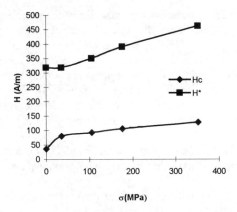

Figure 2. Dependencies of the coercive force H_c and switching field H^* on the tensile stresses σ for the $Co_{67}Mn_9Si_9B_{15}$ AGCW with the radius of the metallic core $R_m = 7.2$ μm and the thickness of the glass cover $t_g = 2.0$ μm

The appearance of the large Barkhausen effect (LBE) in the case of the CoMnSiB AGCW as compared to the CoSiB AGCW which do not present this effect is due to the change in the value of the magnetostriction constant from a negative one ($\lambda \cong -4 \times 10^{-6}$) to a small positive one ($\lambda \cong 10^{-7}$). This modifies the magnetic domain structure from that specific for the negative magnetostrictive AGCW wires (radially magnetized inner core and circumferential magnetized outer shell) [2] to a magnetic structure consisting of a cylindrical inner core with axial magnetization and a thin radially magnetized cylindrical outer shell which is typical for the positive magnetostrictive AGCW [1].

We measured the coercive force H_c, the saturation magnetization M_s, the switching field H^*, and the magnetization at this field M^* of a sample having 15 μm diameter of the metallic core and 2 μm thickness of the glass cover in the as-cast state and after the glass removal, at 400 Hz and 12,000 A/m maximum applied field.

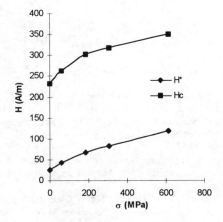

Figure 3. Dependencies of the coercive force H_c and switching field H^* on the tensile stresses σ for the $Co_{67}Mn_9Si_9B_{15}$ wires with the radius of the metallic core $R_m = 7.2$ μm after the glass removal.

The reduced values of the coercive force and also of the switching field as compared to those of the CoSiB AGCW are due to the reduced value of magnetoelastic energy resulting from the smaller absolute value of the magnetostriction constant.

Figure 2 illustrates the stress dependence of the coercive force and switching field for the as-cast AGCW. An external tensile stress applied at the ends of the as-cast wire leads to an increase in the value of the coercive force up to a saturation value (for about 1 GPa). This behavior is similar for the dependence of H* on the applied tensile stress and is due to the cumulative effect of the internal stresses induced during preparation and the applied tensile stresses.

Wires after glass removal also present LBE. The reduced magnetization M^*/M_s remains unchanged this indicating that the volume of the axially magnetized inner core is also very large.

Figure 3 illustrates the stress dependence of the coercive force and switching field for the same wires after glass removal. The glass removal leads to a decrease of the coercive force and switching field of the samples indicating the internal stress relaxation. An applied tensile stress on these samples leads to the increase of the H_c and H*, their values being however smaller than in the case of the glass-covered samples.

4. Conclusions

We investigated the effect of the replacement of 9 at.% Co with Mn on the CoFeSi negative magnetostrictive AGCW and the changes produced by this addition, glass removal and tensile stresses on the magnetic behavior of the wires.

The Mn addition resultin an improvement of the thermal stability by increasing the Curie and crystallization temperatures, the change in sign and absolute value of the magnetostriction constant which leads to the appearance of LBE, and also to the reduction of the coercive force to less than its half value.

The LBE and the values of the reduced magnetization are maintained after glass removal and applied tensile stresses. The glass removal determines a decrease of the coercive force and of the switching field, both quantities being slightly increased by the applied stresses.

References

[1] H. Chiriac,T.A. Óvári and Gh. Pop, Magnetic Behavior of Glass-Covered Wires, *J. Magn. Magn. Mater* **157/158** (1996) 227-228.

[2] H. Chiriac,T.A. Óvári, Gh. Pop and Firuţa Barariu, Magnetic Behavior of Negative and Nearly Zero Magnetostrictive Glass-Covered Wires, *IEEE Transactions on Magnetics* **32** (1996) 4872-4874.

[3] M. Hagiwara and A. Inoue, Production Techniques of Alloy Wires by Rapid Solidification. In H.H. Lieberman (ed.), Rapidly Solidified Alloys: Processes, Structures, Properties, Applications, New York: Marcel Dekker, Inc., 1993, pp 139-155.

[4] K. Narita, J. Yamasaki and H. Fukunaka, Measurement of Saturation Magnetostriction of a Thin Amorphous Ribbon by Means of Small-Angle Magnetization Rotation, *IEEE Transactions on Magnetics* **16** (1980) 435-439.

[5] H. Chiriac,T.A. Óvári, Gh. Pop and Firuţa Barariu, Effect of Glass Removal on the Magnetic Behavior of FeSiB Glass-Covered Wires, *IEEE Transactions on Magnetics* **33** (1997) 782-787.

Non-Linear Electromagnetic Systems
V. Kose and J. Sievert (Eds.)
IOS Press, 1998

Field Sensor Based on the Large Matteucci Effect in FeSiB Amorphous Glass-Covered Wires

Horia CHIRIAC and Catalin-Sandrino MARINESCU

National Institute of Research & Development for Technical Physics, 6600 Iasi 3, Romania

Abstract. Measurement of the Matteucci voltage induced in twisted $Fe_{77.5}Si_{7.5}B_{15}$ amorphous glass-covered wires show the existence of a helical anisotropy component that determines an inclination of easy axis of magnetization in the inner region of the as-cast wires. Applied axial dc field leads to a decrease of the easy axis inclination, determining a decrease of the Matteucci voltage induced in these wires. This fact allows the use of these materials in field sensors. The operating principle, construction, and characteristics of such a device are presented.

1. Introduction

Recently, amorphous glass-covered magnetic wires (AGCMW) have been obtained by the glass-coated melt spinning method, with the diameter of the metallic core ranging between 2 and 25 μm and the thickness of the glass cover ranging between 2 and 15 μm [1]. In these materials high internal stresses are induced during preparation process [2]. The coupling between internal stresses and magnetostriction plays a significant role in the domain structure formation [3].

In the case of $Fe_{77.5}Si_{7.5}B_{15}$ AGCMW, the magnetoelastic coupling between the high positive magnetostriction ($\lambda = 35 \times 10^{-6}$) and internal stresses leads in a first approximation to a domain structure consisting of an inner core with axial magnetization and an outer shell with radial magnetization [4].

However, there is strong evidence that a helical component of the internal stress tensor exists in such wires, leading to an inclination of the easy axis of magnetization in the inner region of amorphous wires from the axial direction toward the azimuthal one [5]. Due to this helical magnetoelastic anisotropy, twisted $Fe_{77.5}Si_{7.5}B_{15}$ AGCMW present a large Matteucci effect that consists of the appearance of a voltage (Matteucci voltage) across the ends of the twisted wire, in presence of an axial alternating field. This effect is strongly improved after glass removal by chemical etching. The magnitude of Matteucci voltage is about several hundreds of mV [5].

The aim of this paper is to present the dependence of Matteucci voltage (V_M) on the axial dc magnetic field for twisted $Fe_{77.5}Si_{7.5}B_{15}$ AGCMW. Based on this dc field sensitivity of V_M, we propose a new sensing device using a twisted $Fe_{77.5}Si_{7.5}B_{15}$ AGCMW that allows a direct and simultaneous determination of the dc field and its direction.

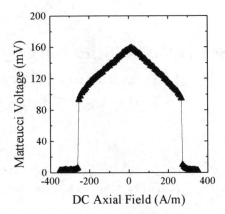

Fig. 1. The dc field dependence of Matteucci voltage for a twisted $Fe_{77.5}Si_{7.5}B_{15}$ AGCMW.

2. Experiment and Results

We prepared $Fe_{77.5}Si_{7.5}B_{15}$ AGCMW by glass-coated melt spinning method having the diameter of the metallic core 16 μm and the thickness of the glass cover 3.65 μm. The Matteucci voltage was determined by measurements of the voltage induced across the ends of the 50 rad/m twisted wire, in a square axial field H having maximum value of 350 A/m at 10 kHz. The measurements were performed on samples having 10 mm in length. The experimental set-up allowed the determination of V_M for samples subjected to an applied axial dc field related to the possible application in field sensors.

Fig. 1 presents the dc field dependence of V_M for a twisted $Fe_{77.5}Si_{7.5}B_{15}$ AGCMW. The applied axial dc field contributes to the decrease of the helical easy axis inclination in the inner region of the metallic core. This fact leads to a decrease of the V_M. The dc field sensitivity of V_M induced in these wires allows the use of these materials in field sensors.

3. Field Sensor. Principle and Construction

The determination of dc field value requires the detection of the V_M magnitude. For this sensor it is important to determine direct - and simultaneous – the dc field value and its direction. With this aim, we analyzed the Matteucci signal for samples subjected to the axial dc field (Fig. 2). From the synchronic diagram, one can observe that the negative alternation of Matteucci signal (V_{M-}) decrease for "positive" dc field direction and positive alternation of Matteucci signal (V_{M+}) decrease for "negative" dc field direction, respectively.

In the following we will analyze how the difference $\Delta V_M = |V_{M+}| - |V_{M-}|$ reflects the presence of the dc field. The Matteucci voltage is due to the flux variation determined by the azimuthal magnetization component M_θ through the longitudinal cross section of the wire in the presence of the alternating axial magnetic field:

$$V_M = l \cdot r_{IC} \cdot \frac{dM_\theta}{dt} \tag{1}$$

where l is length of the wire and r_{IC} is the radius of the inner core.

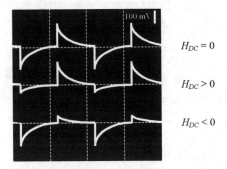

$$H_{DC} = 0$$

$$H_{DC} > 0$$

$$H_{DC} < 0$$

Fig. 2. Synchronic diagram of the Matteucci voltage induced in twisted $Fe_{77.5}Si_{7.5}B_{15}$ AGCMW in the absence of the dc field and dc field with "positive" and "negative" direction, respectively; f=10 kHz, torque angle=50 rad/m, square excitation field 350 A/m, external axial dc field ±250 A/m.

In a first order approximation,

$$V_M = M_\theta \cdot r_{IC} \cdot v_W \tag{2}$$

where v_W is the wall velocity. The Sixtus-Tonks experiment for an $Fe_{77.5}Si_{7.5}B_{15}$ amorphous wire indicated a linear relation between v_W and excitation field H [6].

$$v_W = S(H - H_0) \tag{3}$$

where the coefficient S, related to the wall mobility, is determined by the damping mechanism for the moving wall and the wall length and H_0 is the static critical field.

In presence of axial dc field H_{DC} according to equations (2) and (3)

$$|V_{M\pm}| = M_\theta \cdot r_{IC} \cdot S(H - H_0 \pm H_{DC}), \quad |H_{DC}| < (H-H_0) \tag{4}$$

From equation (4),

$$\Delta V_M = |V_{M=}| - |V_{M-}| = 2 \cdot M_\theta \cdot r_{IC} \cdot S \cdot H_{DC} \tag{5}$$

Equation (5) indicates a linear dependence between ΔV_M and H_{DC} : $\Delta V_M = kH_{DC}$ for $|H_{DC}| < (H-H_0)$, where k is a constant ($k = 2 \cdot M_\theta \cdot r_{IC} \cdot S$). This fact is proved by experiment. Based on this effect, we proposed a new sensing device using a 50 red/m twisted $Fe_{77.5}Si_{7.5}B_{15}$ AGCMW with 16 μm in diameter and 10mm in length, which allows a direct and simultaneous determination of dc field value and its direction by means of the differential-detection technique. The field sensor presented in block-diagram in Fig. 3 consists of a sensitive element, square signal generator (10 kHz) and differential-detection amplifier. The sensor's characteristic (Fig. 4) show a good linearity of response in the range of ±250 A/m.

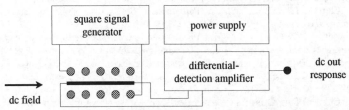

Fig. 3. Block-diagram of the dc field sensor based on the Matteucci effect in twisted FeSiB AGCMW.

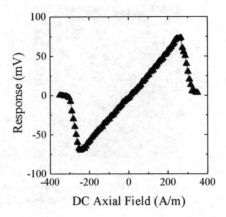

Fig. 4. The response of the field sensor versus the applied axial dc field.

Due to the relatively high value of dc field sensitivity and the good linearity of the Matteucci voltage induced in $Fe_{77.5}Si_{7.5}B_{15}$ AGCMW, a field sensor based on this effect requires very simple electronics. This fact allows the miniaturization and cost reduction of the device.

4. Conclusion

The sensitivity of twisted $Fe_{77.5}Si_{7.5}B_{15}$ amorphous glass-covered magnetic wires to an applied dc field recommends them as sensitive elements in field sensors. In the dc field linearity range of V_M (±250 A/m), the dc signal obtained after differential-detection is proportional to the value of the applied dc field and does not require electronic corrections. The value of the dc field sensitivity of V_M, which is about 280 µV/(A/m), ensures a precision better than 10^{-2} A/m (the voltage measurement precision of differential amplifier is about 2 µV). The advantage of these wires is that they de not require stress-relief annealing like the conventional amorphous wires. It is worthwhile to mention that sensors based on field sensitivity of Matteucci voltage do not require pick-up coils, and the minimum length of the wires can be around 2 mm, that is the minimum length at which the Matteucci effect still appears in such wires.

References

[1]   M. Hagiwara and A. Inoue, Production Techniques of Alloy Wire by Rapid Solidification. In: H. H. Liebermann (ed.), Rapidly Solidified Alloys. Marcel Dekker Inc., New York, 1993, pp. 139-155.
[2]   H. Chiriac, T.A. Ovari, and Gh. Pop, Internal Stress Distribution in Glass-Covered Magnetic Amorphous Wires, *Physical Revue B* **52** (1995) 10104-10113.
[3]   H. Chiriac, T.A. Ovari, Gh. Pop, and F. Barariu, Internal Stresses in Highly Magnetostrictive Glass-Covered Amorphous Wires, *Journal of Magnetism and Magnetic Materials* **160** (1996) 237-238.
[4]   H. Chiriac, T.A. Ovari, and Gh. Pop, Magnetic Behavior of Negative and Zero Magnetostrictive Glass-Covered Amorphous Wires, *Journal of Magnetism and Magnetic Materials* **157/158** (1996) 227-228.
[5]   H. Chiriac, T.A. Ovari, S.C. Marinescu, and V. Nagacevschi, Magnetic Anisotropy in FeSiB Amorphous Glass-Covered Wire, *IEEE Transactions on Magnetics* **32** (1996) 4755-4757.H. Chiriac,
[6]   D. X. Chen, N. M. Dempsey, M. Vasquez, and A. Hernando, Propagating domain Wall Shape and Dynamics in Iron-Rich Amorphous Wires, *IEEE Transactions on Magnetics* **31** (1995) 781-790.

Non-Linear Electromagnetic Systems
V. Kose and J. Sievert (Eds.)
IOS Press, 1998

Magnetic Behavior of Nearly Zero Magnetostrictive CoFeSiB Amorphous Glass-Covered Wires with Cr Addition

Horia CHIRIAC, Tibor-Adrian ÓVÁRI, Gheorghe POP, Firuţa BARARIU
National Institute of R&D for Technical Physics, 47 Mangeron Blvd., 6600 Iaşi, Romania

Ludek KRAUS
Institute of Physics, A.S.C.R.., Na Slovance 2, 18040 Prague 8, The Czech Republic

Abstract. We study the effect of Cr addition on the magnetic behavior of amorphous glass-covered wires of the Co-Fe-Si-B alloy system. Cr addition changes the sign of the magnetostriction constant, leading to the appearance of a bistable magnetic behavior of CoFeCrSiB amorphous glass-covered wires, behavior that is maintained even if the glass cover is removed. The results are explained in terms of the basic interactions that determine the domain structure formation.

1. Introduction

CoFeSiB amorphous glass-covered wires (AGCW) with nearly zero magnetostriction are important for basic and applied research, their magnetic properties being appropriate for sensor applications, especially for those based on the giant magneto-impedance effect. Nearly zero magnetostrictive $Co_{68.15}Fe_{4.35}Si_{12.5}B_{15}$ AGCW have a specific magnetic behavior determined by the still large magnetoelastic coupling between the large internal stresses induced during preparation and their low negative magnetostriction ($\lambda \cong -1 \times 10^{-7}$), which determines a unique domain structure consisting of a radially magnetized inner core and a circumferentially magnetized outer shell [1]. After glass removal, the magnetoelastic anisotropy decreases due to stress relief, and the domain structure is determined by the minimization of the magnetostatic energy, consisting of an axially magnetized inner core and a circumferentially magnetized outer shell [1].

Cr addition generally improves the corrosion resistance of metallic alloys, allowing particularly, at the same time, an easier glass removal in the case of AGCW. On the other hand, a study of its effect on the magnetic behavior of such wires is also required, taking into account their practical importance based on the excellent soft magnetic properties. The aim of this paper is to study the effect of Cr addition on the magnetic properties and behavior of AGCW of the Co-Fe-Si-B alloy system. The found results are discussed starting from the magnetic behavior of nearly zero magnetostrictive $Co_{68.15}Fe_{4.35}Si_{12.5}B_{15}$ AGCW.

2. Experiment

We prepared by glass-coated melt spinning $Co_{67}Fe_4Cr_7Si_8B_{14}$ AGCW with diameters of the metallic wire, Φ_m, of 4 to 24 μm, and the glass cover thickness, t_g, of 9 to 25 μm. For our experimental investigations we have chosen two samples having the same total wire diameter, Φ, of 39 μm, but with different dimensions of the metallic wire and glass cover - the first sample has $\Phi_m = 9$ μm and $t_g = 15$ μm, while the second one has $\Phi_m = 21$ μm and $t_g = 9$ μm, since, according to our previous results related to the influence of the wire dimensions on its magnetic behavior [2], the basic magnetic properties of an AGCW are dependent on the ratio k between the radius of the metallic wire, R_m, and the glass cover thickness, t_g ($k = R_m/t_g$). Thus, we have chosen two wires with the same diameter, but one has $k = 0.3$, while the other has $k = 1.2$, due to the fact that the magnetic properties strongly change if k is smaller or larger than 1 [3]. We performed magnetic measurements by a fluxmetric method on as-cast AGCW and on wires after glass removal with a HF solution. The maximum value of the applied field was 3000 A/m and its frequency was 400 Hz. The same measurements were performed on both AGCW and wires after glass removal subjected to external tensile stresses. The saturation magnetization M_S of the samples was determined with a vibrating sample magnetometer and it was found to be 0.6 T. We performed magnetostriction measurements through the SAMR method [4].

3. Results and discussion

We observed that both $Co_{67}Fe_4Cr_7Si_8B_{14}$ AGCW present a bistable magnetic behavior, i.e. that a large Barkhausen effect (LBE) appears in these wires at relatively small axial fields. LBE consists of a one-step magnetization reversal when the applied field reaches to a threshold value called the switching field, H^*. H^* is 850 A/m for the AGCW with $k = 0.3$, and 300 A/m for the one with $k = 1.2$. The reduced remanence M^*/M_S - the ratio between the remanent magnetization M^* at the switching field and the saturation magnetization M_S is about 0.95 for both wires, value that is similar to that observed in highly positive magnetostrictive AGCW (e.g. $Fe_{77.5}Si_{7.5}B_{15}$ [3]). After glass removal, LBE is maintained, but H^* decreases one order of magnitude, being 87 A/m for the wire with $k = 0.3$, and 47 A/m for the one with $k = 1.2$. M^*/M_S decreases very slightly in both cases, being of about 0.90.

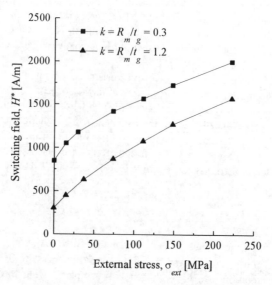

Fig. 1. Dependence of H^* on external tensile stresses, σ_{ext}, for the $Co_{67}Fe_4Cr_7Si_8B_{14}$ AGCW.

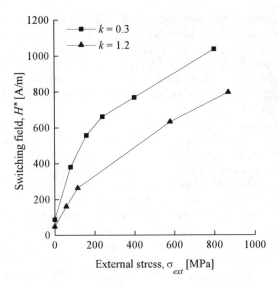

Fig. 2. Dependence of H^* on external tensile stresses, σ_{ext}, for $Co_{67}Fe_4Cr_7Si_8B_{14}$ wires after glass removal.

Fig. 1 shows the dependence of the switching field, H^*, on external tensile stresses, σ_{ext}, applied during measurements, for the $Co_{67}Fe_4Cr_7Si_8B_{14}$ AGCW. Fig. 2 illustrates the same dependence for the $Co_{67}Fe_4Cr_7Si_8B_{14}$ wires after glass removal. One observes that H^* increases with σ_{ext} for both AGCW and wires after glass removal, but at smaller values in the latter case. Similar dependencies were previously reported for $Fe_{77.5}Si_{7.5}B_{15}$ AGCW and wires after glass removal [3, 5].

The magnetic behavior of $Co_{67}Fe_4Cr_7Si_8B_{14}$ AGCW and wires after glass removal is somewhat similar to that of highly positive magnetostrictive ones, especially as concerns the appearance of the LBE and its maintenance after glass removal, but in the latter ones, M^*/M_S decreases after glass removal to about 0.60, and the values of H^* for AGCW are larger than those of CoFeCrSiB ones with similar dimensions. On the other hand, the magnetic behavior of CoFeCrSiB AGCW is different from that of CoFeSiB ones, i.e. the latter do not display LBE until the glass cover is not removed. The characteristics of the LBE in CoFeSiB and CoFeCrSiB wires after glass removal are similar: H^* is several tens of A/m, and M^*/M_S is large - 0.81 for CoFeSiB and 0.90 for CoFeCrSiB [1].

The above remarks allow us to state that the major modification induced by the Cr addition is the change of the magnetostriction constant of the CoFeSiB alloy. Due to the similarity between the magnetic behavior of CoFeCrSiB wires and highly positive magnetostrictive ones, we expect a change in the sign of the magnetostriction constant. Since the characteristics of the LBE in CoFeCrSiB wires after glass removal are very much alike to those of nearly zero magnetostrictive CoFeSiB ones ($\lambda \cong -1 \times 10^{-7}$), we also expect this magnetostriction constant to be small. Indeed, we measured the magnetostriction constant of CoFeCrSiB wires, and we found it to be small and positive: $\lambda \cong +1 \times 10^{-7}$.

This value of the magnetostriction constant explains the observed magnetic behavior of CoFeCrSiB AGCW and wires after glass removal. Thus, in the AGCW, the internal stresses induced during preparation are very large (by the order of several GPa [3]), and their magnetoelastic coupling with the small positive magnetostriction results in a *large magnetoelastic energy*, whose minimization determines the domain structure formation.

Under these circumstances, the domain structure of CoFeCrSiB AGCW is similar to that of highly positive magnetostrictive AGCW: an axially magnetized inner core that occupies almost the entire wire volume, and a very small radially magnetized outer shell [3]. The LBE occurs in the inner core, H^* being large due to the large magnetoelastic anisotropy, and M^*/M_S due to the large volume of the inner core. These explanations are supported by the stress dependence of H^* (Fig. 1), which has a typical shape for highly positive magnetostrictive AGCW.

After glass removal, stress relief occurs [3], and the role of the magnetoelastic energy becomes less important, the domain structure being determined by the minimization of the *magnetostatic energy*, like in the case of CoFeSiB amorphous wires after glass removal [1]. Thus, LBE is maintained in CoFeCrSiB wires after glass removal, with a large M^*/M_S, since the axial inner core is determined by magnetoelastic energy minimization, unlike in the case of highly positive magnetostrictive wires after glass removal. This fact is supported by the similar values of M^*/M_S in CoFeSiB and CoFeCrSiB wires after glass removal. Stress relief explains the decrease of H^*, but its further increase with σ_{ext} (Fig. 2), which is typical for highly positive magnetostrictive wires after glass removal, is due to the increase of the magnetoelastic term contribution when such samples are subjected to external stresses. This behavior is equivalent to the restoration of the initial situation in CoFeSiB wires after glass removal, where LBE is suppressed when the samples are subjected to external tensile stresses. In CoFeCrSiB wires after glass removal, the magnetoelastic energy retakes its initial role of determining the easy axis distribution.

4. Conclusion

Cr addition to the amorphous CoFeSiB alloy, changes its small negative magnetostriction ($\lambda \cong -1 \times 10^{-7}$) to a small positive one ($\lambda \cong +1 \times 10^{-7}$). As a result, CoFeCrSiB amorphous glass-covered wires display large Barkhausen effect with a remanence to saturation ratio of 0.95, at values of the switching field that depend on their dimensions. The effect is maintained after glass removal, with a large remanence (90%), but the switching field decreases by one order of magnitude. The results are explained taking into account the dominant interactions that determine the domain structure formation, and their changes with glass removal and/or applied tensile stresses.

References

[1] H. Chiriac, Gh. Pop, T.A. Óvári and F. Barariu, Magnetic Behavior of Negative and Nearly Zero Magnetostrictive Glass-Covered Amorphous Wires, *IEEE Transactions on Magnetics* **32** (1996) 4872-4874.

[2] H. Chiriac, Gh. Pop, F. Barariu, T.A. Óvári and M. Tomuţ, Magnetization Processes in Amorphous FeSiB Glass-Covered Wires, *Journal of Non-Crystalline Solids* **205-207** (1996) 687-691.

[3] H. Chiriac, T.A. Óvári, Gh. Pop and F. Barariu, Effect of Glass Removal on the Magnetic Behavior of FeSiB Glass-Covered Wires, *IEEE Transactions on Magnetics* **33** (1997) 782-787.

[4] K. Narita, J. Yamasaki and H. Fukunaga, Measurement of Saturation Magnetostriction of a Thin Amorphous Ribbon by Means of Small-Angle Magnetization Rotation, *IEEE Transactions on Magnetics* **16** (1980) 435-439.

[5] H. Chiriac, T.A. Óvári and Gh. Pop, Magnetic Behavior of Glass-Covered Amorphous Wires, *Journal of Magnetism and Magnetic Materials* **157/158** (1996) 227-228.

Non-Linear Electromagnetic Systems
V. Kose and J. Sievert (Eds.)
IOS Press, 1998

Size Effect on the Giant Magneto-Impedance of CoFeSiB Magnetic Amorphous Ribbons and Wires

Horia CHIRIAC, Franco VINAI[#], Tibor-Adrian ÓVÁRI, Cătălin-Sandrino MARINESCU,
Paola TIBERTO[#] and Aldo STANTERO[#]

Natl. Inst. of R&D for Technical Physics, 47 Mangeron Blvd., 6600 Iaşi, Romania
[#]*IEN Galileo Ferraris, Corso Massimo D'Azeglio 42, Torino I-10125, Italy*

Abstract. We report results on the influence of the samples' shape and dimensions on the sensitivity of the giant magneto-impedance effect in amorphous CoFeSiB ribbons and wires. The sensitivity of the effect increases as the samples' cross section area increases. The practical importance of the results is that they allow the choice of appropriate dimensioned materials for applications based on this effect.

1. Introduction

Giant magneto-impedance (GMI) effect consists of sensitive changes in the high frequency impedance of a magnetic conductor with a small magnetic field. This effect was found to be very large - about 10%/Oe - in CoFeSiB amorphous wires and ribbons [1].

The sensitivity of a material from the point of view of the GMI effect strongly depends on the dynamic magnetization processes that take place in it at high frequencies, which are reflected by the transverse permeability behavior. The transverse permeability influences the distribution of the ac current at high frequencies where the skin effect occurs, since the magnetic penetration depth depends on the value of this permeability.

The aim of this paper is to study the influence of the transverse dimensions on the magnitude and sensitivity of the GMI effect in $Co_{68.15}Fe_{4.35}Si_{12.5}B_{15}$ amorphous ribbons and wires with nearly zero magnetostriction ($\lambda \cong -1 \times 10^{-7}$). With this aim, we firstly compared the GMI effect in two amorphous samples - a ribbon shaped one and a wire shaped one - with similar values of the cross section area, and then we studied the sensitivity of the effect in amorphous ribbons with different values of the cross section area. The experimental results have been explained in terms of transverse magnetic permeability dependence on frequency and on the samples' transverse shape and dimensions.

The obtained results are important as regards the practical applications of the GMI effect in magnetic sensors, allowing us to choose suitable shape and dimensions for a material designated for such applications.

2. Results and discussion

In order to study the influence of the sample's shape on the GMI effect, comparative measurements were performed on samples W-1, an amorphous CoFeSiB wire with 113 μm in diameter, and R-1, an amorphous ribbon with 580 μm width and 15 μm thickness, with equal cross section areas ($S = 0.9 \times 10^{-4}$ cm^2). The driving ac current in this case was $I_{RMS} =$

30 mA. The influence of the transverse samples' dimensions on the GMI effect was studied on four ribbon shaped samples with the following dimensions: sample R-2 with a width, w, of 257 μm, and the thickness, t, of 16 μm, R-3 with $w = 220$ μm and $t = 20$ μm, R-4 with $w = 376$ μm and $t = 29$ μm, and R-5 with $w = 516$ μm and $t = 28$ μm. In this case, we kept the current density constant, $j_{RMS} = 3.5 \times 10^6$ A/m^2, in order to eliminate the differences in the impedance behavior induced by different values of the transverse field. GMI measurements were performed in the frequency range 1 - 10 MHz, using a digital oscilloscope coupled with a computer that allowed data acquisition and processing.

Figs. 1 and 2 illustrate the dependence of the impedance, Z, on the value of the axial dc field, H_{dc}, at 1, 5, and 10 MHz, for the samples R-1 and W-1, respectively. At these high frequencies, the transverse permeability $\mu_\perp$ of the ribbon and the circumferential permeability μ_θ of the wire are strongly diminished, and hence, the values of the reactance for both samples will be very small as compared to the ac resistance [2]. Correspondingly, in order to understand better the processes that dominate the high frequency impedance behavior, we make the following assumption, without loss of generality: at high frequencies, where the skin effect is very strong and $\mu_\perp$, μ_θ are very small, the impedance behavior is given only by the ac resistance. The impedance behavior at dc fields over 40 A/m for all three frequencies is similar for the wire and the ribbon. This observation is in agreement with the above hypothesis, if the diminished values of the permeabilities for the two samples are almost equal: $\mu_\perp \approx \mu_\theta$. The difference in the impedance behavior of the two samples appears especially at fields under 40 A/m, where one observes a sharp increase of Z in the case of the ribbon. For the wire, Z also increases in the low field region, but its increase is less important. In the above hypothesis, this behavior suggests a decrease of the magnetic penetration depth, δ_m, due to an increase of the permeabilities $\mu_\perp$ and μ_θ, where $\mu_\perp$ presents a stronger increase, and thus it is more sensitive to changes of H_{dc}. Such behavior of the permeability can be understood by considering the transverse (for the ribbon) and circumferential (for the wire) magnetization processes occurring in these samples at high frequencies. Thus, it is plausible to state that at high frequencies - where domain wall movements are damped - the magnetization proceeds on the transverse and circumferential directions, respectively, by rotations of the magnetic moments. The rotations are favored by small axial fields in both cases, while for high axial fields even the rotational permeabilities are strongly depreciated. $\mu_\perp$ is more sensitive to small axial dc fields since the magnetoelastic anisotropy forces are smaller in the ribbon.

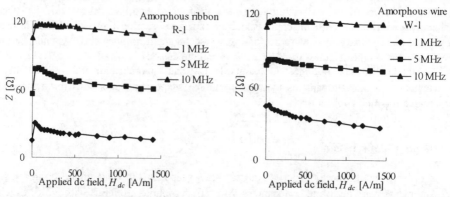

Fig. 1. Field dependence of impedance for sample R-1 (amorphous CoFeSiB ribbon).

Fig. 2. Field dependence of impedance for sample W-1 (amorphous CoFeSiB wire).

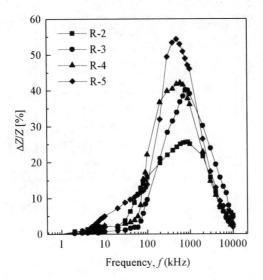

Fig. 3. Frequency dependence of the MI ratio ($\Delta Z/Z$) for the ribbon shaped samples R-2, R-3, R-4, and R-5.

Let us analyze in the following the influence of the samples' transverse dimensions on the sensitivity of the GMI effect in the ribbon shaped samples (R-2, R-3, R-4, and R-5), through the magneto-impedance (MI) ratio defined as $\Delta Z/Z = [Z(H_{dc} = 0) - Z(H_{dc} = 1500$ A/m)]/$Z(H_{dc} = 0)$. This quantity gives us information about the degree of impedance reduction when an axial dc field of 1500 A/m is applied to a sample. Fig. 3 shows the frequency dependence of $\Delta Z/Z$ for all four samples in the frequency range 1 kHz - 10 MHz. The frequency dependencies of $\Delta Z/Z$ have similar shapes up to 10 kHz, where the MI ratios are small; for $f > 10$ kHz, $\Delta Z/Z$ increases for all samples reaching a maximum in the frequency range 500 kHz - 1 MHz; after the maximum, $\Delta Z/Z$ decreases with the increase of f, reaching small values for $f = 10$ MHz.

One observes that the sensitivity of the GMI effect is higher for samples with larger cross section areas. On the other hand, the maximum sensitivity ($\Delta Z/Z)_{max}$, is obtained at lower frequencies for samples with larger cross section areas (Fig. 4). This behavior can be explained by taking into account the role played by the skin effect at high frequencies. Thus, according to the relationship that gives the magnetic penetration depth δ_m at frequencies where the skin effect occurs:

$$\delta_m = (\rho/\pi \cdot f \cdot \mu)^{1/2} \tag{1}$$

where μ is the transverse permeability, ρ the resistivity and f the frequency, the smaller the frequency, the larger is δ_m. Consequently, skin effect appears at lower frequencies in materials with larger cross sections. As a result, the ac resistance of such a sample increases at lower frequencies, while for a sample with a small cross section, it increases at higher frequencies. Clearly, the conditions for the appearance of a sensitive GMI effect for samples with large cross sections are accomplished according to the above considerations at lower frequencies as compared to those with small cross sections.

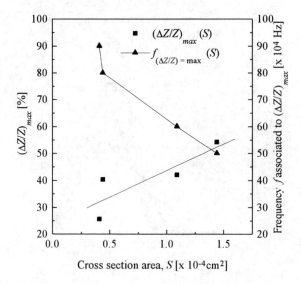

Fig. 4. Dependence of the maximum MI ratio, $(\Delta Z/Z)_{max}$, and of the frequency f associated with it, on the value of the cross section area S, for CoFeSiB amorphous ribbon shaped samples (R-2, R-3, R-4 and R-5).

3. Conclusion

We analyzed the influence of the samples' shape and dimensions on the sensitivity of the giant magneto-impedance effect in amorphous CoFeSiB ribbons and wires. Summarizing, the most important conclusions emerging from our study are:

1. Ribbon shaped samples are more sensitive to small dc magnetic fields, since their transverse permeability is more sensitive to such fields, due to their smaller magnetoelastic anisotropy in this direction, which allows an easier rotation of the magnetic moments by small axial fields.

2. As the samples' cross section area increases, the sensitivity of the GMI effect increases and maximum sensitivity is reached at lower frequencies. This behavior is directly related to the achievement of the frequency conditions at which a strong skin effect occurs.

Acknowledgments

This work was supported by the North Atlantic Treaty Organization under project number SA.12-3-02 (HTECH.LG 950633) 1267 (95) JARC 412.

References

[1] L.V. Panina, K. Mohri, K. Bushida and M. Noda, Giant Magneto-Impedance and Magneto-Inductive Effects in Amorphous Alloys," *Journal of Applied Physics* **76** (1994) 6198-6203.

[2] H. Chiriac, F. Vinai, T. A. Óvári, S. C. Marinescu, F. Barariu and P. Tiberto, Comparative Study of the Giant Magneto-Impedance Effect in CoFeSiB Magnetic Amorphous Ribbons and Wires, *Materials Science and Engineering A* (1997) in press.

Non-Linear Electromagnetic Systems
V. Kose and J. Sievert (Eds.)
IOS Press, 1998

Length Effect on Magneto-Impedance of Coaxial Waveguide with FeCoSiB Amorphous Wire at Microwave Frequencies

L. V. Panina   and   K. Mohri

Department of Electrical Engineering, Nagoya University, Nagoya 464, Japan

A.S. Antonov   and   A.N. Lagarikov

Scientific Center for Applied Problems in Electrodynamics, IVTAN, Russian Academy of Sciences, Moscow 127412, Russia

Abstract. The input impedance of a coaxial wave guide with a CoFeSiB amorphous wire used as a central conductor is measured in a wide frequency range of 1 - 1200 MHz and analyzed on the basis of an exact solution of the corresponding electromagnetic problem. It is demonstrated that for a system several cm long the radiation effects can be essential already at frequencies of 100 MHz, resulting in a resonance-like frequency dependence of the input impedance with a reduced sensitivity to an applied field in comparison with that typical of the wire surface impedance.

1. Introduction.

Since the discovery of giant magneto-impedance (MI) in Co-based amorphous wires, there is an increasing effort to apply this effect to fabrication of various micro-sized magnetic sensors[1-3]. In the case of the device with quick response, the carrier current frequency of the MI element is required to be increased to about 1 GHz to detect signals at 50-100 MHz. However, at such frequencies, the radiation effects of electromagnetic wave reflection and interference become substantial even for the total length of the MI element (including lead wires) of several centimeters. The radiation effects result in MI characteristics depending non-linearly on the length and being quite different compared with those arising from the surface impedance. In the present paper, this problem is considered both theoretically and experimentally, in which MI characteristics are studied for a coaxial waveguide with a near-zero magnetostrictive CoFeSiB amorphous wire replacing the central conductor. This experiment makes it possible to obtain reliable data in a wide frequency range of 1-1200 MHz and to analyze the data using an exact solution of a corresponding electromagnetic problem.

The MI effect is based on the dependence of the surface impedance ζ on the skin depth δ. For frequencies when the skin effect is strong, ζ is described by a square root function of the frequency ω and the permeability μ. For certain magnetic structures (for example, a circular domain structure typical of Co-rich amorphous wires) μ can be very sensitive to an external magnetic field. Then, the MI effect reflecting the field and frequency dependence of μ has to be efficient for frequencies up to the ferromagnetic resonance frequency ω_r, which determines the permeability frequency dispersion. In the case of CoFeSiB amorphous wires, the value of $\omega_r / 2\pi$ is about 200 MHz for zero external field. On the other hand, in many experimental works (see, for example [4,5]) on the MI effect in soft ferromagnetic alloys it is found that the MI ratio has a maximum at frequencies of 1-10 MHz and then quickly decreases becoming very small for frequencies higher than 50-100 MHz. The authors believe that such frequency dependence of MI is caused by the radiation effects, and if the MI element and lead wires are shortened (< 1 cm) the MI ratio reaching a maximum would slightly change with frequencies

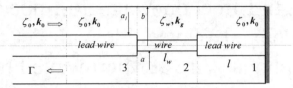

Fig. 1. Coaxial waveguide cell with amorphous wire

$\omega < \omega_r$. At high frequencies the measured impedance corresponds to the input impedance of the system, the   frequency dependence of which is demonstrated to exhibit a characteristic maximum at a frequency depending on a wire length   and a lead wire length, as well as on an external field   applied along the wire.

2. Input impedance of a waveguide system

The input impedance Z_{inp} is deduced from the measurements of the   complex reflection   coefficient, Γ: $Z_{inp} = \varsigma_0 (1 + \Gamma)/(1 - \Gamma)$ where ς_0 is the characteristic impedance of a non-disturbed waveguide. The authors are interested in finding the effect on Z_{inp} of the magnetic properties of the amorphous wire placed in the central section of a waveguide, as shown in Fig. 1. For this purpose, the recurrent formula for a multilayer system [6] can be iteratively used:

$$Z^{(n)}{}_{inp} = \frac{Z_{inp}^{(n-1)} - j\varsigma_c^{(n)} \tan(k^{(n)} l_n)}{\varsigma_c^{(n)} - jZ_{inp}^{(n-1)} \tan(k^{(n)} l_n)} \varsigma_c^{(n)} \tag{1}$$

Equation (1) connects the input impedance $Z_{inp}^{(n)}$ of the Nth layer with that of the (N-1) layer, $\varsigma_c^{(n)}$, $k^{(n)}$, l_n are the characteristic impedance, the propagation constant and the length of the Nth layer, respectively. Here $\varsigma_c^{(1)} = \varsigma_c^{(3)} = \varsigma_0$, $k^{(1)} = k^{(3)} = k_0 = \omega / c$ and $Z_{inp}^{(0)} = 0$. The wave parameters $\varsigma_c^{(2)} = \varsigma_w$ and $k^{(2)} = k_g$ of the wire section, depending on the permeability μ, can be deduced from the solution of Maxwell's equations for an infinitely long waveguide.

The authors' case corresponds to the propagation of a transverse magnetic wave having a circular magnetic field H_φ, and an electric field with radial E_r and longitudinal E_z components, depending on the exponential factor $\exp(-j\omega t + jk_g z)$. From Maxwell's equations it follows that the transverse components H_φ and E_r can be deduced from E_z :

$$E_r = \frac{jk_g}{k_{1,2}^2} \frac{\partial E_z}{\partial r}, \quad H_\varphi = j\frac{\omega}{c} \frac{\varepsilon_{1,2}}{k_{1,2}^2 - k_g^2} \frac{\partial E_z}{\partial r} \tag{2}$$

where indices 1,2 are related to the wire region ($r<a$, a is the wire radius) and to the hollow   region   ($a<r<b$, b is   the   guide   outer   radius),   respectively, $k_1 = (1+j)/\delta$, $\delta = c/\sqrt{2\pi\sigma\omega\mu}$, $k_2 = k_0$; $\varepsilon_1 = j4\pi\sigma/\omega$, $\varepsilon_2 = 1$, σ is the wire conductivity. The longitudinal component E_z is described by the wave equation:

$$\frac{1}{r}\frac{\partial}{\partial r}(r\frac{\partial E_z}{\partial r}) + (k_{1,2}^2 - k_g^2)E_z = 0 \tag{3}$$

The value of k_g can be found from the boundary conditions that E_z and H_φ are continuous at the wire surface ($r=a\pm 0$) and $E_z(b) = 0$. The characteristic impedance ζ_w is represented by the ratio V/I where V is the potential difference between the wire and outer conductor and $I = a\, cH_\varphi(a)/2$ is the current flowing along the wire. ζ_w can be expressed through the wire surface impedance $\varsigma = E_z(a)/H_\varphi(a)$.

This approach allows the authors to find how the wire surface impedance is related to the measured input impedance of the system. At high frequencies ($k^{(n)}l_n \geq$) this relationship depends on a certain way of wire connection in the experimental scheme, in particular, there is a non-linear dependence on the magnetic wire length as well as on the lead wire length. The problem can be reduced to the quasi-static case for $k^{(n)}l_n \ll 1$. In this approximation it is also reasonable to assume that $k_1 a \gg k_g a$, which gives an analytical representation for k_g: $k_g^2 = k_0^2 + j\varsigma/ca\ln(b/a)$. Then, the input impedance is written in the form:

$$Z_{inp} = -j(\omega/c^2)\ln b/a + \varsigma/2ac \tag{4}$$

Expression (4) corresponds to that well known for the complex resistance of a wire, obtained in a quasi-static case [7]. The first term results from the external inductance of the coaxial line $L_e = 2l\ln b/a$, and the second one describes the energy consumption inside the wire.

3. Results and discussion

The input impedance Z_{inp} is calculated using (1-3) and considering that the permeability is due to the magnetization rotation only [8]. Fig. 2 shows the frequency dependence of Z_{inp} for different values of the wire length l_w and the external field H_{ex}. There is a characteristic maximum at a frequency f_m the value of which is lower for a longer wire (or a longer lead wire). For example, in the case of a wire of 0.5 cm long, f_m =900 MHz ($H_{ex} = 2.5\ H_K$, $l_1 = l_3 = l = 2$ cm, H_K is the anisotropy field) where as $f_m =$ 300 MHz for $l_w = 3$ cm. For comparison, in Fig. 2a the frequency dependence of impedance calculated in a quasi-dc approximation (formula (4)) is given, demonstrating a monotone increase with frequency. At frequencies higher than 100 MHz this curve deviates considerably even from that seen for $l_w = 0.5$ cm. Application of the external field also effects the impedance vs. frequency behavior: $H_{ex} < H_K$ decreases f_m, but if $H_{ex} \gg H_K$, the maximum again shifts to higher frequencies. This behavior is associated with the field dependency of the permeability which has a maximum at $H_{ex} \sim H_K$. The analysis made shows that the radiation effects which are responsible for resonance-like impedance vs. frequency behavior eventually result in a decrease of the MI element sensitivity to the field. At GHz frequencies, the MI element length has to be reduced down to 1 cm in order to reach the sensitivity typical of intrinsic surface impedance.

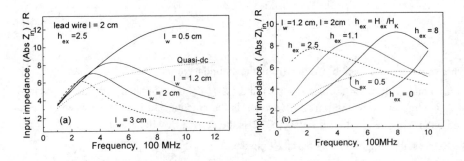

Fig. 2. Input impedance vs. frequency for different wire length in (a) and for different fields in (b). $h_{ex} = H_{ex}/H_K$, R is the dc resistance

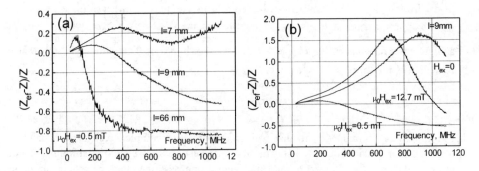

Fig. 3. Experimental plots impedance vs. frequency for different values of the wire length (a) and the applied field (b). Z - measured impedance, Z_{ef} -excluding lead wire effect

Characteristically the same tendency is seen in Fig. 3, where the experimental data are presented. The complex reflection coefficient Γ is measured by a waveguide method [9] in a broad frequency range from 1 MHz to 1.2 GHz. A 50 Ohm coaxial transmission line having the outer diameter of 7 mm with a 30 μm diameter CoFeSiB amorphous wire placed as shown in Fig. 1 is used.

References

[1] K. Mohri, Application of Amorphous Magnetic Wires to Computer Peripherals, *Materials Science and Engineering*. **A185** (1995) 141.
[2] L. V. Panina, K. Mohri, K. Bushida, M. Noda, Giant Magneto-Inductive and Magneto-Impedance Effects in Amorphous Alloys, *J.Appl. Phys.* **76** (1994) 6198.
[3] K. Mohri, L.V. Panina, T. Uchiyama, K. Bushida, Sensitive and Quick-Response Micro Magnetic Sensor Utilizing Magneto-Impedance in Co-Rich Amorphous Wires, *IEEE Trans. Magn.***31** (1995) 1266.
[4] P. Ciureanu et al, *J.Appl. Phys.* **79** (1996) 5136.
[5]   M.Tejedor et al, *Journal of Magnetism and Magn.Mater.* **157-158** (1996) 141.
[6] J. D. Kraus, *Electromagnetics*, New York: MCGraw-Hill . (1984) 404.
[7] L.D. Landau and E.M. Lifshitz, *Electrodynamics of Continuous Media,* Pergamon Press,Oxford (1975) 195 .
[8] L.V. Panina, K. Mohri, T. Uchiyama, K. Bushida, Giant Magneto-Impedance in Co-Rich Amorphous Wires and Films *IEEE Trans. Magn.* **31** (1995) 1249.
[9] A. Antonov, et al, The features of GMI Effect in Amorphous Wires at Microwaves, *Physika A* **241** (1997) 420.

Non-Linear Electromagnetic Systems
V. Kose and J. Sievert (Eds.)
IOS Press, 1998

Lift-off-compensated three Coil Eddy-Current Sensor for Non-contact Temperature Measurement

Frank Röper

Institut für Elektrische Meßtechnik, TU Braunschweig, Hans-Sommer-Str. 66,
38106 Braunschweig, Germany

Abstract. A three coil eddy-current sensor is presented, which allows non-contact temperature measurements. With the help of analytical calculations the principle of lift-off-compensation is discussed. The influence of soft magnetic cores is investigated by FEM-analysis. Finally measurements that agree with the theoretical predictions are presented.

1. Introduction

An eddy-current sensor consists of a coil which is used to create a time-varying electromagnetic field, within which the conducting test material is placed. Eddy-currents induced in the test material tend to oppose and weaken the exciting field. The opposite can be true when using a ferromagnetic test material: the high magnetic permeability results in an increase in flux density. These two effects influence the apparent impedance of the primary coil and the induced voltage in a pick-up coil. The magnitude, phase lag and spatial distribution of the eddy-currents, which determine the output signal, are dependent on the sensor-specimen distance (lift-off) and the electrical conductivity, magnetic permeability and thickness of the test material. Choosing one of these to be the desired measuring property, the others become disturbing influences. On knowing the temperature dependence of the electromagnetic material properties, it is also possible to measure the surface temperature of the test material.

Non-contact temperature measurement is normally done by radiation-pyrometry, but in the case of measuring the surface temperature of a rotating roll the highly reflecting surface represents a serious difficulty. The use of an eddy-current technique allows to circumvent this problem, furthermore eddy-current sensors are not influenced by dust, steam or non-conducting foils covering the roll. The lift-off sensitivity critically affects the temperature measurement, because the lift-off change due to temperature expansion generates a change of the output signal, that lies in the same order of magnitude as the change due to variation of the material properties. Förster [1] reduced the lift-off sensitivity of a single-coil sensor by using a resonant circuit formed of the coil and a capacitor. A sensor, with the phase of the output-signal being relatively insensitive to the lift-off, was presented in [2]. The sensor consists of a large primary and two pick-up coils with different lift-off distances inside the primary coil.

Generally a transformer sensor offers two main advantages vs. a single coil sensor:

- the (temperature-sensitive) resistance of the coil has no influence on the output signal,
- there are more degrees of freedom to adjust the sensor to the measured property.

Here a three coil sensor with one primary and two concentric pick-up coils, that are located parallel to the test objects surface is presented (Fig. 1). The principle of the lift-off suppression is discussed with the help of analytical calculations. The influence of a ferrite-cup-core is investigated by FEM-analysis and experimental verification is given.

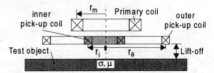

Fig. 1. Geometry of the three coil sensor, number of turns of the primary and pick-up coils: n_{ex}, n_{pi}, respectively

2. Analytical Solution of the air-core system

In [3] the analytical solution for an air-core circular coil of rectangular cross-section placed over a half space with the coil axis perpendicular to the surface is presented. The movement of the test-object is not considered. The magnetic vectorpotential A is expressed in the form of integral-equations. This analytical solutions show that A is determined by the product: conductivity · test-frequency, and not by the two parameters separately. From A any observable electromagnetic induction phenomena can be calculated, for example:

induced voltage in a single-turn pick-up coil: $U_{pi} = j \omega 2 \Pi r_a A[r_a, l_a] = f \cdot$ Function[$\sigma \cdot f$] ,

eddy-currents in the test material: $J(r, z) = -j \omega \sigma A[r, z] = f \cdot$ Function[$\sigma \cdot f$],

normal component of magnetic field: $H_z(r,z) = \dfrac{1}{\mu_0 \mu_r} \dfrac{1}{r} \dfrac{\partial r A(r,z)}{\partial r} =$ Function[$\sigma \cdot f$],

with the angular frequency ω, the imaginary unit j and the radius and lift-off of the pickup coil: r_a, l_a respectively. A *MATHEMATICA* program has been written to calculate the integral-solutions numerically. The ohmic resistance of the coil windings is not incorporated in this solution. In the following the influence of the sample's thickness is neglected (thickness much larger then penetration depth) and only nonmagnetic materials are considered.

Fig. 2 shows the calculated induced voltage in one pick-up coil and in two differentially connected pick-up coils vs. the pick-up coil radius for a typical metallic test material. The output-voltage has its maximum when pick-up and primary coil radii are nearly the same. Further increase of the r_a results in an decrease of the voltage, because of the change of magnetic field orientation. An increase in the lift-off yields an increase in the output voltage, caused by the reduced weakening of the magnetic field due to the eddy-currents. The crossing point of two (U vs. r)-curves for different lift-off indicates the lift-off-compensation: no change in the output-signal with varying lift-off. For the single pick-up coil this crossing point occurs at an unpractical large r_a, when the voltage vanishes. The insertion of the inner pick-up coil shifts the crossing to a lower radius: $r_a = r_{comp}$. The amplitude of U now differs from zero, because of the different phase angles: Phase(U_{ri}) > Phase(U_{rcomp}). This phase shift results from the phase-lag of the eddy-current compared to the exciting currents. The value of the eddy-current's phase lag depends on the material properties [4].

The amplitude normalized to the number of turns and phase angle of the output signal as a function of the lift-off are plotted in Fig. 3. The radii of the three-coil sensor are chosen according to Fig. 2: $r_i = 0.5\ r_m$, $r_a = 2.75\ r_m$, $r_m = 9.75$mm. By placing the sensor at the distance at which the amplitude has its minimum, |U| becomes lift-off independent over a limited distance range. The phase runs through zero at this distance, allowing an easier installation procedure.

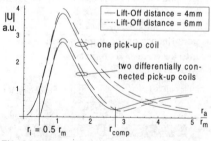

Fig. 2. Output voltage for one and two pick-up coils vs. normalized coil radius. $r_m = 9.75$mm, $f = 1$ kHz, $\sigma = 20$ MS/m, pick-up coils 1.4mm below primary coil

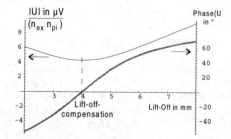

Fig. 3. Amplitude and phase of the output voltage as a function of the lift-off distance, $r_i=0.5\ r_m$, $r_a=2.75\ r_m$, $I_{ex}=1$A, other parameters see Fig. 2.

3. Experimental verification

A simple experimental set-up was build to verify the theoretical predictions. The three coils were wound on PVC-cores and positioned in front of an aluminium alloy plate. With a micrometer-screw the lift-off distance was varied. Amplitude and phase of the differentially connected pick-up coils were measured with an lock-in amplifier.

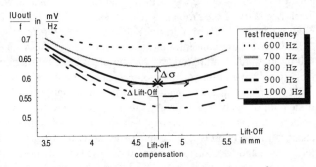

Fig. 4. Normalized output-voltage vs. lift-off for various test frequencies; r_m=5mm, r_i=12.5mm, r_a=36mm, n_{ex}=n_{pi}=320, Ac-Gain=20, I_{ex}=28mA, σ_0 = 15 MS/m

Fig. 4 shows the normalized voltage $|U|/f$ vs. lift-off for different test-frequencies. The minimum of the curves, indicating lift-off-compensation, shifts with varying frequency. A change in the test objects temperature (ΔT) can be approximated by a variation of the test frequency:

$$\frac{|U(\sigma_0 + \Delta\sigma, f_0)|}{f_0} = \frac{|U(\sigma_0, f_0 + \Delta f)|}{f_0 + \Delta f}, \text{ with } 1 + \frac{\Delta\sigma}{\sigma_0} = \frac{1}{1 + \alpha\,\Delta T}.$$

f_0 denotes the actual test frequency, α is the temperature coefficient of the resistivity (α_{Al} = 0.004 1/K) and $\Delta\sigma$ the variation of the conductivity due to the temperature change. The lift-off sensitivity can be read directly from the graph for the chosen f_0.
The following sensitivities are derived from the graphs for f_0 = 800 Hz:

lift-off change = 100 μm: $\dfrac{\Delta|U|}{\Delta Lift-off} = 5.4\,{mV}/{mm}$,

increased temperature: $\dfrac{\Delta|U|}{\Delta T} = 0.97\,{mV}/{K}$, decreased temperature: $\dfrac{\Delta|U|}{\Delta T} = 0.89\,{mV}/{K}$.

The different values for increasing and decreasing temperature indicate, that the sensor has a non-linear temperature characteristic.

4. Finite element modelling

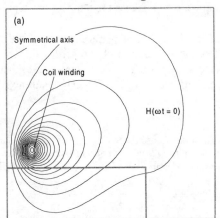
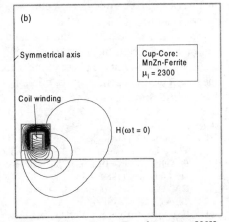

Fig. 5. Field lines of (a) an air-core sensor and (b) a ferrite-cup core sensor. Test frequency: 500Hz, r_m = 9.75mm, lift-off = 4.5 mm, test material: aluminium alloy (σ = 20 MS/m)

The integral-solutions used so far can only handle air-core coils. To investigate the influence of soft magnetic cores a FEM-analysis with *ANSYS* was performed. Because of the axial-symmetry of the problem a two-dimensional, harmonic analysis could be used. A comparison of the calculated field-lines of an air- and a ferrite-core set-up is shown in Fig. 5.

The ferrite-core concentrates a greater portion of the magnetic field under the sensor, resulting in a higher amplitude of the output voltage. Furthermore it avoids interaction of the pick-up coils with other metallic objects or magnetic fields behind the sensor.

The calculations show that the lift-off-compensation works also with the ferrite-cup core, if the radii of the two pick-up coils in front of the primary coil are properly adjusted.

In the following table the output-voltage at lift-off-compensation (l_a = 4mm), the relative changes in the output-signal due to variations of lift-off or electrical conductivity for one ferrite-core and two corresponding air-core sensors are compared.

Table 1. Calculated amplitude and rel. changes of air- and ferrite-core sensors

Parameters: r_m = 9.75mm, f=500Hz, σ = 20 MS/m, Lift-off = 4.5mm	Ferrite-cup-core r_i = 5mm r_a = 16.2 mm	Air-core sensor with same inner pick-up coil, r_a = 25.1 mm	Air-core sensor with same outer pick-up coil, r_i = 7.8 mm
$\|U\|$ in $V/(n_{ex} \cdot n_{pi})$	$0.77 \, 10^{-5}$	$0.30 \, 10^{-5}$	$0.27 \, 10^{-5}$
$\dfrac{\Delta\|U\|}{\|U\|}, \dfrac{\Delta\sigma}{\sigma} = 10\%$	5.5 %	6.2 %	5.1 %
$\dfrac{\Delta\|U\|}{\|U\|}, \Delta Lift - o. = 0.5mm$	1.7 %	1.7 %	1.9 %

The ferrite-core increases the amplitude of the output signal (factor 2.5), but the relative measuring sensitivities are nearly the same for all three sensors. The radius of the outer pick-up coil of the ferrite-core sensor decreases, if r_m and r_i are held constant, making the sensor smaller compared to air-core sensors.

5. Conclusions

A sensor design has been presented, that suppresses the lift-off sensitivity of eddy-current sensors at a distinct lift-off, making non-contact temperature measurement possible. The combination of the two pick-up coil radii, necessary for lift-off-compensation, is dependant on the chosen lift-off distance and test frequency as well as on the material properties. Further work will be carried out to optimise these parameters to enhance the temperature sensitivity.

The fact, that the output voltage is a function of ($\sigma \cdot$ f), offers the possibility to adjust the sensor automatically to different test objects: after the installation of the sensor the frequency is varied until the phase of the output signal crosses zero. This procedure assures that the sensor always operates in the minimum of the lift-off curve, see Fig. 3 and 4, resulting in the best lift-off-compensation.

References
[1]   F. Förster, Theoretische und experimentelle Grundlagen der zerstörungsfreien Werkstoffprüfung mit Wirbelstromverfahren, I Das Tastspulverfahren, *Zeitschrift f. Metallkunde* **43** (1952) 163-171
[2]   P. Keller, A new Technique for Noncontact Temperature Measurement of Rotating Rolls, *Iron and Steel Engineer* (1980) 42-44
[3]   C.V. Dodd *et al.* Solutions to some Eddy Current Problems, *International Journal of Nondestructive Testing* **1** (1969) 29-90
[4]   F. Röper, Analyse von transformatorischen Wirbelstromsensoren zur berührungslosen Messung von Geometriedaten oder Materialparametern, *VDI Berichte* Nr.1255 (1996) 399-404

Non-Linear Electromagnetic Systems
V. Kose and J. Sievert (Eds.)
IOS Press, 1998

Improvement of Signal-to-Noise Ratio on Magnetic Defect Sensor for Wire Rope of Elevator

Y. Ohhira, Y. Hirama, H. Yajima*, H. Wakiwaka*

Hitachi Building Systems Co., Ltd., 4-16-29 Nakagawa Adachi-ku, Tokyo 120, JAPAN
*Faculty of Engineering, Shinshu Univ., 500 Wakasato, Nagano 380, JAPAN

Abstract. A rope tester is a sensor which detects damage in a steel rope. The rope tester for cranes and lifts used successfully. However, it is difficult to detect damage in the rope for elevators with the rope tester for cranes. A big noise is generated, because the rope for elevators is a structure of twisted fine wires.

We developed a rope tester [1]. The signal-to-noise ratio of the rope tester for elevators is discussed in this paper. The signal-to-noise ratio is improved by examining two aspects. They are the structure of the rope tester and the signal processing of output voltage. The improved rope tester has two detection heads and low-pass filter. The signal-to-noise ratio has been improved to nine.

1. Introduction

A rope tester is a sensor which detects damage in a steel rope. Damage in the thick single rope for cranes and lifts has been detected with the rope tester. The ropes for crane and lifts are 20 millimeter in diameter and a single structure. It has been difficult to detect damage in the rope for elevators. Because the rope for elevators has a structure of twisted fine wires, then the performance of high sensitivity and low noise is demanded for the rope tester of elevators. The signal-to-noise ratio of the rope tester for elevators is discussed in this paper;

(1) the structure and the principle operation of the rope tester for elevators,

(2) the method of decreasing noise,

(3) the measurement of the signal-to-noise ratio in the rope for elevators.

2. Structure and principle operation

2.1 Structure and principle operation of the rope tester

Fig. 1 shows the structure of the rope tester. Table 1 shows the specification. Exciting yokes and detection heads have the shape of the character U because of surrounding the rope. The rope between two exciting yokes is magnetized by the field of the exciting coil. The magnetic flux leaks at damaged parts of the rope. Leaked magnetic flux flows into the detection head, when the damaged part approaches the detection head. The change in the magnetic flux is converted into the voltage by the electromagnetic induction law with the detection coil. The induced voltage V_e is given as follows;

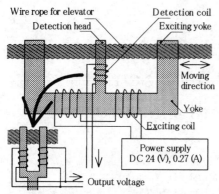

Wire rope for elevator
Detection head
Detection coil
Exciting yoke

Moving direction

Yoke

Exciting coil

Power supply
DC 24 (V), 0.27 (A)

Output voltage

Two detection heads type

Fig.1   Structure of rope tester.

Table 1   Specification of rope tester.

Item	Value
Overall dimensions (mm³)	270 × 85 × 131
Weight (kg)	13
Material of yoke and head	Fe
Exciting coils	
Number of coil windings	3000
Voltage (V)	DC 24
Current (A)	0.27
Detection coils	
Number of coil windings	11200

8 × 21 filler wire Type U
8 strands, 21 wire per strand, 1/2 in, and larger

8 × 19 Warrington
8 strands, 19 wire per strand, 7/16 in, and smaller

Fig.2   Structure of rope for elevator.

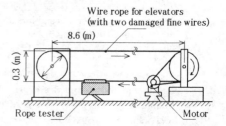

Wire rope for elevators
(with two damaged fine wires)

8.6 (m)

0.3 (m)

Rope tester   Motor

Fig.3   Measuring method.

$$V_e \;=\; -N\frac{\Delta\Phi_c}{\Delta t}, \;\; (V) \tag{1}$$

$$\;=\; -N\frac{\Delta\Phi_c}{\Delta x}v. \;\; (V) \tag{2}$$

Where, V_e : the induced voltage in the detection coils (V),
  N : the number of the detection coil windings,
  Φ_c : magnetic flux through the detection head (Wb),
  t : time (s),
  v : velocity of the rope (m/s).

2.2 Structure of the rope for elevator

Fig. 2 shows the structure of the rope for elevators. The rope for elevator is a strand with fine wire of 1 millimeter in diameter. The performance of high sensitivity is demanded for the rope tester for elevators. But noise is generated at the rope tester, because the surface of the strand is uneven.

3. Improvement of the signal-to-noise ratio

3.1 Cause and decreasing of noise

The two following points are enumerated as a cause of the noise:

(1) the surface of the strand is uneven. The distance between the surface of the rope and the detection head changes because of the movement of the rope. As a result, a strand noise is generated,

(2) an external noise generated by peripheral electric machinery and apparatus.

These noises are decreased by the following methods:
(1) two detection heads and the inverse-parallel connection of two detection coils,
(2) processing with a low-pass filter.

3.2   Inverse-parallel connection of two detection coils

An improved rope tester has two detection heads and two detection coils as Fig. 1. The generated output voltage V_o of the detection coils connected as inverse-parallel is given as follows;

$$V_o = \frac{V_{eA} - V_{eB}}{2}. \ (V) \tag{3}$$

Where, V_o : the output voltage (V),
$\quad\quad\quad V_{eA}$: the induced voltage of the detection coil A (V),
$\quad\quad\quad V_{eB}$: the induced voltage of the detection coil B (V).

If a similar strand noise is generated in each coil, the strand noise is suppressed at the equation. Two detection heads are arranged so that the distance between each detection head and the surface of the rope may become equal. The interval of each detection head is the integral multiple of the pitch of the rope's irregularities.

3.3   Processing with a low-pass filter

The frequency of the strand noise is given as follows;

$$f \ = \ \frac{v}{\tau}. \ (Hz) \tag{4}$$

Where, τ : the pitch of the rope's irregularities $(= 10 \ (mm)) \ (m)$.

When the velocity of the rope is 0.5 meter per second, the frequency of the strand noise becomes 50 Hertz.

The strand noise and the external noise are decreased by a low-pass filter. The low-pass filter of the improved rope tester is a 2nd order active filter. The cut-off frequency of the filter is 34 Hertz.

4.   Measurement

4.1   Measuring method

Fig. 3 shows a measuring method with the rope for elevator. The rope for elevator with two damaged fine wires is moved by a motor. The output voltage of the fixed rope tester is measured.

4.2   Result

Fig. 4 shows the measured output voltage of the single detection head type rope testers and the two detection heads type rope tester. The two detection heads type rope tester was able to detect damage. The noise with no damaged wire was from 5 to 8 millivolts. The output voltage with two damaged wires was 28 millivolts.

Fig. 5 shows the measured output voltage of the rope tester with the low-pass filter and without the filter. The noise has become smaller through the low-pass filter. The signal-to-noise ratio has been improved from 9.5 dB to 19.1 dB by the low-pass filter.

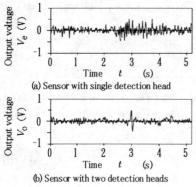

(a) Sensor with single detection head

(b) Sensor with two detection heads

Fig.4   Effect of two detection heads.

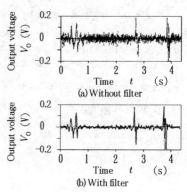

(a) Without filter

(b) With filter

Fig.5   Effect of the low-pass filter.

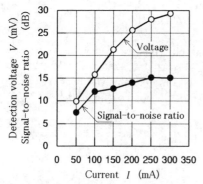

Fig.6   Characteristics of exciting current to detection voltage.

The maximum output voltage of the rope tester with two damaged fine wires is defined as detection voltage. Fig. 6 shows the characteristics of exciting current to detection voltage. The detection voltage has increased proportional to the excitation current. The signal-to-noise ratio has increased proportional to the excitation current because of the increasing detection voltage, too. The signal-to-noise ratio was saturated at 250 milli-ampere.

5.   Conclusion

Three results were obtained;
(1) The structure of the rope tester for elevators was shown. The principle operation was clarified.
(2) In order to decrease the noise, two detection coils of detection head were connected in inverse-parallel. In addition, the output voltage was processed with the low-pass filter.
(3) The signal-to-noise ratio was measured with the rope for elevators that have two damaged wires. It was confirmed that the signal-to-noise ratio has been improved from 9.5 dB to 19.1 dB by the above-mentioned methods.

References

[1] Hirama, Yauti, Ohhohira: Development of rope tester for crane, IEEJ, MAG-82-118 (1988) 9-18.

Non-Linear Electromagnetic Systems
V. Kose and J. Sievert (Eds.)
IOS Press, 1998

Magneto-optical and Electrical Properties of Giant Magnetoresistive $(La,Pr)_{0.7}(Ca,Sr)_{0.3}MnO_3$ Thin Films Prepared by Aerosol MOCVD

Elena A. GAN'SHINA, Oleg Yu. GORBENKO*, Natalia A. BABUSHKINA**, Andrey R. KAUL*, A. G. SMECHOVA, Lubov' M. BELOVA**

Physics Department, Moscow State University, 119899 Moscow, Russia
*Chemistry Department, Moscow State University, 119899 Moscow, Russia
** Kurchatov Institute, 123182 Moscow, Russia

Abstract. $(La,Pr)_{0.7}(Ca,Sr)_{0.3}MnO_3$ thin films were prepared by aerosol MOCVD on the different single crystalline substrates ($LaAlO_3$, $SrTiO_3$, MgO and $ZrO_2(Y_2O_3)$). Spectral dependence of the transversal Kerr effect (TKE) was similar for all compositions in the energy range 1.4-3.5 eV. The highest TKE was observed for $(La_{0.5}Pr_{0.5})_{0.7}Sr_{0.3}MnO_3$ $-33*10^{-3}$ at $\hbar\omega=3.05$ eV. The remarkable feature of $(La_{0.5}Pr_{0.5})_{0.7}Ca_{0.3}MnO_3$ films is the temperature hysteresis of TKE. The appearance of the hysteresis, Curie temperature and maximum TKE was found to change significantly with the variation of the substrate material. The results are compared with the data of the electrical measurements. The highest negative magnetoresistance MR was measured for $(La_{0.5}Pr_{0.5})_{0.7}Ca_{0.3}MnO_3$ on $LaAlO_3$: MR(70K, 3T)=10^6; MR(10K, 1T)>10^8.

1. Introduction

$R_{1-x}M_xMnO_3$ perovskites, where R^{3+} is a rare earth cation, M - doubly charged cation of large ionic radius, with both R and M populating A positions of ABO_3 perovskite lattice, have attracted considerable attention due to the recent discovery of giant negative magnetoresistance (GMR) of such compounds [1,2]. Thin epitaxial films are especially important objects for study owing to the material anisotropy and lack of contribution of random grain boundaries to the sample resistance. Magnetic field, position and movement sensors, reading heads for hard disks with areal density up to 10 Gbit/in.^{2} are perspective areas for GMR materials applications.

As compared to $La_{1-x}Ca_xMnO_3$ and $La_{1-x}Sr_xMnO_3$ (x= 0.2-0.5), revealing one maximum of the resistance near Curie temperature (T_c) and giant magnetoresistance nearby this maximum [3], and to $Pr_{1-x}Ca_xMnO_3$, which is semiconducting down to 4.2 K [4], the solid solutions in the system $(La,Pr)_{0.7}(Ca,Sr)_{0.3}MnO_3$ can possess very complex dependence of the resistance and magnetoresistance in the range from 4.2 K to room temperature. In this work the temperature, magnetic field and spectral dependencies of the transversal Kerr effect (TKE, δ), providing non-destructive diagnostics of the ferromagnetic phase, were measured for $(La,Pr)_{0.7}(Ca,Sr)_{0.3}MnO_3$ thin films prepared by aerosol MOCVD on the different single crystalline substrates.

2. Experimental

Films were deposited on $LaAlO_3$, $SrTiO_3$, MgO and $ZrO_2(Y_2O_3)$ single crystalline substrates with (001) orientation at the temperature 750°C and oxygen partial pressure 3 mbar (total pressure 6 mbar) and annealed at the deposition temperature in the oxygen flow at atmospheric pressure for 1h to fix their oxygen content. XRD with four circle diffractometer Siemens D5000 with secondary graphite monochromator (Cu K_α radiation) was applied to determine phase composition, orientation and lattice parameters. θ-2θ, φ scans and rocking curve measurements were used. Apparent cubic lattice parameter and (001) orientation with 'cube-on-cube' in-plane alignment were found for the films by XRD, except for the films on $ZrO_2(Y_2O_3)$, where (110) orientation of the film was preferable with the diagonal in-plane alignment (the body diagonal of the perovskite cube along the face diagonal of the fluorite cube) [5]. SEM was accomplished by CAMSCAN equipped with EDAX system for quantitative analysis. Electrical resistivity measurements were carried out in a conventional four-point probe configuration. Spectral dependence of TKE was measured in the energy range 1.3-3.8 eV (magnetic field up to $2.8 \cdot 10^5$ A/m).

3. Results and Discussion

Spectral dependence of δ was similar for all compositions in the energy range 1.4-3.5 eV (Fig.1). The maximum δ position fluctuated with the change of the film composition and substrate material up to 0.2 eV. The result, being considered in relation to the crystallographic data [1], means that in the energy range under study magneto-optical activity is mostly due to the transitions in MnO_6 octahedra with the permanent Mn^{4+}/Mn^{3+} ratio in the whole series. δ(H) dependencies show the saturation in magnetic field $> 1.6 \cdot 10^5$ A/m, except for films on $ZrO_2(Y_2O_3)$ (Fig.2). On the other hand, maximum δ value varied about 3 times with the variation of the film composition. The highest δ was observed for $(La_{0.5}Pr_{0.5})_{0.7}Sr_{0.3}MnO_3$: $-33*10^{-3}$ at $h\omega=3.05$ eV and the beam inclination angle $\psi = 67°$. $(La_{0.5}Pr_{0.5})_{0.7}Sr_{0.3}MnO_3$ can be considered as a homogeneous ferromagnetic below T_c (~300 K) without any hysteresis for δ(T) curves.

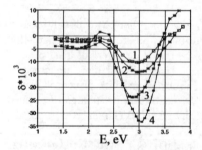
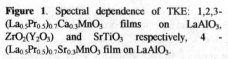
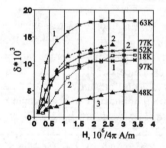

Figure 1. Spectral dependence of TKE: 1,2,3- $(La_{0.5}Pr_{0.5})_{0.7}Ca_{0.3}MnO_3$ films on $LaAlO_3$, $ZrO_2(Y_2O_3)$ and $SrTiO_3$ respectively, 4 - $(La_{0.5}Pr_{0.5})_{0.7}Sr_{0.3}MnO_3$ film on $LaAlO_3$.

Figure 2. Magnetic field dependence of TKE for $(La_{0.5}Pr_{0.5})_{0.7}Ca_{0.3}MnO_3$ films on $SrTiO_3$ (1), $LaAlO_3$ (2) and $ZrO_2(Y_2O_3)$ (3) at the different temperatures.

The TKE variation for other samples is due to the change of the ferromagnetic phase quantity. Besides that, $(La_{0.5}Pr_{0.5})_{0.7}Ca_{0.3}MnO_3$ films revealed significant hysteresis of the δ(T) curves (Fig.3). These results imply a more complex state in the case of $(La_{0.5}Pr_{0.5})_{0.7}Ca_{0.3}MnO_3$. The variation of rare earth or alkaline earth cation is known to produce a deformation of the

MnO$_6$ octahedron array by tilting and rotations of octahedra rather than a deformation of each octahedron [6]. On the other hand, the film - substrate lattice mismatch creates an anisotropic strain in the film which results in the lattice distortion compatible with the cooperative Jahn-Teller effect [7]. The strain is responsible for the difference of δ(T) and resistivity curves measured for the films on LaAlO$_3$ and SrTiO$_3$ (Fig.3,4). Both substrate materials are perovskites and provide the epitaxial growth of (La$_{0.5}$Pr$_{0.5}$)$_{0.7}$Ca$_{0.3}$MnO$_3$, but they produce strain on the reverse direction in the film [5]. The behaviour of the film on SrTiO$_3$ was simpler and corresponded to the paramagnetic - ferromagnetic phase transition above 100 K. A hysteresis of TKE was observed only in the transition area (Fig.3a).

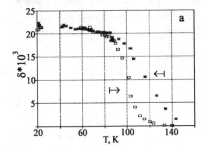

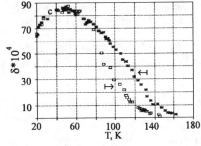

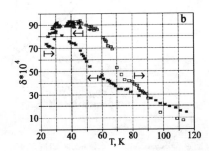

Figure 3. Temperature dependences of TKE of (La$_{0.5}$Pr$_{0.5}$)$_{0.7}$Ca$_{0.3}$MnO$_3$ films on SrTiO$_3$ (a), LaAlO$_3$ (b) and ZrO$_2$(Y$_2$O$_3$) (c) measured during cooling and heating (as shown by arrows).The photon energy 3.0 eV and the amplitude of the magnetic field 4·10^4 A/m were used for measurements in an optical continuous-flow helium cryostat.

The film on LaAlO$_3$ showed hysteresis of δ(T) extended to the lower temperature with inversion of the hysteresis sign at 90 K. Below this temperature the hysteresis was of the opposite sign to that of the film on SrTiO$_3$. The ferromagnetic transition took place at about 30 K lower temperature as compared to the film on SrTiO$_3$. Next, a sharp decrease of δ was observed below 30 K. The difference manifests itself also in the resistivity curves (Fig.4) with maximum resistivity temperature T$_p$ being close to T$_c$ found from the δ(T) curve. (La$_{0.5}$Pr$_{0.5}$)$_{0.7}$Ca$_{0.3}$MnO$_3$ film on LaAlO$_3$ revealed a resistivity hysteresis even in the zero magnetic field. During cooling, the film showed dielectric behaviour. When starting the heating from 4.2 K the low resistance window appeared in the range 20-70 K (Fig.4b). The nature of the antiferromagnetic state below 20 K is not clear: it can be a charge ordered state similar to that found in Pr$_{1-x}$Ca$_x$MnO$_3$ [8] or a two phase state with ferromagnetic clusters embedded in the antiferromagnetic matrix like in the magnetic semiconductors of the EuSe family [9]. Then the applied magnetic field assists the melting of the charge ordered lattice or growth of the ferromagnetic clusters providing the percolation at the critical dimension. Metamagnetic phase transition in the film below 20 K results in the anomalous high negative magnetoresistance (MR= R(0) - R (H))/R(H), where R(0) and R(H) represent resistance in the zero field and applied magnetic field): MR(70K, 3T)=10^6; MR(10K, 1T)>10^8. The film on SrTiO$_3$ revealed weaker negative magnetoresistance: MR(100K, 3T)=10^4.

The $(La_{0.5}Pr_{0.5})_{0.7}Ca_{0.3}MnO_3$ film on $ZrO_2(Y_2O_3)$ possessed much higher electrical resistance than films on the perovskite substrates. As shown by us in [5] the effect is owing to the special domain boundaries in the (110) oriented film. The electrical resistance fluctuated strongly below 100 K indicating complicated percolation behaviour. According to the TKE measurements (Fig.3c) the broad ferromagnetic transition started below 140 K but ferromagnetism decreased again below 40 K (similar to the film on LaAlO₃).

4. Conclusions

The temperature and magnetic field dependencies of TKE in the giant magnetoresistive $(La,Pr)_{0.7}(Ca,Sr)_{0.3}MnO_3$ films firstly studied in the current work are rather complex. Hysteretic phenomena indicate the complicated magnetic structures existing in the wide temperature range. The nature of such structures which can be only guessed now is worthy further study.

This study was supported by the Copernicus Program (grant No. PL 96-4292), BMBF Program (grant No. 13 N 6947/5) and RFBR (grant No. 97-03-32979a).

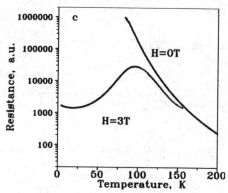

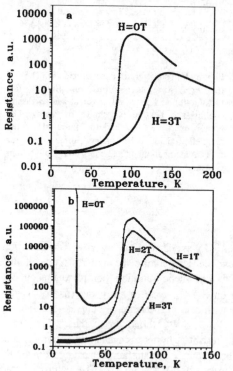

Figure 4. Temperature dependences of the resistivity of $(La_{0.5}Pr_{0.5})_{0.7}Ca_{0.3}MnO_3$ films on SrTiO₃ (a), LaAlO₃ (b) and $ZrO_2(Y_2O_3)$ (c) in the zero magnetic field as well as in the applied magnetic field.

References

[1]   C.N.R.Rao, A.K.Cheetham and R.Mahesh, *Chem. Mater.* **8** (1996) 2421.

[2]   K.Chahara, T.Ohno *et al.*, *Appl. Phys. Lett.*, **63** (1993) 1990.

[3]   P.G.Radaelli, M.Marezio *et al.*, *J. Solid State Chem.* **122** (1996) 444.

[4]   M.R.Lees, J.Barratt *et al.*, *Phys. Rev. B* **52** (1995) 1.

[5]   O.Yu.Gorbenko, N.A.Babushkina *et al.*, *J. de Physique* IV, (1997) to be published in proceedings of EUROCVD - XI.

[6]   H.Y.Hwang, S.W.Cheong *et al.*, *Phys. Rev. Lett.*, **75** (1995) 914.

[7]   A.J.Millis, *Phys. Rev. B*, **53** (1996) 8434.

[8]   Y.Tomioka, A.Asamitsu *et al.*, *Phys. Rev. B* **53** (1996) 1689.

[9]   S.von Molnar and S.Mathfessel, *J. Appl. Phys.* **38** (1967) 959.

Non-Linear Electromagnetic Systems
V. Kose and J. Sievert (Eds.)
IOS Press, 1998

GMI Effect in Amorphous Ribbons with Creep-induced Magnetic Anisotropy

L. KRAUS, M. KNOBEL*, V. HAŠLAR, A. MEDINA*, and F.G. GANDRA*

*Institute of Physics, Academy of Sciences of the Czech Republic, Na Slovance 2,
CZ-18040 Praha 8, Czech Republic*
**Instituto de Fisica "Gleb Wataghin", Universidade Estadual de Campinas (UNICAMP),
P.O.B. 6165 - Campinas 13083-970, SP Brasil*

Abstract. Giant magnetoimpedance effect (GMI) and biased transverse permeability μ_t were investigated in $Co_{67}Fe_4Cr_7Si_8B_{14}$ amorphous ribbons with hard-ribbon-axis magnetic anisotropy induced by stress annealing. The magnitude of effective anisotropy field H_K was varied by changing the stress applied during annealing. Both the biased transverse susceptibility and the GMI show typical peculiarities at $H_0 = H_K$. While μ_t is nearly constant for $H_0 \leq H_K$, it shows a rapid decrease with bias field H_0 larger than H_K. The GMI effect shows remarkable maxima at $H_0 \approx H_K$. The experimental results are compared with the theoretical model based on a simple stripe domain structure. The theory qualitatively well describes the main features of behaviour. The observed magnitudes of GMI are, however, much lower than the calculated ones. Possible origins of this discrepancy are discussed.

1. Introduction

It was observed by Machado et al. [1] that ac impedance of a conductor made of soft magnetic amorphous metal exhibits a strong dependence on external magnetic field. This effect, so called "giant magneto impedance" (GMI), has been attracting attention of many researchers in the last few years because of its prospective applications in sensor elements. Though the principle of the phenomenon is well known - it is based on the classical electrodynamics [2,3], the actual limits of GMI have not been determined (neither theoretically nor practically). The magnitude of GMI is closely related to the circumferential permeability μ_ϕ and its field dependence. The permeability can be effectively controlled by an induced magnetic anisotropy and the related domain structure. For example, weak circumferential anisotropy, produced in Co-rich amorphous wires by current-annealing [4] or by stress-annealing [5], has been reported to be particularly suitable for large GMI. To achieve the best performance of GMI elements better understanding of the phenomenon is needed. Some attempts to calculate GMI theoretically have already been done [6-8]. An exact solution based on the simultaneous evaluation of Maxwell equations and the equation of motion in real domain structures is, however, extremely difficult and is still absent.

The aim of this work was to investigate the influence of creep-induced magnetic anisotropy on GMI effect in a Co-rich amorphous ribbon and to compare the experimental results with the theoretical predictions of a simple model. The ribbon shape of sample was used because it enables direct measurements of transversal permeability. Also the theoretical calculations are somewhat easier for the planar geometry. The creep-induced anisotropy, on the other hand, provides a regular domain structure, which is more appropriate for the theoretical treatment.

2. Theoretical

The typical domain structure of a ribbon with the easy plane perpendicular to the ribbon axis (z) is shown in Fig.1. It is assumed that transverse magnetization takes place by domain wall displacement (Δz) and magnetization rotations (m_1, m_2). The quasistatic transverse permeability μ_t

$$\mu_t = \mu_0\left(1 + \chi_{mov} + \chi_{rot}\right) = \mu_0\left(1 + \frac{2\,\Delta z\,M_s \sin\theta_0}{Dh_{ex}} + \frac{m_{1y} + m_{2y}}{2h_{ex}}\right). \tag{1}$$

is obtained by minimizing the corresponding free energy E with respect to Δz, m_1 and m_2. Here χ_{mov} and χ_{rot} are the domain wall movement and the rotational susceptibilities, respectively. The meaning of other symbols used in Eq.(1) is apparent from the figure. Both χ_{mov} and χ_{rot} contribute at applied dc fields H_0 lower than the anisotropy field H_K. For $H_0 > H_K$, where the domain structure disappears (cos θ_0 = 1), a uniform rotation of magnetization takes place and χ_{rot} is given by

$$\chi_{rot} = M_s/(H_0 - H_K + N_y M_s), \tag{2}$$

where N_y is the transversal demagnetizing factor. The maximum μ_t is obtained for $H_0 = H_K$ and its magnitude is limited only by N_y. If the ribbon is magnetized by a uniform ac field h_{ex} N_y is determined by the sample dimensions. If, however, the sample is magnetized by an axial ac electric current then ac magnetization is circumferential and magnetic charges on the side walls of ribbon are negligibly small. Then the circumferential permeability μ_ϕ = $\mu_t(N_y\to0)$ theoretically goes to infinity for $H_0 = H_K$.

The Eqs. (1) and (2) are not applicable to higher frequencies, where the dynamic effects must be taken into account. The domain wall movements are heavily damped by eddy currents. Then the rigid domain structure ($\Delta z = 0$) is a good approximation. The gyroscopic nature of magnetization rotations is accounted for by the general equation of motion

$$\dot{M}_i = \gamma\,M_i \times H_{eff\,i} - (\alpha/M_s)\,M_i \times \dot{M}_i - (1/\tau)(M_i - M_{i0}) \tag{3}$$

where α is the Gilbert damping parameter, τ the Bloch-Bloembergen relaxation time and the effective field $H_{eff\,i} = -\partial E/\partial M_i$. Two damping terms are used in Eq.(3). As will be shown, the GMI effect is quite sensitive to the choice of particular damping term. The complex impedance of the ribbon can be obtained by the simultaneous solution of Maxwell equations and the equation of motion. Even for the rigid domain structure the exact solution is practically impossible. Therefore an effective medium approximation (assuming a homogeneous permeability) is used. The relative impedance Z/R_{dc} of the ribbon is then

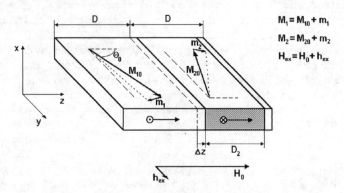

Fig.1. Schematic domain structure

$$Z / R_{dc} = k \, t / 2 \cot(k \, t / 2), \qquad k = (1 + i) / \delta_0 \sqrt{\mu_{eff} / \mu_0} \tag{4}$$

where R_{dc} is the dc resistance, δ_0 the classical skin depth and t the ribbon thickness. The effective dynamic permeability μ_{eff} is calculated from Eq.(3). The theoretical $|Z(H_0)|$ curves show sharp maxima at $H_0 \approx H_K$. Their heights and widths depend on a number of material parameters. The behaviour of $|Z(H_0)|$ at lower frequencies is, however, particularly sensitive to the relaxation time τ. The details of theoretical calculations will be published elsewhere.

3. Experimental

An amorphous ribbon (10 mm $\times$ 22 μm) with the nominal composition $Co_{67}Fe_4Cr_7Si_8B_{14}$ was investigated. Five pieces of the ribbon were stress-annealed for 1 h. at 380°C with the applied stress $\sigma_a = 0, 200, ..., 800$ MPa. 7 cm long samples were then used for the measurements. Anisotropy field H_K and the demagnetizing factor N_y were determined from dc hysteretis loops measured along and transverse to the ribbon axis, respectively. The transverse permeability μ_t as a function of longitudinal bias field H_0 was measured at the frequency 75 Hz using a simple inductive method. The impedance Z of the ribbon was investigated at several frequencies in the range 10 kHz - 120 kHz with the bias field applied either parallel or transverse to the ribbon axis.

4. Results and Discussion

The longitudinal hysteresis loops (Fig.2a) are linear with a sharp knee at the anisotropy field H_K, which corresponds to the theoretical: $\cos \theta_0 = H_0/H_K$. The anisotropy is proportional to the stress σ_a applied during annealing. The domain observation by SEM proved that, with the exception of the vicinity of ribbon edges, the structures are close to the theoretical one. The transversal hysteresis loops of all the samples are practically identical showing a sharp knee at $H_0 \approx 1060$ A/m, from which the demagnetizing factor $N_y = 2.37 \times 10^{-3}$ can be calculated.

The relative change of transverse permeability $\mu_t(H_0)$, measured with $h_{ex} = 12$ A/m, is shown in Fig.2b. As can be seen, $\mu_t(H_0)$ is nearly constant for $|H_0| < H_K$, which indicates that domain wall movement dominates in this field region. The observed hysteresis may be explained by some domain structure changes with applied bias field H_0. When smaller h_{ex} is used for μ_t-measurement then the local minima on $\mu_t(H_0)$ become deeper and the hysteresis increases. The behaviour of $\mu_t(H_0)$ for $|H_0| > H_K$ does not depend on h_{ex} and is well described by the theoretical dependence (Eq.(3)).

The GMI-effect measured at the frequency of 100 kHz with the ac current of 10 mA is shown in Fig.2c. In the central part of $|Z(H_0)|$ curve some hysteresis is also present and sharp maxima can be seen at. The amplitude of GMI increases by about one order of magnitude when the measuring frequency increases from 10 kHz to 120 kHz. With the dc field applied transversally to the ribbon axis some changes of Z were also found at low bias field. But the effect is very weak and seems to be independent of H_K.

Fig.2 clearly shows that the transverse permeability and the GMI-effect are closely related to the transverse magnetic anisotropy. These results are in a good qualitative agreement with the simple theoretical model described above. The theoretical $|Z(H_0)|$ curves also show sharp maxima at $H_0 = H_K$. When, however, the magnitudes of experimental and theoretical GMI curves are compared the agreement becomes much worse. If the Bloch-Bloembergen damping term is neglected ($1/\tau = 0$) the maximum theoretical value $|Z(H_K)|$ is frequency independent and is enormously high. For example, using the known parameters

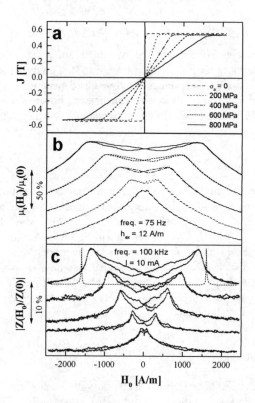

Fig. 2 (a) - longitudinal hysteresis loops, (b) - transverse permeability and (c) - GMI effect in stress annealed amorphous $Co_{67}Fe_4Cr_7Si_8B_{14}$ ribbons.

($t = 22$ μm, $M_s = 4.5 \times 10^5$ A/m, $H_K = 1600$ A/m and $\rho = 1.5 \times 10^{-6}$ Ωm) and a reasonable value $\alpha = 10^{-2}$ for the Gilbert damping parameter (see e.g. [9]) one obtains $|\Delta Z_{max}|/R_{dc} = 32.5$, which is almost three orders higher than the experimental value. By introducing the Bloch-Bloembergen damping term into the equation of motion more realistic frequency dependence of GMI can be obtained. To explain the magnitude of $|\Delta Z_{max}|$ measured at 100 kHz $\tau = 1.3 \times 10^{-9}$ s should be used. This value of relaxation time is, however, unreasonably high and moreover the peak on the theoretical $|Z_0(H_0)|$ dependence remains still too sharp (dashed curve in Fig.2c). The discrepancy between the theory and experiment may be also explained by the dispersion of H_K or local easy axes [7]. Such dispersion could substantially decrease the height and increase the width of GMI peak at low frequencies, but it would be quite ineffective at higher frequencies. To explain the high frequency behaviour of GMI some additional mechanisms (such as exchange effects or the surface roughness) must be taken into account.

References

[1]   F.L.A. Machado. B. Lopes da Silva and E. Montarroyos, J. Appl. Phys. **73** (1993) 6387-6389.
[2]   R.S. Beach and A.E. Berkowiz, Appl. Phys. Lett **64** (1994) 3652-3654.
[3]   L.V. Panina and K. Mohri, Appl. Phys. Lett. **65** (1994) 1189-1191.
[4]   J.L. Costa-Krämer, K.V. Rao, IEEE Trans. Magn. **31** (1995) 1261-1265.
[5]   J. Pokorný and L. Kraus, Sensors and Actuators A **59** (1997) 65-69.
[6]   R.S. Beach, A.E. Berkowiz, J. Appl. Phys. **76** (1994) 6209-62.
[7]   L.V. Panina et al., IEEE Trans. Magn. **31** (1995) 1249-1260.
[8]   T. Morikawa et al., IEEE Trans. Magn. **32** (1996) 4965-4967.
[9]   L. Kraus, Z.Frait and J. Schneider, Phys. Stat. Sol. (a) **63** (1981) 669-675.

Gradient Magnetic Biosensor for Weak Dia- or Paramagnetic Shifts in Magnetic Susceptibility of Porous Beads or Surrounding Fluid

Svetlana NORINA
Magnetism Chair, Moscow State University, Moscow, 119899, Russia
Stanislav RASTOPOV
General Phys.Inst.,Vavilova Str.38,Moscow, 117942, Russia

Abstract. A gradient magnetic biosensor is described as suitable for rapid quantitative detection of single or double macromolecular layers in porous nonmagnetic beads or surrounding fluid. The measurements of capture travelling time or accumulation radius in gradient magnetic field have shown that it is possible to determine the weak dia- or paramagnetic shifts of magnetic susceptibility up to $0.7*10^{-10}$(SI) equall to 0.20 mg/ml of macromolecular amount.

1. Introduction

Numerous biological applications of high gradient magnetic separation (HGMS) achieved mainly for human blood cells [1,2] have shown that it is possible to distinguish between the diamagnetic Fe-zero-spin oxidised states and the paramagnetic Fe-high-spin reduced ones in red blood cells, and the achieved accuracy of magnetic susceptibility determination, equall to 10^{-8} (SI) [1], allowed the detection of corresponding shifts of diamagnetic haemoglobin concentrations of 35.2 mg/ml. The capture and build-up of particles with 5-40 μm sizes on single magnetised wires were investigated theoretically and experimentally [3-5] and the calculations of particles' trajectories on the basis of a combination of magnetic, fluid viscous and gravitational forces under appropriate hydrodynamic parameters of potential (inviscid) flow theory were reported [5]. In the axial case of study the cylindrical symmetry imposed parallel to a wire produces equations as analytical solutions for the particle with the plane of wire magnetisation trajectories. When assuming azimuth angle ϑ of the particle with the plane of wire magnetisation to be zero, then the equations of motion can be integrated, giving the following equation:

$$T= -1/V_m [y-y_0] \qquad (1)$$

where $y=r_a^4/4 - kr_a^2/2 + K^2/\ln(r_a^2+K)/2$, $V_m=2\chi b^2 M_s \mu_0 H_0/9\eta a$, r_a is the normalised co-ordinate; T is the travelling time of the particle from position r_a to r_{a0}; $K=M_s/2\mu_0 H_0$ with M_s being the saturation magnetisation of the wire in an applied magnetic field H_0 perpendicular to the axis of the wire having radius a, χ-the relative magnetic susceptibility of the surrounding fluid, η is the fluid viscosity, b is the particle radius. HGMS- study considers the capture and retention of particles as they move near a wire for the axial flow configuration: the capture radius R_c [4], the accumulation radius R_a [5] and the saturation accumulation radius R_{as} [5].

The aim of the paper is HGMS-study of the capture travelling time and accumulation parameters for the experimental foundation of the gradient magnetic biosensor which can be applied for the screening detection of the small macromolecular changes within porous nonmagnetic beads or in the surrounding fluid.

2. Experimental methods

The successive procedures of the present method can be described by:

-obtaining agarose beads (control 1 and control 2) to immobilise the macromolecular ligands in their meshes; the incubation of beads with the ligand sample or the injection of the sample through affinity mini- column [6];

- obtaining the magnetic parameters [4,5]: capture travelling time T (using HGMS), initial build-up radius R_a, capture radius R_c and saturation accumulation radius R_{as} (using the original portable device based on the fibre optic method and HGMS in parallel, and magnetic 10 μm poles gap surface - for semi-quantitative screenings).

It was found that R_c and R_{as} did not depend on the beads radius b within values 3-12 μm in the case of the vertical wire configuration. Indeed, assuming flow velocity V_0 [5] equall to the gravity settling velocity for nearly spherical beads $V_g=V_o=2\rho gb^2/9\eta$, which is parallel to the vertical wire axis, we obtained :

$$R_{as,c} = A_1(\chi M_s\mu_0H_0/\rho ag)^m \qquad (2)$$

where ρ - density of single bead, A_1=4.05, m=0.33 for R_{as} ; A_1=2, m=1 for R_c [4].

Agarose beads with 4 vol.% of sorbent net ("Pharmacia")   crashed to diameters 3-40 μm are prepared for the ligand immobilisation according to the conventional Br-CN-activated procedure [6].The adjusting optimisation of the magnetic susceptibility values was reached by the shaking of clear beads in 1 mM HCl solution for 5 and 30 min. to obtain control 1 and control 2 , respectively, under 4°C. The immobilised single macromolecular layer was covalently bound to agarose beads. Such ligands were : bovine serum albumen (BSA) and human immunoglobulin G (Ig G) (" Serva"). Double layers were presented by antigen (Ag) - antibody (Ab) pairs: IgG and goat antihuman immunoglobulin G ("Sigma") or prepared as in [6]. The micro-enzyme-linked immunosorbent assay (micro-ELISA) [6] determined the content of the linked protein on the beads to calibrate the suggested method.

3. Results and discussion

In many immunotests demonstrated by Fig.1, the immobilised antigen-antibody complexes on porous agarose beads have shown undoubted difference from control 1, the clear beads having paramagnetic susceptibility   with mean capture time less than 0.2 s. Single layer of antigen bound in 4.5 mg/ml concentration on agarose beads shifted T to 1.0 s, next antibody additions in 0.6-3.2 mg/ml on wet agarose beads have given valuable changes of T up to 10 s and more, with further appearance of diamagnetic properties of beads. These curves can be described by hyperbolic curves corresponding to eq.(1), from which $T = const/\chi b^2$, where χ is proportional to the tested macromolecular amounts on beads.

The collection normalised parameters of the magnetic build-up near magnetised wire (a=27 μm) under diamagnetic accumulation processes with control 2 beads are presented in Fig.2-4. The dependence of the saturation accumulation radius R_{as} via the protein concentration on agarose beads was established (Fig.2) by curve 1, which included the linear part up to $R_{as}=4$; C=1.5 mg/ml and the saturation level was $R_{as}=7$; C=6 mg/ml.

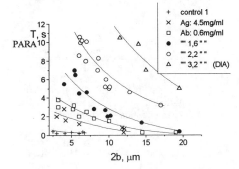

Fig.1. Capture travelling time T of porous agarose beads via their diameters 2b for the different concentrations of linked macromolecules in the meshes of beads: antigen (Ag)- single layer, antigen-antibody (Ag-Ab) - double layers.

Curve 1 (Fig.2) was obtained for different size ranges of beads maintaining the approximation from eq.(2), where the R_{as}-values are independent of dimensional sizes but depend on the relative magnetic susceptibility. As soon as the clear beads of control 2 were weak paramagnetic that is nearly" zero" and R_{as} increases linearly on curve 1 with the macromolecular ligand concentrations the relevant working range of the method becomes apparent. The saturation of R_{as}-curve means the limited immobilising capacity of the

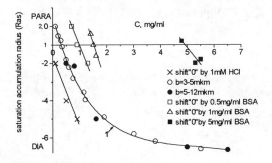

Fig.2. Saturation accumulation radius R_{as} via macromolecular concentrations C on porous agarose beads. Curve 1 (black and white circles) were obtained for different b-size ranges. Shifts of «zero» - value, comparative to curve 1, were obtained by 1mM HCl treatment of clear beads and BSA-addings into liquid.

agarose sorbent under 6-10 mg/ml concentrations of the protein ligands, that is a well known fact in affinity chromatography [6]. From Fig.2 the accuracy of the macromolecular ligand determination consists of as little as 0.20 mg/ml of protein on sorbent beads (Fig.2) or $0.7*10^{-10}$ (SI) for the magnetic susceptibility value, that can be compared with results [1,2] for red blood cells where accuracy was not higher than 10^{-8}(SI).

The initial build-up radius R_a as a function of the time t for the different ligand concentrations on beads is presented in Fig.3, where the separated curves show the high sensitivity of such measurements for the rapid direct detection (within several minutes) of linked macromolecular processes.

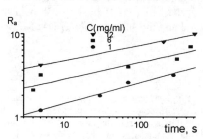

Fig.3. Relative accumulation radius R_a as a function of time T for agarose beads (b=5-7 μm, $\mu_0 H_0$=1T) for different macromolecular (BSA) concentrations on beads.

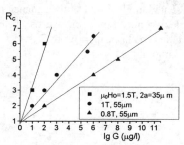

Fig.4. Capture radius R_c of porous agarose beads (b=3-12 μm) as a function of antibody Ig G concentration in tested liquid.

The capture radius R_c is the mean distance from which the beads are caught by gradient magnetic field, and depends strongly on magnetic technical parameters H_0 and a, not on beads radius b, as Fig.4 demonstrates, where R_c, as a function of the macromolecular antibodies concentration in tested liquid, is plotted before the antigen-antibody pairs formation. According to Fig.4, the sensitivity of the antibody detection in the sample is equall to about 1 μg/ml in concentration values, and consists 10 ng in ligand mass terms assuming that the minimal tested volume was 10^{-2} ml of fluid.

The gradient magnetic biosensor made on the basis of numerous experimental results gave the possibility for rapid quantitative detection of very small macromolecular amounts and weak dia- or paramagnetic shifts as one can see in Fig.2 observing shifts "0" by 1 mM HCl treatment of beads and BSA-addings into surrounding fluid.

4. Conclusion

The accuracy of the magnetic susceptibility measurements for porous beads is more than two orders better than HGMS-studies in [1,2]. The fast testing of small magnetic changes in porous sorbent beads or surrounding fluid can be applied in the biotechnological and ecological analysis.

References

[1]   M.Takayasu, N.Duske, S.R.Ash, and F.J.Friedlaender, HGMS Studies of Blood Cell Behavior in Plasma, *IEEE Trans.Magn.***18**(1982)1520-1522.
[2]   A.Shalygin, S.Norina, and E.I.Kondorsky, Behaviour of Erythrocytes in High gradient Magnetic Field, *J.Magnetism and Magnetic Mat.***31-34** (1983)555-556.
[3]   F.J.Friedlaender, M.Takayasu,J.R.Rettig, and C.P.Kentzer,Particle Flow and Collection Process in Single Wire HGMS Studies, *IEEE Trans. Magn.***14**(1978)1158-1164.
[4]   J.H.Watson and A.S.Bahaj, Vortex Capture in High Gradient Magnetic Separators at Moderate Reynolds Number,*IEEE Trans.Magn.***25**(1989)3803-3809.
[5]   F.J.Friedlaender, M.Takayasu, J.B.Rettig, and C.P.Kentzer, Studies of Single Wire Stream Type HGMS, *IEEE Trans.Magn.***14**(1978)404-406.
[6]   P.D.G.Dean, W.S.Johnson and F.A.Middle, Affinity Chromatography, IRL Press Limited, Oxford-Washington DC, (1985)44-154.

Non-Linear Electromagnetic Systems
V. Kose and J. Sievert (Eds.)
IOS Press, 1998

Fluxgate Sensor with Multiple Paths of Flux

Octavian BALTAG and Doina COSTANDACHE
*The National Institute of Research and Development for Technical
Physics, 47, Mangeron Blvd., PO Box 834, 6600 Iasi, Romania*

Abstract. The paper describes the behaviour of a new type of fluxgate sensor
with a new geometry of the core. Operating principle of a fluxgate is based on the
pulse exciting method. The performances and the operating limits are also
presented.

1. Introduction

Many works have been published on the utilization of certain new materials, such
as, for instance, the amorphous magnetic materials [1]. Making use of the non-linear
magnetic properties of the ferromagnetic, the authors considered different methods to
determine the magnetic field magnitude. For example, the same fluxgate can be used in
magnetometers working on the second harmonic of the exciting signal, or in time domain.
The signal coil can be tuned, via a capacitor in the second harmonic, or the fluxgate can
work in impulse mode. The measurement can be made in a simple way by determining the
amplitude and phase of the second harmonic, or using more complex conversion procedures
in time domain [2], [3]. The utilization of the amorphous materials made it possible to
increase the working frequency and to adopt new working principles, such as the
dependence of the width of the impulses from the signal circuit in the magnetic field.
Although most of the authors tend to use materials with new properties we have chosen the
permalloy, a classic material.

The sensor of a fluxgate magnetometer consists of a ferromagnetic material core
which is periodically driven into saturation by means of an exciting current of a frequency f.
One of the problems with the fluxgate magnetometers is that a sensor must be of high
sensibility, low noise and reduced power consumption. The performances of the sensor can
be improved by using a certain exciting regime [4], the same passive elements in the
exciting circuit and tuning the output circuit in the second harmonic of the exciting signal
[5] or by reducing the demagnetizing factor for the measured field [6].

2. Description and operating principle

The fluxgate with multiple paths of flux consist of a permalloy sheet representing
the ferromagnetic core. The thickness of permalloy core is 50 μm; the length of the core is
24 mm. In the ferromagnetic core a number of windows was cut in two parallel rows.
Through these windows the exciting coil passes, which sense of winding alternates from
one window to another, so that on the common paths of flux of two adjacent windows, the
induction should add to each other. By the windows, the edges of the sheet were cut round
so that the segments with a smaller magnetic induction than the saturation should be
excluded. On the median axis, other windows were cut, in order to eliminate from the sheet
the portion which would remain unsaturated due to the deleting of the magnetic inductions

of opposite signs. A solenoidal coil covering the whole core pick up the second harmonic of the exciting current.

Our fluxgate sensor consists of eight paths of flux and can be assimilated to a number of eight toroidal sensors adjoined so that a new geometry of the ferromagnetic core should result (Fig. 1). The fluxgate can be used with sinusoidal or impulse excitation. Since the material reaches the saturation quite soon, the sinusoidal excitation presents the disadvantage of a high power demand and an excessive heating of the excitation winding. After the saturating of the material the complex impedance decreases to the value of its resistive component, which results in an excessive current and an excessive heating of the excitation winding.

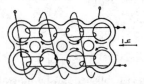

Fig. 1. The ferromagnetic core of the fluxgate sensor.

After the saturation of the material the complex impedance decreases to the value of its resistive component, which results in an excessive increase of the excitation current until the moment of the off saturation condition. And this happens every half period. In order to avoid this shortcoming we used the impulse excitation with a filling factor $t/T = 1/8$.

The fluxgate operation principle is presented in Fig.2. Every element can be considered as a toroidal sensor. Fig. 2 presents the diagrams of the magnetic induction and the voltage induced in the signal coil in two cases, when the magnetic field to be measured is $h>0$ and $h<0$, respectively.

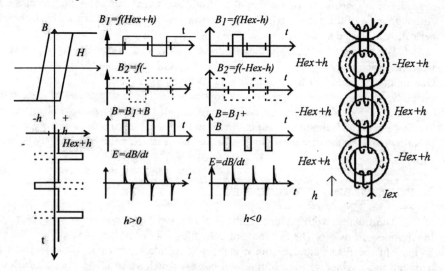

Fig.2. Operating principle of the fluxgate based on the pulse exciting method.

At the same time, the diagram presents the cases when the phase of the exciting magnetic field is $0°$ and $180°$, respectively. This is important, since each of the elementary core rings of the fluxgate is excited in a magnetic field which phase changes by $180°$ when passing from one ring core to another. As it can be noticed the voltage induced in the signal winding depends only on the magnitude and polarity of the external magnetic field *h* and not on the 0 ° or 180 ° phase of the exciting magnetic field. Taking into account all the toroidal elements of the fluxgate, each of these will synchronously contribute to the voltage induced in the signal winding. The two windings are orthogonally coupled, which means that in the absence of the ferromagnetic core the induced voltage is minimal. Phenomenologically, the operation principle of the sensor with multiple paths of flux is not essentially different from that of other magnetic sensors - it doubles the excitation frequency when applying the continuous magnetic field. The magnetic induction in the core in the regions where the inductions due to the excitation and that due to the field to be measured are parallel, is greater than in the regions where they are opposed to each other.

3. Experimental results and performances

The sensor was introduced inside a Helmholtz coils system by means of which the compensation of the local magnetic field is performed, another pair of a Helmholtz coil system being used to apply known fields. The fluxgate transfer characteristic $U_{2f} = f(h)$ is presented in Fig. 3. The two diagrams present the fluxgate response with (Fig. 3.a) and without feedback loop (Fig. 3.b). As it can be noticed, the feedback loop makes the response linear which extends the measuring range. The noise level of the odd harmonics is $U_{pp} = 48$ mV , the resolution with the feedback loop closed is 8 x10^{-6} A/cm; the fluxgate sensitivity is about 0,4 V/(A/cm).

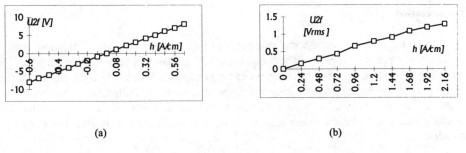

(a)   (b)

Fig.3. The fluxgate transfer characteristic $U_{2f} = f(h)$

The oscillograms in Fig. 4 present the dependence of the second harmonic voltage on the applied field, taking as reference the exciting current. The signal winding is tuned in the second harmonic of the exciting current which improves the signal to noise ratio, by stressing the second harmonic and reducing the amplitudes of the other residual odd and even harmonics. The phase of the second harmonic changes by 180 ° when changing the polarity of the field *h*. The residual signal consisting of the induced odd harmonics are due both to the parasitic coupling between the two windings and to some core non-uniformities. The oscillograms from Fig. 5 represent the signal induced in the fluxgate when the signal winding is not tuned. This case is similar to that presented in Fig. 2. The impulses amplitude depends on the field intensity, their polarity and phase changing by 180 ° according to the positive or negative direction of the field.

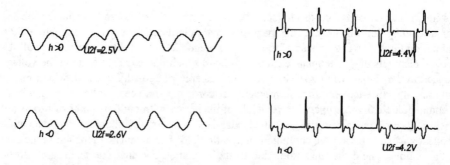

Fig.4. Phase dependence on *h* polarity-tuned   Fig.5. Phase dependence on *h* polarity-untuned

The presence of the resonance in the signal winding enables a 10 times increase of the sensitivity. The fluxgate sensitivity also depends on the excitation conditions, because for the sinusoidal excitation the equivalent quality factor of the signal winding is reduced due to the core being mainly in the saturated state. Under these circumstances, the signal winding is easy to tune, but no significant increase of the second harmonic amplitude is obtained. The impulse excitation is more convenient since the equivalent permeability of the core is higher which results in an increase of the quality factor. When tuning the fluxgate in the second harmonic a non-linear resonance occurs which under certain conditions (high excitation level, high quality factor) can generate oscillations on the second harmonic.

4. Conclusions

The fluxgate sensor with multiple paths of flux shows a sensibility measured outside the non-linear resonant region of about one order higher than that the ring core sensors. The sensibility of our sensor depends on the excitation conditions. When the output circuit is tuned in the second harmonic of the exciting signal the sinusoidal oscillations are generated by the multiple flux sensor having a frequency equals to the double of the excitation frequency. The multiple flux sensor can be determined to work in a steady region by shunting the output circuit.

References

[1]   P. Ripka, Review of fluxgate sensors, Sensors and Actuators A, Vol.33, pp.129, 1992
[2]   M. Lassahn, G.Trenkler, A multi-channel magnetometer for field structure based on time encoded fluxgate sensors, IEEE Trans.on Instr. And Meas., Vol. 42, No 2, pp.635, 1993
[3]   K. Weyand, V. Bosse, Fluxgate magnetometer for low-frequency magnetic electromagnetic compatibility measurements, IEEE Trans. on Instrum. and Meas., Vol. 46, No 2, pp.617-620, 1997
[4]   J. Afanasiev, Fluxgate devices, Energoatom., Leningrad, 1986
[5]   M.H. Acuna, and C.J. Pellerin, A miniature two-axis fluxgate magnetometer, IEEE Trans. Geosci. Electron., Vol. GE-7, pp.256-260, 1969
[6]   F. Primdahl, The fluxgate mechanism, Part I : The gating curves of parallel and orthogonal fluxgates, IEEE Trans. Magn.,vol.6,pp.376,1970

Non-Linear Electromagnetic Systems
V. Kose and J. Sievert (Eds.)
IOS Press, 1998

Evaluation of Magnetostriction Measurement Techniques

P.I. Anderson, A.J. Moses, H.J. Stanbury*

Wolfson Centre for Magnetics Technology, Cardiff School of Engineering, University of Wales
Cardiff, PO Box 917, Cardiff CF2 1XH, Wales, UK
* Orb Electrical Steels Ltd, PO Box 30, Newport, Gwent NP9 0XT, Wales, UK

Abstract. The variation of the magnetostriction of conventional and high permeability electrical steels under stresses of +5MPa to -5MPa is presented. Systems utilising a piezoelectric accelerometer and resistance strain gauges were used to investigate the magnetostriction of standard Epstein strips. Both systems showed the familiar characteristic for both the rolling and transverse directions but with very low harmonic distortion. A significant variation along the length of the strip as well as between materials was observed but high permeability material had a better average characteristic. The localised measurements showed a grain to grain variation in the high permeability material attributable to the grain orientation. The magnetostriction response was also linked to the thickness of the insulating coating via the stored tensile stress. This localised data proved valuable despite the relatively poor performance of the strain gauge system when compared to that of the piezoelectric accelerometer.

1. Introduction

The magnetostrictive behaviour of grain oriented 3% silicon steel is generally accepted to be the main source of transformer noise. Magnetostriction, like other magnetic properties, is particularly sensitive to compressive stress. This may create a problem since transformer laminations are likely to be stressed due to factors such as non-flat laminations, non-uniform clamping and variations in the core temperature.
The objectives of this investigation were to measure the fundamental and harmonic levels of magnetostriction and understand the effects of the insulating coating for a variety of samples of grain oriented silicon steels under applied stress using two techniques.[1]

2. Procedure

Epstein samples cut parallel and perpendicular to the rolling direction were taken from conventional and high permeability grain oriented silicon steel produced by Orb Electrical Steels Ltd.
The basic apparatus used was developed by Stanbury[2]. This measurement system incorporated piezoelectric accelerometers clamped at the ends of a single Epstein strip which was fixed at one end whilst the free end was attached to a thumbscrew stressing mechanism together with load cell measurement facility. The magnetising system included primary and secondary windings across a flux closure yoke with the primary being excited by a power amplifier and sinewave oscillator. The flux density, B, was set up by measuring the voltage across the secondary winding using a mean sensing digital voltmeter. The output from the accelerometers was summed to give a differential result and then filtered and double integrated before passing to a tuneable filter where the frequency of interest was selected.
A second measurement system was constructed to operate in parallel with the first system, utilising the same magnetising and stressing set-up, using commercial strain gauges to measure the

magnetostriction. These gauges were glued to the areas of interest on the strip with epoxy resin and connected to a strain gauge bridge together with a passive gauge on a second strip. The output was filtered using a tuneable amplifier.

Three 120Ω, 20mm gauges were evenly distributed along the conventional grain oriented samples so that any variation of the magnetostriction along the length of the strip could be detected. The large grain size of the high permeability material made it feasible to investigate the grain to grain variation of magnetostriction by careful placement of 120Ω, 5mm gauges. The grains were chosen, using a domain viewer, so as to be large enough to take the gauge and to fall into one of the three categories listed below

- well oriented
- slightly mis-oriented from the rolling direction
- containing supplementary closure domains which suggest an angle of dip of 2-3°.

The samples were weighed so that the induction and stress levels could be calculated before being inserted into the apparatus. The maximum tensile stress (+5MPa) was then applied to the sample and was gradually released as magnetostriction readings were taken. This continued through zero stress and up to the maximum (-5MPa) compressive stress. This process was carried out at a magnetisation frequency of 50Hz for peak inductions of 1.1, 1.5 and 1.7 Tesla.

3. Results

Both measurement systems showed the normal magnetostriction characteristic. Along the rolling direction a very low negative magnetostriction was exhibited when unstressed or under tension. This increased in a complex, non-linear manner under compression. In the samples cut along the transverse direction magnetostriction increased with decreasing tension reaching a plateau before the zero stress condition. This then remained unresponsive to increasing compressive stress in the transverse direction. The harmonics of magnetostriction were found to be below the sensitivity of the measuring equipment (i.e. less than 2×10^{-8}). The magnetostriction was found to vary significantly not only between materials and batches but also within a pack of Epstein samples cut from a single test sheet. The pack results showed the magnetostriction at 5 MPa varied from 0.2 to 2με for the conventional material whilst the high permeability material showed results from 0.15 to 0.5με only. The higher variance in the conventional material is due to the fact that there is a larger range of domain mis-orientation - the primary magnetostriction generator - whereas the orientation in the high permeability material is more tightly controlled.

3.1 Localised Versus Bulk Magnetostriction for Conventional Grain Oriented Steel

The localised magnetostriction measured using the strain gauges showed a variation along the length of the Epstein strip (see figure 1). The results from gauge B, at the centre of the strip, closely follow that of the gauge average which in turn is very similar to the bulk response from the piezoelectric accelerometer measurement system whereas gauges A and C show a significantly different response. This would be expected since the gauges can cover a large number of grains and include representative samples of the whole strip. The results from gauge A (located near the free end of the strip) indicate either a material effect or a physical distortion of the strip caused by the compression. An equipment fault was ruled out by the close agreement between the gauge average and the bulk response. It is thought that the strip actually lifts from the flat base of the windings former and possibly oscillates thus affecting the surface mounted strain gauges.

3.2 Localised Versus Bulk Magnetostriction in High Permeability Grain Oriented Steel

The strain gauges were placed on an Epstein strip of the high permeability material which had its insulating coating chemically removed since changes in the magnetostriction of the coated high permeability material were too small to be measured by the strain gauge technique. Gauge A was attached to a well oriented grain, gauge B to a slightly mis-oriented grain and gauge C to a grain containing a relatively large volume of supplementary closure domains. It was ensured that all the gauges were placed in close proximity to each other in order to eliminate localised stress variations from the grain to grain comparisons.

All of the gauge results exhibit a similar response to that of gauge A on the conventional material in the previous section which was attributed to the buckling of the strip and as such would all be affected by an increased stress around the buckling area having an unknown effect on the grain structure of the material. Despite this, the plots show a predictable response after buckling. It can be seen from figure 2 that the zero crossings for the magnetostriction response occurs in the order B-C-A. The response for the grain measured by gauge B increases first because the domain vectors are initially slightly mis-aligned from the stress axis and are therefore more easily rotated than those measured by gauge A which are well aligned to the stress axis.

The grain measured by gauge C contains domains which are well aligned and supplementary domains in the volume of the sample which are aligned at 90° to the stress axis. The response for gauge C therefore crosses the abscissa at a point between those of the previous two and reaches a maximum before gauge A due to a lower volume fraction of well oriented domains. Grain B exhibits the highest peak magnetostriction at 5MPa, attaining a level of 5 $\mu\varepsilon$. Grains A and C, however, reach a peak value of 3 and 2 $\mu\varepsilon$ respectively before entering into an oscillation. It is unknown whether this oscillation is a result of the strip buckling as described earlier or a material property such as that described by Moses [3]. The oscillation occurs at a compressive stress of 3.5 MPa which, taking into account the shift of the response due to the removal of the 'Magnite S' coating, corresponds to a coated stress of approximately 8.5 MPa. This is very close to the stress where Moses observed the behaviour attributed to an oscillation between two energetically similar domain patterns.

In this case averaging the measurements of the gauges does not produce a close approximation to the bulk measurement. This is due to the fact that the grains were chosen for their size and type and not as a representative sample of the entire grain population of the strip as well as being concentrated in a single region of the strip.

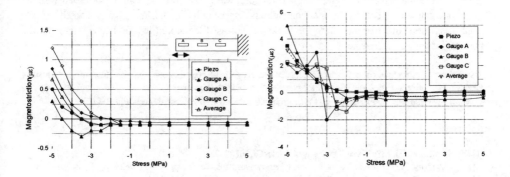

Figure1: Fundamental of Peak Magnetostriction Vs Stress for CGO at 1.5T, 50Hz Using Both Systems

Figure 2: Fundamental of Peak Magnetostriction Vs Stress for De-Coated Hi-B at 1.5T, 50Hz Using Both Systems

4. Comparison of Methods

Despite measuring the same parameter the two systems employed here actually serve entirely different functions. The piezoelectric accelerometer offers a comparatively quick, easy and accurate method of measuring the bulk magnetostriction of the strip and as such is suitable for batch testing of many samples. The strain gauge system, however, should be considered as a research tool for measurement of the grain to grain variation of the magnetostriction but is limited by the long preparation time required for each strip. The sensitivity and uncertainty of the strain gauge system employed here were comparatively poor (the resolution of 0.2 µε being a factor of 10 higher than that of the accelerometer system) but this was largely due to the use of 'off the shelf' measurement systems.

However, both systems produced responses showing near identical trends for the bulk and average magnetostriction which offers further validity to the methods used.

5. Additional Measurements

Additional measurements were taken, using the accelerometer system only, to assess the effects of the insulating coating on the magnetostriction of coated conventional and high permeability grain oriented silicon steels. Measurements taken before and after a chemical pickle to totally remove the coating showed a shift of the characteristic along the stress axis corresponding to a release of tensile stress of between 3 and 5 MPa. The largest shift observed was for the high permeability material which uses an improved coating. The tensile stress imposed by the coating, produced by a combination of differential contraction and line tension, was found to be related to the thickness of the coating as measured by a Fischer Permascope in the conventional material whereby an increased thickness offered increased tension and hence improved magnetostriction characteristics. The high permeability material showed little variation in the magnetostriction characteristic with coating thickness due to the coating being significantly thicker than that on the conventional material.

6. Conclusions

The magnetostriction response measured by the two systems was found to show similar trends although the piezoelectric system provided a far better resolution. For the conventional grain oriented material the strain gauges showed a variation of the magnetostriction characteristic along a single Epstein strip with the gauge average correlating closely with the bulk piezoelectric measurement. For the high permeability material a grain to grain variation was observed which can be correlated with the grain orientation. Additionally the coating, and specifically its thickness, was related to the magnetostriction such that increasing the thickness increased the stored tension hence improving the magnetostriction characteristic.

The authors wish to thank the Works General Manager, Orb Electrical Steels and the Head of the School of Engineering, University of Wales, Cardiff for permission to publish this paper.
This work was partially supported by a TCS programme between Orb Electrical Steels and UWC.

References

[1]   Anderson, P.I, MSc Thesis, University of Wales Cardiff (1996).
[2]   Stanbury, H.J, Journal of Magnetism and Magnetic Materials, Vol 26, pp 47-49 (1982).
[3]   Moses, A.J, Journal of Magnetism and Magnetic Materials, Vol 26, pp 185-186 (1982).

Non-Linear Electromagnetic Systems
V. Kose and J. Sievert (Eds.)
IOS Press, 1998

345

Examination of Making Highly Accurate Index Phase on Magnetic Rotary Encoder

Y.Kikuchi, K.Shiotani, F.Nakamura, H.Wakiwaka, and H.Yamada

Graduate School of Science and Technology, Shinshu University
Wakasato 500, Nagano, 380 Japan

Abstract. This paper deals with considerations of an index phase accuracy on a magnetic rotary encoder with high resolution of 4000 pulses per revolution. There is an index phase to detect the original position per rotation. It is necessary to have a highly accurate index phase as the resolution increases. To obtain the stabilized index phase, a novel output method and the accuracy of the index phase as a magneto-resistive element width variation, are considered by experiments and finite element analyses. The required index phase width of 6.9 μm is obtained according to the arrangement of four magneto-resistive elements and the output wave form processing. As a result, index phase widths from 1.48 μm to 6.30 μm are obtained when the detection gap variation is 30 μm.

1. Introduction

In recent years, all kinds of sensors should be made the high resolution and accuracy because of the progress of control technology. Magnetic rotary encoders (MREs) are used in factory automation applications to detect the rotatory speed and angle of servo motors, because of their suitability for use in difficult environments. The required resolution is more than 2000 pulses per revolution (ppr) in the magnetic drum with a diameter of 35 mm.

Many studies have been made to develop the high resolution MRE by use of multiple magneto-resistive (MR) elements and electric manipulation of the output signals or wave forms[1]. The MRE with a resolution of 4000 ppr has been developed by using a double interpolation method of a sinusoidal MR output voltage wave form[2]. There is an index phase to detect an original position on the drum. The index phase with high resolution and accuracy (i.e., less than $\tau/4$, τ is the periodic time of the resolution of 4000 ppr) is also required.

To obtain the stabilized index phase, investigations have been given to an MR stripe width and a novel output method of the index phase. This paper describes the following: 1) the characteristic of the index output wave form and the processing method of the index phase pulse, 2) the MR output characteristics depending on the element width variation, and 3) an optimal output condition of the index phase pulse.

2. Operation principle of MRE

A structural view and the operation principle of the MRE are shown in Fig. 1. The MRE consists of the magnetic drum, i.e., medium, MR elements, and signal processing circuits. Additionally, a magnetization head is used to investigate the influence of the magnetization factors, which is shown in Fig. 1.

To detect both the rotatory angle and direction, the drum consists of a multi-pole

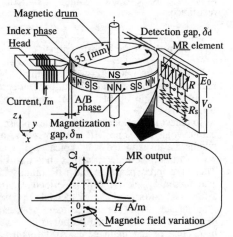

Fig. 1 Structural view and operation principle of MRE ($E_0 = 10$ V, $R_s = 514$ Ω).

Table 1   Magnetic properties and dimensions of magnetization head, drum, and MR element.

Item	Symbol	Value	Unit
Magnetic head material: RM10 (Co—V—Fe alloy)			
Saturated magnetic flux density B_s		1.6	T
Initial relative permeability	μ_i	900	
Magnetic head gap	δ_h	40	μm
Magnetic head track width	w_t	4	mm
Number of turns in coil	N	160	
Magnetization current	I_m	400	mA
Drum material: Isotropic plastic magnet			
Saturated magnetization	M_s	0.172	T
Remanent magnetization	M_r	0.103	T
Coercive force	H_c	151	kA/m
Rectangular ratio	Γ	0.60	
Drum diameter	D	35	mm
Drum rotation speed	v	0.5	s^{-1}
MR element material: Ni (81%) —Fe			
MR element size			
Stripe length	l	3	mm
Stripe width	w	5—40	μm

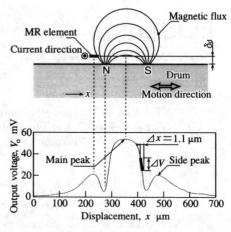

Fig. 2   Output voltage wave form of index phase.

phase, which is the A/B phase. A track with an individual two phase output (i.e., A and B phase to which shifts by 45 degrees) and the index track with the original position per rotation. The A/B and index phases are magnetized by using sine and rectangular wave form currents, respectively. By rotating the magnetized drum, the resistance of the MR element varies as the value of the magnetic field from the drum also turns. It is then possible to detect the rotatory angle and direction by using signal circuits. Moreover, the index phase can be detected in the original position. There are reports that signal circuits also are composed by using several MR elements[3].

The magnetic properties and dimensions of the magnetization head, the drum, and the MR element are shown in Table 1. The MR output voltage is defined at the connected point between the MR element and a fixed resistance. The output voltage of the one stripe MR element can be calculated by eq. (1),

$$V_o = \frac{(R_s/R)\,E_0}{1-(\varDelta R/R)+(R_s/R)} - \frac{(R_s/R)\,E_0}{1+(R_s/R)} \qquad (1)$$

where, V_o is the output voltage, R_s is the fixed resistance, E_0 is the power supply voltage, R is the MR resistance without magnetic field, and $\varDelta R$ is the variation of MR element resistance.

3.   Output voltage of index phase

Note that the magnetic field sensitivity is increased as the MR element width is widened. However, as the spacial resolution is decreased[4], the $\varDelta R/R$ characteristics are saturated to about 3 % in all of the MR element width conditions.

To detect the influence of the index phase, the output voltage of one MR element with varied widths is investigated. A typical MR output wave form of the index phase is shown in Fig. 2. There are three peaks in the MR output wave. The maximal peak in Fig. 2 is defined as a main peak, i.e., output voltage of the index phase. A gradient, i.e., $\varDelta V / \varDelta x$, of the main peak is also defined as the differential output voltage at the half value of the main peak. Additionally, two small peaks at both sides of the main peak are called side peaks. Fig. 3 shows a simulated

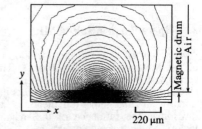

Fig. 3   Distribution of magnetic flux on index phase magnetization by using finite element analysis (2.09×10^{-8} Wb/div.).

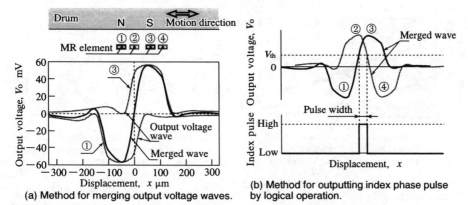

(a) Method for merging output voltage waves.

(b) Method for outputting index phase pulse by logical operation.

Fig. 4   Method for outputting index phase pulse.

result by using a finite element analysis, suggesting that the side peaks are caused by the magnetic flux in the direction along the x axis[5]. The side peak values are better when small. To improve the index phase accuracy of the MRE, the width of the index phase is required to maintain half, (i.e., $\tau/2$, the value of the width of the A/B phase), and it must be stabilized to the variation of the main peak value. For example, for a calculated result at 4000 ppr, the maximal variation value is less than 6.9 μm. The MR output variation causes the *wow* or *jitter* on the drum rotation. The gradient of the main peak, as shown in Fig. 2, requires more than 2 mV/μm (i.e., index pulse width is less than 5 μm).

4.   Characteristics of stabilized index phase

4.1   Novel method for outputting index phase pulse
The magnetization current width of the rectangular wave form is set to 10 μm; even so, the pulse width of the index phase at Δx is more than 100 μm or in other words, the width is too large. Therefore, a novel technique is required to obtain the 6.9 μm pulse width for the necessary condition. The novel method for outputting the index phase pulse is shown in Fig. 4. The principal operations are as follows: as shown in Fig. 4(a), the middle electrical signals have been made by amplifying the differential voltages between the ① and ③ signals or the ② and ④ signals. Then, as shown in Fig. 4(b), the index phase output is made by the logical OR Gate of ①+③ and ②+④ outputs combined with a digital threshold voltage V_{th}, which is the main peak voltage by half. It is found that the index phase pulse width of 6.9 [μm] at the resolution of 4000 ppr can be obtained using the novel technique.

4.2   Gradient characteristics of main peak
To stabilize the index phase pulse, the gradient of the main peak wave form has to be enlarged at the point of the threshold voltage. The gradient of the main peak is defined easily by using eq. (2). The calculated results are indicated with varied MR element width parameters as shown in Fig. 5,

$$\Delta V/\Delta x = \Delta V/(1.1 \times 10^{-6}) \tag{2}$$

where, ΔV is a differential voltage V as Δx is set to 1.1 μm at the main peak voltage by half. The gradient of the MR output wave form is increased as the width of the MR element is increased. However, the gradient seemed to be reduced at a range of less than a detection gap of about 30 μm. However it is experimentally clear (see Fig. 6) that the lowest points between the main and side peaks do not fall down to 0 V in the detection gap range. In other words, it is thought that the spacial resolution is reduced by increasing the MR element width. The magnetic flux through the MR element in the direction of the x axis does not arrive at 0 V. To protect the MR element from the variation of the detection gap, the detection gap must be widened more than 40 μm. Also, to consider the range less than 10 % of the side peak, as shown in Fig. 5, a useful range is indicated by the solid line. In the range, the maximal gradient can be obtained of 2.0 mV/μm as the width of the MR element

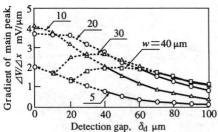

Fig. 5   Characteristics of gradient on
the main peak (w : width of MR element).

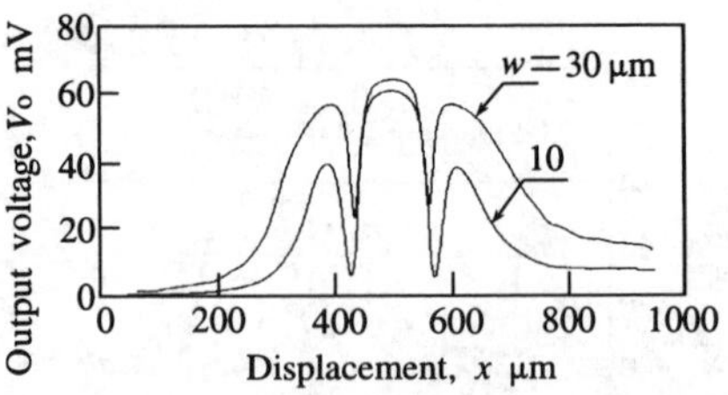

Fig. 6   Wave forms of output voltage according
to the width of the MR element (detection gap
$\delta_d = 10$ μm, w : width of MR element).

width and the detection gap are set to 20 μm and
60 μm, respectively.

4.3   Width simulations of index phase pulse

Influences of the detection condition to the
MR output are found according to the previous
section results. As a next consideration, width
variations of the index pulse are simulated by
using a personal computer as the detection gap
variation produces 30 μm. The index pulse is
merged using four MR output waves which are
obtained from the individual MR stripe elements
by adjusting its situation. Variations of the index

Table 2   Variation of index pulse width.

Width of MR element w μm	Index pulse width w_p μm (Detection gap)	Variation of index pulse width Δw_p μm
10	1.35— 9.96 (70—40 μm)	8.61
20	1.48— 6.30 (80—50 μm)	4.82
40	1.48—12.32 (100—70 μm)	10.84

pulse width are shown in **Table 2**. The variation range is from 1.48 μm to 6.30 μm as the MR
element width is set to 20 μm, which is indicated by a maximal gradient of the main peak.
This condition satisfies the necessary condition of less than 6.9 μm variation on the MRE
with 4000 ppr. Additionally, it was found that a minimal variation of the MR index phase
pulse width can be set to 4.82 μm.

5.   Conclusions

To obtain the stabilized index phase, a novel method and its accuracy for outputting the
index phase as the change in the width of magneto-resistive element, have been considered by
experiments and finite element analyses. It was found that the formation method of the index
phase pulse and the necessary accuracy on the magnetic rotary encoder with 4000 ppr had been
developed. The results are as follows: 1) the formation method of the index phase pulse was
proposed, 2) it was found that the maximal gradient that could be obtained was 2.0 mV/μm as
the width of the MR element width and the detection gap were set at 20 μm and 60 μm,
respectively, and 3) the variation range of the index phase pulse width was from 1.48 μm to
6.30 μm as the MR element width was set to 20 μm. This condition satisfied the necessary
condition of less than 6.9 μm variation. Additionally, a minimal variation of the MR index
phase pulse width could be set to 4.82 μm.

References

[1] Shoichi Kawamata, Tadashi Takahashi, Kunio Miyashita, and Katsuyoshi Tamura, *Proceedings of the 4th
    Sensor Symposium*, (1984)277-280.
[2] Y.Kikuchi, F.Nakamura, Y.Yamamoto, H.Wakiwaka, and H.Yamada, Consideration of a High Resolution of
    Magnetic Rotary Encoder, *IEEE Trans. Magn.*, **32**(1996)4959-4961.
[3] Tadashi Takahashi, Shoichi Kawamata, Kunio Miyashita, and Hiromi Kanai, *Proceeding of the 4th Sensor
    Symposium*, (1984)281-284.
[4] Robert P. Hunt, A Magnetoresistive Readout Transducer, *IEEE Trans., Magn.*, MAG-7, (1971)150-154.
[5] Y.Kikuchi, K.Kusama, Y.Yamamoto, H.Wakiwaka, and H.Yamada, *The 1st International Symposium on
    Linear Drives for Industry Applications, IEE of Japan*, (1995)477-480.

Non-Linear Electromagnetic Systems
V. Kose and J. Sievert (Eds.)
IOS Press, 1998

A Full Passive Fault Current Limiter using the Permanent Magnet for the Burden Critical Equipment

Masayoshi Iwahara, Subhas C. Mukhopadhyay, Sotoshi Yamada and Francis P. Dawson*
*Department of Electrical and Computer Engineering, Kanazawa University,
Kodatsuno 2-40-20, Kanazawa 920, Japan
* Department of Electrical Engineering, University of Toronto, Canada.*

Abstract. The demand for electrical power is continuously increasing. In practice, it is difficult to meet that demand. The ever increasing demand for power puts a heavy burden on the power supply. Faults and other abnormal conditions force high fault currens to flow. This paper describes a passive fault current limiter configured using permanent magnets with high coercive force, high remanent flux-density and ferrite cores with high permeability. The flux distribution inside the PM and ferrite cores has been analyzed using the 3-dimensional finite element method.

1. Introduction

During fault or short circuit conditions a high fault currents flow in a power system. When power semiconductor devices are used this high value of fault current puts a heavy burden on the consumers as these devices are to be selected corresponding to the maximum value of fault current and fault voltage. To avoid that situation some kind of current limiter is used in power systems. The basic characteristics of the Fault Current Limiter (FCL) should be : the voltage drop across it should negligible under normal conditions; the operation from normal to abnormal conditions and vice-versa should be automatic; and it should be able to limit the first peak value of fault current. A passive fault current limiter consisting of permanent magnets (PM) with high coercive force, high remanent flux density and ferrite cores with high permeability, low saturated flux-density compared to that of permanent magnets has already been reported in [1-2]. It does not need any control device to sense the fault current. This device almost met all the requirements of an ideal current limiter. In this paper, using the three dimensional finite element method , the field distribution inside the cores and permanent magnets has been investigated. Field distribution of the system with varying thickness permanent magnets has been studied and the results are presented.

2. Structure of Fault Current Limiter

The configuration of the FCL based on permanent magnets and ferrite cores is shown in Fig.1. It consists of two magnetic devices connected in series and with opposing magnetomotive forces. The magnetic devices are combination of ferrite cores and permanent magnets. The polarity of the connections of the coils is such that at any instant of a.c. cycle, the mmf due to ac source and permanent magnet mmf assists in one core while they oppose in the other.

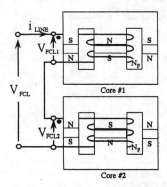

Fig.1 Configuration of FCL.

Table 1 Parameters of core and PM.

Material	Specific Data	Dimension
Ferrite Core	DL-2S (Hitachi Ferrite Co. Ltd.)	Coil Turns, N_F =150
B_s	0.28 T	a = 5.9 mm
H_s	30 A/m	b = 8.2 mm
$\mu_{ru} = \mu_u /\mu_o$	12740	c = 10.1 mm
$\mu_{rs} = \mu_s /\mu_o$	87.6	d = 11.7 mm
Permanent Magnet	NdFeB (EPSON Electric Co. Ltd)	
B_r	1.144 T	
H_c	0.825 MA/m	l_m = 0.5 mm
μ_{rc}	1.103×10^{-6} H/m	

Under normal operation both the cores are saturated, with the result that they offer low saturated impedance of both the coils in series. When a fault occurs, the cores become free from saturation alternately during a cycle and the impedance rises to a higher value, which is the saturated reactance of one coil added to the unsaturated reactance of the other coil, thereby limiting the current. The characteristics of a ferrite core and a permanent magnet are shown in Figs. 2 and 3 respectively. The characteristics of the ferrite core are approximated by two constant lines having different slopes. In the unsaturated zone, the permeability is designated by μ_u , whereas in the saturated zone it is μ_s. The permanent magnet characteristics are approximated by a straight line with constant recoil permeability μ_r.

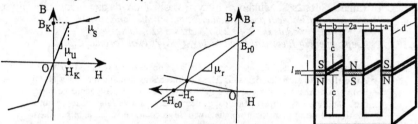

Fig.2 Characteristics of ferrite-core.   Fig.3 PM characteristics.   Fig.4 PM-Ferrite core assembly.

3. Formulation of Finite Element Model

For the analysis, finite element modelling has been carried out for one-fourth of one ferrite core-PM assembly. The ferrite-core-PM assembly is shown in Fig.4, and the dimensions of core, PM and other important parameters are listed in table 1. For modelling field equations are guided by Maxwell's equation and given by (1)

$$\nabla \times \left(\frac{1}{\mu} \nabla \times A \right) = J_0 + \nabla \times M \tag{1}$$

In an alternating field taking the effect of eddy currents in permanent magnets into consideration, the eq(1) can be written as in (2)

$$\nabla \times \left(\frac{1}{\mu} \nabla \times A \right) = J_0 + \nabla \times M - j\omega\sigma A \tag{2}$$

where, A is the magnetic vector potential, J_0 is the external impressed current density, M is the magnetization vector, μ is the permeability, ω is the angular frequency and σ is the electrical conductivity of the permanent magnet material. The elements in the quarter section of the assembly is shown in Fig.5 and the B-H characteristics of the ferrite core is shown in Fig.6. The finite element package JMAG-JVISION[4] has been used for the field analysis. Total elements of 5512 and total nodes of 6480 have been used. The analysis has been carried out in a BPL Fujitsu Vector Parallel processor which takes almost one hour for the solution.

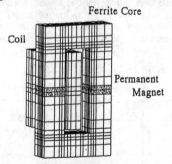

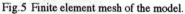

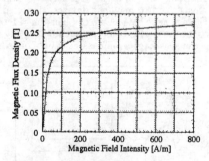

Fig.5 Finite element mesh of the model.   Fig.6 B-H characteristics of ferrite-core
                                                          in FEM model.

4. Results of Field Solution

Using the output magnetic vector potential the fluxlines are plotted and post-processing is carried out. Initially we have taken the thickness of the permanent magnet as 2.2mm. Fig.7a , b and c show the fluxlines inside the ferrite core, coil and permanent magnet respectively. The most important part of this analysis is to show that electrical stress developed due to the flow of eddy currents in the permanent magnet. It is seen from Fig.7 that in the vicinity of the PM and ferrite core, the fluxlines are crowded. Consequently the flux density is higher in the outer portion of the PM and core. This is due to the effect of eddy currents in the PM. Fig.8 shows the eddy current distribution in permanent magnets obtained from field solution. Eddy currents mainly concentrate in the inner portion of the magnets. The field study has also been carried out for varying thickness permanent magnets. Fig.9a and 9b shows the fluxlines in the ferrite core and coil when the permanent magnet thickness is 1.0mm. Fig.10a and 10b shows the fluxlines in the ferrite core and coil when the permanent magnet thickness is 0.5mm. The eddy current density has been reduced with the decrease of PM thickness and also there is a reduction of leakage flux. In addition, with the reduction of PM thickness the range of normal operating current will be less.

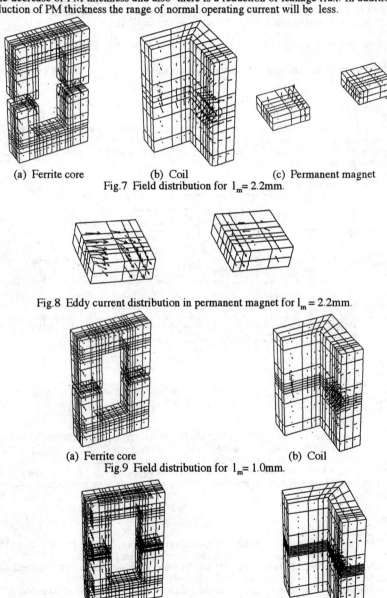

  (a) Ferrite core   (b) Coil   (c) Permanent magnet
Fig.7 Field distribution for l_m= 2.2mm.

Fig.8 Eddy current distribution in permanent magnet for l_m = 2.2mm.

  (a) Ferrite core   (b) Coil
Fig.9 Field distribution for l_m= 1.0mm.

  (a) Ferrite core   (b) Coil
Fig.10 Field distribution for l_m= 0.5mm.

5. Inference of Calculated Results

In this section some of the governing equations which have been used for the analysis of the FCL circuit will be discussed. The basic equation is written in (3) for the circuit shown in Fig.4.

$$U_{mo} = (R_m + R_i)\phi; \quad R_m = \upsilon_m \frac{l_m}{S_i}, \quad R_i = \upsilon_i \frac{l_i}{S_i} \tag{3}$$

where, U_{mo} is the magnetomotive force of the PM corresponding to retentivity, R_i and R_m are the reluctances of the ferrite core and PM respectively. υ_i and υ_m are the reluctivity of the ferrite core and PM respectively. l_i and l_m are the magnetic path length of the ferrite core and PM respectively. S_i is the area of the ferrite core and PM. ϕ is the operating flux through the ferrite core and PM. Assuming the ferrite core operates in saturation, the equation(3) can be written as in (4).

$$\phi_0 = \frac{1}{(R_m + R_i)}\left\{U_{mo} + (R_{iu} - R_{is})\phi_k\right\} \qquad R_{iu} = \upsilon_{iu}\frac{l_i}{S_i}, \quad R_{is} = \upsilon_{is}\frac{l_i}{S_i} \tag{4}$$

In order to guarantee that the core will operate in saturation i.e., $\phi_0 > \phi_k$, the following condition must be satisfied

$$H_{co} \ge (K_x + K_\mu)H_k; \quad H_k = \upsilon_{iu} B_k \qquad K_x = \frac{l_i}{l_m}, \quad K_\mu = \frac{\mu_{ru}}{\mu_{rm}} \tag{5}$$

where H_{co} is the retentivity of PM, B_k and H_k are the flux density and magnetic intensity at knee point of the ferrite core. μ_{rm} and μ_{rm} are the permeability of the ferrite core at unsaturated state and PM respectively. Based on the above discussion a FCL has been fabricated and experimented. Fig.11 shows the waveform under a fault condition.

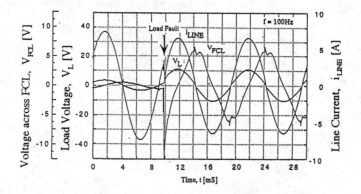

Fig.11 Typical waveform at fault condition.

6. Conclusion

This paper has described a passive fault current limiter consisting of permanent magnets and ferrite cores. The flux distribution inside the cores and permanent magnets have been analyzed using the finite element method taking into consideration the flow of eddy currents in the permanent magnets. The governing equations have also been listed and the design criterion has been discussed. This type of fault current limiter can be used for limiting fault current in practical power system applications.

References

[1]   F.P.Dawson, S.Yamada and M.Iwahara, "Experimental Results for a Two-Material Passive DI/DT Limiter", IEEE Transc. on Magnetics, Vol. 31, No.6, Nov. 1995, pp 3734-3736.
[2]   M.Iwahara, S.Yamada, F.P.Dawson and G.Fillion, "A Passive Current Limiter for Power Semiconductor Protection", IEEE Intermag conference, SanDiego, 1996.
[3]   S.C.Mukhopadhyay, M.Iwahara, S.Yamada and F.P.Dawson, "Consideration of Operation of Passive Fault Current Limiter Including Eddy Current by means of Tableau Approach", Proceeding of National Symposium of Electrical Engineers, Japan, Kyoto, March 26-28, 1997, pp 2-336 to 2-337.
[4]   "JMAG-JVISION" Software Package of Magnetic Field Computation, The Japan Research Institute Limited, 1995.

Non-Linear Electromagnetic Systems
V. Kose and J. Sievert (Eds.)
IOS Press, 1998

New Type of Sensors with Compensation by Magnetic Fluid Force

Radu OLARU* and Corneliu TOCAN**

Technical University of Iasi, Faculty of Electrical Engineering, 53 Mangeron Blvd.,
**BIT Research Centre of Iasi,*
6600 Iasi, Romania

Abstract. This paper presents two new types of sensors with magnetic liquid which operate in conformity with the principle of balancing of the forces, in a closed - feedback loop - measuring system. The tilting sensor with a horizontal pendulum and the pressure sensor with a membrane contain a magnetofluidic arrangement on feedback which produces the compensating forces. Output voltage is quadratic in terms of the input magnitude. In order to obtain a linear behaviour of the measuring system, the nonlinearity introduced by the magnetofluidic arrangement has to be compensated for by means of a function generator (e.g. of the root extractor type).

1. Sensors Operation Principle

The beneficial effects of the negative feedback in the measuring systems with sensors are well known: accuracy, stability, immunity to disturbing factors. Usually, either an electric feedback is used in the measuring circuit, or an electromechanical one with influence on the detecting element of the transducer (e.g. pressure transmitter). In this last case the balancing forces method is applied, but the use of a lever group leads to constructive difficulties and the increase of transducer size and weight.

By using a magnetic liquid (ML), magnetic fluid forces may be developed that would compensate the forces produced by the magnitude to be measured. Fig.1 shows the operation principle of the two non-conventional sensors, a tilt (acceleration) one and a pressure one. The three armatures create a differential capacitive element sensitive to the displacement of

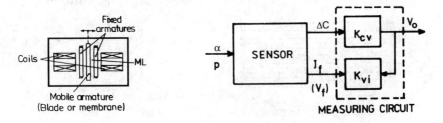

Fig. 1 - Operation principle of the sensors   Fig. 2 - Measuring system

the blade of the tilting sensor pendulum and to the deformation of the pressure sensor membrane, respectively.

The coupling of the magnetic field **H** to the ML with magnetization **M** is given by the force per unit volume **f** as

$$\mathbf{f} = \mu_0(\mathbf{M}.\nabla)\mathbf{H}, \tag{1}$$

where μ_0 is the permeability of the vacuum. The magnetization M is approximately a linear function of the field H, at small values of H, but it approaches a saturation value M_s at sufficiently high field strength. Therefore, with a dc field of moderate amplitude, the force will be quadratic in terms of the dc current producing the magnetic field [1].

The magnetic fluid forces compensate for the forces produced by the unit magnitude to be measured in such a way that, during operation, the mobile armature (Fig.1) is kept in the zero position and the current (or voltage) that feeds the coil towards which the armature is directed depends on the quantity of magnitude to be measured.

The feedback-measuring system of the sensors (Fig.2) contains a block for the conversion ΔC capacity to voltage and its gain, characterised by the transfer factor k_{cv} and a voltage-current convertor on the feedback circuit characterised by factor k_{vi}. A feedback coil with ML will be characterised by factor $k_{if} = f / I_f = k'_{if}I_f = k_{vf}V_o$ (k'_{if}, k_{vf} = const.) if relation (1) is accounted for. For a higher magnitude of k_{cv}, the dependence output-input of the system is

$$V_o^2 \cong k_\alpha\alpha, \text{ and } V_o^2 \cong k_p p, \tag{2}$$

respectively (k_α, k_p = const.).

For a linear behaviour of the measuring system the nonlinearity $f = k'_{if}I_f^2$ has to be compensated by means of a function generator (e.g. root extractor type) contained in k_{vi} block so that $k_{vi}.k_{if}$ = const.

2. Tilt Sensor with a Horizontal Pendulum

The sensor construction with a horizontal pendulum, which can also function as a vertical pendulum with a lower mechanical sensitivity, is shown in Fig. 3 [2]. This detects the tilt angle α produced by its rotation around axis xx'. A petrol base ML with a saturation magnetization $M_s \cong 30$ kA/m and kinematic viscosity $\nu = 7.5 \ 10^{-6}$ m^2/s was used.

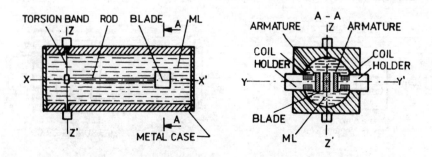

Fig. 3 - Tilt sensor with a horizontal pendulum

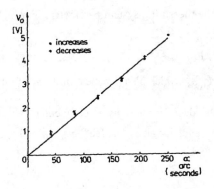

Fig. 4 - Voltage-inclination angle diagram

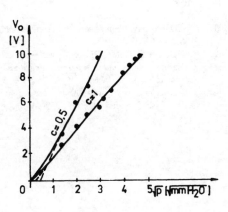

Fig. 5 - Pressure sensor with a membrane

By linearising the feedback circuit a linear voltage-inclination angle dependency is obtained, $V_o = k'_\alpha \alpha$ (Fig. 4), where k'_α is a constant ce depending on the sensor's constructive parameters, on k_{vi} coefficient and the ML caracteristics (physical properties - viscosity, magnetization) [2].

3. Pressure Sensor with a Membrane

Pressure sensors with a single membrane or two detecting membranes can also be made on the principle of forces balancing, in several constructive-functional variants [3, 4]. Such a variant which uses a membrane and a capacitive armature is shown in Fig. 5. The measurement of pressure p is done by keeping the metallic membrane close to its rest position, checked by the armature-metallic membrane capacitive element, by means of the magnetic fluid forces produced when the electromagnet is fed.

The static behaviour of the measuring system is given by the relationship [3]

$$k_1 V_o^2 + V_o = k_2 p \qquad (3)$$

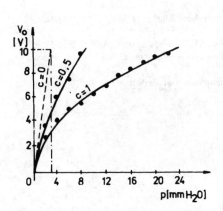

Fig. 6 - Experimental dependence $V_o = F(p)$

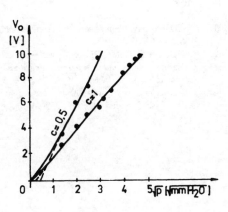

Fig. 7 - Experimental dependence $V_o = F(\sqrt{p})$

where k_1 and k_2 are coefficients depending on the measuring system parameters. For high amplification values, the relationship (3) becomes

$$V_o^2 \cong \frac{k_2}{k_1} p = \frac{1}{k_3} p, \qquad (4)$$

where $k_3 = k_1 / k_2$ is a coefficient proportional to k_{vi}^2 which also depends on the feedback magnetofluidic arrangement.

The graphs in Figs. 6 and 7 represent the dependencies, by experimental values, $V_o = F(p)$ and $V_o = F(\sqrt{p})$, respectively, for several values of feedback in voltage, $c = V_f / V_o$.

4. Conclusions

The sensor with a horizontal pendulum and ML presented here can be used in measuring the low tilts of some surfaces and bodies, from ten up to several hundred of arc seconds. As compared to the conventional sensors with a vertical pendulum, this is simpler to construct, has smaller dimensions and is more robust. At the same time it can be used in measuring vibrations and accelerations.

The sensor with a membrane and ML presented above may be used as follows:
- to measure the pressure p, by introducing in feedback an element meant to compensate nonlinearity, e.g. a root sign extractor;
- to measure the volume flow $Q = k_q \sqrt{p}$, together with a hydraulic resistor, while ensuring high values for the coefficients k_2 and particularly k_1.

The sensor can measure pressures up to several tens of mm H_2O and gases and liquids volume flow by means of a constrictional element which produces a pressure drop of the same magnitude. For instance, for a diaphragm with the orifice diameter d = 27.5 mm, a flow of water of $3 \cdot 10^{-4}$ m^3/s could be measured, corresponding to a pressure drop of about 35 mm H_2O, in a pipe-line with diameter D = 50 mm.

References

[1] P. S. Dubbelday, *Characteristics of a hydro-acustic ferrofluid projector*, J. Magn. Magn. Mater., 39 (1983) 159-161.
[2] R. Olaru and C. Cotae, *Tilt sensor with magnetic liquid*, Sensors and Actuators A 59 (1997) 133-135.
[3] R. Olaru, *Pressure and flow transducer with membranes and magnetic fluid*, IMEKO TC3, Warszawa, Poland, 1995, pp. 261-263.
[4] R. Olaru, *Pressure transducers with balancing magnetofluidic forces*, Buletin I. P. Iasi, Tom XLI (XLV), Fasc. 5, Sectia III, 1995, 939-943.

Rotational Magnetic Flux Sensor with Three Axis Search Coil for Non-Destructive Testing

M. OKA*   and   M. ENOKIZONO**
*Department of Computer and Control Engineering, Oita National College of Technology,
1666 Maki, Oita, 870-01, Japan
**Faculty of Engineering, Oita University, 700 Dannoharu, Oita, 870-11, Japan

Abstract. The conventional magnetic sensors using an AC or DC magnetic field can't estimate beneficial information about the depth or position of a reverse side crack. Accordingly, we are improving an electromagnetic non-destructive testing (NDT) system, i.e. a rotational magnetic flux sensor with a three axis search coil, to estimate it. This improved magnetic sensor can obtain some effective information on the depth or position of the reverse side crack on a thick steel plate. The experimental results show the validity of this improved rotational magnetic sensor. In this paper, we present the construction of the rotational magnetic flux sensor and experimental results.

1. Introduction

An electromagnetic non-destructive testing (NDT) system has been widely utilized for inspection of many materials used in dangerous situations[1]. In order to prevent accidents, it is important to detect and estimate reverse side cracks in the structural components of metallic materials. In recent years, a need has arisen for a more reliable and quantitative detection method for ensuring safety.

The authors have already presented several types of the rotational magnetic flux sensor for detection of a reverse side crack on a thick steel plate[2-4]. These sensors are excited by low frequency currents and have the three-axis search coil. These sensors have some effective points compared with a conventional magnetic sensor using an AC or DC magnetic field. One of them is sensitivity to detect a reverse side crack. However, in these sensor using air cores, signals from a reverse side crack on a thick steel plate were very small because of a weak application of magnetic flux. Therefore, we improved the exciting device of the rotational magnetic flux sensor by using laminated silicon steel cores. We obtained good results from experiments.

2. Magnetic Sensor System

Figure 1(a) shows the construction of the improved rotational magnetic flux sensor. The pair of U-shaped exciting cores (perpendicularly intersecting each other) are made of silicon steel sheets. They generate the rotational magnetic field in the specimen due to two-phase exciting currents with the 90 degree difference. B-coils are wound on each exciting core leg for measurement of the applied magnetic flux density. Figure 1(b) shows the details of the three-axis search coil. We installed it in the center of the exciting cores. It is just in contact with the specimen. It can measure the magnetic leakage flux disarranged by the reverse side crack by decomposing into three perpendicular components. In order to obtain large output signals from the reverse side crack, the x- and y-axis search coil has an amorphous core(7 mm x 7 mm x 1 mm). All search coils are made of 0.04 mm diameter copper wire.

Figure 2 shows the set up of experimental equipment. The D/A converter set in the PC generates two-phase exciting voltages with 90 degree phase difference. They are applied to two pairs of exciting coils. The signals which are disturbed by the reverse side crack are detected by the three-axis search coil. Three signals from the three-axis search coil and two signals from B-coils are introduced to the A/D converter and inputted to the PC, where they undergo numerical integration. We calculate the x-axis magnetic flux density(B_x) using the output signal of the x-axis search coil. B_y and B_z are processed in the same way. The maximum value of B_x(B_{xmax}) is computed from the measured values, and B_{ymax}, B_{zmax}. The angle θ_{Bzmax} which indicates the maximum value of B_z is computed.

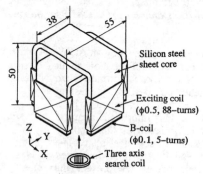

Fig. 1(a)   Construction of the sensor.

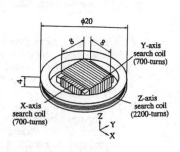

Fig. 1(b)   The details of the search coil.

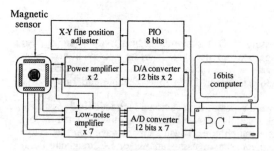

Fig. 2   Experimental equipment.

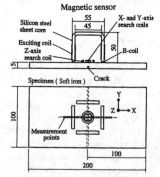

Fig. 3   The arrangement of the sensor and the dimensions of the specimen.

3. Experimental Results and Discussions

Figure 3 shows the magnetic sensor, dimensions of the specimen which is the 5 mm thick steel plate(SS41) and the positions of measurement points. The depth of the crack differ in each specimen at in steps of 1 mm(0, 1, 2, 3, 4 mm). The width of the crack is 0.4 mm. We carried out the experiment under following conditions; We mainly used 2.5 Hz as the exciting frequency in our experiments so as to consider the skin effect. The maximum exciting magnetic flux density is 0.56 T. Under this condition, the measured values of B_{xmax} and B_{ymax} are 0.01 T. The magnetic sensor was moved by the X-Y fine position adjuster at intervals of 1 mm. We carried out the experiment as follows; Firstly, in order to equalize two applied magnetic flux densities, the magnetic sensor is put on the specimen where there is no reverse side crack. At that place, exciting currents are adjusted by the PC so that the Lissajous's figure which is drawn by B_x and B_y becomes nearly a circle. When the Lissajous's figure becomes a circle, we put the magnetic sensor in the first measurement point, and begin measurement.

Figure 4, 5 and 6 show the relationship between the maximum value of each axis magnetic flux density and the sensor position. The change of B_{xmax}, B_{ymax} and B_{zmax} are related the distance between the magnetic sensor and the reverse side crack. When the depth of the crack is great, the change of each axis magnetic flux density is large. The position where B_{xmax} indicates the maximum value is in agreement with the position of the reverse side crack. This magnetic sensor can take an accurate measurement of the reverse side crack. Since the crack crosses at right angles to the x-axis of this magnetic sensor, the change of B_{xmax} is bigger than B_{ymax} near the crack. If we use this characteristic to estimate the direction of the crack, this magnetic sensor can estimate the direction of the crack. δB_{zmax} indicates large values at both side of the crack, and small values on the crack. The position indicating the minimum values of B_{zmax} is in agreement with the position of the crack. But, there is no distinct difference between any of the cracks other than the 4 mm depth crack. Figure 7 shows the relationship between θ_{Bzmax} and the sensor position. θ_{Bzmax} indicates large values at the left-hand side of the crack, and small values at the right-hand side of the crack. We can determine the relative position between the sensor and the crack according to this information. This change is caused by the change of the eddy current in the specimen owing to the relative position between the magnetic sensor and the crack.

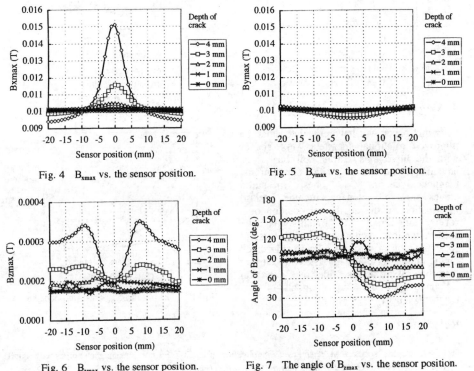

Fig. 4 B_{xmax} vs. the sensor position.

Fig. 5 B_{ymax} vs. the sensor position.

Fig. 6 B_{zmax} vs. the sensor position.

Fig. 7   The angle of B_{zmax} vs. the sensor position.

Figure 8 shows the experimental frequency characteristics of this sensor(The crack depth = 2 mm, the crack width = 0.4 mm.). In this figure, B_{xmax} is standardized using the mean values without a crack as comparison under the same condition. When the exciting frequency is 2.5 Hz, B_{xmax} is bigger than at the three other exciting frequencies. This means that the low exciting frequency is useful for detection of the minor reverse side crack. On other hand, when the exciting frequencies are 2.5 Hz and 5 Hz, B_{xmax} increases near the crack. It decreases near the crack in case of 10 Hz or 20 Hz. This phenomenon is owing to the eddy current near the crack. Magnetic flux generated by this eddy current decreases the magnetic flux passed through the search coil.

Figure 9 shows the same relationship as Fig. 8 with analytical results. B_{xmax} is calculated using the 2D-F.E.M. based A-φ method. We carried out numerical analysis under following conditions; Current density of exciting coils is 4.0×10^6 A/m², conductivity of the specimen 1.3×10^7 S/m, relative permeability of the specimen 300, relative permeability of the amorphous core is 30,000, relative permeability of the exciting core 500. The crack width is 1 mm and the depth is 2 mm. B_{xmax} is the average of the part of the amorphous core. These two figure are in good agreement.

Figure 10 shows the relationship between the B_{xmax} change and the crack depth (the crack width = 0.4 mm.). The B_{xmax} change caused by a reverse side crack is very large at frequency of 2.5 Hz. This means that the low exciting frequency is useful for estimation of a minor reverse side crack on the thick steel plate because of the skin effect.

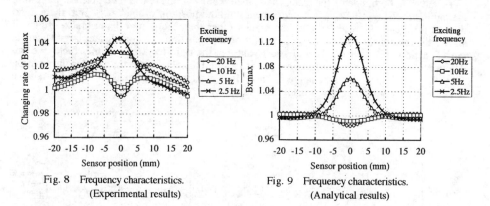

Fig. 8   Frequency characteristics.   Fig. 9   Frequency characteristics.
         (Experimental results)   (Analytical results)

4. Conclusions

In this paper, we have elucidated that the improved rotational magnetic flux sensor with the three axis search coil has some advantages compared with conventional magnetic sensors. (1) These measured values of this sensor include useful information on the depth and position of a reverse side crack. (2) In this magnetic sensor, the low exciting frequency is useful for estimation of a minor reverse side crack on the thick steel plate. (3) The eddy current disturbed by the reverse side crack exerts an influence upon the output signals of this magnetic sensor.

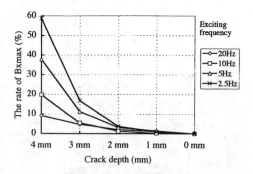

Fig. 10   The rate of B_{xmax} vs. the crack depth.

References

[1]T. Maehata, M. Isobe and M. Nishikawa, "Application of RFECT for non ferromagnetic pipes with small diameter", *Journal o f the Japan Society of Applied Electromagnetics and Mechanics*, vol. 3, No. 2, pp. 11-18,1995

[2] M. Enokizono and S. Nagata, "Non-Destructive Testing with Magnetic Sensor Using Rotational Magnetic Flux", *Journal of the Magnetics Society of Japan*, vol. 15, No. 2, pp. 455-460, 1991

[3] M. Oka and M. Enokizono, "Searching Defect Under Rotating Magnetic Field", *Nonlinear Electromagnetic Systems,(Proceedings of ISEM-Cardiff)*, IOS Press, pp. 756-759, 1996

[4] M. Oka and M. Enokizono, "A Detection of Backside Crack Using Rotational Magnetic Flux Sensor with Search Coils", *IEEE Transactions on Magnetics*, Vol. 32, No. 5, pp. 4968-4970, 1996

Non-Linear Electromagnetic Systems
V. Kose and J. Sievert (Eds.)
IOS Press, 1998

Nondestructive Testing by Differential Type of Rotational Magnetic Sensor

M. Enokizono*, M. Oka**, T. Todaka*, M. Akita*, T. Chady*, Y. Tsuchida*,
R. Sikora***, S. Gratkowski*** and J. Sikora****

Faculty of Engineering, Oita University, 700 Dannoharu, Oita, 870-11, Japan
**Oita National College of Technology, 1666 Maki, Oita, 870-01, Japan*
***Technical University of Szczecin, AL. Piastow, 19, Szczecin 70-310, Poland*
****Warsaw University of Technology, PL. Politechniki I, 00-660 Warsaw, Poland*

Abstract. It is very difficult to estimate a depth, position and shape of a minor reverse side defect because a signal derived from it is very small and noisy. Therefore, we are improving an electromagnetic non-destructive testing (NDT) system such as a rotational magnetic flux sensor and a differential type of eddy current probe to estimate it. We combined two types of the magnetic sensor into a differential type of a rotational magnetic flux sensor. This new magnetic sensor can obtain some effective information for the depth, position and shape of a reverse side defect on a metallic material plate. The experimental results show the validity of this new magnetic sensor.

1. Introduction

From the standpoint of prolonging the service life-time of a structure by replacement of defective parts, it will become quite important to estimate an unknown reverse side defect on a metallic material plate. These studies have been established as Grant-in-Aid for International Scientific Research(Joint Research), which is the scientific and technological cooperation joint project between Japan and Poland. The authors have already presented two non-destructive testing systems, one was driven under rotational magnetic flux[1-2] and the other was a differential type of eddy current probe[3-5]. Then, we combined two types of magnetic sensor into the differential type of rotational magnetic flux sensor. This new sensor which is driven under a rotating magnetic field by a two-phase power source, shows good sensitivity as a differential type of eddy current probe. Experimental results show that this sensor has the possibility of estimating the depth, shape or position of an unknown reverse side defect on a 5 mm thick steel plate. In this paper, we present the construction of this sensor, the principle of detection and fundamental results of numerical analysis and experiments.

2. Magnetic Sensor System

Figure 1(a) shows the construction of the differential type of rotational magnetic flux sensor. Figure 1(b) shows a photograph of this sensor. This sensor consists of one ferrite core, two axial exciting coils and ten pickup coils. Two exciting coils are wound on connective parts of the ferrite core. The rotational magnetic field is generated in the specimen by two-phase exciting currents. B-coils are wound under each exciting coil for measurement of applied the magnetic flux density. Ten pick-up coils are wound on each end of the core leg. They can measure each part of the magnetic flux density which is disarranged by the reverse side defect near the specimen. B-coils are made of 0.1 mm diameter copper wire. Figure 2 shows the principle of detection of this differential type of rotational magnetic flux sensor. Figure 2(a) shows the disposition of pickup coils. Figure 2(b) shows the cross section A. Figure 2(c) shows the connection of each axial pickup coil. The x- and y-axis pickup coil are each connected to four pickup coils, and the z-axis pickup coil is connected to two pickup coils. These connections of each pickup coil are differential vertically and horizontally. When the magnitude and phase of each pickup coil's output voltages(e_x, e_y and e_z) are well-balanced, e_x, e_y and e_z are zero. In order to achieve this condition, we used a comparative specimen(see Fig. 4.) which is of the same material as the specimens. Under this condition, if there is a reverse side defect on the specimen, e_x, e_y and e_z are not zero, because the output voltage of each pickup coil is not equal.

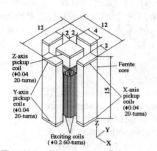

Fig. 1(a)   Construction of the sensor.

Fig. 1(b)   Photograph of the sensor.

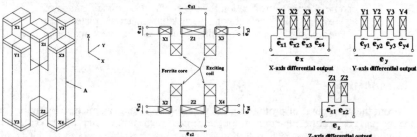

(a) The disposition of pickup coils. (b) The cross section of this sensor.   (c) The connection of pickup coils.
Fig. 2   The principle of detection of the differential type rotational magnetic flux sensor.

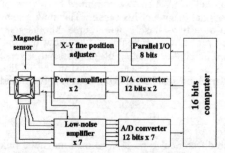

Fig. 3   Experimental equipment.

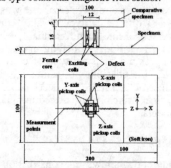

Fig. 4   The arrangement of the sensor and the dimension of the specimen.

Figure 3 shows the arrangement of the experimental equipment. The D/A converter generated two exciting voltages 90 degree out of phase. They were applied to two pairs of exciting coils. Three signals from the three-axis pickup coil which was disturbed by a reverse side defect, and two signals from the B-coils were sampled by the A/D converter and transferred to the PC, where they were undergone numerical integration. We calculated the difference of the x-axis magnetic flux density(δB_x) using e_x. δB_y and δB_z were processed in the same way. The maximum value of $\delta B_x(\delta B_{xmax})$, δB_{ymax} and δB_{zmax} is computed from the measured values. We calculated the maximum value of each axis applied magnetic flux density(B_{xmax} and B_{ymax}) using output voltages of the B-coils. Figure 4 shows the dimensions of the comparative specimen and the 5 mm thick steel plate(SS41) together with the positions of measurement points. The depth of the defect is different in each specimen at a rate of 1 mm(0, 1, 2, 3, 4 mm). The width of the defect is 0.4 mm. The magnetic sensor was moved by the X-Y fine position adjuster in step of 1 mm.

3. Numerical analysis

We carried out numerical analysis using the A-ϕ based Finite Element Method(FEM) under the following conditions. The exciting frequency is 10 Hz. The current density of the exciting coils is 7.5×10^6 A/m^2, and conductivity of the comparative specimen and specimens is 1.3×10^7 S/m. The relative permeability of the comparative and the main specimens are 300, and relative permeability of the ferrite core is 2600. Figure 5(a) and (b) show the distribution of magnetic vector potential (The defect width is 1 mm.). From the results, when there is not a reverse side defect on a specimen, the distribution of magnetic vector potential is almost symmetrical, both vertically and horizontally in Fig. 5(b). However, the size of the comparative specimen is different from the size of the main specimen. There is a little difference between the upper part and the lower part in Fig. 5(b), but, if a reverse side defect exists on the main specimen this difference increases in Fig. 5(a).

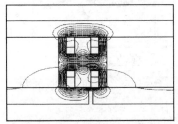 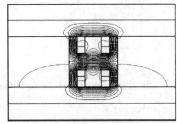

(a) With a 4 mm defect.   (b) without a defect.

Fig. 5   Results for numerical analysis.

4. Experimental Results and Discussions

The experiment was carried out under the following conditions. We used 10 Hz as the exciting frequency so as to consider the skin effect. The maximum exciting magnetic flux density was 0.2 T at connective parts of the ferrite core legs. Since this sensor was a differential type, it was important that e_x, e_y and e_z were zero at a place without a defect. In this experiment, they are not equal to zero because there are minor imbalances of each pickup coil's output voltage. This sensor is very sensitive to lift-off which is the distance between the sensor core and the comparative specimen and the main specimen. The experiments were carried out carefully in order that e_x, e_y and e_z may become nearly zero.

Firstly, in order to equalize the two applied magnetic flux densities, the magnetic sensor is put on the specimen without the defect. At that place, exciting currents are adjusted by the PC so that the two applied magnetic flux densities may become equal. When the two applied magnetic flux densities became equal, we put the magnetic sensor at the first measurement point, and began measurement.

Figure 6 shows the relationship between the axis ratio α(= minor axis / major axis) of Lissajous's figure drawn by measured δB_x and δB_y, and the sensor position [1]. Figure 7, 8 and 9 show the relationship between the maximum value of each axial differential magnetic flux density and the sensor position. δB_{xmax}, δB_{ymax} and δB_{zmax} are set in relation to the distance between the magnetic sensor and the reverse side defect. When the defect is deep, the change of each magnetic flux density along each axis is large. The position where α indicates the minimum value is in agreement with the position of the reverse side defect. This magnetic sensor can detect the position of the reverse side defect. Since the defect crosses at right angles to the x-axis of this magnetic sensor, δB_{xmax} is larger than δB_{ymax} near the defect. If we use this characteristic to estimate the direction of the defect, this magnetic sensor can estimate the direction of the defect. δB_{zmax} indicates large values at the left hand side of the defect, and small values at the right hand side of it. We can determine the relative position between the sensor and the defect by using this information.

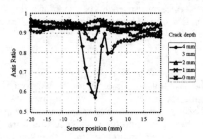

Fig. 6 The axis ratio vs. the sensor position.   Fig. 7 δB_{xmax} vs. the sensor position.

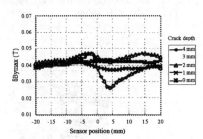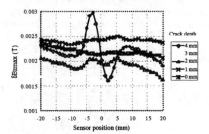

Fig. 8 δB_{ymax} vs. the sensor position.   Fig. 9 δB_{zmax} vs. the sensor position.

5. Conclusions

In this joint project, we combined two types of magnetic sensor into the differential type of rotational magnetic flux sensor. In this paper, we have shown that it has some advantages over the conventional magnetic sensors.
(1) We present a new conceptual magnetic sensor for NDT such as the differential type of rotational magnetic flux sensor.
(2) This sensor has high sensitive characteristics to detect a reverse side defect on the 5 mm thick steel plate on account of the differential type.
(3) Each axis output signal of this sensor includes useful information on the depth, position and shape of the reverse side defect.
(4) This sensor is very sensitive to imbalance of the distance between the core and a specimen.

References

[1] M. Enokizono and S. Nagata; "Non-Destructive Testing with Magnetic Sensor Using Rotational Magnetic Flux", *Journal of the Magnetics Society of Japan*, Vol. 15, No. 2, pp. 455-460,1991
[2] M. Oka and M. Enokizono; "A Detection of backside Defect Using Rotational Magnetic Flux Sensor with Search Coils", *IEEE Transactions on Magnetics*, Vol. 32, No. 5, pp. 4968-4970, 1996
[3] T. Chady, M. Komorowski; "Investigation of Differential Eddy Current Probe", *International Journal No. 5 joint to International Symposium on Theoretical Electrotechnics*, Budapest, Oct. 12, 1994
[4] R. Sikora, M. Komorowski, T. Chady; "A Neural Network Model of Eddy Current Probe", *Electromagnetic Non-Destructive Evaluation*, IOS-press, pp. 231-238, 1997
[5] T. Chady, M. Komorowski, R. Sikora; "Eddy current nondestructive testing of modeled cracks in plane conducting plates", *JSAEM Studies in Applied Electromagnetics*(Vol. 4, 1996)

Advanced Mathematical and
Computational Techniques

Non-Linear Electromagnetic Systems
V. Kose and J. Sievert (Eds.)
IOS Press, 1998

Learning Methods for Fuzzy Systems

Rudolf Kruse and Andreas Nürnberger
Department of Computer Science, University of Magdeburg
Universitätsplatz 2, D-39106 Magdeburg, Germany
Phone : +49.391.67.18706, Fax : +49.391.67.12018
E-Mail: kruse@iik.cs.uni-magdeburg.de

Abstract. Fuzzy systems are currently being used in a wide field of industrial and scientific applications. Therefore, it is necessary to have algorithms which construct and optimize such systems automatically. Since the idea of learning is being studied in other research areas like machine learning and data mining, some of the developed methods have been made available and optimized for the learning process in fuzzy systems. In this paper, we present a short survey of these methods and take a closer look at a special learning approach, the neuro-fuzzy systems.

1. Introduction

At present, fuzzy systems are being used in a wide range of industrial and scientific applications with the main application areas being fuzzy control, data analysis and knowledge based systems. Fuzzy controllers, for instance, model the control strategy of a human expert to control a system for which no mathematical or physical model exists. They employ a set of linguistic rules to describe the human behavior. A rule in the domain of speed control, for example, could have the form:

'*If* the speed is very high and the distance is small, *then* brake very strongly'.

The linguistic rules describe a control surface, which defines an appropriate output value for every vector of input values. Thus, a function is defined which 'fits' the rules.

The major benefits of fuzzy techniques are the convenient method to model technical systems and the good interpretability of the system description by using linguistic rules. However, the implementation of a fuzzy system can be very time consuming because there are no formal methods to determine its parameters (fuzzy sets and fuzzy rules). Therefore it is necessary to have algorithms which can learn fuzzy systems automatically from data.

Since the idea of learning is being studied in other research areas like machine learning and data mining, it is obvious to make the developed methods available for the learning process in fuzzy systems. The most important methods are currently derived from statistics [8], cluster analysis [4] or neural network theory [15]. All methods use sample data (data vectors, observation data) to learn from.

In this paper we present a short introduction to learning methods for fuzzy systems and discuss one approach in more detail: the neuro-fuzzy systems.

2. Fuzzy Systems

Different methods have been developed to automatically construct fuzzy systems from data. Most of them are based on the principle to construct a fuzzy system for function approximation. Such a system consists of r parallel rules. Let $x_1, .. ,x_n$ be n input values, y the output value and $\mu_r^{(1)}, ..., \mu_r^{(n)}$ and v_r fuzzy sets, then the fuzzy rules of a fuzzy system can be defined as:

$$if\ x_1\ is\ \mu_r^{(1)}\ and\ x_2\ is\ \mu_r^{(2)}\ and\ ...\ and\ x_n\ is\ \mu_r^{(n)}\ then\ y_1\ is\ v_r\ .$$

To calculate the output of this system, the output of every rule is computed first. Then, all outputs are combined into a single system output (usually, a fuzzy set). A *crisp* (exact) output value is derived by a defuzzification procedure [8], e.g. the center of gravity of the resulting fuzzy set. A fuzzy system defined in this way approximates an unknown function based on vague samples, which are described by the fuzzy sets. To construct or learn a fuzzy system, the fuzzy rules as well as the membership functions describing the fuzzy sets have to be defined. This can be done by using learning methods like cluster analysis, neuro-fuzzy approaches or a combination of them.

Cluster analysis is a technique for classifying data, i.e. to divide given data into sets of classes or *clusters*. In 'classical' cluster analysis each datum has to be assigned to exactly one class. This strict division is often not applicable to real world problems. Fuzzy cluster analysis relaxes this requirement by allowing gradual memberships. This offers the opportunity to deal with data that belong to more than one class. A survey to fuzzy clustering algorithms is presented in [4].

By using fuzzy clustering methods it is possible to learn fuzzy *if-then* rules from data. An example of learned fuzzy rules, which assigns one output value to two input values, is presented in Figure 1. The data points (depicted in the left top of Figure 1) are connected to the cluster for which they achieve the highest membership value. Every cluster represents a

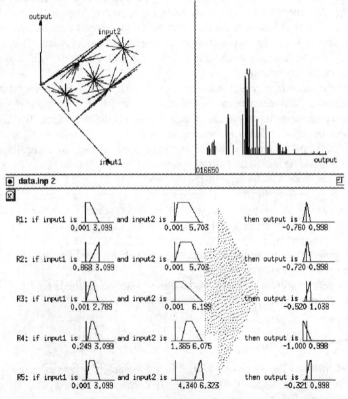

Figure 1. An example of fuzzy cluster analysis created with the tool FCLUSTER

fuzzy *if-then* rule and can be seen as a multidimensional fuzzy set. A fuzzy set in a single dimension is derived by projecting a cluster to one dimension (see the graph in the right top of Figure 1). The presented 5 rules can be obtained by projecting all clusters and smoothing the projection to triangular or trapezoidal fuzzy sets. Using these techniques, for example, fuzzy controllers and fuzzy approximation systems can be constructed [5].

A major drawback of this approach is that a rulebase derived by fuzzy clustering methods is often not easy to interpret, since no restrictions concerning the form of the fuzzy sets can be defined. Further, prior knowledge given by already existing fuzzy rules is not easy to introduce into the learning process. Neuro-fuzzy systems present a solution to these problems. They can be used to learn fuzzy rules and fuzzy sets, but also to optimize fuzzy systems derived by fuzzy clustering algorithms.

3. Neuro-Fuzzy Systems

Neuro-fuzzy systems are fuzzy systems that are trained by a learning algorithm derived from neural network theory. The (heuristical) learning procedure operates on local information, and causes only local changes to the underlying fuzzy system. The learning process is not knowledge based, but data driven.

A neuro-fuzzy system can be viewed as a special 3-layer feedforward neural network. The units in this network use t-norms or t-conorms instead of the activation functions common in neural networks. The first layer represents input variables, the middle (hidden) layer represents fuzzy rules, and the third layer represents output variables. Fuzzy sets are encoded as (fuzzy) connection weights. Some neuro-fuzzy models use more than 3 layers, and encode fuzzy sets as activation functions. It is usually possible to transform these models into a 3-layer architecture. This view of a fuzzy system illustrates the data flow within the system and its parallel nature. However, this neural network view is not a prerequisite for applying a learning procedure, it is merely a convenience.

A neuro-fuzzy system can always (i.e. before, during and after learning) be interpreted as a system of fuzzy rules. It is both possible to create the system out of training data from scratch, and it is possible to initialize it by prior knowledge in form of fuzzy rules.

The learning procedure of a neuro-fuzzy system takes the semantic properties of the underlying fuzzy system into account. This results in constraints on the possible modifications applicable to the system's parameters.

Neuro-fuzzy systems approximate n-dimensional (unknown) functions that are partially given by the training data. The fuzzy rules encoded within the system represent vague samples, and can be viewed as vague prototypes of the training data. A neuro-fuzzy system should not be seen as a kind of (fuzzy) expert system, and it has nothing to do with fuzzy logic in the narrow sense [8]. An approach to fuzzy expert systems is realized by possibilistic networks [3].

Therefore, neuro-fuzzy can be considered as a specific technique to derive a fuzzy system from data, or to enhance it by learning from examples. The exact implementation of the neuro-fuzzy model is of no importance. It is possible to use a neural network to learn certain parameters of a fuzzy system, like using a self-organizing feature map to find fuzzy rules [18] (cooperative models), or to view a fuzzy system as a special neural network, and directly apply a learning algorithm [12] (hybrid models).

An example of a hybrid neuro-fuzzy system is the NEFCON model. It has been originally developed to learn and optimize the rulebase of a fuzzy controller [11], but it can also be seen as a neuro-fuzzy system for function approximation.

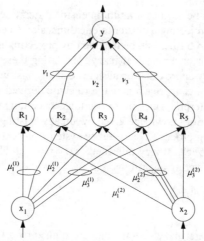

Figure 2. A NEFCON system with two inputs, 5 rules and one output

4. The NEFCON-Model

The NEFCON-Model is based on a three layer generic fuzzy perceptron [13; 15]. Fig. 2 presents an example of a NEFCON system, which describes the structure of a fuzzy system with 5 rules, 2 inputs, and one output. The inner nodes R_1, ..., R_5 represent the rules, the nodes x_1, x_2, and y the input and output values, and $\mu_r^{(i)}$, v_r the fuzzy sets describing the antecedents $A_r^{(i)}$ and conclusions B_r. Rules with the same antecedents use so-called shared weights, which are represented by ellipses in Fig. 2. They ensure the integrity of the rulebase. Node R_1, for example, represents the rule: R_1: if x_1 is $A_1^{(1)}$ and x_2 is $A_1^{(2)}$ then y is B_1.

The learning process of the NEFCON model can be divided into two main phases. The first phase aims at finding an appropriate initial rulebase as soon as possible. If no prior knowledge is available, the rulebase is learned from scratch. If, on the other hand, a manually defined rulebase is already given, the algorithm completes this rulebase. In the second phase, the acquired rulebase is optimized by shifting or modifying the fuzzy sets of the rules. Both phases use a fuzzy error, which describes the quality of the current system state, to learn or to optimize the rulebase. The fuzzy error can be derived by calculating the difference to the desired output (in case of function approximation, if the output is known), or by a linguistic error definition (in case of fuzzy control). In case of a linguistic error description, the linguistic rules are used to describe 'good' and 'bad' situations of the dynamic system, which has to be controlled [15]. So the system is able to learn online, since no input/output pairs must be given. The fuzzy error plays the role of a critic element in reinforcement learning models (e.g. [1; 2]).

5. An Implementation of the NEFCON Model

The NEFCON model has been implemented in different environments. A recent implementation was done under MATLAB/SIMULINK[1] with updated learning methods for the

---

[1] MATLAB/SIMULINK is a simulation tool developed and distributed by 'The Mathworks' Inc., 24 Prime Park Way, Natick, Mass.01760; WWW: http://www.mathworks.com.

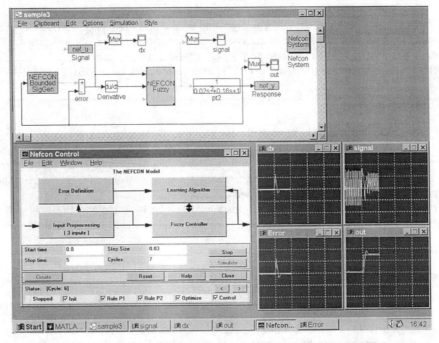

Figure 3. Sample of a development environment under MATLAB/SIMULINK (PT2-System)

development of fuzzy controllers in industrial research applications. The NEFCON learning algorithms [14; 16] learn and optimize the rulebase of a Mamdani-like fuzzy controller [10]. The fuzzy sets of the antecedents and conclusions can be represented by any symmetrical membership function.

The major goal of the implementation of the NEFCON model under MATLAB/SIMULINK was to provide an interactive tool for the construction and optimization of a fuzzy controller. This tool enables the user to include prior knowledge into the fuzzy system, to stop and to resume the learning process at any time, to modify the rulebase and the optimization parameters interactively, and to define the fuzzy error in a convenient way. Figure 3 presents the simulation environment of a sample application. It was created under Microsoft Windows NT 4.0.

The simulation results concerning a conventional PT_2 system (see [9; 6]) during the learning cycles are presented in Figure 4. The learning algorithm started with three equally distributed trapezoidal fuzzy sets for each input variable and five for the output variable. The fuzzy error was described using 'fuzzy intervals' [16]. The corresponding simulation

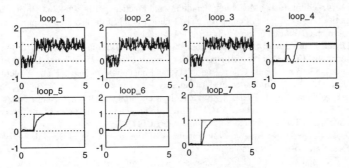

Figure 4. Simulation results for a PT_2 system

environment is shown in Figure 3. The algorithm used a noisy reference signal during rule learning to improve the coverage of the system state space (see cycles 1-3 in Figure 4). The system was able to find an appropriate rulebase within 3 rule learning and 3 optimization cycles (with 167 iteration steps each cycle). The optimized rulebase consists of 25 rules.

6. Conclusion

In this paper we presented some methods of learning fuzzy systems from data. However, the interpretability of learned or optimized fuzzy systems can be sometimes poor when there are no constraints applied to the learning process. Learning methods should be seen as an addition to manually defined and (semantically) clearly structured fuzzy systems. Such systems can be more easily maintained and checked for plausibility.

The tools mentioned in this paper, in particular FCLUSTER and NEFCON, as well as further information concerning fuzzy systems, are available on the Internet at: *http://fuzzy.cs.uni-magdeburg.de.*

References

[1]   Barto, A.G., Sutton R. S., Anderson, C. W. (1983): Neuronlike adaptive elements that can solve difficult learning control problems, IEEE Transactions on Systems, Man and Cybernetics, 13:834-846
[2]   Barto, A.G. (1992): Reinforcement Learning and Adaptive Critic Methods, In [17]
[3]   Gebhardt, Jörg and Kruse, Rudolf (1996): Learning Possibilistic Networks from Data, in D. Fisher, H. Lenz (eds.): Learning from Data, Artificial Intelligence and Statistics 5, Lecture Notes in Statistics 112, 143-153, Springer, New York
[4]   Höppner, Frank; Klawonn, Frank; Kruse, Rudolf (1996): Fuzzy-Clusteranalyse, Computational Intelligence, Friedr. Vieweg & Sohn Verlagsgesellschaft mbH, Braunschweig, Wiesbaden
[5]   Klawonn, Frank and Kruse, Rudolf (1997): Constructing a Fuzzy Controller from Data, In Fuzzy Sets and Systems 85
[6]   Knappe, Heiko (1994): Comparison of Conventional and Fuzzy-Control of Non-Linear Systems, in [7]
[7]   Kruse, Rudolf; Gebhardt, Jörg; Palm, Rainer (Eds.) (1994): Fuzzy Systems in Computer Science, Friedr. Vieweg & Sohn Verlagsgesellschaft mbH, Braunschweig, Wiesbaden
[8]   Kruse, Rudolf; Gebhardt, Jörg, Klawonn, Frank (1994): Foundations of Fuzzy Systems, John Wiley & Sons, Inc., New York, Chichester, et.al.
[9]   Leonhard, Werner (1992): Einführung in die Regelungstechnik, Friedr. Vieweg & Sohn Verlagsgesellschaft mbH, Braunschweig, Wiesbaden
[10] Mamdani, E. H.; Assilian S. (1973): An Experiment in Linguistic Synthesis with a Fuzzy Logic Controller, International Journal of Man-Machine Studies, 7:1-13
[11] Nauck, Detlef and Kruse, Rudolf (1993): A Fuzzy Neural Network Learning Fuzzy Control Rules and Membership Functions by Fuzzy Error Backpropagation, In Proc. IEEE Int. Conf. on Neural Networks 1993, San Francisco
[12] Nauck, Detlef and Kruse, Rudolf (1996): Designing neuro-fuzzy systems through backpropagation, In Witold Pedryz, editor, Fuzzy Modelling: Paradigms and Practice, pages 203-228, Kluwer, Boston
[13] Nauck, Detlef (1994): A Fuzzy Perceptron as a Generic Model for Neuro-Fuzzy Approaches, In Proc. of the 2nd German GI-Workshop Fuzzy-Systeme '94, München
[14] Nauck, Detlef; Kruse, Rudolf; Stellmach, Roland (1995): New Learning Algorithms for the Neuro-Fuzzy Environment NEFCON-I, In Proceedings of Neuro-Fuzzy-Systeme '95, 357-364, Darmstadt
[15] Nauck, Detlef; Klawonn, Frank; Kruse, Rudolf (1997): Foundations of Neuro-Fuzzy Systems, John Wiley & Sons, Inc., New York, Chichester, et.al. (to appear)
[16] Nürnberger, Andreas; Nauck, Detlef; Kruse, Rudolf; Merz, Ludger (1997): A Neuro-Fuzzy Development Tool for Fuzzy Controllers under MATLAB/SIMULINK, to appear in: Proc. of the 5th European Congress on Intelligent Techniques and Soft Computing (EUFIT'97), Aachen
[17] White, D. A., Sofge, D. A., Hrsg. (1992): Handbook of Intelligent Control. Neural, Fuzzy and Adaptive Approaches, Van Nostrand Reinhold, New York
[18] Witold Pedryz and Card, H.C (1992): Linguistic interpretation of self-organizing maps, In Proc. IEEE Int. Conf. On Fuzzy Systems 1992, pages 371-378, San Diego

Non-Linear Electromagnetic Systems
V. Kose and J. Sievert (Eds.)
IOS Press, 1998

Wavelet Methods and Applications in Physics

Peter MAASS

Universität Potsdam, Am Neuen Palais 10, 14415 Potsdam, Germany

Abstract. Wavelet analysis was originally introduced as a tool for analyzing non–stationary signals. Since then wavelet methods have been adapted to a large variety of different applications, see [13].

The aim of this paper is to sketch the contributions of wavelet analysis in physics. These contributions fall in two categories: wavelet models for physical entities and wavelet methods for analyzing measured physical quantities.

1. Introduction

The classical tool in signal processing is Fourier analysis, which is perfectly suited to analyzing stationary signals. Fourier methods have an optimal frequency resolution, however their capability for detecting non–stationary, time–varying structures is limited. The windowed Fourier transform or Gabor transform partially remedies this drawback by convolving the signal with a window function. However the window width is fixed throughout the analysis, i.e. the time resolution is again limited.

Wavelet analysis was designed as a mathematical microscope with (theoretically) arbitrarily sharp time resolution: the wavelet transform of a signal f with respect to the wavelet ψ is defined by

$$L_\psi f(a,b) = |a|^{-1/2} \int_{I\!\!R} f(t)\psi\left(\frac{t-b}{a}\right) dt . \tag{1}$$

The parameter $a \in I\!\!R$ determines the width or scale of the window ψ: as a tends to zero finer and finer details of f are analyzed. The name wavelet transform originated in the early 80's, however this mathematical tool has been rediscovered many times, for a very readable survey see [10].

The basic operations which are used in the construction of the wavelet transform are translations by b and dilations or scalings by a. This results in a decomposition of f with structural components $\psi\left(\frac{t-b}{a}\right)$ which are labelled by a time variable b and a scale/size–variable a. Hence wavelet transforms are particularly suited to applications where the underlying physical or technical model exhibits a multiscale or time–scale structure (rather then a time–frequency structure for which Fourier methods are preferable).

The definition of a wavelet transform (1) can be examined from different mathematical points of view. The success of wavelet methods in signal processing relies on these different interpretations: e.g. (1) can be viewed as a correlation of f with a scaled and translated ψ or as a convolution product, which results in a simultaneous filtering of f with bandpass filters parametrizd by a, or as an asymptotic microscope which reveals points of singularity, edges or discontinuities as $a \to \infty$. Many survey articles and books have been written on wavelet applications in signal processing, see e.g. [14, 13, 16, 12].

The purpose of this paper is to discuss some basic concepts of wavelet methods in physics. These applications fall in two classes:

- Signal analysis of physical measurements,
- Wavelet models in physics.

The first class of applications can be treated with standard signal processing methods. The success relies on a careful choice of the wavelet to be used and on additional tools for interpretating the resulting computations. For example adapted wavelets have been constructed for measurements in electrodynamics [5] or the evaluation of scattering potentials [6]. For an excellent and extensive overview of wavelet applications in physics see [3].

The emphasis of this exposition is on the second class of applications. The construction of wavelet models in physics often relies on the inversion formula for the wavelet transform: If ψ satisfies the admissibility condition ($\hat{\psi}$ denotes the Fourier transform)

$$0 < c_\psi = \int_R \frac{|\hat{\psi}(\omega)|^2}{|\omega|} \, d\omega < \infty \tag{2}$$

then f can be reconstructed from its wavelet transform $L_\psi f(a, b)$ via

$$f(t) = \frac{1}{c_\psi} \int \int L_\psi f(a, b) \frac{1}{|a|^{1/2}} \psi\left(\frac{t-b}{b}\right) \frac{da\,db}{a^2} . \tag{3}$$

Again this reconstruction formula can be viewed differently. First of all it gives a resolution of the identity or a decomposition of f into wavelet coherent states $\psi\left(\frac{t-b}{b}\right)$ which are labelled by the elements of the affine group of translations and dilations. This concept extends the notion of the canonical coherent states, see [1] and leads to a large variety of meaningful extensions and modifications, see [2].

This concept also demonstrates the relationship between Fourier analysis and wavelet theory, which is based on a powerful mathematical construction. From an abstract group theoretic point of view the windowed Fourier transform (leading to the canonical coherent states) and the wavelet transform (leading to wavelet coherent states) stem from the same construction principle of square integrable group representations (Weyl–Heisenberg group and windowed Fourier transform, resp. affine group and wavelet transform). For a description of this approach see [8, 9].

2. Wavelet models in physics

From the beginning applications in physics and mathematical physics have strongly influenced the development of wavelet analysis [10], in fact the foundations of wavelet analysis were laid by theoretical physicists [8]. Accordingly the list of wavelet contributions in various fields of modern physics is rich and long.

Hence the following selection is very much incomplete and rather reflects the authors preferences. A more complete and excellent summary of the interrelations of physics and (continuous) wavelet analysis is presented in [3].

1. Renormalization and quantum field theory: this is one of the oldest applications of wavelet analysis in physics, see [7] and the references given there. The action of

a renormalization group on a function f is based on a local averaging procedure followed by rescaling. Local averaging and rescaling exactly leads to an integral transform as in (1). If f is a function on an integer lattice – as in [7] – then the integral (1) becomes a sum defining the discrete wavelet transform of f. This point of view has brought new insight into phase cell decompositions, spinor quantum electrodynamics and other areas of quantum field theory.

2. Coherent states and localization operators: The interpretation of the wavelet inversion formula (3) as a decomposition of f by affine coherent states (or wavelet coherent states) is the starting point for a natural connection between wavelet analysis and quantum physics. For example a special choice of ψ was used in the investigation of the Schroedinger equation for the hydrogen atom [17], the concept of wavelet coherent states has been generalized and investigated in great detail in [2]. Moreover, introducing localization operators based on weight functions $w(a, b)$ in (3),

$$f(t) = \frac{1}{c_\psi} \int \int w(a,b) \, L_\psi f(a,b) \, \frac{1}{|a|^{1/2}} \psi \left(\frac{t-b}{b} \right) \frac{da \, db}{a^2} \, , \tag{4}$$

leads to a variety of wavelet models in quantum physics, see [15, 17].

3. Wavelet Electromagnetics, see [5]: the applications of wavelet methods in this field are based on two different considerations. First, the set of solutions of Maxwell's equations is invariant under translations and dilations, i.e. the affine group is a subgroup of the full invariance group (conformal group). The standard wavelet transform is constructed from the action of the affine group on functions $f \in L^2(I\!R)$, extending this construction to the action of the conformal group on $L^2(I\!R^n)$ leads to a resolution of the identity, which is tailored for electromagnetic waves. Second, if an electromagnetic field is measured with a device which moves at constant speed, then a simple calculation shows that the measured data is the wavelet transform of the electromagnetic field, where the wavelet corresponds to the impulse response function of the measurement device.

4. Scattering on fractal potentials, see [6]: the mathematical basis for this application differs from the previous examples. Certain physical quantities e.g. turbulent flows, exhibit power laws. Conversely, a quantitive analysis of these laws leads to a characterization of the underlying phenomenon. In [6] the authors consider one–dimensional scattering with a fractal potential. Let q denote the momentum transfer between the incoming and outgoing wave, then the scattered intensity $I(q)$ obeys an averaged power law,

$$\int_{q/3}^{q} I(q') \, dq' \sim q^{1-D} \, ,$$

where D is the fractal dimension of the scattering potential. The wavelet transform is frequently called a mathematical microscope with lense ψ, where the amplification/resolution is determined by the parameter a. Hence –with a properly chosen wavelet– selfsimilar phenomena can be characterized by analyzing the asymptotic behaviour of the wavelet transform as $a \to 0$. In particular the fractal dimension of the scattering potential can be determined in this way by a stable numerical procedure.

5. Radar imaging: Let us consider a signal ψ, which has been emitted at time $t = 0$ and which is reflected from an object moving with velocity y at distance x (x, y denote position and speed in range–Doppler–coordinates). The reflected signal exhibits a frequency shift due to the Doppler effect, hence the time–delayed return signal is of the form $\psi\left(\frac{t-x}{y}\right)$. Considering many reflecting objects, e.g. rain clouds whose distribution in range–Doppler coordinates is given by $D(x, y)$, leads to a return signal

$$f(t) = \int\int D(x,y)\, \psi\left(\frac{t-x}{y}\right)\, dx dy \; ,$$

which –up to a normalzation factor– is identical with (3). Usually one assumes a narrowband signal which allows further simplifications, however the general modelling of radar measurements directly leads to wavelet analysis. This is the starting point for wavelet methods in radar and sonar, see [4, 11].

References

[1] J.R. Klauder and B.S. Skagerstam, Coherent states – applications in physics and mathematical physics. World Scientific, Singapore, 1985.

[2] S.T. Ali, J.–P. Antoine, U.A. Mueller, Coherent states and their generalization, *Reviews Math. Phys.* **7** (1995) 1013–1104.

[3] J.-P. Antoine, Wavelet Methods in Physics, Chapters 1,2, and 8. In: J.C. van den Berg, (ed.), Wavelets in Physics, Cambridge Univ. Press, Cambridge, 1997 (in press).

[4] P. Maass, Wideband approximation and wavelet transform. In: F.A. Gruenbaum *et al.* (eds.), Radar and Sonar, Part II. Springer Verlag, New York, 1992.

[5] G. Kaiser, Space–time–scale analysis of electromagnetic waves. In: Proc. of IEEE–SP Int. Symp. on Time-Frequency and Time–Scale Analysis, Victoria, Canada, 1992

[6] C.-A. Guerin, M. Holschneider, Scattering on fractal measures, *J.Phys.A:Math.Gen.* **29** (1996) 7651–7667.

[7] G. Battle, Wavelets: A renormalization group point of view. In: G. Ruskai *et al.* (eds.), Wavelets and their applications. Jones and Barlett Pub., Boston, 1992.

[8] A. Grossmann *et al.*, Transforms associated to square integrable group representations, *Ann. Inst. H. Poincare* **45** (1986) 293–309.

[9] H.-G. Feichtinger and K.-H. Groechenig, Banach spaces related to integrable group representations and their atomic decomposition, *J. Funct. Anal.* **23** (1992) 244–261.

[10] Y. Meyer, Wavelets: Algorithms and Applications. SIAM, Philadelphia, 1993.

[11] H. Naparst, Radar signal choice and processing for a dense target enviroment. Ph.D. thesis, University of Berkeley, California, 1988.

[12] P. Maass, Wavelet Methods in Signal Processing. In: Proc. Metrology Conf. Berlin (1996)

[13] Ch. Chui (ed.), Series on: Wavelet Analysis and its Applications, vol. I –V. Academic Press, San Diego, 1992–96

[14] I. Daubechies, Ten Lectures on Wavelets. CBMS, SIAM Publ., Philadelphia, 1992.

[15] I. Daubechies and T. Paul, Time–frequency localization operators, *Inverse Problems* **4** (1988) 661–680.

[16] A.K. Louis, P. Maaß, A. Rieder, Wavelets: Eine Einführung in Theorie und Anwendungen. Teubner Verlag, Studienbücher Mathematik, 1994.

[17] T. Paul, Affine coherent states and the radial Schroedinger equation, *J.Math.Phys.* **25** (1994).

Non-Linear Electromagnetic Systems
V. Kose and J. Sievert (Eds.)
IOS Press, 1998

Wavelet Analysis of the Field Data
for Inverse Problems

Iliana MARINOVA

Department of Electrical Apparatus, Technical University of Sofia, Sofia 1756, Bulgaria

Yoshifuru SAITO

College of Engineering, Hosei University, Kajino, Koganei, Tokyo 184, Japan

Abstract. A wide range of inverse problems is ill-posed and usually uses indirect noisy measurements. In this paper, we propose the wavelet approach to reduce the noise from the data used in inverse problems. Taking the discrete wavelet transform to the field data and analyzing the wavelet coefficients we apply a suitable thresholding procedure. The original data with reduced noise are reconstructed by inverse discrete wavelet transform. It is shown that the wavelet approach works well for field data used in inverse problems.

1. Introduction

A wide range of inverse problems in electromagnetics (NDT, identification problems, field synthesis, design problems, etc.) uses the measured data of the field [1-2]. This information is of great importance for the inverse problem solutions. One of the main difficulties in dealing with such problems is the wide variability of the data values that are used. Usually, the field data are much less than needed and in most cases located in different scales. As a result, the inverse problems tend to be ill-posed. In most cases the data are combined with noise that reflect on the solution essentially. Because of that, control and reduction of noises are of great importance in solving the inverse problems under consideration.

In the past few years, wavelet methods have been applied to the multiscale representation and analysis of signals [3-4]. In this paper, we propose an approach based on the wavelet technique to analyze the noise measured data. In order to reduce the noise we take the discrete wavelet transform (DWT) to the data. After detailed analysis of the received wavelet coefficients we apply a suitable thresholding procedure. The data with reduced noise are reconstructed using inverse discrete wavelet transform.

The proposed approach was applied to analyze the field data measured around a transmission cable and used to determine the current density distribution of the cable. It is shown that the wavelet approach works well for field data used in inverse problems.

2. Wavelet analysis of the field data

Let us consider the field data vector Y used for solving inverse problems to be measured at discrete points t_i for $i = 1, \ldots, n$, where $n = 2^J$ for some integer J. The field data are always corrupted by noise, so that the measured data $y(t)$ can be presented by

$$y(t) = y_e(t) + \varepsilon(t), \tag{1}$$

where $\varepsilon(t)$ is a noise at point t.

The discrete version of (1) is

$$y_i = y_e(t_i) + \varepsilon_i. \tag{2}$$

Let measured field data as well as noise variables have independent normal distributions and variance σ^2 is known.

In order to reduce the noise we take the discrete wavelet transform (DWT) to the data. After detailed analysis of the received wavelet coefficients we apply a suitable thresholding procedure to them and invert the wavelet transform to reconstruct the original data with reduced noise.

The DWT is taken to the field data vector Y. Since the whole DWT is a composition of orthogonal linear operations, so the DWT is an orthogonal linear operator.

The DWT consists of applying a wavelet filter coefficient matrix with four coefficients (3) to the data vector Y.

$$
\mathbf{C} = \begin{bmatrix}
c_0 & c_1 & c_2 & c_3 & & & & & \\
c_3 & -c_2 & c_1 & -c_0 & & & & & \\
& & c_0 & c_1 & c_2 & c_3 & & & \\
& & c_3 & -c_2 & c_1 & -c_0 & & & \\
\vdots & & & & \ddots & & & & \\
& & & & & c_0 & c_1 & c_2 & c_3 \\
& & & & & c_3 & -c_2 & c_1 & -c_0 \\
c_2 & c_3 & & & & & & c_0 & c_1 \\
c_1 & -c_0 & & & & & & c_3 & -c_2
\end{bmatrix}
\tag{3}
$$

where

$$c_0 = (1+\sqrt{3})/4\sqrt{2} \qquad\qquad c_1 = (3+\sqrt{3})/4\sqrt{2}$$
$$c_2 = (3-\sqrt{3})/4\sqrt{2} \qquad\qquad c_3 = (1-\sqrt{3})/4\sqrt{2} \tag{4}$$

Here, blank entries signify zeroes.

The wavelet filter coefficient matrix with six, eight, ten, etc. coefficients can be composed in similar way.

The action of the matrix is to perform two related convolutions, then to decimate each of them by half (throw away half the values), and interleave the remaining halves.

For such a characterization to be useful, it must be possible to reconstruct the original data vector Y from its $\frac{n}{2}$ smooth or s-components and its $\frac{n}{2}$ detail or d-components.

That is effected by requiring the matrix (3) to be orthogonal, so that its inverse is just the transposed matrix of the matrix (3).

The procedure is called a pyramidal algorithm.

The transformation matrix acting first on a full column data vector Y, then to the "smooth" vector of length $\frac{n}{2}$, then to the "smooth-smooth" vector of length $\frac{n}{4}$, and so on until only a trivial number of "smooth-...smooth" components (usually 2) remain.

The output of DWT will always be a vector with two S's and all the "detail" components of D's, d's, etc. that were accumulated along the way. A diagram for data vector Y of length $n=8$ make the procedure clear:

$$
Y = \begin{bmatrix} y_1 \\ y_2 \\ y_3 \\ y_4 \\ y_5 \\ y_6 \\ y_7 \\ y_8 \end{bmatrix} \xrightarrow{\ C\ } \begin{bmatrix} s_1 \\ d_1 \\ s_2 \\ d_2 \\ s_3 \\ d_3 \\ s_4 \\ d_4 \end{bmatrix} \xrightarrow{\ P\ } W_8^{(1)}Y = \begin{bmatrix} s_1 \\ s_2 \\ s_3 \\ s_4 \\ d_1 \\ d_2 \\ d_3 \\ d_4 \end{bmatrix} \xrightarrow{\ C'\ } \begin{bmatrix} S_1 \\ D_1 \\ S_2 \\ D_2 \\ d_1 \\ d_2 \\ d_3 \\ d_4 \end{bmatrix} \xrightarrow{\ P'\ } W_8^{(2)}Y = \begin{bmatrix} S_1 \\ S_2 \\ D_1 \\ D_2 \\ d_1 \\ d_2 \\ d_3 \\ d_4 \end{bmatrix}
\tag{5}
$$

The elements in vector Y are sorted by means of the matrix

$$
P_8 = \begin{bmatrix} 1 & 0 & 0 & 0 & 0 & 0 & 0 & 0 \\ 0 & 0 & 1 & 0 & 0 & 0 & 0 & 0 \\ 0 & 0 & 0 & 0 & 1 & 0 & 0 & 0 \\ 0 & 0 & 0 & 0 & 0 & 0 & 1 & 0 \\ 0 & 1 & 0 & 0 & 0 & 0 & 0 & 0 \\ 0 & 0 & 0 & 1 & 0 & 0 & 0 & 0 \\ 0 & 0 & 0 & 0 & 0 & 1 & 0 & 0 \\ 0 & 0 & 0 & 0 & 0 & 0 & 0 & 1 \end{bmatrix}
\tag{6}
$$

Thus the transformation $W_8^{(1)}, W_8^{(2)}, \ldots$ are presented by

$$
W_8^{(1)} = P_8 C_8
$$
$$
W_8^{(2)} = (P_8' C_8')(P_8 C_8)
\tag{7}
$$

where
$$
P_8' = \begin{bmatrix} P_4 & 0 \\ 0 & I_4 \end{bmatrix}, \quad C_8' = \begin{bmatrix} C_4 & 0 \\ 0 & I_4 \end{bmatrix},
\tag{8}
$$

and I_j is the j-th order unit matrix.

The last m-th transformation $W_n^{(m)}$ represents the wavelet spectrum.

A value d_i of any level is termed a "wavelet coefficient" of the original data vector, the final values S_1, S_2 should strictly be called "mother-function coefficients".

The wavelet coefficients are estimated by applying the soft threshold function
$$
\delta_\lambda(x) = sign(x)(|x| - \lambda)
\tag{9}
$$
or the hard threshold function
$$
\delta_\lambda(x) = x \ \text{if} \ |x| > \lambda, \ 0 \ \text{otherwise}
\tag{10}
$$

for some threshold value $\lambda \geq 0$. A "universal threshold" $\lambda = \sigma\sqrt{2\ln n}$ can be used.

To invert the DWT, we reverse the procedure (5) from right to left. Instead of matrix (3) its inverse (in this case - transposed) matrix is used.

So, we apply inverse discrete wavelet transform (IDWT) to the thresholding wavelet coefficients in order to reconstruct the vector Y with noise reduced data.

3. Examples

In order to estimate the current density distributions in a transmission cable using the SPM method the magnetic field data are measured around the cable cross section. Here, we use inversions of numerically simulated experimental data, rather than real ones..

The wavelet analysis of noise corrupted data is presented. The exact field data at the points of measurement are shown in Fig. 1(a). The number of points n is set to 1024, so the applied wavelet transform has ten levels. The field data with different data-to-noise ratio (DNR) are analyzed. Fig. 1(b) and Fig. 1(c) show the field data with DNR=1 and corresponding wavelet spectrum, respectively. The results from the wavelet noise reduction are shown in Fig. 1(d). The relative data errors are calculated before and after wavelet noise reduction. The maximum value of the relative error decreases essentially .

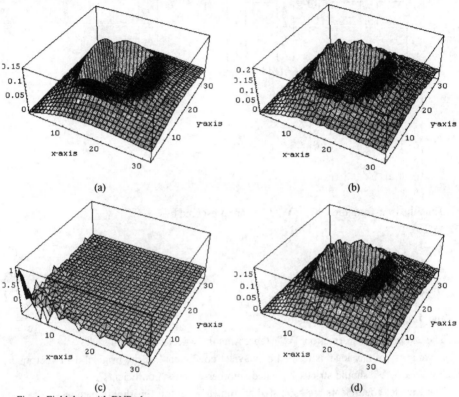

(a)   (b)

(c)   (d)

Fig. 1. Field data with DNR=1.
  (a) Exact field data; (b) Field data with noise; (c) Wavelet spectrum;
  (d) Field data with reduced noise.

4. Conclusion

In this paper, we used wavelet analysis to reduce the noise of the measured field data used in inverse problems. Our proposed wavelet approach has several features that made it quite useful. First, noise problems are analyzed easier in the wavelet domain than in the original domain. Second, the thresholding procedure of the separate treating of wavelet coefficients makes it possible to analyze the data with jumps, spikes and other nonsmooth features. Third, suitable choice of the threshold is very important in the noise reducing process.

The examples show that the wavelet approach works well for field data used in inverse problems.

References
[1]   A.M. Fedotov, Ill-posed problems with random data errors. Nauka, Novosibirsk, 1990.
[2]   A.N. Tikhonov and V.Y. Arsenin, Solutions of Ill-posed Problems. New York: Wiley, 1977.
[3]   Y. Midorikawa et al, A thin film common mode noise filter and its evaluation by wavelet   analysis, *IEEE Trans. on Magnetics*, Vol.32, No.5, pp.5001-5003, September 1996.
[4]   I. Daubeches, Ten Lectures on Wavelets. Philadelphia: PA:Soc. Ind. and Appl. Math., 1992.

Non-Linear Electromagnetic Systems
V. Kose and J. Sievert (Eds.)
IOS Press, 1998

The State Space Description: Mathematical Models and Problems while Processing

Ivan KALCHEV

Technical University of Sofia, Faculty of Automation, BG 1797, Sofia, Bulgaria

Abstract. The paper deals with both the possibilities of modern computer science for mathematical modelling of mechanical structures and the problems which may occur in the algorithmization of these tasks.

1. Object of modelling

A variety of complete mathematical models exists for the analysis of a great many objects under non-destructive testing [1] and other related fields. They themselves present differential equations with partial derivatives. Exact solutions exist only for several partial cases, but in the common case approximating methods are widely known. The numerical solution of such equations in its classical comprehension is reduced to the solution of a set of linear algebraic equations. This task is executed by means of finite-difference methods [2] and by applying the principles of parallelism on a modern computing environment. This usually is an iterative continuous process.

In particular, one of the conventional tests of steel welded joints in our nuclear power plants and in our chemical industry is related to assessment of the ferrite phase content the presence of which increases the sensitivity of the material to the progress of intercrystalline corrosion and the sensitivity to cracking. At the stage of choosing appropriate electrodes for the welding, the tests are regulated by standard specification. After the manufacture of the welded joint, a determination of the ferrite phase in the surface layer (accessible to control) was made by using ferrite meters, designed on electromagnetic, magnetostatic, induction, pond metric, etc. principles. The thermal deformation and radiation effects as well as the contact of the material of the welded joint with corrosive mediums create the necessity of subsequent verifications for the presence of changes in the ferrite phase content, formed on the grains boundary with the purpose of diagnosis of the weld joint state and determination of the weld joint resource.

2. The process of mathematical modelling

As can be seen there are many tasks which may be solved in terms of partial differential equations by means of numerical analysis.

2.1. Basic terminology and postulation

The first step in modelling the investigated structures, subjected to testing is determination and building of a preliminary information abstract graph. This highest level of abstraction gives the main state space description and the properties of the object. The graphs are convenient for both visual perception and logical analysis.

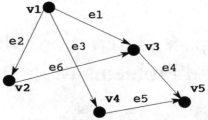

Figure 1. The information graph

Figure 1 shows the information graph $G\{V,E\}$ of a finite state space mechanical structure. Every vertex v_i of the set V interacts with one or several vertices v_j.

The graph edges are defined according to the function $G = G\{V, \Re v\}$, where $\Re$ is the operator defining the functional relationship of these interactions $e = \Re v$.

The second step of the mathematical modelling is the representation of such an information graph as a table of the incidences, shown in Table 1. This is an indispensable condition for further computer processing.

Every vertex of this graph uniquely corresponds to a row. With each edge is associated a column. For each edge $+1$ is written into the cell at the crossing with the row, corresponding to the vertex, from which comes the edge, and -1, where the edge enters. The rest cells are 0. Such matrices very often are large sparse ones. As a rule, the bigger the matrix, the greater the scattering.

The next step is to find suitable topological equations. They are designed both for the vertices and the edges of the equivalent scheme in accordance with concrete objectives. We can use a method based on the application of information, contained in the matrix of the contours and of the sections - M-matrix. As a matter of fact, it represents a logical tree.

An algorithm forming M-matrices (Table 2):

• Each edge, not entered in a logical tree is used for the design of closed contours.

• Each contour is examined in a given direction.

• If the direction of the edges coincides with the direction of the examination $+1$ is written into the cell, -1 in contrary case, and 0 if the edge does not enter this contour .

The system of topological equations, based on the M-matrix can be written as follows:

(1)
$$M.U_e + U_v = 0$$
$$I_e - M^T.I = 0 \quad,$$

where: U_e, U_v are argument vectors, similar to any potential difference,

I_e, I_v are argument vectors, similar to any current.

M^T is transposed M-matrix.

A similarity is possible with the well known Kirchhoff's laws.

Table 1.

Vertices	Edges				
	e1	*e2*	*e3*	*e4*	*e5*
v1	+1	+1	+1	0	0
v2	0	-1	0	0	0
...	...	...	...	...	...
v5	0	0	0	-1	-1

Table 2.

EDGES	e1	e2	e4
e3	-1	+1	0
e5	0	-1	+1

2.2. Topological equations of distributed objects

The basis of this description for further array treatment is the partial differential equation (generalized Poisson's equation), shown in a non explicit form, as follows:

$$(2) \qquad F(x, y, e''_{xx}, e''_{yy}) = 0,$$

where x, y are the Cartesian's state coordinates of each element of the set V, but

the values e are given along the boundary, by imposing n initial Dirichlet's conditions.

If the equation (2) is subjected to a differentiation with respect to $(n \times m)$ space points by means of standard procedures of the finite-difference calculus, a set of linear algebraic equations is obtained.

One of the simplest iteration methods for solution of such a set is the Jacobi's method, where for solving the linear algebraic set $A.v = e$ the next iterations are used:

$$(3) \qquad e_i[k+1] = f\{e_i[k]\}, \qquad i = 1, ..., (n \times m)$$
$$k = 0, 1, ...$$

This method is not so widespread, but all the intrinsic processing problems typical for the principles of the parallelism are inherent to it. That is why, Jacobi's method is convenient when we have to search reserves elsewhere, for example in the process of digitization.

3. Discussion on the processing problems

The problems which may occur while processing, in most cases, are dependent on the experiment design. First of all, there are some questions related to both a suitable space digitization and the determination of the boundary conditions for each homogeneous region. An incorrect algorithmization of the task very often leads to overcharging even modern computers. What do we have to do to decrease the processing time under a requirement for maximal admissible uncertainty in the final results? Experience suggests, that the possibilities of such known numerical methods are practically exhausted. While processing, after a number of iterations, the coefficient of convergence begins to increase itself asymptotically (a phase of saturation is observed), which is the main processing time problem for achieving the necessary treatment accuracy [3]. There is a need to investigate the impact of the degree of parallelism in the used algorithms on the computing convergence to obtain passable results.

Summarizing, the processing problems include:

• the exchange of the differential operator with a finite-difference analogue,

• the reduction of the iteration's number for obtaining a given accuracy.

At first sight, the domain decomposition seems a simple task. But, in practice, very often essential difficulties occur.

For example, under a stride fining away, the finite-difference scheme may not be giving a solution, trending to the exact one.

A second problem is the maximal admissible error of approximation for every space point: $\max \delta = e(h) - e$, where h is the net stride.

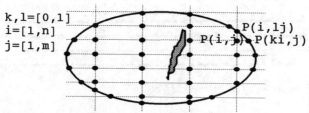

Figure 2. The space digitization

Last, but not least, is the problem of finite-difference scheme stability. This net is stable that does not augment the errors when the initial boundary constraints are translated into the near-by net points for a region with an arbitrary form (Fig. 2.). Moreover, for this interpolation equation (4) still no common approach and the time processing losses are comparable with the time for the full task solution.

(4) $$P(i,j) = F\{P(k.i,j), P(i,l.j)\}, \quad i,j \in \Omega.$$

Especially in the NDT multiple solutions are possible, when the investigated structure is subjected to an electrical or magnetic test for finding out space defects.

For that purpose it is necessary to exclude of parts of the boundary constraints in a given order followed by comparison with the results of numerical calculus.

The non-coincidence will be an indication of presence of mechanical cracks and of a breach in the homogeneity.

Pertaining to iteration reduction we have to bear in mind that the degree of parallelism during the whole calculus process is not constant. A good speedup of a parallel algorithm can be obtained by a suitably chosen coefficient of over relaxation - r :

(5) $$e_{x,y}^{new} = r.e_{x,y}^{*} + (1-r).e_{x,y}^{old},$$
$$1 \le r \le 2.$$

There are reserves also in the preliminary stage of preparation of the linear algebraic set of equations. Like the well known simplex method we can design a reference plan by a suitable choice of a part of the variables as basic ones, or by introducing additional variables to make an artificial basis.

4. Conclusion

While the mathematical modelling is based on the finite-difference calculus by using modern computing environment, it has its own specific problems, the nature of which have been presented in this paper.

References
[1]   I. Kalchev, Mathematical Modelling of Mechanical Structures under Non-Destructive Testing. In: Proc. of the International Welding Technology '96 Symposium, 15-17 May 1996, Istanbul, Turkey.
[2]   I. Kalchev and I. Petrov, Reliability Multiparametrical Optimization Procedure. In: Proc. of the 6th International Conference on Industrial Metrology, 25-27 Oct 1995, Zaragoza, Spain, pp. 602-605.
[3]   I. Kalchev, Fast Parallel Algorithms in Environment Monitoring Systems. In: Proc. of the 3rd International Symposium on Measurement Technology and Intelligent Instruments, Sept. 30 - Oct. 3, 1996, Hayama, Kanagawa Pref., Japan, pp. 143-146.

Non-Linear Electromagnetic Systems
V. Kose and J. Sievert (Eds.)
IOS Press, 1998

The Application of the Extended Hamiltonian Formalism to Analyze a Moving-Conducting Speaker

U. Diemar

Technical University of Ilmenau
Dept. of Theoretical Basics of Electrical Engineering
D-98684 Ilmenau GERMANY

Abstract. The contribution offers a method for the analysis of coupled electro-magneto-mechanical systems which also contain dissipations by means of the "extended" Hamiltonian. The electro-mechanical part of a spherical cap loudspeaker is to be investigated. The acoustical part is not included here. The analysis here sets up the Hamiltonian as well as its derived canonical equations of motion.

1. Introduction

The Lagrangian formalism is based on the evaluation of the energy balance of the system under consideration and supplies the so-called $\{L, D\}$-model, where L, which is a function of the generalized coordinates, their temporal derivation and the time, is the Lagrangian and D, which is a function of the generalized velocity (temporal derivation of the coordinates), is the dissipation function. Hence, the technical problem must be modelled as a variational problem.

The Hamiltonian, which is a function of the generalized coordinates, momenta and the time, must be set up with the help of the Lagrangian and the well known relations for the canonical coordinates resulting from the Lagrangian. Starting with this Hamiltonian and the dissipation function we can establish the canonical equations of motion (see [1],[2]). It can be shown, that the general dissipations are included in one type of the canonical equations (e.g.[3],[4]).

Depending on the technical character of the system the generalized coordinates and the generalized momenta respectively describe specific technical quantities.

2. The application of the extended Hamiltonian formalism

The application of the extended Hamiltonian formalism will be demonstrated on the example of a moving-conducting speaker. The speaker is an electro-mechanical instrument which consists of an electrical and mechanical part. The coupling occurs through the electro-magnetic field.

On the membrane a speaker voice coil is fixed, which moves depending on the strength of the electrical current, which flows through the coil. The oscillation of the membrane generates an acoustical signal.

The state variable or describing quantity of the electrical part of the system is the electrical charge q and the electrical current $\dot{q}$, respectively and the mechanical deflexion z of the speaker voice coil for the mechanical part. The electrical charge q and the mechanical deflexion z are the generalized coordinates q_k of the entire system. The temporal derivation of the generalized coordinate - in this case the electrical current - is the generalized velocity.

In the model the mechanical part consists of the mechanical damping, the spring effect of the membrane and the oscillating mass. The electrical part includes the internal resistance, the inductance of the coil and the voltage source (excitation voltage). The schematic sketch and the equivalent electrical circuit are shown in Figure 1.

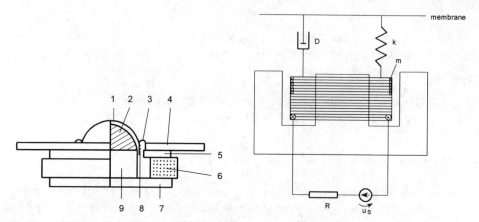

Figure 1: Spherical cap loudspeaker and equivalent electrical circuit

(1 membrane, 2 sound absorbent material, 3 boundary suspension, 4 front panel, 5 pole plate, 6 ferrite magnet, 7 bottom plate, 8 loudspeaker voice coil support, 9 iron core)

For the elements of the scheme the $\mathcal{L}$- and $\mathcal{D}$- terms are well known and are given in Table 1.

The electromechanical coupling is determined by a separate Lagrangian term. To set it up, it is necessary to use the formula (Lorentz-force):

$$d\vec{F} = I(d\vec{l} \times \vec{B}) . \tag{1}$$

The geometry of the example yields for the force $\vec{F}$:

$$\vec{F} = I \cdot B \cdot \pi \cdot d_c \cdot w \cdot \vec{e}_z , \tag{2}$$

with d_c = the diameter of the coil, w = the number of turns, $\vec{e}_z$ = unit vector in force direction, and so the Lagrangian term follows with:

$$\mathcal{L}_{el.-mech.} = B \cdot \pi \cdot d_c \cdot w \cdot z \cdot \dot{q} , \tag{3}$$

Table 1: $\mathcal{L}$- and $\mathcal{D}$-terms of the system elements and the actual values

Elements of the system	$\mathcal{L}$-term	$\mathcal{D}$-term	values
Excitation voltage u_s	$u_s q$		$15V\sin(25000.t)\cos(100000.t)$
Internal resistance R		$\frac{R}{2}\dot{q}^2$	5Ω
Inductance L	$\frac{L}{2}\dot{q}^2$		$90mH$
Total oscillating mass m	$\frac{m}{2}\dot{z}^2$		$0.3g$
Damping constant D		$\frac{D}{2}\dot{z}^2$	$10kg/s$
Spring constant k	$-\frac{k}{2}z^2$		$300N/m$
electro-mechanical coupling	$B\pi d_c wz\dot{q}$		$B = 1.1T$ $d_c = 25mm$ $w = 85$

for the electro-mechanical coupling. To obtain the Lagrangian and dissipation function, it is necessary to sum up all Lagrangian and dissipation terms. The $\{\mathcal{L}, \mathcal{D}\}$-model is given by:

$$\mathcal{L} = u_s q + \frac{m}{2}\dot{z}^2 - \frac{k}{2}z^2 + \frac{L}{2}\dot{q}^2 + B\pi d_c wz\dot{q} ,$$

$$\mathcal{D} = \frac{R}{2}\dot{q}^2 + \frac{D}{2}\dot{z}^2 . \tag{4}$$

To set up the Hamiltonian $\mathcal{H}$, it is necessary to derive the canonical momenta p_k. For the momenta p_k follows:

$$p_q = \frac{\partial \mathcal{L}}{\partial \dot{q}} = L\dot{q} + B\pi d_c wz \quad \rightarrow \quad \dot{q} = \frac{p_q - B\pi d_c wz}{L} ,$$

$$p_z = \frac{\partial \mathcal{L}}{\partial \dot{z}} = m\dot{z} \quad \rightarrow \quad \dot{z} = \frac{p_z}{m} . \tag{5}$$

The Hamiltonian of the system is:

$$\mathcal{H} = \mathcal{H}(p_k, q_k, t) = p_q\dot{q} + p_z\dot{z} - \mathcal{L} ,$$

$$= \frac{(p_q - B\pi d_c wz)^2}{2L} + \frac{p_z^2}{2m} + \frac{k}{2}z^2 - u_s q . \tag{6}$$

The derivations of (6) and (4) yield:

$$\dot{p}_k = -\frac{\partial \mathcal{H}}{\partial q_k} - \frac{\partial \mathcal{D}}{\partial \dot{q}_k} \quad \text{and} \quad \dot{q}_k = \frac{\partial \mathcal{H}}{\partial p_k} , \tag{7}$$

and in the special case:

$$\dot{p}_q = u_s - R\dot{q} \quad \text{and} \quad \dot{p}_z = \frac{p_q B\pi d_c w}{L} - \frac{(B\pi d_c w)^2 z}{L} - kz - D\dot{z} , \tag{8}$$

and

$$\dot{q} = \frac{1}{L}(p_q - B\pi d_c wz) \quad \text{and} \quad \dot{z} = \frac{p_z}{m} . \tag{9}$$

Obviously (8) contains the dissipations of the system. This results in a set of four differential equations of first order. Using the values of Table 1 in (8) and (9) the system is solved for z, the mechanical deflexion of the membrane. With $\tau = 69.3$ ms ;

$\omega_1 = 75000.s^{-1}$ and $\omega_2 = 125000.s^{-1}$ the solution is given by:

$$z(t) = -2.83 \; \mu\text{m} \; e^{-\frac{t}{\tau}} - 1.01 \; \mu\text{m} \; \cos(\omega_1 t) + 0.73 \; \mu\text{m} \; \cos(\omega_2 t)$$
$$+3.81 \; \mu\text{m} \; \sin(\omega_1 t) - 0.69 \; \mu\text{m} \; \sin(\omega_2 t) \; . \tag{10}$$

The transient waveforms of the excitation voltage $u_s(t)$ and the mechanical deflexion of the membrane $z(t)$ are shown in Figure 2. The grafics are generated by using "Mathematica" [6]. The phase retardation of the mechanical deflexion versus the excitation voltage is visible. The amplitude shown in Fig. 2 is in good accordance with the value predicted by analytical formulae, e.g. from [5].

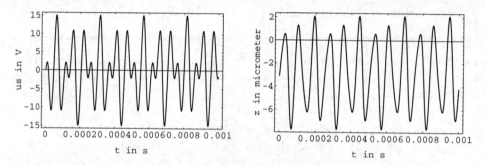

Figure 2: Excitation voltage u_s and mechanical deflexion z of the membrane

3. Conclusion

The possibility of the analysis of an electro-mechanical system and its internal coupling on the basis of an extended Hamiltonian formalism has been demonstrated. The dissipations of the system are included in the canonical equations of motion $\dot{p}_k$. The superiority of the Hamiltonian formalism contrary to other analysis methods is expressed in a unified description of systems, consisting of electrical as well as mechanical quantities.

References

[1] R. Süße, U. Diemar, G. Michel: Theoretische Elektrotechnik, Band II: Netzwerke und Elemente höherer Ordnung. VDI Verlag GmbH, Düsseldorf, 1996

[2] U. Diemar, Analyse und Synthese von Systemen mittels erweiterten Lagrange- und Hamilton-Formalismus unter Einbeziehung von Elementen höherer Ordnung. Ilmenau: PhD thesis, Technische Universität Ilmenau, 1995

[3] M. de Leon and P.R. Rodrigues, Generalized classical mechanics and field theory. North Holland - Amsterdam, New York, Oxford: Elsevier Science Publishers B.V. 1985

[4] J. Baumgarte, "Eine rein Hamiltonsche Formulierung der Mechanik von Systemen mit holonomen Bindungen", Acta Mechanica Vol. 36, No. 3-4, 1980

[5] B. Stark, Lautsprecher-Handbuch; Theorie und Praxis des Boxenbauens. München: Pflaum Verlag, 5. Auflage, 1992

[6] S. Wolfram, Mathematica, Ein System für Mathematik auf dem Computer. Addison-Wesley-Verlag GmbH, 2. Auflage, 1992

Non-Linear Electromagnetic Systems
V. Kose and J. Sievert (Eds.)
IOS Press, 1998

Nonlinear Materials and Adaptive Mesh Generation

Lutz JÄNICKE and Arnulf KOST
Lehrstuhl Allgemeine Elektrotechnik, BTU Cottbus
Postfach 101344, D-03013 Cottbus, Germany

Abstract. Even though the error estimation is derived only for linear problems, adaptive mesh generation can also be applied for nonlinear problems. The convergence of the nonlinear iteration is also stable for the generation of a very inhomogeneous mesh.

1. Introduction

Even though the theory of error estimators (e.g. [1]) does not cover the case of nonlinear materials, the resulting error estimates can be used for adaptive mesh generation. The reliability of the estimated global error on the one hand and the properties of the generated mesh with regard to the convergence on the other hand will be discussed for the case of 2-D nonlinear field problems.

2. Error Estimation

A lot of error estimation schemes have been discussed in recent years. When doing adaptive meshing those schemes which deliver local estimates with still acceptable computer time requirements have to be applied. Normally these local estimators are based on the residuals in the finite element solution.

2.1 Error Indicator

Due to the weak formulation, the finite element solution does not fulfill the differential equation for the magnetic vector potential

$$\nabla \times (\nu \nabla \times \vec{A}) = \vec{J}. \tag{1}$$

While the continuity of $\vec{A}$ preserves the continuity of B_n, the jump of the normal derivatives on the elements' boundaries yield a jump in the tangential component of the field strength $\vec{H}$

$$r_\Gamma = H_{t1} - H_{t2} = \left(\nu_{t1} \frac{\partial A_{t1}}{\partial n} - \nu_{t2} \frac{\partial A_{t2}}{\partial n} \right). \tag{2}$$

This residual (erronous) part can physically be interpreted as a current sheet on the elements' boundaries.

In the case of nonlinear material the reluctivity is realized by a different but constant value for each element. Equation (2) hence has to be evaluated for the reluctivities ν_{t1} and ν_{t2} after the nonlinear solution was obtained. The physical interpretation remains the same.

2.2 Global Error

When applying the 'local error problem' error estimator [1], the local and global error are computed in energy norm. While the estimate seems to be reliable for the linear case, as tested by comparison against known field solutions [2, 3], the computed value for the nonlinear case can only be an 'indicator'. While for the linear case the finite element solution is an approximation computed for the 'exact' problem, the solution for the nonlinear case is obtained for an 'approximated' problem with element-wise constant reluctivity. Hence the estimated error may not be too trustworthy. In order to evaluate the reliability it would be necessary to compare the results to analytically obtained solutions, which are however not available.

3. Nonlinear Process

For the solution of the nonlinear equation system, the Newton-Raphson method is applied

$$\mathbf{x}^{(k+1)} = \mathbf{x}^{(k)} + \Delta \mathbf{x}^{(k)}. \tag{3}$$

The convergence can be judged on by evaluating the relative change in the solution $\mathbf{x}$ $\alpha = |\Delta \mathbf{x}^{(k)}| / |\mathbf{x}^{(k+1)}|$.

4. Sample Problems

As examples a simple test problem and an electrical machine have been chosen.

4.1 Simple Test Problem

Figure 1 shows a simple test problem. This part of an iron core has a sharp inner edge leading to a singularity. The vector potential is given on the boundaries so that the mean flux density is 1.5 T. Since the problem does not take the air around the core into account, the flux cannot go through the air in the saturated region at the sharp corner, so that the singularity is kept.

The homogeneous and the adaptive mesh are shown in Fig. 2 with approx. 5000 nodes and 10000 triangles.

Figure 3 shows the convergence behaviour of the Newton-Raphson process for both meshes in Fig. 2. It can be seen that after the first steps, in which convergence is slow, the behaviour is similar. It should be noted, that the first computation is done for the linear case, which leads to unphysically high flux densities (more than 4 T) in the corner region, leading to the slow convergence at the beginning of the process.

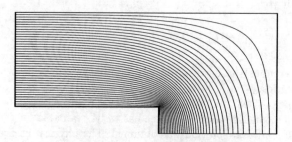

Figure 1: Simple nonlinear test problem

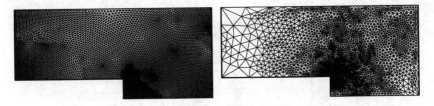

Figure 2: Homogeneous and adaptive mesh

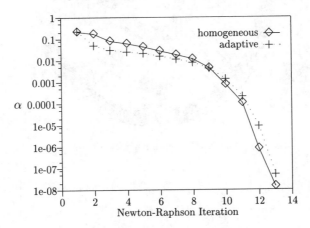

Figure 3: Convergence behaviour for test problem

4.2 SR Motor

Figure 4 shows the cross section of a switched reluctance motor. The magnetic field in the motor has been computed with a homogeneous mesh (≈ 65000 nodes) and an adaptive mesh (≈ 55000 nodes).

The convergence behaviour of the Newton-Raphson process is shown in Fig. 5. In this case the inner edges were smoothed to small circles (according to the real fabrication process) in order to avoid problems with singularities. Nevertheless, highly saturated regions occur due to the sharp contours of the stator and rotor and the thin airgap.

The global accuracy of the solution as given by the error estimator is not completely reliable. Nevertheless the values obtained are $\tilde{\delta}_{\mathrm{hom}} = 0.74$ and $\tilde{\delta}_{\mathrm{adap}} = 0.027$, respectively, for the relative error in energy norm [3]; the refinement procedure and parameters as described for 2-D problems in [3] were applied. While for flux computations the result for the homogenous mesh would be sufficient, torque computations are much more demanding, so that error values as in the adaptive case are necessary.

5. Conclusions

When investigating a number of nonlinear problems calculated using an adaptive mesh generation scheme, especially electrical machines, it turns out that the generated meshes fit the problems even when taking into account the nonlinear behaviour, so that the Newton-Raphson process converges quickly and smoothly. Therefore even when the estimated error cannot be proven to be near to the real error, the error estimation and adaptive refinement turn out to be a suitable tool for nonlinear computations.

Figure 4: Switched reluctance motor

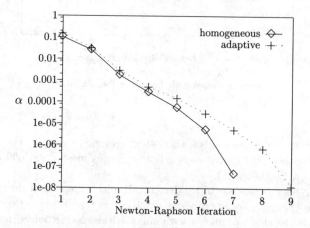

Figure 5: Convergence behaviour for SR motor

References

[1] Randolph E. Bank and Alan Weiser. Some a posteriori error estimators for elliptic partial differential equations. *Mathematics of Computation*, 44(170):283–301, April 1985.

[2] Lutz Jänicke and Arnulf Kost. On the convergence of the finite element method and the reliability of the error estimation. In *8th International Symposium on Theoretical Electrical Engineering, Thessaloniki*, pages 366–369, 1995.

[3] Lutz Jänicke and Arnulf Kost. Error estimation and adaptive mesh generation in the 2d and 3d finite element method. *IEEE Transactions on Magnetics*, 32(3):1334–1337, May 1996.

Non-Linear Electromagnetic Systems
V. Kose and J. Sievert (Eds.)
IOS Press, 1998

Non-linear Magnetostatic Field Analysis by Dual Reciprocity Boundary Element Method

Eugeniusz KURGAN[*]

University of Mining and Metallurgy, Dept. of Electrical Engineering
al. Mickiewicza 30, 30-059 Krakow, Poland

Abstract: This paper presents an application of the dual reciprocity boundary element method to analyse of magnetostatic fields in non-linear materials. The main difficulty in application of the BEM in analysis of magnetostatic fields is, that for non-linear materials a domain discretization is needed to take into account the non-linear effects and distributed currents. Thus the main advantages of the BEM over the FEM are lost. To overcome this difficulty the dual reciprocity boundary element method (DRBEM) is used, which transforms domain integrals in governing integral equations into equivalent boundary integrals. Both total and reduced scalar potentials are used for the formulation of field equations.

1. Introduction

The main difficulty in analysis of magnetostatic problems by the BEM method is, that for non-linear materials a domain discretization is needed to take into account the non-linear effects and distributed currents [6,7]. Thus the main advantages of the BEM over the FEM are lost. To overcome this difficulty the dual reciprocity boundary element method is suggested, which transforms domain integrals, which appear in governing integral equations describing a given magnetostatic problem, into equivalent boundary integrals.

This paper describes the application of the dual reciprocity boundary element method for magnetostatic field calculation using both total and reduced scalar potentials. It is assumed that all permeable media are non-linear, homogeneous and isotropic. At the end an illustrative example is given.

2. Basic Formulations

The solution region for the considered magnetostatic problem consists of two domains: ferromagnetic material region Ω_i with non-linear permeability $\mu_i(|\mathbf{H}|)$, where $\mathbf{H}$ is the field intensity and non-ferromagnetic material region Ω_0 (see [5,6,9,10] for detailed explanations and notations). The basic equations of magnetostatics for this problem are [4,5]:

$$\nabla \times \mathbf{H} = \begin{cases} 0 & \text{in } \Omega_i \cup \Omega_0 \\ \mathbf{J} & \text{in } \Omega_s \end{cases} \tag{1}$$

---

[*] This work was supported by the University of Mining and Metallurgy, under grant 10.120.247.

$$\nabla \cdot \mathbf{B} = 0 \quad \text{in} \quad \Omega = \Omega_i + \Omega_0 + \Omega_s \tag{2}$$

$$\mathbf{B} = \mu \mathbf{H} \begin{cases} \mu = \mu_i(|\mathbf{H}|) & \text{in} \quad \Omega_i \\ \mu = \mu_0 & \text{in} \quad \Omega_0 \cup \Omega_s \end{cases} \tag{3}$$

subject to the open boundary condition at infinity and appropriate boundary conditions at interface Γ_i.

In iron domain Ω_i no currents are assumed and magnetic field can be calculated as $\mathbf{H} = -\nabla \psi$, where ψ is a total scalar potential, which has to satisfy the non-linear differential equation [5,8]

$$\nabla^2 \psi = b \tag{4}$$

where

$$b = -\frac{\nabla \mu_i}{\mu_i} \nabla \psi = -\frac{1}{\mu_i} \left(\frac{\partial \mu_i}{\partial x} \frac{\partial \psi}{\partial x} + \frac{\partial \mu_i}{\partial y} \frac{\partial \psi}{\partial y} \right). \tag{5}$$

Because

$$\frac{\partial \mu_i}{\partial x} = -\frac{1}{|\mathbf{H}|} \frac{\partial \mu_i}{\partial |\mathbf{H}|} \left(\frac{\partial^2 \psi}{\partial x^2} + \frac{\partial^2 \psi}{\partial x \partial y} \right) \tag{6}$$

and similarly for $\dfrac{\partial \mu_i}{\partial y}$, we finally get

$$b = \frac{1}{|\mathbf{H}| \mu_i} \frac{\partial \mu_i}{\partial |\mathbf{H}|} \left[\left(\frac{\partial^2 \psi}{\partial x^2} + \frac{\partial^2 \psi}{\partial x \partial y} \right) \frac{\partial \psi}{\partial x} + \left(\frac{\partial^2 \psi}{\partial x \partial y} + \frac{\partial^2 \psi}{\partial y^2} \right) \frac{\partial \psi}{\partial y} \right] \tag{7}$$

The modulus of the magnetic strength can be calculated as

$$|\mathbf{H}| = \sqrt{\left(\frac{\partial \psi}{\partial x} \right)^2 + \left(\frac{\partial \psi}{\partial y} \right)^2} \tag{8}$$

The boundary integral equation associated with (4) is easy to obtain [5,6,7,10]

$$c_1(\mathbf{r})\psi(\mathbf{r}) + \int_{\Gamma_i} \psi(\mathbf{r}') \frac{\partial G(\mathbf{r},\mathbf{r}')}{\partial n} d\Gamma(\mathbf{r}') = \int_{\Gamma_i} G(\mathbf{r},\mathbf{r}') \frac{\partial \psi(\mathbf{r}')}{\partial n} d\Gamma(\mathbf{r}') + \int_{\Omega_i} b G(\mathbf{r},\mathbf{r}') d\Omega(\mathbf{r}') \tag{9}$$

The second final integral equation can be derived from (1),(2) and (3) and is presented in final form [7,10].

$$c_2(\mathbf{r})\psi(\mathbf{r}) - \int_{\Gamma_i} \psi(\mathbf{r}') \frac{\partial G(\mathbf{r},\mathbf{r}')}{\partial n} d\Gamma(\mathbf{r}') = -\frac{1}{\mu_0} \int_{\Gamma_i} \mu_i(|\mathbf{H}(\mathbf{r}')|) G(\mathbf{r},\mathbf{r}') \frac{\partial \psi(\mathbf{r}')}{\partial n} d\Gamma(\mathbf{r}') + V_s(\mathbf{r}) \tag{10}$$

Exact derivations of the above equations and adequate notations can be found in [5,10]

The current approach uses the dual reciprocity boundary element method to transfer the domain integral in (9) to series of boundary integrals, each one for particular solution of a Laplace equation. Let it be assumed that function $b(\mathbf{r})$ can be in domain Ω and on boundary Γ approximated by the following series expansion [1,2]:

$$b(\mathbf{r}) \cong \sum_{j=1}^{N+L} f_j(\mathbf{r}'_j,\mathbf{r})\alpha_j \tag{11}$$

where N is a number of boundary nodes, L is a number of internal points called also poles, $f_j(\mathbf{r}'_i,\mathbf{r})$ is an approximating function which depends only on problem geometry and is associated with point $\mathbf{r}'_i$. If co-ordinate $\mathbf{r}$ can be chosen in such a way, that for $i \in (1,N)$ overlaps with all boundary collocation points, and for $i \in (N+1,N+L)$ it is any freely chosen point in Ω, then there exist $(N+L)$ values of $f_j(\mathbf{r}'_i,\mathbf{r})$. The coefficients α_j associated with each function f_j are initially unknown. With each function $f_j(\mathbf{r}_i', \mathbf{r})$ there is associated a particular solution $\hat{u}_j$ of Poisson's equation, such that $\nabla^2 \hat{u}_j(\mathbf{r}) = f_j(\mathbf{r}_j', \mathbf{r})$ [3]. Equation (10) in discretized form can be written as (assuming that $b = \mathbf{F} \cdot \alpha$) [9]

$$\mathbf{H}\{\psi\} - \mathbf{G}\left\{\frac{\partial \psi}{\partial n}\right\} = \left(\mathbf{H}\hat{\mathbf{U}} - \mathbf{G}\hat{\mathbf{Q}}\right)\mathbf{F}^{-1}b = \mathbf{S}b \tag{12}$$

where $\mathbf{S} = (\mathbf{H}\hat{\mathbf{U}} - \mathbf{G}\hat{\mathbf{Q}})\mathbf{F}^{-1}$ is now a known matrix. As the functions $f_j(\mathbf{r}_j', \mathbf{r})$ we choose

$$f_j(\mathbf{r}_j', \mathbf{r}) = 1 + r(\mathbf{r}_j', \mathbf{r}) + r(\mathbf{r}_j', \mathbf{r})^2 \tag{13}$$

where $r(\mathbf{r}_j', \mathbf{r})$ is a distance from an approximation point $\mathbf{r}_j'$ to field point $\mathbf{r}$ [2].

Now we have to relate the partial derivatives of ψ on the right side of (8) to the node values of the potential ψ and nodal derivatives $\frac{\partial v_l}{\partial n}$. Assuming approximation $\psi = \mathbf{F}\beta$ we can derive relations between spatial partial derivatives of ψ and ψ alone:

$$\frac{\partial \psi}{\partial x} = \frac{\partial \mathbf{F}}{\partial x}\beta = \frac{\partial \mathbf{F}}{\partial x}\mathbf{F}^{-1}\psi \tag{14}$$

and similarly for other derivatives. Finally, the right side of (4) can be expressed as

$$b = \mathbf{M}\left[\left(\frac{\partial^2 \mathbf{F}}{\partial x^2} + \frac{\partial^2 \mathbf{F}}{\partial x \partial y}\right)\frac{\partial \mathbf{F}}{\partial x} + \left(\frac{\partial^2 \mathbf{F}}{\partial x \partial y} + \frac{\partial^2 \mathbf{F}}{\partial y^2}\right)\frac{\partial \mathbf{F}}{\partial y}\right]\mathbf{F}^{-1}\{\psi\} = \mathbf{R}\{\psi\} \tag{15}$$

where definition of the matrix $\mathbf{R}$ results from a comparison of the left and right sides of (15) and thus is can be assumed at each iteration step as a known value, and

$$\mathbf{M} = diag\left[\frac{1}{|\mathbf{H}|\mu_i(|\mathbf{H}|)} \frac{\partial \mu_i}{\partial |\mathbf{H}|}\right] \tag{16}$$

is a diagonal matrix calculated for all poles. The strength vector $|\mathbf{H}|$ is calculated as

$$|\mathbf{H}| = \sqrt{\left(\frac{\partial \mathbf{F}}{\partial x}\mathbf{F}^{-1}\psi\right)^2 + \left(\frac{\partial \mathbf{F}}{\partial y}\mathbf{F}^{-1}\psi\right)^2} \tag{17}$$

The final form of equation (9) in discretized form is

$$(\mathbf{H} + \mathbf{SR})\{\psi\} = \mathbf{G}\left\{\frac{\partial \psi}{\partial n}\right\} \tag{18}$$

Equation (10) can be discretized in the usual way [1,5].

The solution procedure is now iterative since the coefficients of matrix $\mathbf{R}$ are functions of ψ. First $\mathbf{R} = 0$ is assumed and the Laplace equation is solved to obtain a first estimate of $\{\psi\}$ and $\left\{\frac{\partial \psi}{\partial n}\right\}$. In the next step the value of $|\mathbf{H}|$ in (17) and matrix $\mathbf{R}$ through (15) is to calculate and to solve equation (18) together with the discretized form of (10) to obtain a second approximation of the boundary values. This process is continued until convergence is obtained.

3. Computational Results

After discretization of the coupled boundary integral equations (9) and (10) a system of algebraic equations is obtained. This process is well described in literature [5,6,7] and will not be presented here.

This example presents the calculation of the non-linear magnetic field by the method described in this paper. As an example, a test problem described in [6] is considered.

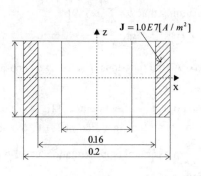

Fig.1. Geometry of the test problem.

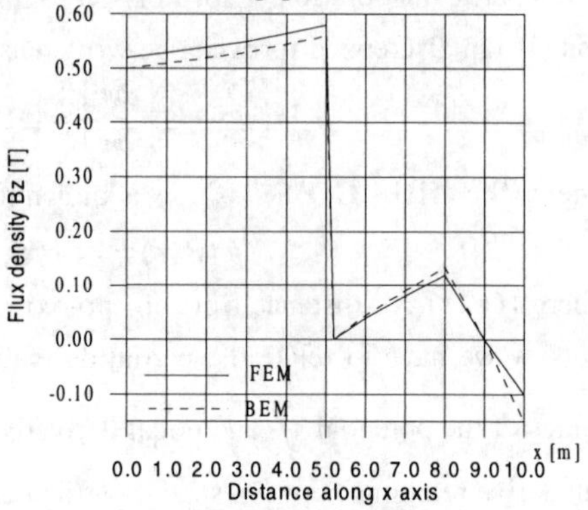

Fig.2. Magnetic flux density along x axis.

An iron cube in the field of a cylindrical coil is shown in Fig. 1. The non-linear behaviour of the material is described by the $B(H)$ characteristic $|\mathbf{B}| = \sqrt{\dfrac{|\mathbf{H}|}{3000}}$ [T]. The surface Γ_i of the iron cube was divided into isoparametric elements of the third order. The current density within the coil was assumed to be $\mathbf{J} = 1 \cdot 10^7 \left[A / m^2 \right]$. In Fig. 2 the dependence of the magnetic flux density B_z from distance x is presented. This relationship is compared with a result obtained from a commercial FEM program. There is generally good agreement between these two calculations. In the present formulation it was not necessary to calculate the gradient of the total scalar potential ψ and the divergence of the magnetic field $\mathbf{H}$ in domain Ω_i, as it has been presented in [6]. This results in simplification of the calculations and the programming effort. The efficiency of this method and related problems were not investigated.

References

[1] C.A. Brebbia, On treatment of domain integrals in boundary elements, in BETECH89, Computational Mechanics Publications, Southampton, 1989.
[2] C.A. Brebbia: On two different methods for transforming domain integrals to boundary, Advances in Boundary Elements, Vol. 1, Springer-Verlag, 1993, pp. 59-72
[3] T. Yamada, L.C. Wrobel, A new approach to magnetic field analysis by the dual reciprocity boundary element method, *Int. Journal for Numer. Methods in Engineering*, pp. 2073-2085, 1993.
[4] A. Kost, Numerische Methoden in der Berechnung elektromagnetischer Felder, Springer Verlag, Berlin, 1994.
[5] H. Igarashi, S. Sakai, T. Nakamura, T. Morinaga, T. Honma, A boundary element analysis of space charge fields in corona device, *IEEE Trans. on Magn.*, vol. 29, pp. 1508-1511, 1993.
[6] W.M. Rucker, Ch. Magele, E. Schlemmer, K.R. Richter, Boundary element analysis of 3-D magnetostatic problems using scalar potentials, *IEEE Trans. on Magn.*, vol. 28, 1099-1102, 1992.
[7] W.M. Rucker, K.R. Richter, Three-dimensional magnetostatic field calculation using boundary element method, *IEEE Trans. on Magn.*, vol. 24, pp. 23-26, 1988.
[8] M. Ayoub, F. Roy, F. Bouillault, A. Razek, Numerical modelling of 3D magnetostatic saturated structures, *IEEE Trans. on Magn.*, vol. 28, pp.1052-1055, 1992.
[9] E. Kurgan, Magnetostatic field calculation in anisotropic media using boundary element method, Proc. Symposium on Electromagnetic Phenomena in Non-linear Circuits, Poznań, 83-88, 1994.
[10] R. Sikora, Teoria pola elektromagnetycznego, Warszawa, WNT, 1985 (in Polish).

Non-Linear Electromagnetic Systems
V. Kose and J. Sievert (Eds.)
IOS Press, 1998

Neural Network Approach to Inverse Problems Solution: Nonlinear Boundary Conditions Identification

E. RATAJEWICZ-MIKOLAJCZAK, J. STARZYNSKI* & J. SIKORA*

*Department of Fundamental Electrical Engineering, Lublin Technical University,
ul. Nadbystrzycka 38a, 20-618 Lublin, Poland.
*Department of Electrical Engineering, Warsaw University of Technology,
ul. Koszykowa 75, 00-662 Warsaw, Poland*

Abstract. An electrical field model of the copper electrodeposition process on a printed circuit board (PCB) coupled with ultrasonic field is presented in this paper. On the basis of measurements, the numerical (FEM) model has been proposed. The inverse problem has been formulated for those two coupled by boundary conditions (BC) fields, making it possible to control the current density and the layer thickness. To solve this problem the artificial neural network will be applied.

1. Introduction

The problems of copper plating and the current distribution into through-holes and ultrasonic intensification of the electrodeposition process are well known and are the subject of many investigations [5]. The result of our measurement experiment indicates that a copper layer in a through-hole grows faster than on the flat surface of a PCB in this process [6]. The same process without ultrasounds gives thinner layer because of a smaller potential gradient in the through-hole and worse mixing conditions. As it is known, current efficiency of the copper electrodeposition process in a pyrophosphate bath is 99%, then the mass of the coated copper and the thickness of its layer depends proportionally on current density on the PCB surface [1].

In order to control the layer thickness by means of the ultrasonic field (which has an influence on current density) the Inverse Problem (IP) for boundary conditions can be formulated. The main goal of this paper is to define the efficient algorithm allowing calculation of certain coefficients of the function describing mixed BC in such a way that current density on the surface of a PCB achieves imposed values. The IP has been formulated making it possible to control the ultrasonic effect on the changes of the electrical field in the model under consideration. To solve this problem efficiently, the artificial neural network approach is proposed in this paper.

2. The numerical model of the system

The model of the system can be created on the basis of simplifying assumptions which have been motivated by the results of the experiments [6,7]. The electrical field at the steady state of the electrodeposition process is a direct current flow field and its structure can be described by a potential function determined by Laplace's equation.

However in the diffusion layer the volumetric charge density is not zero and Laplace's equation inside this layer is not correct. Further, the boundaries of the considered area are surfaces located at a distance from the electrode metal equal to the thickness of the diffusion layer. On the metal - electrolyte phase boundary the dependence of the electrode polarization φ on the current density J is a nonlinear function approximated by the Butler - Volmer equation [1]:

$$J = J_0\left(e^{(1-\beta)F\varphi/RT} - e^{-\beta F\varphi/RT}\right) \tag{1}$$

where J - current density, J_0 - density of the exchange current, β - asymmetry coefficient, F - faraday, φ - overpotential (the difference between electrode potential - one at current flow and the other one at the equilibrium state), R - gas constant, T - absolute temperature.

The graphical representation of this function is shown in Fig.1. Only the cathode current density can be considered in practice within the range of useful current density.Equation (1) is simplified then to the following form:

$$J = J_0 e^{-\beta F\varphi/RT} \tag{2}$$

According to this equation, the curves obtained experimentally (Fig.2) can be approximated by the following dependence:

$$J = e^{a\phi + b} \tag{3}$$

where a and b - coefficients depend on ultrasonic field parameters, ϕ - potential in the considered area.

Fig.2 presents experimentally determined dependencies of current density absolute values on the potential on the metal - electrolite boundary during the process carried out without the ultrasonic field (curve 1) and in this field on 337.7 kPa acoustic pressure (curve 2). At this value of acoustic pressure above the ultrasonic cavitation treshold the effect of build up of the layer in the through - hole is most visible.

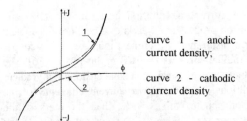

curve 1 - anodic current density;

curve 2 - cathodic current density

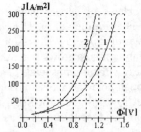

Fig. 1. The graphical representation of the electronation and de-electronation current densities

Fig. 2. The experimetal functions : J - absolute values on the cathode current density, Φ - absolute values on potential of boundary layer

Assuming the above conditions on the metal - elctrolite boundary, the model of this system has been created. The model consists of an unbounded plate with a cylindrical through-hole (cathode) and two parallel plates - anodes (Fig.3). The boundary conditions are shown in Fig.4. In order to solve this problem with nonlinear mixed boundary conditions, the FAT language is used [3]. The adaptive mesh generation is applied to discretize the region due to the non-smooth edge with the 3rd kind BC.

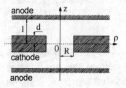

R - through-hole radius,
d - half-width of the cathode,
l - distance between the anode and the cathode center

Fig.3. The model diagram

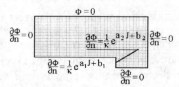

Fig.4. Region under consideration

3. Implementation of the nonlinear mixed BC in the FAT system

The current implementation of nonlinear mixed BC in the FAT system allows nonlinear Neumann boundary conditions to be set in the following form:

$$\kappa \frac{\partial \phi}{\partial n} = f(\phi), \qquad (4)$$

where κ is conductivity of electrolytic bath, Φ is the electrode potential, n denotes direction normal to the boundary and f is a function (3) which is described by the cubic spline.

The current implementation forces the user to specify the initial, linear approximation of the function f, i.e. to specify m and n coefficients of the linearized BC:

$$\kappa \frac{\partial \phi}{\partial n} = m\phi + n, \qquad (5)$$

The nonlinear procedure starts with the application of the condition (5) to the FE model of the problem. This is a well known procedure which leads to modification of the element matrices and right-hand vectors in all elements adjacent to the boundary with mixed BC. After the FEM based equation system is solved the values of the potential at the boundary are obtained and the new BC may be applied according to (4). This nonlinear BC affects the right-hand vector only. Solving the new system we shall obtain the new potential etc.

4. Results of the Forward Problem Solution

The solution has been obtained for the parameters of the electrodeposition process on the physical model of a PCB coated with and without the ultrasonic field [7,8]. The current density vector has been calculated in both cases. The results of calculations with the aid of FAT are presented in Fig.5 and Fig.6.

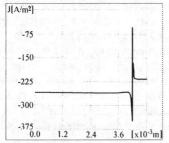
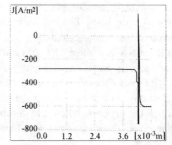

Fig.5. The current density distribution of the cathode without the presence of the ultrasonic field

Fig. 6. The current density distribution of the cathode with the presence of the ultrasonic field

5. The Inverse Problem Solution by Means of the Artificial Neural Network

To enable the control of the thickness of the layer of copper coating in the electrodeposition process, it is necessary to control boundary conditions. To solve such IP formulated for BC the ANN approach was used [2,6]. In order to identify the BC problem the simple feedforward multilayer ANN has been used (see Fig. 7). The data base for ANN training has been generated with the aid of FAT language. The exponential curve (Fig.2) that follows from this phenomenon has been linearized to simplify the problem. Approximating coefficient values have been determined within the range of current dencity 50 - 250 A/m² during the process carried out in the ultrasonic field. The parameters which have been changed during the data base creation procedure are coefficients for the linearized function of the boundary conditions.

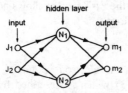

Fig.7. The neural network structure J_1, J_2 - current density, m_1, m_2 - coefficient of linearized function

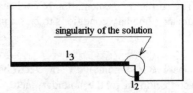

Fig.8. Boundary lines l_2 and l_3 where the design parameters were defined

Due to the singularity of the solution in the corner of the boundary (Fig.8), the authors have restricted the definition of design parameters m_1 and m_2 to the fragments l_2 and l_3 of the boundary line (Fig.8).

In order to learn the ANN the NETTEACH programme has been used [5]. The training procedure have employed 100 element vector which covered the whole solution space for the investigation range. The optimal solution is presented in Fig.9.

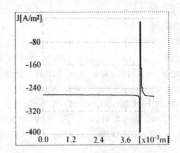

Fig.9. Optimal solution

The calculations proved that network behaves stably giving results with an error not exceeding 1.5%.

6. Conclusions

Neural network application makes it possible to solve a broad class of IP of the Electromagnetic Field Theory, including the IP formulated for mixed boundary conditions. The load of the whole calculation can be taken up by the teaching process of the network. When the ANN is trained, the solution of the IP is fast enough to control the process in real time which was extremely difficult to achieve for the other approaches. In the industrial applications neurocomputing has another significant advantage. The ability to solve IP with the aid of any ordinary commercial FEM packages. The proposed method is very convenient to control the process, where the number of DV (decision variables) is relatively low and the problem demands professional FEM analysis (big dimension that occure in the problem).

References

[1]   Bockris J.O'M., Khan S.U.M.: *Surface electrochemistry - molecular level approach*, Plenum Press, New York, London, 1993.
[2]   Osowski S.: *Algorithm' approach for neural networks* (in Polish), Wydawnictwa Naukowo-Techniczne, Warszawa 1996.
[3]   Korytkowski J., Starzyński J.,: *FAT- Reference manual and user guide, ver.3.2*, IETiME, Warsaw University of Technology, 1995.
[4]   Perusich S.A. et al: *Ultrasonically Induced Cavitation Studies of Electrochemical Passivit and Transport Mechanism*, Journal of the Electrochemical Society, Vol.138, No.3 March 1991, pp 708
[5]   Pesco A.M. and Cheh H.Y.: *The Current Distrbution within Plated Through-Holes*, part I and II, J. Electrochem. Soc., Vol.136, No.2, February 1989, pp.399-414.
[6]   Ratajewicz-Mikołajczak E.: *The Artificial Neural Network Approach for Mixed BC Identification, Archives of Electrotechnics (in print)*.
[7]   Ratajewicz-Mikołajczak E. and Sikora J.: *Ultrasounds influence on the electrical field of the electrodeposition process*, Nonlinear Electromagnetic Systems, A.J.Moses and A.Basak (Eds.). IOS Press, 1996, pp.250-253

Non-Linear Electromagnetic Systems
V. Kose and J. Sievert (Eds.)
IOS Press, 1998

401

Identification of Cracks Using Neural Networks

Ryszard Sikora, Mieczysław Komorowski, Tomasz Chady
Technical University of Szczecin, Al. Piastów 19, 70-310 Szczecin, Poland

Abstract. A system for eddy current nondestructive testing of conducting plates has been developed and described by the authors [3]. This paper presents a neural network model used for crack identification. Only a 2-D surface profile is used. Experimental results are reported.

1. Introduction

Thanks to the present large, high speed computers associated with modern numerical methods (gradient techniques, evolution strategies, artificial neural networks [1], [6] simulated annealing technique, genetic algorithms and other stochastic methods), many of the inverse electromagnetic tasks are now solvable. The use of artificial neural networks is a modern and effective way of solving inverse problems [1]. In this paper a conventional feed-forward neural network is applied to identify the geometry of a surface crack in a conducting plate using NDT.

2. Eddy current gauge

A detailed description of the eddy current testing system has been given in [1]. Figure 1 shows the transducer over a conducting plate with flaws. The defect causes changes in eddy current flow and results in a non zero output signal from the search coil. The signal depends on the flow type and distance between the flaw and the gauge.

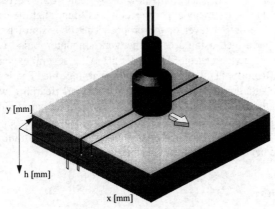

Fig. 1 View of the eddy current probe over the plate with shallow flaws.

3. Neural system for crack identification

It was assumed that the geometry of the flawed plate can be described by one dimensional function. The values of the function correspond to the surface coordinates taken for all probe positions (Fig. 1). The net makes it possible to determine geometrical parameters of a plate from a signal generated by the NDT probe. The input vector contains samples of the output signal from the probe and the net output corresponds to predicted coordinates of the plate surface. A feed-forward network (Fig. 2) was used to solve the problem.

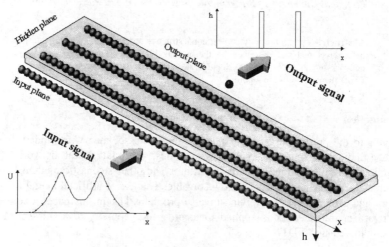

Fig. 2 Neural network used for crack identification

The proposed net contains two layers with connections to the outside world: an input plane which receives a signal from the sensor, and an output neuron which gives the response of the network to a fixed input. The main design parameters of the network are the number of hidden neurons which guarantee the desired degree of accuracy of the approximation. There is no established rule to strictly determine how many neurons are needed to obtain a given degree of accuracy [1]. After many experiments it was found that the net should consist of 150 hidden units. The output layer of the proposed net consists of only one neuron. All calculations were performed using a neural network simulator called *the Stuttgart Neural Network Simulator* on a workstation SUN Sparc 10. After several experiments a „Resilient back-propagation (Rprop)" [1] was selected as the learning rule. The learning process took about 2 hours (300 iterations). The data used for network learning were achieved from measurements and theoretical calculations (FEM). The data set was divided into three parts: training set, validation set and the test set. The learning should be stopped at the minimum of the validation set error. When learning is not stopped, overtraining occurs and the performance of the net on the whole data decreases.

4. Simulation results

In order to demonstrate the advantage of the neural network models developed by the authors, several experiments were carried out. The responses of the net for single cracks having different width are presented in Fig.3, 4 and 5. Some experiments with identification of shallow flaws placed side by side in the plate were also performed. Results of these investigations are presented in Fig.6 and 7.

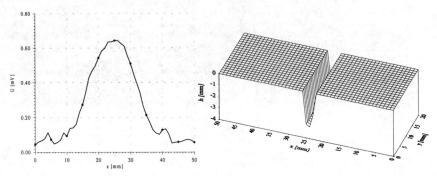

Fig. 3 Output signal from transducer for plate with single flaw and reconstructed shape of the crack; flaw width w = 2 mm; flaw depth h = 3 mm.

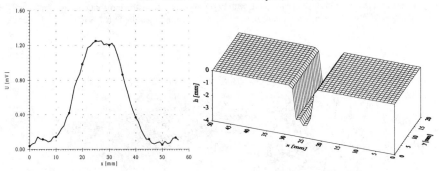

Fig. 4 Output signal from transducer for plate with single flaw and reconstructed shape of the crack; flaw width w = 5 mm; flaw depth h = 3 mm.

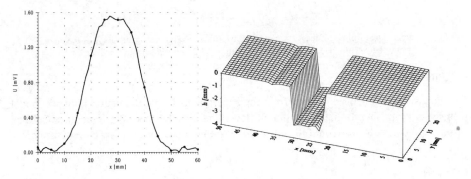

Fig. 5 Output signal from transducer for plate with single flaw and reconstructed shape of the crack. flaw width w = 10 mm; flaw depth h = 3 mm.

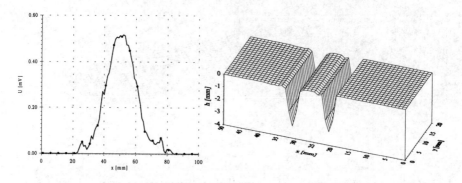

Fig. 6 Output signal from transducer for plate with two flaws and reconstructed shape of the cracks; distance between flaws d = 10 mm; flaw width w = 1.0 mm; flaw depth h = 3 mm.

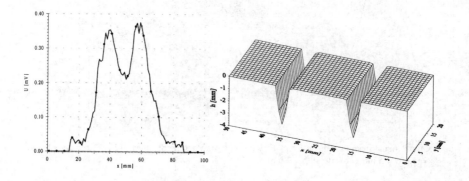

Fig. 7 Output signal from transducer for plate with two flaws and reconstructed shape of the cracks; distance between flaws d = 20 mm; flaw width w = 1.0 mm; flaw depth h = 3 mm.

5. Conclusions

The principal conclusion drawn from the experiments is that the neural network model seems very promising for the evaluation of flaw properties. Further investigations should be performed in order to work out fully a two dimensional model of the eddy current probe.

References

[1]   O. A. Mohammed, D. C. Park, G. F. Uler, „Design Optimization of Electromagnetic Devices Using Artificial Neural Networks", IEEE Trans. Magnetics 28 (5), 1992, pp. 1931-1934.

[2]   S. R. H. Hoole, „Artificial neural networks in the solution of inverse electromagnetic field problems", IEEE Trans. Magnetics 29 (2), 1993, pp. 1931-1934.

[3]   M. Komorowski, T. Chady, „Investigation of Differential Eddy Current Probe, International Journal No. 5 joint to International Symposium on Theoretical Elecrtrotechnics, Budapest 1994, pp. 145-152.

[4]   E. Coccorese, R. Martone, F. C. Morabito, „A Neural Network Approach for the solution of Electric and Magnetic Inverse Problems", IEEE Trans. Magnetics 30 (5), 1994, pp. 2829-2839.

[5]   Riedemiller M., Braun H., „A direct adaptive method for faster backpropagation learning: The RPROP algorithm", IEEE Proc. of the International Conference on Neural Networks 1993.

[6]   J. M. Mann, L. W. Schmear, J. C. Moulder, „ Neural Network Inversion of Uniform-Field Eddy Current Data", Mat. Evaluation, 1991, pp. 34-39.

Non-Linear Electromagnetic Systems
V. Kose and J. Sievert (Eds.)
IOS Press, 1998

A Reconstruction Method of Surface Morphology from an SEM Image with Genetic Algorithms

Tetsuji KODAMA, Kenichi TAJIMA, Xieoyuan LI and Yoshiki UCHIKAWA
Department of Information Electronics, School of Engineering, Nagoya University,
Nagoya 464-01, Japan

Abstract. A novel reconstruction method of surface morphology from a scanning electron microscope (SEM) image is reported. In comparison with conventional stereometric and multiple-detector methods, the system requires only a secondary electron image taken by a conventional SEM for reconstruction of microstructures of a specimen surface. This method is an application of genetic algorithms, which are search algorithms based on the mechanics of natural selection and genetics, to surface morphology in SEM.

1. Introduction

The image in the SEM [1-3] is produced as shown in Fig. 1. The grey level of a pixel on the photographs taken by a conventional SEM is determined by the number of secondary electrons that are collected by the detector system at that pixel position on the specimen. The number of secondary electrons that escape from a pixel on the specimen is determined by the topography of the specimen surface and the atomic number of the elements in the specimen at that pixel position. If both the atomic number of the elements in the specimen and the detector set-up in the specimen chamber are known, the surface slope at a pixel on the specimen can be calculated from the grey level of that pixel position on the photographs. Supposing the topography of the specimen surface is the cause and the grey level of each pixel on the photographs is the effect, the reconstruction of the topography from the grey level can be regarded as the inverse problem, however it is nonlinear and ill-posed. (This sort of inverse problem is known as shape from shading in the field of computer vision.) In this paper, we try to solve this inverse problem with genetic algorithms [4].

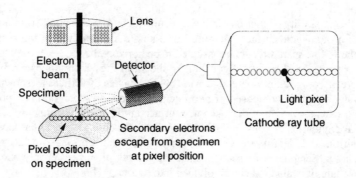

Fig. 1 The image in the scanning electron microscope.

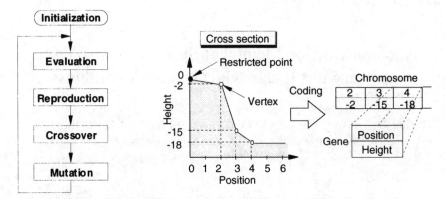

Fig. 2 Main structure of   Fig. 3 Coding.
genetic algorithms.

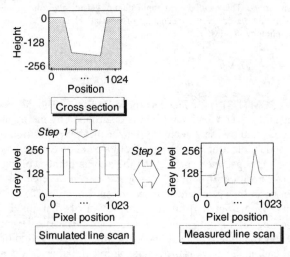

Fig. 4 Procedure of evaluation.

2. Genetic algorithms

To test the capability of this method, we reconstruct a cross section from a calculated line scan of a hemicycle surface composed of a single element, where the surface slope perpendicular to line scan direction is assumed to be horizontal.

The main structure of genetic algorithms is shown in Fig. 2. In this paper, we select a coding where a cross section is coded as a set of coordinates of vertices as shown in Fig. 3. In this coding, a gene is composed of two integers. One represents the pixel position and another represents the height of the specimen surface at that pixel position.

An initial population is constructed from hundreds of chromosomes that have a vertex whose position and height are chosen at random. Each chromosome in that initial population has a triangular cross section, since the number of vertices is only one.

The procedure of evaluation can be divided into two steps (Fig. 4).

Step 1: we first calculate the grey level of each pixel on the line scan of a decoded cross section. The grey level of a pixel is generally given by

$$g \propto \eta \delta I,$$

where η is the collection efficiency by the detector system, δ is the secondary electron yield and I is the primary electron current. In calculation of this grey level, we use some assumptions concerning the secondary electron emission [5]. The secondary electron yield is given by

$$\delta \propto 1/\cos\theta,$$

where θ is an angle of incidence of primary electrons. The energy and the angular distribution of secondary electrons are $\varepsilon/(\varepsilon+\varepsilon_0)^4$ [6] and $\cos\phi$ [5] respectively, where ε represents the initial energy of the secondary electron, ε_0 is a constant depending on surface composition, and ϕ is the initial angle of the secondary electron measured relative to the surface normal.

In these assumptions, calculation of secondary electron yield δ becomes simple, however calculation of collection efficiency η is still complex, since the collection efficiency depends on the shape of cross section and the detector set-up in the specimen chamber. To calculate the collection efficiency, we must simulate secondary electron trajectories in the specimen chamber for various initial energy and initial angles. (Their distributions are given by $\varepsilon/(\varepsilon+\varepsilon_0)^4$ and $\cos\phi$.) In the simulation, we use the detector set-up in the specimen chamber of JEOL JSM-820 (Fig. 5), the electric field in that specimen chamber is calculated by the three-dimensional finite element method, and the secondary electron trajectories are calculated by the fourth order Runge-Kutta method.

Step 2: the line scan calculated at step 1 is then compared with a line scan measured by a conventional SEM. (This is the line scan to be reconstructed.) To calculate the degree of similarity between the two line scans, we select a fitness function

$$f(n) = \frac{1}{\displaystyle\sum_{i=1}^{N-1} w_i \left| g_s(i;n) - g_m(i) \right|},$$

where $g_s(i;n)$ is the grey level of i th pixel on n th simulated line scan, $g_m(i)$ is the grey level of i th pixel on a measured line scan, and $w_i = \sqrt{g_m(i)}$.

For reproduction, we adopt a roulette wheel, where a chromosome is selected with a probability proportional to its fitness function value. Once a chromosome has been selected, a copy of the chromosome is made. This chromosome is then entered into a tentative new population for further genetic operator action. The number of chromosomes in the tentative new population is set to be equal to the number of chromosomes in the initial population.

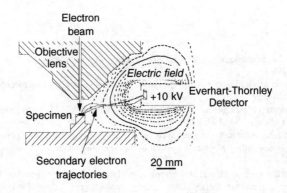

Fig. 5 Simulation result of secondary electron trajectories.

For crossover, we adopt one-point crossover. In this technique, chromosomes in the tentative new population are mated at random as parents. The crossover point is chosen at random from pixel positions of the cross section. The crossover is performed with 0.8 probability except for the chromosome that has the largest fitness value.

A vertex whose position and height are chosen at random is added to a chromosome selected for mutation. The mutation is performed with 0.1 probability.

3. Results and discussions

To remove reconstruction error due to the assumptions concerning the secondary electron emission, we have executed a numerical experiment, where a line scan calculated with the same procedure as step 1 in the evaluation is used as a line scan to be reconstructed. In Fig. 6, maximum and average population fitness are plotted as a function of the number of generations. Each inset shows a reconstructed cross section that has the largest fitness value in that generation. Toward the generation 1000, a form of convergence is observed. This convergent behavior cannot be observed if we fail to apply genetic algorithms to the problem. We can therefore conclude, from these results, that the genetic algorithms we have developed work appropriately as we had expected.

This work was supported by a Naito research grant.

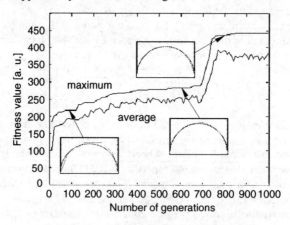

Fig. 6 Maximum and average population fitness.

References

[1] R. E. Lee, Scanning electron microscopy and x-ray microanalysis. P T R Prentice-Hall, Inc., New Jersey, 1993.

[2] T. Oshima *et al.*, Stereomicrography with a scanning electron microscopes, *Photogramm. Engin.* **36** (1970) 874-879.

[3] T. Suganuma, Measurement of surface topography using SEM with two secondary electron detectors, *J. Electron Microsc.* **34** (1985) 328-337.

[4] J. H. Holland, Adaptation in natural and artificial systems. Massachusetts Institute of Technology, 1992.

[5] H. Seiler, Secondary electron emission in the scanning electron microscope, *J. Appl. Phys.* **54** (1983) R1-R18.

[6] M. S. Chung, T. E. Everhart, Simple calculation of energy distribution of low-energy secondary electrons emitted from metals under electron bombardment, *J. Appl. Phys.* **45** (1974) 707-709.

[7] T. Kodama *et al.* , Genetic algorithms as applied to reconstruction of surface morphology from an SEM image. *J. Electron Microsc.* **46** (1997) 215-220.

Non-Linear Electromagnetic Systems
V. Kose and J. Sievert (Eds.)
IOS Press, 1998

Rapid Eddy Current Inversion for 3D Problems

Y. Matsumoto, S. Kojima, H. Komatsu, Z. Badics, F. Nakayasu
Nuclear Fuel Industries Ltd.
950, Ohaza-noda, Kumatori-cho, Sennan-gun, Osaka 590-04, Japan

S. Fukui
Japan Power Engineering and Inspection Co.
Business Court Shin-urayasu Bldg. 9-2, Mihama 1-chome, Urayasu-shi, Chiba, Japan

Abstract. A flaw reconstruction system for eddy current testing is developed by utilizing an inversion scheme based on optimization. Initial flaw estimates for the optimization loop are generated by a signal processing system based on neural networks. In the optimization loop, the signal of a flaw estimate is computed by a fast 3D forward solver that uses a pre-calculated reaction data set constructed by a 3D finite element field simulator before the inversion. The reconstruction examples for different specimen shapes are shown in order to demonstrate the capabilities of the reconstruction system.

1. Introduction

Eddy current testing (ECT) techniques have been widely used for the in-service inspection of steam generator tubing in Pressurized Water Reactor (PWR) atomic power plants. Recently, some flaw reconstruction systems based on inversion schemes that use signal processing techniques are being developed in the nondestructive evaluation field. We are developing a flaw reconstruction system based on optimization which is suitable for the ECT signal data obtained from steam generator tube inspection. In this inversion scheme, two main difficulties arise when we use the finite element method. One is that the forward solution that we need to perform several times during the iteration is very time consuming [1]. The other is that the error function - objective function - generally has several local minimums. In order to avoid these problems, our inversion scheme exploits both a neural network pattern recognition technique and an optimization loop based on using pre-calculated reaction data sets. By using neural networks, the number of necessary forward solutions and its solution time can be reduced. In this paper, as a sequel to [2], we show further examples and investigation results for 3D flaws in tube specimens in order to illustrate the efficiency of the reconstruction scheme.

2. Concept of Flaw Reconstruction

Figure 1 shows the flowchart of the flaw reconstruction system. As a first step, an initial flaw estimate is generated by a neural network system. In the second step, an inversion loop is performed where the signal of the flaw estimate and the corresponding error are calculated. In this step, a pre-calculated reaction data set is incorporated into the forward solver. The error between the measured and calculated data is minimized by changing the shape and/or the material parameters of the flaw. When the calculated data is in good agreement with the measurement, the corresponding shape is probably close to the actual shape.

It is important to estimate a good initial shape in order to avoid being trapped in a local minimum of the error function during the minimization. Good initial shape means that the error value is close enough to the global minimum. Here we estimate the initial flaw shape in real-time by using a neural network for this purpose. The other advantage of utilizing the neural network system is that the number of necessary forward solutions and the calculation time can be reduced.

We have the following set of Maxwell's equations for the arrangement with a flaw [3].

$$\nabla \times \mathbf{E} = -j\omega\mu\mathbf{H},\tag{1}$$

$$\nabla \times \mathbf{H} = \sigma_u\mathbf{E} + \mathbf{P} + \mathbf{J}_S, \qquad\qquad \mathbf{P} = (\sigma - \sigma_u)\mathbf{E} = \eta\mathbf{E}\tag{2}$$

where the current density $\mathbf{J}_S$ denotes the source current density of either coil 1 or coil 2, and the field quantities without suffix – here and later – denote the corresponding fields; σ and σ_u are the electric conductivity with flaws present and absent. $\mathbf{P}$ is an unknown incident *electric current dipole density* and the source of the field distortion due to a flaw (so-called *flaw field*) may be taken to be $\mathbf{P}$ radiating in the unflawed arrangement. η is the conductivity change and is called *flaw function* because it characterizes a flaw.

Probe signal $\mathbf{S}$ is defined as the reaction between the unflawed field of coil 2 and the flaw field generated by coil 1, that is, [4,5]

$$\mathbf{S} = \int_{\Omega_f} \mathbf{E}_2^{(u)} \cdot \mathbf{P}_1 d\tau \tag{3}$$

where Ω_f denotes the flaw region.

We approximate the current dipole density by a set of appropriate basis functions that can describe all allowable flaw functions. These functions are constructed on the basis of rectangular cells. Before the iteration, the region, which is assumed to include the true flaw region, is divided into rectangular cells and the reaction data set is calculated by a finite element simulator. During the inversion, the corresponding coefficient set for a given flaw region is picked up from the pre-calculated data set and a signal is calculated by solving the equation system consisting of these coefficients. This algorithm ensures fast forward calculation and an arbitrary flaw can be modelled by the same pre-calculated data set [2,6].

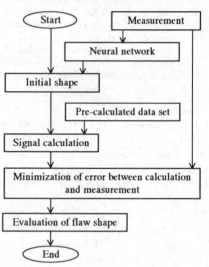

Figure 1. Concept of flaw reconstruction

3. Test arrangements

We use an ECT probe to investigate the efficiency of the reconstruction system. The probe consists of 16 transmitter pancake coils and 32 receiver coils arranged in circumferential direction. Receiver coils have the gap with transmitter coils in axial direction. The specimens are Inconel 600 tubes with an outer diameter of 22.23 mm and a thickness of 1.27 mm. Test signals shown in Table 1 are calculated by a FE simulator. The pitch in the calculation is 0.8 mm in axial and 11.25 degrees in circumferential directions, respectively. Figure 2 shows the structure of a three-layer neural network to estimate the initial flaw shape from the input pattern obtained by parameterization of probe signals. Input parameters based on

Table 1. Parameters of flaw

No	Direc.	Loc.	Length	Width	Depth
1	Circ.	Inner	5 mm	0.2 mm	20 %TW
2	Circ.	Outer	5 mm	0.2 mm	20 %TW
3	Axial	Inner	5 mm	0.2 mm	20 %TW
4	Axial	Outer	5 mm	0.2 mm	20 %TW

phase, amplitude, coefficient of correlation and fourier coefficients of signal are selected to efficiently classify ECT signals and estimate flaw dimensions. As the result of the selection, the number of input node is seven. Output vectors are defined to estimate axial and circumferential length, location and depth of flaws.

Table 2 shows the structure of a pre-calculated data set for both axial and circumferential flaws. It is possible to estimate an axial or circumferential flaw that is within a 10 mm long and 1.5 mm wide area.

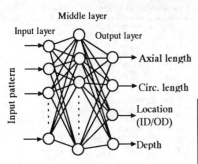

Figure 2. Structure of neural network

Table 2. Structure of pre-calculated data set

Direc-tion	Unit cell	Circ. Data set		Axial data set	
		Cells	Region	Cells	Region
Axial	0.254 mm	6	1.5 mm	40	10 mm
Circ.	0.024 rad	43	10 mm	6	1.5 mm
Depth	0.127 mm	10	100 %TW	10	100 %TW

4. Test results

Figure 3 shows the parameters and the coordinate used to express the rectangular shape flaws for the reconstruction example. (L1+L2: flaw length; L3X2: flaw width; L4: flaw depth.) Four parameters are utilized to describe the target slot in this problem. Table 3 shows their initial estimates by the neural network technique and the final estimated results, and the flaw reconstruction process are displayed in Figure 4. In these four examples, the reconstruction of the circumferential inner flaw takes most iterations. For the axial outer flaw, the iteration number is zero because of the good initial estimate. We can see that each final estimate is very close to the actual flaw shape.

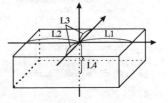

Figure 3. Parameters for reconstruction

Table 3. Parameters of flaw

No	Initial estimate				final estimate			
	Direc.	Loc.	Length	Depth	Loc.	Length	Width	Depth
1	Circ.	Inner	6.7 mm	22 %TW	Inner	5.1 mm	0.25 mm	20 %TW
2	Circ.	Outer	6.5 mm	15 %TW	Outer	5.1 mm	0.25 mm	20 %TW
3	Axial	Inner	5.1 mm	18 %TW	Inner	5.1 mm	0.25 mm	20 %TW
4	Axial	Outer	5.2 mm	20 %TW	Outer	5.1 mm	0.25 mm	20 %TW

5. Conclusion

We have developed a flaw reconstruction system for 3D eddy current testing and carried out a very promising feasibility study. The inversion is performed by first-order optimization where the initial flaw estimate is generated by a neural network system. The fast forward solver in the optimization loop utilizes a pre-calculated reaction data set.

It is demonstrated that this eddy current testing system is very promising for performing flaw reconstruction in tube specimens.

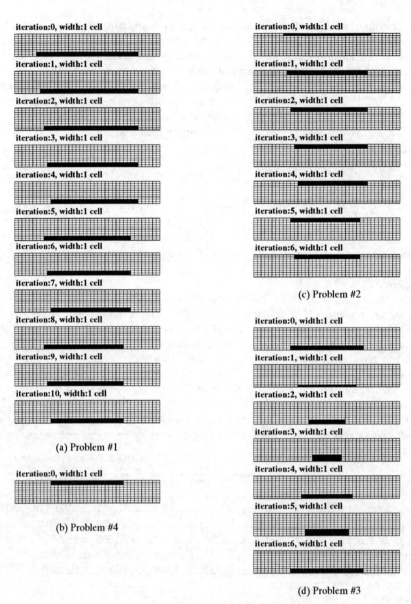

(a) Problem #1

(b) Problem #4

(c) Problem #2

(d) Problem #3

Figure 4. Reconstruction examples for four problems

References
[1]   Z. Badics, Y. Matsumoto, K. Aoki, and F. Nakayasu, "Defect-shape reconstruction in three-dimensional eddy current NDE problems by an optimization technique," *Nonlinear Electromagnetic Systems*, A. J. Moses, A. Basak, Eds., Amsterdam, IOS Press, pp. 780-783, 1995.
[2]   Z. Badics, et al, "Rapid flaw reconstruction scheme for three-dimensional inverse problems in eddy current NDE," E'NDE Tokyo, 2nd International Workshop on Electromagnetic Nondestructive Evaluation, organized by Japan Society of Applied Electromagnetics and Mechanics, Tokyo, Japan, October 28-29, 1996.
[3]   J. R. Bowler, S. A. Jenkins, L. D. Sabbagh, and H. A. Sabbagh, "Eddy-current probe impedance due to a volumetric flaw," *J. Appl. Phys.*, vol. 70, no. 3, pp. 1107-1114, 1991.

[4]   R. F. Harrington, *Time Harmonic Electromagnetic Fields*. New York: McGraw-Hill, 1961.

[5]   Z. Badics, et al, "Accurate probe-response calculation in eddy current NDE by finite element method," *J. Nondestructive Evaluation*, vol. 14, no. 3, 1995.

[6]   Z. Badics, et al, "Inversion scheme based on optimization for 3D eddy current NDE problems," *J. Nondestructive Evaluation*, (submitted)

Non-Linear Electromagnetic Systems
V. Kose and J. Sievert (Eds.)
IOS Press, 1998

An Approximate Solution of Nonlinear Time–Dependent Equations of Electrodiffusion Theory

V.Lebedev, A.Timokhin, V.Ababiy, A.Kolchuzhkin
Tomsk Polytechnic University, 634034, Tomsk, Russia

Abstract. An iterative method has been developed for solution of the Nernst-Planck and Poisson equations of electrodiffusion theory. The solution is used to study the time - dependence of effective diffusion coefficients in plain, cylindrical and spherical geometries.

1. Nernst-Planck and Poisson equations

The system of time–dependent nonlinear partial differential equations Nernst–Planck and Poisson (NPP) [1, 2] is widely used in plasma physics and electrochemistry for description of diffusion-migration phenomena. With i referring to the ionic species, we have:

$$\frac{\partial}{\partial t}c_i(\vec{r},t) = -\operatorname{div}\vec{J}_i(\vec{r},t),$$

where $c_i(\vec{r},t)$ is the local concentration of ions with charge z_i and $\vec{J}(\vec{r},t)$ is the ionic molar flux density vector consisting of two components:

$$\vec{J}_i(\vec{r},t) = -D_i\operatorname{grad}c_i(\vec{r},t) + \frac{F}{RT}z_i c_i(\vec{r},t)\vec{E}(\vec{r},t)$$

caused by gradient of concentration and by motion of ions in electric field $\vec{E}(\vec{r})$. R is the gas constant, F the Faraday constant, T the temperature, D_i ionic diffusion coefficients.

We consider that electric field $\vec{E}$ is the field created by the ions. It is related to c_i by the differential Gauss theorem or Poisson equation:

$$\operatorname{div}\vec{E}(\vec{r},t) = \frac{F}{\varepsilon\varepsilon_0}\sum_i z_i c_i(\vec{r},t).$$

The Poisson equation is very often replaced by the local electroneutrality condition:

$$\sum_i z_i c_i(\vec{r},t) = 0. \tag{1}$$

In this case and for $1:1$ electrolyte the substitution $c = c_1 = c_2$ gives the equation:

$$\frac{\partial}{\partial t}c(\vec{r},t) = D_{\text{eff}}\Delta c(\vec{r},t), \qquad D_{\text{eff}} = \frac{2D_1 D_2}{D_1 + D_2}. \tag{2}$$

It is seen from (2) that in the local electroneutrality approximation both sorts of ions have the same diffusion coefficient D_{eff} rather than D_1 and D_2.

An approximate solution of NPP equations has been obtained in this paper without the electroneutrality condition (1) by the successive approximation method.

Depending on the symmetry of the problem, we use Cartesian x, y, z, cylindrical ρ, ϕ, z or spherical r, θ, ϕ coordinates. In an infinite homogeneous medium the symmetry is determined by the symmetry of initial concentration profiles. For the following, we consider the problems where $c_i(\vec{r}, t)$ depends on a single lateral coordinate and divergence and the Laplace operator can be written in the form: $\text{div} = \dfrac{1}{r^\nu}\dfrac{\partial}{\partial r}r^\nu$, $\Delta = \dfrac{1}{r^\nu}\dfrac{\partial}{\partial r}(r^\nu\dfrac{\partial}{\partial r})$, where ν equals 0, 1 or 2 and r is x, ρ or r for plane, cylindrical or spherical symmetry, respectively. Then for all three geometries NPP equations can be written as

$$\frac{\partial}{\partial t}c_i = \frac{D_i}{r^\nu}\frac{\partial}{\partial r}\left(r^\nu\left(\frac{\partial}{\partial r}c_i - \alpha z_i c_i E\right)\right),\tag{3}$$

$$\frac{1}{r^\nu}\frac{\partial}{\partial r}(r^\nu E) = \frac{1}{\beta}(c_1 - c_2),\tag{4}$$

where $\alpha = \dfrac{F}{RT}$ and $\beta = \dfrac{\varepsilon\varepsilon_0}{F}$.

We use the boundary condition which accounts for mirror symmetry of c_i and the initial condition corresponding to a local instantaneous source:

$$\frac{\partial}{\partial r}c_i(r, t)|_{r=0} = 0,\qquad c_i(r, t)|_{t=0} = \frac{N_\nu}{\omega_\nu r^\nu}\delta(r).$$

Parameter ω_ν equals 2, 2π or 4π for plane, cylindrical and spherical symmetry,

$$N_\nu = \omega_\nu\int_0^\infty r^\nu c_i(r, t)dr\tag{5}$$

is the normalization integral determining the number of ion pairs in the system under consideration, $\delta(r)$ is the Dirac δ - function.

2. Successive approximation method

For solving the system of NPP equations (3), (4) we use an iteration procedure:

$$\frac{\partial}{\partial r}\left(r^\nu\left(\frac{\partial}{\partial r}c_i^{(n+1)} - \alpha z_i c_i^{(n)}E^{(n)}\right)\right) = \frac{r^\nu}{D_i}\frac{\partial}{\partial t}c_i^{(n)}.\tag{6}$$

Integral relation:

$$r^\nu E^{(n)} = \frac{1}{\beta}\int_0^r r^\nu(c_1^{(n)} - c_2^{(n)})dr\tag{7}$$

is used in our calculation instead of (4).

In much the same way as in [3] we suppose that $c_i(r, t) = 0$, for $r \geq d_i(t)$, where $d_i(t)$ is dependent on the time boundary of the diffusion region which has to be determined. In the region $(0 \leq r \leq d_i)$ c_i and $\partial c_i/\partial r$ are continuous functions of r.

As an initial approximation we use the step function

$$c_i^{(0)}(r, t) = \begin{cases} N_\nu\dfrac{\nu + 1}{\omega_\nu d_i^{\nu+1}(t)}, & 0 \leq r \leq d_i \\[2mm] 0, & d_i < r < +\infty. \end{cases}\tag{8}$$

We assume that $D_2 > D_1$ and therefore $d_2 > d_1$. Substitution of (8) into (6), (7) and integration of both sides of (6) over r in interval $(0, r)$ gives the formula for $\partial c_i^{(1)}/\partial r$. This expression has to be integrated over r in the interval (r, d_i) to get formulas for $c_i^{(1)}$:

$$c_1^{(1)} = \frac{1}{D_1}\frac{N_\nu(\nu+1)^2}{\omega_\nu}\frac{\dot{d}_1}{d_1^{\nu+2}}\frac{d_1^2 - r^2}{2} - \alpha\frac{N_\nu^2(\nu+1)}{\omega_\nu^2\beta d_1^{\nu+1}}\left(\frac{1}{d_1^{\nu+1}} - \frac{1}{d_2^{\nu+1}}\right)\frac{d_1^2 - r^2}{2}, \tag{9}$$

$$c_2^{(1)} = \begin{cases} \dfrac{N_\nu(\nu+1)^2}{D_2\omega_\nu}\dfrac{\dot{d}_2}{d_2^{\nu+2}}\dfrac{d_2^2 - r^2}{2} - \dfrac{\alpha N_\nu^2(\nu+1)}{\omega_\nu^2\beta d_2^{\nu+1}}\left(\dfrac{1}{d_1^{\nu+1}} - \dfrac{1}{d_2^{\nu+1}}\right)\dfrac{r^2}{2} + A_\nu, \ 0 \le r \le d_1 \\[2em] \dfrac{N_\nu(\nu+1)^2}{D_2\omega_\nu}\dfrac{\dot{d}_2}{d_2^2}\dfrac{d_2^2 - r^2}{2} + \begin{cases} \dfrac{\alpha N_\nu^2}{\omega_\nu^2\beta}\left(\dfrac{d_2^{1-\nu} - r^{1-\nu}}{1-\nu} - \dfrac{d_2^2 - r^2}{2d_2^{\nu+1}}\right), \nu = 2, 3 \\[1.5em] \dfrac{\alpha N_1^2}{2\pi^2\beta d_2^2}\left(\ln\dfrac{d_2}{r} - \dfrac{d_2^2 - r^2}{2d_2^2}\right), \quad \nu = 1 \end{cases} \ \ d_1 \le r \le d_2 \end{cases}$$

$$A_1 = \frac{\alpha N_1^2}{2\pi^2\beta d_2^2}\ln\left(\frac{d_2}{d_1}\right), \quad A_\nu = \frac{\alpha N_\nu^2(\nu+1)^2}{2\omega_\nu^2\beta d_2^{\nu+1}(\nu-1)}(d_2^{1-\nu} - d_1^{1-\nu}), \quad \nu = 2, 3.$$

Substitution of (9) into the normalization condition (5) gives a nonlinear ordinary differential equation for determination $d_i(t)$:

$$\dot{d}_1(t) = D_1\frac{\nu+3}{d_1}\left(1 + \frac{\alpha N_\nu}{\omega_\nu\beta(\nu+3)}d_1^2\left(\frac{1}{d_1^{\nu+1}} - \frac{1}{d_2^{\nu+1}}\right)\right), \tag{10}$$

$$\dot{d}_2(t) = D_2\frac{\nu+3}{d_2}\left(1 - \frac{\alpha N_\nu(\nu+1)(d_2^2 - d_1^2)}{2\omega_\nu\beta(\nu+3)(d_2^{\nu+1})}\right).$$

The formulas (9) with $\dot{d}_i$ expressed in terms of d_i using (10) can be substituted into (6), (7) to make the next iteration. But here we restrict our consideration to the first approximation.

An analysis of equations (10) shows, that the asymptotic behavior of $d_i(t)$ for plane and spherical symmetry is described by the formulas:

$$d_i^{(s)}(t) = a_i^{(s)}\sqrt{t} + b_i^{(s)}t + o(t^{3/2}) \text{ small } t; \ d_i^{(l)}(t) = a_i^{(l)}\sqrt{t} + b_i^{(l)} + o(t^{-1/2}) \text{ large } t, \tag{11}$$

with a_i, b_i given by the formulas:

$$a_i^{(s)} = \sqrt{6D_i}, \quad a_i^{(l)} = \sqrt{\frac{12D_1 D_2}{D_1 + D_2}},$$

$$b_1^{(s)} = \frac{\alpha N_0}{3\beta}D_1\left(1 - \sqrt{\frac{D_1}{D_2}}\right), \quad b_2^{(s)} = \frac{\alpha N_0}{6\beta}(D_1 - D_2), \tag{12}$$

$$b_1^{(l)} = -\frac{3\alpha N_0}{8\beta}\frac{(D_2 - D_1)(D_2 + 3D_1)}{(D_1 + D_2)^2}, \quad b_2^{(l)} = \frac{3\alpha N_0}{8\beta}\frac{(D_2 - D_1)(3D_2 + D_1)}{(D_1 + D_2)^2}$$

for plane geometry and

$$a_i^{(s)} = \sqrt{\frac{20D_1 D_2}{D_1 + D_2}}, \quad b_1^{(s)} = -b_2^{(s)} = \frac{100}{9}\frac{\pi\beta}{\alpha N_2}\frac{D_1 D_2(D_2 - D_1)(13D_1 - D_2)}{(D_1 + D_2)^3}, \tag{13}$$

$$a_i^{(l)} = \sqrt{10D_i}, \quad b_1^{(l)} = \frac{\alpha N_2}{20\pi\beta}\left(1 - \left(\frac{D_1}{D_2}\right)^{\frac{3}{2}}\right), \quad b_2^{(l)} = \frac{3\alpha N_2}{40\pi\beta}\left(1 - \frac{D_1}{D_2}\right)$$

for spherical geometry. In the case of cylindrical symmetry the system (10) has the analytical solution: $d_i(t) = \sqrt{8D_{i\,\text{eff}}t}$, where

$$D_{1\,\text{eff}} = D_1(2 - D_{2\,\text{eff}}/D_2) \tag{14}$$

and $D_{2\,\text{eff}}$ is the positive solution of the equation

$$D_{2\,\text{eff}}^2 - D_2\left(1 - \frac{\alpha N_1}{8\pi\beta}\left(1 + \frac{D_1}{D_2}\right)\right)D_{2\,\text{eff}} - \frac{\alpha N_1}{4\pi\beta}D_1 D_2 = 0. \tag{15}$$

Substitution of (11)-(13) into (9) and (7) enables the formulas describing the time variation of concentration profiles and the corresponding electric field to be obtained.

3. Effective coefficients of diffusion

If electric field E is not taken into account in equations (10) these equations take the form: $\dot{d}_i\,d_i = (\nu + 3)D_i$. We use a similar formula to define the effective diffusion coefficient in the general case:

$$D_{i\,\text{eff}} = \frac{\dot{d}_i d_i}{\nu + 3} \tag{16}$$

with d_i obeying the differential equations (10). The time - dependence of $D_{i\,\text{eff}}$ is determined by the time - dependence of d_i.

Asymptotic behavior of $D_{i\,\text{eff}}$ can be obtained by substitution of (11)-(13) in (16). It is easy to show that in plane geometry $D_{i\,\text{eff}}(t \to 0) = D_i$ and $D_{i\,\text{eff}}(t \to \infty) = 2D_1D_2/(D_1 + D_2)$. This means that at the beginning of the diffusion process the ion with intrinsically higher diffusion coefficient tends to move ahead. However, this leads to violation of local electroneutrality and the resulting space charge creates an electric field that pulls on the slower ion and retards the faster one until they have the same effective diffusion coefficient in accordance with (2). But in spherical geometry $D_{i\,\text{eff}}(t \to 0) = 2D_1D_2/(D_1 + D_2)$ and $D_{i\,\text{eff}}(t \to \infty) = D_i$, whereas in cylindrical geometry $D_{i\,\text{eff}}$ does not depend on t and is defined by the formulae (14), (15). These analytical results were confirmed by comparison with the data of direct numerical solution of NPP equations. It is possible that the time-dependencies of ionic diffusion coefficients differ from one another because of the difference in lateral variation of electric fields created by charge distributions of different symmetry: the electric field of plane charge distribution does not depend on a distance whereas the fields of cylindrical and spherical charge distributions decrease with a distance in $1/\rho$ and $1/r^2$ fashion.

4. Conclusions

The analysis of the Nernst-Planck and Poisson equations shows that the electric field which is caused by violation of local electroneutrality changes the diffusion rates of ions, and the time-variation of the effective diffusion coefficients depend on the symmetry properties of the problem under consideration.

References

[1] J.S.Newman, Electrochemical Systems, N.J.: Prentice-Hall, Englewood Cliffs, 1973.
[2] R.P.Buck, Kinetics of bulk and interfacial ionic motion: microscopic bases and limits for Nernst-Planck equation applied to membrane system. *J.Mem.Sci.***17**(1984)1-62.
[3] N.M.Beljaev, A.A.Rjadno, Metod teorii teploprovodnosty. M.: Vyshaja Shcola, 1982.

Non-Linear Electromagnetic Systems
V. Kose and J. Sievert (Eds.)
IOS Press, 1998

A New Modeling of Vector Magnetic Properties for Magnetic Field Analysis

Masato ENOKIZONO and Naoya SODA
Dept. of Electrical and Electronic Engineering, Faculty of Engineering,
Oita University, 700 Dannoharu, Oita, 870-11, Japan

Abstract. This paper presents a new modeling of vector magnetic properties in an arbitrary direction considering alternating hysteresis of grain-oriented silicon steel sheets. We define the reluctivities in a new way, calculating from data obtained by two-dimensional magnetic measurement. Moreover, our new expression is introduced into a finite element formulation, and is applied to a simple anisotropic magnetic field problem. As a result, it is shown that our expression is applicable generally to anisotropic problems considering the hysteresis phenomena.

1. Introduction

The study of soft magnetic materials commonly used in rotating machines and three-phase transformers is very important for saving energy. In order to save the energy, we have to grasp correct behaviors of B- and H-vectors in those core materials, and improve the numerical field analysis taking the properties into consideration. Furthermore, the optimum design for the construction of the electrical machines and apparatus must be produced.

Recently, a technique for measuring the magnetic field strength H and the magnetic flux density B as vector quantities has been developed, and the vector magnetic properties, so called two-dimensional magnetic properties, have been made clear for some kinds of magnetic materials[1]. It is therefore necessary to develop the effective expression of the vector properties in the magnetic field analysis. In the numerical approach to investigate the properties, some methods have been proposed. However, the magnetic field analysis considering hysteresis under the alternating flux conditions has not yet been built up. Accordingly, we suggest a new expression considering the alternating hysteresis in finite element formulations. We carry out the magnetic field analysis of a single-phase transformer core model, using this expression. As a result, it is shown that our new method is generally applicable to the anisotropic magnetic field problems.

2. Definition

In our conventional magnetic field analysis, the vector magnetic properties of the arbitrary direction under the alternating flux condition were expressed by reluctivity tensor in the following equation:

$$\begin{Bmatrix} H_x \\ H_y \end{Bmatrix} = \begin{bmatrix} v_x & 0 \\ 0 & v_y \end{bmatrix} \begin{Bmatrix} B_x \\ B_y \end{Bmatrix}, \quad (1) \qquad v_x = f(B, \theta_B) \ , \ \ v_y = f(B, \theta_B). \quad (2)$$

The reluctivity v_x and v_y depends on magnetic flux density B and an inclination angle θ_B as shown in Eq. (2). However, this expression was unable to express the alternating hysteresis [2].

Accordingly, we define the relationship between B- and H-vectors considering the alternating hysteresis as follows:

$$H_x = v_{xr}B_x + v_{xi}\frac{\partial B_x}{\partial t}, \qquad H_y = v_{yr}B_y + v_{yi}\frac{\partial B_y}{\partial t}, \qquad (3)$$

where, v_{xr}, v_{xi}, v_{yr} and v_{yi} are obtained from the measurement data. In this case, the data must be measured under the control of complete sinusoidal flux condition by two-dimensional excitation [3]. Therefore, B_x and B_y are expressed as follows:

$$B_x = B_{sxm}\sin\omega t, \qquad B_y = B_{sym}\sin\omega t, \qquad (4)$$

where, B_{sxm} and B_{sym} are each maximum values of B_x and B_y. We express H_x and H_y including the third higher harmonic by the following equations:

$$\begin{cases} H_x = A_1\cos(\omega t + \alpha_1) + A_2\cos(3\omega t + \alpha_2) \\ H_y = B_1\cos(\omega t + \beta_1) + B_2\cos(3\omega t + \beta_2) \end{cases}, \qquad (5)$$

where, A_1, A_2, B_1, B_2, α_1, α_2, β_1 and β_2 are constants obtained from the measurement data. Thus, the reluctivity coefficients v_{xr}, v_{xi}, v_{yr} and v_{yi} are expressed as follows:

$$\begin{cases} v_{xr} = k_{xr1} + k_{xr2}B_x^2 \quad , \quad v_{xi} = k_{xi1} + k_{xi2}\left(\frac{\partial B_x}{\partial t}\right)^2 \\ v_{yr} = k_{yr1} + k_{yr2}B_y^2 \quad , \quad v_{yi} = k_{yi1} + k_{yi2}\left(\frac{\partial B_y}{\partial t}\right)^2 \end{cases}, \qquad (6)$$

where,

$$\begin{cases} k_{xr1} = -\dfrac{A_1\sin\alpha_1 + 3A_2\sin\alpha_2}{B_{sxm}} \quad , \quad k_{xr2} = \dfrac{4A_2\sin\alpha_2}{B_{sxm}^3} \\[2mm] k_{xi1} = \dfrac{A_1\cos\alpha_1 - 3A_2\cos\alpha_2}{B_{sxm}} \quad , \quad k_{xi2} = \dfrac{4A_2\cos\alpha_2}{(\omega B_{sxm})^3} \\[2mm] k_{yr1} = -\dfrac{B_1\sin\beta_1 + 3B_2\sin\beta_2}{B_{sym}} \quad , \quad k_{yr2} = \dfrac{4B_2\sin\beta_2}{B_{sym}^3} \\[2mm] k_{yi1} = \dfrac{B_1\cos\beta_1 - 3B_2\cos\beta_2}{B_{sym}} \quad , \quad k_{yi2} = \dfrac{4B_2\cos\beta_2}{(\omega B_{sym})^3} \end{cases}. \qquad (7)$$

Figure 1 shows the relationship between the coefficient k_{xr1}, k_{xr2}, k_{xi1}, k_{xi2}, k_{yr1}, k_{yr2}, k_{yi1} and k_{yi2}, the magnetic flux density B and inclination angle θ_B. The specimen used in this study was a grain-oriented steel sheet (23ZDKH90 produced by NSC, 0.23 mm thickness). As space is limited here, we omit the detailed description of the vector magnetic measurement apparatus [3].

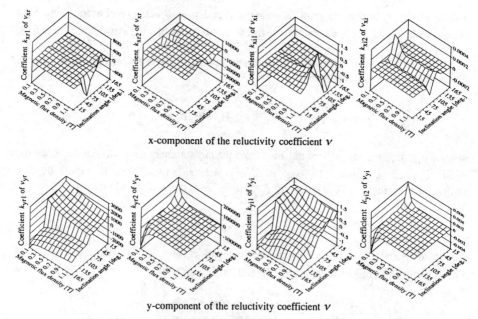

x-component of the reluctivity coefficient ν

y-component of the reluctivity coefficient ν

Fig. 1. Relationship between the coefficients of the reluctivity coefficient ν,
the magnetic flux density and the inclination angle.

Substituting (3) into Maxwell's equations in a two-dimensional quasi-static magnetic field, we can obtain the following equation with the vector potential A $(= A_z)$:

$$\frac{\partial}{\partial x}\left(\nu_{yr}\frac{\partial A}{\partial x}\right) + \frac{\partial}{\partial y}\left(\nu_{xr}\frac{\partial A}{\partial y}\right) + \frac{\partial}{\partial t}\left\{\frac{\partial}{\partial x}\left(\nu_{yi}\frac{\partial A}{\partial x}\right) + \frac{\partial}{\partial y}\left(\nu_{xi}\frac{\partial A}{\partial y}\right)\right\} = -J_0, \qquad (8)$$

where, J_0 is the exciting current density. We can carry out the non-linear magnetic field analysis considering an alternating hysteresis in an arbitrary direction.

3. Results and Discussion

We applied the expression presented here to a single-phase transformer core model under the alternating flux condition. This core model was made by cutting off the unnecessary part from rectangular sheets, as shown in Fig. 2. The core was constructed by the grain-oriented steel sheet, and its number of laminations was equal to 40. Since the core is symmetric, a quarter of the region was analyzed. In this analysis, the leakage flux was neglected.

Figure 3 shows the loci of B- and H-vectors calculated by using our new expression. Figure 4 shows the loci of B- and H-vectors at the corner region of the core. As shown in Fig. 3, the H-vector was especially large at the corner region of the core. Furthermore, it can be seen that an alternating hysteresis occurs as shown in Fig. 4. It was shown that our new method was applicable to the analysis of common electrical machines and apparatus.

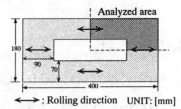

←——→ : Rolling direction   UNIT: [mm]

Fig. 2. The single-phase transformer core model.

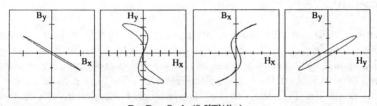

(a) Loci of B-vector   (b) Loci of H-vector

Fig. 3. Distribution of the calculated B- and H-vectors.

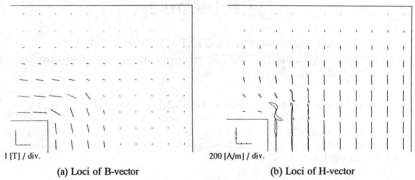

B_x, B_y : Scale (0.5[T]/div.)

H_x, H_y : Scale (100[A/m]/div.)

Fig. 4. Loci of B- and H-vectors at the corner region of the core.

4. Conclusion

In this paper, a new modeling of the vector magnetic properties for non-linear analysis has been presented. We can analyze a single-phase transformer core model considering an alternating hysteresis in an arbitrary direction. Moreover, it can be said that the presented method is very useful to analyze electrical machines and apparatus under the alternating flux condition.

References

[1] M. Enokizono, T. Todaka, S. Kanao and J. Sievert, "Two-dimensional Magnetic Properties of Silicon Steel Sheet Subjected to a Rotating Field", IEEE Trans. on Magnetics, Vol. 29, No. 6, pp.3550-3552, 1993.

[2] M. Enokizono and N. Soda, "Direct Magnetic Loss Analysis by FEM Considering Vector Magnetic Properties", Proceedings of the 11th Conference on the Computation of Electromagnetic Fields, PA2-2, 1997.

[3] M. Enokizono, G. Shirakawa and J. D. Sievert, "Anomalous Anisotropy and Rotational Magnetic Properties of Amorphous Sheet", Journal of Magnetism and Magnetic Materials 112, pp.195-199, 1992.

A Large Deformation Theory of Solids Subjected to Electromagnetic Forces and Its Application to Magnetostatic Problems

Isoharu NISHIGUCHI and Kazuyoshi HASEYAMA

Kanagawa Institute of Technology, 1030 Shimoogono, Atsugi-shi, Kanagawa-ken, Japan

Abstract. A large deformation theory of deformable solids is proposed by the authors and the explicit tangent matrix for the coupled problem is obtained when the electromagnetic force is given as the divergence of a tensor. In the present paper, the magnetoelastic buckling behavior of a ferromagnetic beam plate is analyzed based on the theory.

1. Introduction

When the deflection of a structure under an electromagnetic field is not small, it is impossible to obtain the displacement field and the electromagnetic field independently. Both fields should be solved simultaneously as a coupled problem. For example, for a component of a fusion reactor, the deflection under disruption predicted by a coupled calculation is smaller than the one by an uncoupled calculation. In this case, the deformation has the effect of reducing the electromagnetic force caused by eddy currents in a disruption and this effect is known as magnetic dumping [1]. The buckling of ferromagnetic materials under a magnetic field [2] is another example of coupled problems. This phenomenon cannot be predicted without considering the change of the electromagnetic forces caused by the deformation.

A considerable number of papers have been published to predict these problems. As far as the authors know, however, most of existing theories use plate or shell approximations and elastic material is assumed for the material properties. Though these ad hoc assumptions may have been adequate for previous problems, we believe it is important to treat these problems from a more general point of view when thinking of the expanding field of applications of electromagnetic phenomena in engineering.

A coupled theory [3][4] has therefore been developed by the authors and the features of the theory are as follows:

-This is a large deformation theory based on the Lagrange description and satisfies the condition known as objectivity or material frame-indifference in continuum mechanics [5]. By this, the distinction between deformed and undeformed configuration can be treated rigorously.

-The constitutive equation for the deformation field is not limited to elasticity and general loading path dependency of plasticity or inelasticity as is proposed in [6] can be treated.

In the previous paper [4], we reduced our theory to the inelastic deformation under the magnetostatic field and we found the explicit tangent matrix for the coupled problem when the electromagnetic force is given as the divergence of a tensor such as the Maxwell stress tensor. In the present paper, the magnetoelastic buckling behavior of a ferromagnetic beam plate is analyzed based on the theory.

2. Finite Element Formulation

Weak forms of Maxwell's equations and the balance of momentum, both in a fixed reference configuration constitute the governing equations for our theory [4].

When the updated Lagrangian formulation (ULF) is employed, the rate form of Ampere's law can be expressed as

$$\int_{B(t)} L_v(\mathbf{\eta}) \cdot (\nabla \times \mathbf{w}) dv - \int_{B(t)} \tilde{\imath} \cdot \mathbf{w} dv = 0 \tag{1}$$

where $\mathbf{w}$ is a vector function, $\mathbf{\eta}$ is the magnetic intensity in the co-moving frame, $L_v(\mathbf{\eta})$ is the Lie derivative of $\mathbf{\eta}$ with respect to the velocity field $\mathbf{v}$ and $\tilde{\imath}$ is given by

$$\tilde{\imath} = \frac{1}{J} L_v(J\imath) \tag{2}$$

where $J = \det \mathbf{F}$, $\mathbf{F}$ is the deformation gradient, and $\imath$ the conduction current density.

In order to determine the electromagnetic field, we need a constitutive equation between the magnetic intensity and the magnetic-flux density. A simple equation was proposed in [4] and is employed in our calculations shown later.

Using the magnetic vector potential $\mathbf{a}$ at current configuration as a primary unknown, the incremental equation for the Finite Element Analysis is obtainable:

$$\mathbf{k}^{EM} \Delta \mathbf{a} + \mathbf{l}^{EM} \Delta \mathbf{u} = 0 \tag{3}$$

where $\Delta \mathbf{a}$ and $\Delta \mathbf{u}$ are vectors of nodal incremental values of magnetic vector potential and the displacements, $\mathbf{k}^{EM}$ and $\mathbf{l}^{EM}$ are the corresponding coefficient matrices, respectively. In equation (3), the current term is neglected for simplicity.

As for the linear momentum balance, the following discretized equation was derived in [4] for the ULF.

$$\left(\mathbf{k}^{DS} + \mathbf{k}^{DG} \right) \Delta \mathbf{u} = \Delta \mathbf{f}^E \tag{4}$$

where $\mathbf{k}^{DS}$ is the conventional stiffness matrix and $\mathbf{k}^{DG}$ is the geometric matrix.

The incremental nodal force vector due to electromagnetism, $\Delta \mathbf{f}^E$, can be expressed by

$$\Delta \mathbf{f}^E = -\mathbf{l}^{DA} \Delta \mathbf{a} - \mathbf{k}^{DU} \Delta \mathbf{u} . \tag{5}$$

Note that $\Delta \mathbf{f}^E$ depends not only on the incremental vector potential but also on the incremental displacement.

Equations (4) and (5) can be put in the form

$$\mathbf{k}^D \Delta \mathbf{u} + \mathbf{l}^D \Delta \mathbf{a} = 0 \tag{6}$$

corresponding to equation (44) in [4], where $\mathbf{k}^D = \mathbf{k}^{DS} + \mathbf{k}^{DG} + \mathbf{k}^{DU}$ and $\mathbf{l}^D = \mathbf{l}^{DA}$.

To solve the large deformation under magnetostatic condition, a system of equations can be obtained by assembling all elements whose tangent matrices appear in (3) and (6) with some components of $\Delta \mathbf{a}$ and $\Delta \mathbf{u}$ are prescribed. The unknown components of $\Delta \mathbf{a}$ and $\Delta \mathbf{u}$ can be obtained simultaneously with each increment and the instability may be detected using the global tangent stiffness matrix.

Alternatively, an iterative solution may be employed, e.g.,

$$For \ p = 1, 2, 3, \cdots$$
$$\mathbf{K}^{EM} \Delta \mathbf{a}_{(p)} = -\mathbf{L}^{EM} \Delta \mathbf{u}_{(p-1)} \quad with \ \Delta \mathbf{u}_{(0)} = \mathbf{0} \tag{7}$$
$$\left(\mathbf{K}^{DS} + \mathbf{K}^{DG} \right) \Delta \mathbf{u}_{(p)} = \mathbf{F}^{E}_{(p)} - \mathbf{F}^{E(n-1)} \tag{8}$$

where p is the iteration number, and the $\mathbf{K}$'s and $\mathbf{L}$ are the assembled matrices of $\mathbf{k}$'s and $\mathbf{l}$, respectively. $\mathbf{F}^{E}_{(p)}$ is the nodal magnetic force at iteration number p in the n-th step and $\mathbf{F}^{E(n-1)}$ is the nodal magnetic force at the previous step. These forces are obtainable   by   assembling the contributions from all elements as is proposed in [7].

All matrices   and $\mathbf{F}^{E}_{(p)}$ in this scheme may be calculated with respect to the configuration at $\mathbf{x}^{(n-1)} + \alpha \Delta \mathbf{u}_{(p-1)}$ with the value of the magnetic potential of $\mathbf{a}^{(n-1)} + \alpha \Delta \mathbf{a}_{(p-1)}$ where $\mathbf{x}^{(n-1)}$ is the location at (n-1)-th step and $\alpha = 0.5$ as a natural choice.

3. Numerical Example

To examine the present theory, the elastic buckling   of   a soft ferromagnetic plate in a uniform transverse field has been investigated.   This problem has been analyzed from the 1960's by many   researchers based on the thin plate approximations of the deformation field and some assumptions   on the magnetic field [8]-[11].   We note that such *ad hoc* assumptions are not needed in our analysis and can be applied to the deformation analysis of an arbitrary configuration directly.

The critical magnetic flux B_{cr} of an elastic plate in a homogeneous transverse magnetic field was estimated by Moon and Pao as follows [8]:

$$B_{cr} = \frac{\pi^{3/2}}{\sqrt{192(1 - v^2)}} \left(\frac{t}{L} \right)^{3/2} \mu_0 E \tag{9}$$

where t and L are the thickness and length of the plate, respectively, μ_0 the permeability of vacuum, E the Young's modulus, and v the Poisson's ratio.

Note that this equation was derived under the assumption that the plate was infinitely wide and long.   The theoretical values are 100-200 percent larger than the experimental results [10] and several investigators   have tried to explain this discrepancy [8]-[11].

a) Finite Element Model

The 20-node isoparametric brick elements are used in the analysis.   In the magnetic field analysis, 490 elements and 2576 nodes are used.   The permeability is   set   to be isotropic with $\mu / \mu_0 = 10^4$.

For the deformation field, the plate is modelled by   36 elements and 264 nodes.   The length of the plate, $L = 100 \, mm$, the width $b = 20 \, mm$, and the thickness, $t = 1 \, mm$.   For the constitutive equation hypoelasticity with $E = 195000 \, MPa$ and $v = 0.3$ is assumed. The Jaumann rate is used as the objective stress rate.   The nodes at one end   of the plate are constrained   to simulate a cantilever.

b) Results

The critical magnetic flux of the plate was calculated by the present method. In the analysis, the three-field formulation proposed in [12] and [13] was employed to improve the accuracy of the magnetic flux at the plate surface. The critical magnetic flux was obtained as the value at which the iterations based on the initial configuration become unstable. For the initial configuration, both an undeformed configuration and one with initial imperfection comparative to the displacement by its own weight in the transverse direction were employed. As shown in Table 1, the prediction of the FEM and the one by equation (9) is comparable. On the other hand, the value decreases considerably when the initial imperfection is taken into account in the calculation and this is consistent with the experimental results.

Table 1   Critical magnetic flux

Equation (9)	0.21 T
FEM without imperfection	0.23 T
FEM with imperfection	0.09 T

Acknowledgement

The support of the Grant-in-Aid for Scientific Research (c) from the Japan Ministry of Education, Science, Sports and Culture is acknowledged.

References

[1] Y. Yoshida, K.Miya, K.Demachi, M.Kurokawa, Magnetic damping effects on vibration of conductive shells, *Int. J. Applied Electromagnetics in Materials* 4 (1993) 1-11.

[2] F.C. Moon, Magneto-solid mechanics, John Wiley & Sons: New York, 1984, pp.131-175.

[3] I.Nishiguchi and J. Yamamoto, A large deformation theory of inelasticity subjected to electromagnetic forces. In: T. Honma (ed.), Simulation and design of applied electromagnetic systems. Elsevier Science Publisher B.V., 1994, pp.105-108.

[4] I. Nishiguchi and M. Sasaki, A large deformation theory of solids subject to electromagnetic loads, *IEEE Transactions on Magnetics* 30 (1994) 3272-3275.

[5] C. Truesdell and W. Noll, The non-linear theories of mechanics, in Encyclopedia of Physics, vol. III/3, Springer-Verlag, 1965.

[6] I. Nishiguchi, T.-L. Sham and E. Krempl, A finite deformation theory of viscoplasticity based on overstress: part I - constitutive equations, *ASME Journal of Applied Mechanics* 57 (1990) 548-552.

[7] A. Kameari, Local force calculation in 3D FEM with edge elements, *Int. J. Applied Electromagnetics in Materials*. 3 (1993) 231-240.

[8] F. C. Moon and Yih-Hsing Pao, Magnetoelastic buckling of a thin plate, *ASME Journal of Applied Mechanics* 35 (1968) 53-58.

[9] J. M. Dalrymbple, M. O. Peach   and G. L. Viegelahn, Magnetoelastic buckling of thin magnetically soft plates in cylindrical mode, *ASME Journal of Applied Mechanics* 41 (1974) 145-150.

[10] K. Miya, K. Hara, K. Someya, Experimental and theoretical study on magnetoelastic buckling of a ferromagnetic cantilevered beam-plate, *ASME Journal of Applied Mechanics* 45 (1978) 355-360.

[11] A.A.F. van de Ven, The influence of finite specimen dimensions on the magneto-elastic buckling of a cantilever. In: G.A. Maugin (ed.), The mechanical behavior of electromagnetic solid continua. Elsevier Science Publisher B.V., 1994, pp.421-426.

[12] I. Nishiguchi and H. Matsuzawa, Three-field formulation and its application on magnetostatic problems. In: A.J. Moses and A. Basak (ed.), Nonlinear electromagnetic systems. IOS Press, Amsterdam, 1996, pp.560-563.

[13] I. Nishiguchi, H. Matsuzawa, M. Sasaki, Comparative study of magnetostatics using three-field formulation and conventional vector potential formulations., *IEEE Transactions on Magnetics* 33 (1997) 1243-1246.

Non-Linear Electromagnetic Systems
V. Kose and J. Sievert (Eds.)
IOS Press, 1998

A DC Solution Approach for Large Nonlinear Circuits

Gabriel Stefan POPESCU

Technical University Iasi, Bd. Copou Nr. 11, 6600 Iasi, Romania

Abstract. An event-driven approach to achieve the DC solution of the transistor level modelled VLSI circuits is presented. The quiescent state obtained as the transient solution of an associated circuit with the original one is assimilated with the DC solution. The presented approach bypasses the inefficiencies of classical Newton-Raphson based methods, expensive to use with large circuits, by using an event-oriented way equivalent to the decoupling of the circuit equations.

1. Introduction

The DC solution remains a difficult problem for the transistor level VLSI circuit simulation. This is because of the great complexity of the non-linear equations which must be solved to obtain the circuits solution. Time domain simulation implies solving a nonlinear differential system of equations with the initial solution at time $t = 0$ which is often the DC operating point of the circuit.

Many time domain simulation techniques were developed to seek and exploit the latent character of electronic circuits. These are mostly discrete-event based methods (event-driven methods) [1]-[4], or relaxation based methods [5]. Their main purpose is the reduction of the computational resource consumption during the simulation. Event-driven techniques for simulating the time domain behaviour of large circuits define a set of events and a way to approximate the time step h to each event.

An event-driven algorithm developed for the time domain simulation is the core of the present approach [1]. The DC solution of the circuit is obtained as the overall quiescent state of the transient simulation of an associated circuit. The approach is equivalent with the Newton-Raphson algorithm from the point of view of the convergence proprieties while the computational resources requirements are strongly reduced.

2. The Discrete-Event Approach

In this section a brief presentation of the event-driven approach is made. To explain the integration method the following state equations are used:

$$\dot{x} = A \cdot x + \ddot{u}$$
$$x(0) = x_0 \tag{1}$$

Here the vector $x(t) \in \mathbf{R}^m$ represents the state vector, and $\ddot{u}(t) \in \mathbf{R}^m$ is the equivalent source vector. The matrix $A \in \mathbf{R}^{m \times m}$, where m is the system dimension, and the source vector $\ddot{u}$

constitute the model of the linear(ized) circuit. Using the backward Euler formula (BE) to solve the system of differential equations (1) the problem becomes:

$$(1 - h \cdot A) \cdot x_{n+1} = I_n \; ; \; I_n = x_n + h \cdot \tilde{u}_{n+1} \tag{2}$$

In equation (2) $x_{n+1} = x(t_{n+1})$, $\tilde{u}_{n+1} = \tilde{u}(t_{n+1})$, and h is the time step. The solution $x_{n+1i} \; ; \; i = 1,...,m$ can be asymptotically approximated as:

$$x_{n+1i} = \left\{ I_{ni} + h \cdot \left(\sum_{p=1, p \neq i, a_{ip} \neq 0}^{m} a_{ip} \cdot I_{np} - I_{ni} \cdot \sum_{p=1, p \neq i, a_{ip} \neq 0}^{m} a_{pp} \right) \right\} \cdot \left(1 - h \cdot \sum_{p=1, a_{ip} \neq 0}^{m} a_{pp} \right)^{-1} \; ; \; i = 1,...,m \tag{3}$$

The method is not fundamentally linked with a circuit partition. Even so, the circuit partitioning significantly speeds up the simulation [1] – [4].

The event-driven approach consists of two stages. In the first stage, *scheduling*, the next event is calculated. In the second stage, *rescheduling*, the state of the network corresponding to the scheduled event is calculated. The scheduling and rescheduling stages are performed only for non-quiescent variables as long as new events do not appear in the signal source vector.

The discretization of the amplitude domain of the state variables and the definition of the events associated with the network is made using (3) and some assertion on the network [1]. Equation decoupling can be solved using (3) by associating every state variable x_i with its own time step h_i.

For a given value of a state variable x_{n+1i} at the moment $t_{n+1} = t_n + h_i$ the time step evaluated from (3) is:

$$h_i = \frac{\Delta x_{n+1i}}{\tilde{x}_{ni} + \Delta x_{n+1i} \cdot d_i} \; ; \; i = 1,...,m \tag{4}$$

where $\Delta x_{n+1i} \doteq x_{n+1i} - x_{ni}$, $\tilde{x}_{ni} = \sum_{j=1, a_{ij} \neq 0}^{m} a_{ij} \cdot x_{nj} + \tilde{u}_{n+1i}$ and $d_i \doteq \sum_{j=1, a_{ij} \neq 0}^{m} a_{ij}$. It can be easily shown

that $d_i = \sum_{j=1, a_{ij} \neq 0}^{m} a_{ij} = \sum_{j=1, a_{ij} \neq 0}^{m} p_j$ where p_j are the pole values of a linear subsystem which is built around the equation i augmented with all the equations j for which $a_{ij} \neq 0$. For any circuit which is not an oscillator the equivalent model in the vicinity of quiescence is absolute stable. This implies that $d_i < 0$. During the scheduling process the singularity point $-\frac{\tilde{x}_{ni}}{d_i}$ of (4) gives a clue about the next amplitude step value of the state variable x_i and about the corresponding time step h_i. The time step value corresponding to every state variable are ordered in a event queue. The state variable x_i reaches the quiescent state when the amplitude step $|\Delta x_{n+1i}|$ becomes smaller than a fixed threshold value $|\Delta x_{n+1i \, min}|$.

3. The Associated Circuit

The DC solution of the circuit is calculated using the transient simulation of an associated circuit. The most general formulation of a system of nonlinear differential equations which models a nonlinear dynamic circuit is:

$$C(x(t), u(t))\dot{x}(t) = f(x(t), u(t))$$
$$x(0) = x_0 \tag{5}$$

where $u(t) \in \mathbb{R}^r$ is the independent source vector, $C : \mathbb{R}^m \times \mathbb{R}^r \rightarrow \mathbb{R}^{m \cdot m}$ is such as that $C(x, u)^{-1}$ exists and is uniformly bounded with respect to x, u and $f : \mathbb{R}^m \times \mathbb{R}^r \rightarrow \mathbb{R}^m$ is globally Lipschitz continuous with respect to x for all $u(t) \in \mathbb{R}^r$.

The DC solution of the circuit is the solution of (5) with $\dot{x}(t) = 0$ and $u(t) = U$, where U is the independent signal source vector containing the DC values of it. This yields to:

$$f(x, u) = 0 \tag{6}$$

To exploit the benefits of the event-driven approach, taking into account the previous discussion, (7) can be derived from (5) with C the identity matrix and $u(t) = \begin{cases} \alpha t \ ; \ t \in (0, t_0] \\ U \ ; \ t > t_0 \end{cases}$:

$$\dot{x}(t) = f(x, u)$$
$$x(0) = x_0 \tag{7}$$

were $\alpha \in \mathbb{R}^m$ and $u(t)$ is continuous.

4. The DC Solution as the Quiescent State of the Associated Circuit

The DC solution of the circuit is the quiescent solution of (7) for $t \geq t_0$. Equation (7) can be modified as follows:

$$\ddot{x} = \frac{\partial f(x, u)}{\partial x} \dot{x} + \frac{\partial f(x, u)}{\partial u} \dot{u}$$
$$\dot{x}(0) = 0 \tag{8}$$

Using the event-driven technique to solve (8) with Forward Euler (FE) as the rescheduling formula, let h_i be the next time step. For the non-quiescent state variable the updated value of $\dot{x}_j$ at t_{n+1} is:

$$\dot{x}_{n+1\,j} = \dot{x}_{n\,j} + h_i \left(\frac{\partial f_j(x_n, u_n)}{\partial x} \dot{x}_n + \frac{\partial f_j(x_n, u_n)}{\partial u} \dot{u}_n \right) \ ; \ j \in \{1, ..., m ; \ \dot{x}_j \ non\text{-}quiescent \} \tag{9}$$

Supposing the algorithm converges and it is close to the solution, then $|\dot{x}_{n\,i}| \gg |\dot{x}_{n+1\,i}| \approx 0$. Under this condition, from (9) results:

$$\dot{x}_{n\,i} = -\frac{1}{h_i} \left(\frac{\partial f_i(x_n, u_n)}{\partial x} \right)^{-1} \left(\dot{x}_{n\,i} + \frac{\partial f_i(x_n, u_n)}{\partial u} \dot{u}_n \right) \tag{10}$$

Using this result the value of x_i at t_{n+1} is:

$$x_{n+1\,i} = x_{n\,i} - \left(\frac{\partial f_i(x_n, u_n)}{\partial x} \right)^{-1} \cdot \left(f_i(x_n, u_n) + \frac{\partial f_i(x_n, u_n)}{\partial u} \dot{u}_n \right) \tag{11}$$

For others non-quiescent variables the value at time t_{n+1} is obtained using the BE formula:

$$x_{n+1\,j} = x_{n\,j} + h_i \left\{ \dot{x}_{n\,j} + h_i \left(\frac{\partial f_j(x_n, u_n)}{\partial x} \dot{x}_n + \frac{\partial f_j(x_n, u_n)}{\partial u} \dot{u}_n \right) \right\}; j \in \{1, \ldots, m \; ; \dot{x}_j \, non\text{-}quiescent \;, j \neq i\} \tag{12}$$

For $t > t_0$ (11) becomes:

$$x_{n+1\,i} = x_{n\,i} - \left(\frac{\partial f_i(x_n, u_n)}{\partial x} \right)^{-1} \cdot f_i(x_n, u_n) \tag{13}$$

which is the Newton-Raphson formula. This result proves that, in the vicinity of the quiescence, the presented approach is equivalent to the Newton-Raphson algorithm as far as the convergence is concerned.

5. Conclusions

The presented approach provides the DC solution of the circuit as the solution of a transient simulation. The DC solution is assimilated with the quiescent state of the overall associated circuit.

The approach avoids any complex matrix processing involved by classical algorithms while preserves their convergence performances.

The approach remains an iterative one, for every start of the algorithm at most one solution is found. The differences between multiple starts of the algorithm are the values of the elements of the vector α and the matrix C.

Even if the approach was presented in the context of an particular event-driven algorithm, the idea is not necessarily linked with it. The approach can be reformulated to be used with other event-oriented algorithms in the literature. It provides an equation decoupling solution.

References

[1]   G.S. Popescu, Event Driven Transient Simulation Using Implicit Algorithm, Proc. of The European Simulation Multiconference. ISBN:1-56555-065-X. SCS, 1989, pp. 552 -556.

[2]   A. Devgan and R.A. Rohrer. Adaptively Controlled Explicit Simulation. IEEE Trans. on CAD, vol. 13 (Jun. 1995), pp. 746-762.

[3]   J.T.J. van Eijndhoven, M. T. van Stiphout and H. W. Buurman. Multirate Integration in a Direct Simulation Method. *Proc. of The European Design Automation Conference* (Glasgow, Mar. 12-15 1990), EDAC and IEEE, pp. 306-309.

[4]   C. Viswehwariah and R.A. Rohrer. Piecewise Approximate Circuit Simulation. Digest of Technical Papers (Santa Clara, California, Nov. 5-9 1989). IEEE and ACM, pp. 248-251.

[5]   A.R. Newton and A.L. Sangiovanni-Vincentelli. Relaxation-Based Electrical Simulation. IEEE Trans. on CAD of Integrated Circuits and Systems, vol. 3 (Oct. 1984), pp. 308-340.

Non-Linear Electromagnetic Systems
V. Kose and J. Sievert (Eds.)
IOS Press, 1998

Global Simulation on Free Boundary Dendritic Growth

A. Tanaka

Department of Electrical and Information Engineering

Yamagata University, Yonezawa 992, Japan

Abstract. In this paper, we propose one numerical technique on growing dendrite. Our model is global but rather simple, and has many advantages compared with other models. We obtained many pictures of dendrites and carried out quantitative measurements. For consistency with theory and experiment, we corrected our model and obtained the required results.

1. Introduction

The problem of dendritic growth is a fascinating subject in pattern formation. There are many kinds of beautiful shapes and the mechanism of sidebranching is not yet fully understood. The dominant equation of dendritic growth is a simple diffusion equation, but one of the boundary conditions is given at the interface of the crystal. Hence the boundary condition is itself the function of time. It is called the free boundary problem. Since we know it is hard to find an analytical solution, we must solve such a problem numerically. Several kinds of models have been already proposed, but it is well known that local models are inadequate for complicated dendritic growth[1, 2]. We must, therefore, take a global model it takes a huge amount of computational time[3]. For this reason, we take a simplified global model called the Phase Field Model.

2. Definition of the model

A Phase Field Model of crystal growth has been previously proposed by R. Kobayashi. He succeeded in obtaining several shapes of bulky and dendritic crystal[4]. His scheme has some advantages as compared with others, that is (1) easy expression for the interaction among interfaces and (2) extensibility to higher dimensional simulation.

In the Phase Field Model of 2-dimensional crystal growth, two field variables, a phase field $p(x,t)$ and a temperature field $T(x,t)$ are introduced. The former variable p is an ordering parameter, and $p = 0$ and 1 means liquid and solid states respectively. The thin layer of the interface is expressed by the interval $[0,1]$. The model is derived from the conservation law of enthalpy with the following Ginzburg-Landau type free energy including m as a parameter,

$$\Phi[p; m] = \int \frac{1}{2}\varepsilon^2 |\nabla p|^2 + F(p; m)dx \tag{1}$$

$$F(p; m) = \frac{1}{4}p^4 - \left(\frac{1}{2} - \frac{m}{3}\right)p^3 + \left(\frac{1}{4} - \frac{m}{2}\right)p^2 \tag{2}$$

where ε is a small parameter representing the thickness of the interface and $|m| < 1/2$. The resulting equations are written as follows,

$$\tau\frac{\partial p}{\partial t} = -\frac{\partial}{\partial x}\left(\varepsilon\frac{\partial\varepsilon}{\partial\theta}\frac{\partial p}{\partial y}\right) + \frac{\partial}{\partial y}\left(\varepsilon\frac{\partial\varepsilon}{\partial\theta}\frac{\partial p}{\partial x}\right) + \nabla\cdot(\varepsilon^2\nabla p) + p(1-p)(p - \frac{1}{2} + m) \tag{3}$$

$$\frac{\partial T}{\partial t} = \nabla^2 T + K\frac{\partial p}{\partial t} \tag{4}$$

where K is the dimensionless latent heat. The anisotropy of the crystal is included in ε as follows.

$$\varepsilon = 1 + \delta_1 \cos 4\theta + \delta_2 \cos 8\theta. \tag{5}$$

3. The scaling law from theory and experiment

The scaling relations are main concerns in statistical physics. What determines the scaling of the system? We have to calculate the shapes of dendrite not only qualitatively but also quantitatively. In theoretical study, J. S. Langer and H. Müller-Krumbhaar have obtained the relationship $v\rho^2 \sim$ const. in determining the tip radius of needle crystal. That means we cannot predict the tip radius and tip velocity independently. This relationship has been verified in several experimental systems[5].

Let us go back to the basic equation of the crystal growth. One of the boundary conditions at the interface is as follows,

$$T_{\text{Interface}} = T_{\text{Melting}} \left(1 - \frac{\gamma}{L}\kappa\right). \tag{6}$$

This means that the temperature at the interface is slightly different from that of melting. The difference comes from the surface energy of the crystal and is proportional to the tip curvature. This relationship is well-known and called the Gibbs-Thomson relationship. Since it is derived theoretically near equilibrium, it was uncertain that it holds for dendritic growth. It was, however, verified in experiment using optical techniques[7].

4. Numerical results

In Eq.(5) we introduced a 4-fold anisotropy parameter δ_1 and 8-fold one δ_2 for consistency with experiment and varied them in our simulations. We can obtain many splendid pictures of parabolic and oscillating dendrites. They were summarized in the phase diagram in [6].

We carried out a quantitative estimate of our numerical simulations. At first, we examined the relationship between tip velocity and curvature, and the result is shown in Figure 1. It can be easily seen that our numerical result is quite different from the experimental and theoretical one, that is that the tip radius should be inversely proportional to the square root of the tip velocity.

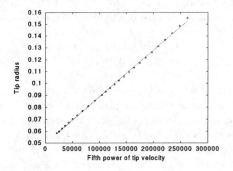

Figure 1: The relationship between tip radius and velocity. It is not consistent with the experimental result $v\rho^2 \sim$ const.

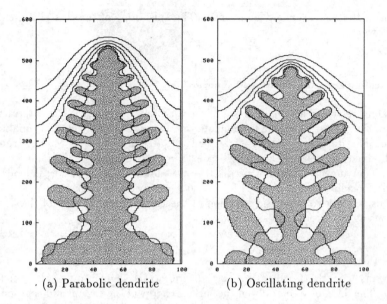

(a) Parabolic dendrite   (b) Oscillating dendrite

Figure 2: Dendrites with isothermal lines. It is clear that some isothermal lines cross the interface of dendrite.

Secondly we examined the temperature field around the dendrite. Parabolic and oscillating dendrites with isothermal lines are shown in Figure 2. Some isothermal lines are clearly shown across the interface in both cases. This means that the temperature at the interface is not constant and different from that of melting. Although we measured the difference as a function of tip curvature, our result was quite inconsistent with the Gibbs-Thomson relationship.

5. Improvement of the model

As mentioned above, since the quantitative estimate of our numerical result is quite different from the one we hoped to obtain, we must correct our model properly. However we tried several kinds of corrections, here we only explain one of them.

Although we already know the anisotropies of the crystal, δ_1, δ_2, are important, we regarded them as constant and the same for the thinness of the layer ε. Here we take them as linear functions of the degree of undercooling as follows.

$$\delta_1 = 0.16 * \Delta T - 0.14, \quad \delta_2 = 0 \tag{7}$$
$$\varepsilon = 0.02 - 0.01 * \Delta T \tag{8}$$

The resulting quantitative relationships are shown in Figure 3.

The straight lines are fitting lines estimated from experiment. Though the former result (a) is quite good, the latter one (b) shows a significant difference. We have to say that our model is still a little inadequate. But the results were rather worse when we made other corrections, e.g. the inclusion of interfacial kinetics etc. We could only obtain similar results to the original model, which were not those required. Therefore this correction is one of the candidates of the best model so far.

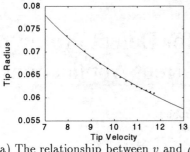

(a) The relationship between v and ρ.

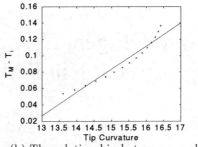

(b) The relationship between κ and ΔT_i.

Figure 3: The quantitative estimates of numerical results. The straight lines are fitting lines obtained from experiment.

6. Summary

In summary, we carried out a numerical simulation on dendritic crystal growth using the Phase Field Model. This is a simplified global model and has many advantages in numerical simulation of dendrite. We estimated our numerical results quantitatively for comparison with theory and experiment. Since we could not obtain consistent results, we corrected our model and obtained desired results. We can say that the anisotropy and the treatment of layer is important for the simulation of dendrite. We will propose a more favorable model in the near future.

Acknowledgements

We thank to T. Yokoyama, R. Kobayashi, Y.Sawada and M.Sano for fruitful discussions and suggestions.

References

[1] R. C. Brower, D. A. Kessler, J. Koplik and H. Levine, "Geometrical Approach to Moving-Interface Dynamics", Phys. Rev. Lett., **51**, 1111 (1983).

[2] E. Ben-Jacob, N. Goldenfeld, J. S. Langer and G. Schon, "Dyanamics of Interfacial Pattern Formation", Phys. Rev. Lett., **51**, 1930 (1983).

[3] Y. Saito, G. Goldbeck-Wood and H. Müller-Krumbhaar, "Dendrite Crystallization: Numerical Study of the One-Sided Model", Phys. Rev. Lett., **58**, 1541 (1987).

[4] R. Kobayashi, "Modeling and Numerical Simulations of Dendritic Crystal Growth", Phisica D, **63**, 410 (1993).

[5] A. Tanaka, Doctoral thesis, Tohoku University (in Japanese)(1991).

[6] A. Tanaka and R. Kobayashi, "Oscillating and Steady Growth of a Dendrite", Sup. Int. J. Appl. Electromag. Mat., 521 (1992).

[7] A. Tanaka and M. Sano, "Measurement of the Kinetic Effect on the Concentration Field of a Growing Dendrite", J. Crystal Growth, **125**, 59 (1993).

Non-Linear Electromagnetic Systems
V. Kose and J. Sievert (Eds.)
IOS Press, 1998

An Intelligent Network for Defect Profile Reconstruction in Eddy Current Applications[1]

Francesco Carlo Morabito

University of Reggio Calabria, Faculty of Engineering

DIMET, Via Graziella, Loc. Feo di Vito, I-89127 Reggio Calabria

Tel.:+39-965 875224, Fax: +39-965 875247, E-Mail: morabito@unirc.it

Abstract. In this paper a network topology which refers to multilayer feed-forward neural networks (FFNN) is proposed to process eddy current testing data. The detection and characterization of a defect in a specimen is carried out by making use of computational intelligence. The fuzzy inference techniques puts constraints on the possible inverse problem solutions. The introduction of wavelet nodes allows to cope with singularities of the mapping by using multiresolution analysis. Fine details of the signals are thus studied with a resolution matched to their scale, in contrast to standard FFNN processing which relies on global (sigmoidal) functions. Although only simple test case is treated here, we claim that the present approach could be useful for more general eddy current non-destructive evaluation (EC-NDE) problems.

1. Introduction

Eddy current testing of metallic materials is based on detecting, characterizing, and sizing defects by processing electromagnetic data picked up by suitable sensors located in the vicinity of the specimen under test. The outcomes of these measurements depend on the geometry of the inspected specimen, on the probing system, and they are affected by various sources of inaccuracies, e.g. the imprecise contour of the defect and the measurement noise. To decide whether there is a defect and, if there is one, to determine its shape and size, can be a difficult task, which typically involves prior knowledge of the problem in terms of human expertise. An EC-NDE problem can be interpreted as an inverse two-step problem, i.e., the recognition of defective specimens and the estimation of the defect configuration starting from the measured effects. The typical laboratory arrangement consists of a conducting body in which eddy currents are induced by a time-varying magnetic field. The presence and the configuration of cracks in the material is revealed by the deformation of the field structure. Computationally expensive iteration procedures are commonly used in order to extract any useful information from raw data. Models based on a network approach are of primary importance in copying with this task [3-4].

In this work, a network approach is proposed in which, within the FFNNs framework, both the potential benefits of fuzzy inference engines and the features of a wavelet decomposition in approximating non-smooth contours are included. The fuzzy logic block will allow to reduce the inherent ill-posedness (in the sense of Hadamard) of the problem [1], while the multiresolution properties of wavelets will be helpful in approximating sharp discontinuities of some field patterns around the specimen [2]. The proposed network model allows us to cope simultaneously with the requirements of practical EC-NDE approaches: fast interpretation of raw data in real time inspection, possibility of introducing prior knowledge about the problem within the basic architecture and without reducing the data throughput, and, finally, the capability of treating signals showing abrupt changes.

[1] This work was partially supported by the INCO-COPERNICUS project of the European Commission.

2. A Test Case

The benchmark problem is a cylindrical perfectly conducting structure simulating steam carrying tubes excited by using the rotating field concept [5]. The aim of the EC-NDE processing is to signal the presence of a defect and to provide a vector representing the defect's coordinates and size, if there is one. The exciting field is generated by a coil system similar to the stator of an induction machine. The rotating field gives rise to an eddy current distribution, possibly altered by the presence of a defect with respect to the same distribution in absence of defect. The detection of the defect is committed to a ring of differential pick-up coils. As the tube is moved through the induction system, the coils detect the presence of the defect. After this rough detection step, a high resolution measurement system provides a finer information about the defect. We will also use the frequency diversity of the exciting signal to discriminate surface from buried cracks.

The database to be used for FFNN training has been generated using the CARIDDI 3-D code. We simulate a 2 m long nonmagnetic tube, with an internal radius of 10 cm and a radial thickness of 1 cm, with resistivity of $1.0 \cdot 10^7$ Ohm·m. The rotating field is generated by superimposing two uniform $4\pi \cdot 10^{-4}$ T magnetic fields perpendicular to each other, with a phase lag of $\pi/2$. The crack is simulated as one or more elements having a higher resistivity. Details of the mesh and of the data base generation can be found in [5]. Fig.1 depicts the EC-NDE example problem.

We simulate the effects of cracks of 25 different sizes in the range $\Delta z \in [0.001, 0.01]$ m, $\Delta \theta \in [0.01, 0.1]$ rad. For each geometry we consider 10 different cases for the radial position of the crack, varying the electrical conductivities σ_1 to σ_5 of the five crack layers for a grand total of 250 possible defect configurations. For each of them we calculate the radial field due to the eddy current patterns in the tube at 60 points (15 rows of 4 pick-up coils at various θ locations) covering the region r=13 mm, $-\pi/12 \leq \theta \leq \pi/12$, $0 \leq z \leq 6$ mm. Ten excitation frequencies are considered for each case corresponding to skin depth varying from 6 mm to 3 cm.

3. The Intelligent Network For Crack Characterization

A hybrid fuzzy-neural procedure has been recently proposed to solve the problem [5]. In contrast to that approach, apart from the introduction of wavelet nodes, the fuzzy system block not only decides about the presence of a defect in the specimen, but, in addition, it yields a first hypothesis about the defect's characteristics. This is carried out by a fuzzy inference system based on the implementation of common sense rules in the network framework [1]. The resulting estimate is used to reduce the ill-posedness of the task, by putting some constraints on the possible solutions of the problem. This constraints implementation technique was already tested by the author for other problems [1]. The fuzzy inference block makes a simple guess about the expected size and location of the crack. The FFNN training is consequently simplified, since this avoid to start the learning procedure from an arbitrary initialization point. Furthermore, the number of hidden nodes may be reduced. The final tuning of the model is left to the wavelet part of the expansion scheme. Another important difference to [5] is that the fuzzy logic block is embedded in a unique network avoiding the need of two different sigmoidal FFNNs for separately processing buried and surface cracks. The final model is then most flexible and may be fine tuned both off and on-line by using an adaptive algorithm. Fig.2 shows the above described network. In EC-NDE applications, the need of simultaneously processing a lot of spatial data to cope with ill-posedness prevents from applying the fuzzy logic approach standalone.

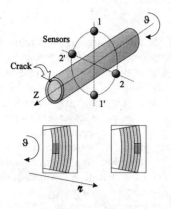

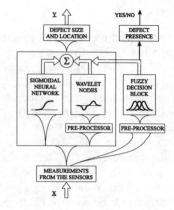

Fig. 1 - The analysed EC-NDE system   Fig. 2 - The proposed intelligent network scheme

Indeed, the number of rules of a fuzzy inference engine model grow exponentially with the number of model inputs. In contrast, FFNNs extract the information about the target from a contextual interpretation of a pattern of measurements. However, the fuzzy approach allows to simplify the search for an appropriate solution by implementing rules. This is what an expert usually makes when he/she expresses judgements about the solution of a problem. He is able of excluding some possibilities based on personal experience. This can hardly be implemented in non-fuzzy processors. Our approach allows us to reduce the size of the data base used for training, since there is no need of examples of what is already known. Furthermore, the extrapolation properties of the mapping are improved, since rules may be used where no observed data are available.

The fuzzy approach also reduces to a concise expansion representation most suitable for a network implementation, by using the Fuzzy Basis Functions concept [1, 4].

A network of wavelets (*WaveNet*) is an adaptive discretization of the continuous inverse wavelet transform: it is, in practice, a single hidden layer neural network whose activation functions are wavelets. Each hidden node implements a function characterized by the prototype wavelet function, properly dilated and shifted. It is possible to initialize the WaveNet on the basis of a regular wavelet lattice. However, the lattice would include a lot of useless wavelets, and, consequently, a number of parameters in excess whose presence complicates the tuning process. In our case, the global coverage of the optimization space is already ensured by sigmoids in the standard section of the hidden layer. Thus, there is no need of adding such levels by wavelets. This means that the dilation levels corresponding to low resolution are supplied by sigmoidal nodes. The number of wavelets corresponding to high dilation coefficients (high resolution) is limited by the fact that the coefficients are about zero everywhere but in correspondence to the mapping discontinuities [2]. This strongly simplifies the wavelet approach. The representation parameters of the wavelets can be fine tuned by using a backpropagation procedure.

The final mapping can be represented by a unique model, as follows: $y = \sum_i w_i f_i (a\underline{x} - b)$ where $i=1, \ldots , N_h$ with N_h number of the hidden nodes, and f_i is a sigmoidal function centered on b and properly scaled, or a fuzzy basis function, or a wavelet function.

4. Results and Conclusions

Each record of the training database includes 120 field measurements corresponding to the 60 different sensor locations at two different frequencies (177 and 4421 Hz) plus 20 field measurements corresponding to a frequency sweep at two different sensor locations.

The outputs are Δz and $\Delta \theta$ of the crack and its radial configuration. The patterns correspond to 10 different radial conductivity distributions (buried and surface cracks). A dimensionality reduction which reduces the input vector size from 140 to 30 is required before to start any learning procedure for reducing the computational burden. For this reason, we introduce a bottleneck linear layer within the inputs and the non-linear hidden layer. The fuzzy system block has just 3 inputs which were selected by preprocessing the measurements by using the concept of a Fuzzy Curve [6]. The rules were extracted by the same fuzzy curves yielding a total of 15 rules. The importance of each rule in the inference process is controlled by an additional parameter. Each input is coded in three possible fuzzy sets with Gaussian membership functions. The Gaussians were centered using a K-means clustering algorithm. Once the design of the fuzzy block is completed, the training of the FFNN is carried out using the backpropagation algorithm. Finally, the mapping is modified by adding 10 wavelet nodes corresponding to a guessed initial dilation coefficient. The wavelet part of the processor is also tuned by using backpropagation. The resulting processor performance has been compared to that of standard FFNNs (140-30-20-3), to systems based only on fuzzy logic, and to wavelet networks (3 pre-processed inputs, 70 hidden nodes corresponding to 5 dilation levels). Tab.1 reports the results of the comparison (the "/" indicates that, in our experimentations, the corresponding output is not recoverable by that approach).

Output Variable	Δz	$\Delta \theta$	Δr
FFNN	14.5	13.6	/
Fuzzy Estimate	16.1	15.8	8.3
FFNN + FuzzyEstimate	9.9	8.7	6.1
Wave Net	17.7	17.5	/
FFNN + FE + WN	6.2	5.6	2.2

Tab.1- Root Mean Squared Percent Error Reported to the Full Scale of Each Variable

We may conclude that the combined model is better than NNs because they are only able to cope with smooth mapping, better than fuzzy systems because the related data base of rules is difficult to obtain and the adaptive tuning of membership functions and parameters is tedious and difficult; and, finally, with respect to a WaveNet based on a wavelet decomposition which is far more expensive in terms of computational burden of the training phase as well as in the number of nodes in the final model.

References

[1] F.C.Morabito and M.Campolo, Ill-Posed Problems in Electromagnetics: Advantages of Neuro-Fuzzy Approaches, *2nd Intl. Symposium on Neuro-Fuzzy Systems*, EPFL, Lausanne, CH, 31 August 1996.

[2] G.Chen, Y.Yoshida, K.Miya, M.Kurokawa, Reconstruction of Defects from the Distribution of Current Vector Potential using Wavelets, *Intl. Journ. of Applied Electr. in Materials* 5 , (1994) 189-199.

[3] E.Coccorese, R.Martone and F.C.Morabito, A Neural Network Approach for the Solution of Electric and Magnetic Inverse Problems, *IEEE Trans. on Magn.*, Vol.30, No.5, (Sept 1994) 2829-2839.

[4] F.C.Morabito, M.Campolo, A novel neural network approach in non-destructive testing, *in Proceedings of the Third Int. Conf. on Comp. in Electr.*, Bath, UK, (10-12 Apr 1996), 364-369, London, UK, Apr '96, IEE Conf. Publ. No.420.

[5] R.Albanese, A.Formisano, R.Martone, C.Morabito, G.Rubinacci, F.Villone, A Method for the Analysis of Metallic Tubes with ECT and Neuro-Fuzzy Processing, paper presented at *IV Workshop on Optimization and Inverse Problems in Electromagnetism*, Brno (June 1996).

[6] F.C. Morabito, M.Versaci, The Use of Fuzzy Curves for the Reconstruction of the Plasma Shape and the Selection of the Magnetic Sensors, presented at the SOFT'96, Lisboa, Portugal.

Non-Linear Electromagnetic Systems
V. Kose and J. Sievert (Eds.)
IOS Press, 1998

Waveform Reconstruction of Impulsive Force and Pressure by Means of Deconvolution

Zdzisław Kaczmarek, Arkadiusz Drobnica
Kielce University of Technology, Al. 1000-lecia P.P. 7, 25-314 Kielce, Poland

Abstract. A method of reconstructing a waveform of impulsive pressure and force, that allows the removal of an effect of dispersion in pressure bar on the accuracy of the reconstruction, is presented. The stabilization of deconvolution based on Tikhonov's regularization technique was accomplished in frequency domain using one-parameter regularization filters. The reconstruction results of the impulsive pressure produced by an electrical discharge in water obtained by using the described technique are presented.

1. Introduction

For measurements of impulsive force and pressure an elastic cylindrical bar is widely utilized as a mechanical transducer. The measured quantities affect the end of the bar and generate elastic stress waves travelling through the bar as longitudinal waves. Taking into account the contact effects, reflection effects and process of waves interference, it is possible to fix a central segment of the bar in which the elastic wave of stress represents the signal of measured quantity. If the wavelengths of stress waves λ_{min} in the bar are sufficiently long in comparison with the bar diameter a ($a/\lambda_{min}<0.1$), it may be assumed that the central segment of the bar, which is established according to above-mentioned principles, is a non-dispersive mechanical transducer (non-distorting). However, if the measured quantity contains the frequency components which wavelengths λ_{min} are of the same order of magnitude as the bar diameter, these components propagate along the bar with different phase velocity. In this case it is assumed that the bar is a dispersive transducer. The dispersion phenomenon causes distortion of the stress wave propagating down the bar [1]. This paper presents a method for correction of dispersion in the mechanical transducer to improve the accuracy of the impulsive force and pressure waveform reconstruction. The process of measured quantity reconstruction was carried out in the frequency domain making use of the FFT techniques. The stabilization of the process was accomplished in frequency domain using regularization filters. In order to construct two types of regularization filters, the Tikhonov's regularization method was used [2]. This method was utilized to reconstruct the waveform of impulsive pressure produced by an electrical discharge in water. The results of the measurements are presented in the subsequent part of the paper.

2. Method principle

The essence of the method that aims at reconstruction of an input function incidenting the end face of the bar is the utilization of empirically determined dynamic characteristics of selected segments of the bar in the form of impulse response or transmittance. This technique uses output signals of two strain gauges located at different points on the lateral surface of the bar (*Fig. 1*). This method is based on the assumption that the bar is made of homogeneous material and the distance between these two gauges is equal with the distance between the first one and the end of the bar. The above-mentioned segments are modelled as linear systems of the same dynamic characteristics.

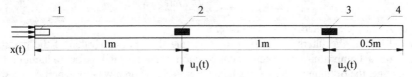

Fig. 1. *Impulsive pressure and force sensor: 1 - hypothetical gauge; 2, 3 - strain gauges; 4 - elastic bar.*

With all these assumptions, the output signal $u_0(t)$ of the hypothetical gauge located at the end of the bar, being a measure of input function incidenting the end of the bar $x(t)$, can be expressed as

$$u_0(t) = u_1(t) * u_1(t) * u_2^{-1}(t) \qquad (1)$$

Where "*" denotes the convolution operation, $u_1(t)$ and $u_2(t)$ are the output signals of the 1st gauge and the 2nd gauge respectively, $u_2^{-1}(t)$ is an inverse function of $u_2(t)$ in the sense that

$$u_2^{-1}(t) * u_2(t) = \delta(t) \qquad (2)$$

where $\delta(t)$ is a delta function. The relationship (1) in the frequency domain can be written as

$$U_0(j\omega) = \frac{U_1(j\omega) \cdot U_1(j\omega)}{U_2(j\omega)} \qquad (3)$$

where $U_0(j\omega)$, $U_1(j\omega)$ and $U_2(j\omega)$ denote the Fourier transforms of $u_0(t)$, $u_1(t)$ and $u_2(t)$ respectively. Using the inverse Fourier transform applied to (3), the input function $u_0(t)$ takes the form

$$u_0(t) = \frac{1}{2\pi} \int_{-\infty}^{\infty} \frac{U_1(j\omega) \cdot U_1(j\omega)}{U_2(j\omega)} \cdot e^{j\omega t} d\omega \qquad (4)$$

The determination of quantity $u_0(t)$ from either (1) or (4), is a task mathematically classified as an ill-posed problem and leads to an unstable solution. The instability is caused by data errors in measurements of $u_1(t)$ and $u_2(t)$ and that the frequency band of signal $u_1(t)$ is wider than the one of signal $u_2(t)$. For this class of problem it is not possible to determine the exact solution $u_0(t)$. Instead, in practice, an approximate solution in the form of an estimator $\tilde{u}_0(t)$ is sought with the aid of the frequency-domain regularization technique [3].

3. Deconvolution regularization

A stable estimator $\tilde{u}_0(t)$ of the reconstructing wave is achieved by multiplying the element of integration in the relationship (4) by a regularization function $R(\omega)$, physically represented by amplitude characteristic of the regularization filter. Hence, the estimate $\tilde{u}_0(t)$ can be written in the form

$$\tilde{u}_0(t) = \frac{1}{2\pi} \int_{-\infty}^{+\infty} \frac{U_1(j\omega) \cdot U_1(j\omega)}{U_2(j\omega)} \cdot R(\omega) \cdot e^{j\omega t} d\omega \qquad (5)$$

For waveform reconstruction, the Tikhonov's regularization method was used [2]. Two types of regularization filters were constructed. The frequency characteristics of the filters are as follows

$$R_1(\omega) = \frac{|U_2(j\omega)|^2}{|U_2(j\omega)|^2 + \lambda} \tag{6}$$

$$R_2(\omega) = \frac{|U_2(j\omega)|^2}{|U_2(j\omega)|^2 + \gamma\omega^4} \tag{7}$$

where λ and γ are regularization parameters.

For the first of the filters, a criterion for selecting the optimum value of the parameter is based on a compromise between deconvolution accuracy and noise reduction. For the second one, the optimum value of this parameter is chosen in a way to minimize the standard deviation in an imaginary part of the deconvolved waveform [4].

4. Experiment results

The results of the measurements of the impulsive pressure produced by an electrical discharge in water, obtained by using the technique discussed in the previous sections, are presented below. The measured waves obtained from strain gauges installed at points 1m and 2m from the bar end with their phase characteristics are shown in *Figs. 2-4*. Reconstructed waves of the measured pressure incidenting the bar end, before and after the regularization process are shown in *Figs. 5-7*. The amplitude characteristics of the utilized regularization filters are presented in *Figs. 8-9*. From the obtained results it follows that the peak amplitude of the pressure wave incidenting the end of the bar is about 91% higher than that acquired from the gauge located at 1m distance from the end of the bar. The width of the reconstructed pressure pulse is 5 μs and the width of the pulse acquired from the 1st gauge is 10 μs.

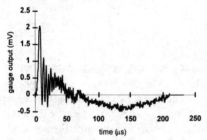

Fig. 2. Strain wave observed at 1m from the bar end.

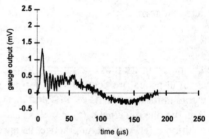

Fig. 3. Strain wave observed at 2m from the bar end.

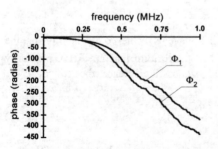

Fig. 4. Unwrapped phase spectras for the waves illustrated in figures 2 and 3 (ϕ_1 and ϕ_2 respectively).

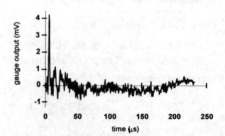

Fig. 5. Reconstructed strain wave before the regularization process.

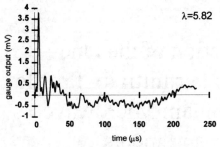

Fig. 6. *Reconstructed strain wave after the regularization with the use of filter (6).*

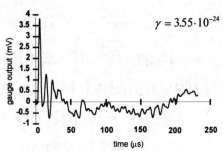

Fig. 7. *Reconstructed strain wave after the regularization with the use of filter (7).*

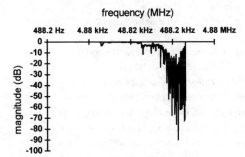

Fig. 8. *Amplitude characteristic of regularization filter (6) for $\lambda = 5.82$.*

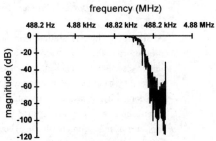

Fig. 9. *Amplitude characteristic of regularization filter (7) for $\gamma = 3.55 \cdot 10^{-24}$.*

5. Conclusion

The results of experiments illustrated above confirm that the proposed reconstruction method with a suitable regularization filter allows the removal of an effect of dispersion in the mechanical transducer. The method permits the application of the dispersive transducer in measurements of impulsive mechanical quantities.

References

[1] A. Gorham, A numerical method for the correction of dispersion in pressure bar signal, *J. Phys. E. Sci. Instrum.,* Vol. 16, 1983, pp. 477-479.
[2] Tikhonov A. N., Arsenin V. Y., Solution of Ill-Posed Problems, New York, Wiley, 1977.
[3] Riad, The Deconvolution Problem: An Overview, *Proc. IEEE,* Vol. 74, No. 1, January 1986, pp. 82-85.
[4] Guillaume, J. C. Bizeul, Mesure de la réponse impulsionnelle d'un milieu de transmission: application aux fibres optiques, *Ann. Télécommunic.,* Vol. 36, No. 3-4, Mars-Avril, 1981, pp. 179-186.

Non-Linear Electromagnetic Systems
V. Kose and J. Sievert (Eds.)
IOS Press, 1998

Iterative Reconstruction of the One-Dimensional Dielectric Permittivity Profile Using Frequency Domain Microwave Reflection Measurements

Valeri A. MIKHNEV

Institute of Applied Physics, Skoriny 16, 220072 Minsk, Belarus

Abstract. An inverse scattering method to reconstruct the dielectric permittivity profile from the reflection coefficient data based on the Newton-Kantorovich iterative scheme is proposed. A numerical instability of the solution is considerably reduced by handling the inverse problem in terms of an optical path length instead of geometrical distance. The initial estimate for the optimization procedure is calculated using another reconstruction method based on a minimax criterion for the modulus of reflection coefficient. The accuracy and the robustness of the approach are demonstrated for a discontinuous dielectric profile of high contrast. It is shown that both the permittivity and the conductivity profiles can be reconstructed if the total thickness of the inhomogeneous slab and the background permittivity are known.

1. Introduction

The penetration of microwaves into various materials gives the active microwave imaging a large potential for industrial and medical applications. This relatively new imaging technique aims to derive the spatial distribution of the dielectric permittivity from the knowledge of the scattered field outside the region under investigation. Various inverse scattering theories have been proposed and discussed in scientific literature.

However, despite a large number of investigations carried out in the field of microwave imaging, a very limited amount of acceptable realistic results has been obtained until recently. Even in the simplest case of one-dimensional inhomogeneity, accurate and reliable reconstructions are performed only for smooth profiles or profiles of low contrast [1]-[2].

In this paper, a new two-step inversion approach valid for the reconstruction of discontinuous profiles of high contrast without using *a priori* information is presented.

2. Inversion theory

Consider the reflection of a normally incident plane wave from an inhomogeneously layered medium of unknown complex permittivity $q(k,x) = \varepsilon(x) - j\eta_0\sigma(x)/k$, where k is the free space wavenumber, η_0 is the characteristic impedance of free space, $\varepsilon(x)$ and $\sigma(x)$ denote medium permittivity and conductivity profiles, respectively. The problem of interest is to find the complex permittivity from a knowledge of the reflection coefficient given for a set of discrete frequencies. The complex reflection coefficient $r(k,x)$ satisfies the Riccati nonlinear differential equation [1]:

$$\frac{dr(k,x)}{dx} = 2jk\sqrt{q(k,x)}r(k,x) + \frac{1-r^2(k,x)}{4q(k,x)}\frac{dq(k,x)}{dx}.$$ (1)

This equation is nonlinear, and its solution cannot be written in closed form with subsequent inversion. Hence, an accurate determination of the complex permittivity distribution requires utilization of some optimization approach. In this work, the reconstruction is accomplished using the Newton-Kantorovich iterative technique[2] applied to the Riccati equation rewritten in terms of the optical path length. Accordingly, the inversion process can be summarized as follows:

step 1   introducing a new variable, the optical path length $t = \int_0^x \sqrt{\varepsilon(x')}dx'$ (2)

step 2   solution   of   a   forward   problem   for   the   initial   profile $q^{(0)}(k,t) = \varepsilon^{(0)}(t) - j\eta_0\sigma^{(0)}(t)/k$ using (1) rewritten in terms of t. Computation of a discrepancy between calculated and given reflection coefficient data $\Delta r(k,0)$

step 3   derivation of the linear integral equation relating a small change of the complex permittivity profile $\Delta q(k,t) = \Delta\varepsilon(t) - j\eta_0\Delta\sigma(t)/k$ and the corresponding change of the reflection coefficient. First-order estimation of the functions $\Delta\varepsilon(t)$, $\Delta\sigma(t)$ by solving the integral equation

step 4   updating the profile function $q^{(1)}(k,t) = q^{(0)}(k,t) + \Delta q(k,t)$

step 5   go to step 2 as long as $\Delta r(k,0)$ is larger than an acceptable error

step 6   otherwise, stop the iterative procedure and return to the geometrical distance in the final profile using (2).

A comparison of reconstructions performed using the described approach and its counterpart derived with the use of a usual spatial coordinate, showed apparent advantage of the presented formulation with respect to the convergence and the stability of the solution. Moreover, the method retains its robustness for discontinuous profiles of high contrast. This can be attributed to a better accuracy of the linearized integral equation in step 3 in case the optical path length rather than the usual geometrical distance is used.

Despite the good convergence of the solution in the proposed approach, quite a lot of computation time can be saved by a careful selection of the initial profile used as a starting point for the iterative procedure. Usually, the initial estimate is taken from *a priori* knowledge about the object under investigation. Nevertheless, at least for low-loss media the initial profile can be calculated from the given reflection coefficient data using another inverse scattering method employing a successive reconstruction of dielectric interfaces and layers.

This approach is based on the behavior of the maximum of modulus of the reflection coefficient in the frequency band of operation. It can be shown, that an addition of one more dielectric interface to a lossless step-like profile results in increasing the maximum of the modulus of the reflection coefficient in an infinite frequency range, and the increase is the larger, the higher is the installed step. Hence, the principle of discrete reconstruction can be formulated by inverting this statement: the parameters of the layers in an unknown multilayered slab are to be chosen one after another in order to minimize the maximum of the modulus of the reflection coefficient for a remaining region of the slab.

The discrete reconstruction method is very fast and yields true inversion of simple one- and two-layered profiles. As far as more complicated multilayered structures or continuous profile functions are concerned, the reconstruction is approximate due to the frequency of maximal reflection can be situated outside the given frequency band. Nevertheless, dominant features of the profile, or its general view, are always reconstructed quite correctly. Thus, the discrete reconstruction method creates good starting conditions for the optimization approach described above saving a lot of computation time and avoiding the use of any *a priori* information.

3. Numerical examples

The capabilities of the proposed approach are demonstrated for highly contrasted multilayered profiles using simulated reflection coefficient data calculated in the frequency range of 0.5 to 15 GHz. Fig.1 shows the reconstruction of a four-layered lossless slab on a substrate, i.e. the total number of unknowns is 9, and the permittivities of the layers vary from 1 to 15.

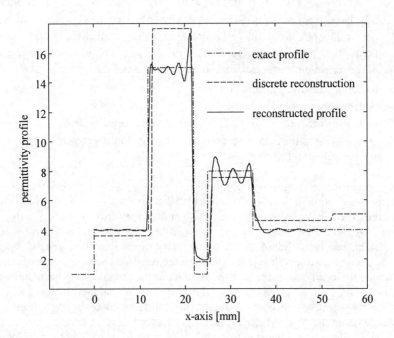

Fig.1. Reconstruction of a lossless discontinuous profile.

First, the discrete reconstruction method is applied. Although the accuracy of the recon-struction is not high, a discrepancy between the exact profile and the reconstructed one is not too large, as seen from Fig.1. For this reason, the result is used as a starting point for the Newton-Kantorovich optimization method, which completes the reconstruction in a few iterations. The quality of the reconstruction is good everywhere except in the vicinity of the interface between the second and the third layer. To estimate the robustness of the approach, a random signal uniformly distributed over the interval [-0.02 , +0.02] was added to the real and imaginary parts of the reflection coefficients at all frequencies in the given frequency band. Even for such a strong noise the solution of the inverse problem was found to retain a good stability. It is also worth noting that an air gap (the third layer) is reconstructed quite well though its width is considerably less than the shortest wavelength used in the input data.

Fig.2 presents a two-layered profile on a substrate where the second layer is lossy.

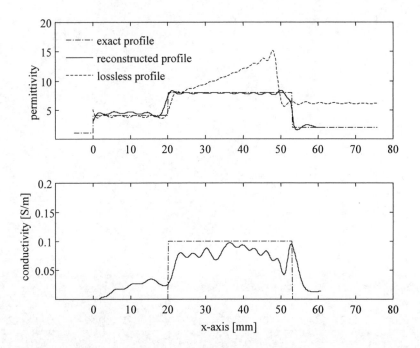

Fig.2. Reconstruction of both the permittivity and the conductivity profiles.

First, an attempt was undertaken to find a lossless profile using the given reflection coefficients. As a result, the lossless profile shown in Fig.2 has almost exactly the same reflection coefficient data in the given frequency band. Therefore, the reflection measurements at normal incidence do not provide sufficient information to obtain both the permittivity and the conductivity profiles. To correct the situation, a total thickness of the inhomogeneous slab and a background permittivity are assumed to be known. Under these constraints imposed on the solution of the inverse problem, the method yields accurate reconstruction of the permittivity profile and qualitative reconstruction of the conductivity profile as shown in Fig.2.

References

[1] T.J. Cui and C.H. Liang, Reconstruction of the permittivity profile of an inhomogeneous medium using an equivalent network method, *IEEE Trans. Antennas Propagat.* **AP-41,** 1993, 1719-1726.

[2] N.P. Zhuk and D.O. Batrakov, Inverse scattering problem in the polarization parameters domain for isotropic layered media: solution via Newton-Kantorovich iterative technique, *J. Electromagn.Waves Appl.* **8,**1994, 759-779.

Material Properties, Design and Modeling

Non-Linear Electromagnetic Systems
V. Kose and J. Sievert (Eds.)
IOS Press, 1998

A Fixed Point Iteration by *H* Scheme in the Solution of Hysteretic Electromagnetic Field Problems

Oriano BOTTAUSCIO (°), Mario CHIAMPI (*),
Carlo RAGUSA (°) and Maurizio REPETTO (*)

(°) IEN Galileo Ferraris, C. M. d'Azeglio 42, I-10125 Torino, Italy
(*) Dip. Ing. Elettrica Ind. - Politecnico di Torino, C. Duca degli Abruzzi 24,
I-10129 Torino, Italy

Abstract. The paper presents a Finite Element formulation for the analysis of the magnetic and electric field distribution in ferromagnetic sheets under periodic supply excitations. The field equations are written in terms of magnetic vector potential in the harmonic domain. The Fixed Point iteration by *H*-scheme is employed for handling magnetic nonlinearities. The hysteresis is modelled through the Preisach model including moving and dynamic effects. The proposed approach is applied to the analysis of a NO 2.5%wt FeSi lamination and the numerical results are found in satisfactory agreement with the experiments.

1. Introduction

The analysis of ferromagnetic losses in machine cores requires a preliminary accurate estimation of the electromagnetic behaviour of laminations. In particular, the electric and magnetic field distribution have to be predicted within a sheet through which a known magnetic flux waveform is imposed. This analysis can be conveniently performed by a mathematical model of the electromagnetic phenomena, which also includes the hysteretic behaviour of the material. The Classical Preisach Model (CPM) of hysteresis is proved to be an efficient tool for such a purpose; however, this model imposes some properties (congruency and rate independence) which are usually not verified in actual ferromagnetic materials. The Moving Preisach Model (MPM) and Dynamic Preisach Model (DPM) have been developed in order to override these limits [1-3]. However, these improved models are more complex and, when applied inside electromagnetic field computation, impose severe requirements on the numerical implementation.

Different formulations have been proposed in literature to analyse electromagnetic field problems inside ferromagnetic laminations using an improved Preisach model. The proposed approaches include the effect of the domain wall motion (DPM), discretizing the time evolution through a time stepping technique and analyzing the transient evolution of magnetic and electric field quantities until the steady state solution is reached [4-5]. Alternative approaches have been proposed by the authors developing different formulations based on Finite Element method performing field computations in the harmonic domain [6-9]. In this paper, the methodology is generalized including the MPM, in addition to DPM, to better reproduce the electromagnetic behaviour under low flux excitations. This goal is reached by formulating the field problem in terms of magnetic vector potential and by applying the Fixed Point technique for the treatment of nonlinearity in the modified version (iteration by *H*-scheme) proposed by Hantila in [10].

After a description of MPM and DPM implementation (Sect. 2), the field formulation is detailed in Sect. 3 together with the main feature of the numerical procedure. Finally, Sect. 4 presents some numerical results, developed on a NO 2.5%wt FeSi lamination under different supply conditions.

2. Implementation of Moving and Dynamic Preisach models

The Preisach model is nowadays the most widespread hysteresis model because it generally well reproduces the hysteretic behaviour for a large number of ferromagnetic materials and, in addition, it is suitable for introduction into numerical field computational schemes. However, comparisons with experiments show that the Classical Preisach Model (CPM) does not account for some particular phenomena of the magnetisation process. In particular, the CPM imposes congruency property of minor loops, which is not observed in actual material. Then, the CPM considers the magnetisation process independent of the supply frequency, neglecting the dynamic effects of domain wall motion (responsible for excess losses). In order to improve the description of material behaviour, several modifications to the CPM have been proposed, giving rise in particular to the Moving Preisach Model (MPM) [1, 2] and the Dynamic Preisach Model (DPM) [3].

The DPM establishes a time evolution of magnetisation of the Preisach operators and allows them to assume intermediate values between positive and negative saturation. As a consequence, the state of the material can no longer be defined by a staircase boundary separating positive and negative operators, as in the CPM, but the magnetisation of any single operator has to be stored. The number of elementary operators obviously affects the response of the model: a great number of operators ensures smooth responses but, at the same time, increases the computational cost; on the contrary, a small number of operators reduces the accuracy of the hysteretic behaviour. To overcome these problems, the authors have introduced modified elementary operators, associating with each of them a portion of Preisach plane [9]. Following this approach, the generic elementary operator centred in (α_o, β_o) has two intervals of influence: $\alpha_1 < \alpha_o < \alpha_2$ and $\beta_1 < \beta_o < \beta_2$; the magnetisation begins to increase when the applied field H is greater than α_1; for $\alpha_1 < H < \alpha_2$ the operator is partially activated, while for $H > \alpha_2$ the behaviour is the one of the operator concentrated in α_o. The differential equations governing the evolution of magnetisation for the positive switching is so modified:

$$\frac{dM}{dt} = 0 \qquad \text{for } H < \alpha_1$$

$$\frac{dM}{dt} = \int_{\alpha_1}^{H} k_d (H - h') w(h') dh' \quad \text{for } \alpha_1 \leq H \leq \alpha_2 \qquad (1)$$

$$\frac{dM}{dt} = k_d (H - \alpha_o) \qquad \text{for } H > \alpha_2$$

where k_d is the dynamic constant and w is a weighting function, which is piecewise linear in $\alpha_1 \div \alpha_2$ and zero elsewhere (Fig. 1). A similar behaviour is imposed for decreasing excitations.

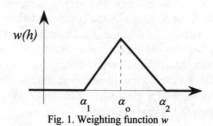

Fig. 1. Weighting function w

The use of such operators, associated with a portion of Preisach plane, allows a very efficient modelling of dynamic behaviour, reducing the number of operators without affecting the accuracy of computational results.

The implementation of the DPM in a periodic field problem requires that at least an initial state of the Preisach triangle is known. This difficulty is overcome by applying the computed H waveform, starting from the demagnetised state and analysing the subsequent transient until the periodic steady state solution is reached.

To remove the congruency property of the CPM, the MPM has been introduced in literature. Following this approach, the applied field H is replaced with an effective field H_p by a mean field correction:

$$H_p = H + f(M) \tag{2}$$

where $f(M)$ is a suitable function of the magnetisation, which introduces a feedback in the hysteresis model, increasing the computational burden. In this paper, we assume, as proposed in [11], the following correction function:

$$f(M) = k_1 M + k_3 M^3 \tag{3}$$

Of course, in order to apply the hysteresis model previously discussed, distribution function $P(\alpha,\beta)$ and moving parameters (k_1 and k_3) have to be preliminarily identified. This identification procedure is based on the method proposed by Kadar [12]. The technique is generalized to include the estimation of moving parameters through an optimization procedure which tries to minimize the discrepancies between experimental and reconstructed minor cycles [13].

Once the Preisach distribution function is known, the operators of DPM, above described, are placed in the Preisach plane. The operators are concentrated in the regions where the gradient of function $P(\alpha,\beta)$ is greater, so that accurate and smooth predictions can be obtained also using a reduced number of operators. Nevertheless, experience has shown that their number cannot be reduced below a certain limit; in the considered cases about 3000 operators have been used with satisfactory results.

3. Field formulation and numerical technique

The aim of the present paper is the analysis of the electromagnetic field within an infinite lamination supplied by a known periodic magnetic flux. This analysis includes eddy currents (skin effect) and requires the solution of a 1D electromagnetic field problem.

The electromagnetic field problem can be formulated assuming either the magnetic field H or the vector potential A ($B = curl A$) as unknown. The first choice makes the use of the Preisach model easy, but requires that the supply conditions are imposed through an integral constraint. The use of the vector potential allows an easy imposition of the supply flux through non-homogeneous Dirichlet boundary conditions and, in addition, it is also suitable for 2D problems; on the contrary, such a formulation requires an inverse application of the Preisach model which is very difficult when the DPM is used. This difficulty is removed by employing the Fixed Point (FP) iterative method following the scheme proposed in [10]. The electromagnetic field problem is then formulated in terms of magnetic vector potential A:

$$\nabla \times \left[\zeta (\nabla \times A) \right] = -\sigma \frac{\partial A}{\partial t} + \frac{\sigma}{S} \int_\Omega \frac{\partial A}{\partial t} d\Omega \tag{4}$$

where function $\zeta(B,P)$ is the nonlinear hysteretic relationship between $B(P,t)$ and $H(P,t)$. Following FP technique, $H=\zeta(B,P)$ is split as $\mathbf{H}(P) = \mathbf{R}(\mathbf{H}, P) + v_{FP} \cdot \mathbf{B}(P)$, being reluctivity v_{FP} a suitable constant and $\mathbf{R}(H,P)$ the residual nonlinearity.

Under periodic field excitation, the linearized problem can be developed in the harmonic domain; the field equation for harmonic n, at the generic iteration i, reduces to:

$$\nu_{FP}\nabla \times \nabla \times A_i^{(n)} = -j\omega n\sigma A_i^{(n)} + j\omega n\frac{\sigma}{S}\int_\Omega A_i^{(n)}d\Omega - \nabla \times R_{i-1}^{(n)} \quad (5)$$

where $A_i^{(n)}$ and $R_{i-1}^{(n)}$ are phasors and ω is the fundamental angular frequency.

The problem is then solved using the Finite Element Method (FEM) following the iterative scheme here reported:
1) the electromagnetic field problem is solved in the harmonic domain obtaining, for each element, the harmonic spectrum of magnetic flux density B_i at the considered i-th iteration;
2) for each element the waveform of B_i is given by the inverse Fast Fourier Transform (FFT);
3) an estimate of the magnetic field is obtained by FP relationship $H_i = R_{i-1} + \nu_{FP}\cdot B_i$;
4) the effective field for the Preisach Model is deduced using H_i and $M_i = B_i - \mu_o H_i$ for the mean field correction ($H_p = H_i + k_1 M_i + k_1 M_i^3$);
5) the waveform of M_i^* is computed using the DPM and B_i^* is evaluated;
6) the residual term for the next iteration is computed: $R_i = H_i(P) - \nu_{FP}\cdot B_i^*(P)$;
7) the harmonic spectrum of R_i is obtained using the FFT;
8) the process is repeated starting from step 1) until convergence is reached.

The numerical results are then elaborated by a post-processor module which provides the main numerical and graphical quantities. In particular, the dynamic hysteresis cycle and the total losses are evaluated enabling a comparison with experiments. In addition a separation of losses can be obtained computing the eddy current and the dynamic hysteresis losses.

4. Numerical results

A preliminary step has been devoted to the investigation of the convergence properties and the efficiency of the proposed approach which employs the iterative FP technique in the H-iteration version. For such a purpose, a ferromagnetic material has been analyzed, under the same supply conditions, comparing the new procedure with a model, previously developed, based on a field formulation in terms of H with an additional integral constraint

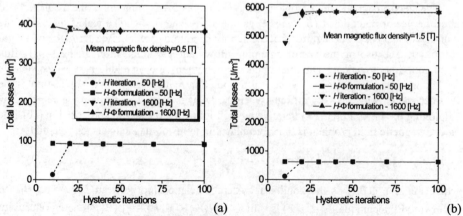

Fig. 2 - Convergence rate for H iteration scheme and H-Φ formulation:
a) low magnetic flux density (0.5 T), b) high magnetic flux density (1.5 T)

(*H-Φ* formulation) [6]. The tests have been performed for a set of frequencies and flux values, comparing the computed losses and flux distributions. The same results have been found for both models; moreover, the convergence rate is very similar, as shown by the diagrams of Figs. 2a and 2b, which present the evolution of computed magnetic losses during the iterative process.

The proposed model has been successively applied to the analysis of an industrial NO 2.5%wt FeSi 0.65 mm thick lamination whose conductivity σ is $2.366 \cdot 10^6$ S/m.

The identification process has shown that the significant mean field effects can be satisfactorily taken into account by including a linear and a cubic term whose values have been estimated by an optimization procedure. Since the microscopic analysis of the material was not available, the constant k_d of the dynamic model has been numerically evaluated by comparing the computational results with dynamic cycles measured at different frequency and flux density values.

The analysis has been performed for frequencies ranging from 50 Hz to 400 Hz and for flux values from 0.325 mWb (0.5 T average induction value) to 0.975 mWb (1.5 T average induction value). Some dynamic hysteresis cycles computed by the model are compared with the experimental ones in Fig. 3, showing that the agreement is more than satisfactory. The effect of moving components is evidenced in Fig. 4a which presents the results obtained in the case of low magnetic flux; one of the cycles has been computed identifying the material neglecting the parameters k_1 and k_3 of moving correction, while the other is obtained taking into account their presence. The comparison with the experimental loop clearly shows that the moving correction sensibly improves the accuracy of the prediction even if some discrepancies are found between measurements and computation. The effect of the moving component is also evident by the analysis of the skin effect in the lamination. Fig. 4b shows the distribution of peak value of magnetic flux density along the lamination with and without the moving terms. As can be seen, the simulation carried out considering moving terms gives rise to a lower skin effect in the lamination.

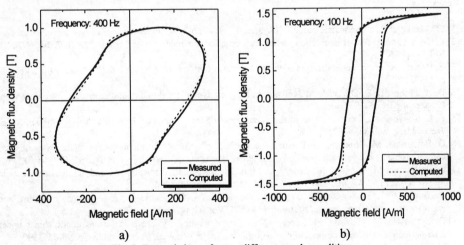

a)   b)

Fig. 3 -Dynamic loops for two different supply conditions:
a) mean magnetic flux density of 1 T at 400 Hz   b) mean magnetic flux density of 1.5 T at 100 Hz

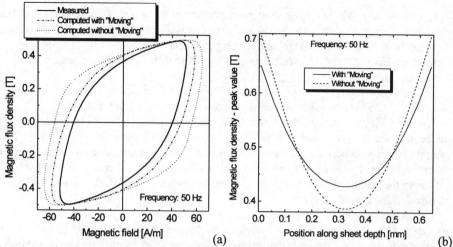

Fig. 4 -Effect of moving correction: a) dynamic loop for mean magnetic flux density of 0.5 T at 50 Hz, b) distribution of peak magnetic flux density versus thickness at 0.5 T, 50 Hz

5. Conclusions

The paper has presented a FE model for the analysis of the electromagnetic field inside a lamination supplied by periodic magnetic flux. The method employs the Preisach model, including the contribution of dynamic and moving effects and the FP technique in the H scheme version. The model has been applied to the analysis of a NO 2.5%wt FeSi lamination, showing a satisfactory agreement between computation and experiments. The results provide evidence of the important role of moving terms, particularly at low flux excitations.

References

[1] I.D.Mayergoyz, Mathematical models of hysteresis, New York, Springler-Verlag, 1991.
[2] E. Della Torre, Effect of interaction on the magnetisation of single domain particles, *IEEE Trans. AUDIO*, Vol. 14, p. 86, 1966.
[3] G. Bertotti, Dynamic generalization of the scalar Preisach model of hysteresis, *IEEE. Trans. Mag.*, Vol. 28, p. 2599, 1992.
[4] D. A. Philips, L. R. Dupré, J. A. Melkebeek, Magnetodynamic field computation using a rate dependent Preisach model, *IEEE. Trans. Mag.*, Vol. 30, p. 4377, 1994.
[5] L. L. Rouve et al., Determination of the parameter k of the generalized Dynamic Preisach model, *IEEE. Trans. Mag.*, Vol. 32, p. 4219, 1996.
[6] O. Bottauscio, M. Chiampi, M. Repetto, Comparison of different formulations for the analysis of eddy current problems in hysteretic media, *Proc. 3rd Int. Workshop on EMF*, Liege (Belgium), 1996, p. 285.
[7] O. Bottauscio, M. Chiampi, M. Repetto, A Finite Element solution for periodic eddy current problems in hysteretic media, *J. Magn. Magnetic Materials.*, Vol. 160, p. 96, 1996.
[8] V. Basso et al., Power losses in magnetic laminations with hysteresis: finite element modeling and experimental validation, *Journal of Applied Physics*, Vol. 81, pp. 5606, 1997.
[9] O. Bottauscio, M. Chiampi, M. Repetto, Modelling dynamical hysteresis in ferromagnetic sheets under time-periodic supply condizions, *Appl. Comp. Electr. Society Journal*, Vol. 12, no. 2, pp. 90-95, 1997
[10] I.F. Hantila, A method for solving stationary magnetic field in non-linear media, *Rev. Roum. Sci. Techn. - Electrotechn. et Energ.*, Vol. 20, 1975, p. 397.
[11] V. Basso et al., Preisach model study of the connection between magnetic and microstructural properties of soft magnetic materials, *IEEE Trans. MAG*, Vol. 31, p. 4000, 1995.
[12] G. Kadar, The Preisach function of ferromagnetic hysteresis, *J. Appl. Phys.* 61, p. 4013, 1987.
[13] V. Basso et al., Comparison of identification procedures for the Preisach model distribution, to appear on: Proc. 8th ISEM Conf.

Non-Linear Electromagnetic Systems
V. Kose and J. Sievert (Eds.)
IOS Press, 1998

High Speed Rotating HT$_c$SC Magnetic Bearing of New Topology

W.-R. Canders, H. May and R. Palka

Institut für Elektrische Maschinen, Antriebe und Bahnen,
Technische Universität Braunschweig, Hans Sommer-Str. 66
38106 Braunschweig, Germany

Abstract: The paper presents a study on inherently stable contact free magnetic bearings. The required forces are created by the interaction of monolithic High Temperature Super Conductors (HT$_c$SC) and Permanent Magnets (PM). The proposed new topology of the magnetic bearing as a truncated cone enables an optimal adaptation to a wide range of applications.

To reduce the amount of applied HT$_c$SC and PM material, excitation elements with flux concentrating iron poles and optimised pole pitches are introduced. The influences of the air gap and the pole pitch on the force and stiffness are given in detail including the effect of limited current densities associated with real monolithic superconductors.

Furthermore the effects of different freezing conditions (transition to superconductivity) on the performance of the magnetic bearings are expressed.

1. Introduction

Contactless acting magnetic bearings are most suitable to restrict the motion of high speed rotating shafts to one degree of freedom (rotation). If monolithic HTSC's interact with PM's such contact free bearings are inherently stable (Earnshaw's-theorem [1]) and thus avoid complex control, costs and electro-magnetic compatibility (EMC) problems of conventional active magnetic bearings [2,3]. As these bearings are free from contaminating lubricants they can be incorporated directly within chemical and other processes. The favourable operation of HTSC-PM bearings (HMB) has been demonstrated successfully in a wide range of applications and has been published in many papers [4,5,6].

For machines with horizontal shafts (Fig. 1) bearings with radial load capacity are necessary and axial bearings are used for machines with vertical shafts respectively (Fig. 2). Both types of bearings are more or less inherently stable in r- and z-directions but it might be useful to combine axial and radial bearings with adaptable properties.

The state of the PM's and the HTSC's during transition to superconductivity characterises the freezing modes. Distinction can be made between Zero Field Frozen (ZFF) and Field Frozen (FF) procedures. Although the ZFF mode leads to max. repulsive forces by the PM's approaching the HTSC's this mode cannot be used in bearing applications as no lateral stability can be observed. Furthermore it is nearly impossible to achieve these freezing conditions within a machine.

Related to the operational clearance between PM and HTSC (air gap) three practicable modes of transition to superconductivity can be distinguished. The Maximum Field Frozen (MFF) mode is characterised by a minimum clearance during the transition. Furthermore at the Operational Field Frozen (OFF) mode the transition state corresponds to the operational

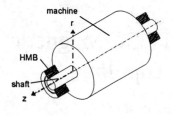

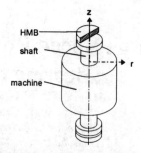

Fig. 1: Support arrangement with horizontal shaft   Fig. 2: Support arrangement with vertical shaft

clearance. If the transition state differs only marginally from that of the operational clearance the mode is defined as Operational Field Freezing with an offset (OFFo).

In an earlier study [8] the values of repelling forces and stiffnesses are given for the ZFF mode dependent on the operational clearance. As these are the highest values these data give an impression of the feasibility of HTSC-PM bearings.

2. Normal- and lateral forces of HTSC-PM bearings at FF mode

Although the shape of all bearings used for rotating shafts are axis symmetric a periodic cartesian model (Fig. 3) was used to calculate both the normal and lateral forces and stiffnesses with an extended Finite Element program. This simplification is permissible as all dimensions are large related to the magnetic gap. For this comparative study an operational clearance between PM and HTSC of g=1.5 mm together with a pole pitch of τ_p=11.3 mm were chosen. For the OFF operation mode the HTSC was frozen with a distance of g=1.5 mm above the excitation system (Fig. 3) and displaced in a lateral and vertical direction.

The resulting normal and lateral forces are shown in Fig. 4 and 5 respectively As this mode exhibits no forces in any direction in the freezing position a disadvantageous displacement has to be accepted to compensate the weight or other external loads. This results in a restricted radial or axial movement of the bearing.

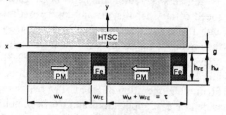

Fig. 3: Calculation model for HMB's
in a flux concentrating arrangement

The topology of the axial bearing of Fig. 2 features a special freezing procedure leading to prestressing forces in the nominal position. To assure this the HTSC in the upper bearing is MF-frozen and the lower HTSC is frozen in an offset position (OFFo) with an enlarged air gap. With the assumed nominal air gap of g=1.5mm the max. displacements are ±1.5 mm in the axial direction and the OFFo position of the HTSC is 3 mm above the surface of the PM's.

The normal forces of the upper bearing (MFF) are shown in Fig. 4 as a function of the normal displacement g and the lateral forces of the lower bearing (OFFo) are shown in Fig. 5 for lateral displacements in the nominal air gap state of g=1.5 mm.

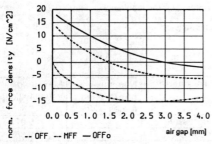

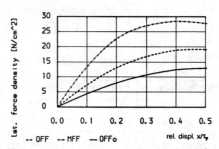

Fig. 4: Normal force density as a function of normal displacement for different freezing modes.

OFF: freezing clearance = 1.5 mm
MFF: freezing clearance ≈ 0 mm
OFFo: freezing clearance = 3.0 mm

Fig. 5: Lateral force density at g=1.5mm as a function of lateral displacement for different freezing modes.

OFF: freezing clearance = 1.5 mm
MFF: freezing clearance ≈ 0 mm
OFFo: freezing clearance = 3.0 mm

3. Comparison of the characteristic data of different bearing topologies

With the knowledge of these specific forces the actual performances of an axial and a radial bearing with the same basic dimensions of the excitation unit (h_M=20mm, τ_p=11.3mm) (Fig. 1 and 2) were calculated. The major determining data for these bearings are shown in Table 1. The radial bearing exhibits superior data with respect to both the stiffness and operational position and the load capacity but due to the symmetry no preferred axial position exists. Therefore the compensation of static loads by prestressing the bearings is not possible. To compensate this disadvantage but using the favourable data of the radial bearing a truncated conic bearing (Fig. 6 and 7) was introduced and calculated (Table 1). The excellent data of this topology result first off all from the combination of the performances of the axial and radial bearing and secondly from the decreased magnetic air gap associated with the conical design. If the mechanical clearance is 1.5 mm in this case with a conic angle of 45° the magnetic air gap is reduced to approximately 1.06 mm.

All calculations have been carried out with the presumption of Meissner state (critical current density $J_c \Rightarrow \infty$) of the HTSC's leading to max. forces. The effect of the finite J_c have been examined earlier [7] with the result that the normal forces are reduced to 65% assuming a J_c of 500 A/cm² and to 40% in the case of a J_c of 100 A/cm².

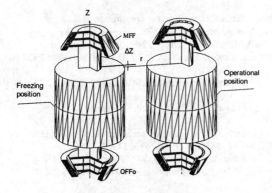

Fig. 6: Conical HTSC-PM bearing
Left: freezing position,
upper bearing MFF, lower bearing OFFo
Right: operational position

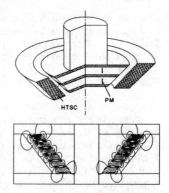

Fig. 7: Conical HTSC-PM bearing
Drawing and cross sectional view
of the field distribution (OFF)

Table 1: Characteristic data of different bearing topologies

Bearings of equivalent active surface and magnet masses:

Mechanical clearance $g_{mech} = 1.5$ mm

Flux concentrating arrangement of magnets, pole pitch $\tau_p = 11.3$ mm

Bearing Type	Freezing Mode	Stiffnesses at nom. Position [kN/mm]		Characteristic	Nominal prestressing forces [kN]	Max. load capacity [kN]	
Radial Bearing	OFF	c_{ax}	= 2.14	degressive		F_{ax}	=5.92
0°		c_{rad}	= 0.93	progressive		F_{rad}	=1.56
Axial Bearing	OFF	c_{ax}	= 1.85	progressive		F_{ax}	=4.61
90°		c_{rad}	= 0.43	degressive		F_{rad}	=1.18
Axial Bearing,	upper bearing	c_{ax}	= 1.18	progressive	$F_{ax,p} = 3.8$	F_{ax}	=4.14
90°	MFF	c_{rad}	= 0.77	degressive		F_{rad}	=1.75
prestressed by	lower bearing	c_{ax}	= 1.93	progressive	$F_{ax,p} = 2.03$	F_{ax}	=2.04
axial displacem.	OFFo	c_{rad}	= 0.25	degressive		F_{rad}	=0.79
Conical bearing	OFF	c_{ax}	= 2.0	progressive		F_{ax}	=5.27
45°		c_{rad}	= 0.68	nearly linear		F_{rad}	=1.37
Conical bearing	upper bearing	c_{ax}	= 2.55	degressive	$F_{ax,p} = 5.1$	F_{ax}	=6.2
45°	MFF	c_{rad}	= 0.68	progressive		F_{rad}	=2.1
prestressed by	lower bearing	c_{ax}	= 1.64	progressive	$F_{ax,p} = 3.1$	F_{ax}	=6.8
axial displacem.	OFFo	c_{rad}	= 0.6	nearly linear.		F_{rad}	=1.8

4. Conclusions

In contrast to contactless active magnetic bearings all topologies of superconducting bearings exhibit stiffness and load capacity in both the radial and the axial directions. Due to the relative small stiffness low critical speeds will have to be expected and the relative high load capacities will allow the design of bearings of utilizable size. The characteristic data of a bearing depend not only on the geometric design but also on the mode of transition to superconductivity (freezing mode) thus giving an additional degree of freedom for the layout of the bearing. The bearing with the conical geometry shows the most preferable combination of properties with respect to stiffness, spring characteristic, preloading capability and load capacity. Further studies are necessary for the optimisation of the geometry and the investigation of the dynamic properties of superconducting bearings.

References

[1] Earnshaw, S.: On the nature of the molecular forces which regulate the constitution of the luminiferous ether. Trans. Cambridge Philosophical Society (1839), Vol. 7, Part 1, 97-112
[2] Budig, P.-K.: Aktive magnetische Lager- eine Übersicht über theoretische Grundlagen, Aufbau, Wirkungsweise, Entwurf und spezielle Probleme. Elektrie 42 (1988) Teil 1: H.10, S. 365-370, Teil2, H 11 S. 424-428
[3] Habermann, H.; Liard, G.: Aktive Magnetlagerung in Turbomaschinen, SKF-Kugell.-Zeits. 195 S. 16-20
[4] Moon, F.C.; Chang, P-Z.: High -speed rotation of magnets on high Tc superconducting bearings. Applied Physics Letter, 56(4) (1990) 397-399.
[5] Marinescu, M.; Marinescu, N.; Tenbrink, J.; Krauth, H.: Passive Axial Stabilization of a Magnetic Radial Bearing by Superconductors. IEEE Trans. On MAG, Vol. 25, No 5, 32267
[6] Fuchs, G.; Stoye, P.; Staiger, Th.; Krabbes, G.; Schätzle. P.; Gawalek, W.; Görnert, P.; Gladun, A.: Melt Textured YBCO Samples for Trapped Field Magnets and Levitation Bearings. Proc. Appl. Superc. Conf. 25th-30th August 96, Pittsburg.
[7] Canders, W.-R.; May, H.; Palka, R.: Identification of the Current Density Distribution of Monolithic Superconductors. ISTET'97, Palermo, June 1997
[8] Canders, W.-R.; May, H.; Palka, R.: Topology and Performance of Superconducting Magnetic Bearings. ISTET'97, Palermo, June 1997

Non-Linear Electromagnetic Systems
V. Kose and J. Sievert (Eds.)
IOS Press, 1998

Eddy Current Loss and Dynamic Hysteresis Loss in Electrical Steel Sheet under Two Dimensional Measuring Conditions

M. Birkfeld and K.A. Hempel

Institut für Werkstoffe der Elektrotechnik, Aachen University of Technology, Templergraben 55, 52056 Aachen, Germany

Abstract. The total power loss of electrical steel sheet under two dimensional excitation can be measured directly. The separation of the total power loss into its components, eddy current loss, hysteresis loss and the so called anomalous loss, depends on some assumptions concerning the physical behaviour of electrical steel sheet and is not accessible by direct measurements. A mathematical description of the non–linear and anisotropic behaviour of electrical steel sheet under two dimensional excitation including the field penetration into the bulk of the material allows the direct calculation of the loss components for arbitrary two dimensional flux conditions. The results of this calculation are compared to those obtained from a loss separation by a frequency variation.

1. Two Dimensional Excitation

The components of the mean value of the magnetic flux density $\vec{B}$ in the cross section of the steel sheet sample are controlled to vary sinusoidally with time t at frequency f and the angular velocity $\omega = 2\pi f$. $\vec{B}$ generally has the form of an ellipse:

$$\vec{B}(t) \;=\; \hat{B}\exp(j\omega t)\begin{pmatrix} +\cos\vartheta & -\sin\vartheta \\ +\sin\vartheta & +\cos\vartheta \end{pmatrix}\begin{pmatrix} 1 \\ -jra \end{pmatrix}, \qquad j^2 = -1 \qquad (1)$$

$\hat{B}$ is the amplitude of the long axis of the ellipse of the flux density vector rotating in the mathematically positive ($r = +1$) or negative sense ($r = -1$), respectively. The amplitude of the short axis is $a\hat{B}$ and $0 \le a \le 1$. The long axis of the ellipse is inclined by the inclination angle ϑ towards the rolling direction of the sample. In this paper, only the special case $\vartheta \equiv 0$ is considered for convenience.

Due to the non–linear behaviour of electrical steel sheet, this kind of excitation leads to higher harmonics of the magnetic field strength $\vec{H}$, although the components of the magnetic flux density $\vec{B}$ are restricted to the fundamental harmonic.

2. The Reluctance Tensor

The reluctance tensor $\overset{\leftrightarrow}{\nu}$ is the link between the harmonics of the magnetic field strength $\vec{H}$ and the magnetic flux density $\vec{B}$ in the sample. The cartesian coordinate system $\Sigma_{(x,y,z)}$ is chosen in such a way, that the sample lies parallel to the (x,y)–plane and that its origin corresponds with the center of the sample. The measurements are performed in the system of main axes [1], the relation between $\vec{H}$ and $\vec{B}$ for every set of coordinates $\vec{r} = (x, y, z)$ can be written in the following form:

$$\overset{\leftrightarrow}{\nu}\,\vec{B}(\vec{r},t)\;=\;\sum_{n>0}\left\{\begin{pmatrix} \nu_x^{(n)} & 0 \\ 0 & \nu_y^{(n)} \end{pmatrix}\exp(j(n-1)\omega t)\right\}\vec{B}(\vec{r},t)\;=\;\mu_0\vec{H}(\vec{r},t). \qquad (2)$$

$\nu_x^{(n)}$ and $\nu_y^{(n)}$ denote the complex reluctances of the n-th order in x- and y-direction, respectively, combining the fundamental harmonic of $\vec{B}$ with the n-th harmonic of $\vec{H}$. The anisotropic behaviour of the steel sheet sample is considered by the absolute values $|\nu_x^{(n)}|$ and $|\nu_y^{(n)}|$, the lag angle in time and space between the vectors of the magnetic field strength and the magnetic flux density is considered by the arguments $\arg \nu_x^{(n)}$ and $\arg \nu_y^{(n)}$ of the reluctances [1][4]. The reluctance tensor represents the magnetic behaviour of a steel sheet due to grains, the polycrystalline structure and due to the domain configuration inside the sample for a given two dimensional excitation.

3. Solution of the Differential Equation of Field Penetration

Due to the eddy currents flowing in the sample of conductivity σ, the following differential equation must be solved to calculate the field distribution in the sample:

$$\nabla \times \nabla \times \vec{B}(\vec{r},t)\;=\;-\sigma\mu_0\,\overset{\leftrightarrow}{\nu}^{-1}\frac{\partial}{\partial t}\vec{B}(\vec{r},t)$$

The boundary conditions are given by the average value of $\vec{B}$ (equ. (1)) in the cross section of the sample with thickness d. Due to the construction of the magnetizing yoke system [3], the flux distribution in the plane of the sample can be assumed to be homogenous in an infinitesimal small layer of thickness dz:

$$\vec{B}(\vec{r},t)\;=\;\vec{B}(z,t)$$

The solution [1]

$$\vec{B}(z,t)\;=\;\begin{pmatrix} C_1\coth(s_x z) & 0 \\ 0 & C_2\coth(s_y z) \end{pmatrix}\vec{B}(t)$$

with constants C_1, C_2, depends in a non-linear way on z and on the elements of the reluctance tensor. The inverse penetration depths s_x and s_y are defined as follows:

$$s_x\;=\;\sqrt{\frac{j\omega\sigma\mu_0}{\nu_x^{(1)}}}\quad\text{and}\quad s_y\;=\;\sqrt{\frac{j\omega\sigma\mu_0}{\nu_y^{(1)}}}. \qquad (3)$$

Only first order elements of the reluctance tensor are involved in the solution of the differential equation, because $\vec{B}$ contains only the fundamental harmonic.

4. Total Loss, Eddy Current Loss and Dynamic Hysteresis Loss

The specific total loss p can be measured directly and can be calculated from the electrical field strength $\vec{E} = \nabla \times \vec{B}$ and magnetical field strength $\mu_0\vec{H} = \overset{\leftrightarrow}{\nu}\,\vec{B}$ (equ. (2)) at the surface of the sample ($z = d/2$) [1][4] (ρ mass density):

$$p\;=\;\Re\left\{\frac{1}{m}\oint_A\left(\vec{E}\times\vec{H}^*\right)\big|_{z=d/2}\,d\vec{A}\right\}$$

$$=\;\Re\left\{\frac{-j\omega\hat{B}^2 d/2}{2\rho\mu_0}\left[\left(s_x\nu_x^{(1)}\coth(s_x d/2)\right)^* + (ra)^2\left(s_y\nu_y^{(1)}\coth(s_y d/2)\right)^*\right]\right\}$$

The eddy current loss p_E must be calculated considering the z–dependence of the electrical field strength $\vec{E}$ in the cross section of the sample (A surface area of the sample):

$$p_E = \frac{1}{m} \int_0^{d/2} \sigma A |\vec{E}(z,t)|^2 \, dz = \frac{1}{16} \frac{\sigma}{\rho} \left(\omega \hat{B} d\right)^2 \left(C_x + (ra)^2 C_y\right)$$

with constants

$$C_x := \frac{\frac{\sinh(s_{x,r}d)}{s_{x,r}d} - \frac{\sin(s_{x,i}d)}{s_{x,i}d}}{\left(\sinh(s_{x,r}d/2)\cos(s_{x,i}d/2)\right)^2 + \left(\cosh(s_{x,r}d/2)\sin(s_{x,i}d/2)\right)^2}$$

$$C_y := \frac{\frac{\sinh(s_{y,r}d)}{s_{y,r}d} - \frac{\sin(s_{y,i}d)}{s_{y,i}d}}{\left(\sinh(s_{y,r}d/2)\cos(s_{y,i}d/2)\right)^2 + \left(\cosh(s_{y,r}d/2)\sin(s_{y,i}d/2)\right)^2}$$

The eddy current loss for a homogenous flux distribution evaluates to:

$$p_{E,hom} := \frac{1}{24} \frac{\sigma}{\rho} \left(\omega \hat{B} d\right)^2 \left(1 + (ra)^2\right) \qquad (4)$$

The value of p_E approaches that of $p_{E,hom}$ in the case of negligible inverse penetration depths (equ. (3)) compared to the thickness d of the sample. The dynamic hysteresis loss is defined as the total power loss in case of vanishing conductivity:

$$p_H := p(\sigma = 0) = \frac{\omega \hat{B}^2}{2\rho\mu_0} \left[\nu_{x,i}^{(1)} + (ra)^2 \nu_{y,i}^{(1)}\right]$$

The reluctances used to calculate the dynamic hysteresis loss are those investigated from measurements of a conducting sample [4], considering the domain configuration under dynamic excitation conditions. As a consequence, the dynamic hysteresis loss p_H comprises both static hysteresis loss and the anomalous loss caused by the domain wall motion, but no contribution of the Ohmic loss due to the eddy currents.

5. Loss Separation

In order to discuss the proposed loss separation by calculating the field distribution inside the sample, we compare these results to those obtained by means of a well known and widely accepted method [2], that makes use of a frequency variation:

$$\text{Statical hysteresis loss:} \quad p_{H,stat} = c_H f$$

$$\text{Anomalous loss:} \quad p_A = c_A f^{3/2}$$

$$\text{Eddy current loss:} \quad p_{E,hom} = c_E f^2$$

With equ. (4):

$$c_E = \frac{1}{24} \frac{\sigma}{\rho} \left(2\pi \hat{B} d\right)^2 \left(1 + (ra)^2\right)$$

The coefficient c_H can be investigated from extrapolation of the cyclical magnetic work p/f at very low frequencies. Fitting of c_A optimally to the measured data of the total power loss p in the frequency range from 10 Hz to 1 kHz yields:

$$p = p_H + p_E \overset{!}{=} c_H f + c_A f^{3/2} + c_E f^2 = p_{H,stat} + p_A + p_{E,hom}$$

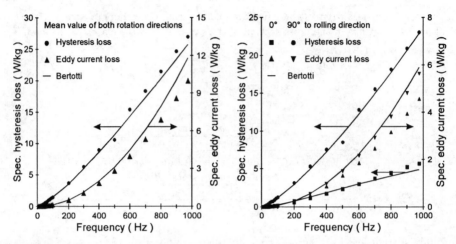

Bild 1: Eddy current loss p_E and dynamic hysteresis loss p_H calculated by means of the reluctance tensor and the corresponding values investigated by Bertotti's method for an oriented silicon iron steel sheet. Measurements are performed under rotational (left) and alternating (right) flux conditions at 0.35 T and 50 Hz, respectively.

From Fig. 1 we can conclude, that

$$c_H f + c_A f^{3/2} \;<\; p_H \qquad \text{and} \qquad c_E f^2 \;>\; p_E$$

The loss components sum up to the specific total power loss p with both methods of loss separation. But the components must be compared very carefully, because the physical assumptions are different in both methods.

6. Discussion

With the mathematical model of the reluctance tensor the field distribution inside an electrically conducting and magnetically anisotropic silicon iron steel sheet can be calculated. The eddy current loss calculated including the field penetration is smaller than that in the case of homogenous fields, because the lower the reluctance of the sample, the less is the field penetration into the bulk. This can be seen very clearly on the right hand side of Fig. 1. In order to fit the measured data of the total power loss, c_A is chosen in such a way, that the sum of statical hysteresis loss and anomalous loss is smaller than the calculated dynamic hysteresis loss p_H.

References

[1] M. Birkfeld, K.A. Hempel, "An Extended Description of the Behaviour of Steel Sheet under Rotational Flux Conditions", J.Magn.Magn.Mater. Vol.160 (1996) 17–18

[2] G. Bertotti, "Physical Interpretation of Eddy Current Losses in Ferromagnetic Materials: I Theoretical Considerations, II Analysis of Experimental Results", J.Appl.Phys., Vol. 57, No. 6 (1985) 2110–2126

[3] W. Salz, "A Two–Dimensional Measuring Equipment for Electrical Steel," IEEE Trans. Mag. Vol. 30 No. 3 (1994) 1253 –1257

[4] M. Birkfeld, "Investigation of the Permeability Tensor of Electrical Steel Sheet," IEEE Trans. Mag. (Submitted for publication)

Non-Linear Electromagnetic Systems
V. Kose and J. Sievert (Eds.)
IOS Press, 1998

Computationally Efficient Vector Hysteresis Model Using Flux Density as Known Variable

A. J. Bergqvist, S. A. Lundgren, S. G. Engdahl

Electric Power Eng., Royal Institute of Technology, SE–10044 Stockholm, Sweden

Abstract. A vector hysteresis model labelled the nonlinear vector Ishlinskiĭ model is described and compared with experiments on SiFe. Agreement is generally quite good though some modifications may be needed to achieve high accuracy for oriented materials. The model is simple and quite fast for arbitrary processes in any number of dimensions and adjustable hysteretic parameters are few and easy to determine.

1. Introduction

Devices such as transformers and electrical machines can be more accurately described at a design stage if hysteresis effects are incorporated in computer simulations. This can be done by coupling the macroscopic Maxwell equations to constitutive laws between field intensity $\vec{H}$ and flux density $\vec{B}$ that account for hysteresis. A hysteresis model should be capable of dealing with the various types of phenomena that can emerge in real applications. The local flux density will often vary in direction as well as size and, because of nonlinearity, will generally have a more or less irregular waveform. Hence, a model needs to be applicable for arbitrary vectorial processes. It should be capable of representing fundamental features of magnetic hysteresis such as: the net loss over a closed cycle $\oint \vec{H} \cdot d\vec{B}$ is always non–negative; as the material approaches saturation, the influence of past magnetic history is gradually erased; minor loops are observed to close or almost close after the first cycle; there is rotational hysteresis, meaning that if $\vec{B}$ is rotating in the plane, $\vec{H}$ also rotates and on average leads $\vec{B}$.

To be of practical value, a model needs to be computationally fast and economical in the sense of employing few variables to represent the influence of past history. Also, it should be easy to determine adjustable material parameters from experiments.

Several approaches to magnetic vector hysteresis modeling have been proposed over the years, the most widely known probably being the Stoner–Wohlfarth model [1] and vector Preisach models [2]. The former has a sound physical basis but is not well suited to soft magnetic materials and is in several ways demanding to use. Vector Preisach models are general enough to give accurate predictions but determination of material parameters requires extensive measurements and calculations and computation times tend to increase exponentially with the number of dimensions.

In this work a model is presented which is an attempt to reconcile these considerations. It may be seen as a modification of the Ishlinskiĭ model [3, 4] which was originally proposed as a model of mechanical hysteresis. While the approach has not received much attention in magnetics it has numerous attractive features including computational simplicity and speed in any number of dimensions and the capability to easily reproduce all the properties of magnetic hysteresis listed above. The current formulation, which will be called the nonlinear vector Ishlinskiĭ model, also employs

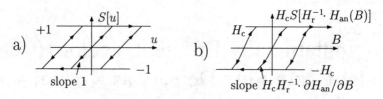

Figure 1: a) Stop element in one dimension. b) Scaled stop element.

only a small number of adjustable hysteretic parameters. Unlike the models mentioned above, it takes $\vec{B}$ as the known variable and $\vec{H}$ as the calculated which is often the desired computational order, a common example being when solving a problem with respect to the magnetic vector potential.

2. The Model

For soft materials, hysteresis phenomena are often neglected altogether and the relationship between $\vec{H}$ and $\vec{B}$ is expressed as $\vec{H} \approx \vec{H}_{\mathrm{an}}(\vec{B})$. $\vec{H}_{\mathrm{an}}(\vec{B})$ is called the anhysteretic curve and can be nonlinear or anisotropic or both and is the single–valued curve the medium would exhibit if the magnetization process could be carried out reversibly. This will due to hysteresis give some deviation from the actual field value. The difference is denoted here by $\vec{H}_{\mathrm{hyst}}$ and will depend on the trajectory path the $\vec{B}$–vector has gone through in time. We write

$$\vec{H}[\vec{B}] = \vec{H}_{\mathrm{an}}(\vec{B}) + \vec{H}_{\mathrm{hyst}}[\vec{B}] \tag{1}$$

The square brackets signify a dependence on the past history of the argument.

To express the hysteresis part, we employ the so–called *stop* hysteresis element S which in its simplest form can be defined in the following manner [4]. Given two vector variables $\vec{u}(t)$ and $\vec{w}(t)$ of arbitrary physical nature with $\vec{w}$ depending on $\vec{u}$, we state that if $\vec{w} = S[\vec{u}]$, then $d\vec{w} = d\vec{u}$ except that $|\vec{w}|$ may not exceed 1. That is,

$$\vec{w} = S[\vec{u}] \stackrel{\mathrm{def}}{\Longrightarrow} \vec{w} + d\vec{w} = \frac{\vec{w} + d\vec{u}}{\max(|\vec{w} + d\vec{u}|, 1)}. \tag{2}$$

This formula is also a useful approximation in numerical computations with $d\vec{u}$, $d\vec{w}$ replaced by $\Delta\vec{u}$, $\Delta\vec{w}$. It is drawn in the one–dimensional case in Fig. 1a. It also has vectorial hysteresis properties. For instance, if $\vec{u}$ rotates with amplitude > 1, $\vec{w}$ will rotate with amplitude 1 and lead $\vec{u}$ by some angle that depends on $|\vec{u}|$.

The resulting relationship between $\vec{u}$ and $\vec{w}$ has some schematic resemblance to the relationship between $\vec{B}$ and $\vec{H}_{\mathrm{hyst}}$ with some appropriate scaling functions $\vec{B} = \vec{\varphi}(\vec{u})$, $\vec{H}_{\mathrm{hyst}} = \vec{\psi}(\vec{w})$. For instance, with the more specific form $\vec{H}_{\mathrm{hyst}}[\vec{B}] = \overset{\leftrightarrow}{H}_{\mathrm{c}} \cdot S[\overset{\leftrightarrow}{H}_{\mathrm{r}}^{-1} \cdot \vec{H}_{\mathrm{an}}(\vec{B})]$, the coercivity in different directions is given by the tensor $\overset{\leftrightarrow}{H}_{\mathrm{c}}$, whereas the reversible reluctivity $\partial\vec{H}_{\mathrm{hyst}}/\partial\vec{B}$ is $\overset{\leftrightarrow}{H}_{\mathrm{c}}\overset{\leftrightarrow}{H}_{\mathrm{r}}^{-1} \cdot \partial\vec{H}_{\mathrm{an}}/\partial\vec{B}$ as illustrated in Fig. 1b.

While this should in many instances be a significant improvement over neglecting the hysteretic part altogether, minor loops are much simpler than for real materials. To express more complex minor loop nesting, we take a continuous distribution of such elements with $\overset{\leftrightarrow}{H}_{\mathrm{r}}^{-1}$ scaled by some value λ and $\overset{\leftrightarrow}{H}_{\mathrm{c}}$ scaled by $\zeta(\lambda)\,d\lambda$ and express $\vec{H}_{\mathrm{hyst}}$ as the integral

$$\vec{H}_{\mathrm{hyst}}[\vec{B}] = \overset{\leftrightarrow}{H}_{\mathrm{c}} \cdot \int_0^\infty \zeta(\lambda)\, S[\lambda\overset{\leftrightarrow}{H}_{\mathrm{r}}^{-1} \cdot \vec{H}_{\mathrm{an}}(\vec{B})]\, d\lambda \tag{3}$$

The major loop half–width is $\overset{\leftrightarrow}{H}_c \int_0^\infty \zeta(\lambda)\, d\lambda$, so it is natural to impose a normalization $\int_0^\infty \zeta(\lambda)\, d\lambda = 1$. Also, the reversible slope $\partial \vec{H}_{\text{hyst}}/\partial \vec{B}$ is $\overset{\leftrightarrow}{H}_c \overset{\leftrightarrow}{H}_r^{-1} \cdot \partial \vec{H}_{\text{an}}/\partial \vec{B} \int_0^\infty \lambda \zeta(\lambda)\, d\lambda$, prompting a second normalization $\int_0^\infty \lambda \zeta(\lambda)\, d\lambda = 1$. Thus the width of (infinitely) large loops and the slope of (infinitesimally) small loops are then independent of $\zeta(\lambda)$.

In numerical calculations the integral is approximated by a sum as

$$\vec{H}_{\text{hyst}}[\vec{B}] = \overset{\leftrightarrow}{H}_c \cdot \sum_{j=1}^{N} \zeta_j\, S[\lambda_j \overset{\leftrightarrow}{H}_r^{-1} \cdot \vec{H}_{\text{an}}(\vec{B})] \tag{4}$$

where $\zeta_j \sim \zeta(\lambda_j)\Delta\lambda_j$. Typically we use in the order of $N \approx 10$ terms.

The resulting model exhibits the main features of magnetic hysteresis listed in the introduction such as closure of minor loops, rotational hysteresis and erasure of past magnetic history near saturation. It can also be proven that $\oint \vec{H} \cdot d\vec{B} \geq 0$ for all cyclic processes if $\overset{\leftrightarrow}{H}_c$, $\overset{\leftrightarrow}{H}_r$ and $\nabla_{\vec{B}} \vec{H}_{\text{an}}(\vec{B})$ are all positive semidefinite.

3. Parameter determination

The adjustable parameters are on the one hand the anhysteretic curve $\vec{H}_{\text{an}}(\vec{B})$ and on the other the hysteretic parameters $\overset{\leftrightarrow}{H}_c$, $\overset{\leftrightarrow}{H}_r$ and $\zeta(\lambda)$. The determination of $\vec{H}_{\text{an}}(\vec{B})$ is a problem not unique to this model and will not be discussed in detail here.

As noted above, the components of $\overset{\leftrightarrow}{H}_c$ simply correspond to the coercivities in main directions and are thus very easy to determine from experiments. It may be remarked though that since for real materials, the loop width is not necessarily constant, it may be more robust to set $\overset{\leftrightarrow}{H}_c$ components to the average half–width of major loops. Similarly, since the reversible reluctivity is $(\partial \vec{H}_{\text{an}}/\partial \vec{B} + \vec{H}_c \overset{\leftrightarrow}{H}_r^{-1} \cdot \partial \vec{H}_{\text{an}}/\partial \vec{B})$, $\overset{\leftrightarrow}{H}_r$ can be determined from the slope of small minor loops in main directions. Very small and very large loops are independent of $\zeta(\lambda)$ as mentioned previously. However, loosely speaking, areas of medium–sized loops decrease as the standard deviation of $\zeta(\lambda)$ increases. Although an expression relating $\zeta(\lambda)$ to for instance the virgin curve can be derived, it appears that using $\zeta(\lambda) = e^{-\lambda}$ often works well enough, leaving no adjustable parameters for $\zeta(\lambda)$. In the discretized case, we can take $\zeta_j = 1/N$ and $\lambda_j = N \ln(2N/(2N - 2j + 1))/ \sum_i \ln(2N/(2N - 2i + 1))$ for $j = 1, ..., N$.

4. Results

Comparisons with quasistatic (in the range 5–10 Hz) experiments were made for non–oriented and oriented SiFe. Alternating losses in rolling (Q_{RD}) and transversal (Q_{TD}) directions and rotational losses (Q_{rot}) were considered as well as some B, H–curves in main directions. Measurements were done with a rotational single sheet tester [5]. For the sake of testing the model, the anhysteretic curve was assumed to have the form $\vec{H}_{\text{an}}(\vec{B}) = \nabla_{\vec{B}} F_{\text{an}}(\vec{B})$ with

$$F_{\text{an}}(\vec{B}) = \left(A_0^2 + A_1^2 B_x^2 + A_2^2 B_y^2 + A_3^2 B_x^2 B_y^2\right)^{1/2} + A_4 \exp[(A_5 B_x^2 + A_6 B_y^2 + A_7 B_x^2 B_y^2)^6] \tag{5}$$

with A_i, $i = 0, ..., 7$ adjustable. While not very accurate, it will serve here to illustrate the hysteresis model. Results are shown in Figs. 2 and 3. For the non–oriented sample, agreement is generally very good, except that Q_{rot} decreases near saturation in calculations but not in measurements. In this respect, typical literature results tend to agree with the model. For the oriented material certain deviation is observed, especially for Q_{TD} and Q_{rot}. One clear source for these errors is that the measured major loop in TD has variable width, unlike the calculation. Agreement could therefore be improved

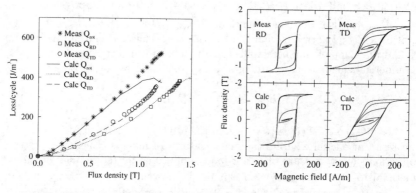

Figure 2: Measured and calculated losses and B, H–curves for non–oriented SiFe. $H_c = (69, 0; 0, 87)$ A^2/m^2; $H_r = (0.69, 0; 0, 7.8)$ A/m; $A_0 = 252$ Jm^{-3}, $(A_1, A_2) = (40, 140)$ Am^{-1}, $A_3 = 0$, $A_4 = 8.3$ Jm^{-3}, $(A_5, A_6) = (0.48, 0.69)$ T^{-2}, $A_7 = 0$.

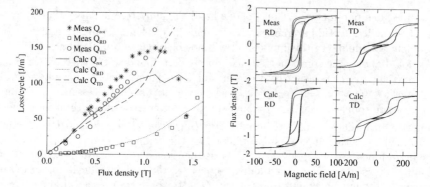

Figure 3: Measured and calculated losses and B, H–curves for oriented SiFe. $H_c = (12, 0; 0, 35)$ A^2/m^2; $H_r = (0.72, 0; 0, 5.2)$ A/m; $A_0 = 40.5$ Jm^{-3}, $(A_1, A_2) = (10, 135)$ Am^{-1}, $A_3 = 10$ Am^{-1}T^{-2}, $A_4 = 8.3$ Jm^{-3}, $(A_5, A_6) = (0.33, 0.62)$ T^{-2}, $A_7 = 0.55$ T^{-4}.

by generalizing the model to allow $\overset{\leftrightarrow}{H}_c$ to be a function of $\vec{B}$. Another problem is the unstable behaviour in (5) for high $\vec{B}$–values in the hard direction which gives the somewhat irregular curve form for the calculated Q_{rot} near saturation.

References

[1] E. C. Stoner and E. P. Wohlfarth, "A mechanism of magnetic hysteresis in heterogeneous alloys," *Trans. R. Soc. London*, vol. 240, p. 599, 1948.

[2] I. D. Mayergoyz, *Mathematical models of hysteresis*. New York: Springer, 1991.

[3] A. Y. Ishlinskiĭ, "General theory of plasticity with linear strain–hardening (russian)," *Ukrain. Matem. Zhurnal*, vol. 6, p. 314, 1954.

[4] A. Visintin, *Differential models of hysteresis*. Berlin: Springer, 1994.

[5] A. Lundgren, A. Bergqvist, and G. Engdahl, "A system for dynamic measurements of magnetomechanical properties of arbitrarily excited silicon-iron sheets," in *Nonlinear electromagnetic systems* (A. J. Moses and A. Basak, eds.), (Amsterdam), pp. 528 – 531, IOS Press, 1996.

Non-Linear Electromagnetic Systems
V. Kose and J. Sievert (Eds.)
IOS Press, 1998

The Moving Vector Preisach Model Applied To Metal Evaporated Tapes

Florence OSSART and Fabienne CORTIAL *
LEG (CNRS UMR 5529 INPG/UJF) - BP 46 - 38402 St Martin d'Hères - France
* LETI/DMITEC - CEA/G - 17, rue des Martyrs - 38054 Grenoble Cedex 9 - France

Abstract. A moving vector Preisach model is presented, which includes the basic features of particulate recording media: uniaxial anisotropy, switching and rotation of the magnetization, dependence of interaction on magnetization. The model is adapted to the very peculiar anisotropy of metal evaporated tapes and comparison with experimental data proves its accuracy.

1. Introduction

Magnetic hysteresis is a vector phenomenon including reversible and irreversible phenomena and it should be modeled as such for the simulation of magnetic recording applications. For some years now, the Preisach and Stoner-Wohlfarth models have been combined to build a phenomenological vector hysteresis model reaching a good compromise between accuracy and computational efficiency [1,2,3]. An excellent agreement with various experimental data was found for metal particles and barium ferrite tapes and the model was used for magnetic recording simulation [4]. The present work continues the investigation of the model's performances presented in [1], by studying the case of metal evaporated tapes. The peculiar anisotropy of this new kind of recording media [5] requires the model to be adapted. Comparison with experimental data is carried out for very different measurements: major loops; δM curves in both longitudinal and transverse directions; magnetization under a rotating field. The fit with experimental data is still very good, although the microstructure of this material, and hence the magnetization processes, are very different from those of particulate media.

2. The 3D moving vector Preisach model

2.1. Principle

The moving vector model is a combination of the Preisach and Stoner-Wohlfarth models. The two-state scalar hysteresis operator of the Preisach model, characterized by its coercive field hc and interaction field hi, is replaced by a vector hysteresis operator. Its behavior is given by the Stoner-Wohlfarth model (uniaxial anisotropy) and includes rotation and switching of the magnetization (Fig.1). The rotation of the magnetization is a purely reversible process, while the switching is an irreversible one.

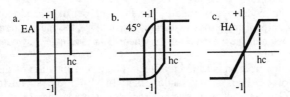

Fig. 1. Stoner-Wohlfarth elementary operator
Major loops calculated for a field along the easy axis (a), at 45° degrees (b) and along the hard axis (c)

The easy axes of the Stoner-Wohlfarth operators are distributed around the orientation direction of the medium. Each operator, called $SW_{hc,hi,\alpha,\beta}$, is characterized by its coercive and interaction fields hc and hi, and its easy axis direction given by the spherical coordinates α and β. The Preisach interaction field hi is assumed to act along the easy axis of each operator. At this stage, the model $m(H)$ is given by (1) :

$$m(\mathbf{H}) = \sum_{\text{operators}} SW_{hc,hi,\alpha,\beta}(\mathbf{H}) \qquad (1)$$

A mean interaction field proportional to the magnetization is added to the applied field. This defines the moving model $M(H)$, given by the self-consistent equation (2), in which k is the mean interaction factor.

$$\mathbf{M(H)} = \mathbf{m}\,(\,\mathbf{H} + k.\mathbf{M(H)}\,) \qquad (2)$$

The mean interaction field is intended to model the dependence of the interaction on the magnetization. The dominant kind of interaction in the material is experimentally estimated by the δM curve, which compares the remanent magnetization after application and subsequent removal of a field H for two different initial states (Fig. 2). The δM curve gives an indication of the relative stability of the ac-demagnetized and saturated states. It is usually negative, but it can be positive in some materials, revealing collective magnetization mechanisms [1,2,3].

It has been shown that the Preisach interaction field distribution leads to negative δM curves, while a positive mean interaction field can change the shape of the δM curve and make it positive, thereby allowing materials with positive interaction to be modeled.

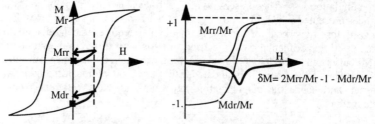

Fig. 2. Definition of the remanence curves: Mdr, Mrr and δM

Due to the manufacturing process, the anisotropy axis of metal evaporated tapes points out of the plane, at an angle between 20° and 40°. As the layer is so thin (less than 0.2 microns) there is an important demagnetizing field, modifying the effective field in the material. To model this behavior, the orientation direction of the model is tilted out of the plane at an angle given by experimental data, and a self-consistent computation is carried out to determine the equilibrium field. This process is similar to the one described above for the moving model, except that only the perpendicular component is concerned.

2.2 Parameters of the model and experimental data needed

The behavior of the model is governed by the Preisach density function p(hc,hi), a factor q adjusting the SW particle anisotropy (truncated astroid [1]), the mean interaction factor k, and the angular dispersion of the easy axes $g(\beta)$, where β is the angle between the easy axis and the orientation direction.

The Preisach density function is factored as the product of a function of the coercive field hc and a function of the Preisach interaction field hi. Both the coercivity and interaction distributions are Gaussians, characterized by the parameters *hc_center*, σ_hc and σ_hi . The angular dispersion is also a Gaussian characterized by its width σ_β and centered around the anisotropy direction β_ani .

The experimental data required to determine those parameters are the major and δM curves in the longitudinal direction. They are used in the trials and errors procedure presented in [1]. We use interpolation procedures to speed up the search of the parameters, but "know-how" is still very helpful in this task, which is the major weakness of the model.

3. Results and discussion

The results reported hereafter show the very good behavior of the model for metal evaporated tapes. Table I summarizes the main experimental characteristics of the medium and the values of the model parameters. S* is a parameter proportional to the slope of the major loop at the coercive field and important for recording performances. The values previously found for metal particles and barium ferrite are also reported to emphasize the difference between the materials.

	Experimental data					Parameters of the model				
	$\mu_0.H_c$ (T)	S Mr/Ms	S*	δM	βani (°)	σ_β (°)	$\mu_0.h_{c\_center}$ (T)	$\mu_0.\sigma\_h_c$ (T)	$\mu_0.\sigma\_h_i$ (T)	q	k
MP	0.152	0.81	0.70	>0	0	28	0.152	0.042	0.016	0.29	0.43
BaFe	0.129	0.66	0.80	>>0	0	56	0.138	0.041	0.012	0.20	0.94
MEv	0.107	0.75	0.43	<<0	27	21	0.129	0.028	0.017	0.02	1.25

Table I. Sets of parameters used for metal particles (MP), BaFe and metal evaporated (MEv) tapes

Figs. 3 and 4 show the major loop and δM curve in the longitudinal and transverse directions. No adjustment of any kind was made to fit experimental data in the transverse direction. Only in-plane measurements were carried out because perpendicular measures are very difficult due to the demagnetizing field.

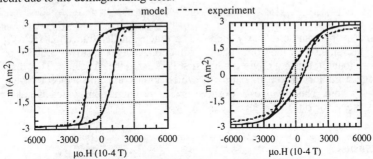

Fig. 3 : Major loops - left : longitudinal - right : transverse direction

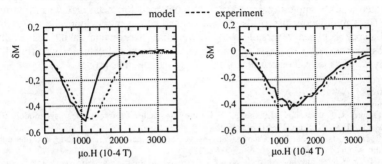

Fig.4 : δM curves - left : longitudinal - right : transverse direction

Fig.5 shows the magnetization tip trajectory when a 360° rotating field of constant magnitude is applied to the material initially saturated to the right. Fig. 6 displays the corresponding torque. The anisotropy is very well reproduced at low and high fields, when mainly rotation is involved. A discrepancy appears when the applied field is close to the coercivity, which is not surprising if one considers the complexity of the phenomena involved.

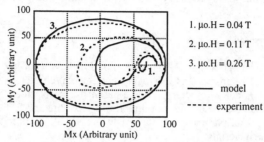

Fig.5 : Magnetization tip trajectory for an anti-clockwise rotating field

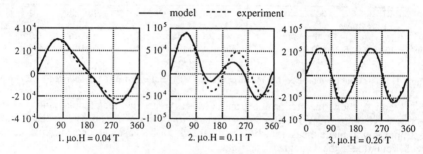

Fig.6 : Torque for a 360° rotating applied field

Compared to materials previously modeled, a much smaller truncation factor q is found, meaning that one is very far from coherent rotation. A larger positive feedback parameter is needed, although this is not obvious when first looking at the δM curve. This may be related to the microstructure of the medium, in which grains are much closer together than in particulate media. This involves a positive exchange interaction, which is balanced by the negative interaction due to the demagnetizing field.

4. Conclusion

The moving vector Preisach model has been adapted to the peculiar anisotropy of metal evaporated tapes, completing previous results obtained for metal particle and barium ferrite tapes. The simultaneous use of the δM curve and of the major loop allows us to find the right balance between reversible and irreversible phenomena taking place during the hysteresis process. A very good agreement with experimental behavior was found, which is the only way to assess the accuracy of this phenomenological model.

References

[1] F. Ossart, R. Davidson, S.H. Charap, A 3D moving vector Preisach hysteresis model, IEEE Trans. Magn., vol. 31, N°3, pp. 1785-1788, 1995.

[2] F. Ossart, R. Davidson, S.H. Charap, Moving vector Preisach hysteresis model and δM curves, IEEE Trans. Magn. , vol.30, N°6 , pp. 4260-4262, 1994

[3] H.A.J. Cramer, A moving Preisach vector hysteresis model for magnetic recording media, J. Magn. Mat., vol. 88, pp. 194-204, 1990.

[4] R. Davidson, S.H. Charap, Combined vector hysteresis models and applications, IEEE Trans. Magn., vol.32, N°5 , pp. 4198-4203, 1996

[5] H.J. Richter, R.J. Veitch, H. Hibst, H. Schildberg, Distribution profile of magnetization in ME tapes, IEEE Trans. Magn. , vol. 29, N°6 , pp. 3754-3756, 1993

Non-Linear Electromagnetic Systems
V. Kose and J. Sievert (Eds.)
IOS Press, 1998

Use of Stochastic Optimization Procedures in the Identification of the Generalized Moving Preisach Model

Bernhard BRANDSTÄTTER (^), Christian MAGELE (^)
Maurizio REPETTO (*) and Carlo RAGUSA (°)

(^) IGTE, Technische Universität Graz, Graz, Austria
(*) Dipartimento di Ingegneria Elettrica Industriale, Politecnico di Torino, Torino, Italy
(°) Istituto Elettrotecnico Nazionale "Galileo Ferraris", Torino, Italy

Abstract. The paper presents an identification method for the Generalized Moving Preisach Model (GMPM) by means of stochastic optimization procedures. The use of the GMPM can overcome some limitations of the Classical Preisach Model, but it lacks of established identification procedure. The use of a factorization of the Preisach Distribution Function and an analytical definition of the resulting one dimensional function allows one to reconstruct BH loops as a function of few parameters. The identification can thus be reduced to a best fit problem where optimization procedures have been widely used. Stochastic optimization routines are global optimizers so that they guarantee convergence to the global minimum of the function avoiding spurious solutions tied to local minima of the cost function. Results obtained by different procedures are shown and discussed.

1. Introduction

The Classical Preisach Model (CPM) is a powerful tool for the simulation of the hysteretic behaviour of ferromagnetic materials. However, as it is well known, not all the features of ferromagnetic hysteresis are well represented by CPM, congruency of minor loops, for instance, which is one of the main properties of CPM, is not usually found in ferromagnetic materials. Several modifications have been proposed to overcome this limitation adding a magnetization feedback to the CPM creating the well known Moving Preisach Model (MPM) [1]. It is proposed that the Generalized Moving Preisach Model (GMPM) [2] will add a more complex feedback to the CPM by using some physical insight of the magnetization process.

Although these models follow better the material behaviour, traditional identification procedures, developed for the CPM, cannot be directly used. Optimization procedures automatically looking for the minimum of a function can be used to identify the model, minimizing, for instance, the rms. error between measured and simulated hysteresis loops. The choice of optimization algorithm is very wide including classical deterministic algorithms (Newton, gradient etc.) and stochastic ones (Genetic Algorithms, Evolution Strategies, Simulated Annealing). Global optimization algorithms are preferred here because, despite their usually high computational cost, they can work on functions where no information can be gathered "a priori", like presence of many local minima, not differentiable points etc..

In the following a brief description of the GMPM used will be given together with an outline of the Optimization algorithms used; the results obtained will be discussed.

2. Generalized Preisach Model and identification

In order to override the congruency property of CPM many actions have been proposed. Usually the insertion of a feedback action acting between the applied magnetic field H_a and the magnetization M is employed. The effective magnetic field acting on the magnetic material H_e is considered to be given by:

$$H_e = H_a + f(M) \tag{1}$$

where f is a function of the material magnetization.

Function f should be determined by considering the physical phenomenon inside the material. MPM tries to accomplish this goal by adding to the applied field a linear term of M. A more thorough approach has been proposed by Bertotti et al. in [2]. By reasoning on the different contributes of the magnetization process to the free Gibbs energy, it is in fact possible to introduce a correction of the applied field as:

$$H_e = H_a + k_1 M + k_3 M^3 \tag{2}$$

where constants k_1 and k_3 are dependent on the material.

Using GMPM the identification problem requires one to find the usual Preisach Distribution Function (PDF) $\mu(\alpha,\beta)$, which rules the distribution in the Preisach plane of rectangular operators and, in addition, the two constants k_1, k_3. Following the approach proposed in [2] the factorization of the PDF can be performed by means of a Lorentzian function, PDF is thus expressed by:

$$\mu(\alpha,\beta) = \lambda(\alpha)\lambda(-\beta) \quad \text{with} \quad \lambda(\alpha) = N / \left(e^2 + (\alpha - d_0)^2\right) \tag{3}$$

where N is a normalization factor, d_0 is the lorentzian center and e is one half of the emivalue-width.

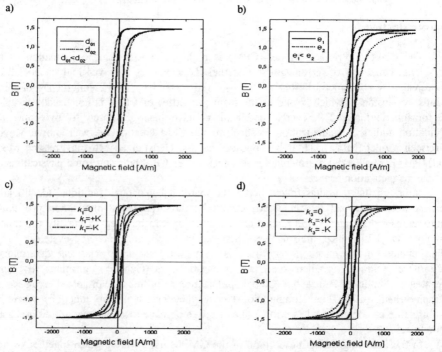

Fig. 1. Effects of the four parameters d_o, e, k_1, k_3 on the shape of the hysteresis loop. a) variation of lorentzian center; b) variation of emivalue-width; c) variation of k_1; d) variation of k_3.

The identification problem can be seen as a best fit problem of the experimental material characteristics as a function of the applied field and of the 4 parameters: d_0, e, k_1 and k_3.

$$M = GMPM(H_a, d_0, e, k_1, k_3) \tag{4}$$

In Fig. 1 the effects of the modifications obtained in the hysteresis loops by changing the four parameters one by one are shown.

3. Genetic algorithms

In nature, evolution can be taken as an example of a very efficient adaptation process of living organisms to their environment. Thus it is very likely to use nature analogous problem solving strategies to achieve an optimal adaptation of the given system.

The initialization procedure creates at random a population of solutions. At first, an individual is produced randomly, and its feasibility is checked with respect to the constraints. If it is not feasible, the algorithm adds a penalty term to its fitness value. The process is repeated until the number of the individuals in the population equals the specified population size, usually of 21. The selection is a process to choose some individuals of high fitness for breeding. Here the commonly used roulette wheel selection [3] is adopted. The operators used for genotypic represented chromosomes (floating point) are quite different from the classical ones, but, because of the intuitive similarities, we will divide them into the standard classes, mutation and crossover. Instead of using the classic coding by means of the binary system an arithmetical crossover is defined as a linear combination of two vectors: if s_w^t and s_v^t are to be crossed, the resulting offsprings are

$$s_v^{t+1} = a\, s_w^t + (1-a)s_v^t \quad \text{and} \quad s_w^{t+1} = a\, s_v^t + (1-a)s_w^t \tag{5}$$

The mutation tries to produce an offspring in the most promising direction. A mutation direction is given by the vector which is the difference between the best and the second best individual. A vector set up randomly around this mutation vector is added to the selected one.

It should be outlined, that this mutation is not dependent on any progress-parameter of the optimization strategy (e.g. no stepsize as it is used in evolutionary strategies) and is therefore a typical genetic zeroth order operator. The algorithm stops if no improvement is obtained for 500 iterations.

The genetic operators are iteratively applied corresponding to their probabilities: crossover is performed with a probability of 53%; mutation is carried out with a probability of 7% while the rest of the individuals are reproduced. These probability values are the main differences beteween Evolutationary Strategy and GA, in fact, in the first, mutation plays a central role, while in GA reproduction and recombination are the most important genetic operators. These optimal values have been located by a meta-GA where a combination of a stochastic and deterministic optimization strategy is employed. A deterministic optimization algorithm can be used to optimize the performance of GA changing its parameters. This is very costly but can increase convergence speed up to a factor of 100.

4. Results of the optimized identification

The proposed procedure has been implemented using Genetic Algorithm as optimizer. Other optimization runs were performed using Simulated Annealing and Evolutionary Strategy to assess the quality of the obtained results. The procedures were tested using a 3% wt SiFe NO material where 3 experimental symmetric loops were available at 0.5, 1 and 1.5 T, corresponding to peak values of magnetic field respectively of 43, 108, 1860 A/m. The maximum relative rms. error between computed and measured cycles was used as the cost function of the optimization. Each evaluation of the cost function required the computation of the integral of the PDF and the reconstruction of three hysteresis loops, this process took few CPU seconds on a Pentium 100 PC. Results obtained are shown in Table I while Fig. 2 reports measured and computed loops at 1 T, at the beginning and at the end of the procedure.

Table I
Results obtained with the three optimization procedures

	Lorentzian center A/m	Lorentzian mvw A/m	k_1 A/Tm	k_3 A/T^{3}m	iterations	final rms. value mT
Genetic Algorithm	27.00	28.64	6.63	-26.41	2385	6.67
Evolutionary Strategy	27.57	28.56	6.23	-24.69	8000	8.41
Simulated Annealing	26.58	28.97	7.08	-26.97	6041	7.53

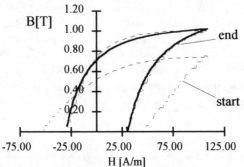

Fig. 2. Measured (solid line) and simulated (dashed line) hysteresis loop for the FeSi 3%wt material. Starting and final loops are shown.

5. Conclusions

The proposed identification procedure has shown the feasibility of the use of optimization algorithms in the determination of the parameters needed to evaluate GMPM with Lorentzian function. All the procedures tested have shown a good convergence speed to the optimal point and a good reproducibility of the result. As a matter of fact, an exploration of the cost function in the vicinity of the minimum showed that this point lies in a shallow valley where only one minimum point was present. This fact can suggest that, for this class of materials, deterministic optimization algorithms can be used as well, reducing then the computational cost of the identification procedure. The results obtained have also shown the good fitting obtained by means of the Lorentzian function which simplifies greatly the identification process.

References
[1] E. DellaTorre, *IEEE Trans. AUDIO*, Vol. 14, P. 86, 1966.
[2] V.Basso, G.Bertotti, A Infortuna and M.Pasquale, "Preisach model study of the connection between magnetic and microstructural properties of soft magnetic materials", *IEEE Trans. Mag.* vol.31, pp.4000 (1995).
[3] A. Gottvald, K. Preis, C. Magele, O.Biro, A. Savini, "Global optimization methods for computational electromagnetics", *IEEE Trans. Mag*, Vol. 28, No. 2, Mar. 1992, pp. 1537-1540.
[4] M. Marchesi, G. Molinari, M. Repetto, "Global optimization for discrete magnetostatic problems", *IEEE Trans. Mag*, Vol. 29, No. 2, Mar. 1993, pp. 1779-1782.
[5] Z. Michaliewicz, *Genetic Algorithms + Data Structures = Evolution Programs*, Springer Verlag, Berlin.

Non-Linear Electromagnetic Systems
V. Kose and J. Sievert (Eds.)
IOS Press, 1998

The State Dependent Reversible Magnetisation in Preisach Type Models

Mihai CERCHEZ, Laurenţiu STOLERIU, Alexandru STANCU
"Alexandru Ioan Cuza" University, Faculty of Physics, IASI, 6600, Romania

Abstract. In the scalar Preisach models the reversible part of magnetisation is computed in different ways. In the Complete Moving Hysteresis Model one is taking into account a distributed state dependent reversible magnetisation. An important problem of this model is that is always considered the fact that all pseudo-particles have the same squareness that is equal to the squareness of the ensemble. The Mixed Model takes into account a distributed squareness that is computed from physical considerations. The analysis of the squareness distribution for oriented and non oriented media is performed.

1. Introduction

The aim of this work is to study the hypothesis, used in the Preisach scalar models [1-4], related to the reversible part of magnetisation.

2. The reversible magnetisation in Preisach phenomenological models

The magnetisation of a ferromagnetic particle system has two component parts: an irreversible part and a reversible part. The classical Preisach model computes only the irreversible part of magnetisation and the apparent reversible magnetisation [5]; the reversible part of magnetisation is not taken into account. In certain models, different methods of computing the reversible part of magnetisation were proposed. In the scalar Preisach models there are two methods:

a) In the Generalised Preisach Model [1] one describes the reversible part of the magnetisation using step operators displaced along the $H_\alpha = H_\beta$ line in the Preisach plane, where H_α and H_β are the particle's switching fields. The reversible magnetisation depends only on the field value and not on the position that the particle has in the Preisach plane. The reversible magnetisation is zero when the applied field is zero. If M_S is the saturation of magnetisation and S is the squareness of the sample, then the reversible part of magnetisation is given by a function that has to be even and it must asymptotically approach the value $M_S(1-S)$ when the applied field is high enough.

The VD1 model (Vajda, Della Torre) [2] computes the reversible part of magnetisation also using step operators but in a moving model. The reversible magnetisation depends on the total magnetisation of the sample. It is presumed that the value of the moving parameter of the reversible part can be different from the value of the moving parameter of the irreversible part.

b) A more realistic way to take into account the reversible part of magnetisation is the state dependent reversible magnetisation. The first state dependent Preisach moving model was developed by Della Torre, Oti and Kadar (DOK model [3]). The hysteresis loop of the particle is decomposed into a rectangular loop that describes the irreversible part of the magnetisation and a non-linear function that describes the reversible part of magnetisation. The value of the reversible part depends on the irreversible part of magnetisation but not on the position of the magnetic particles in the Preisach plane. The irreversible magnetisation is computed using a moving Preisach model

In the Complete Moving Hysteresis (CMH) Model [4] a pseudo-particle is associated with all the real particles corresponding to a point in the Preisach plane with the same critical and interaction fields. One decomposes the hysteresis loop of a pseudo-particle into a fully reversible part that is given by a function and rectangular hysteresis loop that describes the irreversible part of magnetisation. The reversible magnetisation depends on the total magnetisation of the sample and on the interaction field that the pseudo-particle sees. The reversible magnetisation also depends on the magnetic state of the pseudo-particle. In the CMH model one presumes that all the pseudo-particles have the same squareness that is equal with the squareness of the ensemble.

3. The reversible magnetisation in the Mixed Model

In the Mixed Model [6], [7] a mean hysteresis loop of all particles corresponding to a small area in the Preisach plane is computed and the equivalent particle having this loop is called the pseudo-particle. To compute the magnetic moment of the pseudo particle, the Preisach plane is divided into uneven zones of the surface $\delta H_c \delta H_i$ that satisfy the condition that the saturation magnetic moment of the particles corresponding to each zone is the same for all zones, H_c and H_i being the critical field and the interaction field respectively. The critical field of a particle as a function of its anisotropy field H_k and orientation θ is computed using (1).

$$H_c = H_k/g(\theta), \text{ with} \tag{1}$$

$$g(\theta) = \left[(\sin\theta)^{2/3} + \xi(\cos\theta)^{2/3} \right]^{2/3} \tag{2}$$

where ξ is a parameter. If $\xi = 1$ then the relationship (1) describes the Stoner-Wohlfarth case. If $\xi = 1.5$ one obtains a more realistic case in concordance with micromagnetic calculations and measurements on single ferromagnetic particles [9]. Figure 1 presents $H_c(\theta)$ curves for three values of the ξ parameter.

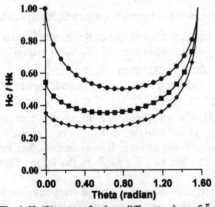

Fig. 1 $H_c(\theta)$ curves for three different values of ξ :

● $\xi = 1$, ■ $\xi = 1.5$, ◆ $\xi = 2$

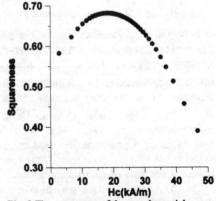

Fig. 2 The squareness of the pseudo particles as a function of the critical fields in the Mixed Model for a floppy disk sample

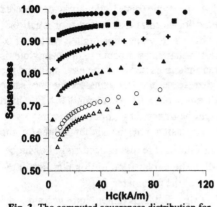

Fig. 3 The computed squareness distribution for log-normal distributed anisotropy fields
● $\theta_\sigma = 10°$, ■ $\theta_\sigma = 20°$, ✚ $\theta_\sigma = 30°$,
▲ $\theta_\sigma = 45°$, O $\theta_\sigma = 90°$, △ $\theta_\sigma = 180°$, $\xi = 1$

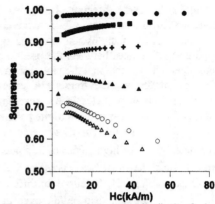

Fig. 4 The computed squareness distribution for log-normal distributed anisotropy fields
● $\theta_\sigma = 10°$, ■ $\theta_\sigma = 20°$, ✚ $\theta_\sigma = 30°$,
▲ $\theta_\sigma = 45°$, O $\theta_\sigma = 90°$, △ $\theta_\sigma = 180°$, $\xi = 1.5$

Using the distributions over the orientations, anisotropy and interactions and the dependence $H_c(\theta)$ one computes for each pseudo-particle the saturation magnetic moment and remanent magnetic moment. In this way the squareness of the pseudo-particle is obtained. Figure 2 presents the squarenesses of the pseudo-particles as a function of the critical fields for a non-oriented system of ferromagnetic particles. In this case the distribution over the orientations was considered uniform and the distribution over the anisotropies was considered Gaussian. The squarenesses of the pseudo particles are very different as a function of the critical field. The presumption that all pseudo-particles have the same squareness may lead to important errors for non-oriented media. In fact the particles with high squarenesses are the particles with small angles between the anisotropy axis and the applied field. If the parameter ξ increases, the number of particles with big squarenesses decreases at high values of the critical field. Figures 3 and 4 show the behaviour of the computed squareness distribution for an oriented medium, for different standard deviations of the angle distribution near the direction of the anisotropy. The field is

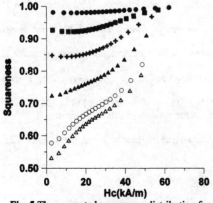

Fig. 5 The computed squareness distribution for Gaussian distributed anisotropy fields
● $\theta_\sigma = 10°$, ■ $\theta_\sigma = 20°$, ✚ $\theta_\sigma = 30°$,
▲ $\theta_\sigma = 45°$, O $\theta_\sigma = 90°$, △ $\theta_\sigma = 180°$, $\xi = 1$

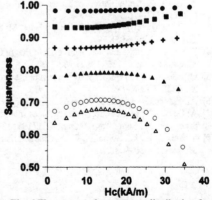

Fig. 6 The computed squareness distribution for Gaussian distributed anisotropy fields
● $\theta_\sigma = 10°$, ■ $\theta_\sigma = 20°$, ✚ $\theta_\sigma = 30°$,
▲ $\theta_\sigma = 45°$, O $\theta_\sigma = 90°$, △ $\theta_\sigma = 180°$, $\xi = 1.5$

applied in the same direction. In figures 3 and 4 the angle distribution was considered Gaussian and the anisotropy fields were considered log-normal distributed. For small θ_σ , the squarenesses of the pseudo-particles are near the same value. For higher values of θ_σ important differences between the pseudo-particles' squarenesses occur. This behaviour is consistent with the fact that the particles have different angles between the easy axis and the direction of the applied field. It is expected that some models considering the same squareness for all particles in the ensemble will work only for very small θ_σ. For high values of θ_σ the determination of the parameters will be affected by important errors.

If $\xi=1$, all the curves are continuously increasing but for higher values of the parameter ξ and for high values of θ_σ a maximum of the squareness is occurring. For higher critical fields, the squareness decreases very rapidly. The explanation of this fact lies in the modification of the $H_c(\theta)$ curve for different values of the parameter ξ .

If a Gaussian distribution over the anisotropy fields is taken into account, the behaviour of the squareness distribution is quite different. Figures 5 and 6 present the computed squareness curves as a function of the critical field for different values of θ_σ and for two values of the parameter ξ . For very small standard deviations of the angle distribution the presumption that all the pseudo-particles have the same squareness is a fairly good approximation.

4. Conclusions

The distribution of squareness depends mainly on the orientations of the easy axis of the particles and on the value of the parameter ξ . For high values of θ_σ the same squareness for all particles is no longer a good approximation and the squareness distribution must be taken into account. Due to its physical hypothesis, the squareness distribution is naturally integrated in the Mixed Model.

References

[1] I. D. Mayergoyz, G. Friedman, Generalised Preisach Model of Hysteresis, *IEEE Trans. Mag.* **24** (1988) 212-217.
[2] F. Vajda, E. Della Torre, Measurements of output-dependent Preisach functions, *IEEE Trans. Magn* **27** (1991) 4757-4762.
[3] E. Della Torre, J. Oti, G. Kadar, Preisach Modelling and Reversible Magnetization, *IEEE Trans. Magn.* **26** (1991) 3052-3058.
[4] F. Vajda, et. al., Analysis of reversible magnetization-dependent Preisach models for recording media, *J. Magn. Mag. Mat.* **115** (1992) 187-189.
[5] O. Benda, To the question of the reversible processes in the Preisach model, *Elect. Eng. J. Slovak Acad. Sc.* **6** (1991) 186-191.
[6] Al. Stancu, et.al., Mixed-Type Models of Hysteresis, *J. Magn. Magn. Mater* **150** (1995) 124-130
[7] Al Stancu, Numerical implementation of mixed-type models for recording media magnetization processes, *J. Magn. Magn. Mater.* **155** (1996) 22-24.
[8] Y. Luo, J.G. Zhu, Switching Field Characteristics of Individual Iron Particles by MFM, *IEEE Trans. Mag.* **30** (1994) 4080-4082.

Non-Linear Electromagnetic Systems
V. Kose and J. Sievert (Eds.)
IOS Press, 1998

Experimental Verification of a Preisach-Type Model of Magnetic Hysteresis

János FÜZI, Elena HELEREA, Danut OLTEANU
Electrical Eng. Dpt., "Transilvania" Univ., Politehnicii 1-3, 2200 Brasov, Romania

Amalia IVÁNYI
Electromagnetic Theory Dpt., Technical University, H-1521 Budapest, Hungary

Helmut PFÜTZNER
Bioelectricity and Magnetism Dpt., University of Technology, Vienna, Austria

Abstract. The classical Preisach model is constructed for grain-oriented and non-oriented electrical sheets in rolling and transverse directions based on measurements performed at very low frequency on a single-sheet tester. The surfaces yielding the Everett integrals defined on the Preisach triangles are plotted. The behaviour of the sheets at higher frequencies is computed - by eddy current simulation in time domain, but ignoring the effects of domain wall dynamics - as well as measured on the same single-sheet tester and the results compared to illustrate the effect of the adopted simplifying assumptions.

1. Construction of the model

The classical Preisach model [1] has been identified by measuring at very low frequency - 1 Hz - the first order reversal curves related to the major hysteresis loop (Fig. 1). The measurements have been performed on a single-sheet tester [2].

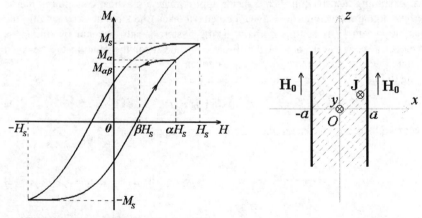

Fig. 1. Measurements for identification of the classical Preisach-model.

Fig. 2. Geometrical configuration

Two kinds of electrical steel sheet have been considered:

- laser scribed, grain-oriented: 27ZDKH95, thickness (Fig. 2) $2a = 0.27$ mm, produced by Nippon Steel.
- non-oriented: STABOCOR 250-50 A, thickness $2a = 0.5$ mm, produced by EBG.

Measurements have been performed in both rolling and transverse directions. The Everett surfaces constructed based on the experimentally obtained data are plotted on Fig. 3.

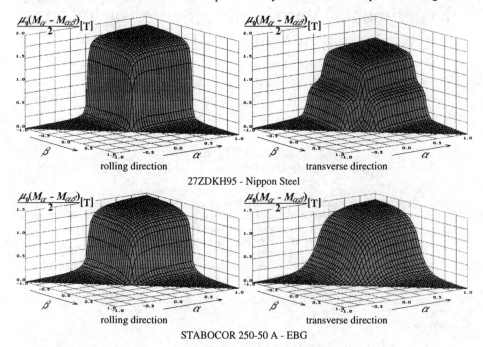

rolling direction   transverse direction

27ZDKH95 - Nippon Steel

rolling direction   transverse direction

STABOCOR 250-50 A - EBG

Fig. 3. Everett integral surfaces defined on the Preisach triangle

The classical Preisach model provides the static magnetic behaviour of the material of the sheets. Assuming that the considered sheets approximate a classical continuum, that is ignoring their domain structure, their dynamic behaviour can be simulated by solving the field equation system [3,4] in time domain, taking hysteresis into account by running a Preisach model at every grid point of the discretizing mesh for the relationship between magnetic field intensity and flux density at a local level.

Referring to Fig. 2, in a wide and long thin sheet we have:

$$\mathbf{H} = \mathbf{k}\, H(x,t) \quad ; \quad \mathbf{J} = \mathbf{j}\, J(x,t). \tag{1}$$

Then Maxwell's equations lead to the differential equation:

$$\frac{\partial H}{\partial t} = \frac{\dfrac{\partial^2 H}{\partial x^2}}{\mu_0 \sigma \left(1 + \dfrac{\partial M}{\partial H}\right)} \quad ; \quad x \in [0,a]\,,\ t \geq 0\,, \tag{2}$$

with the boundary conditions:

$$\frac{\partial H}{\partial x}(0,t) = 0 \quad ; \quad t \geq 0\,. \tag{3}$$
$$H(a,t) = H_0(t)$$

The non-linear term $(\partial M / \partial H)$ in the denominator of Eq. 2 is evaluated by the classical Preisach model. The input signal is the magnetic field intensity at the surface of the sheet

$(H_0(t))$. If the wave form of the output is given (e.g. sinusoidal total flux in the cores of power transformers), then the input is iteratively modified so that the output has the given value at each time-step. If neither the input, nor the output are given, then the larger problem has to be solved, considering also the characteristics of the supply circuits and coupling the circuit analysis with field computation in the ferromagnetic cores of the coils [5].

The average flux density over the cross-section of the sheet

$$B(t) = \frac{1}{a} \int_0^a \mu_0 \big[H(x, t) + M(H(x, t)) \big] dx \qquad (4)$$

is the only magnetic quantity (besides the field intensity at the surface) that can be measured.

2. Results

Measurements have been performed in dynamic operation at 50, 100 and 250 Hz. The length and width of the sheets were much larger (0.5 m) than their thickness. The same regimes have also been simulated, with the input signal (second Eq. 3) equal to the measured values of the field intensity at the surface with respect to time. Some of the resulting magnetizing loops are plotted on Figs. 4 - 6, where s stands for the static hysteresis loop, c for the computed loop, m for the measured dynamic magnetizing loops (average flux density over the cross-section versus the field intensity at the surface).

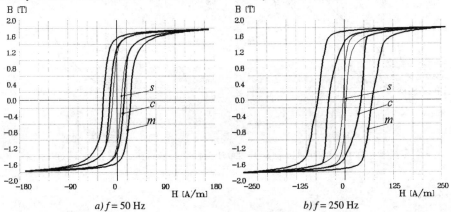

a) $f = 50$ Hz   b) $f = 250$ Hz

Fig. 4. Measured and computed loops for 27ZDKH95 magnetized in rolling direction

3. Conclusions

The advantages of the model (it can be constructed based on static measurements, not influenced by geometry, source frequency or wave-form, nevertheless taking into account their effects in dynamic operation) are obvious and it gives accurate results for materials that can be considered homogeneous (thick medium-carbon steel parts with tiny, randomly oriented domains).

In the case of electrical sheets however, especially grain-oriented sheets, the effects of domain-wall dynamics are too important to be ignored [6,7]. A hysteresis model that takes them into account is required to allow simulation with improved accuracy [8,9].

The slight anisotropy of the STABOCOR sheet (Fig. 3) indicates the presence of a definite domain structure in this non-oriented sheet too. The difference between numerical (taking into account the classical eddy currents only) and measured results also yields anomalous losses [6,7].

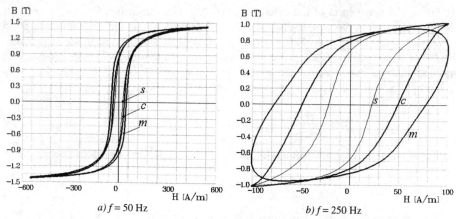

a) f = 50 Hz   b) f = 250 Hz

Fig. 5. Measured and computed loops for STABOCOR 250-50 A magnetized in rolling direction

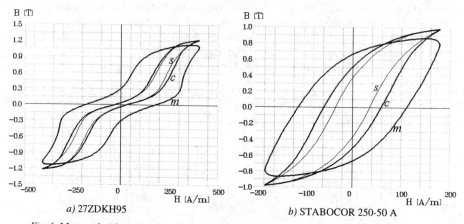

a) 27ZDKH95   b) STABOCOR 250-50 A

Fig. 6. Measured and computed loops at f = 250 Hz for sheets magnetized in transverse direction

References

[1] I.D. Mayergoyz, Mathematical Models of Hysteresis. Springer-Verlag, New-York, 1991.
[2] H. Pfützner, P. Schönhuber, On the Problem of the Field Detection for Single Sheet Testers, *IEEE Magn.* **27** (2) (1991) 778-785.
[3] R.M. Del Vecchio, Computation of Losses in Nonoriented Electrical Steels from a Classical Point of View, *J. Appl. Phys.* **53** (11) (1982) 8281-8286.
[4] J. Füzi, Eddy Currents in Ferromagnetic Sheets Taking Magnetic Hysteresis Nonlinearities into Account, *Per. Polyt. Budapest, Ser. El. Eng.* **39** (2) (1995) 131-143.
[5] J. Füzi, Coils with Ferromagnetic Core in Dynamic Operation Taking Magnetic Hysteresis into Account, *ACES Journ.* Monterey USA, **12** (2) (1997) 96-101.
[6] J.E.L. Bishop, Mechanisms Underlying Eddy Current Loss in Silicon Iron and Other Materials, *Anales de Fisica, Ser. B* **86** (1990) 208-213.
[7] H. Pfützner, P. Schönhuber, B. Erbil, G. Harasko, T. Klinger, Problems of Loss Separation for Crystalline and Consolidated Amorphous Soft Magnetic Materials, *IEEE Magn.* **27** (3) (1991) 3426-3432.
[8] G. Bertotti, Dynamic Generalization of the Scalar Preisach Model of Hysteresis, *IEEE Magn.* **28** (5) (1992) 2599-2601.
[9] L-L. Rouve, F. Ossart, T. Waeckerle, A. Kedouc-Lebouc, Magnetic Flux and Losses Computation in Electrical Laminations, *IEEE Magn.* **32** (5) (1996) 4219-4221.

Non-Linear Electromagnetic Systems
V. Kose and J. Sievert (Eds.)
IOS Press, 1998

Construction of Dynamic Preisach Model
Based on Simulation

János FÜZI

Electrical Eng. Dpt., "Transilvania" Univ., Politehnicii 1-3, 2200 Brasov, Romania

Abstract. The dynamic Preisach model is identified for a ferromagnetic plate by simulation of dynamic behaviour based on the classical Preisach model and computation of electromagnetic field diffusion. The time constants are computed along the major loop and the first-order reversal curves during the process of relaxation of the output (magnetic flux through the cross-section of the core) to its static value subsequent to input (magnetic field intensity on the surface of the core) variations. The obtained model is used to simulate dynamic behaviour of the plate without field computation iterations, thus saving a considerable amount of computer time and especially of required memory.

1. Computation of electromagnetic field diffusion in ferromagnetic media

The classical Preisach-model [1] is a numerical model of static hysteresis, relatively easy to implement and, completed with adequate field computing methods, applicable to various problems involving ferromagnetic media. Coupled with electrical circuit equations, the resulting code can be an efficient and accurate tool of analysis for circuits involving coils with ferromagnetic cores [2].

Evaluating in Maxwell's equations:

$$\nabla \times \mathbf{H} = \mathbf{J}$$

$$\nabla \times \mathbf{E} = -\frac{\partial \mathbf{B}}{\partial t} \qquad \mathbf{J} = \sigma \mathbf{E} \qquad , \qquad (1)$$

$$\nabla \cdot \mathbf{B} = 0 \qquad \mathbf{B} = \mu_0 [\mathbf{H} + \mathbf{M}(H)]$$

the magnetization vector $\mathbf{M}$ by means of a static hysteresis model, the dynamic operation can be simulated. If the field intensity vector does not change its direction, a scalar model is sufficient to describe the relationship between field intensity and magnetization. Using the classical Preisach model accurate results can be expected in case of materials in which most of the dynamic effects are due to classical eddy currents and the effects of domain-wall motion and magnetic viscosity can be ignored [3].

Considering a wide and long ferromagnetic plate (Fig. 1) with thickness $2a$ and symmetric source field along the two surfaces, Maxwell's equations lead to the differential equation:

$$\frac{\partial H}{\partial t} = \frac{\dfrac{\partial^2 H}{\partial x^2}}{\mu_0 \sigma \left(1 + \dfrac{\partial M}{\partial H}\right)} \qquad ; \qquad x \in [0, a] \ , \ t \geq 0 . \qquad (2)$$

In the boundary conditions:

$$\frac{\partial H}{\partial x}(0, t) = 0$$
$$H(a, t) = H_0(t)$$
$$; \quad t \geq 0 \tag{3}$$

$H_0(t)$, the source field at the surface of the plate is either given - the natural case, as in most hysteresis models the field intensity is the input and magnetization the output quantity - or has to be determined from a condition conditioned by the output - the technical case, as in power transformers the flux through the core is sinusoidal. While in the first case the solution of Eq. 2 is obtained straightforwardly by a time-stepping method, in the second case the input has to be adjusted in an iterative way until the given output is obtained within a prescribed error. The hysteresis is by nature a history dependent phenomenon. During the iterative process within one time-step, the Preisach configuration at the beginning of the step has to be stored as a starting situation, so that artificial history is not introduced.

For a plate with $2a = 10$ mm, resistivity $1/\sigma = 48 \cdot 10^{-8}$ Ω m and static magnetic characteristic plotted in Fig. 2, the wave-forms of magnetic field intensity, flux density and eddy current density at different depths in the plate can be seen in Fig. 3.a for sinusoidal source-field intensity and Fig. 3.b for sinusoidal average flux density (sinusoidal total flux) over the thickness respectively (frequency: $f = 50$ Hz). The curves are plotted for 20 equal divisions of plate half thickness (steps of 0.25 mm).

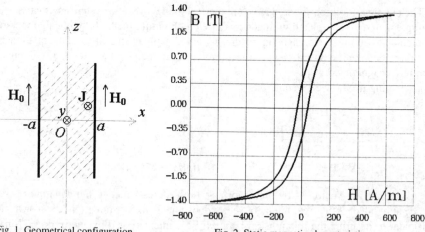

Fig. 1. Geometrical configuration   Fig. 2. Static magnetic characteristic

The computation of time-varying electromagnetic field inside the cores usually requires a large memory and slows down the simulation very much. Once the classical Preisach-model is identified [4] for a certain material, the dynamic behaviour of cores of different geometries in various regimes can be obtained by simulation [5], not needing further experimental measurements.

2. Identification of the dynamic model

A possible way to combine the advantageous features of this approach with that of a dynamic hysteresis model (lower memory size and less computer time required) is to construct the dynamic model by simulation of the otherwise difficult and expensive measurements to be performed for its identification.

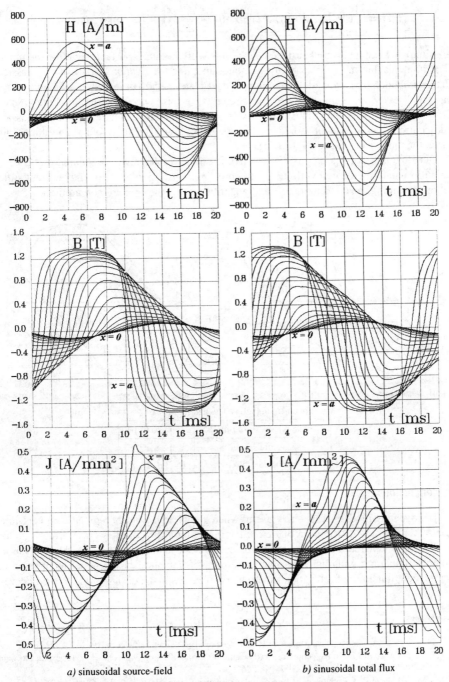

a) sinusoidal source-field   b) sinusoidal total flux

Fig. 3. Wave-form of field quantities at 50 Hz

The relaxation processes subsequent to input variations along the major loop and the first-order reversal curves are simulated and the corresponding time constants determined. Thus the measurements required for identification of the coefficient of the second term of the hysteresis model [1]:

$$f(t) = \iint_{\alpha \geq \beta} \mu_0(\alpha, \beta) \, \hat{\gamma}_{\alpha\beta} \, u(t) \, d\alpha \, d\beta \; + \frac{df}{dt} \iint_{\alpha \geq \beta} \mu_1(\alpha, \beta) \, \hat{\gamma}_{\alpha\beta} \, u(t) \, d\alpha \, d\beta \qquad (4)$$

are simulated and an Everett surface constructed with the time constants to allow the numerical implementation of this part of the model, just like the Everett surface of the classical Preisach model. The dynamic behaviour of steel plates at 50 Hz in the considered regimes can be observed in Fig. 4. The effect of source wave form can be observed, namely that if saturation is reached, then sinusoidal flux condition leads to considerably smaller classical eddy current losses than sinusoidal surface field (narrower dynamic loop). The thinner the plate, the more accentuated the difference.

Once the model for a core is constructed, it can be implemented in the analysis code of circuits containing the coil with the given core - not needing to perform eddy current simulation, providing the distribution of field quantities inside the core are not required.

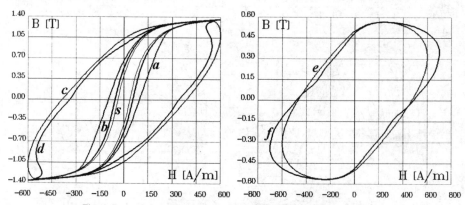

Fig. 4. Dynamic magnetizing loops obtained for the cases (s - static loop):

plate thickness 2a [mm]	1	4	10
sinusoidal field	a	c	e
sinusoidal flux	b	d	f

3. Conclusions

The dynamic model is especially useful when global behaviour of a given core is of interest rather than the distribution of field quantities within the core (e.g. skin effect). Its use allows important reduction of required memory and computing time in the analysis of circuits containing coils with ferromagnetic cores. Simulation of the experiment required for the determination of relaxation times can be effective in the design process of coil cores.

References

[1] I.D. Mayergoyz, Mathematical Models of Hysteresis. Springer-Verlag, New-York, 1991.
[2] J. Füzi, Coils with Ferromagnetic Core in Dynamic Operation Taking Magnetic Hysteresis into Account, Proc. 7th IGTE Symposium, Graz (1996) 460-465.
[3] R.M. Del Vecchio, Computation of Losses in Nonoriented Electrical Steels from a Classical Point of View, *J. Appl. Phys.* **53** (11) (1982) 8281-8286.
[4] J. Füzi, Parameter Identification in Preisach Model to Fit Major Loop Data, Dig. 4th Japan-Hungary Joint Seminar, Fukuyama, Japan (1996) 35-38.
[5] D.A. Philips, L.R. Dupre, Macroscopic Fields in Ferromagnetic Laminations Taking into Account Hysteresis and Eddy Current Effects, *J. Magn. Magn. Mat.* **160** (1996) 5-10.

Non-Linear Electromagnetic Systems
V. Kose and J. Sievert (Eds.)
IOS Press, 1998

Comparison of Identification Procedures for the Preisach Model Distribution

Vittorio BASSO (°), Giorgio BERTOTTI (°), Oriano BOTTAUSCIO (°),
Mario CHIAMPI (*), Daniela CHIARABAGLIO (°), Massimo PASQUALE (°),
Maurizio REPETTO (*) and Carlo SASSO (°)

(°) IEN Galileo Ferraris, INFM and GNSM-CNR C. M. d'Azeglio 42, I-10125 Torino, Italy
(*) Dip. Ing. Elettrica Ind. - Politecnico di Torino, C. Duca degli Abruzzi 24,
I-10129 Torino, Italy

Abstract. The paper presents different techniques for the identification of the Preisach model of hysteresis. Two identification procedures assume a numerical Preisach distribution, while the third one assumes an analytical distribution. The three methods are applied to the analysis of a NO 2.5% wt SiFe electrical steel.

1. Introduction

Recent studies have shown that the Preisach model of hysteresis can be successfully used to predict the static and dynamic hysteresis properties of soft magnetic materials [1,2]. This hysteresis modeling procedure has been also applied to extend the capabilities of Finite Element packages [3]. In the Preisach model the magnetization M, sum of irreversible (M_{irr}) and reversible (M_{rev}) contributions, is given as the superposition of elementary bistable loops $\Phi(\alpha,\beta)$ with up and down switching fields α and β:

$$M = M_{irr} + M_{rev} = \iint_{\alpha>\beta} d\alpha\, d\beta \left(P_{irr}(\alpha,\beta) + P_{rev}(\alpha)\delta(\alpha-\beta) \right)\Phi(\alpha,\beta) \qquad (1)$$

where δ is the Dirac delta function. $P_{irr}(\alpha,\beta)$ and $P_{rev}(\alpha)$ are the irreversible and reversible parts of the Preisach distribution (PD), that is the weight function of the elementary contributions. Furthermore, for a correct description of hysteresis curves, the effective field H, that drives the magnetization process, is related to the applied field H_a by a mean field relation $H=H_a+k_1M$.

The identification procedure requires the determination of $P_{rev}(\alpha)$, $P_{irr}(\alpha,\beta)$ and mean field parameter k_1. Mayergoyz solved the identification problem for $k_1=0$ (Classical Preisach model), using a large amount of measured data [4]. Further complexity arises with the presence of the mean field: the determination of its parameter is still an open problem, strongly connected with the PD shape. Moreover, a reduction of input data needed for the identification is also expected, by making some assumptions, justified on physical basis for a given class of magnetic materials [1].

In this paper the authors present three identification methods and discuss their merits and drawbacks when applied to a NO 2.5% wt SiFe electrical steel core. For all the proposed approaches, the identification of the reversible contribution $P_{rev}(\alpha)$ is performed by evaluating the slope at turning points of some minor loops. As regards the irreversible distribution, three different methodologies are compared. One method, based on the one described by Mayergoyz [4], requires the analysis of a set of minor loop measurements from which the PD is numerically calculated. A second identification method, obtained by a generalization of the one described by Kadar [5], can be used if the factorization $P_{irr}(\alpha,\beta) =\varphi(\alpha)\varphi(-\beta)$ holds true; in this case $P_{irr}(\alpha,\beta)$ is obtained from the full hysteresis loop. Finally, a procedure assuming the previous factorization and a Lorentzian curve for $\varphi(\alpha)$ is employed. In this case the coefficients of the Lorentzian function are derived from the shape of two measured hysteresis curves. For all the methods the mean field effects are taken into account and the k_1 value is obtained by an optimization procedure that is able to find the best agreement between measured loops and model predictions in a wide range of peak inductions (Fig. 1). In the next sections the three methods are described and applied to the prediction of static behavior of magnetic cores, comparing the computed results with the experimental ones.

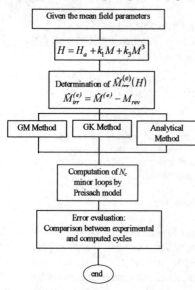

Fig. 1 - Scheme of the procedures

2. Generalized Mayergoyz (GM) method

As outlined by Mayergoyz, the set of experimental data needed for the identification is a set of first order hysteresis curves. Although it was originally shown in the case of return branches, it can be formulated also for symmetric minor loops $M(H,H_p)$, where H is the current field and H_p the peak field. The expression of the PD is:

$$P_{irr}(H,H_p) = -\frac{1}{2}\frac{\partial^2 M(H,H_p)}{\partial H \partial H_p} \qquad (2)$$

The main problem is that the PD must be positive, in order to represent energy dissipating loops, while it can sometimes be experimentally found to be negative. We have therefore developed a suitable algorithm that numerically calculates the distribution from Eq.(2), using experimental data for $M(H,H_p)$, and eliminates possible negative values of distribution. Performing this operation for a range of k_1 values, we have verified that there is a threshold value k_0, such that for the region $k_1 > k_0$ a good agreement between experimental and calculated loops is found. In performing the reconstruction process from the experimental minor loops, we systematically choose the value k_0. Measurements have been performed on non-oriented 2.5% wt SiFe. Two different sets of minor loops: 20 loops up to 1 T and 20 loops at high fields, up to 1.5 T, were performed by sequentially increasing the peak fields.

3. Generalised Kadar (GK) method

The identification procedure proposed by Kadar computes the PD by using the experimental limit cycle [5]. This method is based on a factorization of the PD given as $P_{irr}(\alpha,\beta) = \varphi(\alpha)\varphi(-\beta)$. In this case the first derivative along a limit cycle becomes:

$$\frac{\partial M_{irr}}{\partial H} = \varphi(H)\int_{-\infty}^{H} d\beta \varphi(-\beta) \qquad (3)$$

If the descendent part of the limit cycle is sampled in N points, we have N values of the first derivative of Eq.(3). Taking the N values of the sampled function $\varphi_i = \varphi(\alpha_i)$ as unknowns, the integration can be carried out numerically and equated to the corresponding experimental variation of the magnetisation. The results of this procedure is a nonlinear system of equations whose j-th row is given by:

$$\frac{\Delta M_{irr,j}}{\Delta h_j} = \varphi_{N-j}\left(\sum_{i=1}^{j}\varphi_i \Delta h_i\right) \qquad (4)$$

where Δh_j is the experimental field step, $\Delta M_{irr,j}$ is the experimental variation of the irreversible magnetization. The nonlinear system is solved by means of a successive substitution technique which usually converges in 50-100 iterations.

In this case the mean field relation was taken in a more general form as $H=H_a+k_1M + k_3M^3$. Function φ was calculated from the limit cycle by solving the system of Eq.(4); the values of k_1 and k_3 are determined by a pattern search optimisation algorithm, to minimize the differences between experimental and calculated minor loops of various peak inductions. The optimisation procedure generally estimates the mean field parameters in less then 100 steps.

4. Analytical method

The analytical approach is based on the factorization $P_{irr}(\alpha,\beta) =\varphi(\alpha)\varphi(-\beta)$ and $\varphi(\alpha)$ is assumed to be Lorentzian:

$$\varphi(\alpha) = \sqrt{M_{irr}^{(s)}N_{irr}} \left/ \left[1+\left(\frac{\alpha-d_0}{H_0} \right)^2 \right] \right. \tag{5}$$

where H_0, d_0, $M_{irr}^{(s)}$ are parameters to be identified, and $N_{irr}=2/(\pi H_0^2(\pi+2 \arctan(d_0/H_0)))$ is the normalization factor. With the factorized distribution the expression of the first derivative of hysteresis curves is simplified as in Eq.(3). The derivatives of lorenzian $\varphi(\alpha)$, and also minor loops and the first magnetization curve, can be written as analytical functions. Hence we perform nonlinear fit of two measured curves (numerical first derivative of the first magnetization curve and a minor loop) with the predictions of the lorenzian distribution. The value of the three parameters involved is found by a fast least square algorithm. This procedure can be iterated with different values of the mean field parameter k_1, in order to find a value that minimizes the difference between experimental curves and model predictions.

5. Discussion of results

The proposed methods are applied to the analysis of a non-oriented 2.5%wt SiFe electrical steel core. For the GM procedure, the mean field parameter k_1 was found to be equal to 10 A/(Tm). As regards the GK method, the mean field parameters given by the optimization procedure were: $k_1 = 28$ A/(Tm) and $k_3 = 28.5$ A/(T^{3}m). For the analytical investigation, the parameters found were: mean field $k_1 = 25$ A/(Tm); irreversible part $H_0= 34$ A/m; $d_0 = 46$ A/m; $M_{irr}^{(s)} = 1.3$ T.

The hysteretic cycles were reconstructed for different values of imposed magnetic field. Fig. 2 shows the computed cycles in comparison with the experimental ones for a low and a high magnetization value, respectively. The analysis of these diagrams evidences that all the proposed methods provide good recontructions of the experimental cycles. In particular, the GM method reproduces better the small cycles, while the GK method gives good results for cycles near saturation. The analytical procedure provides satisfactory reconstructions, even if more discrepancies with respect to the numerical ones are found; however, the merit of this last procedure is that the identification of the material is given by few parameters.

The mean field effect is evident in Fig. 2a) which also presents the cycle computed by the GK method imposing the value of parameters k_1 and k_3 equal to zero. In this case the cycle is completely different from the experimental one, showing that the Moving Preisach model is required.

Finally, function $\varphi(\alpha)$ obtained by the GK method is compared in Fig. 3 with the analytical Lorentzian one, showing that the two curves are very similar. The normalized up switching field α varies in the Preisach plane from -0.5 to 0.5, but in the diagram only the interval $-0.1 \div 0.1$ is reported. Fig. 3 also shows how the mean field effect influences the shape of function $\varphi(\alpha)$. This result seems to confirm that the introduction of the mean field effect is not due to the choice of a Lorentzian function; in fact, the Lorentzian-like shape of $\varphi(\alpha)$ is also found using the GK method which does not impose constraints on the PD shape.

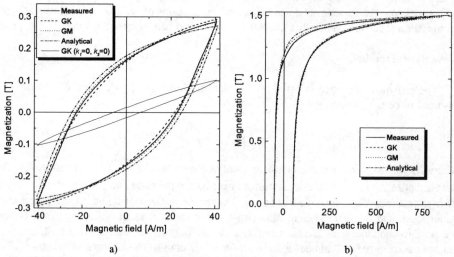

a)   b)

Fig. 2 - Measured and computed hysteresis cycles: a) low magnetization; b) high magnetization.

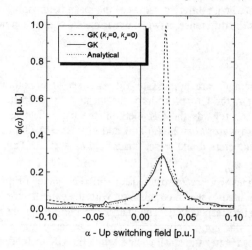

Fig. 3 - Normalized function $\varphi(\alpha)$ for generalized Kadar and analytical approach.

References

[1] V. Basso et al., *IEEE Trans. Magn.*, 31 (1995), pp. 4000-4005.
[2] L.L. Rouve, Th. Waeckerle, A. Kedous-Lebouc, *IEEE Trans. Magn.*, 31 (1995), pp. 3557-3559.
[3] O. Bottauscio, M. Chiampi, M. Repetto, *J. Mag. and Mag. Mat.* 160 (1996), pp. 96-97.
[4] I.D. Mayergoyz, Mathematical Model of Hysteresis. Springer-Verlag, New York, 1991.
[5] G. Kadar, *J. Appl. Phys.* 61 (1987), pp. 4013-4015.

Non-Linear Electromagnetic Systems
V. Kose and J. Sievert (Eds.)
IOS Press, 1998

The Influence of Production Parameters in Optimising Magnetic Materials

Turgut MEYDAN and Hakan KOCKAR

Wolfson Centre for Magnetics Technology, University of Wales, Cardiff, UK

Abstract. In recent years, there have been increasing efforts to design and develop magnetic thin-films and multilayer materials due to their potential applications. It is feasible to produce thin films and multilayers using a novel Rotating Cryostat (RC) technique. The first step of this investigation is to deposit thin iron films in order to systematically understand the deposition process and parameters. Thus an orthogonal design technique for process optimisation in the thin iron film production by the RC system is applied, and the influence of the input parameters over the output functions is investigated and reported here.

1. Introduction

Magnetic thin-film devices, for example thin-film inductors and micro-sensors, are of interest in engineering applications. Soft magnetic thin films with a large magnetostriction can aid to the development of miniaturised magnetoelastic transducers and actuators. In order to achieve this goal, a novel Rotating Cryostat (RC) technique has been used to produce magnetic materials. To understand the system and its characteristics, initially powdered iron was vaporised using a resistively heated tungsten furnace on to a liquid nitrogen cooled polyimide substrate. In this paper, the systematic analysis of the magnetic material production has been accomplished using an orthogonal design technique for production process optimisation. The method is based on studying the relationship of input parameters and output functions, and gives the maximum information with the least number of experiments.

2. Experimental

This method [1] involves finding the setting for each input parameter that optimises the output(s) of the process. The simplest method of process optimisation is the one dimensional research. The other extreme is a full dimensional research exemplified by factorial designs. A number of fractional factorial design approaches have been developed in which certain subsets of the full dimensional search are used to optimise the process. One such approach is orthogonal design. Orthogonal design was developed by Taguchi [1]. Currently, more than thirty orthogonal tables have been derived and developed using linear algebra and appropriate statistical data analysis techniques. Applications of orthogonal design cover all facets of science and technology development. It has a broad application areas such as; optimisation of thin film deposition equipment, plasma etching, photoresist processing, and optical stepper development. The method is based on studying the relationship between process and input parameters and their corresponding output functions by selecting certain representative combinations of the input parameter level settings. The maximum amount of information can be obtained using the least number of experiments, by following the orthogonal tables. In this case, 4 input parameters each with 3 level settings would require only 9 experimental runs, as opposed to 81 runs needed to achieve the optimised condition in the full multidimensional space. It is the essence of the orthogonality that allows statistical analysis of data that, in effect, fills in the "blanks" in the full factorial analysis that are not run as experiments in the orthogonal design. Therefore, the first order correlation between the level settings of the input parameters and the output function values can be explicitly obtained without the necessity of completing the much larger number of experiments required by the full factorial analysis.

The thin iron films were deposited from iron vapour on to a polyimide (Kapton™) substrate. The substrate was attached to the drum of the RC at ambient temperature, which was later filled with liquid nitrogen and rotated at 2000 rpm. Iron powder (99.0 % pure, 1 to 450 micron in diameter) was vaporised from a resistively heated tungsten furnace [2]. The RC system was operated for an hour for each experiment. The thickness uniformity of samples was calculated from the thickness versus distance across the film measured from one edge. The uniformity is expressed as a percentage and is given by ((x/total width of film)*100%), where x refers to film width over which the thickness varies within ± 5 %. The value of x was first established from the film profile. In the case of run 5, x=6.14 mm, therefore the uniformity = (6.14/20)*100, which is 30.7%. The deposition rate was calculated from the average thickness value divided by run-time. For example, for run 5, the deposition rate is 0.0120 nm/s.

3. Results and Discussion

Orthogonal design process has been applied to optimise the thin iron film produced by a RC technique. In this system, it is possible to vary the furnace orifice shape (f. shape), the mass of the iron powder, the power of furnace, and the gap between substrate and source. Therefore, these four variables serve as the input parameters. If we start with a baseline process using (2*4) mm^2 f. shape, iron powder of 90 mg, 355 Watt furnace power, and 24 mm gap, a level variation using one higher and one lower level was listed, yielding 3 level settings for each input parameter. In this experimental set-up, four input parameters, each with three level settings, fit the orthogonal matrix $L_9 3^4$ shown in Table 1 (The numbers in brackets in the table indicate the level settings). The output functions are the uniformity across the film and the deposition rate. Table 1 shows not only the results for each of 9 experiments required by the $L_9 3^4$ matrix but also two extra experiments labelled 1' and 1". These two extra experimental runs were carried out under the identical condition to run 1 for repeatability analysis.

Input Parameters					Output Functions	
Run	F. Shape (mm^2)	Mass (mg)	Power (Watt)	Gap (mm)	Uniformity (%)	Deposition Rate (nm/s)
1	(1) 4*4	(1) 60	(1) 330	(1) 20	U1=13.9	D1=0.057
2	(1) 4*4	(2) 90	(2) 355	(2) 24	U2=44.3	D2=0.013
3	(1) 4*4	(3) 120	(3) 380	(3) 28	U3=31.7	D3=0.075
4	(2) 2*4	(1) 60	(2) 355	(3) 28	U4=12.9	D4=0.010
5	(2) 2*4	(2) 90	(3) 380	(1) 20	U5=30.7	D5=0.012
1'	(1) 4*4	(1) 60	(1) 330	(1) 20	U1'=38.1	D1'=0.052
6	(2) 2*4	(3) 120	(1) 330	(2) 24	U6=55.4	D6=0.163
7	(3) 4*2	(1) 60	(3) 380	(2) 24	U7=25.8	D7=0.024
8	(3) 4*2	(2) 90	(1) 330	(3) 28	U8=18.8	D8=0.094
9	(3) 4*2	(3) 120	(2) 355	(1) 20	U9=25.8	D9=0.152
1"	(1) 4*4	(1) 60	(1) 330	(1) 20	U1"=33.2	D1"=0.077

Table 1. The Orthogonal Table Designed $L_9 3^4$ (input parameters) on the left and corresponding output functions (uniformity and deposition rate) on the right.

The conditions chosen for run 1 were set to level setting 1 for each of the input parameters, i.e., (4*4) mm² f. shape, 60 mg iron, 330 Watt of furnace power, and 20 mm gap, see Table 1. These conditions resulted in a uniformity of 13.9 % and a deposition rate of 0.057 nm/s. Similarly, the data for each of the other experimental conditions and the corresponding results are presented in Table 1.

The first order data analysis was applied. The output function averages (arithmetic means) of uniformity and deposition rate for each level setting and for each input parameter are determined. In the case of uniformity, the uniformity average for (4*4) mm² f.shape setting 1 (runs 1, 2, 3) is given by the average of U1(13.9 %), U2(44.3 %), and U3(31.7 %). This is designated as Us₁ and is 34.8 %. Similarly the uniformity average for f. shape setting 2, (2*4) mm², is given by the average of the uniformity for the experiments 4, 5, and 6 is Us₂=33.0 %. The uniformity average for f. shape setting 3 is Us₃=23.4 %, see Figure 1.Each level setting of the f. shape, the orthogonal property of the matrix correlates the settings for mass, power, and gap so that the effect of each of these parameters tends to cancel out. It can be seen that the first order effect of changing the f. shape (4*4) mm², (2*4) mm², and (4*2) mm² is to decrease the uniformity 34.8 %, 33.0 % and 23.4 % respectively.

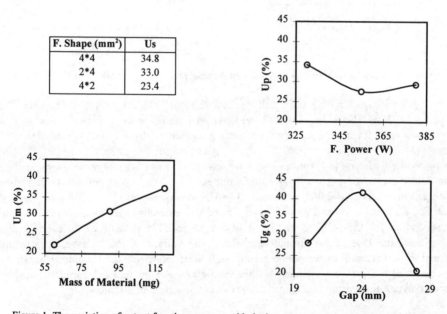

F. Shape (mm²)	Us
4*4	34.8
2*4	33.0
4*2	23.4

Figure 1. The variation of output function averages with the input parameter level settings for uniformity.

The effect of input parameters over the uniformity of the film can be shown by plotting the orthogonal matrix against the level averages as a function of level setting in Figure 1. The f. shape and the mass of the material both have a uniform trend, and by extending the level settings beyond those chosen for the original matrix, it would be possible to achieve even greater uniformity. In practice, however, due to the limitation of shape and volume of the furnace, those values used in the experimental system cannot be exceeded. In the case of furnace power, 355 Watt represents the optimised furnace power settings, and a greater rate could be achieved by altering the power setting to more than 380 Watt or less than

330 Watt. However less than 330 Watt can not initiate the evaporation process or more than 380 Watt melts the pouch. In the case of gap between furnace and substrate, the best uniformity is obtained at 24 mm.

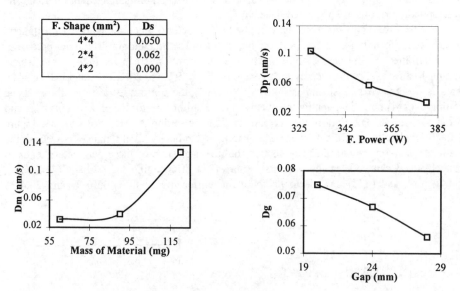

F. Shape (mm²)	Ds
4*4	0.050
2*4	0.062
4*2	0.090

Figure 2. The output function averages a function of the input parameter level settings for deposition rate.

In this investigation, the uniformity is not the sole subject of the discussion, the deposition rate is also important in developing an optimal process. Figure 2 shows that deposition rate, for setting 1, Ds_1=0.05 nm/s is calculated by the average of D1(0.057 nm/s), D2(0.013 nm/s), and D3(0.075 nm/s). It can be seen that the first order effect of changing the f. shape (4*4) mm², (2*4) mm², and (4*2) mm² is to increase the deposition rate 0.05 nm/s, 0.062 nm/s and 0.09 nm/s, respectively. The deposition rate averages for setting 2 and 3 is Ds_2=0.062 nm/s and Ds_3=0.09 nm/s, respectively. F. shape, mass of material, furnace power and gap all have a trend. By extending the level settings beyond those chosen for the original matrix, it would be possible to achieve an even greater deposition rate. Due to the limitation of shape and volume of the furnace, we can not exceed those values. Furnace power setting, 330 Watt represents the optimised value. As discussed in the previous paragraph, this setting can not be reduced further. The best deposition rate for the gap is obtained at the minimum setting of 20 mm.

4. Conclusion

Magnetic thin-films have been produced by a novel Rotating Cryostat. The effect of input parameters on the output functions was investigated by plotting the output function averages against input parameter level settings. The first order data analysis which is presented here is sufficient for most engineering applications involving process optimisation and characterisation. The results of orthogonal analysis indicate that the uniformity across the film was not improved greatly but by using the chosen control parameters, the optimum deposition rate was obtained.

References
[1]   G. Taguchi, "Experimental Designs", 3rd ed., vol. 2, Maruzen Publishing Co., Tokyo, Japan, 1977.
[2]   H. Kockar, et. al., Production and characterisation of thin-films produced by a novel Rotating Cryostat Technique, Non-linear Electromagnetic Systems, IOS press, 1996, pp. 458-461.

Non-Linear Electromagnetic Systems
V. Kose and J. Sievert (Eds.)
IOS Press, 1998

Temperature Dependence of Magnetic Hysteresis of Soft Ferrites

J. G. Zhu, H. Y. Lu, V. S. Ramsden, and K. Tran
Faculty of Engineering, University of Technology, Sydney
P.O. Box 123, Broadway, NSW 2007, Australia

Abstract. This paper presents the experimental measurement and numerical simulation of the temperature dependence of magnetic hysteresis of soft ferrites. The experimental results show that the saturation flux density and the saturation field intensity reduce linearly when the temperature increases. It was also found that the normalised limiting hysteresis loops (B/B_s against H/H_s) at different temperatures were identical. The classical scalar Preisach model was modified to simulate the hysteresis loops at different temperatures. The predicted and experimental results are in substantial agreement.

1. Introduction

In modern power electronic systems, soft ferrites are commonly used as the core materials of electromagnetic devices such as inductors and transformers for high frequency and high power applications because of the advantageous characteristics of high magnetic permeability and low eddy current loss at high excitation frequencies. The magnetic properties of soft ferrites, however, are sensitive to the temperature rise [1-3]. The saturation flux density at $100^{\circ}C$, for example, may be only about half of the value at $20^{\circ}C$ depending on the type of soft ferrite. This causes difficulties in performance prediction and design of electromagnetic devices used in power electronic systems especially at high switching frequencies. Although the significant thermal effects on the magnetic properties of soft ferrites have been observed for a long time, this problem has only received general attention recently [1,4] due to the need for high switching frequency and high power electromagnetic devices in switching mode power supplies. This paper reports the experimental measurement and theoretical modelling of the temperature dependence of magnetic hysteresis of soft ferrites. Section 2 describes the measurement of hysteresis loops of soft ferrites at different temperatures. In section 3, it is shown that the normalised limiting hysteresis loops (B/B_s against H/H_s) at different temperatures are identical. In section 4, the hysteresis loops at different temperatures are simulated using a modified Preisach model and compared with experiments.

2. Experimental Study

Fig.1 illustrates the testing system for experimental study of thermal effects on magnetic hysteresis, where a commercial computer data acquisition system, AMLAB, was used for function generation and data acquisition. For the convenience of measurement, ring samples were used. On the sample, two coils of equal number of turns were wound uniformly. One was used to excite the core and the other for picking up the flux density B signal. The magnetic field intensity H signal was obtained from the voltage across a 1 Ω standard resistor (R in Fig.1). The sample was placed in the middle of an environment chamber, which could maintain the temperature constant over a wide range from −100°C to 200°C. Table 1 lists the dimensions of the ring samples for the study and the number of turns of the excitation and sensing coils, where N30, N27, and N67 samples were supplied by Siemens, and 4C65 and 3C85 by Philips. Fig.2 presents the limiting hysteresis loops measured at different temperatures ranging from 20°C to 120°C.

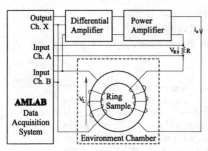

Fig.1 System measuring magnetic hysteresis

Table 1 Samples for testing

Samples	Volume (cm³)	Density (Kg/m³)	D (mm)	d (mm)	h (mm)	N
N30	3.29	4778.88	25.81	15.14	10.10	67
N27	8.59	4665.76	35.88	22.91	14.91	87
N67	8.60	4767.75	35.86	22.87	14.96	73
4C65	1.65	4881.04	23.02	14.19	6.99	67
3C85	0.37	4790.59	12.46	7.60	5.06	48

Note: D and d are the outer and inner diameters, h the height of the ring sample, respectively. N is the number of turns of the excitation and sensing coils.

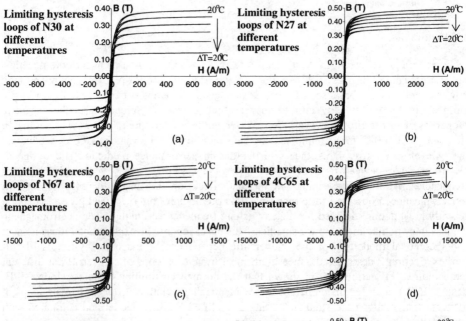

3. Normalised Loops

In the above experiment, the largest loop measured for a sample at a given temperature was chosen as the limiting loop. The saturation point, however, was not reached. When a sample is completely saturated, the permeability should be μ_o since all domain walls disappear and the magnetic moment of the resultant single domain is aligned with the excitation field.

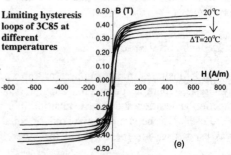

Fig.2 Limiting loops at different temperatures

The saturation point can be found by extending the reversible part of the measured limiting loop by

$$B = \mu_o\left[H + M_{anh}(H)\right] \tag{1}$$

where

$$M_{anh}(H) = M_s\left[\coth\left(\frac{H}{a}\right) - \frac{a}{H}\right] \tag{2}$$

is an expression for the anhysteretic magnetisation curve proposed by Jiles and Atherton [5], M_s the saturation magnetisation, and a is a constant. B_s and H_s can then be determined by finding the point at which $dM/dH = 0.001$, or $\mu = 1.001\mu_o$ (should be $\mu = \mu_o$, but $1.001\mu_o$ is chosen for numerical convenience) on the extended curve. Fig.3 illustrates the curve fitting of the N30 sample at different temperatures. Fig.4 plots B_s and H_s against the temperature. It can be seen that B_s and H_s reduce linearly when the temperature increases. In their data books, the manufacturers of soft ferrites generally supply the limiting loops at 20°C and 100°C. Therefore, $B_s(T)$ and $H_s(T)$, where T is the temperature, can be easily determined in practice.

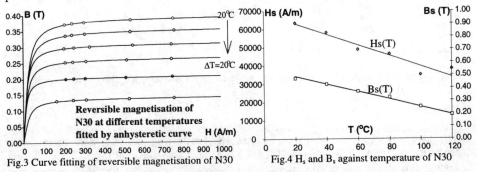

Fig.3 Curve fitting of reversible magnetisation of N30   Fig.4 H_s and B_s against temperature of N30

Fig.5 plots the normalised limiting loops of the N30 sample at different temperatures, which were obtained by dividing the B and H values of the limiting loops in Fig.2(a) by B_s and H_s, respectively. It is shown that these normalised limiting loops are identical.

4. Simulation

The classical scalar Preisach model of magnetic hysteresis with a novel and practical parameter identification method [6] was modified to simulate the magnetic hysteresis of soft ferrites at different temperatures. Fig.6 is the block diagram illustrating the structure of the model with the thermal effects incorporated, where $B_u = B/B_s$ and $H_u = H/H_s$. Fig.7 compares the theoretical and experimental results. As shown, they are in substantial agreement.

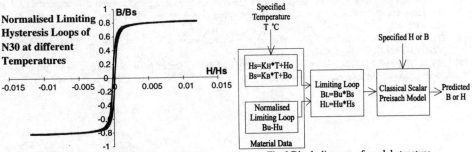

Fig.5 Normalised limiting loops of sample N30   Fig.6 Block diagram of model structure

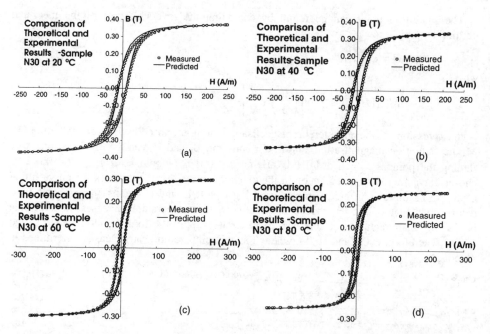

Fig.7 Comparison of theoretical and experimental results of the N30 sample at different temperatures

5. Conclusion

The magnetic hysteresis loops of various soft ferrite ring samples were measured at different temperatures ranging from 20°C to 120°C. The experimental results show that the saturation flux density and field intensity reduce linearly when the temperature increases. This is very useful for modelling, since manufacturers generally supply limiting loops at 20°C and 100°C in their data books. The classical scalar Preisach model was modified to simulate the hysteresis loops at different temperatures. The theoretical and experimental results are in substantial agreement.

References

[1] J.G. Zhu, "Numerical modelling of magnetic materials for computer aided design of electromagnetic devices", Ph.D Thesis, University of Technology, Sydney, Australia, July 1994

[2] E.C. Snelling, "Soft ferrites, properties and applications", Butterworths & Co. (Publishers) Ltd., Second Edition 1988

[3] SIEMENS, "Ferrites and Accessories, data book 1990/1991"

[4] P. Tenant, J.J. Rousseau, and L. Zegadi, "Hysteresis modelling taking into account the temperature", The 6th European Conference on Power Electronics and Applications, 19-21 Sept. 1995, Sevilla, Spain, pp1.001-1.006

[5] D.C. Jiles and D.L. Atherton, "Theory of ferromagnetic hysteresis", Journal of Magnetism and Magnetic Materials, Vol.61, 1986, pp48-60

[6] S.Y.R. Hui and J.G. Zhu, "Numerical modelling and simulation of hysteresis effects in magnetic cores using transmission line modelling and Preisach theory", IEE Proceedings, Part-B, Vol.142, No.1, Jan. 1995, pp57-62

Non-Linear Electromagnetic Systems
V. Kose and J. Sievert (Eds.)
IOS Press, 1998

Simulation of Networks Coupled by a Hysteretic Magnetic Component

M. Pasquale, G. Bertotti, ^S.Brandonisio, *C. Guida

IEN Galileo Ferraris, INFM and GNSM-CNR,C.so M.d'Azeglio 42,Torino 10125, Italy
^ABB Elettrocondutture, Viale dell'Industria 18, Vittuone (MI) I-20010, Italy
*ABB Ricerca SpA, Viale Edison 50, Sesto S. Giovanni (MI) I-20099, Italy

Abstract. In this paper we show that a generic secondary linear network, coupled to an ideal primary current generator by a nonlinear hysteretic component can be described by a simple loaded transformer circuit with a generic $Z(\omega)$ secondary load. This circuit can be described using the Dynamic Preisach Model, that permits the calculation of the secondary current flowing through the $Z(\omega)$ load in a range of dynamic conditions. The secondary current is correctly calculated with no free parameters, taking into account both hysteretic and dynamic contributions produced by the combined non-linearity of the magnetic core and the secondary circuit. Convergence of the numerical procedure under such conditions was obtained by means of an iterative feedback with digital filtering.

1. Introduction

It has been shown [1,2] that the dynamic Preisach model (DPM) can be very helpful in the description and prediction of the magnetic properties of soft materials in a wide range of conditions. Several recent papers have discussed the connection between the hysteresis behavior and the irreversible [$p(h_c,h_u)$] and reversible [$p_{rev}(h_u)$] Preisach distribution functions, where h_c and h_u are the coercive field and the interaction field associated with each Preisach elementary loop [3]. These distributions, whose shape depends on magnetization processes and microstructure, can be directly derived from the measurement of a set of static minor hysteresis loops. Finally one dynamic loop measurement made under defined applied field or induction conditions is required to determine the DPM parameter physically characterizing the correlation length of excess loss processes. Once the DPM description of the materials is defined, the behavior of various transformer circuits containing cores can be predicted in a wide range of conditions. The case of a loaded transformer with a purely resistive secondary load has already been discussed [4]. In this paper we show that the general case of a linear secondary network coupled to the primary by a single hysteretic magnetic core can be correctly simulated by DPM with no free parameters. This result can be achieved when the primary circuit can be assimilated to an ideal current generator of a transformer system, with a secondary circuit of effective impedance $Z(\omega)$. We also show that the convergence of this modeling scheme can be obtained through an iterative procedure based on the description of the mutual interaction between the primary and the secondary linear networks through the hysteretic component. As a practical example, the case of a NiFe permalloy core has been studied in a differential circuit breaker configuration, with a capacitive secondary network connected in parallel to a relay. The equivalent impedance $Z(\omega)$ of the secondary network and the relay was obtained

combining the single components described by linear models. A linear model of the relay
was obtained and checked in the working conditions. Secondary voltages and currents
were obtained by experiment and then compared by simulation with different primary
current waveshapes. An example of the comparisons performed between simulated and
calculated secondary voltages is presented in Fig. 3 with an asymmetric 50 Hz current.

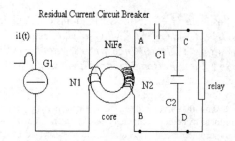

Fig. 1 Primary ideal current generator i_1(t)coupled to the secondary circuit through a VAC ultraperm
NiFe core. The secondary circuit contains typical components for a Residual Current Circuit breaker,
with a total impedance $Z(\omega)$. V_{AB} and V_{CD} secondary voltages were measured and simulated using the
Dynamic Preisach model in different excitation conditions. Results are presented with $N_1/N_2=0.125$.

2. DPM Circuit Solution

The behavior of the transformer circuit described in Figure 1 has been calculated using
DPM, which is a model where all the hysteresis properties are derived in terms of the
internal field H acting inside the core. In DPM H is required to be a periodic function of
time monotone in each half cycle. The connection between the internal field H and the
applied field H_a (proportional to joint effect of the primary and secondary currents) can
be written as H = H_a + k I - f(dI/dt), where k is a constant describing internal
interactions in the mean field approximation, I is the magnetization and f(dI/dt) is a term
taking into account both dynamic magnetization processes and secondary circuit effects.
DPM, using a set of geometric, structural and circuit parameters takes care of the
calculation of f(dI/dt). These contributions increase the lag between magnetization and
excitation current with increasing frequency and introduce distortions in the H
waveform. The effect of a purely resistive secondary circuit can be precisely calculated
in the time domain taking into account the excitation reduction caused by a current flow
in the secondary [4]:

$$H_a(t) = \frac{N_1}{L}i_1(t) - \frac{N_2^2 S}{LR}\left(\frac{dI}{dt}\right), \tag{1}$$

where H_a is the effective magnetic field acting on the core, i_1 and N_1 are the primary
current and windings, L and S are the magnetic path length and sample cross-section,
N_2 is the number of secondary windings, R is the purely resistive load, and I is the
magnetization. In the case of a generic load $Z(\omega)$ in place of R, with an ideal periodic
(also non-sinusoidal) excitation i_1(t), one can use Fourier analysis to calculate the effect
of the secondary circuit on the field H_a. Once all frequency components are determined
H_a(t) and H(t) are reconstructed by inverse fast Fourier transforms, then applying DPM

to calculate the dynamic hysteresis loop I[H(t)]. The new calculated value of I(t) can then be used to determine a new $H'_a(t)$ waveform, and the comparison between $H'_a(t)$ and its desired value, known a priori, can be used iteratively to obtain a solution. This procedure was applied to the calculation of the response of the loaded transformer circuit to different periodic current waveforms $i_1(t)$, and each case was solved using this feedback procedure, where the H waveform applied to DPM at each iteration step is progressively adjusted on the basis of the calculated dI/dt waveform and the load $Z(\omega)$ to obtain an H_a waveform consistent with the ideal primary and the secondary load conditions.

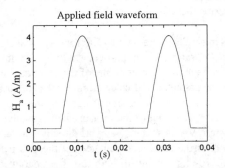

Fig. 2 Asymmetric primary excitation waveform at 50 Hz, generated by truncating the negative part of a sinusoidal signal. This waveform can be used for testing the operating conditions of Residual Current Circuit Breakers.

2.1 Simulation and feedback

1) In the simulation process the initial H(t) can be defined in terms of the applied field waveform $H_a(t)$ and an assumed magnetization waveform I(t), using H(t) = $H_a(t)$ + k I(t) - f(dI/dt). The effect of the secondary load $Z(\omega)$ on H_a is calculated using Fourier transform.
2) Once H(t) has been defined by inverse Fourier transforms of all relevant quantities, magnetization I(t) and its time derivative can be calculated by means of DPM.
3) An iterative procedure is applied to obtain a converging solution: a new applied field waveform $H'_a(t)$ is calculated as a function of the internal field H(t) and the calculated magnetization I(t) using $H'_a(t)$= H(t) - k I(t) + f(dI/dt).
4) $H'_a(t)$ is compared with the reference applied field $H_a(t)$. The feedback procedure is based on the correction of the amplitude and time behavior of H(t) so that $H'_a(t) \approx H_a(t)$ [5,6].
5) The non-linear behavior of the different circuit components and their strong coupling affect the convergence behavior, often inducing some higher frequency oscillations. These problems were overcome by reducing the amplitude and bandwidth of the H waveform corrections.

3. Results and Conclusions

We show that the behavior of a secondary linear network connected to an ideal generator on the primary by a single magnetic core can be simulated using DPM; as an example we present the study of a differential breaker circuit containing a VAC Ultraperm NiFe core. Several steps were needed to accomplish a correct simulation:

1) The model of the core was obtained through a set of 20 static minor loops, analyzed to calculate the irreversible $[p(h_c,h_u)]$ and reversible $[p_{rev}(h_u)]$ Preisach distribution functions.
2) The dynamic characteristics of the core were obtained through the measurement of a few dynamic loops at frequencies of 50 Hz and 400 Hz.
3) The secondary network was defined calculating the equivalent impedance $Z(\omega)$ obtained using the nominal capacitance of C_1 and C_2 and a combination of linear components to describe the relay characteristics.

The circuit was experimentally tested with different H_a waveforms applied to the primary and different N_1/N_2 ratios, and in particular with the excitation shown in Figure 2, sampling the output voltage in different points of the secondary circuit. A comparison is presented between measurements and calculations obtained across the A,B and C,D terminals shown in Figure 1. A good agreement between measurements and modelling can be observed in Figure 3, referring to the measured and calculated secondary voltage V_{AB}, and in Figure 4 where V_{CD} is shown. V_{CD} was directly computed from V_{AB} using a routine taking in to account the relative impedances of the two parallel branches of the secondary circuit. The results, obtained with no free parameters, show that the properties of the magnetic core and the circuit can be correctly described by the Dynamic Preisach Model using an iterative feedback procedure. Future work will be directed to the optimization of the feedback procedure and the study of DPM as an external module for circuit simulation packages.

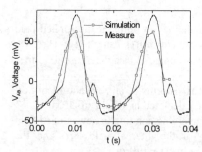
Fig.3 Secondary V_{AB} voltage measured across $Z(\omega)$ (line) and DPM simulated (lines+points) at 50 Hz with the asymmetric excitation waveform shown in Fig.2. V_{AB} is generated by the total secondary current.

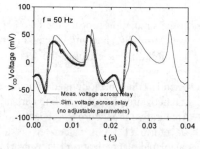
Fig.4 Secondary V_{CD} voltage across the relay measured (line) and calculated (lines+points) from the V_{AB} waveform taking into account the linear modeling of the secondary circuit branches.

References

[1] Bobbio et al., IEEE Trans. Magn., (30), pp.3367-3370, 1994.
[2] V. Basso, G. Bertotti, A.Infortuna and M. Pasquale, IEEE Trans.Magn., (31),p.4000, 1996.
[3] I. D. Mayergoyz "Mathematical Models of Hysteresis", Springer-Verlag, 1991.
[4] M. Pasquale and G. Bertotti, IEEE Trans. Magn., (32), pp. 4231-4233, 1996
[5] G. Bertotti, E.Ferrara,F. Fiorillo, and M. Pasquale, J.Appl.Phys., 73 (10), pp. 5375, 1993.
[6] A.Heitbrink, H.D. Storzer, Nonlin. Electromag. Syst. IOS Press 1996, pp. 580-584.

Non-Linear Electromagnetic Systems
V. Kose and J. Sievert (Eds.)
IOS Press, 1998

503

Simulation of Non-Linear Inductor Circuits in Phenomenological Models

Alexandru STANCU, Petru ANDREI, Ovidiu CALTUN
"Alexandru Ioan Cuza" University, Faculty of Physics, IASI, 6600, Romania

Abstract. Jiles-Atherton, Hodgdon and Preisach models are applied to describe the dynamics of a non-linear RLC circuit comprising a linear resistor and a linear capacitor in series with a non-linear inductance with a MnZn soft ferrite core. The experimental waveform voltage across the capacitor, initially charged and then discharged on RL circuit is compared with the calculated one. A good agreement between the experimental and simulated data is observed for the curves which are used in the identification, but it is not satisfactory when one changes the magnetic field rate.

1. Introduction

To simulate the magnetic components used in the electronic equipment, many attempts to fit experimental and theoretical data were made. In phenomenological modelling one tries to find mathematical functions that fit the experimental data. These models, characterised as mathematical models by Hodgdon [1] and Jiles [2], are usually very efficient in simulating electrical circuits. In this paper we are comparing the results of the magnetisation process simulations made with three models: Jiles-Atherton [2-3], Hodgdon [1,4] and Preisach [5-6].

2. Experimental set-up

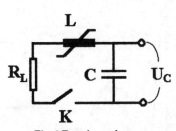

Fig. 1 Experimental set-up

The magnetisation processes are studied as the energy is dissipated in a RLC series circuit. First, the capacitor C is charged by a DC power supply to U_c voltage value. Then, the K switch is turned on and the capacitor is discharged on the RL circuit. The voltage waveforms across the capacitor C are digitised by a storage oscilloscope and transferred by a data acquisition block to a computer.

The magnetic core is a soft MnZn ferrite toroidal core (commercial A-7); the inner radius is r_1=14mm, the outer radius is r_2=26mm, and the height is h=20mm; the number of turns is N=48. The background resistance associated with the windings is R_L=0.24Ω. One can approximately calculate the inductance as a function of the magnetic field (H), magnetisation (M) and their variation rate ($\dot{H}$ and $\dot{M}$) by:

$$L\left(H, M, \dot{H}, \dot{M}\right) = \frac{L_0}{\mu_0}\,\mu\left(H, M, \dot{H}, \dot{M}\right) = \frac{N^2 h}{2\pi}\,\mu\left(H, M, \dot{H}, \dot{M}\right)\ln\frac{r_2}{r_1}, \qquad (1)$$

where the magnetic permeability, μ, is a function of H, M, $\dot{H}$ and $\dot{M}$; L_0 is the inductance of the coil without core.

3. Jiles-Atherton (J-AM) and Hodgdon model (HM)

In the Jiles-Atherton model the magnetic susceptibility is a function of the magnetisation M and the applied magnetic field H. In this paper we have used the model's equation presented in [7]. First, the model parameters are calculated using the method presented in [8], but then, they are slightly modified to improuve the agreement between the simulated and measured waveform across $C=32\mu F$. We have obtained: $M_s= 5.3 \times 10^4$ A/m, $a=40$ A/m, $c=0.1$, $k=3.0$ A/m, $\alpha=4.0 \times 10^{-4}$. The calculated voltage across C versus time is represented by the solid line in Fig. 2 and Fig. 3 for $C=32\mu F$ and for $C=44\mu F$, respectively. The magnetic field versus time is represented in Fig. 4.

In the HM we have used for the identification problem a slightly modified method from that described in [1,4]. Even if the Hodgdon identification method seems to be simpler it needs many experimental points on the major loop; moreover, "it does not insure that these points are on the major loop" [1]. In order to avoid these inconveniences we will find the parameters A_3 and A_4, using a more complicated method. By integrating equation (7) from [1], one may show that the coercive field, H_c, and the remanent induction, B_r must satisfy the equations:

$$\left(\frac{H_c \exp(\alpha B_r)}{A_1 \, tg(A_2 B_r)} - 1\right) \int_{B_r}^{B_{cl}} \exp\left(-\frac{A_4 s}{B_{cl} - s} - \alpha s\right) ds - \int_0^{B_r} \exp\left(-\frac{A_4 s}{B_{cl} - s} - \alpha s\right) ds = 0 \, (2)$$

$$A_3 = -H_c \Big/ \int_0^{B_{cl}} \exp\left(-\frac{A_4 |s|}{B_{cl} - s} - \alpha s\right) ds. \tag{3}$$

The implicit equation (2) can be used to calculate A_4, and the equation (3) to calculate A_3, provided that all of the other parameters are known. Equation (2) has only one solution if

$$H_c > \frac{A_1 \, tg\left(A_2 B_r\right)}{1 - \exp\left[-\alpha\left(B_{cl} - B_r\right)\right]} \cong A_1 \, tg\left(A_2 B_r\right).$$ Unlike the HM, one does not need to know the

permeability at the coercivity μ_c, and remanence μ_r, when one calculates the A_3 and A_4

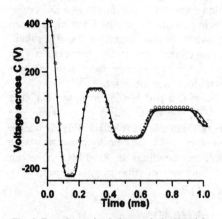

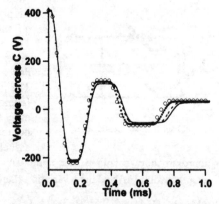

Fig. 2 Experimental (circles) and simulated in J-AM (solid line), HM (dashed line) DPM (points) waveform across the capacitor for $C=32\mu F$

Fig. 3 Experimental (circles) and simulated in J-AM (solid line), HM (dashed line) DPM (points) waveform across the capacitor for $C=44\mu F$

parameters. In order to find the dynamic parameters we have used equation (7) from [4]. Thus, we have obtained the following set of parameters: A_1=193m/A, A_2=20.6T^{-1}, A_3= 0.105, A_4=0.0233, B_{cf}=0.0754T, B_{bp}=0.073T, $\dot{B}_1$=10^{3}T/s, c_1=8x10^{-5}s/T, c_2=4x10^{-5}s/T, μ_s=μ_0, μ_{an}=200μ_0, α=11.3T^{-1} which is determined for C=32μF. The simulated voltage across C is represented in Figs. 2 and 3 by the dashed line for C=32μF and for C=44μF respectively. The magnetic field calculated by using the Hodgdon model is represented in Fig. 4 by the dashed line.

4. Dynamic Preisach model (DPM)

The next DPM is developed from the idea suggested by Pasquale and Bertotti in [8] but, in addition, it has a moving term and a reversible part. Starting from a physical point of view it is shown that at each point of the measured loop, the instantaneous excess field due to the dynamic processes can be approximated with a term proportional to the square root of the total magnetisation rate. Denoting this field by $g(\dot{M})$ and taking into account [5] a moving term $f(M)$, as well, the effective field is given by $H_{eff} = H + f(M) - g(\dot{M})$, where H is the applied field and M is the magnetisation of the sample. The Preisach distribution function (P) is assumed to be the product of a log-normal distribution in the coercive field and a Gaussian distribution in the interaction field. The standard deviation of the interaction field is $H_{\sigma i}$, of the coercive field is σ_c, the mean coercive field is H_0 and it's saturation value is M_{sirr}. The reversible distribution (R) is taken over the first bisetrix and is assumed to be an exponential distribution whose saturation magnetisation is M_{srev} and fitting parameter γ [5-6]. The saturation value of magnetisation is $M_s = M_{sirr} + M_{srev}$. By deriving the magnetisation with respect to the applied magnetic field one obtains:

$$\chi(H, M, \dot{M}) = \frac{dM}{dH} = (I+J)\left[1 + f'(M)\frac{dM}{dH} - g'(\dot{M})\frac{d\dot{M}}{dH}\right], \qquad (4)$$

where we have introduced the notations:

$$I = 2\int_H^{H_{Neff}} P(h, H_{eff})\, dh \text{ if } \dot{H} > 0, \quad I = 2\int_{H_{Neff}}^H P(H_{eff}, h)\, dh \text{ if } \dot{H} < 0 \text{ and } J = 2\, R(H_{eff}) \quad (5)$$

(H_{Neff} is the last extreme value of the effective field in the Preisach plane). But

$$\frac{d\dot{M}(H)}{dH} = \frac{d}{dH}(\chi\,\dot{H}) = \chi\frac{\ddot{H}}{\dot{H}} + \frac{d\chi}{dt} \text{ and, consequently:}$$

$$\frac{d\chi}{dt} = \frac{1}{g(\dot{M})} + \frac{\left[f'(M)\dot{H} - g'(\dot{M})\ddot{H}\right](I+J) - \dot{H}}{(I+J)\,g(\dot{M})\,\dot{H}}\,\chi. \qquad (6)$$

The second derivative of the magnetic field can be obtained by deriving with respect to time the Kirchhoff's law written for the RLC circuit as

$$\ddot{H} = \frac{d^2H}{dt^2} = \frac{RH + nU_c}{L_0(\chi+1)^2}\frac{d\chi}{dt} - \frac{RC\dot{H} + H}{CL_0(\chi+1)}, \text{ where } U_c \text{ is the voltage across } C \text{ and } n \text{ is the}$$

number of windings per unit length.

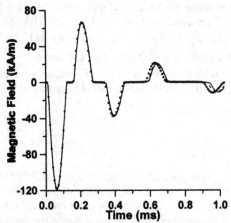

Fig. 4 Calculated magnetic field in J-AM (solid line), HM (dashed line) and DPM (points) for $C=32\mu F$.

In our simulations we have taken a linear moving term $f(M)=\alpha M$ and $g(\dot{M}) = \text{sgn}(\dot{M})\beta\sqrt{|\dot{M}|}$, where α and β are tow fit parameters. These parameters are found when the simulated voltage waveform across the capacitor remains very close to the experimental one. Thus, one finds $M_{sirr}=0.28 \times 10^5$A/m, $M_{srev}=0.2 \times 105$ A/m, $\gamma=0.4$, $\sigma_c=0.72$, $H_a=85$A/m, $H_0=10$A/m, $\alpha=4 \times 10^{-4}$, $\beta=3.5 \times 10^{-2}$s$^{1/2}$A$^{1/2}$m$^{-1/2}$ (determined for $C=32\mu F$). The simulated voltage across C and the magnetic field are represented with points in Figs. 2-3 and in Fig. 4, respectively. Even if the agreement is better for the DPM than for the J-AM or HM, the DPM has some major deficiencies: it does not give an equation for the magnetic susceptibility but for it's first derivative, the time of simulation is approximately a hundred times longer than those required by the J-AM the HM.

5. Conclusions

We have simulated the RLC circuit with three models: Jiles-Atherton, Hodgdon and a modified dynamical Preisach model. We have observed that there is a good agreement between the experimental and simulated data for the curves which are used in the identification ($C=32\mu F$), but it is not satisfactory when one changes the capacitance ($C=44\mu F$). Thus, if one changes too much the values of the components which compose the electronic circuit, one has to change the models' parameters. Even if the Preisach model seems to be better, it needs a longer processing time. It follows that one has to employ a simple model (Hodgdon or Jiles-Atherton models) for usual and fast applications but a more elaborate model (like DPM) for applications requiring more accuracy.

References

[1] M.L. Hodgdon, Mathematical theory and calculations of magnetic hysteresis curves, *IEEE Trans. on Magn.*, vol. 24, no. 6 (1988) pp. 3120-3122.
[2] D.C. Jiles, and Z. Gao, Modelling of hysteresis in magnetic materials, *First Int. Workshop on Simulation of Magnetisation Processes* Wien (1995).
[3] D.C. Jiles and D.L. Atherton, Theory of ferromagnetic hysteresis, *J. Appl. Phys.*, vol. 55, no. 6 (1984) pp. 2115-2120.
[4] M.L. Hodgdon, and C.D. Boley, Model and simulation of hysteresis in magnetic cores, *IEEE Trans. on Magn.*, vol. 25, no. 5 (1989) pp. 3922-3924.
[5] I.D. Mayergoyz, Mathematical models of hysteresis, *IEEE Trans. Magn.*, vol. 22, no. 5 (1986) pp. 603-608.
[6] Al. Stancu, C. Papusoi, Field dependence of the remanent magnetization for thin films containing ferromagnetic particles, *J. Magn. Magn. Mater.*, vol. 134 (1995) pp. 190-194.
[7] J.H.B. Deane, Modelling the dynamics of non-linear inductor circuits, *IEEE Trans. on Magn.*, vol. 30, no. 5 (1994) pp. 2795-2801.
[8] M. Pasquale, and G. Bertotti, Application of the dynamic Preisach model to the simulation of circuits coupled by soft magnetic cores, *IEEE Trans. on Magn.*, vol. 32, no. 5 (1996) pp. 4231-4233.

Non-Linear Electromagnetic Systems
V. Kose and J. Sievert (Eds.)
IOS Press, 1998

Power Losses of Grain Oriented Fe-Si Sheets using Nonsinusoidal Waveforms

B. Weidenfeller and W. Riehemann

IWW TU Clausthal, Agricolastr. 6, D-38678 Clausthal-Zellerfeld, Germany

Abstact. Investigations of power losses in grain oriented silicon iron steels are usually carried out with controlled sinusoidal induction waveforms. Losses appearing under nonsinusoidal induction waveforms cannot be predicted in this way. The aim of this paper is to compare power losses in grain oriented silicon iron steel sheets obtained for different induction signal forms. The power losses were measured at various frequencies, polarizations and induction waveforms, which were characterized by the form factor. Sinusoidal, triangular and rectangular inductions and sinusoidal magnetic fields could be realized. For the investigated frequencies (10–500 Hz), polarizations (1.4–1.7 T) and form factors (1–2.5) it was found that the anomaly factors of the dynamic losses versus frequency times square of the anomaly factor can be described with one curve. For grain oriented silicon iron sheets this results in a simple relationship that enables the comparison of power losses for different induction signals characterized by their form factors.

1. Introduction

Investigations of power losses in soft magnetic materials and in particular power losses in grain oriented silicon iron steel sheets are generally performed at controlled sinusoidal induction signals at test frequencies of 50 Hz and 60 Hz, respectively. However in many practical applications magnetic materials are producing nonsinusoidal induction signals. Power losses measured under trapezoidal induction signals were described in literature [1-3]. Nevertheless, the comparison between power losses appearing in soft magnetic materials for different induction waveforms is generally complex. The aim of this paper is to give a simple formula which enables the comparison of power losses obtained for different induction signals in commercial grain oriented silicon steels.

2. Experimental

Power losses in grain oriented Fe3.2wt.%Si steel sheets were measured in the frequency range 10–500 Hz at amplitudes of polarizations 1.4–1.7 T with different induction waveforms. The ORSI H silicon iron sheets (EBG Gelsenkirchen, Germany) had a width of 30 mm, a length of 305 mm and a thickness of 230 μm. They were measured in untreated conditions and after laser scribing. Eight sheets were used as one Epstein set. $B_8 = 1.945 \pm .003$ T. Measurements were carried out using an Epstein frame with a digital hysteresis recorder system which is described in more detail in [4]. The magnetic losses were measured utilizing a programmable amplifier under sinusoidal, triangular and rectangular induction waveforms and under controlled sinusoidal magnetic field waveform, which leads to induction waveforms varying with frequency and amplitude of polarization. The waveforms of the induced signals $U(t)$ (t:time, T:signal period) were characterized by the numerically calculated form factor k which is defined by

$$k = \frac{\sqrt{\frac{1}{T} \int\limits_0^T U^2(t)\, dt}}{\frac{1}{T} \int\limits_0^T |\, U(t)\, |\, dt} \tag{1}$$

In our investigations k was calculated numerically from the induced and digital voltage $U_i(t_i)$. The form factor varies from $k = 1$ (rectangular) over $k \approx 1.111$ (sinusoidal) and $k \approx 1.155$ (triangular) up to $k = 2.458$ in the case of a controlled sinusoidal magnetizing field and laser scribed sheets.

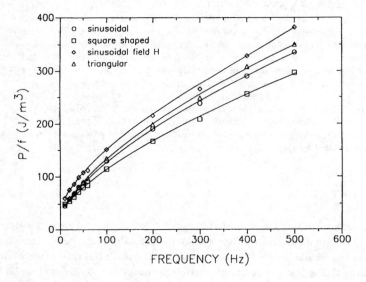

Figure 1: Frequency dependent cyclical magnetic work in grain oriented silicon iron steel sheets measured under different induction waveforms.

3. Results und Discussion

Fig. 1 shows the cyclical magnetic work P/f of laser scribed silicon iron sheets versus frequency for different form factors. As could be expected, the losses increase with increasing frequency and form factor. The dependency of the losses on the form factor can be clarified by separation of the losses. The frequency dependent power loss $P(f)$ can be divided into the frequency independent hysteresis loss P_h which was estimated by extrapolation of power loss $P(f \to 0)$ and the frequency dependent dynamical loss $P_{dyn}(f)$. $P_{dyn}(f)$ can be expressed by the classical losses $P_{cl}(f)$ and the anomaly factor $\eta(f)$.

$$P(f) = P_h + P_{dyn}(f) \tag{2}$$
$$P_{dyn}(f) = \eta(f)P_{cl}(f) \tag{3}$$
$$P_{cl}(f) = (2kdJ_0f)^2/(3\rho) \tag{4}$$

where d is the sheet thickness, J_0 the amplitude of polarization, f the frequency and ρ the electrical resistivity. The classical part $P_{cl}(f)$ of the dynamical losses P_{dyn} depends on the square of the form factor of the induced signal. Comparison of the measured dynamical losses occuring for different waveforms shows that not only the classical losses but also the anomaly factor depend on the form factor.

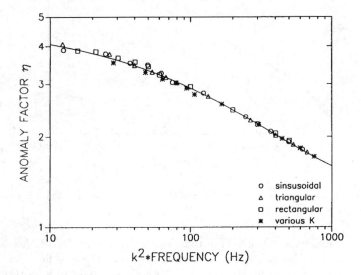

Figure 2: Anomaly factors which were determined from the measured losses versus $k^2 f$ for all measured losses in Fig. 1.

Fig. 2 shows a plot of the anomaly factors which were determined from the measured losses in fig. 1 versus $k^2 f$ in semilogarithmic scale. As can be seen in this plot all points lie nearly perfectly on one graph, which can be approximated by a straight line with a slope -0.25 for frequencies higher than 50 Hz with good accuracy. Therefore Eqs. 2 and 3 yield $P_{dyn}/f \sim k^{1.5}$. Using Eqs. 1-3 this leads to a simple empirical formula for the investigated silicon iron sheets, which allows the comparision of measured power losses in the investigated material in the mentioned frequency range above 50 Hz.

$$P_{rec}(f) = P_h + \left(\frac{k_{rec}}{k}\right)^{1.5} \cdot P_{dyn}(f) \tag{5}$$

In Eq. 5, k is the form factor of the measured induced signal, k_{rec} is the form factor of a rectangular induction voltage ($k_{rec} \equiv 1$) and P_{rec} is the power loss which would be measured in the case of a rectangular induction voltage. If the power losses are to be compared with other forms of induced signals, k_{rec} must be replaced by the form factor of the corresponding signal.

In Fig. 3 the calculated losses reduced to a rectangular induction signal are plotted versus frequency. They were calculated with the aid of Eq. 5 using the frequency and form factor dependent power losses shown in Fig. 1. The calculated values for different measured induction waveforms are in good agreement, and lie in the experimental scatter of the point. The curve is calculated using Eqs. 1-4. Each of the curves shown in Fig. 1 is calculated with the knowledge of static hysteresis loss and the approximated linear frequency dependence of the anomaly factor using Eqs. 1-4. Therefore, the only

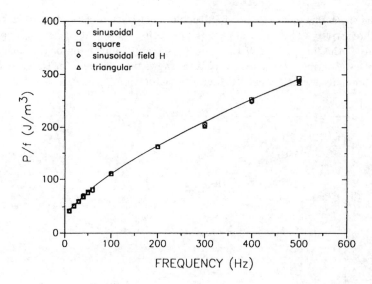

Figure 3: Frequency dependent power losses shown in Fig. 1 reduced to square shaped induction signal ($k = 1$) by Eq. 5.

varying parameter for the calculation of these curves is the form factor. An exception is the curve which fits the values measured with sinusoidal magnetic field, due to variations of the form factors for each measured point. In Fig. 1 the curve fits these points too because the different form factors for each point have been considered.

4. Conclusions

Comparison of power losses in grain oriented electrical steels measured for different forms of the induced signal is possible if their hysteresis loss, dynamic losses and form factors are known.

Acknowledgements

The authors wish to thank Dr. M. Hastenrath, EBG, Gelsenkirchen, for providing us with grain oriented silicon iron sheets. This work was supported by Deutsche Forschungsgemeinschaft.

References

[1] F. Fiorillo, C. Appino, and M. Barisoni, Anales de Fisica B **86** 238 (1990)
[2] T. Barradi, and A. Mailfert, J. Magn. Magn. Mat. **112** 6 (1992)
[3] S. Takada, and T. Sasaki, IEEE Trans. Magn. **28** 2784 (1992)
[4] M. Pott-Langemeyer, W. Riehemann, and W. Heye, Anal. Fisica B **86**, 232 (1990)

Non-Linear Electromagnetic Systems
V. Kose and J. Sievert (Eds.)
IOS Press, 1998

Dependence of ac Magnetic Properties of Electrical Steel due to the Phase Angle of the Higher Harmonic Induction

Derac Son[*], Eun Kyung Kim[*], and J.D. Sievert[**]

[*]Han Nam Univ., Ojung Dong 133 Taejon City, 300-791 Korea
[**]Phys.-Tech. Bundesanstalt, Braunschweig D-38116, Germany

Abstract. We have measured ac magnetic properties of non-oriented silicon steel under harmonic frequencies ranging from the 3^{rd}(180 Hz) to 9^{th}(540 Hz) and harmonic amplitude components from 10% to 50% of the total amplitude (B_{max} =1.5 T). From the experiment, it is found that, if the magnetic induction waveform has above the 9th harmonic frequency components of the magnetic induction, the core losses only depended on the harmonic amplitude component, but if harmonic frequency becomes lower than the 9th harmonic frequency, the core losses depend on the phase angle and the harmonic amplitude, and the phase angle should be considered in the design of electric machines.

1. Introduction

In electrical machines, higher harmonic frequency components of magnetic induction are always generated during ac magnetization of magnetic cores, and analysis for higher harmonic frequency components of the magnetic induction are necessary. Core loss has been analyzed using the superposition principle of higher harmonic frequency components of magnetic induction without their phase angles or with phase angles only for single harmonic frequency components[1-3].

In this work, we have analyzed the ac hysteresis loop properties depending on the phase angle of the given single higher harmonic frequency component of magnetic induction

2. Construction of Measuring System

For the generation of magnetic induction including higher harmonic frequency components with different phase angles, we have constructed a double yoke type single sheet core loss tester in which the specimen size is 8 cm x 8 cm. It includes an arbitrary waveform synthesizer(hp1445) and a 2-channel transient recorder (hp1429), which were controlled by VXI-bus of the VXI system(hp75000), and a feed-back system for the waveform of the induced voltage from the secondary winding to be the same as the output voltage of the arbitrary waveform synthesizer[4]. A waveform was numerically programmed by the instrument controller (hp745) and the waveform data were transferred to the arbitrary waveform synthesizer via IEEE-488bus. Magnetic induction $B(t)$ and magnetic field strength $H(t)$ were digitized using the 2-channel transient recorder and data were transferred to the instrument controller via IEEE-488 bus.

3. Results and discussions

Fig. 1 shows ac hysteresis loops of non-oriented silicon steel at a magnetizing frequency of 60 Hz, including a 3rd harmonic frequency component (180 Hz) of magnetic induction which has a phase angle varying from $0°$ to $180°$ with $90°$ steps and harmonic amplitude 20% of the total amplitude of the magnetic induction. Fig. 2 shows ac hysteresis loops for the same condition as that in Fig. 1 except for the 5th harmonic frequency component of magnetic induction. From the experiment, we can see that ac hysteresis loops and core loss are strongly dependent on the phase angle of the harmonic frequency component of magnetic induction. If the 3rd harmonic frequency component of magnetic induction is affected near the coercive field strength region of ac hysteresis loop, the core loss could be decreased, but near saturation magnetic induction of the ac hysteresis loop, the core loss was increased. This effect was increased remarkably, when the harmonic amplitude component of magnetic induction was increased. Fig. 3 shows the change of ac hysteresis loops based on phase angle changes from $0°$ to $180°$ for the case of the 9th harmonic frequency component of magnetic induction and the harmonic amplitude was 20% of the total amplitude of the magnetic induction. In the 9th harmonic case, the ac hysteresis loop shapes vary with different phase angles but the corresponding core loss changes were very small.

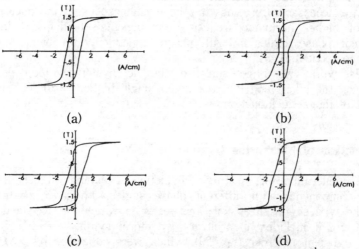

Fig. 1. Ac hysteresis loops dependence on the phase angle of the 3rd harmonic frequency component of magnetic induction (20% of B_{max}=1.5 T); a) without higher harmonic, b) phase angle of $0°$, c) phase angle of $90°$, d) phase angle of $180°$.

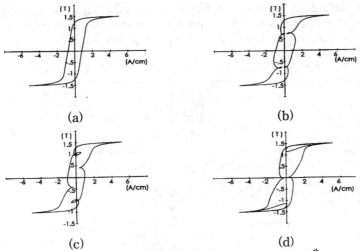

Fig. 2. Ac hysteresis loops dependence on the phase angle of the 5th harmonic frequency component of magnetic induction (20% of B_{max}=1.5 T); a) without higher harmonic, b) phase angle of $0°$, c) phase angle of $90°$, d) phase angle of $180°$.

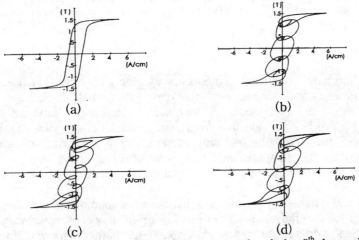

Fig. 3. Ac hysteresis loops depend on the phase angle of the 9th harmonic frequency component of magnetic induction (20% of B_{max}=1.5 T); a) without higher harmonic, b) phase angle of $0°$, c) phase angle of $90°$, d) phase angle of $180°$.

Fig. 4 shows core loss dependence on the phase angle of the harmonic frequency component of magnetic induction under different harmonic frequencies from the 3^{rd}(180 Hz) to 9^{th}(540 Hz) and harmonic amplitude component from 10% to 50% of the total amplitude (B_{max}=1.5 T). From Fig. 5, we can see that core losses strongly depend on the phase angle and harmonic amplitude in the lower harmonic frequency component of magnetic induction. However as the harmonic frequency becomes higher, the core losses becomes a weak function of the phase angle while strongly dependent on harmonic amplitude. From this experimental result, if waveforms of magnetic induction in electric machines have higher harmonic   frequency components of magnetic induction, phase angles should be considered when frequency components are below the 9th harmonic frequency.

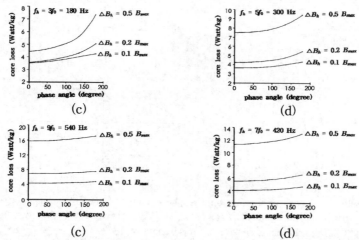

Fig. 4. Core loss depends on the phase angle of the higher harmonic frequency components of magnetic induction; a) for 3rd harmonic, b) for 5th harmonic, c) for 7th harmonic, and d) for 9th harmonic case.

4. Conclusion

Higher harmonic frequency components of magnetic induction are always generated in electric machines and the core loss   analysis for the higher harmonic frequency components of magnetic induction is important in the design of high efficiency electric machines. From the experimental results, we can see that core losses only depend on the harmonic amplitude component when the magnetic induction waveform has above the 9th harmonic frequency components of magnetic induction. As the harmonic frequency becomes lower than the 9th harmonic frequency, the core losses depend on the phase angle and the harmonic amplitude. Thus, the phase angle should be carefully considered in the design of electric machines.

For further work, an analytical description for higher harmonic frequency components of magnetic induction is recommended, but it is very difficult due to the nonlinear and hysteresis properties of magnetic materials.

References

[1] H. L. Schenk et al., Iron Losses in Induction Motor Teeth, *IEEE Trans. on Magn.*, MAG–17, 3385 (1981).
[2] T. Sasaki et al., Magnetic Losses of Electrical Iron Sheets under Ac Magnetization Superimposed with Higher Harmonics, *IEEE Trans. on Magn. in Japan*, Vol. 7 64 (1992).
[3] F. Fiorillo et al., Power Losses under Sinusoidal, Trapezoidal and Distorted Induction Waveform, *IEEE Trans. on Magn.*, MAG–26, 2904 (1990).
[4] D. Son et al., Core Loss Measurements Including Higher Harmonics of Magnetic Induction in Electrical Steel, *J. of Magn. and Mag. Mat.* Vol. 160, 65 (1996).

Non-Linear Electromagnetic Systems
V. Kose and J. Sievert (Eds.)
IOS Press, 1998

Numeric Modeling of the Electromagnetic and Thermal Coupled Fields in the Inductive Heating of Cylindrical Half-Finished Products for Energy Consumption Optimization

Teodor LEUCA

University of Oradea, 5 Armatei Române st., 3700 ORADEA. ROMÂNIA

Abstract. In this paper, based on a non-linear model for the quasistationary electromagnetic and thermal coupled fields we make the numeric modeling using the finite element method for the induction heating of cylindrical half-finished products made of magnetic steel, using the professional software package Flux 2D. The results of the numeric analysis for a given application allow the correlation of the specific parameters of the heating with the basic technological requirements: obtaining a given temperature within a specified time and a homogenous heating for the entire volume.

1. Introduction

In [4] we presented a non-linear model for the coupled quasistationary electromagnetic and thermal fields, specific to the process of inductive heating of magnetic steels. There are also other models in [1,2,3,4,5] for the coupled electromagnetic and thermal fields in the eddy-current case. The Finite Element Method (FEM) [6] lends itself to numeric approaches to the electrical engineering problems and the use of professional software packages. Flux2D is just one example of such a package.

2. Electromagnetic and thermal coupled fields model

For a ferromagnetic domain with volume V_Σ and material properties: $\mu(H,\vartheta)$ and $\rho(\vartheta)$, the quasistationary electromagnetic field satisfies:

$$\nabla \cdot \overline{B} = 0 ; \qquad \nabla \times \overline{H} = \overline{J} \qquad\qquad (1)$$

$$\nabla \times \overline{E} = -\frac{\partial \overline{B}}{\partial t} ; \qquad \overline{B} = \mu(H,\vartheta)\overline{H} ; \qquad \overline{J} = \frac{1}{\rho(\vartheta)}\overline{E}$$

Considering the ferromagnetic domain as a half-finished cylinder made of magnetic steel within a coil excited in AC the magnetic potential complex vector, $\overline{A}$, satisfies:

$$j\omega\mu_0 \frac{1}{\rho(\vartheta)}\overline{A} + \nabla \times \left[\frac{1}{\mu_r(H,\vartheta)}(\nabla \times \overline{A})\right] = \mu_0 \overline{J}$$

$$\qquad\qquad (2)$$

$$\overline{B}=\nabla \times \overline{A} ; \quad \nabla \cdot \overline{A}=0 ; \quad \overline{A}(0,0,A_z), \text{ cu } A_z = \text{ct on the boundary (Dirichlet condition).}$$

The thermal field , ϑ, satisfies:

$$-\nabla\lambda\cdot\nabla\vartheta - \lambda\nabla^2\vartheta + \gamma c\frac{\partial\vartheta}{\partial t} = J^2\cdot\rho(\vartheta)$$

$$\lambda\frac{\partial\vartheta}{\partial n} + \alpha(\vartheta - \vartheta_a) + \varepsilon(\vartheta^4 - \vartheta_a^4) = 0$$

(3)

with: $\lambda = \lambda(\vartheta)$ and : $c = c(\vartheta)$

The coupling between the fields is made by the second term from the first equation in system (3) and through the thermal field, ϑ.

3. Numerical application. Results.

We consider a half-finished cylinder of magnetic steel with a diameter of 50 mm and a length of 200 mm being heated in a solenoid with a length of 250 mm and internal diameter of 90 mm from the environmental temperature $\vartheta_i=20^0C$, to the milling temperature, $\vartheta_f=1050^0C$. The material properties are:

- the resistivity varies with the temperature according to the relationship :
  $\rho(\vartheta)=0.2(1+4\cdot10^{-3}\vartheta)$ $[\Omega\cdot mm^2/m]$;

- magnetic permeability μ depends on the magnetization status, for temperatures below Curie point, $\vartheta_C=760^0C$; afterwards the material becomes non-magnetic with $\mu=\mu_0$. The magnetization curve is fixed based on the model provided by FLUX 2D, which considers the saturated induction, $B_S=2.1$ T and the initial relative permeability $\mu_r=200$

- density , $\gamma=7800$ Kg/m^3

- specific heat is dependent on the temperature, according to:
  $c(\vartheta)=300(1+0.67\cdot10^{-3}\vartheta)$ $[J/Kg\ ^0C]$;

- thermal conductivity dependent on the temperature, according to the relationship :
  $\lambda(\vartheta)=47(1-0.25\cdot10^{-3}\vartheta)$ $[W/Kg\ ^0C]$;

During the heating the sample loses heat through its front and sides. This transfer is defined by the value $\alpha=20$ W/m^2 ^0C of the transmitivity and by the emission factor $\varepsilon=0.5$. The numeric modeling of the electro-thermal interaction specific for the induction heating was done using the magneto-thermal model of FLUX 2D, for three values of the frequency f: 1000 Hz, 2500 Hz, and 8000 Hz, and for ranges of the solenation and of the current density J determined by preliminary tests. The lower limit of the solenation interval is that which cannot provide the final temperature imposed ϑ_f and the upper limit is given by imposing an upper bound on the difference between the maximum and the minimum temperatures at the end of the process.

The results presented here are:

• Figures 1 and 2 show the magnetic field lines at the beginning and at the end of the

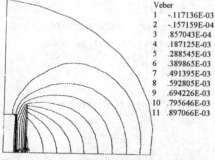

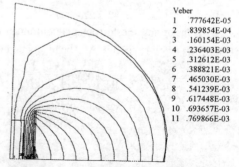

Figure 1. Magnetic field spectrum at the beginning of the heating

Figure 2. Magnetic field spectrum at the end of the heating

process considering f=1000 Hz and J=30 A/mm².

• The distribution of the thermal field at the end of the heating and the isotherms are presented in Figures 3 and 4 for f=1000 Hz and J=30 A/mm². The figures 5 and 6 are the same for f=2500 Hz and J=32,05 A/mm²

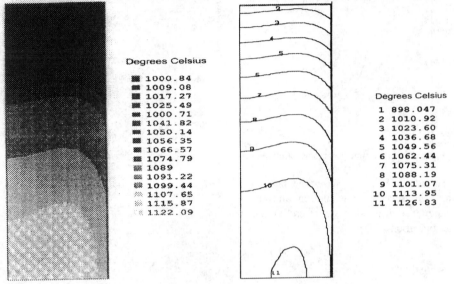

Figure 3. Distribution of the thermal field at the end of the heating; f=1000 Hz; J=30A/mm²

Figure 4. Isotherms; f=1000 Hz; J=30A/mm²

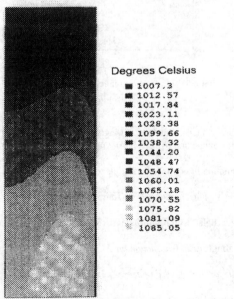
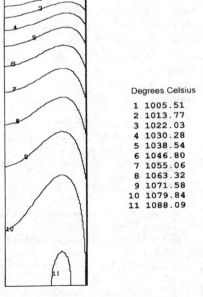

Figure 5. Distribution of the thermal field at the end of the heating; f=2500 Hz; J=32.05A/mm²

Figure 6. Isotherms; f=2500 Hz; J=32.05A/mm²

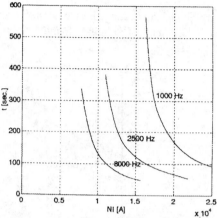

Figure 7. Heating time versus solenation (end of heating)

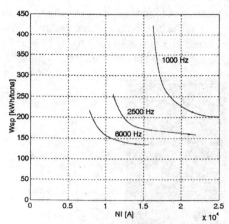

Figure 8. Specific energy consumption versus solenation (end of heating)

• Figures 7, 8 and 9 show the heating time, the specific consumption and the temperature dispersion at the end versus the solenation

4. Conclusions

The space-time distribution of the field values offers a more thorough analysis of the complex induction heating process. The analysis of the dependencies in Figures 7,8 and 9, offers important information on correlating the technological requirements with the electrical parameters of the heating equipment.

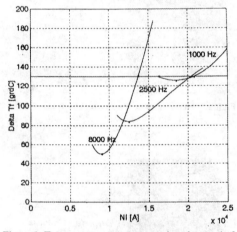

Figure 9. Temperature range versus solenation (end of heating)

References

[1]   Vlatko Čingoski, Akihiro Namera, Kazufumi Kaneda, Analysis of Magneto - Thermal Coupled Problem Involving Moving Eddy - Current Conductors, *IEEE Transaction on Magnetics*, Vol. 32, No. 3, May 1996, pp. 1042 -1045

[2]   H. Hedia, Sinusoidal magnetic field computation in nonlinear materials, *IEEE Transaction on Magnetics, November 1995*, Vol. 31, p. 3527-3529

[3]   T. Kirubarajan, P.R.P. Hoole, S.R.H. Hoole, K. Kiridharan, *IEEE Transaction on Magnetics*, May 1995, Vol. 31, p. 1956-1959.

[4]   T. Leuca, Câmpul electromagnetic şi termic cuplat. Curenţii turbionari, Editura Mediamira, Cluj-Napoca, 1996

[5]   E.J.W. Maten, Simulation of inductive heating, *IEEE Transaction on Magnetics*, Vol. 28, March 1992

[6]   P.P. Silvester, R.L. Ferrari, *Finite Elements for Electrical Engineers*, *Cambridge* University Press, 1994.

[7]   *** , Flux2D, *User's Guide*, Cedrat, Grenoble, 1994.

Non-Linear Electromagnetic Systems
V. Kose and J. Sievert (Eds.)
IOS Press, 1998

Sub–Sonic Domain Wall Dynamics of Yttrium Orthoferrite

Y. S. Didosyan[1], H. Hauser[2], V. Y. Barash[1] and P. L. Fulmek[2]

[1] Russian Institute of Metrological Service,
Ozernaya 46, Moscow 119361, Russian Federation

[2] Institut für Werkstoffe der Elektrotechnik, Technische Universität Wien,
Gußhausstraße 27 – 29, A-1040 Wien, Austria

Abstract. The nonlinear character of the subsonic domain wall dynamics in yttrium orthoferrite is confirmed by numerous sets of measured dependencies of the domain wall velocity v on the magnetic field H. Results of the work are summarized by histograms. About 20 maxima in the velocity range 200–1100 m/s are seen at the histograms. The nonlinearities may be due to the newly discovered lines of zero magnetization in orthoferrites.

1. Introduction

Investigations of the domain wall (DW) dynamics are very significant for the study of ferromagnetism, of the interaction between different sublattices, of magnetization processes, etc. [1]. They are of importance for various practical applications of ferromagnets, including the development of methods of magnetic field (electric current) measurements by means of the measurements of the DW velocity. The DW velocity in orthoferrites is extremely high, cosmic values up to 20 km/s can be reached [2]. Orthoferrites represent canted ferromagnets, i.e. the magnetic moments of the sublattices are almost opposite to each other and, respectively, the resultant magnetic moments are very weak: $M_s = 8$ kA/m.

The fact that the DW velocity in orthoferrites exceeds the speed of sound has provided rich opportunities for the investigations of the DW dynamics, especially for the study of the interaction betweeen the moving DW and a phonon subsystem. It has been found that there are zones of instability of the DW velocity. When the DW velocity reaches the transversal sound velocity it does not any more gradually grow with the increase of the driving magnetic field but rather remains on this speed level or "jumps" to another velocity level corresponding to the longitudinal sound velocity. The excess of the magnetic field energy gained while the velocity remains on the lower level is spent on the emission of phonons [2]. At lower velocities the dependence of the DW velocity on the magnetic field was considered to be linear, until recently. This fact has been established in the classical work [4] and was confirmed by numerous works, both theoretically and experimentally [1–3]. Using a new high resolution DW velocity measurement method, based on the dark field techniques, nonlinearities in the $v(H)$ dependence at subsound velocities, in the range of 200–1100 m/s, have been found [5]. In this paper the studies of the $v(H)$ dependence in this range are continued.

2. Experimental techniques

The sample is a plate of yttrium orthoferrite cut perpendicular to the optical axis. The optical axes of yttrium orthoferrite lie in the bc plane and at the wavelength 0.63μm

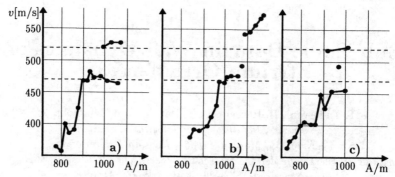

Figure 1: Dependencies of the DW velocity on the external magnetic field measured by using one slit.

at angles of $\pm 52°$ with the axis of weak ferromagnetism (crystallographic c axis); the specific Faraday rotation at this wavelength equals $-2900°/\text{cm}$. The plate's thickness is 100 μm. Perpendicular to the surfaces of the sample a quadrupole magnetic field H_e with a gradient $k = 7000$ kA/m^2 along the a axis was applied. This field created a single rectilinear 180° DW in the sample. To put the DW in motion a pulsed magnetic field h of two small coils, attached to both surfaces of the sample, was used. An additional constant uniform magnetic field H_0, shifting the magnetic field distribution, acted on the sample, too. The efficiency of the light diffraction by a single domain wall in orthoferrites is rather high. Therefore, it is possible to obtain the dark field image of the DW in a form of the bright strip of light. The motion of the strip is measured by means of slits made in the opaque screen [6]. In the process of the DW motion the dark field image passes through a slit and illuminates a photodetector placed behind the screen. The exact moment of the maximum photocurrent is measured.

The measurement of high DW velocities is performed by using two slits. The moments of illumination of the first and of the second slits are measured. The distance between the slits and the difference between these moments give the velocity of the DW. For measurements of low velocities only one slit was used. The moment of the slit's illumination was measured for a given bias field H. Then the value of the bias field was slightly changed by ΔH. This corresponds to changing the initial position of the DW by $\Delta x = \Delta H/k$. The moment of the slit's illumination was measured again. The difference of the moments gives the time interval in which the distance Δx was passed. The values Δx can be made as small as several μm, thus ensuring the high resolution of the velocity measurements.

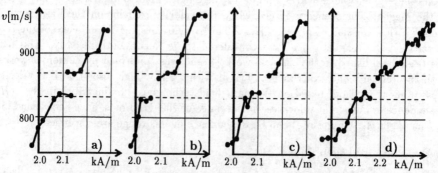

Figure 2: Dependencies of the DW velocity on the external magnetic field measured by using two slits.

3. Experimental results and discussion

Numerous sets of the DW velocity measurements have been performed in the range of 200–1100 m/s. The most of measurements were performed by means of one slit; velocities higher than 600 m/s were measured also by means of two slits. The distance between the slits was 1.0 mm. In the plane of the sample this corresponds to the distance 35 μm. In a number of cases the levels of the velocity constancy — shelves — in the $v(H)$ dependence have been observed.

Figure 1 shows three examples of such curves measured by means of one slit in different places of the sample. The dotted lines indicate two obvious levels of these curves: 470 and 520 m/s. Figure 2 presents four examples of the curves measured by means of two slits. The distance between the slits was 1 mm. The curves a), b) and d) were measured in different places of the sample. Curves b) and c) were measured in the same place of the sample but at different values of the bias field: the difference between the bias fields was equal to 3.2 A/m. Here two velocity levels can be found: 840 and 870 m/s.

The shown shelves, as well as similar shelves at other velocities, do not always appear at the measured $v(H)$ dependencies. Usually only some of them are observed. To summarize the results of the measurements the histogram Fig. 3 can be used. The histogram comprises data of 1230 points at the $v(H)$ dependencies measured by means of one slit and 590 points measured by means of two slits. The horizontal axis shows the velocity of the DW. The total number N of experiments resulting in velocities in the range $v \pm 10$ m/s, among all sets of measurements has been counted. Additional smoothening of the $N(v)$ dependence was performed by averaging in threes adjacent points. The histogram is divided in two parts. Part a) shows the results of the velocity measurements in the range 200–640 m/s, obtained by means of one slit. Part b) shows results of measurements in the range 700–1100 m/s and includes also data obtained by means of two slits. More than 20 maxima separated by intervals of about 40 m/s can be found at the histogram. The positions of the maxima are close to the shelves observed at the $v(H)$ dependencies. Earlier, a sequence of more then ten shelves has been found in orthoferrites in the range between the longitudinal sound velocity and the limiting velocity. The explanation of these levels was given in terms of generation of surface magnons localized near the DW at certain DW velocities [5]. This is based on the assumption of an interaction of the DW with hypothetical nonuniformities proportional to the distance between adjacent levels. But until now the existence of these

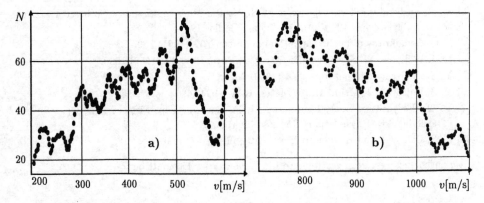

Figure 3: Histogram of the measured $v(H)$ dependencies using a) one slit, b) two slits.

nonuniformities has not been justified. It may be assumed that the shelves in the low and high velocity ranges have a common origin and the whole $v(H)$ dependence has a discrete character. Lines of zero magnetization in orthoferrites have been proposed [7]. They represent the part of the DW separating regions with different polarities in the DW, in analogy to Bloch lines. The lines possess a large energy and it may be assumed that they cause the nonlinearities of the DW dynamics.

By measuring the positions of the DW it is possible to determine the value of the acting magnetic field (electric current). For slowly changing magnetic fields it is sufficient to determine the final, equilibrium DW position. Measurements of short pulses should be possible due to the high DW velocities in orthoferrites, but require the exact knowledge of the relation between the DW velocity and the acting field. Further investigations of the nonlinear DW dynamics will show whether this relation can be established.

4. Conclusions

Numerous measurements of the DW velocity by the dark field method confirm the nonlinear character of the DW dynamics in the sub–sound range. The measurements have been performed by means of one and two slits, at various distances and in various places of the sample. The results of measurements are summarized in a form of histogram showing about 20 maxima separated by intervals about 40 m/s. The positions of the maxima are close to the shelves observed at the $v(H)$ dependencies. Taking into account nonlinearities found earlier in the range of supersonic DW velocities we may assume that the whole dependence of the DW velocity on an external magnetic field has a nonlinear character. Recently Bloch lines – or more precisely lines of zero magnetization – have been discovered in yttrium orthoferrite [8]. One may suppose that generation and motion of these lines is responsible for the nonlinearities in the DW dynamics. Further investigations of the DW dynamics at different thicknesses of samples, at different temperatures and in the higher velocities range may contribute to the understanding of the nature of this effect. These investigations could be also of importance for the development of new methods of magnetic field measurements based on the measurements of the DW velocity.

References

[1] A. P. Malozemoff and J. C. Slonczewski, *Magnetic Domain Walls in Bubble Materials*. Academic Press, New York, 1979.

[2] V. G. Bar'yakhtar et al., *Dynamics of Topological Magnetic Solitons*. Springer Tracts on Modern Physics vol. 129 (Springer–Verlag, Berlin, 1994).

[3] S. O. Demokritov et al., J. Magn. Magn. Mater. **102** (1991) 339.

[4] H. J. Williams, W. Shockley, and C. Kittel, Phys. Rev. **80** (1950) 1090.

[5] Y. S. Didosyan, J. Appl. Phys. **73** (1993) 6828.

[6] Y. S. Didosyan, J. Magn. Magn. Mater. **133**, (1994) 425.

[7] V. G. Bar'yachtar, B. A. Ivanov and M. V. Chetkin, Usp. Fiz. Nauk **146** (1985) 417. [Transl. in: Sov. Phys.–Usp. (USA) **28** (1985) 563.]

[8] Y. S. Didosyan and H. Hauser, Phys. Lett., in press (1998).

The authors express deep gratitude to I.R. Yavorsky for his contribution.

Non-Linear Electromagnetic Systems
V. Kose and J. Sievert (Eds.)
IOS Press, 1998

Non-Linear Dynamics in Magnetostrictive Materials

A.P Holden[*], J. Kennedy[*], N. Holliday[*], D.G. Lord[*], A.G. Jenner[†], P.R. Johnson[†] and H. Ahlers[‡]

[*]*Department of Physics, Joule Laboratory, University of Salford, Salford, M5 4WT, UK,*
[†]*Department of Applied Physics, University of Hull, Hull, HU6 7RX, UK, and*
[‡]*P.T.B., Braunschweig, Germany*

Abstract. Domain populations' and hence magnetisation processes' critical fields have been deduced from measurements of the transverse magnetisation and torque in twinned crystals of Terfenol-D as functions of applied field and crystal orientation. Complementary Barkhausen noise spectra have been obtained in-conjunction with the magnetisation measurements. Critical fields calculated by modelling of the transition processes are in agreement with measured values.

1. Introduction

The ternary cubic Laves phase material Terfenol-D, (typically $Tb_{0.3}Dy_{0.7}Fe_{1.95}$) is characterised by high magnetostrains, typically 1000-1600 ppm, large magnetomechanical strain coefficients, d_{33} ~200 nm/A, and large magnetomechanical coupling coefficients, k_{33} ~0.7, thus making it of significant technological importance for actuator and transducer applications such as sonar and active vibration control [1,2]. The important bulk magnetostriction arises from a combination of low magnetocrystalline anisotropy and a high single crystal magnetostriction coefficient, λ_{111}. The growth of this magnetostrain with increasing applied field is often characterised by discontinuous changes at critical field strengths [3]. Models of the magnetisation processes in this material indicate that these sharp changes arise from non-linear magnetisation reorientations [3,4].

The material available commercially has a growth direction of [112] with dendritic plates running parallel to this growth direction. These plates often contain crystallographic twin boundaries with the plane of both plate and twin generally being the (111) orthogonal to the growth front plane. The material has <111> easy directions of magnetisation at room temperature with the magnetic anisotropy compensation temperature for this material (dependent on composition) around 275 K [5]. It is the anisotropy and hence the domain populations, which determine the appearance of the strain evolution whether this is by domain wall motion or by domain magnetisation rotation.

From measurements of the transverse magnetisation in twinned crystals as a function of applied field, and crystal orientation, magnetic domain populations involved in the magnetisation processes are deduced. Barkhausen noise spectra also provide information complimentary to that obtained through magnetisation measurements and as such are used to elucidate the discontinuous changes at critical field strengths [3].

2. Experimental

All samples used in this work were spark cut and prepared from free-float zoned 8 mm diameter rods supplied by Edge Technologies Inc. Ames, Iowa. For rotating sample and torque magnetometry measurements, (110) and (112) plane discs were cut 3 mm in diameter and 1 mm thick. Bar samples, 50 mm in length and 5 mm by 5 mm square cross-section, for use in transverse magnetisation measurements, were prepared with the principal axis parallel to the [112] growth axis and the faces mutually orthogonal and parallel to the (111) and (110) planes.

Torque curves were obtained for the specimens from anti-clockwise rotation of a 650 kAm^{-1} magnetic field from a torque magnetometer based on the Aldenkamp transducer design over a temperature range 100 K to 280 K [6].

Transverse magnetisation information was collected from the bar specimens by having co-axial dc and ac magnetic fields along the sample's principal axis. Two pairs of pick-up coils mutually orthogonal to each other were placed around the specimen to detect the transverse magnetisation changes in both the [111] and [110] directions. One coil, with its associated air-flux compensation coil, positioned co-axial with the bar's principle [112] axis was used to detect susceptibility parallel to the applied fields. A lock-in amplifier was used to detect all induced signals in-phase with the ac field frequency.

Further anisotropy measurements and complimentary Barkhausen noise spectra were obtained on the disc samples utilising a rotating sample magnetometer. Samples were rotated with constant frequency inside a set of pickup coils placed between the poles of an electromagnet. The pickup coils were arranged such that their sensing axes were transverse to the applied dc field direction. The signals from the sensing coils were amplified on one channel of a digital oscilloscope. The second channel of the oscilloscope was used to observe the intensity of the Barkhausen noise of the measured signal after amplification, filtering and rectifying. The filtering and rectification were implemented utilising a flat response bandpass filter set between 300 Hz and 200 kHz and a full wave rectification circuit. Data was taken for a variety of rotational frequencies (10 Hz - 100 Hz) and for dc applied fields up to 700 kAm^{-1}. The data were stored allowing further processing.

The majority of measurements reported were obtained at zero applied stress and at room temperature.

3. Results and discussions

The angle of the magnetisation with respect to crystal axes used in general torque expressions is related with respect to the <100> direction in the {110} plane and the <110> direction in the {112} plane. Therefore it is necessary to make these directions the zero degree point in the experimental graphs. The expressions for the anisotropy constants for (112) in terms of the Fourier coefficients (A_n, B_n), for terms up to the sixth harmonic, are:

$$K_1 = 24/7[(A_4^2 + B_4^2)^{0.5} - 5K_2/144] \qquad (1)$$

$$K_2 = -576/25 [A_6^2 + B_6^2]^{0.5} \qquad (2)$$

The anisotropy constant K_1 was by analysis of the torque curves corrected for shear due to the misalignment of magnetisation from both (110) and (112) discs and were in good agreement having values respectively of -1.15×10^5 Jm^{-3} and -1.13×10^5 Jm^{-3}.

The values for the anisotropy constant K_2 from both sets of discs, however were not in as a good agreement as those for K_1. The (110) discs measurements yielding a K_2 value of 5.05×10^4 Jm^{-3} and the (112) discs a value of -8.68×10^4 Jm^{-3}.

Torque versus temperature curve analysis of a (112) plane disc is shown in Figure 1. Zero torque indicates an effective zero anisotropy due to the spin reorientation to the <111> easy axes at the higher temperatures. It can be seen from Figure 1 that this transition temperature is around 210 K. This is at least 60 degrees below the normally accepted transition temperature, though it strongly supports recent Lorentz TE microscopy evidence [8].

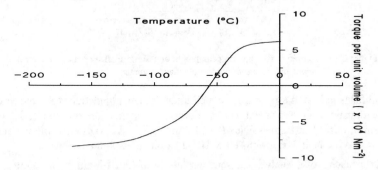

Figure 1 Torque versus temperature plot for a (110) plane disc. Zero torque indicates transition   from (100) to (111) easy axes.

Modelling of the transition processes based on domain populations between the various <111> easy directions and the applied field direction has been carried out [8]. The modelling has indicated the important critical field strengths involved in the transfer of domain populations between the various easy directions during the magnetisation process. These critical fields are given in Table 1 and the angles of the magnetisation for each axis group relative to the field direction in Table 2.

Table 1 Modelled magnetisation processes in Terfenol-D indicating specimen/ field critical fields for domain population redistribution

Axis Group	Zero Field Population	Critical Low Fields Am^{-1}	Critical Medium Fields kAm^{-1}	Critical High Fields kAm^{-1}
I$^+$	1/3	11 DW	7.4, 10.2, 11.0 DW/DR	48.5 DR
I$^-$	0	-	244.0 DW/DR	15.0 DR
II	1/3	11 DW	20.1 DW/DR	
III$^-$	1/3	11 DW	1.6, 11.0, 37.6 DW/DR	16.0 DR
III$^+$	0	-	4.4 DW/DR	48.5 DR
F$^+$	0	-		

Table 2 Angle of <111> to [112] direction for axis groups used in modelling process.

Axis Group	Angle (degrees)
I$^-$	118
I$^+$	62
II	90
III$^-$	160.5
III	19.5
F$^+$	0

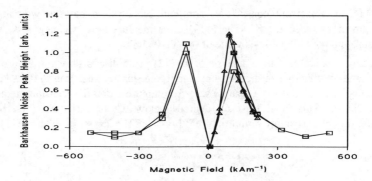

Figure 2 An example of Barkhausen peak height emissions against applied field for a (112) plane disc in a rotating sample magnetometer.

The critical fields indicated are for transfers of initial domain populations to higher order axis groups by domain wall movement (DW), domain magnetisation rotation (DR) or by a combination of both. Quite small fields (~ 11 Am^{-1}) allow the transfer of the initial domain populations into the next axis groups (II & III$^+$) by domain wall movement.

Barkhausen noise analysis of samples has been carried out in a rotating sample magnetometer. An example of the variation of Barkhausen noise emission peaks with applied field is shown in Figure 3. The emissions occur over a range of field strengths due to the domains having different orientations with respect to the field axis and hence different critical field strengths for rotation. The Barkhausen emission decay with increasing higher field strengths appear, as expected, in the region 40 to 100 kAm^{-1} which coincide with the onset of higher field processes predicted by the modelling.

4. Conclusions

Anisotropy constants have been derived from (110) and, for the first time, from (112) specimens. No influence on the (112) from the twin density is expected. An <100> to <111> easy axes transition temperature of 210 K has been observed. Domain redistribution critical fields have been calculated by preliminary modelling. Calculated critical fields in agreement with fields from transverse magnetisation and Barkhausen measurements [8].

References

[1]   R.D. Greenough, I.M. Reed and M.P. Schulze, Chapter 8, "Advances in Actuators" Eds. A.P. Dorey and J.H. Moore, IoP Pub. (1995).

[2]   A.G. Jenner, R.D. Greenough, D. Allwood and A.J. Wilkinson, J. Appl. Phys. **76** (10), 7160 (1994).

[3]   A.E. Clark, J.P. Teter and O.D. McMasters, J. App. Phys. **63**, 3910 (1988).

[4]   D.C. Jiles and J.B. Thoelke, IEEE Trans. Mag. **27**, 5352 (1991).

[5]   M. Al-Jiboory and D.G. Lord, IEEE Trans. Mag., **26**, 2583 (1990).

[6]   A.A. Aldenkamp, C.P. Marks, and H. Zijlstra, Rev. Sci. Instr., **31**, 544 (1960).

[7]   A.P. Holden, D.G. Lord, and P.J. Grundy, J. Appl. Phys., **79**, 4650 (1996).

[8]   D.G. Lord, Private communication (1996).

Non-Linear Electromagnetic Systems
V. Kose and J. Sievert (Eds.)
IOS Press, 1998

Modelling of Nonlinear Magnetic Material using an Effective Permeability – Comparison between Different Formulations

Dieter LEDERER, Hajime IGARASHI*, Arnulf KOST**

Institut für El. Energietechnik, TU Berlin, Einsteinufer 11, 10587 Berlin, Germany
Faculty of Eng., Hokkaido University, Kita 13, Nishi 8, Kita-ku, Sapporo 060, Japan
**Lehrstuhl Allg. Elektrotech., BTU Cottbus, Postf. 101344, 03013 Cottbus, Germany*

Abstract. The calculation of shielding arrangements with nonlinear material and sinusoidal excitation is investigated. The material is modelled using an "effective permeability" which allows a sinusoidal calculation neglecting higher harmonics. Different formulations are taken into account. The calculation results are compared with measurement results. It is concluded that the formulations based on the preservation of the rms-value of H or B are more appropriate for the calculation of shielding arrangements than those based on the preservation of energy.

1. Introduction

In recent years discussion about the effects of electromagnetic fields on technical devices has increased significantly due to their risen sensitivity to electromagnetic fields. Devices like computer screens, electron microscopes or pacemakers may be disturbed by magnetic fields exceeding certain values.

One of the main sources of disturbing magnetic fields are the power supply facilities like power cables or low-voltage distribution devices. The produced magnetic field can be reduced by changing the phase-sequence or the geometrical arrangement of the conductors or by shielding the field source with ferromagnetic material.

The calculation of such a shielding arrangement is investigated in this paper. As the excitation is sinusoidal and the shielding material is nonlinear, the "effective permeability" material model is used, based on which a finite element calculation with complex phasors, neglecting higher harmonics, is performed. Different ways of obtaining the effective permeability from the measured magnetization curve of the material are taken into account [2–7]. The results are compared with measurement results.

In most reported cases the method is applied to the calculation of machines and is often related to the consideration of saturation effects and the calculation of losses. Here the method is used for the shielding of power cables where – under normal conditions (no short circuit, no overload) – the state of saturation is not reached.

2. Formulation

Two-dimensional eddy current problems are considered, for which the relationship

$$\nabla \times \left(\frac{1}{\mu_{\text{eff}}} \nabla \times \underline{\vec{A}} \right) + j\omega\kappa\underline{\vec{A}} = \underline{\vec{J}}_0 \tag{1}$$

is derived from Maxwell's equations. $\underline{\vec{A}} = \underline{A}\,\vec{e}_z$ is the complex vector potential, $\underline{\vec{J}}_0 = \underline{J}_0\,\vec{e}_z$ is the exciting current density and μ_{eff} is the effective permeability obtained by the material model described in the next section. Applying Galerkin's method with the shape functions α, one gets

$$\sum_{k=1}^{n} \left(\int_{\Omega} \frac{1}{\mu_{\text{eff}}} \nabla\alpha_i \nabla\alpha_k \, d\Omega + j\omega \int_{\Omega} \kappa\,\alpha_i\alpha_k \, d\Omega \right) \underline{A}_k - \int_{\Omega} \alpha_i \underline{J}_0 \, d\Omega = 0 \tag{2}$$

with $i = 1, \ldots, n$ for the z-component of the vector potential; n is the number of nodes. These equations are nonlinear due to the dependence of μ_{eff} on the nodal values of the vector potential. For their solution the complex Newton-Raphson method is used [1].

3. Material Modelling

The usage of an effective permeability for the calculation of nonlinear problems with sinusoidal excitation is well-known and various possibilities for its determination have been reported [2–7].

The effective permeability is calculated from the measured magnetization curve under the assumption of either sinusoidal magnetic field strength [2,3,5], or sinusoidal magnetic flux density [4–7]. Of course, in general neither will be sinusoidal; these assumptions are approximations.

Further, a quantity has to be chosen, based on physical considerations, whose value will be preserved in the transformation process. Three different quantities are chosen in this paper: the rms-value of H or B [3,5], the energy density imparted during one period [6] and the mean energy density over one period [2,7]. This leads to the following definitions of the effective permeability:

H sinusoidal: $H(t) = \hat{H}\sin\omega t$ 　　　　 B sinusoidal: $B(t) = \hat{B}\sin\omega t$

$$\mu_{\text{h}1} = \frac{1}{\hat{H}} \left(\frac{2}{T} \int_0^T [B(H)]^2 \, dt \right)^{1/2} \qquad \mu_{\text{b}1} = \hat{B} \left(\frac{2}{T} \int_0^T [H(B)]^2 \, dt \right)^{-1/2}$$

$$\mu_{\text{h}2} = \frac{2}{\hat{H}^2} \int_0^{B(\hat{H})} H \, dB \qquad \mu_{\text{b}2} = \frac{\hat{B}^2}{2} \left(\int_0^{\hat{B}} H \, dB \right)^{-1}$$

$$\mu_{\text{h}3} = \frac{4}{\hat{H}^2 T} \int_0^T \int_0^{B[H(t)]} H \, dB \, dt \qquad \mu_{\text{b}3} = \frac{\hat{B}^2 T}{4} \left(\int_0^T \int_0^{B(t)} H \, dB \, dt \right)^{-1}$$

For all formulations the relationship $B_{\text{rms}} = \mu_{\text{eff}} H_{\text{rms}}$ holds. Figures 1 and 2 show the measured magnetization curve of constructional steel St 37 and the transformed curves obtained by using the effective permeabilities $\mu_{\text{h}1}$, $\mu_{\text{b}1}$ and $\mu_{\text{b}2}$, $\mu_{\text{b}3}$, respectively. For the given magnetization curve, the permeabilities $\mu_{\text{h}2}$ and $\mu_{\text{h}3}$ lead to non-monotonous transformed curves; thus they could not be used.

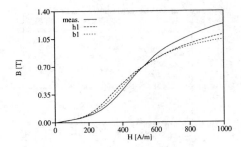

Fig. 1: Magnetization curves: measured, μ_{h1}, μ_{b1}

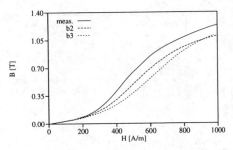

Fig. 2: Magnetization curves: measured, μ_{b2}, μ_{b3}

4. Application

4.1 Problem Description

The analyzed shielding arrangement is shown in Fig. 3: a steel plate over a double line. The plate material is constructional steel St 37 with conductivity $\kappa = 6.41 \times 10^6$ S/m. The dimensions are: $a = 250$ mm, $b = 1390$ mm, $d = 3$ mm, $h = 100$ mm. The exciting current is sinusoidal with the frequency $f = 50$ Hz and the rms-value $I = 900$ A.

4.2 Results and Discussion

Figures 4 and 5 show the measured and calculated rms-values of the total magnetic flux density at $y = 5$ cm above one half of the plate. The curves in Fig. 4 were calculated using the formulations μ_{h1} and μ_{b1} whereas for those in Fig. 5 μ_{b2} and μ_{b3} were used.

Comparing the curves, it is not questionable that the results calculated with the permeabilities μ_{h1} and μ_{b1} are closer to the measurement results than those calculated with the permeabilities μ_{b2} and μ_{b3}. The difference between the curves calculated with μ_{h1} and μ_{b1} is negligible. Thus it can be concluded that for the present type of problem – shielding arrangement with ferromagnetic material, no saturation – the effective permeabilities μ_{h1} and μ_{b1}, i.e. the preservation of the rms-value of H or B in the transformation of the magnetization curve, are more appropriate than the other ones.

The skin depth of the steel plate is in the range of 0.8–1.6 mm, depending on the local value of the permeability. The computed maximum values of the magnetic flux density inside the plate are listed in Table 1. It is obvious that saturation is not reached; however, the material characteristics are significantly nonlinear in this range (see Figs. 1 and 2). Consequently a linear calculation would not be possible, even for such low values of the field quantities.

The data of the calculation process (FEM, 2D) are shown in Table 2: the computation time and the need of memory are considerably low.

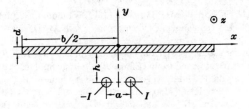

Fig. 3: Steel plate over double line

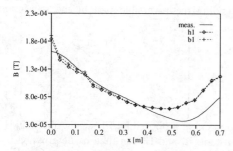 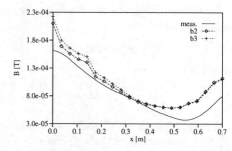

Fig. 4: Magnetic flux density 5 cm above the plate: measured, μ_{h1}, μ_{b1}

Fig. 5: Magnetic flux density 5 cm above the plate: measured, μ_{b2}, μ_{b3}

In the middle of the plate, the calculated and measured curves agree well, whereas near the edge of the plate, they disagree. Two reasons can be found for this result: at first it has to be mentioned that the effective permeability method is a fast and simple method, which neglects higher harmonics; further, hysteresis losses were not taken into account. The measured B-H-characteristics show significant hysteresis loops even for low values of the magnetic flux density. Consequently, the total agreement between measured and calculated curves could not be expected.

In order to achieve higher accuracy the present problem could be calculated by e.g. time step methods which take into account the full time dependence and do not need strong approximations of the material characteristics. The computation time and need of memory, however, would be much higher than those of the present method.

Summarizing, the effective permeability method is an appropriate tool to judge the efficiency of shielding arrangements. In spite of the approximation of the material characteristics reasonable results are obtained. This follows from the limited influence of higher harmonics, in which case the assumption of sinusoidal fieldstrength and flux density, related by the effective permeability, is possible.

Table 1: Maximum values of the flux density

Permeability	μ_{h1}	μ_{b1}	μ_{b2}	μ_{b3}
B_{max} (rms)/T	0.31	0.30	0.27	0.25

Table 2: Data of the calculation process

Elements	Time	Memory	Computer
28672	220 s	50 MB	HP C180

References

[1] D. Lederer, H. Igarashi and A. Kost, The Newton-Raphson Method for Complex Equation Systems, *ACES Journal*, **12** (1997) 113–116

[2] N.A. Demerdash and D.H. Gillott, A new Approach for Determination of Eddy Current and Flux Penetration in Nonlinear Ferromagnetic Materials, *IEEE Trans. Magn.*, **10** (1974) 682–685

[3] J.D. Lavers, Finite Element Solution of Nonlinear Two Dimensional TE-Mode Eddy Current Problems, *IEEE Trans. Magn.*, **19** (1983) 2201–2203

[4] Y. Du Terrail, J.C. Sabonnadière, P. Masse and J.L. Coulomb, Nonlinear Complex Finite Elements Analysis of Electromagnetic Field in Steady-State AC Devices, *IEEE Trans. Magn.*, **20** (1984) 549–552

[5] J. Luomi, A. Niemenmaa and A. Arkkio, On the Use of Effective Reluctivities in Magnetic Field Analysis of Induction Motors fed from a Sinusoidal Voltage Source, *Proc. ICEM* (1986) 706–709

[6] H. Hedia, J.-F. Remacle, P. Dular, A. Nicolet, A. Genon and W. Legros, A Sinusoidal Magnetic Field Computation in Nonlinear Materials, *IEEE Trans. Magn.*, **31** (1995) 3527–3529

[7] C. Guérin and G. Meunier, 3D Non-Linear Eddy-Current Problems Taking Into Account Magnetic Saturation Using Magnetodynamic Finite Elements, *Proc. ISEM* (1995)

Non-Linear Electromagnetic Systems
V. Kose and J. Sievert (Eds.)
IOS Press, 1998

Non-Linear Basic Circuits for Modelling Energetic Systems

W.G. Büntig and W. Vogt

Technical University of Ilmenau
Dept. of Theoretical Basics of Electrical Engineering and Dept. of Mathematics
D-98684 Ilmenau GERMANY

Abstract. An energetic system is very complex because of a large number of sub-systems in three wires. A linear description of sub-systems with hysteresis and magnetic saturation of iron is allowed only in some cases. To estimate the influence of the parameters on the system behavior is a different task. A few simple basic models are necessary at first to describe the principal properties of parts of energetic systems. By using these models one can examine a more complex and more realistic system.

1. Basic Circuit - Model 1

Between the wires of the energy supply the current is very small. The Model of the basic circuit - Model 1 - can therefore be found.

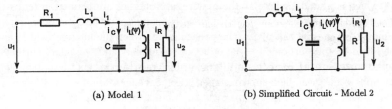

(a) Model 1　　　　　(b) Simplified Circuit - Model 2

Figure 1: Basic Circuit

$$u_1 = u_2 + L_1 \frac{di_1}{dt} + R_1 i_1 \tag{1}$$

The meaning of the elements in Fig. 1 is: R_1-Copper-resistance of the power supply and the magnetic circuit, L_1-Inductance of the power supply and leakage inductance of the magnetic circuit, C-Interturn capacitance and/or outer parallel capacitance $i_L(\psi)$-Characteristic function for describing the non-linear iron-core, and R-Load resistance and/or resistance for modelling the hysteresis behavior. Approximation of $i_L(\psi)$ is possible by using an incomplete power polynomial of the form

$$i_L(\psi) = a^* \psi + b^* \psi^n . \tag{2}$$

By normalizing the model Equation 1 with the abbreviations:

$$\tau = \omega t; \quad U_r = \omega \Psi_r; \quad Z_r = \frac{U_r}{I_r}; \quad X_L = \omega L_1; \quad X_C = \frac{1}{\omega C}; \quad a^* = a \frac{I_r}{\Psi_r}; \quad b^* = b \frac{I_r}{\Psi_r^n}; \quad x = \frac{\psi}{\Psi_r}. \tag{3}$$

By assuming an excitation force

$$u_1 = \hat{U}_1 \sin(\omega t) \tag{4}$$

we can find an equation for the normalized flux

$$\frac{d^3x}{d\tau^3} + \delta\frac{d^2x}{d\tau^2} + \alpha(1 + \beta x^{n-1})\frac{dx}{d\tau} + \Delta_1 x + \Delta_2 x^n = \hat{\Gamma}\sin(\tau) \tag{5}$$

with the short cut:

$$\delta = \frac{X_C}{R} + \frac{R_1}{X_L}; \quad \alpha = \frac{X_C}{Z_r}a + \frac{R_1}{X_L}\frac{X_C}{R} + \frac{X_L}{X_C}; \quad \beta = \frac{nb\frac{X_C}{Z_r}}{\alpha} \tag{6}$$

$$\hat{\Gamma} = \frac{\hat{U}_1}{U_r}\frac{X_C}{X_L}; \quad \Delta_1 = \frac{R_1}{Z_r}\frac{X_C}{X_L}a; \quad \Delta_2 = \frac{R_1}{Z_r}\frac{X_C}{X_L}b$$

2. Simplified Basic Circuit - Model 2

If we assume that $R_1 \approx 0$ we find the simplified model of the basic circuit - Model 2:

$$\frac{d^2x}{d\tau^2} + \delta\frac{dx}{d\tau} + \alpha x + \gamma x^n = -\hat{\Gamma}\cos\tau \tag{7}$$

with the abbreviations

$$\hat{\Gamma} = \frac{\hat{U}_1}{U_r}\frac{X_C}{X_L}; \quad \delta = \frac{X_C}{R}; \quad \alpha = \frac{X_C}{Z_r}a + \frac{X_C}{X_L}; \quad \gamma = b\frac{X_C}{Z_r}. \tag{8}$$

In the case that capacitance C is in resonance with the non-linear inductance it is

$$Z_r = \frac{U_r}{I_r} = \frac{1}{\omega C} = X_C \tag{9}$$

The parameter values for the non-linear normalized magnetization curve are taken from [2] $n = 3; \ a = 0,037; \ b = 0,963$ or $n = 9; \ a = 0,25; \ b = 0,75$. In the normal load case of the system $\delta = 0,47$ (nominal-load) for the normal idling case $\delta = 0,14$. In this normal working area of the system and for higher load the simple Model 2 can be used. If $\delta < 0,1$ we have to use the more complicated Model 1.

3. Approximation of Stable and Unstable Manifolds

Models 1 and 2 can be rewritten as first order systems

$$\frac{dx}{d\tau} = f(\tau, x), \quad f : \mathbb{R} \times \mathbb{R}^n \to \mathbb{R}^n, \tag{10}$$

where f is smooth and periodic in τ with period $T_0 = 2\pi$. The Poincaré map (stroboscopic map) $P : \mathbb{R}^n \to \mathbb{R}^n$ of the flow φ_τ defined by

$$P(x_0) = \varphi_{\tau_0+T}(x_0), \quad T = k * T_0, \ k \in \mathbb{N}, \tag{11}$$

transforms periodic solutions into fixed points x^* of the diffeomorphism P. For hyperbolic fixed points x^* of P the Stable Manifold Theorem guarantees the existence of stable and unstable manifolds $W^s(x^*)$, $W^u(x^*)$ tangential to the eigenspaces E^s and E^u. These manifolds can be approximated numerically in the case of dimension $n = 2$ by the following:

Algorithm 1 - Open Invariant Curves

Step 1 Fixed point analysis: Computation of a fixed point x^* of saddle type and of its multipliers and the normed eigenvectors for E^s and E^u

Step 2 Starting procedure on $W^u(x^*)$: Selection of p starting points x_i near x^* on E^u in geometrical order and computation of their images $P(x_i)$

Step 3 Insertion of mapping points on $W^u(x^*)$: Successive interpolation of further points until the manifold is approximated by "sufficiently dense" mapping points

Step 4 Stopping criteria: Stopping the algorithm if the points x_i converge to another (stable) fixed point or escape from a given region

Step 5 Approximation of $W^s(x^*)$: Reversion of time τ yields $W^s(x^*)$ as the unstable manifold of the time-reversed system

Application of Algorithm 1 to Model 2

We demonstrate the method by an analysis of Model 2 Equation (7) where the parameters are

$n = 9; \delta = 0,15; \alpha = 0,65; \gamma = 0,75; T = 4\pi$.

- For $\hat{\Gamma} = 2,08$ the fixed point $P_2 = (-1,32525; -0,73182)$ of saddle type has non-intersecting invariant manifolds (see Fig.2(a)), whereas the fixed points P_{21} and P_{22} are stable. By increasing the amplitude $\hat{\Gamma}$ a period doubling occurs. For $\hat{\Gamma} = 2,15$ the fixed points P_{21} and P_{22} are of saddle type.

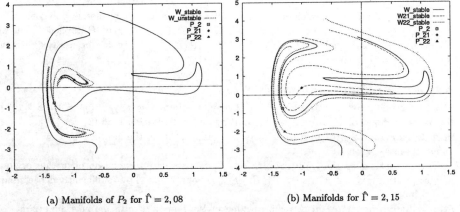

(a) Manifolds of P_2 for $\hat{\Gamma} = 2,08$ (b) Manifolds for $\hat{\Gamma} = 2,15$

Figure 2: Invariant manifolds and fixed points

- After continuation of the 3 fixed points up to $\hat{\Gamma} = 2,15$ we now approximate (see Fig.2(b)) their stable manifolds $W^s(P_2)$, $W^s(P_{21})$, $W^s(P_{22})$. The curve $W^s(P_2)$ as a separatrix divides the regions of the other invariant manifolds.

- In Fig.3(a) we restrict the method to the fixed point $P_{21} = (-1,01019; 0,37511)$. Here the invariant curves $W^s(P_{21})$ and $W^u(P_{21})$ intersect transversally at homoclinic points $P_H \neq P_{21}$. Due to the Smale-Birkhoff Theorem (see [6]) the Poincaré-map has an imbedded horseshoe-like map.

- This observation also holds for greater values of the amplitude $\hat{\Gamma}$. Fig. 3(b) shows a zooming for $\hat{\Gamma} = 3,0$ in the neighbourhood of the fixed point P_{21} where transversal homoclinic points are displayed.

In [1] the occurrence of a strange attractor for parameter values $\hat{\Gamma} \geq 2,15$ is verified by a bifurcation analysis.

4. Quasi-periodic Responses and Invariant Tori

For study of Model 1 Equation (5) the parameter-values $\alpha = 0,687; \beta = 9,85; n = 9; \Delta_1 = 0,025; \Delta_2 = 0,075$ are chosen. If the parameter δ is decreasing to $0,1$, the fixed point P_2 undergoes an Andronov-Hopf bifurcation and an asymptotically stable invariant curve γ of the Poincaré map occurs.

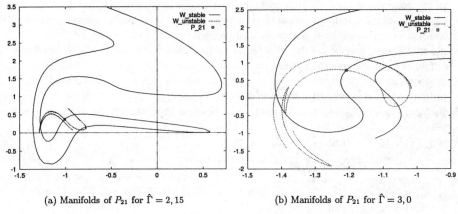

(a) Manifolds of P_{21} for $\hat{\Gamma} = 2,15$ (b) Manifolds of P_{21} for $\hat{\Gamma} = 3,0$

Figure 3: Intersection of invariant manifolds

One of the numerical approaches (see [4], [5]) approximates γ by closed polygons $\Pi = \Pi(\{x_i\}_{i=1}^N) \subset R^n$ with N points $x_1, x_2, \ldots, x_N$. The Poincaré map P applied (in parallel) to these N points yields the "better" polygon $\Pi(\{Px_i\}_{i=1}^N)$ and can be iterateduntil the equation $P\gamma = \gamma$ is solved.

Algorithm 2 - Closed Invariant Curves

Step 1 Starting procedure: Selection of a suitable starting polygon $\gamma_0 = \Pi(\{x_i\}_{i=1}^N)$ in the neighbourhood of an unstable fixed point x^* and initialization k := 0.

Step 2 Computation of mapping points: Computation of the images Px_i, $i = 1..N$, by numerical integration

Step 3 Insertion of mapping points on γ_k: Successive interpolation of further original points x_i until the manifold $\gamma_{k+1} = \Pi(\{Px_i\}_{i=1}^N)$ is approximated by "sufficiently dense" mapping points

Step 4 Stopping criteria: After projection of the points x_i onto the new approximation γ_{k+1} the error can be estimated. Stopping the algorithm if the polygon γ_k converges to γ

This method works in the given cartesian coordinates and it needs no special transformations.

5. Conclusion

These and other results of the authors show that the presented models and numerical methods are well suited for modelling real energetic systems.

References

[1] E.S. PHILIPPOW, W.G. BÜNTIG. *Analyse nichtlinearer dynamischer Systeme der Elektrotechnik.* Carl Hanser Verlag München Wien, 1992.

[2] E.S. PHILIPPOW. *Der ferromagnetische Spannungsstabilisator.* Leipzig, Geest & Portig K.G., 1968.

[3] E.S. PHILIPPOW, W.G. BÜNTIG, O.M. GUERRERO. *Besondere Effekte bei der Untersuchung von Reihenferroresonanzkreisen mit asymmetrischen nichtlinearen Magnetisierungskennlinien.* Proc. Int. Symp. Electromag. Theory, Bratislava, *Elec. Eng. J.*, Vol. 42, No.3, 1991, pp. 115 - 124.

[4] H. BERNET, W. VOGT. *Anwendung finiter Differenzenverfahren zur direkten Bestimmung invarianter Tori.* Z. Angew. Mathemathik u. Mech. Vol. 74, 1994, T577 - T579.

[5] F. NEDWAL, W. VOGT. *Zur numerischen Approximation geschlossener Invarianzkurven von Poincaré-Abbildungen.* TU Ilmenau, Preprint No. M 2/1996.

[6] T.S. PARKER, L.O. CHUA. *Practical Numerical Algorithms for Chaotic Systems.* Springer Verlag New York, 1989.

Non-Linear Electromagnetic Systems
V. Kose and J. Sievert (Eds.)
IOS Press, 1998

Magnetization Process and Statistical Domain Behaviour

Hans HAUSER and Paul L. FULMEK

Institut für Werkstoffe der Elektrotechnik, Technische Universität Wien,
Gußhausstraße 27-29, A-1040 Vienna, Austria

Abstract. Statistical domains are the simplified representation of magnetic domains and are characterized by their volume fractions and the direction of their magnetization. The energetic model of ferromagnetic hysteresis calculates the magnetic state of anisotropic materials by minimizing the total energy function for all statistical domains. Magnetocrystalline anisotropy, shape anisotropy and strain energy give the easy directions for the domains. All reversible contributions of the domain wall movements are described by a reversible field strength. The domain walls move reversibly until an individual Barkhausen jump occurs. From this point the irreversible losses start to accumulate. At a point of field reversal the magnetic state of the material is classified by changing the parameters of the irreversible energy in dependence of the so far covered irreversible jumps. The physical constants of this model are derived from anisotropic energy contributions, initial susceptibility, coercivity, and saturation magnetization.

1. Introduction

The energetic model of ferromagnetic hysteresis calculates the magnetic state of anisotropic ferromagnetic materials by minimizing the total energy function

$$E_T = E_H + E_M \ . \tag{1}$$

The energy of the material

$$E_M = E_c + E_s + E_d + E_R + E_I \tag{2}$$

is divided into magnetocrystalline energy E_c, strain energy E_s, energy of the demagnetizing field E_d, into reversible (E_R) — described by statistical domain behaviour — and irreversible (E_I) terms [1, 2]. The statistical domains are characterized by their volume fraction v_i ($\sum_{i=1}^{n} v_i = 1$) and the direction $\vec{n}_i$ of their magnetization. There are n easy directions, mainly due to magnetocrystalline and strain anisotropy.

The total magnetization of the ferromagnetic material is:

$$\vec{M} = M_s \sum v_i \vec{n}_i \ . \tag{3}$$

2. Domain Wall Displacements and Magnetization Reversal

The reversible energy E_R is interpreted as the energy of the magnetization in a reversible field H_R [1]. This field is given by:

$$H_R = -h \left(\left(\prod_i \frac{v_i^{v_i}}{w_i^{v_i}} \right)^a - 1 \right) , \qquad (4)$$

with the adaptive constant h. The initial occupation probabilities w_i ($\sum_i w_i = 1$) for the easy directions $\vec{n}_i$ represent the real domain structure of the material at the demagnetized state ($v_i = w_i$). The applied field H causes reversible domain wall displacements s_i until an individual Barkhausen jump starting position s_A is reached. The probability density $p_a(s_A)$ is assumed to be decreasing with increasing s_A:

$$p_a = \frac{K_a}{K_c} \exp\left[-\frac{K_a}{K_c} s_A \right] \qquad (5)$$

(adaptive constant: K_a). K_c describes the influence of the total magnetic state on p_a at points of magnetization reversal; $K_c = 1$ for the initial magnetization curve. The energy loss E_l of a single domain wall during an irreversible Barkhausen jump (wall friction: K_r) is increasing with increasing jumping distance s_i:

$$E_l = K_r \left[s_i - s_A - \frac{K_c}{K_a}(K_c - 1) \right] , \qquad (6)$$

K_c depends on the covered displacements $s_i = |(v_i - v_{i,0})/w_i|$ with respect to the starting domain distribution $v_{i,0}$, at a point of field reversal:

$$K_c^{\text{new}} = 2 - K_c^{\text{old}} \exp\left[-\frac{K_a}{K_c^{\text{old}}} s_i \right] \quad \text{with} \quad v_{i,0}^{\text{new}} = v_i . \qquad (7)$$

This follows from the condition of steadiness at these points. The variation of the irreversible energy with the movement of 180°–domain walls is [3]:

$$\frac{\partial E_I}{\partial v_i} = K_r \left(1 - K_{c,i} \exp\left[-\frac{K_a}{K_{c,i}} \left| \frac{v_i - v_{i,0}}{w_i} \right| \right] \right) . \qquad (8)$$

3. Calculation of Magnetization Curves

Stable points of the magnetization curve are points with minimum total energy, i.e. a state of vanishing total variation:

$$dE_T = \sum_{i=1}^{n} (\partial E_T / \partial v_i) dv_i + (\partial E_T / \partial \vec{n}_i) d\vec{n}_i = 0 . \qquad (9)$$

Due to the magnetocrystalline anisotropy and the crystallite–orientation in Goss–texture steel sheets, there exist two pairs of equivalent easy directions in the sheet plane. Therefore, a simplification of the energy expressions for magnetization processes in the symmetric [001]– and [110]–direction can be used to determine the parameters of the model for FeSi–steel with Goss–texture. In the symmetric directions the magnetization stays always parallel to the applied field in the sheet plane. The calculations result in a parametric expression with a Lagrange multiplier Λ. The magnetization curve is described by:

$$v_i = \frac{w_i \exp[\alpha_{iH}\Lambda]}{\sum_i w_k \exp[\alpha_{kH}\Lambda]} \tag{10}$$

$$M = M_s \sum_i v_i \alpha_{iH} \tag{11}$$

$$H_a = h\left(\left(\prod_i \frac{v_i^{v_i}}{w_i^{v_i}}\right)^a - 1\right) + \frac{K_r}{\mu_0 M_s}\sum_i \left|\frac{1}{\alpha_{iH}}\left(1 - K_{c,i}\exp\left[\frac{-K_a}{K_{c,i}}\Big|\frac{v_i - v_{i,0}}{w_i}\Big|\right]\right)\right| \tag{12}$$

The $\alpha_{i,H}$ are the cosines of the angles of the applied field H_a with the easy directions $\vec{n}_i$. The physical constants of this model are derived from anisotropic energy contributions, initial susceptibility, coercivity, saturation magnetization and saturating field strength. The value of saturation magnetization M_s, of w_1/w_3 and the parameters K_a and K_r are obtained by measuring the magnetization curves parallel and perpendicular to the rolling direction (Fig. 1a). Three values have to be measured: the saturation field strength $H_{S[001]}$ and the value of M_k, i.e. the point of intersection of the magnetization curve with the straight line $M/M_s = 1 - (1/H_{S[001]}) \cdot H$ in the rolling direction, and the field strength $H_{S[110]}$ necessary to reach $M_s(1/\sqrt{2})$ in the [110]-direction.

$$a = \frac{\log\left[(1 - M_k)(H_{S[001]}/H_{S[110]})^{(1-M_k)}\right]}{\log\left[M_k^{M_k}((1 - M_k)/2)^{(1-M_k)}\right]} \tag{13}$$

$$\frac{w_1}{w_3} = 2\left(\frac{H_{S[110]}}{H_{S[001]}}\right)^{(1/a)} \tag{14}$$

$$h = H_{S[001]} \cdot w_1^a = H_{S[001]} \cdot \left(2 + 4\frac{w_3}{w_1}\right)^{-a} = \tag{15}$$

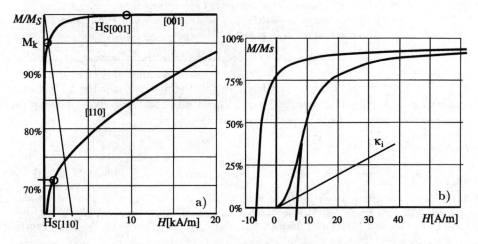

Figure 1: Measurement of magnetization curves of FeSi–steel with Goss–texture. Strong fields a), weak fields b).

Measurements at weak fields (Fig. 1b) in the rolling direction give values for the initial susceptibility and the coercivity. With these values the parameters of the irreversible energy are determined:

$$K_r = H_c \cdot \mu_0 M_s \tag{16}$$

$$K_a \;=\; \frac{2w_1 M_s}{H_c \kappa_i} \tag{17}$$

With the equations from above, the measured values:

$$M_s = 1.6 \text{ MA/m}, \; H_c = 6.5 \text{ A/m}, \; \kappa_i = 17300,$$

$$M_k = 0.95, \; H_{S[001]} = 9.9 \text{ kA/m}, \; H_{S[110]} = 1.1 \text{ kA/m}$$

give the complete set of parameters for the energetic model of ferromagnetic hysteresis:

$$a = 12.4, \; h = 111 \cdot 10^{-6} \text{ A/m}, \; K_a = 6.47, \; K_r = 13.1 \text{ J/m}^3, \; w_1/w_3 = 1.67.$$

Figure 2a shows the calculated curve for Goss–texture steel sheets in [001]–direction. The figure displays the virgin curve, the hysteresis loop, and two asymmetric loops from $M/M_s = -0.5 \ldots 1$ and from $M/M_s = 0.5 \ldots 1$. Figure 2b displays the results of the calculation in [001], [110] and in the 45°–direction with respect to the rolling direction and measured points at strong fields.

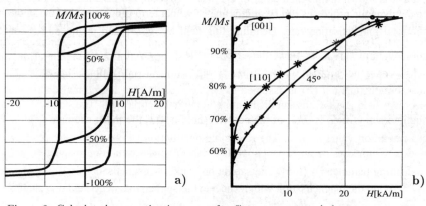

Figure 2: Calculated magnetization curve for Goss–texture steel sheets.
a) Virgin curve, hysteresis curve, symmetric and asymmetric minor loops at weak fields,
b) Calculation and measured points at strong fields.

Furthermore, the model can be extended to describe the magnetic behaviour of isotropic materials by considering an arbitrary angle of the applied field with the easy directions $\vec{n}_i$. The results are comparable to established hysteresis models for isotropic magnetization, e.g. [6]. The approach shows a good agreement with the magnetization curves of both soft and hard magnetic materials [3,4,5].

References

[1] H. Hauser, *Energetic Model of Ferromagnetic Hysteresis*, J. Appl. Phys. **75**, (1994) 2584–2597.
[2] H. Hauser, *Energetic Model of Ferromagnetic Hysteresis 2: Magnetization Calculations of (110)[001] FeSi Sheets by Statistic Domain Behavior*, J. Appl. Phys. **77**, (1995) 2625–2633.
[3] P. L. Fulmek and H. Hauser, *Magnetization Reversal in an Energetic Hysteresis Model*, J. Magn. Magn. Mater. **160**, (1996) 35–37.
[4] P. L. Fulmek and H. Hauser, *Magnetization Calculations of (110)[001] FeSi–Sheets in Different Directions by Statistic Domain Behaviour*, J. Magn. Magn. Mater. **157/158**, (1996) 361–362.
[5] H. Hauser and P. L. Fulmek, *Hysteresis Calculations by Statistical Behaviour of Particles of High Density*, J. Magn. Magn. Mater. **155**, (1996) 34–36
[6] F. Ossart and G. Meunier, *Comparison Between Various Hysteresis Models and Experimental Data*, IEEE Trans. Magn. **26**, (1990) 1342.

Non-Linear Electromagnetic Systems
V. Kose and J. Sievert (Eds.)
IOS Press, 1998

Magnetic Properties of Barium Ferrite Thin Films Prepared at High Deposition Rates

R.K. Kotnala, H.M. Bhatnagar and P.C. Kothari

National Physical laboratory
New Delhi-110012, India

Abstract: In this study, the influence of applied RF power on the crystallographic characteristics and magnetic properties of barium ferrite thin films has been described. Deposition rate increased from 1 nm/min to 5 nm/ min with increase of applied power from 100 W to 400 W. Thin films with small $v \ominus_{50}$ of 2 to 3° could be prepared at high deposition rates. Even at high rates of deposition such as 5 nm/min deterioration in magnetic properties and crystallographic characteristics of the barium ferrite thin films was not apparent. The Faraday rotation measurement for the films prepared at higher power density showed greater improvement of magneto-optical properties at shorter wavelengths.

1. Introduction

To achieve high recording density of the order of 10^8 bits/cm^2, the recording media should have excellent surface smoothness, chemical stability and mechanical strength as well as moderate hard magnetism. Magneto-plumbite type of hexagonal barium ferrite films, whose c-axis of crystallite is well oriented perpendicularly to the film plane, are suitable for magneto-optical recording media [1],[2]. Recording density depends on c-axis dispersion angle $v \ominus_{50}$ for ultra high recording media [3]. In this paper, the influences of applied rf power on the crystallographic characteristics and magnetic properties have been described. For the purpose of magneto-optical applications, the Faraday rotation measurements were made on the films.

2. Experiment

Very simple rf diode sputtering equipment was used to prepare barium ferrite thin films in pure argon gas atmosphere. The substrates used for the film growth were (111) oriented $Gd_3Ga_5O_{12}$ (GGG) single crystal of 30 mm x 20 mm in size which were heated at 580°C during sputtering, ceramic target was used with stoichiometric barium ferrite compositions of $Ba_{1.14}Fe_{12}O_{19}$. Initially, vacuum chamber was evacuated down to 0.133 mPa pressure, and then sputtering was performed at 6.67 Pa pressure. Magnetic properties such as coercive force (H_c) and saturation magnetisation (M_s) of the films in perpendicular direction were measured by a Vibrating Sample Magnetometer.

The crystal structure and composition of deposited films were determined by X-ray diffractometry and electron probe micro-analysis respectively. Faraday rotation of the films was measured at room temperature by the polarization modulation method using a monochromator.

3. Results and Discussion

In a conventional rf diode sputtering, the applied power (P_{in}) varied from 100 W to 400 W for different samples maintaining the substrate temperature at 550 °C and 570 °C. Table 1 shows saturation magnetisation and coercivity for a few samples prepared at different rf power.

Table 1 : Characteristics of barium ferrite thin films at different rf power.

Sample No.	RF Power input,P_{in} (Watts)	Substrate temp. (°C)	Saturation Magnetisation (kA/m)		Coercivity (kA/m)	
			$M_s(\perp)$	$M_s(\parallel)$	$H_c(\perp)$	$H_c(\parallel)$
BaF-22	200	550	156.25	174.0	110	160
BaF-27	250	550	280.4	196.3	110	110
BaF-46	300	550	413	200	80	110
BaF-48	350	550	424.24	217.2	56	56
BaF-53	390	570	393.9	205	56	56

The perpendicular anisotropy of the barium ferrite thin films can be estimated by measuring the dependence of c-axis dispersion angle $\nabla\Theta_{50}$ and c-axis orientation factor f_c of the films on incident power. The value of f_c near 1.0 shows that the c-axis is totally oriented towards the perpendicular direction to the plane of the film which has been the case of our many samples. Interestingly, the orientation of our samples towards the perpendicular direction is greater with an increase in incident rf power and $\nabla\Theta_{50}$ ranges between 2 to 3° even at higher P_{in} values as shown in Fig.1. The preferred orientation along the c-axis with the presence of (006), (008), (200) and (220) lines have been confirmed by X-ray diffraction.

Fig.2 shows the variation of M_s (perpendicular) and H_c (perpendicular) of the deposited films with the increase in rf power. The significant dependence of M_s on P_{in} has been observed. As incident rf power increases, the value of M_s increases and ultimately maximum value of M_s for barium ferrite obtained from P_{in} equals 300 W and onwards. However, H_c of the films decreases gradually from 120 kA to 48 kA/m with increase of incident rf power. It has been also observed that increase in P_{in} increases the squareness ratio of the B-H loop plotted for the films.

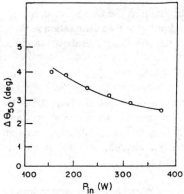

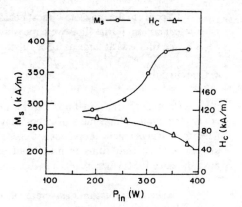

Fig. 1. Variation in $\nabla\theta_{50}$ with increase in rf input power.

Fig. 2. Variation in saturation magnetisation and coercivity with input rf power.

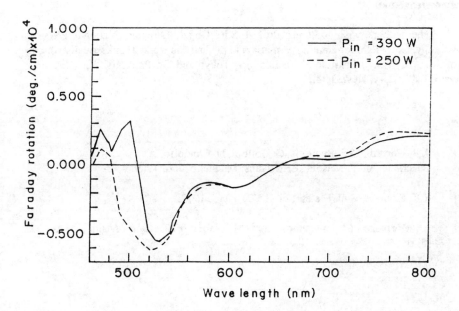

Fig. 3. Faraday rotation versus incident light wavelength.

To confirm the magneto-optical behaviour of the thin film, Faraday rotation spectra of different barium ferrite films were measured. The peak value of Faraday rotation in $Ba_{0.96}$ Fe_{12} O_{19} thin film was as large as 0.3×10^4 deg/cm at 500 nm wavelength as shown in Fig.3.

4. Conclusions

Barium ferrite thin films can be prepared for magneto-optical applications by sputtering even at high rates of deposition such as 5 nm/minute not showing apparent deterioration in magnetic properties and crystallographic characteristics. The Faraday rotation angle for such thin films is maximum at a shorter wavelength, 500 nm, which is advantageous for further increase in recording bit densities.

The Faraday rotation measurement for the films prepared at higher power density showed better improvement of magneto-optical properties at shorter wavelengths.

Acknowledgement

The authors wish to thank Prof. E.S.R. Gopal, Director, National Physical Laboratory, New Delhi for granting permission to publish this work. Thanks are also due to Prof. M. Abe, Tokyo Institute of Technology, Tokyo and Dr. Pran Kishan, Solid State Physics Laboratory, New Delhi.

References

[1].   A.R. Corradi, D.E. Speliotis, G. Bottoni, D. Candolfo, A.
       Ceechetto and F. Masoli, IEEE Trans. Magnet,p4066, 1989.

[2].   R.K. Kotnala, Bull. Mater. Sci., 15(2), p149, 1992.

[3].   A. Morisako, M. matsumoto and M. Naoe; IEEE trans. Magn., **MAG-22**,
       (5),p1146, 1986.

Non-Linear Electromagnetic Systems
V. Kose and J. Sievert (Eds.)
IOS Press, 1998

Magnetic Anisotropy of Amorphous and Polycrystalline Tb-Dy-Fe Films

Jens Gleitzmann

Institut für Elektrische Messtechnik, TU Braunschweig

Hans-Sommer-Str. 66, 38106 Braunschweig, Germany

Abstract. Amorphous and polycrystalline $(Tb_{0.27}Dy_{0.73})_x Fe_{1-x}$ films have been prepared by rf magnetron sputtering. Their composition, structure and magnetic properties were investigated in relation to their preparation conditions. The necessity of a grain orientation is obvious. The influence of film stress on the magnetization orientation in those textured films has been studied.

1. Introduction

Rare-earth-transition metal $(RE - TM)$ compounds, especially the $RE - Fe_2$ cubic Laves phase can exhibit large magnetostriction values at room temperature. This is particulary the case for single crystalline $TbFe_2$ and $DyFe_2$. The Tb containing compound achieves a magnetostriction of $\lambda = 2.5 \cdot 10^{-3}$. Large magnetostriction is very useful for actuators and sensors applications. The applied magnetic field changes the magnetization direction and thus changes the dimensions of the material. On the other hand, one can use the inverse magnetostrictive effect to control the magnetization orientation by applying stress on the material. However, these high magnetostriction values are associated with very large magnetocrystalline anisotropies and large fields are required to realize the full strains. This is a significant disadvantage for their application in magnetoelastic devices. Fortunately, the first anisotropy constants of $TbFe_2$ $(K_1 = -7.6 \cdot 10^6 \, J/m^3)$ and $DyFe_2$ $(K_1 = +2.1 \cdot 10^6 \, J/m^3)$ are opposite in sign and therefore the use of an alloy consisting of Tb, Dy and Fe will reduce the magnetocrystalline energy. The anisotropy constant K_1 of the pseudo-binary alloy $(Tb_{0.27}Dy_{0.73})Fe_2$, called TERFENOL-D is only $-0.06 \cdot 10^6 \, J/m^3$, which leads to higher permeability and lower coercivity.

Compared to the saturation of magnetostriction of single crystalline $Tb_{0.27}Dy_{0.73}Fe_2$ bulk material (about $1.64 \cdot 10^{-3}$) [1] both amorphous and polycrystalline films exhibit only low values. Amorphous films show a magnetostriction up to $0.4 \cdot 10^{-3}$ at an applied field of $1\,T$ and also polycrystalline films reach only magnetostriction values up to $0.75 \cdot 10^{-3}$ at the same field [1].

2. Experimental

R.f. sputtering experiments were performed with a 7.5 cm magnetron cathode mounted in a deposition chamber (Leybold UNIVEX 300) with a target-substrate distance of 5 cm. The films were deposited on silicon and CuBe substrates using a polycrystalline $(Tb_{0.27}Dy_{0.73})Fe_2$ target (FEREDYN EUROPE). Before each deposition, the target was sputtered for 30 min. This presputter treatment is necessary to clean the target and to guarantee reproducible results. The Ar-sputtering pressure was varied between 1 Pa and 10 Pa and the deposition rate between 7 nm/min and 50 nm/min. The substrate could be heated up to 450 °C during the sputter deposition. The thickness of the films varied in the range from 0.5 μm to 8 μm. The chemical composition was determined by wavelength-dispersive x-ray microanalysis (WDX) and depth profiling was obtained using secondary ion mass spectroscopy (SIMS). The crystallographic structure was analysed by x-ray diffraction. The polarization of the films at room temperature as well as the temperature dependence of the polarization in the range from 5 K to 320 K was measured by a SQUID magnetometer with a maximum external field of 5.5 T.

3. Results and Discussion

3.1 Structure and composition

The variation of the sputtering parameters (Ar-pressure and rf-power) has an influence on the chemical composition. The iron content in the films decreases with increasing sputtering pressure. In contrast the ratio of the rare earth metals Tb : Dy = 0.27 : 0.73 is almost independent of the pressure. X-ray diffraction spectrums of films which are deposited at different substrate temperature show good stability of the amorphous state up to 300 °C. At higher temperature partially crystalline films are formed and the Laves phase ($REFe_2$) begins to crystallize.

3.2 Magnetic properties

Fig. 1 shows the polarization hysteresis loops of an amorphous and polycrystalline film with identical composition. The measurements were made by applying a field parallel and perpendicular to the film plane at room temperature.

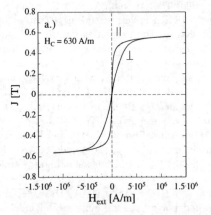 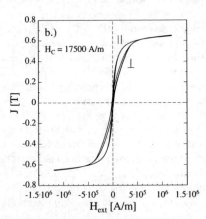

Fig. 1: Polarization vs external field applied parallel ∥ and perpendicular ⊥ to the plane for a) an amorphous film and b) a partially crystalline film

The saturation polarization of the polycrystalline film is higher than the corresponding amorphous composition. This is due to the stronger magnetic coupling in the ordered, crystalline compounds, which is reflected in the higher Curie temperature of about 350 °C.

The amorphous films are characterized by a low Curie temperature ($T_c \approx 135$ °C) and a coercivity of about 630 A/m. In contrast the partially polycrystalline films exhibit distinct higher H_c values up to 18000 A/m, which are in good agreement to data reported by other authors [1]. The large coercivity is likely caused by pinning of the domain walls in the presence of randomly oriented $REFe_2$ grains.

The orientation of the magnetic easy axis is stongly influenced by mechanical film stress.

The difference in thermal expansion coefficients of film ($\alpha_{TbDyFe} = 12\,\mathrm{ppm/K}$) and substrates ($\alpha_{Si} = 4\,\mathrm{ppm/K}$, $\alpha_{CuBe} = 17\,\mathrm{ppm/K}$) leads to planar thermal stresses when the samples are cooled down from sputtering temperature to room temperature. These stresses are tensile on silicon and compressive on CuBe substrates.

3.3 Consequences

The low Curie temperature of amorphous films causes major problems since a strong temperature dependence of magnetostriction is expected for a practical temperature range from $-30\,°C$ to $90\,°C$. The application of polycrystalline films with randomly oriented grains is problematic due to their large coercivity. This disadvantage and the highly anisotropic magnetostriction $\lambda_{111}/\lambda_{100} \gg 1$ at room temperature reveal the immense importance of (111) grain orientation in order to achieve high magnetostriction combined with low H_c values in this material.

It is obvious that the magnetization and magnetostriction is strongly influenced by the orientation of the easy axis. The orientation might be changed by film stress. Therefore, we have studied the effect of in-plane strain in (111) oriented (TbDy)Fe$_2$ films on the magnetization orientation.

3.4 Magnetic anisotropy

Magnetocrystalline anisotropy, shape anisotropy and magnetoelastic energy density were calculated to determine the magnetization in (111) textured (Tb$_{0.27}$Dy$_{0.73}$)Fe$_2$ thin films. For cubic crystals such as the (REFe$_2$) Laves phase the magnetocrystalline anisotropy can be expressed in terms of the direction cosines ($\alpha_1, \alpha_2, \alpha_3$) of the internal magnetization with respect to the three cube edges:

$$F_{mc} = K_1 \cdot (\alpha_1^2 \alpha_2^2 + \alpha_2^2 \alpha_3^2 + \alpha_3^2 \alpha_1^2), \tag{1}$$

where K_1 is the first magnetic anisotropy constant of $-0.06 \cdot 10^6\,\mathrm{J/m^3}$.
Using the coordinate system in Fig. 2, the magnetocrystalline anisotropy energy density can be written as:

$$F_{mc} = K_1 \cdot (\frac{1}{3}\cos^4(\Theta) + \frac{1}{4}\sin^4(\Theta) - \frac{\sqrt{2}}{3}\cos(3\Phi)\cos(\Theta)\sin^3(\Theta)). \tag{2}$$

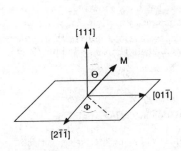

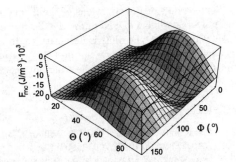

Fig. 2: The coordinate system

Fig. 3: Magnetocrystalline anisotropy energy as a function of Φ and Θ

For a thin plate with a demagnetizing factor $N = 1$ normal to the plane, the shape anisotropy is expressed by

$$F_s = \frac{1}{2}\,\mu_0\,M_s^2\,\cos^2(\Theta),\tag{3}$$

where M_s is the saturation magnetization of $0.8 * 10^6$ A/m. For $\lambda_{111} \gg \lambda_{100}$ and equi-biaxial stress, the magnetoelastic energy density F_{me} can be written as [2]:

$$F_{me} = \lambda_{111}\,\epsilon_{11}\left(\frac{c_{44}(c_{44} + 2c_{12})}{c_{11} + 2c_{12} + 4c_{44}}\right)\cos^2(\Theta),\tag{4}$$

where λ_{111} is the magnetostrictive constant of $1.64 \cdot 10^{-3}$ and ϵ_{11} is the in-plane strain. The c_{ij} are the elements of the stiffness matrix and were measured from single crystalline $Tb_{0.3}Dy_{0.7}Fe_2$ by Clark [3]. The Φ and Θ dependence of the magnetocrystalline anisotropy is plotted in Fig. 3. The energy surface was calculated from $\Theta = 0°$ to $90°$ and $\Phi = 0°$ to $160°$. The surface has two energy minima (two easy axis) in the selected plot range, one along the out-of-plane [111] direction and the other one along the nearly in-plane [11$\bar{1}$] direction at $\Theta = 70.5°$ and $\Phi = 60°$. If the shape anisotropy is taken into account, the magnetization will rotate to the direction at $\Theta = 88°$ and $\Phi = 60°$. The total of F_{mc}, F_s and F_{me} as a function of Θ at $\Phi = 60°$ for various strains is shown in Fig. 4.

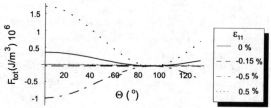

Fig. 4 The total anisotropy energy density for various in-plane strains ϵ_{11}

A tensile strain of $0.5\,\%$ makes the magnetization lie nearly in the film plane. A compressive strain of $-0.15\,\%$ and greater leads to a rotation of magnetization into the out-of-plane direction. Therefore a modest compressive strain will induce the desired out-of-plane magnetization in (111) textured $(Tb_{0.27}Dy_{0.73})Fe_2$ films.

4. Conclusions

Some results of investigations of the magnetic properties of sputter deposited amorphous and polycrystalline TbDyFe films have been presented. It has been shown that a (111) grain orientation in TbDyFe films is necessary to obtain high magnetostriction combined with low coercivity values. In order to obtain (111) textured $(Tb_{0.27}Dy_{0.73})Fe_2$ films with compressive film stress, substrates with higher thermal expansion coefficients than $\alpha = 12\,\text{ppm/K}$ should be used.

References

[1]   P.J. Grundy, P.G. Lord and P.I. Williams, Magnetostriction in *TbDyFe* thin films, *J.Appl.Phys 76*, 10, p. 7003 (1994)

[2]   C.T. Wang, R.M. Osgood et al., Epitaxial growth of (111) *TbFe₂* by sputter deposition, *Mat.Res.Soc.Symp.Proc.Vol.384* p. 79 (1995)

[3]   A.E. Clark, *Magnetostrictive rare earth − Fe₂ compounds* in *Ferromagnetic Materials*, edited by E.P. Wohlfahrt, North Holland, Amsterdam, p. 531 (1980)

Non-Linear Electromagnetic Systems
V. Kose and J. Sievert (Eds.)
IOS Press, 1998

Fundamental Frequency Influence on Harmonic Content of Circular Barkhausen Effect Induced Signal

Ovidiu CALTUN[(1)], Leonard SPINU[(1,2)], Petru ANDREI[(1)], Alexandru STANCU[(1)]

(1) "Alexandru Ioan Cuza" University, Faculty of Physics, 6600, IASI, Romania
(2) Laboratoire du Magnetisme et d'Optique, CNRS-Universite de Versailles-S[t] Quentin, 45
av. des Etats Unis, 78035 Versailles-Cedex, France

Abstract. The circular Barkhausen effect induced signal and its dependencies on
stress and torsion, on DC and AC perpendicular field intensities have been intensely
studied on crystalline and amorphous ferromagnetic wires. The samples are wires
coaxial with a pick-up coil. The waveforms of the signal depend on the
magnetostrictive response of the sample, on the AC and DC perpendicular magnetic
fields, on the frequency, and on the stress and torsion. The induced signal in the
pick-up coil is taken over by an amplifier and analysed by a spectrum analyser. The
DC output signal on a digital voltmeter block is proportional to frequency harmonic
coefficients. The signal is stored and processed by the computer. One observes a
different signal behaviour depending on nickel atoms contents and different
behaviours depending on the fundamental frequency.

1. Introduction

A new approach of Procopiu's effect [1], suggested by experimental studies of
Mohri [2] and Rothenstein [3] have proved the complexity of the induced signal in the
magnetoinductive effect. In this paper is detailed the influence of the fundamental
frequency on the harmonic content.

A spectrum analysis of the signal induced in a coaxial pick-up coil by a NiFe wire
carried by alternating current is made up by a spectrum analyser. The harmonic content
depends on Ni atoms concentration (magnetostrictive response of the sample), on the AC
magnetic field (amplitude of the current), and on longitudinal stress and torsion. For the
sake of simplicity the coherent rotation model [4] can be used to simulate the signal. The
influence of the fundamental frequency on the harmonic content of the induced signal
observed in our experiment cannot be described in this model. A dynamic more complex
model and the influences of torsion, dc parallel magnetic field, thermal and mechanical
history of the samples will be the subject of a future paper.

2. Experimental Set-up

The samples, coaxial with the pick-up coil, are Ni or NiFe alloy wires and are
labeled PN (100% Ni), PP (75% Ni 25% Fe) and P (25% Ni, 75% Fe) [4].

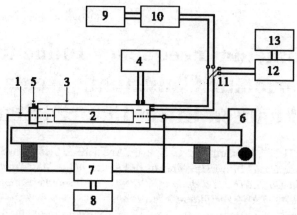

Fig. 1. Experimental set-up. (1) Sample; (2) Pick-up coil; (3) Solenoid; (4) DC Current supply; (5) Mechanical devices for torsion; (6) Mechanical devices for stress; (7) Power amplifier; (8) AC Current supply with variable frequency; (9) Spectrum analyser; (10) Amplifier; (11) Switch; (12) Digital oscilloscope; (13) Computer.

A solenoid coil, coaxial with the system, serves to apply a DC magnetic field, like in the schematic diagram of the experimental set-up, Fig. 1. Two mechanical devices were designed to apply and to measure the torsion and stress. A voltage supply, with variable frequency and a power amplifier ensures 0...2A alternating current through the sample wires, which correspond to a 0...1kA/m magnetic field. The induced signal in the pick-up coil is taken over by an amplifier and analysed by a spectrum analyser. The output signal on the DC digital voltmeter block is proportional to the amplitude of frequency harmonic coefficients. In the second operating mode the oscilloscope stores the signal. The signal is numerically analysed and the amplitudes of harmonics are calculated by computer.

3. Results

The global signal induced by the alternating current of three frequencies, having an amplitude of 0.80 A, without tension and without DC magnetic field, for P, PP and PN samples are presented in Figs. 2, 3 and 4, respectively. One observes different signal behaviours for different Ni contents.

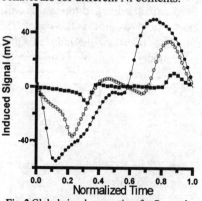

Fig. 2 Global signal versus time for P sample
(■) 100Hz; (o) 500Hz; (•) 1000Hz.

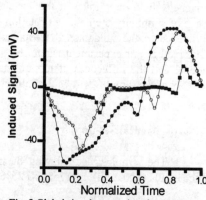

Fig. 3 Global signal versus time for PP sample
(■) 100Hz; (o) 500Hz; (•) 1000Hz.

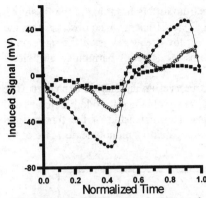

Fig. 4 Global signal versus time for PN sample
(■) 100Hz; (o) 500Hz; (•) 1000Hz.

The amplitudes of the induced signals for all samples normally rise with rising frequency. For the P sample one observes a minor change in the shape of the waveform of the induced signals. In the case of the PP sample the shapes of the induced signal change slowly if the frequency rises. For the PN sample the shapes of induced signals bear important change.

The harmonic contents analyses of the signal proves the presence of odd and even coefficients [4] at low and high frequencies. Both experimental and numerical spectral analysis are consistent. In the Figs. 5, 6 and 7 one presents spectrums for three frequencies, 100, 500 and 1000Hz, for P, PP and PN respectively.

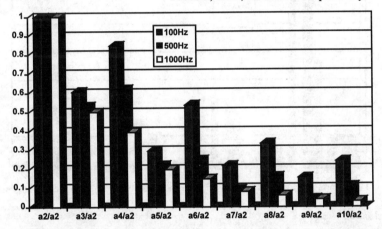

Fig. 5 Normalised spectral coefficients versus normalised frequency for the P sample

- For the P sample the normalised coefficient of the even superior harmonics a_{2n} exceed the odd one a_{2n+1}, (Fig. 5). The normalised coefficients for all superior harmonics are sub-unit and decrease with increasing of fundamental frequency. The relative decrease is more important with rising harmonic order.

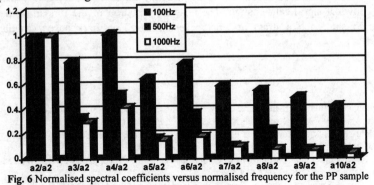

Fig. 6 Normalised spectral coefficients versus normalised frequency for the PP sample

    - For the PP sample, permalloy wire, the spectrum differs from that of the P sample. The values of the a_{2n} coefficients exceed that of the a_{2n-1} coefficients. One observes that the amplitude of even and odd normalised coefficients for low frequency have the same order of magnitude. If the fundamental frequency increases the superior harmonic coefficients drops.

    - For the PN sample, pure nickel wire, one observes an anomalous behaviour. The maximum values of the normalised coefficients (Fig. 7) are obtained for odd harmonics and the normalised amplitude rises for high frequencies. The increasing of the fundamental frequency bears to a monotone decrease of the ratio a_{2n}/a_2 and to a monotone decrease of the ratio a_{2n+1}/a_2.

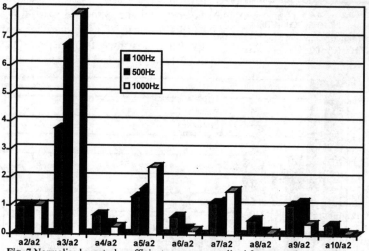

Fig. 7 Normalised spectral coefficients versus normalised frequency for the PN sample

4. Conclusions

The experimental result proves the complex harmonic content of a circular Barkhausen effect induced signal and the influence of the fundamental frequency to the spectrum content. The Fourier coefficients depend on Ni atoms concentration. The behaviour of a_k/a_2 is more meaningful than that of the global signal. Even the waveform of the global signal apparently rest unchanged at a slow variation of the fundamental frequency the harmonic content proves a significant change. The possible variation of the fundamental frequency must be considered if a stress or tension sensor based on Barkhausen circular effect is designed.

References

[1] L. Spinu, O. Caltun, Al. Stancu, Frequency Harmonic Contents Analysis of Induced signal in Procopiu's Effect, An. Univ. "Al. I. Cuza", tom XLI-XLII, fasc.2, Fiz. Sol. (1995) 47-58.
[2] K. Mohri et al., Magnetoinductive effect in amorphous wires, IEEE Trans. on Magn., vol. 28, no. 5 (1992) 3150 3152.
[3] B.F. Rothenstein et al., Contributions to the mechanism of the Procopiu and inverse Wiedemann effects, Phys. Stat. Sol., vol. 19 (1967) 613-622.
[4] O. Caltun, L. Spinu, Al. Stancu, Tension and torsion magnetic sensors based on frequency harmonic content analysis of induced signal in perpendicular field, Sensors and Actuators A, 59 (1997) 142-146.

Non-Linear Electromagnetic Systems
V. Kose and J. Sievert (Eds.)
IOS Press, 1998

Effect of Heat Treatment on the Magnetization Changes of Amorphous Ribbon under the Stress

Y. Iwami*, T. Yamasaki, T. Shimizu, M. Ohkita and A. Saito
*Jomo Kenyatta University, P.O.Box 62000, Nairobi Kenya
Tottori University, Koyama, Tottori 680 Japan

Abstract   The effects of tensile stress and heat treatment on the magnetization properties of amorphous ribbon with negative magnetostriction constant have been investigated. The tension-magnetization curve for annealed samples under constant magnetic field shows peculiar characteristics at small values of magnetic field. To reveal the mechanism of this phenomena, the surface layer of the sample was removed and analyzed.

1. Introduction

Amorphous magnetic materials do not possess crystalline anisotropy and, therefore, induced magnetic anisotropy plays an important part in them. As a consequence, the magnetic characteristic change due to stress manifest themselves more distinctly in these materials.[1] Amorphous materials also have the excellent mechanical properties, which make these materials applicable as a large stress sensor.[2]   In this paper, the effects of stress and the heat treatment on the magnetization of Co base amorphous ribbon are investigated. The tension-magnetization curve under constant magnetic field shows the stress insensitive region in which magnetization does not change due to application of tension, and the magnetization increase for the heat treated sample due to application of tension in spite of a negative magnetostriction constant. The region of stress insensitivity expands with increase of heat treatment temperature.   To reveal the mechanism of the appearance of stress insensitivity, the surface of the ribbon was analyzed by X-ray-Photoelectron-Spectroscope and confirmed as non-crystallized.   The surface layer of the ribbon was removed to compare the characteristics with the sample as heat treated.

2. Sample and Experiment

The ribbon samples used in this investigation are METGLAS 2714A with negative magnetostriction constant. The saturation magnetization, Curie temperature and crystallization temperature of the samples are 0.55T, 205 C and 550 C, respectively. The width, thickness and length of the ribbons are 25 mm, 20um and 400mm, respectively. The tensile stress is applied to the sample smoothly by using a tension tester. The tensile stress was detected by a load cell and fed into the X terminal of XY recorder. Surface Co element was analyzed by the XPS method and crystallization was checked by X-ray analysis. Samples were heat treated in a magnetic field to clarify the mechanism of sensitivity of magnetization.

3. Results and discussion

Figure 1 shows the loci of magnetization for a non-annealed sample under a dc magnetic field of 2.0 A/m due to application of tensile stress up to 100 MPa.   Point s indicate the magnetization at the magnetic field of 2.0 A/m. Point t indicate the magnetization at the tensile stress of 100 MPa and the point e indicate the magnetization under the condition of stress removed.   The magnetization change value s-e is observed only in the initial stress cycle and denoted as an irreversible magnetization change[3], and the magnetization change value e-t measured under the repeated application of stress is denoted as reversible magnetization change. The reason for the appearance of irreversible magnetization change and the reversible magnetization change is directly related to the domain wall structure and domain wall motion.[4]   The magnetoelastic energy induced by applied stress is given as;

$$E = -(3/2) \lambda_s \sigma \cos^2 \theta$$

where λ_s is the saturation magnetostriction constant, σ is the applied tensile stress with positive sign or compressive stress with negative sign and θ is the angle between stress and magnetization direction in which the magnetization vector lies. This equation explains that the application of tension makes the stress direction hard direction for negative magnetostrictive materials. If we apply tension to negative magnetostrictive materials under constant magnetic field, the magnetization must decrease. This is shown clearly in figure 1, in which the magnetization decreases due to application of tensile stress, as expected.

Figure 2 shows the magnetization change due to tension for the sample as annealed at 310 C for 1 hour.   The insensitive region with stress appear up to 500 MPa. This effect is increased by the increase of heat treatment temperature.

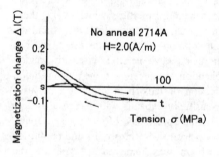

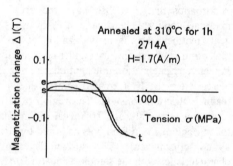

Fig.1.   The loci of magnetization due to tensile stress under magnetic field of 2.0 A/m for non-annealed sample.

Fig. 2.   The loci of magnetization due to tensile stress under magnetic field of 1.7 A/m for annealed sample.

Figure 3 shows the magnetization change due to tension for $2 \mu m$ at surface removed sample and annealing under the same condition as figure 2. Here the insensitive effect disappears and magnetization decreases rapidly by the application of tension. These results shows the appearance of an insensitive region with stress is due to the influence of surface layer of the ribbon.

Figure 4 shows the XPS intensity of Co at the surface with respect to the binding energy for non-annealed and sample annealed at 310°C. Peaks of Co and oxidized Co exist, however there is no meaningful difference between annealed and non-annealed sample.

Figure 5 shows the XPS intensity of Co at removed surface layer with respect to the binding energy for non-annealed and annealed samples. Both results are the same. From these results, it can be stated that decrease of magnetization due to tension in spite of negative magnetostriction constant is actually caused by the surface layer, but it is not due to the oxidization of surface layer by heat treatment.

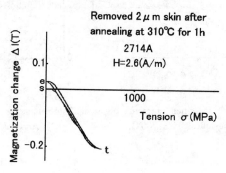

Fig. 3. The loci of magnetization for the sample after 2 μ m surface removal at a field of 2.6 A/m due to tensile stress.

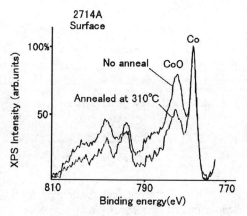

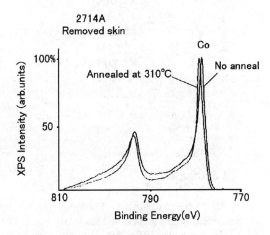

Fig. 4. The XPS intensity of Co at the surface with respect to binding energy for non-annealed and annealed sample at 310 °C.

Fig. 5. The XPS intensity of Co at removed surface with respect to binding energy for non-annealed and annealed sample at 310°C.

Figure 6 and figure 7 shows the X-ray analysis of surface layers for non-annealed and after annealing at 300°C, respectively. It can be judged that no crystallization occurs due to the 300 °C heat treatment, and the effect of surface layer is not caused by the crystallization.

Figure 8 shows the effect of magnetic field anneal in the parallel direction. Here, again an insensitive effect appeared with respect to tensile stress. This may be due to the sticking effect of domain walls due to magnetic field annealing.

Figure 9 shows the appearance of a peculiar stress-magnetization effect due to heat treatment under perpendicular application of magnetic field, for the magnetization increase with respect to application of tensile stress in spite of a negative magnetostriction constant.

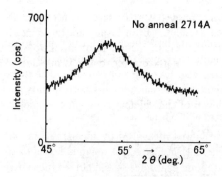

Fig.6. X-ray analysis for non-annealed
sample and one annealed at 300℃.

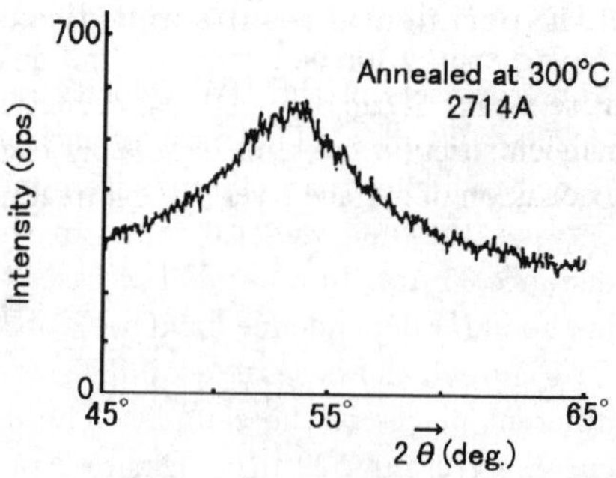

Fig. 7 Example for X-ray analysis of
recrystallized sample.

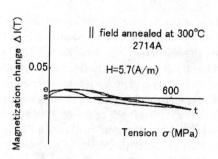

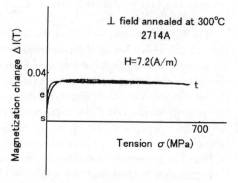

Fig.8 Magnetization change due to tension
with parallel field anneal at 300℃.

Fig. 9 Magnetization change due to tension
with perpendicular field anneal at 300℃.

4. Conclusion

An insensitive region of magnetization change has been observed and this region is extended by application of heat treatment. The phenomena that the magnetization increased or does not changed by the application of tensile stress appeared under constant magnetic field in spite of the negative magnetostriction constant of sample.

It was confirmed by XPS and X ray analysis that the mentioned effects were not due to the surface oxidation or crystallization, and it was also confirmed by surface layer removal that the positive magnetostriction effect lies in the surface layer of sample. The induced anisotropy effects were also investigated by applying magnetic field annealing.

Further investigation is needed, for example anisotropy study or magnetic domain observation, to reveal the mechanism of the stress effect.

References

[1] A. Saito and K. Yamamoto : Journal of Magnetics of Japan, Vol.16, No.2, p.205, 1992

[2] A. Saito et. Al.: Journal of Mag. and Mag. Materials, Vol.113, p.627, 1993

[3] A. Saito et. Al.: Journal of Mag, and Mag. Materials, Vol.112, p.41, 1992

[4] A. Saito and K. Yamamoto: Trans. IEE of Japan, Vol.14-A, No.7/8, p.535. 1995.

Non-Linear Electromagnetic Systems
V. Kose and J. Sievert (Eds.)
IOS Press, 1998

Dynamic Domain Structure and Non-Linear Core Loss Behavior in Grain-Oriented 3% Si-Fe

Yu. N. DRAGOSHANSKII and A. Yu. DRAGOSHANSKII

Inst. of Metal Physics, 620219 Ekaterinburg, Russia

Abstract. The principal mechanism involved in the spontaneous refinement of the domain structure in magnetically soft ferromagnets during dynamic magnetization reversal is defined. The mechanism is traceable to the weakening of hysteresis phenomena and to the inhomogeneity of the effective field in the specimen (notably, from the longitudinal grain-dimension gradient, or structural inhomogeneities). These factors give rise to a continuous translational displacement of domain walls in addition to their vibrational motion under the action of an alternating field. This in turn leads to an increase in the number of domains and to non-linear core loss behavior.

1. Introduction

Grain-oriented Fe-3% Si alloy with rather high magnetic induction at saturation (2,04 T), high maximum permeability (0,1 H/m), low quasi-static losses (not more that 10-20% from total electromagnetic losses) and low magnetostriction is commonly used in rotating electrical generators and power transformers. Its superior magnetic properties in the rolling direction are due to the development of a sharp (110)[001] Goss-texture. The sharpness of (110)[001] crystallographical texture leads to the increase of the portion of 180-degree domain walls displacements and magnetic permeability. However, that leads to increasing of size crystals, domain structure transpolicrystalline. This in turn leads to an increase in the stripe domain width and eddy-current losses.

This state of affairs justifies the search for new ways and means of material control that would reduce domain size and magnetic loss.

The necessary initial step in the investigations, that constitute the objective of this work, was to elucidate the mechanism responsible for the dynamic decrease in domain spacing, non-linear core loss dependence on the magnetic field frequency (from an analysis of domain behavior during dynamic magnetization reversal) and the formation of a method for narrow-domain state stabilisation.

2. Experimental

The behavior of the domain structure in various regimes of magnetization reversal was investigated on single-crystal and polycrystalline specimens of Fe-3% Si electrical-sheet steel in the form of a disk, a triangle, and a rectangle. The surface orientation was (110)[001] or nearly so; the linear dimensions of the specimens were about 100 mm, and their thickness, about 0.2 mm. For domain observation, the specimens were first mechanically polished and annealed in a vacuum at a temperature of 1100° C for 2 h. The 180° stripelike principal domains and the edge 90° flux-closure domains were revealed by using the magnetooptic Kerr effect and the stroboscopic method of observation. The alternating magnetic field applied along the texture axis of the specimens was varied from 0.05 to 10 kHz in frequency so as to achieve a maximum amplitude of 1.0-1.7 T for the sinusoidal magnetic induction in the specimens. The magnetic loss along the texture axis of the stripe specimens was measured in a closed magnetic circuit.

3. Results and Discussion

In the case of sheet specimens Fe-3% Si alloy with the (110)[001] orientation of the surface, the demagnetized state of the grains corresponds to the main domain structure consisting of 180° stripe domains of width D, the magnetic flux of which closes at the edges of the specimen via a system of flux-closure C-domains.

For this kind of domain structure, it has been experimentally established that its spacing decreases with an increase in the frequency of the applied alternating magnetic field [1]. Theoretically, if one follows the assumption that magnetization reversal in a ferromagnetic sheet of length L and thickness h minimized entropy production, and if one takes into account the energy dissipated by the disappearance and appearance of domain walls ($E_\gamma \sim \gamma/D$ erg/cm³), the edge flux-closure 90° domain structure [$E_\sigma \sim (E_{100}\lambda^2_{100}/2)(D/2L)$ erg/cm³], and the eddy currents within the regions of domain wall displacement ($E_{eddy} \sim B^2_m fhD/\rho$ erg/cm³), then for the domain width averaged over a cycle, one obtains

$$D_{dyn} \sim [L\gamma\rho/(E_{100}\lambda^2_{100}\rho + B^2_m Lfh)]^{1/2},$$

where γ is the density of the domain wall energy, E_{100} is Young's moduls, λ_{100} is the magnetostriction constant, B_m is the amplitude of magnetic induction, and ρ is the resistivity of the material. This expression shows that, for a given amplitude B_m of alternating induction, an increase in the alternating field frequency has a significant effect on domain width, beginning from some threshold frequency f, at which the second term in the denominator becomes comparable with or greater than the first term. An increase in the magnetization reversal frequency must cause a reduction in domain width.

One of the sources that serve to make the hypothesized and experimentally observed relationship $D \sim 1/f^{1/2}$ a reality was first observed (using the magnetooptic Kerr effect and an LV-04 lapsed-time photographic unit) on silicon iron crystals with the [001] easy axis making an angle $\beta \sim 2^\circ$ with the surface of the specimen. In such a case, the spacing of the dynamic domain structure decreases owing to the growth, coalescence and transformation of individual flux-closure droplike domains into principal domains [2].

In the range of relatively low frequency remagnetization, the frequency dependence of the domain spacing can be explained qualitatively by including the increase in domain-wall area due to the domain-wall bowing with phase shift and different amplitudes of oscillation of individual points of the domain wall [3].

The continuous translational motion of domain walls [4], which is another dynamic feature in the behavior of the domain structure, plays a more significant role in domain refinement. This form of wall domain displacement (as in a disk, a triangle and a rectangular stripe), observed along with their vibrational motion in step with alternations of the field, is characteristic of only dynamic magnetization reversal. At a certain level of "vibration", it enables domain walls to overcome local potential barriers in the crystal. Under the circumstances, with hysteresis phenomena weakened and with the mobility of domain walls increased, the nonhomogeneity of the effective magnetic field in the specimen shows up quite noticeably. It can arise from structural inhomogeneities in the ferromagnet and from the longitudinal dimension gradient of the crystal, leading to the demagnetizing field gradient.

Because of the effective field gradient, every domain wall finds itself in stronger fields when it moves from the position of equilibrium to one side and in weaker fields when it moves in the opposite direction. Tllis leads to a continuous translational motion of all domain walls.

The continuous translational motion of domain walls first results in the densification of the original domains and then it gives rise to new principal domains and to a core loss decrease (Table 1; P_{calc} by [5]).

Table 1. The electromagnetic loss, number and width domains of Fe-3% Si alloy under influence of dynamic domain movement

Dynamic domain movement	N domain	D mm	$P_{1/50}$ W/kg	$P^{calc}_{1/50}$ W/kg
no	5	1,00	0,57	0,64
yes	6	0,83	0,51	0,56
yes	7	0,71	0,46	0,48

It shows a steady stabilization of the narrow-domain state in Fe-Si alloys after a thermomagnetic treatment (TMT) in elevated-frequency fields [4].

The core loss reduction can be increased by the combination of the local laser radiation, tensile stress coating and the TMT. Recent measurements [6] have shown for optimal Goss-oriented Fe 3 % Si steel ribbons that the complex treatments described above allow reducing the electromagnetic losses up to 30-40 % under various conditions of remagnetization.

4. Conclusions

It is established that reduction of domain space, eddy-current and total electromagnetic losses of Fe-Si alloy in elevated-frequency magnetic fields results from continuous translational displacement of domain walls. The motion of domain walls leads to an increase in the number of domains.

The refinement of domains increases with magnetization reversal frequency, which leads to non-linear electromagnetic loss behavior.

A method is proposed for thermomagnetic treatment in elevated-frequency (50Hz-10kHz) alternating magnetic fields, ensuring the stabilisation of the narrow-domain state and a steady reduction in electromagnetic loss in materials for electrical engineering applications.

The study was supported by the Russian Fand for Fundamental Research (Project 96-02-16000).

References

[1] G. Krose and B. Passon, Die Abhangigkeit der Anfangspermeabilitat Kornorientier Eisen-Silizium-Bleche von der Entmagnetisier Ungsfrequenz, *Z. Angew. Phys.* **23** (1967) 157-160.

[2] Yu.N. Dragoshanskii et al., Dependence of Electromagnetic Loss in Silicon Iron Single Crystals on the Crystallographic Orientation of Their Surface, *Fiz. Met. Melalloved.* **34** (1972) 987-994.

[3] Yu.N. Dragoshanskii et al., Influence of Domain Walls Bowing on the Electromagnetic Losses in Single Fe-Si Crystals, *Fiz. Met. Metalloved.* **39** (1975) 519-523.

[4] Yu. N. Dragoshanskii, Physical Mechanisms Involved in the Dynamic Refinement of the Domain Structure in Electrical-Engineering Materials, *Fiz. Met. Metalloved.* **77** (1994) 53-65.

[5] R.H. Pry and C.P. Bean, Calculating of the Energy Loss in Magnetic Sheet Materials Using a Domain Model, *J. Appl. Phys.* **29** (1958) 532-534.

[6] Yu. N. Dragoshanskii et al., New Methods of Magnetical Anisotropy Optimization in Ribbons for Electrical-Engineering, *Steel* **3** (1996) 58-61.

Non-Linear Electromagnetic Systems
V. Kose and J. Sievert (Eds.)
IOS Press, 1998

Induction Signals of Surface Treated
Electrical Steels

M. Kügler and W. Riehemann

IWW TU Clausthal, Agricolastr. 6, D-38678 Clausthal-Zellerfeld, Germany

Abstact. The power losses of grain oriented iron silicon steels can be reduced by surface treatments such as scratching, laser scribing or related techniques. Usually this loss reduction is related to a domain refinement. An explanation for this frequency dependent improvement is the increase of the effective number of activated domain walls for different frequencies or excess fields. Various techniques of surface treatment can lead to the same calculated number of effectively moving domain walls at a given excess field but cause different power losses. Therefore investigations in the shape of the induction voltages were made. Induced signals with one and two maxima were found to be dependent on the surface treatment and amplitude of polarisation.

1. Introduction

The loss reduction of grain oriented iron silicon steels by scratching or related techniques is usually explained with a refinement of main domains [1]. The increase in the number of moving domain walls n_{eff} with increasing frequency of magnetisation can be described using a simple magnetization model of Williams et al. [2]. For grain oriented silicon steel this was done by Bertotti et al. [3]. For laser scribed grain oriented silicon steel the change of n_{eff} with frequency or excess field shows interesting details of the magnetizing process [4] but cannot explain improvements for more complex magnetisation curves.

2. Experimental

The investigated material was grain oriented Fe3.2wt.%Si steel sheets coated with fosterite. It was supplied by EBG (Gelsenkirchen, Germany) in the form of Epstein strips (l=305 mm, b=30 mm, d=230 μm). The measurements were performed using an Epstein frame with a digital hysteresis recorder system sampling the voltages of the induction coil and of a shunt switched in series with the field coils. Sinusoidal or triangular magnetic fields at frequencies ranging from 0.05 Hz to 1000 Hz and amplitudes of polarisation ranging from 1.4 T to 1.7 T were used. Coated and uncoated samples in a scratched and unscratched state were investigated. The samples discussed here were prepared by mechanical scratching orthogonal to the easy magnetic axis. The scratches were 70 μm wide, 3 μm deep and had a distance of 5 mm.

3. Results and discussion

Neglecting the interaction and bowing of moving domain walls their number can be calculated by

$$n_{eff} = \frac{H_\omega}{H_{exc}} \tag{1}$$

where H_ω is the field, which is necessary for the magnetizing process by one plain 180° domain wall only and H_{exc} is the measured excess field. Following Williams et al. [2]:

$$H_\omega = \frac{4 D_g A_p \dot{J}}{\pi^3 \rho} \tag{2}$$

and

$$H_{exc}(f) = H_d(f) - H_c \tag{3}$$

In these equations A_p represents the cross section of the sample, ρ the specific resistivity, D_g an geometric factor and $\dot{J}$ is the rate of polarisation. $H_d(f)$ is the measured dynamic coercive field depending on frequency f and H_c is the frequency independent coercive force which was determined by linear extrapolation of the measured dynamic coercivities to zero frequency.

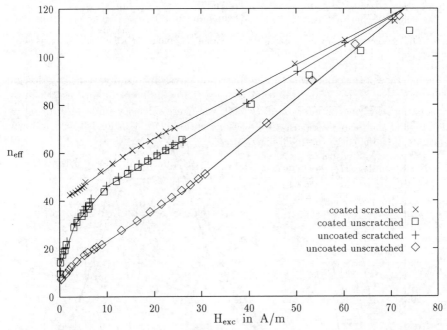

Figure 1: Simultaneously moving number of domain walls of typical samples with different surface treatments. $J_0 = 1.4$ T

Figure 1 shows the dependence of n_{eff} on the excess field which results from the different magnetization frequency of a coated and uncoated sample before and after they were scratched. One of them was coated with fosterite, the other not. The polarization amplitude was 1.4 T. Surface treatment and coating increases n_{eff} as expected. At higher excess fields this effect decreases. In each curve several linear sections are found. These sections indicate the existence of different types of magnetisation carriers which

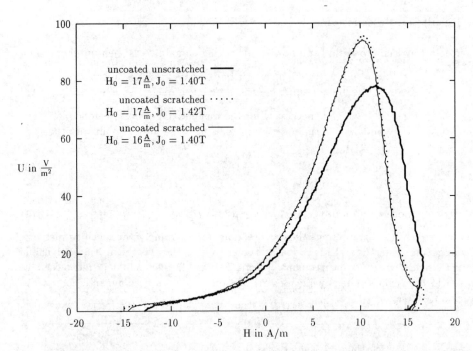

Figure 2: Induction voltages of an uncoated sample with and without surface treatment versus applied field (triangular, f = 2 Hz).

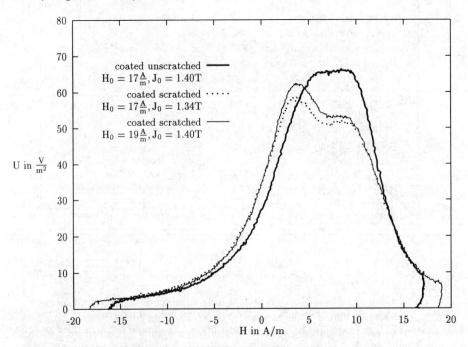

Figure 3: Induction voltages of a coated sample with and without surface treatment versus applied field (triangular, f = 2 Hz).

successively fade out at certain excess fields where the slope of the curve changes as described by Weidenfeller et al. [4]. Though the coated unscratched and the uncoated scatched sample in Fig. 1 show the same behaviour, the uncoated scatched sample produces 4%–9% lower power losses (0 A/m < H_{exc} < 80 A/m) than the coated untreated one.

Figure 2 shows the induction voltage versus applied field during magnetisation of an uncoated unscratched sample at a polarisation amplitude of 1.4 T. The magnetizing field with a frequency of 2 Hz was triangular. After scratching, the sample reaches a somewhat higher polarisation with the same amplitude of field (here: 17 A/m). The third curve shows the induction voltages of the sample in the scratched state at a polarisation of 1.4 T. All curves show just one maximum.

Figure 3 shows the induction voltages of a coated sample measured under the same conditions. In the unscratched case also only one maximum is found although it is broader compared to uncoated specimens. After scratching, *two* maxima occur. The first one is higher than the second and increases more than the second with increasing polarisations. Other measurements have shown that these maxima are more pronounced when the scratches are deeper or have less distances from each other. The effect is stronger at lower magnetizing frequencies and vanishes at magnetizing frequencies higher than 10 Hz ($H_{exc} \simeq 7\ A/m$).

The occurence of the second maximum supports the results of Weidenfeller et al. [4] that different magnetizing carriers are available at different excess fields. Nevertheless the model which leads to Fig. 1 can no longer explain certain power loss improvements by commonly applied techniques of domain refinement. The simple magnetisation model of Williams et al. [2] implies that the whole magnetisation takes place at a particular field.

4. Conclusion

The determination of the effective number of moving domain walls using the simple model of William et al. [2] fails for the description of power loss improvement in case of complex induction signals with two maxima.

References

[1] T. Iuchi, S. Yamaguchi, T. Ichiyama, M. Nakamura, T. Ishimoto and K. Kuroki J., Laser processing for reducing core loss of grain oriented silicon steel, Appl. Phys. **53** (1982) 2410-2412.
[2] H.J. Williams, W. Shockley, and C. Kittel, Studies of the propagation velocity of a ferromagnetic domain boundary, Phys. Rev. **80** (1950) 1090-1094.
[3] G. Bertotti, Physical interpretation of eddy current losses in ferromagnetic materials, J. Appl. Phys. **57** (1985) 2110-2117.
[4] B. Weidenfeller, W. Riehemann, Domain refinement and domain wall activation of surface treated Fe–Si sheets, J. Magn. Magn. Mater. **160** (1996) 136-138.

Non-Linear Electromagnetic Systems
V. Kose and J. Sievert (Eds.)
IOS Press, 1998

Power Losses in Electrical Steel Sheets at Power Frequency with Odd Higher Harmonics of Low Order

Shunji TAKADA, Osamu ARAI, and Tadashi SASAKI
Faculty of Engineering, Gifu University, Yanagido 1-1, Gifu, 501-11 Japan

Abstract. Core losses in electrical steel sheets have been measured with magnetizing wave forms composed of the superposition of 50 Hz and one of its odd higher harmonics of low oder: the 3rd, 5, 7, and 9th harmonics. All the results have been presented, but discussion has been limited to the case without minor loops. The superposition of the 9th harmonic yields a similar effect to that of higher harmonics of high order, whose phase does not change the proportion of losses between the fundamental and the harmonic component.

1. Introduction

The induced voltage of a magnetic device is liable to contain odd higher harmonics of low order because of the non-linear relation between the magnetic flux-density (B) and the magnetic field (H). Therefore, some attempts have been made to understand the influence of the superposition of odd higher harmonics of low order on magnetic properties of electrical steel sheets [1][2].

The problem has increased as the exciting voltage wave form changes from sinusoidal to artificial such as PWM. We previously tested the influence of the superposition of higher harmonics of high order on magnetic losses in electrical steel sheets, and reported that the proportion of losses between the fundamental and the harmonic component changes not with the phase but with the amplitude of the harmonic, and that the higher harmonic magnetization is mainly assoociated with motion of supplementary domains [3].

This report presents the effect of the superposition of odd higher harmonics of low order on magnetic losses in electrical steel sheets. In this case, the maximum flux density and the shape of ac hysteresis loops significantly change not only with amplitude but also with phase of the superposed harmonic.

2. Experimental

The specimens are a toroidal core of conventional oriented Si-Fe sheet (Z) with 110 mm outer diameter and 90 mm inner diameter and a stacked ring core of non-oriented electrical steel sheet (S) with 200 mm outer diameter and 100 mm inner diameter.

Magnetization has been strictly controlled using a digital measuring system so as to be able to express the magnetic flux density $B(t)$ as follows;

$$B(t) = B_1 \sin \omega t + B_n \sin(n\omega t + \theta_n),$$

where, $\omega = 2\pi f$, $f = 50$ Hz, $n = 3, 5, 7$, or 9, θ_n = every 0.1π between 0 and π, B_1 is the amplitude of the fundamental flux density and takes the value of 0.5, 1.0, and 1.5 T for the Z-specimen and 1.4 T instead of 1.5 T for the S-specimen, B_n is the amplitude of the nth

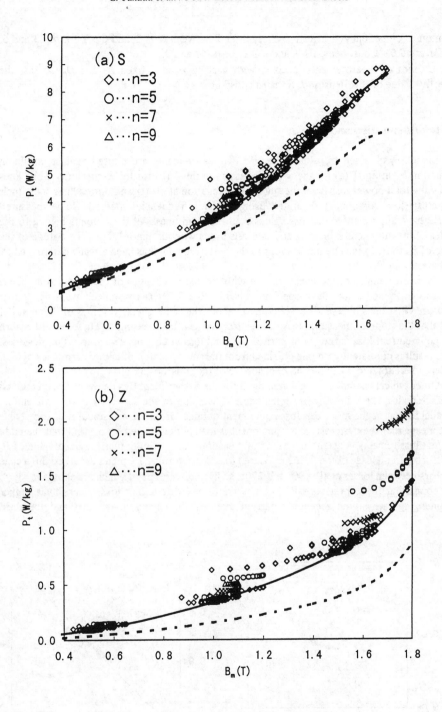

Fig. 1   Characteristics of P_t vs. B_m

harmonic component and has steps every 0.05 T between 0.05 T and 0.4 T for $n = 3$ and 5, and every 0.01 T between 0.01 T and 0.1 T for $n = 7$ and 9.

A series of measurements is taken with varying θ_n at a fixed B_1 and B_n, so that the effective value of flux density B_e remains constant.

3. Results and Discussion

Losses P_t obtained from all the series of measurements are plotted against maximum induction B_m in fig. 1 (a) for the S-specimen and in fig. 1 (b) for the Z-specimen. Solid lines show the total power loss from the sine wave excitation at 50 Hz and dotted lines show their hysteresis loss component separated using the two frequencies method. Some measured points have apparently lower loss values than the solid lines. All these points belong to the series of measurement exhibiting ac hysteresis loops without minor loops. The criteria of the ratio of B_n to B_1 to include minor loops in the hysteresis loop has been already determined by Nakata et al. [1].

Typical examples of the loss variation with B_m due to changes of the superposing phase are shown in fig. 2 for the Z-specimen at $B_1 = 0.5$ T. It is recognized from fig. 2 that measured loss lines intersect the solid line at about the same effective values of flux density, and that the slope of measured loss curves decreases with increasing n. These result because the eddy current loss is the major part of the total loss in this material at low flux densities. This result is consistent with one obtained from the superposition of higher harmonics of high order, where total loss exhibits no dependence of the superposing phase [3].

Other typical examples are given in fig. 3 for the S-specimen. Beside the above mentioned characteristics, the following are remarkable. The slope of the superposed losses is almost parallel to that of the hysteresis loss component obtained from the sinusoidal excitation. This fact means that the hysteresis loss component is mainly determined by B_m, and that the eddy current loss remains constant in a series of measurements, through which a fixed value of B_e is assured. In the case of $n = 3$, the measured loss curve forms a loop and a marked difference appears between losses at $\theta_3 = 90°$ and $270°$, ac hysteresis loops of which are shown in fig. 4. By comparing these hysteresis loops, it is confirmed that higher loss occurs by a higher magnetizing rate in the ascending part of magnetization lying in the first and the third

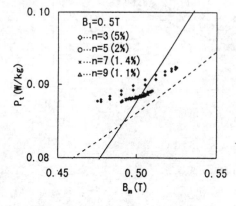

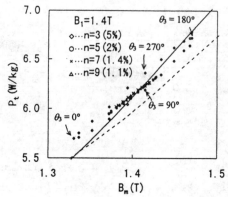

Fig. 2 P_t vs. B_m (Z-specimen)   Fig. 3 P_t vs. B_m (S-specimen)

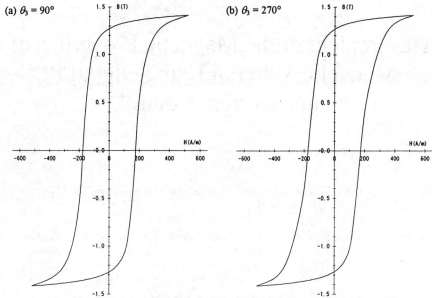

Fig. 4   Examples of dynamic hysteresis loop. (S-specimen, B_1 = 1.4 T, B_3 = 0.07 T)

quadrants. This shows that the hysteresis loss is changed not only with B_m but with the magnetizing rate.

In cases of magnetization accompanying minor loops, magnetic losses always exceed the solid line and the B_m dependence of losses tends to decrease. The detailed results and discussion about these cases as well as the superposing effect, which means the ratio of loss between the fundamental and the harmonic component, will be reported later.

4. Conclusion

Magnetic losses in electrical steel sheets at power frequency with odd higher harmonics of low order have been presented and discussion about the case without minor loop has been made. The results are summarized as follows. (1) The loss ratio of the harmonic component to the fundamental component varies with θ_n. The higher the order of harmonics the less variation of the ratio. (2) When B_n is sufficiently small, the loss vs. B_m curve obtained by the superposed excitation intersects that obtained by the sinusoidal excitation at the same effective value of flux density B_e. (3) Superposing the third harmonic, the total loss vs. B_m curve forms a loop, which is remarkable in non-oriented sheet. This is because the difference in magnetizing rate appears in the ascending part of magnetization. Thus, the hysteresis loss component changes not only with B_m but also with the magnetizing rate.

References

[1]   T. Nakata, Y. Ishihara, and M. Nakano, J. IEEJ, 90 (1970) 115
[2]   A. J. Moses and H. Shirkoohi, Physica Scripta, 39, (1989) 523
[3]   T. Sasaki, S. Saiki, and S. Takada, J. Magn. Soc. Japan, 15, (1991) 271

Non-Linear Electromagnetic Systems
V. Kose and J. Sievert (Eds.)
IOS Press, 1998

Measurement of the Magnetic Properties of Soft Magnetic Material Using Digital Real-Time Current Control

N. Grote*, P. Denke*, M. Rademacher*, R. Kipscholl*, M. Bongards**, S. Siebert*

*DR. BROCKHAUS MESSTECHNIK GmbH & Co. KG, Gustav-Adolf-Str. 4, D-58507 Lüdenscheid
**FACHHOCHSCHULE KÖLN, Abt. Gummersbach, Am Sandberg 1, D-51643 Gummersbach*

Abstract. A device is described which uses a complete digital real-time PI control circuit to control the primary current for reaching a sinusoidal secondary voltage. The nominal value for the secondary voltage and the setpoints for the primary current are produced by software computation and a D/A converter. The complete control circuit has been realized on a personal computer. It is, however, not an iterative control system. Each digitally created setpoint for the actual primary current will be calculated from the digitalized deviation between nominal and actual value of the secondary voltage. The necessary scanning ratio for the measurement, the calculation and the creation of a new actual setpoint is up to 25 kHz. Additionally an online digital offset correction has been implemented. The principle of the digital real-time PI control and its integration into the equipment is shown. Results of measurements on Epstein frames of electrical steel sheets and strips which have been measured and evaluated with this device are presented.

1. Introduction

The measurement of the magnetic properties of soft magnetic materials, especially of electrical steel sheets and strips by means of an Epstein frame are described in the IEC 404-2 [1].

For the measurements of the specific total loss the specific apparent power and the R.M.S. value of the exciting current, the wave form of the secondary voltage should be sinusoidal with a form factor of 1.111 ± 1 % to measure magnetic materials under identical conditions. This ensures the required reproducibility for the comparison of the measuring results which have been received with different measuring devices.

To reach this form factor value most equipment for these measurements uses an analogous based negative feedback to control the primary current. Usually the secondary voltage of the measuring system - which corresponds to dB/dt - is used for the feedback.

A device has been developed which uses a complete digital real-time PI control circuit to control the primary current for reaching a sinusoidal secondary voltage. The nominal setpoint for the secondary voltage and the setpoints for the primary current are produced by software computation and a D/A converter. The complete control circuit has been realized on a personal computer.

2. General Description

The electrical steel sheet is measured in an Epstein frame which - together with the magnetic material - works like a transformer. This transformer will be driven with an alternating current on the primary side. This current is proportional to the magnetizing field. The integral of the secondary voltage represents the polarisation of the material and will be displayed versus the exciting current so that a typical hysteresis loop is created.

If this process is uncontrolled the shape of the secondary voltage is not sinusoidal, as required by the international standard IEC 404-2. This is caused by the non-linear transmission behaviour of the magnetic material used. To reach a sinusoidal shaped voltage it is necessary to control the primary current.

The described digital real-time PI control circuit uses the following algorithm for the calculation of each setpoint:

$$y(k) = y(k-1) - K_p \cdot \left[x_d(k) \cdot \left(1 + \frac{T_0}{2 \cdot T_i} \right) - x_d(k-1) \cdot \left(1 - \frac{T_0}{2 \cdot T_i} \right) \right] \qquad (1)$$

Each setpoint $y(k)$ will be calculated according to the formula (1) from the former setpoint $y(k-1)$ and the control deviations $x_d(k)$ and $x_d(k-1)$ between the nominal and the actual value under consideration of the amplification factor K_p, the scanning time T_0 and the integration time T_i.

In the above described transformer no direct current is transmitted so that an eventually existing offset in the primary signal of the transformer cannot be recognized on the secondary voltage by the controller. In order to take this offset into consideration the offset will be measured on the primary side and used for the calculation of the control deviation $x_d(k)$.

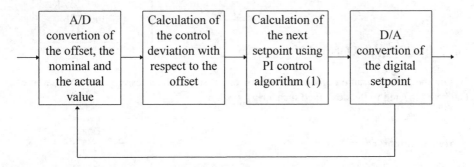

Fig. 1. Principle of the computation of the setpoints

3. Measuring Setup

The digital real-time PI control circuit (see Fig. 2) consists of a personal computer (Pentium 133 Mhz) with an A/D and D/A converter inside. The personal computer calculates each actual setpoint.

With this construction a scanning ratio for the measurement, the calculation and the creation of a new setpoint up to 25 kHz has been realized. The resolution of the A/D- and D/A conversion is 0.05 % in all measuring ranges.

The measurements of the electrical steel sheet take place in an Epstein frame which is built according to the IEC 404-2. The number of the primary and the secondary windings are bozh 700.

Furthermore the construction consists of a power amplifier which supplies a maximum voltage of ± 110 V and ± 40 A and a processor controlled measurement evaluation system with a measuring resolution of 0.05 % in all measuring ranges.

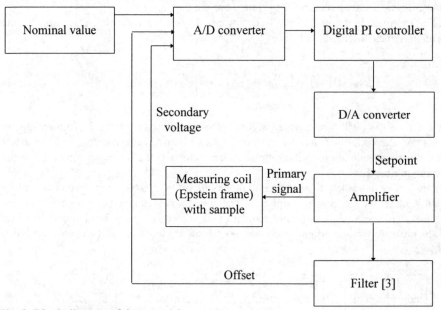

Fig. 2. Block diagram of the measuring set-up

Beside the Epstein frame other coil systems, e.g. a single sheet tester [2] or ring samples, can be connected to the described measuring system.

4. Measurements

In this section results are shown which have been achieved with the above described measuring system. An Epstein sample made from non-oriented material was used. The measurements had been made with a frequency of 50 Hz and a polarisation of 1.5 T. Figure 3 shows the polarisation (integrated secondary voltage) without any control circuit (curve C1) and with digital real-time PI control (curve C2). The same is shown in Figure 4 for the field strength (primary current) of the measurement. In Figures 5 and 6 the hysteresis loops for waveforms with (Fig. 6) and without (Fig. 5) PI control are presented. The form factor ff can be taken as a reference to show how the secondary voltage corresponds to a exact sinusoid. In the IEC 404-2 a form factor ff = 1,111 ± 1% is required. The measurements have shown that this value can be reached with the digital real-time PI control circuit.

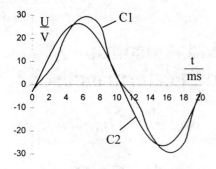

Fig. 3. Secondary voltaged without (C1) and with (C2) primary current control

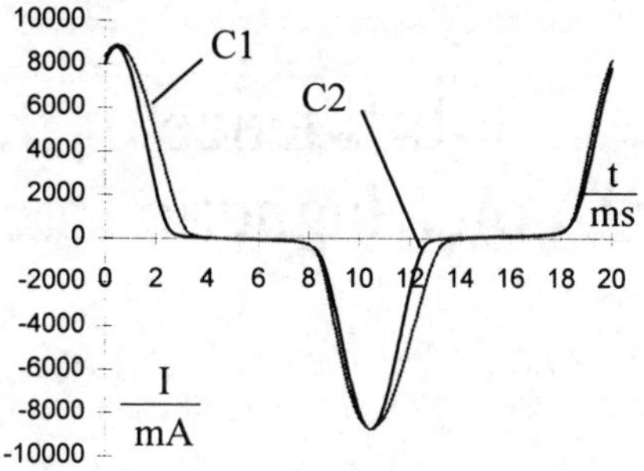

Fig. 4. Primary current without (C1) and with (C2) primary current control

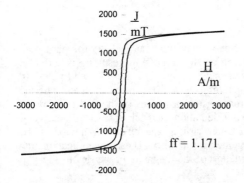

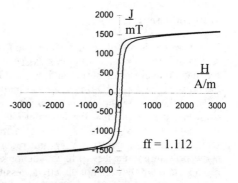

Fig. 5. Hysteresis loop without primary current control

Fig. 6. Hysteresis loop with primary current control

5. Conclusions

The digital real-time PI control circuit is suitable for measurements of electrical steel sheets according to the international standard IEC 404-4. Due to the actual reachable scanning ratio of 25 kHz this control principle can be realized for measuring frequencies up to 100 Hz. An enlargement of this frequency range can be effected by increasing the scanning ratio of the control circuit.

The considered algorithm (1) enables a real-time control even at a much higher scanning ratio and permits maintainig the optimum form factor in the used frequency range $< \pm 1$ %.

References
[1]   IEC 404-2
[2]   IEC 404-3
[3]   B.D.O. Anderson, J.B. Moore: Optimal Filtering, Prentice-Hall, Inc. New Jersey, 1979.
[4]   O. Föllinger: Nichtlineare Regelung, Oldenbourg Verlag München–Wien, 1970.
[5]   H. Ahlers, J. Sievert, PTB Mit. 94, 1984.

Non-Linear Electromagnetic Systems
V. Kose and J. Sievert (Eds.)
IOS Press, 1998

Influence of the Head Anisotropy
on Magnetic Recording Performances

Florence OSSART and Valentin IONITA*

LEG (CNRS UMR 5539 INPG/UJF) - BP 46 - 38402 St Martin d'Hères - France
* Univ."POLITEHNICA" Bucuresti - Spl.Indep. 313 - 77206 Bucuresti - Romania

Abstract. Two structures of thin film heads are modeled and their behaviors are compared. The horizontal head is shown to have much better reading performances than the vertical head, which is due to the different magnetic anisotropy in the poles. This confirms the need to have a better knowledge of the domain distribution and the resulting magnetic properties in the poles.

1. Introduction

Thin film heads are more and more used, for video applications as well as for hard disk recording. Various head structures exist, depending on the processing, resulting in different magnetic domains distribution and hence different magnetic properties in the head poles. In the presented work, a 2D finite element model including hysteresis is used to investigate the effect of the pole anisotropy on the writing and reading performances of the two main types of heads. It is found that the apparent easy axis orientation, vertical in vertical heads and horizontal in horizontal heads, leads to a much better writing and reading behavior of the second head, especially for very thin recording media (50 nm). These theoretical results show how important it is to have a better knowledge of the magnetic properties in the gap region, although they are not directly accessible.

2. The finite element model

To simulate the recording process, one needs to compute the field created by the writing head and the resulting magnetization in the recorded medium as it moves under the head. The more important region of the recorded pattern is the transition between opposite bits, which must be as small as possible to achieve high densities. This requires a strong coercivity of the recording medium, but also a writing field as sharp as possible, with a high field gradient around the coercive field of the medium [1].

The finite element method allows the geometry and magnetic properties of the head to be taken into account, as well as those of the recording media. The global behavior of the head requires a 3D finite element computation, but because the writing process itself is very local, a 2D finite element model can be used to compute the recorded magnetization on a much smaller scale. Eddy currents are neglected and hysteresis in the medium is accounted for by a scalar analytical model [2]. The scalar magnetic potential is used and a modified fixed-point method insures the convergence of this non linear computation [3]. Tests not reported here prove the validity of this local model.

Fig.1 shows the local geometry and the boundary conditions of the two modeled heads. The magnetomotive force is created by a potential difference instead of the writing current. This potential difference is adjusted so as to create the same field as the one computed by the 3D model. As the medium is very thin, its magnetization is assumed to be in the x-direction and to depend only on the Hx component. This assumption may no longer hold at very high densities.

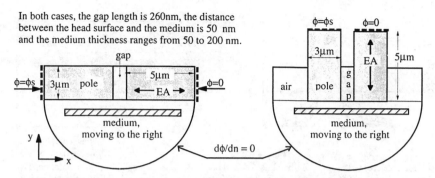

Fig.1.a : Horizontal head   Fig.1.b : Vertical head

The domain structure (resulting from the process and shape effects) leads to an anisotropic magnetic permeability, with an apparent easy axis horizontal for the horizontal head and vertical for the vertical head. The exact value of the permeability in each direction is not well known; we consider here an extreme case and assume the relative permeability to be 500 in the easy axis direction and only 1 along the hard axis.

A very refined mesh (26×25 nm2 second order rectangular elements in the medium and the gap region) insures a good accuracy, especially in the transition region, moving with the medium. The motion of the medium is modeled as a succession of magnetostatic problems, history (local past field extrema) and hysteresis being propagated along the mesh as the medium is moving.

The reading process is modeled using the reciprocity principle [1], assuming a linear behavior of the head and no magnetization change. The reading voltage is given by (1) :

$$V(xo) = \frac{\mu_0\ w\ V}{I_w} \int_{medium} Hx(x,y) . \frac{dMx}{dx} (x-xo)\ dx\ dy \qquad (1)$$

where $M(x,y)$ is the recorded magnetization being read, $H(x,y)$ is the writing field that the reading head would create if fed by the writing current I_w, w is the track width and V is the medium velocity. If a single perfect transition (step function) is assumed, the readback voltage is a pulse, the shape of which is given by the average value of $Hx(x,y)$ in the medium thickness. The narrower this pulse is, the higher density the head can read.

3. Comparison between the vertical and the horizontal head

The first indication of the head performances is given by the writing field. A steep field gradient dHx/dx around the coercive field will write a sharp transition between two opposite bits, provided the recording media has a coercive field strong enough. On the other hand a head with a narrow writing field will read high densities. These goals (sharp field gradient and narrow field) are usually contradictory and a good compromise must be met, depending on the application.

The driving force ϕs was calculated so that both heads produce the same writing field magnitude at $y = -0.1\mu m$ ($\mu_0.Hmax = 0.4$ T, result of a 3D computation with the whole horizontal head and no medium) . Because the permeability of the vertical head is so low in the writing direction, a higher value is required for ϕs (0.26 instead of 0.19 for the horizontal head), meaning that the head has a lower efficiency.

The writing field computed for the two heads (Fig.2) indicates that the horizontal head will both write and read better that the vertical one close to the head surface, but the field vanishes more rapidly and the advantage of the horizontal head may no longer be obvious for thick recording media, and especially for particulate tapes.

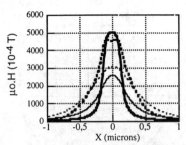

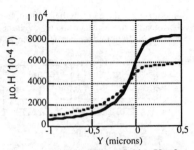

Fig.2.a : Writing field at y = y0
- solid line: horizontal head   • dashed line: vertical head
- thick line : y0 = - 0.05 μm   • thin line : y0 = - 0.20 μm

Fig.2.b : Writing field at X = 0
- solid line: horizontal head
- dashed line: vertical head

Computations were also performed for isotropic permeabilities, showing firstly that the shape anisotropy does not contribute to the difference between the heads and secondly that only the vertical head is sensitive to the pole magnetic anisotropy.

Writing and reading on various commercially available hard disks was simulated (see properties in Table I). The bit length is 0.78 micron, corresponding to 32 kfci (kilo flux changes per inch) and a frequency of 17 MHz if the linear velocity of the medium is 13 m/s.

Tape	μo.Hc (T)	dM/dH at H=Hc	μo.Hsat (T)	μo.Mr (T)	S=Mr/Ms	Thickness (μm)
KOMAG	0.17	24	0.29	0.43	0.75	0.05
FUJI	0.17	32	0.24	0.53	0.89	0,2
CDD	0.23	29	0.32	0.41	0.82	0.05;0.1;0.2

Table I : Properties of the hard disks used in the simulation

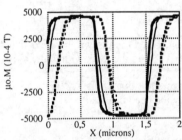

Fig. 3.a: Magnetization in the disk KOMAG
solid : horizontal head - dashed : vertical head
thick : y=-0.05 micron - thin : y=-0.1 micron

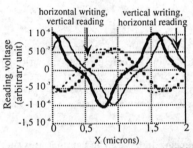

Fig.3.b : Reading voltage for the disk KOMAG
solid : horizontal head - dashed : vertical

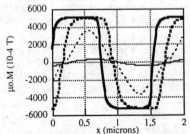

Fig. 4.a: Magnetization in the disk FUJI
solid : horizontal head - dashed : vertical head
thick : y=-0.05 micron - thin : y=-0.25 micron

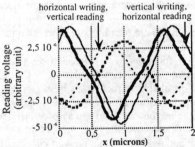

Fig.4.b : Reading voltage for the disk FUJI
solid : horizontal head - dashed : vertical

Figs. 3.and 4 show the magnetization in the media and the readback voltage (arbitrary unit) for the disks KOMAG and FUJI, which have the same coercive field.

The voltage magnitude is much higher for the horizontal head and the pulse shape, still apparent, indicates that higher densities can easily be achieved, which is not the case for the vertical head. The horizontal -resp.vertical- head reading transitions written by the vertical -resp.horizontal- head were also modeled, showing that the good performance of the horizontal is mainly due to a better reading. The good writing performance is limited to the medium suface and at a depth of 0.2 micron, the medium depth is not written at all. In this case, a vertical writing head and a horizontal reading head is the best combination.

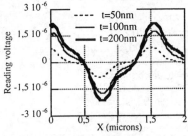

Fig.5.a : Disk CDD with various thickness   Fig.5.b : Disk CDD with various thickness
Readback voltage for the horizontal head   Readback voltage for the vertical head

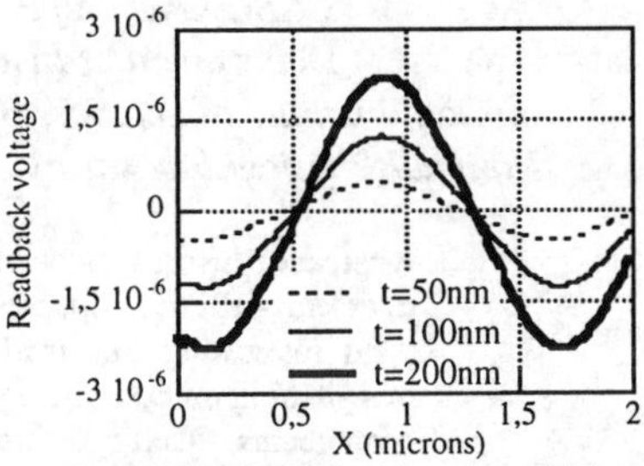

Fig.5 investigates the effect of the media thickness for the CDD disk. As foreseen, the difference between the two heads diminishes as the thickness is increased. The writing field of the horizontal head rapidly vanishes when the distance to the gap increases and the head will not be able to saturate the recording medium very deeply. This effect is more pronounced with the CDD disk because of its higher coercivity.

The relatively bad reading performance of the vertical head is explained by a magnetic gap larger than the physical one because of the bad permeability in the horizontal direction. This difference between the two heads was clearly measured, with a much better behavior of the horizontal head at high densities and no clear explanation of this phenomenon.

4. Conclusion

The presented results show the better behavior of the horizontal head, mostly at reading. During writing, the poles are close to saturation and the anisotropy will not be as pronounced as in the present study, but this is not expected to significantly change the written magnetization. On the contrary, during reading the induction in the head is very low and then anisotropy is shown to have an important effect on the reading performances.

Since the actual permeability of the poles is not precisely known, these results must be considered as trends. More simulations and comparison with experiments are required to refine these results, and the effective permeability should be linked to the domain .

References

[1] H.N. Bertram, Fundamentals of the magnetic recording process, Proc. IEEE, vol. 74, N°11, pp.1494-1512, 1986.
[2] F.Cortial, F.Ossart, J.B.Albertini, M.Aïd, An improved analytical scalar hysteresis model and its implementation in magnetic recording modeling by the finite element method, IEEE Trans. Magn., vol.33, pp1592-1595, 1997
[3] F. Ossart, V. Ionita, Etude de la convergence de la méthode du point fixe pour calculer le champ magnétique dans les matériaux hystérétiques, The European Physical Journal, Applied Physics, *in press*, 1998
[4] J.P.Lazzari, P.Deroux-Dauphin, A new thin film generation IC head, IEEE Trans. Magn. 25, N°5, Sept.89, pp.3190-3192

Non-Linear Electromagnetic Systems
V. Kose and J. Sievert (Eds.)
IOS Press, 1998

Non-Linear Electrical Response Of Chalcogenide Glasses : Memory State Phenomena

S.ASOKAN and E.S.R.GOPAL[*]

Department of Instrumentation,
Indian, Institute of Science, Bangalore 560012, India
National Physical laboratory, New Delhi 110012, India[*]

Abstract : Investigations on the switching behaviour of arsenic-tellurium glasses with Ge or Al additives, yield interesting information about the dependence of switching on network rigidity, co-ordination of the constituents, glass transition & ambient temperature and glass forming ability.

1. Introduction

Chalcogenide glasses exhibit non-linear I-V characteristics and current controlled electrical switching from a high resistance OFF state to a low resistance ON state at high fields [1]. Electrical switching in these materials can be memory or threshold type. Memory glasses stay in the ON state, even if the current is reduced to zero. Threshold samples, on the other hand, switch back to the OFF state, if the current is reduced below a holding value. The memory switching in glassy chalcogenides is associated with the formation of a conducting crystalline channel and the threshold switching occurs when the charged defect states are filled by field induced carriers [1]. This paper deals with the dependence of electrical switching of bulk chalcogenide glasses on other material properties.

2. Experimental

The switching studies reported here were undertaken in a custom made PC based set-up [2]. Samples of about 0.2 mm thickness were mounted between flat plate bottom and point contact top electrodes. A constant current was passed through the sample and the voltage developed across it, was measured.

3. The composition dependence of Electrical Switching of Chalcogenide glasses-the influence of Network topology

$Ge_{7.5}As_xTe_{92.5-x}$ glasses ($15 \leq x \leq 60$) exhibit non-linear electrical behaviour and memory switching (Figure 1). The inset in Figure 1 shows the composition dependence of memory switching fields (E_t) of $Ge_{7.5}As_xTe_{92.5-x}$ glasses. It can be seen that E_t increases linearly with x in the range $15 \leq x \leq 25$. At x = 25, a distinct change in slope is observed and E_t continues to increase in the range $25 < x \leq 50$. There is a reversal in trend at x = 50 and E_t exhibits a minimum at the composition x = 52.5.

The composition dependence of electrical switching fields of chalcogenide glasses is determined by two topological effects, namely Rigidity Percolation and chemical

Ordering [3,4]. Rigidity Percolation refers to the transformation of the material from a floppy polymeric glass to a rigid amorphous solid when lower coordinated chalcogen atoms are replaced by higher coordinated additives. With increasing rigidity, the network becomes less flexible and structural reorganisation becomes more difficult. Consequently, memory switching fields can increase with increasing network rigidity. Considering the above aspects, an increase in E_t with x and a slope change (from a lower to a higher value) across the percolation threshold, can be expected. In the $Ge_{7.5}As_xTe_{92.5-x}$ system, the composition x=25 corresponds to the Rigidity Percolation threshold and it can be seen from Fig.1 (inset) that such a slope change is indeed observed in E_t at this composition.

In any system, the stoichiometric glass is considered to be maximally ordered [3]. Memory switching fields should be the least in any glassy system for the stoichiometric composition (known as the chemical threshold), as the energy barrier and the driving force required for the crystallization of the stoichiometric glass are the lowest. In the $Ge_{7.5}As_x Te_{92.5-x}$ system, the composition x=52.5 corresponds to the chemical threshold and the minimum in E_t is clearly observed at this threshold composition (Figure 1-inset).

4. The dependence of switching behaviour on the coordination of the constituents

The non-linear I-V and switching characteristics of $Al_xAs_{40-x}Te_{60}$ ($5 \leq x \leq 20$) glasses are shown in Figure 2, which indicates that $Al_xAs_{40-x}Te_{60}$ samples of lower Al content exhibit memory and those with higher Al content show threshold switching.

The change in the switching behaviour of $Al_xAs_{40-x}Te_{60}$ glasses is intimately connected with the co-ordination of aluminium. In the glassy state, Al can reside in three different environments, having 4-fold, 5-fold and 6-fold co-ordinations respectively [5]. NMR studies indicate that in $Al_xAs_{40-x}Te_{60}$ glasses, at lower Al proportions, ($x \approx 10$), the ratio between the 4 fold and 6 fold coordinated Al atoms is about 30 %. With increasing x, the number of 4-fold coordinated Al decreases and it becomes zero around x=15 [5]. It is clear that in $Al_xAs_{40-x}Te_{60}$ glasses which exhibit threshold switching, Al is predominantly 6-fold coordinated. The increase in the co-ordination of Al can lead to a higher connectivity of the glass. In samples with higher structural connectivity, memory switching which requires structural reorganisation, becomes more difficult. Hence, these samples exhibit threshold behaviour.

5. Temperature Dependence of Switching

I-V studies on $Ge_{10}As_{45}Te_{45}$ memory glasses at different temperatures, indicate that the characteristics becomes broader and the switching more sluggish at high temperatures (Figure 3). The switching voltages decrease with increasing temperature, which is expected as the crystallization of the sample is more favourable at higher temperature. A model based on the configurational free energy [6], suggests the following relationship between the switching voltage (V_c), glass transition temperature (T_g) and the ambient temperature (T) :

$$V_c^2 = C_1 \exp [C_2 k (T_g - T)/kT_g] \qquad ..... (1)$$

where C_1 and C_2 are constants and k is the Boltzman constant. The variation of log (V_c^2) with $(T_g-T)/T_g$ of $Ge_{10}As_{45}Te_{45}$ glasses (Figure 4) indicates that the temperature dependence of switching voltages of these samples obeys the suggested relationship.

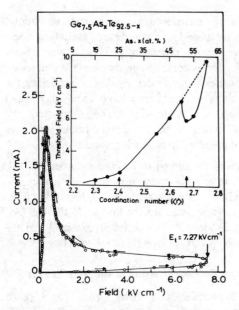

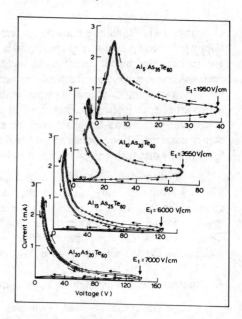

Figure 1 : I-V Characteristics of Ge$_{7.5}$As$_x$Te$_{92.5-x}$ glasses; Inset shows the composition dependence of switching fields.

Figure 2 : I-V characteristics of Al$_x$As$_{40-x}$Te$_{60}$ glasses.

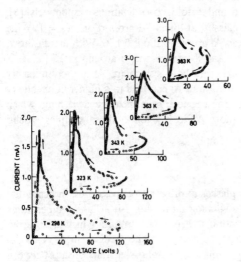

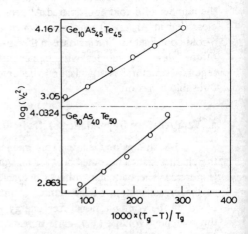

Figure 3 : I-V characteristics of Ge$_{10}$As$_{45}$Te$_{45}$ glasses at different temperatures

Figure 4 : Variation of Log (V$_c^2$) with (T$_g$-T)/(T$_g$) of Ge$_{10}$As$_{45}$Te$_{45}$ glasses.

6. Resettability of the memory state and Glass Forming Ability of Chalcogenides

Chalcogenide glasses, in which the memory state can be resetted by a current pulse, find applications in Read Mostly Memories (RMMs). The memory resetting involves local melting of the conducting crystalline channel formed during switching and its re-solidification as a glass. A direct relationship has been found between the memory resettability and the Glass Forming Ability (GFA) of chalcogenides. The GFA of a glass is given by

$$GFA = (T_c-T_g)/(T_m-T_c) \qquad ..... (2)$$

Table 1 shows the values of the GFA and the magnitudes of the resetting pulse, for $Ge_{10}As_xTe_{90-x}$ ($35 \leq x \leq 50$) memory glasses. It is clear from Table 1 that for samples with higher GFA, the resetting is easier. It is also found that the degradation after repeated resetting is higher for glasses with lower GFA; Samples with higher GFA are found to be stable with less than 5% variation in V_c at more than 10^3 cycles [6].

Table 1

COMPOSITION	GFA	RESETTING PULSE
$Ge_{10}As_{35}Te_{55}$	0.529	20 mA; 50 μSec
$Ge_{10}As_{40}Te_{50}$	0.390	20 mA; 50 μ Sec
$Ge_{10}As_{45}Te_{45}$	3.90	10 mA; 30μ Sec
$Ge_{10}As_{50}Te_{50}$	3.13	10 mA; 30 μ Sec

7. Conclusion

The composition dependence of electrical switching in arsenic-tellurium glasses with Ge or Al additives, has been found to be modulated by two topological effects namely *rigidity percolation* and *chemical ordering*. The co-ordination of the constituents has been found to play a crucial role in the type of switching (*memory/threshold*) exhibited by these samples. Further, a clear relationship has been established between the thermal properties and the electrical switching behaviour of this group of chalcogenide glasses.

Acknowledgements

The authors thank Dr.S.Prakash, Mr.S.Murugavel and Mr.R.Aravinda Narayanan for their help.

References

[1] H.Fritzsche, Switching and Memory in Amorphous Semiconductors : J.Tauc (ed.). Amorphous and Liquid Semiconductors, Plenum Press, New York, 1974, pp.313-359.
[2] R.Chatterjee, K.V.Acharya, S.Asokan and S.S.K.Titus, A PC Based System For Studying Current-Controlled Electrical Switching in Solids, Rev. Sci. Instrum., 65 (1994) 2382-2387.
[3] S.Asokan, M.V.N.Prasad, G.Parthasarathy and E.S.R.Gopal, Mechanical and Chemical Thresholds in IV-VI Chalcogenide Glasses, Phys. Rev. Letts., 62 (1989) 808-810.
[4] A.K.Varshneya, A.N.Sreram and D.R.Swiler, A Review of Average Co-ordination Number Concept in Multi-Component Chalcogenide Glass Systems, Phys. Chem. Glasses, 34 (1993) 179-193.
[5] S.Murugavel and S.Asokan, Solid State NMR Studies on Al based Chalcogenide Glasses, Phys. Rev.B. (submitted)
[6] S.Prakash, S.Asokan and D.B Ghare, Electrical Switching Behaviour of Semiconducting Aluminium Telluride Glasses, Semicond. Sci. Tech. 9, (1994) 1484-1488.

Non-Linear Electromagnetic Systems
V. Kose and J. Sievert (Eds.)
IOS Press, 1998

Domain Structure Destabilization of Fe73,5Cu1Nb3Si13,5B9 Nanocrystalline Alloy

V.V. Shulika, A.P. Potapov

Institute of Metal Physics U.D.RAS GSP-170, Ekaterinburg, Russia

Abstract. A new method of domain structure destabilization - thermomagnetic treatment in a high frequency field - is applied to Fe73,5Cu1Nb3Si13,5B9 nanocrystalline alloy.

1. Introduction

It is known, that a lot of ferromagnetics, cooled at a certain temperature interval, being in multido-main state are affected by local thermomagnetic treatment (TMT). Appearance of local induced magnetic anisotropy results in domain structure destabilization and deterioration of magnetic properties.

Such influence is especially strong in the nanocrystalline soft magnetic alloys in which resulting effective constant of magnetic crystallographic anisotropy is averaged to zero. The destabilizated domain structure may be obtained by quenching from temperature above the Curie point [1] or by TMT in a magnetic field [2].

A new method of domain structure destabilization is developed by the authors, it is TMT in the high frequency magnetic field [2]. The principal idea of this method consists in that the reversal magnetisation in micro-domains of thin ribbons with a frequency more than 50 kHz is carried out by means of inhomogeneous rotation of magnetization. If the specimen annealing is performed upon such reversal magnetization, anisotropy is not induced, domain structure destabilization takes place as at TMT in rotational magnetic field [3].

As shown in [3], TMT in the 80 kHz magnetic field of amorphous alloys leads to the improvement of magnetic properties. It is of interest to learn the opportunity of using high frequency TMT for the domain structure destabilisation of nanocrystalline soft magnetic materials.

2. Experimental

The amorphous ribbons of Fe73,5Cu1Nb3Si13,5B9 were obtained by melt quenching on the rotating disk (ribbon thicknesses are 20-25 μ, width - 10 mm). Toroidal specimens were wound from these ribbons with an outer diameter 30 mm, inner diameter 25 mm.

In the first series of experiments for obtaining nanocrystalline structure samples were annealed at 540° C for 1 h in vacuum. Further annealing in a longitudional magnetic field at 540° C for 0,5 h and cooling in the field to room temperature with the rate 200° C/h were carried out. Some samples were subjected to quenching in water from Curie point (570° C).

In the second series of experiments magnetic field was applied to the sample during the alloy transition from the amorphous state to the nanocrystalline one (heating to 540° C, keeping for 0,5 h, cooling in the field with rate 200° C/h). It was compared effectiveness of TMT in the samples accord-

ing to 1 version and the 2 one. The magnetic properties after TMT in high frequency (80kHz), alternative (50Hz), and direct magnetic fields were compared.

Static hysteresis loops, magnetic losses, initial magnetic permeability, were mearsured. Magnetic
losses were measured at the frequency of reversal magnetization 20 kHz and induction 0,2 T.

3. Results and Discussion

The results of measurements are presented in Fig. 1-3 and Table 1. Static hysteresis loops of samples
after TMT in direct magnetic field are shown in Fig. 1. The first sample was preliminary annealed at the
temperature of 540° C. Annealing at 540° C and TMT were combined for the second sample. It is seen
that effectiveness of TMT is higher for the second sample, it has a more rectangular hysteresis loop after
annealing in magnetic field relative to the first sample. Therefore TMT in subsequent experiments are
performed according to the version 2: annealing and TMT are carried out at the same time.

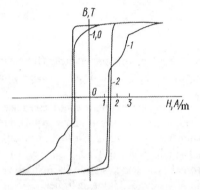

Fig. 1. The hysteresis loops of samples after annealing at 540° C and
subsequent TMT in a direct field (1), annealing at the same
temperature and simultaneous TMT in a direct field (2).

Static hysteresis loops after annealing in high frequency (80 kHz)-1, alternative (50 kHz)-2 and
direct magnetic field-3 are compared in Fig.2. Their magnetic properties (initial permeability μ_0, maximum permeability μ_{max} rectangular coefficient B_r/B_m and magnetic losses P) are summerized in
Table 1. It is necessary to emphasize that TMT was performed from the temperature of alloy transition
from amorphous state to nanocrystalline one. As TMT in high frequency magnetic field leads to nonrectangular hysteresis loop, to higher initial permeability and lower coercivity and magnetic losses in
comparison with TMT in direct or alternative (50 Hz) magnetic fields.

Table 1. The magnetic properties of Fe73,5Cu1Nb3Si13,5B9 nanocrystalline alloy after TMT

Treatment	μ_o*10^{-3}	$\mu_{max}*10^6$	H_c, A/m	B_r/B_m	$P_{0,2/20000}$, W/kg
H, f=80 kHz	53	0,75	0,8	0,68	5,5
H, f=50 Hz	37	1,20	0,8	0,96	8,0
H, f=0	10	1,00	1,2	0,98	11,0

Fig.3 shows the static hysteresis loops of two samples: 1-after annealing at 540° C and subsiquent quenching in water from the Curie point 2-after TMT in high frequency magnetic field. It follows from the comparison of the hysteresis loops that TMT in 80 kHz magnetic field results in a higher remanent induction and lower coercivity, than after quenching in water. The magnetic losses $P_{0,2/20000}$ after TMT in 80 kHz magnetic field are four times lower than those after quenching in water.

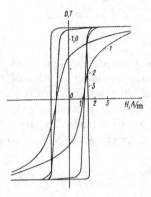

Fig. 2. The hysteresis loops of samples in high frequency (80kHz)-1,
alternative (50Hz)-2, and direct-3 magnetic fields.

The results of the performed investigations confirm conclusion made in [4], that in the nanocrystalline FeCuNbSiB alloy induced anisotropy obtains maximum equilibrium value if magnetic field was applied during the alloy transition from amorphous state to the nanocrystalline one.

The effect of TMT depends on the magnetic field frequency. The influence of TMT in 80 kHz magnetic field on magnetic properties of nanocrystalline FeCuNbSiB alloy essentially differs from the influence of TMT in direct and alternative 50 Hz magnetic fields.

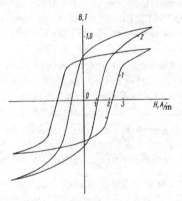

Fig. 3. The hysteresis loops of samples after quenching in water (1)
and high frequency TMT (2).

Apparently, this is due to the following reasons. In high frequency 80 kHz field reversal magnetization is realised by inhomogeneous rotation of magnetization. That is why the uniaxial anisotropy is not induced at TMT in such a field because of the lack of preferable direction of magnetization and a domain structure destabilization takes place. It leads to a nonrectangular hysteresis loop with the B_r/B_m relation equal to 0,68, and therefore to the increase of initial permeability and the decrease of magnetic losses eddy-current component.

The uniaxial anisotropy is induced at TMT in direct magnetic field and hysteresis loop becomes rectangular (curve 3, Fig.2) with a high value of B_r/B_m. It reduces the initial permeability and makes the process of reversal magnetization jump-formed, which leads to sharp growth of eddy-current component of magnetic losses. The reversal magnetization in a nanocrystalline alloy at TMT in alternative 50 Hz field takes place by the domain structure movement. TMT in alternative field also leads to the development of uniaxial anisotropy, because the process of its inducing is an even effect, but its value is lower than after TMT in direct field. Evidently, one can account for the above-mentioned higher values of initial magnetic permeability and lower magnetic losses after TMT in alternative 50 Hz field against TMT in direct field.

The traditional method of domain structure destabilization, namely, quenching in water from temperature above the Curie point in the investigated alloy does not give optimal magnetic properties (curve 1, Fig.3). The sample after quenching has increased values of coercivity, magnetic losses and low values of the initial and maximum magnetic permeability. Evidently, it is specified by the fakt that the nanocrystalline Fe73,5Cu1Nb3Si13,5B9 alloy contains α-Fe-Si grains, having magnetostriction $\lambda \sim 20*10^{-6}$. As a result of quenching the large internal stresses appear in a sample, which lead to deterioration of both static and dynamic magnetic properties.

4. Conclusions

It is established that domain structure destabilization of Fe73,5Cu1Nb3Si13,5B9 nanocrystalline alloy at TMT in a high frequency magnetic field results in maximum decrease of magnetic losses and permeability increase, if high frequency magnetic field is applied to a specimen during the alloy transition from an amorphous state to a nanocrystalline one.

The TMT in a high frequency magnetic field may be used for treatment of magnetic cores produced from the nanocrystalline soft magnetic materials.

References

[1]   A.A.Glazer, V.V. Shulika, A.P.Potapov, Influence of Induced Magnetic Anisotropy upon Static and Dynamic Magnetic Properties of Amorphous Soft Magnetic Alloys with Different Magnetostriction, *Fiz. Met. Metalloved.* **78** (1994) 45-51.

[2]   F.N. Dunaev, N.S. Malev, Thermomagnetic Treatment of 66-Permalloy and Fe-Si Steel in Rotational Magnetic Field, *Fiz. Met. Metalloved.* **20** (1965) 935-937.

[3]   A.A. Glazer, V.V. Shulika, A.P. Potapov, Domain Structure Destabilization of Amorphous Alloys by Thermomagnetic Treatment in High Frequency Field, *DAN* **324** (1992) 1191-1193.

[4]   G.Herzer, Magnetic Field Induced Anisotropy in Nanocrystalline Fe-Cu-Nb-Si-B alloy, *JMMM* **133** (1994) 248-250.

Non-Linear Electromagnetic Systems
V. Kose and J. Sievert (Eds.)
IOS Press, 1998

DESIGN OF A MAGNETICALLY DRIVEN SHAPE MEMORY ALLOY Ni$_2$MnGa

Alexander N. VASIL'EV, Alexei D. BOZHKO, Vladimir V. KHOVAILO
Low Temperature Physics Department, Physics Faculty, Moscow State University, Moscow
119899, Russia
Junji TANI, Toshiyuki TAKAGI, Shunji SUZUKI,
Institute of Fluid Science, Tohoku University
Katahira 2-1-1, Aobaku, Sendai 980-77, Japan
Minoru MATSUMOTO
Institute of Advanced Material Processing, Tohoku University
Katahira 2-1-1, Aobaku, Sendai 980-77, Japan

Abstract. An Heusler-type alloy Ni$_2$MnGa exhibits well defined shape memory properties in a ferromagnetic state, which means that the martensitic transition temperature is lower than the Curie point of this material. The change of composition allows these characteristic temperatures to move towards each other. To study this behaviour the measurements of specific heat, ac magnetic susceptibility and dc resistivity were made.

1. Introduction

Some representatives of the Heusler alloys family are known to exhibit a crystallographically reversible, thermoelastic martensitic transformation resulting in the shape memory effect. This effect is observed usually in the following way: the sample is in the low-temperature martensitic phase when it is deformed and when the stress is removed it will regain its original shape when heated. The process of regaining the original shape is associated with the reverse transformation of the deformed martensitic phase to the higher temperature parent phase. These alloys can be trained by appropriate stress and thermal cycling, so that a two-way shape memory will be obtained. Once this conditioning has been achieved, a sample will spontaneously "bend" when the parent transforms into martensite, and "unbend" to the initial shape during the reverse transformation [1].

2. Shape memory effect in magnetic alloys

In most cases the shape memory alloys are *non-magnetic* and the options to influence their shape and dimensions are restricted to stress and temperature. In Mn-containing Heusler alloys however the indirect exchange interaction between magnetic atoms results in ferromagnetism. It opens the possibility to influence the shape and dimensions of *magnetic* shape memory alloys by an external magnetic field in addition to stress and temperature [2].

This new possibility can be realized either through magnetostriction associated with the reversible rotation of the magnetization vector within the ferromagnetic domains or through spontaneous magnetostriction at the magnetic transition. The most promising way to achieve a substantial effect of magnetic field on shape and dimensions of a ferromagnetic shape memory alloy seems to merge the temperatures of structural and magnetic transitions. In this case the application of a magnetic field would result in martensitic transformation. For a two-way trained shape memory alloy it will lead to a magnetically driven "bending" and "unbending" of the sample

3. Ferromagnetic shape memory alloy Ni$_2$MnGa

The intermetallic compound Ni$_2$MnGa is a ferromagnetic Heusler alloy with L2$_1$ structure. A martensitic phase transition from a cubic to a tetragonal structure takes place on cooling below T_M = 202 K. This phase transition is hysteretic, but reversible on heating and thus the alloy shows the shape memory effect. The magnetic moments of about 4μ_B in this compound are mainly confined to the Mn sites whereas the small Ni-moments disappear at the Curie temperature T_C = 376 K. There is a strong influence of the lattice parameter (and hence of the chemical composition) on the structural properties of the alloys [3].

4. Samples

The ingots of specimens of six compositions were prepared by arc melting high-purity (99.99%) elements and subsequent homogenization of the ingot material by annealing at 1100 K for 9 days. The ingots were quenched in ice water. The composition of the Ni$_{2(1+\delta)}$Mn$_{1-2\delta}$Ga samples is characterized by Ni excess 2δ in the range 0 < 2δ < 0.19. The partial substitution of Mn by Ni resulting in negative chemical pressure leads both to the increase of martensitic transition temperature and to the decrease of ferromagnetic transition temperature. This was evidenced by X-ray diffraction at room temperature and by the measurements of resistivity, magnetic susceptibility and specific heat.

5. Experimental results

The crystal structure of the specimens at room temperature was determined by X-ray diffraction using CuK$_\alpha$ radiation. The high temperature phase of specimens with $\delta \leq 0.05$ had a cubic structure of Heusler type. The lattice parameter a = 5.796 Å for the stoichiometric composition increases with the increase of Ni content. The low temperature phase of the specimens with $\delta > 0.05$ at room temperature had a tetragonal structure.

Both transition temperature and transformation heat of the samples were measured by the differential scanning calorimetry (DSC). The peaks accompanying the martensitic and the magnetic transformation were observed, as shown in Fig. 1, during heating. DSC was not able to resolve both transitions in the case of those temperatures close to each other due to the relatively large width of the peaks observed.

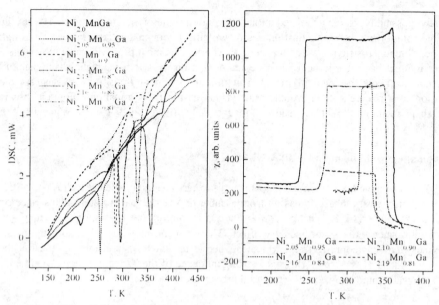

Fig. 1 Differential scanning calorimetry of $Ni_{2(1-\delta)}Mn_{1-\delta}Ga$ specific heat at heating.

Fig.2 Magnetic susceptibility vs temperature dependence $Ni_{2(1-\delta)}Mn_{1-\delta}Ga$.

The temperature dependencies of the ac magnetic susceptibility χ, as shown in Fig. 2, exhibit very sharp changes at martensitic and magnetic transitions. The martensitic phase transition from a cubic to a tetragonal structure is indicated by the drastic drop of χ due to the magnetocrystalline anisotropy of the tetragonal phase. The Curie temperature is indicated by a significant decrease of χ at transition to paramagnetic state.

The dc resistivity measurements provide simple and effective tools to detect both structural and magnetic transitions. As shown in Fig. 3, at the temperature of martensitic transition from a cubic to a tetragonal structure the resistivity exhibits a pronounced jump-like behaviour, while at the Curie temperature the change of the slope takes place. This allows the resolution of martensitic and magnetic transitions even if their temperatures merge as in the samples with highest Ni concentration. Below the martensitic transition temperature the resistivity possesses a marked hysteretic behaviour in the samples with $\delta \geq 0.065$.

The results of all these measurements correspond to each other and allow the establishment of the general tendency of martensitic temperature T_M to increase and Curie temperature T_C to decrease with the Ni excess in the Heusler type alloy Ni_2MnGa. The composition dependencies of T_M and T_C which are the best fit for specific heat, magnetic susceptibility and resistivity measurements are shown in Fig. 4. It appears that these characteristic temperatures of $Ni_{2(1-\delta)}Mn_{1-2\delta}Ga$ will merge at Ni excess $\delta = 0.09 \div 0.10$. In the samples of this composition range the profiles of the observed singularities in physical properties measured change qualitatively.

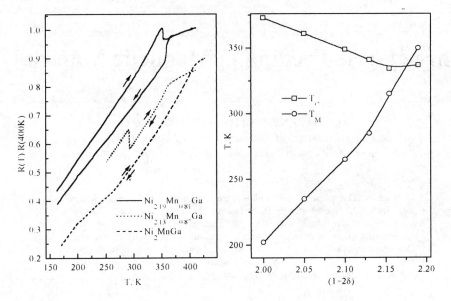

Fig. 3 R(T) dependencies of Ni$_{2.1}$Mn$_{0.9}$Ga at heating and cooling. The curves are shifted with respect to each other for clarity.

Fig. 4 Dependencies of Curie temperature T$_C$ and martensitic transformation temperature T$_M$ on Ni content 2(1+δ).

6. Conclusion

In conclusion, it has been shown that the cubic phase of Ni$_{2(1+\delta)}$Mn$_{1-2\delta}$Ga can be suppressed by a certain Ni excess δ substituting Mn. Due to the increase of the lattice parameter with Ni excess, the Curie temperature of this alloy decreases at this substitution. The temperatures of martensitic and magnetic transitions merge at Ni excess δ = 0.09 ÷ 0.10. To determine the right composition of a magnetically driven shape memory alloy is only the first step to realize this possibility. Further study is required to overcome the brittleness of this compound and to establish the relationship between martensitic and magnetic domain structures.

This work formed part of the Russian Foundation for Basic Research Grant-in-Aid No. 96-02-19755 and A.D.B. has been supported by the Japan Society for the Promotion of Science fellowship (N PS - 95677).

References

[1] C.M. Wayman. Some Applications of Shape-Memory Alloys. Journal of Metals. June 1980. v.2, 129-137.
[2] A.N. Vasil'ev et al.. Magnetoelastic Interaction in the Martensitic Transformation in an Ni2MnGa Single Crystal. Sov. Phys. JETP 82 (1996) 524-526.
[3] S. Wirth. A. Leithe-Jasper, A.N. Vasil'ev and J.M.D. Coey. Structural and Magnetic Properties of Ni2MnGa. Submitted to J. Magn. Magn. Mat. 1997.

Non-Linear Electromagnetic Systems
V. Kose and J. Sievert (Eds.)
IOS Press, 1998

Three Phase Feeding in Magnetic Material

Amália IVÁNYI[1] and Helmut PFÜTZNER[2]
[1] Technical University of Budapest, H-1521 Budapest, Hungary
[2] Vienna University of Technology, A-1040 Vienna, Austria

Abstract. A 2D approach to the nonlinear electromagnetic field in a single sheet tester with a hexagonal sample is evaluated. The feeding rotational flux is a symmetrical three phase excitation of positive sequence. The nonlinear characteristic of the anisotropic magnetic material is represented by the scalar Jiles–Atherton hysteresis operator. The numerical analysis of the electromagnetic field on the basis of the combined global variational and *R*–functions methods is evaluated and the results are plotted in figures.

1. Introduction

For testing the magnetic property of anisotropic material a single sheet tester with a rotational feeding field has been developed. Several papers have been published on the experimental and the theoretical investigations evaluated for the rectangular shape of sample under two-phase feeding field [1], [2]. A hexagonal shape of model under three-phase rotational flux feeding has been introduced and measured by Pfützner et al [3]. In this paper, a 2D field model for the determination of the nonlinear electromagnetic field in the single sheet tester with a hexagonal sample is presented. The numerical field analysis, based on the global variational method is introduced in the time domain to study the effect of the eddy currents and the magnetic hysteresis. In the model, the feedback interaction of the sample on the laminated magnetic poles is disregarded, and the boundary conditions under prescribed flux feeding are satisfied by the aid of the *R*–functions [4], [5].

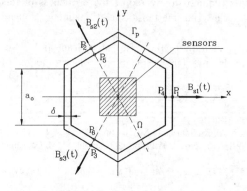

Fig. 1. Model for the single sheet tester

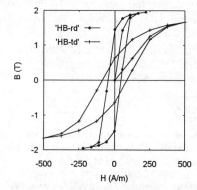

Fig. 2. Simulated hysteresis characteristics

In the arrangement plotted in *Fig. 1* the profile parameter of the hexagonal plate is $a_0 = 80\ mm$, the air gap between the magnetic poles and the sample is $\delta = 0.5\ mm$. The conductivity of the material is assumed to be $\sigma = 10^8/48\ S/m$. The sensor coils have length $\ell_0 = 50\ mm$. For the numerical field analysis the investigated region Ω is closed by the surface of the magnetic poles Γ_p, where the prescribed flux of the three-phase rotational feeding system has only normal components.

In the model of the anisotropic material, at zero cutting angle, the rolling direction is selected to coincidence with the x-axis, the transverse direction is the y-axis. On the basis of the measured data [4], the stationary hysteresis characteristics parallel to the rolling (rd) and the transverse (td) directions are simulated with the scalar Jiles–Atherton hysteresis operator [6], [7] $M = \sum M_i e_i$, $M_i = \mathcal{H}\{H_i, M_i\}$, $i = rd / td$, and the results are plotted in *Fig. 2*.

2. The 2D Field Model

As the magnetic measurement is performed under rotational magnetic induction, the source flux of the magnetic poles is generated by a symmetrical three-phase excitation of positive sequence, $f = 50\ Hz$. For the 2D simulation of the electromagnetic field in the investigated region $\Omega \cup \Gamma_p$, a z-directed magnetic vector potential is introduced in the x,y plane $B = \nabla \times A$, $A = A(x,y,t)e_z$. The source field A_s of the poles and the induced component A_i of the vector potential $A = A_s + A_i$, generated by the interaction of the eddy currents and the nonlinear field of the anisotropic material yields the solution of the nonlinear diffusion equation

$$\nabla \times (\nabla \times (A_s + A_i)) + \mu_0 \sigma (\dot{A}_s + \dot{A}_i) = \nabla \times (\mu_0 M). \tag{1}$$

The source component of the vector potential is a known function describing the flux with homogeneous distribution on the magnetic poles $A_s(x,y,t) = e_z \sum B_{si}(t) w_i(x,y)$, $i = 1,2,3$, where $B_{si}(t)$ is the source flux density on the poles and the functions $w_1(x,y) = y$, $w_2(x,y) = (-\sqrt{3}x - y)/2$ and $w_3(x,y) = (\sqrt{3}x - y)/2$ prove the circular polarised rotational flux. The unknown A_i component of the formulation, having zero value along the magnetic poles $A_i|_{\Gamma_p} = 0$, is approximated by a space dependent entire function set with time dependent unknown coefficients $A_i(x,y) = \mathbf{F}^T(x,y,e_z)\mathbf{a}(t)$. As the magnetic field has no tangential component along the magnetic poles $(\nabla \times A) \times n|_{\Gamma_p} = 0$, thus on the basis of the R–functions method [4], [5] both terms of the magnetic vector potential are modified as $A^* = A + w((\nabla \times A) \times n)$, where $w(x,y)$ is an R-function, describing the boundary surface Γ_p of the region Ω under the conditions

$$w(P) \begin{cases} > 0, & if\ P \in \Omega, \\ = 0, & if\ P \in \Gamma_p, \\ < 0, & if\ P \notin \Omega \cup \Gamma_P, \end{cases} \quad \left|\frac{\partial w}{\partial n}\right|_{\Gamma_P} = 1, \quad and \quad n = -\nabla w. \tag{2}$$

To determine the magnetic field in the examined region the solution of the nonlinear differential equation can be evaluated by the global variational method. For each time step the extreme value of the functional results in an ordinary differential equation for the unknown coefficients

$$\mathcal{F} = \int_{\Omega} (\tfrac{1}{2}\nabla \times (A_s^* + A_i^*) - \mu_0 M)(\nabla \times (A_s^* + A_i^*) + \mu_0 \sigma (\dot{A}_s^* + \dot{A}_i^*)(A_s^* + A_i^*))d\Omega, \tag{3}$$

$$\mathbf{K}_1 \mathbf{a} + \mu_0 \sigma \mathbf{K}_2 \dot{\mathbf{a}} = -\mathbf{G}_1(\nabla \times A_s^*) - \mu_0 \sigma \mathbf{G}_2(\dot{A}_s^*) + \mathbf{G}_3(\mu_0 M). \tag{4}$$

3. Realization of the Solution

(i) First the distribution of the source field of the poles is determined without a sample in between the poles. The magnetic field is determined for $B_{so} = 0.25\ T$ phase amplitude, under the conditions $\sigma = 0$, $M = 0$. In *Fig. 3* the distribution of the imposed flux in the empty single sheet tester is plotted. In *Fig. 3.a* the locus of the flux density at the centre of the magnetic poles, in points P_1, P_2 and P_3 is plotted with the ones at the points δ distance from them, in points P_4, P_5 and P_6. From the figure it can be seen that on the poles the magnetic flux density has only normal component and with the distance the tangential component is increasing. In *Fig. 3.b* the locus of the flux distribution against the distance from the magnetic poles is plotted. The figure proves that about 5~6 *mm* distance from the poles the flux distribution becomes homogeneous.

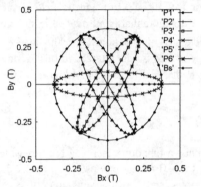

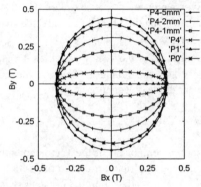

Fig. 3.a. Flux density at and near to the poles   Fig. 3.b. Flux density with distance from the pole

(ii) In second case a metal sample with conductivity $\sigma = 10^8/48\ S/m$ is placed in the single sheet tester and fed by a magnetic field of phase amplitude $B_{so} = 0.25\ T$. The simulation is evaluated under the condition $\sigma \neq 0$, $M = 0$. In *Fig. 4 (left)* the eddy current effect on the flux can be seen at 1 *mm*, 2 *mm* and 5 *mm* distances from the edge of the conducting sample. In *Fig. 4 (right)* the variation of the locus of the flux density at $x = 0$, 50, 55, 60 *mm* in the conducting sample is plotted.

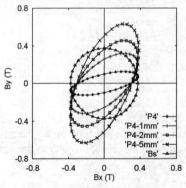

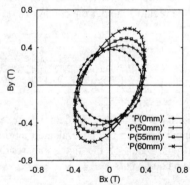

Fig. 4. The flux distribution in the single sheet tester for nonmagnetic material

From the figures it can be seen that due to the eddy currents the magnetic field is increasing about 50~60 *mm* distance from the centre of the sample. At the same time, as in the model the flux of the magnetic poles is fixed (no interaction from the sample to the poles is considered) the magnetic field cannot leave the sample, the poles are working as a magnetic wall. In the central region of the sample the magnetic field proves to have a quasi homogeneous distribution up to 4 *cm* radius.

(iii) Finally the effect of the magnetic material is proved, $\sigma \neq 0$, $M \neq 0$. According to the increased permeability of the material the effect of the z-directed eddy currents becomes dominant, resulting high field at the edge regions of the sample. About 5 *mm* distance from the edge of the sample the material proves saturation with $B \sim 1.8\ T$ as seen in *Fig. 5 (left)*. At the same time according to the strong skin effect there is no magnetic field in the central region of the material as seen in *Fig. 5 (right)*.

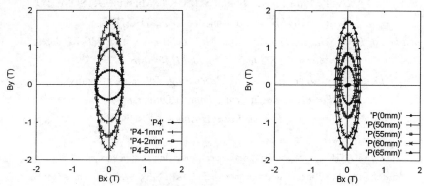

Fig. 5. The flux distribution in the single sheet tester for magnetic material

4. Conclusion

The numerical investigations prove that the single sheet tester with a hexagonal sample is a novel arrangement for measuring the magnetic anisotropy. However, as the arrangement has rotational symmetry, its response is sensitive to the geometry in x,y coordinates. The numerical simulation needs high accuracy to prove the symmetry.

However, the 2D model of the single sheet tester gives information about the distribution of the magnetic field. However it shows deficiency in simulation of the closed loop of the eddy currents and results in high tangential flux near to the border of the sample. Thus, the development of a 3D model is required to describe the nonlinear field in the sample.

Acknowledgement The research work is part of the Project in cooperation with the Wienna University of Technology, sponsored by the Hungarian Ministry of Education and Culture, Pr.No. MKM 790/1996.

References

[1] Enokizono, M. ed.: Two-Dimensional Magnetic Measurement and its Properties, JSAEM Studies in Applied Electromagnetics, vol.1. 1993. Japan,

[2] Moses, A.J., Basak, A. ed: Nonlinear Electromagnetic Systems, Studies in Applied Electromagnetics and Mechanics, vol.10. IOS Press, 1996.

[3] Hasenzagl, A., Wiesner, B., Pfützner, H.: Novel 3-Phase Excited Single Sheet Tester for Rotational Magnetization, J. Magn. Mag. Mat. vol. 160, pp.180-182. 1996.

[4] Rvachev, V.L., Sheiko, T.I.: R–functions in boundary value problems in mechanics, Applied Mechanics Reviews, vol.48. 1995. pp.151-188.

[5] Ivanyi, A.: R–functions in Electromagnetism, Periodica Polytechnica, Technical Report, TUB-TR-93-EE08, Technical University of Budapest, 1993. Budapest,

[6] Jiles, D.: Introduction to Magnetism and Magnetic Materials, Chapmann & Hall, 1995.

[7] Ivanyi, A.: Hysteresis Models in Electromagnetic Computation, Akadémiai Kiadó, Budapest, 1997.

An Application of
Shape Memory Alloy Plate to Flow Control
Using All-Round Shape Memory Effect

Toshiyuki TAKAGI[a], Junji TANI[a], Shunji SUZUKI[b], Minoru MATSUMOTO[c]

Seiji CHONAN[d] and Yoshikatsu TANAHASHI[e]

[a]*Institute of Fluid Science, Tohoku University, Aoba-ku, Sendai 980-77, Japan*

[b]*Graduate Student of Tohoku University*

[c]*Institute for Advanced Materials Processing, Tohoku University*

[d]*Faculty of Engineering, Tohoku University*

[e]*Tohoku Kosai Hospital*

Abstract. This paper describes a new approach to a flow control valve using SMA plates with the all-round shape memory effect (ARSME). As the temperature becomes higher than the transformation temperature, fluid can automatically flow because of the shape change of the plates. The relationship between temperature and flow rate was investigated experimentally. Temperature rises were caused by electromagnetic inductive heating.

1. Introduction

Shape memory alloys (SMAs) exhibit useful behaviors like the shape memory effect (SME) in which a substantial fraction of residual strain is recovered by heating, and pseudoelasticity (PE) in which residual strain is recovered by unloading alone. These characteristics are caused by the martensitic transformation and its reverse transformation. TiNi SMA can show the all-round shape memory effect (ARSME) if the stoichiometry and heat treatment are specially arranged[1]. If a SMA plate is heat-treated on arc shape, it is flat at room temperature and arc shape at high temperature.

Both Ti and Ni have good bio-compatibility, and TiNi SMAs with one way SME have been used for medical research like medical clips and orthodontic braces. Recently an artificial urethral valve for the treatment of urinary incontinence was developed by Chonan et al.[2]. Since artificial urethra valves and anals should be used everyday in a human body, induced heating is better than direct electrical heating to obtain the temperature rise. In order to estimate the performance of the valve it is necessary to evaluate the temperature change and phase transformation of the SMA valve with induced heating by numerical analysis.

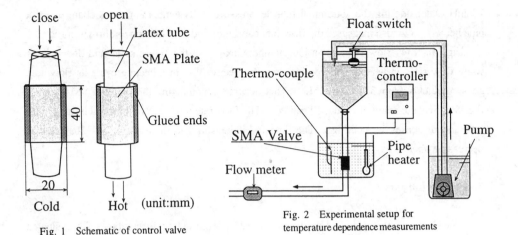

Fig. 1   Schematic of control valve

Fig. 2   Experimental setup for
temperature dependence measurements

2. Experiment

The SMA plate used here is Ti-51at%Ni with a length of 40mm, a width of 20mm, and a thickness of 0.25mm.   First we annealed it after keeping 1073K for 2 hours and then removed the oxide-surface.   Next we fixed it to the outside of a aluminum pipe (outer diameter of 25mm) in circular shape and annealed it after keeping 673K for 25 hours.

Transformation temperatures of the specimen were measured by a differential scanning calorimetry (DSC) without stress.   In this paper, we use a shape change caused by the R (rhombohedral) phase transformation, which appears between the A phase and the M phase. The shape is flat at room temperature and arc shape at high temperature.

Fig. 1 shows a flow control part of an induced heating valve.   This flow control part is made of two ARSME plates and a latex tube inserted between these plates.   When these plates are heated, their shapes change into arc shape each other.   Hence fluid flows automatically.   To evaluate the efficiency of the valve, we measured the flow rate at various temperatures, as shown in Fig. 2.   We used water for this experiment, with pressure kept

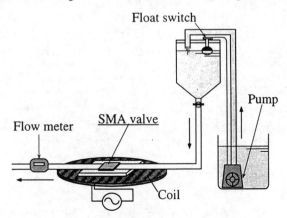

Fig. 3   Experimental setup for induced heating measurement

70cmH$_2$O considering the abdominal muscle pressure.   Its temperature was changed by a pipe heater.   We also measured the flow rate caused by induced heating as shown in Fig. 3.

Fig. 4 shows flow rates at various temperatures.   In the figure solid and dotted lines show flow rates in heating and cooling steps respectively.   In a heating cycle its flow rate goes up quickly from 320K to 335K.   This occurs with changing the shape of SMA plates. Fig. 5 shows the result of induced heating.   The flow rate increases by induced heating with time and decreases by cooling.   The time for increasing the flow rate needs 20s and for decreasing about 15s.

3. FEM Analysis

The shape change of an SMA valve is caused by a phase transformation.   We evaluated temperature rises of the SMA plate with different coil shapes. The shape of a specimen was kept 5mm width, 10mm length, and 0.25 mm thickness considering the inner diameter of human urethra, which is smaller than the testpiece used in the experiment.   The distance from SMA plate to a coil is determined to be 30mm considering the application to human bodies.   An analytical model is a quarter one using symmetry conditions.   Finite elements used here are eight nodes isoparametric 3D elements.   The number of elements and nodes are 3608 and 4416 respectively.   The conductivity of the SMA is $1 \times 10^6 \, S/m$.   Current density of a coil is determined $3.13 \times 10^6 \, A/m^2$ from the measurement of magnetic flux density.

Fig. 6 shows a result of an electromagneto-thermal analysis.   The solid line shows the results from 0s to 100s and obtained temperature rise was only about 3K.   It is difficult to heat SMA plates by this induced heating condition.   Hence we changed current density and frequency from the experimental condition of $3.13 \times 10^6 \, A/m^2$ to $5 \times 10^6 \, A/m^2$ and from

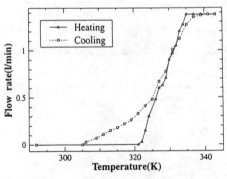

Fig. 4   Relationship between flow rate and temperature

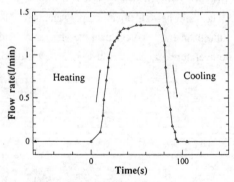

Fig. 5   Relationship between flow rate and time

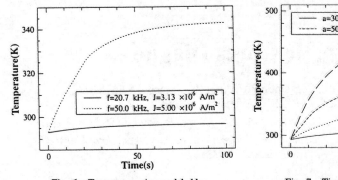

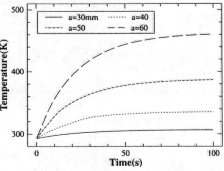

Fig. 6   Temperature rise modeled by
induced heating

Fig. 7   Time dependence of an SMA plate
temperature with various coil diameters

20.7kHz to 50kHz.   The dotted line shows the analytical results.   Temperature rose to 343K and this case can heat the SMA plate to the shape change temperature region.

Next, we investigated the temperature rise of the SMA plate when the shape of coil is changed.   The width of a coil changes from 30mm to 60mm and the distance is kept 30mm. The temperature rise became larger with the wider coil as shown in Fig. 7.   From this analysis we can make out that the shape of coil has large influence on the temperature rise.

4. Summary

A new approach to a flow control valve using SMA plates with ARSME was proposed. The new valve was fabricated and its flow performance was demonstrated.   A small hysteresis flow control valve using ARSME plates was successfully manufactured. Practical heating method was also discussed by changing values of current density and frequency, and the shape of a coil by using finite element calculation.

We are grateful to Dr. K. Yamauchi of Tokin Co. for supplying us SMA plates. We also acknowledge Prof. R. Kainuma of the Department of Materials Science, Tohoku University for his help in DSC measurement.

References

[1]   R. Kainuma et al., The Mechanism of the All-round Shape Memory Effect in a Ni-Rich TiNi Alloy, *Proc. of the Intern. Conf. on Martensitic Trans. JIM*, (1986) pp.717-722.

[2]   S. Chonan et al., Development of an Artificial Valve Using SMA Actuators, *Smart Mater. Struct.*, 6 (1997) pp.410-414.

Non-Linear Electromagnetic Systems
V. Kose and J. Sievert (Eds.)
IOS Press, 1998

Magnetic Cores Diagnosis

Ovidiu CALTUN[1], Constantin PAPUSOI[1], Alexandru STANCU[1],
Petru ANDREI[1], Wilhelm KAPPEL[2]

[1] *"Alexandru Ioan Cuza" University, Faculty of Physics, 6600, IASI, ROMANIA*
[2] *Research Institute for Electrotechnics, Magnetic Materials Dept., Splai Unirii 313*
BUCHAREST, ROMANIA

Abstract. One presents a dynamic testing method of magnetic properties of magnetic cores. A complex installation serves to diagnose the magnetic integrity of the magnetic samples by displaying the hysteresis loops. The first operation mode allows the study of the magnetisation process at high frequency and intense magnetic field in a RLC circuit. A capacitor, initially charged is discharged on circuit by a switch. A data acquisition system transfers the experimental results pointed out by the voltage signals across C and across a sensing current resistor R. In the second operation mode, the primary coil is carried by an ac current having different frequencies. The magnetisation processes at low frequencies and low magnetic fields are displayed by integrating with respect to the time the signal induced in the secondary coil and plotting with respect to the magnetic field, proportional to the voltage across a sensing current resistor. The hysteresis loops for various samples in the same experimental conditions are compared with standard hysteresis loops stored by computer. If the experimental results present differences greater than an admissible standard deviation, the magnetic core tested is rejected.

1. Introduction

In order to diagnose the magnetic integrity of magnetic cores, like MnZn soft ferrites, a complex installation was designed [1]. Magnetisation processes have formed the subject of many theoretical review papers. [2,3] The magnetic properties of diverse soft ferrite magnetic cores and their behaviours in different experimental conditions have been the subject of thousands of papers. Many attempts have been made to fit experimental and theoretical data. In the phenomenological models one tries to find mathematical functions that fit the experimental data. For each sample, belonging to a category, the same set of parameters is used corresponding to the experimental conditions to display the experimental and calculated hysteresis loops. Simultaneously a standard core hysteresis loop stored by the computer is displayed. If any difference is observed the magnetic core is re-tested and if the disagreements remain the core is rejected.

2. Experimental set-up

The experimental installation and the two operating modes are presented in Fig. 1. The soft ferrite magnetic cores are toroidal in shape. In the first operating mode the sample is the magnetic core of the inductance L, a component part of the RLC series circuit. If the switch K_1 is turned off the capacitor C is charged by the DC power supply at V_0 voltage value. If K_1 is turned on and K_2 is turned off the capacitor is discharged to the RL primary circuit. The voltage waveform across the capacitor C (K_3 turned off) and the voltage across

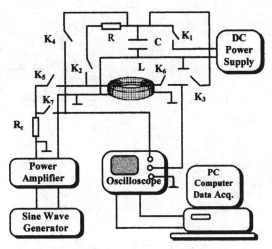

Fig. 1 Experimental set-up

R (K_4 turned off) are digitised by the storage oscilloscope HM 205-3 and transferred by the data acquisition block to the computer.

In the second operating mode the switch K_5 is turned off and the RL primary circuit carries an AC current with variable frequency (a sine wave generator RFT 51035 and a power amplifier RFT LV-103 is used). The signal induced in the secondary coil, proportional to the magnetic flux variation, $U_L = L\dot{H}/n$, and the voltage signal across the sensing current resistor R_C are captured (K_6 and K_7 are turned off) and digitised by the storage oscilloscope. By integrating with respect to the time of the magnetic flux rate and plotting it versus field one obtains the hysteresis loop. Other magnetisation curves can be calculated [3].

In this work the experimental results are limited only to a single category of samples. These are MnZn soft magnetic ferrite toroidal cores (commercial A-7) having inner radius, $r_1=23$ mm, outer radius $r_2=35$ mm, and height $h=12.5$ mm respectively. The number of primary and secondary coils is equal, 56 turns. The background resistance associated with the windings is $R=0.32\Omega$. One can approximately calculate the inductance as a function of magnetic field (H), magnetisation (M) and their variation rate ($\dot{H}$ and $\dot{M}$) by

$$L\left(H, M, \dot{H}, \dot{M}\right) = \frac{L_0}{\mu_0} \mu\left(H, M, \dot{H}, \dot{M}\right) = \frac{N^2 h}{2\pi} \mu\left(H, M, \dot{H}, \dot{M}\right) \ln\frac{r_2}{r_1}, \qquad (1)$$

where the magnetic permeability, μ, is a function of H, M, $\dot{H}$ and $\dot{M}$, L_0 is the inductance of the coil without a core. If one neglects the deplacement current, Ampere's law gives

$$H = nI = \frac{L_0}{NS\mu_0} I = \frac{N}{2\pi(r_2 - r_1)} \ln\frac{r_2}{r_1} I. \qquad (2)$$

3. Experimental Results

In the first operating mode the capacitor was 48µF and was charged to $V_0=300$V. The voltage waveform across the capacitor and across the resistor are presented in Fig. 2 and Fig. 3, respectively.

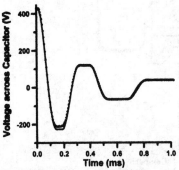

Fig. 2 Voltage across the capacitor
(———) simulated, (•) experimental

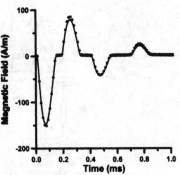

Fig. 3 Voltage across the resistor
(———) simulated, (•) experimental

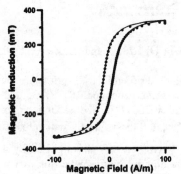

Fig. 4 Hysteresis loops for the standard core
(———) simulated, (•) experimental

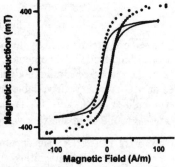

Fig. 5 Hysteresis loops for diagnosed core
(———) standard core, (•) diagnosed core

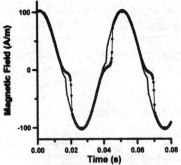

Fig. 6 Voltage across the sensing current
(———) simulated, (•) experimental

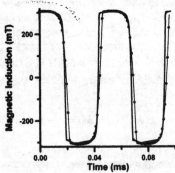

Fig. 7 Magnetisation versus time
(———) simulated, (•) diagnosed core

Any difference in the shape of experimental curves is analysed. Figs. 4 and 5 present a comparison between a sample having very close magnetic properties and dissimilar magnetic properties from those of the standard core.

The voltage waveform across the sensing current resistor and the waveform of magnetisation, obtained by integrating the induced voltage in the secondary coil in the second operating mode for the standard core are presented in Fig. 6 and Fig. 7 respectively. The optimum set of parameters [3], chosen to fit the Jiles-Atherton model experimental curves for the standard core, is used to calculate the standard hysteresis loop. Fig. 8 and Fig.

9 present the experimental and simulated hysteresis loops obtained in the second operating mode at v_f=20Hz at V=25V (value of the AC amplitude of the electromotive force). Fig. 8 presents the hysteresis loop of the sample having close and Fig. 9 dissimilar magnetic properties to the standard core.

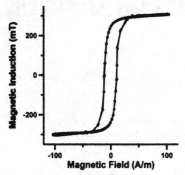

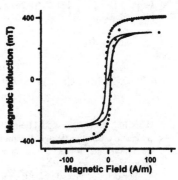

Fig. 8 Hysteresis loops for the standard core   Fig. 9 Hysteresis loops for diagnosed core
(———) simulated, (•) experimental   (———) standard core, (•) diagnosed core

In all the experimental hysteresis loops saturation is attained but the variation rate of the magnetic field is different; consequently, the power losses and the coercive fields are different. In the Jiles-Atherton model to display the hysteresis loops in different experimental conditions it is necessary to choose carefully the model's parameters. So, the parameters used in the modelization of the RLC circuit at high frequency and high value of the applied magnetic field: M_s=6.0x10^{4}A/m, a=100A/m, c=0.12, k=10.3A/m, α=4.2x10^{-4} differ from those used in the modelization of the RL circuit at low frequency and low value of the applied magnetic field: M_s=3.2x10^{5}A/m, a=41.82A/m, c=0.133, k=39.27A/m, α=3.63x10^{-4}.

4. Conclusions

Despite the intricacy of the modelization of the magnetisation processes our work proves that the Jiles-Atherton model can be used in typical conditions to calculate the hysteresis loops, using parameters carefully chosen. The experimental and simulated waveform across the components of the circuit for the standard and tested core allow the diagnosis of the core magnetic properties. If the magnetic properties of the core under the test differ from expected then it is rejected.

References

[1] Al. Stancu, O. Caltun, P. Andrei, Mixed models of hysteresis in magnetic cores, *Proceeding of 7th International Conference on Ferrite*, Bordeaux, 1996.

[2] D.C. Jiles and D.L. Atherton, Theory of ferromagnetic hysteresis, J. Magn. and Magn. Mater., vol. 61, (1986) 48-60.

[3] O. Caltun, C. Papusoi et al., Magnetization Processes in Soft MnZn Ferrite, Anal. Univ. Iasi, tom XLI-XLII, fasc.2, Solid State Phys., (1995) 37-46.

Homogeneity of Magnetic Properties of FINEMET Before and After Heat Treatment

D. Ramin and W. Riehemann

IWW, TU Clausthal, Agricolastr. 6, 38678 Clausthal-Zellerfeld, Germany

Abstract. Previous investigations of FINEMET have shown serious variations of the magnetic properties though the amorphous material has been heat treated identically. Structural fluctuations in the as quenched amorphous ribbon were assumed to be the main cause for this effect.

For detailed investigation the magnetic properties of various narrow samples cut out of an amorphous $Fe_{73.5}Si_{13.5}Cu_1Nb_3B_9$-ribbon have been measured using a digital data recorder before and after nanocrystallisation under identical conditions. Neighbouring specimens were compared in the lateral as well as in the longitudinal direction regarding their power losses and the magnetic coercitivities at polarisations between 0.6 T and 1 T in the frequency range from 3.2 Hz to 20 kHz.

Correlation of the measurements before and after nanocrystallisation indicate the influence of structural differences in the amorphous material on the magnetic properties of FINEMET.

1. Introduction

Investigations of the properties of a new material often require that measurements be performed on small samples. In this case, it is essential that the sample's properties represent that of the whole material, otherwise the obtained results cannot be attributed to the overall properties of the material itself. Inhomogeneities may result in serious deviations of a measurement from an averaged value especially if small sample sizes are employed. Previous experiments investigating the influence of surface treatment of FINEMET [1-4] on the cycle losses have shown that the measured values of different specimens were distributed over a wide range, although identical treatments were applied. This was observed for frequencies ranging from 3.2 Hz to 20 kHz at polarisations between 0.6 T and 1 T.

2. Experimental

Specimens 110 mm long and 3 mm wide were cut from an amorphous melt spun $Fe_{73.5}Si_{13.5}Cu_1Nb_3B_9$-ribbon of 14.9 mm width, 21 µm thickness, and a length of several meters. The ribbon was supplied by Vacuumschmelze, Hanau. During the following preparation process it was ensured that the sample's position in the tape could be restored in order to relate the magnetic properties of the specimen to their original position. In the amorphous state the magnetisation losses were obtained by measuring hysteresis loops at polarisations ranging from 0.3 T to 1 T and frequencies between 3.2 Hz and 20 kHz. This was done by an automated computer controlled system of a synthesizer, a linear amplifier, a coil system, and an analogue-digital data recorder. The coil system consists of several coils

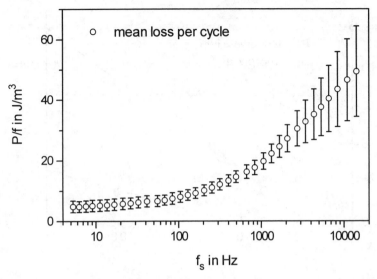

Fig. 1: Mean value of total cycle losses at a polarisation of 0.8 T for 13 samples of FINEMET annealed at 580° C for 1 hour plotted vs. the frequency f_s reduced to a sinusoidal waveform [5]. The error bars indicate the standard deviation of a single sample.

of different winding numbers to ensure optimal measuring conditions over a wide range of frequencies. The earth's magnetic field was compensated by modified Helmholtz coils and the sample's temperature was stabilised at 30° C. After collecting all necessary data about the amorphous samples they were sealed under vacuum in quartz tubes. A heat treatment of 580° C for one hour was applied to the specimens followed by quenching into cold water. Finally the magnetisation losses were measured analogously to the measurements of the amorphous material.

3. Results

Plotting the magnetic losses of FINEMET versus the frequency, as shown in fig. 1, for a polarisation of 0.8 T a high deviation of losses of a single sample from the mean value has been detected at all frequencies. This behaviour was observed in previous investigations after various preparation techniques, for example heat treatments between 560° C and 590° C and different quenching procedures.

Although the magnetic properties of the amorphous material and the nanocrystallised one are generally quite different, it was found that the total losses of the material in either state differ only by a small factor ranging from two to four, compared to the magnetic field necessary for the appropriate polarisations. This field can be a hundred times higher for the amorphous phase relative to FINEMET. As illustrated in figure 2 there is a good linear dependence between the polarisation and the total cycle losses both before and after nanocrystallisation. This relationship was observed at all measured frequencies and holds for most specimens.

Plotting the total losses of the amorphous samples at a frequency of 250 Hz and a polarisation of 0.8 T against the specimen's original position on the ribbon, it can be seen

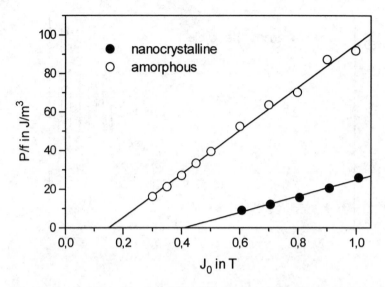

Fig. 2: Total cycle losses of a single sample of $Fe_{73.5}Si_{13.5}Cu_1Nb_3B_9$ before and after a heat treatment of 580° C for one hour at a frequency of 250 Hz plotted vs. the maximum polarisation.

that the losses vary gradually from sample to sample (fig. 3, left side). The change of losses along the longitudinal direction which corresponds to the melt spinning direction is much smaller than in the perpendicular direction. This behaviour is also observed in FINEMET (Fig. 3, right side) but it is not as distinctive as in the amorphous state, which might be caused by a combination of effects arising from variations of the nanocrystallisation process.

It can be noticed that the cycle losses of FINEMET correlate with those in the amorphous phase especially if they are plotted in their consecutive order as shown in figure 4. This dependence cannot be described by a simple continuous function but might be better understood if all acquired data and other information i.e. saturation magnetisation or sample thickness are taken into account.

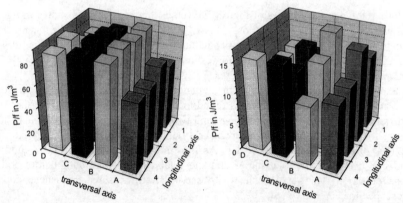

Fig 3: Total cycle losses of $Fe_{73.5}Si_{13.5}Cu_1Nb_3B_9$ at 250 Hz and a maximum polarisation of 0.8 T before (left) and after (right) heat treatment at 580° C for an hour according to the samples original position on the melt spun ribbon. Unit length: longitudinal 110 mm, transversal 3 mm.

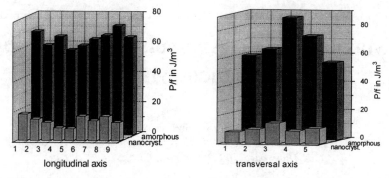

Fig 4: Total cycle losses of $Fe_{73.5}Si_{13.5}Cu_1Nb_3B_9$ at 250 Hz and a maximum polarisation of 0.8 T before and after heat treatment at 580° C for an hour according to the samples original position on the melt spun ribbon. The longitudinal axis corresponds to the direction of melt spinning. Unit length: longitudinal 110 mm, transversal 3 mm.

4. Conclusions

According to the variation of the sample's location in the melt spun ribbon it was found that the magnetic losses of $Fe_{73.5}Si_{13.5}Cu_1Nb_3B_9$ before and after nanocrystallisation vary gradually with position in a similar manner. Although the losses before and after nanocrytallisation are strongly dependent on each other this cannot be described by a simple continuous function. Future experiments will be made using tapes of different thickness and surface structure. Further efforts will be made to find a better way of correlating the magnetic properties of the amorphous phase to those of FINEMET.

Acknowledgements

The authors wish to thank Dr. G. Herzer and Dr. H.R. Hilzinger, Vacuumschmelze, Hanau for providing us with amorphous $Fe_{73.5}Si_{13.5}Cu_1Nb_3B_9$-tapes. This work was supported by Deutsche Forschungsgemeinschaft.

References

[1]   Y. Yoshizawa, S. Oguma, and K. Yamauchi, New Fe-based soft magnetic alloy composed of ultrafine grain structure, *Appl. Phys.* **64** (1988) 6044-6064

[2]   G. Herzer and H. Warlimont, Nanocrystalline soft magnetic materials by partial crystallization of amorphous alloys, *Nanostructured Materials*, **1** (1992) 263-268

[3]   C. Wittwer and W. Riehemann, Dynamic losses in nanocrystalline finemet for various annealing temperatures, *J. Magn. Magn. Mat.*, **133** (1994) 287-290

[4]   D. Ramin and W. Riehemann, Loss Improvement in Finemet by Surface Modification, *Mat. Sci. Eng.*, accepted for publication, 1996

[5]   B. Weidenfeller and W. Riehemann, Power Loss in Grain Oriented Fe-Si Sheets under Non-Sinusoidal Induction Signal, this volume, 1997

602
Non-Linear Electromagnetic Systems
V. Kose and J. Sievert (Eds.)
IOS Press, 1998

Magnetic Field Analysis of Anisotropic Permanent Magnets

Masato ENOKIZONO and Tsuyoshi TSUZAKI

*Dept. of Electrical and Electronic Engineering, Faculty of Engineering,
Oita University, 700 Dannoharu, Oita, 870-11, Japan*

Abstract. The numerical analysis of the magnetization situation in permanent magnets was carried out, using the improved finite element method with the VMSW model. It was clarified that the magnetization affects the performance of the permanent magnet motor. Moreover, a few magnetizer models were originated and the optimal distribution of magnetization was examined.

1. Introduction

In general, magnetic properties of permanent magnets are explained by means of the magnetization vector in materials. It is therefore important to consider the distribution of magnetization in magnets when we carry out analysis of electrical machinery using permanent magnets. However in the case of the magnet has a strong anisotropy, it is difficult to measure the magnetization in the hard-direction, and to express the value as vector quantity. To solve the such problem, it is significant to established an analytical method for the magnetizing situation of the anisotropic hard magnetic material, and also to find out distribution of magnetization in permanent magnets by using numerical methods. The improved finite element method was used with the VMSW (Variable Magnetization and Stoner-Wohlfarth) model[1] to carried out the numerical analysis of the magnetization.

2. Method

2.1 Calculation of initial magnetization

A general spheroidal magnetic body in a uniform external field H_0 is shown as in Fig. 1. When the magnetization M is induced by an external field H_0, the volumetric energy can be written as,

$$U = \frac{M^2}{2\mu_0}\left\{N_a + (N_b - N_a)\sin^2(\theta - \phi)\right\} + \frac{M^2}{2}\left(\frac{\cos^2\theta}{\chi_e} + \frac{\sin^2\theta}{\chi_h}\right) - MH_0\cos(\Omega - \theta) , \tag{1}$$

where χ_e and χ_h are the magnetic susceptibilities in the direction of easy axis and hard axis. Both N_a and N_b ($N_b > N_a$) are demagnetizing factors in the a-axis and b-axis, and Ω, θ, ϕ, are defined as shown in Fig. 1. The detailed description of the calculation of demagnetizing factor [1] is omitted here. The principal axes of the demagnetizing field are named the a-axis and the b-axis, and N_a and N_b are the eigenvalues of the tensor.

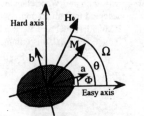

Fig. 1 Definition of notations.

The angles are chosen so that the a-axis makes an angle ϕ and H_0 makes an angle Ω with the e-axis (easy axis).

The total energy is minimized by the conditions,

$$\frac{\partial U}{\partial M} = 0 , \quad \frac{\partial U}{\partial \theta} = 0 .$$ (2)

The following equations can be obtained by (1) and (2)

$$\left\{ N_a + (N_b - N_a)\sin^2(\theta - \phi) \right\} \frac{M}{\mu_0} + \left(\frac{\cos^2\theta}{\chi_e} + \frac{\sin^2\theta}{\chi_h} \right) M - H_0 \cos(\Omega - \theta) = 0$$ (3)

$$\frac{1}{2\mu_0} (N_b - N_a) M \sin 2(\theta - \phi) + \frac{1}{2} \left(\frac{1}{\chi_h} - \frac{1}{\chi_e} \right) M \sin 2\theta - H_0 \sin(\Omega - \theta) = 0 .$$ (4)

These two equations form a simultaneous equation, which can be solved by utilizing two curves, M_e-H_e and M_h-H_h curves. These curves are the initial magnetizing curves. Therefore, initial magnetizing process can be analyzed in the above way. Equation (3) is the Stoner-Wohlfarth equation with two anisotropic fields: the shape magnetic anisotropic field and the intrinsic anisotropic field.

2.2 Calculation of residual magnetization

As the magnetizing field decreases, both M and θ decrease. Finally, when the magnetizing field approaches zero, M and θ become M_r and θ_r, as shown in Fig.2. The residual angle θ_r is the angle between the residual magnetization M_r and the e-axis. It occurs when the direction of H_0 is not parallel to the e-axis as shown in Fig.2. The relationship between θ and θ_r can be expressed as follows:

$$\tan\theta_r = \left(\frac{1}{R_s(e)} - 1 \right) \tan\theta ,$$ (5)

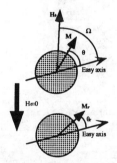

Fig. 2 Occurrence of residual angle.

where $R_s(e)$ is the rectangular ratio of magnetization curve in the easy direction. θ_r can be obtained by Equation (5). When the residual angle occurs, we assume that the e-axis of the representative particle inclines to the applied field direction by θ_r. The residual magnetization is given by the following equations.

$$M_r(\theta_r) = M_m R_s(\theta_r) / \cos(\omega - \theta_r - \phi)$$ (6)

$$\frac{1}{2} H_a(M_m) \sin 2\phi - H_m \sin(\omega - \theta_r - \phi) = 0$$ (7)

with

$$R_s(\theta_r) = \sqrt{R_s^2(e)\cos^2\theta_r + R_s^2(h)\sin^2\theta_r}$$ (8)

$$H_a(M_m) = \frac{M_m}{M_s} H_a(M_s) ,$$ (9)

where, R_s is the degree of particle alignment and $R_s(h)$ is in the hard direction. The value of M_r can be obtained by using the Storner-Wohlfarth equation of the demagnetizing curve. In the way described above, the magnetizing process should be analyzed at all elements. As a result, the final distribution of the magnetization in the magnetized magnet can be derived, and the magnetic field produced by the magnet can be calculated.

3. Results and Discussion

Three kinds of magnetizer models are shown in Figs.3(a), (b) and (c). Model-1, Model-2 and Model-3 have 4 poles, 6 poles and 8 poles, respectively. The magnetic material is a Sr-ferrite bonded magnet, $M_s = 0.289[T]$, $R_s(e) = 0.781$, $R_s(h) = 0.318$, the anisotropic field $H_a(M_m) = 1.6/\mu_0$ [A/m]. The achieved magnetization are shown in Fig.4.   q is the inclination angle from the x-axis (Fig.4). The distribution of the magnetization vectors after the demagnetizing process in the permanent magnet is shown in Fig.5.   The initial magnetizing curve used in this analysis is shown in Fig.6.   By using these results, a permanent magnet motor is shown in Fig.7 was analyzed. The flux distribution in the permanent magnet motor are shown in Fig.8.   Figs.8(a), (b) and (c) shows the result of using Model-1, Model-2 and Model-3, respectively. The waveforms of the cogging torque are shown in Fig.9.

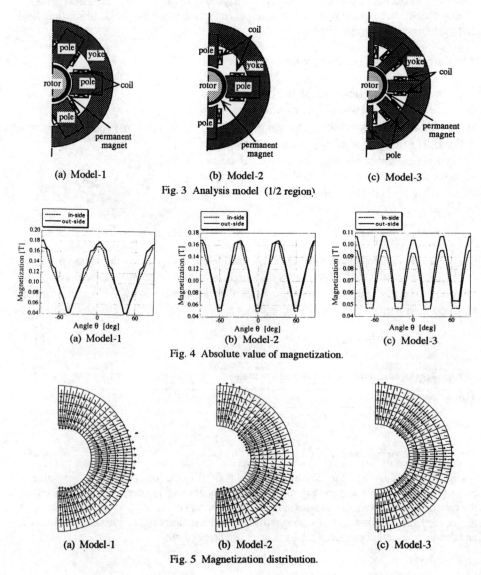

(a) Model-1   (b) Model-2   (c) Model-3

Fig. 3 Analysis model (1/2 region)

(a) Model-1   (b) Model-2   (c) Model-3

Fig. 4 Absolute value of magnetization.

(a) Model-1   (b) Model-2   (c) Model-3

Fig. 5 Magnetization distribution.

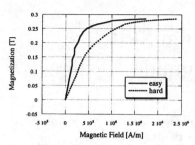

Fig. 6 Initial magnetizing curve.

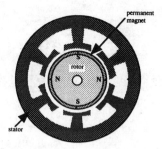

Fig. 7 Permanent magnet motor.

(a) Model-1   (b) Model-2   (c) Model-3

Fig. 8 Flux distributions in permanent magnet motor.

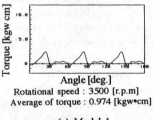
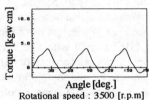

Rotational speed : 3500 [r.p.m]
Average of torque : 0.974 [kgw•cm]

Rotational speed : 3500 [r.p.m]
Average of torque : 1.863 [kgw•cm]

Rotational speed : 3500 [r.p.m]
Average of torque : 0.797 [kgw•cm]

(a) Model-1   (b) Model-2   (c) Model-3

Fig. 9 Wavefoms of the cogging torque.

4. Conclusions

The modified finite element method with the VMSW model can be applied to anisotropic permanent magnets, the transition of the cogging torque, which is calculated by using the magnetization distribution obtained from the analysis of the different magnetizing process were presented.

References
[1] M. Enokizono, K. Matsumura, and F Mohri, "Magnetic Field Analysis of Permanent Magnet Problems by Finite Element Method,", Proceedings of the Seventh Biennial IEEE Conference on Electromagnetic Field Computation (1996)

Non-Linear Electromagnetic Systems
V. Kose and J. Sievert (Eds.)
IOS Press, 1998

Analog Behavioural SPICE Macromodelling of Magnetic Hysteresis

Adrian MAXIM

Department of Electronics, Technical University "Gh. Asachi" Iasi 6600 Romania

Danielle ANDREU, Jacques BOUCHER

Department of Electronics, E.N.S.E.E.I.H. - I.N.P. University Toulouse 31071 France

Abstract. This paper presents a new method of magnetic hysteresis SPICE macromodelling, based on an electric equivalent circuit of electromagnetic laws. The B-H anhysteretic non-linear curve is piece-wise-linear approximated with a "look-up table" controlled source, and the hysteresis is introduced with a current source, that shifts the anhysteretic curve with the coercive force to either side of the B axis. Using the Analog Behavioural Modelling facilities of SPICE simulators, an accurate and highly computationally efficient model was developed.

1. Introduction

In the last decade, many mathematical hysteresis models have been developed but, unfortunately, few of them can be implemented in SPICE like electronic simulators [1]. Moreover, the currently used SPICE models are not general, as they give a good agreement with the experimental curves only for a determined family of magnetic materials.

Three methods of magnetic components SPICE modelling exist: using the linear or non-linear intrinsic models, developing an electrical equivalent circuit for the electromagnetic laws, and writing a C code model subroutine, that describes magnetic component behaviour.

Many of the currently used versions of SPICE simulator include the popular Jiles-Atherton model as an intrinsic model [2, 3]. It fits well with ferromagnetic materials behaviour, but cannot describe the abrupt B-H curves of amorphous materials. Moreover, in the highly non-linear applications long analysis times result and sometimes convergence problems appear.

The second widely used method of non-linear magnetic components SPICE modelling is to produce a structural equivalent electric circuit for the electromagnetic laws. Its main advantage is that the resulting model is made up of only standard SPICE elements and therefore it is portable in all SPICE 2G6 compatible simulators.

The simplest structural macromodel was obtained by adding two polynomial current sources in parallel with an ideal inductor, in order to introduce the hysteretic behaviour [4].

Another structural macromodel was introduced in Intusoft SPICE [5]. It approximates the anhysteretic B-H curve with two linear branches, generated with a piece-wise-linear resistive circuit. The hysteresis is introduced with two diode depletion capacitances, that shift the anhysteretic curve, due to the capacitor's charging and discharging processes. Although this macromodel assures a low analysis time, it gives a rather poor accuracy and introduces errors through using a false dependence of coercive force on the flux density level. Also, it cannot simulate a magnetic core working only in the saturation regime.

The C code models of magnetic components are faster and more flexible than the structural macromodels, but they require an open SPICE simulator and a high skill of C programming. Moreover, the C code modelling has not been yet standardised and thus the resulting models are not portable. The most representative C code models are : the Tabrizi model [6], implemented in the SPICE PLUS simulator, that uses two hyperbolas to model the B-H loop and the Chan model [7], implemented in the DSPICE simulator, that uses also two hyperbolas to model the B-H loop, but which includes the rate and temperature dependencies.

A simple and efficient way to develop an accurate model of magnetic hysteresis is to use the Analog Behavioural Modelling (ABM) facilities of modern SPICE like simulators [2].

The ABM enables the direct implementation of algebraic and differential equations with "in line equation" controlled sources and the piece-wise-linear approximation of non-linear dependencies with "look-up table" controlled sources. Thus, the electromagnetic laws can be modelled with non-linear controlled voltage and current sources, that form a behavioural macromodel representation of the magnetic component.

The ABM method is general enough as it allows the piece-wise-linear approximation of the B-H loop and hence it simulates virtually all kinds of magnetic materials. Another great advantage is that the ABM macromodels keep the circuit representation of the model, which is easier to be understood and modified by the user, than the C code models.

As amorphous magnetic materials are increasingly used, a new SPICE macromodel for magnetic components, that can simulate abrupt B-H hysteresis curves must be developed.

2. Magnetic hysteresis behavioural macromodel description

This new SPICE macromodel of hysteresis is based on an electric dual of magnetic quantities. Thus, the electromagnetic laws are modelled with an electric equivalent circuit, that can be analysed with SPICE type simulators. The detailed behavioural macromodel of magnetic hysteresis is presented in Fig.1.

Faraday's electromagnetic equation is modelled with the Gfaraday "in line equation" controlled current source, that calculates the integral of the input terminal voltage (V(in)):

$$v = \frac{d}{dt} \iint_S B dS \rightarrow B = \frac{1}{N * AREA} \int v(t) dt \quad \rightarrow \quad Gfaraday = SDT(Vin) / (N * AREA) \quad (1)$$

where AREA is the mean magnetic cross-section and N is the windings turns ratio.

This current source delivers a unity resistance (RBtemp=1Ω) and gives a corresponding voltage (V(Btesla)) in Volts, equal to the flux density in Tesla. Defining the TC1 linear and TC2 quadratic thermal coefficients of RBtemp resistor, the V(RBtemp) voltage describes the parabolic ambient temperature dependence (Tamb) of the magnetic flux density :

$$B(Tamb) = V(RBtemp) = Gfaraday \cdot [1 + TC1 \cdot (Tamb - Tnom) + TC2 \cdot (Tamb - Tnom)^2] \quad (2)$$

The non-linear anhysteretic B-H curve is modelled with GB_Hanhyst "look-up table" ABM source [2], that performs a piece-wise-linear approximation of the experimental curve. The major advantage of this new behavioural macromodel is that it assures a point-by-point description of the B-H curve, and thus it can describe any kind of magnetic material.

To introduce the hysteresis, the anhysteretic B-H curve must be shifted with the coercive force (Hc) to either side of the B axis, for rising and respectively falling flux density values. The EBrise_fall "in line equation" source [2] computes the derivative of the flux density and the Ghyst "If-Then-Else" source [2] gives a current equal to Hc coercive force for rising flux density and equal to -Hc for falling flux density values.

The current delivered by Ghyst "If-Then-Else" source is injected together with the GB_Hanhyst current in the RHtemp unity resistor. Thus, the voltage V(RHtemp)=V(Ha_m), that is equal in Volts to the magnetic field intensity in A/m, exhibits an hysteretic variation. In the same way, the TC1 linear and TC2 quadratic thermal coefficients of the RHtemp resistor model the parabolic field intensity dependence on the ambient temperature :

$$H(Tamb) = (GB\_Hanhyst + Ghyst) \cdot (1 + TC1 \cdot (Tamb - Tnom) + TC2 \cdot (Tamb - Tnom)^2] \quad (3)$$

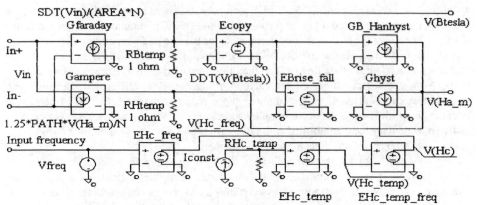

Fig. 1. New ABM non-linear SPICE macromodel of magnetic hysteresis

Finally, the Ampere law is modelled with the Gampere "in line equation" ABM source, that imposes the current through the input winding :

$$N * I = \oint_l Hdl \quad \rightarrow \quad Gampere = I = \frac{0.4 * PATH * \pi}{N} * H \tag{4}$$

where PATH is the mean magnetic path length.

An important feature of a magnetic component macromodel is the ability to model accurately the frequency dependence of the B-H hysteresis curve. The classic Jiles-Atherton model accounts for the frequency dependence of the coercive force (Hc), but it gives a non-realistic frequency dependence of the linear regime permeability (μLIN), when simulating amorphous magnetic materials.

The frequency of the input signal is delivered to the macromodel as a parameter, that is transformed in a voltage by a Vfreq independent source. This voltage is used to control the EHc_freq "look-up table" ABM voltage source, that performs a piece-wise-linear approximation of the experimental coercive force versus the frequency characteristic.

The coercive force exhibits a non-linear temperature dependence, that cannot be linear or parabolic approximated. To model such a dependence, firstly a linear temperature dependence is generated with a constant current source Iconst=1A, that delivers the RHc_temp unity resistor, with TC1=1. This voltage is further used to control the EHc_temp "look-up table" voltage source, that approximates piece-wise-linear the experimental coercive force versus the temperature characteristic. Finally, the V(Hc) voltage, that includes both the frequency and the temperature dependencies, is used by the Ghyst source to generate the hysteresis loops.

3. Simulation results

Using the PSPICE 7.1 circuit simulator, the proposed ABM non-linear core macromodel was compared in terms of accuracy and analysis time with currently used SPICE core models.

Fig. 2 presents the simulated and the experimental temperature dependencies of the normalised magnetic flux density and field intensity, for the orthonol amorphous material [8]. Fig. 3 presents the simulated

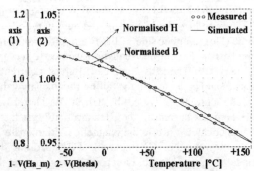

Fig. 2. The simulated and experimental temperature dependencies of the magnetic field intensity and flux density

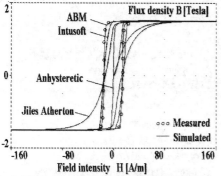

Fig. 3. The B-H loops obtained with classical and new ABM SPICE core models

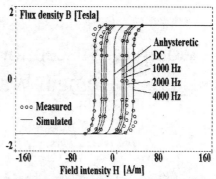

Fig. 4. The B-H loops obtained with the ABM macromodel for different input signal frequencies

B-H curves, obtained with the new ABM macromodel, the Jiles-Atherton model and the Intusoft structural macromodel. Fig. 4 shows the simulated and the experimental B-H hysteresis curves for different frequencies of the input signal.

It can been observed that this new ABM macromodel greatly increases the simulation accuracy of amorphous magnetic materials, with a comparable analysis time.

4. Conclusions

This new behavioural macromodelling method performs a piece-wise-linear approximation of the hysteretic B-H curve, instead of the empirical equations' fitting used in existing models. Therefore the model is able to simulate any kind of magnetic material. Moreover, the data sheets' parameters (e.g. Hc, Br, Path, Area, etc.) are directly specified as model parameters and thus the parameters extraction algorithm was eliminated.

The behavioural macromodel is made up only of low level SPICE devices, which leads to a high computational efficiency, with no convergence problems.

Second order effects such as temperature variations of magnetic field intensity and flux density, as well as the frequency dependence of the B-H hysteresis loop have been included.

The present behavioural macromodel simulates only the major B-H loops. The minor loops can be further modelled by considering more sophisticated electromagnetic differential equations, that can be easily implemented in modern SPICE simulators with the SDT time integrator and DDT time derivative predefined ABM functions.

This ABM method of magnetic components' modelling is portable and can be used in all the SPICE simulators that have ABM facilities : PSPICE, IsSPICE, Berkeley SPICE3, etc.

References

[1] M. Takach, P. Lauritzen, Survey of magnetic core models , Proceeding of IEEE PESC'95, pp. 560-566.

[2] *** Microsim PSPICE A/D reference manual, Microsim Corporation, October 1996.

[3] D. Jiles, D. Atherton, Theory of ferromagnetic hysteresis, Journal of Magnetism and Magnetic Materials, no. 61, 1986, pp. 48-60.

[4] D. Pei, P. Lauritzen, A computer model of magnetic saturation and hysteresis for use on SPICE 2 , IEEE Trans. on Power Electronics, vol. 1, no. 2, April 1986, pp. 101-110.

[5] I. Meares, C. Hymowitz, Improved SPICE model simulates transformer's physical process, EDN August 1993, pp. 105-110.

[6] M Tabrizi, The nonlinear magnetic core model used in SPICE Plus, Proceeding of APEC'87, pp. 32-36.

[7] J. Chan, A. Vladimirescu, X. Gao, P. Leibmann, J. Valainis, Non-linear transformer model for circuit simulation, IEEE Trans. on Computer Aided Design, vol. 10, no. 4, April 1991, pp. 476-482.

[8] *** Amorphous material cores data sheets, Magnetics Inc. 1995, pp. 41-54.

Complex Susceptibilities Associated with Domain Wall Relaxation

C.St. Bădescu, Margareta Ignat*, V.Stancu**

Inst. of Tech. Phys., 47 Mangeron Blvd., Iaşi 6600, Romania
*Dept. of Phys., "Al.I.Cuza" Univ., 11 Copou Blvd., 6600 Iaşi, Romania
**Dept. of Phys., "Gh.Asachi" Techn. Univ., 71 Mangeron Blvd., 6600 Iaşi, Romania

Abstract. In this work we analyze the power losses by domain wall motion damping in ferromagnetic samples of low quality factors Q. The model used takes into account the structure of the domain wall, which can be modified by a magnetic field transverse to the anisotropy axis, and it allows a comparison between viscous and eddy current damping. The spectrum of the dissipated power is obtained for different values of the fields and the power of the main harmonic is analysed as a function of the transverse field's intensity and orientation, in circumstances of missing precession.

1. Introduction

When the displacements of a domain wall (DW) in a periodic coercive field are small, the response of the system to an ac magnetic field parallel to the anisotropy axis is the solution of a second order differential equation for a damped harmonic oscillator [1]. The dissipation can be deduced from the complex susceptibility of the system. When the displacements of the DW are large, not only the oscillations are nonlinear, but also the system must be described by two dynamic variables: the position of the wall q and the azimuth Ψ of the magnetization at the DW level. Instead of a susceptibility associated with a DW movement with the frequency of the field, we will deal with a large spectrum of susceptibilities associated with different Fourier components of the nonlinear oscillations.

The case of ferromagnetic films with perpendicular anisotropy and characterized by Q factors very close to 1 is treated here, for which the models used for bubble materials [1] are not suitable. We consider an appropriate model [2], which is an extention of [3] for including ac fields, coercive fields and eddy current damping. The latter was discussed in [4] and [5] in connection with the chaotic motion of a DW, but in a theoretical frame which could not include the transverse fields. The model [2] is used for obtaining temporal series of q, which lead to the susceptibility spectrum and then to the dissipated powers for different applied field values. The change in the spectrum under the action of a transverse field and the competition between the viscous damping and that by eddy currents are discussed.

2. The model

We consider an uniaxial ferromagnet characterized by saturation magnetization M_S, anisotropy constant K, and exchange constant A. The quality factor and the anisotropy field are, respectively, $Q = 2K / \mu_0 M_S^2$ and $H_K = 2K / \mu_0 M_S$. We choose a reference system such that the planar DW moves along axis x. The DW is driven by an ac field $H_{//} \sin \omega_0 t$ which lies along the anisotropy axis. It is damped due to both the viscosity, characterized by the Gilbert constant α, and the eddy-currents, described by a geometrical constant a [6]. The DW moves in a periodic coercive field $H_c \sin(2\pi q / \ell)$. A continuous magnetic field $H_\perp$ is applied crosswise to the anisotropy axis, under an angle Ψ_H

relative to the DW. This field modifies the structure of the DW [3], producing a localized perturbation of the magnetization, which will make an azimuth Ψ with the DW, and it also determines the magnetization inside a magnetic domain to be oriented under a nonzero angle θ_0 or $\pi - \theta_0$ with respect to the anisotropy axis. The motion equations are derived in detail in [2]. They are:

$$
\left\{
\begin{aligned}
\dot{q} &= \frac{1}{(1 + \alpha^2 f_0 r_0)\sqrt{n_0 / r_0} + \alpha f_0 a}\left[\alpha f_0 h_{t//} + \left(\frac{f_0 \sin 2(\Psi + \Psi_H)}{2Q} + \right.\right. \\
&\qquad \left.\left. g_0\left(h_\perp \sin\Psi - \frac{\sin\theta_0 \sin\Psi_H}{Q}\cos(\Psi + \Psi_H)\right)\right)\right] \\
\dot{\psi} &= \frac{1}{(1 + \alpha^2 f_0 r_0)\sqrt{n_0 / r_0} + \alpha f_0 a}\left[\sqrt{n_0 / r_0}\,h_{t//} - (\alpha\sqrt{n_0 r_0} + a)\left(\frac{f_0 \sin 2(\Psi + \Psi_H)}{2Q} + \right.\right. \\
&\qquad \left.\left. g_0\left(h_\perp \sin\Psi - \frac{\sin\theta_0 \sin\Psi_H}{Q}\cos(\Psi + \Psi_H)\right)\right)\right]
\end{aligned}
\right. \tag{1}
$$

where a system of units was used in which $\Delta_B = \sqrt{K/A} = 1$ and $\omega_A = \gamma H_K = 1$, for the sake of generality. All fields are given in normalized values, $h = H / H_K$. Functions r_0, g_0, f_0, n_0 are:

$$
\left\{
\begin{aligned}
f_0 &= \left[\pi \sin^2\theta_0 + 2\cos^2\theta_0 + \sin\theta_0 \cos\theta_0 (\pi - 2\theta_0)\right] / 2\cos\theta_0 \\
g_0 &= \left[\pi \sin\theta_0 + \cos\theta_0 (\pi - 2\theta_0)\right] / 2\cos\theta_0 \\
r_0 &= \left[2\cos\theta_0 - \sin\theta_0 (\pi - 2\theta_0)\right] / 2\cos^2\theta_0 \\
n_0 &= \left[1 + \sin^2(\Psi + \Psi_H)/Q\right]f_0 - 2g_0\left[h_\perp \cos\Psi + \sin\theta_0 \sin\Psi_H \sin(\Psi + \Psi_H)/Q\right] + \\
&\quad + \pi/2\,tg\theta_0\left(2h_\perp + \sin\theta_0 \sin^2\Psi_H - \sin\theta_0\right)
\end{aligned}
\right. \tag{2}
$$

We consider a region of unit volume consisting of two domains, which at $t = 0$ lay from $-L/2$ to 0, respectively from 0 to $L/2$ on the x axis, at this moment the total magnetic moment being 0. The displacement q produces a total magnetic moment $M(t) = 2q(t)M_S \cos\theta_0 / L$. We work with the quantity LM, independent of L, with the following Fourier representation:

$$
LM(t) = \sum_{n=0,N-1} \frac{1}{N} M_L(\omega_n)\exp(-i\omega_n t) \tag{3}
$$

with $\omega_n = 2\pi n / N\delta$, δ the sampling time, and N the Nyquist critical number. We may use only the positive frequency part of (3), twice of which we denote by $LM_+(t)$:

$$
LM(t) = \mathrm{Re}(LM_+(t)) = \mathrm{Re}\left(\sum_{n=1,N-1} \frac{1}{N} H_0 \chi_L(\omega_n)\exp(-i\omega_n t)\right), \tag{4}
$$

where $\chi_L(\omega_n)$ is the susceptibility of the considered material region, $\chi_L(\omega_n) = 2M_L(\omega_n)/H_0$. Denoting by $\overline{w}_H = \mu_0 H_K^2 / 4$ the energy density of a magnetic ac field of strength H_K, we obtain for the normalized spectral component of the complex power:

$$
P_{L+}(\omega_n) = \frac{1}{\overline{w}_H}\int_0^{T_0} \frac{1}{T_0} H_+^*(t)M_+(\omega_n)d(\exp(-i\omega_n t)) = -4\omega_n h_{t//}^2 \chi_L \frac{\sin\pi(1 - \omega_n/\omega_0)}{\pi(1 - \omega_n/\omega_0)}\exp(-i\pi(1 - \omega_n/\omega_0)), \tag{5}
$$

in which the plus sign denotes the positive frequency part, and $H_+(t) = iH_0 \exp(-i\omega_0 t)$. The dissipated power in the unit volume of length L is the imaginary part of this expression.

3. Results

We used $h_C = 0.1$ as a typical value, $Q = 1.01$ as slightly different from 1, which is a singularity for the motion equations, and the normalized $\omega_0 = 10^{-4}$ corresponding to the MHz range.

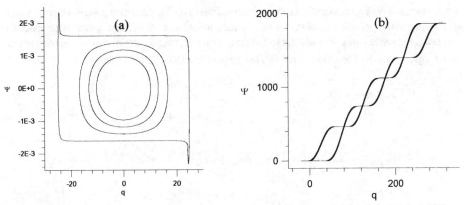

Fig.1 Phase space trajectories to 0 transverse field, for a) applied fields 0.06, 0.073, 0.086, 0.099 from inside to outside, and b) applied field value 0.105.

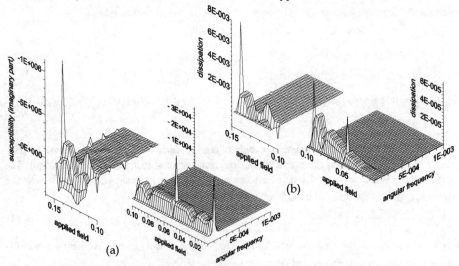

Fig.2 Spectra of a) Imaginary part of the susceptibility, and b) Dissipated power, for transverse field 0.8 parallel to the wall, Gilbert constant 0.01 and a=0.01.

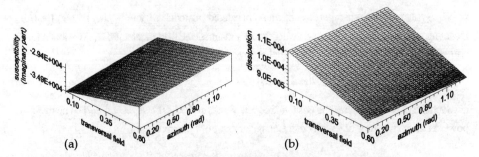

Fig.3 a) First harmonic amplitude of the susceptibility and b) First harmonic amplitude of the dissipated power for applied field 0.09.

The Nyquist frequency used in the Fast Fourier Transforms of q was $v_N = 50v_0$. All spectra are concentrated below $10 v_0$. The oscillations have completely different forms in the case in which the drive field is smaller than the coercive field from those in which the drive field is higher than h_C. This can be seen in Figure 1, which shows space phase trajectories for the drive field 0.06, 0.073, 0.86, 0.099 (a) and 0.105 (b), and in Figure 2, which shows the spectra of the imaginary part of the susceptibility and the dissipated power as functions of the drive field amplitude at constant $h_\perp$. One may see in Figure 1 that, while the precession does not occur below h_C and the oscillations are limited to the potential well, strong precessions appear immediately above h_C: the DW travels long distances and the magnetization rotates many times in the same direction. The precession does not appear at the Walker field $h_W = 2\alpha/Q = 0.0198$, which is smaller than h_C, thus the considerations from the stationary circumstances [3] do not apply here. The dissipation is fundamentaly different in the two ranges of $h_{//}$. Below h_C both the spectrum of the imaginary part of the susceptibility and that of the dissipated power are concentrated around the first harmonic (Fig. 2). Above h_C the spectra have more components, of much higher amplitudes - with two magnitude orders - so the description must be made in therms of a set of susceptibllities. We have observed the same behavior for any combination of $h_\perp$ and Ψ_H values.

In Figures 3(a) and 3(b) the first harmonic's variation versus $h_\perp$ and Ψ_H was followed for the imaginary part of the susceptibility and for the dissipation, just below h_C ($h_{//} = 0.09$). The dissipation decreases when increasing $h_\perp$, thus $h_\perp$ has a stabilizing effect, and it also decreases with Ψ_H. Simulations proved that this harmonic is dominant, the second harmonic being four orders of magnitude smaller.

The response is determined by the drive frequency ω_0, for this eigenfrequencies of the DW not being excited for the displacement oscillations, but only for the precession, when $h_{//} > h_C$. Some times, peaks appear like those in Figure 2, but these are isolated and are not a characteristic of the ferromagnet.

Simulations were also performed for different values of a, from α to 100α. The same border appeared between the two regimes. The damping by eddy currents increased the dissipation up to the order -2, in direct proportionality with a. At high conductivity this second mechanism is dominant. The response of the system is of the same kind: harmonics of the drive field are present and no eigenfrequencies are excited for the wall displacement.

In brief, the dynamics of a DW in a low anisotropy ferromagnet is a complex phenomenon and the dissipation is not due only to the DW displacements when the conductivity is small. The present model is an appropiate one for studying the power losses when the DW structure is modified by a transverse field.

References

[1] A.P.Malozemoff, J.C.Slonczewski, Magnetic Domain Walls in Bubble Materials, Academic Press, New York, 1979.
[2] C. Şt. Bădescu, Margareta Ignat, N.Rezlescu - A General Model for Power Dissipation by Domain Wall Movement in Thin Ferromagnetic Films, submitted to Z.Phys.B
[3] V.L.Sobolev, H.L.Huang, S.C.Chen, Domain Wall Dynamics in the Presence of an External Magnetic Field Normal to the Anisotropy Axis, JMMM. 147(1995) 284-298.
[4] H.Okuno, Y.Sugitany, Bifurcation Diagram and Large Energy Dissipation Caused by Chaotic Domain Wall Motion, IEEE Trans. Magn., 30(6) (1994) 4305-4307.
[5] C.Şt.Bădescu, M.Ignat, S.Oprişan, On the Chaotic Oscillations of Bloch Walls and Their Control, Chaos, Solitons and Fractals, 8(1) (1997) 33-43.
[6] H.J.Williams, W.Shockley, C.Kittel, Studies of the Propagation Velocity of a Ferromagnetic Domain Boundary, Phys. Rev. 80(6) (1950) 1090-1094.

FEM and BEM Calculations

Non-Linear Electromagnetic Systems
V. Kose and J. Sievert (Eds.)
IOS Press, 1998

Vector Control Simulation of Synchronous Reluctance Motor Including Saturation by Finite Element Method

*Jung-Ho LEE, *Sang-Baeck YOON, **Jung-Chul KIM and *Dong-Seok HYUN

Dept. of Electrical Engineering, Hanyang University, Seoul 133-791, KOREA
**LG Electronics Co., Living System Lab., Seoul 153-023, KOREA*

Abstract. This study investigates the vector control characteristics of a Synchronous Reluctance Motor (SynRM), with segmental rotor structure, using the finite element method which uses a moving mesh technique. The focus of this paper is general property and speed response characteristic of SynRM under magnetic non-linearity. Extensive experimental work has been conducted to verify the accuracy of the proposed analysis method.

1. Introduction

The influence of the saturation cannot be neglected since the high-speed operation and precision torque control of Synchronous Reluctance Motor (SynRM) are required. The q-axis usually shows minimal saturation for high saliency ratio, but the high permeance d-axis usually shows significant saturation when the machine is being operated at rated operations[1][2]. In the case of d-axis excitation, the saturation is the combined effect of saturation in the stator yoke, stator teeth, and rotor rib, and can reduce the inductance L_d, which has an effect on the torque, power factor and efficiency of SynRM, by as much as 50%. Therefore it is difficult to expect a good performance of SynRM without consideration of the above defects.

Fortunately, the advantages of FEM unlike voltage equation modeling are primarily their ability to model the complicated internal structure within a SynRM and ability to model magnetic saturation to a high degree of accuracy.

This paper presents the vector control characteristics analysis method of a SynRM with magnetic non-linearity using the 2-D FEM considered rotor motion.

Extensive experimental work has been conducted to verify the accuracy of the analysis.

2. Analysis Model

The geometry of the stator and rotor structure is shown in Fig.1. This rotor structure was chosen according to the proposed rotor structure of Vagatii[3], without any attempt at optimization, and because of its simplicity. The stator of Fig.1 is that of a standard 400 W, four-pole induction machine with a sinusoidally distributed double-layer winding.

Fig. 1 Components of SynRM

3. Voltage Source Finite Element Formulation With Rotor Motion

When the moving coordinate system is used, the governing equation in 2D is given as follows.

$$\frac{\partial}{\partial x}\frac{1}{\mu}(\frac{\partial A_z}{\partial x}) + \frac{\partial}{\partial y}\frac{1}{\mu}(\frac{\partial A_z}{\partial y}) = -J_z \tag{1}$$

Where, A_z : z component of magnetic vector potential J_z :   current density
        μ :   permeability

The circuit equation is written as

$$\{V\} = [R]\{I\} + [L_0]\frac{d}{dt}\{I\} + \{E\} \tag{2}$$

Where, $\{E\}$: E.M.F. vector in the winding $\{V\}$: supply voltage vector
        $\{I\}$: phase current vector

To solve (1), we used the Galerkin finite element method. For the time differentiation in (1) and (2), time stepping method is used with backward difference formula. Coupling (1) and (2), the governing equation is given as follows.

$$\left[\begin{bmatrix} \frac{1}{\mu}S & -N \\ 0 & R \end{bmatrix} + \frac{1}{\Delta t}\begin{bmatrix} 0 & 0 \\ LG^T & L \end{bmatrix}\right]\begin{Bmatrix} A \\ I \end{Bmatrix}_t = \frac{1}{\Delta t}\begin{bmatrix} 0 & 0 \\ LG^T & L \end{bmatrix}\begin{Bmatrix} A \\ I \end{Bmatrix}_{t-\Delta t} + \begin{Bmatrix} 0 \\ V \end{Bmatrix}_t \tag{3}$$

4. Simulation and experimental results

The torque acting on the SynRM at each time is calculated by the line integral of the Maxwell stress tensor. In order to improve the calculation accuracy, the air gap part is divided into three layers. The torque is calculated at each layer and averaged[4][5]. A direct convergence method, which is not diverged at any initial value and has the advantage of converging more quickly than the Newton-Raphson method, is used to provide a saturated field solution. An indirect field-oriented control algorithm is applied to the proposed analysis model for dynamic characteristics. This technique is illustrated in Fig.2.

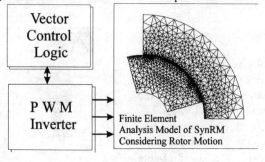

Vector
Control
Logic

P W M
Inverter

Finite Element
Analysis Model of SynRM
Considering Rotor Motion

Fig. 2 Scheme of simulation

The proposed analysis method is applied to the step speed command (500rpm) in vector control logic part. Fig.3 (A) ,(B) and Fig.4(A), (B) shows the speed, A-phase current and A, B-phase currents response of simulation and experiment. Experimental results closely match those obtained in the simulation results of the proposed method. Speed response of experiment, however, is slower than those of simulation because iron core loss reduces torque. Various investigations on the iron core loss consideration in the proposed method is underway, the result will be discussed in the near future.

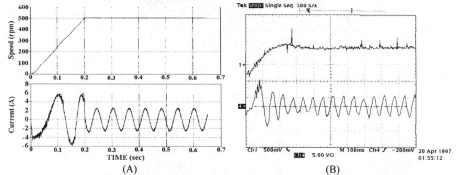

(A)   (B)

Fig.3 Speed and A-phase current response of simulation(A) and experiment(B)

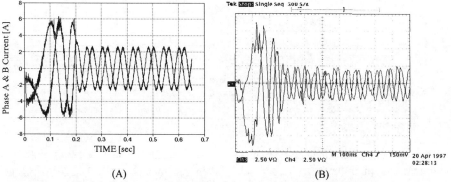

(A)   (B)

Fig.4 A, B-phase current response of simulation(A) and experiment(B)

Fig.5 shows flux plots of SynRM with rotor motion. This represents partial saturation on the stator teeth and rotor rib nearby the entrance and exit of flux barrier (1.8 - 2.1 T) at synchronous speed. Fig.6, Fig.7 represent the D-, Q- flux response and Ld, Lq inductance. It is known that Ld variation due to PWM and cogging torque of proposed SynRM (it is not skewed) is larger than Lq.

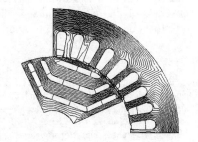

Fig. 5   Flux plots of SynRM with rotor motion

These large variations of Ld is a cause of torque pulsation as shown in Fig.8. In Fig.8 (A) the torque response is calculated by eq. $\lambda_d i_q - \lambda_q i_d$ and (B) is calculated by the Maxwell stress tensor method applied in this paper.

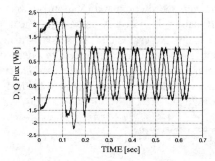

Fig.6 D-,Q-Flux response of simulation

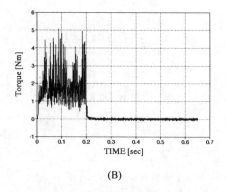

Fig.7 Ld, Lq parameters response of simulation

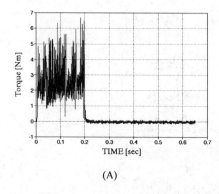

(A)   (B)

Fig.8 Torque response calculated by eq.($\lambda_d i_q - \lambda_q i_d$)(A)
and maxwell stress tensor method(B)

5. Conclusion

A dynamic characteristic analysis method has been proposed, which is suited to machines with magnetic non-linearity, in particular when parameters are affected by magnetic saturation.

Computer simulation and extensive experimental result for the SynRM show the effectiveness of the proposed method. The proposed method can be used as a numerical analysis tool to improve design procedure and for further control scheme development.

References

[1] R.E. Betz, R. Lagerquist, M. Jovanovic, T.J.E. Miller and R.H. Middleton, "Control of Sychronous Reluctance Machines ," IEEE Trans. on IAS, Vol. 29, No.6, pp.1110-1121, Nov.1993.

[2] Longya Xu, Jiping Yao,"A Compensated Vector Control Scheme of A Synchronous Reluctance Motor Including Saturation and Iron Losses," IEEE Trans. Ind. Appl., Vol. 28, No. 6, pp. 1330-1338, 1992.

[3] A. Vagati , G. Franceschini, I. Marongiu, G.P. Troglia: "Design Criteria of High Performance Synchronous Reluctance Motors", IEEE- IAS Annual Meeting, Houston, U.S.A., October 1992.

[4] Dal-Ho Im and Chang-Eob Kim, "Finite element force calculation of linear induction motor taking into account of the movement," IEEE Trans. Magn. , Vol.30, No.5, pp.3495-3498,1994.

[5] D.H. Im, B.I. Kwon, J.H. Lee, "Dynamic Analysis of Induction Motor with Tapped stator Winding for Vector Control using FEM," ISEM'95 pp. 140-143, Cardiff, Wales, UK, 17-20 September, 1995.

Non-Linear Electromagnetic Systems
V. Kose and J. Sievert (Eds.)
IOS Press, 1998

Shape Optimization of a 400W AC Servo Motor Using FEM with Advanced Evolution Strategy

Jei-Hoon Baek , Pan-Seok Shin
Dept. of Electrical Engineering, Hong-Ik University
Youngi-gun Chochiwon Chungnam 339-701, Rep. of Korea

Abstract. This paper presents an optimization method to find a shape of the permanent magnet and slot of the motor for minimization of cogging torque and maximization of efficiency. The magnetic field computation is done by Finite Element Method(FEM) and the torque calculation is performed by the virtual work method. For the shape optimization, multi-objective program technique and advanced Evolution Strategy(ES) are adopted. Simulation result has brought an optimized shape of the magnet and the slot of the motor. The performance of the optimized motor is compared with those of the initial model.

1. Introduction

AC servo motors with permanent magnet have many advantages like good controllability, compactness, easy maintenance, long life times and others. Especially, the closed slot type of the stator has merits of higher filling factor of coil and higher flux density in the air gap. Recently, applications of these kinds of motors are credibly increased in industries. In addition, the industries require the motor to be more compact and higher efficiency. So, this paper presents an optimized method of AC servo motor in the viewpoint of reduction of cogging torque and increase of efficiency.

For the magnetic field analysis of the motor, 2-dimensional FEM is employed. Virtual work method is also used to calculate magnetic forces and cogging torque of the model. For the shape optimization of the rotor surface and stator slot, multi objective program and advanced evolution strategy are adopted. The method is reported to have an excellent convergence characteristics regardless of local minimum[1],[2]. Multiobjective functions are chosen to minimize the cogging torque and to maximize the efficiency of the motor. For this study, a 400 W AC servo motor is selected as a simulation model, which consists of 9 poles on the stator and 8 poles of rotor with rare-earth permanent magnet. The main rates of the motor are speed of 3000 rpm, torque of 13 $kg \cdot cm$ and input voltage of 200 V.

2. Numerical Analysis of the Magnetic Field

To analyze a 2-dimensional magnetic field, a conventional FEM is used. The magnetostatic governing equation of the field can be described from Maxwell's equation and the magnetic vector potential A as follows;

$$\nabla \times (\nu \nabla \times A) = J_0 + \nabla \times \nu_0 M \qquad (1)$$

where ν_0 is reluctivity of material, J_0 is input current density and M is magnetization of permanent magnet. For finite element formulation, Galerkin's weighted residual method is applied to eq.(1). After discrete the week form of the eq.(1) using 2nd order triangular elements and their shape functions, a system matrix equation is obtained and solved for A. The flux density B is then calculated by the definition of $B = \nabla \times A$.

The torque on an object is found as the derivative of the magnetic coenergy with respect to position at constant current. For given set of current sources and materials, the magnetic co-energy, W_1', can be computed. If the object is moved by $\Delta\theta$ in the direction of the torque, the magnetic field could be resolved with the same currents and the coenergy, W_2', could be calculated. Then, the torque can be computed as;

$$T_\theta = \frac{W_2' - W_1'}{\Delta\theta} \tag{2}$$

In order to apply the virtual work method to compute torque, magnetic stored energy W_{mag} is calculated as follows.

$$W_{mag} = \int \int \int_v \frac{1}{2} B \cdot H dv = \frac{1}{2\mu}[A]^T[S][A] \tag{3}$$

where [A] is vector potential and [S] is stiffness matrix obtained from eq. (1). It is well known that the energy functional could be directly differentiated with respect to a virtual displacement. Thus, to compute torque is to differentiate eq.(3) with respect to θ. In this case, A is constant due to the requirement of constant flux linkages. If the region under consideration is linear, then $\frac{\partial \nu}{\partial \theta} = 0$, the expression for torque becomes

$$T_{\theta_e} = \frac{\partial W_{mag}}{\partial \theta} = -\frac{\nu_e}{2}[A]^T\left[\frac{\partial S_e}{\partial \theta}\right][A] \tag{4}$$

Then total torque can be computed by summing up the contributions of all the elements of the problem.

3. Multi-objective Programming and Optimization Using (1+1) ES

3.1. Multi-objective Programming

The general multi-objecive problem with p objectives is defined as follows:

maximize $Z(x_1, x_2, \cdots, x_n)$
$$= [Z_1(x_1, x_2, \ldots, x_n), Z_2(x_1, x_2, \ldots, x_n), \ldots, Z_p(x_1, x_2, \ldots, x_n)] \tag{5}$$
subject to $(x_1, x_2, \ldots, x_n) \in F_d$ (6)

where Z is objective function, x_i is design variable and F_d is constraints of design variables. The multi-objective problem can be converted to a single-objective problem by using constrain method[4] ;

maximize $Z_h(x_1, x_2, \ldots, x_n)$ (7)
subject to $(x_1, x_2, \ldots, x_n) \in F_d$ (8)
$Z_k(x_1, x_2, \ldots, x_n) \geq L_k$ $(k=1, 2, \ldots, h-1, h+1, \ldots, p)$ (9)

where Z_h, objective solution, is chosen for maximization and L_k is constraint of k-th objective function.

The procedure of the multi-objective program is as follows.

<step 1> Construct a payoff table
<step 2> Convert multi-objective programming problem to constrained
          problem such as eq.(9).
<step 3> The n_k and M_k from step 1 represent a range for objective k in
          the noninferior set : $n_k \leq Z_k \leq M_k$
          (M_k : maximum value, n_k : minimum value)
<step 4> Solve the constrained problem set up in step 2 for the L_k.

$$L_k = n_k + [t/(r-1)](M_k - n_k), \quad t = 0,1,2,\ldots,(r-1)$$

(r is the number of different value for L_k)

In this paper, the objective functions are cogging torque, Tc, and efficiency, η. In the program, the cogging torque Z_1 is minimized and the efficiency Z_2 is maximized. The objective functions of Z_1 and Z_2 are expressed by

$$Z_1 = T_c = \frac{W(\theta_m + \Delta\theta_m) - W(\theta_m)}{\Delta\theta_m}, \quad Z_2 = \eta = \frac{output}{output + totall\ loss} \qquad (10)$$

To find an optimum shape of the motor, 2 constraints are given in the program; the cogging torque is under $1\ kg \cdot cm$ and the efficiency is over 90%.

3.2. (1+1) Evolution Strategy (ES)

ES is one of the stochastic optimization methods, which is combined with genetic algorithm and simulated annealing. In the simplest (1+1) ES scheme, there are three basic steps; reproduction, selection and annealing[3]. Fig.1 shows a flow chart of the (1+1) ES scheme and the procedure can be explained by the following 6 steps.

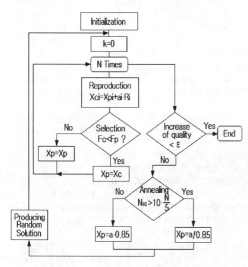

Fig 1. Flow chart of the (1+1) ES

Step 1 is initialization of the program to select more superior first parent variable(X_p). In this step, FEM is used to calculate the initial objective function of torque and efficiency. Step 2 is reproduction of offsprings ($X_{ci} = X_{Pi} + a_i \cdot R_i$, , a is mutation step width) and step 3 is selecting of superior solutions by the 'fitness of nature'. Step 4 is annealing to adjust the mutation step width by the criteria of N_{10}. Conventionally, the number 0.85 is used in this step. Step 5 is producing of random solution and comparison in the whole evolution range. Step 6 is terminating. If termination value were satisfied, the procedure would be ended. Otherwise, the procedure goes back to the step 2 of reproduction.

4. Simulation Results

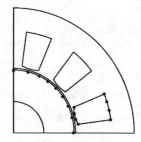

Fig 2. Initial shape of motor

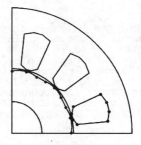

Fig 3. Optimized shape of motor

Fig.2 shows the initial shape of the simulation model which has 8 variables on the slot of the stator and 9 variables on the magnet of the rotor. After performed the (1+1) ES program with FEM, the final optimized shape of the slot and the magnet are obtained as shown in Fig.3.

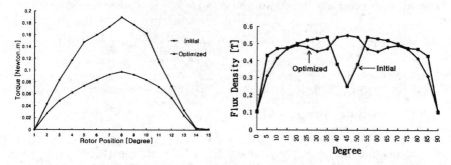

Fig 4. Cogging Torque   Fig 5. Air-gap Flux Density

In order to compare the ripple of cogging torque for the 2 models, torques are calculated at the different position of the rotor over the half slot pitch(15 degrees). As shown in Fig. 4, the cogging torque of the final model is reduced from $1.86\,kg\cdot cm$ to $0.95\,kg\cdot cm$. As shown in Fig.5, the distribution of the air-gap flux density of the optimized model is much smoother than those of the initial, which implies that the smoothness of the flux density distribution in the air gap make cogging torque decrease. Each efficiency of the model is calculated by eq.(10). The efficiency of the optimized model is also improved from 87 % to 90.6 %.

5. Conclusions

This paper presents a shape optimization algorithm for the closed-slot type 400 W AC servo motor using a multi-objective ES program with FEM to minimize cogging torque and to maximize efficiency of the motor. The field analysis is performed by FEM and the shape optimization is used by the multi-objective program under the condition of given constraints. Also, advanced evolution strategy is employed to find noninferior solution. Under the constraints of cogging torque under $1.0\,kg\cdot cm$ and efficiency over 90 %, the final results are fairly good improvement of the motor; the cogging torque is reduced by 46 % and the efficiency is increased by 3.6 %, respectively. If the multi-objective shape optimization algorithm would be improved for more objectives and expanded to 3-dimensional problems, it would be good tools for industry application to the optional design of all kinds of actuators.

References

[1] R.E.Hnitsch and K.Hameyer, Rotating Actuator with Bonded Permanent Magnets. Eight International symposium on Magnetic Anisotropy and Coercivity Rare-Earth Transition Metal Alloys. September 15, 1994.
[2] J.S. Chun and Hyun-Kyo Jung, Shape Optimization od Closed Slot Type Permanent Magnet Motors for Cogging Torque Reduction using Evolution Strategy. IEEE : Transactions on Mag. VOL.33, No.2, March 1997. pp. 1912-1915
[3] Manfred Kasper, Shape Optimization by Evolution Strategy. IEEE : Transactions on Magnetics, vol .28, no.2, March 1992, pp. 1556-1560
[4] Jared L. Cohon, Multiobjective Programming and Planning. Academic Press, 1978.

Non-Linear Electromagnetic Systems
V. Kose and J. Sievert (Eds.)
IOS Press, 1998

Finite Element Implementation of an Internal Variable Magneto-Elastic Hysteresis Model

Cédric GOURDIN, Laurent HIRSINGER & René BILLARDON

Laboratoire de Mécanique et Technologie, E.N.S. Cachan / C.N.R.S. / Université Paris 6

61, Avenue du Président Wilson, 94 235 Cachan Cedex France

Abstract. Recent progress in numerical simulations suggests the use of numerical tools to tackle electrical engineering problems. An accurate treatment of such problems requires a realistic modelling of the complex couplings between the different phenomena involved in the real thermo-magneto-mechanical behaviour of electrical machines. In this paper, non-linear irreversible magnetic constitutive equations coupled with elasticity are given. This internal variable hysteresis model which takes into account the mechanical state of the material has been implemented in the finite element package CASTEM 2000. The ultimate aim of this paper is to illustrate the effect of eddy currents and mechanical stresses on the 2-D magnetic field in an infinite conductive plate.

1. Magneto-elastic Constitutive Equations

Different models have already been proposed in the literature to couple magnetic and magnetostriction hystereses as well as the stress effect on both types of hysteresis [1-2]. Based on the thermodynamics of continuous media the model used in this paper enables the description of the magneto-elastic behaviour of a material [3].

The magnetisation process is the result of independent non-linear reversible and non-linear irreversible mechanisms: the former corresponds to the so-called anhysteretic curve M_{an} and the latter to the pinning of the domain walls on various obstacles. In the proposed model, the magnetic field H is partitioned into reversible anhysteretic and irreversible parts denoted by H_{an} and H_i respectively, so that:

$$H = H_{an} + H_i \tag{1}$$

$$M = M_{an}(H - H_i) \tag{2}$$

The magnetic field is also responsible for the magnetostriction deformation mechanism. It may be assumed that total strains ε are the sum of pure mechanical strains ε^m, pure magnetostriction strains ε^μ and thermal expansion strains ε^{th} so that when small strain and linear elasticity assumptions are made, [4]

$$\varepsilon = \varepsilon^m + \varepsilon^\mu + \varepsilon^{th} = \mathbb{E}_{(T)}^{-1} : \sigma + \varepsilon^\mu(H,H_i,T) + \varepsilon^{th}(T) \tag{3}$$

where $\mathbb{E}$, σ and T respectively denote the elasticity moduli tensor, the stress tensor and temperature.

The specific free enthalpy Ψ used as a state potential may be written as a function of temperature, stress tensor, magnetic field, and internal variable H_i:

$$\rho\Psi = \rho\Psi(\sigma, H, H_i, T) = \rho\Psi^{\mu m}(\sigma, H, H_i, T) + \rho\Psi^{\mu r}(H, T) + \rho\Psi^{\mu i}(H, H_i, T) \tag{4}$$

where ρ denotes the mass density. The first term $\rho\Psi^{\mu m}$ is chosen as:

$$\rho\Psi^{\mu m} = -\frac{1}{2}\sigma : \mathbb{E}_{(T)}^{-1} : \sigma - \left(\varepsilon^\mu(H,H_i,T) + \varepsilon^{th}(T)\right) : \sigma \tag{5}$$

A state uncoupling is granted between reversible $\Psi^{\mu r}$ and irreversible $\Psi^{\mu i}$ parts of the magnetic state.

The mechanical and magnetic state laws are derived from this state potential as:

$$\cdot = -\frac{\partial(\rho\,\Psi^{\mu m})}{\partial \cdot} \tag{6}$$

$$\mu_0\,\mathbf{M} = -\frac{\partial(\rho\Psi^{\mu r}+\rho\Psi^{\mu i})}{\partial\mathbf{H}} + \frac{\partial \cdot\mu}{\partial\mathbf{H}}(\mathbf{H},\mathbf{H_i},T) : \cdot = \mu_0\,\mathbf{M_{an}}(\mathbf{H}-\mathbf{H_i}) + \frac{\partial \cdot\mu}{\partial\mathbf{H}}(\mathbf{H},\mathbf{H_i},T) : \cdot\tag{7}$$

$$\mu_0\,\mathbf{M_i} = -\frac{\partial(\rho\,\Psi^{\mu i})}{\partial\mathbf{H_i}} + \frac{\partial \cdot\mu}{\partial\mathbf{H_i}}(\mathbf{H},\mathbf{H_i},T) : \cdot \tag{8}$$

where μ_0, $\mathbf{M}$ and $\mathbf{M_i}$ respectively denote the vacuum permeability and the variables associated with $\mathbf{H}$ and $\mathbf{H_i}$. Mechanical state law (6) corresponds to equation (3).

The dissipative mechanisms responsible for the magnetic hysteresis are linked to the irreversible part $\mathbf{H_i}$ of the magnetic field, since the Clausius Duhem inequality is :

$$\mathcal{D} = -\rho\dot\Psi - \cdot : \cdot \quad -\mu_0\,\mathbf{M}\cdot\dot{\mathbf{H}} = \mu_0\,\mathbf{M_i}\cdot\dot{\mathbf{H_i}} \geq 0 \tag{9}$$

For isotropic magnetic hysteresis model, the evolution law of the internal variable $\mathbf{H_i}$, which must be chosen such that this inequality is satisfied, and the anhysteretic curve $\mathbf{M_{an}}$ have the following expressions:

$$\dot{\mathbf{H_i}} = f(H, H_i)\,\dot{\mathbf{H}} = \frac{\chi_{a0} - \chi_0}{\chi_{a0}}\left(1 - \frac{H_i}{H_c}\,\mathrm{sign}(\dot H)\right)\dot{\mathbf{H}} \tag{10}$$

$$M = M_{an}(H - H_i) = \frac{2\,M_s}{\pi}\,\mathrm{Arctan}\left(\frac{\pi}{2}\,\chi_{a0}\,\frac{H - H_i}{M_s}\right) \tag{11}$$

where, M, H and H_i respectively denote $\mathbf{M}$, $\mathbf{H}$ and $\mathbf{H_i}$ moduli whereas, constants χ_0, χ_{a0}, M_s and H_c denote material parameters respectively related to initial slopes of hysteretic and anhysteretic curves, saturation magnetisation and coercive magnetic field.

In the following for sake of simplicity, the coupling between magnetic hysteresis and elasticity is taken into account through the dependence of these parameters on elastic stresses. These parameters have been identified for an industrial Fe-3%Si alloy and subjected to mechanical tension applied in the same direction as the magnetic field (cf. Figures 1-2).

2. Numerical Implementation of the Isotropic Magnetic Hysteresis Model

The constitutive equations presented above have been implemented in the finite element software package CASTEM 2000 by using a vector potential formulation and an analogy with a non-linear heat transfer analysis [4].

The global problem consists in solving the equilibrium equations with corresponding boundary conditions, and the local problem consists in the integration of the constitutive

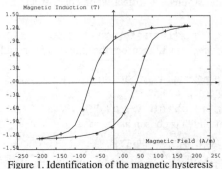

Figure 1. Identification of the magnetic hysteresis
model parameters without applied stress
for Fe-3%Si (lines: experiment, points: model):
M_s=1114200A/m, H_c=95A/m, χ_{a0}=35466, χ_0=18087.

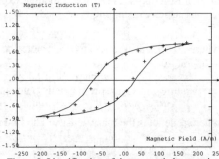

Figure 2. Identification of the magnetic hysteresis
model parameters for an elastic stress (100 Mpa)
for Fe-3%Si (lines: experiment, points: model):
M_s=810000A/m, H_c=95A/m, χ_{a0}=12900, χ_0=6575.

equations. The global non-linear problems are iteratively solved by using a θ-method with a consistent tangent operator scheme [4-5]. The local non-linear balance laws, as detailed in the following, are iteratively solved by applying a θ-method and a pure Newton algorithm.

Since the values H_n and H_{in} of the magnetic field are known at the beginning of each time increment Δt and the total induction increment ΔB is given, the local integration algorithm allows H_{n+1} and H_{in+1} to be determined at each integration point.

The variables must satisfy the constitutive equations:

$$\dot{H}_i = f(H, H_i)\, \dot{H} \tag{12}$$

$$B = \mu_0(H + M) \qquad \text{with,} \qquad M = M_{an}(H - H_i) \tag{13}$$

The evolution law (12) is decretized in the incremental form:

$$\Delta H_i = f(H_{n+\theta}, H_{in+\theta})\, \Delta H \tag{14}$$

where, $\Delta H = H_{n+1} - H_n$
and, $(\ )_{n+\theta} = (1 - \theta)(\ )_n + \theta(\ )_{n+1}$, with $0 \le \theta \le 1$.

At each t_{n+1}, variables H_{n+1} and H_{in+1} must satisfy equations (13) and (14), or the following system:

$$g(H,H_i)_{n+1} = \Delta H_i - f(H_{n+\theta}, H_{in+\theta})\, \Delta H = 0 \tag{15}$$

$$h(H,H_i)_{n+1} = B_{n+1} - \mu_0 \{ H_{n+1} + M_{an}(H_{n+1} - H_{in+1}) \} = 0 \tag{16}$$

This set of non-linear equations is solved iteratively by a pure Newton method:

$$g(H,H_i)_{n+1}^s + \left(\frac{\partial g}{\partial H}\right)_{n+1}^s C_H + \left(\frac{\partial g}{\partial H_i}\right)_{n+1}^s C_{H_i} = 0 \tag{17}$$

$$h(H,H_i)_{n+1}^s + \left(\frac{\partial h}{\partial H}\right)_{n+1}^s C_H + \left(\frac{\partial h}{\partial H_i}\right)_{n+1}^s C_{H_i} = 0 \tag{18}$$

where C_H and C_{H_i} denote "corrections to H and H_i":

$$C_H = (H_{n+1}^{s+1} - H_{n+1}^s) \qquad C_{Hi} = (H_{in+1}^{s+1} - H_{in+1}^s) \tag{19}$$

The iterations are stopped if the following tests are verified:

$$\| g(H,H_i)_{n+1}^s \| \le Tol_g \qquad \text{and} \qquad \| h(H,H_i)_{n+1}^s \| \le Tol_h \tag{20}$$

3. Application to an Infinite Conductive Plate

The final aim of the above-described numerical tools is to predict the magnetic losses in electro-mechanical devices.

The example analysed in this paper corresponds to an infinite conductive plate subjected to a magnetic field and a mechanical traction in the same direction as the plate plane. A symmetric sinusoidal magnetic field with a maximum value of 200A/m is applied to the plate. This analysis enables the study of the influence of eddy currents on the average values B_m of the magnetic field through the plate thickness e,

$$B_m = \frac{1}{e} \int_{-e/2}^{e/2} B\, dz \tag{21}$$

Plots given in figures 3 and 4 correspond to 2-D transient electro-magnetic analysis. They correspond to stabilized hysteresis loops, which occur after the fourth magnetic field excitation period, between B_m and the magnetic field measured at the surface H_{surf}.

Figure 3 corresponds to B_m versus H_{surf} plots for different frequencies of the applied magnetic field with zero stress applied.

Figure 4 corresponds to the same plots, the plate being subjected to a stress approximately equal to a quarter of the yield stress of the material.

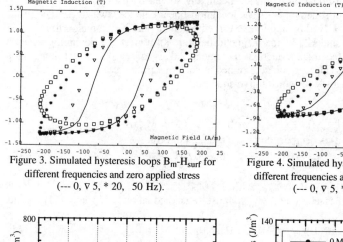

Figure 3. Simulated hysteresis loops B_m-H_{surf} for different frequencies and zero applied stress (--- 0, ▽ 5, * 20, 50 Hz).

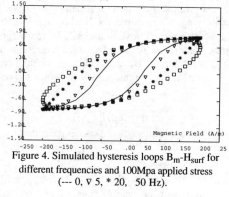

Figure 4. Simulated hysteresis loops B_m-H_{surf} for different frequencies and 100Mpa applied stress (--- 0, ▽ 5, * 20, 50 Hz).

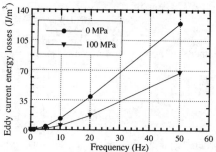

Figure 5. Total energy losses per cycles versus frequency for different stresses and H_{max}=200A/m, and for Fe-3%Si.

Figure 6. Eddy current energy losses per cycles versus frequency for different stresses and H_{max}=200A/m, and for Fe-3%Si.

The total energy dissipated per cycle can be computed by integration of the previous hysteretic loops. These results are gathered in Figure 5 as total energy losses plotted as a function of the applied magnetic field frequency.

These magnetic losses can be easily divided into two main parts: first, pure hysteresis losses which are stress dependent and second, eddy currents losses which are strongly dependent on the geometry. The latter, $E_{eddy\ current}$, are plotted in Figure 6 as a function of the applied magnetic field frequency and correspond to the following integrations:

$$E_{eddy\ current} = \frac{1}{T\ e} \int_{-e/2}^{e/2} \left(\int_{1\ cycle} - j\ dA \right) dz \qquad (22)$$

where T, j and A respectively denote the period, the eddy current and the vector potential.

References

[1]   D.C. Jiles, Theory of the magnetomechanical effect, *J. Phys. D: Applied Physics*, **28** (1995), 1537-1546.

[2]   M. Sabir, Constitutive relations for magnetomechanical hysteresis in ferromagnetic materials, *Int. J. Engng. Sci.*, 33, **9** (1995) 1233-1249.

[3]   G. Barbier, Proposition d'un modèle de couplage magnéto-mécanique pour les ferromagnétiques doux, Rapport de DEA de Mécanique, Université Paris 6, LMT Cachan, 1995.

[4]   L. Hirsinger *et al.*, Static and dynamic 2-D magnetic analysis with an internal variables hysteresis model, Numerical Methods in Engineering '96, 2nd ECCOMAS Proc., Wiley, 1996, pp. 318-324.

[5]   A. Benallal *et al.*, An integration algorithm and the corresponding consistent tangent operator for fully coupled elastoplastic and damage equations, *Communications in Applied Numerical Methods*, **4** (1988) 731-740.

Non-Linear Electromagnetic Systems
V. Kose and J. Sievert (Eds.)
IOS Press, 1998

Finite Element Analysis of
Superconducting Couplers

S.-A. Zhou and H. Hesselbom

Ericsson Components AB, S-164 81 Kista-Stockholm, Sweden

Abstract. A finite element method is introduced to analyse electromagnetic properties of superconducting couplers of arbitrary cross sectional geometry. Basic formulation of a multi-superconductor system is first given, including the normal skin effect, the Meissner effect, as well as the effect of kinetic energy of superelectrons. These effects can be of importance in affecting properties of superconducting interconnects and couplers operating at high frequencies. In the formulation, Galerkin's method is applied to the integrodifferential equations derived for the field problem of superconducting couplers in conjunction with the usual triangular finite element method. With the aid of the finite element formulation, the field problem of superconducting couplers is solved in terms of the magnetic vector potential in the frequency domain. The numerical results are then used to calculate line parameters, such as the mutual inductance, mutual capacitance, resistance and conductance of the equivalent coupled transmission lines based on a quasi-TEM approximation. Both symmetric and asymmetric coupled structures can be analysed. By using the coupled mode approach and the line parameters obtained from the finite element code, properties of superconducting directional couplers are studied numerically. The approach introduced here is quite general, and can equally be used to analyse quantitatively cross-talks among superconducting interconnects, which may be of interest in superconducting electronic circuit designs.

1. Formulation of multi-superconducting strip lines

AC losses may exist in superconductors at finite temperatures, especially when they are operating at high frequencies. When power is absorbed in a superconductor, there must be a component of the Poynting vector directed into the superconductor. In general, not only transverse electromagnetic field components, but also longitudinal components of both electric and magnetic fields may exist inside the superconductor. According to Maxwell's theory, a longitudinal component of magnetic field will have associated with the transverse currents on the superconductor. Since these transverse currents arise only because of the perturbation of the TEM mode into a mode with longitudinal field components, they are small in comparison with the longitudinal currents. In the case of the transverse current is negligible, we may assume that the current in superconductors of a superconducting strip line along z-axis flows only in the z (longitudinal) direction, and the EM field is independent of z-coordinate at the quasi-TEM approximation [1]. Thus, introducing a vector potential A, we may find that A has only a z-component, which satisfy the following equation:

$$-\nabla^2 A_z(x, y) = \mu_o J_z(x, y) \tag{1}$$

where J_z is the current density inside superconductors along z-axis direction and μ_o is the permeability in vacuum. In the dielectric medium (or free space) denoted by S_D, surrounding the straight superconductors, we have the field equation: $\nabla^2 A_z = 0$. Thus, inside superconductors, the electric field in time-harmonic fields ($\sim \exp(i\omega t)$) can be expressed by

$$E_z = -i\omega A_z - \frac{\partial V}{\partial z} \tag{2}$$

where V is the electric scalar potential and ω is the radian frequency of the field.

By the classical two-fluid model [2], we write the total current density $J_z = J_z^{(s)} + J_z^{(n)}$, with the normal current density $J_z^{(n)} = \sigma_n E_z$, following Ohm's law, and the superconducting current density $J_z^{(s)} = E_z/(i\mu_0\omega\lambda_L^2)$, according to London's theory. Here, σ_n denotes the normal conductivity of the superconductor, and λ_L is the London penetration depth. Thus, by introducing a complex conductivity σ by $\sigma = \sigma_n - i/(\mu_0\omega\lambda_L^2)$, Eq.(1) can be expressed as

$$-\nabla^2 A_z + i\omega\mu_0\sigma A_z = -\mu_0\sigma\frac{\partial V}{\partial z} \tag{3}$$

For a uniform straight superconductor, the gradient of the electric potential V may be assumed to be constant along the superconductor. Thus, for a system of multi-superconductor strip lines, we have the following set of equations:

$$-\nabla^2 A_z + i\omega\mu_0\sigma_k A_z - \frac{i\omega\mu_0\sigma_k}{S_k}\int_{S_k} A_z dS = \frac{\mu_0}{S_k}I_k \quad in \ S_k \tag{4}$$

with S_k being the cross sectional area of the kth superconductor ($k=1,2,...,Q$), and I_k the total current flowing in the kth superconductor, given by

$$I_k = -S_k\sigma_k\frac{\partial V}{\partial z} - \int_{S_k} i\sigma_k\omega A_z dS \tag{5}$$

which may be considered as the sum of the source current (defined by $I_s = -S_k\sigma_k\partial V/\partial z$) and the total induced or eddy current flowing inside the superconductor. Eq.(4) is an integro-differential equation for the magnetic vector potential component A_z. To introduce a finite element formulation, we may apply the Galerkin integral form:

$$\int_{S_D} (\nabla w)\cdot\nabla A_z dS + \sum_{k=1}^{Q}\int_{S_k}\left[(\nabla w)\cdot\nabla A_z + i\omega\mu_0\sigma_k\left(A_z - \frac{\eta_k}{S_k}\int_{S_k} A_z dS\right)w - \frac{\mu_0}{S_k}I_k w\right]dS = f_b \tag{6}$$

in which we have introduced the parameter η_k for the convenience of numerical calculation. The parameter η_k equals one ($\eta_k=1$) for the kth superconductor carrying non-zero source current (I_k is specified). If the kth superconductor carrying no source current, we set $\eta_k=0$ and $I_k=0$ in the equation. In such a case, the total current flowing in the superconductor is equal to the induced (eddy) current in the absence of the source current. In eq.(6), f_b is given by

$$f_b = \int_\Gamma w(\nabla A_z)\cdot n d\Gamma = \int_\Gamma w\frac{\partial A_z}{\partial n}d\Gamma \tag{7}$$

where w is a testing function, and n is the outward unit vector normal to the outer boundary Γ of the system domain. For a homogeneous boundary condition (*i.e.*, if the strips are shielded by a magnetic material of high permeability (noting: $H\times n=\partial A_z/\partial n$)), we have $f_b=0$. For open boundary value problems, the region may be bounded by an approximate Dirichlet or Neumann boundary, or the infinite element approach could be used.

By the well-known finite element method, we may discretise the domain of interest, and approximate the unknown function A_z within each element. If, for instance, linear triangular elements are used, the unknown function A_z within each element can be approximate by

$$A_z^e(x,y) = \sum_{j=1}^{3} N_j^e(x,y)A_{zj}^e \tag{8}$$

where A_{zj}^e (j=1,2,3) denote the node potential, and N_j^e are the interpolation or expansion functions. Taking $w=N_i^e$ (i=1,2,3) as the testing function in each element, we may then obtain a set of linear algebraic equations for the determination of discrete values of the magnetic potential component A_z at all nodes of the domain of interest.

For a system of Q superconductors with M signal lines ($M<Q$), the ac resistance matrix R and the inductance matrix L may be obtained by computing, in general, the magnetic potential distributions for $M(M+1)/2$ independent excitation states. However, for a system of having certain symmetry, the number of computations can be reduced. In general, the matrices R and L can be determined respectively by the following relations:

$$\sum_{i,j=1}^{M} R_{ij}I_iI_j^* = \sum_{k=1}^{Q} \int_{S_k} \frac{1}{\sigma_k^{(n)}}|J_z^{(n)}|^2 dS \tag{9}$$

and

$$\sum_{i,j=1}^{M} L_{ij}I_iI_j^* = \sum_{k=1}^{Q} \left[Re\left\{ \int_{S_k} A_z J_z^* dS \right\} + \int_{S_k} \mu_o \lambda_k^2 |J_z^{(s)}|^2 dS \right] \tag{10}$$

where the asterisk (*) denotes complex conjugate. $\sigma_k^{(n)}$ and λ_k are respectively the normal conductivity and the London penetration depth of the kth superconductor. It can be seen that only the normal conduction current components $J_z^{(n)}$ causes the ac loss of the superconductors, and the second term on the right hand side of eq.(10) expresses the kinetic inductance per unit length of the superconductor.

2. Analysis of superconducting directional couplers

This section introduces a coupled transmission-line model [3] for analysing properties of superconducting microstrip couplers (ex., shown in figure (a)). In this model, the behavior of two coupled transmission lines is described by the following set of differential equations:

$$\frac{dV}{dz} = -(R+i\omega L)\cdot I \tag{11}$$

$$\frac{dI}{dz} = -(G+i\omega C)\cdot V \tag{12}$$

where $V=(V_1, V_2)$ and $I=(I_1, I_2)$ are the voltage vector and the current vector with V_1, I_1 and V_2, I_2 being the voltage and current on the 1st and 2nd microstrip transmission lines respectively. R, L, G, and C are respectively the resistance (Ω/m), the inductance (H/m), the conductance (1/Ω·m), and the capacitance (F/m) matrix of the coupled lines, which can be determined by field analyses based on the above finite-element method together the well-known approach [4] for the capacitance and conductance matrices.

Introducing a complex propagation constant γ by

$$V = V_o e^{-\gamma z} \quad and \quad I = I_o e^{-\gamma z} \tag{13}$$

we may arrive at the following eigen-value equation:

$$det[(R+i\omega L)\cdot(G+i\omega C)-\gamma^2\delta] = 0 \tag{14}$$

in which δ_{ij} is the Kronecker delta matrix. This equation gives, in general, four possible values of the complex propagation constant $\gamma=\alpha+i\beta$, in which α is the attenuation constant and β is the real propagation constant. Each of values of the complex propagation constant corresponds to a distinct mode of propagation. These are the incident and reflected waves for an even mode, where the voltages as well as the currents on the two lines are in phase, and the incident and reflected waves of an odd mode, where the voltages as well as the currents on the two lines are out of phase with each other. When the two coupled lines are identical (symmetric), then the two lines have voltages of equal magnitude and currents of equal magnitude. The general solution of eqs.(11) and (12) can be obtained by

$$\begin{bmatrix} V(z) \\ I(z) \end{bmatrix} = \sum_{k=1}^{4} B_k F_k e^{-\gamma_k z} \tag{15}$$

where B_k (k=1,2,3,4) are four unknown constants to be determined by proper boundary conditions. F_k is the eigen-vector (voltage and current) corresponding to the kth eigen-value γ_k (k=1,2,3,4) (*i.e.*, the four components of the eigen-vector F_k correspond to $F_k(1)=V_{1k}$, $F_k(2)=V_{2k}$, $F_k(3)=I_{1k}$, and $F_k(4)=I_{2k}$ respectively).

Illustratively, let us consider a superconducting directional coupler, shown in figure (*a*) and (*b*), where the load is assumed to be at $z=L$ and the generator at $z=0$. Thus, we have the following boundary conditions: $I_1(0)Z_0+V_1(0)=2V$ at port 1; $I_2(0)Z_0+V_2(0)=0$ at port 2; $I_2(L)Z_0=V_2(L)$ at port 3; and $I_1(L)Z_0=V_1(L)$ at port 4, where Z_o denotes the matching impedance, usually taken to be 50 Ω. By applying the boundary conditions, the four unknown constants B_k (k=1,2,3,4) can be got from eq.(15). Thus, the properties of the superconducting directional coupler can be determined. In general, the coupling coefficient C_{dB} of the coupler can be defined by $C_{dB}=10\times\log_{10}(P_1/P_2)$ in dB, and the directivity D_{dB} of the coupler by $D_{dB}=10\times\log_{10}(P_2/P_3)$. The isolation of the coupler is numerically the sum of C_{dB} and D_{dB}, and the insertion loss Π_{dB} of the coupler is given by $\Pi_{dB}=10\times\log_{10}(P_1/P_4)$, where P_1, P_2, P_3, and P_4 denote respectively the power at port 1, port 2, port 3 and port 4 of the directional coupler, shown in figure (*b*). Some numerical analyses are carried out for 3dB directional couplers in either superconducting or normal state at different frequencies [5].

3. Conclusions

A finite-element method is developed for analysing superconducting as well as normal electrical couplers. Numerical results indicate that superconducting directional couplers may have higher directivity and are less frequency-sensitive than normal conducting ones due to their lower resistance and nearly frequency-independent inductance.

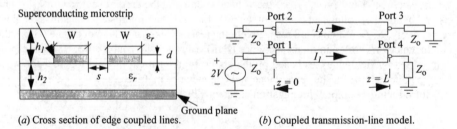

(*a*) Cross section of edge coupled lines.   (*b*) Coupled transmission-line model.

References

[1]   K. Konrad, Integrodifferential Finite Element Formulation of Two-Dimensional Steady-State Skin Effect Problems, IEEE Trans. Mag., MAG-**18** (1982) 284-292.
[2]   S.-A. Zhou, Electrodynamic Theory of Superconductors. Peter Peregrinus Ltd, London, 1952.
[3]   D.H. Schrader, Microstrip Circuit Analysis. Prentice Hall PTR, New Jersey, 1995.
[4]   R.E. Collin, Foundations for microwave Engineering. McGraw-Hill, New York, 1995.
[5]   S.-A. Zhou and H. Hesselbom, Electromagnetic Simulation of Electrical Couplers. Ericsson Technical Report, EKA/SI 97:009, Kista-Stockholm (1997).

Non-Linear Electromagnetic Systems
V. Kose and J. Sievert (Eds.)
IOS Press, 1998

Dynamic Finite Element Analysis with a New Sliding Element and Composite Mesh Scheme

Takashi TODAKA and Masato ENOKIZONO

*Dept. of Electrical and Electronic Engineering, Faculty of Engineering,
Oita University, 700 Dannoharu, Oita 870-11, Japan*

Abstract. This paper presents a composite mesh scheme with a new sliding element. The stationary mesh without a moving object and the shifting mesh including the object are usually used in the conventional schemes. In this case, because the initial field greatly differs from the real field, many iterative calculations are needed to obtain the solution. In our method, the moving object is also considered in the stationary mesh with sliding elements to improve the initial field. Therefore the computing time can be considerably reduced in comparison with that of the conventional schemes. Results show the usefulness of this method in dynamic eddy current problems.

1. Introduction

In the conventional remesh method, because modifications of the finite element distributions influence the error distribution of the numerical results in the considered region, it is necessary to use many finite elements to avoid such variations. In addition, when the moving part has acceleration, the displacement for a period becomes an unknown quantity and many iterative calculations are needed to obtain the static solution. For the above reasons, the reports on transient magnetic field analysis considering movement of actual electrical machinery and apparatus are very few.

In this paper, we introduce a special sliding element into an axisymmetric finite element method to simulate a dynamic problem of a magnetic hammer model. The element has two-material parts: the one is the rod material and the other is its outer air region. The movement of the rod can be treated as a modification of the positional data in the sliding elements, which are distributed near the moving rod boundary. The elemental coefficient matrices can be obtained approximately with a linear interpolation without additional nodes. Modifying the interpolation parameters with a composite mesh scheme, the movement and the magnetic field can be calculated accurately.

In addition, to avoid oscillating solutions in the convection-diffusion equation, an upwind interpolation function that has been suggested in the field of the finite analytic method, is introduced into the formulation [1]. Because the elemental coefficient matrices can be calculated analytically in this case, we can achieve simplicity of the upwinding formulation and accurate calculation of the motional term.

2. Formulation

The governing equation and the circuit equation in the axisymmetric eddy current problems considering movement of a conductive material, are written as follows:

$$\frac{v}{r}\frac{\partial}{\partial r}\left(r\frac{\partial A}{\partial r}\right) + v\frac{\partial^2 A}{\partial z^2} = \sigma\frac{\partial A}{\partial t} + v_z\frac{\partial A}{\partial z} - J_0 \quad (1) \qquad\qquad V = IR + \frac{d}{dt}\oint_C A\,dl \quad (2)$$

where, A is the θ-directional component of the magnetic vector potential, J_0 the exciting current density, v the magnetic resistivity, σ the conductivity, v_z the velocity of a moving conductor in z-direction, V the terminal voltage, I the exciting current, and R the resistance of an exciting coil. The finite element equations can be expressed as follows:

$$M_{ij}\frac{A_j^{n+1} - A_j^n}{\Delta t} + (G_{ij} + V_{ij})A_j^n + C_i I = 0 \quad (3) \qquad V = IR + H_k\frac{A_k^{n+1} - A_k^n}{\Delta t} \quad (4)$$

where, Δt is the time step width. Combining (4) with (3), the exciting current can be treated as an unknown value, thus the impedance change can be also simulated directly.

2. 1 Upwind Techniques

Figure 1 shows the definition of the rectangular element. In the upwinding approximation of the motional term V_{ij}, we use the following upwind interpolation function [1].

$$L_i = \frac{\eta_i\left(e^{\kappa\eta} - e^{-\kappa\eta_i}\right)}{4\sinh\kappa}(1 + \xi_i\xi) \quad (5)$$

$$\kappa = a\,\mu\sigma\,v_z \quad (6)$$

$$\begin{aligned}
i=1: &\quad \xi_i = 1, \quad \eta_i = 1\\
i=2: &\quad \xi_i = -1, \quad \eta_i = 1\\
i=3: &\quad \xi_i = -1, \quad \eta_i = -1\\
i=4: &\quad \xi_i = 1, \quad \eta_i = -1
\end{aligned} \right\} \quad (7)$$

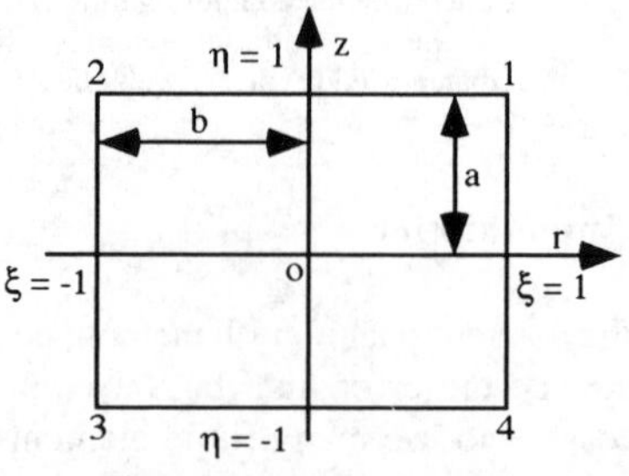

Fig. 1 Rectangular element.

where, κ is the half value of the cell Peclet-number in z-direction. The motional term V_{ij} can be calculated analytically as follows:

$$V_{ij} = \int_{\Omega_e} v_z N_i\frac{\partial L_j}{\partial x}r\,drdz = \frac{b\sigma v_z\eta_j}{4}\left\{r_0\left(1 + \frac{\xi_i\xi_j}{3}\right) + b\frac{\xi_j + \xi_i}{3}\right\}\left\{1 + \eta_i\left(\coth\kappa - \frac{1}{\kappa}\right)\right\} \quad (8)$$

where, N_i is the linear interpolation function for the rectangular element. Because the upwind interpolation function corresponds to N_i in the limit when κ approaches zero, Eq. (6) can be applied to the problems in low Peclet-number ranges as it is.

2. 2 Sliding Element

Two meshes, namely, the stationary mesh without a moving object and the shifting mesh including the object, are usually used in the conventional composite mesh schemes [2, 3]. In this case, because the initial field greatly differs from the real field, many iterative calculations for each mesh are needed to obtain the numerical solution. In our method, the moving object is also considered in the stationary mesh as shown in Fig. 2 to improve the initial field. Therefore the elements indicated with the mark "s" are including the two parts that have a different material constant and velocity.

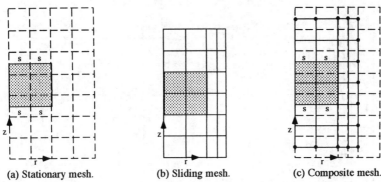

(a) Stationary mesh.   (b) Sliding mesh.   (c) Composite mesh.

Fig. 2 Definition of composite meshes.

To analyze such models, the additional nodes (5 and 6) are usually distributed as shown in Fig. 3. However, we can remove the nodes by using the following interpolation relationships:

$$A_5 = \frac{A_2\alpha + A_3\beta}{\alpha + \beta}, \quad A_6 = \frac{A_1\alpha + A_4\beta}{\alpha + \beta} \tag{9}$$

where, A_k ($k = 1 - 6$) is the vector potential at each point and A_1 through A_4 are the nodal potentials. For example, the elemental coefficient matrices can be obtained as follows:

$$\left[G_{ij}^1\right]\begin{Bmatrix} A_1 \\ A_2 \\ A_5 \\ A_6 \end{Bmatrix} + \left[G_{ij}^2\right]\begin{Bmatrix} A_6 \\ A_5 \\ A_3 \\ A_4 \end{Bmatrix} = \left[G_{ij}\right]\begin{Bmatrix} A_1 \\ A_2 \\ A_3 \\ A_4 \end{Bmatrix} \tag{10}$$

$$\begin{aligned} G_{i1} &= G_{i1}^1 + \phi_1\left(G_{i4}^1 + G_{i1}^2\right) \\ G_{i2} &= G_{i2}^1 + \phi_1\left(G_{i3}^1 + G_{i2}^2\right) \\ G_{i3} &= G_{i3}^2 + \phi_2\left(G_{i2}^2 + G_{i3}^1\right) \\ G_{i4} &= G_{i4}^2 + \phi_2\left(G_{i1}^2 + G_{i4}^1\right) \end{aligned} \tag{11}$$

$$\phi_1 = \frac{\alpha}{\alpha + \beta}, \quad \phi_2 = \frac{\beta}{\alpha + \beta} \tag{12}$$

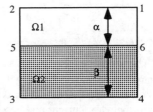

Fig. 3 Sliding element.

where, the superscripts (*1* and *2*) show the regions (Ω_1 and Ω_2).

Firstly, the potential distribution on the stationary mesh is calculated with the initial interpolation parameters given by the rod position. Then the boundary conditions at the nodes indicated by the black symbol in the sliding mesh are obtained with a linear interpolation. Secondly, the potential distribution on the sliding mesh is calculated under the conditions, and the parameters (α and β) can be renewed with the relationship (9). Modifying the interpolation parameters in the same way, the movement of the rod and the transient magnetic field distribution can be calculated.

3. Results and discussion

Figs. 4 and 5 show a simplified magnetic hammer model and the movement of the rod, respectively. Fig. 6 shows the flux distributions and the rod positions at each time step. The moving simulation was carried out with the step-by-step method.

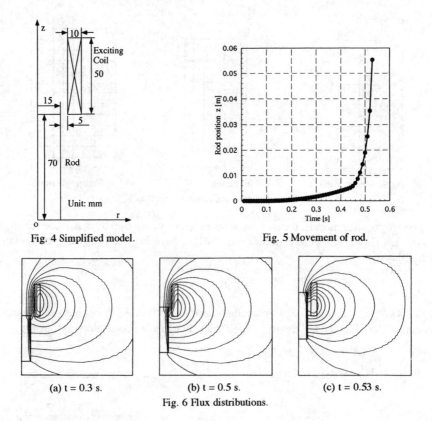

Fig. 4 Simplified model.   Fig. 5 Movement of rod.

(a) t = 0.3 s.   (b) t = 0.5 s.   (c) t = 0.53 s.
Fig. 6 Flux distributions.

The conditions used in this analysis were: the time step width Δt = 0.01 s, the terminal voltage V = 50 V, and the weight of rod m = 0.4 kg. The terminal voltage was supplied to the coil as a step function at t = 0. As shown in Fig. 4, the rod was attracted slowly at first due to the eddy current and accelerated rapidly past about 0.45 s. The number of the iterative calculations for the two meshes at each time was within five times.

4. Conclusion

In this paper, we have presented the composite mesh scheme with the sliding element and the upwinding technique. The example showed usefulness of this method in the dynamic magnetic field analyses. The advantage of this method is that the computing time can be considerably reduced in comparison with that of the conventional schemes.

References

[1] T. Tanahashi and Y. Oki, New Theory of Hybrid-Upwind Technique and Discrete Del Operator for Finite-Element Method, Comp. Fluid Dyn., Vol. 7, pp. 3-14, 1996.
[2] J. L. Steger, On the Use of Composite Grid Schemes in Computational Aerodynamics, J. Computer Methods in Applied Mechanics and Engineering, Vol. 64, pp. 301-320, 1987.
[3] T. Todaka, et al., Analysis of Magnetic Field Problems including Velocity Term with UFEM, Paper of the 17th Symp. on Comp. Electrical and Electric Engineering., SST of Japan, I-6, pp. 31-34,1996.

Non-Linear Electromagnetic Systems
V. Kose and J. Sievert (Eds.)
IOS Press, 1998

Development of the Special Element for the Laminated Core for the Finite Element Method

Masato ENOKIZONO, Takashi TODAKA and Hiroki TAKIMIZU
Dept. of Electrical and Electronic Engineering, Faculty of Engineering,
Oita University, 700 Dannoharu, Oita, 870-11, Japan

Abstract. This paper presents a new special finite element to analyze laminated cores. This special element is based on the one order rectangular element, and has two regions including the steel sheet and surface coating for insulation. We applied this special element considering surface coating to a simplified magnetic field problem. As a result, it is shown that the special element is applicable to the model including laminated cores.

1. Introduction

Generally, silicon steel sheet has a surface coating for insulation. In conventional finite element analysis of electromagnetic problems including laminated silicon steel sheets, the magnetic materials are usually modeled as a non-laminated core or yoke, because a great number of elements have to be distributed in the region. Actually, the silicon steel sheets are each insulated and laminated with a stacking factor. It is therefore important to develop the model of a laminated core considering the stacking factor to analyze practical electrical machinery and apparatus. It is also desirable that the model should be simple with a reduction in the number of elements.

To solve this problem, a simplified model considering the insulator is developed. In the examples of verification of this element, the results with the special elements are compared with ones of conventional finite elements. The results show the applicability of this element to magnetic field problems including laminated cores.

2. Formulation for the two-dimensional problem

Generally, Poisson's equation for the two-dimensional electrostatic problems is written as follows:

$$\frac{\partial}{\partial x}\left(v_y \frac{\partial A}{\partial x}\right) + \frac{\partial}{\partial y}\left(v_x \frac{\partial A}{\partial y}\right) = -J, \tag{1}$$

where, v is the magnetic reluctivity, A is the magnetic vector potential, J the exciting current density. The Garelkin's equation for the two-dimensional problems is written as follows:

$$\iint_{S^{(e)}}\left(v_y \frac{\partial N_i}{\partial x}\frac{\partial N_j}{\partial x} + v_x \frac{\partial N_i}{\partial y}\frac{\partial N_j}{\partial y}\right)A_j dxdy = \int_{S^{(e)}} N_i J dxdy, \tag{2}$$

where, Nk is the shape function of the rectangular element, and $S^{(e)}$ the region of each element.

Figure 1 shows the special element considering the insulator region. $S^{(e)}$ is the steel region, $\Omega^{(e)}$ is the insulator region, and δ the thickness of the insulator. A_1 through A_6 indicate the nodal value of the vector potential. When the thickness of the insulator is very thin in comparison with the one of the steel, A_5 and A_6 can be approximated as,

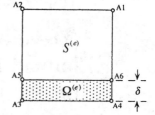

Fig. 1 Special element considering the insulator.

$$A_5=A_3, \ A_6=A_4. \qquad (2)$$

Therefore the elemental coefficient matrix M_{ij} can be written as,

$$\left[M_{ij}\right]\begin{Bmatrix}A_1\\A_2\\A_3\\A_4\end{Bmatrix}=\left[S_{ij}\right]\begin{Bmatrix}A_1\\A_2\\A_3\\A_4\end{Bmatrix}+\left[Q_{ij}\right]\begin{Bmatrix}A_3\\A_4\end{Bmatrix}, \qquad (3)$$

where, S_{ij} and Q_{ij} are the matrix in $S^{(e)}$ and $\Omega^{(e)}$, respectively. We can therefore take account of the insulator region without additional nodes: A_5 and A_6. In the insulator region, Q_{ij} can be expressed as follows:

$$Q_{ij}=\delta\int_\Gamma v\frac{\partial L_i}{\partial x}\frac{\partial L_j}{\partial x}d\Gamma \qquad (4)$$

where, $L_k(k=1,2)$ is the linear shape function and given by

$$L_1=\frac{1}{2}(1-\xi), \ L_2=\frac{1}{2}(1+\xi) \qquad (5)$$

3. Numerical Results

Figure 2 shows the analyzed model for the laminated core. A laminated core is constituted of the steel and insulator (the thickness of the steel: 3 mm, relative permeability: 2000, the thickness of the insulator: 0.03 mm, relative permeability: 1). In this analysis, because we formulate the special element as shown in Figure 1., the insulator is considered only on the under part of the steel. But we can consider over and under parts of the steel by laminating the special elements. The number of laminations is 20, the current density of the exciting coil is $3\times10^6 \ A/m^2$. In this analysis, we used the one order triangular element as the conventional method. Figures 3(a) and (b), show the distribution of the magnetic flux, Figures 4(a) ,(b) and (c), show the value at the vector potential of evaluated positions calculated by the conventional method and the new method. Table 1 shows the number of nodes and elements used in the analysis by the conventional and the new method.

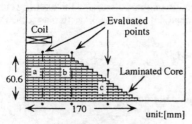

Fig.2 Analyzed model.

Table1 Number of nodes and elements.

	Number of nodes	Number of elements
Conventional method	5895	5720
Special element	1395	1320

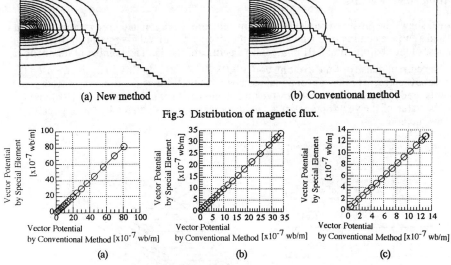

(a) New method   (b) Conventional method

Fig.3 Distribution of magnetic flux.

Fig.4 Distribution of vector potential.

4. Formulation for the three-dimensional problem

Generally, the Garelkin's equation for three-dimensional problems is written as follows:

$$v \int_{V^{(e)}} \left(\frac{\partial N_i}{\partial x} \frac{\partial N_j}{\partial x} + \frac{\partial N_i}{\partial x} \frac{\partial N_j}{\partial x} + \frac{\partial N_i}{\partial x} \frac{\partial N_j}{\partial x} \right) dV = \int_{V^{(e)}} N_i J dV ,$$ (6)

where, v is the magnetic reluctivity, N_k the shape function of the 3D element, J the exciting current density, $V^{(e)}$ the volume of each element, and A the magnetic vector potential. Figure 5 shows a special element considering the insulator. When the thickness of the insulator is very thin in comparison with the steel, A_9, A_{10}, A_{11} and A_{12} can be approximated as

$$A_9 = A_1, \quad A_{10} = A_2, \quad A_{11} = A_3, \quad A_{12} = A_4.$$ (7)

We can calculate the coefficient matrix for the three-dimensional problem as well as for the two-dimensional. The coefficient matrix S_{ij} in the region of the steel and Q_{ij} in the region of the insulator is shown as follows:

$$S_{ij} = v \int_{V^{(e)}} \left(\frac{\partial N_i}{\partial x} \frac{\partial N_j}{\partial x} + \frac{\partial N_i}{\partial y} \frac{\partial N_j}{\partial y} + \frac{\partial N_i}{\partial z} \frac{\partial N_j}{\partial z} \right) dV , \quad Q_{ij} = \delta \int_S v \sum_{j=1}^{4} \left(\frac{\partial M_i}{\partial x} \frac{\partial M_j}{\partial x} + \frac{\partial M_i}{\partial y} \frac{\partial M_j}{\partial y} \right) dS d\Omega ,$$

(8)   (9)

where, M_k is the shape function of the rectangular element, δ the thickness of the insulator.

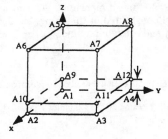

Fig.5 Special element considering the insulator.

5. Numerical Results

Figure 6 shows the analyzed model for a laminated core. A laminated core is constituted of steel and a surface coating film (thickness of the steel: 1.5 mm, relative permeability: 3000, thickness of the coating film: 0.01 mm, relative permeability: 1). The number of laminations is 20, the current density of the exciting coil is $1.0 \times 10^6 \, A/m^2$. In this analysis, we used the one order 3D element as the conventional method. Table 2 shows the number of nodes and elements used in the analysis by the conventional and new method. Figures 7(a), (b) and (c) show the value of the vector potential at evaluated positions calculated by the conventional and new methods.

Table 2 Number of nodes and element.

	Number of nodes	Number of elements
Conventional method	29791	20181
Special element	27000	18000

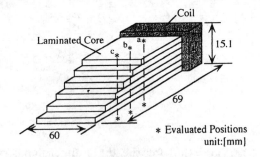

Fig.6 Analyzed model.

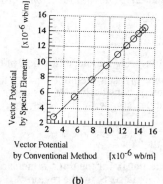

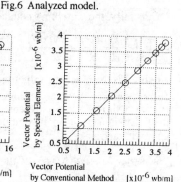

Fig.7 Comparison of the vector potential.

6. Conclusion

In this paper, we applied a special element considering the insulator to a simplified electromagnetic model for two and three dimensions. As a result, we found that the special element is useful for the model constituted of the laminated cores, and that it is possible to save a large amount of memory and calculation time.

References

[1] T.Nakata and N.Takahashi, The Finite Element Method in Electrical Engineering, Morikita-syutsupan, Tokyo, Japan, 1986.

[2] W.Hisaichirou, H.Miyamoto, Y.Yamada, Y.Yamamoto and T.Kawai, Finite Element Method Handbook, Baihuukan, Tokyo, Japan, 1981.

Non-Linear Electromagnetic Systems
V. Kose and J. Sievert (Eds.)
IOS Press, 1998

Control Parameter Calculation of Synchronous Reluctance Motor Using FEM and Preisach Modeling

*Jung-Ho LEE, *Sang-Baeck YOON, *Yeon-Ho SON, **Jung-Chul KIM and *Dong-Seok HYUN
Dept. of Electrical Engineering, Hanyang University, Seoul 133-791, KOREA
**LG Electronics Co., Living System Lab., Seoul 153-023, KOREA*

Abstract. It is necessary to calculate the exact parameters considering iron loss and saturation for a precise control characteristics analysis. This paper presents the calculation of the control parameters of a synchronous reluctance motor (SynRM) using the 2-D finite element method (FEM) and Preisach's model. Preisach's model is adopted for analysis of a SynRM with hysteresis characteristics. An iterative method, which does not diverge at any initial value and has the advantage of converging more quickly than the Newton-Raphson method, is used to provide a saturated field solution.

1. Introduction

The effects of saturation and iron losses are often an important issue in the performance of synchronous reluctance motors.[1]

The saturation effect in the d-axis of the rotor is very different from that of the q-axis because the nature of the magnetic paths is different. In the case of d-axis excitation, the saturation is the combined effect of saturation in the stator yoke, stator teeth, and rotor ribs, and can reduce the L_d inductance by as much as 50%.[2] In high speed applications iron losses can become the major cause of power dissipation. Therefore, whereas in other kinds of machines a rough estimation of iron losses can be accepted, their importance in SynRM justifies a greater effort in calculating them more precisely. Finite element methods are used primarily because of their ability to model the complicated internal structure within a SynRM and their ability to model magnetic saturation to a high degree of accuracy.

Preisach's model, which allows fast and accurate prediction of iron losses, is adopted in the procedure to provide a nonlinear solution .

In this paper, a coupled finite element analysis and Preisach model for a Synchronous reluctance motor (SynRM) is presented, and control parameters are calculated under the combined effects of saturation and iron losses.

2. Finite Element Formulation of Analysis Model

Considerable attention has been paid in the past to improving the rotor design of SynRM[3]. This paper's rotor structure was chosen according to the proposed rotor structure of Vagatii [1], without any attempt at optimization, and because of its simplicity.

In the proposed SynRM, displacement currents and eddy currents are negligible. Maxwell's equations can be written as

$$\nabla \times \vec{H} = \vec{J}_0, \qquad \nabla \cdot \vec{B} = 0, \qquad \vec{B} = \mu_0 \vec{H} + \vec{M}$$

where, $\vec{M}$ is the magnetization with respect to the magnetic intensity $\vec{H}$. Other symbols have the usual meaning.

The magnetic vector potential $\vec{A}$ and the equivalent magnetizing current $\vec{J}_m$ are expressed as follows

$$\vec{B} = \nabla \times \vec{A} , \quad \vec{J}_m = \frac{1}{\mu_0}\nabla \times \vec{M}$$

The Governing equation, derived from upper equations, is given by

$$\frac{1}{\mu_0}(\nabla \times \nabla \times \vec{A}) = \vec{J}_0 + \vec{J}_m$$

The vector potential is given as

$$A(x,y) = \sum_{i=1}^{n} N_i(x,y)A_i$$

where, n ; A number of unknown nodes, N_i ; Shape function for each element
Using the Galerkin method, the system matrix can be expressed as

$$\{K\}\{A\} = \{F\}$$

where, $\{K\}$; The finite element coefficient matrix,

$\quad\quad \{F\}$; The forcing matrix with the excitation current and the magnetization.

3. Application of Preisach's Model

One of the reasons for the recent interest in Preisach's model is the need to account for hysteresis effects of materials in numerical field computations.[4] In the Preisach model in this paper, the M-H relation is used instead of the B-H relation. Use of the M-H characteristic is shown to guarantee iteration stability because the slope approaches unity as μ tends to infinity. The Preisach theory describes the hysteresis of a magnetic material via an infinite set of magnetic dipoles, which have rectangular hysteresis loops, as shown in Fig.1. The magnetization M, including the magnetic field H, is expressed as

$$M = \int_S \mu(\alpha,\beta)\gamma_{\alpha\beta}(H)d\alpha d\beta$$
$$= \int_{S^+} \mu(\alpha,\beta)d\alpha d\beta - \int_{S^-} \mu(\alpha,\beta)d\alpha d\beta$$

where, S is the triangular region $H_{sat} \geq \alpha \geq \beta \geq -H_{sat}$ on the (α,β) plane in the Fig.2 (known as the Preisach's diagram), H_{sat} is the saturation magnetic field strength, α and β are the magnetic field strengths in the increasingly positive and negative directions, $\mu(\alpha,\beta)$ is the distribution function of the dipoles, $\mu(\alpha,\beta) = \mu(-\alpha,-\beta)$, and $\mu(\alpha,\beta) = 0$ if $(\alpha,\beta) \notin S$, $\gamma_{\alpha\beta}(H) = 1$ on the S^+, and $\gamma_{\alpha\beta}(H) = -1$ on the S^-. The interface between S^+ and S^- .is determined by the history and the present state of magnetization. Through certain function transforms, the area integration can be related to the hysteresis loops.

A more convenient treatment of this model is to substitute an Everett plane for the Preisach one. The relation of Preisach's plane $\int_S \mu(\alpha,\beta)\gamma_{\alpha\beta}(H)d\alpha d\beta$ and the Everett one $E(\alpha,\beta)$ is given as

$$E(\alpha,\beta) = \int_S \mu(\alpha,\beta)\gamma_{\alpha\beta}(H)d\alpha d\beta$$

In the Everett plane, the distributions of H have Gaussian ones.[5]

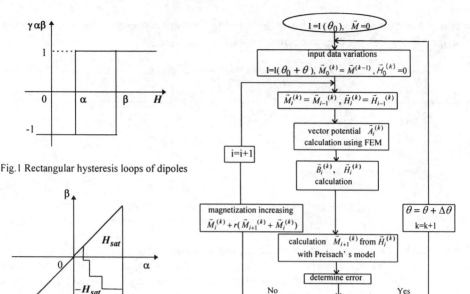

Fig.1 Rectangular hysteresis loops of dipoles

Fig. 2 Typical Preisach diagram   Fig.3 Flowchart of FEM using Preisach's model

4. Results of simulation

The flow chart for the proposed analysis method is shown in Fig. 1. It is assumed that the d-axis current is applied to the proposed the SynRM and the rotor rotates synchronously to the d axis current. $\theta_0 (\pi/3)$ is the initial angle to align the electrical d-axis with the mechanical d-axis. The q-axis simulation uses the same procedure as the d-axis one. Flux distribution for d-axis and q-axis excitation, for the SynRM with a conventional stator having 36 slots and four poles, are shown in Fig.4.

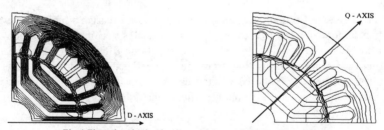

Fig.4 Flux plots in the d axis excitation and the q axis excitation

Fig.4 shows that the saturation effect in the d-axis is very different from that of the q-axis. As shown in Fig.4, the saturation in the SynRM having segmental rotor type is particularly significant in the rotor ribs and the stator teeth. This saturation can reduce the Ld inductance, which can have an effect on the high torque density, power factor and efficiency of SynRM.

The iron losses in the stator are of similar origin to those in a conventional induction machine stator. The rotor would have almost zero loss since the spatial flux wave in a SynRM is rotating at the same angular velocity as the rotor, and therefore the rotor flux wave is constant. Fig.5 shows the hysteresis loops in the stator teeth according to input current variations(1[A]-4[A]). It is clear from Fig.4,5 that the iron core losses and saturation in the d-axis are larger than those for the q-axis.

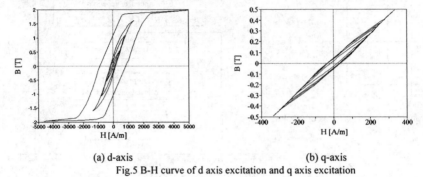

(a) d-axis   (b) q-axis

Fig.5 B-H curve of d axis excitation and q axis excitation

Fig 6 shows the d-axis inductance(Ld) and q-axis inductance(Lq) for d-axis excitation and q-axis excitation respectively. Ld and Lq decrease according as the excitation current increases. Particularly, these indicate the large Ld variation according to angle, which leads to reluctance torque ripple.

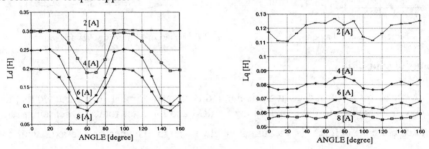

Fig.6 Ld and Lq at each angle

5. Conclusion

The paper presents a FE nonlinear analysis coupled to a Preisach's model for a SynRM. The method is good for iron losses and saturation analysis, compared to a conventional nonlinear analysis. The iteration times decrease about one fifth, whilst control parameters can be calculated more precisely than from the conventional method because this method considers iron losses. The proposed method can be used as a numerical analysis tool to improve the design procedure of nonlinear fields including iron core loss and magnetic saturation

References

[1] A.Vagati and T.A. Lipo, " Synchronous Reluctance motors and Drives A new Alternative," Oct. 2 1994 at the 1994 IEEE IAS annual meeting Denver Co.

[2] R.E.Betz, R. Lagerquist, M. Jovanovic, T.J.E. Miller, R.H.Middleton, "Control of Synchronous Reluctance Machines," IEEE Trans. on IAS, Vol.29, No.6, Nov. 1993.

[3] A.Vagati and G.Franceschini, I. Marongiu, G.P. Troglia: "Design Criteria of High performance Synchronous Reluctance Motors," IEEE-IAS Annual Meeting, Houston, U.S.A., October 1992.

[4] Isaak D. Mayeroyz, "Mathematical models of Hysteresis," IEEE Trans. In Magnetics, Vol. MAG-22, No.5, pp.603-608 Sept. 1986

[5] R.M. Del Vecchio "An Effect Procedure for Modeling Complex Hysteresis Process in Ferromagnetic Materials" IEEE Trans on Magnetics, Vol MAG-16, No. 5, Sept. 1980

Non-Linear Electromagnetic Systems
V. Kose and J. Sievert (Eds.)
IOS Press, 1998

Analysis of Dynamic Characteristics of Permanent Magnet Linear Synchronous Motor Using FEM

B. I. Kwon[a], S. H. Rhyu[a], K. I. Woo[a] and S. C. Park[b]

[a] *Dept. of Electrical Engineering, Hanyang University, Ansan 425-791, Korea*
[b] *Research Institute of Engineering & Technology, Ansan 425-791, Korea*

Abstract This paper deals with the dynamic characteristic analysis of a permanent magnet linear synchronous motor with the secondary aluminium sheet, using time - stepped finite element method (FEM). The conductivity of the secondary aluminium sheet and solid back-iron is taken into account in the electromagnetic field analysis. As a result, we can investigate dynamic speed characteristics as well as some electromagnetic performances such as thrust, attractive force and input power, etc.

1. Introduction

The use of a powerful permanent magnet material such as NdFeB alloy has allowed significant improvements on the performance of electrical machines. Especially, permanent magnet linear synchronous motor could obtain high thrust and linear speed without mechanical energy conversion device, which results in the wide spread applications such as in linear elevator or tool machines.

Some researchers have analyzed the steady-state characteristics of permanent magnet linear synchronous motors [1,2,3]. Static thrust has been analyzed by 2-dimensional FEM [1,2], and the effect of damper winding has also been investigated by 3-dimensional FEM followed by a sequence of experiments [3]. However, previous works didn't analyze the dynamic characteristics based on transient magnetic field.

This paper deals with the analysis of dynamic characteristics of a permanent magnet linear synchronous motor with aluminium sheet, which has damper winding effect. Because asynchronous speed of the motor during transient period induces the eddy currents on the secondary conductive sheet and solid back-iron, these phenomena are considered. Also, in simulation results, thrust, attractive force, speed of the mover and input power are shown under the drive condition of constant input voltage and frequency.

2. Finite Element Analysis

Magnetic vector potential A is in the z-direction only provided that all currents flow in the z-direction only. The governing equation for the analysis model of synchronous motor with aluminium sheet as shown in Fig. 1 is expressed as

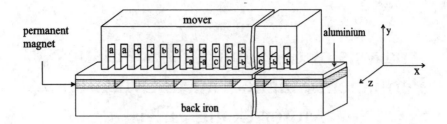

Fig. 1. Analysis model of permanent magnet linear synchronous motor
with the secondary aluminium sheet.

$$\frac{1}{\mu}\frac{\partial^2 A}{\partial x^2} + \frac{1}{\mu}\frac{\partial^2 A}{\partial y^2} = -J_o + \sigma\frac{\partial A}{\partial t} - \frac{1}{\mu_r}\left(\frac{\partial M_{ry}}{\partial x} - \frac{\partial M_{rx}}{\partial y}\right),$$ (1)

where A is vector potential in the z-direction, M_{ry} and M_{rx} are remanent magnetization in the
y- and x- direction respectively, and σ is the conductivity of the secondary. Because the
moving coordinate system is utilized, the emf in Eq. (1) has time derivative component only
[4].

The pole pitch of analysis model is 40[mm], the stack height is 60[mm], the number of
pole is 6, series turns per phase is 216[turns], and aluminium sheet depth is 0.5[mm].

Using the Galerkin's method introducing shape function N_j into weighted function, we can
obtain the Eq. (2) in one element.

$$I_{je} = \int_{S^e}\frac{1}{\mu}\sum_{i=1}^{3}\left(\frac{\partial N_{ie}}{\partial x}\frac{\partial N_{je}}{\partial x} + \frac{\partial N_{ie}}{\partial y}\frac{\partial N_{je}}{\partial y}\right)A_{ie}dxdy - \int_{S^e}\frac{1}{\mu_r}\left(M_{rx}^e\frac{\partial N_{je}}{\partial y} - M_{ry}^e\frac{\partial N_{je}}{\partial x}\right)dxdy$$

$$- \int_{S^e}J_0 N_{je}dxdy + \sigma\frac{\partial}{\partial t}\int_{S^e}\sum_{i=1}^{3}N_{ie}N_{je}A_{ie}dxdy \qquad (j=1,2,3).$$ (2)

Integrating Eq. (2) over all elements leads to Eq. (3), where unknown variables are vector
potentials at nodes and phase currents,

$$\begin{bmatrix}[S] & -[C]\end{bmatrix}\begin{bmatrix}[A]\\[I]\end{bmatrix} + \frac{\partial}{\partial t}\begin{bmatrix}[T][0]\end{bmatrix}\begin{bmatrix}[A]\\[I]\end{bmatrix} = [G].$$ (3)

In Eq. (3), [S] is a coefficient matrix related to node positions and permeabilities, [C] and
[T] are coefficient matrixes for coil current density and eddy current density respectively,
and [G] is a driving matrix corresponding to equivalent magnetization current density. Eq.
(3) is combined with circuit equations so that

$$\frac{d}{dt}[\Psi] + [L_0]\frac{d}{dt}[I] + [R][I] = [V],$$ (4)

where [V] : voltage vector, [I] : phase current vector, [R] : phase resistance vector, $[L_0]$: the
primary end winding leakeage inductance vector, [Ψ] : magnetic flux linkage vector.
Because [Ψ] can be expressed as vector potential A, Eq. (4) becomes

$$h_{eff}[C]^T\frac{\partial}{\partial t}[A] + [L_0]\frac{d}{dt}[I] + [R][I] = [V],$$ (5)

where h_{eff} is the effective stack height. Therefore, the system matrix in Eq. (6) could be
obtained by using Eqs. (3), (5) and backward difference method in the time derivative
approximation.

$$\left[[S] + \frac{[T]}{\Delta t} \quad -[C] \atop -[C]^T \quad -\frac{[L_0] + \Delta t[R]}{h_{eff}} \right] \left[[A] \atop [I] \right]_{t+\Delta t} = \left[\frac{[T]^T}{\Delta t} \quad [0] \atop -[C]^T \quad -\frac{[L_0]}{h_{eff}} \right] \left[[A] \atop [I] \right]_t + \left[[G] \atop \frac{\Delta t}{h_{eff}}[V] \right]_{t+\Delta t} \quad (6)$$

3. Simulation results

The flow chart of this analysis is illustrated in Fig. 2. Analysis of the static magnetic field set up by permanent magnet only is carried out to satisfy initial condition for applying backward difference approximation in Eq. (6). Thrust F_x and attractive force F_y are calculated by Maxwell stress tensor method given by

$$F_x = \frac{1}{\mu_0} \int h_{eff} \left[\{ (B_x^2 + B_y^2) n_x + 2n_y B_x B_y \}/2 \right] dL, \quad (7)$$

$$F_y = \frac{1}{\mu_0} \int h_{eff} \left[\{ (B_y^2 + B_x^2) n_y + 2n_x B_x B_y \}/2 \right] dL, \quad (8)$$

where n_x and n_y are x and y component of unit normal vector on integral path respectively. The speed of mover, v, could be obtained by solving the equation of motion,

$$F_x = (M + M') \frac{dv}{dt} + F_l, \quad (9)$$

where M is the primary mass, F_l is load thrust(20N), and M' is the equivalent mass due to attractive force. Fig. 3 shows magnetic flux plots according to the movement determined by Eq. (9), when the supplied voltage is 7[V] and the frequency is 2[Hz].

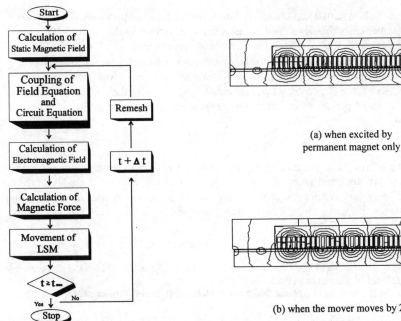

(a) when excited by permanent magnet only

(b) when the mover moves by 2.6 [mm]

Fig. 2. The flow chart of dynamic analysis

Fig. 3. Some flux plots

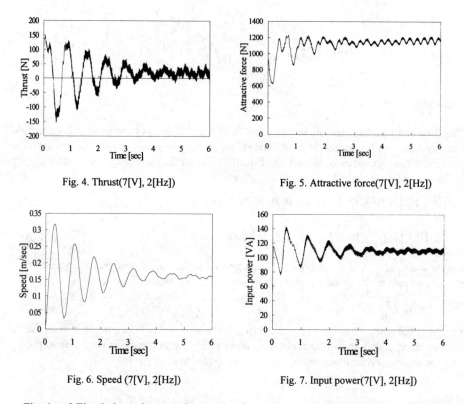

Fig. 4. Thrust(7[V], 2[Hz])

Fig. 5. Attractive force(7[V], 2[Hz])

Fig. 6. Speed (7[V], 2[Hz])

Fig. 7. Input power(7[V], 2[Hz])

Fig. 4 and Fig. 5 show thrust and attractive force characteristic respectively. The attractive force is much larger than the developed thrust values as shown in the figures. The speed characteristic is shown in Fig. 6. The mover arrives at synchronous speed, 0.16[m/sec] after about 5 [sec]. The settling time reaching to steady-state synchronous speed is somewhat long because of excessive attractive force. However this attractive force values are the resultants reduced by repulsive force due to eddy currents induced on the secondary aluminium sheet in the asynchronous speed period. Also, input power characteristic is shown in Fig. 7.

4. Conclusion

Dynamic characteristics of permanent magnet linear synchronous motor with secondary aluminium sheet are analyzed using the time-stepped FEM considering the eddy currents induced on the secondary conductive sheet during the asynchronous speed. As a result, the instantaneous thrust, attractive force, speed of the mover and input power are investigated.

References

[1] Tsutomu Mizuno, Hajime Yamada, Magnetic Circuit of a Linear Synchronous Motor with Permanent Magnets, IEEE Trans. on Magnetics 5 (1992) 3027-3029.
[2] Gieras Jacek F., et al., Analysis of a Linear Synchronous Motors with Buried Permanent Magnets, Proc. of LDIA (1995) 323-326.
[3] Takashi Onuki, yushi Kamiya, Jun Enomoto, et al., Secondary Conduction of Linear Synchronous Motor with Divided Permanent Magnets, T. IEE Japan, 116-D, 1 (1996) 88-93
[4] K.Muramatsu, T.Nakata, N.Takahashi and K.Fujiwara, Comparison of Coordinate Systems for Eddy Current Analysis in Moving Conductors, IEEE Trans. on Magnetics 2 (1992) 1186-1189.

Non-Linear Electromagnetic Systems
V. Kose and J. Sievert (Eds.)
IOS Press, 1998

Analysis of Magnetic Field Problems by Using Upwind-FEM with Discrete Derivative Operator

Masato ENOKIZONO, Takashi TODAKA and Masafumi NOMURA

*Dept. of Electrical and Electronic Engineering, Faculty of Engineering,
Oita University, 700 Dannoharu, 870-11 Oita, Japan*

Abstract. This paper presents an application of the finite element method with an upwind interpolation function and its adjoint interpolation function for magnetic field problems. This method is based on the finite analytic method developed in the field of the finite difference method. Furthermore a discrete derivative operator is used in discretizing the finite element equation to simplify the formulation. In this paper, we demonstrate the effectiveness of usage of these techniques in eddy current problems.

1. Introduction

Travelling magnetic field problems are governed by the convection-diffusion equation. When such problems are analyzed by using the standard Galerkin finite element method, it is well known that unnecessary oscillating solutions appear depending on the Pechlet number. To avoid this inconvenience, several upwinding techniques are suggested in the field of the computational hydrodynamics [1-2]. However, reports on stabilization of the oscillations in eddy current problems are very few.

Recently, an upwind interpolation function is proposed in the field of the finite analytic method, and introduced into the field of the finite element method for hydraulic calculations by Tanahashi [3-4]. This interpolation function can be applied to a low Pechlet number problem as it is. Furthermore a discrete derivative operator defined as an elemental average of the gradient of the interpolation function is also suggested in the same field [4-5]. This operator allows easy coding and reduction of memory storage in the computer. In this paper, we apply these techniques to a travelling magnetic field problem and compare the results with the conventional ones obtained by the standard Galerkin finite element method. The example shows the effectiveness of this method.

2. Formulation

The governing equation of a two-dimensional eddy current problem that includes a conductive object in motion, is expressed as the following advection-diffusion equation:

$$v \nabla^2 A = \sigma \left(\frac{\partial A}{\partial t} + \frac{\partial \phi}{\partial z} + V_x \frac{\partial A}{\partial x} + V_y \frac{\partial A}{\partial y} \right) - J_0 , \tag{1}$$

where, A is the z-directional component of the magnetic vector potential, J_0 the exciting current density, v the magnetic resistivity, σ the conductivity, and ϕ the electric scalar potential. V_x and V_y are the components of the velocity in x- and y-direction, respectively. In the two-dimensional problems, the value of $(\partial \phi / \partial z)$ becomes constant in each conductor:

$$\frac{\partial \phi}{\partial z} = \frac{- \int_\Omega \left(\frac{\partial A}{\partial t} + V_x \frac{\partial A}{\partial x} + V_y \frac{\partial A}{\partial y} \right) d\Omega}{S}, \tag{2}$$

where, S is the cross-sectional area of the conductor. The finite element equation of Eq. (1)

in a static field can be expressed as follows:

$$K_{ij} A_j + \sigma C_i \left(\frac{\partial \phi}{\partial z}\right) = C_i J_0 ,\tag{3}$$

where, K_{ij} or C_i is the coefficient matrix. Combining Eq. (2) with Eq. (3), we can carry out the magnetic field analysis.

The numerical solution obtained by the discretized equation with the standard Galerkin method (GFEM), will oscillate between the adjacent nodes depending on the physical and dimensional parameter, that is, the cell Pechlet number (the cell magnetic Reynolds number). To avoid the oscillation solutions, we use the following upwind interpolation function L_i and its adjoint upwind interpolation function L_i^* for the tetragonal element (Fig. 1) [3-4].

$$L_i = \frac{\xi_i\left(e^{\lambda\xi} - e^{-\lambda\xi_i}\right)}{e^\lambda - e^{-\lambda}} \cdot \frac{\eta_i\left(e^{\kappa\eta} - e^{-\kappa\eta_i}\right)}{e^\kappa - e^{-\kappa}} , \quad L_i^* = \frac{\xi_i\left(e^{\lambda\xi_i} - e^{-\lambda\xi}\right)}{e^\lambda - e^{-\lambda}} \cdot \frac{\eta_i\left(e^{\kappa\eta_i} - e^{-\kappa\eta}\right)}{e^\kappa - e^{-\kappa}} ,\tag{4}$$

$$\lambda = \frac{\sigma V_x \Delta_x}{2v}, \quad \kappa = \frac{\sigma V_y \Delta_y}{2v} ,\tag{5}$$

where, λ or κ is the half value of the cell Pechlet number in each direction. Δ_x and Δ_y are the mean lengths of the tetragonal element in x- and y-direction, respectively. The upwind interpolation function can be derived from the static one-dimensional convection-diffusion equation. The function L_i corresponds to the linear interpolation function in the limit when λ and κ approach zero. In this case the coefficient matrices are written as follows:

$$K_{ij} = v \int_\Omega \left(\frac{\partial L_i^*}{\partial x}\frac{\partial L_j}{\partial x} + \frac{\partial L_i^*}{\partial y}\frac{\partial L_j}{\partial y}\right) d\Omega + \sigma \int_\Omega \left(V_x L_i^* \frac{\partial L_j}{\partial x} + V_y L_i^* \frac{\partial L_j}{\partial y}\right) d\Omega , \quad C_i = \int_\Omega L_i^* d\Omega .\tag{6}$$

The coefficient matrices can be simplified in the following section using the idea of elemental average, that is, the discrete derivative operator.

3. Discrete Derivative Operator

Here, we consider two smooth functions f and g. The elemental average of the function f is defined by Eq. (7). Also the elemental average of the product of f and g can be approximated with Eq. (8) in a sufficiently small element.

$$\langle f \rangle_e = \frac{1}{\Omega_e} \int_{\Omega_e} f \, d\Omega , \quad (7) \qquad \langle fg \rangle_e = \frac{1}{\Omega_e} \int_{\Omega_e} fg \, d\Omega \approx \frac{1}{\Omega_e} \int_{\Omega_e} \langle f \rangle_e \, g \, d\Omega = \langle f \rangle_e \langle g \rangle_e ,\tag{8}$$

where, Ω_e is the area of the tetragonal element. In the same way, we define the discrete derivative operator as an elemental average of the gradient of the interpolation function L_i. We call this operator "the discrete del operator." The adjoint operator is also defined for L_i^*.

$$\nabla_i = \frac{1}{\Omega_e} \int_{\Omega_e} \nabla L_i \, d\Omega , \quad \nabla_i^* = \frac{1}{\Omega_e} \int_{\Omega_e} \nabla L_i^* \, d\Omega .\tag{9}$$

Both of the discrete del operators have two components similar to the vector gradient operator: $\nabla_i = \left\{ (\nabla_i)_x, (\nabla_i)_y \right\}, \ \nabla_i^* = \left\{ (\nabla_i^*)_x, (\nabla_i^*)_y \right\}$. In order to obtain the discrete del operators, we need the inverse Jacobian matrix:

$$[J_{ij}]^{-1} = \frac{1}{J}[G_{ij}], \quad J = det\,(J_{ij}),\tag{10}$$

where, G_{ij} is the cofactor of the Jacobian matrix. Each component of the discrete del operator can be written with this cofactor as follows:

$$(\nabla_i)_x = \frac{1}{\Omega_e}\int_{\Omega_e}\left(G_{11}\frac{\partial L_i}{\partial \xi} + G_{12}\frac{\partial L_i}{\partial \eta}\right)d\Omega', \qquad (\nabla_i)_y = \frac{1}{\Omega_e}\int_{\Omega_e}\left(G_{21}\frac{\partial L_i}{\partial \xi} + G_{22}\frac{\partial L_i}{\partial \eta}\right)d\Omega'. \qquad (11)$$

where, $d\Omega' = d\xi\,d\eta$. Hence the discrete del operator can be rewritten with the approximation of Eq. (8):

$$(\nabla_i)_x = \frac{4}{\Omega_e}\left\{\langle G_{11}\rangle_e\left\langle\frac{\partial L_i}{\partial \xi}\right\rangle_e + \langle G_{12}\rangle_e\left\langle\frac{\partial L_i}{\partial \eta}\right\rangle_e\right\}, \qquad (\nabla_i)_y = \frac{4}{\Omega_e}\left\{\langle G_{21}\rangle_e\left\langle\frac{\partial L_i}{\partial \xi}\right\rangle_e + \langle G_{22}\rangle_e\left\langle\frac{\partial L_i}{\partial \eta}\right\rangle_e\right\}. \qquad (12)$$

The cofactors and the elemental averages of the derivative of the interpolation function in Eq. (12) are given by

$$\langle G_{11}\rangle_e = (y_1 + y_2 - y_3 - y_4)/4, \qquad \langle G_{12}\rangle_e = -(y_1 - y_2 - y_3 + y_4)/4,$$

$$\langle G_{21}\rangle_e = -(x_1 + x_2 - x_3 - x_4)/4, \qquad \langle G_{22}\rangle_e = (x_1 - x_2 - x_3 + x_4)/4, \qquad (13)$$

$$\left\langle\frac{\partial L_i}{\partial \xi}\right\rangle_e = \frac{\xi_i\,\eta_i}{4}\left(\frac{1}{\kappa} - \frac{e^{-\kappa\eta_i}}{sinh\,\kappa}\right), \qquad \left\langle\frac{\partial L_i}{\partial \eta}\right\rangle_e = \frac{\xi_i\,\eta_i}{4}\left(\frac{1}{\lambda} - \frac{e^{-\lambda\eta_i}}{sinh\,\lambda}\right). \qquad (14)$$

The similar expression can be also obtained for the adjoint operator. The coefficient matrices, K_{ij} and C_i are therefore rewritten with these elemental averages as follows:

$$K_{ij} = \nu\,\Omega_e\left\{(\nabla_i^*)_x(\nabla_j)_x + (\nabla_i^*)_y(\nabla_i)_y\right\} + \sigma\,\Omega_e\left\{V_x\,\langle L_i^*\rangle_e(\nabla_j)_x + V_y\,\langle L_i^*\rangle_e(\nabla_j)_y\right\}, \qquad (15)$$

$$C_i = \Omega_e\,\langle L_i^*\rangle_e, \qquad (16)$$

$$\left\langle\frac{\partial L_i^*}{\partial \xi}\right\rangle_e = -\left\langle\frac{\partial L_i}{\partial \xi}\right\rangle_e, \qquad \left\langle\frac{\partial L_i^*}{\partial \eta}\right\rangle_e = -\left\langle\frac{\partial L_i}{\partial \eta}\right\rangle_e, \qquad \langle L_i^*\rangle_e = \frac{\xi_i\,\eta_i}{4}\left(\frac{1}{\lambda} - \frac{e^{\lambda\eta_i}}{sinh\,\lambda}\right)\left(\frac{1}{\kappa} - \frac{e^{\kappa\eta_i}}{sinh\,\kappa}\right). \qquad (17)$$

In the case of the expression (6), generally we have to use a numerical integration formula to obtain the coefficient matrices. Using the discrete del operators as mentioned in the above, those matrices can be calculated directly. As a result, the computation time can be reduced in comparison with the conventional one.

4. Results and Discussion

The method presented here was applied to a simple inductor model as shown in Fig. 2 to validate the accuracy of this scheme. In this model, we assumed that the thin conductive plate was sliding in x-direction at a fixed velocity. The plate was placed inside the gap of the electromagnet. The martial constants assumed in the aluminum plate are: the permeability $\mu = 4\pi \times 10^{-7}$ H/m, the conductivity $\sigma = 2.32 \times 10^7$ S/m. Fig. 3 shows the flux distributions obtained by using the standard Galerkin method.

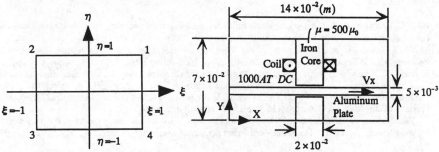

Fig. 1. Tetragonal element.   Fig. 2. A simple inductor model.

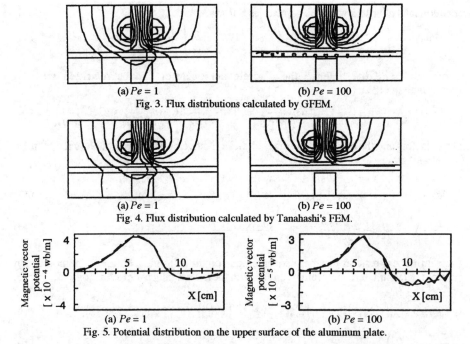

(a) $Pe = 1$ (b) $Pe = 100$
Fig. 3. Flux distributions calculated by GFEM.

(a) $Pe = 1$ (b) $Pe = 100$
Fig. 4. Flux distribution calculated by Tanahashi's FEM.

(a) $Pe = 1$ (b) $Pe = 100$
Fig. 5. Potential distribution on the upper surface of the aluminum plate.

The region under consideration was divided into 924 tetragonal elements. The maximum mean length of the elements in x-direction was 5 mm, thus the cell Pechlet number became 1 when the velocity V_x was equal to 6.85 m/s. That was the limit to obtain a stable solution with this arrangement of the elements as shown in Fig. 3 (a). In the case of Pe = 100 (Fig. 3 (b)), the oscillation solution occurred in the conductive plate region. Fig. 4 shows the flux distributions calculated with our method. The oscillation solution was successfully suppressed as shown in Fig. 4 (b). Fig. 5 shows the comparison of the calculated values of the magnetic vector potential on the upper surface of the aluminum plate.

5. Conclusion

In this paper, we applied Tanahashi's upwind scheme and discrete method to the magnetic field problem. This method was very effective in the performance of stable analysis for the high and low Pechlet number problems. In addition, the method using the discrete del operator enabled easy coding and reduction of memory storage in the computer. The advantages of this method are that the computing time can be reduced and the motional term can be calculated accurately in comparison with those of the conventional methods. Extension of this method in three-dimensional problems is also possible.

References

[1] E. K. C. Chan and S. Williamson, "Factors Influence the Need for Upwinding in Two-Dimensional Field Calculation," IEEE Trans. on Magn., Vol. 28, No. 2, pp. 1611-1614, 1992.
[2] M. Ito, T. Takahashi, and M. Odamura, "Up-Wind Finite Element Solution of Travelling Magnetic Field Problem," IEEE Trans. on Magn., Vol. 28, No. 2, pp. 1605-1610, 1992.
[3] E. Inada, K. Yamashita, and T. Tanahashi, "New Hybrid-Streamline-Upwind Finite Element Method in a Dual Space, Trans. of Japan Society of Mechanical Engineers, Vol. B62, No. 597, pp. 1685-1692, 1996.
[4] T. Tanahashi and Y. Oki, " New Theory of Hybrid-Upwind Technique and Discrete Del Operator for Finite Element Method," Comp. Fluid Dyn., Vol. 7, pp. 3-14, 1996.
[5] T. Tanahashi and T. Yamamoto, "Discrete Del Operator Using Streamline-Upwind Shape Function for Mixed-Element Method," Transactions of Japan Society of Mechanical Engineers, Vol. B62, No. 599, pp. 2660-2667, 1996.

Non-Linear Electromagnetic Systems
V. Kose and J. Sievert (Eds.)
IOS Press, 1998

A Numerical Method for Thin Structures: An FEM-BEM Hybridization for the Analysis of Ferromagnetic Layers

H. Igarashi, D. Lederer*, T. Honma and A. Kost*

Faculty of Engineering, Hokkaido University, Japan
*Institut für Elektrische Energietechnik, TU Berlin, Einsteinufer 11, D-10587, Germany

Abstract. This paper describes a numerical method for analysis of electromagnetic fields around thin ferromagnetic layers with time-dependent source currents. In this method, the electromagnetic fields inside the thin layer are assumed to obey the one dimensional magnetic-diffusion equation while the field in air around the layer is assumed to be a quasi-static magnetic field in full dimensions. The nonlinearity in the magnetic characteristics of the ferromagnetic material is taken into account. The interior and exterior fields are analyzed by finite and boundary element methods, respectively. The present method gives results similar to experimental results for a shielding thin plate.

1. Introduction

Electromagnetic fields with voluminous material have been well analyzed by means of the finite element method (FEM) or boundary element method (BEM), even when the material has nonlinear electromagnetic properties. However, when the thickness of the material is extremely thin compared to its overall size, there is a significant difference between the scale of the spatial change in the electromagnetic fields in the direction of the thickness and of that in the transverse direction. When such a field is analyzed by the FEM, the difference between element sizes in both directions must be within a reasonable extent to get a well conditioned matrix. This results, however, in a huge number of unknowns with unnecessary discretization in the transverse direction. Instead, if very thin elements are employed to avoid this difficulty, then the resultant matrix will become ill-conditioned and consequently the iterative solution of the matrix system becomes time-consuming. This problem would also undermine BEM.

When the material is assumed to possess linear permeability, the above difficulty can be overcome by introducing the impedance boundary condition (IBC) on the surface of the material [1-3]. In this method the electromagnetic field inside the material is analytically evaluated provided that it obeys the one dimensional equation considering its significant spatial change in the direction of thickness. This approximation leads to the IBC which describes the relationship between the field quantities on both sides of the material surface. The outer field in air is then numerically determined under the IBC. The authors have shown that the IBC can be expressed in terms of the vector potential, and this leads to a suitable formulation for analysis. It is, however, difficult to analyze fields including material with nonlinear permeability by using the IBC.

In this paper, we describe a numerical method for analysis of thin material with nonlinear permeability. In this method the electromagnetic field inside the material is analyzed by means of one dimension FEM to get the boundary conditions on the surface of the material. Further the BEM is employed to determine the quasi-static magnetic field around the material under the boundary conditions which the FEM gave. These processes are repeated iteratively to get a self-consistent solution. The results for a magnetic shielding system will be discussed and compared them with experimental results.

2. Formulation

Let us consider a closed thin layer Γ with thickness d immersed in a time-varying magnetic field (see Fig. 1). The layer Γ will consist of an air layer as well as ferromagnetic layer.

The thickness of Γ is assumed to be sufficiently thin compared to its overall size, so that the spatial change in the field in the direction of thickness would be dominant compared to the changes in the tangential directions. This assumption leads to the one-dimensional equation for the vector potential as

$$\sigma \frac{\partial A}{\partial t} = \frac{\partial}{\partial n}\left(\frac{1}{\mu}\frac{\partial A}{\partial n}\right) + \frac{\partial M}{\partial n},\tag{1}$$

where A is the z-component of the vector potential, n the local coordinate whose direction is parallel to the direction of thickness, μ the permeability, σ conductivity, and M is a tangential component of the magnetization vector. The weighted residual expression of (1) with the backward Euler method can be written by

$$\int_0^d wA^k dn + \frac{\Delta t}{\sigma}\int_0^d \frac{1}{\mu}\frac{\partial w}{\partial n}\frac{\partial A^k}{\partial n}dn + \frac{\Delta t}{\sigma}\int_0^d \frac{\partial w}{\partial n}M^k dn = \int_0^d wA^{k-1}dn,\tag{2}$$

where w is a weighting function. Discretizing (2) by FEM, we solve it under prescribed boundary conditions for A_i^k on the two sides of Γ, belonging to Ω_1 and Ω_2, which correspond to $n = 0, d$, respectively. Since M and μ depend on A^k due to their nonlinearlity, the Newton-Raphson scheme is employed for the solution of the FE equations. The normal derivatives $\partial A_i^k/\partial n$ obtained by the FEM process are used as boundary conditions for the next BEM process. The boundary values are connected across the air-material interface by considering that A is continuous there while $1/\mu_0(\partial A/\partial n)_{air} = 1/\mu(\partial A/\partial n)_{material} + M$. The magnetic field with source current J_s around the material obeys Ampere's law

$$\nabla^2 A = -\mu_0 J_s.\tag{3}$$

The integral equations corresponding to (3) at time step k can be written in the form

$$\eta_i A_i^k = \pm\int_\Gamma \frac{\partial A_i^k}{\partial n}G ds \mp \int_\Gamma \frac{\partial G}{\partial n}A_i^k ds + \mu_0\int_\Gamma G J_{si}^k ds,\tag{4}$$

where η_i are constants coming from the Cauchy singularity in the integrals, G is the two-dimensional free space Green function, upper and lower signs correspond to the regions Ω_1 and Ω_2 respectively, and suffix i represents the region Ω_i. Equation (4) is solved for the vector potential A_i^k by BEM under the boundary conditions for $\partial A_i^k/\partial n$. The resultant A_i^k are used as boundary conditions for the next FEM process. The above mentioned FE-BE processes are repeated until the results converge at each time step.

3. Results for a shielding problem

The present method is applied to the analysis of a shielding problem, whose schematic is shown in Fig. 2. The anti-parallel currents below a ferromagnetic thin plate generates magnetic fields at 50 [Hz]. The measured value of σ is $6.41\times10^6[1/\Omega m]$. Although the plate has finite length, 2.87 [m], in the direction parallel to the currents, this system is assumed to be two dimensional. Note that the ratio of the thickness of the plate to the width is less than 0.002.

We first analyze this problem assuming the linear permeability, $\mu = 1000\mu_0$, to test the validity of the present method. In the analysis, 60 constant boundary elements and 10×20 second-order finite elements are used. The number of time steps in one period is taken to be 20. This problem is also solved by a usual FEM based on $j\omega$-formulation with adaptive mesh generation to allow comparison of both results. The magnitude of magnetic induction B at $y=0.05$ [m] is plotted in Fig. 3. The agreement between these results seems satisfactory. Since the skin depth, evaluated to be 0.89 [mm], for this case is less than one third of the thickness of the plate, the result is thought to be significantly influenced by eddy currents. The vector plot of the magnetic field obtained by the present method is shown in Fig. 4.

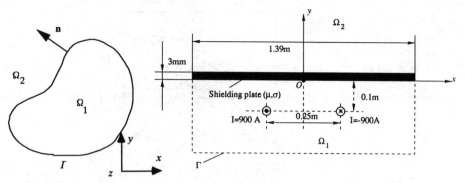

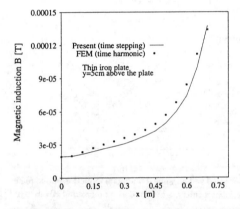

Fig. 1 Thin closed layer

Fig. 2 Shielding plate above anti-parallel currents

Fig. 3 Magnetic induction above plate (linear permeability)

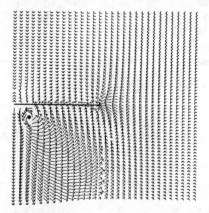

Fig. 4 Magnetic field distribution around plate

The present method is now applied to the nonlinear problem. The measured BH characteristics of the ferromagnetic plate are shown in Fig. 5. We compute the fields for two cases; in the first case the BH curve is assumed to be the initial curve without hysteresis, and in second case the full hysteresis is taken into account. In the latter case the measured BH data are

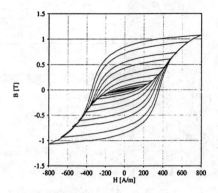

Fig. 5 Measured BH characteristics

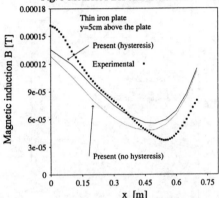

Fig. 6 Measured and computed magnetic induction above plate.
Computational results are obtained using measured BH data in Fig. 5.

interpolated to obtain curves between the measured hysteresis curves. The results are shown
in Fig. 6. In contrast to the result for linear permeability shown in Fig. 3, the measured
magnetic induction has a minimum near the edge of the plate, i.e., $x=0.695$ [m]. Both
computed results also indicate this property, although they are not very close to the measured
data. This discrepancy may be caused by the anisotropy of the material, or numerical errors,
or three dimensional effects.

4. Conclusion

This paper has described a numerical method for analysis of electromagnetic fields around
thin ferromagnetic material. The present method gives the spatial distribution of magnetic
induction for a shielding system, which agrees well with that obtained by conventional
FEM under the assumption of linear permeability. The magnetic fields are also computed
by the present method using the measured BH data with hysteresis. There are still discrepancies
between the computed and measured magnetic field distributions although those quantitative
tendencies are similar.

References
[1] L. Krähenbühl and D. Muller, Thin layers in electrical engineering. Example of shell models in analysing
    eddy-currents by boundary and finite element methods, *IEEE Trans. Mag.*, **29** (1993) 1450-1455.
[2] I.D. Mayergoyz and G. Bedrosian, On calculation of 3-D eddy currents in conducting and magnetic shells,
    IEEE Trans. Mag., **31** (1995) 1319-1324.
[3] H. Igarashi and T. Honma, An analysis of thin magnetic materials using hypersingular integral equations,
    IEEE Trans. Mag., **32** (1996) 682-685.
[4] H. Igarashi, A. Kost and T. Honma, Impedance boundary condition for vector potentials on thin layers
    and its applications to integral equations, *Proc. NUMELEC97*, Lyon France (1997) 6-7, extended version
    appears in *European Physical Journal, Applied Phys.*, **1**, 1998.

Non-Linear Electromagnetic Systems
V. Kose and J. Sievert (Eds.)
IOS Press, 1998

3-D Heat Analysis of Motors Taking into Account Eddy Current Loss Using Finite Element Method

Yoshihiro KAWASE, Koji SUWA and Shokichi ITO*
Department of Electronics and Computer Engineering, Gifu University
Yanagido 1-1, Gifu, 501-11, Japan
*Department of Electronics, Fukuoka Institute of Technology
Wajirohigashi 3-30-1, Higashi-Ku, Fukuoka, 811-02, Japan

Abstract. This paper describes a newly developed heat analysis system by the 3-D finite element method taking into account the heat generated by the eddy current loss which is analyzed by the dynamic magnetic field analysis using the 2-D finite element method. The new method is applied to obtain the temperature distributions in a permanent-magnet motor taking into account the eddy current loss in the permanent magnets. It is shown that our method is very useful for the optimum design of motors.

1. Introduction

Motors are widely used for machine tools and robotics. 3-D heat analysis is very important for the optimum design of motors. In order to obtain the exact temperature distributions, the heat generated by the eddy current loss as well as the coil loss should be considered. Therefore, we have developed a heat analysis system by the 3-D finite element method taking into account the heat generated by the eddy current loss which is calculated by the dynamic magnetic field analysis using the 2-D finite element method. Our method is applied to obtain the temperature distributions in a permanent-magnet motor taking into account the eddy currents in the permanent magnets. It is clear that our method to obtain temperature distributions is very useful.

2. Method of Numerical Analysis

The fundamental equation of the magnetic field taking into account the eddy currents can be written in terms of the magnetic vector potential A as follows [1] :

$$\mathrm{rot}(v\mathrm{rot}A) = J_0 + J_e + \mathrm{rot}\,v_0 M \quad , \quad J_e = -\sigma\left(\frac{\partial A}{\partial t} + \mathrm{grad}\phi\right) \tag{1}$$

where v is the reluctivity, J_0 is the current density, J_e is the eddy current density, ϕ is the electric scalar potential, σ is the conductivity, v_0 is the reluctivity of vacuum and M is the magnetization of the permanent magnet.

The eddy current loss W_{ed} is calculated by the following equation:

$$W_{ed} = \frac{1}{\tau/2} \int_{t}^{t+\tau/2} \left\{ \int_{V_e} \frac{(J_e^t)^2}{\sigma} dV \right\} dt \qquad (2)$$

where τ is the cycle and V_e is the region of the conductor.

The fundamental equation of the heat conduction is given as follows:

$$\frac{\partial}{\partial x}\left(K_{xx} \frac{\partial \theta}{\partial x} \right) + \frac{\partial}{\partial y}\left(K_{yy} \frac{\partial \theta}{\partial y} \right) + \frac{\partial}{\partial z}\left(K_{zz} \frac{\partial \theta}{\partial z} \right) + Q = dc \frac{\partial \theta}{\partial t} \qquad (3)$$

where K_{xx}, K_{yy} and K_{zz} are the thermal conductivity in x, y and z direction respectively, θ is the temperature, Q is the heat generated by the eddy current loss W_{ed} as well as the copper loss in coils, d is the material density, and c is the specific heat.

3. Analyzed Model

Figure 1 shows the analyzed model of a permanent-magnet motor [2]. The length of the armature is the same as that of the stator. The rotational direction of the armature is counterclockwise. The magnetization of the magnets is 0.9 Tesla. Figure 2 (a) and (b) show the 2-D finite element mesh for the dynamic magnetic field analysis and the 3-D finite element mesh for thermal analysis respectively.

The 2-D dynamic magnetic fields are analyzed in two conditions. One is when the armature is rotating at synchronous speed (synchronous speed analysis), and the other is when the armature is locked (locked-rotor analysis).

The 3-D temperature distributions are calculated by (3). When the armature is rotating, the thermal conductivity in (3) is assumed to be 10.0 W/m°C because of the stirring of the air between the armature and the stator. When the armature to be is locked, the thermal conductivity is assumed to be 0.025 W/m°C as well as 10.0 W/m°C to clarify the effects of the thermal conductivity.

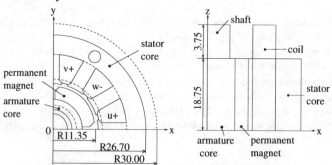

Figure 1. Analyzed model of a permanent-magnet motor.

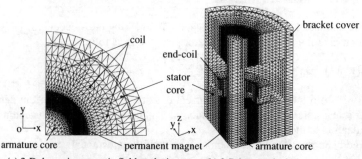

(a) 2-D dynamic magnetic field analysis   (b) 3-D heat analysis

Figure 2. Finite element meshes.

4. Results and Discussion

Figure 3 (a) and (b) show the examples of the flux distributions when the armature is rotating at synchronous speed (synchronous speed analysis) and the armature is locked (locked-rotor analysis) respectively.

Figure 4 shows the eddy current loss distributions in the permanent magnet. The eddy current loss of the locked-rotor analysis is much larger than that of the synchronous speed analysis.

Figure 5 shows examples of the temperature distributions by the heat analysis. The heat generated by the eddy current loss in the permanent magnet is small compared with the copper loss in the coils, because the size of this motor is small. Therefore, the effect of the eddy current loss in the permanent magnet on the temperature distributions is small in this model.

Figure 6 shows the time variations of the temperature in the permanent magnet. The effects of thermal conductivity are larger than that of the effects of eddy current loss in the permanent magnet shown in Figure 4.

Table 1 shows discretization data and CPU time.

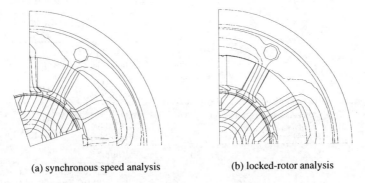

(a) synchronous speed analysis   (b) locked-rotor analysis

Figure 3. Flux distributions.

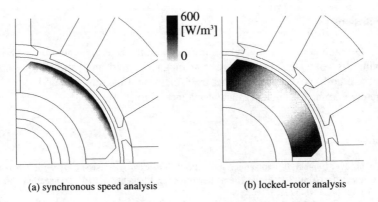

(a) synchronous speed analysis   (b) locked-rotor analysis

Figure 4. Distributions of eddy current loss in permanent magnet.

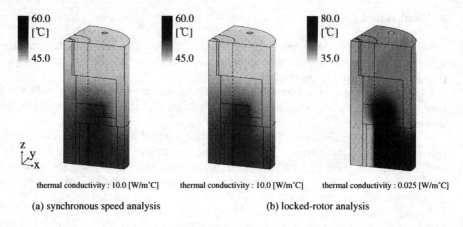

thermal conductivity : 10.0 [W/m°C]   thermal conductivity : 10.0 [W/m°C]   thermal conductivity : 0.025 [W/m°C]

(a) synchronous speed analysis   (b) locked-rotor analysis

Figure 5. Temperature distributions by heat analysis(time=5min.).

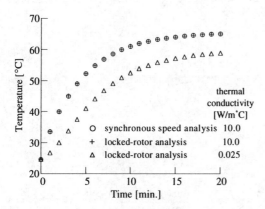

Figure 6. Time variations of temperature
in permanent magnet.

Table 1. Discretization data and CPU time.

	2-D dynamic field analysis	3-D heat analysis
number of elements	3167	370539
number of unknowns	1574	65280
number of time steps	360	20
CPU time [min.]	12.7	47.8

computer used : SUN Microsystems
SPARC station 20 model HS21

5. Conclusions

The 3-D heat analysis method taking into account the heat generated by the eddy current loss, which is calculated by the dynamic magnetic field analysis using the 2-D finite element method, is developed. The usefulness of this new method is in obtaining temperature distributions in a permanent-magnet motor.

References

[1] Y.Kawase, O.Miyatani, T.Yamaguchi and S.Ito, "Numerical Analysis of Dynamic Characteristics of Electromagnets Using 3-D Finite Element with Edge Elements", IEEE Transactions on Magnetics, vol.30, no.5, pp. 3248-3251, 1994.
[2] Y.Kawase et al., "3-D Finite Element Analysis of Permanent-Magnet Motor Excited from Square Pulse Voltage Source", IEEE Transactions on Magnetics, vol.32, no.3, pp. 1537-1540, 1996.

Non-Linear Electromagnetic Systems
V. Kose and J. Sievert (Eds.)
IOS Press, 1998

A 3D BEM Analysis of Power Frequency Electromagnetic Fields in Low Conductivity Bodies

Oriano BOTTAUSCIO (°), Mario CHIAMPI (*) and Maurizio REPETTO (*)

(°) IEN Galileo Ferraris, C. M. d'Azeglio 42, I-10125 Torino, Italy
(*) Dip. Ing. Elettrica Ind. - Politecnico di Torino, C. Duca degli Abruzzi 24,
I-10129 Torino, Italy

Abstract. The paper presents a numerical approach for the evaluation of induced currents in living organisms as a consequence of the exposure to ELF electric and magnetic fields. The numerical model is based on an integral formulation of the electromagnetic field problem derived from the Green vector theorem. The problem is solved by the Boundary Element Method, taking advantage of the features of this technique, particularly apt to the solution of linear open boundary field problems. After a description of the problem formulation, the implementation of the method is described and an application is presented.

1. Introduction

The analysis of the interaction between living organisms and the extremely low frequency (ELF) electromagnetic field nowadays holds a lot of interest, because of the large debate in the scientific world about the possible risk consequent to the exposure of human beings to these fields. An important topic in these research activities is the evaluation of the intensity of the currents induced inside the bodies by the contemporary presence of electric and magnetic fields of given intensity. This analysis is performed either by an experimental approach, using physical models of human beings and measuring the induced currents, or by a numerical model simulating the exposure to ELF fields [1-5]. The second approach enables the analysis of different exposure conditions and, in addition, it allows a deeper investigation of the influence of the biological tissue characteristics on induced current density.

In this paper the authors present a numerical approach able to analyse the electromagnetic field distribution in low conduction bodies, as biological tissues or human bodies, consequent to the presence of applied electric and magnetic fields. This model is based on a 3D electromagnetic field computation, applying the Boundary Element Method (BEM) which gives great advantages in the study of linear open boundary field problems. After a description of the adopted numerical formulation, the numerical implementation of the model is discussed and then an application is presented, in order to illustrate the main features and powerfulness of the method.

2. Electromagnetic field problem and computational code

The evaluation of induced currents in low conductivity bodies exposed to ELF electromagnetic fields requires the solution of the Maxwell equations in their complete form:

Research partially supported by ENEL - Centro di Ricerca Elettrica (Italy)

$$curlE = -j\omega B \qquad\qquad curlH = j\omega D + J + J_s$$
$$divB = 0 \qquad\qquad\qquad divD = \rho \qquad\qquad (1)$$

where J and J_s respectively represents the induced and the imposed (magnetic field sources) current densities, ρ is the volume electric charge (electric field sources) and ω is the angular frequency of the field sources. The electromagnetic behaviour of the materials is assumed linear, that is $J = \sigma E$, $D = \varepsilon E$, $B = \mu H$.

The field formulation is developed applying the vector Green theorem [6, 7]:

$$\xi E\left(P_s\right) = -\int_\Omega (n \cdot E)\nabla\psi dS - \int_\Omega (n \times E) \times \nabla\psi dS + \int_\Omega j\omega\psi (n \times B) dS - j\omega\mu \int_V \psi J_s dV + \frac{1}{\varepsilon}\int_V \rho\nabla\psi dV$$
$$(2)$$

In this equation the unknown electric field E in point P_s inside a volume V is expressed as a function of surface integrals extended to boundary Ω of volume V and to volume integrals including the field sources; coefficient ξ is ½ if P_s lays on the boundary Ω and 1 elsewhere. The Green function ψ is given by $\psi = exp(-jkr)/4\pi r$, where k depends on material characteristics ($k^2 = \mu(\varepsilon - j\sigma/\omega)\omega^2$) and r is the distance between the *source point* and the *computational point*.

If the conductivity of the materials is low, as in biological tissue ($\sigma=0.01\div1$ S/m), the induced currents do not sensibly affect the local value of the magnetic flux density B, which, consequently, only depends on the imposed currents:

$$B\left(P_s\right) = \mu\int_V \psi J_s \times \nabla\psi dV \qquad\qquad (3)$$

Thus, the surface integrals involving B in eqn. (2) are known quantities and the only unknown is the electric field E.

Applying the BEM, the bodies surfaces are discretized into triangular elements, over which vector E is assumed to be constant. Considering the generic element i, which separates the volumes (a) and (b), the integral equations (2) for volume (a) become:

$$\xi E_i^{(a)} \sum_{m=1}^{M^{(a)}} \left(n_m^{(a)} \cdot E_m^{(a)}\right) \int_{\Omega^{(a)}} \nabla\psi dS + \sum_{m=1}^{M^{(a)}} \left(n_m^{(a)} \times E_m^{(a)}\right) \times \int_{\Omega^{(a)}} \nabla\psi dS =$$
$$= j\omega \sum_{m=1}^{M^{(a)}} \left(n_m^{(a)} \times B_m^{(a)}\right) \int_{\Omega^{(a)}} \psi dS - j\omega\mu \int_{V^{(a)}} \psi J_s dV + \frac{1}{\varepsilon}\int_{V^{(a)}} \rho\nabla\psi dV$$
$$(4)$$

where $M^{(a)}$ is the number of triangular elements into which $\Omega^{(a)}$ is subdivided. In addition to (4), we introduce the interface conditions on normal and tangential components of E

$$E_i^{(a)} \cdot n_i^{(a)} = \frac{\varepsilon^{(b)} - j\left(\sigma^{(b)}/\omega\right)}{\varepsilon^{(a)} - j\left(\sigma^{(a)}/\omega\right)} E_i^{(b)} \cdot n_i^{(b)} \qquad \begin{aligned} E_i^{(a)} \cdot \tau_{1i}^{(a)} &= E_i^{(b)} \cdot \tau_{1i}^{(b)} \\ E_i^{(a)} \cdot \tau_{2i}^{(a)} &= E_i^{(b)} \cdot \tau_{2i}^{(b)} \end{aligned} \qquad (5)$$

where n_i, τ_{1i} and τ_{2i} represents the normal and tangential unit vectors for triangle i.

The unknowns of the problem are then represented by one normal component ($E_{ni}^{(a)}$ or $E_{ni}^{(b)}$) and two tangential components ($E_{t_1i}^{(a)}, E_{t_2i}^{(a)}$) for each triangle. When only magnetic field sources are present an additional approximation can be introduced to limit the number of unknowns. In fact, since the conductivity of the air surrounding the human model is zero, the normal component of electric field on the triangles of the external surface is zero, so that they can be removed as unknowns and only $E_{t_1i}^{(a)}$ and $E_{t_2i}^{(a)}$ have to be computed.

The application of BEM approximation reduces the terms including the surface integrals to $\int_\Omega \nabla\psi dS$ and $\int_\Omega \psi dS$ which have to be computed by numerical integration. These terms

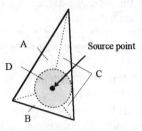

A
D
Source point
C
B

Fig. 1 - Reduction of singularity

must be carefully computed, mainly when the source point is near the considered triangular element (distance r in the Green function tends to zero). To achieve a good accuracy without increasing dramatically the computational burden, an adaptive integration algorithm, based on Kronrod scheme [8], has been worked out. Following this scheme, the number of integration points is increased up to obtain a stated precision in the result, starting from a limited number of them. If the convergence of the procedure is not reached, the triangle is subdivided into four subtriangles and the scheme is repeated for each of them.

When the source point is included into the considered triangular element, a singularity of the Green function appears because distance r becomes zero. The singularity is automatically removed in the integral of ψ by the presence of the infinitesimal surface dS. For the computation of the integral of Green function gradient, the triangle is divided into four subelements (Fig. 1): the contribution of the circular subelement (D) having centre in the source point vanishes due to symmetry considerations and only the integrals over A, B, and C have to be evaluated.

An essential feature of the code for BEM analysis is the possibility of describing and discretizing complex structures as those representing the morphology of human bodies. A specific pre-processing module has been tailored to this purpose; it enables the description and discretization of the human body by a set of horizontal sections which are automatically connected to form the external surface. The volume elements, which represent the field sources, are described by a set of 8 or 20 vertix hexahedra, arranged to build different field sources; the more common structures (solenoids, bars, coils, spheres, etc.) have been previously prepared and stored in library, from which they can be recalled and adjusted. Finally, the pre-processor module assigns the electromagnetic characteristics of the materials which constitute the body.

The solver module prepares the matrix and solves the algebraic system of equations to obtain the values of the unknowns. This phase represents the heaviest one from a computational point of view, taking into account the great number of matrix element which have to be evaluated and the successive solution of the large linear system. The numerical results are then elaborated by a specific post-processor which enables the estimation and the graphical representation of the field quantities at each point of the domain. In particular, it evaluates the induced current density inside each portion of the body and provides a map of intensity in order to illustrate the more critical regions.

The results obtained by the numerical procedure have been preliminarily validated by comparison with analytical solutions available in the case of simple geometries, such as spheres, spheroids and cylinders; this comparison has also given an estimate of the computational accuracy [9].

3. Application of the model

The numerical procedure has been applied to the analysis of typical exposure conditions. In particular, we have focused the attention here on the evaluation of the induced currents in a human body model as a consequence of the exposure to the electric and magnetic fields which can be experienced beneath a HV transmission line (380 kV) under maximum load conditions (1500 A current). Under these conditions, the field intensities on the model are about 2.6 kV/m for the electric field (directed mainly along the vertical axis) and 18 μT for the magnetic flux density (having a direction normal to the human thorax).

The model of a human body is sketched in Fig. 2. The presence of the main internal organs has been simulated by suitable volumes, having a stated conductivity value. Following the data available in literature, we assume for the heart and brain regions σ = 0.7 S/m, for the intestines σ = 0.03 S/m, while for the lung and liver σ = 0.1 S/m. The mesh of the external surface of the body and the internal organs has been generated by the pre-processor module using about 2000 triangular elements. The results of the simulation are presented in Fig. 2, where a map of the induced current density values is reported. The maximum value of induced currents, which is found in the heart and in the brain, is about 0.16 mA/m^2.

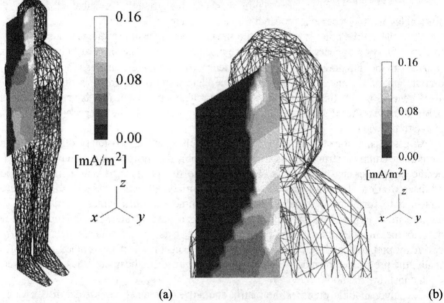

(a)   (b)

Fig. 2 - Current density induced in human model by exposure to electric and magnetic field beneath a HV transmission line (380 kV, 1500 A) - (values in mA/m^2): a) complete view, b) brain and heart regions

References

[1] A. Chiba et al., Application of finite element method to analysis of induced current densities inside human model exposed to 60 Hz electric field, *IEEE Trans. PAS*, Vol 103, no. 7, 1984, pp. 1895.

[2] A. Chiba et al., A method for numerical determination of induced current density in human model exposed to power frequency electric field, *Proc. of the 8th ISH*, Yokohama (Japan), Aug. 1993, pp. 535.

[3] P. J. Dimbylow, Finite difference calculation of current density in a homogenous model of a man exposed to extremely low frequency electric fields, *Bioelectromagnetics*, Vol. 8, 1987, pp. 355.

[4] A. Bossavit, A theoretical approach to the question of biological effects of low frequency fields, *IEEE Trans. MAG*, Vol. 29, no. 2, 1993, pp. 1399.

[5] P. Baraton et al., Three dimensional computation of the electric fields induced in a human body by magnetic fields, *Proc. of the 8th ISH*, Yokohama (Japan), Aug. 1993, pp. 517.

[6] J. A. Stratton, Electromagnetic theory, New York: McGraw-Hill, 1941.

[7] T. Misaki, H. Tsuboi, Techniques for boundary element analysis of three-dimensional eddy current distribution, *IEEE Trans. MAG*, Vol. 24, no. 1, 1988, pp. 146.

[8] R. Piessens et al., Quadpack, Springer-Verlag, Berlin, 1983.

[9] O. Bottauscio, G. Crotti, A numerical method for the evaluation of induced currents in human models by ELF electromagnetic fields, *Proc. 3rd Int. Workshop on EMF*, Liège (Belgium), May 1996, pp.141.

Non-Linear Electromagnetic Systems
V. Kose and J. Sievert (Eds.)
IOS Press, 1998

The Performance Analysis of a DC Motor
with Skewed Slots by the 2-D FEM

T.Ohba, K.Harada, Y.Ishihara and T.Todaka
Department of Electrical Engineering, Doshisha University
Kyotanabe 610-03, Japan

Abstract. This paper presents a technique to calculate the dynamic characteristics of a motor with skewed slots by using the 2-D Finite Element Method (FEM). The technique is employed in the performance analysis of the brushless DC servomotor with skewed magnets. The results calculated by the proposed technique agree well with the measured values.

1. Introduction

It is well known that skewed slots is effective in improving the characteristics of a motor. In general, it is necessary to use the 3-D FEM to analyze motors with skewed slots, and the static characteristics have been analyzed by using the 3-D FEM[1]. However, as a matter of fact, the present conditions, such as memory capacity in computer and calculation time, make it difficult and impractical to calculate dynamic characteristics by using the 3-D FEM. For that reason, the authors have proposed a technique that applies the 2-D FEM to analyze in a shaded pole motor with skewed slots which has one slot pitch by modifying the rotor circuit equations[2].

This paper describes a technique to calculate the dynamic performances of motors with skewed slots which have any slot pitch by using the 2-D FEM. In this calculation, the skewed effects are considered by dividing a motor into segments. This technique is applied to the analysis of the brushless DC servomotor with skewed magnets. Comparison of the calculated results and the measured values proves the validity of the proposed technique. The influences of the number of segments in this technique are also investigated.

2. Procedure to consider skewing

Figure 1(a) shows continuous skewed magnets model. In this technique, continuous skewed magnets model is divided into n segments and is considered as step skewed ones shown in Fig.1(b). It is assumed that the flux distribution of each segment is uniform in the axial direction for calculation by the 2-D FEM. In Fig.1(b), when the rotor is revolving, the whole induced voltage E_s is given by

$$E_s = E_1 + E_2 + E_3 + \cdots + E_n \tag{1}$$

where E_1, E_2, E_3, $\cdots$, and E_n are the induced voltage of each segment respectively.

The whole induced voltage E_s is coupled with the circuit equation derived from the equivalent circuit of a DC motor as shown in Fig.2. The circuit equation considering the skewed effects is given as follows :

$$V = E_s + RI + L\frac{dI}{dt} \tag{2}$$

where V is the supplied voltage, R is the resistance, I is the circuit current, and L is the leakage reactance. Solving Eq.(2) enables the 2-D FEM to take the skewed effects into account.

In the proposed technique, if a motor is divided into n segments, it is necessary to prepare n meshes for each segment. The mesh of each segment is prepared by rotating the rotor mesh of the first segment according to the adjusted skew pitch of each segment, and then the meshes of the all segments are considered as one mesh, which is applied to the calculation. Preparing the mesh in this technique needs less experience than that in the 3-D FEM, and this is very practical.

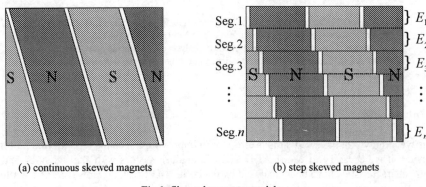

(a) continuous skewed magnets   (b) step skewed magnets

Fig.1. Skewed magnets models.

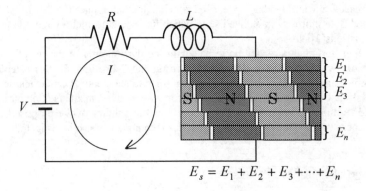

$$E_s = E_1 + E_2 + E_3 + \cdots + E_n$$

Fig.2. Coupling E_s with circuit equation.

3. Analyzed model

Figure 3 shows the analyzed model of a 3-phase, 4-pole, brushless DC servomotor with skewed magnets. The rotor is composed of magnets divided into 4 segments in the axial direction as shown in Fig.3(b). The length of each segment is 12.5 mm in the axial direction. The skew angle of each segment is 10 deg. in the rotary direction, and the total skew angle is 10 deg. $\times$ 3 = 30 deg., and consequently the rotor is skewed by one slot pitch.

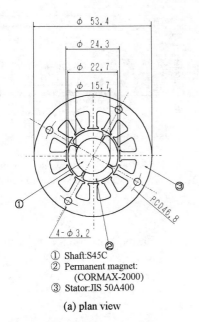

① Shaft:S45C
② Permanent magnet:
   (CORMAX-2000)
③ Stator:JIS 50A400

(a) plan view

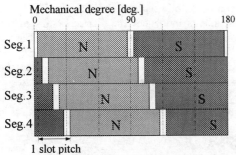

(b) front view of rotor magnets

Fig.3. Analyzed model.

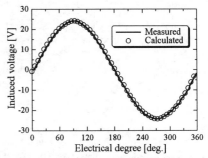

Fig.4. Waveforms of induced voltage (600 rpm).

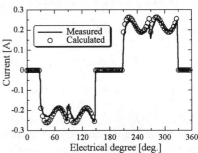

Fig.5. Waveforms of current (900 rpm).

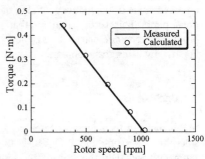

Fig.6. Speed - average torque characteristic.

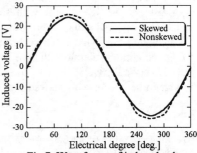

Fig.7. Waveforms of induced voltage
(skewed vs. nonskewed, 600 rpm).

4. Results

4.1 Comparison of calculated results and measured values

Figure 4 shows the waveforms of the induced voltage in the case of the generator when the rotor speed is 600 rpm. Figure 5 shows the waveforms of the current in the U-phase winding when the rotor speed is 900 rpm, and Fig.6 shows the average torque characteristic versus the rotor speed. The calculated results are in excellent agreement with the measured values, and therefore the validity of the proposed technique is confirmed.

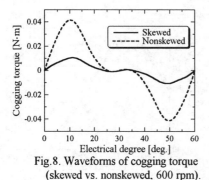

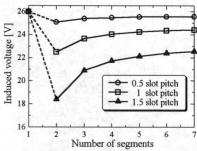

Fig.8. Waveforms of cogging torque (skewed vs. nonskewed, 600 rpm).

Fig.9. Relationship between the number of segments and induced voltage.

4.2 Investigation of the skewed effects

Figures 7 and 8, which compare the skewed results with the nonskewed ones, show the waveforms of the induced voltage and the cogging torques respectively. The skewed effects reduce the harmonic components of the induced voltage and the cogging torque.

4.3 Influences of the number of segments

Although the rotor of the analyzed model is composed of 4 segments as shown in Fig.3, it is investigated how changing the number of segments influences the characteristics of the motor. Figure 9 shows the relationship between the number of segments and the effective value of the induced voltage in the generator when the rotor speed is 600 rpm. In Fig.9, the skew pitch is given by 0.5, 1 and 1.5 slot pitch. As the number of segments increases, the effective value of the induced voltage of each segment converges on the value defined by the skew pitch. This result suggests that giving the appropriate number of segments in the proposed technique makes it possible to calculate the continuous skewed model as shown in Fig.1(a) by using this technique.

5. Conclusions

In this paper, a technique is presented to calculate motors with skewed slots which have any slot pitch by using the 2-D FEM. In the proposed technique, a motor is divided into segments, and then the sum of the induced voltage of each segment is coupled with the circuit equation, so that it becomes possible for the 2-D FEM to take the skewed effects into account. The utility and the validity of the proposed technique are shown by the analysis of the brushless DC servomotor with step skewed magnets. If the appropriate number of segments is selected in this technique, the continuous skewed model can be analyzed.

Calculating the dynamic characteristics by using the 3-D FEM is a very hard task in present conditions, therefore the proposed technique to analyze a skewed model by the 2-D FEM is very effective. The proposed technique can be applied not only to a DC motor but also any other motor with skewed slots, such as an induction motor.

References

[1] H.Kometani, M.Inoue and S.Sakabe, "3-Dimensional Magnetic Force Calculation for Permanent-magnet Synchronous Motor", *Papers of Combined Technical Meeting on Rotating Machinery*, RM-94-39, 1994.
[2] T.Matsubara, Y.Ishihara, S.Kitamura and Y.Inoue, "Magnetic field analysis in shaded pole motor taking skewed slot effects into account", *IEEE Trans. Magn.*, Vol.31, No.3, pp.1916-1919, 1995.

Non-Linear Electromagnetic Systems
V. Kose and J. Sievert (Eds.)
IOS Press, 1998

Vector Control Scheme of Synchronous Reluctance Motor Considering Iron Loss

*Jung-Ho LEE, *Kil-Hwan KIM, **Jung-Chul KIM, and *Dong-Seok HYUN
*Dept. of Electrical Engineering, Hanyang University, Seoul 133-791, KOREA
**LG Electronics Co., Living System Lab., Seoul 153-023, KOREA

Abstract. This paper proposes a new method, for the Synchronous Reluctance Motor (SynRM), which selects appropriate stator d,q-axis currents that the influence of iron core loss on the developed torque can be minimized, and suggests the algorithm which compensates the iron core loss effect for better control characteristics, in the high speed range.

1. Introduction

The general principles of vector control are derived under the assumptions that the motor parameters are constant, and that the magnetizing flux saturation and iron loss are not considered, but such an idealized situation is not met when precision torque characteristics are needed.

Some papers which discussed the influence of iron loss on vector control have been presented. They discussed the influence on the behaviour of the induction motor [1][2] and the SynRM[3]. The reason for nonlinear torque characteristic is that input stator terminal d,q-axis currents are not the exact ones directly used for the flux and torque control due to d,q-axis additional coupling caused by iron loss.

In this paper, the torque estimation error is investigated quantitatively when the SynRM including iron core loss is vector-controlled directly by using the input stator current (i_{ds}, i_{qs}).

It is found that the torque estimation error can be minimized if the input d,q-axis currents ratio (i_{qs}^* / i_{ds}^*) is selected properly with the proposed method. Computer simulations and experiments are performed by using the selected d,q-axis currents ratio (i_{qs}^* / i_{ds}^*) to evaluate the proposed method.

2. Modeling of SynRM Including Iron Core Loss

The d,q-axis equivalent circuits of the SynRM are shown in Fig.1, where R_m is the equivalent resistance to represent the iron loss. As can be seen R_m allows the input stator current be diverted away from the torque-producing branch of the machine model, therefore this makes inevitable torque estimation error because the stator current components i_{dm}, i_{qm} which are directly related to the torque production are not equal to the input stator current components i_{ds}, i_{qs} and contain i_{qm}, i_{dm} respectively. Therefore this additional mutual coupling between two components makes it difficult to control torque precisely.

This phenomenon is shown in the simulation results of Fig.2.

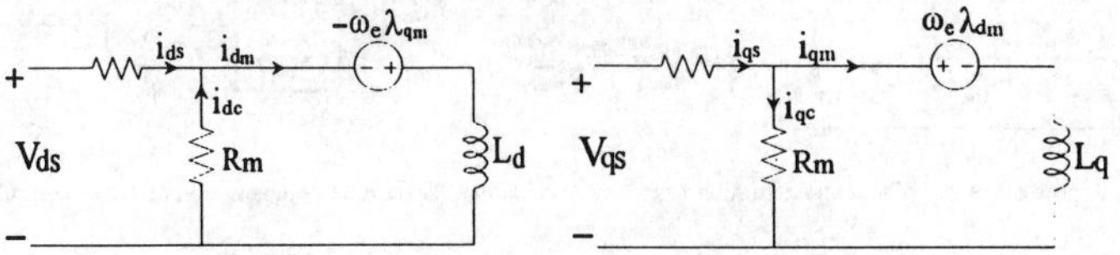

(a) d-axis   (b) q-axis

Fig.1. d-q axis equivalent circuit including iron loss

Voltage and torque equation in the synchronous reference frame are,

$$v_{ds} = R_s i_{ds} + L_d \frac{di_{dm}}{dt} - \omega_e \lambda_{qm} \qquad v_{qs} = R_s i_{qs} + L_q \frac{di_{qm}}{dt} + \omega_e \lambda_{dm}$$

$$i_{ds} = i_{dm} - i_{dc}, \qquad i_{qs} = i_{qm} + i_{qc} \tag{1}$$

$$T_e = \tfrac{3}{2}\tfrac{P}{2}(L_d - L_q)i_{dm}i_{qm} = \tfrac{3}{2}\tfrac{P}{2}(L_d - L_q)(i_{ds} + i_{dc})(i_{qs} - i_{qc}) \tag{2}$$

and in the steady state,

$$i_{dc} = \frac{\omega_e \lambda_{qm}}{R_m} = \frac{\omega_e L_q i_{qm}}{R_m} \qquad i_{qc} = \frac{\omega_e \lambda_{dm}}{R_m} = \frac{\omega_e L_d i_{dm}}{R_m} \tag{3}$$

3. Error on Torque Estimation and Discussion

To analyze the amount of torque estimation error quantitatively when the iron loss is considered and not considered in the steady state, $\%e_{torque}$ is defined as follows:

$$\%e_{torque} = \frac{T_{e-iron} - T_e}{T_{e-iron}} \times 100 \ [\%] \tag{4}$$

Where, $\qquad T_e = \frac{3}{2}\frac{P}{2}(L_d - L_q)i_{ds}i_{qs}$ and $T_{e-iron} = \frac{3}{2}\frac{P}{2}(L_d - L_q)i_{dm}i_{qm}$ (5)

T_e and T_{e-iron} are developed torque when iron loss is neglected and considered respectively.

Substituting (5) into (4), it follows

$$\%e_{torque} = \left(1 - \frac{i_{ds}i_{qs}}{i_{dm}i_{qm}}\right) \times 100 \ [\%] \tag{6}$$

By substituting (1) and (3) into (6), the following with ω_e can be obtained.

$$\%e_{torque} = \left[1 - \cfrac{1}{\left[1\big/\left(1 + \frac{\omega_e^2 L_d L_q}{R_m^2}\right)\right]^2 \left[\frac{\omega_e L_d}{R_m}\frac{i_{ds}}{i_{qs}} + \frac{\omega_e L_q}{R_m}\frac{i_{qs}}{i_{ds}}\left(1 - \frac{\omega_e^2 L_d L_q}{R_m^2}\right)\right]} \right] \times 100$$

$$i_{ds}i_{qs} = \text{constant} \tag{7}$$

Fig.4. shows the percentage torque error ($\%e_{torque}$) for a variety of i_{ds} and ω_e under constant torque command 2.17[Nm]. Fig.4 indicates that it is possible to find out i_{ds} and i_{qs} reference values at which torque response is less affected by iron loss averaged .

In (7), the denominator can be reduced as follows.

$$\frac{\omega_e L_d L_q}{R_m} \approx 0, -L_d \frac{i_{ds}}{i_{qs}} + L_q \frac{i_{qs}}{i_{ds}} = K \quad (8) \ denominator = \left[1\big/\left(1 + \frac{\omega_e^2 L_d L_q}{R_m^2}\right)\right]^2 \left(\frac{\omega_e}{R_m}\right)\left(K + \frac{R_m}{\omega_e}\right) \tag{9}$$

If ω_e[rad/sec] is below rated speed range, $\qquad 1\big/\left(1 + \frac{\omega_e^2 L_d L_q}{R_m^2}\right) \approx 1 \tag{10}$

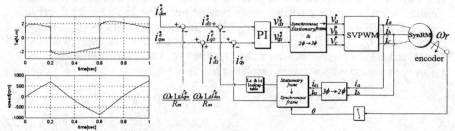

Fig.2. Torque and speed response without compensation   Fig.3 Block diagram for compensation of vector controlled SynRM

(9) can be approximated by substituting (8), (10) into (7) as follows,

$$(9) = 1^2 \cdot \left(\frac{\omega_e}{R_m}\right) \cdot K + 1 \tag{11}$$

Accordingly, %e$_{torque}$ can approach 0 in (7) if (11) is approximately equal to 1, and following relationships are obtained.

$$\therefore \ K = 0, \ -L_d \frac{i_{ds}}{i_{qs}} + L_q \frac{i_{qs}}{i_{ds}} = 0, \ \sqrt{\frac{L_d}{L_q}} = \frac{i_{qs}}{i_{ds}} \tag{12} \qquad i_{ds} \cdot i_{qs} = \text{constant} \tag{13}$$

From the above equations (12), (13), it is possible to calculate stator currents i_{ds}, i_{qs} which have less iron loss effect on torque when the given torque is constant. Simulation results are shown in Fig.8(b) when the stator current components have been selected by the proposed method. Here i_{ds}, and i_{qs} references were calculated as 1.6[A] and 2.5[A] respectively under the torque command 2.17[Nm]. Simulation results show improved torque characteristic compared to Fig.2 and these results are comparable to Fig. 8(a) performed by compensation algorithm shown in Fig. 3.

4. Simulation and Experimental Results

To prove the propriety of the proposed method, the TMS320C31 DSP control unit and Power Circuits are equipped as shown in Fig. 5. Motor parameters are identified using FEM and Preisach Modeling and are listed in Table1. For the purpose of comparison, torque reference 2.17[Nm] was made by applying different stator d,q axis current components 2[A],2[A] and 1.6[A], 2.5[A] which is based on proposed method (Fig. 7). Torque characteristics are compared by observing each speed response. As shown in the experimental results (Figs. 9, 10) nonlinear torque characteristic can be improved when reference currents for certain torque are determined by using proposed current ratio with (12) and (13).

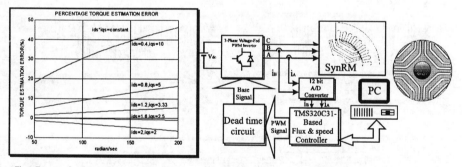

Fig.4 .Torque estimation error versus speed and i$_{ds}$ Fig. 5. Experimental system configuration   Fig.6. Rotor structure

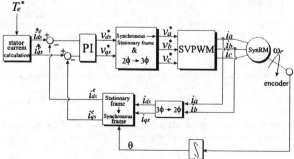

Fig. 7. Block diagram with proposed method

5. Conclusion

In this paper, torque estimation error is investigated quantitatively when the SynRM including iron core loss is vector-controlled directly by using input stator current (i_{ds}, i_{qs}). The torque estimation error due to iron loss can be minimized if the input d,q-axis reference currents are selected properly with the proposed method. And the simulation results also show that proposed method has good performance comparable to the system configuration in which simple iron loss compensation routine is added. Simulation and experimental results clarified that the theory described in this paper is reasonable.

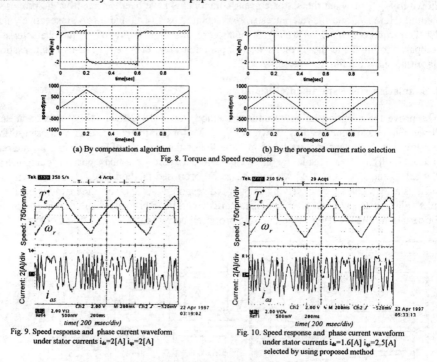

(a) By compensation algorithm   (b) By the proposed current ratio selection

Fig. 8. Torque and Speed responses

Fig. 9. Speed response and phase current waveform under stator currents i_{ds}=2[A] i_{qs}=2[A]

Fig. 10. Speed response and phase current waveform under stator currents i_{ds}=1.6[A] i_{qs}=2.5[A] selected by using proposed method

Table1: Specifications and parameters of the SynRM

Rotor type	Segmental Rotor	Stator phase resistance	2.82 Ω
Power rating	450 W	Number of phase	3 phases
Air gap	0.4 mm	Poles	4
Rotor length	55mm	L_d	0.3025 [H]
Rm	147 Ω	L_q	0.1220 [H]

References

[1] E.Levi, "Impact of iron loss on behaviour of vector controlled induction machines", *IEEE IAS Annual Meeting, Denver Colorado*, pp74-80, October 1994.
[2] G.O. Garcia, J.A. Santisteban, and S.D. Brignone,"Iron loss influence on a field-oriented controller", *IEEE IECON Conf. Rec*, pp633-638, 1994.
[3] Longya Xu, and Jiping Yao, "A compensated vector control scheme of a synchronous reluctance motor including saturation and iron losses", *IEEE Tran. on Ind. Appl.*, vol. 28, No.6, pp1330-1338, November/December 1992.

Magnetic Devices, Levitation

Non-Linear Electromagnetic Systems
V. Kose and J. Sievert (Eds.)
IOS Press, 1998

Yamanashi Maglev Test Line and Electromagnetic Characteristics of Superconducting Maglev

Shunsuke   FUJIWARA
Railway Technical Research Institute
2-8-38, Hikari-cho, Kokubunji-shi, Tokyo,185, Japan

Abstract. A new test line of superconducting maglev has been constructed in Yamanashi Prefecture, and the running test has begun. A priority section has a length of 18.4 km, partially with double tracks. The first train set consists of 3 cars which form an articulated bogie system. The second train set is under construction. Two groups of power converters are installed, because two trains are operated at the same time. The capacities of the power converters are 38MVA and 20MVA respectively. Both levitation and propulsion coils are arranged on the vertical surfaces of the guideway. The levitation and propulsion system has some distinctive features. One is the characteristic of correlation between lateral and rolling motion, which affects lateral stability at low speed. Another feature occurs when a coil on board or on the ground loses its function. These characteristics are described briefly.

1. Introduction

The superconducting maglev system was proposed by J.R.Powell and G.R.Danby. In Japan it has been under development for the past 25 years. So far running tests have been done with MLU002N vehicle on the Miyazaki Test Track. Now a new test line is opening in Yamanashi Prefecture [1] [2] this spring. This test line is designed by the project team formed by Railway Technical Research Institute, Central Japan Railway Company and Japan Railway Construction Public Corporation. Running tests began on a priority section which has a length of 18.4 km. From now on a series of running tests will be done for a period of three years. This paper presents an outline of the Yamanashi Test Line, and then presents some distinctive characteristics of this EDS system.

2. Outline of the Yamanashi Test Line

2.1 Ground facility

The priority section of the Yamanashi Test Line is 18.4 km long, partially with double tracks. It has tunnel sections which account for about 80 % of the whole line with a radius of curvature 8,000m and a cant angle of 10 degrees. Its maximum gradient is 4 %. The distance between the centers of tracks is 5.8m. The cross section of the tunnels

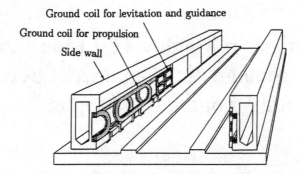

Ground coil for levitation and guidance

Ground coil for propulsion

Side wall

Figure 1: Configuration of the guideway and arrangement of the coils

is about 20 % larger than that of the Shinkansen for the purpose of reducing aerodrag and micro pressure wave. The guideway has a U type cross section and adopts the combined levitation and guidance system. At main installations there are a test center, a platform, turnouts, a power converter station and a car depot.

Two groups of power converters are installed, because two trains are operated at the same time according to the test plan. One of the power converters has a capacity of 38MVA, the other has that of 20MVA [3]. The maximum frequency of the inverter is 56.6Hz which corresponds to the vehicle speed of 550km/h. The train is fed its traction power by three converters sequentially on the feeding section of about 450m.

The configuration of the guideway and the arrangement of the ground coils are shown in Figure 1. There are two kinds of ground coils which are attached to the side wall of the guideway [4]. One is the armature coil of the linear synchronous motor for propulsion. Aluminum winding is molded out of epoxy resin to ensure its electric insulation. Its arranging pitch is 0.9m and in double layer for the purpose of reducing the spacial harmonic magnetic field. At the end of the guideway beam or panel, a special short end coil is installed to prevent the coil from spanning two beams. The other is the combined levitation and guidance coil. A coil is composed of two-unit rectangular windings connected in a figure eight shape. Two sets of aluminum windings are molded in a lump from GFRP which is chosen to ensure a high mechanical strength. They are arranged in 0.9m pitch (the coil has 0.45m pitch).

2.2 Train set

The first train set consisting of 3 cars is used for the running test. The train set is an articulated bogie system and has a magnetic shielding structure [5] [6]. Figure 2 shows an outside view of the vehicle. The length of the car is 21.6m for middle car and 28m for the head car. The width of the car body is 2.9m and that of the bogie is 3.22m, and the height of the car is 3.28m. The cross sectional area of the car is about 8.9m². The bogies are arranged with a 21.6m distance. Each bogie has 8 superconducting coils with a rated magnetomotive force of 700kA. Four superconducting coils are installed in a cryostat, and on-board refrigeration is provided above the coils in alignment with a liquid helium tank and liquid nitrogen tank. The dimensions of the magnet are 5.4m in length and 1.17m in height. The dimensions of the coil is shown in Table 1. The braking

Figure 2: Outside view of the train

Table 1: Coil dimensions

superconducting coil	number	4 /in a cryostat
	length × width	1.07 × 0.5 m (race track shape)
	pole pitch	1.35m
	mmf	700 kA
levitation coil	length × width	0.35 × 0.34m (unit rectangular)
	pitch	0.45m
propulsion coil	length × width	1.42 × 0.6 m (race track shape)
	number of turns	8, 10
	pitch	0.9m

of the train is usually made by the regenerative braking system. As a back-up brake system, four types of brakes are provided. They are dynamic brake, coil shortening brake, aerodynamic brake and wheel disc brake.

The second train set is now under construction. Although the main specifications of the second train set are almost the same as those of the first train set, the following differences are noted. The bogies adopt a resiliently-mounting method of SCM to improve ride quality; the inductive power collection system is adopted; furthermore the length of the middle car is 2.7m longer than that of the first train set.

2.3 The test schedule

The total check of the system before starting on the running test has already been completed, and the running test has started using the first train set. The running test is planned to continue for three years. To confirm the total technical system and its influence on the environment, the running speed will be raised up to the maximum target of 550 km/h. Using double tracks, the characteristics of two trains passing each other will be confirmed at the maximum relative speed of 1000 km/h.

3. Electromagnetic characteristics

The steady state electromagnetic characteristics of EDS has been well studied and the results are reflected in the design of coils. This superconducting maglev system has

The steady state electromagnetic characteristics of EDS has been well studied and the results are reflected in the design of coils. This superconducting maglev system has some distinctive features. One of these features is a correlation between lateral motion and rolling motion. If a bogie generates a lateral displacement, it will induce a rolling moment because the levitation force of the superconducting magnet which approaches the ground coils grows large. The correlation increases with the increase of vertical displacement of the bogie, and it is zero when the vertical displacement is zero. This effect reduces the equivalent lateral stiffness, and affects a lateral stability at low speed. Division of the guidance current so as to generate opposite rolling moments improves this characteristic.

This system uses flux cancelling to reduce power loss, by adopting a symmetric coil arrangement and circuit. Therefore when this structure changes, a large force may occur. If the superconducting coil loses its stored energy, the guidance system loses its flux balance in the lateral circuit, and a large lateral force occurs. If the guideway coil is short-circuited or open-circuited, the system also generates a particular force. If an armature coil is short-circuited, a large current is induced in the coil because of strong coupling between the armature coil and on board superconducting coils. These phenomena have been analyzed and tested at the Miyazaki Test Track. We confirmed the safety of the system by tests in three cases, where a superconducting coil was de-energized, an armature coil was short-circuited, and three levitation coils were removed.

4. Concluding Remarks

The development of the superconducting maglev system will enter into a new phase upon beginning the running test on a new test line which will yield many important data. The ground transportation system is gradually advancing as a highspeed service, with the total safety and environmental condition being confirmed. The maglev system is one of the promising candidates for future high speed transportation. The development of the maglev system has been subsidized by the Japanese Ministry of Transport.

References

[1] H.Nakashima and A.Seki, The Status of the Technical Development for the Yamanashi Maglev Test Line, The 14th International Conference on Magnetically Levitated Systems and Linear Drives, pp.31-35, Nov. 1995

[2] H.Nakashima, The Status and the Plan of the Yamanashi Test Line R & D, Quarterly Report of RTRI, Vol.37, No.2, pp.59-62, 1996

[3] I.Kawaguchi, H.Ikeda, S.Kaga and H.Ohtsuki, Power Supply System in Yamanashi Maglev Test Line, ibid. pp.90-93

[4] K.Sawada, H.Tsuruga and T.Iida, Development of Ground Coils for Yamanashi Maglev Test Line, ibid. pp.99-103

[5] K.Takao, M.Yoshimura, N.Tagawa, Y.Matsudaira, A.Inoue and S.Hosaka, Development of the Superconducting Maglev Vehicles (MLX01 type) on the Yamanashi Test Line, ibid. pp.63-70

[6] H.Seino, H.Oshima, H.Takizawa, K.Watanabe, K.Kato and H.Yoshioka, The Maglev Bogie System Developments for the Yamanashi Test Line, ibid. pp.78-83

Non-Linear Electromagnetic Systems
V. Kose and J. Sievert (Eds.)
IOS Press, 1998

Development and Application of Magnetic High Speed Trains in the Federal Republic of Germany

Dieter ROGG

Dornier SystemConsult GmbH, Friedrichshafen, Germany

Abstract: When, in 2005 the first passengers will hover safely and comfortably at 450 km/hr. on the route between Hamburg and Berlin, the Transrapid can look back to more than 30 years of history: since 1970, the German Federal Ministry of Research and Technology has been promoting the development of high-speed ground transportation systems whose levitation, guidance, and propulsion forces are transferred by electromagnetic fields without contact. The objective of this development is an attractive, efficient and economical high-speed ground transportation system in the velocity range from 300 to 500 km/hr. for the transportation of passengers and high-quality freight. The system is destined for operation in the Federal Republic of Germany and in Europe; it is also planned to be exported to oversea countries. With low negative effects on the environment, the system is projected to relieve the roads and the air space in highly congested traffic corridors. In combination with the traditional railways, the system is intended to increase the common share in the transportation market. It shall contribute to ease urgent traffic problems and to keep ecological damage low.

1. Description of the System

The high speed transportation sytem Transrapid is a track-bound train system, capable of operation at speeds of 400 to 500 km/hr. Its characteristics are shown in Table 1.

Characteristics of the Transrapid

- High speed tracked transport system for passengers and containerised goods
- Speed range   300 to 500 km/hr.
- Stations in connection with railway stations, airports, urban traffic centers and highways
- Trains consisting of   2 to 10 sections for 170 to 1000 passengers
- Capacity   5 to 35 million passengers per year

Table 1: Transrapid, general features of the system

The Transrapid vehicles are supported by separate, controlled electromagnets for levitation and guidance, arranged at both sides alongside the vehicle. A safe-life power supply for the on board energy is provided by linear generators, integrated into the poles of the levitation magnets. At low speeds the on board power supply is buffered by batteries. High redundancies, failure tolerant behavior and on-line diagnostics guarantee safe hovering [1]. The structure and equipment of the cabins correspond to modern railway coaches.

A long stator synchronous linear motor propels the vehicles. The energy supply is provided via substations along the track at a distance of some 50 km with variable voltage and frequency. The primary of the motor consists of a three phase cable winding, fixed in stator packets, which are mounted on both sides of the guideway, beneath the upper, table-shaped part of the guideway beams. The motor primary is divided in sections of about 1 to 4 km; only sections, where the vehicles are located, are connected to the substations. The stator packets in the guideway also serve as levitation rails, the levitation magnets as exciting magnets for the linear synchronous motor. High reliability is reached by redundant configurations of the power supply and cable winding.

The automatic train control consists of a fully automated communication and control system. It maintains the trainset speed within safe limits and provides a safe and unobstructed travel path. The system relies mainly on microprocessors which are designed and verified with fail-safe, fail-active and fail-tolerant techniques.

For the route planning various types of guideway structures are available with regard to cost efficiency and environmental concerns:

- single or double track elevated guideways using steel or prestressed concrete beams mounted on piers
- different single or double track at-grade structures for tunnels, cuttings, bridges and sections where the track runs in parallel to existing railway tracks.

Bending switches have been developed and tested for turnout speeds of 200 km/hr..

2. The Development Program

The today's "state of the art" is the result of more than 25 years of thorough development. The program was started in 1970 and is sponsored by the German Ministry for Research and Technology. After demand and feasibility studies the development and testing of the key components took place. The research program covered the whole variety of possible techniques: air cushion technique, permanent magnetic levitation, electrodynamic levitation, different short- and long stator propulsions. In 1978, after comprehensive evaluations, the technology of the Transrapid system - that means electromagnetic suspension and synchronous long stator populsion was selected for the further development.

3. Emsland Transrapid Test Facility

After the system concept was decided upon, the definition phase for a Transrapid test facility on the river Ems in the northwest of Germany was initiated in 1978 for trials on a fully operational scale. The construction of the first section of the guideway of 20 km in length, consisting of a 12 km long straight track and an 8 km long turning loop was started in 1979. In 1983, the vehicle Transrapid 06 was subjected to riding tests; in 1984, the experimental vehicle achieved a speed of more than 300 km/hr. After the commissioning of the second turning loop with another 10 km in length in December 1988, the design speed of 400 km/hr. was exceeded. 412 km/hr. were measured, at that time representing a world record for passenger maglev vehicles. This value was further improved by the Transrapid 07 to 435 km/hr. in December 1989 and to 450 km/hr. in June 1993. In routine test operation, a riding profile with a maximum speed of 420 km/hr. is usual today.

The concept of the Emsland Transrapid test facility - allowing both high-speed and endurance tests - proved to be successful. Changes in the basic concepts were necessary only in a few cases. By the end of 1996, more than 400,000 km were covered in total; the longest non-stop ride was almost 1700 km.

Fig: 1 Test facility Emsland and vehicle TR07

On account of the experience gained on the Emsland test track and the technical progress, additionally improved solutions could be put into practice, for instance in the second generation of the vehicle and the operations control technique as well as propulsion and guideway components. Also in 1991 the "operational readiness" could be confirmed by the experts of the German Railway after intensive examinations. They stated, "that the legal prerequisites for planning procedures were fulfilled and that the overall system as well as the subsystems include no system- or safety risks".

In the meantime the technology has been further improved and now the licensing procedure for application of the Transrapid is going on. For the operation of the Transrapid as a public means of transport comprehensive safety proofs in a multistage process and licensing on a legal basis are necessary. An important stage is the type-licensing of the preproduction models of the vehicles, the propulsion, controlling and signalling as well as of different guideway designs. After the completion of this type-licensing at the Transrapid Test Facility in the years 1997 to 1999 the manufacturing of the series-models of all components of the first application line can be carried out according to the time schedule and without technical risks.

4. Development Program Participants

The following functions of participating companies are particularly worth mentioning: The Federal Ministry of Education, Science, Research and Technology supports the development with financial contributions; Dornier SystemConsult assists the ministry in planning and controlling the program.

"Versuchs- und Planungsgesellschaft für Magnetbahnsysteme mbH, MPV" (Test and Planning Corporation for Maglev Systems), an affiliate of the German Railway and Lufthansa, is responsible for the operation of the Emsland Transrapid test facility.

Thyssen Transrapid System GmbH is charged with the technological optimization of the system, the vehicle, the guideway and the operations facilities. Propulsion and operations control technology are work shares of Siemens AG. Other companies are involved, for instance ADtranz and Stahlbau Lavis.

Safety assessments are conducted by different Technical Control Boards.

Additional basic research is being performed at university institutes, especially in Braunschweig but also in Aachen, Berlin, and Hannover.

5. Complexity of the Development Task

The development period of nearly twenty years starting with the system selection has to be compared with the volume and the degree of difficulty of the tasks to be fulfilled. The goal was to develop a complete new transportation system. In many cases, no examples were available on which the activities could have been based. Very often, the state of the art was not sufficient so that basic new developments were necessary. For instance, the levitation and guidance system of the TR 07 is required to guide a 50 m long and 100 t heavy vehicle on the Emsland test track at 450 km/hr. at a distance of approximately 10 mm from the levitation stators and guidance rails. In order to minimize the guideway costs - which represent a dominant part of the system costs - considerable deflections of the guideway beams under load, thermal deformations and, also manufacturing and mounting inaccuracies had to be tolerated. Moreover, external loads, such as centrifugal and aerodynamical forces at 450 km/hr. have to be mastered. Permanent levitation and the non-contact capability between the magnet and the stator have to be proved to be safe. It was possible to prove that all these problems could be solved with the TR 07.

6. Application Profile of the Transrapid

Examples of special features are as follows:

The Transrapid has no wheeled undercarriage. The levitation and guidance forces are generated by controlled electromagnets and are transmitted without contact and without the riding noises caused by wheels, excluding the risk of derailment: mechanical systems subject to wear are replaced by electronics.

The Transrapid vehicles are not equipped with locomotives with voluminous and heavy transformers, inverters, and switching devices. In the Transrapid system, these components are located on the guidewayside: the complete vehicle length with the exception of the driver cabins can be fully used for passenger seats.

The forces to be transferred from the propulsion system are not limited by friction factors which diminish when the speed increases: the propulsive forces are made available by the linear motor without contact. Moreover, the motor in the guideway can be designed with high power in acceleration and inclination sections and cost saving with less power in sections of settled speed.

The Transrapid has no mechanical, friction-depending brakes. The linear propulsion acts as a non-contact service brake, while the non-contact eddy-current brake is used as a safety brake: also in this case, mechanical systems subject to wear are replaced by electronics.

The Transrapid's energy transmission to the vehicle is not ensured by contact wire and power collectors which would be subject to wear and susceptible to failures and, at the same time, would increase the required tunnel cross-sections and generate aerodynamic sound difficult to shield at high speeds.

The Transrapid can be provided completely with an aerodynamic fairing, even in the bottom area of the levitation and guidance system: this decreases the required energy and the aerodynamic sound emission.

These special features of the Transrapid result in quite a series of characteristics which make this system particularly suitable for an economically efficient and ecologically acceptable application at very high speeds from 400 to 500 km/hr. Here are some examples:

- Relatively light vehicles with much space for payload, low track loads and good energy utilization.
- Short headways and short waiting times for passengers.
- High acceleration capability makes relatively short stop intervals possible without much loss of time.
- Relatively small horizontal radii and high climbing capability up to 10 %, thus flexible, cost-saving adaptation of the route to the terrain, minimizing the number of cost-intensive tunnels, cuttings, high bridges, and other decisive changes to the landscape.
- Little wear to vehicles and guideway, thus low maintenance and repair costs, high degree of availability, 24-hour operation possible.
- Emission values for sound and vibrations are relatively low, in particular in the medium and lower speed range when the train approaches or leaves the stations in the towns.

A high-speed ground system striving at the velocity of 400 km/hr. and higher must have all the characteristics mentioned. Only then is it possible to keep the construction and operating cost within the necessary frame in order to be economically acceptable considering the traffic volume which can be expected in practice.

The Transrapid can be employed as a superimposed high-speed long-distance transportation system in congested corridors, closely interlaced with the railway which provides the feeder service or subsequent transportation for passengers via its more close-meshed network. Central or peripheral stations which, in particular cases, have to be built anew are used as connection points. Other important connection points are airports, highway junctions and suburban traffic centers.

This means that the Transrapid is not intended to compete with the railway but to cooperate with it in order that, in a synergetic interaction, important additional segments of the traffic market can be opened up.

Another field of application for short distances is the express connection between neighboring population centers as well as the connection to and between airports.

The description of the Transrapid characteristics in this paragraph is based on documented and reproducible measuring results gained on the Transrapid Test Facility.

7. Application of the Transrapid in the Federal Republic of Germany, Transrapid-Line Berlin - Hamburg

Despite all these advantages, it is quite understandable that it is not easy for such a completely novel transportation system like the Transrapid to gain a foothold in the relatively well developed traffic infrastructure of Germany. The situation changed after the German reunification. In the course of the elaboration of the first transportation master plan for the whole of Germany it was necessary to complement the traffic routes, which up to that date were oriented mainly in north-south direction, with connections in east-west direction.

In this context, various Transrapid routes were investigated and assessed, taking into account aspects of national economy and operational economy, under the management of the Ministries of Research and Transportation. The connection between Germany's two largest towns, Berlin and Hamburg, turned out be the best suited route.

Therefore the German government decided to put the Berlin-Hamburg route into practice together with industrial companies. Important data on the Berlin-Hamburg project are given in Table 2. The start of operation is scheduled for 2005. The planning work has been going on since 1994. All the legal regulations which are necessary have been passed by the parliaments.

The responsibilities are agreed upon as follows: the Government makes a loan of 6,100 million DM available for the investment costs of the guideway. Vehicles and the other operational equipment will be financed by an industrial consortium of the companies ADtranz, Siemens and Thyssen with 500 million DM risk capital and 3,200 million DM from the capital market. The Deutsche Bahn AG is responsible for the construction and the operation, repays the loan to the government and pays a rental to the financing consortium.

Transrapid Berlin - Hamburg

Distance	292 km
Maximum operational speed	450 km/h
Travelling time	< 60 min
Number of train runs per day and direction	50
Trip frequency	15 min
Train length (four/five sections)	100/125 m
Train capacity	340/450 seats
Traffic volume	11,4 to 15,2 million passengers per year
Traffic performance	2.600 to 3.500 million passenger kilometers per year

Table 2: Transrapid route Berlin - Hamburg

As the planning of the Berlin - Hamburg line proceeds, the interest in the Transrapid system increases.

In the long run, the Transrapid will be able to offer attractive travel times to the passenger traffic between European metropolises. The integration of Europe presupposes means of transportation such as the Transrapid with sufficient capacity and an operating speed commensurate with the distances, at economical investment and operating costs and with acceptable effects on the environment.

The establishment of large-scale maglev connections in Germany, Europe, and throughout the world can only be made step by step as is the case with other traffic infrastructures. Permanently available industrial development and delivery capacities are prerequisites. The necessary basis will be created by implementing the Transrapid route Berlin - Hamburg.

References

[1] Miller L.,u.a., "Magnetschnellbahn Transrapid, Systemdarstellung", Oldenburg-Verlag, Elektrische Bahnen 93 (1995) 7

[2] Rossberg Ralf Roman, "Radlos in die Zukunft", Orell Füssli Verlag Zürich und Schwäbisch Hall 1983, ISBN 3 280 01503 0

Non-Linear Electromagnetic Systems
V. Kose and J. Sievert (Eds.)
IOS Press, 1998

Nonlinear Modelling of Three Phase Transformers

Z.Q. Wu, G.H. Shirkoohi, J.Z. Cao and Jan Sikora*

School of EEIE, South Bank University, London SE1 0AA, UK

* *Technical University of Warsaw, Poland*

Abstract. This paper presents a new method to determine the nonlinear (saturation and hysteresis) three phase transformer model by taking into account the instantaneous saturation curve, the power losses at different flux densities supplied by the producer of the steel core, the geometry dimensions of the core, the flux balance equations in the core section joint points and the magnetic force balance equations in the magnetic loops in the core structure. This model was used in more accurate calculation of open conductor overvoltage of the three phase transformer.

1. Introduction

When one or two of the supply conductors to an unloaded or very lightly loaded ungrounded power transformer are switched or interrupted, leaving a transformer coil energizing through the capacitive coupling with the other phases, overvoltage will occur at the interrupted phases, the ultimate of this condition would be ferroresonance. The fundamental factor to the open conductor overvoltage is the nonlinear magnetic behaviour (i.e. saturation and hysteresis) of the transformer.

The open conductor overvoltage has been observed and presented in many papers; however, the magnetization characteristics of the transformer adopted when simulating the overvoltage are only suitable for single phase power transformers in most of the papers [1]. Although a few papers deal with three phase power transformers, they are not connected with the nonlinear hysteresis in detail [2], [3].

This paper presents a new method to determine the nonlinear (saturation and hysteresis) three phase transformer model by taking into account the instantaneous saturation curve, the power losses at different flux densities supplied by the producer of the steel core, the geometry dimensions of the core, the flux balance equations in the core section joint points and the magnetic force balance equations in the magnetic loops in the core structure. This model was used in more accurate calculation of open conductor overvoltage.

2. The Dynamic Hysteresis Expression by Differential Equation

The dynamic hysteresis phenomenon can be modelled by the following differential equation:

$$H = a_1 B + a_2 B^{n2} + a_3 B^{n3} + a_4 \frac{dB}{dt} \tag{1}$$

where a_1, a_2 and a_3 can be determined by fitting the corresponding magnetisation characteristics of the core; B is the magnetic flux density; a_4 is the so called dynamic hysteresis coefficient, it is a nonlinear function of the flux densities and can be modelled in piecewise linearity and polynomial function.

3. The Determination of a_4 by Piecewise Linearity Method

A relationship or curve between the dynamic hysteresis coefficient and the instantaneous value of the magnetic flux density can be imagined. Assume that this nonlinear relationship can be modelled in piecewise linearity (Fig.1). There are J segments in the imagined curve, B is the absolute value of magnetic flux density and $B_{m,i}$ is the assumed ith segment point value of the flux density in the imagined curve of the dynamic hysteresis coefficient vs the instantaneous value of the magnetic flux density. There are also J separate dynamic hysteresis coefficients $a_{4,1}$, $a_{4,2}$, ..., $a_{4,j}$ which define the instantaneous dynamic hysteresis coefficient nonlinearity.

We can calculate an integral, which is the hysteresis loop area. After calculating the integral, furthermore the power loss (including hysteresis, eddy current and anomalous losses), we can summarize the segment dynamic hysteresis coefficients as follows:

$$a_{4,1} = W(B_{m,1}) / [f V \; \omega \; B_{m,1}{}^2 \; \pi] \tag{2}$$

$$a_{4,2} = 2 \; [W(B_{m,2})/(2fV) - f(1)] / [\omega \; B_{m,2}{}^2 (2\beta_1 - \sin 2\beta_1)] \tag{3}$$

$$\beta_1 = \arccos B_{m,1}/B_{m,2}$$

$$\cdots$$

$$a_{4,k} = 2[W(B_{m,k})/2fV - f(1) - ... - f(k-1)] /[\omega B_{m,k}{}^2 (2 \; \beta_{k-1} - \sin 2\beta_{k-1})] \tag{4}$$

$$\beta_i = \arccos B_{m,i}/B_{m,k}$$

$$i=1, 2, ..., k-1; \; k=2, ..., j.$$

where $\omega = 2\pi f$ is the frequency; V is the volume of the core section; $W(B_{m,k})$ is the power loss of the core section at the peak flux density of $B_{m,k}$;

$$f(1) = a_{4,1} \; \omega \; B_{m,k}{}^2 [\pi/2 - \beta_1 - \sin 2\beta_1 / 2] \; ;$$

$$\cdots$$

$$f(k-1) = a_{4,k-1} \; \omega \; B_{m,k}{}^2 [\beta_{k-2} - \beta_{k-1} - (\sin 2\beta_{k-2} - \sin 2\beta_{k-1})/2] \; .$$

4. The Polynomial Approximation of a_4

In this approach, the coefficient a_4 can be modelled in the following polynomial function

$$a_4 = a_{40} + a_{4q} B^q \tag{5}$$

where $q=2,4,6, ...$, B is the instantaneous value of flux density in the core section.

We can calculate the power loss under the above hysteresis expression. For $q=2$, we have

$$W(B_m) = fV\pi\omega (B_m{}^2 a_{40} + 0.25*B_m{}^4 a_{4q}) \tag{6}$$

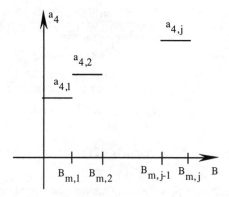

Fig.1 The imagined instantaneous nonlinearity of the dynamic hysteresis coefficient with the magnetic flux density in piecewise linearity.

where B_m is the peak value of the flux density; $\omega = 2\pi f$ is the power frequency; V is the volume of the core.

According to the power losses of the steel core at different flux densities, we can determine the coefficients a_{40} and a_{4q} by least squares curve fitting method.

5. The Three Phase Transformer Model

According to the nonlinear H - B relationship (1), magnetic fields in the sections can be written as follows:

$$H_a = a_1 B_a + a_2 B_a{}^{n2} + a_3 B_a{}^{n3} + a_4 dB_a/dt \tag{7}$$

$$H_b = a_1 B_b + a_2 B_b{}^{n2} + a_3 B_b{}^{n3} + a_4 dB_b/dt \tag{8}$$

$$H_c = a_1 B_c + a_2 B_c{}^{n2} + a_3 B_c{}^{n3} + a_4 dB_c/dt \tag{9}$$

According to Equations (7, 8, 9) and the geometry dimensions of the core, we can constitute a three phase power transformer model by combining the flux balance equations in the core section joint points, the magnetic force balance equations in the magnetic loops in the core structure and the voltage and current balancing equations at the terminals of the transformer.

6. Modelling Results

The core geometry of the three phase three limb power transformer and the specific power losses for the electrical steel cores at different flux densities considered in the case study is taken from [4]. The dynamic hysteresis coefficient was calculated for the commercial grain oriented (CGO), high permeability grain oriented (HiB) and laser scribed high permeability grain-oriented (ZDKH) silicon iron materials in their respective practical model cores. The piecewise linearized dynamic coefficients are presented in Table 1.

In the polynomial function approach, for CGO, HiB and ZDKH core transformers, the nonlinear dynamic hysteresis can be modelled by Equations (10), (11) and (12), respectively.

$$a_4 = 6.16*10^{-2} + 1.15*10^{-2}B^4 \tag{10}$$

$$a_4 = 6.26*10^{-2} + 7.32*10^{-3}B^2 \tag{11}$$

$$a_4 = 5.84*10^{-2} + 2.8*10^{-3}B^4 \tag{12}$$

Table 1 The global instantaneous dynamic hysteresis coefficients predicted from the identical three phase three limb transformer core constructed from the three materials for various instantaneous induction ranges.

	B≤ 1.0	1.0 ≤ B ≤1.3	B≥ 1.3
CGO	0.0651	0.0725	0.1329
HiB	0.0651	0.0725	0.0914
ZDKH	0.0574	0.069	0.084

By comparing the results of Equations (10-12) and Table 1, it has been found that they are in good agreement.

By the measured B-H characteristics of CGO core material and the linear least squares fit procedure, from a number of possible combinations, the anhysteresis B-H characteristics was fitted as

$$H = 11.1515B + 1.2151B^7 + 3.2*10^{-7}B^{33} \tag{13}$$

7. The Open Conductor Overvoltage Study

The model transformer is fed through a cable, the transformer's windings are star - star connected with no grounding, and the secondary winding is opened. Phase A of the transformer is connected to the supply source, while phases B and C are disconnected, hence, the capacitance to ground of the disconnected conductors appears in series with the transformer magnetising inductance. The transformer terminal voltages were simulated. According to these simulations, the terminal voltages of the open phases are affected by the dynamic hysteresis coefficient (as well as the anhysteresis magnetisation curve). Critical capacitance of the equivalent shunt capacitor can be calculated by simulation. If the critical voltage at the open phases is 125% of the rated voltage, then, the critical capacitance for constant hysteresis coefficient is 0.28 μF, while the critical capacitance for variable dynamic hysteresis coefficient is 0.24 μF. The difference for these two cases is 14.3%.

8. Conclusions

This paper presents a good nonlinear (saturation and hysteresis) model for a three phase power transformer. For the nonlinear hysteresis modelling, two methods are presented. The results obtained by both methods are in good agreement. This model was used to more accurately study the open conductor overvoltage (ferroresonance) of a three phase transformer. Studies show that the open conductor overvoltage (ferroresonance) is very sensitive to the nonlinear magnetization saturation curve and nonlinear hysteresis damping.

References

[1]   T. Tran-Quoc and L. Pierrat, " An Efficient Non Linear Transformer Model and its Application to Ferroresonance Study", IEEE Trans on Magnetics, Vol.31, No.3, May 1995, pp. 2060-2063.

[2]   X.S. Chen and P. Neudorfer, " Digital model for transient studies of a three phase five legged transformer", Proc. IEE, Vol. 139, No. 4, July 1992, pp.351-358

[3]   D. Dolinar, J. Pihler and B. Grcar, " Dynamic model of a three phase power transformer", IEEE Trans. on Power Delivery, Vol.8, No.4, October 1993, pp.1811-1819.

[4]   A. Sakaida, PhD Thesis, University of Wales, Cardiff, 1986.

Non-Linear Electromagnetic Systems
V. Kose and J. Sievert (Eds.)
IOS Press, 1998

Model of a Five-Limb Amorphous Core Used in a Three-Phase Distribution Transformer

Zbigniew GACEK[1], Marian SOIŃSKI[2], Rafał SOSIŃSKI[1]

[1]*Silesian Technical University, ul. Krzywoustego 2, 44-100 Gliwice, Poland*
[2]*Technical University of Częstochowa, Al. Armii Krajowej 17, 42-200 Częstochowa, Poland*

Abstract. The paper presents a mathematical model of a five-limb wound core as well as its experimental verification. The object of the experiment is the five-limb wound core of amorphous ribbon designed for construction of three-phase distribution transformers with a power rating of 160 kVA. This paper contains the results of computer simulation as well as the experimental results for the "star" with neutral point connection of primary windings. The tests made it possible to determine the propagation of magnetic flux and measure magnetic induction. The model has been suggested as: a vectorial diagram of magnetic flux appearing in a core, a block diagram and a system of equations.

1. Introduction

The manufacture of distribution transformers with increased efficiency requires the use of energy-saving magnetic materials. Due to its low losses an amorphous ribbon fulfils this criterion. The core of the amorphous ribbon is a wound core, cut through in order to place primary and secondary transformer winding. The mass production of amorphous ribbon caused more frequent use of this material in single-phase distribution transformers. Single-phase distribution transformers with amorphous cores are commonly used in the United States, Japan and India. It is not possible, however, for these transformers to be widespread in Europe because of a different type of distribution network used here. However, single-phase amorphous cores have already been used for the construction of five-limb magnetic circuits in three-phase transformers. The most popular, for technological reasons, is the construction of a five-limb wound core set up by four similar single-phase cores.

The construction of a wound core of a three-phase transformer, being different from that of a stacked core, requires another method of calculation of flux and magnetic induction distribution.

The aim of the paper is to propose a mathematical model of a five-limb amorphous core to be employed in making a three-phase transformer. The model has been suggested as: a vectorial diagram of magnetic flux appearing in a core, a block diagram and a system of equations.

2. Mathematical Model

Experimental research proved the existence of specific propagation of magnetic flux in a five-limb wound core of amorphous ribbon which is completely different from the propagation of flux in a five-limb stacked core (Fig.1.) [1]. Therefore, electromagnetic calculation based on a model of a stacked core, known from the literature (e.g. [2, 3]), cannot be applied here.

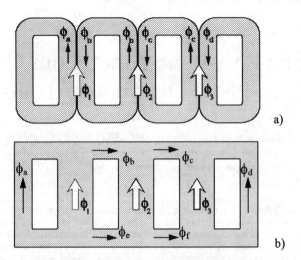

Fig.1. Propagation of magnetic flux in: a) a wound core, b) a stacked core

The specific propagation of magnetic flux in a five-limb wound core is due mainly to its construction. It consists of four single-phase wound cores put side by side (Fig.1a). Such a construction of the core is responsible for the fact that the cores standing side by side do not touch each other with their whole surface. There is a tiny air-gap between them. This air-gap has magnetic resistance big enough, as compared to the resistance of the core, to prevent the flux flow between each constituent core. It means that an individual flux flows in each core, i.e. there are four independent magnetic circuits. The flux in one of the main limbs of the core inducing the voltage in the secondary winding is the sum of a vector of two independent fluxes that flow through the adjoining cores e.g. ϕ_a and ϕ_b are phase shifted in relation to each other by 120°. Then one of these fluxes (e.g. ϕ_a) is shifted in relation to the flux in the main limb (ϕ_1) by 60° and the other (ϕ_b) -60° so that the flux in the main limb should be the sum of constituent fluxes (ϕ_a, ϕ_b). The same characteristics refer to other limbs and constituent cores.

The fluxes in the main limb should be in phase with forcing fluxes (ϕ_R, ϕ_S, ϕ_T) only when:
- we assume a linear characteristic of core magnetisation,
- magnetising windings are connected into a "star" with neutral point connection,
- constituent cores have the same magnetisation characteristic.

On the grounds of this above assumption a vectorial diagram of fluxes in a core (Fig.2) has been made. It shows magnetic flux propagation in a core and the reciprocal phase shifts of fluxes appearing in different parts of a magnetic circuit.

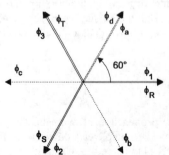

The magnetic fluxes: ϕ_a, ϕ_b, ϕ_c and ϕ_d are fluxes following through single cores: a, b, c and d. The fluxes ϕ_1, ϕ_2 and ϕ_3 are ones in the limbs of a core, being the sum of fluxes in constituent cores of each limb (e.g. $\phi_1 = \phi_a + \phi_b$). The fluxes ϕ_R, ϕ_S and ϕ_T are ones forced by the primary windings of a transformer.

A linear characteristic of core magnetisation has been assumed in further consideration in order to make easier the description of phenomena occurring in the core. It has to be emphasised, that the proposed model will be true in the case of the non-linear core characteristic.

Fig. 2. Vectorial diagram of magnetic fluxes in a five-limb wound core.

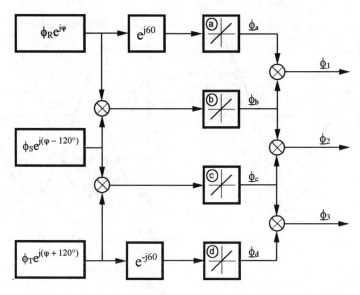

Fig.3. Block diagram of five-limb wound core model

On the grounds of the above assumptions and the vectorial diagram of magnetic flux the block diagram of a core model has been produced. The diagram is presented in Fig.3. Blocks a, b, c and d represent a magnetisation characteristic of each constituent core. Both the block diagram the vectorial one are described by equations which refer to the behaviour of magnetic fluxes in the whole core:

$$
\begin{cases}
\phi_1(t) = \phi_R e^{j\phi} \cdot e^{j60^\circ} \cdot F_a(B,H,t) + (\phi_R e^{j\phi} + \phi_S e^{j(\phi-120^\circ)}) \cdot F_b(B,H,t) \\
\phi_2(t) = (\phi_R e^{j\phi} + \phi_S e^{j(\phi-120^\circ)}) \cdot F_b(B,H,t) + (\phi_S e^{j(\phi-120^\circ)} + \phi_T e^{j(\phi+120^\circ)}) \cdot F_c(B,H,t) \quad (1) \\
\phi_3(t) = \phi_T e^{j(\phi+120^\circ)} \cdot e^{-j60^\circ} \cdot F_d(B,H,t) + (\phi_S e^{j(\phi-120^\circ)} + \phi_T e^{j(\phi+120^\circ)}) \cdot F_c(B,H,t)
\end{cases}
$$

where ϕ – time shift of a forcing flux ϕ_R, $F_a(B,H,t)$, $F_b(B,H,t)$, $F_c(B,H,t)$ and $F_d(B,H,t)$ - functions describing magnetisations curves of each constituent core: a, b, c and d.

In order to apply the above model to a real core one should determine real magnetisation curves of each constituent core, and then describe them by known non-linear functions [4].

3. Experimental Verification

Experimental tests have been carried out to verify the proposed model. The object of the experiment was a five-limb wound magnetic circuit of three-phase distribution transformers with a power rating of 160 kVA. The tests made it possible to determine the propagation of magnetic flux and measure magnetic induction as well as the deformations of voltage induced in the secondary windings. The tests were carried out for the "star" with neutral point connection of the primary windings. The measurements were taken with a digital oscilloscope with a memory. The results of measurements and calculations of fluxes in time are presented in Fig.4.

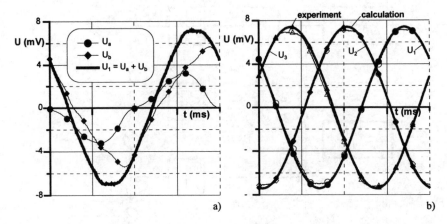

Fig.4. Fluxes in time: a) experimental data, b) comparison between experiment and calculations

For example, real voltages in time in measuring windings referred to fluxes ϕ_a and ϕ_b (Ua and Ub) as well as voltage in time in a secondary winding on limb "1" (U1 refers to ϕ_1) are presented in Fig.4a. Comparisons between measured and calculated voltages in time in the secondary transformer windings are presented in Fig.4b. These voltages refer to fluxes in the main limbs of five-limb cores. In order to calculate real fluxes in time, real voltages in time should be integrated.

On the grounds of formula (1) a class of non-linear functions was examined up to the fifth order inclusive, but the application of a linear function gives satisfying results as compared to experimental data results (Fig.4b).

4. Summary

The proposed mathematical model of a five-limb amorphous core is a simplified one. Its aim is to show the propagation of magnetic flux in a real core. The theoretical basis for the model was the assumption that magnetic flux flows only in a single core and does not penetrate into other cores. This means that four independent magnetic circuits arise. The above regularity has been experimentally proved.

Despite the occurrence of non-linear phenomena in the core, their linearization gave satisfying results. The proposed model may serve as the basis for an analogous three-limb wound core model in the future.

Since the assumptions for the model are universal they can be used for making alternative models for other schemes of connection of primary windings ("delta" and "star" with ungrounded neutral point) as well as for geometric configurations of wound cores (e.g. a wound core of a single phase transformer set up by two single-phase cores).

References

[1]   S. Sieradzki, *Structural and Technical Conditions of the Oil Distribution Transformer Construction with Five-Limb Amorphous Core*, PhD Thesis, Technical Univ. of Wrocław, Poland, 1997

[2]   J. L. Michel, J. Poittevin and G. Seyrwing: *Representation of the Transformer in the Electrical Network. Modelling from PSPICE Software* - Proc. of 19-th Sem. - SPETO, pp.183-198, Gliwice 1996, Poland

[3]   A. Kozłowska, H. Pabis: *Analysis of Flux Distribution in 3-Phase 5-Limb Transformer Cores under Overfluxing Conditions*. Technical University of Szczecin, Report Nr 37, Szczecin 1975, Poland

[4]   D. Jiles: *Introduction to Magnetism and Magnetic Materials*, London, 1989

Non-Linear Electromagnetic Systems
V. Kose and J. Sievert (Eds.)
IOS Press, 1998

Control of High-Tc Superconducting Radial Bearings

Kosuke NAGAYA, Kouichi KANNO and Nobuyuki HAYASHI

Department of Mechanical Engineering, Gunma University
Kiryu, Gunma 376, JAPAN

Abstract: The high-Tc superconducting levitation system is stable in both radial and axial directions without contacts, so it has been applied to bearings of an electric energy storage flywheel. The present article discusses a method for passing through critical speeds by using magnetic bearings as actuators. The analytical results for the system are first obtained, then a control method is presented. Numerical calculations are carried out for a typical problem, and it is ascertained that the present system and control method are applicable to the high-Tc superconducting bearings.

1 Introduction

Recently, large size bulk superconductors have been produced, and a number of applications have been presented on fly wheels, magnetic bearings and dampers. For the fly wheel, electrical energy is stored by transferring electrical energy to mechanical energy, so as to have a high efficiency, with a significantly high rotating speed of the wheel. This implies that the wheel should pass through some critical speeds. The present article discusses a magnetic bearing using a high-Tc superconductor. A system of radial magnetic bearings consisting of a super conducting disc and two radial magnetic bearings is presented. In the system, the superconducting disc is used for levitating the fly wheel in both radial and thrust directions by using pinning forces in both directions. To pass through the critical speeds, usual magnetic bearings are used. The magnetic bearings are used for controlling the vibrations of the wheel only when increasing or decreasing the speed. When the rotating speed becomes steady after passing through critical speeds, the control is cut off. Then the wheel rotates steadily without control. The analytical expressions for designing the controller of the magnetic bearings are obtained based on our previous analysis and modeling. In this system, the control energy should be small as just mentioned, so a method of control of the magnetic bearings is presented in which the sum of areas of the frequency response curve is taken as a cost function.

2 Analysis

Figure 1 shows the geometry of the flywheel system given in this paper. The shaft with the flywheel is levitated by the pinning force between the circular high-Tc superconductor and a circular ring shaped permanent magnet without contact. The shaft is rotated by the motor without bearings attached to the top of the shaft. The advantage of this system is to support the shaft by using only one superconductor. The equations of motion of the flywheel with the shaft are

$$m\frac{d^2x}{dt^2} + Kx = F_x, \qquad m\frac{d^2y}{dt^2} + Ky = F_y$$

$$J\frac{d^2\theta_y}{dt^2} - J_p\Omega\frac{d\theta_y}{dt} + K'\theta_y = T_y$$

$$J\frac{d^2\theta_x}{dt^2} + J_p\Omega\frac{d\theta_x}{dt} + K'\theta_x = T_x$$

$$\text{(1)}$$

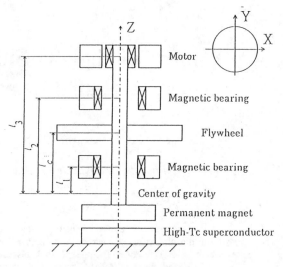

Fig.1 Analytical model

where m is the mass, J the moment of inertia and J_p the polar moment of inertia. The superconducting levitation system has a nonlinear relationship between the restoring force and the amplitude of the displacement of the shaft. But in this system, the vibration displacement is significantly small (less than 100 μ m), so the restoring force can be assumed to be linear with respect to the amplitude. The other important phenomenon for the system is that the levitation force varies with the vibration frequency of the shaft. This means that the levitation force is a function of the rotating speed of the shaft. By reference to our previous report[1,2], the spring constants can be written in the following forms:

$$K = a\left\{1 + b\left(1 - e^{-c\Omega}\right)\right\}$$

for the translation motion, and

$$K' = a'\left\{1 + b'\left(1 - e^{-c'\Omega}\right)\right\}$$

for the rotational motion, where Ω is the angular velocity of the shaft. There are the following relationships between the transitional displacements x_1, x_2, y_1, y_2 and rotational displacements:

$$x_1 = x - l_1\theta_y, \quad x_2 = x + l_2\theta_y, \quad y_1 = y - l_1\theta_x, \quad y_2 = y + l_2\theta_x \quad \text{(2)}$$

where the subscript 1 denotes the location at the upper electromagnetic actuator (magnetic bearing) and 2 the lower bearing, and where l_1 is the length between the lower actuator and the center of gravity, and l_2 the length between the upper actuator and the center of gravity. We have also the following relationships:

$$\theta_x = (x_2 - x_1)/(l_1 + l_2), \quad \theta_y = (y_2 - y_1)/(l_1 + l_2) \quad \text{(3)}$$

The forces F and the torque T at the actuators are

$$F_x = F_{x1} + F_{x2} + f_x + f_{cx}, \quad F_y = F_{y1} + F_{y2} + f_y + f_{cy}$$

$$T_y = -l_1F_{x1} + l_2F_{x2} + f_xl_3 + f_{cx}l_c, \quad T_x = -l_1F_{y1} + l_2F_{y2} + f_yl_3 + f_{cy}l_c \quad \text{(4)}$$

where $f_{x\sim y}$ is the force given by the motor, $f_{cx\sim cy}$ the unbalance force of the flywheel, l_3 the length between the rotor of the motor and the center of gravity, and l_c the length between the flywheel and the center of gravity. The drag force of the actuator is dependent on the air gap of the actuator (between the ferrite ring and the magnetic bearing), but when the displacement of the shaft is small, it can be written as

$$F_{x1\sim y2} = -kl_{x1\sim y2} \quad \text{(5)}$$

where k is the constant. The drag force and centrifugal forces are

$$f_x = f_0 \cos(\Omega t + \alpha), f_y = f_0 \sin(\Omega t + \alpha), f_{cx} = me\Omega^2 \cos\Omega t, f_{cy} = me\Omega^2 \sin\Omega t \quad (6)$$

where e is the eccentricity of the wheel.

3 Control Method

To control the shaft vibration, three control methods are combined. The first is the cross feedback control which eliminates the Corioli's force given by Matusita et al[3], second is the inertia force cancellation technique given by the present authors[4], and third is the feedback control with frequency weight.

(a) When we consider the feedback control with the frequency weight, the differential equations for the control currents are shown to be as follows with consideration of the Cross feedback control[3]:

$$c_{11}\ddot{I}_{x1} + c_{12}\dot{I}_{x1} + c_{13}I_{x1} = g_{x1}\theta_x , \quad c_{21}\ddot{I}_{x2} + c_{22}\dot{I}_{x2} + c_{23}I_{x2} = g_{x2}\theta_x$$

$$d_{11}\ddot{I}_{y1} + d_{12}\dot{I}_{y1} + d_{13}I_{y1} = -g_{y1}\theta_y, d_{21}\ddot{I}_{y2} + d_{22}\dot{I}_{y2} + d_{23}I_{y2} = -g_{y2}\theta_y \quad (7)$$

(b) The inertia force cancellation has a considerable effect on reducing the vibrations [4], and the equations become

$$c'_{11}\ddot{I}'_{x1} + c'_{12}\dot{I}'_{x1} + c'_{13}I'_{x1} = g'_{x1}\ddot{x}_1 , \quad c'_{21}\ddot{I}'_{x2} + c'_{22}\dot{I}'_{x2} + c'_{23}I'_{x2} = g'_{x2}\ddot{x}_2$$

$$d'_{11}\ddot{I}'_{y1} + d'_{12}\dot{I}'_{y1} + d'_{13}I'_{y1} = g'_{y1}\ddot{y}_1, \quad d'_{21}\ddot{I}'_{y2} + d'_{22}\dot{I}'_{y2} + d'_{23}I'_{y2} = g'_{y2}\ddot{y}_2 \quad (8)$$

(c) To reduce peak values, the feedback of the velocity and the displacement is valid:

$$c''_{11}\ddot{I}''_{x1} + c''_{12}\dot{I}''_{x1} + c''_{13}I''_{x1} = g''_{x1}\dot{x}_1 + g'''_{x1}x_1 , \quad c''_{21}\ddot{I}''_{x2} + c''_{22}\dot{I}''_{x2} + c''_{23}I''_{x2} = g''_{x2}\dot{x}_2 + g'''_{x2}x_2$$

$$d''_{11}\ddot{I}''_{y1} + d''_{12}\dot{I}''_{y1} + d''_{13}I''_{y1} = g''_{y1}\dot{y}_1 + g'''_{y1}y_1, \quad d''_{21}\ddot{I}''_{y2} + d''_{22}\dot{I}''_{y2} + d''_{23}I''_{y2} = g''_{y2}\dot{y}_2 + g'''_{y2}y_2 \quad (9)$$

where I is the control current, the subscripts x and y show the directions of x and y, and $c_{11} \sim g'''_{y2}$ are the coefficients of the controller. Substituting Eqs.(2) through (9) into Eq.(1) yields the equations under the control. There are many coefficients in the controller, so it is difficult to obtain optimal values directly. To have the optimal controller, the following cost function is considered:

$$J = \int_0^p \left\{ w_1\left(x_1^2 + x_2^2 + y_1^2 + y_2^2\right) + w_2\left(I_{x1}^2 + I_{x2}^2 + I_{y1}^2 + I_{y2}^2\right) \right\} d\Omega \quad (10)$$

where w_1 and w_2 are the weights. The above equation illustrates the cost function is constructed by both the displacements of the shaft and the control currents. This implies that the vibrations of the shaft become minimum with minimum control energy over the whole frequency range when the cost function becomes minimum.

Table 1 denotes the dimensions of the shaft, flywheel and the superconducting levitation system. By using the numerical values, the optimal coefficients of the controller are calculated. Those values are given in Table 2. The comparison between the frequency response of the displacement with the present control (Authors' control 1) and that without control are depicted in Fig.2. The result for the optimal PD control without cross feedback and inertia force cancellation (Authors' control 2) is also given in the figure. The resonance peaks are reduced in both the present method and the optimal PD control. Since the control current is included in the cost function, the control currents are also restrained within 10A.

4 Conclusion

A superconducting levitation system of an electrical energy storage flywheel system has been presented . A method for passing through the critical speeds has been presented in which the radial magnetic bearings are applied as actuators. It is clarified that the present method and analysis are valid for controlling vibrations of the system.

Table 1 Dimensions of the
superconducting bearing

$m[kg]$	2.00
$J[kg \cdot m^2]$	1.24×10^{-4}
$J_p[kg \cdot m^2]$	2.00×10^{-4}
$l_1[m]$	2.25×10^{-2}
$l_2[m]$	6.75×10^{-2}
$l_3[m]$	2.02×10^{-1}
$l_c[m]$	2.25×10^{-2}
$e[m]$	1.00×10^{-5}
$k[N/A]$	1.00
$f_0[N]$	1.00
$a[N/m]$	5.60×10^4
$a'[N \cdot m/rad]$	5.17×10^1
b	5.00×10^{-5}
b'	1.00×10^{-1}
c	3.60×10^8
c'	5.00×10^9

Table 2 Optimal coefficients of the controller

g_{x1}	9.36×10^{-1}	g''_{y2}	9.79×10^{-1}
g'_{x1}	1.24×10^0	g'''_{y2}	9.54×10^{-1}
g''_{x1}	1.00×10^0	c_{11}	7.75×10^{-1}
g'''_{x1}	1.00×10^0	c_{12}	1.00×10^0
g_{x2}	1.29×10^2	c_{13}	1.00×10^0
g'_{x2}	1.30×10^2	c_{21}	-2.11×10^{-3}
g''_{x2}	9.29×10^{-1}	c_{22}	2.00×10^{-1}
g'''_{x2}	9.98×10^{-1}	c_{23}	1.00×10^0
g_{y1}	1.77×10^{-4}	d_{11}	1.03×10^0
g'_{y1}	2.05×10^3	d_{12}	9.98×10^{-1}
g''_{y1}	2.00×10^0	d_{13}	1.00×10^0
g'''_{y1}	1.00×10^0	d_{21}	-1.72×10^{-3}
g_{y2}	9.97×10^{-1}	d_{22}	8.44×10^{-2}
g'_{y2}	1.30×10^2	d_{23}	-9.96×10^{-1}

References

[1] K.Nagaya, *IEEE Trans. on Magnetics*, Vol.32 (1996), 445-452.
[2] K.Nagaya and S. Shuto, *IEEE Trans. on Magnetics*, Vol.32 (1996), 1888-1896.
[3] O. Matsusita et al. Int. Conf. Rotordyn. *JSME & IFTOMM, Tokyo* (1986), 421-426.
[4] K. Nagaya, *Jour. Sound & Vib.* 184-2 (1995), 185-194.

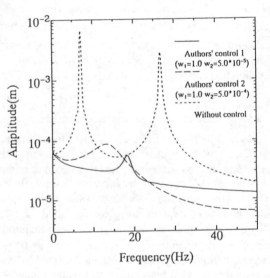

Fig.2 Frequency response of the displacements of the shaft at the upper magnetic bearing.

Non-Linear Electromagnetic Systems
V. Kose and J. Sievert (Eds.)
IOS Press, 1998

Fundamental Study of a Linear Induction Motor for Propulsion and Levitation of a Steel Plate

Hitoshi HAYASHIYA, Hiroyuki OHSAKI, Eisuke MASADA

Department of Electrical Engineering, Faculty of Engineering,
The University of Tokyo. 7-3-1, Hongo. Bunkyo-ku. Tokyo 113, JAPAN

Abstract. To accomplish requirements regarding quality improvement of the products in the steel making process, a non-contacting conveyance system for a steel plate has been investigated. In this paper, some non-contacting conveyance methods are discussed, assuming the conveyance of a steel plate by means a practical approach.

At the beginning, a non-contacting supporting system by electromagnets is shown. In accordance with this supporting system, three propulsion systems, namely, a system with LIMs, a system using reluctance force by electromagnets, and a system giving a tilt to produce thrust force, are discussed. The advantages and the disadvantages of these methods are emphasized in order to outline the possible overall system.

1. Introduction

In the steel making process, there is always the demand to improve the quality of the product. To realize such a requirement, a non-contacting conveyance system for a steel plate has been considered. In our previous studies, a conveyance system with electromagnets for levitation and linear induction motors for propulsion has been investigated[1]. Instead of this system, the conveyance system using not only the thrust force but also the normal force of a SLIM is proposed to simplify the total system. Figure 1 shows the concept of a combined lift and propulsion system for a steel plate conveyance.

In this paper, three non-contacting conveyance methods by magnetic force are investigated and compared. The first one is a system with SLIMs as shown in Fig.1. The second one is a system which uses reluctance force acting on the edge of a steel plate. The last one is a system giving a tilt to a steel plate to produce propulsion force.

2. Investigation of Non-Contacting Conveyance Methods of a Steel Plate

2.1. Levitation System by Electromagnets

Prior to comparing some propulsion systems for a non-contacting steel plate conveyance, a non-contacting supporting system by electromagnets was designed.

Fig. 1 Combined lift and propulsion system for a steel plate conveyance

To support a steel plate by electromagnets, small electromagnets have the advantage of a better controllable magnetic force. It is desirable, however, to use big electromagnets to support a heavy steel plate. Regarding its shape, an electromagnet with a large aspect ratio can avoid magnetic saturation in a steel plate[2]. On the contrary, a large aspect ratio causes the increase of energy consumption and the deterioration of controllability.

By giving constraints for the flux density in the steel plate of 1 T and a current density for electromagnets of 3 A/mm^2, the supporting system by electromagnets was designed as shown in Table 1. A thick steel plate with 10 m length, 1 m width and 50 mm thickness and a thin steel plate with 3 m length, 1 m width and 2 mm thickness are assumed here. The gap length was 10 mm.

N_m refers to the number of electromagnets arranged above a steel plate. The dimensions of an electromagnet, a and b are defined as shown in Fig. 2. The occupation ratio R_o refers to the ratio of the occupation area by electromagnets to the surface area of a steel plate.

In the following sections, propulsion systems added to this levitation system by electromagnets are investigated. In every estimation, the size of the steel plate was assumed to be 10 m length, 1 m width and 50 mm thickness, and the gap length was assumed to be 10mm.

Table 1   Configurations of supporting systems

—	Thick plate	Thin plate
Number of magnets N_m	14	16
Dimension of magnets a	102 mm	35 mm
Dimension of magnets b	52 mm	118 mm
Occupation ratio R_o	5.2 %	7.9 %

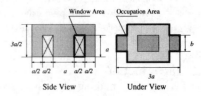

Fig. 2   Definition of dimensions of electromagnets

2.2. Propulsion system by LIMs

The secondary of a LIM for a steel plate conveyance consists of solid iron and this structure produces strong normal forces between the LIM and the steel plate. If it is possible to control the normal force of a LIM and use this force to support a steel plate, the number of electromagnets can be reduced and the total system will be simplified. On the other hand, there is iron saturation near the surface of the steel plate because of the skin effect. To produce strong normal forces to support a heavy steel plate, a LIM with long pole pitch is preferable, however, it causes more magnetic saturation and deteriorates the efficiency of the LIM.

Figure 3 shows the normal force and the thrust force characteristics of a LIM, designed to produce strong normal forces, having the following dimensions;

Length(L):1180 mm   Width(h):120 mm   Pole pitch(τ):240 mm
Slot pitch(t_s):40 mm   Slot width(w_s):20 mm   Slot depth(d_s):40 mm
Phase No.(m):3   Air gap(d):10 mm   Slip Frequency(sf):10 Hz

The two dimensional finite element method (FEM) was used in the analysis and the repetition of the FEM analysis was performed to take magnetic saturation into account. The permeability of each element was modified according to the flux density of the element in every iteration and the flux density in the secondary iron was restricted to less than 1.5 T. The relative permeability of iron was chosen to be 1000 for the beginning.

Because of the magnetic saturation, the normal force of the LIM was decreased and it was necessary to use twenty LIMs to support the total mass of a steel plate with 10 m length, 1 m width and 50 mm thickness even when the primary current density is 3 A/mm². In such a system, the occupation ratio by LIMs on the steel plate is 29.6 %, much higher than that of the supporting system by electromagnets.

2.3. Propulsion system using reluctance force

When the steel plate is supported by electromagnets as shown in Section 2.1, the reluctance force acts on the steel plate by exciting electromagnets near the edge.

Figure 4 shows the force characteristics acting on a steel plate with 50mm thickness. The electromagnet for the thick steel plate conveyance in Table 1 was used in calculation and the x axis represents the distance of the steel plate from one side of the electromagnet. Figure 5 shows the flux distributions in selected positions.

The forces in Fig. 4 are values for one electromagnet with 50 mm thickness, and the steel plate is driven with about 150 N intermittently when the electromagnets are arranged in two lines.

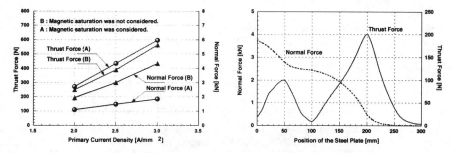

Fig. 3 Normal force and thrust force characteristics of the LIM

Fig. 4 Reluctance force characteristics

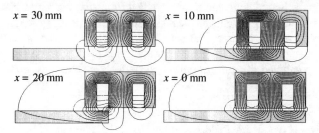

Fig. 5 Flux distribution at the edge of a steel plate

2.4. Propulsion system by giving a tilt

The method to use the gravitational force by giving a tilt to an object has been studied in the application for the conveyance system in a clean room and to produce a guidance force for a non-contacting steel plate conveyance[3].

The thrust force by giving a tilt to a steel plate is calculated in Eq. 1. Here, M is the mass of the steel plate and θ_p is the pitching angle. When the gap difference between each edge is 10 mm, the thrust force is 38.5 N.

$$F_t = Mg\tan\theta_p \approx Mg\theta_p \quad \text{N} \tag{1}$$

Because it is difficult to give a sufficiently large tilt to a long steel plate under fixed electromagnets in this application, the thrust force is very weak compared with other methods. The advantage of this method is, however, the possibility to produce a continuous thrust force and to control it almost independently of the levitation force.

3. Comparison of Non-Contacting Conveyance Methods for a Steel Plate

Fig. 6 shows the features of each method. In this figure, the strength and the controllability of the force have a direct influence on controlling the position and velocity of a steel plate. Therefore, the propulsion system with LIMs is preferable at the beginning and at the end of the conveyance line to strictly control the position of a steel plate and to realize good acceleration and deceleration. To keep the movement of a steel plate after the acceleration, the propulsion system by giving a tilt to the plate is practically sufficient.

In the future, the controllability of the normal force of a LIM will be investigated. If it is possible to control it, the number of electromagnets to support a steel plate will be reduced and the disadvantages of this method in Fig. 6 will be reduced.

	Equipment for Propulsion	Strength of Thrust Force	Continuity of Thrust Force	Controllability of Thrust Force	Independence from Levitation
LIM	○	◉	○	◉	△
Reluctance	◉	○	△	×	×
Tilting	◉	△	◉	○	○

◉ : Excellent
○ : Good
△ : Little
× : No

Fig. 6   Comparison of the non-contacting conveyance methods

4. Conclusions

Non-contacting propulsion methods by magnetic forces for a steel plate conveyance system were investigated and compared. The propulsion system with LIMs has an advantage in producing a strong and controllable thrust force. On the contrary, the method by giving a tilt to a steel plate can easily produce a constant propulsion force.

In the future, the controllability of the normal force of a LIM will be studied and the feasibility of a combined lift and propulsion system for a steel plate conveyance will be evaluated.

References

[1] H.Hayashiya, N.Araki, J.E.Paddison, H.Ohsaki and E.Masada. "Magnetic Levitation of a Flexible Steel Plate with a Vibration Suppressing Magnet". *IEEE Transactions on Magnetics*, Vol.32, No.5, pp.5067-5069, 1996

[2] H.Osabe, M.Watada, S.Torii and D.Ebihara. "The Effect on the Multipolar Electromagnet for the Levitation of Thin Iron Plate". *Proceedings of MAGLEV'95*, pp.145-150, 1995

[3] M.Morishita and M.Akashi. "A Slant Guide Technique for Magnetically Levitated Steel Plate". *National Convention Record IEEJ*, Vol.5, pp.181-182, 1997 (*in Japanese*)

Non-Linear Electromagnetic Systems
V. Kose and J. Sievert (Eds.)
IOS Press, 1998

Magnetic Levitation without Airgap Sensor

Th. Friedrich, G. Henneberger

Institut für Elektrische Maschinen, RWTH Aachen, 52056 Aachen, Germany

Abstract. The paper presents different methods of stabilizing the suspension magnet of a magnetically levitated conveyance system without an airgap sensor.

1. Introduction

This paper describes three different methods of stabilizing a hybrid excited magnetic levitation system without using an airgap sensor. It is part of a contactless conveyance system (Fig. 1(a)) [1]. The studies have been carried out for one magnet. A special device which can suppress horizontal motion has been built and was used in our investigations (Fig. 1(b)).

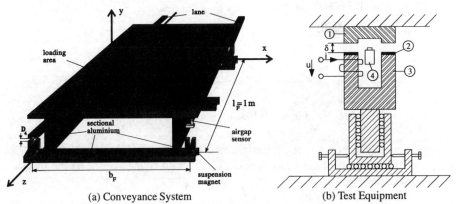

(a) Conveyance System   (b) Test Equipment

Figure 1: Suspension Magnet as Part of the Conveyance System

The suspension magnet consists of laminated U-cores ③ with 2 mm thick NdFeB magnets ②. Each coil has got 100 turns and they are connected in series. The rail ① consists of solid steel. The airgap sensor ④ is usually needed to stabilize the levitation system [2].

2. Differential Equations

2.1 Mechanical Equations

The equation of the motion of the magnet is as follows:

$$m\ddot{\delta} = mg - F_M(\Theta_L, \delta). \tag{1}$$

δ is the value of the airgap, m represents the mass of the magnet together with its load, g is the acceleration of gravity and F_M is the magnetic force. The force depends on the ampere turns Θ_L and on the airgap δ. In order to linearize this differential equation one has to calculate a tylor-series, which is truncated after the first term:

$$F_M \approx F_M|_{\Theta_0,\delta_0} + (\Theta_L - \Theta_0)\frac{\partial F_M}{\partial \Theta}|_{\Theta_0,\delta_0} + (\delta - \delta_0)\frac{\partial F_M}{\partial \delta}|_{\Theta_0,\delta_0}. \qquad (2)$$

The magnetic force F_M of the magnet has been calculated using a three-dimensional Finite Element analysis. The interpolation of the computed values and the determination of the partial derivatives were done using a bicubic spline-interpolation [3].

2.2 Electromagnetic equations

In order to describe the eddy current losses, the magnetic levitation system can be compared with a short-circuited transformer. The primary winding represents the coils of the levitation system. The eddy current paths in the rail form the transformers secondary winding. In practise the leakage inductances can be neglected (cf. Fig. 2(a)).

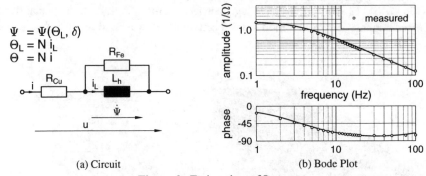

(a) Circuit   (b) Bode Plot

Figure 2: Estimation of Losses

The relationship between the impedance of the system and the frequency is used to determine the elements of the equivalent circuit diagram. Fig. 2(b) shows the measured and the calculated impedance [6].

The voltage equation is $u = R_{Cu}i + \dot{\Psi}$.
The flux $\Psi(\Theta_L, \delta)$ and its derivatives are also computed using a three-dimensional Finite Element method.

Now the transfer function of the system can be calculated. Fig. 3(a) shows the bode-plot of the transfer function with the voltage as the input and the airgap as the output. The attenuation is about -70 dB at a frequency of 100 Hz. The transfer function of the sensorless system, where the voltage is the input and the current is the output, is presented in Fig. 3(b). In this case an attenuation of -70 dB is reached at a frequency over 10 kHz. This indicates, that it is more difficult to stabilize the sensorless system than the system with an airgap sensor [6].

3. Sensorless Control

The equation of the voltage is the main equation of all conrollers without a sensor for the airgap:

$$u = R_{Cu} \cdot i + \dot{\Psi} \quad \text{with} \quad \dot{\Psi} = \frac{\partial \Psi}{\partial \Theta_L} \cdot \dot{\Theta}_L + \frac{\partial \Psi}{\partial \delta} \cdot \dot{\delta} = u_\delta + u_\Theta. \qquad (3)$$

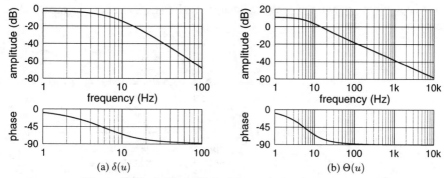

(a) $\delta(u)$ (b) $\Theta(u)$

Figure 3: Bode Plots of the Magnetic Levitation System

Investigations on three different methods for the sensorless control were done.

Theoretically the airgap can be evaluated by analyzing the dependency between the airgap and the inductivity $\partial\Psi/\partial\Theta_L$. In this case the voltage u_δ has to be neglected [6].

Another approach is to stabilize the electrical part of the whole system. In this case the voltage is the input and the current is the output. One possible controller is a robust H_∞-controller.

A state observer can also be used instead of an airgap sensor. One gets a robust controller concerning the noise of measurement and of the system using a Kalman Filter [5].

4. Simulation and Measurements

Fig. 4(a) presents the current together with the switched voltage for a certain airgap. Experiments have shown, that the rise time of the current varies only a little with the airgap. Thus it is not practicable to stabilize the magnetic levitation system using the rise time instead of an airgap sensor.

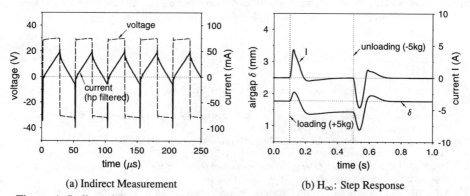

(a) Indirect Measurement   (b) H_∞: Step Response

Figure 4: Indirect Measurement of the Airgap and Step Response of a H_∞-controller

Fig. 4(b) shows the simulation results using a H_∞-controller, loading 5 kg at 0.1 s and taking it away at 0.5 s. The differential equations are solved by a 4th order Runge-Kutta method taking into account the nonlinear characteristics (F_M, Ψ) of the magnets and the noise of the measurement.

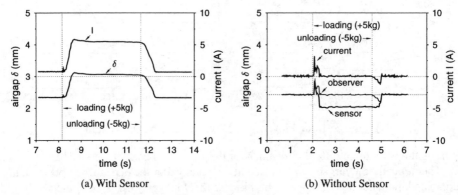

(a) With Sensor   (b) Without Sensor

Figure 5: Load: 5 kg

Experimental results with a load of 5 kg are presented in Fig. 5. Fig. 5(a) shows the value of the airgap and the current using a state controller without observer. The current increases with the additional load. Fig. 5(b) shows the behaviour of the current and the airgap using a Kalman-Filter instead of a sensor. The airgap sensor is only used to lift the carrier up from the lower stop. Afterwards the controller with the observer is used to stabilize the system. In this case the current is adjusted automatically. No current will flow in the coils after the recovery time.

5. Conclusions

This article points out that a magnetic levitation system with hybrid excitation can be stabilized without an airgap sensor, using an observer. The airgap sensor is still used to lift the carrier up from the lower stop. An indirect measurement does not lead to a practicable sensorless controller.

References

[1] Reuber, Chr.: *Berührungsloses Transportsystem mit Synchronlinearantrieb.* Dr.-Thesis, RWTH-Aachen: 1996.

[2] Rödder, D.: *Berechnung und Auslegung der berührungslosen Lagerung eines Förderfahrzeuges mit Hybridmagneten.* Dr.-Thesis, RWTH-Aachen: 1994.

[3] Friedrich, Th.; Henneberger, G.; Ress, Chr.: *Sensorless Magnetically Levitated System with Reduced Observer.* 14th Int. Conf. on Magnetically Levitated Systems. MAGLEV 95, Bremen: 1995.

[4] Friedrich, Th.; Henneberger, G.: *Optimization of an observer for the airgap of a hybrid excited magnetic levitation system.* Electrimacs, St.-Nazaire 1996.

[5] Friedrich, Th.; Kahlen, Kl.; Henneberger, G.: *Einsatz eines Beobachters für den Luftspalt einer hybridmagnetischen Schwebeanordnung.* Archiv für Elektrotechnik, Band 79, Nummer 6, Berlin 1996.

[6] Friedrich, Th.: *Sensorlose magnetische Lagerung für ein Förderfahrzeug.* Dr.-Thesis, RWTH-Aachen: 1997.

Non-Linear Electromagnetic Systems
V. Kose and J. Sievert (Eds.)
IOS Press, 1998

The Levitation System for a Magnetic Non-contact Transportation Device of Thin Steel Plates

T. ISHIWATARI, M. WATADA, S. TORII and D. EBIHARA

Musashi Institute of Technology, 1-28-1, Tamazutsumi, Setagayaku, Tokyo, 158 Japan

Abstract Thin steel plates are needed to be transported without cracks on the surface and the unevenness of lubricating oil in the field of producing steel-ware. The non-contact transportation of thin steel plates by magnetic levitation is one of the solutions. The authors propose the levitation system with two desired values to prevent transverse waves from propagating along the plate. It is found that the system possesses a wide stable range for the disturbance of the elastic force by simulations. The result indicates the possibility of the stable levitation by proposed system.

1. Introduction

The body of a motorcar should appear to be of high quality. Therefore, two problems exist for the requirement. First, the surface of the thin steel plate used as the material of the body must be kept smooth without cracks for the high quality coat. Second, the lubricating oil coated on the surface must be even in the area where the thin steel plate is pressed. The contact between the plate and the device of transportation causes problems. The technology of levitation by magnetic attractive control may enable the thin steel plate to be transported without contact. However, the technology to levitate the thin steel plate is not advanced enough for practical use.

The thin steel plates bend more easily than the thick steel plates. If the control system for the levitation of the thin steel plate is designed based on the stiff body model, the limit of stability is low, because the thin steel plate vibrates easily. The robust control systems for disturbance forces are needed for levitation of thin plates. The design method of the control system is widely studied, for example, as the design based on the robust control theory in recent years[1][2]. However, the optimum composition of the levitation system is not yet known. We have been studying the arrangement of the electromagnets and the control method , so that the levitation systems are not influenced by the elastic force from the vibration of plates, with the system contained in a number of suspended points. The elastic vibration arises from the transverse waves propagating along the sheet.

In this paper, we propose a control method that a group of some electromagnets has additional function to suppress the propagation. This system is called the levitation system with two desired values. It is expected that the respective electromagnets absorbs the elastic energy and one magnet-group suppresses the propagation of the transverse waves. We investigate the effect of the levitation system with two desired values using computer simulation.

2. Proposition of the levitation System with two desired values

The attractive force is independently controlled in the normal system of the magnetic levitation with some electromagnets. The usual system to levitate the thin steel plate is shown in Fig.1. All magnets in the system are controlled based on the fixed desired value. In fact, however, the state parameters in each control system are not independent variables, because the stress by the elasticity of the bent plate acts on each system. If the relative systems such as the levitation system of the thin steel plate control each magnet independently, some transverse waves of the plate easily cause the control systems to resonate.

The proposed system is shown in

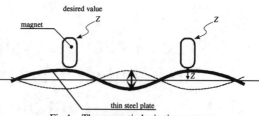

Fig.1   The magnetic levitation system with independent control

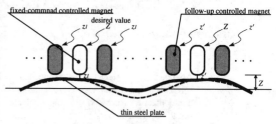

Fig.2   The magnetic levitation system with proposed control

Fig.2. The block diagram of the system is shown in Fig.3. The magnets of the several blocks suspend a thin steel plate. One block consists of one fixed-command controlled magnet and $2 \times n$ follow-up controlled magnets. The desired value of the fixed-command controlled magnet is a constantly steady parameter. The desired value of the two follow-up controlled magnets placed nearest the fixed-command controlled magnet are state parameters of the fixed-command controlled magnet. All of the follow-up controlled magnets in one block controlled by the cascade connection. They have the relationship of master and slave each other. The proposed system is called the levitation system with two desired values. The relation of the electromagnet as master and slave can flats the thin steel plates under one magnet-group, and causes the elastic energy reduces in the thin plate.

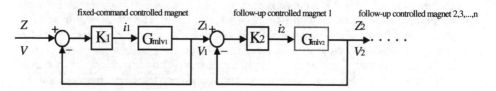

Fig.3   The block diagram of the levitation system with two desired values

3. Effect of the levitation system with two desired values

The bend motion of thin steel plate is formulated as non-linear differential equation of high dimension. The simulation by the strict equation model is difficult. One the other hand, the bend motion of thin steel plate is represented by three thick plates connected with springs and dampers. The model that three levitation system involves is shown in Fig.4. At present, the thick plate (20mm thickness) levitates experimentally in our laboratory. The property of the simulator for one levitation system is confirmed.

The levitation system with two desired values could be compared with the independent levitation system by simulation. The difference between both systems is only the desired value. The comparison of the gap response for the disturbance is shown in Fig.5. The disturbance force of 2N acts on the plate below the magnet I from 0.1s to 0.2s. Elastic coefficient k=2300N/m and damping coefficient c=115Ns/m decide the quality of the material modeled as the thin steel plate. Fig.5(a) shows the gap response of the magnet III by the independent levitation system, Fig.5(b) and Fig.5(c) shows the gap response of one by the levitation system with two desired values in case that the desired values of two follow-up controlled magnets are have not the relationship as master and slave and have one, respectively.

In Fig.5, the characteristics of the system (b) and (c) are both convergent, though the one of system (a) is divergent. When damping coefficient c decreases to 105Ns/m, the comparison of the characteristics of the system (b) and system (c) is shown in Fig.6. In the figure, system (b) is divergent , though system (a) is still convergent.

It is found that the levitation system with two desired values that the follow-up controlled magnets have the relationship as master and slave possesses the widest stable range for the disturbance of the elastic force in the three systems, and the effect gives the levitation control a robustness.

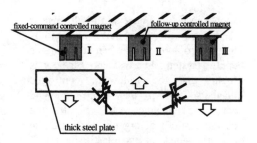

Fig.4   The model of three control systems

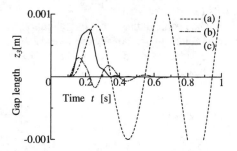

Fig.5   The gap response of three control systems
((b) is convergent)

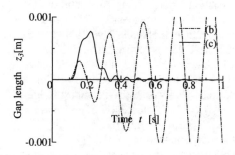

Fig.6   The gap response of three control systems
((b) is divergent)

4. Resonance of the control by the levitation system with two desired values

It is considered that the good relativity in the controlled variables of each system prevents transverse waves from propagating along the plate. The levitation system with two desired values possesses a wider stable range than the independent system. Though, being designed in error, it is possible that the control systems of some follow-up controlled magnets in the proposed system causes the resonance, respectively. The characteristic in this case could be confirmed by simulation.

When the disturbing force of sine wave acts on the plate below the magnet I, the gap response of the magnet III is simulated. Elastic coefficient k=2300N/m, and damping coefficient c=115Ns/m are set, and the disturbance force of 5N acts from 1s continuously. When the frequency of the disturbance force is 2Hz, 8Hz, and 15Hz, the gap responses are shown in Fig.7, Fig.8, and Fig.9. 8Hz is the resonant frequency that is obtained by the closed loop of the velocity in one control system. In Fig.7 and Fig.9, it is found that the follow-up controlled magnets follow the fixed-command controlled magnet. In Fig.8, the characteristic is divergent. 8Hz is not the eigenfrequency of the elastic model. It shows that the characteristic frequency of resonance exists in the levitation system with two desired values.

The results indicate that the levitation system with two desired values needs to be designed so as to prevent the resonance.

5. Conclusion

The vibration of the thin steel plate may be regarded as the propagation of some transverse waves overlapped. If the levitation system has the ability to prevent the transverse wave from propagating, the stable levitation of the thin steel plate is realized. The simulation with the simple model of the elastic plate shows that the levitation system with two desired values possesses the widest stable range for the disturbance of the elastic force. The result offers one possibility for the stable levitation of thin plates. The system should be designed based on the relationship between the arrangement of the magnet and the propagation of the transverse wave, so as to prevent the resonance.

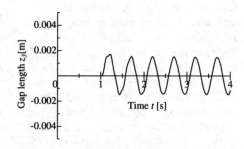

Fig.7   The gap response for the sine-wave disturbance
(f =2Hz)

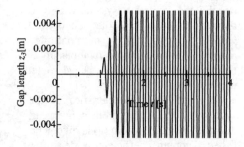

Fig.8   The gap response for the sine-wave disturbance
(f =8Hz)

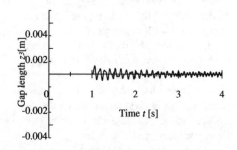

Fig.9   The gap response for the sine-wave disturbance
(f =15Hz)

The characteristics of the levitation vary extensively with the network of some magnets. The effect of the proposed system should be confirmed experimentally.

References

[1] Muraguchi Y:"CONTROL SYSTEM DESIGN APPLIED AN H ∞ ROBUST THEORY FOR MAGNETICALLY LEVITATED THIN IRON PLATE", Japan Industry Applications Society Conference, International Sessions of 8th Annual Conference, LD-8, pp.193-196(1994)

[2] Kawaguchi Y, Nakagawa T: "Transportation Technique for Magnetically Levitated Thin Iron Plates", IEEE TRANSACTIONS ON MAGNETICS (INTERMAG'93), VOL.29, NO.6, pp.2974-2976(1993)

Non-Linear Electromagnetic Systems
V. Kose and J. Sievert (Eds.)
IOS Press, 1998

Influence of Coil Quench on the Electromagnetic Force of the Superconducting Magnetically Levitated Train

S. OHASHI, Y.HIRANE, T.HIKIHARA*, H.OHSAKI**, E. MASADA**

Kansai University, 3-3-35, Yamate-cho, Suita, Osaka 564, Japan
* Kyoto University, Yoshida-honmachi, Sakyo, Kyoto 606-01, Japan
**The University of Tokyo, 7-3-1, Hongo, Bunkyo, Tokyo 113, Japan

Abstract. A running simulation program for a superconducting(SC) magnetically levitated train (JR Maglev system) has been developed. The train motion for the case of the SC coil quenching was calculated. The levitation and guidance force of the SC coil next to the quenched one became about twice and seven times as large as those in the normal state respectively. Maximum lateral displacement of the quenched bogie edge was about 60% of the lateral air gap. The influence of the SC coil quenching on the other bogies was shown to be very small.

1. Introduction

A superconducting (SC) magnetically levitated transportation system (JR Maglev system) has been developed in Japan, and the new test line has been constructed in Yamanashi prefecture[1]. Figure 1 shows a cross-section of the system, and Fig. 2 shows the arrangement of ground coils and SC coils. In this system, a SC linear synchronous motor is used for propulsion, and a SC side wall electrodynamic suspension system is used for levitation and guidance. This levitation and guidance system is a non-linear one, and numerical analysis is used to verify its running characteristics. However running characteristics under various disturbances are not clearly understood. We have developed a running simulation program that includes the propulsion, the levitation system and the mechanical coupling between a bogie and a cabin. Using this program, the three-dimensional movement of the Maglev train in the case of SC coil quenching has been investigated.

2. Analysis method

The calculation consists of three parts[2]: the EDS, the LSM and the secondary suspension part. The virtual displacement method is used to calculate the levitation and guidance forces. The current induced in the levitation coil needs to be calculated

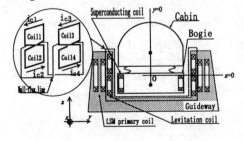

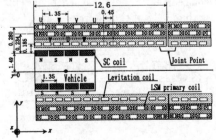

Figure 1: Cross-section of the super-conducting Maglev system

Figure 2: Arrangement of ground coils for levitation and SC coils

as follows: the EDS system is given as an air-core coil system, and modeled as electric circuits. Then solving the electric circuit equations, we can calculate the current of the levitation coils when the bogie passes them. For the LSM system, we calculate the flux generated by the LSM primary coils. Combining these with the current of the SC coils, the thrust force by LSM is given. The motions of the bogies and cabins are calculated by putting these electromagnetic forces and the force of the secondary suspension into the motion equations. Iterating these procedures, the transient motion of the train is given. Figure 3 shows the train model. The train consists of three cabins and four bogies. Four mechanical springs and dampers are used to connect a bogie and a cabin.

3. Results

The horizontal center of the guideway is located at y=0, and the vertical center of the LSM armature coils at z=0. In the analysis, the current of the quenched SC coil i_{sq} was obtained from the following equation:

$$i_{sq}(t) = I_0 \times (1.0 + t/T) \times e^{-t/T} \quad T = 0.501sec \tag{1}$$

where I_0 is the SC coil current when quenching occurs. This equation is derived from the experimental results at the Railway Technical Research Institute(RTRI)[2]. Figure 4 shows the arrangement of the quenched SC coil. From the numerical analysis of the bogie model, the influence of the quenching on the bogie position is largest when the first one (SC1) is quenched. From the results of the train model[3], the influence on the train position is largest in the case of the bogie between the cabins quenching. Therefore the running simulation in the case of SC1 of the second bogie quenching is shown. The levitation force balances the train weight at a bogie position of (y, z)=(0,-0.039m) for the running speed v=140m/s. The initial positions for calculation are set there. The train starts running in x direction and keeps these positions for a period of 0.045sec.

Figure 5 shows the average position of the train from t=2.8 to 3.8 sec. In this period, the train finished the transient motion caused by the quenching. In the figures, $CM(M=1-3)$ is the Mth cabin, $BN(N=1-4)$ is the Nth bogie. The solid lines show the cross over lines between the horizontal plane of the bogies including their center of gravity and the side plane of the bogie. The dotted lines show those of the cabins. The upper figures show the displacement of the left side of the vehicle and the lower ones show that of the right side in the running direction. Figure 5(a) shows lateral displacement ($x - y$ plane). After SC 1 quenched, the thrust force of B2 became small. Therefore B2 was pushed by C2 and B3 behind, and sandwiched between C1 and C2. Then the lateral displacement and yaw angle of B2 became larger than that in the one bogie model[2]. The positions of other bogies and cabins were little changed. Lateral displacement of the quenched bogie (B2) edge reached about 0.07m. As this value is less than 60% of the lateral air gap, the train can run without touching the sidewall

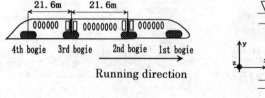

Figure 3: Train model

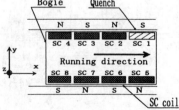

Figure 4: Arrangement of the quenched SC coil

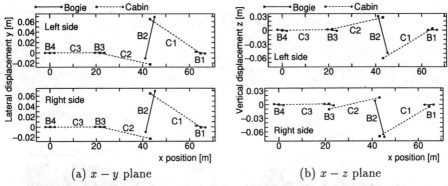

(a) $x - y$ plane   (b) $x - z$ plane

Figure 5: Average position of the train after SC coil quenching

of the guideway. Figure 5(b) shows vertical displacement ($x - z$ plane). The ordinate shows vertical displacement from the initial positions. Due to the SC coil quenching, the left side position of B2 and C2 became higher than those of the right side. As in the case of the lateral displacement, B2 was pushed by C2 and B3 behind, and sandwiched between C1 and C2. Therefore the vertical displacement and pitch angle of B2 became larger than that in the one bogie model. The vertical positions of other bogies and cabins were also little changed. The vertical displacement of B2 edge reached about -0.06m, and the total vertical displacement of B2 edge from $z=0$ became about -0.0994m. Although the train can run without touching the bottom of the guideway, the vertical gap became very small. The lateral and vertical displacement of the bogie next to the quenched one, was about 1% of the air gap. The electromagnetic spring of the EDS system is stronger than the mechanical spring of the secondary suspension. Therefore the mechanical spring absorbed the influence of the quenched bogie. In this case, the train can run without touching the side wall of the guideway.

Next the electromagnetic force applied to the quenched bogie is shown. The influence was most pronounced in the SC2 which is next to the quenched one. Figure 6 shows the levitation force. The small 300Hz oscillation in figure 6 is caused by the levitation coil pitch. The average force of SC2 became about 2.4 times as large as in the normal state. Figure 7 shows the guidance force. The average force of SC2 reached -40kN which was about seven times as large as in the normal state. As SC2 is close to the levitation coils on the side wall, the amplitude of the oscillation (300Hz) became 14kN. Because of the bogie displacement, a large guidance force was also applied to SC5. Figure 8 shows the thrust force. After SC 1 quenching, the bogie moved to the quenched side. On the

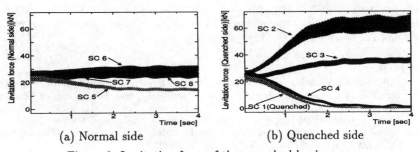

(a) Normal side   (b) Quenched side

Figure 6: Levitation force of the quenched bogie

normal side, the lateral distance of the SC5-7 from the side wall became large, and
the thrust force of these SC coils decreased. Because of the rotational movement of
the bogie, the lateral distance of the SC8 became small. As a result, only the thrust
force of the SC 8 increased. On the quenched side, the thrust force of SC2-4 increased.
The thrust force on the normal side decreased 12.7%, that of the quenched side about
18.5%. The total force of the quenched bogie was 15.6% smaller than the normal state.

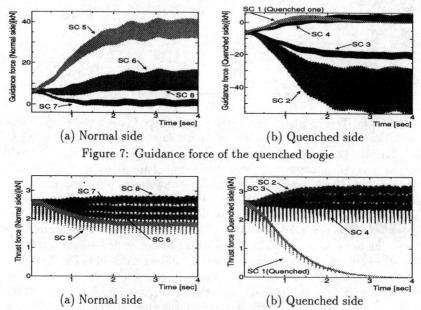

(a) Normal side   (b) Quenched side

Figure 7: Guidance force of the quenched bogie

(a) Normal side   (b) Quenched side

Figure 8: Thrust force of the quenched bogie (LSM)

4. Conclusion

A three-dimensional running simulation of the Maglev train in the case of SC coil
quenching has been undertaken. The levitation and guidance force of the SC coil next
to the quenched one became about twice and seven times as large as those in the
normal state respectively. This may cause the problem with SC coil stability. The
electromagnetic spring of the EDS system is stronger than the mechanical spring of the
secondary suspension. Therefore the mechanical spring absorbed the influence of the
quenched bogie. As a result, the influence of the SC coil quenching was only seen in
the cabins connected to the quenched bogie. In this case, the train can run without
touching the side wall of the guideway. We are planning to include the active damper
system to produce more stability good riding comfort.

References
[1]   H.Nakasima and A.Seki, The Status of the Technical Development for the Yamanashi Maglev Test
      Line, The 14th International Conference on Magnetically Levitated Systems and Linear Drives
      (Maglev'95) (1995) 31-35
[2]   S.Ohashi, K.Higashi H.Ohsaki and E.Masada, Influence of Magnet Quench on Running Charac-
      teristics of the Superconducting Magnetically Levitated Bogie, The 8th International Symposium
      on Superconductivity (ISS'95), Vol.2, (1995) 1339-1342
[3]   S.Ohashi, K.Higashi H.Ohsaki and E.Masada, Influence of Magnet Quench on the Superconduct-
      ing Magnetically Levitated Bogie, IEEE Trans. on Magnetics, Vol.32, No.5, (1996) 5046-5048

Non-Linear Electromagnetic Systems
V. Kose and J. Sievert (Eds.)
IOS Press, 1998

A Dynamical Model of High Speed Rotor Suspended by High-Tc Superconducting Bearing

Takashi HIKIHARA and Yoshisuke UEDA
School of Electrical Engineering, Kyoto University
Yoshida-honmachi, Sakyo, Kyoto, 606-01 JAPAN

Abstract A dynamical model of a high speed rotor suspended by a high-Tc superconducting bearing is discussed. The bearing has the hysteretic characteristics in the force-displacement relation induced by the magnetic characteristics of the high-Tc superconductor. The model shows the possibility that the hysteretic characteristics cause the drift of levitation under the gyroscopic motion of the rotor.

1. Introduction

Since the discovery of a high-Tc superconductor (HTSC), many applications have been proposed. Among these, the superconducting magnetic bearing is one of the most possible applications and has been studied in many institutes and companies [1-3]. It has a possibility of the non-contact bearing without any control. The dynamics of the high speed rotor suspended by the magnetic bearing exhibits gyroscopic motion. The high-Tc superconducting bearing has a coupled structure of the HTSC and the rare earth permanent magnets. The force-displacement relation between the HTSC and the magnets has the hysteretic characteristics, which cause the complex nonlinear vibration.

The authors have studied the force-displacement relation between the superconductor and the permanent magnet experimentally. Based on the result, a model which can represent the dynamic force-displacement relationship has been proposed by authors [4-6]. It can be applied in the differential equations of the system which includes the high-Tc superconducting bearing. In this paper, the dynamics of the high speed rotor suspended by the high-Tc superconducting bearing is discussed numerically based on the analysis of a dynamic model with respect to the hysteretic force relationship.

2. System model

The high speed rotor suspended by the high-Tc superconducting bearing has a structure as shown in Fig. 1. The rotor is suspended by the flux pinning effect of the high-Tc superconductor (HTSC). The dynamics of the rotor is given by Euler's equation with the nonlinear hysteretic force between a permanent magnet (PM) and a HTSC as follows [7];

$$\frac{dI_x \omega_x}{dt} = (I_y - I_z)\omega_y \omega_z = M_x \qquad \omega_x = \dot{\varphi}\cos\theta\cos\psi + \dot{\theta}\sin\psi$$

$$\frac{dI_y \omega_y}{dt} = (I_z - I_x)\omega_z \omega_x = M_y \quad (1) \qquad \omega_y = \dot{\theta}\cos\psi - \dot{\varphi}\cos\theta\sin\psi \qquad (2)$$

$$\frac{dI_z \omega_z}{dt} = (I_x - I_y)\omega_x \omega_y = M_x \qquad \omega_z = \dot{\psi} + \dot{\varphi}\sin\theta$$

where I_i denotes the inertia of rotor, ω_i the angular velocity, M_i the external moment and (φ, θ, ψ) the Euler coordinate system. Under the condition $I_x = I_y = I$ and $I_z = J$ (which implies the shape of a cylindrical body), Eq. (1), the moment relationship are given by [8];

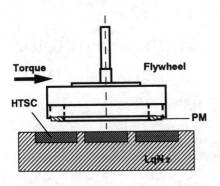

Fig.1 Flywheel rotor supported
by HTSC magnetic bearing.

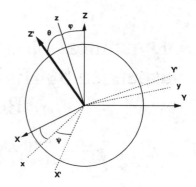

Fig.2 Coordinate system

$$I(1 + A \sin \nu\psi)(\ddot{\varphi}\cos\theta - \dot{\theta}\dot{\varphi}\sin\theta) + J\dot{\theta}(\dot{\psi} + \dot{\varphi}\sin\theta) = M_x = h_x$$

$$I(\ddot{\theta} + \dot{\varphi}^2\sin\theta\cos\theta) - J\dot{\varphi}\cos\theta(\dot{\psi} + \dot{\varphi}\sin\theta) = M_y = h_y \qquad (3)$$

$$\dot{\psi} = \Omega - \dot{\varphi}\sin\theta$$

On the other hand, the center of mass (x_0, y_0, z_0) has the relationship;

$$m\ddot{x}_0 = -k(x_0 + \theta z_p)$$

$$m\ddot{y}_0 = -k(y_0 - \varphi z_p) \qquad (4)$$

$$m\ddot{z}_0 = f_{lev}(x, y, z, \varphi, \theta, \psi) - mg$$

where f_{lev} denotes the levitation force working on the rotor and z_p the center of horizontal force. Here the moment h_x and h_y have the hysteretic characteristics caused by the hysteretic force-displacement relationship between the HTSC and the PM. They can be approximately represented by

$$h_x \cong A_1\varphi F_z + A_2\varphi^2 F_x$$

$$h_y \cong A_1\theta F_z + A_2\theta^2 F_y \qquad (5)$$

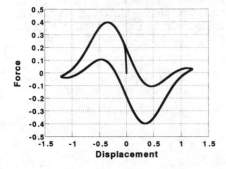

Fig.3 Lateral force-displacement
hysteresis.

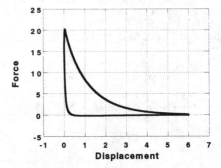

Fig.4 Vertical force-displacement
hysteresis.

We have already obtained a model of the force-displacement relationship in the HTSC and PM system [4-6] as for F_z and F_x;

$$\dot{F}_z = -\gamma\{F_z - f_z(z,\dot{z})\}$$
$$f_z(z,\dot{z}) = f_{z1}(z)\{1 + f_{z2}(\dot{z})\}$$
$$f_{z1}(z) = F_0 e^{-\beta z}$$

$$\dot{F}_x = -\gamma\{F_x - f_x(x,\dot{x},z)\}$$
$$f_x(x,\dot{x},z) = f_{x0}(x)\{f_{x1}(x,z) - f_{x2}(\dot{x},z)\}$$
$$f_{x0}(x) = e^{-hx^2}$$
$$f_{x1}(x,z) = f_{z1}(z)(x^3 - x)$$

$$f_{z2}(\dot{z}) = \begin{cases} -\mu_1 - \alpha_1\dot{z}, & \varepsilon \leq \dot{z}, \\ -\dot{z}(\mu_1 + \mu_2)/2\varepsilon, & -\varepsilon \leq \dot{z} < \varepsilon, \\ \mu_2 - \alpha_2\dot{z}, & \dot{z} < -\varepsilon. \end{cases}$$

$$f_{x2}(\dot{x},z) = \begin{cases} -\xi_1 e^{-\beta z}, & \varepsilon \leq \dot{x}, \\ -(\xi_1/\varepsilon)\dot{x}e^{-\beta z}, & -\varepsilon \leq \dot{x} < \varepsilon, \\ \xi_1 e^{-\beta z}, & \dot{x} < -\varepsilon. \end{cases}$$

$$\ldots\ldots\ldots(6)$$

$$\ldots\ldots(7)$$

These models of hysteresis can represent the shape of hysteretic characteristics as shown in Figs. 3 and 4. When the angles φ and θ are small, the torque is given by the first term in Eq. (5).

3. Simulated results

In our experimental results[4,6], it has been experimentally found that the external vibration causes the drift of the levitated magnet upon the HTSC bearing. Recently, it was confirmed that the external vibration caused the gap decay of the flywheel bearing by Coombs and Campbell [9]. Moreover, we have obtained the result which showed the drastic gap drift of the flywheel suspended by the HTSC bearing under free revolution [10]. The drift appeared at the revolution of the mechanical resonance in the flywheel system. However, we will not discuss the experimental results in this paper. Here, we will focus on the nonlinear dynamics of the flywheel rotor and the drift caused by the mechanical vibration of the flywheel rotor.

3.1 Nonlinear dynamics.   The flywheel system described by the above differential equations is nonlinear both in the force-displacement relationship and in the rigid body dynamic relationship depending on the angular factor. Therefore, the imbalance of the rotor and the initial position make the dynamics complicated. The rotor dynamics without any contact becomes a Hamiltonian system. However, the hysteretic characteristics of the HTSC bearing causes the dissipation. One of the simulated dynamics is shown in Fig. 5.   The parameters of the rotor were set as follows:

$$I = 6.690 \times 10^{-4} \ [kg \cdot m^2], \qquad J = 1.24 \times 10^{-3} \ [kg \cdot m^2]$$
$$k = 950 \ [N/m], \qquad m = 0.7854 \ [kg],$$
$$z_p = -8.86 \times 10^{-3} \ [m] \qquad A_1 = 2020 \ [Nm/rad]$$

The parameters in the hysteretic function were set as follows:

$$\gamma = 0.1, F_0 = 10.0, \beta = 0.8, \mu_1 = -1.2, \mu_2 = -0.2,$$
$$\varepsilon = 0.01, \alpha_1 = -2.0, \alpha_2 = 0.0$$

The initial position of the rotor was at $(\varphi, \theta, \psi) = (0.001, 0, 0)$ and the velocity at $(\dot{\varphi}, \dot{\theta}, \dot{\psi}) = (0,0,0)$.   At a speed of 36,000 [rpm], the trajectory of the dynamics is as shown in Fig.5a. When we get the Poincaré slice at $-\sigma \leq \dot{\varphi} < \sigma$ in $(\varphi, \dot{\varphi}, \theta, \dot{\theta})$ space, the projection of the discrete points show the distribution as shown in Fig. 5b , where $\sigma = 0.2$.   The Poincaré points stay in a closed region but do not show any structure. It implies the hysteretic characteristics make the system a non-Hamiltonian. Fig. 5c shows one of the moment and angle relationship. The torque shows several minor loops in the major loop. However, the results should be examined based on the experimental results.

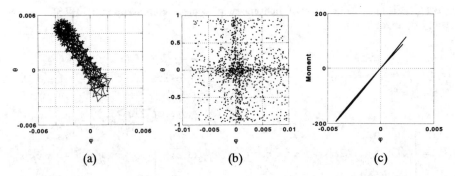

(a)   (b)   (c)

Fig.5   Simulated results, (a) trajectory, (b) Poincaré map, and (c) moment-angle relation.

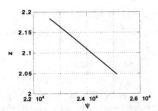

Fig.6   Drift of levitation.

3.2 Levitation drift.   At speed, the self oscillation occurs as shown in Fig.5. The vertical position of the rotor drifts as shown in Fig. 6. The initial gap between the HTSC and the PM is set at 2.0. Here the vertical position of the center of mass is given. The decay of the gap coincides with the experimental results which have been already obtained [8]. The simulated result shows the possibility of the levitation drift being caused by the hysteretic force-displacement relationship.

4. Concluding remarks

We have obtained a dynamic model of the rotor suspended by the HTSC magnetic bearing considering the force-displacement hysteretic relationship between the HTSC and the PM. The model shows the agreement with the experimental results.

References

[1] F.C.Moon and P-Z.Chang, "High-speed rotation of magnets on high Tc superconducting bearings", Applied Physics Letter, 56(4) (1990) 397 - 399.
[2] R.Takahata and H.Ueyama, "Characterization of superconducting magnetic bearings (vibration damping and hysteresis of magnetic force in superconductor), Proc. of the 4th Int. Symp. on Supercon. (1991) 1089 - 1092.
[3] H.Takaichi and et. al., " The application of bulk YBaCuO for a practical superconducting magnetic bearing", Proc. of the 3rd Int. Symp. on Magnetic Bearings (1992) 307 --316.
[4] T.Hikihara and F.C.Moon, "Chaotic levitated motion of a magnet supported by superconductor", Physics Letters A, 191 (1994) 279 - 284.
[5] T.Hikihara and F.C.Moon, " Levitation drift of a magnet supported by a high-Tc superconductor under vibration", Physica C, 250 (1995) 121 - 127.
[6] T.Hikihara and G.Isozumi, "Modeling of lateral force-displacement hysteresis caused by local flux pinning", Physica C, 270 (1996) 68 - 74.
[7] D.T.Greenwood, Principles of dynamics, 2nd Ed. (Prentice Hall, 1988).
[8] R.D.Bourke, NASA CR-108 (1969).
[9] T.A.Coombs and A.M.Campbell, "Gap decay in superconducting magnetic bearings under the influence of vibration", Physica C, 256 (1996) 298-301.
[10] T.Hikihara, M.Adachi, S.Ohashi, Y.Hirane and Y.Ueda, "Levitation drift of flywheel and HTSC bearing system caused by mechanical resonance", Physica C (in press).

Non-Linear Electromagnetic Systems
V. Kose and J. Sievert (Eds.)
IOS Press, 1998

A Hysteresis Hardening Effect for Levitation Force in HTSC Magnetic Bearing

Y. Luo, Y. Yoshida, K. Miya and H.Higasa*

*Nuclear Engineering Research Laboratory, Faculty of Engineering,
The University of Tokyo, Tokai, Ibaraki 319-11, Japan
Shikoku Research Institute, Inc., Takamatsu, 761-01, Japan

Abstract. This paper describes a *Hysteresis Hardening Effect* in HTSC magnetic bearings and its contribution to the improvement of the magnetic force relaxation. It was found that the magnetic force relaxation can be improved by using the hysteresis property of the supercurrent, and the extent of this improvement depends on the hysteresis current distribution. Theoretical explanation of the effect was conducted and verified by the numerical evaluation of force decay. Possible engineering applications were also discussed.

1. Introduction

In the design of high temperature superconducting (HTSC) magnetic bearings, some electromagnetic phenomena due to the mixed state of type-II superconductors must be taken into account. Flux creep is especially responsible for the time decay of the shielding current and it indirectly causes the decay of the magnetic force between the two main elements of the bearing: permanent magnets(rotor) and superconducting bulks(stator).

Besides the improvement of material properties, some researchers have suggested that magnetizing the superconductor at a higher temperature than the operating temperature can decrease the probability of the flux motion[1] and restrain the flux creep. However, for this method, a complicated cryogenic system is necessary and this will result in the unreasonable increase in the cost of the HTSC magnetic bearing.

It was reported[2] that the relaxations of the levitated heights in an axial bearing show different behaviors, when the magnet is driven descending towards HTSC or with a descending-ascending procedure (*e.g.*, the decay of levitated height in the latter case becomes much smaller than in the former). Concerning the force decays[3], such a phenomenon was also observed by oscillating the superconducting specimen.

To reflect this knowledge in the design of the HTSC magnetic bearing, a numerical prediction of the relaxation properties therefore becomes an important issue.

In the present paper, a numerical evaluation of the relaxation properties was performed, on the configuration of a practically radial-type HTSC magnetic bearing. A theoretical explanation was also given to clarify the physical mechanism of the above mentioned phenomenon (which can be referred as *Hysteresis Hardening Effect*). The conclusions were that the magnetic force relaxation can be improved by using the hysteresis property of the supercurrent, and that the extent of this improvement depends on the hysteresis current distribution.

Discussions were conducted, based on the numerical results and taking into account the operating time of the HTSC magnetic bearing.

2. Theory

In a superconductor where shielding currents are present, a critical state is generally established, based on the equilibrium between the pinning force and the Lorentz force acting on the fluxoids. In practical experiments, however, it was found that, even under the force equilibrium, a measurable voltage appears, leading to a small flux motion. It is considered that, due to the thermal fluctuation, fluxoids jump out of the pinning center with a certain probability and generate an average flux motion. This phenomenon (called *flux creep*) which leads to the decay of the shielding current, was generally negligible in a conventional superconductor but becomes an important issue in type-II superconductors, because of the relatively higher operating temperature. Furthermore, in some engineering applications, like HTSC magnetic bearings operated in a broad time range, flux creep may induce the decay of the levitation force or of the distance between the superconductors and the permanent magnets.

A well-known description of flux creep phenomena, the Anderson and Kim's theory[4][5], can be summarized as follows. The probability of the jumps of fluxoids over the pinning wall is expressed as $\sim exp(-U/k_B\Theta)$, where U denotes the pinning potential, k_B, the Boltzmann constant and Θ, the isolated temperature. Considering the existence of current in the superconductor, the effective pinning potential can be written as:

$$U_+ = U_0(1 - \frac{J}{J_c}),\tag{1}$$

along the direction of Lorentz force, and

$$U_- = U_0(1 + \frac{J}{J_c})\tag{2}$$

in the opposite direction. U_0 denotes the pinning potential in the case without transformed current in the superconductor, J the current density and J_c the critical current density. By using the net potential in the pinning center, the probability of the flux jump is given by

$$\Omega = exp(-U_+/k_B\Theta) - exp(-U_-/k_B\Theta).\tag{3}$$

The velocity of the flux motion becomes

$$v = \Omega d\{\exp\left(-\frac{U_+}{k_B\Theta}\right) - \exp\left(-\frac{U_-}{k_B\Theta}\right)\}\tag{4}$$

$$= 2\Omega d\exp\left(-\frac{U_0}{k_B\Theta}\right)\sinh\left(\frac{U_0}{k_B\Theta}\frac{J}{J_c}\right),\tag{5}$$

where d is the distance of one jump. If the resistivity of flux creep is defined as:

$$\rho_c = B\Omega d/J_c,\tag{6}$$

by using Maxwell's equation we finally obtain

$$E = vB = 2\rho_c J_c\sinh\left(\frac{U_0}{k_B\Theta}\frac{J}{J_c}\right)\exp\left(-\frac{U_0}{k_B\Theta}\right).\tag{7}$$

Substituting this constitutive relation in a numerical formulation of Faraday's law where the current potential and the electric field are connected as[6]

$$\mu_0\dot{T} + \frac{\mu_0}{4\pi}\mathbf{n}\cdot\int\dot{T}\nabla'\frac{1}{R}dS = -\dot{B}_0 - \mathbf{n}\cdot\nabla\times\mathbf{E},\tag{8}$$

the relaxation of the critical current can be evaluated quantitatively. In e.q.(8), the scalar T is reduced from the current vector potential $\mathbf{T}$ using a thin plate approximation, by considering the anisotropic property of the type-II superconductor. R is the distance from source point to field point and $\mathbf{E}$ is the field induced by the motion of the fluxoids.

3. Numerical evaluations

The numerical approach applied in this paper for computing the supercurrent distribution is based on the current vector potential method expressed by eq.(8). For numerical analysis, the critical current density of YBCO bulk sample was $1.0 \times 10^8 A/m^2$, the pinning potential was chosen from the B-dependence[7] of YBCO as 0.15eV.

The numerical evaluations were performed for a practical radial-type HTSC magnetic bearing. The bearing consists of a permanent magnet array of three annular-shaped Pr magnets, and a tube-shaped HTSC assembled with eight YBCO bulks. The details of the arrangement are shown in Fig.1.

The magnetic force which sustains the levitation arises when a relative displacement is given between the magnets and the HTSCs, along the direction parallel to the axis of the bearing. The magnets were supposed to be displaced with both a descending procedure and an ascending one. During these two procedures, the displacements are held at various values, in order to investigate the behavior of the force decay.

4. Results and discussions

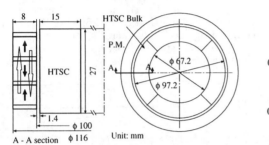

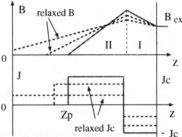

Figure 1 Configuration of a radial-type magnetic bearing

Figure 2 Profiles of the hysteresis current, the flux density and their relaxations

The displacement-force curves obtained from experiment and numerical calculation are in agreement as shown in Fig.3. After this validation, the force decay was investigated by holding the displacements at various values.

The results depicted in Fig.4 clearly show the different behaviors for the two operating procedures, i.e., on the ascending curve, the force relaxation seems to be remarkably reduced compared with that on the descending curve. This improvement is considered owing to the hysteresis current distribution which depends on the history of the external field variation. Though in the configuration of present HTSC magnetic bearing the field of the magnet array exhibit a complicated profile, in a local position of HTSC, however, it can be approximately considered that the external field varies monotonically in the descending process but has an alternative variation in the descending-ascending process. This fact allows us to use a half-space superconductor exposed in an uniform field parallel to its surface to discuss the results mentioned above, for simplicity. We assume that the external field is varied with an increasing-decreasing procedure. In the increasing process, the current arises to compensate the penetration of the increasing external field within the penetration depth region, but in the decreasing case a new layer of current, induced by the decreasing field, appears near the surface. Here, obviously the currents possess opposite directions in the two layers. By using the Bean model, the profiles of the magnetic density and the supercurrent were illustrated in Fig.2. In the inner layer II (see Fig.2), the current remains in the initial direction, whereas in the outer layer I, it takes a reversed direction, leading to opposite magnetic forces. Thus entire force can be obtained by the superposition of the both forces

$$F_{total} = F_{II} - F_I, \qquad F_I, F_{II} > 0. \tag{9}$$

By denoting the force variations in the two layers during a certain time range as $\triangle F_I, \triangle F_{II} > 0$, the relaxed entire force becomes

$$F'_{total} = (F_{II} - \triangle F_{II}) - (F_I - \triangle F_I) = F_{total} - (\triangle F_{II} - \triangle F_I). \tag{10}$$

The force relaxation is determined by integrating the product of the current decay and the external field density over the supercurrent region. Neglecting the dependence of the current relaxation on the magnetic field, the force decay is mainly determined by (i)the volume of current region and on (ii)the external field in this region.

Generally, in a practical system like the HTSC magnetic bearing, since the external field decreases exponentially from the surface to the interior of the HTSC, the force decay per unit volume in the outer layer is larger than that in the inner layer, under the existence of a hysteresis current profile. This is considered the reason of the improvement of the force decay because of that the decay of the reversed force in the outer layer can play a role of increasing the totally magnetic force. In addition the second term on the extreme right of eq.(10) may take a positive or negative sign, which depends on the proportion of the volumes of the region I and II. Obviously, the larger the volume of the region I, the more remarkable is the *Hysteresis Hardening Effect*. In an extreme example(see case a1 in Fig.4), the force *increases* owing to the flux creep.

In conclusions, it was found that the force relaxation in a HTSC magnetic bearing can be improved under a hysteresis current profile. However, the extent of the improvement effect depends on the volume of the outer current layer and this results in a reduced initial levitation force. Accordingly, it is required in the design of a commercial HTSC magnetic bearing (a)to control the relaxation of levitation by applying the above mentioned effect under an allowance decay during the whole operating time range and (b)to obtain a maximum levitation force simultaneously. This becomes a further issue concerning the optimization of the HTSC magnetic bearing.

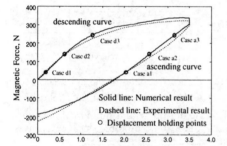

Figure 3 Levitation properties of the radial-type magnetic bearing

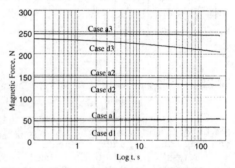

Figure 4 *Hysteresis Hardening Effect*, in cases d1~d3 the displacement was held on the descending curve and in cases a1~a3 on the ascending curve.

References

[1] Morita. et al., Proceedings of the 46th Cryogenic Engineering Conference, A 1-39 (1991) in Japanese

[2] M. Okano, Proceedings of the 49th Cryogenic Engineering Conference, A 2-12 (1993) in Japanese

[3] T. Hikihara and F.C. Moon, Physica C 250 (1995) 121

[4] P.W.Anderson, *Phys. Rev. Letters* 9,(1962) 309

[5] P.W.Anderson and Y.B.Kim, *Rev. Mod. Phys.* 36,(1964) 39

[6] Y. Yoshida, M. Uesaka and K. Miya, *IEEE Trans.of Magn.* Vol. 30, No.5,(1994) 3503

[7] K. Yamafuji, *Proceedings of the ICTPS'90*,(1990) 110

Non-Linear Electromagnetic Systems
V. Kose and J. Sievert (Eds.)
IOS Press, 1998

Modelling of Hysteresis and Anisotropy of High Temperature Superconductors for Levitating Applications

István VAJDA [a], József LUKÁCS [a], László MOHÁCSI [b]

[a] Dept. of Electrical Machines and Drives, Technical University of Budapest,
Egry József utca 18, H-1111 Budapest, Hungary

[b] Silex ltd, H-1751 Budapest, POB 93, Hungary

Abstract. For the numerical calculation of the magnetic field and forces between HTSC materials and permanent magnets an algorithm and a 2D finite difference computer code has been developed. The basic ideas of the physical modelling are discussed.

1. Introduction

The magnetically levitated passive HTSC bearings are among the most possible and promising applications of HTSC materials. For the numerical calculation of the magnetic field and forces between HTSC materials and permanent magnets the critical state model, including the Bean-, the Kim- and several other models [1–3] are used commonly. The authors have applied the alternative approach to treat the superconductor as a special kind of magnetic material rather than a special kind of conducting material. A computer code called "WinTer" has also been developed by the authors of this paper. For more details see [4].

2. General: The Ways of Treating the Superconductor

There are two possible ways of describing the magnetic properties of superconductors.
(1) Ascribe magnetic behavior to circulating (screening) currents.
(2) Describe the superconductor as a magnetic material having magnetization.

2.1. First case: Screening Currents

The corresponding equations for the first case are as follows:

$$B = \mu_0 H, \qquad M = 0$$
$$\nabla \times H = j, \qquad j = j_c \tag{1}$$

For the solution of the above equations the spatial current distribution should be known *a priori*. If one has ideal or, at least, near–to–ideal samples, one may be able to determine the current paths. If, however, the sample is far from being ideal, then the determination of the current paths is principally impossible. For this reason the alternative approach described below has been chosen.

2.2. Second case: Magnetization

The corresponding equations for the second case are as follows:

$$B = \mu_0 H + M$$
$$\nabla \times H = 0 \tag{2}$$

In this case it is assumed that *there are no circulating currents*. On the other hand the magnetization curve of the superconductor should be available as a function of H.

3. The Magnetization Curve

The magnetization curve can be measured on samples, thus, e.g. on bulks. The magnetization curves measured on bulk samples are not suitable for the development of the algorithm. These curves, among others, are highly dependent on the forms and geometries of the samples as well as on the distribution of the external magnetic field. As a result they cannot be used for general purposes, like, e.g. for the calculation of arbitrarily shaped samples placed in inhomogenous magnetic fields.

The generality of the problem treatment necessitates that the algorithm be based on the magnetization curve of a 'negligibly' small sample measured in a homogenous magnetic field as the input characteristics. As a consequence, the material properties of the superconductor should be homogenous for its each elementary volume to have the same magnetization curves.

The material is divided into 'negligibly' small elementary volumes (EV) having a width of *unit*, and the field distribution will be built up as the resultant and *not as the sum* of the fields generated by the permanent magnets and by all the EVs due to the nonlinearity of the problem.

Having the input characteristics, the algorithm calculates the average magnetization of the sample with extended dimensions (or bulk) depending on the configuration or arrangement of permanent magnets, superconductors, etc. In particular, the algorithm calculates the average magnetization of the sample simulating measuring conditions, i.e. a sample placed in a homogenous magnetic field swapped between given values.

The measurement of $M(B)$ is to be carried out, e.g., on a slab having a width of $2b$. The algorithm subdivides the sample into EVs having the same dimensions which are called "*units*". The measured curve can be used *directly* as input to the calculations if:

$$unit = 2b \tag{3}$$

If, however, one wants to use a different *unit*, he can recalculate the measured curve according to the model that has been chosen.

4.   Determination of $M(B)$ from $M(H)$

In this case H is the field generated by external currents (or permanent magnets) and as such it is the same both in the presence and the absence of any superconducting parts. Thus, when the superconductor is treated as a magnetic material, the knowledge of the H field distribution is not enough for the calculation of the magnetic field distribution *within* the superconductor. To avoid this difficulty, the H field is replaced by the B field by applying the following equations according to [5]:

$$H = \mu_0^{-1}(B - M)$$
$$\nabla \times (B - M) = 0 \tag{4}$$

The idea for the derivation of the $M(B)$ relationship is as follows: Remove the element in question (virtually thus leaving a hole in the bulk). Calculate the field in the hole generated by all permanent magnets and by the rest of the EVs. Assume that the field in the hole is homogenous due to the 'negligibly' small dimensions of the removed piece. Thus the conditions of a small piece of a superconductor placed in a homogenous magnetic field (flux density) in the air are created and the $M(B)$ curve can be applied, where B is the *external flux density*. Further we will use the designation B_{ext}, where

$$B_{ext} = \mu_0 H \tag{5}$$

The *internal flux density* B_{int} as a function of the magnetization can be expressed as

$$B_{int} = B_{ext} + M \tag{6}$$

The relationship is shown in Fig.1 for the first magnetization curve.

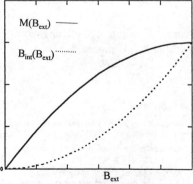

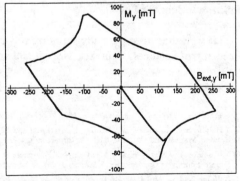

Fig.1 Magnetization and internal flux density as a function of the external magnetic field (calculated).

Fig.2 Hysteresis loop of a slab placed in a homogenous external magnetic field (calculated).

This way the $M(H) \to M(B_{ext})$ 'conversion' is completed by equaling $\mu_0 H = B_{ext}$. Then B_{ext} can be calculated in each EV and M can be determined from the $M(B_{ext})$ curve as shown in Fig. 2 for a slab placed in a homogenous magnetic field.

5. Determination of the Vector of the Resultant Flux Density Considering Non-linearity

First, the vector potential distribution is calculated. Given the vector potential distribution, the components of the magnetization M_x and M_y can be calculated at each spatial point. The components of the flux density B_x and B_y can be determined from the hysteresis curve. The vectorial sum of the components gives B_{sum}. The magnitude of the resultant flux density is obtained by cutting B_{sum} with the circle of B_{max} (if $B_{sum} > B_{max}$).

The first magnetization curve is described by a concave parabolic function. We have assumed that all the internal loops starting from any point of the magnetization curve may also be described by parabolic functions: in a decreasing magnetic field the working point of an arbitrary elementary volume moves on a convex parabola while in an increasing magnetic field the working point moves on a concave parabola.

For the consideration of the anisotropy, the material should be placed so that the crystal lattice axes be aligned with the corresponding axes of the coordinate system. Otherwise the calculation is carried out in a similar way as it has been described above.

6. Conclusions

(1) The method of the description of the superconductor as a special kind of magnetic material has been developed.
(2) The magnetic behaviour of a bulk superconductor material has been explained on the assumption that the bulk is a complex of elementary superconducting volumes.
(3) A method and an algorithm has been developed for the determination of the resultant magnetic field considering the nonlinear and anisotropic magnetization characteristics of the superconductor.

Acknowledgement

This work has been partially supported by National Scientific Fund of the Hungarian Academy of Sciences, Contract. No. 022532 and by the Silex ltd, Hungary.

References

[1] F.C.Moon, *Superconducting Levitation*. New York, etc.: John Wiley and Sons (1994).

[2] T.H.Johansen, H.Bratsberg, A.B.Riise, H.Mestl and A.T.Skjeltorp, Measurement and Model Calculations of Forces Between Magnet and Granular High–T_c Superconductor. *Applied Superconductivity* **2**, 535–548 (1994).

[3] T. Coombs, D.A.Cardwell and A.M.Campbell, Dynamic Properties of Superconducting Magnetic Bearing. Applied Superconductivity Conference (25–30 July 1996), to be published in Trans. on Applied Superconductivity.

[4] I. Vajda and L. Mohácsi, Advanced Hysteresis Model for Levitating Applications of HTSC Materials. Applied Superconductivity Conference (26–31 August 1996), to be published in IEEE Trans. on Applied Superconductivity.

[5] K. Simonyi, *Electromagnetic Theory*. Budapest: Akadémiai Kiadó (1973), In Hungarian.

Non-Linear Electromagnetic Systems
V. Kose and J. Sievert (Eds.)
IOS Press, 1998

Flux-Concentration Type Electromagnetic Pump for Transferring Sodium

Sotoshi YAMADA, Masayoshi IWAHARA, and Takayoshi MINAMIZONO
Laboratory of Magnetic Field Control and Applications,
Faculty of Engineering, Kanazawa University,
2-40-20 Kodatuno, Kanazawa 920, Japan

Abstract. The proposed flux-concentration type electromagnetic pump is one of the devices that uses the flux-concentration effect depending on the positive aspect of eddy currents. The electromagnetic pump can overcome the barrier of capacity and efficiency that the conventional machines hold. This paper described the calculated and experimental results on the in-sodium tests. The experimental results show that the maximum efficiency was about 11%, which is double that of the conventional type small-sized electromagnetic pump. For a higher-capacity device, we discuss the new structure of our EMP and its characteristics.

1. Introduction

Liquid sodium is used as the coolant material in the cooling system of the fast breeder reactors (FBR). The material is chemically so active that the total system including the pump apparatus requires the highest mechanical reliability. In this aspect, an electromagnetic pump can move liquid metal under airtight conditions, so it has a valid reason for use in the FBR system. However, the electromagnetic pump has never been produced with a large capacity because of efficiency and scale compared with mechanical pump. To overcome the engineering barrier, we have proposed a "Flux-Concentration Type Electromagnetic Pump" that employs the flux-concentration effect using eddy currents[1],[2].

Eddy currents induced in a conducting material have not only the negative attribute of eddy-current losses but also the positive attribute of magnetic shielding and flux concentration. The flux-concentration type electromagnetic pump (EMP) is one of the attractive devices that employ the flux-concentration effect by eddy currents. The key idea of this electromagnetic pump is to decrease leakage flux by the shielding effect and concentrating the flux into the gap for liquid sodium.

We have designed a small scale EMP and examined the characteristics on the testing sodium loop. This paper describes the in-sodium test results of the EMP and the comparison with the characteristics of the conventional EMP. We also discuss the improved structure of the higher capacitor EMP.

2. Flux-Concentration Type Electromagnetic Pump

Fig.1 shows the flux-concentration type electromagnetic pump. The EMP is basically of an annular-linear type in structure. When a three-phase source is applied, travelling electro-magnetic field is generated in the duct for liquid metal. The conductive plate inserted between exciting coils in each slot is a key element of the device. The shapes of the plate are shown in

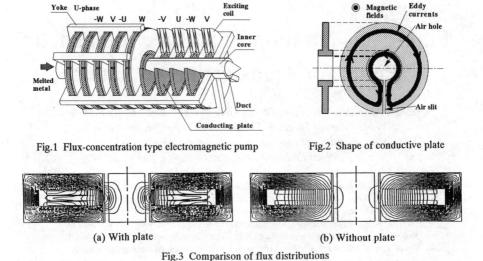

Fig.1 Flux-concentration type electromagnetic pump   Fig.2 Shape of conductive plate

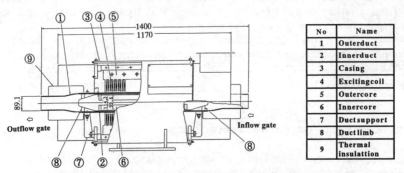

(a) With plate   (b) Without plate

Fig.3 Comparison of flux distributions

Fig.2. The copper-made plate has a unique shape with a r-direction slit and acts as the magnetic shield for leakage flux between the yokes. If external alternating magnetic fields are applied perpendicularly, eddy currents are induced as shown by the broad line. The slit enables eddy currents to flow around an air hole. The eddy currents suppress magnetic flux passing into the conducting plate. Fig.3 shows the difference in calculated flux-distributions on one slot with and without a conductive plate. These results show clearly that the leakage flux between the yokes is suppressed and the amount of the flux through the duct increases.

3. Characteristics of In-Sodium Test

Fig.4 shows the outline of the EMP designed for liquid sodium. The measurement was carried out in the circulating system for liquid sodium supported by the Power Reactor and Nuclear Fuel Development Corporation. An input voltage has kept constant in these

No	Name
1	Outerduct
2	Innerduct
3	Casing
4	Excitingcoil
5	Outercore
6	Innercore
7	Ductsupport
8	Ductlimb
9	Thermal insulattion

Fig.4 Flux-concentration type electromagnetic pump for liquid sodium

experiments and the temperature of sodium was 200°C. The conductivity of sodium is 7.4x10⁶ S/m at this temperature.

   Fig.5 shows the comparisons of pressure vs. flow rate and efficiency characteristics between the experimental and calculated results. It was shown by the experiments that the maximum flow rate and pump head of the pump were about 520 ℓ/min and 2.6 kg/cm² respectively under the input voltage of 400 V at 50 Hz. The maximum efficiency was about 9.5%. In calculation, the FEM analysis was applied to the axisymmetrical model with a range of one pole pitch[2]. For simplicity of analysis, the molten metal is assumed to flow with a constant velocity and the liquid friction between molten metal and the duct is not considered. Then a slight decrease of pressure can be observed on the calculated result unlike the experimental results.

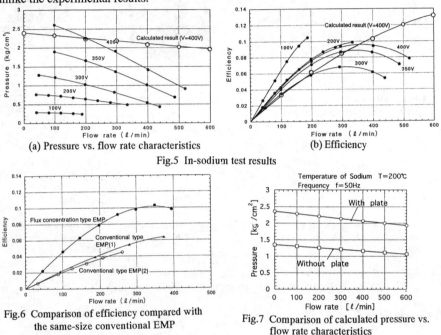

(a) Pressure vs. flow rate characteristics   (b) Efficiency

Fig.5 In-sodium test results

Fig.6 Comparison of efficiency compared with the same-size conventional EMP

Fig.7 Comparison of calculated pressure vs. flow rate characteristics

   Fig.6 shows the efficiency characteristics compared with the same-sized conventional electromagnetic pump. The flux-concentration type EMP has twice the efficiency. Fig.7 shows the comparison of pressure vs. flow rate characteristics by calculation. The result suggests that the in-sodium testing EMP has 1.8 times higher pressure than the conventional one.

4. Structure of Higher Capacity EMP

For the higher-capacity EMP, we proposed the flux-concentration type EMP that has double the excitation in structure. Fig.8 shows the structure of the model. The double excitation means that the excitation coils are arranged both inside and outside the yoke. As the characteristics have been calculated by using the FEM analysis the structure can decrease the leakage flux and is more efficient than the previous type. Fig.9 shows the comparison of the flux distributions. Fig.10 shows the efficiency of pressure and flow rate characteristics. Both the effects of the conductive plate and double excitation can improve the characteristics remarkably.

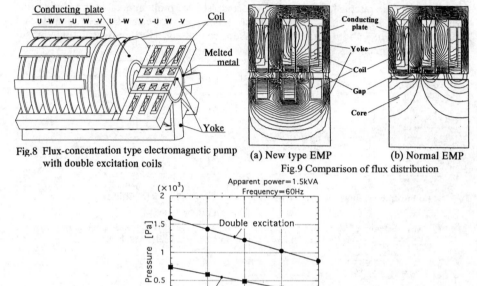

Fig.8   Flux-concentration type electromagnetic pump with double excitation coils

(a) New type EMP   (b) Normal EMP

Fig.9 Comparison of flux distribution

Fig.10 Calculated pressure vs. flow rate characteristics

5. Conclusions

We confirmed the advantage of the flux-concentration type EMP by measured and calculated results. The in-sodium characteristic testings give greater confidence that the magnetic shielding effect by eddy currents is effective in the EMP. However, the power losses in the conductive plate, caused by eddy currents, give the additional problem of coil cooling and structure. The design of the actual scale EMP needs the calculation of heat and liquid flow.

References

[1] K.Bessho, and et al, Characteristics of a new flux concentration type electromagnetic pump for fast breeder reactor, *Proc. of ISEM-Sendai,* Elsevier Publisher (1992) 79-82.
[2] S.Yamada, H.Mamada, K.Tsuyama, W.Iwahara, and K.Bessho, Dynamic performances of flux-concentration type electromagnetic pump using U-alloy as liquid metal, *JSAEM Studies in Applied Electromagnetics 3, Applied Electromagnetic in Materials,* Elsevier Publisher (1994) 285-295.

Non-Linear Electromagnetic Systems
V. Kose and J. Sievert (Eds.)
IOS Press, 1998

Effects of Radiation Damage on Core Loss of Conducting Ferromagnetics

Eugeniusz KLUGMANN

Department of Solid-State Electronics, Technical University of Gdańsk,
ul.G.Narutowicza 11/12, 80-952 Gdańsk, Poland,

Abstract. Owing to the non-homogeneous distribution of radiation defects and losses in the core, a conception of local resistivity is discussed. It is shown that nitrogen ion implantation in cobalt leads to decrease of the eddy-current loss and increase of residual losses.

1.Introduction

Various attempts have been made to reduce core loss in crystalline magnetic materials, such as stress inducing coating, surface scratching, laser scribing [1] or ion implantation [2].

The structural defects created by those methods increase a lattice parameter giving rise to compressive stress, which in turn reduce the 180 degree main domain wall spacing, reducing the core loss. Apart from extended crystal defects, the implantation damage of a crystal includes point defects decreasing the average domain wall mobility (magnetic permeability μ) and increasing the electrical resistivity ρ, thus reducing the heat generated in a core per cycle of magnetization of a period T:

$$Q = \int_V \left(\oint \vec{H} \cdot d\vec{B} + \rho(\vec{r}) \int_0^T \left| \vec{j}\,(\vec{r},t) \right|^2 dt \right) dV \qquad (1)$$

where $\vec{j} = \dfrac{\vec{E}}{\rho}$ is the eddy-current density.,

Consider the local resistivity:

$$\rho(\vec{r}) = \rho_0(\vec{r}) + \rho_T + \Delta\rho_T \qquad (2)$$

in a small physical volume dV of a core.

In this formula: $\rho_0(\vec{r})$ - the irradiation dose dependent residual resistivity,

ρ_T - the conduction electron-phonon scattering resistivity,

$\Delta\rho_T = A\rho_0 \langle\varepsilon^2\rangle$ - an additional temperature dependent resistivity due to the effects of point defects on the thermal vibrations of a crystal [3],

     A - a constant,

     $\langle\varepsilon^2\rangle$ - denotes the mean-square thermal deformation in the vicinity of a lattice defect (impurity atom).

The aim of this work is to analyse the influence of radiation defects produced by ion implantation on the core loss.

2. Experimental Procedure

Polycrystalline cobalt fabricated from powder-metallurgy with average grain size of 60 μm in a form of wire (D=0.5mm) has been preannealed under the vacuum of 10^{-5} Pa (4h,1300K) for stress relief. Samples were irradiated with 80 keV nitrogen ions at the target temperature of T<320K to integrated dose 10^{20} N^{+}/m^{2} in two opposite directions to the nape. The loss angle of a demagnetized sample was estimated before and after ion implantation, using an automated Maxwell-bridge measuring system [4] at sinusoidal magnetizing field at the fixed temperature of 303 K.

The loss angle $\tan\delta = \dfrac{\Delta R}{2\pi f L}$ of a demagnetized cobalt sample was estimated before and after irradiation by means of complex permeability measurements.

3. Results and Analysis

Fig.1 shows the total loss angle-versus-frequency characteristics of annealed (——) and a nitrogen ion implanted (---) cobalt sample in magnetizing field of the amplitude of 24 and 48 A/m.

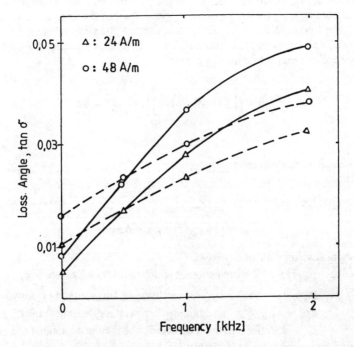

Fig.1.Loss angle versus frequency characteristics of annealed (——) and nitrogen implanted (---) cobalt, measured in ac magnetizing field of amplitude 24 A/m (Δ) and 48A/m (O).

These characteristics may be divided into:
-   static hysteresis loss angle (at frequency f = 0),
-   "classical" eddy current loss angle (linear in frequency) [2]:

$$\tan \delta_e = \pi^2 r^2 \frac{\mu}{\rho} f \cdot 10^{-9} \tag{3}$$

-   non-linear versus frequency excess loss angle related to domain wall dynamics and microstructure.

This means that the loss angle of eddy currents is determined by the sum of losses due to "classical" and "domain" eddy currents. The observed non-linear characteristics can be clarified by the skin effect [7,8] and microcurrents surrounding moving domain walls and producing local inhomogeneous magnetic fields bending the surface of the main 180° domain walls of cobalt. A peculiarity of cobalt is the large uniaxial magnetocrystalline anisotropy, therefore the measured permeability and the loss angle are largely conditioned by the 180° domain wall dynamic. The size of domain wall bending increases with frequency and amplitude of magnetizing field influencing the curvature of the characteristics and leads to the reduction of losses as seen in Fig.1. The static hysteresis losses of ion implanted samples are slightly higher because of the increase with irradiation dose residual losses, as has been shown earlier [2,5].

Fig.2 shows the variation of total loss angle over a magnetizing field amplitude range of 8÷48 A/m, at f=1kHz, before (——), and after (-----) nitrogen ion implantation of cobalt. The reduction of the loss angle at this field range is 10÷24%.

In an ion implanted sample the defects distribution is not uniform (with defects "condensation" near the surface of a sample) therefore resistivities $\rho(\bar{r})$ and $\Delta\rho_T$ are space dependent. Particularly dislocations and grain boundaries created or excited by the ion implantation can promote propagation of implanted damage and improve changes in the radial - loss distribution of the sample.

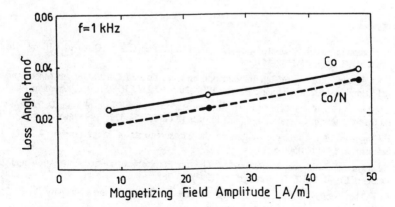

Fig.2. Loss angle versus magnetizing field amplitude at 1 kHz before (——) and after (---) nitrogen ion implantation.

The local residual resistivity depends on various radiation defects concentration $C_i(\bar{r})$ [6]:

$$\rho_0(\bar{r}) = \frac{m\upsilon_F}{e^2} \sum_i C_i(\bar{r}) \sigma_i \tag{4}$$

where: υ_F - the Fermi velocity of electrons scattering with point defects,
    m - effective mass of an electron,
    e - its electric charge,
    σ_i - electron scattering cross-section from i-type defects.

If the space and time dependence of the current density $j(\bar{r},t)$ and resistivity $\rho(\bar{r})$ is known, then the power loss per unit volume is

$$P_e = <\left|\vec{j}(\bar{r},t)\right|^2> <\rho(\bar{r})> \tag{5}$$

where the angular brackets indicate averaged expression.

From (2), (4) and (5) it may be concluded that point defects and impurity atoms acts as centres for nonelastic scattering of conduction electrons leading to the increase of resistivity and to the diminishment of flow of "classical" eddy currents and the eddy currents localized at domain walls, influencing the core loss.

4. Conclusion

In conclusion, the core loss of conducting magnetic material can be reduced by means of radiation defects or disorder using e.g. non-contact techniques such as laser irradiation or ion implantation.

References

[1] M.Nakamura et al.,Characteristics of Laser Irradiated Grain, Oriented Silicon Steel, IEEE Trans.MAG 18 (1982)1508-1510.

[2] E.Klugmann, Influence of Ion Implantation on Core Loss of Cobalt, In: A.J.Moses and A.Basak (ed.),Nonlinear Electromagnetic Systems.ISBN:90 5199 2513.IOS Press, Amsterdam 1996.

[3] A.A.Maradudin, Theoretical and Experimental Aspects of the Effects of Point Defects and Disorder on the Vibration of Crystals, Solid State Physics 18 (1966)273-420; 19 (1966)1-134.

[4] E.Klugmann, An Automatic System for the Measurements of Magnetic After-Effects, J.Phys.E. Sci.Instrum 13 (1980)500-503.

[5] E.Klugmann, Influence of Ion Implantation on Permeability Disaccommodation and Magnetic Losses of Cobalt, IEEE Trans.Magn. 30 (1994)763-765.

[6] M.W.Thompson, Defects and Radiation Damage in Metals. At the University Press, Cambridge 1969.

[7] E.Klugmann, H.J. Blythe, Influence of the Skin-Effect on the High-Temperature Magnetic Relaxation Spectrum of Cobalt, Archiwum Nauki o Materialach 9 (1988) 83-90.

[8] E.W. Lee, Eddy-Current Effects in Rectangular Ferromagnetic Cores, Proc.IEEE 107C (1960) 257-264.

Non-Linear Electromagnetic Systems
V. Kose and J. Sievert (Eds.)
IOS Press, 1998

Forces Exerted on a Particle and its Trajectory in the Vicinity of a Separator Matrix Element

Ryszard GOLEMAN, Zbigniew ZŁONKIEWICZ
Lublin Technical University, ul. Nadbystrzycka 38A, 20-618 Lublin, Poland

Abstract. The paper discusses the influence of an elliptical cylinder ferromagnetic element that is placed in a uniform magnetic field on a paramagnetic particle moving along a suspension flux. The mathematical model consisting of particle movement equations taking into account the magnetic force, the force of medium dynamic resistance and the gravitational force allows evaluation of the effectiveness of the magnetic separation process depending on the following factors: magnetic flux density, flow velocity, matrix element dimensions, particle properties and medium properties. The paper presents examples of particle trajectory in the vicinity of the matrix element, distribution of forces along these trajectories and the capture zone.

1. Introduction

In matrix separators, the nonuniformity of the magnetic field is obtained from matrices which are differently constructed in dependence on the type of separation. The matrices can be in the form of grates, grids or screens made of flat bars forming a coil. For the separation of very fine particles, the matrices are made of a number of fine irregular fibres. To obtain a relatively high magnetic force, some appropriate ratio between the collector radius and the particle radius should be maintained [1,2,5]. The paper discusses the influence of a ferromagnetic element in the form of an indefinitely long elliptic cylinder of given semiaxes lengths placed in a uniform magnetic field on a sphere paramagnetic particle moving in a suspension flux in relation to this element. The consideration is based on the analysis of three forces: magnetic, gravitational and hydrodynamic (dynamic resistance of the medium).

2. Forces acting on a particle

The force of nonuniform magnetic field that acts on a particle should be the dominant property in the process of phase separation. This is a necessary condition to acquire successful separation. The magnetic force per unit volume can be expressed in the following way:

$$\vec{f}_m = \tfrac{1}{2}\mu_0\left(\chi_c - \chi_o\right)\cdot \operatorname{grad}\left(H^2\right) \tag{1}$$

where: χ_c, χ_o - magnetic susceptibility of a particle and medium, respectively, μ_0 - vacuum magnetic permeability, H - absolute value of magnetic field intensity vector.

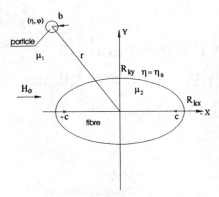

Fig.1. Collector of elliptical cross-section used for investigations

The magnetic field in separators is irrotational and potential. Assuming that the collector axis is on „z" axis, the magnetic scalar potential in the elliptic-cylindrical coordinates is the function of η and φ and is described by Laplace's equation. Solving this equation and assuming that the magnetic field is uniform at a distance from a matrix element:

$$\vec{H} = H_o \cdot \vec{i}_X. \tag{2}$$

one can indicate [3,4] that the magnetic field intensity vector outside the elliptic cylinder has two components H_η and H_φ and the magnetic force per particle unit volume can be expressed:

$$\vec{f}_m = \tfrac{1}{2}\mu_0\left(\chi_c - \chi_o\right)\frac{H_0^2}{c} \cdot \frac{K}{\left(\mathrm{ch}^2\eta - \cos^2\varphi\right)^{\frac{5}{2}}}\left\{-\vec{i}_\eta \cdot \left[2\mathrm{ch}^2\eta \cdot \cos 2\varphi - \right.\right.$$

$$\left.\left. - K\left(1 - e^{-2\eta}\right)\cdot\cos 2\varphi + 2\left(K + 1\right)\cos^2\varphi\right] - \vec{i}_\varphi \cdot \left[K \cdot e^{-2\eta} + \mathrm{sh}2\eta\right]\cdot\sin 2\varphi\right\} \tag{3}$$

The parameter K describes the influence of magnetic properties of the matrix element and its transverse dimensions on magnetic force values

$$K = \frac{R_{ky}}{R_{kx}}\left(\frac{\mu_2}{\mu_1} - I\right)\left(1 + \frac{R_{ky}}{R_{kx}}\cdot\frac{\mu_2}{\mu_1}\right)^{-1}\left(1 - \frac{R_{ky}}{R_{kx}}\right)^{-1} \tag{4}$$

where: μ_1, μ_2 - magnetic permeabilities of a collector and medium.

The dynamic resistance of the medium is a reaction force from the medium that surrounds this particle. This force is described by Stoke's equation when small particles are:

$$\vec{F}_d = 6\cdot\pi\cdot\eta\cdot b\cdot\left(\vec{\vartheta} - \vec{\vartheta}_c\right) \tag{5}$$

where: η - dynamic coefficient of viscosity, b -particle radius, $\vec{\vartheta}$ - medium velocity vector, $\vec{\vartheta}_c$ - particle velocity vector.

The medium velocity is determined through the consideration of potential flow of an incompressible medium described by a velocity potential. The medium velocity vector is expressed as this potential gradient and the components of the vector outside the cylinder have the form [4]:

$$\begin{bmatrix}\vartheta_\eta \\ \vartheta_\varphi\end{bmatrix} = \frac{\vartheta_0}{\left(\mathrm{ch}\eta_0 - \mathrm{sh}\eta_0\right)\sqrt{\mathrm{ch}^2\eta - \cos^2\varphi}}\begin{bmatrix}\left(\mathrm{ch}\eta_0\mathrm{sh}\eta - \mathrm{sh}\eta_0\mathrm{ch}\eta\right)\cos\varphi \\ -\left(\mathrm{ch}\eta_0\mathrm{ch}\eta - \mathrm{sh}\eta_0\mathrm{sh}\eta\right)\sin\varphi\end{bmatrix} \tag{6}$$

These components of the magnetic force and of the medium velocity vector can be converted from the elliptic-cylindrical coordinates to the rectangular coordinates according to the equation:

$$\begin{bmatrix} W_x \\ W_y \end{bmatrix} = \frac{1}{\sqrt{(ch^2\eta - \cos^2\varphi)}} \begin{bmatrix} sh\eta \cdot \cos\varphi - ch\eta \cdot \sin\varphi \\ ch\eta \cdot \sin\varphi + sh\eta \cdot \cos\varphi \end{bmatrix} \begin{bmatrix} W_\eta \\ W_\varphi \end{bmatrix} \qquad (7)$$

The gravitational force is constant in the whole separator. This force is reduced by the upthrust from the medium and can be expressed:

$$\vec{F}_g = (\rho_c - \rho_o) \cdot V_c \cdot \vec{g} \qquad (8)$$

where: ρ_c - particle density, ρ_o - medium density, V_c - particle volume, $\vec{g}$ - acceleration of gravity vector.

3. Equations of motion and trajectories of a particle

It is necessary to solve the system of differential equations in the system of Cartesian coordinates for the horizontal component and for the vertical one to determine a particle trajectory:

$$\frac{d\vartheta_x}{dt} = \frac{F_{mx} + F_{gx} + F_{dx}}{m}, \quad \frac{d\vartheta_y}{dt} = \frac{F_{my} + F_{gy} + F_{dy}}{m}, \quad \frac{ds_x}{dt} = \vartheta_x, \quad \frac{ds_y}{dt} = \vartheta_y$$

$$(9)$$

where: $\vartheta_{cx}, \vartheta_{cy}$ - particle velocity vector components of $\vec{\vartheta}_c$; s_x, s_y - the components of a distance $\vec{s}$ covered by a particle; F_{mx}, F_{my} - magnetic force components of $\vec{F}_m$; F_{gx}, F_{gy} - gravitational force components of $\vec{F}_g$; F_{dx}, F_{dy} - medium dynamic resistance components of $\vec{F}_d$; m - particle mass.

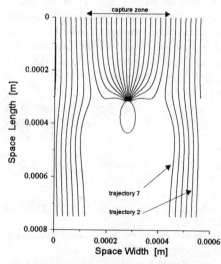

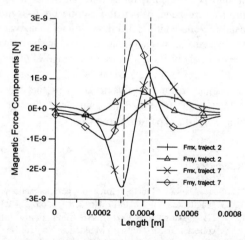

Fig.2. Particle trajectories for an elliptical fibre (dimensions in meters); $R_{kx}=6\cdot10^{-5}$ m, $R_{ky}=3\cdot10^{-5}$ m.

Fig.3. Magnetic force components vs. current position of a particle on noncaptured trajectories (2 and 7), at elliptical fibre cross-section (trajectory 2 and 7 on the right in Fig.2); vertical lines show collector position.

Magnetic flux density - 0.5 T, fluid velocity - 0.05 m/s, particle radius - 7.5×10^{-6} m, particle/fluid susceptibility - 0 / 0.009, particle/fluid density - 5×10^3 / 10^3 kg/m³, fluid kinematic viscosity - 0.001 N/m·s

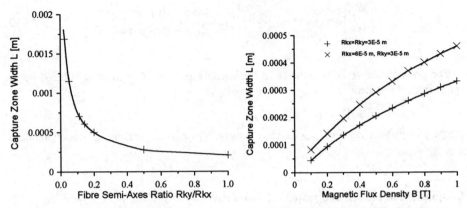

Fig.4. Capture zone width vs. the ratio of collector semiaxes

Fig. 5. Capture zone width vs. magnetic flux density

Particle radius - 7.5×10^{-6} m, fluid velocity - 0.05 m/s, particle/fluid susceptibility - 0 / 0.009, particle/fluid density - 5×10^3 / 10^3 kg/m³, fluid kinematic viscosity - 0.001 N/m·s

Runge-Kutty's procedure of the fourth order has been applied to solve the system of equations (9). The integration step has been selected automatically. Figs. 3-5 present computational results.

4. Conclusions

The collector of an elliptical cross-section is more effective than the one of circular cross-section because its capture zone width is longer at each magnetic flux density value. It follows from simulations that the flatter the collector cross-section becomes, the bigger the capture zone. When the elliptic cross-section is ten times thinner (R_{kx}/R_{ky}=1-10) the capture zone width is 3.6 times longer. This width is changed almost linearly together with the increase of magnetic flux density. In the case of an elliptic collector (R_{kx}/R_{ky}=2) the zone increases about six fold within the range of magnetic flux density between 0.1 T to 1 T. This width decreases rapidly when velocity increases. At matrix elliptic elements and at the range of velocity from 0.01 m/s to 0.1 m/s the capture zone is reduced from 4.88×10^{-4} m to 2.22×10^{-4} m. It should be mentioned that at low medium flow velocity the operating efficiency of a separator is reduced.

References

[1]   A. Aharoni, Tracting force on paramagnetic particles in magnetic separators. IEEE Trans. on Magnetics, Vol. MAG-12, No. 3, 1976, pp. 234-235.

[2]   L.P. Davies and R Gerber, 2-D Simulation of ultra-fine particle capture by a single-wire magnetic collector. IEEE Trans. on Magnetics, Vol. 26, No. 5, 1990, pp. 1867-1869.

[3]   R. Goleman and Z. Złonkiewicz, Magnetic force acting on a particle in matrix separator with fibres of an elliptical cross-section. Anales de Fisica, Serie B, Vol.86, No.1, 1990, pp. 40-42.

[4]   R. Goleman and Z. Złonkiewicz, Modelling of distributions of forces acting on a particle in the surrounding of an elliptic matrix element in magnetic separators, 20-th Seminar of Fundamentals of Electrotechnics and Circuit Theory, Gliwice-Ustroń, 1997, (in polish).

[5]   F.E. Luborsky and B.J. Drummond, High gradient magnetic separation: theory versus experiment. IEEE Trans. on Magnetics, Vol. MAG-11, No. 6, 1975, pp. 1696-1700.

Non-Linear Electromagnetic Systems
V. Kose and J. Sievert (Eds.)
IOS Press, 1998

Electric Field Calculation for Precipitators

Tamás BARBARICS, Amália IVÁNYI
Department of Electromagnetic Theory, Technical University of Budapest
Budapest, Hungary, H-1521

Abstract. This paper deals with the calculation of the electric field for electric precipitators. The numerical results are determined by using the R-functions method. The movement of the charged particles is described on the basis of the Navier Stokes equations and the Coulomb forces.

1. Description of the Problem

Prevention of environmental pollution requires techniques for cleaning up the flue gas in the coal-boiler. In general, electrical precipitators control the emission of the particles. These devices permit the removal of flying ash, avoiding their emission into the atmosphere. The electrically charged ash is collected by the grounded plates under the action of the electric field. The ash is charged by the ions formed through the corona effect on the high voltage emitting electrodes. During the precipitation the particles are charged by the corona discharge on the high voltage electrode, and the charged particles are going to move in the direction of the grounded electrode. Since the electric field is strongly affected by the space charges, and vica versa, in the electric precipitators, they must be self-consistently determined by an appropriate numerical technique.

Several kinds of numerical methods, such as the finite difference method, the finite element method, the charge simulations have been used to resolve the problem [1,2].

At the present time, the precipitators usually work with impulse excitation. The flying particles are charged by the corona discharge and during the impulse excitation the current density and the number of the emitted charged particles are increased [3]. The value of the potential on the high voltage electrode has an influence on the emitted charges, so it has to be known exactly. In an attempt to avoid this problem, the R-functions method is applied. This method ensures that the solution proves the correct value of the potential at every point of the electrodes with prescribed Dirichlet-type boundaries for any selected approximation function. The possibility of modelling the electrodes, not only in a cylindrical shape but in any artificial forms, is the main advantage of the R-function method.

The Ritz method, applying in the global variational method, provides a solution in semi-analytical form at any point in the examined region. Combining the global variational method with the R-functions, the solution of the differential equations fulfilling the prescribed boundary conditions can be determined exactly.

In the application of the finite element and finite difference methods simulation of the homogeneous Neumann-type boundary conditions can be fulfilled either as forced or as natural conditions. In finite element or finite difference analysis, it is not easy to take into account the forced Neumann-type boundary conditions. To solve this problem, the number of unknowns must be increased by extending the examined region. By using the R-functions method, this problem can be eased when the solution includes the prescribed Neumann-type boundary conditions as well. Applying this last method, the solution can be determined without an increased number of unknowns.

A numerical method based on the global variational method, combined with the R-functions is applied in this paper to determine the three dimensional electrical field of an electrical precipitator.

2. Basic Equations

The electric field and the charge density can be derived from Maxwell equations by introducing the scalar potential.

$$\nabla E = - \, e \, n \, / \, \varepsilon_0, \qquad E = - \, \nabla\varphi(r,t), \qquad n = n_+ - n_- - n_e \tag{1}$$

where E is the electric field intensity, φ the electric scalar potential, ε_0 the permittivity of the free space, n_+ and n_- are the numbers of the positive and the negative ions, n_e is the number of the electrons and μ the mobility of ions. These equations yield the expressions for the current density of the ions and the electrons as follows:

$$e \, \partial n_+/\partial t = - \, \nabla J_+ + e \, n_+^{\, *}, \quad e \, \partial n_-/\partial t = - \, \nabla J_- + e \, n_-^{\, *}, \quad e \, \partial n_e/\partial t = - \, \nabla J_e + e \, n_e^{\, *} \tag{2}$$

$$n_+^{\, *} = \alpha \, J_e \qquad\qquad n_-^{\, *} = \eta \, J_e \qquad\qquad n_e^{\, *} = e \, n_+$$

where J_+ and J_- are the current density of the positive and the negative ions, J_e is the current density of the free electrons, n_+, n_- and n_e the number of the positive, negative ions and the electrons originating in the region Ω, α is the shock ionisation and η is the absorbtion. There are only two prescribed Dirichlet type boundary conditions, the first is on the grounded electrode, where φ_1 is equal to zero, and the other one on the high voltage electrode, where φ_2 is equal to $\varphi_o(t)$.

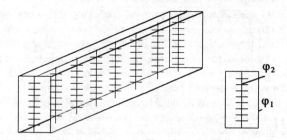

Fig. 1. The model and its cross-section

The cross-section of the three-dimensional electrostatic precipitator considered in the numerical simulation can be seen in *Fig.1*. The Dirichlet-type boundary conditions are prescribed on Γ_{D1} and Γ_{D2} as

$$\varphi\Big|_{\Gamma_{D1}} = 0 \, , \qquad\qquad \varphi\Big|_{\Gamma_{D2}} = \varphi_2 \tag{3}$$

where Γ_{D1} is the grounded electrode, and Γ_{D2} models the high voltage electrode with the potential of φ_2.

For the model on the boundaries Γ_{N1} and Γ_{N2} the homogeneous Neumann-type boundary conditions are fulfilled as forced Neumann-type boundary conditions on Γ_{N1} and Γ_{N2}.

$$\left.\frac{\partial\varphi}{\partial n}\right|_{\Gamma_{N1}} = 0, \quad \left.\frac{\partial\varphi}{\partial n}\right|_{\Gamma_{N2}} = 0 \tag{4}$$

To solve the Laplace-Poisson equation, the global variational method is applied, where the solution of the differential equation is the potential function, ensuring the minimal value for energy related functional

$$W(\varphi)=1/2 \int_{\Omega} (\varepsilon\, \nabla\varphi\nabla\varphi - 2\,\rho\,\varphi)\, d\,\Omega. \tag{5}$$

which can be solved by using the Ritz method in a bounded region [4].

3. Solution of Field Equations

Because of the changes of the potential along the high voltage electrode the numbers of the electric particles (n_+, n_-, n_e) will be changed during the time, as well as the current densities ($\mathbf{J_+}$, $\mathbf{J_-}$, $\mathbf{J_e}$).

In the first step to find the solution, the numbers of the electric particles in the region are selected to be zero. The number of the electrons on the high voltage electrode are increasing due to the excitation as

$$n_e^{**} = e \int u(t)\, dt\, /\, W_{out}, \tag{6}$$

where W_{out} is the leaking energy. The field does not change until there is no electron in the region Ω. When the potential is high enough to produce an electron, the calculation can be started. The potential has to be determined due to the change in the numbers of the electrical particles and the current density [5].

4. Movement of Charged Particles

The velocity vector of the flue-gas particles has only one component in the cross-section of the precipitator in the region before the electrodes if the flow is laminar. The emitted ions attach to these particles and the charged particles going to have new velocity components because of the Coulomb force $\mathbf{F} = q\,\mathbf{E}$, where q is the charge of the particles. This force makes changes in the orbits of the flying particles, they are going to move in the direction of the grounded electrode, and they are going to stick on the metal wall, because of their charge.

The movement of the particles can be determined from the Euler equation [6] as

$$\frac{\partial\mathbf{v}}{\partial t} + \mathbf{v}\,\nabla\mathbf{v} - \mathbf{g} + \frac{1}{\rho}\nabla\mathbf{x}\,\mathbf{p} + \mathbf{a_e} = 0, \tag{7}$$

where $\mathbf{v}$ is the velocity, $\mathbf{g}$ is the gravitation, ρ is the density and $\mathbf{p}$ is the pressure of the flue-gas and $\mathbf{a_e}$ is the acceleration from the Coulomb force as $\mathbf{a_e} = q\,\mathbf{E}\,/\,m_p$.

From (7) the Navier Stokes equation (8) can be evaluated by taking into account the frictional forces

$$\frac{\partial v}{\partial t} + v \nabla v - g + \frac{1}{\rho} \nabla x \, p - v \Delta v + a_c = 0 \tag{8}$$

5. Results and Conclusion

For this investigation, the field equations and the model were examined by the unit step excitation, because in this case the calculated results can be checked by measured data (*Fig.2*) [1, 3]. In this figure the current density on the grounded wall are plotted under different distances (10 or 20 cm) between the IEH type high voltage electrodes.

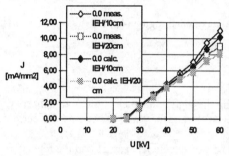

Fig. 2. Measured and calculated values of the current
density for different electrodes

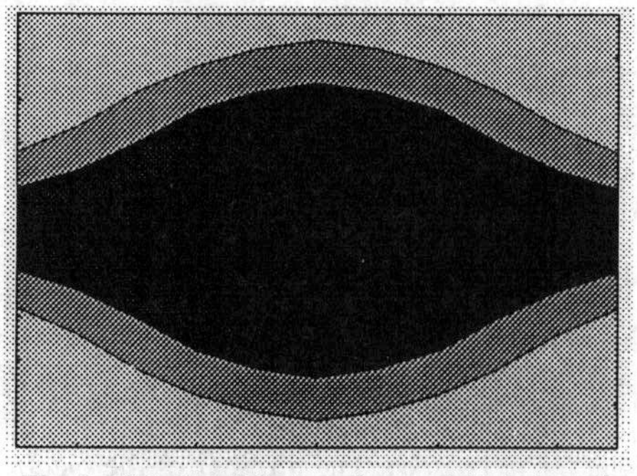

Fig. 3. Calculated dust on the precipitator's wall

Fig. 3 shows the width of calculated dust on the precipitator's grounded wall. Most of the flue-gas particles attach to the wall opposite to the high voltage electrode, where the current density has the highest value.

In conclusion it can be stated that the electric field determined in a three-dimensional model, proove the movement of the charged particles and the value of the current density.

A more accurate model of the problem can be established in the future by taking into account the effect of the back corona discharge.

References

[1] S. Cristina, G. Dinelli, M. Feliziani: Numerical Computation of Corona Space Charge and V-I Characteristic in dc Electrostatic Precipitator, IEEE Trans. on Ind. App., Vol.27., 1991. pp.147-153.

[2] J.L. Davis, J.F. Hoburg: Wire-Duct Precipitator Field and Charge Computation Using the Finite Element and Characteristics Method, Journal of Electrostatics, Vol.14., 1983. pp.187-199.

[3] T. Barbarics, A. Iványi: Determination of the Emitted Current in an Electrical Precipitator, Proceedings of 7th International IGTE Symposium on Numerical Field Calculation in Electrical Engineering, Sept., 1996, pp.83-87.

[4] T. Barbarics, H. Igarashi, A. Iványi, T. Honma: Determination of the Electric Field and the Space Charges of a Precipitetor Using the R-functions and the Method of Characteristics. Journal of Electrostatics 38, (1996) pp.269-282

[5] T. Barbarics, A. Ivanyi: Discharge of Impulse Series in Presence of Space Charges, Presentation on the VI. International Conference on Electrostatic Precipitation, Budapest, 18-21. June, 1996.

[6] I. Varga: Áramlástan (in Hungarian), Text Book, Technical University of Budapest, Sept., 1993.

Non-Linear Electromagnetic Systems
V. Kose and J. Sievert (Eds.)
IOS Press, 1998

Design and Development of a Low Cost Repulsive Type Magnetic Bearing and its H$^\infty$ Control

S.C.Mukhopadhyay, T.Ohji, M.Iwahara, S.Yamada and F.Matsumura
Faculty of Engineering, Kanazawa University,
2-40-20 Kodatsuno, Kanazawa - 920, Japan.

Abstract. This paper reports on the design and development of a low cost repulsive type magnetic bearing in which the radial bearing section has been achieved by the utilization of repulsive forces acting between the stator and rotor permanent magnets. A new type of permanent magnet configuration has been analyzed by the finite element method to obtain better stiffness. For axial stabilization an H$^\infty$ controller has been configured around a digital signal processor to stabilize the bearing system. A prototype system is designed and experiments are carried out.

1. Introduction

Passive magnetic bearings are not widely accepted by industries for commercial applications because of the absence of adjustable damping and stiffness characteristics. Active magnetic bearings(AMB) have a good reputation with industries but it is a five-axis control problem and the need for electromagnets and a complicated control circuit. It is becoming difficult to meet current machinery requirements such as compactness, lightweight and energy saving features. Different types of passive magnetic bearings and couplings using PMs have been discussed in [1]-[2]. It is not possible to implement fully passive magnetic bearing using only PMs or constant current electromagnets. This paper discusses the development of a magnetic bearing system which is a combination of PMs and controlled electromagnets. The stability of the magnetic bearing system is strongly influenced by the characteristics and configuration of the permanent magnets in the bearing system. It reports a new type of PM configuration which has been analyzed by the finite element method to obtain better stiffness. The improved stiffness characteristic helps to reduce the effect of radial disturbance on the system. For the axial stabilization an H$^\infty$ controller has been configured around a digital signal processor to stabilize the bearing system. A prototype system has been designed and developed in our laboratory and experiments are carried out.

2. Design and Development of Bearing System

The rotor is levitated by the repulsive forces acting between the stator and rotor permanent magnets. The analysis of repulsive forces and stiffnesses have been carried out by the 2-D finite element method. Depending on the system requirement, utilizing finite element analysis an optimum magnet configuration is obtained which is a trade-off between the repulsive force and stiffness. To attenuate the effect of radial disturbance higher radial stiffness is desirable. In order to improve radial stiffness a small piece of stator permanent magnet is placed at the top of the rotor. The characteristics of repulsive force and stiffness along the vertical axis is shown in Figs. 1 and 2 for the configuration with and without upper stator magnet respectively. The permanent magnet configuration is shown in Fig.3. The upper portion of the stator magnet is very important for radial disturbance attenuation. Using this magnet configuration the bearing system has been developed as shown in Fig.4. The stator permanent magnet is made of Nd-Fe-B magnet and the rotor is of Strontium-Ferrite magnet. The total mass of the rotor is 8kg. Gap sensors are used to give the information of the gap between the electromagnet and the rotor.

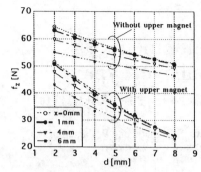

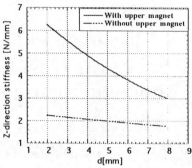

Fig.1 Repulsive force along vertical axis.   Fig.2 Stiffness along vertical axis.

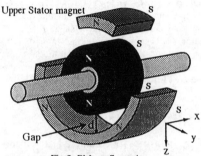

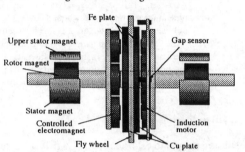

Fig.3 PM configuration.   Fig.4 Configuration of magnetic bearing system.

3. System Modeling and Control

The system has been modeled in state space form, the detailed of which is given in [3] and [4]. The magnetic bearing system developed here is a single axis controlled problem. The acting forces are the attractive forces of the electromagnets and the repulsive forces of permanent magnets. The four electromagnets for the axial control are connected in series. The governing equations are listed in (1) and (2).

$$m\frac{d^2x}{dt^2} = 2\left(S + \frac{\sum_i F_i}{W}\right)x + 2\frac{\sum_i F_i}{I}i \quad (1) \qquad L\frac{di}{dt} = e - Ri \quad (2)$$

where m is the mass of the rotor, S is a constant obtained from the repulsive force characteristics of the permanent magnet, F_i is the attractive force of ith electromagnet on the rotor, W is the nominal gap between the rotor and the electromagnet, I is the nominal current of the electromagnet, L and r are the inductance and resistance of the electromagnet. x is the gap displacement from the nominal value and i is the current deviation from the nominal value. The gap-displacement, x and its derivative and the current deviation, i, are taken as state variables. Using the values of different parameters the simplified state space representation is given by (3) and the output equation in (4), the details of parameters are given in [3].

$$\frac{d}{dt}\begin{bmatrix} x \\ \dot{x} \\ i \end{bmatrix} = \begin{bmatrix} 0 & 1 & 0 \\ 5885 & 0 & 2.654 \\ 0 & 0 & -60.5 \end{bmatrix}\begin{bmatrix} x \\ \dot{x} \\ i \end{bmatrix} + \begin{bmatrix} 0 \\ 0 \\ 5.165 \end{bmatrix}e \quad (3) \qquad y = \begin{bmatrix} 1 & 0 & 0 \end{bmatrix}\begin{bmatrix} x \\ \dot{x} \\ i \end{bmatrix} \quad (4)$$

The system is controllable and observable. The transfer functions of this system are given by (5).

$$G(s) = \frac{13.7079}{(s - 76.7138)(s + 76.7138)(s + 60.5)} \tag{5}$$

A simplified H^∞ controller has been designed for the magnetic bearing system. The details of the H^∞ controller have been described in [5]. The H^∞ norm specification is given by

$$\left\| \begin{matrix} \gamma W_1(s) & S(s) \\ W_2(s) & T(s) \end{matrix} \right\|_\infty < 1 \tag{6}$$

where $W_1(s)$ and $W_2(s)$ are desired weighting functions, $\gamma > 0$ is an adjusting scalar parameter, $S(s)$ is the sensitivity function and $T(s)$ is the complementary sensitivity function. In this study the weighting functions $W_1(s)$ and $W_2(s)$ are chosen as

$$W_1(s) = \frac{1.329}{1 + (s / 2\pi \cdot 0.016)} \tag{7}$$

$$W_2(s) = 1.5 \times 10^{-6} \cdot \left[1 + \frac{s}{2\pi \cdot 0.00012} \right]\left[1 + \frac{s}{2\pi \cdot 4.78} \right]\left[1 + \frac{s}{2\pi \cdot 55.7} \right] \tag{8}$$

and γ is 13.6.

The bode plot of the sensitivity S with $\gamma^{-1}W_1^{-1}$ and the complementary sensitivity T with W_2^{-1} are shown in Figs. 5 and 6 respectively. In Figs. 5 and 6, the sensitivity S approaches to $\gamma^{-1}W_1^{-1}$ at low frequencies and the complementary sensitivity T approaches W_2^{-1} at high frequencies. These are essentially based on the remarkable all-pass property in the H^∞ theory. Using the above weighting functions and using Robust control toolbox MATLAB, the H^∞ controller has been designed and the system has been simulated. Fig. 7 shows the simulated response characteristics at step input of the system.

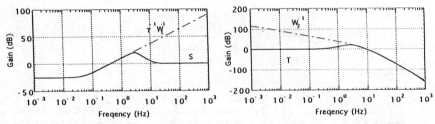

Fig.5 Sensitivity function.   Fig.6 Complementary sensitivity function.

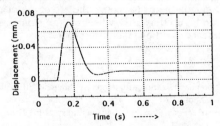

Fig.7 Simulated response characteristics.

4. Controller implementation

The controller has been configured around a digital signal processor as shown in Fig.8. The position of the gap-sensor is used as input and goes to the controller through the A-D port of the DSP. The DSP outputs through D-A port to the power-amplifier for adjusting the current of the electromagnet.

5. Experimental Results

With the help of the controller, the rotor has been stabilized. Fig.9 shows the rotor oscillation at a steady state along x axis. When a disturbance is applied along the vertical axis, the responses of the rotor are shown in Figs. 10a and 10b respectively. In Fig.10a the characteristic is less stiff, being without the stator permanent magnet configuration having less stiffness. Fig.10b corresponds to higher stiffness and has been achieved by properly configuring the upper stator magnet. It clearly shows that the effect of disturbance has been attenuated to almost one third of its original value.

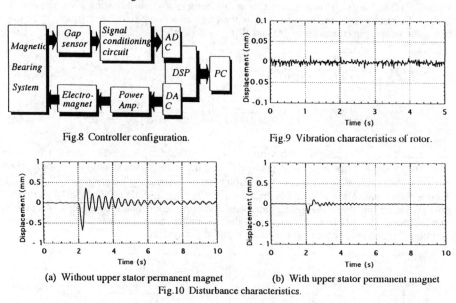

Fig.8 Controller configuration.　　　　　Fig.9 Vibration characteristics of rotor.

(a) Without upper stator permanent magnet　　　(b) With upper stator permanent magnet

Fig.10 Disturbance characteristics.

6. Conclusion

This paper has reported the development of a repulsive type magnetic bearing system and its simplified H^∞ control. The application of the H^∞ controller makes the system robust resulting in a low cost, high performance magnetic bearing system. Permanent magnet configuration has been carried out employing finite element analysis. Proper placement of the stator permanent magnet helps to reduce the effect the of radial disturbance of the system. The system can be used as a fly-wheel energy storage system or for other such applications in the presence of little radial disturbance.

References

[1]   J.P.Yonnet, "Passive Magnetic Bearing with Permanent Magnets", IEEE Transc. on Magnetics, Vol-Mag 14, No. 5, pp 803-805, Sept. 1978.

[2]   J.P.Yonnet, "Permanent Magnet Bearings and Couplings", IEEE Transc. on Magnetics, Vol-Mag 17, No. 1, pp 1169-1173, Jan. 1981.

[3]   S.C.Mukhopadhyay, T.Ohji, M.Iwahara, S.Yamada and F.Matsumura, "A New Repulsive Type Magnetic Bearing - Modeling and Control", Proc. of IEEE PEDS Conference, Singapore, pp 12-18, May 26-29, 1997.

[4]   F.Matsumura and H.Kobayashi, "Fundamental Equations for Horizontal Shaft Magnetic Bearing and its Control System Design", JIEE, Vol. 101C, No.6, pp. 137-144, 1981.

[5]   M.Fujita, F.Matsumura and M.Shimizu, "H^∞ Robust Control Design for a Magnetic Suspension System", Proceeding of 2nd ISMB, pp. 349-356, July 1990, Tokyo, Japan.

[6]   M.A.Franchek, "Selecting the Performance Weights for the μ and H^∞ Synthesis Methods for SISO Regulating Systems", Measurement and Control. Vol. 118, pp. 126-131, March 1996.

Non-Linear Electromagnetic Systems
V. Kose and J. Sievert (Eds.)
IOS Press, 1998

Comparison of Permanent Magnet Excited High Speed Synchronous Machines with Internal and External Rotor

Wolf-Rüdiger CANDERS,   Helmut MOSEBACH
Institut für elektrische Maschinen, Antriebe und Bahnen,
Technische Universität Braunschweig, Postfach 3329,
D 38023 Braunschweig, Germany

Abstract. The paper deals with a comparative discussion of typical design features of high speed permanent magnet excited synchronous machines with internal and external rotor. After a discussion of characteristic design restrictions, a 20,000 rpm machine with a circumferential speed of 200 m/s is treated in detail. The torque calculation is based on a new analytical method. For the selected example, the internal rotor machine shows advantages with respect to length, iron, magnet and copper masses, whereas the external rotor machine is favourable concerning the overall diameter.

1. Introduction

Permanent magnet excited synchronous machines (PMSM) for high speeds are usually designed as cylindrically shaped machines with the magnets arranged at the surface of the rotor iron (surface magnets) providing a fairly uniform mechanical stress distribution in the rotor as compared to machines with "buried magnets" for flux concentration.

High speed PMSM are applied in a large variety of drives such as machines for advanced flywheel energy storage systems and applications in turbo machinery [1,2,3,4].

2. Basic Structures

2.1 Internal Rotor Machine

The internal rotor machine is the classical solution most commonly found in all types of electrical machinery (Fig. 1).

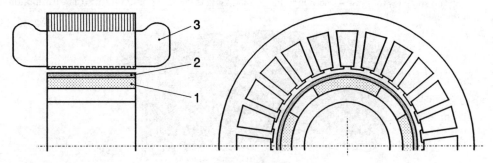

Figure 1 Internal rotor machine
1 ... permanent magnet
2 ... binding
3 ... stator winding

The bearing situation is conventional. The machine shows fairly simple rotor dynamics.

Mechanical problems may arise from the large centrifugal forces exerted on the magnets which are usually glued to the rotor yoke surface. The necessary mechanical strength is most frequently achieved with a can or binding around the magnets. The thickness of the usually nonmagnetic binding adds to the overall magnetic air gap, thus deteriorating the flux utilization of the magnet material.

2.2 External Rotor Machine

The external rotor is less frequently found in electrical machines and requires an arrangement where the rotor embraces the stator (Fig.2). Usually a single-sided bearing is adopted here being somewhat detrimental to the rotor dynamics. The magnets are housed at the inner surface of the rotor yoke and are ideally supported with increasing speed.

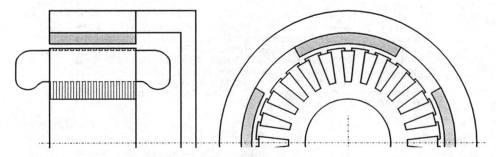

Figure 2 External rotor machine

3. Mechanical and Thermal Design Restrictions

The design of the two machine types encounters typical mechanical and thermal restrictions. The resulting interdisciplinary aspects have to be duely accounted for by an appropriate simultaneous consideration including the magnetic circuit as discussed later.

In the internal rotor machine the thickness of the binding, preferably made of carbon fiber, must be large enough to support the magnets and to limit the maximum stress in the rotor yoke which may here be dimensioned mainly following magnetic design aspects.

In the external rotor machine the rotor yoke itself must provide the necessary mechanical strength to sustain its own centrifugal forces and those of the magnets, as well as the required magnetic properties. This imposes restrictions on the selection of the yoke material. The determination of the rotor yoke stress distribution is best achieved by applying a numerical FE program such as ANSYS. As shown in Fig.3, the magnets cause a non-uniform stress distribution in the yoke.

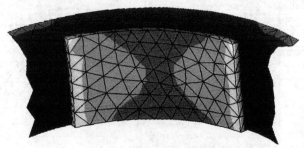

Figure 3 Rotor stress distribution

The thermal restrictions refer to the necessity to remove various loss contributions. These are the usual winding power loss, the hysteresis and eddy current power loss in the stator iron, and finally the gas friction loss. Due to the high frequency and the large peripheral speed, iron loss and gas friction are in the same order of magnitude as the copper losses.

The heat is preferably removed from the machine by an indirect water cooling jacket applied at the respective free surface of the stator back iron.

In the internal rotor machine, the heat management is facilitated by the increasing cross sectional area of the stator yoke towards the outer rim. The external rotor machine is more critical in this respect, an obvious consequence of the reduced inner cross sectional area.

It is essential to find a reasonable compromise between the yoke height to be bridged by the heat flow and the iron loss increasing with the yoke flux density.

The gas friction power loss comprises a contribution in the air gap [5] as well as a term occurring at the free rotor yoke surface. The latter contribution may be considered small in the internal rotor machine or even eliminated by the shaft. In the external rotor machine, it represents together with the air gap friction the main loss portion. In the air gap, the gas friction power loss depends mainly on the surface roughness of the outer part and is thus larger in the internal rotor machine with its slotted outer stator.

4. Specialities of the Magnetic Circuit

The magnetic circuit of both high speed PMSM is strongly influenced by the large magnetically effective gap and the intense curvature of the active surfaces. Both features are much more pronounced than in conventional ac machines and suggest the application of a 2-dimensional solution of the field equations, written in polar coordinates, as proposed in [6].

The radial dependence of the available cross sectional areas imposes various influences on the slot design and the achievable current loading. Due to the tapered shape of the slots, the internal rotor machine features more favourable conditions to establish large current loadings than in the external rotor machine, where it is found difficult to house enough copper in the slots, getting smaller with increasing rated speed. Leakage flux considerations or geometrical restrictions prevent the desired compensation of the reduced slot area by simply making them deeper. The situation is still aggravated by the flux concentration of the large no-load field of the magnets towards the stator surface. To maintain a constant tooth width as compared to the internal rotor design, it might be necessary to reduce the magnet height and/or to increase artificially the mechanical clearance which is also of benefit for the unbalanced magnetic pull and a reduction of the armature reaction.

5. Design Example

For a comparative and exemplary design study, a speed of 20,000 rpm was chosen at a peripheral speed of 200 m/s at the stator surrface. The machines are assumed to rotate in air at the normal atmospheric pressure. With 4 poles, the fundamental frequency reaches 667 Hz requiring the application of quite thin iron laminations. Therefore, a 0.20 mm iron sheet was assumed. Further common features are the same teeth and yoke flux density, a constant tooth width, and magnets with a remanence flux density of 1.0 T at a relative permeability of $\mu_r = 1.05$. The pole arc is 60 degrees.

The geometry of the two designs is summarized in Table 1 and illustrated true to scale in Fig.4.

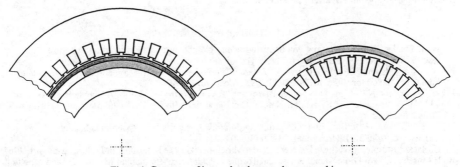

Figure 4 Geometry of internal and external rotor machine

Table 1 Geometry

Dimension	Internal rotor	External rotor
External / internal diameter	290 / 122 mm	259 / 88 mm
Height of stator / rotor yoke	33 / 24 mm	34 / 25 mm
Air gap / magnet height	1.0 / 7.5 mm	3.0 / 5.5 mm
Thickness of binding	2.5 mm	-
Slot height / tooth width	16 / 8.5 mm	18 / 8.5 mm

The specific utilization of the materials may be concluded from Table 2.

Table 2 Specific quantities

Physical quantity	Internal rotor	External rotor
Tooth flux density	1.65 T	1.65 T
Stator yoke flux density	1.20 T	1.20 T
Rotor yoke flux density	1.60 T	1.60 T
Current density	15 A/mm²	12.9 A/mm²
Surface current density	583 A/cm	380 A/cm
Average force density	27.4 kN/m²	18.8 kN/m²
Iron loss per unit length	22.4 kW/m	13.4 kW/m
Copper loss per unit length	14.0 kW/m	7.8 kW/m
Gas friction per unit length	7.2 kW/m	15.0 kW/m
Overall temperature rise	115 K	125 K

The result of the comparison is listed in Table 3 expressing the external rotor machine data in percent of the internal rotor machine with the same total output power.

Table 3 External rotor machine relative data

Quantity	Per cent of int. rotor machine of same power		
Iron length	146	Rotor inertial moment	483
Overall diameter	89	Iron power loss	87
Iron mass	126	Copper power loss	81
Copper mass	111	Gas friction power loss	302
Magnet mass	123		

6. Conclusion

Summarizing the results, the external rotor machine shows advantages concerning the overall diameter, and the iron and copper power loss. The other features are superior in the classical internal rotor machine. Especially the additional requirement of the costly magnet material in the external rotor machine is a result commonly not expected and is mainly caused by the extended length. It may be concluded that the external rotor machine is mainly preferable in applications with strict diameter constraints and in cases where the integration into the overall system is followed by additional advantageous properties.

References
[1] W.-R.Canders, Zur Berechnung von Schwungradenergiespeichern aus Faserverbundwerkstoff mit elektrischem Energiewandler. Doct. Thesis Technical University of Braunschweig, 1982
[2] W.-R.Canders and K.Reuter, Integrated System of Turboexpander with Magnetic Bearings and High Frequency Generator. Proc. MBMDDGS 1992, Alexandria, VA
[3] W.-R.Canders, N.Ueffing, U.Schrader-Hausmann, R.Larsonneur, MTG 400: A Magnetically Levitated 400 kW Turbo Generator System for Natural Gas Expansion. 4th Int. Symp. on Magnetic Bearings 1994, Zürich, Schweiz
[4] W.-R. Canders, N.Ueffing, Energiegewinnung durch Nutzung des Druckpotentials bei der Erdgasexpansion. 3R International, 33-4/5 (1994)
[5] U.Mack, Luftreibungsverluste bei elektrischen Maschinen kleiner Baugröße. PhD thesis Stuttgart, 1967
[6] H.Mosebach, Two-Dimensional Analytical Torque Calculation of Permanent Magnet Excited Actuators and Machines with a Large Air Gap. Proc. ISEM 1997, Braunschweig

Non-Linear Electromagnetic Systems
V. Kose and J. Sievert (Eds.)
IOS Press, 1998

Airgap Magnetic Field Evaluation in Permanent Magnet Motors Taking into Account the Slot Presence

Raul RABINOVICI

Department of Electrical and Computer Engineering
Ben- Gurion University of The Negev
P. O. Box 653, 84105 Beer- Sheva, Israel

Abstract. The paper presents a simple approximate way to evaluate the magnetic field distribution in the airgap of a permanent magnet motor. The influence of the stator slots on the field distribution is taken into account. Results of finite numerical analyses, that support the approach, are shown.

1. Introduction

The electric, and the magnetic behaviors of a permanent magnet motor (PMM) depend largely on the values of the magnetic flux density in its airgap [1]. The average magnetic permeance of a tooth- slot combination can be considered by relatively simple methods, e. g. Carter coefficients [2]. As a result, the stator surface for magnetic calculations can be considered as a smooth one [3,4]. The local influence of the stator slots on airgap magnetic field can be neglected in most of the design, and evaluation problems. However, there are cases where this simplification cannot be more applied. One of them is the calculation of the eddy current losses in metallic cans that support the rotor permanent magnets against centrifugal forces. Generally, the evaluation of the magnetic field distribution in the airgap of PMMs, while the slot presence is taken into account, is performed by numerical finite element methods [5]. Moreover, several simplified models that simplify the exact permanent magnet- slot configuration were proposed [6,7]. As a result, an analytical solution is obtained. By using a close formula for the magnetic field distribution, the calculations could be done much faster than by numerical finite element methods.

The paper presents also an approximate way to estimate the magnetic field distribution in the PMM airgap with a slotted stator. An important advantage of the present approach is that slots of any form could be considered. Moreover, the stator slots could be close one to the other, while their influence on the magnetic field distribution interferes.

2. Approach Description

To calculate the magnetic field distribution in the airgap of a PMM, while taking into account the stator slot presence, the real configuration is separated in two partial configurations, much simpler to analyze than the complex initial one. The magnetic field produced by the permanent magnets is evaluated in a smooth slotless stator structure. The slot contribution, is estimated in a configuration without permanent magnets. The actual field value is obtained by multiplying the two waveforms that were obtained in the partial configurations. The slot influence is estimated by calculating the distribution of the magnetic flux density between two opposite equipotential surfaces. One of them is smooth, and corresponds to the rotor surface. The other, that corresponds to the stator surface, contains the slots. The motor ferromagnetic core is presumed to be of infinite magnetic permeability, while the permanent magnets are of a magnetic permeability equal to that of air. In the explanation that follows, we refer to the radial component of the magnetic field. However, it is the determining one in the PMM operation.

The present approach lacks a theoretical justification. However, it could be introduced intuitively. At every airgap point, of radius r and angle θ, the magnetic flux density can be calculated as the ratio between a local magneto- motive force δF, a local magnetic reluctance δR, and a small surface δA related to δR :

$$B(r,\theta) = (\delta F / \delta R) / \delta A \qquad (1)$$

The local magnetic reluctance δR, and surface δA are the same for all the airgap points, where the magnetic field distribution is calculated. The local magneto- motive force depends on the magnetization of the rotor permanent magnets. These factors are taken into consideration by the magnetic analysis of the partial configuration with permanent magnets, and a smooth stator.

However, the local magneto- motive force depends also on the manner in which the total magneto- motive force of the permanent magnets is spread with along the airgap magnetic circuit of the airgap, between the surfaces of the rotor, and stator. Its occurrence is different if the airgap region is found under a stator slot, or not. This factor is taken into account by the analysis of the second partial configuration, with stator slots, but without the rotor permanent magnets.

Therefore, magnetic flux density at an airgap point could be evaluated by:

$$B(r,\theta) = B_1(r,\theta) * [B_{21}(r,\theta) / B_{22}(r)] \qquad (2)$$

B_1 is the magnetic flux density at that point, in the first partial configuration. B_{21} is the magnetic flux density at that point, in the second partial configuration. B_{22} is the magnetic flux density at that point, between the two equipotential surfaces of the second partial configuration, but both of them are now smooth ones, without slots. B_{22} is a reference magnetic field density. It depends only on the value of the radius at the point where the magnetic flux density is evaluated. B_{22} can be obtained very easily in an analytical mode, in a one dimensional (radius) analysis. It can be also taken from the magnetic analysis of the second partial configuration, in an airgap section where the stator slot influence is not perceptible. e. g. the airgap section under the axis of a stator tooth.

3. Results and Discussion

The present approach was verified by FEM numerical tools. The tested configurations are equivalent to a developed model of PMM. The relative magnetic permeability of the permanent magnets is one. The stator and rotor cores are of infinite magnetic permeability.

Fig. 1 exemplifies how the present method works. Fig. 1a shows the complete configuration, with both rotor permanent magnets and stator slots. Fig. 1b shows a partial configuration, with only permanent magnets, but with a smooth surface stator. Fig. 1c shows the second partial configuration, with only stator slots, but without permanent magnets.

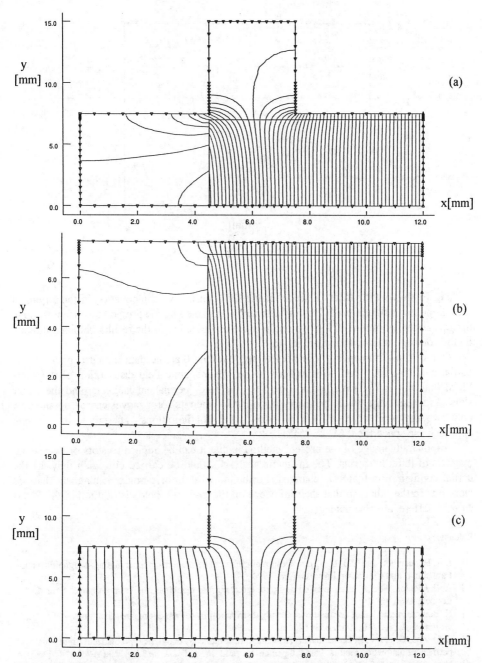

Fig. 1. Magnetic field distribution in the airgap of a PMM: a) the real configuration, with both permanent magnets and stator slots; b) the first partial configuration, with only permanent magnets; c) the second partial configuration, with only stator slots; d) comparative results of the airgap magnetic flux density (y-component). The airgap thickness is 0.5 mm, the magnet thickness is 7 mm, while the slot opening is 3mm. y=7.5 mm: —— complete analysis, o present method; y=7.1 mm: ---- complete analysis, x present method.

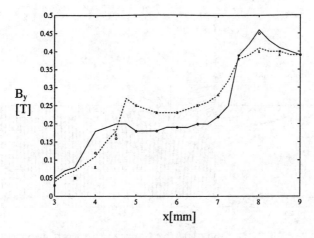

Fig. 1d

Fig. 1d shows the values of the magnetic flux density (y- components) in the airgap of the complete configuration of Fig. 1a. The results obtained by the present approach, through the partial configurations of Fig. 1b and c, were compared with the results obtained directly on the complete configuration of Fig. 1a.

One of the advantages of the present approach is that the data obtained once on the partial configurations, can be employed to get the magnetic field distribution of the same PMM but with any relative position between the rotor permanent magnets, and the stator slots. As a result, the whole behavior of the PMM when its rotor moves can be obtained by only two magnetic analyses, of the two partial configurations. Therefore, the present approach can speed the CAD of PMMs.

Another advantage of the present method is that it can be applied to slots or permanent magnets of different forms. The magnetic analysis should be carried out again only for the partial configuration that was changed. Furthermore, slots or permanent magnets that are close one to the other, so that their influence on the magnetic field distribution in the PMM airgap interfere, are also manageable.

References

1] J. R. Hendershot, and T. J. E. Miller, Design of brushless permanent- magnet motors, Magna Physics Publishing, and Clarendon Press, Oxford, 1994

2] A. E. Clayton, The performance and design of direct current machines, Sir Isaac Pitman & Sons, Ltd., London, 1966, pp. 1- 52.

3] N. Boules, Prediction of no- load flux density distribution in permanent magnet machines, *IEEE Trans. on Ind. Appl.*, IA- 21 (1985) 633- 643.

4] Z. Q. Zhu, D. Howe, E. Bolte, and B. Ackerman, Instantaneous magnetic field distribution in brushless permanent magnet dc motors. Part I: Open circuit field, *IEEE Trans. on Mag.*, 29 (1993) 124- 135

5] N. A. Demerdash, T. W. Nehl, F. A. Fouad, and A. A. Arkadan, Analysis of the   magnetic field in rotating- armature electronically commutated DC machines by finite elements, *IEEE Trans. on Power Apparatus and Systems*, PAS- 103 (1984) 2223- 2231

6] N. Boules, Impact of slot harmonics on losses of high- speed permanent magnet machines with a magnet retaining ring, *Elec. Mach. And Electromech.*, 6 (1981) 527- 539

7] Z. Q. Zhu, and D. Howe, Instantaneous magnetic field distribution in brushless permanent magnet dc motors. Part III: Effect of stator slotting, *IEEE Trans. on Mag.*, 29 (1993) 143- 151

Non-Linear Electromagnetic Systems
V. Kose and J. Sievert (Eds.)
IOS Press, 1998

Development of 50kg Cold Crucible Levitation Melting Furnace

K.Kainuma[a] , T.Take[a], M.Fujita[a] and S.Hayashi[b]

[a] Fuji Electric Corporate Research and Development Ltd. , 5520 Suzuka Mie 513, Japan
[b] Fuji Electric Co., Ltd., 5520 Suzuka Mie 513, Japan

Abstract. A cold crucible levitation melting furnace is a useful apparatus for melting metal without contaminating it. The molten metal is levitated in the crucible by strongly induced electromagnetic forces. Using this apparatus, we succeeded in melting 50kg of cast iron and tapping it through a hole in the bottom of the crucible. During the development of the apparatus, we investigated eddy current distribution in order to discover the amount and the distribution of magnetic force and heat loss. We calculated heat transfer and fluid flow problems to evaluate the cooling ability of the crucible. Furthermore, we analyzed free boundary problem for a mass of molten metal.

1.Introduction

The contamination of molten metal from a crucible made of refractory materials and the reaction which occurs between the molten metal and the crucible have made it difficult to melt pure reactive metals. For this reason, new methods of melting metal have been studied energetically. Cold Crucible Levitation Melting (CCLM) has recently attracted special attention, because of its unique characteristics: ① It allows metals to be melted without contamination; ② Molten metals become uniform in composition because they are strongly stirred in the crucible; ③ Metals with high melting points can be melted with it; ④ A crucible can be used repeatedly. However, this superior method has been applied only in research, because until recently it has been possible to melt only a few kilograms of metals with a CCLM furnace. The large CCLM furnace we have developed can melt 50kg of cast iron and can be tapped from a hole in the bottom of the crucible. This apparatus can be applied to practical casting processes. When designing a large CCLM furnace, evaluating the cooling ability of a crucible and estimating the shape of the molten metal mass are important problems. These problems can be solved numerically with a computer.

2.Principle of CCLM

Figure 1 shows the configuration of the CCLM furnace. A copper crucible is constructed of many segments, each of which is water-cooled internally. High frequency currents flow through the coils. Eddy currents are induced in the crucible and metal to be melted. The metal in the crucible is levitated with electromagnetic repulsion forces. The eddy current generates Joule heat in the metal, and the metal melts. To realize stable melting conditions, we have designed a CCLM furnace with a duplex inverter configuration[1)2)]. The upper coil, in which the current frequency is several tens of kilohertz, is used mainly to heat metals and the lower one, in which current frequency is several kilo hertz, is used mainly to levitate molten metal[3].

3.Design of Large CCLM

3.1 Eddy Current Analysis

In order to evaluate the amount and the distribution of electromagnetic force and heat loss, and in order to optimize crucible and coil structures, we have analyzed eddy currents in the crucible and coils numerically. In a CCLM furnace, magnetic flux leaks over a wide area and eddy current depth is shallow compared with the thickness of conductor segments. We used an integro-differential method of analysis with a thin plate element with current vector potential[4]. The integro-differential equation for the normal component of the electric vector potential $\mathbf{T}$ is described as follows:

$$\frac{1}{\sigma}\nabla^2 T = \frac{j\omega\mu h}{4\pi}\iint_s \frac{\{\nabla\times(\mathbf{n}T)\}\times\mathbf{r}\cdot\mathbf{n}}{r^3}ds + j\omega \mathbf{B}s\cdot\mathbf{n} \qquad (1)$$

where T is the normal component of $\mathbf{T}$, $\mathbf{r}$ is the vector from the source point to the field point, $\mathbf{B}s$ is the density of the magnetic flux generated by the external source, h is the thickness of the active parts, s is the surface of the conducting bodies and $\mathbf{n}$ is a normal vector on the conductor surface. The value chosen for h is the same value as that for skin depth. The density $\mathbf{J}$ of the eddy current flowing in each element, the density(Joule heat) W of the power loss in each element and the electromagnetic force $\mathbf{F}$ sustained in each element are calculated as follows:

$$\mathbf{J} = \nabla\times(\mathbf{n}T) \qquad (2)$$

$$W = \frac{h}{\sigma}\left|\nabla\times(\mathbf{n}T)\right|^2 \qquad (3)$$

$$\mathbf{F} = \{\nabla\times(\mathbf{n}T)\}\times\mathbf{B} \qquad (4)$$

Figure 2 shows equi-potential lines on segments of the crucible. Eddy currents flow along these lines. We evaluated the

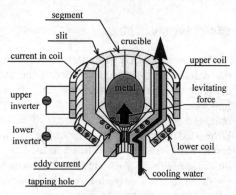

Fig.1   Principle of CCLM Furnace

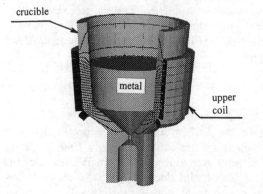
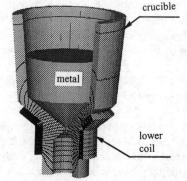

(a)Current Flowing through upper coil   (b)Current Flowing through lower coil

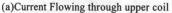

Fig.2   Eddy Current Distribution

power loss in a segment and the levitation force exerted on the molten metal using this result and equations (2) and (3).

3.2 Thermal Analysis

Our research indicates that a large CCLM furnace induces a great heat loss especially around the hole in its bottom. Thus cooling there will be a very important problem. We analyzed heat transfer in the crucible to discover the temperature distribution in the segments in order to optimize the crucible cooling structure. In our thermal analysis, the distribution of the heat source is calculated from equation (3) using a 3-dimensional eddy current analysis. The cooling water fluid problem was calculated at the same time to consider the distribution of the heat transfer coefficient between the crucible and cooling water.

Figure 3 shows the temperature distribution and the flow of cooling water on a cross section of the segment near the bottom of the crucible. The arrows plotted in the figure indicate the direction and velocity of cooling water.

3.3 Estimation of Shape of Molten Metal Mass

The stability of the molten metal is influenced by the induced magnetic forces, which depend on the shape of the crucible and the electrical parameters of the operating coils. To understand the conditions necessary for levitation and tapping, we calculated the free boundary of the molten metal. In this analysis, we developed a coupled analysis in which magnetic fields and the free boundary of the molten metal are considered. The electromagnetic force exerted on the molten metal was calculated using a 2-dimensional finite element method (FEM). The free boundary of the molten metal was calculated by the volume fraction method (VOF). In this method, we calculated the fluid volume percentage for every element. The equi-ratio line (the ratio is 50%) represents the shape of the molten metal mass. To discover the shape of the molten metal mass and the conditions necessary for levitating and tapping, we calculated the magnetic force and free boundary alternately until the shape of the molten metal mass become invariable when compared with the shape calculated in the previous step. Figure 4 shows the results obtained in the free boundary calculation. These results show that it is possible to levitate the molten metal mass and to tap the molten metal from a hole in the bottom of the crucible by controlling the power applied to the upper and lower coils.

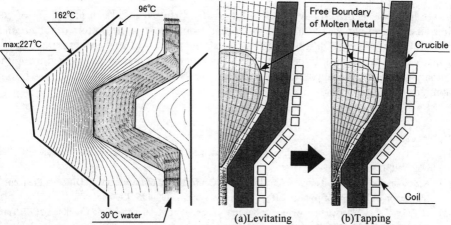

Fig.3   Temperature and water-flow
distribution around the bottom
hole of the crucible

Fig.4   The results of free boundary calculations
of molten metal simulation

4.Experimental Results Obtained with Large CCLM Developed

We constructed a prototype apparatus based on the analysis described above and conducted melting and tapping experiments. During levitation melting, the power supply of the upper inverter was 400kW and that of the lower one was 200kW. 50kg of cast iron were melted in 5 minutes. Figure 5 shows the melting conditions of 50kg of cast iron in the Large CCLM. During the melting experiment, we confirmed that the molten metal mass did not contact the crucible and that it melted stably. Figure 6 shows molten cast iron being tapped from the bottom of a crucible. Molten metal tapping was carried out by balancing narrowing forces, levitation forces and gravity. Tapping time can be controlled by controlling the power supply of the upper and lower coils. The tapping time is 16-27 seconds when 40kg of cast iron are tapped.

Fig.5   Melting of 50kg of cast iron   Fig.6   Tapping of 50kg of cast iron

5.Conclusion

We have succeeded in developing a large CCLM furnace that can melt 50kg of cast iron in only 5 minutes, and that allows tapping from a hole in the bottom of the crucible in about 20 seconds. In designing this furnace, the numerical analyses presented in this paper were a useful tool. Experiments in which metal was melted and tapped using this furnace show that this apparatus has reached a level of development that allows it to be applied in industrial fields.

References

[1]   A.Fukuzawa, et al.,: "Cold Crucible Type Levitation Melting by Supplying Two Frequencies", EPM Nagoya ISIJ, 172-177, 1994

[2]   K.Sakuraya, et al.,: "Levitation Melting by Duplex Configuration of Inverters with Different Frequency", Simulation and Design of Applied Electromagnetics Systems, 483-486, 1994

[3]   H.Tadano, et al.,: "Levitation Melting of Several Kilograms of Metal with a Cold Crucible", IEEE Trans. on Magn.Vol30(6),4740-4742, 1994

[4]   H.Tsuboi, et al.,: "Three-Dimensional Eddy Current Analysis of Eddy Current Analysis of Induction Heating in Cold Crucibles", IEEE Trans. on Magn. Vol.29(2), pp1574-1577, 1993

Non-Linear Electromagnetic Systems
V. Kose and J. Sievert (Eds.)
IOS Press, 1998

Design of a High-Frequency Excitation Coil Used for Electromagnetic Hotplate

M. Enokizono*, T. Todaka*, H. Nakamura* and R. Fukuda**

*Dept. of Electrical and Electronic Engineering, Faculty of Engineering,
Oita University, 700 Dannoharu, Oita 870-11, Japan
**FUJIMAK Corporation,
246-1 Komaki, Shishibu, Koga, Kasuya, Fukuoka 811-31,Japan

Abstract. This paper deals with the design of a high frequency induction coil shape of an electromagnetic hotplate using the three-dimensional finite element method. In the case of heating the conventional exciting coils, the heating condition is not uniform. Therefore, we improve the exciting coil shape for uniform heating by means of numerical analysis and experiment.

1. Introduction

The high frequency induction heating system is widely used in high frequency induction baking, welding of pipes and dissolution of metals and so on. Many applications of this system in electromagnetic ranges are currently being carried out because of saving energy, keeping high-efficiency and preventing a serious fire. The induction systems are still designed using the results of experiments based on experiences of uniform heating or local heating. Therefore, it is necessary to introduce numerical approaches into the designing to investigate eddy current and temperature distribution in detail.

This paper deals with the design of a high frequency induction coil shape of a Chinese pan heater using the three-dimensional finite element method (T-method) [1]. In the case of gas heating, the center of pan can be effectively heated. On the contrary, in the case of heating with the conventional exciting coils, the eddy current becomes smaller near the center of a pan. It is therefore necessary to design the eddy current paths by optimizing the coil shape. The investigations were carried out for various shapes of windings using three-dimensional analysis. In this paper, the eddy current, the heat source and the temperature distributions in the Chinese pan heater for some typical exciting coil shapes are presented.

2. Formulation

Generally, when the displacement currents can be negligible, the Maxwell's equations for the quasi-static magnetic field can be written as follows:

$$\nabla \times H = J ,\tag{1}$$

$$\nabla \times E = -\frac{\partial}{\partial t}\left(B_0 + B_e\right),\tag{2}$$

$$\nabla \cdot B = 0, \qquad \nabla \cdot J = 0,\tag{3}$$

where H is the magnetic field intensity, J is the current density, E is the electric field intensity, B_0 is the magnetic flux density due to the source currents and B_e is the magnetic flux density due to the eddy currents. The constitutive relationships are given by $B = \mu H$ and $J = \sigma E$ (μ : the permeability, σ : the conductivity).

Substituting the current vector potential T, which is defined by $J = \nabla \times T$, into eq. (2), we can obtain the following equation.

$$\nabla \times \nabla \times T = -\sigma \frac{\partial}{\partial t}\left(B_0 + B_e\right). \tag{4}$$

On the other hand, Helmholtz's law for any vector function A is expressed as:

$$A = \frac{1}{4\pi}[\int_V (\nabla' \times A) \times \nabla' \frac{1}{R} dV' + \int_V (\nabla' \cdot A)\nabla' \frac{1}{R} dV'$$

$$-\int_S n \times A \times \nabla' \frac{1}{R} dS' - \int_S (n \cdot A)\nabla' \frac{1}{R} dS'. \tag{5}$$

Therefore, the current vector potential T can be expressed as the following equation, with the gauge condition $\nabla \cdot T = 0$ and the boundary condition $n \times T = 0$.

$$T = \frac{1}{4\pi}[\int_V (\nabla' \times T) \times \nabla' \frac{1}{R} dV' - \int_S (n \cdot T)\nabla' \frac{1}{R} dS']. \tag{6}$$

Thus the following governing equation was obtained by eqs. (4) and (6) and Biot-Savart's law:

$$\nabla \times \nabla \times T + \mu\sigma \frac{\partial}{\partial t} T + \frac{\mu}{4\pi} \frac{\partial}{\partial t} \int_S Tn \nabla' \frac{1}{R} dS' = -\sigma \frac{\partial}{\partial t} B_0, \tag{7}$$

where,

$$B_0 = \frac{\mu}{4\pi} \int_V J \times \nabla'' \frac{1}{R} dV''. \tag{8}$$

The finite element equation with the Galerkin's method can be written as:

$$\int_V Ni \cdot \left(\nabla \times \nabla \times T + \mu\sigma \frac{\partial}{\partial t} T + \frac{\mu}{4\pi} \frac{\partial}{\partial t} \int_S Tn \nabla' \frac{1}{R} dS' + \sigma \frac{\partial}{\partial t} B_0 \right) dV = 0,$$

$$\int_V Ni \cdot \nabla \times \nabla \times T dV + \mu\sigma \frac{\partial}{\partial t} \int_V Ni \cdot T dV + \frac{\mu}{4\pi} \int_V Ni \frac{\partial}{\partial t} \int_S Tn \nabla' \frac{1}{R} dS' dV$$

$$= -\sigma \frac{\partial}{\partial t} \int_V Ni \cdot B_0 dV. \tag{9}$$

Where Ni is the linear interpolation function. The matrix form of eq. (9) can be written as:

$$[A]\{T\} + \frac{\partial}{\partial t}[M_1]\{T\} + \frac{\partial}{\partial t}[M_2]\{T\} = \frac{\partial}{\partial t}\{B_0\}. \tag{10}$$

3. Results and Discussion

Figure 1 shows a high frequency induction heating model. The conditions assumed in this analysis were: the permeability $\mu = 4\pi \times 10^{-7}$ H/m, the conductivity $\sigma = 1 \times 10^7$ S/m, the exciting current $I_0 = 8$ A, the exciting frequency f = 20 kHz.

We have investigated various shapes of windings. As space is limited, we show the results for only one typical example. Figure 2 shows the exciting coil shapes. (a) is the conventional type. (b) and (c) are newly investigated. Figures 3, 4 and 5 show the distribution of the eddy current density at $\omega t = \pi / 2$, and the power loss in the Chinese pan for each coil.

As shown in Fig. 3, it was evident that the eddy currents flowing at the center of the pan were very small and the center part could not be heated properly. It is therefore necessary to design the optimizing coil shape for the eddy current paths. Figures 4 and 5 show the results obtained for Coil-B and -C. In the case of Coil-B, because the large eddy currents were flowing only at the center of pan, the center part could be heated and the other part could not be heated. It was realized that the Coil-B is not good for the Chines pan heater. For Coil-C, the flowing area of the large-eddy-currents was widened. The loss distribution shows that it is possible to heat the Chinese pan uniformly by using this coil.

Figure 6 shows the temperature distributions in the Chinese pan for Coil-A, -B and -C. As shown in Fig. 5, the Chinese pan is not heated uniformly for Coil-A and -B. It can be observed that Coil-C heats the center of the Chinese pan uniformly.

Fig.1 High frequency induction heating model

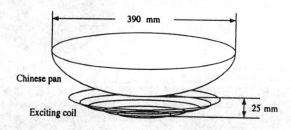

(a) Coil-A   (b) Coil-B   (c) Coil-C
Fig.2 Exciting coil shapes.

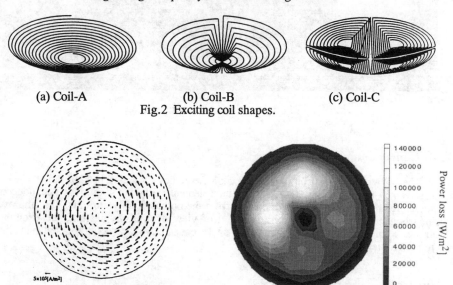

(a) Eddy current density   (b) Power loss
Fig.3 Distributions of eddy current density and power loss for Coil-A.

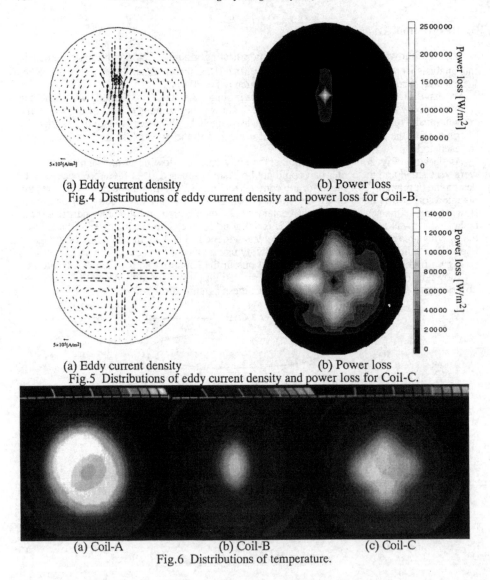

(a) Eddy current density   (b) Power loss
Fig.4 Distributions of eddy current density and power loss for Coil-B.

(a) Eddy current density   (b) Power loss
Fig.5 Distributions of eddy current density and power loss for Coil-C.

(a) Coil-A   (b) Coil-B   (c) Coil-C
Fig.6 Distributions of temperature.

4. Conclusion

The eddy current distributions of the high frequency induction heating model have been analyzed by using the three-dimensional finite element method (T-method). The induced heat source distributions were also calculated from the obtained eddy currents. To improve the induction heating apparatus from the point of view of uniform heating, the exciting coil shapes were investigated. Consideration of the constant voltage condition still remains as a future problem.

Reference

[1] K.Miyata, T.Sugiura, H.Hasizume and K.Miya, "Fast and accurate technique for 3D eddy current analysis using T method", Paper of technical meeting on static machine and rotational machine, IEE of Japan, SA-87-39, RM-87-76, 1987.

Non-Linear Electromagnetic Systems
V. Kose and J. Sievert (Eds.)
IOS Press, 1998

Numerical Computation of the Electromagnetic, Fluid Flow and Temperature Fields in an Induction Furnace

Dan SOREA

University "Transilvania", B-dul Eroilor 29, Braşov - 2200, Romania

Abstract. A mathematical simulation has been developed for the electromagnetic, fluid flow and temperature fields in a crucible induction furnace. The fluid flow field was represented by writing the axisymmetric turbulent Navier-Stokes equations, containing the electromagnetic body-force term. The electromagnetic field was calculated by using numerical methods. The k-ε model was employed for evaluating the turbulent viscosity, and the resultant differential equations were solved numerically, by the finite-difference method.

The computed results were found to be in good agreement with measurements reported in other works in this field regarding the flow pattern, mean velocity, and turbulence parameters.

1. Introduction

In recent years the interest in the theoretical study of molten metals in electrical induction furnaces has continuously increased.

The electric current in a coil which surrounds the crucible (Fig. 1) produces eddy currents in molten metal. In the presence of the magnetic field, these currents generate an electromagnetic body-force field, which is the cause of the turbulent agitation inside the furnace. The turbulent flow remains an unsolved problem. The k-ε model (1974) seems to be the most appropriate for that purpose. The temperature field is described by the enthalpy equation.

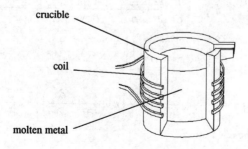

Fig. 1. Crucible induction furnace

2. Mathematical model

The simplifying assumptions used in mathematical modeling are the following:
- the symmetry in cylindrical coordinate system allows the neglection of the derivates of all quantities with respect to φ coordinate.
- the magnetic vector potential and the current density have a single component in the direction of φ.
- the magnetic induction, the electromagnetic force, the fluid flow velocities have components only in direction of r and z axis.

2.1. The electromagnetic field

The equation of φ component of the complex magnetic vector potential is[1]:

$$\Delta \underline{A} = \mu_0 \sigma \left(j\omega \underline{A} - u_z \underline{B}_r + u_r \underline{B}_z \right) - \mu_0 \underline{J}_i .$$

Solving this equation, we shall have the following:
- the magnetic induction components:

$$\underline{B}_r = -\frac{\partial \underline{A}}{\partial z}, \quad \underline{B}_z = \frac{1}{r}\frac{\partial}{\partial r}(r\underline{A}),$$

- the eddy current density in molten metal:

$$\underline{J} = \sigma \left(-j\omega \underline{A} + u_z \underline{B}_r - u_r \underline{B}_z \right),$$

- the components of the electromagnetic force:

$$f_r = \mathrm{Re}\left\{ \underline{J} \cdot \underline{B}_z^* \right\}, \quad f_z = -\mathrm{Re}\left\{ \underline{J} \cdot \underline{B}_r^* \right\}.$$

2.2. The fluid flow field and the temperature field

The governing equations of these fields are[2,4]:
- the continuity equation:

$$\frac{1}{r} \cdot \frac{\partial}{\partial r}(ru_r) + \frac{\partial u_z}{\partial z} = 0,$$

- the r momentum equation:

$$\frac{\partial u_r}{\partial t} + \frac{1}{r} \cdot \frac{\partial}{\partial r}(r u_r u_r) + \frac{\partial}{\partial z}(u_z u_r) = -\frac{1}{\rho} \cdot \frac{\partial p}{\partial r} + \frac{f_r}{\rho} + \frac{2}{r} \cdot \frac{\partial}{\partial r}\left(r \nu \frac{\partial u_r}{\partial r} \right) - \frac{2\nu u_r}{r^2}$$

$$+ \frac{\partial}{\partial z}\left[\nu \left(\frac{\partial u_r}{\partial z} + \frac{\partial u_z}{\partial r} \right) \right],$$

- the z momentum equation:

$$\frac{\partial u_z}{\partial t} + \frac{1}{r} \cdot \frac{\partial}{\partial r}(r u_r u_z) + \frac{\partial}{\partial z}(u_z u_z) = -\frac{1}{\rho} \cdot \frac{\partial p}{\partial z} + \frac{f_z}{\rho} + \frac{1}{r} \cdot \frac{\partial}{\partial r}\left[r \nu \left(\frac{\partial u_r}{\partial z} + \frac{\partial u_z}{\partial r} \right) \right] + 2\frac{\partial}{\partial z}\left(\nu \frac{\partial u_z}{\partial z} \right),$$

where the viscosity is given by the equations:

$$\nu = \nu_l + \nu_t, \quad \nu_t = c_\mu \cdot \frac{k^2}{\varepsilon}, \quad \text{where } c_\mu = 0.09.$$

In these equations, v_l is the kinematic laminar viscosity, k represents the kinetic energy of turbulence and ε its rate of dissipation. The k, ε and temperature[3] equations may be written in a single form as:

$$\frac{\partial \Phi}{\partial t} + \frac{1}{r} \cdot \frac{\partial}{\partial r}(r u_r \Phi) + \frac{\partial}{\partial z}(u_z \Phi) = \frac{1}{r} \cdot \frac{\partial}{\partial r}\left(r \frac{v}{\sigma_\Phi} \cdot \frac{\partial \Phi}{\partial r}\right) + \frac{\partial}{\partial z}\left[\frac{v}{\sigma_\Phi} \frac{\partial \Phi}{\partial z}\right] + S_\Phi,$$

where:

$$S_k = \frac{G}{\rho} - \varepsilon, \quad S_\varepsilon = \frac{\varepsilon}{k} \cdot \left(c_1 \frac{G}{\rho} - c_2 \varepsilon\right), \quad S_T = \frac{J^2}{c_p \sigma \rho},$$

$$G = \rho v_t \left\{ 2\left[\left(\frac{\partial u_r}{\partial r}\right)^2 + \left(\frac{\partial u_z}{\partial z}\right)^2 + \left(\frac{u_r}{r}\right)^2\right] + \left(\frac{\partial u_r}{\partial z} + \frac{\partial u_z}{\partial r}\right)^2 \right\},$$

$$\sigma_k = 1, \quad \sigma_\varepsilon = 1.3, \quad \sigma_T = 0.9, \quad c_1 = 1.44, \quad c_2 = 1.92.$$

3. Discretization equations

For solving the equations the finite-differences method was used. The general explicit form of a discretizated equation for a grid point P, by applying the pseudotransient method is[1]:

$$\Phi_P^{new} = \Phi_P + \Delta t \cdot \left(\sum a_{nb} \cdot \Phi_{nb} - a_P \cdot \Phi_P + b_P \right).$$

In these iterative calculations the resulting velocities do not satisfy the continuity equation. These velocities and the pressure must be corrected simultaneously. For that correction the SOLA algorithm was preferred, which evaluates the corrections for the pressure and velocities based on the velocity divergence at a given time level and then repeats the process until the divergence becomes less than a minimal value (theoretical equal to zero).

In order to determine the free surface shape, the pressure of an exterior upper grid point in the middle of the furnace is determined by linear extrapolation and the lower pressure grid points shall be eliminated in flow calculation; the flow computation shall begin again with the new shape[1].

4. Results

The computation of a laboratory furnace model containing Wood-metal was made. The radius of the crucible is $r = 0.1$ m and the height of the molten metal 0.2 m. The coil current frequency is $f = 250$ Hz, and the current density in the inductor $J_i = 10$ A/mm^2.

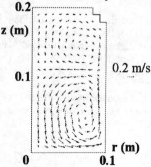

Fig. 2. Fluid flow pattern

The fluid flow pattern is shown in Fig. 2. The kinetic energy of turbulence may be seen in Fig. 3. Due to the motion, the temperature is practically constant, with small peaks in domains of reduced velocities.

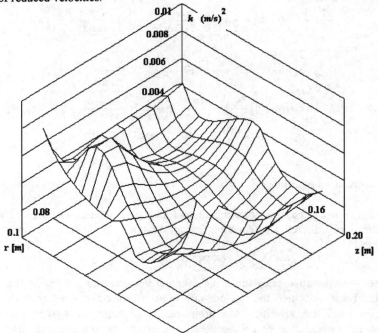

Fig. 3. The kinetic energy of turbulence in an induction furnace

5. Conclusions

The results of the computation were found to be in good agreement with the results given in other works in the field [2,4]. A small difference appears in the flow field, namely the velocity values in the lower recirculation loop are greater than those in the upper loop.

References

[1]   O. Peşteanu, Simulation of Turbulent Molten Metal Flow in Electromagnetic Field. Fortschritt-Berichte VDI, Reihe 7, No. 237. ISBN 3-18-143707-7. VDI-Verlag, Düsseldorf, 1994.

[2]   E. Baake, Grenzleistungs-und Aufkohlungsverhalten von Induktions-Tiegelöfen. Fortschritt-Berichte VDI, Reihe 19, No. 74. ISBN 3-18-307419-2. VDI-Verlag, Hannover, 1993.

[3]   W.J. Mynkowicz, et al., Handbook of Numerical Heat Transfer. ISBN 0 471 83093 3. John Wiley & Sons New York, 1988, p. 247.

[4]   N. El-Kaddah, J. Szekely, The Turbulent Recirculating Flow Field in a Coreless Induction Furnace, a Comparison of Theoretical Predictions with Measurements. J. Fluid Mech. 133 (1983) pp. 37-46.

Non-Linear Electromagnetic Systems
V. Kose and J. Sievert (Eds.)
IOS Press, 1998

Methods for the Optimization of the Electrical Efficiency of Inductive Heating Devices

S. DAPPEN and G. HENNEBERGER
Institut für Elektrische Maschinen, RWTH Aachen, 52056 Aachen, Germany

Abstract. This paper presents two FE optimization approaches for the electrical efficiency of an induction furnace. An adaptive axisymmetric time harmonic calculation is used for the solution of the eddy current field problem. This is coupled with the globally convergent genetic algorithm for the optimization of the cross sectional shape of the turn of the inductor coil. A gradient technique based on the sensitivity analysis is taken for the shape optimization of the whole inductor coil.

1. Introduction

The electrical efficiency η in inductive heating applications is in most cases significantly smaller than in other electrical devices. This is usually combined with a large amount of energy necessary for heating and melting metals. As a consequence the optimization of the efficiency is one of the basic demands of the designing process. Although an improvement in the efficiency has been examined over many years a direct coupling of numerical field calculations with advanced optimization techniques was still missing.

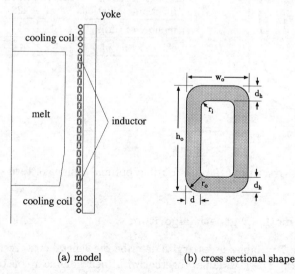

(a) model   (b) cross sectional shape

Figure 1: model of induction furnace and design variables of the shape optimization

All field computations in this paper have been realized using an axisymmetric time harmonic FE solver. Combined with an adaptive mesh generation the computational error of the efficiency was reduced to less than 0.2 %. The FE models contain about 12000 fieldnodes after 2 adaption steps.

In recent time stochastic methods have become very popular due to their flexible application to different design problems. They can be used with standard field calculation packages and the implementation of (nonlinear) boundary conditions is easy. Moreover the global optimum will reliably be found. The major drawback is the large number of field calculations needed to find the optimum.

On the other hand gradient based methods can reduce the computational amount significantly especially when dealing with a large number of design variables. As these methods are not globally convergent their application has to be handled with special care.

In the following sections the application to a practical induction heating problem will be shown. The examined induction furnace is displayed in Fig. 1a. The inductor is an inverter fed copper coil. It is flanked by two passive stainless steel cooling coils. Each turn is made from a solid hollow copper section (Fig. 1b). The radial position of the inductor is fixed by the crucible and the yoke.

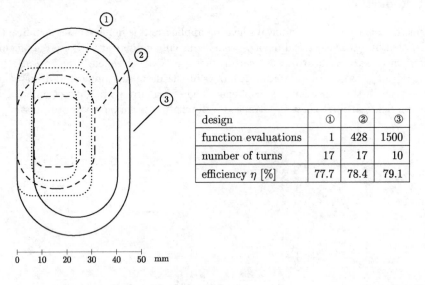

design	①	②	③
function evaluations	1	428	1500
number of turns	17	17	10
efficiency η [%]	77.7	78.4	79.1

Figure 2: results of GA optimization: optimal designs of turn shape

2. Stochastic method: genetic algorithm

Fig. 1b shows the 7 variables necessary to describe the general cross sectional shape of the inductor's turn. These variables together with their nonlinear constraints produce the design space of the optimization problem. The number of turns is chosen according to the given power supply and has been examined separately.

For the optimization two genetic algorithms (GA) have been developed. Both have a population size of 30 and a steady-state reproduction scheme. The first uses Gray-encoded chromosomes with the classical operators one-point-crossover ($p_c = 70 \ldots 50 \%$)

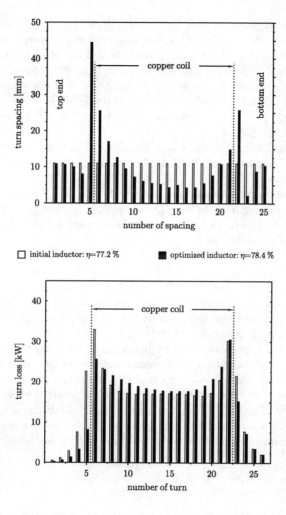

Figure 3: result of the spacing variation

and bit mutation ($p_m = 2\%$). The second is based on real number encoded chromosomes employing the operators average crossover, uniform crossover, mutation and local search [1] with varying weigths. In fact both algorithms show a similar convergence behaviour.

The results of two optimization runs for different numbers of turns are given in Fig. 2. A large number of function evaluations was necessary to obtain the optimized designs ② and ③. The comparison to the initial design ① shows only a small benefit to the efficiency.

Several other optimizations give similar results [2].

3. Gradient method: conjugate gradient

In this section a variation of the vertical position of each of the 26 turns (Fig. 1a) is allowed while keeping constant the total length of the coil. Through this a small increase in the electrical efficiency can be observed as reported in [3]. The design variables are the 25 spacings between the turns. With this number of variables an efficient computation of the gradient of the electrical efficiency becomes necessary. This is done by the combination of the sensitivity analysis and the adjoint variable method. The former is a method to gain the gradient information by differentiating the element interpolation functions. One characteristic of the method is that one solution of a linear set of equations is necessary for each component of the gradient. Combined with the adjoint variable method the amount can be reduced significantly to only one matrix solution for the total gradient [4].

To obtain a global optimum by a gradient method the design space has to be unimodal. For this reason several optimization runs have been performed starting from different initial designs.

The initial and optimized spacing distribution after 62 field computations and the corresponding losses displayed in Fig. 3. The spacings at both ends of the inductor increase whereas the middle turns move together. The losses in the end turns decrease and the loss distribution is flattened.

4. Conclusions

A shape optimization of the inductor turn is realized by a genetic algorithm. The shape is described by 7 design variables. Here several hundreds of adaptive field solutions are necessary to obtain the optimal shape of the turn.

In a second section a gradient based optimization of the 25 spacings between the turns of the inductor coil is carried out. The small number of necessary field solutions qualifies the method as a powerful and fast tool for the design optimization process.

Both methods seem to have found the global optimum but they show only little effect on the electrical efficiency. With the commonly used coil structure of solid copper turns no further benefit to the efficiency is possible. Additional results and informations concerning the optimization method and the field calculations can be taken from [2].

References

[1] L. Davis. *Handbook of Genetic Algorithms.* van Nostrand Reinhold, New York, 1991.

[2] S. Dappen. *Numerische Verfahren zur Berechnung und Optimierung von Induktionstiegelöfen.* Dissertation, RWTH Aachen, Shaker-Verlag, Aachen, 1997.

[3] J. Zgraja, A. Eggers. Calculating the electrical efficiency of inductor coils as a function of design and operating parameters. *Elektrowärme International,* vol. 48, no. 3, pp. 107–114, 1990.

[4] S. Dappen, G. Henneberger. A Sensitivity approach for the Optimization of Loss Efficiencies. *IEEE Transactions on Magnetics,* vol. 33, no. 2, pp. 1836-1839, 1997.

Non-Linear Electromagnetic Systems
V. Kose and J. Sievert (Eds.)
IOS Press, 1998

Two-Dimensional Analytical Torque Calculation of Permanent Magnet Excited Actuators and Machines with a Large Air Gap

Helmut MOSEBACH
Institut für elektrische Maschinen, Antriebe und Bahnen,
Technische Universität Braunschweig, Postfach 3329,
D 38023 Braunschweig, Germany

Abstract. The calculation is based on the vector potential method in polar coordinates and is especially suitable to account for the effects of large magnetic air gaps. The torque is calculated on the rotor side evaluating the interaction of the armature field and the fictitious magnet currents located at the magnet edges, yielding a closed analytical expression. The proposed technique represents an excellent tool in the design and the optimization of PM excited devices.

1. Introduction

Actuators and machines with permanent magnets arranged at the air gap (surface magnets) have a rather large magnetically effective gap. It is composed of the magnet height, and the mechanical clearance. A further contribution may arise from an eventual binding of the magnets and a protecting shield in the interspace between the stationary and the moving part. As a consequence, there are field components of considerable magnitude in circumferential direction, weakening the usable permanent magnet flux. Conventional analytical design methods rely frequently on the validity of a cartesian machine model and on the assumption of a small air gap and fail in the cases discussed here.

The prediction of torque in electromechanical devices is usually based on energy considerations or an evaluation of the Lorentz forces [1,2]. With the Lorentz force method, the "no-load"-field is determined, which is then evaluated at the stator surface interacting there with the stator currents. Permanent magnet systems are sometimes modelled by a Fourier series for the coercivity [3], or by a representation with edge surface currents [4], without leaving the force calculation at the stationary part. In the paper, a direct torque calculation at the magnet edges is presented.

2. Two-Dimensional Analysis

2.1 Physical Model

The basic structure of the electromechanical devices discussed here is shown schematically in Fig.1, showing one pole of an exemplary arrangement with p=3 pairs of poles. It comprises a stationary external part (the stator) carrying a polyphase ac winding, and a movable secondary (the rotor), rotating in the bore of the stator. There are two regions describing magnet and air gap.

The coils of the stator winding are housed in slots and may be represented by an equivalent infinitesimally thin axially directed current sheet $K_z(\varphi,t)$ on the stator surface. The iron is considered to be smooth and to have infinite permeability. The influence of the real slot openings and the finite iron permeability may be accounted for by Carter's coefficient and an additional air gap.

The permanent magnets have an alternating polarity and extend over a fraction α of the complete pole arc π/p.

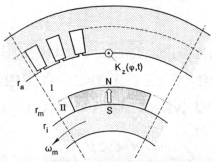

Figure 1 Physical model

It is assumed that the interspaces between adjacent magnets have the same permeability as the magnet material itself, thus providing uniform material properties over the φ-coordinate as a precondition for the separation of variables. With modern magnet materials and relative permeabilities of < 1.2, the induced error turns out to be very small.

2.2 Magnet Representation and Torque Calculation

Radially magnetized sector magnets as depicted in Fig.2a are applied. For the analytical approach, they can be represented by a fictitious surface current density K_m located at the edges of the magnet (Fig.2b). This current per unit length is given by

$$K_m = B_{rem} / \mu_m \tag{1}$$

with B_{rem} the remanent flux density, and μ_m the magnet permeability.

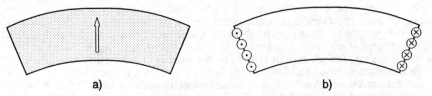

a)   b)

Figure 2 Magnet representation by surface currents

In contrast to the classical torque calculation, the torque is here determined on the rotor side applying the mutual interaction of the armature field with the magnet surface current density. For the torque production, the radial component $B_r(r)$ of the armature field is to be taken. The torque contribution of a single magnet pole edge is then obtained from

$$T_{pole\ edge} = l K_m \int_{r_i}^{r_i+h_m} B_r(r) r \, dr \tag{2}$$

where the Lorentz force element $B_r\, l\, K_m\, dr$ is multiplied with the respective radius r to yield the torque element dT to be integrated. The integral (2) can be solved easily, and yields a closed analytical expression. The overall torque of the device is then simply obtained by summing up the appropriate pole edge contributions.

The first step for the torque calculation is therefore the determination of the armature field. It is sufficient to solve the field equations for a single sinusoidal stator current loading wave. Winding harmonics or non-sinusoidal currents may be treated by Fourier analysis and simple superposition, without requiring a coupled simultaneous consideration.

2.3 Vector Potential Approach

The analytical treatment of the field problem is based most suitably on the vector potential method. To solve the Laplace equation in polar cordinates, the vector potential $V_z(r,\varphi,t)$ and the stator current sheet $K_z(\varphi,t)$ are expressed in terms of rotating waves,

$$V(r,\varphi,t) = Re\{\hat{V}(r)\,e^{j(\omega t - p\varphi)}\}$$
$$K_z(\varphi,t) = Re\{\hat{K}_z\,e^{j(\omega t - p\varphi)}\}$$

(3)

This approach allows the separation of the variable r from the variables φ and t, yielding an ordinary differential equation for V(r). Suitable types of solutions are r/r_i to the $\pm$ p-th power. The flux density components are derived from B=-rotV and are

$$\hat{B}_r(r) = j\mu_0\hat{K}_z[C(\tfrac{r}{r_i})^{p-1} + D(\tfrac{r}{r_i})^{-p-1}]$$
$$\hat{B}_\varphi(r) = \mu_0\hat{K}_z[C(\tfrac{r}{r_i})^{p-1} - D(\tfrac{r}{r_i})^{-p-1}]$$

(4)

where C and D are dimensionless integration constants to be determined by appropriate boundary conditions (BC). There are 4 BC for an equal number of unknowns, leading to a linear system of 4 equations.

For a more generalized representation of the result it is appropriate to define a reference value T_o for the torque

$$T_0 = 4r_a^2 l\hat{K}_z B_{rem}sin(\alpha\pi/2)$$

(5)

T_o is the hypothetical torque of a machine with the same stator bore diameter under idealized conditions. For machines with $p>1$ the torque is

$$\frac{T}{T_0} = \frac{2p(\frac{r_a}{r_i})^{p-1}}{p^2-1}\cdot\frac{(p-1)[(\frac{r_m}{r_i})^{p+1}-1]-(p+1)[(\frac{r_m}{r_i})^{-p+1}-1]}{(\mu_{rm}+1)[(\frac{r_a}{r_i})^{2p}-1]+(\mu_{rm}-1)[(\frac{r_a}{r_m})^{2p}-(\frac{r_m}{r_i})^{2p}]}$$

(6)

3. Comparison with One-Dimensional Analysis

For comparison, a simple one-dimensional cartesian analysis is applied. The respective reference torque T_1 is derived from T_o by taking into account the finite air gap,

$$T_1 = h_m/(h_m + \mu_{rm}\delta)T_0$$

(7)

The resulting torque ratio is shown in Fig.3 vs. r_a/r_i for different numbers of pairs of poles. For the magnets, $\mu_{rm}=1.05$ and a magnet height equalling the air gap was assumed.

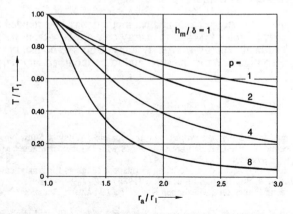

Figure 3 Comparison with idealized one-dimensional solution

The simple one-dimensional analysis produces reasonable results only for radii ratios close to one. With growing magnet height and air gap, and with increasing number of pairs of poles, the curvature of the active surfaces as well as the tangential flux components are felt more intensely making the exact two-dimensional analysis mandatory.

4. Example

For a critical validation it is expedient to choose an example with rather extreme geometrical dimensions, and to compare the results with those obtained from a numerical 2D-treatment considered to yield the "true values". The test example is an actuator of an autoclave measuring system exerting a controllable torque on the movable part. The magnetic field has to bridge the nonmagnetic wall of the cylindrical container as well as the thickness of the heating fluid layer inside the container. The geometric and magnetic situation is illustrated by Fig.4. The heating fluid region and the nonmagnetic wall are not shown. The field plot is the result of a Finite-Difference calculation using a 53*18 grid.

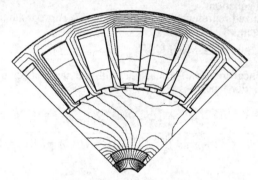

Figure 4   Actuator field plot

With only the fundamental current sheet wave, and neglecting the slot openings and the finite iron permeability, the analytically calculated torque is only 1.7 % larger than the numerical value indicating a very good agreement. Taking further into account Carter's coefficient, the analytical result matches perfectly with the numerical value, or lies at least within the error bounds which may be reasonably expected from the numerical method.

5. Further Applications of Method

The basic principle of the suggested analytical method may be extended to synchronous machines with an external rotor [5], and to energy converters without rotor magnet back iron as discussed for kinetic energy storage systems. Only the boundary conditions have to be adapted to the configuration in question and to be modified appropriately.

References

[1]   H.H.Woodson and J.R.Melcher, Electromechanical Dynamics. Wiley & Sons, New York, 1968
[2]   D.Platt and S.Geetha, Torque Calculation of Machines with Permanent Magnet Materials. Electric Machines and Power Systems 24 (1996), 393-415
[3]   K.Yoshida and H.Weh, A Method of Modeling Permanent Magnets for Analytical Approach to Electrical Machinery. Archiv für Elektrotechnik 5 (1985), 229-239
[4]   A.C.Smith, Magnetic Forces on a Misaligned Rotor of a PM Linear Actuator. Proc. ICEM 1990, Boston, 1076-1081
[5]   W.-R.Canders and H.Mosebach, Comparison of Permanent Magnet Excited High Speed Synchronous Machines with Internal and External Rotor. Proc. ISEM 1997, Braunschweig

Non-Linear Electromagnetic Systems
V. Kose and J. Sievert (Eds.)
IOS Press, 1998

Numerical Simulation of Brushless Synchronous Generator

S. Nagata[a] and M. Enokizono[b]
[a]Faculty of Engineering, Miyazaki University
1-1 Gakuenkibanadai-nishi, Miyazaki, 889-21, JAPAN
[b]Faculty of Engineering, Oita University
700 Dannoharu, Oiita, 870-11, JAPAN

Abstract
This paper presents the electromagnetic field analysis including the
motional electromotive force in a rotating electric machine. In this
analysis, the boundary element method is applied to the synchronous
generator. The method takes the effect of the relative motion between
the field winding and the armature winding into account.

Introduction

Magnetic field analysis of a brushless synchronous generator is presented. This kind of
machine has armature windings which are connected to the diodes. The static magnetic
field generated by additional field windings induces speed electromotive force in the closed
armature windings. Connected diodes rectify alternating currents, then they can provide
d.c. armature currents without brushes.

Numerical methods, such as the finite element method or the boundary element method,
have become established tools for electric machine analysis. However, the application of
conventional methods to the electric generator is difficult.

The reasons are as follows:

(1) The current induced by the relative motion has not been possible to determine.

(2) The governing equation including the motional electromotive force becomes a
convection-diffusion equation. It is well-known that the high Peclet number produces
unstable solutions.

(3) When using FEM or FDM, the mesh subdivisions should be modified as the rotor
moves.

In a rotating machine, the effect of the relative motion is of considerable importance in
some practical application. In this paper, the boundary element method taking the external
power source [1] is developed so that the motion induced current can be determined. This
method is applied to the synchronous generator, and clarifies the behavior of the magnetic
field when a load is applied.

Additionally, we expand the boundary element method to apply to a brushless synchronous
generator. The functioning of this apparatus depends on the nonlinear characteristics of the
diodes connected to the armature windings. It is necessary to analyze the apparatus taking
into account not only speed electromotive force but also nonlinear electronics circuits.
With this method, the behavior of the magnetic field due to the armature reaction and the
contributions of diode rectifiers can be clarified.

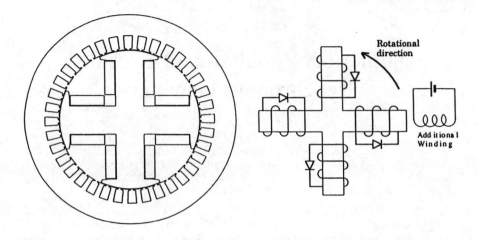

Fig.1 Model of brushless synchronous generator

Formulation

In the stationary coordinate system Maxwell's equations are always valid and expressed as follows:

$$\nabla \times \mathbf{E} = -\frac{\partial \mathbf{B}}{\partial t}$$
$$\nabla \times \mathbf{H} = \mathbf{J}$$
$$\nabla \cdot \mathbf{B} = 0$$
$$\nabla \cdot \mathbf{D} = 0$$

$$(1)$$

Where $\mathbf{E}$ is the electric field intensity, $\mathbf{B}$ the magnetic flux density, $\mathbf{H}$ the magnetic field intensity, $\mathbf{J}$ the current density, and $\mathbf{D}$ the electric flux density. It is assumed that the electric charges and the displacement currents are negligible.

The Galilean space-time transformation may be applied to Eq. (1) to give Maxwell's equations in a moving coordinate system [2]:

$$\nabla' \times (\mathbf{E} + v \times \mathbf{B}) = -\frac{\partial \mathbf{B}}{\partial t}$$
$$\nabla' \times \mathbf{H} = \mathbf{J}$$
$$\nabla' \cdot \mathbf{B} = 0$$
$$\nabla' \cdot \mathbf{D} = 0$$

$$(2)$$

where primes denote quantities expressed in the moving coordinate system, and v is the velocity of a point in the moving coordinate system (x', y', z', t'). The quantities $\mathbf{H}, \mathbf{B}, \mathbf{D}$ and $\mathbf{J}$ are not altered in both coordinate systems, however, only the electric field intensity vector E is modified as:

$$\mathbf{E}' = \mathbf{E} + v \times \mathbf{B} \tag{3}$$

In two-dimensional analysis, substituting B=rot A into Eq. (3) yields,

$$\mathbf{E}' = -\frac{\partial \mathbf{A}'}{\partial t'} - grad\varphi' - v_x \frac{\partial \mathbf{A}'}{\partial x'} - v_y \frac{\partial \mathbf{A}'}{\partial y'} \tag{4}$$

where A is the magnetic vector potential, φ is the scalar potential, v_x and v_y are x- and y-directional components of the velocity, respectively. On the other hand, in the stationary coordinate system, the electric field intensity E is

$$\mathbf{E} = \frac{\partial \mathbf{A}}{\partial t} - grad\varphi \tag{5}$$

In the case of a generator, the electric field vector intensity is induced in the windings. The terminal voltage U_t of the windings is expressed as follows:

$$U_T = -n_W \frac{d\phi}{dt} = n_W \int_l \mathbf{E} \cdot d\mathbf{l} \tag{6}$$

where n_w is the number of turns of winding, Φ is the interlinkage flux, and l is the length around the winding. Assuming the windings are connected to the external electric circuit as shown in Figure 2, the circuit equations are obtained as follows:

(a) In a stationary winding:

$$U_1 = n_W \int_l \frac{dA_{i1}}{dt} dl + R_1 I_1 \tag{7}$$

(b) In a moving winding:

$$U_2 = n_W \int_l \left(\frac{dA_{i2}}{dt} + v_x \frac{\partial \mathbf{A}_{i2}}{\partial x} + v_y \frac{\partial \mathbf{A}_{i2}}{\partial y} \right) dl + R_2 I_2 + U_{diode} \tag{8}$$

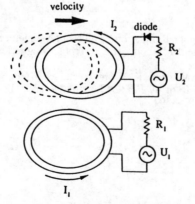

Fig.2 Relative motion of windings.

The subscript i means the value at the winding, U is the voltage of the external power source, R the resistance, I the current and U_{diode} the terminal voltage of the connected diode. When the rotation is assumed to be around the z-axis at an angular frequency w_r, v_x and v_y are given by:

$$v_x = -\omega_r y, \quad v_y = \omega_r x \tag{9}$$

The boundary integral equation including induced current is expressed as follows:

$$C_i A_i + \int_\Gamma q^* A d\Gamma = \int_\Gamma A^* q d\Gamma + \mu_0 \frac{n_w}{S} \int_\Omega A^* I d\Omega \tag{10}$$

In order to determine the unknown current I, the circuit equation including the external electric circuit should be combined with the field equation (10).

In general, A and its derivatives are unknown. However, they can be given by the boundary integral equations

$$A_i = \frac{1}{C_i}\left(-\int_\Gamma q^* A d\Gamma + \int_\Gamma A^* q d\Gamma + \mu_0 \frac{n_w}{S}\int_\Omega A^* I d\Omega \right)$$

$$\frac{\partial A_i}{\partial x_i} = \frac{1}{C_i}\left(-\frac{\partial C_i}{\partial x_i} A_i - \int_\Gamma \frac{\partial q^*}{\partial x_i} A d\Gamma + \int_\Gamma \frac{\partial A^*}{\partial x_i} q d\Gamma + \mu_0 \frac{n_w I}{S}\int_\Omega \frac{\partial A^*}{\partial x_i} d\Omega \right) \tag{11}$$

$$\frac{\partial A_i}{\partial y_i} = \frac{1}{C_i}\left(-\frac{\partial C_i}{\partial y_i} A_i - \int_\Gamma \frac{\partial q^*}{\partial y_i} A d\Gamma + \int_\Gamma \frac{\partial A^*}{\partial y_i} q d\Gamma + \mu_0 \frac{n_w I}{S}\int_\Omega \frac{\partial A^*}{\partial y_i} d\Omega \right)$$

After the discretizing process, the field equations (10)(11) and equations (7)(8) are combined to solve the equations.

Conclusions

In this paper, the boundary element method has been applied to the brushless synchronous generator. In order to analyze the induced current in the armature windings, the boundary element method including the motion induced field was developed. This techniques is based on the boundary element method taking the external power source into account.

References

[1] M. Enokizono and S. Nagata, Simulation Analysis of Magnetic Sensor for Nondestractive Testing by Boundary Element Method, IEEE, Trans. Magnetics, Vol. 26, No.2 (1990), pp.877-880
[2] M. Enokiono, T. Todaka, M. Aoki, K. Yoshioka and M. Wada, Analysis of Characteristics of Single Phase Induction Motor by Finite Element Method, IEEE Trans. on Magnetics, Vol. 23, No.5 (1987), pp.3302-3304

Miscellaneous

Non-Linear Electromagnetic Systems
V. Kose and J. Sievert (Eds.)
IOS Press, 1998

Contact Bounce Phenomena in an Electrical Switching System

K.Hamana, T.Sugiura, K.Suzuki*, M.Yoshizawa

Department of Mechanical Engineering, Faculty of Science and Technology, Keio University, 3-14-1 Hiyoshi, Kouhoku-ku, Yokohama-shi, Kanagawa, 223, Japan
*Fuji Electric Corporate Research and Development, Ltd, 1, Fuji-machi, Hino-shi, Tokyo, 191, Japan

Abstract   An electrical contact bounce usually occurs in an electrical switching system. It is accompanied by impact vibration between traveling and fixed contacts of the system. In this research, we examine such a mechanical bounce. First a feature of the contact bounce phenomena was confirmed from experimental results. This feature is clarified by theoretical analysis of dynamics including electromagnetic force. Moreover we discuss this mechanical bounce from the viewpoint of nonlinear dynamics.

1   Introduction

It is well-known that an electrical contact bounce usually occurs in an electrical switching system. So research into the bounce is indispensable[1]~[3] because it causes defective working of the system. These phenomena are also very interesting as the basic problem is associated with both magneto-solid interactions and impact oscillations. In particular, it is one of the important problems to which modern nonlinear dynamics of the bouncing ball is applicable. This research investigates impact between traveling and fixed contacts of the system, and discusses a feature of its mechanical bounce process from a viewpoint of nonlinear dynamics[4] associated with a multibody system in the electromagnetic field.

2   A Feature of Electrical Contact Bounce Phenomena

The electrical switching system consists of a traveling actuator, a fixed magnet, a fixed spring, two fixed contacts and a traveling contact which is in connection with a traveling actuator through a traveling spring, as shown in Fig.1. The traveling contact is assumed to be a rigid body with a concentrated mass at both ends. The fixed contacts are assumed to be flexible cantilevered beams with a concentrated mass end.

When the electrical circuit is closed, the traveling actuator is attracted to a fixed magnet by electromagnetic force. Then the traveling contact is driven by the actuator through the traveling spring. The traveling contact makes several inelastic impacts with the fixed contacts. Then transient current flows across the traveling and fixed contacts. After the impacts, they oscillate together, and the steady-state current passes through the contacts.

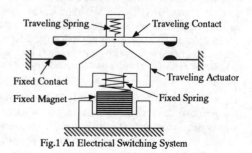

Fig.1 An Electrical Switching System

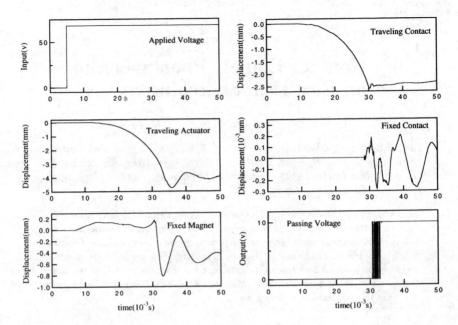

Fig.2 Experimental Results

3 Experimental Results

Experiments were carried out to confirm the contact bounce phenomena in the electrical switching system by applying low voltage to observe dynamic behavior of them. Fig.2 shows results of measuring the applied voltage V_{in}(volt), the displacements of the traveling actuator y_a(mm), the fixed magnet y_m(mm), the traveling contact y_t(mm) and the one-side fixed contact y_f($\times 10^{-3}$mm) and the voltage flowing the contact circuit V_{out}(volt).

The results show several impacts between the traveling and fixed contacts. And contact bounce phenomena are confirmed from the figure of the voltage passing through the contacts.

4 Theoretical Discussion

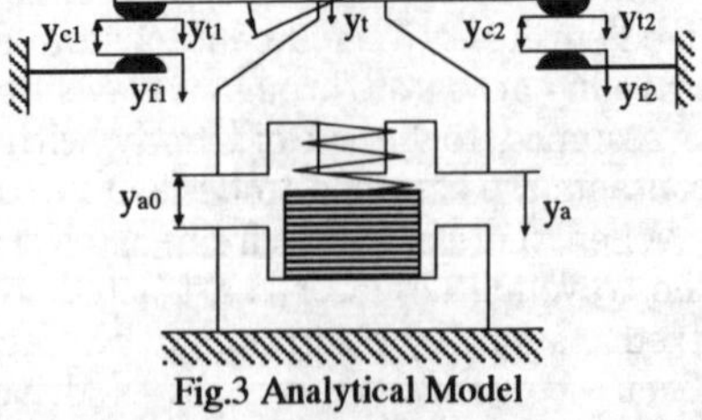

In this section, we discuss the contact bounce phenomena theoretically. The analytical model for the electrical switching system is shown in Fig.3. We derive the governing equations of the magnetic flux density B, motion of the traveling actuator, two fixed contacts and the traveling contact. Then we

Fig.3 Analytical Model

put these equations into nondimensional form, where nondimensional parameters are as follows: the natural period of the fixed contact $T = l_f^2\sqrt{\rho_f A_f / E_f I_f}$ for time, $B_0 = TV_0/NA_m$ for magnetic flux density, applied voltage V_0 for voltage, the mass of the point of the traveling contact m_t for mass, the length of the traveling contact l_t for x-axis displacement, the initial distance between the traveling and the fixed contact in side 1 y_{c1} for y-axis displacement. We also transpose the displacements of the fixed contacts: $y_{fi}(x,t) = \Phi_f(x)Y_{fi}(t)$, then using the eigenfunction expansion of the fixed contacts, the ordinary differential equations are obtained as follows:

$$\dot{B} = -a_0\{a_1(1 - a_2 y_a) + 1\}B + V, \tag{1}$$

$$(2 + M_b + M_a)\ddot{y}_a = PB^2 - k_{s2}(y_a + z_{s2}), \tag{2}$$

$$\ddot{Y}_{f1} + 2\mu_{fs}\dot{Y}_{f1} + \Omega_f^2 Y_{f1} = \gamma_f H_1 \sum_{n=1}^{\infty} \delta(t - \tau_{1n}), \tag{3}$$

$$\ddot{Y}_{f2} + 2\mu_{fs}\dot{Y}_{f2} + \Omega_f^2 Y_{f2} = \gamma_f H_2 \sum_{n=1}^{\infty} \delta(t - \tau_{2n}), \tag{4}$$

$$(2 + M_b)\ddot{y}_t = k_{s1}(y_a - y_t + z_{s1}) - H_1 \sum_{n=1}^{\infty} \delta(t - \tau_{1n}) - H_2 \sum_{n=1}^{\infty} \delta(t - \tau_{2n}), \tag{5}$$

$$(2 + M_b)\ddot{\theta} = -\frac{6}{L_t D}H_1 \sum_{n=1}^{\infty} \delta(t - \tau_{1n}) + \frac{6}{L_t D}H_2 \sum_{n=1}^{\infty} \delta(t - \tau_{2n}), \tag{6}$$

where

$$a_0 = \frac{RT}{N^2} \int \frac{dl}{\mu_0 \mu_r A_m(l)} \,, \quad a_1 = \frac{2y_{a0}}{\mu_0 \mu_r} \frac{1}{\int \frac{dl}{\mu_0 \mu_r A_m(l)}} \,, \quad a_2 = \frac{y_{c1}}{y_{a0}} \,. \tag{7}$$

and

$$P = \frac{T^2}{m_t y_{c1}} \frac{B_0^2 A_m}{2\mu_0} \,, \quad \mu_{fs} = \frac{\mu_f}{M_{fs}} \,, \quad \Omega_f = \frac{\omega_f}{\sqrt{M_{fs}}} \,, \quad \gamma_f = \frac{M}{M_{fs}} \Phi_f(1) \,, \tag{8}$$

$$D = \frac{l_f}{y_{c1}} \,, \quad \left(2\mu_f = \frac{c_f T}{\rho_f A_f} \,, \quad M = \frac{m_t}{\rho_f A_f l_f} \,, \quad M_{fs} = 1 + M M_f \Phi_f^2(1)\right).$$

R is the resistance of the electrical circuit, N refers to turns of the excitation coil, A_m is the area of the electromagnet pole face, y_{a0} and y_a are the initial distance between the traveling actuator and the fixed magnet and the displacement of the traveling actuator respectively. M_b and M_a are the mass of the body of the traveling contact and the traveling actuator respectively. k_{s1} and k_{s2} are the spring constant of the traveling and fixed spring respectively. z_{s1} and z_{s2} are the initial compressed length of the traveling spring and fixed spring respectively. ρ_f is the density; A_f and I_f are the area and moment of inertia of the cross section; E_f is Young's modulus; c_f is the damping coefficient; m_f is the mass of the point of the fixed contact. y_{fi} is the displacement of the fixed contact. y_{c2} is the initial distance between the traveling and fixed contacts in side 2. h_i is the impulse of impacts between the traveling and fixed contacts. y_t is the displacement of the center of gravity; θ is the angle; J_t is the moment of the inertia. Therefore the displacement of the points of the traveling contact is expressed as $y_{t1} = y_t + l_t \theta/2$ and $y_{t2} = y_t - l_t \theta/2$.

During the semielastic impact between the traveling and fixed contacts, the approximate formula regarding the velocities before and after, is assumed using the constant coefficient of restitution, in addition to the conservation law of the momentum.

Numerical results are obtained by solving the above equations. Fig.4 shows the time histories of the displacement of the traveling actuator and the points of traveling and fixed contact for the symmetrical fixed contacts($y_{c2} = 1$). Fig.5 shows the time histories for the asymmetrical fixed contacts ($y_{c2} = 0.8$).

From Fig.4, we can see a few impacts between the traveling and fixed contacts. When $t = 185$, they are in contact and oscillate together. In Fig.5, y_{t1} sticks to y_{f1} before y_{t2} sticks to y_{f2}. We can consider that the steady-state current flows through the contacts from $t = 202$.

These phenomena are shown as the impact map of the traveling contact from the viewpoint of nonlinear dynamics[5], as shown in Fig.6. We only show the impact map for the symmetrical fixed contacts and the conservative system, i.e., the coefficient of restitution $e = 1$ and the damping coefficient of fixed contact $c_f = 0$. Fig.7 shows the impact map for a more flexible fixed contact, i.e., $\Omega_f \times 0.1$.

In Fig.6, plots exist within an oval boundary. This means the energy supplied from the impact is limited. In Fig.7, plots are on an oval. This means that both contacts vibrate periodically.

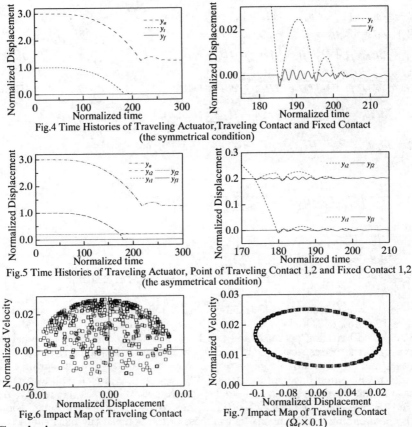

Fig.4 Time Histories of Traveling Actuator,Traveling Contact and Fixed Contact
(the symmetrical condition)

Fig.5 Time Histories of Traveling Actuator, Point of Traveling Contact 1,2 and Fixed Contact 1,2
(the asymmetrical condition)

Fig.6 Impact Map of Traveling Contact

Fig.7 Impact Map of Traveling Contact
($\Omega_f \times 0.1$)

5 Conclusion

In this research, the contact bounce phenomena in an actual electrical switching system were observed by experiments. They have been discussed theoretically by using not only the time histories of the displacement of the traveling and fixed contacts but also their impact maps. As a main result, the global properties of the bounce phenomena have been clarified for the typical contacts. The theoretical approach presented here is applicable to analysis of bounce phenomena of various contacts.

Furthermore it is expected to explain the stability of the bounce phenomena by investigating the characteristic value of the impact map.

References

[1] P.Barkan, A Study of the Contact Bounce Phenomenon, *IEEE Transactions on Power Apparatus and Systems* **PAS-86**-2 (1967) 231-240.
[2] D.J.Mapps, John W.McBride, G.Roberts, Peter J.White, Design Optimization of Rocker-Switch Dynamics to Reduce Pivot Bounce and the Effect of Load Current in Modifying Bounce Characteristics, *IEEE Transactions on Components, Hybrids, and Manufacturing Technology* **CHMT-9**-3 (1986) 258-264.
[3] John W.McBride, An Experimental Investigation of Contact Bounce in Medium Duty Contacts , *IEEE Transactions on Conponents, Hybrids, and Manufacturing Technology* 14-2 (1991) 319-326.
[4] Albert C.J.Luo and Ray P.S. Han, The Dynamics of a Bouncing Ball with a Sinusoidally Vibrating Table Revisited, Nonlinear Dynamics 10, 1996.
[5] T.M. Mello and N.B. Tufillaro, Strange attractors of a bouncing ball, *Am.J.Phys.* **55**-4 (1987) 316-320.

Non-Linear Electromagnetic Systems
V. Kose and J. Sievert (Eds.)
IOS Press, 1998

Contribution to the Calculation of Molten Metal Flow in Electromagnetic Channels

Ovidiu PEŞTEANU

"Transilvania" University of Braşov, 29 B-dul Eroilor, RO-2200 Braşov, Romania

Abstract. The electromagnetic and hydraulic calculations of the channel conveying molten metals is presented, by using a version of the k-ε turbulence model with nonisotropic viscosity. Finally, several distributions of electromagnetic and flow fields quantities determined for a channel transporting Al alloys are given.

1. Introduction

The electromagnetic channel is composed of a three-phase flat inductor situated under an inclined refractory-lined channel. Electromagnetic forces are exerted by interaction between currents induced in the molten metal and the travelling magnetic field, conveying the metal upwards in the channel. The advantages of inductive transportation are: electrically adjustable flow rate, precise dosing into moulds and purification of the conveyed metal [1].

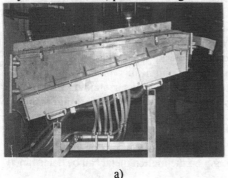

a)   b)

Fig. 1. Industrial electromagnetic channels for conveying molten: a) Al alloys b) cast iron.

2. Electromagnetic Calculation

The inductor and the channel lateral steel plates are replaced by the current sheet cs and the hatched ferromagnetic walls of infinite permeability represented only on the left side of Fig. 2a, due to the symmetry. For the rectangular domains free of currents $-H < y < 0$ and $y > d$ in Fig. 2a, the magnetic scalar potential is obtained using the variables-separation method [2], while in the domain $0 < y < d$ containing molten metal, the complex magnetic vector potential can be calculated by applying the finite-difference method for the equations

$$\Delta \underline{A} = - \mu_0 \underline{J} = \mu_0 \sigma (j \omega \underline{A} - \underline{E} + \operatorname{grad} \underline{V}), \quad \Delta \underline{V} = \operatorname{div} \underline{E}, \qquad (1)$$

where $\underline{E} = \mathbf{v} \times \underline{B}$ and $\underline{V}$ is an electrical potential determined inside the molten metal.

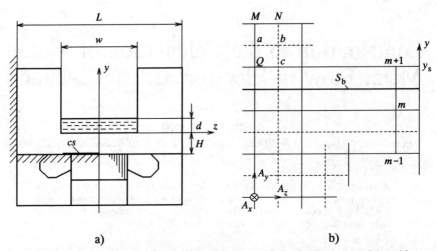

a)   b)

Fig. 2 a) Computational model in the channel transverse section b) Upper part of the grid.

Considering that for the travelling electromagnetic field, to the derivative with respect to x of a quantity there corresponds the phasor multiplication by $-j\pi/\tau_p$ (τ_p being the length of pole pitch), a two-dimensional grid will be used. As in flow calculation, the components A_y and A_z of the magnetic vector potential are determined for points of the staggered grid indicated in Fig. 2b by arrows, because thus they satisfy numerically more accurately the Coulomb gauge div $\mathbf{A} = 0$ and yield more precise values of the magnetic induction. At the molten metal boundary, e.g. its upper surface S_b (Fig. 2b) the following will apply [2]:
- the continuity of the tangential components of the magnetic induction computed by the values of the magnetic scalar potential, namely B_z for the grid point a and B_x for b yields the boundary values of the tangential components A_x and A_z for the grid points M and N;
- the boundary value of the normal component A_y for point a must be determined by the Coulomb gauge applied in discretized form for a volume element surrounding the point Q;
- at $y=y_s$ the component B_y obtained by means of the scalar potential expression results in a Fourier series depending on z of period $2L$; its coefficients are calculated numerically using the values B_y computed by the potential $\mathbf{A}$ for the staggered grid points c at $y=y_s$.
- the condition $J_y = 0$ on S_b yields the boundary value of potential V for grid point Q.
After field calculations the force density $\mathbf{f}$ acting on the molten metal can be determined.

3. Flow Calculation

Owing to nonlinearity, interlinkage and variable source terms of the flow equations, the steady flow will be determined as the final solution of a theoretical transient flow. The equations of the unidirectional flow with $\mathbf{v} = u\mathbf{i}$ in long channels can be written as

$$\frac{\partial \Phi}{\partial t} = \frac{1}{c_\Phi} \left[\frac{\partial}{\partial y} \left(\nu_{ty} \frac{\partial \Phi}{\partial y} \right) + \frac{\partial}{\partial z} \left(\nu_{tz} \frac{\partial \Phi}{\partial z} \right) \right] + S_\Phi . \qquad (2)$$

In the streamwise momentum equation: $\Phi = u$, $c_u = 1$, $S_u = f_x/\rho$, ρ being the metal mass density and allowing for damping at the free surface of the velocity normal fluctuations [3]:

$$\nu_{ty} = \frac{C}{C + 1.5\varphi} \frac{C}{C + 2\varphi} \nu_t, \quad \nu_{tz} = \frac{C + 2.5\varphi}{C + 2\varphi} \nu_t, \quad \nu_t = c_\mu \frac{k^2}{\varepsilon} . \qquad (3)$$

In Eq. (3) the damping function φ depends on the distance Y to the free surface by [3]:

$$\varphi = \frac{c_3}{(c_4 + Y/l)^2}, \quad l = \frac{c_\mu^{0.75} k^{1.5}}{\kappa \varepsilon}. \tag{4}$$

The equations of the turbulence kinetic energy k and its dissipation rate ε result by replacing in (2) $\Phi = k$, $S_k = G - \varepsilon$ and $\Phi = \varepsilon$, $S_\varepsilon = (c_1 G - c_2 \varepsilon)\varepsilon/k$, respectively, where [3]

$$G = \nu_{ty}\left(\frac{\partial u}{\partial y}\right)^2 + \nu_{tz}\left(\frac{\partial u}{\partial z}\right)^2. \tag{5}$$

The used model constants are [3,4]: $\kappa = 0.4$, $c_\mu = 0.09$, $c_k = 1$, $c_\varepsilon = 1.3$, $c_1 = 1.44$, $c_2 = 1.92$, $C = 1.5$, $c_3 = 0.1$, $c_4 = 0.16$. For a near-wall point P, with a wall distance n_P, the wall shear stress τ_w and the boundary value of ε are determined by [2,4]:

$$\tau_w = \rho U_{\tau P}^2 \frac{u_P}{|u_P|}, \quad U_{\tau P} = \frac{\kappa |u_P|}{\ln(30 n_P / h_r)}, \quad \varepsilon_P = \frac{c_\mu^{0.75} k_P^{1.5}}{\kappa n_P}, \tag{6}$$

h_r being the roughness height of the channel surface [5]. At the free surface of Al alloys, taking into consideration the oxide thin film of zero velocity, in the momentum equation of a near-film grid point P, the film shear stress τ_f is calculated as for smooth walls

$$\tau_f = \rho U_{\tau P}^2 \frac{u_P}{|u_P|}, \quad U_{\tau P} = \frac{\kappa |u_P|}{\ln(9 U_{\tau P} Y_P / \nu)}, \tag{7}$$

where Y_P is the film distance of P and ν the kinematic laminar viscosity. All other boundary conditions, inclusively on the y axis are zero normal gradients of variables.

The general explicit discretization equation of (2) for a grid point P can be written as

$$\Phi_P^{new} - \Phi_P = \Delta t \left(\sum a_{nb} \Phi_{nb} - a_P \Phi_P + S_c\right), \quad a_P = \sum a_{nb} - S_P, \tag{8}$$

where the summation is to be performed for all neighbour grid points of P denoted by the subscript nb; S_P represents the always negative coefficient of Φ_P in the linearized formula of the discretized source term for the point P [4]: $S_{\Phi P} = S_c + S_P \Phi_P$. Applying Brauer's theorem [6] for the error matrix of Eq. (8) yields for its eigenvalues λ

$$|\lambda - 1 + \Delta t\, a_P| \leq \Delta t \sum |a_{nb}|. \tag{9}$$

From $|\lambda| < 1$ results the approximative formula of the stability limited time step:

$$\Delta t = \frac{K}{\max\{a_P\}}, \tag{10}$$

where $\max\{\ \}$ denotes the greatest value from the last iteration, i.e. last time level, while $K = 0.8...0.9$ for the first tens of iterations and then $K = 1$.

The small cross-stream velocity components can be also computed, thus resulting a pressure field further used for determination of the free surface form. For a selected metal height d in the middle of the channel (Fig. 2a) the pressure for the upper exterior point $m+1$ near the y axis (Fig. 2b) is calculated by linear extrapolation: $p_{m+1} = 2p_m - p_{m-1}$ and the grid points with pressures smaller than p_{m+1} are eliminated in flow calculation [2]. Successive calculations of the electromagnetic field and turbulent flow are performed until sufficient convergence is reached.

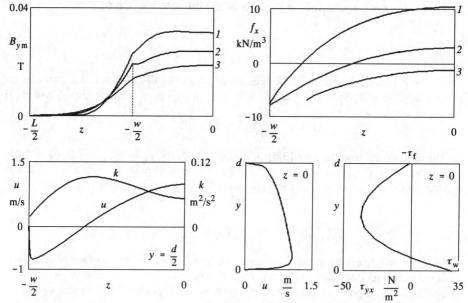

Fig. 3. Calculated distributions of the electromagnetic and flow fields quantities; $\tau_{yx} = \rho \nu_{ty} \dfrac{\partial u}{\partial y}$.

4. Results

Calculations were performed for the channel shown in Fig. 1a using the following data
- inductor: 50 Hz, amplitude of the line current density $7 \cdot 10^4$ A/m, $\tau_p = 0.156$ m;
- channel: inclination angle $18°$, $w = 0.175$ m, $H = 0.043$ m, $L = 0.4$ m, $h_r = 0.001$ m;
- Al alloy: $\sigma = 3.57 \cdot 10^6$ S/m, $\rho = 2400$ kg/m^3, $\nu = 0.56 \cdot 10^{-6}$ m^2/s, $d = 0.025$ m.

In Fig. 3 curves 1, 2, 3 of the amplitude B_{ym} and the density force component f_x are determined for the points of the first near-bottom grid line with $y = 0.00125$ m, the central line at $y = d/2$ and the upper last near-film grid line with $y = 0.02375$ m, respectively. Curve 3 of f_x is interrupted near the lateral wall because the corresponding grid points were eliminated in determination of the free surface form.

Using by means of Eqs. (3) a reduced viscosity ν_{ty} near the surface film, leads to small values of stress τ_f, thus explaining that the oxide film does not tear even at high velocities.

References

[1]   O. Peșteanu, Contribution to the Calculation of Electromagnetic Channels for Molten Metal Transportation, *European Transactions on Electrical Power Engineering* 2 (1992) No. 6, pp. 397-405.

[2]   O. Peșteanu, Simulation of Turbulent Molten Metal Flow in Electromagnetic Field. Fortschritt-Berichte VDI, Reihe 7, No. 237. ISBN: 3 18 143707 7. VDI-Verlag, Düsseldorf, 1994.

[3]   D. Naot, W. Rodi, Calculation of Secondary Currents in Channel Flow, *Journal of the Hydraulics Division*, ASCE 108 (1982) No. HY8, pp. 948-968.

[4]   W. J. Minkowycz *et al.*, Handbook of Numerical Heat Transfer. ISBN: 0 471 83093 3. John Wiley & Sons, New York, 1988, p. 225 a. pp. 281-283.

[5]   G. Preißler and G. Bollrich: Technische Hydromechanik. Vol. 1. Orderno.: 561 754 9. VEB Verlag für Bauwesen, Berlin, 1980, pp. 185-195 a. 277.

[6]   G. D. Smith, Numerische Lösung von partiellen Differentialgleichungen. ISBN: 3 528 08296 8. Vieweg Verlag, Braunschweig, 1970, pp. 95-100.

Non-Linear Electromagnetic Systems
V. Kose and J. Sievert (Eds.)
IOS Press, 1998

Compound Material for Electromagnetic Shielding Boxes

K. Weyand

Physikalisch Technische Bundesanstalt, Bundesallee 100, D-38116 Braunschweig, Germany

Yi Zhang , H. Bousack

Forschungszentrum Jülich GmbH (KFA), Postbox 1913, D-52428, Jülich, Germany

Abstract. A compound material based on aluminum and amorphous, soft magnetic alloys has been developed by means of gluing techniques which are commonly used in aerospace applications to produce metallic or fibrous laminates. It is well suited for the construction of high-quality containers shielding against electromagnetic fields from dc to some 100 kHz. Its magnetic properties can be modified over a wide range by varying the number and thickness of the laminate. The magnetic properties are not affected by mechanical processes such as turning, cutting or drilling. The material has also proved to be suitable for applications together with HTS-SQUIDs at 77 K.

1. Introduction

The shielding factor S_M - which is the ratio of the magnetic flux density B_o outside the container to B_i inside it - of a spherical or cylindrical container with an inner diameter D_c and made from magnetizable material is proportional to the ratio Q_{WD} of the wall thickness WT over the diameter D_c, and proportional to the permeability of the wall material μ_{rel} : $S_M \sim \mu_{rel} WT / D_c$. In order to achieve high shielding factors it is state-of-the-art practice to construct multiple-shell containers. If the distance between each shell is sufficient, the total shielding factor is found by multiplying the shielding factors of the single shells: $S_{M,tot} = S_{M1} \bullet S_{M2} \bullet \ldots\ldots S_{MN}$, where N is the number of shells. In ac fields, theory predicts a rising shielding factor for these materials as for any metal. This additional shielding effect is due to eddy currents and increases with electrical conductivity σ and frequency f [1] : $S_{EC} \sim f, \sigma$. However, in experiments, a decrease of S is observed in most cases. It is caused by quite different effects which may be material-dependent, or which are construction-dependent such as unavoidable opening or imperfect closing [2].

Containers made from crystalline Ni-Fe alloys and intended for shielding magnetic dc and low-frequency ac fields ($f < 1$kHz) have the advantage of good soft magnetic properties. However, when they are subjected to mechanical stress an elaborate annealing process is necessary afterwards. Amorphous alloys based on Fe, Ni and Co with permeability values in the order of $\mu_{rel} = 50,000 - 100,000$ can be used without additional treatment if their magnetostriction is low. But they are available only as thin strips less than 0.03 mm thick and with a width of about 50 mm [3],[4]. Bearing in mind the three effects of electromagnetic shielding mentioned above, we have designed a compound material based on aluminum and amorphous Ni-Fe-Co foils - where the aluminum separates the magnetic shells - which provides a high shielding factor independent of, or even increasing with, frequency, and which does not need any annealing.

2. Laminated wall structure

Using an algorithm given by Wadey [5], quite different lamination structures have been analyzed in order to find the limit at which lamination becomes pointless [6]. The limit depends on the permeability and on the Q_{WD} ratio of the container: with decreasing permeability the distance between the magnetic shells must be increased in order to produce a significant contribution to the total shielding factor. For permeability values $\mu_{rel} \geq 25,000$ or large Q_{WD} ratios, the optimum shielding factor is achieved if the distance between the magnetic shells is almost equal to their thickness. With a decreasing number of shells or decreasing Q_{WD} ratio, the optimum tends to

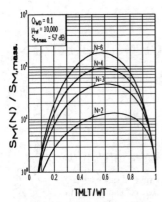

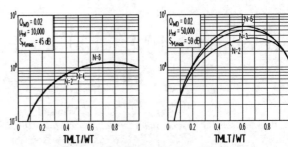

Fig.1   Shielding factor of containers with different laminated walls related to that with massive wall versus the total thickness $TMLT$ of N magnetic shells over wall thickness WT.

smaller distances with more magnetic material being required. Fig.1 shows the calculated shielding factor of different laminated walls versus the total amount of magnetic material in the wall. The wall parameters are noted in the diagram insets. With magnetic material of μ_{rel} = 10,000, for a thick wall about 10 % material is needed to achieve the values of the corresponding massive wall while about 50 % is required for a thin wall, where the grade of lamination does not have any effect and where S_M does not raise. In the case of highly permeable material, it obviously makes sense to laminate even a thin wall. To achieve the high optimum shielding factors given in Fig.1, it is vital that each shell is completely enclosed. Transitions from one shell to the next at a corner, for example, in particular where a cylindrical container is closed by a plate, will significantly reduce the shielding factor.

3. Container with laminated walls

Epoxy-bonded material from which the prototype containers had been fabricated [7] proved to be unsuitable for low temperature applications. Together with the necessity to produce a more uniform layer structure, this has led to a modified bonding procedure. Before being fixed together, amorphous material foils and aluminum foils are coated on both sides with a film of hot adhesive 10 µm thick which becomes active at a temperature above 100 °C. From the coated foils, cube-shaped boxes with a wall structure and dimensions as shown in Figs. 2a,b and the cylindrical container in Fig.2c designed for use with an HTS-SQUID were manufactured in the same way: the layers are wound on a plexiglass core, fixed on the outside by means of a brass clamp and then heated at 130 °C for about 3 hours in a hot-air oven without any special starting procedure or temperature program being observed. Due to pressure on the wall resulting from the different expansion coefficients of core and clamp, a mean adhesive film thickness of about 8 µm is achieved when the heating process is finished. Afterwards the core can easily be removed from the bonded material by cooling the whole arrangement to a few degrees below room tem-

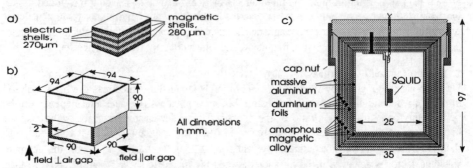

Fig.2a,b: Cube-shaped shielding box and structure of its walls. Fig.2c: Sectional view of a cylindrical box, designed for low temperature application.

perature. The final wall of the cube-shaped boxes contains about 40 % magnetic material, 40 % aluminum and 20 % adhesive, and has a fairly uniform layer structure. Before starting the bonding process on the cylindrical box, the layer arrangement was inserted into a well-fitting aluminum tube with a 1 mm thick wall for outside clamping, and another aluminum cylinder was inserted after removal of the plexiglass core to increase the eddy current shielding in the low frequency range. The multiple layer end plates are structured in the same way as the cylinder wall.

4. Measurement results
a) Room temperature measurements
Using a Helmholtz coil for field generation and an induction coil magnetometer for field measurement, the shielding factor versus frequency of two boxes with the dimensions given in Fig. 2 was compared with that of an Mu metal box of nearly equal dimensions, - curves "Mu" in Fig. 3. The "Vi" curves were measured on a box, the four magnetic shells of which are made of Vitrovac 6025X, whereas those denoted by "Ni" are for a box with the two outermost magnetic shells made of an NiFeB alloy which has a higher saturation magnetization but lower permeability. For the measurements shown in diagram 3a, the field was applied transversely to the air gap remaining between the box parts; diagrams 3b and 3c show the results with the field parallel to the air gap - see also Fig. 2b. All boxes yield higher shielding factors with the air gap and the field in parallel, because fewer flux lines can enter the box undistorted, but, due to the interrupted current path, eddy current shielding occurs at a higher frequency. A comparison of diagrams 3a and 3b shows that the difference in shielding factors of the three boxes observed below 10 Hz is partially dependent on different air gaps - with box "Ni" having the largest air gap. The curvature observed with composite material in the frequency range around 100 Hz also strongly depends on the air gap and may vary if the box has been dismounted. Above 20 Hz where the shielding factor of the Mu metal box decreases by about 10 dB/decade, and with increased flux density, boxes made from the composite material achieve considerably better results.

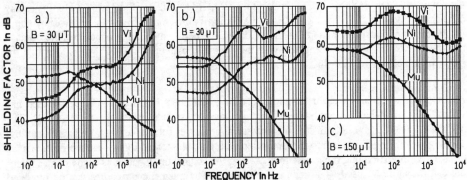

Fig.3 Shielding factor of cube shaped boxes versus frequency; (Vi) magnetic shells made from Vitrovac; (Ni) the two outermost magnetic shells made from an NiFeB alloy; (Mu) Mu metal box; 3a: field ⊥ air gap; 3b,3c: field ∥ air gap.

b) Low temperature measurements
When a shielding box is to be used with a SQUID, it is more important to consider its noise contribution than its shielding factor. Measurements of noise and shielding factor were carried out with an HTS-rf SQUID with a field sensitivity of about 100 fT/√Hz using 150 MHz electronics. The results obtained with the box shown in Fig. 2c were compared with those for two double-shelled Cryoperm boxes with about the same dimensions but of different charges, designed and annealed for high permeability at 80 K by Vacuumschmelze[8].The boxes with the SQUID inside were fixed in a fiber glass cryostat filled with liquid N_2. The composite box shows a field noise of 250 fT/√Hz in the frequency range above 10 Hz - Fig.4, plot "a". This is more than twice that obtained with the Cryoperm boxes; plots "b1" and "b2" in Fig.4. The assumption that the inner-

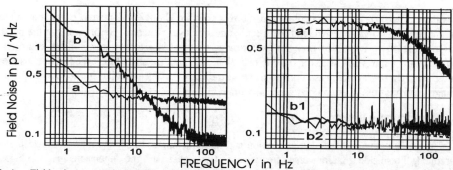

Fig.4   Field noise measured with HTS-rf SQUID versus frequency; lefthand diagram: SQUID inside a composite material box; a)with innermost aluminum cylinder; b) without aluminum cylinder; righthand diagram: SQUID inside a double-shell Cryoperm box; a1) aluminum cylinder inserted; b1),b2) different charges, without aluminum cylinder.

most aluminum cylinder is the source of the higher noise was checked by inserting an aluminum inlet into a Cryoperm box. The noise value of the arrangement 'Cryoperm box plus aluminum inlet' is shown in Fig.4, plot "a1".The noise contribution was found to be in the order of 750 fT/√Hz. Then the innermost aluminum cylinder of the composite box was removed. Fig.4, plot "b", shows the noise measurement with the modified composite box. The noise above 15 Hz is clearly reduced, and above 50 Hz it is lower than with the Cryoperm boxes. It turned out, however, that the 1/f noise without innermost aluminum cylinder is remarkably enhanced. The 1/f noise is obviously produced in the magnetic shells and it was partially screened before the aluminum tube was removed. It may be caused by the movement of the foils or flux jumps from foil to foil inside a shell, but most probably it is due to flux jumps from shell to shell at the corners of the box. Measurements of the shielding factor with the field transverse to the box axis show that the Cryoperm shielding factor is higher by about 45 dB at 1 Hz and by 15 dB at 200 Hz. Especially at lower frequencies the shielding factor of the composite material is found to be lower by about 40 dB than previously estimated. The large difference between evaluated and measured values may be caused by a lower permeability of the magnetic material at 80 K than is specified or, again, by imperfectly closed shells.

5. Conclusion
A composite material for electromagnetic shielding containers has been developed, which has several important features and some advantages over crystalline soft magnetic materials: The magnetic properties are not affected by mechanical processes such as turning, drilling or cutting. Turning and milling transverse to the lamination with titanium-coated tools is as unproblematic as drilling and thread-cutting transverse to and along the lamination. Containers made of this material can therefore be used as structural components. The magnetic properties can be modified over a wide range by varying the number and thickness of the laminate. With the shielding factor identical, the weight is considerably reduced in comparison with boxes made from massive soft magnetic material.

References

[1]   Kaden,H., "Wirbelströme und Schirmung in der Nachrichtentechnik",Springer, München, 1959
[2]   Mager,A.J., "Magnetic Shields",*IEEE Trans.Magn.*,**MAG-6**, 1970, pp.67-75
[3]   Vacuumschmelze GmbH, Amorphe Metalle VITROVAC®, *Data Sheet PV-006*, Hanau, 1989
[4]   Allied Corp., *Data Sheet* "MetGLAS Electromagnetic Alloys", Parsippany, N.J. USA, 1981
[5]   Wadey,W.G., "Magnetic Shielding with Multiple Cylindrical Shells",*Rev.Sci.Instr.*,**27**, 1956, pp.910-916
[6]   Weyand,K., "Verbundwerkstoff für elektromagnetische Abschirmungen", *Metall*, **47**,1993, pp.359-360
[7]   Vacuumschmelze GmbH, "Ringbandkerne aus VITROVAC", *Data Sheet VC-004*, Hanau, 1984
[8]   Vacuumschmelze GmbH: "CRYOPERM® 10", Data Sheet PW-004, Hanau, 1992
®   registered trademarks of VACUUMSCHMELZE GmbH, Hanau, Germany

Non-Linear Electromagnetic Systems
V. Kose and J. Sievert (Eds.)
IOS Press, 1998

Bistable Switching Cell Based on a Thin Semiconductor Film

Serghei L. GAIVAN and Petr I. KHADZHI
Institute of Applied Physics, Academy of Sciences of Moldova
Strada Academiei 5, 2028 Kishinev, Moldova

Abstract. A theoretical investigation is reported of steady-state non-linear transmission of resonant laser radiation by a thin semiconductor film under the conditions of two-pulse generation of excitons and biexcitons. The equations of state are derived for the description of bistable behavior of the amplitudes of the transmitted pulses and of the quasiparticle concentration, depending on the parameters of the exciting fields. The criteria of existence of an optical bistability are determined.

It is known [1] that a wide spectrum of nonlinear optical effects, including cavity-free optical bistability, can appear in a thin film consisting of two-level atoms. A medium of three-level particles provides even more extensive opportunities for the control of the transmission of incident radiation.

We shall report a theoretical study of resonatorless optical bistability observed in the transmission by a thin semiconductor film (TSF) under the conditions of generation of coherent excitons and biexcitons by photons belonging to two different pulses. We shall consider simultaneously two quantum transitions (two channels), in the exciton range of the spectrum and in the region of the M-band. The M luminescence band is the result of optical exciton-biexciton conversion and is shifted towards longer wavelengths relative to the exciton absorbtion band by an amount equal to the biexciton binding energy [2]. In CuCl-type crystals, the biexciton binding energy is of the order of 40 meV, so that a transition in the M-band region is characterized by a considerable detuning of a resonance from a transition in the exciton range of the spectrum. Moreover, the giant oscillator strength of the exciton-biexciton conversion process [2] may favor the appearance of optical nonlinearities even at moderate levels of excitation of a crystal.

We shall assume that a TSF, located in a vacuum, is exposed to two normally incident monochromatic laser pulses with the electric field envelopes E_{01} and E_{02} varying slowly with time and with the photon frequencies ω_1 and ω_2, respectively. We shall also postulate that the film thickness L is considerably less than the incident radiation wavelength. The interaction of the radiation fields with the excitons and biexcitons in a semiconductor can be described by the Hamiltonian [3]

$$H = -\hbar g(a^+ E_1^+ + a E_1^-) - \hbar\sigma(a^+ b E_2^- + a b^+ E_2^+), \qquad (1)$$

where a and b are the amplitudes of the exciton and biexciton waves, respectively; E_i^+ and E_i^- are, respectively, the positive- and negative-frequency components of the electric field of the i-th pulse ($i = 1, 2$); g is the constant of the interaction excitons with the field; σ is the constant of the optical exciton-biexciton conversion process.

The Hamiltonian (1) yields readily the Heisenberg (material) equations for the amplitudes a and b:

$$i\dot{a} = \omega_0 a - i\gamma_e a - g E_1^+ - \sigma b E_2^-, \quad i\dot{b} = \Omega_0 b - i\gamma_m b - \sigma a E_2^+, \qquad (2)$$

where ω_0 and Ω_0 are the self-frequencies of the exciton and biexciton states, respectively; γ_e and γ_m are the phenomenologically introduced damping constants of the exciton and biexciton modes, respectively.

Our task is to determine the amplitudes of the fields of the pulses transmitted by the film, E_1 and E_2, when the amplitudes of the pulses incident on the TSF, E_{01} and E_{02}, are given. We find from the Hamiltonian (1) and from the boundary conditions postulating constancy of the tangential components of the fields at the film-vacuum interface that

$$E_1^+ = E_{01}\exp(i\omega_1 t) + i\alpha_1 ga, \quad E_2^+ = E_{02}\exp(i\omega_2 t) + i\alpha_2\sigma a^+ b, \tag{3}$$

where $\alpha_1 = 2\pi\hbar\omega_1 L/c$; $\alpha_2 = 2\pi\hbar\omega_2 L/c$. The set of equations (2) and (3) solves the problem formulated above.

Next, we shall introduce dimensionless quantities $Y_1 = \sigma E_1/\gamma$, $Y_2 = \sigma E_2/\gamma$, $X_1 = \sigma E_{01}/\gamma$, $X_2 = \sigma E_{02}/\gamma$, $C_1 = \alpha_1 g^2/\gamma$, $C_2 = \alpha_2 g^2/\gamma$. For simplicity, we assume here that $\gamma_e = \gamma_m = \gamma$. Equating to zero the derivatives $\dot{a}$ and $\dot{b}$, under the conditions of exact resonance, $\omega_1 = \omega_0$ and $\omega_1 + \omega_2 = \Omega_0$, we obtain the following steady-state equations of state:

$$X_1^2 = n\left[1 + C_1 + \frac{X_2^2}{(1 + C_2 n)^2}\right]^2, \quad N = \frac{X_2^2}{(1 + C_2 n)^2}n, \tag{4}$$

$$X_1 = Y_1\left(1 + \frac{C_1}{1 + Y_2^2}\right), \quad X_2 = Y_2\left(1 + \frac{C_2 Y_1^2}{(1 + Y_2^2)^2}\right), \tag{5}$$

where $n = (\sigma/g)^2|a|^2$ and $N = (\sigma/g)^2|b|^2$ are the normalized densities of coherent excitons and biexcitons, respectively.

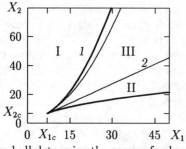

Fig. 1: Bifurcation curves in the parametric space (X_1, X_2) calculated for $C_1 = C_2 = 10$.

We shall determine the range of values of the pump amplitudes X_1 and X_2 in which the densities of excitons n and biexcitons N, as well as the amplitudes of transmitted pulses Y_1 and Y_2, can be multivalued functions of X_1 and X_2. The boundaries of saddles in the parametric space (X_1, X_2) are defined by the following system of parametric equations:

$$X_1^2 = n\left[\frac{4C_2 n(1 + C_1)}{3C_2 n - 1}\right]^2, \quad X_2^2 = \frac{(1 + C_1)(1 + C_2 n)^3}{3C_2 n - 1}, \tag{6}$$

(curve *1* in Fig. 1). We shall now consider the stability of our steady-state solutions. A system of parametric equations

$$X_1^2 = n\left[\frac{(2 + C_1)(1 + C_2 n)}{C_2 n}\right]^2, \quad X_2^2 = \frac{(2 + C_1 + C_2 n)(1 + C_2 n)^2}{C_2 n}, \tag{7}$$

determines the stability boundary of nonsaddle singularities (curve *2* in Fig. 1). The resultant bifurcation curves split the parametric space (X_1, X_2) into three regions. In

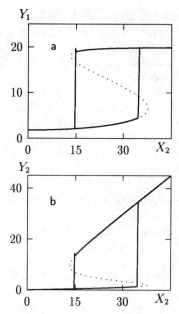

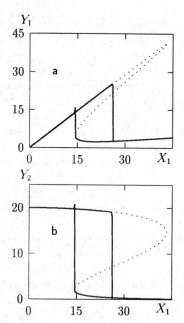

Fig. 2: Hysteretic dependencies of the amplitude Y_1 of the output radiation in the exciton part of spectrum (a) and of the amplitude Y_2 of the output radiation in the M-band region (b) on the pump amplitude X_2 in the M-band region, calculated for $X_1 = 20$ and $C_1 = C_2 = 10$.

Fig. 3: Hysteretic dependencies of the amplitude Y_1 of the output radiation in the exciton part of spectrum (a) and of the amplitude Y_2 of the output radiation in the M-band region (b) on the pump amplitude X_1 in the exciton range of spectrum, calculated for $X_2 = 20$ and $C_1 = C_2 = 10$.

region I there is one singularity, namely, a stable focus; in region II there are three singularities, one of which is stable and the other two unstable; finally, in region III there are two stable solutions and a saddle-type singularity.

Let us now consider possible types of hysteretic dependencies. We shall begin with the behavior of the functions $Y_1(X_2)$ and $Y_2(X_2)$ at a fixed value of X_1. The dotted curves in Fig. 2 represent expressions (5) and the solid lines obtained by numerical solution of a relevant system of equations on the assumption that $X_1 = 20$ and that X_2 is a triangular pulse with a maximum normalized amplitude $X_{20} = 45$ and of width $500\gamma^{-1}$. The values of the pump amplitude X_2 in the M-band region, where switching takes place from one branch to another, can be found graphically from Fig. 1. The behavior of the function $Y_1(X_2)$ and $Y_2(X_2)$ has a simple physical explanation. At $X_2 = 0$ the transitions occur only in the exciton range of spectrum (first channel) and this determines the output amplitudes $Y_1 = X_1/(1 + C_1)$ and $Y_2 = 0$. An increase in the pump amplitude X_2 in the M-band region results in conversion of an increasing proportion of excitons into biexcitons and, at the of switching to the upper branch, the number of excitons in the TSF falls to zero and the biexciton density rises strongly. When X_2 is increased still further, the exciton density remains zero and the biexciton density falls monotonically to zero in the range $X_2 \gg X_1$. This saturation effect makes the film transparent and this occurs in both channels.

Fig. 3 illustrates the case which is the reverse of that just discussed: the dependencies $Y_1(X_1)$ and $Y_2(X_1)$ are calculated numerically (solid lines) and analytically (dotted curves) on the basis of equations (5) assuming that $X_2 = 20$ and that X_1 varies adiabatically slowly from 0 to 45. It is evident from Fig. 3 that for $X_1 = 0$ the output signal Y_2 is equal to X_2, whereas in the absence of excitons the incident pulse in the M-band region is transmitted by the film under total transparency conditions. So long as $X_1 < X_2$, all the excitons generated in the TSF are converted into biexcitons. Therefore, as X_1 increased, the amplitude Y_2 falls very slowly, whereas Y_1 rises almost linearly. A further increase in X_1 results in an abrupt jump in the exciton density and a reduction in the biexciton density, the role of saturation becomes much weaker, and the behavior of the output radiation amplitudes is governed solely by transitions in the exciton range of spectrum. In the limit $X_1 \gg X_2$, we obtain $Y_2 = 0$, $Y_1 \approx X_1/(1 + C_1)$.

It follows from the above that the TSF can operate efficiently as a bistable integrated-optical switching cell when excitons are generated resonantly in the film and are simultaneously converted into biexcitons by photons from two different pulses. Figures 2 and 3 made it clear that the output of one channel is governed by pumping not only of this channel, but also of the other channel. This presents an opportunity for controlling bistable behavior of the first beam by the second.

We shall conclude with estimates of the relevant quantities. In the case of CuCl, we have $\gamma \sim 10^{11}$ s^{-1}, $g^2 = 3 \times 10^{46}$ J^{-1}s^{-2}, $\sigma^2 = 3 \times 10^{22}$ J^{-1}s^{-2}m^3, $\omega_1 \approx \omega_2 = 5 \times 10^{15}$ s^{-1} [3]. However, if $L = 10^{-8}$ m, we have $C_1 \approx C_2 \sim 10$. These values of the parameters C_1 and C_2 correspond to the critical values of the normalized amplitudes $X_{1c} \approx X_{2c} \approx X_{cr} = 7$. The pump intensities at which resonatorless optical bistability can occur should be of the order of $I = c\gamma^2 X_{cr}^2/8\pi\sigma^2 \sim 2 \times 10^8$ W/m^2. Using computer simulations we found that the switching times from one steady state to another are no more than 10 ps.

References

[1] V. Ben-Aryeh, C. M. Bowden, and J. C. Englund, Intrinsic optical bistability in collections of spatially distributed two-level atoms, *Physical Review* **A34** (1986) 3917-3926.

[2] P. I. Khadzhi, Kinetics of Exciton and Biexciton Recombination Radiation in Semiconductors. Ştiinţa, Kishinev, 1977.

[3] P. I. Khadzhi, Nonlinear Optical Processes in System of Excitons and Biexcitons in Semiconductors. Ştiinţa, Kishinev, 1984.

Non-Linear Electromagnetic Systems
V. Kose and J. Sievert (Eds.)
IOS Press, 1998

Characterization of Video Recording Heads

Helmut FRAIS–KÖLBL and Hans HAUSER
*Institut für Werkstoffe der Elektrotechnik, Technische Universität Wien,
Gußhausstraße 27–29, A–1040 Vienna, Austria*

Abstract. Measuring the magnetization curve of the assembled video–head leads to the detection of remanence and the prevention of un–intended tape erasing. The temperature distributions caused by eddy currents have also been visualized by high resolution thermography. The described measuring methods allow conclusions for the optimization of many parameters of the amorphous CoZrNb sandwich heads.

1. Introduction

To optimize the layer structure of the magnetic circuit and the preparation parameters of sandwich video head layers, several methods have been investigated. Figure 1 shows these heads schematically. The track width is 32 μm and the gap is about 30 nm. The development of amorphous magnetic materials gives an alternative to ferrites, which have been used for the magnetic circuit of recording/replay–heads in video recorders. The advantages of amorphous magnetic materials are the higher saturation flux density and the higher permeability. Compared to the ferrites the low specific resistance and therefore the eddy current loss is the main disadvantage [1].

The superior magnetic properties of amorphous $Co_{89,1}Zr_{7,9}Nb_{2,4}$–layers result mainly from annealing the sputtered film in magnetic fields of various directions. Thereby an induced uniaxial anisotropy and a domain structure refinement is caused. Already during the sputtering process a magnetic field can cause the corresponding orientation of the easy directions and the domain structure.

2. Experimental

The determination of $M(H)$–curves usually takes place by inductive methods. These methods give the average flux density B over the area of the measuring coil versus

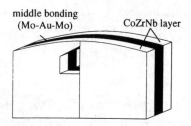

Fig. 1. Sandwich video head with amorphous CoZrNb layers and insulating ZrO_2 layers on a $CaTiO_3$ substrate

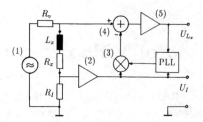

Fig. 2. Experimental set–up

the applied field H_a. These results depend on the geometrical arrangement. The $M(H)$–curve, magnetization versus the true inner field, can be obtained easily if the demagnetization factor N and the effective areas of the coil A_c and specimen A_s, or the factor $k_f = A_s/A_c$, respectively, are determined:

$$M = \frac{B - \mu_0 k_f H_a}{\mu_0(1 - N)} \tag{1}$$

$$H_i = \frac{\mu_0 k_f H_a - N B}{\mu_0 k_f(1 - N)} \tag{2}$$

Figure 2 shows the experimental set–up: Driven by the signal generator (1) with a resistance R_v, the magnetization curve of the complete magnetic circuit of the sandwich video head can be determined by subtracting (4) the voltage due to the resistance R_x of the pick–up–winding (inductance L_x) of the head from the total voltage drop over the winding. A signal U_I proportional (resistance R_I) to the current through the coil of the video head is amplified (2) and multiplied (3) with the output of a phase locked loop (PLL) and subtracted from the voltage on the coil. The PLL is controlled by the amplified (5) voltage U_{L_x} on the pick–up coil and the current through the coil. The voltage U_{L_x} is integrated with respect to the time and the result is proportional to the flux in the magnetic circuit. This automatic resistance–compensation allows measurements at very low frequencies (from 5 MHz down to 1 kHz). A similar method has already been applied successfully to the quality control of small ferromagnetic parts [2].

The distribution of the eddy currents in the area of the head gap can be determined by thermography. The local heating of the laminated layer causes stationary temperature (T) profiles. For known thermal properties of the material (thermal conductivity λ) these profiles can be determined by a high–resolution infrared microscope and analyzed by numerical methods, e.g. solving the equation

$$d\Phi = -\lambda \, \nabla T \cdot dA \tag{3}$$

for the heat current Φ through the area A. Much experience has been gained in different fields of thermal simulation [5], which are now exploited for eddy–current analyzation.

3. Results and Discussion

Figure 3 shows the resulting magnetization curves of two different sandwich video heads and a ferrite head at a frequency of 1 MHz. The difference between the unlaminated head and the sandwich video head is evident. Both the latter structure and the ferrite head have much lower hysteresis losses, but the ferrite has also a lower flux density.

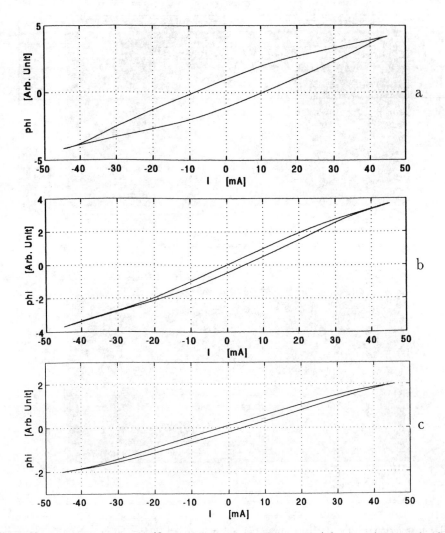

Fig. 3. Hysteresis measurement (flux versus magnetization current) for an unlaminated video head (a) a laminated sandwich head (b), and a ferrite head (c)

AGEMA 9000 thermography equipment has been used to determine the eddy current distribution. The spatial resolution was 17 μm/pixel, the temperature resolution was 0.1° K. The video heads were supplied with a sinusoidal voltage of 10 V peak at a frequency of 5 MHz in order to increase both losses and temperatures.

Figure 4a shows the temperature distribution for an unlaminated video head at 150 mA (Hysteresis shown in Fig. 3a). According to eddy–current losses, the temperature in the vicinity of the gap (highest flux density) is much higher than the windings´ temperature. Figure 4b) shows the result for the laminated sandwich head at 25 mA. Due to very low losses (see Fig. 3b) the only temperature maximum is located within the windings. The ferrite head showed no significant temperature increase.

The measuring methods as described above allow conclusions for the optimization of the following parameters: sputtering rate, anisotropy field, parameters for anneal-

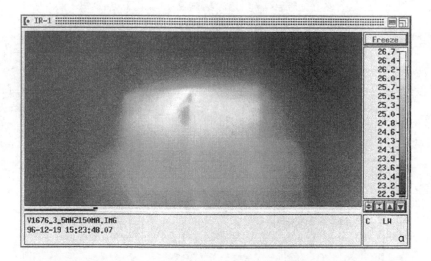

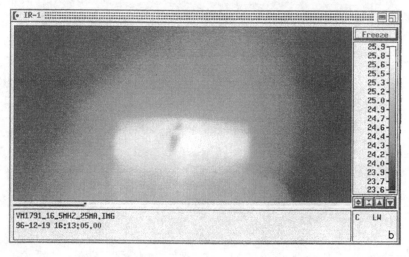

Fig. 4. Thermography of an unlaminated video head at 10 V, 150 mA, 5 MHz (a), and a laminated sandwich head at 10 V, 25 mA, 5 MHz (b)

ing in the magnetic field, mechanical deformation by substrate materials, geometry of the lamination and the air gap, etc. Measuring the magnetization curve of the assembled video–head leads — beneath general electromagnetic characterization — to the detection of remanence and the prevention of unintended tape erasing.

References

[1] H. Takano et al., IEEE Trans. Magn. **31** (1995), 2651.
[2] H. Hauser, Sensors and Actuators A **46** (1995), 588.
[3] H. Hauser and P. Fulmek, Nondestr. Test. Eval. **10** (1993), 359.
[4] H. Hauser, F. Haberl, J. Hochreiter, and M. Gaugitsch, Appl. Phys. Lett. **64** (1994), 2448.
[5] J. Nicolics and G. Hobler, Compel Trans. **13** (1994), 845.

Non-Linear Electromagnetic Systems
V. Kose and J. Sievert (Eds.)
IOS Press, 1998

Anharmonic Effects at Some Solid Surfaces Induced by Polar Impurities[*]

Jerzy KAPELEWSKI
Institute of Electronics Materials Technology,
133 Wólczyńska St., 01-919 Warsaw, Poland.

Abstract. A coherent way of modelling near-surface impurity-induced changes in electromechanical properties of complex crystal materials is demonstrated. The approach is formulated in terms of the pseudocontinuum lattice theory, and illustrated by application to some vibronic and off-center impurity systems. The potential of such systems for producing local polarisation-related phenomena of practical interest is discussed.

1. Introduction

Impurities, as introduced to the crystal, can affect material parameters in various ways, depending on the strength and a character of their influence on the lattice potential. Those situated at off-center positions occupy one of several possible sites determined by local symmetry, between which either tunnelling or jumps are allowed. That is why the impurity - defect configurations are not frozen but can undergo a reorientation under external or internal fields, producing microscopic changes in their crystal surrounding which, at high enough concentration of impurities, may lead to local phase changes of either ferroelectric or ferroelastic type. In a number of cases these kinds of changes can be modified in a controllable way. If the centres are of Jahn-Teller (JT) type, the corresponding changes essentially refer to mechanical parameters such as the elasticity tensor, whereas for the electrical ones (e.g. the dielectric susceptibility) a mechanism influencing dipole configuration is generally required (this is exemplified by the pseudo-JT version of the vibronic coupling). A common peculiar feature of the phenomena in consideration is an inherently microscopic character of their conditioning. It resides in an essential role of configurational symmetry of ligands in the near vicinity of the defect centre (with a frequent predominance of one of the irreducible representation (irr. repr.) of the crystal point group involved as it is the case for vibronic-like phenomena) as well as of anharmonic effects associated with relatively large values of microdisplacements in that region. These circumstances provide a framework for constructing any tractable and realistic model system of the properties in question. The discreteness of the material may, basically, be accounted for through introducing a nonlocality to the description of microscopic material characteristics. This can be formally performed by making use of a pseudocontinuum-type approach constructed appropriately to the existing size parameter as determining the basic volume element [1, 2]. The latter is usually taken in relation to the unit cell of the crystal or to its extended variety consisting of a group of the cells of a determined macroscopic symmetry. When treating point defects or more complex point irregularities, the natural

---

[*] This work was supported by the Committee of Scientific Research.

choice for the volume element is the region surrounding the defect centre with high enough magnitude of the dominant microdisplacements (with respect to that at an asymptotic distance). The mentioned pseudocontinuum approach, in its conventional form, constitutes in fact a method of interpolation based on the Shannon - Kotelnikov theory as applied to a periodic crystal lattice; the eventual microstructure being then treated in terms of Cosserat micromoments used to simulate relative intracell modes of microdisplacements. In a series of papers [see, e.g., 3, 4, 5], we have developed an alternative method, especially well suited to treat the defect-induced microdistortion in complex crystals. The method is derived from the microscopic lattice statics considerations and involves expansion of physical quantities in terms of locally symmetrized superpositions of polarisation vectors which span the appropriate subspace of the host lattice vibration modes. Such an approach allows one to express basic microdisplacement and force array characteristics in the framework a single lattice space composed of a set of the volume elements (each considered to be an integral whole). Owing to its flexibility and inherent ability to separate out particular local symmetries of microdistortion, the method has proved to be very useful in modelling distortion-related phenomena in crystalline materials, especially those associated with point defects having some internal structure. For greater clarity of the treatment presented below, all the indices (including those for irr. repr. involved), will be, hereafter, omitted. In the integrals over the phonon density and related functions the shorthand notation $\int dQ$ is used for $\sum_{\Gamma\gamma} dQ_{\Gamma\gamma}$.

2. An example of a pseudo-JT system

Following the general approach to model the defect - induced distortion as given in [5] we assume the local adiabatic potential for the system of pseudo - JT centres in the form:

$$W(\mathbf{l}) = \left[\Delta\sigma_z + vQ^{(\mathbf{l})}\sigma_x\right]c(\mathbf{l}) + \frac{1}{2}\sum K_{\textit{eff}}(\mathbf{l},\mathbf{l}')Q(\mathbf{l})Q(\mathbf{l}')c(\mathbf{l})c(\mathbf{l}') \qquad (2.1)$$

where c(l) stands for the defect occupation number, σ_i for the Pauli matrices and $K_{\textit{eff}}(\mathbf{l}, \mathbf{l}')$ for the effective harmonic interaction between particular volume elements, allowing for the coupling with the macroscopic strain tensor of the lattice [3]. The existence of the first term in (2.1) is known to be possible only for an odd type irr. repr. corresponding to the dipole - type microdistortion. Then the system free energy can be described by means of the formula:

$$\mathcal{F}(\mathbf{l}) = U(\mathbf{l}) - \beta^{-1}c(\mathbf{l})\ln Z(\mathbf{l}) \qquad (2.2)$$

where $$U(\mathbf{l}) = <W(\mathbf{l})>, \quad \beta^{-1} = kT, \qquad (2.3)$$

with k being the Boltzmann constant,
stands the ensemble-averaged form of (2.1), and Z(l) - for the respective statistical sum. For the case of double - well centres, from (2.2) we obtain :

$$\mathcal{F}(\mathbf{l}) = \frac{1}{2}\sum K_{\textit{eff}}(\mathbf{l},\mathbf{l}')Q(\mathbf{l})Q(\mathbf{l}')c(\mathbf{l})c(\mathbf{l}') - \beta^{-1}\ln\left\{2\left[1 + \text{ch}\left(\beta^{-1}\sqrt{\Delta^2 + v^2Q^2(\mathbf{l})}\right)\right]\right\} \qquad (2.4)$$

Denoting for particular lattice planes :

$$<Q(\mathbf{l})> = Q(n), \; <c(\mathbf{l})> = c(n), \; <c(\mathbf{l})K_{\textit{eff}}(\mathbf{l},\mathbf{l}')c(\mathbf{l}')> = K_{\textit{eff}}(n,n'), \; \mathbf{l} = (\mathbf{l}_\parallel, n) \qquad (2.5)$$

the equilibrium condition reads

$$\frac{\partial F}{\partial Q(n)} = -\frac{v^2 Q(n)c(n)}{\sqrt{\Delta^2 + v^2 Q(n)}} \text{th}\left[\frac{\beta}{2}\sqrt{\Delta^2 + v^2 Q(n)}\right] + \sum K_{\text{eff}}(n,n')Q(n') = 0 \qquad (2.6)$$

In the limit of small values of Q it leads to the secular equation:

$$\det\left(\hat{A}\right) = 0 \qquad (2.7)$$

with
$$\mathbf{A}_{n,n'} = -\frac{v^2}{\Delta}c(n)\text{th}\left(\frac{\beta}{2}\Delta\right)\delta_{n,n'} + \mathbf{K}_{\text{eff}}(n,n') \qquad (2.8)$$

from which the local critical temperature for the defect system in consideration, can be evaluated. Note that the values of $c(n)$ as defined in (2.5) describe the concentration of the impurities being considered. Its dependence of the distance *(n)* from the surface can be, basically, controlled according to the doping technique used.

3. The case with defect - induced local anharmonicity

Let us now consider the case with local defect-induced anharmonicity, restricting ourselves to small displacements up to the fourth order in Q, of the central cell region. In view of the centrosymmetric character of the "lattice" (irrespective of the actual microscopic symmetry of the structure in consideration) the adiabatic quasiharmonic potential of the system does not consist of terms with odd powers of Q, resulting in the form,

$$U(\mathbf{l}) = U_\varepsilon[Q(\mathbf{l})] + \frac{1}{2}\lambda\varepsilon^2(\mathbf{l}), \qquad (3.1)$$

with
$$U_\varepsilon[Q(L)] = c(\mathbf{l})\left\{\left[\frac{A}{2}Q^2(\mathbf{l}) + \frac{B}{4}Q^4(\mathbf{l})\right] + \frac{1}{4}\sum_{\mathbf{l}'} c(\mathbf{l})c(\mathbf{l}')\mathbf{K}(\mathbf{l},\mathbf{l}')[Q(\mathbf{l}) - Q(\mathbf{l}')]^2 + h\varepsilon(\mathbf{l})Q(\mathbf{l})\right\}$$

where the parameter A originates from contribution to the harmonic force constants from the volume element containing defect (together with "self-interaction" term K(**l**, **l'**)), ε for the locally symmetrized strain tensor, B for the strength of the local anharmonic interactions, and h for the piezoelastic tensor. For any of the activated portions of the crystal, the molecular field (m.f.) potential can then be written:

$$U_{m.f.} = \frac{1}{2}\sum_n N_d(n)c(n)Q(\mathbf{l})\sum_{n'} K(n,n')Q(n') + \sum_{\mathbf{l}} c(\mathbf{l})W_0[Q(\mathbf{l})] \qquad (3.2)$$

with $N_d(n)$ being the number of the active centres in the n-th lattice plane, and

$$W_0[Q(\mathbf{l})] = U(\mathbf{l}) - Q(\mathbf{l})F(n) \qquad (3.3)$$

where the average local force $F(n)$ is given by the static distortion, according to the formula:

$$F(n) = \sum_{n'}[K(n,n')Q(n') + h(n,n')\varepsilon(n')] \qquad (3.4)$$

For the local entropy contribution to the density of the free energy, as taken at the n-th lattice plane , we thus obtain : $\qquad (3.5)$

$$TS(n) = T\ln\int dQ \exp\{-\beta U(Q) + \beta QF(n)\}\left[1 - \beta\frac{B}{4}Q^4\right] = \frac{[F(n)]^2}{2A}\left(1 - \frac{3BT}{A}\right) + \frac{B}{4A^4}[F(n)]^4$$

where TS^0 (n), being the term independent of $<Q>$, becomes

$$TS^0(n) = T \ln \int dQ \exp\left[1 - \frac{\beta A Q^2}{2}\right]\left[1 - \frac{B}{4}Q^4\right] = \frac{2\pi T}{\beta A}\frac{\pi}{A}\left(1 - \frac{15B}{4(\beta A)^2}\right) \qquad (3.6)$$

The free energy density is then given by:

$$\mathcal{F}(n) = \frac{1}{2}Q(n)\sum_{n'}K(n,n')Q(n') - TS(n) - TS^0(n) \qquad (3.7)$$

With the equilibrium condition $\frac{\partial \mathcal{F}}{dQ} = 0$ it leads to the equation

$$F(n)\left[1 - \frac{K(n,n)}{A}\left(1 - \frac{3BT}{A^2}\right)\right] + \frac{BK(n,n)}{A^4}[F(n)]^3 = \sum_{n'}h(n,n')\varepsilon(n') \qquad (3.8)$$

This is a fundamental set of equations from which, by means of (3.4), the full set of $Q(n)$ can be evaluated. The distinctive features of the particular solution of Eq. (3.8) can readily be seen, when restricting itself to the two limiting cases corresponding respectively to that of small values of Q and vanishing values of piezoelastic coupling (h → 0). For the first one the nonlinear term in (3.8) can be disregarded, to give :

$$Q(n) = \alpha(T,n)\sum_{n'}N(n,n')\varepsilon(n') = \alpha(T,n)\sum_{n'}M(n,n')\sigma(n') \qquad (3.9)$$

where $M = NC^{-1}$, $N = K^{-1}h$, $\alpha(T,n) = \frac{\gamma(T,N)}{1 - \gamma(T,N)}$, $\gamma(T,N) = \frac{K(n,n)}{A}\left(1 - \frac{3BT}{A^2}\right)$

with σ being the near-surface stress tensor.
In the second case we obtain

$$Q(n) = 3A[T_c(n) - T]K^{-1}(n,n) \qquad (3.10)$$

where the parameter $T_c(n) = \dfrac{A^2}{3BK(n,n)}$

can be interpreted as a local critical temperature of the system.

The local anharmonicity as described above, is known to be representative for off-center impurity-defects. The, then dominant, axial symmetry of microdistortion and, consequently, that of Q, results, in analogy with the previous case, in a near-surface polarisation field, which is determined here by the foregoing formulae (Eqs. 3.4 and 3.8).

The phenomena treated above are of considerable practical interest in many technological applications. It is especially profitable in surface acoustic waves (SAW) devices technology for providing a useful technique to modify the surface acoustic impedance of the substrate surfaces, as well as its temperature dependence, through an appropriate adjusting the impurity content and implantation conditions. Such investigations are now under way in our group.

References

[1] A. S. Maugin, Nonlocal theories or gradient-type theories: a matter of convenience?, Arch. Mech. 31.01.15, 1979.
[2] E. Kröner, B. K. Datta, Nichtlocale elastostatic Ableitung aus der Gittertheorie, Zs. f. Phys., 196,203,1966.
[3] J. Kapelewski, A one-lattice formulation of crystal micro-elasticity. I. The crystals with a microstructure, Int. J. Engn. Sci., 27,9,1077,1989.
[4] J. Kapelewski, A defect-induced, near-surface ferroelasticity, Ferroelectrics, 81,157,1121 (1988);
[5] J. Kapelewski, Nonlocal modelling in some electroelastic materials, in: Electromechanical Interactions in Deformable Solids and Structures, ed. Y. Yamamoto, K. Miya, Elsevier, p. 335, 1987.

Non-Linear Electromagnetic Systems
V. Kose and J. Sievert (Eds.)
IOS Press, 1998

A Theoretical and Experimental Investigation of a High Permeability Parallel Flux Parametric Oscillator

D. Delic and D. Kearney

School of Physics and Electronic Systems Engineering, University of South Australia
Ingle Farm 5095, Australia

Abstract. The paper presents two experimentally verified lumped parameter models of a parallel flux parametric oscillator constructed from two high permeability amorphous magnetic cores. The previously published literature on the analysis of this type of system has focussed on low permeability materials. The behavioural dynamics of the system has been examined and three different stable forms of oscillations have been discovered, and confirmed by experiment. Conditions for these various modes of oscillations to occur have been mathematically analysed. And it can be shown that a simple linear analysis can be used to predict the value of capacitance required for parametric oscillations.

1. Introduction

The phenomena of parametric oscillations occurs in many physical systems. The principle underlying its mechanism could be generally stated as : *If the parameter of a storage element in a oscillatory system changes periodical at twice the natural frequency f of the system, then energy is imparted to the system causing it to oscillate at frequency f* . In this paper first a simple non-linear parametric equivalent circuit model is analysed. Next a parallel flux current driven circuit is presented containing a high permeable and highly non-linear amorphous magnetic B/H curve characteristic. Techniques are presented to find the value of C required to sustain parametric oscillations. The results are confirmed experimentally. The purpose of characterising this system, has significance in establishing the feasibility of developing a new very low current RCD magnetic sensor.

2. Simple Non-Linear Parametric Resonant Circuit

The classic electric circuit demonstrating parametric oscillations is shown in Figure 1. In this case the inductance parameter has to vary at twice the frequency the capacitor is tuned to. This kind of oscillations can be generally described by a second-order linear differential equation with periodically time varying coefficients of the Hill's or Mathieu's [1] type.

$$\frac{d^2v}{dt^2} + \frac{1}{R_L C}\frac{dv}{dt} + \frac{v}{L(t)C} = 0, \quad L(t) = L_o + \hat{L}_1 \cos(2\omega t)$$

$$\frac{d^2y}{dz^2} + \left(a - b\cos(2z)\right)y = 0, \quad a = \frac{\omega_o^2}{\omega^2} - \frac{k^2}{4\omega^2},$$

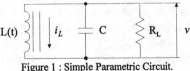

$$L(t) \quad i_L \quad C \quad R_L \quad v$$

Figure 1 : Simple Parametric Circuit.

$k = \dfrac{1}{R_L C}$, $\quad \omega_o^2 = \dfrac{1}{L_o C}$, $\quad b = \dfrac{m\omega_o^2}{2\omega^2}$, $\quad z = \omega t$. $\quad m$ is the modulation index, L_o is the quiescent value of the inductance, $\hat{L}_1$ is the peak amplitude of the change, ω_o is the resonant frequency of the circuit for $m \to 0$. For parametric oscillations to exist, considering a loss-less circuit $k = 0$, then the two conditions $\omega_o^2 = \omega^2$ and $m = \dfrac{\hat{L}_1}{L_o} > 0$ must be met.

3. Current Driven Parallel Flux Circuit

Two equivalent circuit realisations of the previous parametric circuit is shown in Figure 2 & 3 respectively. In this work the rational-fraction approximation to the B/H curve was used. For further details of our model see [4,6].

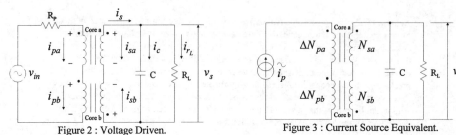

Figure 2 : Voltage Driven.   Figure 3 : Current Source Equivalent.

Here, v_{in} represents a primary sinusoidal excitation voltage, i_p is the primary sinusoidal independent current source, i_s is the secondary current, v_s is the output secondary voltage, the coil resistance (not shown) $R_s = 0$, and load resistor $R_L \sim \infty$. Toroidal cores a and b are equivalent with the same magnetisation characteristic, physical geometry and primary and secondary turns.

This section provides a concise explanation of the parametric oscillation phenomena in these circuits. Because the secondary windings of both cores are connected in phase opposition, any alternating energy source injected into the primary of both cores will induce a voltage on each secondary of equal magnitude, hence the net voltage measured across the secondary of both cores under ideal conditions will be zero; they will cancel out. In practise however because the cores will not be perfectly matched a small secondary current flows. Under certain conditions dependent on the magnitude of the input excitation (which sets up the bias position on the curve), B/H curve properties (permeability and degree of non-linearity), and finally capacitance (which sets the phasing of the secondary and primary current relationship respectively), the interaction of currents i_p & i_s push and pull each of the cores continuously into and out of saturation, hence changing respective instantaneous inductances. Each core contributes a disproportionate amount to the net instantaneous inductance seen at the output. This could be due to unequal core parameters, injected electrical disturbances or noise on the secondary side, or differing remanent flux offsets residing in each core. The result however is a vestigial inductance signal oscillating at double the frequency of the primary excitation signal. Once the perceived net secondary inductance of the secondary of the transformer is made to modulate at twice the natural frequency of the capacitive tuned circuit, parametric amplification and oscillations result. Previous studies [2,3,5] on low permeability cores have concentrated on obtaining the correct value of C using non-linear analysis of the equivalent circuit. However experimental results reported in this paper show that there is no need to solve a non-linear parametric differential equation to obtain this value of C for higher permeability materials. It has been found both using simulation and experiment the value of C that maximises the output secondary voltage for a small signal analysis is very close to the best choice to sustain non-linear parametric oscillations. In the next section a brief outline of this method is given. For a more detailed study see [6].

4. Theoretical and Experimental Results

This section shows how to map the parallel flux circuit to the simple non-linear parametric model in Figure 2. For Figure 3, the equation describing the primary circuit serves no more then to establish the magnetic flux points in corresponding cores. The secondary equation is a second order non-linear differential. From this it can be shown [6] that for parametric excitation we require $L_s(i_s) = N_s A_c B_s / i_s = L(t) = L_o + \hat{L}_1 \cos(2\omega t)$, where $L(t)$ is the time varying inductance defined as the perceived secondary inductance $L_s(i_s)$, which is a function of secondary current i_s. This expression is purely dependent on the existence of a secondary current to set up the double frequency inductance conditions for parametric oscillations. It is then possible to find the value of C required for parametric oscillations [6].

Another approach using a linear approximations to the B/H curve is recognising that for *small* primary current driving force excitation we can assume the core characteristics are linear. If we represent each core by its own second order linear differential equation we find that tuning for each core and hence the whole system is determined by : $C = \dfrac{L_m m_{lin}}{2N_s^2 A_c \omega_o^2}$

where m_{lin} is equal to the inverse of the absolute permeability μ for the magnetic material.

To *fully* explore the behaviour of the system a numerical study of the non-linear model (Fig. 2) has been used. The non-autonomous non-linear system of differential equations are as follows.

$$\frac{dB_p}{dt} = \frac{v_{in}}{N_p A_c} \sin \omega t - \frac{R_p L_m}{2N_p^2 A_c} \zeta_p(B_p, B_s, t) B_p$$

$$\frac{dx}{dt} = -\frac{1}{RC} x - \frac{L_m \zeta_s(B_p, B_s, t)}{2N_s^2 A_c C} B_s$$

$$\frac{dB_s}{dt} = x$$

Note : B_p & B_s are the additive and subtractive magnetic flux components seen in both cores, while $\zeta_p(B_p, B_s, t)$ and $\zeta_s(B_p, B_s, t)$ represent the non-linear functions [6] relating B_p & B_s to the primary and secondary side respectively. The test sample used in this paper has the following specifications : Core material - Vitrovac 6025F, 16 mm x 12 mm x 6 mm, physical parameters N_p = 150 t, N_s = 500 t, A_c =3e-6 m^2, L_m =0.0511 m. Test parameters for all simulations and experiments are $\hat{v}_{in}$ =26 V, ω=100π, R_L =1 MΩ. The results of these simulations are shown in Figure 4, the error tolerance is 1e-6, and initial conditions are $\dot{B}_{s(0)} = 0$, $B_{p(0)} = 0$, $B_{s(0)} = 0.1$. The experimental results are shown in Figure 5. For parametric mode 1 (w1/m2.2) R_p =35 kΩ, C=0.22 uF, parametric mode 2 (w2/m1.2) R_p = 5 kΩ, C=1.6 uF and subharmonic mode (w3/ch2) R_p =10 kΩ, C=5 uF.

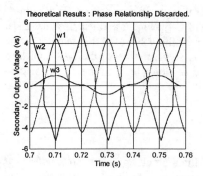

Theoretical Results : Phase Relationship Discarded.

Figure 4 : Simulated Results.

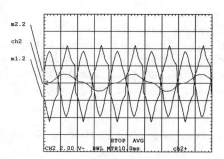

Figure 5 : Experimental Results.
Time base = 10 ms, Vertical axis : v_s.

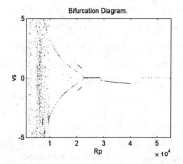

Bifurcation Diagram.

Figure 6 : N_p =150, N_s =500, C=0.22uF,
$\hat{v}_{in}$ =26V. 200 periods, first 150 ignored.

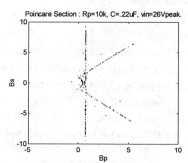

Poincare Section : Rp=10k, C=.22uF, vin=26Vpeak.

Figure 7 : 600 periods shown.
B_p - additive flux, B_s - subtractive flux.

5. Discussion and Conclusion

For large values of R_L the output voltage of the circuit exhibits basically five modes of behaviour depending on the magnitude of the forced excitation: linear tuning, parametric oscillations (Fig. 4 : {w1, w2}) & (Fig. 5 : {m1.2, m2.2}), quasi-periodic oscillations, sub-harmonic oscillations (Fig 4 : w3) & (Fig. 5 : ch2) and chaotic or abnormal behaviour (Figures 6 & 7). Figure 6 illustrates that parametric oscillations are sustained when 32 kΩ < R_p < 42 kΩ and that the oscillator displays the classical period-doubling route to chaos. At the onset of instability corresponding to R_p = 10 kΩ, the Poincaré section generated (Figure 7) shows this chaotic behaviour and structure.

References

[1] N. Minorsky, "Nonlinear Oscillations", Princeton, N. J. : D. Van Nostrand Company, 1962.
[2] Z. H. Meiksin, "Parallel-Flux Parametric Voltage Regulator and Comparison with Orthogonal-Flux Parametric Voltage Regulator," IEEE Trans. Ind. Appl., vol. IA-10, no. 3, pp. 428-430, May/June 1974.
[3] T. Kitagawa, "Almost-Periodic Oscillation and Generation of Chaos in a Parametric Excitation Circuit Having an External Force," Electron. Comm. Japan, pt. 1, vol. 71, no. 9, pp. 27-34, 1988.
[4] F. T. Widger, "Representation of Magnetisation Curves Over Extensive Range by Rational-Fraction Approximations," Proc. IEE, vol. 116, no. 1, pp. 156-16 Jan. 1969.
[5] F. Tatsuta and M. Tadokoro, "Chaotic Behaviour in Parametric Oscillations," Trans. of the Institute of Electrical Engineers of Japan, vol. 114, pt D, no. 2, pp. 157-64, Feb. 1994.
[6] D. Delic, "A High Permeability Parametric Oscillator", University of South Australia Research Report.

Non-Linear Electromagnetic Systems
V. Kose and J. Sievert (Eds.)
IOS Press, 1998

Magnetic Shielding in Alternating Field by Grain Oriented Silicon Steel Sheets

Yasuo OKAZAKI

Dep. of Electrical & Electronic Engineering, Gifu University,
Yanagido 1-1, Gifu, 501-11, Japan

Abstract. Shielding effectiveness in an alternating magnetic field was studied experimentally and analytically on both conventional grain oriented silicon steel GO and high permeability grain oriented silicon steel HIB. Magnetic shielding for an alternating field was more complex than that for a static field due to frequency dependent permeability and eddy currents. Analytical results by 3-D FEM showed discrepancy when compared to experimental results.

1. Introduction

Magnetic shielding technology has been gaining in importance along with the advancement of scientific technology and is inevitable for electromagnetic apparatus. Recently, an alternating magnetic field at around commercial power frequency, which is defined as extremely low frequency, has become a crucial subject of the electromagnetic environment. For magnetic shielding performed by ferromagnetic materials, shielding effectiveness in a dc or static field has been clarified theoretically and experimentally based on the high permeability of ferromagnetic materials. Magnetic shielding for higher frequency over 10 kHz is considered to be carried out effectively by metallic magnetic materials due to low electric resistivity. Magnetic shielding for an alternating field of 50-60 Hz would be more complex than that for a static or higher frequency field due to frequency dependent permeability and eddy currents[1-3].

The author has reported excellent dc or static shielding effectiveness on grain oriented silicon steels of conventional steel (GO) and high permeability (HIB) compared with high permeability 78Ni-Fe (Permalloy) considering their cost and construction of shield[4]. Grain oriented silicon steel has extremely high permeability in the direction of easy magnetization <100> axes ; on the other hand, other directions show relatively lower permeability. Accordingly, for large shields GO or HIB sheets were constructed to shielding panels laminated with <100> axes alternately perpendicular. The author has also shown that shielding effectiveness in the <100> direction of grain oriented silicon steels decreases with the increase of frequency from 100 Hz to 20 kHz in cylindrical shields [2]. Analytical results from theoretical formula and FEM simulation showed a complicated effect of ac magnetic field frequency [3].

Here, alternating magnetic shielding effectiveness on two grain oriented silicon steels GO and HIB in relatively large 450 mm cubic box shields is shown experimentally at the magnetic field frequency of 50 Hz. 3-D FEM analysis by the A-φ method is performed to simulate the above experimental results.

2. Experiment

Two sheets of each shielding material of 0.23 mm thick HIB (23P) and 0.35 mm thick GO (35G) were laminated into panels with their easy magnetization <100> axes alternately perpendicular. The ac permeability of 35G and 23P at 50 Hz measured by the Epstein method in the rolling direction <100> are shown in Fig.1 as m-H curves. The permeability of 23P was shown to be about 30% higher than that of 35G up to the maximum. An alternating magnetic field at 50 Hz were produced by a set of Hemholtz coils of 1.8 m diameter and applied to shielding boxes which were set so as to face one plane perpendicular to the field in the center of the Helmholz coils. The field was virtually parallel within a 500 mm cubic space. A shield of a cubic box of 450 mm length constructed by laminated panels was set in the coils which imposed ac fields of 5 to 100×10^{-7} T at 50 Hz. The shielding field was measured by a three axial gauss meter at the center of the boxes. The shielding effectiveness S was calculated as $S(dB) = 20$ $log(Be/Bi)$ where Be and Bi are the external and internal field of the shield, respectively.

The experimental results are shown in Table 1. S (dB) increased with the external field strength till 100×10^{-7} T for both 35G and 23P shields which would correspond to the increase of ac permeability of 23P and 35G till 100×10^{-7} T as shown in Fig.1. A shield of 35G showed higher shielding effectiveness than that of 23P due to greater thickness. Table 1 shows the effectiveness, $S1$ and $S2$ at $Be = 25$ and 100×10^{-7} T, respectively. $S*1$ and $S*2$ in Table 1 for 23P are calculated to be $t = 0.35$ mm, according to the equation[5], $S (Be/Bi) = 1 + a \mu t / D$ (a : constant, μ : permeability, t : thickness, and D : length of a shield). Table 1 indicates higher permeability of 23P did not work so effectively at lower field strength. Considering 30% higher μ of 23P than 35G, $S*1$ and $S*2$ should have been 14 dB and 17 dB, respectively.

Table 1 Experimental shielding effectiveness

material	thickness	$S1$	$S2$	$S*1$	$S*2$
	(mm)	(dB)		(dB)	
35G	0.70	13	15	-	-
23P	0.46	10	14	13	17

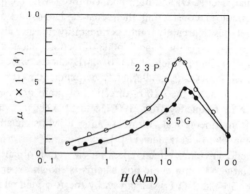

Fig.1. ac permeability for the shielding materials

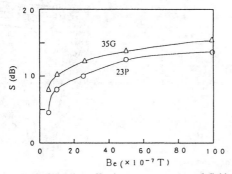

Fig.2. Shielding effectiveness to ac external field

3. 3-D Field Analysis

3-D field analysis was carried out by the A-ϕ method for the shielding boxes of GO and HIB panels respectively, which simulated the above experiments. Calculations were done in the case of external fields of 25 (S1) and 100 (S2)$\times 10^{-7}$ T at 50 Hz. Governing equations for the 3-D finite element method are as follows;

$$B = \operatorname{rot} A \qquad (1)$$

$$E = - dA/dt - \operatorname{grad} \phi \qquad (2)$$

$$\operatorname{rot} \mu^{-1} \operatorname{rot} A = Jo - \sigma (dA/dt + \operatorname{grad} \phi) \qquad (3)$$

$$- \operatorname{div} \sigma (dA/dt + \operatorname{grad} \phi) = 0 \qquad (4)$$

Input data of ac permeability chosen in the rolling direction (RD) of <100>, transverse direction (TD) <110> and thickness direction <110> at 25 and 100×10^{-7} T , are shown in Table 2. The permeability in the thickness direction <110> was considered to be the same as that of the transverse direction. Electric resistivity of both steels is also shown in Table 2. The shielding boxes are constructed so as to make the <100> directions of the sheets circular; Fig. 3 shows the arrangement of a shield and the <100> directions of each laminated sheet of panels, an inner layer (a) and an outer layer (b).

Total nodes for the analysis were 5065 to make a hexahedron element mesh of a shield and its adjacent space. A sheet was divided into three layers in the thickness direction and the inside and outside spaces of the shield box were divided into nine geometric ratios. Fig.4 shows the 3-D FEM hexahedron mesh. Exciting field, Be =25, 100×10^{-7} T , was applied from x-axis direction parallel to one edge of the shield to simulate the above experiment. .

Table 2.   Input data for 3-D analysis of permeability at Be=25, 100×10^{-7} T

Material	RD		TD		Resistivity
(Be)	25	100	25	100	(Ωm)
35G	14000	28000	100	100	48 x10^{-8}
23O	20000	41000	100	100	50 x10^{-8}

The results of shielding effectiveness are shown in Table 3 which were calculated at the center of the inside space of the shields. The magnetic flux density By, and Bz were considerably lower than Bx, being the direction of the applied field, Be.

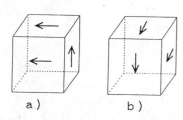

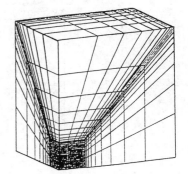

Fig.3. <100> arrangement of sheet panels
a) inner shell, b)outer shell of a shield

Fig.4   3-D FEM hexahedron mesh

Table 3.   Shielding effectiveness at the center of the inside by 3-D analysis

Material	Thickness (mm)	Be	Bx ($\times 10^{-7}$ T)	By	Bz	S (dB)
35G	0.70	25	1.2	0.05	0.01	21.6
		100	4.6	0.01	0.03	21.6
23P	0.46	25	1.3	0.03	0.01	19.5
		100	5.1	0.10	0.02	19.5

The shielding effectiveness in an alternating magnetic field obtained by 3-D FEM analysis was much higher than the effectiveness obtained by experiments as shown in Table 1. The difference of the shielding effectiveness between the two results of the experiment and the analysis in Table 1 and 3 were considerably larger than the previous results of static shielding[2] which showed good agreement between the experimental and analytical data. These results reaffirm the difficulties of calculation of shielding effectiveness in alternating field described previously[3]. The main reasons for the discrepancy could be pointed out 1) no consideration of air gap at joints of each shielding panel, 2) material property of constant permeability, 3) neglect of multilayer effect of laminated panels. It may be suggested that the gap at joints of the cubic edges should be small enough in the construction process of shields in experiments and taking the gap into consideration for the 3-D analysis as input data.

4. Conclusion

The shielding effectiveness in an alternative field of 50 Hz obtained by experiments and 3-D FEM analysis showed discrepancy in relatively large shields of the grain oriented silicon steel panels. Shielding effectiveness should be studied more at higher frequency up to 10 kHz by experiment and analysis to understand the roles of frequency dependent permeability and eddy current in shielding materials.

References
[1]Y.Okazaki and K. Ueno,J.Magn.Magn.Matr.,112(1992)192-194
[2]Y.Okazaki and H.Mogi,Matr.Sci.& Engnr.,AISI/A182(1994)1374-1377
[3]Y.Okazaki, JSAEM, vol.3, pp387-394(1995).
[4]Y.Okazaki and M.Fujikura, JSAEM, Nonlinear Electromag. Sys. JSAEM10, ISO Press 644-647(1996)
[5]V.O.Kelha, IEEE TransMag.,VOL.MAG-18,No.1, 260-270(1982)

Non-Linear Electromagnetic Systems
V. Kose and J. Sievert (Eds.)
IOS Press, 1998

Acoustically Excited
Electromagnetic Wave in Metal

Klára ČÁPOVÁ and Ivo ČÁP

University of Žilina, Faculty of Electrical Engineering, 010 26 Žilina, Slovakia

The present paper deals with an effect of generation of electromagnetic field inside the metal medium caused by mechanical vibrations of its electrodynamic structure in presence of an external constant magnetic field. The excited electromagnetic field can be emitted from the metal surface under proper conditions into the surrounding space. The mentioned effect can be used for the detection of mechanical processes in metal. The theoretical analysis and some experimental results are presented.

1. Introduction

The metal medium can be taken into account as a structure of an ionic lattice and a gas of electrons. The mechanical excitation of the medium is connected mainly with motion of the elastically bounded lattice. At room temperature and without any external electromagnetic field the ionic motion is completely screened by the electron gas, so that no corresponding electric polarization occurs. At very low temperatures and at a high frequency of mechanical excitation the mean free path of electrons can be comparable with the wavelength of the elastic wave and the screening of the lattice motion by the electron gas is perturbed. Similarly, the perturbation of the screening occurs without respect to the temperature in presence of a magnetic field, which influences the motion of ions and electrons in a different way due to their different masses. In both mentioned cases the distributions of positively and negatively charged particles are not the same and the resulting charge distribution gives rise to a corresponding electromagnetic field in the metal medium. In the case of its proper polarization the induced electromagnetic field is radiated from the metal surface and it can be detected, e.g. by a wire coil.

Similar effects connected with the excitation of electrodynamic structure of the metal were investigated from the opposite point of view as mechanisms of electromagnetic generation of acoustic waves in metals, e.g. [1], [2], [3] and [4].

2. Theoretical analysis

The case of an isotropic conducting medium at room temperature and at RF frequency of the exciting process is considered. Under these conditions the medium is taken as an electromagnetically sensitive elastic continuum.

The electromagnetic field in the medium is described by the wave equation

$$rot\ rot\ \boldsymbol{E}(\boldsymbol{r},t) + \varepsilon\mu\ \frac{\partial^2 \boldsymbol{E}(\boldsymbol{r},t)}{\partial t^2} = -\frac{\partial \boldsymbol{J}(\boldsymbol{r},t)}{\partial t} \quad , \text{ where } \quad \boldsymbol{J} = \boldsymbol{J}_{electron} + \boldsymbol{J}_{ion} \quad , \quad (1)$$

where $E(r,t)$ is the electric intensity, $J(r,t)$ is the total current density, ε and μ are the permittivity and the permeability of the medium. The total current density contains both components which correspond to the motion of electrons and ions

$$J = \sigma . \left[E - \frac{m}{e\tau} \frac{\partial \xi}{\partial t} + \frac{2 E_F}{3e} \, grad \, div \, \xi \right] + ne \frac{\partial \xi}{\partial t} \quad , \tag{2}$$

where σ is the electric conductivity tensor, m and $(-e)$ are the mass and the charge of an electron, τ is the relaxation time of electrons, E_F is the Fermi-energy of electrons, ne is the charge density of the ion-lattice, and ξ is the ionic displacement.

The mechanical motion of the elastic medium is described by the equation

$$c_l \, grad \, div \, \xi(r,t) - c_t \, rot \, rot \, \xi(r,t) - \rho \frac{\partial^2 \xi(r,t)}{\partial t^2} = f \quad , \tag{3}$$

where c_l and c_t are the longitudinal and shear elastic constants, ρ is the mass density of the medium, f is the density of the total force acting on the ion-lattice. It consists of the direct electromagnetic force and of the collision drag force of electrons

$$f = ne \left(E + \frac{\partial \xi}{\partial t} \times B \right) - \frac{nm}{\tau} \left(\frac{1}{ne} J_e + \frac{\partial \xi}{\partial t} \right) \quad , \tag{4}$$

where B is the resulting magnetic induction and J_e is the electron-current density (see the first term in (2)).

The interaction of mechanical and electromagnetic processes in the medium is represented by right side terms (2) and (4) in the wave equations (1) and (3). If we neglect the effect of the time-dependent component of the internal magnetic field, the equations (1) and (3) are linear and without any external field the mechanical and electromagnetic processes are independent. The interaction of both processes rises with increasing of the external constant magnetic field. If we take the time-dependent internal magnetic field into account, the non-linear interaction occurs.

The linear interaction of mechanical and electromagnetic processes in the presence of an external constant magnetic field can be studied for the elementary time-harmonic plane waves. After selecting the first-harmonic terms the Laplace images of the space-dependent wave equations have the form

$$\left[i\omega ne \left(I - \frac{\sigma_c}{\sigma_o} \right) + \frac{2 E_F}{3e} (\sigma_c . q)q \right] \cdot \xi^*(q) + \frac{1}{i\omega\mu} \left[(q^2 - \omega^2 \mu\varepsilon) I - qq + i\omega\mu \, \sigma_c \right] \cdot E^*(q) = -J_o^* $$

$$\tag{5}$$

$$\left[-c_t q^2 I - (c_l - c_t) qq + \rho\omega^2 I - \frac{2}{3} n E_F \left(\frac{\sigma_c}{\sigma_o} \cdot q \right) q \right] \cdot \xi^*(q) + ne \left(I - \frac{\sigma_c}{\sigma_o} \right) \cdot E^*(q) = -\rho \, \Phi_o^* \quad ,$$

where I is the identity tensor, q is the wave vector, ω is the angular frequency, σ_c is the magneto-conductivity tensor taking only the constant magnetic field into account and σ_0 is the DC conductivity of the medium. The terms J_o^* and Φ_o^* represent the boundary conditions and physically the mechanical and electrical sources of excitation. In the case of mechanically excited process $J_o^* = 0$.

The set of resulting equations can be analysed for different polarizations of elastic waves and for different orientations of the external magnetic field. From the experimental point of view we are looking for the generated electromagnetic waves with transverse polarization, which are radiated from the metal surface and can be detected. Both significant shear and longitudinal polarizations of the exciting mechanical vibrations $\Phi_o^* = [\Phi_{ox}^*, 0, \Phi_{oz}^*]$ together with $q = [0,0,q]$ were investigated.

In the case of longitudinal orientation of the magnetic field $B_c = [0,0,B_c]$

$$\frac{\sigma_c}{\sigma_o} = \begin{bmatrix} 1 & -\omega_c\tau & 0 \\ \omega_c\tau & 1 & 0 \\ 0 & 0 & 1 \end{bmatrix} \qquad \frac{\sigma_c}{\sigma_o} \cdot qq = \begin{bmatrix} 0 & 0 & 0 \\ 0 & 0 & 0 \\ 0 & 0 & q^2 \end{bmatrix} , \qquad (6)$$

where $\omega_c = e\,B_c\,/\,m$ is the cyclotron angular frequency of the electron.
The set of the equations (5) become a form

$$\left(c_i q^2 - \rho\omega^2\right)u_x^* - ne\left(\omega_c\tau\right)E_y^* = \rho\,\Phi_{ox}^*$$

$$\left(c_i q^2 - \rho\omega^2\right)u_y^* + ne\left(\omega_c\tau\right)E_x^* = 0$$

$$\left(c_l^* q^2 - \rho\omega^2\right)\xi_z^* = \rho\,\Phi_{oz}^*$$

$$\left(i\omega\mu\right)\left(i\omega ne\right)\left(\omega_c\tau\right)\xi_y^* + \left(q^2 + i\omega\mu\sigma_o\right)E_x^* - \left(i\omega\mu\sigma_o\right)\left(\omega_c\tau\right)E_y^* = 0 \qquad (7)$$

$$\left(i\omega\mu\right)\left(i\omega ne\right)\left(\omega_c\tau\right)\xi_x^* - \left(q^2 + i\omega\mu\sigma_o\right)E_y^* - \left(i\omega\mu\sigma_o\right)\left(\omega_c\tau\right)E_x^* = 0$$

$$(2/3)nE_F\,q^2\,\xi_z^* + ne\,E_z^* = 0$$

The resulting electromagnetic wave has components

$$E_y^* = \frac{\left(i\omega\mu\right)\left(i\omega ne\right)}{\left(c_i q^2 - \rho\omega^2\right)\left(q^2 + i\omega\mu\sigma_o\right)}\left(\omega_c\tau\right)\rho\,\Phi_{ox}^*$$

$$E_x^* = \frac{\left(i\omega ne\right)}{\sigma_o}\left(\omega_c\tau\right)^2\frac{\left(i\omega\mu\sigma_o\right)^2}{\left(c_i q^2 - \rho\omega^2\right)\left(q^2 + i\omega\mu\sigma_o\right)^2}\rho\,\Phi_{ox}^* \qquad (8)$$

$$E_z^* = -\frac{2}{3}\frac{E_F}{e}q^2\frac{1}{\left(c_l^* q^2 - \rho\omega^2\right)}\rho\,\Phi_{oz}^*$$

The dominant component is given by E_y^* which is proportional to the $\omega_c\tau \ll 1$. The second component E_x^* is proportional to $(\omega_c\tau)^2$ and it can be neglected. The component E_z^* exists only inside the medium and cannot be detected by an external detector. The time-space dependence of the dominant wave is

$$E_y(z,t) = -\frac{i\omega\,B_c}{1-\dfrac{i\omega}{s_t^2\,\mu\sigma_o}}\,\xi_{ox}\left\{exp\left[i\left(\frac{\omega}{s_t}z - \omega t\right)\right] - exp\left[i\left(-\sqrt{i\omega\mu\sigma_o}\,z - \omega t\right)\right]\right\}.$$

The first elastic mode is weakly damped and accompanies the elastic wave with the propagation velocity $s_t = \sqrt{c_t/\rho}$. The second mode is strongly damped with the normal electromagnetic penetration depth of $\delta = \sqrt{2/\left(\omega\mu\sigma_o\right)}$.

In the case of transverse orientation of the external magnetic field $B_c = [B_c,0,0]$ we obtain in a similar way the dominant electromagnetic wave given by the component

$$E_y(z,t) = -\frac{i\omega\,B_c}{1-\dfrac{i\omega}{s_l^2\,\mu\sigma_o}}\,\xi_{oz}\left\{exp\left[i\left(\frac{\omega}{s_l}z - \omega t\right)\right] - exp\left[i\left(-\sqrt{i\omega\mu\sigma_o}\,z - \omega t\right)\right]\right\}.$$

The electromagnetic field is radiated from the metal surface and can be detected by a proper detecting system which is able to detect and distinguish the polarization of the elastic wave in the medium.

3. Experimental results

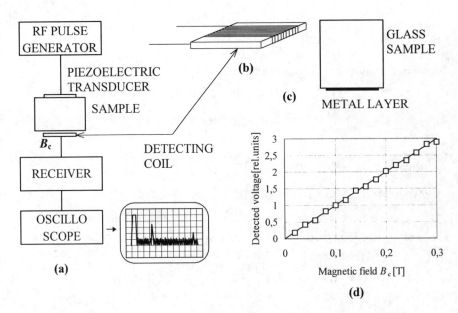

Fig. 1. Experimental arrangement, (a) experimental set-up, (b) detecting coil, (c) sample,
(d) measured dependence.

We have proved the theoretical results under different conditions. An RF pulse system with the frequency of about 10 MHz, pulse length of 1 μs and pulse power of 1 kW was used, Fig. 1(a). The shear or longitudinal elastic waves were generated by means of piezoelectric transducers. The radiated EM field was detected by means of the flat coil, Fig. 1(b), and the constant magnetic field was oriented according to the theory. The voltage detected by the coil was proved to be proportional to the magnitude of the magnetic induction B_c, Fig. 1(d), and be independent on the distance between the coil and the sample surface up to about 0.5 mm. A heterodyne receiver with the gain of 60 dB was used. According to the theory the delay of detected pulses corresponded to the elastic waves velocities in the samples. It proved that the detected EM field was generated by mechanical vibrations of the metal. The same results were obtained in cases of metallic samples as well as of glass ones with a thin metallic layer of a depth $d > \delta$ bound to its surface, Fig. 1(c). The mentioned effect was used both in the non-destructive testing of metallic samples, e.g. of rails, and in the investigation of crystal anisotropy of materials.

References

[1] E. Kartheuser, L. R. Ram Mohan and S. Rodriguez, Theory of electromagnetic generation of acoustic waves in metal. *Advances in Physics*, vol. 35 (1986), No. 5, pp. 423 - 505.

[2] I. Čáp, Direct Electromagnetic 2nd Harmonic Ultrasound Generation in Metals. *Acta Physica Slovaca* 32 (1982), No. 1, pp. 77 - 83.

[3] I. E. Aronov and V. L. Falko, Electromagnetic Generation of Sound in Metals in a Magnetic Field. *Physics Reports* 221 (1992), No. 2-3, pp. 81 - 166.

[4] I. Čáp and K. Čápová, Electromagnetic Wave Transition through a Metal Layer due to an Acoustic Excitation. *Interdisciplinary Applied Electromagnetics*, Brno 1996, pp. 156 - 164.

Non-Linear Electromagnetic Systems
V. Kose and J. Sievert (Eds.)
IOS Press, 1998

"Unusual" Electron Transport in Ultra-Pure Metals: a Possible Way to Low-Temperature Electronic Devices

V.V. Marchenkov [a,b,c,*], V.E. Startsev [a,c], A.N. Cherepanov [a], and H.W. Weber [b]
a Institute of Metal Physics, 620219 Ekaterinburg, Russia
b Atominstitut der Österreichischen Universitäten, A-1020 Wien, Austria
c International Laboratory of High Magnetic Fields and Low Temperatures, 53-529 Wroclaw, Poland

Abstract. Dc and ac measurements of the resistivity and of the current-voltage characteristics in ultra-pure tungsten, molybdenum and rhenium single crystals were made at low temperatures and high magnetic fields. "Unusual" magnetotransport in the crystals was observed, which can be used for making non-linear devices.

1. Introduction

Ultra-pure metal single crystals with resistivity ratios of up to $10^4 \div 10^5$ have been obtained during the past few years. The electron mean free path in these metals can reach $1 \div 10$ mm at helium temperatures and an electron can move without any collisions over a distance comparable to the sample size. In a magnetic field, the electron trajectories in such crystals are formed according to the topology of the Fermi surface of the metal used and the orbit sizes are determined by the magnitude and the direction of the magnetic field. For example, in metals with a closed Fermi surface, high magnetic fields lead to a "localized state", i.e. an electron moves along a closed path in the plane perpendicular to the magnetic field [1]. In this case the magnetoresistivity of compensated metals may increase by a few million times with magnetic field and reach $r_{xx} = 2 \cdot 10^{-3}$ Ohm cm [2], which is equal to the resitivity of semiconductors (Fig. 1). If an electron moves along an open path, the "current state" occurs and the field dependence of the magnetoresistivity is very weak [1]. The transition from one state to the other can be induced by
- applying a magnetic field;
- the influence of temperature;
- scattering of the conduction electrons by the sample surface.

In the first case, the magnetic breakdown phenomenon [1] can occur as a result of quantum tunneling of the conduction electrons from one Fermi surface sheet to another.In the second case, the temperature breakdown [3] results from intersheet electron-phonon scattering. Finally, quasi-open trajectories can appear near the sample surface in real space as a result

* Permanent address: Institute of Metal Physics, Kovalevskaya Str. 18, 620219 Ekaterinburg, Russia
E-mail: vvm@fmm.e-burg.su

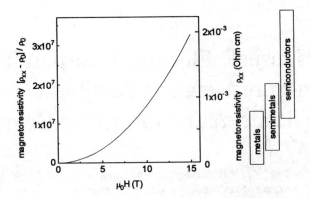

Fig. 1. Field dependence of the magnetoresistivity of a tungsten crystal at T=4.2 K.

of strong electron-surface scattering. This leads to a concentration of the dc current near the crystal surface. This phenomenon is called the static skin effect (SSE) [4]. In all cases, the type of the electron orbits is changed substantially in momentum or real space. Consequently the electron transport properties are changed as well. Besides, under the SSE the current density near the sample surface can be in $10^3 \div 10^4$ times higher than in the bulk of the crystal and can reach $10^5 \div 10^6$ A/cm^2 [2,5,6]. These non-linear effects in metals clearly go beyond traditional ideas [1].

The above phenomena show that the electron motion and the current flow in metals can be controlled by magnetic field and temperature and by taking the peculiarities of electron-surface scattering into account. It is very important to note that the electron concentration is about 10^{10} el./cm^3 in vacuum tubes, about $10^{14} \div 10^{18}$ el./cm^3 in semiconductors, and that the concentration of the conduction electrons in metals is $10^{22} \div 10^{23}$ el./cm^3. Therefore, the new non-linear effects in pure metals can be employed for applications in high-current low temperature electronic devices.

2. Experimental

Dc and ac measurements of the resistivity and of the current-voltage characteristics (CVC) in ultra-pure tungsten, molybdenum and rhenium single crystals were made in the temperature range from 2 to 60 K and in magnetic fields up to 15 T. Tungsten crystals with resistivity ratios (RRR) of up to 100.000 as well as molybdenum and rhenium crystals with RRR's of up to 30.000 were used. The mean free path of the conduction electrons was about 4 mm in tungsten and about $0.6 \div 1.0$ mm in molybdenum and rhenium at T=4.2 K. The measurements were carried out on rectangular plates and Corbino disks. In the Corbino disk geometry, the magnetic field is directed perpendicular to the disk plane and the electric current flows outward from the center of the disk to its edges [7].

3. Experimental results

The electron-surface interaction in compensated metals with a closed Fermi surface can lead to the static skin effect, i.e. a concentration of the dc current near the sample

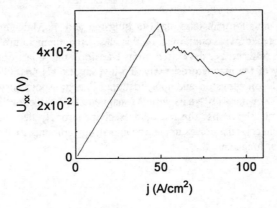

Fig. 2. Typical current-voltage characteristics of a tungsten crystal under SSE conditions.

surface [4,5]. Our experiments on tungsten show that the SSE occurs in plate-like samples, but is absent in the Corbino disk geometry. Under SSE, the current density near the sample surface can reach $10^5 \div 10^6$ A/cm^2, although the mean current density in the crystal is about $10^1 \div 10^2$ A/cm^2.

According to Refs. [6,8,9], non-linear effects occur in metals, even when the current densities are more than 10^5 A/cm^2. Figure 2 shows the CVC of a tungsten crystal under SSE conditions. One can see that a non-linear CVC of the S-type appears at a current density of about 50 A/cm^2. An analysis of the experimetal and literature data allows us to conclude that non-linearities are caused by the SSE and can be due to the effect of phonon generation [10,11].

Another "unusual" effect occurs in these crystals under SSE. A dc voltage is observed at the potential leads, when an ac current of low frequency (10^1-10^3 Hz) flows through the sample (Fig. 3). This dc voltage increases quadratically with magnetic field and disappears either with temperature or with the introduction of impurities into the sample, i.e. when the SSE is absent.

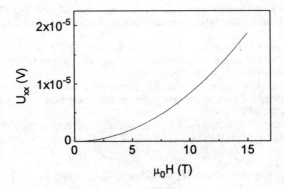

Fig. 3. Dc-voltage at the potential leads of a tungsten crystal under SSE. Ac current I = 0.1 A , f = 600 Hz.

A very interesting situation can take place in tungsten and molybdenum for H ‖ <110> under temperature breakdown conditions. In this case, the temperature breakdown may result from intersheet transfer, i.e. the closed orbits transform into quasi-open ones. This can lead to an anisotropy of the magnetoresistivity. To investigate this possibility, the following experiment was made on tungsten and molybdenum. The transport current was directed either along the <100> or the <110> axis with the magnetic field direction H ‖ <110> in both cases. Our experiments show that the magnetoresistivity strongly depends on the current direction in these cubic crystals. For example, the anisotropy amounts to about 83% at T=16 K in a field of 14 T for tungsten.

4. Conclusions

1. It was demonstrated that the type of the electron trajectories in metals can be changed by the magnitude and the direction of the magnetic field and by temperature. It was shown that the current flow in metal crystals is determined by the form and the size of the samples.

2. Non-linear current-voltage characteristics were observed under the static skin effect, where the electric current density near the sample surface is more than 10^5 A/cm^2.

3. A dc voltage was also observed on the potential leads, when an ac current flowed through the sample under the static skin effect.

The observed "unusual" electron transport properties of ultra-pure metals can be used for making high-current non-linear devices for low temperature metallic electronics.

References

[1]   I.M. Lifshitz, M.Ya. Azbel', M.I. Kaganov, Electron Theory of Metals. Consultants Bureau, New York, 1973.

[2]   V.E. Startsev, V.V. Marchenkov, and A.N. Cherepanov, Electronic Transport Properties for the Ultra-Pure Single Crystals of the Transition Metals under Conditions of the Strong Static Skin Effect at High Current Densities. In: P.M. Oppeneer and J. Kübler (ed.), Physics of Transition Metals, World Scientific, Singapore, 1993, pp. 232-238.

[3]   V.V. Marchenkov, H.W. Weber, A.N. Cherepanov, and V.E. Startsev, Experimental Verification and Quantitative Analysis of the Temperature (Phonon) Breakdown Phenomenon in the High-Field Magnetoresistivity of Compensated Metlas, J. Low Temp. Phys. 102 (1996) 133-155.

[4]   V.G. Peschanskii and M.Ya. Azbel', Magnetoresistivity of Semimetals, Sov. Phys. JETP 28 (1969) 1045-1055.

[5]   A.N. Cherepanov, V.V. Marchenkov, V.E. Startsev et al., High-Field Galvanomagnetic Properties of Compensated Metals under Electron-Surface and Intersheet Electron-Phonon Scattering (Tungsten), J. Low Temp. Phys. 80 (1990) 135-151.

[6]   V.V. Marchenkov, A.N. Cherepanov, and V.E. Startsev, Nonlinear Current Voltage Characteristics of Magnetoresistance of Tungsten Single Crystals in Static Skin Effect, The Physics of Metals and Metallography 73 (1992) 132-140.

[7]   V.V. Marchenkov, A.N. Cherepanov, V.E. Startsev et al., Temperature Breakdown Phenomenon in Tungsten Single Crystals at High Magnetic Fields, J. Low Temp. Phys. 98 (1995) 425-451.

[8]   A.A. Slutskin and A.M. Kadigrobov, Electric Domains in Metals at Low Tempertures, Sov. Phys. JETP Letters 28 (1978) 219-222.

[9]   V.V. Boiko, Yu. F. Podrezov, and N.P. Klimova, Electric Field Domains in Metals at Low Temperatures, Sov. Phys. JETP Letters 35 (1982) 524-526.

[10]   L. Esaki, New Phenomenon in Magnetoresistance of Bismuth at Low Temperatures, Phys. Rev. Letters 8 (1962) 4-7.

[11]   Yu.A. Bogod, Properties of Bismuth under Conditions of Elastic Wave Generation, Sov. J. Low Temp. Phys. 8 (1982) 787-829.

Non-Linear Electromagnetic Systems
V. Kose and J. Sievert (Eds.)
IOS Press, 1998

Generation of High Frequency Voltages

Andrzej WAC-WŁODARCZYK
Lublin Technical University, Faculty of Electrical Engineering
ul. Nadbystrzycka 38a, 20-618 Lublin, Poland

Abstract. This paper describes a hybrid system to convert frequency. This system is aimed to fill the gap in electric energy sources which generate voltages below 100 kHz and supply high power. The idea of the system is the assumption of cascade combination of two systems: a semiconductor frequency converter and a magnetic frequency multiplier. This paper describes the process of construction and tests of a system model. It can be found that the operation is correct and there are promising perspectives.

1. Introduction

The increase of demand for electric energy and thus its reasonable use forces the interest in high frequency sources. They are among the most important fields in power electronics which has also shown a dynamic development.

Supplying systems of inductive heating and high-speed drive units require electric energy sources of frequency from the mains one (50 Hz) to several MHz and power of 10 W - 1 MW. High frequency sources, which have been used in most cases so far, are the following:
- thyristor frequency converters within the range: 100 Hz to 10 kHz and power of 1 MW, but high power limits thyristors' frequencies to 1 kHz [5],
- magnetic frequency multipliers within the range: 100 Hz to 450 Hz and power of a few MW,
- frequency motor converters up to 10 kHz and power of a few MW,
- high frequency valve generators, frequency: 50 kHz to MHz and power of several hundred kW.

The above classification indicates a distinct gap in the range from a few to tens of kHz and power above several tens of kW. The system, presented in this paper, is the optional attempt to fill this gap. It consists of a cascade combination of two basic units: a semiconductor frequency converter and a magnetic frequency multiplier (m.f.m.). For the first time the idea of this system was presented in Japan in 1983 [2, 5] and now, it is presented in Fig. 1.

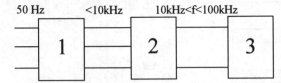

50 Hz	<10kHz	10kHz<f<100kHz

1 - semiconductor frequency converter (inverter),
2 - magnetic frequency multiplier (m.f.m.),
3 - receiver.

Fig. 1. Flow chart of the described hybrid system to convert frequency

A semiconductor inverter of a three-phase output converts the mains frequency of input voltage into the range from a few to 10 kHz. Then, this voltage supplies the m.f.m. that multiplies the frequency from 3 to 9 times depending on the type of the magnetic multiplier.

However, not so long ago m.f.m. applications in the above frequency were not possible, the recent development of magnetic material processing has brought these magnetic multipliers again to the group of modern advanced converters. For this reason, and due to numerous advantages such as simplicity of construction, reliable operation, symmetric single-phase loads and relatively low maintenance and investment cost, they predominate over the other systems in many applications.

2. Selection of Inverter System

The shape and values of the output voltage of a current inverter depend mainly on the load. For this reason, the inverter is not a suitable supplier of m.f.m. Moreover, the DC sources in inverter systems are mains converters which contain chokes of high inductance. The chokes are selected so as to reduce input current pulsation. It can be found that current inverters are not suitable to supply m.f.m. of frequency regulation within the wide range.

Thyristor voltage inverters are not considered here either as they are even more complex systems and need additional commutation circuits. Neither do, so called, series (resonance) thyristors nor series-parallel ones seem to be suitable in our case because their frequency can be changed only in a very small range and they need constant load parameters.

Following the above consideration a transistor voltage inverter can be selected. Its construction is very simple and it is a very reliable device within the power range 20 - 30 kW [1, 6]. In the case when the device operates correctly, it is necessary to consider thyristor systems of inverters when one wants to widen this power range.

The physical model comprises a three-phase voltage inverter with a power terminal LGXT 1050, Power Compact jsc. [3]. This is a compact unit comprising the whole three-phase system of six transistors with their backward diodes and insulated impulse generators, all in one insulated housing. The generators release transistor gates. Moreover this model is equipped with a three-phase diode bridge-rectifier which enables supply to the inverter from a three-phase AC mains. It is always possible to supply the inverter from a DC source without a rectifier. Its control unit comprises an integrated modulator (HEF 4752, Philips [4]) using pulse width modulation (PWM).

3. Selection of Core Material for Magnetic Unit System

Particularly careful selection of core material is needed because of the special range of its operation within a few kHz. The main criterion of this selection is tendency to minimize losses of a particular magnetic material which determines saturation density.

Table 1. Properties of selected magnetic materials at B = 1T and f = 50Hz

Material	B_s T	B_r T	H_C A/m	$\Delta P_{50Hz, 1T}$ W/kg
Fe-Si(ET-5)	2.0	1.7	12.2	1.20
Permaloy Ni79Fe17Mo4	0.7	0.5	1.2	0.07*
Metglas 2605 SC	1.61	1.42	3.2	0.15
Metglas 2605 Co	1.80	1.60	4.0	0.20
Metglas 2714 A	0.55	0.40	0.4	0.01*
Metglas 2826 MB	0.88	0.50	0.4	0.09*
Metglas 2605 SM	1.30	0.80	2.6	0.15
Vitrovac 4040F	0.80	0.65	1.0	0.15*
Vitrovac 6025F	0.55	0.45	0.4	0.02*

* at saturation induction B_s

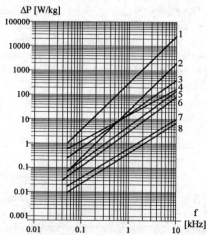

Fig. 3. Power loss ΔP vs. frequency in different materials at saturation density B_s, where: 1. silicon sheet ET-5; 2. permaloy Ni79Fe17Mo4; 3. Metglas 2605 SC; 4. Metglas 2605 S3; 5. Metglas 2826 MB; 6. Vitrovac 4040F; 7. Vitrovac 6025F; 8. Metglas 2714 A

The saturation of magnetic flux density is not high. The need to reduce the magnetizing current points out to the materials of rectangular magnetizing characteristics. The cascade systems constructed in Japan used so called supermaloy [2] or ferrite [6]. However, it seems that an amorphous magnetic materials are more convenient and more competitive.

Unit power loss in amorphous magnetic materials at 50 Hz is sometimes one order lower than in ferrosilicon sheets. Amorphous tapes are thin and have three times higher resistivity. Basic properties of magnetic materials, essential to m.f.m. operation, are listed in Table 1, whereas Fig. 3 illustrates power loss in selected materials at different frequencies.

The data are taken from catalogues indicated in [7] and the loss computations carried out in accordance to the equations are also presented in [7].

Power losses in some magnetic materials do not exceed 10 W/kg at saturation induction and 10 kHz which can be found in the figure. There is a conclusion that such magnetic cores can work with natural cooling.

4. Construction and Testing of the System Model

According to the idea the system was constructed using models of two basic units, which are the inverter and m.f.m.

Basic technical parameters of the inverter are the following:
- collector - emitter max. voltage - 1000 V,
- collector continuous current at 85°C - 27 A,
- frequency range 60 - 1140 Hz,
- dimensions 236×221×98 mm.

Upper limit of the frequency range can be even higher by using a more complex integrated control system instead of the modulator HEF 4752.

M.f.m. has been constructed as a nontupler (Fig. 4).

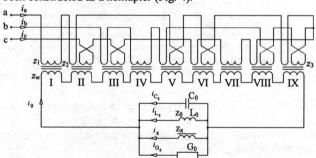

Fig. 4. Magnetic frequency nontupler system scheme

It comprises 9 identical cores made of an amorphous material - Vitrovac 6025F manufactured by Vacuumschmelze, Germany. Each core is in the form of a fine amorphous tape, thickness 25 μm and its dimension are: outer diameter - 25 mm, inner diameter - 16 mm, height - 10 mm. The following numbers of turns have been established (Fig. 4): $z_1 = 37$, $z_2 = 14$, $z_3 = 28$, $z_w = 47$ on the basis of turn proportions between the primary winding turns and small core windows. The windings of each core have been wound on its whole perimeter and made of copper enamelled leads of the following diameters: $d_1 = 0.7$ mm - the primary windings and $d_2 = 0.6$ mm - the secondary windings.

Measurements and observations of waveforms at no-load, short circuit and load states of the multiplier have been carried out at three different frequencies of input voltage: 1, 2 and 3 kHz (output frequencies: 9, 18 and 27 kHz, respectively). This multiplier operates properly when the supplying voltage threshold value is exceeded. The nontupler core is then introduced to the saturation state. The voltages that are illustrated in Fig. 5 at the above frequencies, exceed 17 V, 40 V and 60 V, respectively.

U$_{20}$ [V]

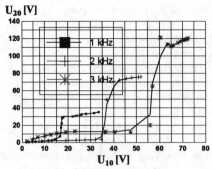

U$_{10}$ [V]

Fig. 5. Voltage characteristics at no-load

U$_2$ [V]

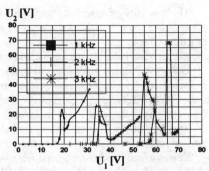

U$_1$ [V]

Fig. 6. Output voltage vs supplying voltage of the
nontupler at capacitive load C = 100nF

Some unstable operation and hysteresis output voltage characteristics vs. supplying voltage (Fig. 6) can be observed at capacitive load in accordance with theoretical results obtained by means of the mathematical model of the multiplier [7]. The instabilities can be observed only at capacitive loads and follow ferroresonance phenomena in the system. It can be found that the device operates properly which can be proved by numerous oscillograms presented in [7] and Fig. 7.

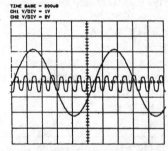

Fig. 7. Oscillograms of the phase voltage, which supplies the nontupler, and the secondary voltage U$_9$ at
f = 1 kHz and resistance - capacitive load of R = 100Ω; C = 47nF; U$_1$ = 27,1V; U$_9$ = 8V; I$_1$ = 0.5A

5. Conclusions

Temporal availability of only small amorphous cores causes some inconvenience as inverters are much more capable. The constructed model of magnetic frequency nontupler is the first approach, to the author's knowledge, to the application of metal glass cores in m.f.m. and it can be useful study for construction of larger physical models.

The laboratory tests proved that these cores do not heat too much and the output signal has a regular shape in oscillograms. The above can lead to the conclusion that the operation of the two stage hybrid system to convert frequency is successful.

References

[1] R. Barlik, M. Nowak, Thyristhor technique, (in Polish), WNT, Warszawa, 1993.
[2] K. Bessho, S. Yamada, High frequency power supply composed of a combination of a thyristor inverter and magnetic frequency multiplier, Fifth International Telecommunication Energy Conference, IEEE CH 1855-6, 1983, pp. 570-573.
[3] POWER COMPACT SA Catalogue, Power Unit of Transistor Bridge Inverter LGXT 1050, Chemin de Marget, France, 1991.
[4] Integrated Numerical Systems Catalogue CMOS series HE 4000, B-Philips, part 4, 1983.
[5] H. Suzuki, T. Aikawa, M. Tadokoro, Operating characteristics of hybrid inverter system, IEEE Translation Journal of Magnetics in Japan, TJMJ - 1, No 5, 1988, pp. 600-602.
[6] H. Tunia, M.P. Kaźmierkowski, Electric drive automatics, (in Polish), PWN, Warszawa, 1987.
[7] A. Wac-Włodarczyk, Hybrid systems to convert frequency, (Research report - not published, in Polish) PB 930/S5, Contract No 8 8211 92 03, Lublin Technical University, 1994.

Non-Linear Electromagnetic Systems
V. Kose and J. Sievert (Eds.)
IOS Press, 1998

Magnetic Ordering of Water Molecules Under Magnetic Flux Densities of up to 14T

Masakazu IWASAKA and Shoogo UENO
*Department of Biomedical Engineering, Graduate School of Medicine, University of Tokyo,
Tokyo 113, Japan*

Abstract. In this study, we investigated the effects of strong magnetic flux densities of up to 14T on the thermal agitation of water and ethanol. We measured the Raman spectrum and Rayleigh scattering of water-ethanol system. The Raman spectra of the water and ethanol in the range of 450– 700nm were not affected by an 8T magnetic field. However, the Rayleigh scattering intensity decreased significantly during 8T magnetic field exposures. We also investigated the near-infrared spectrum of water. The peak wavelength shifted from 1930nm to 1932-1933nm under a 14T magnetic field. There is a possibility that the formation of hydrogen bonds is enhanced and the thermal agitation of water molecules is decreased under strong magnetic flux densities of up to 14T.

1. Introduction

Water, an important material for living systems, is a diamagnetic material. We have observed the phenomenon that the surface of water was parted by magnetic fields and the bottom of the water chamber appeared when the water was exposed to gradient magnetic flux densities up to 8 T [1][2].

The so called "Moses' effect" can be explained by the diamagnetic property of water. Since water is diamagnetic, when a magnetic force acting on water reaches a high enough value it presses back the water. The visible phenomenon of bulk water under high gradient magnetic fields are very drastic.

Also, it is important to understand the mechanism of the interaction of a homogeneous magnetic field with diamagnetic molecules. Magnetic orientation of fibrin polymers [3] is due to the magnetic torque rotation that is induced under a homogeneous magnetic field.

In this study, we focus on the interaction of a homogeneous magnetic field with a structure of water molecules. We investigate the effects of strong magnetic flux densities of up to 14T on the thermal agitation and light scattering properties of water and ethanol.

2. Methods

We measured the Raman spectrum and Rayleigh scattering of water and ethanol. Raman spectrum and Rayleigh scattering of water and ethanol were measured using an Argon laser (488 nm), as shown in Fig. 1. We used a horizontal superconducting magnet 700 mm long with a bore 100 mm in diameter. The magnet produced 8 T at its center. The measurements were carried out in 10mm optical pathway. To examine the contribution of the water molecules that bound in the surface of quartz cell, a Rayleigh scattering measurement was also carried out in 0.5mm optical pathway.

In addition, we investigated the near-infrared spectrum of water molecules using a horizontal type of superconducting magnet which produces 14 T at its center. We measured the near-infrared spectrum of water in the range of 1930nm to 2000nm that originated from the three species of water molecules [4], as shown in Fig. 2. Each of the spectra was obtained from 7 measurements. The optical path length of a cell was 0.2mm. The spectral resolution of NIR spectrophotometer was 0.2nm.

3. Results and discussion

No effects of magnetic flux densities of up to 8 T on the Raman spectrum of ethanol and water were observed in the range of 450 nm – 700 nm, as shown in Fig. 3. A peak observed at 580 nm corresponds to the O–H vibration. The result also indicates that magnetic flux densities of up to 8 T have no effect on O–H vibration of molecules.

On the other hand, Rayleigh scattering intensity in 10mm optical path length decreased significantly during 8T magnetic field exposures, as shown in left figures of Fig. 4.

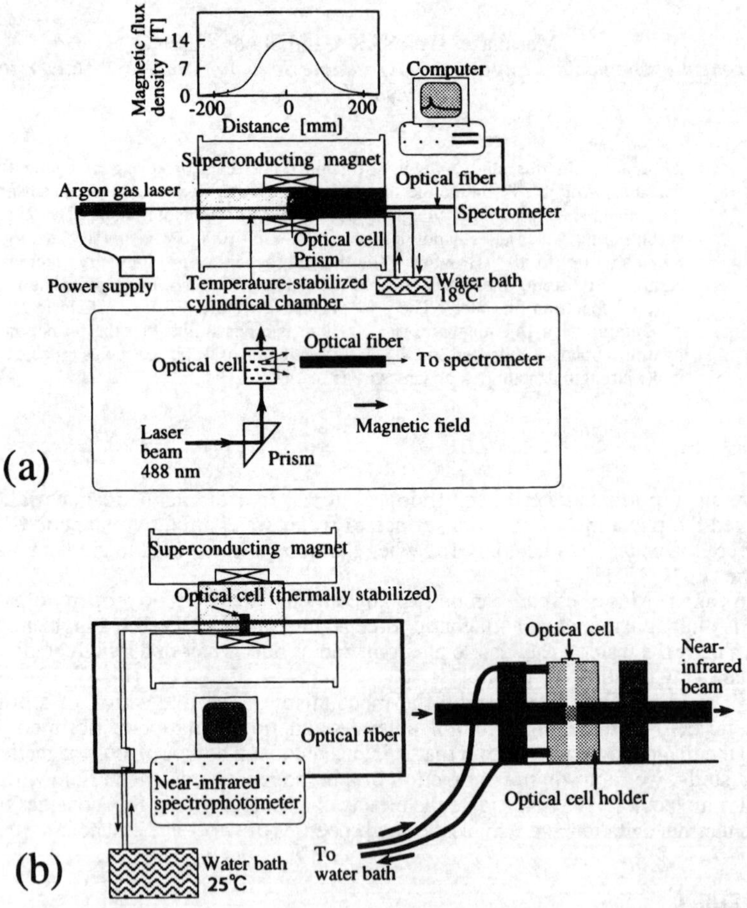

Fig. 1 Experimental set-up for the Raman spectrum and Rayleigh scattering measurements of water and ethanol.

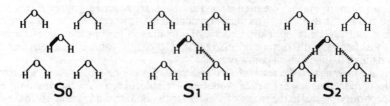

Fig. 2 Three species of water molecules, classified by the association of hydrogen bonds [4].

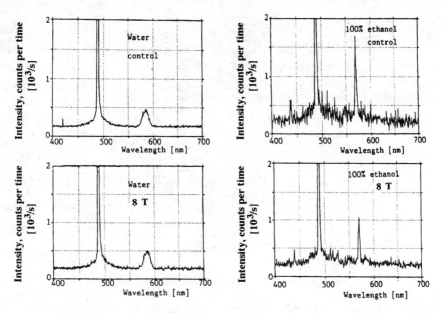

Fig. 3 Raman spectrum of water and ethanol in the range of 450 nm – 700 nm.

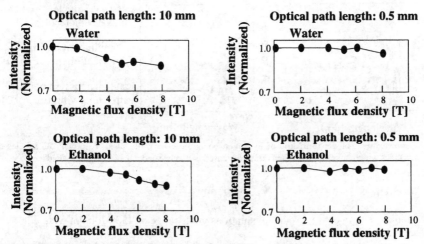

Fig. 4 Rayleigh scattering intensity during 8 T magnetic field exposures.

However, significant change was not observed in 0.5mm optical path length. The contribution of the molecules that bound in the surface of quartz cell was negligible. The decrease of Rayleigh scattering intensity of both water and ethanol was due to bulk solution. In the case of the near-infrared spectrum of water molecules, the peak wavelength of water shifted from 1930 nm to 1932~1933 nm under a 14T magnetic field, as shown in Fig. 5. We obtained the mean value and standard deviation from seven experiments. The results indicate that the species of water-molecules which have two hydrogen bonds increased in number under a 14T magnetic field (Fig. 6).

There is a possibility that the formation of hydrogen bonds is enhanced and thermal agitation of water molecules decreased under strong magnetic flux densities of up to 14T.

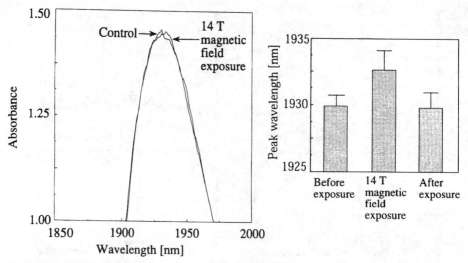

Fig. 5 The near-infrared spectrum of water molecules under a 14T magnetic field.

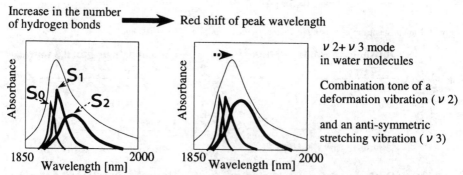

Fig. 6   Enhancement of the formation of hydrogen bonds in water molecules. A model: the species of water-molecules which have two hydrogen bonds increased in number under a 14T magnetic field.

4. Conclusions

1. The Raman spectra of water-ethanol solutions in the range of 450– 700 nm were not affected by an 8 T magnetic field.
2. The Rayleigh scattering intensity of water-ethanol solutions decreased significantly during 8 T magnetic field exposures.
3. The peak wavelength in the near-infrared spectrum of water shifted from 1930 nm to 1932-1933 nm under a 14 T magnetic field.

References

[1] S. Ueno and M. Iwasaka: Properties of Diamagnetic Fluid in High Gradient Magnetic Fields, Journal of Applied Physics, 75(10), pp. 7177-7179, 1994
[2] S. Ueno and M. Iwasaka: Parting of Water by Magnetic Fields, IEEE Transactions on Magnetics, 30, 6, pp.4698-4700, 1994
[3] J. Torbet, M. Freyssinet, and G. Hudry-Clergeon: Oriented Fibrin Gels Formed by Polymerization in Strong Magnetic Fields, Nature, Vol. 289, pp. 91-93 , 1981
[4] M. Kato, Y. Taniguchi, S. Sawamura, and K. Suzuki, Physics and Chemistry of Ice, ed. by N.Maeno and T. Hondoh, 83 (Hokkaido University Press), 1992

Non-Linear Electromagnetic Systems
V. Kose and J. Sievert (Eds.)
IOS Press, 1998

Hydration of Sugar Under Intense Magnetic Fields

Masakazu IWASAKA, Shoogo UENO and Ichiro OTSUKA*
*Department of Biomedical Engineering, Graduate School of Medicine, University of Tokyo,
Tokyo 113, Japan*
*Department of Chemistry, Ohu University, Fukushima 963, Japan

Abstract. In this study, we investigated the effects of strong magnetic fields of up to 14 T on the hydration of a sugar. We measured a near-infrared absorbance at 958nm under a magnetic field at 14 T to investigate the hydration of solutes in aqueous solution. When the magnetic fields were changed from 0 T to 14 T, the absorbance of 1 mol of D-glucose solution increased 40% – 50%. However, when we measured the absorbance of distilled water at 958nm under magnetic fields in the range of 0 T - 14 T, no change in the absorbance was observed. The results indicate that the hydration of glucose changed under magnetic fields of up to 14 T.

1. Introduction

It is important to know whether or not magnetic fields affect the property of water molecules. As water is an essential material in the living body, knowledge of the behavior of water under magnetic fields provides information for discussion on the effect of magnetic fields on living systems.

Solutions of biological materials such as proteins and amino acids are mostly diamagnetic, and their magnetic susceptibility is small compared to paramagnetic inorganic materials.

Recently, some studies visualized such weak diamagnetism by observing liquid surfaces under intense magnetic fields. In past studies, we have observed the phenomenon that the surface of water was parted by magnetic fields of up to 8 T [1][2].

We have reported the effect of magnetic fields up to 8 T on the enzymatic reaction of glucose oxidase with D-glucose using an electrochemical technique [3]. We observed that electric currents decreased 10% - 20% after magnetic field exposure for 1 to 2 hours. We conjecture that the diffusion process of glucose in aqueous solution, into the pore of the electrode, appeared to be inhibited by the magnetic field.

In this study, we report the effects of a magnetic field at 14 T on the hydration of sugars.

2. Methods

We measured a near-infrared absorbance under a magnetic field at 14 T to investigate the hydration of solutes in aqueous solution. The hydration of D-glucose was evaluated by the absorbance of solutes at 958nm[4]. We measured the absorbance of 1 mole of D-glucose solution and water.

We used a near-infrared spectrophotometer which has an external optical cell box in a superconducting magnet, as shown in Fig. 1. Two optical fibers connected the external optical cell with the spectrophotometer. The external optical cell was fixed at the center of the magnet's bore, and the measurement was carried out under a static condition.

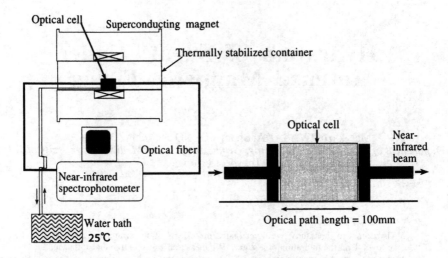

Fig. 1 Experimental set-up for near-infrared absorbance measurement under magnetic fields
of up to 14 T.

3. Results and Discussion

When the magnetic fields were changed from 0 T to 14 T, the absorbance of 1 mole of D-glucose solution increased 40% – 50%, as shown in Fig. 2. The absorbance returned to the same level when the magnetic field was off. This phenomenon was reversible and reproducible.

However, when we measured the absorbance at 958nm of distilled water under magnetic fields in the range of 0-14 T, no change in the absorbance was observed, as shown in Fig. 3. The results indicate that the hydration of glucose changed under magnetic fields of up to 14 T.

The reported effect of a magnetic field on the diffusion of saccharides through a porous membrane was obtained at 1.1 T [5]. It is possible for a strong magnetic field to affect the hydration of sugar.

4. Conclusion

The near-infrared absorbance of 1 mole of D-glucose solution at 958nm under a magnetic field at 14 T increased compared to the absorbance of solution at 0 T.
No change was observed in the near-infrared absorbance of water.
The results indicate that the hydration of glucose changed under magnetic fields of up to 14 T.

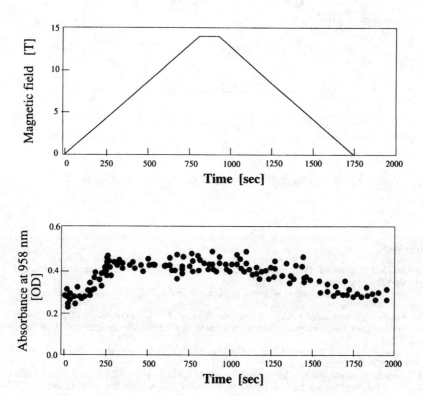

Fig. 2   Time course of change in the near-infrared absorbance of 1 mole of D-glucose solution. The magnetic field of the superconducting magnet was changed between 0 T and 14 T.

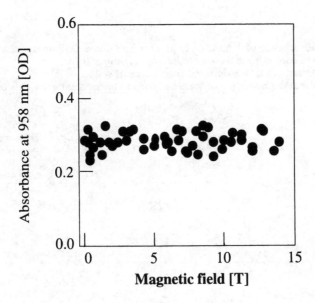

Fig. 3 Near-infrared absorbance of water under magnetic fields of up to 14 T.

References

[1] S. Ueno and M. Iwasaka: Properties of Diamagnetic Fluid in High Gradient Magnetic Fields, Journal of Applied Physics, 75(10), pp. 7177-7179, 1994
[2] S. Ueno and M. Iwasaka :Parting of Water by Magnetic Fields, IEEE Transactions on Magnetics, 30, 6, pp.4698-4700, 1994
[3] M. Yaoita, M. Iwasaka, and S. Ueno: Enzymatic Reaction of Glucose Oxidase in Magnetic Fields up to 8T, in *Nonlinear Electromagnetic Systems*, Studies in Applied Electromagnetics and Mechanics, Vol. 10, IOS Press, 1996
[4] J. L. Hollenberg and J. B. Ifft; Hydration Numbers by Near-Infrared Spectrophotometry. 1. Amino Acids, J. Phys. Chem., 86, pp.1938-1941, 1982
[5] J. Lielmezs, V. Atwal, and H. Aleman: Magnetic Field Effect on Free Diffusion of Selected Saccharides in Aqueous Solution Through an Inert Porous Membrane, J. Electrochem. Soc., 137,12, pp.3809-3814, 1990

Non-Linear Electromagnetic Systems
V. Kose and J. Sievert (Eds.)
IOS Press, 1998

The Influence of Structure Defects on the Damage Threshold of Transparent Dielectrics

V. A. Feodorov, I. V. Ushakov, V. P. Shelohvostov

Tambov State University, 392622 Tambov, Russian Federation

Abstract. The influence of a macrocrack on the damage threshold of transparent dielectrics is investigated. The probability of laser-induced macrocrack growth was established. The sketch of a relationship between the damage threshold and time of cracks' existence is shown. The mechanism of laser-induced macrocrack growth is discussed.

1. Introduction

Presence of cracks will reduce the damage threshold of transparent dielectrics due to: 1. interference phenomena [1]; 2. absorption of impurities, environmental contamination etc. [1,2]; 3. emission of electrons by growing cracks [3].

The purpose of the investigation was to establish: 1. the influence of one macrocrack on the damage threshold and mechanical strength of crystals characterised by different optical, mechanical and other properties; 2. the reasons for and nature of laser-induced crack growth.

2. Experimental procedure

The materials used in the present work were: 1. $CaCO_3$ single crystals: a) $CaCO_3$ with impurity content $> 10^{-2}$ %, the size of the biggest (particulate inclusion) PI were about 0,1 mm; b) $CaCO_3$ with impurity content $< 10^{-2}$ %, without big PI; 2. LiF single crystals with impurity content $< 10^{-3}$ %; 3. NaCl single crystals a) NaCl alloyed by Cr^{3+} 10^{-2} %; b) NaCl alloyed by Cr^{3+} 10^{-3} %; c) NaCl with impurity content $< 10^{-3}$ %.

The size of the samples was about 15x30x8 mm. The cracks were driven in along the (010) plane. Their lengths were about 8 mm. A pulse laser GOS 1001 with wavelength 1060 nm and $\tau \approx 1$ msec., was used. The energy of pulses was changed from 200 to 500 j. and the radiated area from 1 to 40 mm^2. The energy of the first pulse was $0.7E_{th}$ (E_{th} - the threshold energy). The output energy of each successive pulse was increased by 10%, till failure of the sample. Ageing of the surface was modelled by heating the calcite crystals (at $T \approx 973$ K the surface layers begin to decompose). At the temperature samples of calcite were radiated once by a pulse of $E > E_{th}$.

3. Results

1. Crystals of calcite (group 1a) were destroyed due to the formation of laser-induced cracks at PI. The laser-induced cracks were often associated with a former macrocrack.

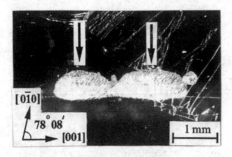

Fig. 1. Damaged area on exit surface of sample of calcite with a macrocrack. It is extending along the crack due to: 1. natural birefringence, its corresponding parts are indicated by arrows; 2. reflection of light from the plane of the macrocrack, and reduction of the radiated area in *[0̄10]* direction.

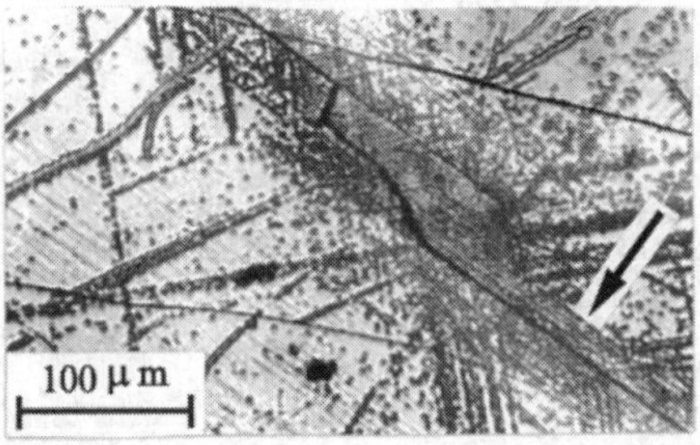

Fig. 2. The association of a laser-induced crack with a former macrocrack in LiF single crystal. The former crack is indicated by an arrow. There is a dislocation rosette at its tip.

Samples of group 1b were damaged at the entrance or exit surfaces (80%). Very often at the exit surface of samples with a former macrocrack, the damaged area was stretched along the crack (Fig 1).

The samples of group 1a and 1b (heated to 973 K) were damaged at entrance and exit (100%). The threshold of laser-induced damage was decreased significantly due to decomposition of surface layers of calcite (the threshold of laser-induced damaged of decomposed calcite is very low).

2. Crystals of LiF were destroyed due to: 1. growth of the former macrocrack; 2. initiation at PI laser-induced cracks (they were usually associated with the former macrocrack) (Fig 2).

As a rule all significant destructions in the irradiated area interacted with a former macrocrack. At the same time, the former macrocrack might grow up and destroy the crystal without any observed destruction.

3. The statistics of laser-induced destruction of NaCl and $CaCO_3$ single crystals, with a former macrocrack and without one is shown in Table 1, where P - the probability of activation of the crack growth after laser-induced crystal failure.

4. Discussion

According to the relationship between I and T (Fig. 3) (where I is the damage threshold, T is the time of crack existence) the discussion is divided into three parts. The relationship discussed below is valid for the crystal with the impurity content mentioned above. For transparent dielectrics with very low (or high) impurity content the relationship will be different.

TABLE 1. Statistics of failure of samples with and without a former macrocrack.

Number of group of crystals	The former macrocrack grew up and destroyed sample.	The former macrocrack grew up (2-3 mm).	Crystal destroyed without growth of former crack	P
NaCl - 3a	78,6 %	7,14 %	14,28 %	0,88
NaCl - 3b	80 %	---------	20 %	0,8
NaCl - 3c	37,5 %	37,5 %	25 %	0,7
$CaCO_3$ - 1a	60 %	---------	40 %	0,6

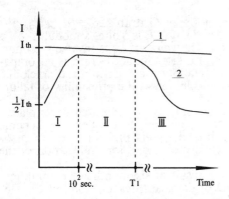

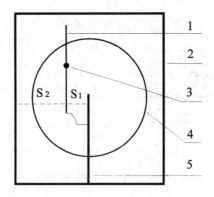

Fig. 3. The relationship between the damage threshold and the time of crack existence. I, II, III - periods of crack existence. 1 - damage threshold of sample without cracks, 2 - damage threshold of samples with cracks. T_1 - after the time natural ageing of crack surfaces determines the value of damage threshold.

Fig. 4. Scheme of the former macrocrack growth, due to it association with the laser-induced crack. 1 - the laser induced crack, 2 - crystal, 3 - particulate inclusion, 4 - the radiated area, 5 - the former macrocrack.

1. The growth of cracks in crystals is accompanied by emission of electrons. As a result the damage threshold for such crystals will be low [3]. Free electrons may stay on fresh surfaces for hours and even days [4], though they may be the cause of damage during the time of growth of the crack and for the first minutes after its formation (Fig. 3) (during the time their concentration is high enough [5]).

2. In the present work crystals were radiated in $\approx$ 5 min. after samples preparation and crack formation. The emission of electrons and relaxation of high mechanical stresses was over in five minutes. At the same time, the effect of natural ageing did not decrease the damage threshold yet, as the process demands more time [2]. In the second period (Fig. 3) the crack decreases the damage threshold and mechanical strength due to: 1. interference phenomena; 2. association of laser-induced defects with the former macrocrack.

A sample with a macrocrack was damaged at the exit surface at a lower power level than the sample without a crack. One cause of it is the interference of incident light, light reflected from the exit surface and from the surface of the macrocrack. Many other factors of low damage threshold of materials with microcracks and pores are described in [1].

Heating of a PI situated in front of the crack tip (Fig. 4) will initiate cracks in the bulk material. They will be extending along {100} planes. In the case of the situation of a laser-induced crack at some distance from the plane of the former crack, the laser-induce crack may change the plane of the propagation from (010) to {100} or {110} (if $S_1 < 0.5 S_2$ [6]). Fig. 4 illustrates such a case.

For the case of crack initiation at the PI, which was situated in another place, the mechanism of their association and growth will be analogous to that discussed above.

There is a stress interval $\sigma_n > \sigma > \sigma_v$ (σ_n - elastic stress, σ_v - critical stress) inside which crystal failure resulted from activation of the growth of the former macrocrack. If $\sigma >> \sigma_v$ crystal failure resulted from initiation at PI through cracks (as a rule these cracks interacted with the former crack). On the basis of nonlinearity increasing stress at PI after exceeding critical light intensity [7] and narrow stress interval $\sigma_n > \sigma > \sigma_v$, it may be concluded that stresses initiated at PI were either $\sigma << \sigma_n$ or $\sigma >> \sigma_v$. So we may explain that: 1. as a rule, crack growth was accompanied by sufficient failure of the specimen; 2. for all groups of samples, both NaCl and $CaCO_3$ the probability of activation of crack growth was about 80%; 3. absence of a strong tendency of decreasing strength of crystals by a macrocrack (see Table 1).

The crack growth is governed by the equation of Griffith $\delta \geq \delta_{cr} = (\alpha \gamma E / L_{max})^{1/2}$ [8], where α is coefficient, γ is efficient surface energy, E is Young's modulus, δ_{cr} is minimal stress required for crack propagation. The value of δ_{cr} is inversely proportion to the length of the crack L_{max}. Growth of a laser-induced crack is accompanied by relaxation of mechanical stresses. Association of a laser-induced crack and a former macrocrack lowers the value of δ_{cr} and increases the probability of failure of the crystal.

Under laser radiation PI burned up. Cracks and cavities appeared in its place . Under radiation by constant energy laser pulses, cracks and cavities may become stable [7]. Presence of a former macrocrack provides great probability of the association of defects with it. So a macrocrack creates obstacles for stabilisation of the laser-induced defects.

3. According to data presented in [2], as a result of six months of natural ageing of a polished surface 2-3 times lowering of damage threshold took place. Decreasing of the damage threshold of heated samples of calcite (in the present work) resulted from decomposition of the surface layer of calcite.

Thus contact between surfaces of cracks and environmental contamination lowers the damage threshold. The process depends on temperature and the presence of active elements. This factor determines the damage threshold during the last period of a crack's existence (Fig. 3).

5. Conclusions

1. Emission of electrons by growing cracks and fresh surfaces determines the damage threshold during the first period of crack existence.

2. Interference phenomena and growth of a crack due to its association with laser-induced defects determines the damage threshold during the second period of crack existence.

3. Environmental contamination of the crack surfaces due to airborne particles and absorption of impurities determines the damage threshold during the third period of crack existence.

Acknowledgement: the authors wish to acknowledge the sponsors of the 8th international symposium ISEM-Braunschweig and RFBR. Due to their financial support the authors managed to participate in the symposium.

References

[1] A. J. Glass and A. H. Guenther, Laser Induced Damage of Optical Elements - a Status Report, *Applied Optics* **12** (1973) 637-649.
[2] Bilibin S. V. *et al.,* An Effect of Optical Processing Parameters and Natural Ageing Process on the Depth of the Damage Layer and Beam Strength of the Surface of KCl Single Crystals, *Kvantovaya electronica* **9** (1982) 1912-1915.
[3] V. N. Smirnov, The decreasing of the threshold of laser-induced damage near the surface of the deforming crystal, *Technical Physics Letters,* **14** (1988) 316-321.
[4] J. Kramer, Untersuchungen mit dem Geiger - Spitzenzahler an bestrahlten Kristallen, *Zeitschrift Physics,* **129** (1951) 34-44.
[5] B. V. Derjagin and M. S. Mezik, Contribution of Electrical Forces into the process of cleaving of mica, *Solid State Physics* **1** (1959) 1521-1528.
[6] J. J. Gilman, Atomic Mechanism of Fracture. Moscow, Metallurgy, 1963. pp. 220-250.
[7] V. P. Krutjakova and V. N. Smirnov, Fluorescence of Alkali-Halide Crystals Under Irradiation of Laser Pulse with l=10.6 μm, *Journal of Technical Physics* **48** (1978) 844-852.
[8] V. I. Vladimiriv, Physical nature of metal destruction. Moscow, Metallurgy, 1984. p. 280.

Non-Linear Electromagnetic Systems
V. Kose and J. Sievert (Eds.)
IOS Press, 1998

835

Solution of Maxwell's Equations in a Nonlinear Complex-Permitivity Dielectric

J.P. DUCREUX, G. MARQUES

Electricite de France, Direction des Etudes et Recherches, 1 Avenue Général de Gaulle,
92141 Clamart, France

Abstract - In the study of electrical devices such as lightning conductors, the electric part modelling with Maxwell's equations leads to a nonlinear complex problem. This paper describes the partial differential equation resolution with a finite element method coupled to a specific nonlinear algorithm. This method is especially suited for devices where a permitivity is described by a nonlinear complex function.

1. Introduction

The electrical devices conception and analysis force engineers to search for local information such as magnetic induction, electric field or current density. Hence Maxwell's equations have to be solved and a very efficient way is provided by the finite element method. Besides some dielectric permitivity depends on the applied electric field. Thus, the above mentioned equations become nonlinear. This study is concerned with solving time harmonic problems using a finite element formulation coupled with nonlinear algorithms [1].

We investigate here a Fixed-Point algorithm and a Newton-Raphson algorithm. The first algorithm can be straightforwardly used. But the second algorithm must be adapted because the unknowns here are complex numbers. We implement the Fixed-Point and the Newton-Raphson algorithms in a finite element software. We compare the convergence and the accuracy of the two methods by solving several cases.

2. Hypotheses and equations

Maxwell's equations are characterised by Equation (1) with the implicit description of the material behaviour in Equation (2), where σ is the conductivity and ε is the permitivity.

$$\operatorname{rot}\vec{E} = -\frac{\partial \vec{B}}{\partial t} \qquad (1); \qquad\qquad \operatorname{rot}\vec{H} = \vec{J} + \frac{\partial \vec{D}}{\partial t} \qquad (2)$$

$$\vec{J} = \sigma \vec{E} \qquad (3); \qquad\qquad \vec{D} = \varepsilon \vec{E} \qquad (4)$$

We introduce here the magnetic vector potential $\vec{A}$ ($\vec{B} = \operatorname{rot}\vec{A}$). Substituting this definition in Equation (1) defines the electric scalar potential V :

$$\vec{E} = -\frac{\partial \vec{A}}{\partial t} - \operatorname{grad} V \qquad (5)$$

The Equation (2) divergence leads to :

$$\text{div}\left(\vec{J} + \frac{\partial \vec{D}}{\partial t}\right) = 0 \tag{6}$$

For a harmonic solution in terms of angular velocity ω, the time derivative can be replaced by the complex operation $j\omega$ ($j^2 = -1$). Equations (3), and (4) modify also the last equation (6):

$$j\omega \, \text{div}\,[(\sigma + j\omega\varepsilon)\vec{A}] + \text{div}\,[(\sigma + j\omega\varepsilon)\text{grad}\,V] = 0 \tag{7}$$

The first part of Equation (7) can be written :

$$\text{div}\,[(\sigma + j\omega\varepsilon)\vec{A}] = (\sigma + j\omega\varepsilon)\text{div}\,\vec{A} + \vec{A}.\text{grad}(\sigma + j\omega\varepsilon) \tag{8}$$

We are only interested here in 2D problems. In these cases the magnetic vector potential reduces to only one component, which is perpendicular to the study plane (i.e $\vec{A} = (0,0,A_z(x,y))$). On the other hand, the gradient of conductivity and permittivity is in the study plane ($\sigma = \sigma(x,y), \varepsilon = \varepsilon(x,y)$). With these remarks and Coulomb's gauge ($\text{div}\,\vec{A} = 0$), only the second part of Equation (7) remains :

$$\text{div}((\sigma + j\omega\varepsilon).\text{grad}\,V) = 0 \tag{9}$$

3. Formulation and methods

First of all Equation (9) is projected on a space of functions $(N_i)_{1 \le i \le N}$ that we'll define later and then integrated on the whole domain Ω, we obtain :

$$\forall i \quad 1 \le i \le N \quad \iiint_\Omega N_i \cdot \text{div}((\sigma + j\omega\varepsilon).\text{grad}\,V).d\Omega = 0 \tag{10}$$

Then with Ostrogradsky's theorem applied to Equation (10) :

$$\forall i \; 1 \le i \le N \; -\iiint_\Omega \text{grad}\,N_i.(\sigma + j\omega\varepsilon).\text{grad}\,V.d\Omega + \oiint_S N_i.(\sigma + j\omega\varepsilon).\text{grad}\,V.d\vec{S} = 0 \tag{11}$$

(N_i) functions are chosen in order that $V = \displaystyle\sum_{i=1}^{N} N_i.V_i$ ((V_i) are V nodes values). So Equation (11) is transformed into Equation (12) :

$$M(V).V = F \tag{12}$$

where : $\quad V = (V_j)_{1 \le j \le N}; M = (M_{ij})_{\substack{1 \le i \le N \\ 1 \le j \le N}} = \iiint_\Omega \text{grad}\,N_i.(\sigma + j\omega\varepsilon).\text{grad}\,N_j.d\Omega \quad$ and

$F = (F_i)_{1 \le i \le N} = \oiint_S N_i.(\sigma + j\omega\varepsilon).\text{grad}\,V d\vec{S}$. Boundary conditions are accounted for into the matrix M.

4. Nonlinear algorithms

4.1 Fixed-Point Method

This algorithm involves calculating iteratively the result given by Equation (13), until the difference between V^k and V^{k+1} is smaller than an accuracy criterion :

$$V^{k+1} = [M(V^k)]^{-1}.F = f(V^k) \tag{13}$$

4.2 Newton-Raphson method

A problem exists in that Equation (12) is not differentiable with respect to the complex variable V_j. Real and complex parts have been separated to obtain real unknowns ($V_j^{\mathcal{R}}$ and $V_j^{\mathcal{I}}$). As a material law depends on $|\vec{E}|^2$ ($|\vec{E}|^2 = E_x \cdot \overline{E}_x + E_y \cdot \overline{E}_y$) in first approximation, we may find the derivative of Equation (12) versus this variable. Then the Jacobian is [2]:

$$[J] = \iiint_\Omega \begin{bmatrix} [\sigma \operatorname{grad} N_i \cdot \operatorname{grad} N_j] & [-\omega\varepsilon \operatorname{grad} N_i \cdot \operatorname{grad} N_j] \\ [\omega\varepsilon \operatorname{grad} N_i \cdot \operatorname{grad} N_j] & [\sigma \operatorname{grad} N_i \cdot \operatorname{grad} N_j] \end{bmatrix}$$

$$+2 \begin{bmatrix} [\operatorname{grad} N_i \cdot \vec{E}^{\mathcal{R}} \dfrac{\partial\sigma}{\partial|\vec{E}|^2} \operatorname{grad} N_j \cdot \vec{E}^{\mathcal{R}}] & [\operatorname{grad} N_i \cdot \vec{E}^{\mathcal{R}} \dfrac{\partial\sigma}{\partial|\vec{E}|^2} \operatorname{grad} N_j \cdot \vec{E}^{\mathcal{I}}] \\ [\operatorname{grad} N_i \cdot \vec{E}^{\mathcal{I}} \dfrac{\partial\sigma}{\partial|\vec{E}|^2} \operatorname{grad} N_j \cdot \vec{E}^{\mathcal{R}}] & [\operatorname{grad} N_i \cdot \vec{E}^{\mathcal{I}} \dfrac{\partial\sigma}{\partial|\vec{E}|^2} \operatorname{grad} N_j \cdot \vec{E}^{\mathcal{I}}] \end{bmatrix}$$

$$+2 \begin{bmatrix} [-\omega \operatorname{grad} N_i \cdot \vec{E}^{\mathcal{I}} \dfrac{\partial\varepsilon}{\partial|\vec{E}|^2} \operatorname{grad} N_j \cdot \vec{E}^{\mathcal{R}}] & [-\omega \operatorname{grad} N_i \cdot \vec{E}^{\mathcal{I}} \dfrac{\partial\varepsilon}{\partial|\vec{E}|^2} \operatorname{grad} N_j \cdot \vec{E}^{\mathcal{I}}] \\ [\omega \operatorname{grad} N_i \cdot \vec{E}^{\mathcal{R}} \dfrac{\partial\varepsilon}{\partial|\vec{E}|^2} \operatorname{grad} N_j \cdot \vec{E}^{\mathcal{R}}] & [\omega \operatorname{grad} N_i \cdot \vec{E}^{\mathcal{R}} \dfrac{\partial\varepsilon}{\partial|\vec{E}|^2} \operatorname{grad} N_j \cdot \vec{E}^{\mathcal{I}}] \end{bmatrix} d\Omega$$

$$(14)$$

The obtained asymmetrical Jacobian will increase the computation time. Hence we have chosen to remain in complex arithmetic with an approximation of the Jacobian. The tangential matrix is approximated by :

$$\forall (i;j) \in \{1;N\}^2 \quad M_{t\ ij} = M_{ij} + 2.(\iiint_\Omega \frac{\partial(\sigma + j\omega\varepsilon)}{\partial|\vec{E}|^2} [\operatorname{grad} Ni \cdot \vec{E}^{\mathcal{R}}][\operatorname{grad} N_j \cdot \vec{E}^{\mathcal{R}}] d\Omega)$$

$$+ 2.(\iiint_\Omega \frac{\partial(\sigma + j\omega\varepsilon)}{\partial|\vec{E}|^2} [\operatorname{grad} Ni \cdot \vec{E}^{\mathcal{I}}][\operatorname{grad} N_j \cdot \vec{E}^{\mathcal{I}}] d\Omega)$$

$$(15)$$

This means that the other terms in the actual gradient matrix are neglected [3].

5. Implementation and Algorithm tests

5.1 Theoritical cases

To control all the parameters, we have written a fortran code. We have used a simple square geometry. The domain was meshed in square linear elements. To compare the two algorithms in terms of accuracy and convergence speed, we used first cases with analytical solution as given by Equation (16) :

$$\forall (x,y) \in \mathfrak{R}^2 \quad V(x,y) = e^{x + j.\sqrt{2}.y} \quad \sigma + j\omega\varepsilon = \sqrt{\frac{\partial V}{\partial x} \cdot \frac{\partial \overline{V}}{\partial x} + \frac{\partial V}{\partial y} \cdot \frac{\partial \overline{V}}{\partial y}} \quad (16)$$

In comparison with the analytical solution the Fixed-Point software gives accurate results while the Newton - Raphson accuracy is not better than 1%. We have used another analogous test :

$$\forall (x,y) \in \Re^2 \quad V(x,y) = e^{x+j\cdot\sqrt{\frac{5}{3}}\cdot y} \quad \sigma + j\omega\varepsilon = \sqrt[3]{\frac{\partial V}{\partial x}\cdot\frac{\partial \overline{V}}{\partial x} + \frac{\partial V}{\partial y}\cdot\frac{\partial \overline{V}}{\partial y}} \quad (17)$$

The N.R accuracy is improved. However it must be noticed that for all theoretical cases, the Fixed-Point method was faster and more accurate than the Newton-Raphson method. Fixed-Point errors are due to numerical noise. It takes 110 F.P. iterations vs 163 for the Newton-Raphson method to achieve the same accuracy.

5.2 Physical cases

We have first modeled a plane condensor with various dielectric materials : permitivity dependent on the square modulus of $\vec{E}$, several shape functions to test the derivative terms rigidity in the Newton-Raphson formulation, and so on... For each test, the calculated solution was exact for both methods.

Then we have modeled a simple device as shown in Figure (1) (Horizontal faces have Dirichlet boundary conditions and vertical ones have Neuman boundary conditions):

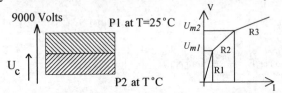

Figure 1 : geometry and characteristic of the problem.

Neither of these two algorithms converged. Their error criterion seemed, after a few iterations, to swing between two values. For the Newton-Raphson method, this drawback can be cancelled with the introduction of an under relaxation factor. The results given by this modified Newton-Raphson method are good.

6. Conclusion

Two methods to solve nonlinear complex equations have been described in this paper. The Point Fixed algorithm, although simpler and faster for theoretical cases, was unable to solve all kind of problems.

Hence a Newton-Raphson algorithm, which separated real and imaginary parts, was built with an under relaxation factor. This method seems more reliable to model industrial devices with strong permitivity nonlinearity. Moreover, the lightning conductor model is now performed with this method.

References

[1]   Y. Du Terrail and J.-C. Saboonadière, « Nonlinear Complex Finite Elements Analysis of Electromagnetic Field in Steady-State AC Devices » IEEE Transaction on Magnetics, vol. 20, pp. 549-552, July 1984.

[2]   G. Dhatt and G. Thouzot, « Une présentation de la méthode des éléments finis » Collection Université de Compiègne, Maloine S.A. Editeur

[3]   A. G. Jack, B. C. Mecrow, « Methods for magnetically nonlinear problems involving significant hysteresis and eddy currents », IEEE Transaction on Magnetics, vol. 26, pp. 424-429, March 1990.

Non-Linear Electromagnetic Systems
V. Kose and J. Sievert (Eds.)
IOS Press, 1998

Nonlinear Effects in Antennae and the Methods of their Analysis

Y.S. SHIFRIN and A.I. LUCHANINOV
Kharkov State Technical University of Radioelectronics
Lenin Avenue, Kharkov, 310726, Ukraine

Abstract. The main nonlinear effects in antennae (NEA), methods of their analysis and the developed theory applications are treated in a brief form. The paper is largely based on the authors' results.

1. NEA general characteristics

The interest in the NEA investigation is related, firstly, to the introduction into practice of antennae with nonlinear elements (ANE) such as antennae - rectifiers, multipliers, mixers, etc. Secondly, the interest is due to the fact that NEA can noticeably complicate the EMC problem of radioengineering systems. The NEA can be due to both the nonlinear elements (NE) specifically built in the antenna and parasitic nonlinearities stipulated either by its construction (for examples, contacts in reflector antennae) or by an unfavorable operation regime of its active elements (for example, amplifiers in the active phased antenna arrays - APAA). Apart from the nonlinearity nature its presence manifests itself in the appearance of new spectral components in the antenna response and its characteristics' dependence on the input signal level. Appearance of new spectral components is useful in a number of cases. Thus, it forms the basis of functioning of the above mentioned new types of ANE. At the same time it manifests itself in the emergence of spurious (usually "harmful") radiation on the basic signal harmonic or combinational frequencies. The latter situation can take place both in multi-frequency (transmitting) and single-frequency (transmitting or receiving) antennae with nonlinearities, if the latter are affected by a noise signal along with the basic signal. If the combinational frequencies get into the receiving passband the formation of accessory reception channels in the receiving antennae is feasible.

The fact that antenna characteristics depend on the input signal level dictates the necessity of the NEA computations with various input signal levels. It can also manifest itself in deterioration of antenna parameters on the basic frequency. Thus, if the signals amplitude limitation takes place in the amplifiers of the receiving APAA, then it results in the widening of the antenna radiation pattern (RP) and in the side lobe rise.

It should be emphasized that the degree of manifestation of both useful and harmful NEA depends on the concrete antenna construction, the places of inclusion of the elements with nonlinear characteristics, the input signal, etc. Hence, NEA correct analysis requires an integrated approach, where it is necessary to consider characteristics of all the ANE parts and also its excitation conditions.

2. ANE theoretical analysis methods

The aim of analysis is to define the spectral composition of the ANE response versus the character of external action. It is supported to characterize the latter by the input actions vector whose components describe ANE excitation both by an electromagnetic field from the outer space and by other exterior sources (exterior generators). The ANE response is characterized by the output parameters vector whose components are the values describing the antenna connection with space and exterior devices (receivers and generators).

At present an approach based on the ANE state concept is mainly used for the ANE analysis. This state is defined by the values of a number of state variables depending on the input action and the ANE scheme parameters. The corresponding dependencies are described by the state equations (SE).

A sufficiently general method for deriving SE and subsequent ANE analysis was offered by the authors in [1-3]. The present work proposes a further refinement of this method significantly widening its possibilities. The generalized ANE scheme (Fig.1), as in [1-3] is the initial one, where the antenna is presented as a joint circuits of three linear multiports (LM) and one nonlinear multiport (NM).

The linear multiport LM-1 describes the ANE radiator system. It is characterized by the scattering matrix $\hat{S}(\omega)$ connecting the vectors of the incident $\mathbf{a}''(\omega)\rangle$ and reflected $\mathbf{b}''(\omega)\rangle$ waves on the radiators system inputs (section $\beta - \beta$) and the vector of the incident $\mathbf{u}'_{in}(\omega)\rangle$ and reflected $\mathbf{u}'_r(\omega)\rangle$ (section $\delta - \delta$) "spherical waves in the free space" [4].

The linear multiport LM-2 describes the load and is characterized by the scattering matrix $\mathbf{S}_L(\omega)$ connecting the vectors of the incident $\mathbf{a}'(\omega)\rangle$ and reflected $\mathbf{b}'(\omega)\rangle$ waves on its inputs. The vector $\mathbf{b}_0(\omega)\rangle$ characterizing the waves excited by the sources in the LM-2 is also shown on the Fig.1.

The differences in the method proposed in this work from that described in [1-3] consist in the modes of describing the nonlinear multiport characteristics and the linear multiport LM-3 connected with it. Thus, in [1, 2] the NM input condition was described (as it is accepted in NE models used in the electron circuits analysis) by the vectors of currents $\mathbf{i}_{NL}(t)\rangle$ and voltages $\mathbf{u}_{NL}(t)\rangle$ connected by nonlinear operators $\widetilde{G}$ or $\widetilde{R}$ ($\mathbf{i}_{NL}(t)\rangle = \widetilde{G}\{\mathbf{u}_{NL}(t)\rangle\}$ or $\mathbf{u}_{NL}(t)\rangle = \widetilde{R}\{\mathbf{i}_{NL}(t)\rangle\}$). Recently, owing to the "advancement" of NE to the microwave region, development of new NE types and their models, the NM input operation began to be described in terms of the incident $\mathbf{a}_{NL}(t)\rangle$ and reflected waves $\mathbf{b}_{NL}(t)\rangle$ connected by the nonlinear operator $\widetilde{S}$ $\left(\mathbf{b}_{NL}(t)\rangle = \widetilde{S}\{\mathbf{a}_{NL}(t)\rangle\}\right)$.

The given method modification proposes to use simultaneously all three methods in its description for characterizing NM. To this end, a common operator $\Re\{\cdot\}$ uniting $\widetilde{G}\{\cdot\}, \widetilde{R}\{\cdot\}$

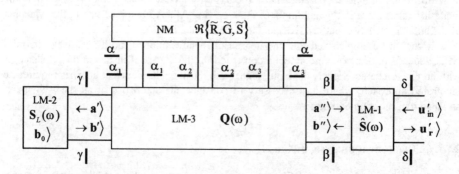

Fig. 1.

and $\widetilde{S}\{\}$ is introduced in the treatment, it transfers the input action on the NM: the vector $x_{NL}^{\alpha}(t)\rangle = (i_{NL}^{\alpha_1}(t)\rangle, u_{NL}^{\alpha_2}(t)\rangle, a_{NL}^{\alpha_3}(t)\rangle)_T$ into its response - vector $y_{NL}^{\alpha}(t)\rangle = (u_{NL}^{\alpha_1}(t)\rangle,$ $i_{NL}^{\alpha_2}(t)\rangle, b_{NL}^{\alpha_3}(t)\rangle)_T$. Here T index means the operation of transposition. In this case the NM inputs (section $\alpha - \alpha$, Fig. 1) are divided into three groups: $\alpha_1 - \alpha_1$, $\alpha_2 - \alpha_2$ and $\alpha_3 - \alpha_3$, supposing that in the first two of them the regimes are described by currents and voltages, the connection between them is given by the operators $\widetilde{G}\{\}$ and $\widetilde{R}\{\}$, respectively. In the section $\alpha_3 - \alpha_3$ it is described by the incident and reflected waves coupled by the operator $\widetilde{S}\{\}$.

Generalization in the NM description cited above requires an alternative (as compared with that accepted in [1-3]) characterization of the multiport LM-3. It is reasonable to characterize the operation of its input connected with the nonlinear multiport ($\alpha - \alpha$ section) by the vectors of the same type as those characterizing the nonlinear multiport inputs operations. For this purpose it is convenient to use the mixed matrix $Q(\omega)$ connecting the vectors $x^{\alpha}(\omega)\rangle$, $a^{\beta}(\omega)\rangle$, $a^{\gamma}(\omega)\rangle$, on the one hand, with the vectors $y^{\alpha}(\omega)\rangle$, $b^{\beta}(\omega)\rangle$, $b^{\gamma}(\omega)\rangle$, on the other hand, defined by the following relationship:

$$\begin{pmatrix} y^{\alpha}\rangle \\ b^{\beta}\rangle \\ b^{\gamma}\rangle \end{pmatrix} = \begin{pmatrix} Q_{\alpha\alpha} & Q_{\alpha\beta} & Q_{\alpha\gamma} \\ Q_{\beta\alpha} & Q_{\beta\beta} & Q_{\beta\gamma} \\ Q_{\gamma\alpha} & Q_{\gamma\beta} & Q_{\gamma\gamma} \end{pmatrix} \begin{pmatrix} x^{\alpha}\rangle \\ a^{\beta}\rangle \\ a^{\gamma}\rangle \end{pmatrix}.$$

The proposed new method of NM and LM-3 description allows the development of a common algorithm and complex programs suitable for analyzing the ANE containing NE at rather complicated models which can be dissimilar in different parts of radio band, and depending on the required ANE analysis depth.

Determination of LM matrix $\mathfrak{R}\{\}$ operator, input vectors $u'_{in}(\omega)\rangle$ and $b_0(\omega)\rangle$ for the specific ANE, selection of the state variables vector, derivation of the state equations using conditions of NM and LM connections in $\alpha - \alpha$ section form the first ANE analysis stage.

As a state variable either $x^{\alpha}(\omega)\rangle$ vector or $y^{\alpha}(\omega)\rangle$ vector can be chosen. The choice of either of two indicated vectors is absolutely equivalent because, firstly, either of them unambiguously defines the output operation in $\alpha - \alpha$ section and, secondly, knowing either of them it is possible to define the whole ANE scheme operation, i.e. its operation in $\beta - \beta$, $\gamma - \gamma$ and $\delta - \delta$ sections.

Thus, if $x^{\alpha}(\omega)\rangle$ vector is chosen as a state variables vector then the SE system assumes the form for steady conditions with periodic or nearly periodic excitation on frequencies $\{\omega_0, \omega_1, ..., \omega_q\}$

$$\sum_{n=-N}^{N} \delta_n \{r_h\}^{-1} x^{\alpha}(v_n)\rangle e^{jv_n t} - \mathfrak{R}\left\{ \sum_{n=-N}^{N} \delta_n \{r_h\} Q_{\alpha\alpha}^{\Sigma} x^{\alpha}(v_n)\rangle e^{jv_n t} + \Psi(t) \right\} = 0, \quad (1)$$

where δ_n - is the Kroeneker symbol; $\{r_h\}$ is the normalizing matrix [4], $Q_{\alpha\alpha}^{\Sigma}$ is the mixed matrix characterizing the whole linear subscheme of ANE (with allowance for the multiports LM-1 and LM-2 connected to LM-3); $\Psi(t)\rangle$ is a vector of external actions converted to sections $\alpha - \alpha$. The values v_n characterize the possible combinations of input actions:

$$v_n = m_{0n}\omega_0 + m_{1n}\omega_1 + ... + m_{qn}\omega_q; \quad m_{in} = 0, \pm 1, \pm 2, ...$$

In general, a real device response is of interest not at all the frequencies v_n ($-\infty < n < \infty$), but at some limited frequencies set. Therefore we confine ourselves to the values $|n| \leq N$ in the relationship (1).

The SE system solution presents the second stage of ANE analysis.

This solution can be obtained numerically [1, 2] or analytically [3]. In the first case the state variables are presented by the Fourier series. Then the initial system of non-linear SE is transformed to a more simple system of nonlinear equations for the harmonic components of the state variable components being solved further with some iteration methods.

The analytical method of SE solution described in [3] is based on the Volterra series application in the matrix formulation. Practically this method results in the ANE structural model design. An essential property of this method is that the increase of the external action level results in the ANE structural model increase, the structural models of the lower level remaining undisturbed.

The last, third stage of the ANE general method consists in the determination of the system output parameters vectors $\mathbf{u}'_r(v_n)\rangle$ and $\mathbf{a}'(v_n)\rangle$ by the state variables vector $\mathbf{x}^\alpha(\omega)\rangle$ calculated in the second stage and the known input actions vectors $\mathbf{u}'_{in}(\omega_n)\rangle$ and $\mathbf{b}_0(\omega_n)\rangle$. After $\mathbf{u}'_r(v_n)\rangle$ and $\mathbf{a}'(v_n)\rangle$ calculation by the formulae given in [4], all the external parameters required for the given ANE type are defined: its RP, directive gain, load power, etc. The parameters should be defined not only at the external action frequencies but at all the new frequencies caused by nonlinearities, with various levels of the input action, separately for transmitting and receiving antennae (due to ANE non-reciprocity).

The ANE analysis method specified above in a brief form was presented in more detail in [1-3]. Application of this method of the state variables generalized vector $\mathbf{x}^\alpha(\omega)\rangle$ to a new version results naturally in the generalization of the ANE analysis procedure described in [1-3], of its different stages. In particular, it is a combination of the approaches to the SE solution proposed in [1,2] and [3].

3. Conclusion

In conclusion it should be noted that the ANE analysis method developed by the authors has been applied to a number of important problems. Among them there are: rectennas investigation [5], calculation of the scattering cross-section of the objects with nonlinear elements, investigation of nonlinear effects in APAA [1], analysis of antennae with distributed nonlinearities [3], etc.

References

[1] Y.S.Shifrin, A.I.Luchaninov. "Nonlinear effects in active phased arrays", Radiotekhnika i Electronika, vol. 39, No. 7, 1994 (in Russian).

[2] Y.S.Shifrin, A.I.Luchaninov "Antennae with nonlinear elements". Chapter 10 in "Handbook of antennae technique. Vol. 1". Edited by L.D. Bakhrah and E.G. Zelkin. Radiotekhnika Press, Moscow, 1997 (in Russian).

[3] Y.S.Shifrin, A.I.Luchaninov. "State-of-the-art of antennae with nonlinear elements theory", Radioelectronika, vol. 39, No. 9-10, 1997, p. 4-16 (in Russian).

[4] Sazonov. "Principles of antenna array matrix theory", in: Collection on Applied Electrodynamics, Vyssh. Shk., Moscow, No. 6, 1988, p. 111-162 (in Russian).

[5] Y.S.Shifrin, A.I.Luchaninov, V.M. Shokalo, A.A. Shcherbina. "Rectennas", Actes des Journees Internationales de Nice sur les Antennes - JINA-94, novembre 1994.

Non-Linear Electromagnetic Systems
V. Kose and J. Sievert (Eds.)
IOS Press, 1998

Nonlinear Lateral Vibration of Three-Phase Cables Enclosed by a Cylindrical Shell

H.Kawamoto, S.Hiraishi, T.Sugiura, M.Yoshizawa

Department of Mechanical Engineering, Faculty of Science and Technology, Keio University, 3-14-1 Hiyoshi, Kouhoku-ku, Yokohama, Kanagawa, 223, Japan

Abstract This research discusses analytically nonlinear vibrations of conducting cables enclosed by a cylindrical conductor. We show analytical evaluation of the electromagnetic forces acting on a cable enclosed by a shell and on three-phase cables. Those forces have been estimated by introducing a dimensionless parameter. According to this estimation, the force due to the interaction between the cables is dominant over the one due to the induced current in the shell. Nonlinear dynamics analysis of the cables predicts two types of their lateral vibrations: forced vibration and parametric excitation. In particular, the latter leads to nonplanar vibration.

1 Introduction

In new power-transmission systems using SF6 gas insulation or superconducting cables, there are very interesting problems related to the dynamics of cables, from the viewpoint of nonlinear dynamics associated with magneto-solid interactions. As a fundamental study of these problems, we discuss analytically nonlinear lateral vibrations of three-phase cables enclosed by a conducting shell[1,2].

2 Estimation of Electromagnetic Forces Acting on a Cable

Three-phase cables under consideration are symmetrically enclosed in a conducting cylindrical shell as shown in Fig.1. The three cables are placed at a distance R_0 from the central axis of the shell which is a nonmagnetic conductor of inner radius R, thickness d_p, conductivity σ_p and magnetic permeability μ_0. Compared with R, the length l of the cables is sufficiently long and their radius is negligible.

In this model, the electromagnetic force acting on each cable consists of two components, i.e. the force due to the induced current in the conducting shell and the force due to the currents in the other two cables.

When an alternating current I of amplitude I_0 and frequency N is carried in a cable as shown in Fig.2, an electromagnetic force caused by the interaction between I and the induced current in the conducting shell acts on the cable. As a first approximation with respect to δ/R (< 1), this force F_s is expressed as follows[3]:

$$F_s \approx \frac{\mu_0^2 I_0^2 N \sigma_p d_p}{4\pi R} \delta \sin 2Nt \tag{1}$$

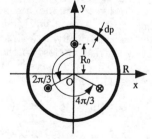

Fig.1 Three-Phase Cables Enclosed by a Conducting Shell

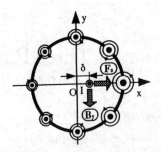

Fig.2 A Cable Inside the Shell

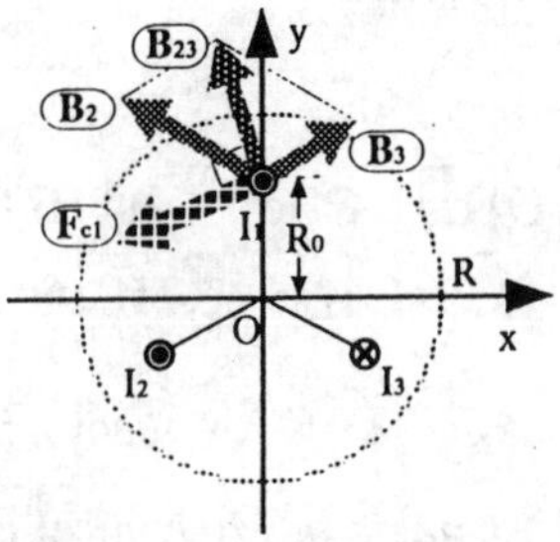

Fig.3 Three Cables Set on the Apex
of a Equilateral Triangle

Three cables, carrying three-phase currents, are set on the apex of a equilateral
triangle at a distance R_0 from the center of x-y orthogonal coordinates.

The electromagnetic forces $\mathbf{F_{c1}}$ acting on Cable 1 are caused by the interaction
between the current $I_0 \sin Nt$ in Cable 1 and $I_0 \sin(Nt + 2\pi/3)$ in Cable 2, and by the
interaction between the current in Cable 1 and $I_0 \sin(Nt + 4\pi/3)$ in Cable 3, as shown
in Fig.3. These forces are expressed analytically as,

$$F_{c1x} = \frac{\mu_0 I_0^2}{12\pi R_0} \sin Nt \left[-3\cos Nt + (\sin Nt/R_0)u_1 + (3\cos Nt/R_0)v_1 \right.$$
$$\left. + (\sin(Nt + 2\pi/3)/R_0)(u_2 - \sqrt{3}v_2) + (\sin(Nt + 4\pi/3)/R_0)(u_3 + \sqrt{3}v_3)\right] \quad (2)$$

$$F_{c1y} = \frac{\mu_0 I_0^2}{12\pi R_0} \sin Nt \left[3\sin Nt + (3\cos Nt/R_0)u_1 - (\sin Nt/R_0)v_1 \right.$$
$$\left. - (\sin(Nt + 2\pi/3)/R_0)(\sqrt{3}u_2 + v_2) + (\sin(Nt + 4\pi/3)/R_0)(\sqrt{3}u_3 - v_3)\right] \quad (3)$$

where F_{c1x} and F_{c1y} are the x and y component of $\mathbf{F_{c1}}$, respectively. u_i and v_i are the
x and y component of the dynamic displacement of the cable $i(i = 1, 2, 3)$. The force
acting on the cable 2 and 3 are expressed in the same manner from the symmetry.

Using Eqs.(1) $\sim$ (3), we obtain a dimensionless parameter,

$$\frac{F_s}{F_c} \approx \frac{\mu_0 N \sigma_p d_p R_0^2}{R} \quad (4)$$

to estimate the electromagnetic force due to the induced current in the conducting
shell. With parameters of a practical system, $\mu_0 = 4\pi \times 10^{-7}$, $N = 80\pi$, $\sigma_p = 5.91 \times 10^7$, $d_p = 5 \times 10^{-4}$, $R_0 = 4 \times 10^{-3}$ and $R = 6 \times 10^{-3}$, the magnitude of $\mathbf{F_s}$ is
10^{-2} and small compared with that of $\mathbf{F_c}$.

Thus we consider only the vibration of the cable caused by the electromagnetic
force due to the currents in the other two cables, in the following section.

However it is interesting to discuss vibration
of a cable due to the induced current in a conducting shell. Lateral vibration of the cable is
excited parametrically if the frequency of the alternating current is close to the natural one of
the cables. The equations of motion of the cable
are expressed as follows:

$$\ddot{\xi} + (1 - \epsilon \hat{\alpha}_1 \nu \sin 2\nu t)\xi = -\epsilon \hat{\chi}\xi(\xi^2 + \eta^2) - 2\epsilon \hat{\mu}\dot{\xi} \quad (5)$$

$$\ddot{\eta} + (1 - \epsilon \hat{\alpha}_1 \nu \sin 2\nu t)\eta = -\epsilon \hat{\chi}\eta(\xi^2 + \eta^2) - 2\epsilon \hat{\mu}\dot{\eta} \quad (6)$$

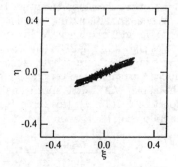

Numerical results of Eqs.(5) and (6) are shown in
Fig.4. Approximate analytical solutions can also
be obtained by the method of multiple scales[4].

Fig.4 Trajectory of Planar Motion
(Numerical Simulation, $\nu = 1.02$)

3   Derivation of the Equations Governing Vibrations of the Cables

We consider the motion of three-phase cables with density ρ, cross section A, length l, Young's modulus E, damping ratio c, fixed at both ends with initial tension T_0. Introducing dimensionless variables with the asterisk, $u_i^* = u_i/R_0$, $v_i^* = v_i/R_0$, $z^* = z\pi/l$, $t^* = t(\pi/l)\sqrt{T_0/\rho A}$ and dimensionless parameters, $\mu = cl/2\pi\sqrt{\rho A T_0}$, $\chi = EAR_0^2\pi^2/4T_0l^2$, $\alpha_1 = \mu_0 I_0^2 l^2/12\pi^3 T_0 R_0^2$, $\nu = N(l/\pi)\sqrt{\rho A/T_0}$, we obtain the equations governing nonplanar vibrations of the cables in the dimensionless form as follows:

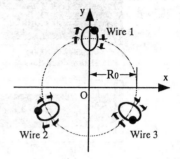

Fig.5 Mutual Relation of Motion

$$\frac{\partial^2 u_{ij}}{\partial t^2} + 2\mu\frac{\partial u_{ij}}{\partial t} = \left\{1 + \chi\int_0^\pi \left[\left(\frac{\partial u_{ij}}{\partial z}\right)^2 + \left(\frac{\partial u_{ik}}{\partial z}\right)^2\right] dz\right\}\frac{\partial^2 u_{ij}}{\partial z^2} + \alpha_1 f_{ij}, \tag{7}$$

$$u_{ij}(0,t) = u_{ij}(\pi,t) = 0, \quad (i = 1, 2, 3 \quad j = x, y \quad k = y, x).$$

where $\alpha_1 f_{ij}$ is the dimensionless j-component of the electromagnetic forces acting on the cable i. The asterisks are omitted in Eq.(7) and henceforward. Substituting an assumption $u_{ij}(z,t) = [\xi_i(t), \eta_i(t)]\sin z + \cdots$ into Eq.(7), we obtain simultaneous differential equations as follows[4]:

$$\ddot{\xi} + 2\mu\dot{\xi} + \xi = -\chi\xi(\xi^2 + \eta^2) + \frac{3}{2}\alpha_1\left[-\frac{4}{\pi}\sin 2\nu t + \xi - \cos 2\nu t\,\xi + \sin 2\nu t\,\eta\right] \tag{8}$$

$$\ddot{\eta} + 2\mu\dot{\eta} + \eta = -\chi\eta(\xi^2 + \eta^2) + \frac{3}{2}\alpha_1\left[\frac{4}{\pi} - \frac{4}{\pi}\cos 2\nu t - \eta + \sin 2\nu t\,\xi + \cos 2\nu t\,\eta\right] \tag{9}$$

The differential equations of other cables are expressed in the same manner from the symmetry shown in Fig.5.

4   Numerical Results and Discussion

It is analytically predicted that the cables are excited parametrically or forcibly by the electromagnetic forces.

Forced vibration of the cables is due to the term of the electromagnetic forces $\sin 2\nu t$ and $\cos 2\nu t$, if the frequency of the alternating current is close to half of the natural one of the cables. By theoretical analysis based on the method of multiple scales, the amplitudes a of this vibration can be obtained from the following equation, $\frac{1}{a^2} = \left(\frac{4\mu}{3\alpha_1}\right)^2 + \left(\frac{8(\nu-1)}{3\alpha_1} - \frac{\chi a^2}{2\alpha_1} + \epsilon\right)^2$. Analytical prediction of the vibration can be confirmed by numerical integration as shown in Fig.6.

Parametric excitation of the cable occurs owing to the terms of the electromagnetic forces $\cos 2\nu t\,\xi$ and $\sin 2\nu t\,\eta$, if the frequency of the alternating current is close to the natural one of the cables. Theoretical analysis estimates that the order of the amplitude of this vibration is equal to that of $\sqrt{\alpha_1/\chi}$, the square of which is the ratio of the electromagnetic force to the nonlinear term. Moreover nonplanar vibration of a cable occurs due to the terms, $\sin 2\nu t\,\eta$ and $\cos 2\nu t\,\xi$, which are the forces normal to the displacement of the cable. Dynamics of the cable stated above can be comfirmed by numerical integration, as shown in Fig.7.

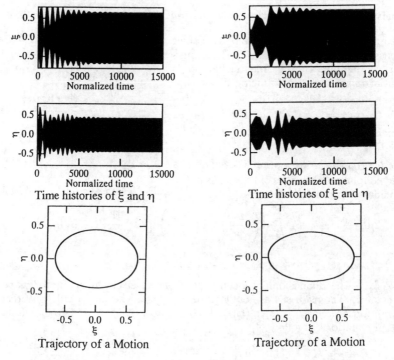

Fig.6 Forced Vibration ($\nu = 0.5$)   Fig.7 Parametric Excitation ($\nu = 1.003$)
(Numerical Simulation)   (Numerical Simulation)

5 Conclusions

In this research, we have investigated theoretically nonlinear lateral vibrations of three-phase current-carrying cables in a conducting shell. Conclusions are summarized as follows.

1. The electromagnetic forces acting on a cable due to the other two cables, $\mathbf{F_c}$, are sufficiently large compared with the one due to the induced current in the shell, $\mathbf{F_s}$. This fact can be explained by introducing a dimensionless parameter which indicates the ratio of $\mathbf{F_c}$ and $\mathbf{F_s}$.

2. The nonlinear lateral vibrations of the three-phase cables due to $\mathbf{F_c}$ are of two types, i.e. the parametric excitation and forced vibration. In particular, parametric excitation leads to nonplanar vibration of the cables.

Acknowledgment. The authors would like to express their gratitude to Hitachi-Cable.Co. for financial support, Dr.A.Bossavit for providing them with valuable references, Mr.S.Shimokawa and Mr.T.Kawaguchi, a graduate student and an undergraduate student of Keio Univ. for their assistance in calculation.

References

[1] M.N.Zervas and E.E.Kriezis, Integral for the field and the forces in a system of conducting cylindrical shell: a general approach, IEE PROCEEDINGS, Vol. 134, Pt.B, No.5, pp.269-275,September 1987.
[2] A.Safigianni and Prof.D.Tsanakas, Short-circuit electromagnetic forces in three-phase gas insulated systems with aluminium enclosure, IEE PROCEEDINGS, Vol.133, Pt.B, No.5, pp.331-340, September 1986.
[3] Moon.F.C., *Magneto-Solid Mechanics*, John-Wiley & Sons, Inc., 1984.
[4] A.H.Nayfeh and D.T.Mook, *Nonlinear Oscillations*, Wiley, 1979.

Non-Linear Electromagnetic Systems
V. Kose and J. Sievert (Eds.)
IOS Press, 1998

Partially Relaxed MHD Equilibria in Gun-Injected Tokamak

Atsushi KAMITANI* and Soichiro IKUNO**

* *Yamagata University, Johnan 4-3-16, Yonezawa, Yamagata 992, Japan*
***University of Tsukuba, Tennoudai 1-1-1, Tsukuba, Ibaraki 305, Japan*

Abstract. The method for producing the gun-injected tokamak is to insert a center conductor into the flux conserver and to apply an electric current I_s externally. Equilibrium configurations of the gun-injected toroidal plasma are determined numerically and their stability is investigated by use of the Mercier criterion. The results of computations show that the equilibrium changes its state from spheromak through ultra-low q to tokamak with the increase of I_s. In addition, the Mercier limit of the spheromak decreases to zero as I_s is increased, whereas that of the tokamak becomes larger with the increase of I_s.

1. Introduction

The gun-injected tokamak is a tight-aspect-ratio toroidal plasma that has many potential advantages as a fusion reactor: large elongation, high beta ratio, large plasma current and simplicity of the metallic vessel. In addition, the gun-injected tokamak has a possibility of translation of a plasma, which allows the formation chamber to be separated from the burn chamber and, therefore, enables two toroidal plasmas to be merged into one plasma with a larger current. On account of these advantages, the gun-injected tokamak has recently attracted great attention as an alternative to the conventional fusion device.

The method for producing the gun-injected tokamak is to use the Marshall gun. After production in the gun, the plasma is ejected into the metallic vessel which is called a flux conserver (FC). Although this method is very similar to that for the spheromak formation, the only difference is the center conductor inserted along the symmetry axis of the FC. In this center conductor, the current is flowing to generate the toroidal field. The experiments by this method have been used in the HIST[1] and the "rodmak[2]" experiments.

Although there is a growing interest in the gun-injected tokamak, further experimental and numerical investigations are required to distinguish the gun-injected tokamak from other toroidal fusion plasmas. The purpose of the present study is to numerically determine the MHD equilibrium configurations in the HIST device and to investigate the effect of the center-conductor current on their stability.

2. MHD Stability and Critical Beta Ratio of Tokamak Plasma in Spherical FC

In this section, we investigate the MHD stability of the plasma confined in the spherical FC. The center conductor of radius R_i is inserted into the FC and the current I_s is forced to flow in the conductor. This poloidal current, I_s, generates the toroidal flux in the plasma. Throughout the present paper, the value of R_i/L is fixed as 0.114, where L denotes the radius of the spherical FC. In Fig. 1, we show the model of the FC that is used in this section.

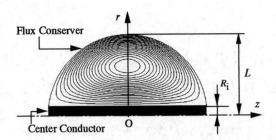

Fig. 1. The equilibrium configuration with $\beta_0 = \sigma = 0$ and $I_s/I_p = 0.2$.

As is well known, the equilibrium configurations of an axially symmetric plasma can be determined by solving the Grad-Shafranov equation:

$$-\hat{L}\psi = \chi(z, r)\left(r^2\frac{\mathrm{d}p}{\mathrm{d}\psi} + \frac{1}{2}\frac{\mathrm{d}I^2}{\mathrm{d}\psi}\right) ,$$ (1)

where $p(\psi)$ and $I(\psi)$ denote the pressure and the toroidal magnetic field function, respectively, and ψ is a flux function. Further $\chi(z, r)$ denotes the characteristic function which takes unity in the plasma region and vanishes elsewhere. The boundary conditions for Eq. (1) are assumed as $\psi = 0$ on the surface of the FC and the center conductor. As $p(\psi)$ and $I(\psi)$, we assume

$$\frac{\mathrm{d}p}{\mathrm{d}\psi} = \frac{\beta_0}{(RL)^2}\left[\lambda\,(1 + \sigma) + \frac{I_s\,L}{2\,\pi\,\psi_{\text{axis}}}\right]^2 (\psi_{\text{axis}} - \psi) ,$$ (2)

$$I(\psi) = \frac{\lambda\,\psi_{\text{axis}}}{L}(\Psi + \sigma\,\Psi^2) + \frac{I_s}{2\,\pi} ,$$ (3)

where ψ_{axis} denotes a value of ψ on the magnetic axis and $\Psi \equiv \psi/\psi_{\text{axis}}$. In addition, β_0, λ, and σ are constants and R is the r-coordinate of the magnetic axis for the case of $\beta_0 = \sigma = I_s = 0$. The similar assumption for $I(\psi)$ is used by Browning et al.[3] although they leave the pressure effect out of consideration. Under the above assumptions, Eq. (1) and the boundary conditions constitute the nonlinear Eigenvalue problem with Eigenvalue λ. This problem is solved numerically by means of the algorithm of the nonlinear Eigenvalue problem[4]. A typical example of equilibrium configurations is shown in Fig. 1.

Next, we investigate the stability of equilibrium configurations by use of the Mercier criterion. The Mercier criterion is actually tested on magnetic surfaces $\psi = i\,\psi_{\text{axis}}/20$ ($i = 1, 2, \cdots, 19$). Let us evaluate the maximum value of the beta ratio, β_{max}, below which the Mercier criterion is fulfilled on every magnetic surface. In Fig. 2, we show the dependence of β_{max} on I_s/I_p for various values of σ. Here I_p denotes the total plasma current. As I_s/I_p is increased, β_{max}

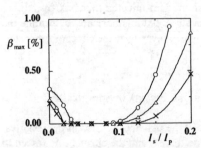

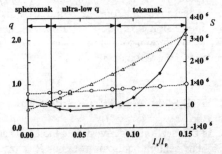

Fig. 2. Maximum beta ratio β_{max} as functions of I_s/I_p. $\bigcirc$: $\sigma = 0.0$, $\triangle$: $\sigma = 0.3$, $\times$: $\sigma = 0.5$.

Fig. 3. Values of q_s, q_{axis} and S for equilibria with the maximum beta ratios and $\sigma = 0.0$. The symbols $\triangle$ and $\bigcirc$ denote the values of q_s and q_{axis}, respectively, and the solid curve indicates dependence of S on I_s/I_p.

decreases monotonously until it vanishes. After β_{max} remains almost zero for $0.02 < I_s/I_p <$ 0.08, it increases with the increase of I_s/I_p for $I_s/I_p > 0.08$. This behavior of β_{max} can be explained from the standpoint of the change of states. As is well known, the safety factor of the spheromak takes the minimum on the plasma surface and increases monotonously toward the magnetic axis. In contrast, the safety factor of the tokamak behaves inversely. In order to characterize the safety factor profile, let us define the shear product by

$$S \equiv \underset{0 \le \Psi \le 1}{\text{Max}}\left(\frac{dq}{d\Psi}\right) \cdot \underset{0 \le \Psi \le 1}{\text{Min}}\left(\frac{dq}{d\Psi}\right) . \tag{4}$$

The equilibrium configuration with $S > 0$ corresponds to tokamak and spheromak because the safety factor changes monotonously with Ψ. Which of the states the configuration becomes is dependent on the sign of $q_{axis} - q_s$. Here q_{axis} and q_s denote the values of the safety factor on the magnetic axis and on the plasma surface, respectively. On the other hand, for the equilibrium configuration with $S < 0$, there exists at least one magnetic surface with zero shear in the plasma region and the Mercier criterion is violated there. Therefore, this equilibrium becomes unstable against localized perturbations. For equilibrium configurations that give β_{max}, values of q_{min}, q_{axis} and S are calculated and are plotted as functions of I_s/I_p in Fig. 3. We see from this figure that the equilibrium configuration changes its state from spheromak through ultra-low q to tokamak with the increase of I_s/I_p.

3. MHD Equilibria of HIST Plasma

The flux conserver actually installed in the HIST device is joined to the helicity injector that is used for amplifying the plasma current externally. The divertor bias coil is placed inside the helicity injector as shown in Fig. 4. The coil is covered with the shielding wall so that the plasma may not be in contact with the coil.

Since the divertor bias coil is turned on long before the plasma is injected into the FC, the poloidal field generated by the coil extends all over the space. On the other hand, the life of the plasma is sufficiently short, as compared with the skin time of the FC and the center conductor, that the magnetic field generated by the plasma current can penetrate neither inside the center conductor nor outside the FC. By taking these circumstances into account, the equilibrium configurations in the HIST device are determined by use of the same assumptions on $p(\psi)$ and $I(\psi)$ as used in the previous section. Although the nonlinear Eigenvalue problem becomes complicated due to such complex boundary conditions, it can be solved successfully by means of the combination of FDM and BEM[4].

In Fig. 4, we show an equilibrium configuration of the force-free state. The difference between the equilibria obtained in the previous section and those in the HIST device is that X point can exist inside the FC of the HIST device. Here X point denotes a point at which the poloidal field vanishes. Even if no X point can be observed in the FC, all magnetic field lines do not close themselves in the plasma region. This is why the stability analysis cannot be

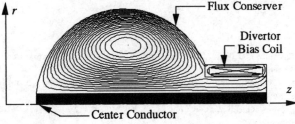

Fig. 4. An equilibrium configuration in the HIST device for the case of $\beta_0 = \sigma = 0$, $I_s/I_p = 0.15$, and $I_D/I_p = 0.05$. Here I_D denotes a coil current.

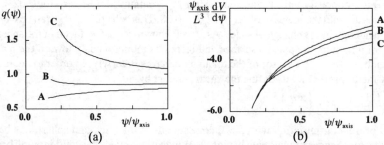

(a)   (b)

Fig. 5. (a) Safety factor $q(\psi)$ and (b) specific volume $dV/d\psi$ for equilibrium configurations with $\beta_0 = \sigma = 0$ and $I_D/I_p = 0.05$. A: $I_s/I_p = 0.00$, B: $I_s/I_p = 0.03$ and C: $I_s/I_p = 0.10$.

performed by using the Mercier criterion. However, surface quantities can be calculated on the magnetic surface between the separatrix and the magnetic axis. Thus let us evaluate the safety factor $q(\psi)$ and the specific volume $dV/d\psi$ inside the separatrix and investigate the stability of the HIST plasma against localized perturbations. In Figs. 5(a) and 5(b), we show the profiles of $q(\psi)$ and $dV/d\psi$ for the force-free states. These figures indicate that the configuration changes its state from spheromak through ultra-low q to tokamak with the increase of I_s/I_p. This tendency bears a close resemblance to the state transition with the increase of I_D/I_p. Further, we see from Fig. 5(b) that the specific volume increases monotonously from the separatrix to the magnetic axis regardless of the values of I_s/I_p. Thus the stabilization by the average-minimum B cannot be expected for the equilibrium in the HIST device. In addition, since the ultra-low q configuration has a region with no shear, it becomes unstable against localized perturbations.

4. Conclusions

We have determined the equilibria of the gun-injected toroidal plasma and have investigated their stability by using the Mercier criterion. Conclusions obtained in the present study are summarized as follows.

1) Equilibria of the tight-aspect-ratio toroidal plasma can be classified into three types of configurations: spheromak, ultra-low q and tokamak.

2) As I_s/I_p is increased, the equilibrium changes its state from spheromak through ultra-low q to tokamak.

3) As long as the equilibrium is a spheromak, β_{max} decreases with the increase of I_s/I_p. The reason for this is explained as follows. As I_s/I_p is increased, q_s increases more rapidly than q_{axis} and, therefore, the magnetic shear becomes smaller near the plasma surface. In contrast, β_{max} increases remarkably with I_s/I_p while the equilibrium becomes a tokamak. This is because shear stabilization is enhanced near the plasma surface with the increase of I_s/I_p.

References

[1] M. Nagata, N. Fukumoto, H. Haruoka, S. Kano, K. Kuramoto and T. Uyama: *Proc. ICCP 96, Nagoya, 1996* (to be published).
[2] P. K. Browning, G. Cunningham, R. Duck, S. J. Gee, K. J. Gibson, D. A. Kitson, R. Martin and M. G. Rusbridge: Phys. Rev. Lett. **68** (1992) 1722.
[3] P. K. Browning, J. R. Clegg, R. C. Duck and M. G. Rusbridge: Plasma Phys. Control. Fusion **35** (1993) 1563.
[4] A. Kamitani, T. Kanki, M. Nagata and T. Uyama: *Advanced Computational Electromagnetics*, ed. T. Honma (IOS Press, Amsterdam, 1995) p. 373.

Non-Linear Electromagnetic Systems
V. Kose and J. Sievert (Eds.)
IOS Press, 1998

Microscopic Analysis of Arcs between Relay Contacts

Hans HAUSER, Heinz HOMOLKA, and Manuela FRANZ
*Institut für Werkstoffe der Elektrotechnik, Technische Universität Wien,
Gußhausstraße 27–29, A–1040 Vienna, Austria.*

Abstract. Using a high speed CCD camera with reduced resolution, the dynamic behaviour of arcs between miniature relay contacts of different types has been studied under dc-current rating and inductive/ohmic load. If the moving contact has a negative potential with respect to the non–operative contact, the arcs can burn permanently with higher probability and are capable of destroying the relay thermically.

1. Introduction

For quality control, miniature power relays for printed circuit boards (specified for a switching capacity of 250V/8A ac) have been tested for dc power rating of the contacts. These components find wide applications in industrial electronics. Reliability problems arise, if even a small dc current in an inductive/ohmic circuit has to be switched off. Modern contact materials [1] as sintered Ag/ZnO or Ag/SnO compounds did not fully satisfy the needs to solve these problems.

2. Experimental

The test circuit (see Fig. 1) consists of a constant voltage source ($U = 110$V) and an ohmic/inductive load (S_1, S_2, S_3, S_4 : $R/L = 690\Omega/1.5$H). Driven by a low frequency

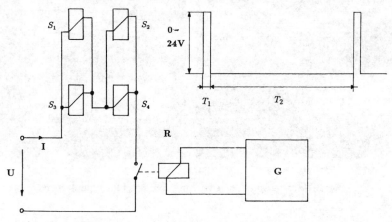

Fig. 1. Schematical test circuit and duty cycle

Fig. 2. Open and closed relay contacts in the image plane of the microscope

generator G ($T_1 = 250$ms, $T_2 = 4750$ms), the sample relay R connects the source with the load once per five seconds. Because a protective circuit (e.g. recovery diode) has been omitted intentionally, the energy $LI^2/2$ is released in the arc between the contacts. During the steady–state condition the current amounts to $I = 160$mA.

The contacts of the relay are located in the image plane of a microscope objective (see Fig. 2). The ocular is replaced by a high speed CCD–camera of reduced resolution. A computer processes the video signals of the camera. The mechanical switching behaviour has been also monitored by measuring the magnetic flux in the core by a compensation method [2].

3. Results and Discussion

The video analysis shows that both the intensity and the burning time of the arc is dependent on the polarity of the contacts in a nonlinear manner. If the moving contact has a negative potential with respect to the non–operative contact, the probability of a permanent arc burning is significantly higher (25%) than in the other case (5%). These arcs can even burn over to the contact springs.

Figure 3 shows burning arcs between the contacts of a miniature power relay of type A. The moving contact has a negative potential with respect to the other one.

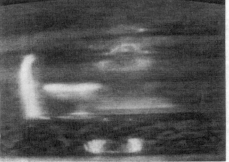

Fig. 3. Arcs between the switching contacts of a type A relay

Figure 4 is a superposition of five switching cycles, recorded at relay type A with high resolution. The small sparks between the contacts just after the beginning of the opening are caused by burning contact material during the comparatively slow contact movement. The magnetizing coil current I_G (and therefore the opening force) increases with the time t after applying the output voltage of the generator G as

$$I_G = \frac{U_G}{R_c} \left(1 - \exp\left[-t\frac{R_c}{L_c}\right]\right) , \tag{1}$$

where R_c and L_c are the resistance and inductance of the relay coil, respectively. For this relay type, the contact movement is in a small range proportional to the force.

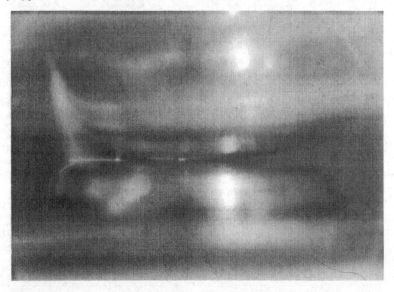

Fig. 4. Superposition of arcs between the switching contacts of a type A relay

Figure 5 illustrates the effect for a relay of type B at positive potential of the moving contact. Figure 6 shows arcs for a type C relay at negative moving contact potential.

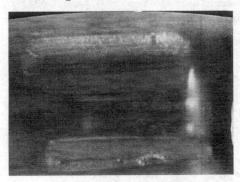 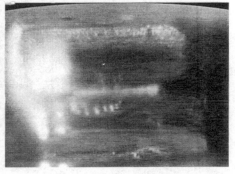

Fig. 5. Arcs between the switching contacts of a type B relay

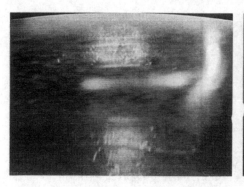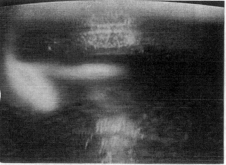

Fig. 6. Arcs between the switching contacts of a type C relay

4. Conclusions

The effects reported above are able to reduce the reliability or even to destroy the miniature relay by thermical overload. The analysis of different types of relays show that these drop outs can be minimized by mechanical design improvements. These are, for instance, a thicker contact spring (to increase the heat capacity, at least in the vicinity of the contacts), longer contact pieces (to allow longer arc distances, burning on the contact material), and the selection of contact materials with a low micro–migration (material transfer) rate [3] (to reduce the danger of short–circuiting).

Acknowledgments

The authors are grateful to Prof. G. Fasching and Dr. P. Matyas † for making these investigations possible and to Paul L. Fulmek for friendly discussions. Financial support was provided by Schrack Components GmbH.

References

[1] H. Hauser, Kontaktwerkstoffe. In: G. Fasching, Werkstoffe für die Elektrotechnik, 3rd Edition. ISBN: 3-211-82610-6. Springer–Verlag, Wien/New York, 1994, pp. 517–524.

[2] H. Hauser, Automatic Quality Control of Small Relays and Their Magnetic Parts, *Sensors and Actuators A* **46** (1995) 588–592.

[3] R. Holm, Electric Contacts, 4th Edition. Springer–Verlag, Berlin/Heidelberg/New York, 1967.

Non-Linear Electromagnetic Systems
V. Kose and J. Sievert (Eds.)
IOS Press, 1998

Magnetic Field Effects on Wave Motion of a Magnetic Liquid

M. OHABA[1], M. TOMOMORI[1], T. SAWADA[2], S. SUDO[3] and T. TANAHASHI[2]

[1] Department of Control System Engineering, Toin University of Yokohama
1614 Kurogane-cho, Aoba-ku, Yokohama, 225-0025 Japan
[2] Department of Mechanical Engineering, Keio University
3-14-1 Hiyoshi, Kouhoku-ku, Yokohama, 223-0061 Japan
[3] Department of Mechanical Engineering, Iwaki Meisei University
5-5-1 Iino, Chuoudai, Iwaki, 970-8044 Japan

Abstract. The wave motion of a magnetic liquid in a cylindrical container subject to a non-uniform magnetic field is examined. The theoretical surface response is firstly analyzed, and then the experimental results of the magnetic field effects on the wave are described. The surface displacement is in quantitative agreement with the theory. The nonlinear swirling phenomenon is made stronger due to the increase in the applied magnetic field and can be controlled using the shift of the resonance frequency in a non-uniform magnetic field.

1. Introduction

Wave motion of a liquid in an accelerated container is an important engineering problem in liquid storage tanks and aerospace vehicles [1]. For example, dynamic responses such as nonlinear wave motion in an accelerated cylindrical container, that is, the swirling behavior, are difficult to analyze because of the complexity of the strong nonlinear phenomena. In our study, the apparent gravitational acceleration on a magnetic liquid can be changed by increasing/decreasing an applied magnetic field gradient using the magnetic body force [2]. Thus, it is possible to terrestrially simulate the nonlinear liquid characteristics in the accelerated container using this method. In a previous study [3] we clarified the effect of a non-uniform magnetic field on the subharmonic oscillations of a magnetic liquid.

2. Theoretical Analysis

Consider the oscillation of a magnetic liquid of height h in a cylindrical container of radius $r = a$ subject to a non-uniform magnetic field $H(r, \theta, z, t)$ along the z-axis, as shown in Fig. 1. When a lateral vibration $y_f = Y_0 e^{i\omega t}$ is applied to the container we derive the resonance frequency and surface displacement from linear theory using a cylindrical coordinate system (r, θ, z), where Y_0 is the forcing amplitude and ω the forcing angular frequency. If the magnetic liquid is inviscid and incompressible, its flow is irrotational, and velocity potential $\phi\,(r, \theta, z, t)$ will satisfy the Laplace equation

$$\left(\frac{\partial^2}{\partial r^2} + \frac{1}{r}\frac{\partial}{\partial r} + \frac{1}{r^2}\frac{\partial^2}{\partial \theta^2} + \frac{\partial^2}{\partial z^2} \right)\phi = 0, \tag{1}$$

with the following conditions at the bottom of and the wall of the container.

$$\frac{\partial \phi}{\partial z} = 0 \quad at \ z = -h, \quad \frac{\partial \phi}{\partial r} = i\omega Y_0 \cos\theta\, e^{i\omega t} \quad at \ r = a. \tag{2}$$

Hence if $\eta\,(r, \theta, t)$ denotes the displacement of the surface, the linearized dynamic boundary condition with Maxwell's stress and surface tension is given by

$$\rho\frac{\partial \phi}{\partial t} + \rho g\eta - \frac{1}{2}\mu_0\left(\chi_m H^2 + M^2\right) - \sigma\left(\frac{1}{r}\frac{\partial \eta}{\partial r} + \frac{\partial^2 \eta}{\partial r^2} + \frac{1}{r^2}\frac{\partial^2 \eta}{\partial \theta^2}\right) = 0 \quad at \ z = 0, \tag{3}$$

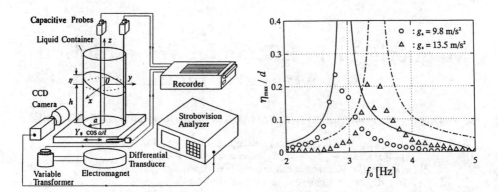

Fig. 1 Experimental apparatus   Fig. 2 Surface response of the first mode

where ρ is the liquid density, g the gravitational acceleration, σ the surface tension, μ_0 the magnetic permeability of vacuum, χ_m the initial magnetic susceptibility, and M the magnetization. Applying Lagrange derivative D/Dt to Eq.(3), and using the linearized kinematical boundary condition $\partial\phi/\partial z = \partial\eta/\partial t$, the surface dynamic condition can be written by

$$\frac{\partial^2\phi}{\partial t^2} + g\frac{\partial\phi}{\partial z} - \frac{\sigma}{\rho}\left(\frac{1}{r}\frac{\partial^2\phi}{\partial r\partial z} + \frac{\partial^3\phi}{\partial r^2\partial z} + \frac{1}{r^2}\frac{\partial^3\phi}{\partial\theta^2\partial z}\right) - (1+\chi_m)\frac{\mu_0 M}{\rho}\frac{\partial H}{\partial z}\frac{\partial\phi}{\partial z} = 0 \quad at\ z = 0. \tag{4}$$

From Eqs.(1) and (2), the general solution of the velocity potential may be

$$\phi(r,\theta,z,t) = \sum_{m=1}^{\infty} A_m e^{i\omega t}\left[i\omega Y_0 r + J_1\left(\xi_m r/a\right)\right]\cos\theta\cosh\xi_m(z+h)/a, \tag{5}$$

where A_m is an unknown coefficient, $J_1(r)$ the first order Bessel function of the first kind and ξ_m the m th root of the equation $\dfrac{d}{dr}J_1\left(\xi_m r/a\right) = 0$ at $r = a$. Determining A_m to satisfy Eq.(4), we get the velocity potential $\phi(r,\theta,z,t)$ as

$$\phi(r,\theta,z,t) = aY_0\omega\left[\frac{r}{a} + 2\sum_{m=1}^{\infty}\frac{\zeta_m^2}{1-\zeta_m^2}\frac{J_1\left(\xi_m r/a\right)\cosh\xi_m(z+h)/a}{(\xi_m^2-1)J_1\left(\xi_m\right)\cosh\left(\xi_m h/a\right)}\right]\cos\theta\cos\omega t, \tag{6}$$

where ζ_m, the resonance circular frequency ω_m, and the apparent gravitational acceleration g_a are given

$$\zeta_m = \frac{\omega}{\omega_m}, \quad \omega_m^2 = \frac{g_a\xi_m}{a}\tanh\left(\xi_m h/a\right), \quad g_a = g + \frac{\xi_m\sigma}{\rho a} - (1+\chi_m)\frac{\mu_0 M}{\rho}\frac{\partial H}{\partial z}. \tag{7}$$

Finally, the surface displacement $\eta(r,\theta,t)$ is then determined

$$\eta(r,\theta,t) = \frac{aY_0\omega^2}{g_a}\left[\frac{r}{a} + 2\sum_{k=1}^{\infty}\frac{\zeta_m^2}{1-\zeta_m^2}\frac{J_1\left(\xi_m r/a\right)}{(\xi_m^2-1)J_1\left(\xi_m\right)}\right]\cos\theta\sin\omega t. \tag{8}$$

3. Experimental Apparatus and Procedure

Figure 1 shows the experimental apparatus. An electrodynamic exciter is used to generate a sinusoidal excitation of the container. The liquid container has a height 230 mm and a inside diameter of 90 mm. The surface displacement is measured using Teflon capacitive probes which are fastened vertically in the container at a distance 5 mm from the wall. One of these probes is positioned along the forcing axis y (point P_y), and the other (point P_x) makes a right angle to the forcing axis. The forcing displacement is measured by a differential transducer. An electromagnet mounted under the container produces a vertical magnetic field, which is variable by varying the coil voltage. A water-based magnetic liquid is used and the liquid depth is $h = 20$ mm.

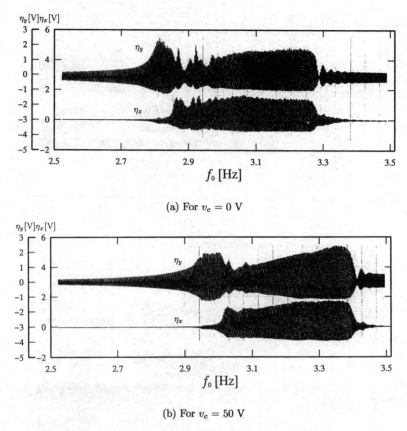

(a) For $v_c = 0$ V

(b) For $v_c = 50$ V

Fig. 3 Surface response of first mode for different values of coil voltage v_c

4. Experimental Results and Discussion

Figure 2 shows the maximum surface displacement of the mode (1,1) for $g_a = 9.8$ m/s^2 and 13.5 m/s^2 and $Y_0 = 2$ mm. In Fig. 2, f_0 is the forcing frequency, η_{max} the maximum surface displacement at P_y, d the diameter of the container and the solid and dash-dotted curves the theoretical results. As g_a increases, the resonance frequency increases and the surface response shifts upwards to the higher frequency range. The experimental results are in good agreement with the theoretical predictions.

Figure 3(a) and 3(b) show the liquid surface responses of the mode (1,1) as the forcing frequency is slowly increased from $f_0 = 2.0$ Hz to 3.8 Hz, where the forcing amplitude is $Y_0 = 2.0$ mm, η_x and η_y the output voltages from the capacitive probes at the point P_x and P_y, respectively. The appearance η_x indicates the onset of the swirling. Compared with the results in Fig. 3(a), in the case of the coil voltage $v_c = 50$ V, the resonance frequency range shifts to a higher frequency range and the maximum displacement of the swirling motion increases, as shown in Fig. 3(b). That is, increasing the apparent gravitational acceleration strengthens the nonlinear swirling motion.

Figures 4(a) and 4(b) show the mode (1,1) transient surface responses η_x, η_y for different coil voltages $v_c = 50$ V and 90 V. Before applying the magnetic field, the container is firstly vibrated by the forcing vibration y_f ($f_0 = 3.0$ Hz and $Y_0 = 2.0$ mm) and the mode (1,1) standing wave occurs. Thereafter, rotating wave motion (swirling) begins and grows along the liquid container wall. In Fig. 4(a), after the coil voltage $v_c = 50$ V is applied at the time $t = 23$ s, surface beating motions in η_x and η_y are observed. In Fig. 4(b), after $v_c = 90$ V at $t = 24$ s, the swirling wave displacement η_x quickly turns into small. That is, the swirling motion is suppressed by the applied magnetic field. Thus the

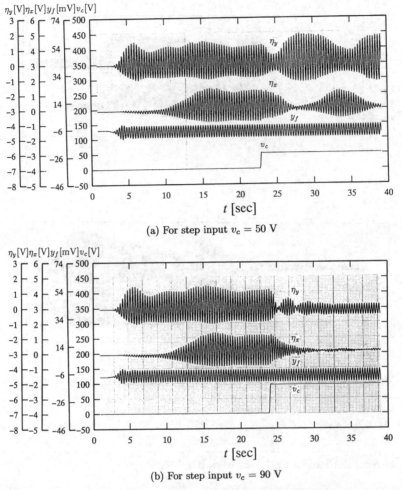

(a) For step input $v_c = 50$ V

(b) For step input $v_c = 90$ V

Fig. 4 Transient surface response for step input of different values of v_c at $f_0 = 3.0$ Hz

increase in g_a caused by the magnetic field gradient shifts the resonance to a higher frequency and ceases the swirling motion. This response suggests the control of a nonlinear liquid wave motion such as swirling using a non-uniform magnetic field.

5. Conclusion

We examined the dynamic surface response of a magnetic liquid subject to a non-uniform magnetic field. The results are as follows: (1) the resonance frequency increases with an increase in magnetic field, (2) the nonlinear swirling is stronger under increased apparent gravitational acceleration, and (3) it may be possible to control the swirling wave motion using the shift of the resonance frequency.

References

[1] H. N. Abramson, The Dynamic Behavior of Liquids in Moving Containers, NASA SP-106 (1966).
[2] R. E. Rosensweig, Ferrohydrodynamics, Cambridge University Press, (1985).
[3] M. Ohaba and S. Sudo, Liquid Surface Behavior of a Magnetic Liquid in a Container Subjected to Magnetic Field and Vertical Vibration, J. Magn. Magn. Mate., 149, (1995), 38-41.

Non-Linear Electromagnetic Systems
V. Kose and J. Sievert (Eds.)
IOS Press, 1998

Behavior of a Magnetic Fluid Drop at an Air–Magnetic Fluid Interface

Seiichi SUDO[a], Noriaki WAKAMATSU[a],
Toshiaki IKOHAGI[b] and Kazunari KATAGIRI[b]

[a] *Dept. of Mechanical Engineering, Iwaki Meisei University, Iwaki 970, Japan*
[b] *Institute of Fluid Science, Tohoku University, Sendai 980, Japan*

Abstract. This paper describes a study of the behavior of some kinds of liquid drops floating on an air–liquid interface under non–magnetic and magnetic fields. A high speed video camera system revealed the effect of magnetic fields applied tangentially to the interface upon a floating magnetic fluid drop at the air–magnetic fluid interface.

1. Introduction

When a container holding water is vibrated at a natural frequency of the fluid–container system, the free surface is disintegrated and spray particles are ejected [1]. The impaction, bouncing, and coalescence of small water drops on plane and curved air–water interfaces have been investigated [2]. When the cylindrical container holding a magnetic fluid is vibrated at the natural frequency of the fluid–container system, liquid drops floating on the free surface are formed [3]. Some effects of applied magnetic fields on fluid surface responses and disappearance of drops on the fluid surface in the laterally vibrated cylindrical container were studied [3]. However, the effect of magnetic fields upon the floating magnetic fluid drop at an air–magnetic fluid interface have been unsolved until now.

In the present report, the behavior of a magnetic fluid drop floating on an air–magnetic fluid interface under non–magnetic and magnetic fields was studied theoretically and experimentally. It was found that the floating time of the magnetic drop on the interface decreased with the magnetic field strength.

2. Experimental apparatus and procedures

A schematic diagram of the experimental apparatus is shown in Fig.1. Liquid drops were formed with a liquid reservoir. The liquid drops fell on a plane surface of liquid contained in a glass dish. The height of fall was adjusted to form the floating drop on the air–liquid interface. The behavior of the liquid drop on the interface was observed with two digital high speed cameras. The drop floating phenomena were analysed three–dimensionally by using a motion grabber system and a personal computer. In this experiment, three different liquids were used; ethanol, water and magnetic fluids

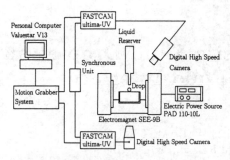

Fig 1 Schematic diagram of experimental arrangement

Table. 1 Physical properties of test liquids at 20°C

sample liquid	density (kg/m^3)	viscosity (Pa·S)	surface tension (N/m)
water	998.2	0.0010	0.0728
ethanol	789.0	0.0012	0.0228
magnetic fluid	1385.0	0.0141	0.0294

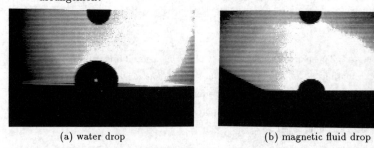

(a) water drop   (b) magnetic fluid drop

Fig. 2 Photographs of floating liquid droplet on the interface

(water–based ferricolloid W–35). A list of the physical properties of the test liquids is given in Table 1. Tests were conducted under applied magnetic fields by using a water–cooled electromagnet. The direction of the applied magnetic field was parallel to the interface.

3. Experimental results and discussion

When the liquid drop was impacted on the plane liquid surface at appropriate velocities (low velocities), the liquid drop floating on the interface was formed. Typical photographs of floating liquid droplets are shown in Fig.2. The floating period of the liquid droplet is not long. Figure 3 shows the time variation of horizontal diameter of liquid drops, D_y. The stairlike change of D_y for a water drop shows the formation of a smaller secondary droplet due to the coalescence between the liquid surface and drop. In the case of a magnetic fluid drop, the secondary drop formation was not observed. The initial drop was observed to oscillate after bouncing (at $t=0.01\sim 0.05$s in Fig.3b). It can be seen from Fig. 3b that the magnetic fluid drop floats on the surface for 0.2s. Under the influence of the magnetic field applied tangentially to the interface, the magnetic fluid drop was observed to coalesce directly with a plane magnetic fluid surface. In the next paragraph, we consider the experimental results theoretically.

4. Theoretical considerations

We consider the moving drop at the interface as shown in Fig.4. Both the drop and the interface are distorted by the impact of drop, and a thin film is formed between the drop and the interface. We may consider the spherical coordinates (r, θ, ϕ) as being

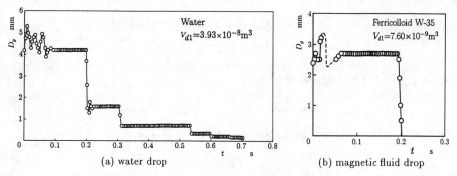

(a) water drop   (b) magnetic fluid drop

Fig. 3 Time variation of horizontal diameter of liquid drop

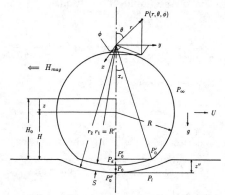

Fig. 4 Coordinate system and symbols

moved with the velocity U. The velocity components in the inside of the drop, v_r, v_θ, v_ϕ, are described by

$$\left.\begin{array}{l} v_r = v_r'(r) + U \sin\theta \sin\phi \\ v_\theta = v_\theta'(r,\ \theta) + U \cos\theta \sin\phi \\ v_\phi = U \cos\phi \end{array}\right\} \tag{1}$$

The velocity v_r' can be omitted as a small value compared with the other values. Navier–Stokes equations and the equation of continuity are described by

$$-\rho_g \frac{1}{r}(v_\theta'^2 + v_\theta U \cos\theta \sin\theta) = -\frac{\partial P}{\partial r} + \mu_\ell \frac{1}{r^2}\frac{\partial^2}{\partial r^2}(r^2 v_r') \tag{2}$$

$$\frac{1}{2}\rho_g \frac{\partial}{\partial\theta}(v_\theta' + U \cos\theta \sin\phi)^2 = -\frac{\partial P}{\partial\theta} + \mu_\ell \frac{1}{r}\frac{\partial}{\partial r}\left(r^2 \frac{\partial v_\theta'}{\partial r}\right) \tag{3}$$

$$\frac{1}{r}\frac{\partial}{\partial r}(r^2 v_r') + \frac{1}{\sin\theta}\frac{\partial}{\partial\theta}(\sin\theta v_\theta') = 0 \tag{4}$$

It is assumed that v_θ' may be expressed as follows:

$$v_\theta'(r,\ \theta) = K_1(r)\cot\frac{\theta}{2} \tag{5}$$

From Eq.(5), we obtain the following form:

$$\frac{\partial^2 P}{\partial r\partial\theta} = \frac{\partial}{\partial r}\left\{\frac{\mu_\ell}{r}\frac{\partial}{\partial r}\left(r^2\frac{\partial K_1(r)}{\partial r}\right)\right\}\cot\frac{\theta}{2} = 0 \tag{6}$$

where, the small inertia term of v_θ' was neglected. We obtain K_1 from Eq.(6). From

$v'_\theta=0$ at $r = r_1$ and $r = r_2 = r_1 + S$, the velocity distribution is expressed as follows:

$$v'_\theta = C_1 \frac{(r - r_1)(r - r_2)}{r} \cot \frac{\theta}{2} \tag{7}$$

Substituting Eq.(7) into Eq.(3), we have:

$$P_{r=r_1} - P_0 = 2\mu_\ell C_1 \log(\sin \frac{\theta}{2}) + \frac{1}{2}\rho_g U^2 \sin^2 \theta \sin^2 \phi \tag{8}$$

If we put $P = P_\infty$ at $\theta = \theta_c$, we get the following equation:

$$P_0 - P_\infty = \frac{\sigma}{R} \tag{9}$$

where σ is the surface tension of the liquid. From Eq.(9), we can get C_1. Hence, we write:

$$P_{r=r_1} - P_\infty = \frac{\sigma}{R} \left\{ 1 - \frac{\log(\sin \frac{\theta}{2})}{\log(\sin \frac{\theta_c}{2})} \right\} + \frac{1}{2}\rho_g U^2 \sin^2 \phi \sin^2 \theta_c \left\{ \frac{\sin^2 \theta}{\sin^2 \theta_c} - \frac{\log(\sin \frac{\theta}{2})}{\log(\sin \frac{\theta_c}{2})} \right\} \tag{10}$$

If we put $x = \pi - \theta$, then the repulsive force F_f is written as follows:

$$F_f = \int_0^{x_c} \int_0^{2\pi} (P_{r=r_1} - P_\infty) \cos x (r_1^2 \sin x) d\phi dx \tag{11}$$

Expanding $x_c = \pi - \theta_c$ in a series, and neglecting the small terms, F_f is expressed as follows:

$$F_f \cong \frac{\pi}{2} r_1^2 \sin^2 x_c \frac{\sigma}{R}(1 + 0.15 \sin^2 x_c) \tag{12}$$

This force supports the drop on the interface.

On the other hand, a magnetic fluid is magnetized under the applied fields. Therefore, the magnetic force has an effect on the floating drop phenomena. We consider the magnetizable drop–liquid system under the applied fields H_{mag} as shown in Fig.4. It is a complicated system, so to simplify, it is assumed that a small magnet with magnetic dipole moment P_m is located at distance R from a disk magnet with radius R_ℓ and magnetic dipole moment τ per unit area. Then the force acted upon the small magnet is described by

$$F_m = -\frac{P_m \tau R_\ell^2}{2\mu_0} \frac{3R}{(R_\ell^2 + R^2)^{\frac{5}{2}}} \tag{13}$$

where μ_0 is the permeability of vacuum. Therefore, the total replusive force F is described by

$$F = F_f + F_m = \frac{\pi}{2} r_1^2 \sin^2 x_c \frac{\sigma}{R}(1 + 0.15 \sin^2 x_c) - \frac{P_m \tau R_\ell^2}{2\mu_0} \frac{3R}{(R_\ell^2 + R^2)^{\frac{5}{2}}} \tag{14}$$

The second term of Eq.(14) speeds up the coalescence under an applied magnetic field.

References

[1]   S. Sudo, H. Hashimoto, J. Tani and M. Chiba, Dynamic Behavior of a Liquid in a Cylindrical Container Subject to Horizontal Vibration, *Bulletin of JSME* **29–254** (1986) 2455–2461.

[2]   O. W. Jayaratne and B. J. Mason, The Coalescence and Bouncing of Water Drops at an Air/Water Interface, *Proc. R. Soc. Lond.* **A280** (1964) 545–565.

[3]   S. Sudo, H. Hashimoto, A. Ikeda and K. Katagiri, Some Studies of Magnetic Liquid Sloshing, *J. Magn. Magn. Mat.* **65** (1987) 219–222.

Non-Linear Electromagnetic Systems
V. Kose and J. Sievert (Eds.)
IOS Press, 1998   863

Axisymmetric Oscillations of Magnetic Liquid Columns in a Magnetic Field

Motoyoshi OHABA[1], Tatsuo SAWADA[2], Masatsugu Tomomori[1],
Seiichi SUDO[3] and Takahiko TANAHASHI[2]

[1] *Department of Control System Engineering, Toin University of Yokohama*
1614 Kurogane-cho, Aoba-ku, Yokohama 225-0025, Japan
[2] *Department of Mechanical Engineering, Keio University*
3-14-1 Hiyoshi, Kouhoku-ku, Yokohama 223-8522, Japan
[3] *Department of Mechanical Engineering, Iwaki Meisei University*
5-5-1 Iino, Chuoudai, Iwaki 970-8044, Japan

Abstract. A magnetic liquid column formed in a uniform magnetic field is vibrated axisymmetrically. The dynamic characteristics of the axisymmetric mode are theoretically and experimentally examined. The stable range is enhanced by increasing the magnetic Bond number. The experimental interfacial response and resonant frequency of the liquid column show quantitative agreement with the theoretical prediction. Further, the dynamic disintegration is observed at the lowest resonant point.

1. Introduction

The Floating Zone Technique [1] is used in the crystal growth of semiconductors such as silicon. In this Technique, a liquid column is supported by surface tension alone and, as such, vibrational disturbances may cause the interface to oscillate. In a microgravity environment, this interfacial oscillation cannot be neglected. The Neutral Buoyancy Method, in which a liquid column is formed in another immiscible liquid having the same density, is typically used to perform terrestrial simulations to study the stability and resonant frequency of a liquid column. However, in order to precisely examine the dynamic behavior of liquid columns this method is not suitable because of the dynamic influences of the surrounding liquid [2]. For this reason we propose a new simulation method, that is, the Magnetic Liquid Method, as an alternative method for the Neutral Buoyancy Method. In a previous study [3, 4], we showed that the Magnetic Liquid Method could be used to clearly estimate the resonant frequency of a liquid column's lateral oscillations. Further, we have analyzed the axisymmetric oscillation of a magnetic liquid column in a uniform magnetic field [5].

In the present paper, a magnetic liquid column subjected to excitation in a vertical direction (i.e. axisymmetrically) is studied both theoretically and experimentally. In particular, the magnetic field effects on the stability range and the dynamic interfacial behavior are examined.

2. Theoretical Interfacial Response

Consider a liquid column with radius a and height h formed between two non-magnetic parallel plates under a uniform magnetic field H_0. We assume that the magnetic liquid is inviscid and incompressible. When the liquid column begins vibrating from a quiescent state, the flow is irrotational, and the velocity potential can be defined. Defining the vibration in the vertical direction by $Z_0 e^{i\omega t}$, the velocity potential $\phi(r, z, t)$ is described by Eq.(1),

$$\phi(r, z, t) = \sum_{n=1}^{\infty} \frac{(-1)^{n-1} 4Z_0 \omega^3 I_0(\kappa r)}{\kappa^2 h (\omega_{0n}{}^2 - \omega^2) I_0(\kappa a)} \sin \kappa z \sin \omega t, \tag{1}$$

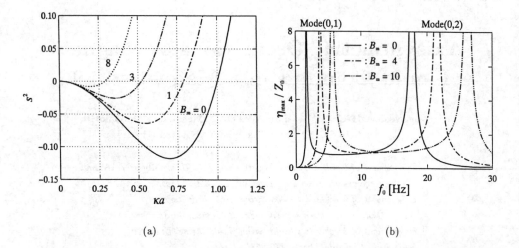

(a)   (b)

Fig. 1 Theoretical analysis of an axisymmetrically oscillating magnetic liquid column for various magnetic Bond numbers B_m: (a) dispersion relation; (b) interfacial frequency response.

where Z_0 is the excitation amplitude, ω the excitation circular frequency, $I_\nu(x)$ the modified Bessel function of the first kind of order ν, $\kappa = (2n-1)\pi/h$ the wave number in the axial direction, ω_{0n} the axisymmetric resonant circular frequency of the mode $(0, n)$, n the axial mode number and h the column height. In considering the interfacial oscillation, the magnetic perturbation potential concerning the surrounding gas is determined by the interfacial magnetic field conditions

$$\psi(r, z, t) = \sum_{n=1}^{\infty} \frac{4(-1)^n Z_0 \omega^2 (\mu_1 - \mu_2) H_0 I_0(\kappa a) I_1(\kappa a) K_0(\kappa r) \cos \kappa z \, e^{i\omega t}}{(2n-1)\pi (\omega_{0n}^2 - \omega^2) [\mu_1 I_1(\kappa a) K_0(\kappa a) + \mu_2 I_0(\kappa a) K_1(\kappa a)]}, \tag{2}$$

where μ_1, μ_2 are the permeabilities of the liquid column and surrounding gas respectively and $K_\nu(x)$ is the modified Bessel function of the second kind of order ν. Solving for ω_{0n}^2 gives

$$\omega_{0n}^2 = -\frac{\sigma \kappa I_1(\kappa a)}{\rho a^2 I_0(\kappa a)} \left[1 - a^2 \kappa^2\right] + \frac{\kappa^2 (\mu_1 - \mu_2)^2 H_0^2 I_1(\kappa a) K_0(\kappa a)}{\rho [\mu_1 I_1(\kappa a) K_0(\kappa a) + \mu_2 I_0(\kappa a) K_1(\kappa a)]}, \tag{3}$$

where σ is the surface tension ρ the density. Using both the linearized kinematical boundary condition and Eq.(1), we get the axisymmetric interfacial displacement $\eta(z, t)$

$$\eta(z, t) = \sum_{n=1}^{\infty} \frac{(-1)^{n-1} 4 Z_0 \omega^2 I_1(\kappa a)}{\kappa h (\omega_{0n}^2 - \omega^2) I_0(\kappa a)} \sin \kappa z \cos \omega t. \tag{4}$$

Figures 1(a) and 1(b) show the theoretical analysis of the dynamic characteristics of the magnetic liquid column for axisymmetric mode $(0,1)$ oscillations. Figure 1(a) shows the dispersion relation for various magnetic Bond numbers B_m. B_m is defined by $B_m = (\mu_1 - \mu_2) a H_0^2 / 2\sigma$, which is the dimensionless quantity of the magnetic force opposing the surface tension. In Fig. 1(a), κa is a dimensionless wave number and s dimensionless resonant circular frequency $s = (\rho a^3/\sigma)^{1/2} \omega_{on}$. The physical values used are: $\mu_1/\mu_2 = 1.5$, $\mu_2 = 4\pi \times 10^{-7}$ N/A², $\rho = 1390$ kg/m³ and $\sigma = 0.026$ N/m. In the range $s^2 \leq 0$ the liquid column becomes unstable and breaks, however, as shown in Fig. 1(a), the stable range $s^2 > 0$ is enhanced by increasing the magnetic Bond number B_m.

Figure 1(b) shows the dimensionless maximum interfacial displacement η_{max}/Z_0 of the mode $(0,1)$ for various magnetic Bond numbers B_m. For purposes of calculation, the dimensionless excitation amplitude is $Z_0/h = 0.05$ and the column aspect ratio defined by $A = 2a/h$ is 0.8. The maximum interfacial displacement η_{max} increases with the excitation frequency f_0 until the mode $(0,1)$ resonant

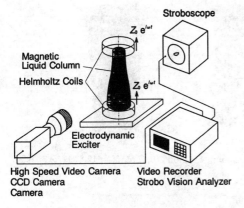

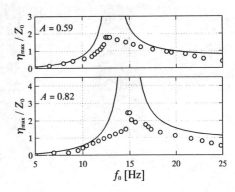

Fig. 2 Experimental apparatus

Fig. 3 Interfacial frequency response

point is reached, and decreases, then increases again at the mode (0,2) resonant point. The resonant range for the mode (0,2) is wider than that of the mode (0,1). When B_m is increased, the resonant point shifts to a higher frequency and the resonant range also expands.

3. Experimental Apparatus and Procedure

Figure 2 shows the experimental apparatus. A magnetic liquid column of height $h = 9.78$ mm is formed between Helmholtz coils generating a uniform magnetic field. The Helmholtz coils are supported on an electrodynamic exciter table and, after formation of the liquid column, are subjected to a sinusoidal wave excitation in the vertical direction. The column's oscillating behavior is observed as the frequency is gradually increased. In particular, a CCD camera, a stroboscope, high speed video recorder and a synchronized strobo-vision analyzer are used to observe the oscillating column. This system provides slow motion pictures of the column for analysis. We used a kerosene-based magnetic liquid having specific gravity 1.390; coefficient of viscosity 10.5 mPa; surface tension 0.026. The frequency showing the maximum displacement at the interface is defined as the resonant frequency.

4. Experimental Results and Discussion

In Figure 3, the theoretical and experimental mode (0,1) interfacial response is shown for $H_0 = 38.3$ kA/m, $Z_0 = 0.30$ mm, and $A = 0.59, 0.82$. The horizontal-axis shows the excitation frequency f_0 and the vertical-axis shows the dimensionless maximum interfacial displacement η_{max}/Z_0. At $A = 0.59$, there is clear agreement with the theoretical curves. However, at $A = 0.82$, variance from the theory increases, perhaps due to the viscous dissipation and dynamic clustering phenomenon. However, in general, since for $A = 0.59$, resonant frequency $f_r = 12.4$ Hz, and for $A = 0.82$, $f_r = 14.9$ Hz, it appears that the resonant frequency increases with the aspect ratio, which agrees well with the theory.

Figure 4 shows photographs of the lowest mode (0,1) standing wave in the range of the resonant point. These interfacial configurations correspond to each phase $\pi/2$ advance. The experimental conditions are: applied magnetic field $H_0 = 39.8$ kA/m, excitation frequency $f_0 = 12.8$ Hz, aspect ratio $A = 0.58$, and excitation amplitude $Z_0 = 0.60$ mm. Anti-nodes are observed at both column ends and it is clear that there is a node at almost the center of the column interface.

Figure 5 shows a sequence of photographs when resonant instability appears at $f_0 = 10.2$ Hz for $A = 0.25$, $H_0 = 38.1$ kA/m, and $Z_0 = 0.60$. At the excitation frequency of $f_0 = 10.2$ Hz, near the resonant point, a sudden increase appears in the liquid interface, and, simultaneously, the mode

Fig. 4 Sequence photographs of oscillating liquid column for mode (0,1) at $f_0 = 12.8$ Hz, $A = 0.58$.

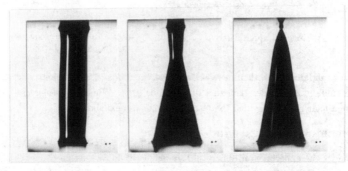

Fig. 5 Photographs showing resonant instability of liquid column at $f_0 = 10.2$ Hz, $A = 0.25$.

(0,1) standing wave quickly changes into a progressive wave. Somewhere in this resonant range the minimum diameter of the column's upper end becomes zero, and the column's upper face shows a slight separation response. Because of this dynamic disintegration for an axisymmetric mode oscillation involving a non-magnetic liquid, it is essential to carefully control the collapse of the liquid column.

5. Conclusion

Through the use of terrestrial simulation, we have studied the dynamic characteristics of a liquid column such as that used in the Floating Zone Technique. In particular, we observed the dynamic interfacial response and nonlinear resonant instability of an axisymmetrically oscillating magnetic liquid column in a uniform magnetic field. The experimental interfacial response agrees well with the theoretical prediction, further confirming the validity of the Magnetic Liquid Method.

References

[1] P. H. Keck, M. Green and M. L. Polk, Shapes of Floating Zones between Solid Rods, J. Appl. Phys. 24, (1953), 1479-1481.
[2] I. Martine, J. M. Haynes and D. LangSbein, Fluid Statics and Capillary. In: H. U. Walter (ed.), Fluid Sciences and Materials Science in Space, Springer-Verlag, Berlin, (1987), 70-75.
[3] M. Ohaba, et al., Vibrational Characteristics of a MagneticLiquid Bridge Subject to a Collinear Magnetic Field. In: A. J. Moses and A. Basak (ed.), NonlinearElectromagnetic Systems, IOS Press, Amsterdam, (1996), 410-413.
[4] M. Ohaba, et al., Magnetic Field Effects on a Laterally Vibrating Magnetic Liquid Column, Transport Phenomena in Material Processing and Manufacturing, ASME HTD-Vol. 336/FED-Vol. 240, (1996), 207-213.
[5] M. Ohaba, et al., Axisymmetric Surface Oscillations of a Magnetic Liquid Column, JSME ICFE-97-203, (1997), 263-268.

Non-Linear Electromagnetic Systems
V. Kose and J. Sievert (Eds.)
IOS Press, 1998

Rotating Regulator with Electrorheological Fluid

K. Shimada, M. Iwabuchi, S. Kamiyama*,

T. Fujita**, H. Nishida*** and K. Okui

Dept. of Mechanical Intellectual Systems Engineering,

Toyama Univ., 3190 Gofuku, Toyama 930, Japan

Inst. of Fluid Science, Tohoku Univ.,

2-1-1 Katahira, Aoba-ku, Sendai 980-77, Japan

**Akita Univ., Dept. of Geosciences,*

Mining Engineering and Materials Processing

1-1 Tegatagakuen-cho, Akita 010, Japan

***Ishikawa Polytechnic College,*

I-45-1, Yuigaoka, Anamizu, Ishikawa 927, Japan

Abstract. Experimental research on a rotating regulator using electro-rheological fluid (ERF) has been carried out. A colloidal suspension ERF was used. Experimental data of the torque, the current density and the electric power for a constant given number of revolutions of the disk from the time of applying the electric field were measured, and then compared with one another on the D.C. electric field strength and the gap between the electrodes. The experimental data depend strongly on experimental conditions, both qualitatively and quantitatively.

1. Introduction

An ERF is an intelligent fluid which is reacted upon by a supplied electric field. By using the enhancement of the apparent viscosity of the ERF in the case of applying D.C. or A.C. electric field, the number of revolutions of the rotating disk in the ERF can be controlled. This device has also the role of not only regulating the angular velocity but also braking. The concept of the device was first proposed by Winslow [1]. However, the experimental confirmation has not yet been investigated precisely. We use a colloidal suspension ERF and by attempting to solve many problems of the suspension ERF, the engineering application of ERF are expected.

The torque and current density for a constant given number of revolutions of the disk from the time of applying D.C. electric field are measured. They are compared with one another on the electric field strength and the gap between the electrodes.

2. Experimental Investigation

The rotating disk is connected between the torque motor and the torque-tachometer. The disk and ERF is settled in the casing which is held constant at temperature. Between the disk and the casing, the D.C. electric field strength E is applied with the gap G.

We use the colloidal suspension ERF which has been improved by one author of this paper. It has smectite particles of nm order size in a silicone oil base which has 20% mass concentration.

3. Results and Discussion

Figure 1 shows the characteristics between the torque T and the number of revolutions N. The dotted line is power input from the torque motor to the device.

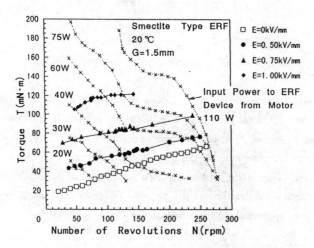

Fig.1 The characteristics between the torque and the number of revolutions.

The motion of the rotating disk is varied along the dotted line of the input power of the torque motor into the device by supplying the electric field. The regulation of the number of revolutions of the disk can be controlled by the electric field.

The characteristics as shown in Fig.1 are varied by the factors of the gap G and the voltage V of the supplied electric field. Therefore, we use the typical shear rate D_o which is defined by the angular velocity at the edge of the disk. Figure 2 shows relationship between the torque and the shear rate.

The experimental data can be arranged by the D_o even if the factors of G and V are different in the device.

Figure 3 shows the relationship between the current density I and the shear rate D_o.

In the current density, the influence of the gap is remarkable compared to the torque (Fig. 2). Because the current density becomes smaller at the constant shear rate and the same voltage when the gap is larger.

On the other hand, when increasing the shear rate, the current density decreases at the same gap.

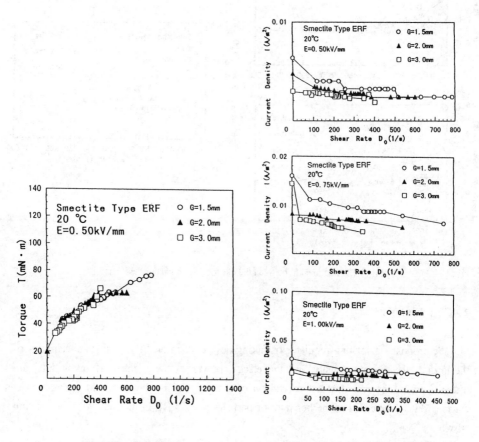

Fig.2 The relationship between the torque and the shear rate.

Fig.3 The relationship between the current density and the shear rate.

We want to know the required quantitativeness of electric power supplied to the device by the D.C. electric power supply. Figure 4 shows the relationship between the electric power W and the shear rate D_o.

The electric power needs to be greater at the constant shear rate and the same voltage as smaller gap.

On the other hand, with increasing shear rate, the electric power is needed to be smaller at the same gap. However, at that time the number of revolutions of the rotating disk is smaller. Therefore, if we expect to get the same quantitative decreasing number of revolutions at the same gap, more electric power is needed.

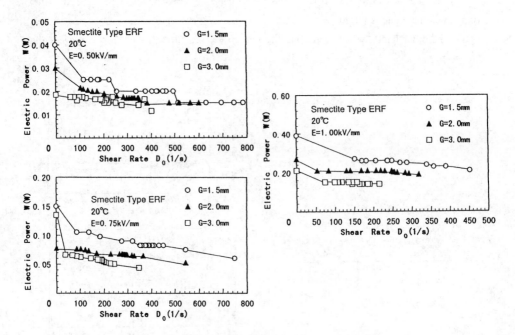

Fig.4 The relationship between the electric power and the shear rate.

4. Conclusions

We proposed a rotating regulator or raking device using an ERF. The characteristics of the torque, the current density and the electric power to the number of revolutions are different under the experimental conditions of the gap and the electric field strength. The cause is due to the effect of the behavior of particles in the ERF influenced from the rotating flow.

Reference
[1] W. M. Winslow, Induced fibration of suspensions, *J. Appl. Phys.*, **20** (1949) 1137-1141.

Non-Linear Electromagnetic Systems
V. Kose and J. Sievert (Eds.)
IOS Press, 1998

Lateral Vibration of a Magnetic Fluid Column

Tatsuo SAWADA[1], Yasuhiro FUJIWARA[2], Nobuaki OMURA[1],
Motoyoshi OHABA[3] and Takahiko TANAHASHI[1]

1) Department of Mechanical Engineering, Keio University
   3-14-1 Hiyoshi, Kohoku-ku, Yokohama 223-8522, Japan
2) Toyoda Automatic Loom Works, Ltd.
   2-1 Toyoda-cho, Kariya 448-0848, Japan
3) Department of Control System Engineering, Toin University of Yokohama
   1614 Kurogane-cho, Aoba-ku, Yokohama 225-8502, Japan

Abstract. Dynamic behavior of a magnetic fluid column which is formed
between two electromagnets is investigated. An electrodynamic shaker is used
to generate a lateral sinusoidal vibration. Experiments are carried out to clarify
the influence of magnetic field intensity, column height and fluid volume on the
vibration of a magnetic fluid column. A theoretical approach is also taken
using linearized wave theory. The first resonant frequency of displacement of
the free surface increases with the magnetic field intensity, and decreases with
the radius and height of the magnetic fluid column. These results are in good
agreement with theoretical results.

1. Introduction

With the advances of space engineering technology, the floating zone melting technique for material processing has been developed. This method is widely used in crystal
growth and refining of high-melting-point material. A large effort has been devoted
to the study of the stability of liquid bridges[1,2]. In recent years, the vibration of
cylindrical liquid bridges under a micro-gravity environment has received attention in
manufacturing processes in space. In order to support these projects, several investigations have been reported on the oscillations of liquid bridges[3,4]. There are only a
few experimental studies, mainly for a very small liquid bridge[5,6], because it is very
difficult to obtain a long liquid bridge on the earth.

Recently, we have carried out some experimental studies in order to perform terrestrial simulation of a liquid column[7,8]. We have used a magnetic fluid as a test liquid.
When a magnetic field is applied to a sample of magnetic fluid, the magnetic particles
in the fluid tend to remain rigidly aligned with the direction of the orienting field. One
of the most interesting results is that a micro-gravity environment is realized on the
earth using a magnetic fluid. If a magnetic force is applied to a magnetic fluid in the
direction in which the gravity is cancelled, there exists no body force in the magnetic
fluid.

In the present paper, the cylindrical magnetic liquid bridge which is formed between
a pair of electromagnets is studied. Effects of magnetic field, column length and fluid
volume on the dynamic characteristics of a magnetic liquid bridge column subject to
a lateral oscillation are investigated.

2. Experiment

Figure 1 illustrates a schematic diagram of the experimental apparatus. Each device number indicates the following: 1: function synthesizer, 2: power amplifier, 3: oscilloscope, 4: CCD camera, 5: stroboscope, 6: electrodynamic shaker, 7: strobo vision analyzer, 8: temperature control unit, 9: displacement transducer and 10: electromagnet. The main system consists of two electromagnets, an electrodynamic shaker, a cooling device and a displacement measuring instrument. Each electromagnet is 10 mm in inner diameter and approximately 90 mm in outer diameter. These electromagnets are arranged coaxially and vertically, and the distance is changeable. When a magnetic fluid is put between two electromagnets, the vertical cylindrical magnetic fluid column is formed between these electromagnets as shown in Fig. 2. A small thin thermostatic bath is glued to each electromagnet to keep the temperature of the magnetic fluid constant. Each thermostatic bath is filled with water at a constant temperature, 20 °C, which is supplied by a temperature control unit. The temperature of circulating water is controlled to an accuracy of 0.1 °C. An electrodynamic shaker is used to generate a sinusoidal excitation. Its amplitude and frequency are controlled by a function synthesizer. The amplitude of oscillation is $\eta_0 = 0.1$ mm for all experiments. The surface displacement of a liquid column is measured by an optical technique which uses a CCD camera, a stroboscope and a strobo-vision analyzer.

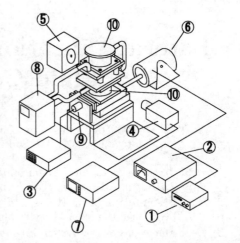

Fig. 1 Experimental apparatus.

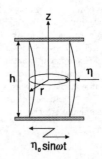

Fig. 2 Magnetic fluid column and coordinate.

The magnetic fluid is a water-based 24 % weight concentration of fine magnetite particles Fe_3O_4. Its kinematic viscosity, density and intensity of saturated magnetization are $\nu = 16.75 \times 10^{-6}$ m²/s, $\rho = 1.40 \times 10^3$ kg/m³ and $M_S = 47.5$ mT at 20 °C, respectively.

3. Results and Discussion

The ratio of the peak amplitude η of the free surface oscillation at $z = 0$ to the amplitude of the electrodynamic shaker is represented as a function of the forced frequency f in Figs. 3 and 4. Here, V is the volume of a magnetic fluid column, h the distance between two electromagnets and I the supplied current to each electromagnet. When the supplied current varies from 1.0 A to 1.6 A, the averaged magnetic force acting on the center of the fluid column is 3.2 to 7.7 times the gravitaional force. Theoretical results obtained by linearized wave theory are also shown in these figures. As the forcing frequency increases, the surface displacement also increases until the

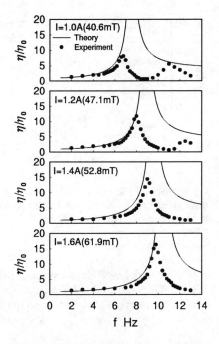

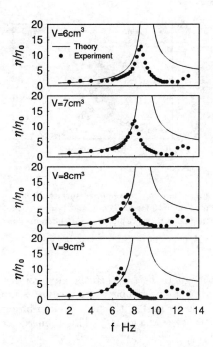

Fig. 3 Frequency response of free surface displacement at $z=0$ for $h=21\,\mathrm{mm}$ and $V=7\,\mathrm{cm}^3$ for different supplied currents.

Fig. 4 Frequency response of free surface displacement at $z=0$ for $h=21\,\mathrm{mm}$ and $I=1.2\,\mathrm{A}$ for different fluid volumes.

column surface is intensively vibrated near the first resonant frequency. Over the first resonant frequency, the surface disturbance is repressed. The first resonant frequency shifts to a higher frequency region as the magnetic field intensity increases or the fluid volume decreases. As can be observed, experimental data show the same trends as theoretical results except for the high frequency region over the first resonant frequency. Since the theoretical calculation is obtained by a linear analysis, differences between experimental and theoretical results increase with the forced frequency. The second resonant frequency is observed in some experimental data.

Dynamic behavior of a magnetic fluid column is influenced by gravity, surface tension, forced vibration and magnetic force. From dimensional analysis these effects are expressed by the following dimensionless parameters:

$$Fr = \frac{U^2}{Lg}, \qquad We = \frac{\rho U^2 L}{\sigma}, \qquad Rp = \frac{\mu_0 \chi_m H^2}{\rho U^2} \qquad (1)$$

where μ_0, χ_m, σ and g is the magnetic permeability of vacuum, the magnetic susceptibility, the surface tension and the gravitational acceleration, respectively. L, U and H are the reference quantities of length, velocity and magnetic field, respectively. Fr and We are the ordinal Froude and the Weber numbers. Rp is an additional parameters for a magnetic fluid and is the dimensionless ratio of magnetic forces to inertial forces. It is called the magnetic pressure number. In order to investigate the magnetic effect on the dynamic characteristic of a magnetic fluid column, Figs. 5 and 6 are illustrated. Here $U = \eta_0 \omega$ and the averaged radius of a magnetic fluid column is used for L. Various experimental data shown in Figs. 3 and 4 are rearranged in order by magnetic pressure number in Figs. 5 and 6. This indicates that the magnetic pressure number is a very useful parameter to predict the resonant frequency of a magnetic fluid column under

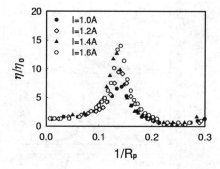

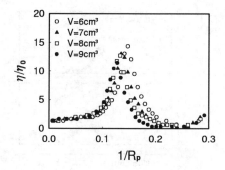

Fig. 5 Relation between free surface displacement at $z=0$ for $h=22$ mm and $V=8$ cm^3 and magnetic pressure number for different supplied currents.

Fig. 6 Relation between free surface displacement at $z=0$ for $h=22$ mm and $I=1.4$ A and magnetic pressure number for different fluid volumes.

a magnetic field. When the fluid volume is changed, the rearranged data are scattered especially near the resonant region as shown in Fig. 6. When the fluid volume increases, the lower part of the fluid column becomes thicker in comparison with the upper part because of the gravity. This shape deformation produces scattering of the rearranged data.

4. Conclusion

An experimental study of the dynamics of a magnetic fluid column is presented. It is found that the first resonant frequency increases with magnetic field intensity and decreases with fluid volume. The change of the first resonant frequency depends on the magnetic pressure number. Theoretical results are in good agreement with experimental ones under the first resonant frequency.

References

[1] R.D.Gillette and D.C.Dyson, Stability of Fluid Interfaces of Revolution between Equal Solid Circular Plates, Chem.Eng., 2 (1971), 44–54.

[2] D.Langbein, Stability of Liquid Bridges between Parallel Plates, Microgravity Sci.Technol., 5 (1992), 2–11.

[3] J.M.Perales and J.Meseguer, Theoretical and Experimental Study of the Vibration of Axisymmetric Viscous Liquid Bridges, Phys.Fluids, A-4 (1992), 110–1130.

[4] W.Eidel and H.F.Bauer, Non-linear Oscillations of an Inviscid Liquid Column under Zero-gravity, Ingenieur-Archiv, 58 (1988), 276–284.

[5] D.J.Mollot et al., Nonlinear Dynamics of Capillary Bridges: Experiments, J.Fluid Mech., 255 (1993), 411–435.

[6] S.Ahrens, Experiments on Oscillations of Small Liquid Bridges, Microgravity Sci.Technol., 7 (1994), 2–5.

[7] M.Ohaba et al., Magnetic Field Effects on a Laterally Vibrating Magnetic Liquid Column, Transport Phenomena in Materials Processing and Manufacturing, ASME HTD-336/FED-240 (1996), 206–213.

[8] M.Ohaba et al., Vibrational Characteristics of a Magnetic Liquid Bridge Subject to a Collinear Magnetic Field. In: A.J.Moses and A.Basak(eds.), Nonlinear Electromagnetic Systems, IOS Press, Amsterdam, (1996), 410–413.

Non-Linear Electromagnetic Systems
V. Kose and J. Sievert (Eds.)
IOS Press, 1998

Investigation on Vibration Characteristics of Magnetic Fluid Active Damper

Shinichi KAMIYAMA, Shuji KAWABE*, Tadashi YAMANE,
Tadamasa OYAMA, and Kazuyuki UENO

Institute of Fluid Science, Tohoku University,
2-1-1, Katahira, Aoba-ku, Sendai, 980-77, Japan
*Japan Nuclear Fuel Ltd., 2-2-2, Uchisaiwai, Chiyoda-ku, Tokyo, 100, Japan

Abstract. Study on the control of vibration of a spring-mass system attached to a piston immersed in a magnetic fluid is conducted to develop a magnetic fluid active damper. It is clarified that the vibration of the piston is effectively suppressed by the action of a magnetic force controlled by an oscillating non-uniform magnetic field applied to the magnetic fluid.

1. Introduction

There have been several investigations for the development of magnetic fluid viscous dampers in which the viscosity of the magnetic fluid is controlled by the applied magnetic field [1-3]. However, the development of an active damper has not yet been encoureged because of the weakness of the magnetic force acting on the magnetic fluid. Recently it is expected that an increase in the magnetic force will become possible by the preparation of a magnetic fluid having a high quality [4]. Therefore it is very important to clarify the dynamic characteristics of a magnetic fluid active damper.

In the present paper, the control of vibration of a simple spring-mass system attached to a piston immersed in a magnetic fluid is studied theoretically and experimentally as a basic study of a magnetic fluid active damper.

2. Theoretical Analysis

2.1 Equation of piston motion

Let us consider the oscillating motion of a piston immersed in the magnetic fluid as shown in Fig.1. The piston which is connected to a weight is supported by a cylinder with a spring. The cylinder is oscillated by a vibration exciter with amplitude Δu and frequency ω. Then the oscillation of the piston is induced.

The governing equation of the piston motion is given as follows:

$$
\begin{aligned}
(m + m_A)\frac{d^2 z}{dt^2} &+ (C + C_e)\frac{dz}{dt} + kz \\
&= m_A \frac{d^2 u}{dt^2} + (C + C_e)\frac{du}{dt} + ku - \pi r_0^2 \left\{ \int \frac{\partial p'}{\partial z_r}dz_r + \int \mu_0 M dH \right\}
\end{aligned}
\tag{1}
$$

where $z_r = z - u = \Delta z_r \sin(\omega t - \phi)$, $u = \Delta u \sin \omega t$, $m = m' + \pi r_0^2 \ell \rho'$, m_A:added mass, $p' = p + \rho g z - \int \mu_0 M dH$, C:viscous damping coefficient, C_e:equivalent damping coefficient due to Coulomb friction force, k:spring constant, r_0:radius of piston, ℓ:length of piston, H:magnetic field strength, M:magnetization, m':mass of weight, ρ':density of piston, ρ:fluid density, g:acceleration of gravity.

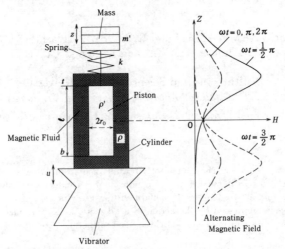

Figure 1   Analytical model of magnetic fluid active damper

The condition of complete suppression of piston oscillation is given with the right hand side of eq.(1)= 0 as follows.

$$\pi r_0^2 \int \mu_0 M dH = F_m \sin \omega t \tag{2}$$

where

$$\left.\begin{array}{l} F_m = \sqrt{(-m_A \Delta u \omega^2 + k\Delta u - \pi r_0^2 \ell p_2)^2 + \{(C + C_e)\Delta u \omega - \pi r_0^2 \ell p_1\}^2} \\[2mm] \tan \phi = \dfrac{(C + C_e)\Delta u \omega - \pi r_0^2 \ell p_1}{-m_A \Delta u \omega^2 + k\Delta u - \pi r_0^2 \ell p_2} \\[2mm] \dfrac{\partial p'}{\partial z_r} = p_1 \cos \omega t + p_2 \sin \omega t \end{array}\right\} \tag{3}$$

2.2 Method of Damping

If we put $F_v \sin \omega_v t$ as the force for driving piston oscillation, $-F_m \sin(\omega_m t + \psi)$ as the force for suppression and $F_m = F_v$ as the condition of control, then the total force acting on the piston F is given by

$$F = 2F_v \cos \frac{(\omega_m + \omega_v)t + \psi}{2} \sin \frac{(\omega_m - \omega_v)t - \psi}{2} \tag{4}$$

We set the most effective suppression at time t_0 during the time interval $[t_0 - t_1, t_0 + t_1]$, namely,

$$\left.\begin{array}{l} F(t_0) = 0 \\[2mm] F(t_0 - t_1) = -F(t_0 + t_1) \end{array}\right\} \tag{5}$$

Then, the conditions of oscillation suppression are obtained as

$$
\left.\begin{array}{l}
t_0 = \dfrac{n\pi}{\omega_v} \qquad (n = 0, 1, 2, 3, \cdots) \\[3mm]
\psi = \left(1 - \dfrac{\omega_m}{\omega_v}\right) n\pi
\end{array}\right\} \tag{6}
$$

The successive change of ψ is made with

$$
\left.\begin{array}{l}
t_0 = 1.5 + 3i \\[2mm]
n = 2f_v(1.5 + 3i) \qquad (i = 0, 1, 2, \cdots)
\end{array}\right\} \tag{7}
$$

3. Experimental Study

3.1 Experimental Apparatus and Procedure

Figure 2 shows an experimental apparatus for studying the damping of a piston immersed in the magnetic fluid. The radii of the piston and cylinder are 9 mm and 11 mm respectively. The magnetic fluid used here is hydrocarbon based magnetic fluid with 50 percent mass concentration of magnetite particles. The displacement of the piston and the vibration exciter is measured by an optical method (laser displacement meter).

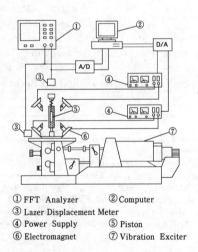

① FFT Analyzer   ② Computer
③ Lazer Displacement Meter
④ Power Supply   ⑤ Piston
⑥ Electromagnet   ⑦ Vibration Exciter

Figure 2   Experimental apparatus

3.2 Test Result and Discussion

Figure 3 shows the test results of suppression of piston oscillation in the case of $f = 2$, 4 and 6 Hz for $m' = 76.73 \times 10^{-3}$ kg, $k = 70$ N/m which has a resonance frequency of $f = 4$ Hz. It is clear that although the reduction of oscillation amplitude is not complete for $f_v = 2$ Hz $[f_v < f_r$ (resonance frequency)], stable control of suppression is possible in the cases of $f = 4$ and 6 Hz.

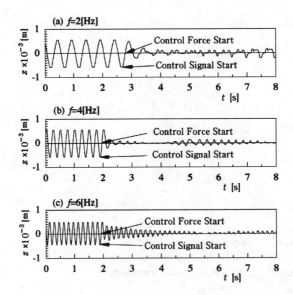

Figure 3   Damping of piston oscillation

4. Conclusion

The obtained results are summarized as follows:

(1) One method for control of a piston oscillation immersed in a magnetic fluid is proposed by utilizing a non-uniform applied magnetic field.

(2) In the case of low frequency, Coulomb friction force strongly influences the piston motion and disturbs complete suppression. However, for increased frequency, the viscous damping force exceeds the Coulomb force in the piston motion and stable control of piston oscillation becomes possible.

References

[1] K. Raj and R. Moskowitz, A Review of Damping Applications of Ferrofluids, *IEEE Trans. Magn.*, **MAG-16** (1980) 358-363.

[2] K. Nakatsuka *et al.*, Damper Application of Mangetic Fluid for a Vibration Isolating Table, *J. Magn. Magn. Mater.*, **65** (1987) 359-362.

[3] S. Kamiyama and K. Shimada, Basic Study on Dynamic Characteristics of Magnetic Fluid Viscous Damper, *Trans. Japan Soc. Mech. Engrs.*, Ser.B, **57** (1991) 4111-4115.

[4] I. Nakatani *et al.*, Iron-nitride Magnetic Fluids Prepared by Vapor-liquid Reaction and their Magnetic Properties, *J. Magn. Magn. Mater.*, **122** (1993) 10-14.

Non-Linear Electromagnetic Systems
V. Kose and J. Sievert (Eds.)
IOS Press, 1998

Light-Induced Quasi-Static and Dynamic Instability of Magnetization in Magnets

A.Kabychenkov

Institute of Radio Engineering and Electronics of Russian Academy of Sciences,
Mokhovaya st. 11, GSP-3, Moscow, 103907, Russia

Abstract. The influence of a light wave in the magnet transparency band on static and dynamic characteristics of the magnet is considered. The phase diagrams of magnetization in the light field are plotted. Light induced changes of domain boundaries and structures are determined. Conditions for parametric excitation of spin waves by a biharmonic light field are found. Light wave instability at a spontaneous phase transition is investigated. Two-parametric magnetooptical solitons are examined.

A light wave (LW) interacts with the magnetization of sublattice M^n through the magnetic component of the light field (LF) (magnetodipol (MD) interaction) and the electric component e of LF (magnetoelectric (ME) interaction) [1]. Linear MD and ME interactions are very small at optical frequencies because these are a few orders bigger than the precession frequencies of M^n. Besides, the latter only exists in crystals with no centre of symmetry. Therefore, the LF-M^n interaction is chiefly a nonlinear ME interaction.

As a result of this interaction an additional magnetization M^{nL} and an effective magnetic field H^{nL} are created in the magnet [2,3]. The characteristics of M^{nL} and H^{nL} depend on both LF parameters and the crystal magnetic structure. A quasimonochromatic LW produces static homogeneous or inhomogeneous M^{nL} and H^{nL}. A discrete spectrum LF besides producing static components gives rise to time-dependent components of M^{nL} and H^{nL} alternating at the difference frequencies of LF components as well. These frequencies lie in a broad band limited by the spectral widths of the components on the lower end and the light frequency on the upper end. If the frequency of the light-induced (LI) field coincides with the frequency of natural oscillations in the magnetic subsystem LI ferromagnetic (FM), antiferromagnetic (AFM), domain boundary (DB) or domain structure (DS) resonances may occur. This forms a basis for active nonlinear optical spectroscopy of magnetics. In addition to the above mentioned low frequency LI magnetization and magnetic field LW produces components of very small amplitudes with multiple and sum frequencies of the LF spectral components.

H^{nL} may be presented as a sum of exchange, anisotropy and magnetic fields. The latter is only produced by the circularly polarized component of LF, while the former exists also at linear polarization. The LI magnetic fields affect a magnetic subsystem as the intrinsic effective magnetic field does. H^{nL} is small since it is quadratic in e and the magnetooptical (MO) constants are small. For instance, at light intensities $I \sim 1 MW/cm^2$ in bithmuth-containing ferritic garnets (BFG) of $(CdBi)_3(FeALGa)_5O_{16}$ type or at $I \sim 0,1 MW/cm^2$ in EuO, EuSe rare earth magnets (REM) H^{nL} is about 1Oe. For this reason, the LF effect is considerable near the points of static and dynamic instabilities, where susceptibility is anomalously large. LI fields can have advantages over fields produced by permanent magnets or current-carrying conductors. They are localized within the light beam, they do not produce interference and, depending on the polarization, they act like an external magnetic field or uniaxial tension (compression) and may be of very short duration.

LW also creates an electric polarization at the optical frequencies of LF proportional to $\mathbf{M}^\eta$. The nonlinear susceptibility due to $\mathbf{M}^\eta$ is of the order of 10^{-2} for the exemplified BFG and 10^{-1} for REM. A polarization at combination frequencies in the optical range, which is proportional to the amplitude of small oscillations of $\mathbf{M}^\eta$ is also produced.

Thus, in general, LW propagation in a magnet is a self-consistent problem. Nevertheless, there are extreme cases, when the influence of LW on $\mathbf{M}^\eta$ and vice versa can be ignored. The first case is the classic MO effect, for example the Faraday or Foht effect [1]. Characteristic of this case are weak light intensities I and long optical paths l, required to observe changes in LW phase or polarization due to $\mathbf{M}^\eta$. The second case is the inverse MO effect, i.e., optomagnetic (OM) effects [2,3]. High I and short l, with the LW polarization almost intact are typical. The range of nonlinear MO effects lies between these extreme cases. Here LW instabilities are typical due to varying $\mathbf{e}$ - varying $\mathbf{M}^\eta$ interaction.

This paper concerns chiefly OM effects, in particular, LI phase transitions (PT), LI changes in DB and DS, parametrically generated spin waves (SW). Nonlinear MO effects, in particular, MO instabilities in LW and MO solitons also are considered. The combination scattering effects are not discussed.

1. General Equations

The energy density of a multisublattice magnet in a discrete spectrum LF can be represented by a sum of the nonmagnetized crystal energy, magnetodipol energy, internal energy of the magnetic subsystem contaning homogeneous and inhomogeneous exchange, anisotropy energy and LF-magnetic interaction energy:

$$W = W^0 + W^H + W^M + W^L, \quad W^H = -\mathbf{M}(\mathbf{H} - \tfrac{1}{2}\mathbf{H}_d) - \tfrac{1}{2}\chi_{ij}H_iH_j, \quad W^M = W^{hex} + W^{iex} + W^a,$$

$$W^{hex} = \tfrac{1}{2}A^{\lambda\eta}\mathbf{M}^\lambda\mathbf{M}^\eta + \tfrac{1}{4}B^{\lambda\eta\tau\rho}(\mathbf{M}^\lambda\mathbf{M}^\eta)(\mathbf{M}^\tau\mathbf{M}^\rho), \quad W^{iex} = \tfrac{1}{2}a_{ij}^{\lambda\eta}(\partial\mathbf{M}^\lambda/\partial x_i)(\partial\mathbf{M}^\eta/\partial x_j),$$

$$W^a = \tfrac{1}{2}K_{ij}^{\lambda\eta}M_i^\lambda M_j^\eta + \tfrac{1}{4}K_{ijkl}^{\lambda\eta\tau\rho}M_i^\lambda M_j^\eta M_k^\tau M_l^\rho, \quad W^L = -\varepsilon_{ij}^{\mu\nu}e_{ij}^{\mu\nu}, \tag{1}$$

where $\mathbf{H}=\mathbf{H}_0+\mathbf{H}_d$, $\mathbf{H}$, $\mathbf{H}_0$ and $\mathbf{H}_d$ are the internal, external and demagnetizing fields, $\mathbf{M}$ and χ are the total spontaneous magnetization and paramagnetic susceptibility, $\varepsilon_{ij}^{\mu\nu}(\omega_\mu,\omega_\nu,\mathbf{M}^\eta,\partial\mathbf{M}^\eta/\partial\mathbf{x},\mathbf{H})$ is the permittivity tensor, $e_{ij}^{\mu\nu} = \overline{e}_{ij}^{\mu\nu}\exp(i\omega_{\mu\nu}t)$ is the LF intensity tensor, $\overline{e}_{ij}^{\mu\nu} = e_{0i}^{\mu*}(\omega_\mu)e_{0j}^\nu(\omega_\nu)/16\pi$, ω_μ and $\mathbf{e}_0^\mu$ are the frequency and complex amplitude of the LF spectral components $\mathbf{e}^\mu = Re\,\mathbf{e}_0^\mu\,exp(i\omega_\mu t)$, $\omega_{\mu\nu}=\omega_\mu-\omega_\nu$.

The dynamics of $\mathbf{M}^\eta$ is determined by the equation

$$\dot{\mathbf{M}}^\eta = g^\eta[\mathbf{M}^\eta\widetilde{\mathbf{H}}^\eta] + (1/\tau_2^\eta)\,\widetilde{\mathbf{H}}^\eta - (1/\tau_1^\eta|\mathbf{M}^\eta|^2)[\mathbf{M}^\eta[\mathbf{M}^\eta\widetilde{\mathbf{H}}^\eta]] \quad . \tag{2}$$

In (2) g^η is the η-sublattice gyromagnetic ratio. The effective magnetic field is

$$\widetilde{\mathbf{H}}^\eta = \mathbf{H} + \widetilde{\mathbf{H}}^{\eta M} + \widetilde{\mathbf{H}}^{\eta L}, \quad \widetilde{\mathbf{H}}^{\eta M} = -\delta W^M/\delta\mathbf{M}^\eta, \quad \widetilde{\mathbf{H}}^{\eta L} = \hat{e}^{\mu\nu}\delta\hat{\varepsilon}^{\mu\nu}/\delta\mathbf{M}^\eta. \tag{3}$$

Assuming $\varepsilon_{ij}^{\mu\nu} = \overline{\varepsilon}_{ij}^{\mu\nu} + ie_{ijk}\alpha_{kn}^{\mu\nu\eta}M_n^\eta + \beta_{ijkn}^{\mu\nu\eta\lambda}M_k^\eta M_n^\lambda + \gamma_{ijklmn}^{\mu\nu\eta\lambda}\partial M_k^\eta/\partial x_l\,\partial M_m^\lambda/\partial x_n$, where e_{ijk} is the Levi-Civita tensor, we find $H_i^{\eta L} = H_i^{\eta L hex} + H_i^{\eta L iex} + H_i^{\eta L a} + H_i^{\eta L h}$, $H_i^{\eta L hex} = A^{\lambda\eta L}M_i^\lambda$ and $H_i^{\eta L iex} = a_{klmi}^{\lambda\eta L}\partial^2 M_m^\lambda/\partial x_k\partial x_l$ are LI homogeneous and inhomogeneous exchange fields, $A^{\lambda\eta L} = \tfrac{2}{3}e_{kl}^{\mu\nu}\beta_{klrr}^{\mu\nu\lambda\eta}$, $a_{klmi}^{\lambda\eta L} = 2e_{rt}^{\mu\nu}\gamma_{rtklmi}^{\mu\nu\lambda\eta}$, $H_i^{\eta L a} = K_{ij}^{\lambda\eta L}M_j^\lambda$, $K_{ij}^{\lambda\eta L} = 2e_{kl}^{\mu\nu}(\beta_{klij}^{\mu\nu\lambda\eta} - \tfrac{1}{3}\beta_{kirr}^{\mu\nu\lambda\eta}\delta_{ij})$ is the constant of LI anisotropy, δ_{ij} is the Kroneker tensor, $H_i^{\eta Lh} = g_k^{\mu\nu}\alpha_{ki}^{\mu\nu\eta}$ is LI magnetic field, $g_k^{\mu\nu} = ie_{ij}^{\mu\nu}e_{ijk}$. τ_1^η and τ_2^η are the relaxation times of $\mathbf{M}^\eta$.

The LF is determined by the equations

$$\Delta\mathbf{e}_0^\mu - grad\,div\,\mathbf{e}_0^\mu + c^{-2}(\omega_\nu^2\hat{\varepsilon}^{\mu\nu}\mathbf{e}_0^\nu - i\hat{\varepsilon}'^{\mu\nu}\dot{\mathbf{e}}_0^\nu) = 0, \quad div\,\hat{\varepsilon}^{\mu\nu}\mathbf{e}_0^\nu = 0, \tag{4}$$

where c is the velocity of light, $\hat{\varepsilon}'^{\mu\nu} = \partial\omega_\nu^2\hat{\varepsilon}(\omega_\mu,\omega_\nu)/\partial\omega_\nu$. Below the general equations (1)-(4) are applied to concrete problems.

2. Reorientation of the Magnetic Moment Vector in the Light Field

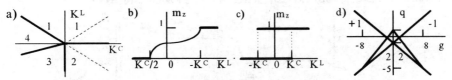

Fig.1. Phase diagram for the magnet in a linearly polarized LF (a), changes in $\mathbf{m}$ in LF at $K^m<0$ (b), $K^m>0$ (c); phase diagram for the case of circularly polarized LF for $K^m>0$ (d).

Let us consider a cubic magnet at low temperature (when $|\mathbf{M}^\eta|$ is conserved) in a LF linearly polarized along the [001] edge at $\mathbf{H}=0$. In this simplest case (1) is

$$W = W^a + W^L \ , \ W^a = K^m(m_x^2 m_y^2 + m_x^2 m_z^2 + m_y^2 m_z^2) \ , \ W^L = -K^L m_z^2 \ , \tag{5}$$

where $\mathbf{m} = \mathbf{M}/|\mathbf{M}|$ ($\mathbf{m}=\mathbf{L}/|\mathbf{L}|$, $\mathbf{L}=\mathbf{M}^1-\mathbf{M}^2$ for AFM), K^m is the magnetic anisotropy constant, $K^L = \beta e_z^{\mu\mu}$ is the constant of LI uniaxial anisotropy with axis parallel to $\mathbf{e}$, $\beta = \beta^{\mu\mu}$, $\beta^{\mu\nu} = (\beta_{zzzz}^{\mu\nu\eta\eta} - \beta_{zzyy}^{\mu\nu\eta\eta})|\mathbf{M}^\eta|^2$. The stationary states and their ranges of stability are

1) $m_z^2 = 1, K^L > -K^m$;   2) $m_z = 0, m_{x,y}^2 = 1 - m_{y,x}^2 = 1, K^L < K^m, K^m > 0$;

3) $m_z = 0, m_x^2 = m_y^2 = \frac{1}{2}, K^L > \frac{1}{2}K^m, K^m < 0$; $\qquad\qquad\qquad\qquad$ (6)

4) $m_z^2 = (1 - 2K^L / K^m)/3, m_x^2 = m_y^2 = (1 - m_z^2)/2, \frac{1}{2}K^m < K^L < -K^m, K^m < 0$.

In the collinear phases (CP) 1,2 and 3 $\mathbf{m}$ is oriented along the [001], [100] or [010] and [110] axes, respectively. In the angle phase (AP) 4 $\mathbf{m}$ lies on the (110) plane. The phase diagram (PD) is presented in Fig.1a. Transitions 1-4 and 3-4 are second-order LI spin-reorientation PT. At the transition $\mathbf{m}$ changes continuously (Fig.1b). Transitions 1-2 and 2-3 are LI first-order PT with and without hysteresis, respectively. There is a discontinuity in $\mathbf{m}$ on the instability lines (IL) (dashed lines) (Fig.1a,c).

Now let a circularly polarized LW be propagating along [001] $\parallel \mathbf{h}=\mathbf{H}/|\mathbf{M}|$. Then (1) is

$$W = W^H + W^a + W^L \ , \ W^H = -M_z H \ , \ W^L = K^L m_z^2 - H^L M_z \ , \tag{7}$$

where $H^L = 2\alpha_{xx}^{\mu\mu\eta}|\mathbf{M}^\eta|e_{xx}^{\mu\mu}$. LI anisotropy with the axis lying along the light beam and a magnetic field directed along or opposite $\mathbf{h}$ for right or left hand polarization are created in this case. The stationary states are given by

1) $m_z^2 = 1$; 2) $4m_z^3 - 2(1+q)m_z + g = 0, m_x m_y = 0$; 3) $3m_z^3 - (1-2q)m_z - g = 0, m_x^2 = m_y^2$,

where $q = K^L/|K^m|$, $g = |\mathbf{M}|(H + H^L)/|K^m|$. Stability of the states is investigated within the framework of the catastrophe theory. The PD is shown in Fig.1d. The point $(q,g)=(1,0)$ is a critical point. If a state trajectory in PD lies above it, then $\mathbf{m}$ changes continuously between $m_z=\pm 1$ and if a trajectory lies underneath it, $\mathbf{m}$ jumps.

3.Influence of the Light Field on Domain Boundaries and Domain Structures

Now we consider an inhomogeneously magnetized crystal at $\mathbf{e}\|[110]$. Then (1) is

$$W = W^{iex} + W^a + W^L, W^{iex} = \frac{1}{2}a(\partial \mathbf{m}/\partial x_i)^2, W^L = \frac{1}{2}a^L(\partial \mathbf{m}/\partial x_i)^2 - K^L(m_x^2 + m_y^2) - K_b^L 2m_x m_y,$$

where $a = a_{ii}^{\eta\eta}|\mathbf{M}^\eta|^2$, $a^L = 4\gamma_{iiiii}^{\mu\mu\eta\eta}|\mathbf{M}^\eta|^2 e_{xx}^{\mu\mu}$, $K_b^L = 2\beta_{xyxy}^{\mu\mu\eta\eta}|\mathbf{M}^\eta|^2 e_{xx}^{\mu\mu}$ is the constant of the LI basis anisotropy. $\mathbf{m}$ is close to that for homogeneous states in the areas far from DB. These states are CP 1)$\mathbf{m}\|[001]$, 2)$\mathbf{m}\|[110]$, 3)$\mathbf{m}\|[\bar{1}10]$ and AP 4) $m_z=0$, $m_{x,y}^2 = (v/k)^2$,

5) $m_z^2 = [1 + 2(u+v)/k]/3$, $m_x=m_y$ and 6) $m_z^2 = [1 - 2(u-v)/k]/3$, $m_x=-m_y$ ($u = K^L/|K_b^L|$, $k = K^m/|K_b^L|$, $v = K_b^L/|K_b^L|$) with $\mathbf{m}$ turning on (001),($\bar{1}10$) and (110) planes. Phases 3,6(2,5) are unstable at $K_b^L>0(K_b^L<0)$. PD for $K_b^L>0$ is shown in Fig.2a. Cross sections of PD give $\mathbf{m}$ versus K^L or K^m dependence (Fig.2b,c).

Bloch DB in CP 1 is $m_z^2 = 1/[1 + \rho_1^{-1} sh^{-2}(\xi/\delta_1)]$, where ρ_1 shows how far 1 is from its IL, $\delta_1 = (a_t/k_t)^{1/2}$ is the DB thickness, $a_t = a + a^L$, $k_t = 2(|K_b^L|/|\mathbf{M}|^2)(k-u-v)$. In a FM plate of (001) type at $k<0$ the period of a [110] oriented DS is $d=d_0 F$, where $F\sim 1$ at

$m_z \sim 1$ and $F \ll 1$ at $m_z \ll 1$, $d_0 = [2^{1/2}(k_t a_t)^{1/2} l / \sigma]^{1/2}$, $\sigma \sim 1,7$, l is the plate thickness. Relative change $\Delta d / d \sim 3\beta I / K^m c\bar{\varepsilon}^{1/2} \sim 10^{-2}$ at $I \sim 1 MW/cm^2$ for BFG.

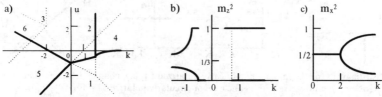

Fig.2. Phase diagram for the magnet in LF at $e\|[110]$ (a), change in **m** in LF (b), (c).

4. Generation of the Spin Waves by the Discrete Spectrum Light Field

For ground state $m\|h\|z\|[001]$ and biharmonic LF with $e^\mu\|[001]$ and wave vectors $k_\mu\|[100]$ equations (2) take the form

$$\omega_M^{-1}\ddot{m}_{x,y} + \rho_{1,2}m_{x,y} + a_k m''_{y,x} - [h_{2,1} + 2\,Re\,h^{La}\,exp(i\psi)]m_{y,x} = \mp k_{1,2}(m_{x,y}^2 + \tfrac{5}{3}\rho_{1,2}m_{y,x}^2)m_{y,x}, \quad (8)$$

where $\omega_M = g^\eta|M^\eta|$, $\rho_\varsigma = h_\varsigma / \omega_M \tau_1$, $\varsigma = 1,2$, $h_\varsigma = h + 2k^m + k^L + 4\pi\delta_{1\varsigma}$, $k^{m,L} = K^{m,L}/|M|^2$, $a_k = ak_{\mu\nu}^2$, $h^{La} = \beta^{\mu\nu}\bar{e}_{zz}^{\mu\nu}$ is the LI anisotropy field amplitude, $\psi = \omega_{\mu\nu}t - k_{\mu\nu}x$, $k_\varsigma = 3k^m + 2\pi\delta_{2\varsigma}$, $p_1 = 5/3$, $p_2 = 1 + 2/[3 + (2\pi/k^m)]$. Trajectories of **m** found from (8) are ellipsoids $(|m_x|^2 / m_{x0}^2) + (|m_y|^2 / m_{y0}^2) = 1$ with semiaxes $m_{i0} = A_i(|h^{La}|^2 - |h_{th}^{La}|^2)^{1/2}$, h_{th}^{La} is the threshold of the parametric excitation of SW. The ellipses are compressed by the dynamic $\tilde{h}_d$ along x. Approximately $I_{th} \approx (c\omega_{\mu\nu}|M^\eta|/4\pi\omega_M\beta^{\mu\nu})\Delta H^\eta$, ΔH^η is FMR width, $m_0 \approx 8\pi(\beta^{\mu\nu}I_{th}/3c|M|^2)[(I - I_{th})/I_{th}k^m h]^{1/2}$. In BFG at $I - I_{th} \approx I_{th} \approx 1 MW/cm^2$ $m_0 \approx 10^{-2}$. The behaviour of m_0 at the threshold is similar to that of the order parameter at PT.

5. Instability of a Light Wave in the Magnet. Magnetooptic Solitons.

Let a quasimonochromatic linearly polarized LW be propagating in an AFM at Neel point. Then $\varepsilon_L \equiv \varepsilon_{zz}^{\eta\eta} = \bar{\varepsilon}_{zz}^{\eta\eta} + \beta L_z^2$. In magnetoquasistatic approach $L_z^2 = (L_z^{(0)})^2 + b|e_z|^2$, $(L_z^{(0)})^2 = -A^{\eta\eta}/B$ is the equilibrium AFM moment, $B = B^{\eta\eta\eta\eta}$, $b = \beta/8\pi B$. In this case a solution of the equation (4) can be presented in the form $e_{0z}^\mu = e_0\,exp(-ik_\mu x)$, $k_\mu^2 = \omega_\mu^2\varepsilon_{L0}/c^2$, $\varepsilon_{L0} = \varepsilon_L(L^{(0)})$. The amplitude satisfies the nonlinear Schrödinger equation

$$i(\dot{e}_0 + v_g e_0') - \tfrac{1}{2}\tilde{a}\Delta e_0 - \tilde{b}|e_0|^2 e_0 = 0, \quad (9)$$

where $v_g = 2c\omega\varepsilon_{L0}^{1/2}/\tilde{\varepsilon}$ is the group velocity, $\tilde{\varepsilon} = \partial\omega^2\varepsilon_{L0}/\partial\omega$; $\tilde{a} = 2c^2/\tilde{\varepsilon}$ and $\tilde{b} = \beta\beta'/8\pi B\tilde{\varepsilon}$, $\beta' = \partial\omega\beta/\partial\omega$ determine the dispersion and nonlinearity of the magnet. It follows from (9) that LW will have modulation instability at $\beta\beta' > 0$ in an infinite magnet. Thresholds of MO instability arise because of the finite both l and transverse dimension of the light beam d. The longitudinal instability threshold $I_{th\|} = c(\lambda/l)^2 B/\beta\beta'$, λ is the wavelength. Assuming $\beta L^2 \approx 0,1$, $\lambda \approx 1\mu m$, $BL^4 \approx 10^8 erg/cm^3$, $l \approx 0,1 cm$ we find $I_{th\|} \approx 1 MW/cm^2$.

Evolution of the instability results in two component MO solitons. The **e**-component is a radio soliton, the **L**-component is a video soliton $L_z^2 - (L_z^{(0)})^2 = bC^2 sch^2[|\tilde{b}/2\tilde{a}|^{1/2} C(x - v_0 t)]$, C is a constant, v_0 is the soliton velocity. These nontopological solitons are stable due to compensation of the dispersion by nonlinearity both arising thanks to **L**. Evolution of the transverse instability results in MO self-focusing or collapse.

This work is partly supported by the RFBR, project 96-02-16082-a

References
[1].L.D.Landau and E.M.Lifshitz, Electrodynamics of continuous media. Oxford. 1960
[2].P.S.Pershan, I.P.van der Ziel, L.D.Malmstrom, Phys.Rev. **143**, 574(1966)
[3].A.F.Kabychenkov, Zh.Eksp.Teor.Fiz. **100**,1219(1991)

Non-Linear Electromagnetic Systems
V. Kose and J. Sievert (Eds.)
IOS Press, 1998

Numerical Calculations of Polarization Ellipses and Poynting-Vectors in Absorber Lined EMC Test Chambers

Th. SCHRADER, K. MÜNTER, M. SPITZER and J. GLIMM
Physikalisch-Technische Bundesanstalt (PTB)
Electromagnetic Fields Lab.
38116 Braunschweig, Germany

Abstract
With today's usual equipment, the results of EMC susceptibility tests can vary considerably, depending on the actual field generator. To investigate these inconsistencies, a research project was performed at PTB. It was shown by measurements and numerical calculations, that the surface current distribution on a conductive device describes the real test severity and is itself defined by the superposition of all incoming waves. Therefore it is necessary to verify the field structures in detail, especially the polarization properties. The computer models used for the small PTB anechoic room and the measurement set-up are described. Furthermore, the empty field distributions, the polarization ellipses, and the Poynting vectors in some typical field generators are calculated. The results are presented and compared with measured data.

1. Introduction

In an ideal free space environment a good linearly polarized antenna should produce a well defined field at the location of the DUT, because there is only one wave propagating, and any scattered waves from the DUT vanish and can be neglected. In commonly used field generators the situation is different. Even without a DUT the "empty field strength" at any point is the superposition of the incident wave and all scattered waves from the field generator itself. In a semi-anechoic chamber (as required in most test standards) the direct wave and the wave reflected at the metal floor give the two dominant contributions, in addition there may be scattered waves from the DUT [1], if the field generator has insufficient attenuation. Therefore, discrepancies for the test severity are possible, although the field strength measured with an isotropic sensor is the same.

2. Model for Numerical Calculations

Field calculations were performed with a commercial software package [2]. For this program the volume of interest is discretized into a 3-dimensional orthogonal grid, and the boundary conditions and material parameters have to be defined. After this geometric model is established, a special matrix algorithm based on the Finite-Integration-Method calculates an approximate numerical solution of Maxwells' equations by a recursive single time-step-technique.

With this numerical method it is necessary to describe the material distribution of the investigated object in great detail with correct material parameters. In the geometric model the mesh width has to be adequately chosen, especially in regions where gradients and divergent fields are expected. Besides that, staircase-like approximations of borders and edges must be avoided. For the transformed solutions in the frequency domain there is a correlation between the upper frequency limit, the time step and the mesh width. For good accuracy the mesh width should be less than one tenth of the wavelength at the highest frequency.

Under these hardware constraints it is impossible to define details of the absorbers inside an anechoic chamber of practical dimensions, therefore a model using layers with fitted reflection parameter conditions [3] was employed. A 2+1-layered model from the literature describing the material properties of pyramidal absorbing cones with 30 cm length was found sufficient. As a source of electromagnetic fields a logarithmic-periodic antenna was chosen. To represent details of its actual geometry, the mesh had to be partially refined to 1 cm grid length. With the field excitation and the absorbers adequately described, the fields in two models of anechoic chambers (one with a metallic reflecting groundplane, the other fully anechoic) were calculated up to 600 MHz with proper spatial resolution.

3. Measurement Set-up

To obtain field structure information a system measuring the amplitude and phase is needed. A sensor system developed at PTB [4] was modified to measure magnetic fields and surface currents with a flat frequency response up to 1 GHz. Together with a complex network analyzer a suitable system for high-speed measurements with a large dynamic range is formed. The measurement uncertainty for the complete system (sensor, network analyzer, power amplifier, cables) is 18 % in amplitude and 9 degrees in phase for the used coverage factor k = 2 [5].

4. Evaluation of Measurement and Numerical Data

In a general time-harmonic field the electric and magnetic field vectors rotate in separate planes, with their ends tracing an ellipse. These electric and magnetic ''polarization ellipses'' are not necessarily perpendicular, but may be oriented arbitrarily in space. Such a complicated structure as to be investigated here, is often found in the near fields around a radiator or scattering object. Orientation and shape of the polarization ellipses give good insight into the field structure at a location. In this investigation all necessary amplitude and phase data versus frequency of all field components are available, for convenience they are referenced to a coordinate system (x,y,z) defined by the main axes of the test set-up (e.g. the room).

The first structure information is obtained by calculating the main axes of the polarization ellipses from the amplitude and phase informations. Orientation and shape of the ellipses are easily imagined from these vectors. Another important quantity is the power flux density, given by the instantaneous Poynting-Vector [6], defined by the cross product equation

$$\underline{\vec{S}} = \underline{\vec{E}} \times \underline{\vec{H}}, \tag{1}$$

In oscillating fields the end of the instantaneous Poynting-vector is also tracing an ellipse, but (in contrast to the polarization ellipses) its movement is not restricted to a plane. Depending on the field conditions this Poynting-vector-ellipse (PVE) could be orientated arbitrarily in space. The instantaneous Poynting vector can therefore be separated into one time-invariant component $\vec{S}^R$ (describing the mean flow of energy, often referred to as the time-averaged Poynting vector) and an oscillating vector $\vec{S}^X$ with a time average of zero.

$$\vec{S}(t) = \vec{S}^R + \vec{S}^X(t) \tag{2}$$

Table 1.1: Fully-Anechoic Chamber, calculated Polarization Ellipses

f in MHz	E_{max} in %	E_{min} in %	$\vartheta_{Form,E}$	H_{max} in %	H_{min} in %	$\vartheta_{Form,H}$
100	100	0,2	0,002	101,7	3,7	0,04
200	100	0,9	0,009	100,3	9,7	0,1
300	100	12,8	0,13	100,9	50,4	0,5
400	100	2,2	0,02	100,9	10,4	0,1
500	100	0,05	0,0	100,3	10,2	0,1
600	100	4,0	0,04	100,3	11,3	0,11

Table 1.2: Poynting-Vector, calculated   measured

$|S|/S_0$	θ in °	Sx/S_0	Sy/S_0	Sz/S_0	$|S|/S_0$	θ in °	Sx/S_0	Sy/S_0	Sz/S_0
4589	1,0	920	-132	4494	5580	4,2	-4559	450	3186
4946	0,9	433	-150	4925	5543	0,3	-2250	-276	5059
5121	6,4	909	-505	5014	5162	1,8	702	-430	5096
4251	4,1	-292	-62	4241	5107	0,4	262	-27	5101
4044	4,7	622	-74	3996	4990	0,2	-26	-450	4970
4251	4,3	598	-61	4208	5031	0,5	15	-224	5026

Table 2.1: **Semi-Anechoic-Chamber**, measured Polarization Ellipses

f in MHz	E_{max} in %	E_{min} in %	$\vartheta_{Form,E}$	H_{max} in %	H_{min} in %	$\vartheta_{Form,H}$
100	116,3	40,2	0,35	100,3	2,5	0,025
200	113,1	7,1	0,06	100,2	4,0	0,04
300	101,2	26,4	0,26	100,5	9,6	0,1
400	100,9	11,9	0,12	100,1	16,4	0,16
500	100	7,7	0,08	101,2	2,5	0,025
600	100,9	12,4	0,12	100,3	3,4	0,034

Table 3.1: „GTEM-1500" cell, measured Polarization Ellipses

f in MHz	E_{max} in %	E_{min} in %	$\vartheta_{Form,E}$	H_{max} in %	H_{min} in %	$\vartheta_{Form,H}$
100	100,8	2,1	0,021	100,1	5,9	0,06
200	101,5	15,2	0,15	100,1	5,4	0,05
300	100,9	15,1	0,15	104,3	6,1	0,06
400	100,4	5,3	0,05	100,5	9,9	0,09
500	100,4	18,5	0,18	100,1	2,2	0,02
600	100,0	20,5	0,2	100,3	2,5	0,025

Table 3.2: Measured Poynting-Vector

$|S|/S_0$	θ in °	Sx/S_0	Sy/S_0	Sz/S_0
5047	0,1	-646	228	5001
5063	0,9	-765	190	5001
5045	1,9	-521	997	4918
5036	0,3	-441	356	5004
5018	0,3	-431	215	4995
5008	0,3	-152	329	4995

The x components of these vectors are calculated from the E and H amplitude- and phase measurements as follows:

$$\vec{S}_x^{\,R} = \frac{1}{2}\left(\left|E_y H_z\right|\cos\!\left(\phi_y^E - \phi_z^H\right) - \left|E_z H_y\right|\cos\!\left(\phi_z^E - \phi_y^H\right)\right) \tag{3a}$$

$$\vec{S}_x^{\,X}(t) = \frac{1}{2}\left(\left|E_y H_z\right|\cos\!\left(2\omega t + \phi_y^E + \phi_z^H\right) - \left|E_z H_y\right|\cos\!\left(2\omega t + \phi_z^E + \phi_y^H\right)\right), \tag{3b}$$

y and z components are obtained by cyclic replacement of the lower x,y,z indices.

In the following tables the lengths of the main axes of the polarization ellipses are listed, normalized to the desired field component (here E_x and H_y). The three spatial components S_x, S_y, S_z of the time-invariant term $\vec{S}^R$ and the maximum angle between the extreme positions of the instantaneous vector are specified as structure informations about the energy flow. In a fully anechoic chamber (Table 1.1) the E ellipse shows good linear polarization, with some elliptical polarization of the H field at 300 MHz. The calculated data show an energy flow (Table 1.2) mainly into the z direction as expected, but the measured performance below 200 MHz is obviously degraded. At these low frequencies the calculation model assumptions (absorbers?) seem to be inadequate. In a semi-anechoic chamber the measured polarization ellipses (Table 2.1) indicate a pronounced elliptical polarization at the lowest frequency and around 300 MHz, with the worst frequency of the magnetic field at about 400 MHz. For comparison Table 3.1 and 3.2 present some data measured in a large "GTEM cell". Here the magnetic field is linearly polarized, while the electric field indicates some problems between 200 and 300 MHz and at the higher frequencies. The energy flow in that cell shows only minor distortions.

Conclusion

In different test facilities with complicated field structures, the use of isotropic field sensors leads to discrepancies and should therefore be avoided, if comparable EMC test results must be obtained.

References

[1]   Th. Schrader, Vergleich von Feldgeneratoren für EMV-Prüfungen,
       Ph.D. Thesis, Technical University (TU), Braunschweig, Germany, 1997
[2]   The MAFIA Collaboration, Users' Guide MAFIA 3.22, CST GmbH, Darmstadt
[3]   Ch. Bornkessel, Analyse und Optimierung der elektrodynamischen Eigen-
       schaften von EMV-Absorberkammern durch numerische Feldberechnung,

Ph.D. Thesis, Technical University (T.H.), Karlsruhe, 1993

[4]   J. Glimm, Th. Schrader, K. Münter, M. Spitzer, A New Direct-Measuring Field Sensor up to 1 GHz with an Analog Fibre-Optical Link -Design, Traceable Calibration, and Results, 11th Symposium on EMC, Zurich, 1995

[5]   European Coorporation for Accreditation of Laboratories, R-2: Expression of the Uncertainty of Measurement in Calibration, 1997

[6]   H.G. Booker, Energy in Electromagnetism, Peregrinus Ltd., New York, 1982

Excitonic Non-Linearity of Semiconductors at High Level of Laser Excitation

P.I. Khadzhi, D.V. Tkachenko

Transnistrian State-Corporative University, 25 October str., 128, Tiraspol,
Moldova, 3300

Abstract. Field-induced are predicted bistable behaviour and abrupt changes of the dispersive and absorptive susceptibilities of semiconductors in the exciton range of the spectrum at a high level of excitation taking into account the exciton-photon and elastic exciton-exciton interactions.

1. Introduction

Electromagnetically induced changes in the optical properties of semiconductors at a high level of laser excitation has attracted great attention in recent years due to its possible practical applications. The field-induced transparency, optical Stark-effect, blue and red shifts of the atomic and exciton resonances and other nonlinear optical effects have proved to be very important and a number of experiments have demonstrated these effects. A very large high-energy shift of the lowest exciton resonance in GaAs multiple-quantum-well structures is observed [1] during irradiation of the sample with femtosecond laser pulses. The effect was explained in terms of excitons "dressed" by photons. The experiments and theoretical description of the optical Stark-effect in the exction range of the spectrum of semiconductors are presented in [2,3]. The possibilitiy of media with a very large refractive index has also been demonstrated [4]. In [5] it was shown that field-induced transparency changes the dispersive characteristics and the possibility of probe gain due to the presence of the pump laser exist in the system of high-density excitons modelled as a system of anharmonic oscillators. A similar problem was discussed in [6] for the system of Bose-condensed excitons in the presence of pump and probe fields.

2. Results

In this paper we present the main results of an investigation of the bistable and spectral peculiarities of the dispersive and absorptive susceptibilities of semiconductors in the exciton range of the spectrum taking into account the exciton-photon and elastic exciton-exciton interactions. We assume a strong (pump) electromagnetic wave with amplitude E_o and frequency ω_e and a weak (test) wave with amplitude E and frequency ω incident on the crystal. The strong electromagnetic wave propagating in the crystal excites the high density coherent excitons with the self-frequency ω_o. The Hamiltonian of the system of excitons interacting with the external fields in the frame rotating with the frequency ω_e of the strong field has the form

$$H = -\hbar\Delta\alpha^+\alpha + \hbar v\alpha^+\alpha^+\alpha\,\alpha + \hbar g(E_o^+\alpha^+ + \alpha\,E^-) - \tag{1}$$

$$- \hbar g(\alpha^+ E^+ e^{-\iota\delta t} + \alpha\,E^- e^{\iota\delta t}),$$

where $\Delta = \omega_e - \omega_o$ and $\delta = \omega - \omega_e$ are the detunings, α is the amplitude of the exciton wave, g is the exciton-photon coupling constant and v is the elastic exciton-exciton interaction constant. Then the semiclassical equation for the exciton wave amplitude α is

$$i\frac{\partial}{\partial t}\alpha = -\Delta\alpha - i\gamma\alpha - g\,E_o^+ - g\,E^+ e^{-\iota\delta t} + v|\alpha|^2\alpha, \tag{2}$$

where γ is the phenomenological damping constant of the exciton. Then calculating the response of the exciton system to all orders in E_o and to the first order in E it is easy to obtain the following expressions for the dispersive X' and absorptive X'' susceptibilities of the crystal:

$$X'/X_o = \left[x - 2z + z^2\cdot(\delta_o - \Delta_o + 2z)/(1 + (\delta_o - \Delta_o + 2z)^2)\right]\cdot F^{-1}, \tag{3}$$

$$X''/X_o = \left[1 - z^2/(1 + (\delta_o - \Delta_o + 2z)^2)\right]\cdot F^{-1}, \tag{4}$$

$$F = \left(1 - \frac{z^2}{1 + (\delta_o - \Delta_o + 2z)^2}\right)^2 + \left(x - 2z + z^2\frac{\delta_o - \Delta_o + 2z}{1 + (\delta_o - \Delta_o + 2z)^2}\right)^2, \tag{5}$$

where $X_o = \hbar g^2/\gamma$, $\Delta_o = \Delta/\gamma$, $\delta_o = \delta/\gamma$, $x = \Delta_o + \delta_o = (\omega - \omega_o)/\gamma$, $z = v n_o/\gamma$, n_o is the density of coherent excitons, excited by the strong field, which is expressed by the nonlinear equation

$$z\left[(\Delta_o - z)^2 + 1\right] = f^2. \tag{6}$$

Here $f = E_o/E_s$, $E_s^3 = \gamma^3/(v\,g^2)$. It follows from (6), that the density of coherent excitons z shows the bistable behaviour depending on f for fixed values.

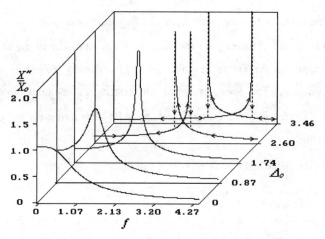

Fig.1 Dependence of the absorptive susceptibility on the level of excitation f and on the laser detuning Δ_o

of Δ_o, if $\Delta_o^2 > 3$ and depending on Δ_o for fixed values of f, if $f > f_{cr} = (4/3)^{3/4}$. It is known [7], that the susceptibilities under the action of only the pump pulse depend on Δ_o and f. Taking into consideration this conclusion we have investigated the behaviour of the real X' and imaginary X'' parts of susceptibility depending on the pump detuning Δ_o and the test detuning δ_o as well as on the level of the pump excitation f.

3. Discussion

In Fig.1 the absorptive susceptibility X'' is plotted as a function of the pump amplitude f for different values of the pump detuning Δ_o and a fixed value of the probe detuning $\delta_o = 0$. The absorption at first increases with increasing pump amplitude f, reaches its maximum, then decreases, if the pump detuning $\Delta_o^2 < 3$. But if $\Delta_o^2 > 3$ the function X'' shows bistable behaviour as a function of the pump amplitude f, and jumps at several values of f, when the field amplitude f increases or decreases. This effect is due to the exciton-exciton interaction, which gives rise to exciton resonance shifting. X'' versus x are plotted correspondingly in Fig.2 and Fig.3 for $\Delta_o = 0$ and different values of the pump amplitude f for the cases of increasing (Fig.2) and decreasing (Fig.3) f. We can see that probe pulse gain appears on the long-wave tail of the function $X''(x)$, which is located in the spectral region $2(\Delta_o - z) - \sqrt{z^2 - 1} < x < 2(\Delta_o - z) + \sqrt{z^2 - 1}$. The width of this region depends on the frequency and amplitude of the pump pulse. The physical reason for the gain effect is due to the processes of paarwise leaving of the excitons which are excited by the strong laser field from the coherent state, and their passing to the long-wave and short-wave ranges of the polariton branches with respect to the frequency of the strong field. The radiative recombination of these excitons is the main reason for the dumping suppression effect and hence for the gain band formation. The absorption band $X''(x)$ shows a blue shift, when the pump amplitude f increases. It is important to note, that the probe absorption and gain bands undergo abrupt spectral changes, namely, blue and red shifts, when the pump amplitude f approaches the critical values, which are consistent with the positions of the jumps of the function $z(f)$. It is evident, that the function $X'(x)$ undergoes similar changes.

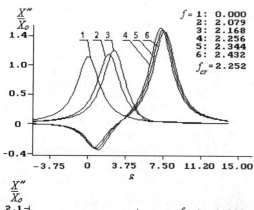

$$f = \begin{array}{ll} 1: & 0.000 \\ 2: & 2.079 \\ 3: & 2.168 \\ 4: & 2.256 \\ 5: & 2.344 \\ 6: & 2.432 \end{array}$$

$f_{cr} = 2.252$

Fig.2 Dependence of X'' on probe δ_o detuning for the case increasing f.

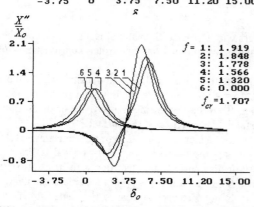

$$f = \begin{array}{ll} 1: & 1.919 \\ 2: & 1.848 \\ 3: & 1.778 \\ 4: & 1.566 \\ 5: & 1.320 \\ 6: & 0.000 \end{array}$$

$f_{cr} = 1.707$

Fig.3 Dependence of X'' on probe δ_o detuning for the case decreasing f.

References

[1]   A.Mysyrowicz, D.Hulin, et al., *Phys. Rev. Lett.*, **56**, (1986) 2747-2751.

[2]   M. Combescot, *Phys. Reports* **221** (1992) 167-249

[3]   D. Frohlich et al., *Phys. Stat. Sol.* (B) **173** (1992) 83-89.

[4]   M.O. Scully, *Phys. Rev. Lett.* **67** (1991) 1855-1858.

[5]   G.S. Agarwal, *Phys. Rev.* **A51** (1995) R2711-R2714.

[6]   S.A. Moskalenko, V.G. Pavlov, *JETP* **111** (1997) 6.

[7]   P.I. Khadzhi, Y.D. Slavov, *Ukr. Fiz. Jurn.* **33** (1988) 824-827.

Non-Linear Electromagnetic Systems
V. Kose and J. Sievert (Eds.)
IOS Press, 1998

Method for Prepolarizing Ferroelectric Particles

G. Randolf, H. Frais-Kölbl, and H. Hauser
Institut für Werkstoffe der Elektrotechnik, Technische Universität Wien
A-1040 Gusshausstrasse 27-29, AUSTRIA

Abstract. A new manufacturing process of ferroelectric capacitors has been developed and is currently investigated. The main motivation is to increase the electric energy density of the dielectric matter. Powdery ferroelectric particles are mixed up with a thermoplastic powder and are pressed under vacuum and high temperature. The permittivity of the capacitor can be varied in a large range by choosing the ferroelectric-powder-to-substrate-powder-ratio. Furthermore, the pill shaped sample is prepolarized by a high electrical bias field that is aligned parallel to the surface of the sample. The electrical field that is applied in the semiliquid state of the sample during the manufacturing process is able to align the ferroelectric particles. After freezing this configuration the dipoles remain in a state of low total crystal energy. If a charging field is applied to the sample perpendicular to the sample surface the momentum of the dipoles is rotated by the maximum angle in a reversible process. Therefore a capacitor can be produced having an expected energy density of 10^6 - 10^7 J/m^3. We observed a reduction of hysteresis losses of about 5 % in our first experiments using a bias field of 300 kV/m. In the currently used configuration we use a bias field of 1000 kV/m and we expect an according decrease of hysteresis losses.

1. Introduction

The optimization of the design of capacitors with ferroelectric material leads to considerable energy density. In the following the technological possibilities of further improvements of dielectric energy storage are discussed.

In conventional capacitors using a high permittive material like high density BaTiO$_3$ most of the energy is stored in the small gap between the solid bulk material and the electrodes. This gap originates, besides the contacting, from the decreasing permittivity in the surface layer caused by the breakdown of the crystal field interaction at the surface.

The permittivity of the surface region cannot be increased because of the vanishing crystal fields. As the thickness of the surface layer is in the range of several 10 crystal layers, the thickness of the bulk is in the range of 10μm to improve the electrode-dielectric ratio but turns a moderate voltage to a high electrical field strength [1]. Trying to increase the energy density by increasing the bulk permittivity is not successful because the material will be saturated immediately at low applied fields. The energy is then stored in the small gap and therefore, the overall energy density decreases.

2. Energy Storing in the Bulk Material

In order to increase the energy density of a capacitor it is necessary to raise the energy stored in the large bulk material. Therefore an optimization of permittivity-to-saturation-polarization-ratio is required.

As shown in Figure 1 the energy that can be stored in the material has its maximum in case (b) where the polarization has its saturation at the breakdown voltage of the dielectric

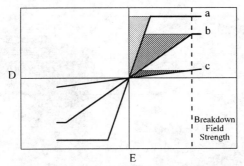

Figure 1 Energy stored in various hysteresis curves

material. There is also no advantage, having both a high permittivity and a high saturation polarization together, as long as their ratio is not optimized (curve a). A reduction of the permittivity of material (a) causes a shearing of the hysteresis curve and an increase of the energy density.

In order to obtain the maximum energy density, Equation (1) should be valid, where P_s is the spontaneous polarization, ε_0 and $\varepsilon_{r(E)}$ the permittivity of vacuum and of the material (depending on E) respectively, and E the applied field, that should be close to the breakdown field.

$$P_s = E \varepsilon_0 (\varepsilon_r - 1) \tag{1}$$

3. Experimental Set-up

The samples are in shape of discs with an area of 10 cm^2 and a thickness of 1.5 mm to 2.5 mm and have been produced under several, defined conditions. Temperatures at 180°C are necessary to melt the polymer powder and to break the spontaneous polarization of the ferroelectric powder. We used BaTiO$_3$ because of its reasonable polarization to permittivity ratio, a Curie-temperature of $T_C = 120$°C and an easy resourcing.

A mechanical pressure of about 5 MPa and a vacuum of 100 Pa cause a homogeneous sample with low gas inclusion. Finally an electrical bias field of 1 MV/m is applied in the plane of the sample. To combine all these conditions a very special press had to be constructed as shown in Figure 2. The press plates have to be made from fire bricks, which is an electrical insulating but heat resistant material. The guiding system and the power drive are asymmetrically placed behind the electrode with ground potential to avoid distortion of the electric field. The sample mold is made of solid teflon and has a teflon tube to flood the sample with cooled oil after the pressing process to freeze the aligned dipoles while the bias field is still applied.

4. Set-up of Measuring Instruments

For the measurements the samples are contacted with liquid conductive silver in a well defined area. The sample and the sample holder are dipped in insulating oil to prevent creep current on the surface. A sinusoidal voltage up to 6 kV with a frequency of 5 to 10 mHz is applied. The voltage and the charging current are shown by a Hewlett Packard storage oscilloscope as check up but stored by a PCMCIA measurement device and a portable PC as shown in Figure 3. The use of developed software makes it possible to observe either voltage and current vs. time or voltage vs. electrical charge (integration of the current). Several parameters of the sample and the measurement adjustment are stored in a header of the data file so that the conditions of the measurement can be reproduced.

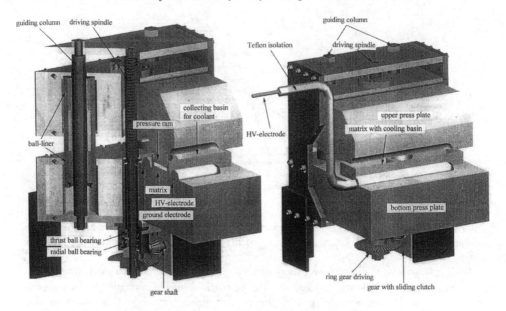

Figure 2 Mechanical components of the pressing device

A further computer code enables automatic evaluation of the data file, by using the information in the file header on dimension of the sample and the instrument settings. Corrections and filtering of the data file are also made by the computer code. From the resulting D vs. E hysteresis loop the stored energy and the energy losses can be calculated.

5. Results and Discussion

In order to show the effect of prepolarization separately one conventional sintered sample has been used. Firstly, the original non polarized, polycrystalline $BaTiO_3$ (volume density 95%) specimen was measured under various voltages up to 4.5 kV with a frequency of 0.1 Hz. Afterwards the sample was heated up to 160° C and an electrical bias field of 800 kV/m was applied. The sample was left for about 20 minutes under these conditions to allow the dipoles to align themselves in the direction of the field. Then the sample was cooled down quickly to 60° C in the electrical field. A liquid conductive silver film was applied and the sample was measured a second time under the same conditions. These results are shown in Fig. 4.

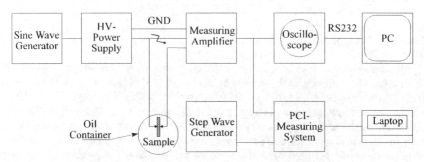

Figure 3 Set-up of measuring instruments

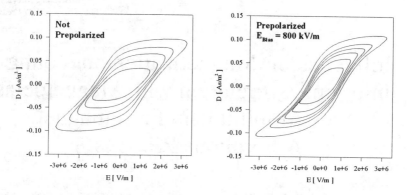

Figure 4 Sample measured with and without an electrical bias field during the manufacturing process. Because of the dielectric breakdown it is not possible to yield the saturation polarization

To reduce the permittivity of the material the density of the dielectric material has to be varied. Due to the porosity occurring in a sintered sample it has to be drenched with an insulating liquid to avoid current leakage. By embedding the dielectric material in polymer substrate the ratio of high permittive to low permittive material can be varied and the result is a compact sample with reduced overall permittivity.

6. Conclusion

These results show the usefulness of prepolarization during the manufacturing process of ferroelectric capacitors.

The efficiency of an energy storing device using ferroelectric material can be considerably increased by reducing the hysteresis losses. For a high energy density several parameters like grain size and a fine tuned ratio between spontaneous polarization and permittivity are important.

These optimizations lead to a capacitive energy storing device with low losses, an ultra quick charging/discharging ability and a high cycle stability. A conventional electro-chemical battery can be combined with the capacitive device and a DC/DC - converter. Thus, the performance and the durability of the battery will be increased.

$BaTiO_3$ is of course not the most suitable dielectric material conceivable. Instead of pure $BaTiO_3$ powder a crystal mixture of $PbTiO_3$ and $Pb(Zn_{1/3}Nb_{2/3})O_3$ has to be used to reach an energy density of $W = 10^6 - 10^7$ J/m^3. The spontaneous polarization of this material is $P_S = 0.52$ As/m^2 and the permittivity is about $\varepsilon_{r(0)} = 5000$ at room temperature. The Curie temperature $T_C = 180\,°C$ [2] is just as suitable but the material is difficult to produce and so our test series are made with $BaTiO_3$ for the present.

Acknowledgments

We want to express our gratitude to Prof. G. M. Fasching, Dir. Peschl and H. Frais-Kölbl for making these investigations possible. Financial support was provided by the "Forschungsförderungsfonds für die gewerbliche Wirtschaft" under Grant No. 3/11705.

References

[1]   G. Randolf, H. Frais-Kölbl, H. Hauser: Optimization of Capacitive Energy Storage. Int. Symp Dielectric and Insulating Systems in Electrical Engeneering, Bratislava (27.-28. November 1995)
[2]   J. Kuwata. K. Uchino, S. Nomura: Jpn. J. Appl. Phys. 21 (1982),1298

Influence of Magnetization on Young Modulus and Correlation with Magnetoelastic Waves Amplitude in Ferromagnetic Amorphous Ribbons

Amleto D'AGOSTINO, Roberto GERMANO, Vincenzo IANNOTTI, Luciano LANOTTE
Istituto Nazionale per la Fisica della Materia, Dipartimento di Scienze Fisiche, Facoltà di Ingegneria dell'Università di Napoli "Federico II", p.le V.Tecchio 80, 80125 Napoli, Italia.

Abstract. This paper deals with the effect of a small (~ 100 A/m) inhomogeneity of magnetic transverse polarization on the spontaneous vibrational mode of $Fe_{80}B_{13}Si_4C_3$ ribbon samples. The local polarizing field is limited to a region of one square centimeter. The magnetoelastic resonant waves are both excited and detected by means of direct and inverse magnetostrictive effect. It is shown that a change of transverse magnetization in a small portion of a ribbon sample can produce a change of longitudinal resonance frequency, namely of magnetoelastic wave velocity and Young Modulus. By placing a local magnetizing field at different points along the vibrating sample, a direct correlation can be shown between the measured Young Modulus and the longitudinal magnetoelastic wave amplitude. In particular, the maxima of magnetoelastic response are related to the minima of Young Modulus. The theoretical interpretation of the effect enables to predict very interesting developments in applications for amplitude modulation signals.

1. Introduction

In previous work, the effect of a longitudinal magnetic field inhomogeneity on the amplitude of resonant magnetoelastic vibrations was shown [1]. This paper represents a more extended investigation by considering different locations of a local transverse magnetizing field H_L on the resonant magnetoelastic wave amplitude V_0. The change of the resonance frequency V_0 due to H_L was particularly considered and a clear correlation between Young Modulus at constant field E_{II} and the amplitude V_0 was shown by experimental evidence. Since a minimum of V_0 means also a maximum of the magnetoelastic coupling coefficient k which governs the exchange between magnetic and elastic energy, the results indicate that V_0 is proportional to k, which is a coherent relationship under the experimental conditions used.

2. Procedure and Theory of the Experiment

The experimental arrangement, both to generate and to detect resonant standing waves in a magnetic ribbon, is shown in Fig.1. In the coil A the current, governed by a variable frequency generator, produces an alternate magnetizing field:

$$He = Heo \, z \, \cos(2\pi v \, t + \varphi)$$ (1)

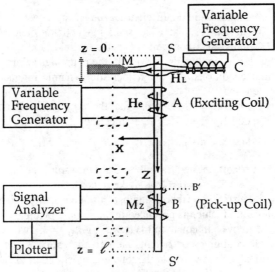

Fig. 1 Experimental apparatus.

whose amplitude Heo can be changed independently of the frequency v and where z is the versor of the z axis. Coil B detects the electromotive force which is induced by the changes of the sample longitudinal magnetization intensity. The coils are shielded on the external surface so that they do not influence each other by mutual induction or interact with other external systems. As a consequence of the experimental arrangement the signal induced in coil B can be written as:

$$V(t) = -\mu_o \left(\frac{\partial}{\partial t} \sum_{i=1}^{N} S_i M_i(t) \right) = -\mu_o \left(\frac{NS}{L} \right) \frac{\partial}{\partial t} \int_{Lo}^{(Lo+L)} M_z(z,t) \, dz$$ (2)

where N is the turn number of coil B, $S=S_i$ is the sample cross section, $M_i(t) = M_z(z,t)$ is the magnetization intensity in the i-th turn of the coil B which has coordinate z, μ_o is the vacuum permeability and L_o is the z coordinate at the end B_1 of coil B (see Fig.1).

The ribbons of amorphous metal used were 22 cm in length, with a cross section of $50\mu m$ x 3 mm, and they were produced by rapid quenching in the composition $Fe_{80}B_{13}Si_4C_3$. Before experimental investigation, all the samples were heat treated by furnace annealing in inert atmosphere for two hours at 400 °C in order to produce the structural relaxation which improves magnetoelastic coupling [2].

Generally the signal induced in B has a negligible amplitude due to the magnetization produced in B by the magnetic field applied in A (the distance between A and B [nearest end] is about 12 cm). But, if He frequency is tuned to the frequency of the longitudinal spontaneous vibrational mode of the ribbon sample, vo, the V(t) signal has a high amplitude because a resonant standing wave, that has both an elastic component and a magnetic one, is produced. We concentrated the experimental investigation on the fundamental longitudinal mode which originates from the magnetoelastic coupling described by means of the classical state equations, which relate the magnetic components along the z axis (Fig.1), Mz and Hz, with the elastic components along the same axis, σ_{zz} and ε_{zz} (σ = stress, ε = strain) [3]:

$$\varepsilon_{zz} = g_{zz}|_{\sigma} \cdot M_z + C_{zz}^{-1}|_H \cdot \sigma_{zz}$$

$$M_z = \chi_{zz}|_{\varepsilon} \cdot H_z + e_{zz}|_H \cdot \varepsilon_{zz} \tag{3}$$

where $g_{zz}|_{\sigma}$ is the magnetoelastic coefficient $\partial\varepsilon_{zz}/\partial M_z$ at constant stress, $\varepsilon_{zz}|_H$ is the magnetoelastic coefficient $\partial M_z/\partial\varepsilon_{zz}$ at a constant magnetizing field, $C_{zz}|_H$ is the elastic coefficient (Young Modulus) at a constant magnetizing field, and $\chi_{zz}|_{\varepsilon}$ is the susceptivity coefficient at constant strain. In equations (3) only the couplings zz are considered because, under our experimental conditions, the magnetoelastic dynamic interactions xz and yz are neither produced nor detected.

From equations (3) it is easy to recognize that the magnetoelastic coupling between M_z and ε_{zz} gives a signal in coil B, where $H_z = 0$, by means of the term

$$M(z,t) = e_{zz}|_H \cdot \varepsilon_{zz}(z,t) = e_{zz}|_H \cdot g_{zz}|_{\sigma} \cdot M_z \tag{4}$$

Now, for standing conditions and free vibration of the sample ends, it is logical to assume
$$M_z = M_{zo} \sin k^*vt \, \cos k^*z \tag{5}$$
where $k^*=2\pi/l_o$, $l_o = 2l$, l, $(2/3)l$, l = ribbon length, v = wave velocity and M_{zo} is the amplitude of the alternate magnetization which is produced in the sample by the exciting field, therefore it is $M_{zo} \propto \chi_{zz}|_{\varepsilon} H_{eo}$. Moreover, in equation (4) $(\varepsilon_{zz}|_H)(g_{zz}|_{\sigma}) = k^2$ where k^2 is the square of the magnetomechanical coupling coefficient, namely the ratio of the transferred energy (magnetic into mechanical or vice versa) to the total energy stored in the magnetized material.

Considering Eqs. (4) and (5), and that for maximum amplitude $v = 1 \, v|_H$ while $k^2 = 1-(C_{zz}|_H /C_{zz}|_B)$, where the symbols $|_H$ and $|_B$ refer to a constant magnetizing field and a constant magnetic induction, respectively, one finds that the amplitude of the induced signal (2) is:

$$V_0(t) \propto \left(1 - \frac{C_{zz}|_H}{C_{zz}|_B}\right) \cdot C_{zz}|_H^{1/2} \cdot \chi_{zz}|_{\varepsilon} \cdot H_{eo} = \left(1 - \frac{v|_H^2}{v|_B^2}\right) \cdot v|_H \cdot \chi_{zz}|_{\varepsilon} \cdot H_{eo} \tag{6}$$

where it was also used: $C_{zz} = v^2 \, 4 \, l^2 \, \rho \tag{7}$

3. Results and Discussion

From equation (6) it is evident that any physical action which produces changes of resonance frequency (and consequently of C_{zz}) is accompanied by V_0 changes.

Under any experimental condition it is easy to evaluate the resonance frequency $v|_H$ by the Fourier analysis of the induced signal. Equation (7) enables the calculation of $C_{zz}|_H$, while the exciting signal can be tuned to $v|_H$ and V_0 measured from the pick-up coil output.

The $C_{zz}|_B$ value is practically obtained by the resonance frequency at magnetic saturation. In the first set of measurements the intrinsic conditions of the ribbon were changed by a static field applied by a magnet M (Fig.1) in the x direction (transverse to the sample plane zy) with a H_{Lz} component of 120 A/m in a sample piece of 1 cm, while the H_{Ly} and H_{Lx} components were practically zero. In Fig.2 the $C_{zz}|_H$ behaviour is compared to the measured curve of V_0 when the magnetic perturbation H_{Lz} was displaced along the whole sample from S (z=0) to S' (z=l) (Fig.1). It is evident that the qualitative behaviour predicted by equation (6) is verified; in particular $C_{zz}|_H$ minima give V_0 maxima ($\chi_{zz}|_H$ changes can be considered negligible because $H_{Lz}=0$). Naturally one expects that $C_{zz}|_H$ variations should be also dependent on the H_{Lx} intensity. Therefore in the second set of measurements we maintained a fixed position of the magnet M, but we changed the H_{Lx} intensity by overimposing a variable transverse field by means of coil C (Fig.1). This coil produces an alternate field $H_{Lx}' = H_{Lx} \cos(2\pi v't)$ so that the

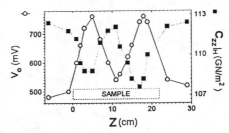

Fig. 2 Magnetoelastic wave amplitude vs. position z of the transversal H_{Lx} field.

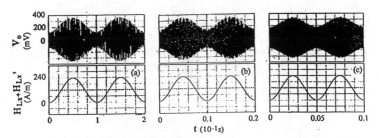

Fig.3 Modulation of Vo by H_{Lx}' at frequency (a) 10 Hz, (b) 100 Hz, (c) 200 Hz.

total local field H_L has a variable value in the range from 0 to $2H_{Lx}$. The response due to three different levels of v' is shown in Fig.3 which confirms the predicted modulation of Vo: when $H_L=H_{Lx}+H_{Lx}'$ increases, $C_{zz}|_H$ decreases and Vo increases; when H_L decreases, $C_{zz}|_H$ increases and Vo has a minimum for $H_L=0$.

It is apparent that the dynamic response to H_L decreases by increasing the frequency v'. This is in agreement with the expected viscoelastic behaviour of the amorphous material investigated here.

4. Conclusions

A local transverse magnetic polarization can influence longitudinal elastic coefficient and, therefore, it changes resonance frequency Vo $|_{II}$; the consequence of this effect on the longitudinal magnetoelastic wave amplitude Vo was verified under new practical conditions.

A theoretical relationship between Vo and Vo $|_{II}$ (and consequently Young Modulus $C_{zz}|_{II}$) was demonstrated to be effective. The evidence that the interaction of Vo with $C_{zz}|_H$ exists in variable transverse local magnetization, was experimentally shown. This result gives indications for the construction of a magnetoelastic wave transmission line which can be modulated in amplitude by a low local field, within the limit of low frequency (< 1 kHz).

References

[1] L. Lanotte, V. Iannotti and R. Bruzzese, Effects of Local Magnetization Changes on Resonant Magnetoelastic Waves, *J.Appl.Electr.Mech.* **6** (1995) 155-163

[2] T. Egami, Magnetic Amorphous Alloys: Physics and Technological Applications, *Rep. Prog. Phys.* **47** (1984) 1601-1725

[3] E. Olsen, Applied Magnetism, Phil.Tech.Libr.ed., The Netherlands (1966), chap. 11

Author Index